Precalculus
Real Mathematics, Real People
Alternate Edition
Sixth Edition

Ron Larson
The Pennsylvania State University, The Behrend College

With the assistance of David C. Falvo
The Pennsylvania State University, The Behrend College

BROOKS/COLE
CENGAGE Learning™

Australia • Brazil • Japan • Korea • Mexico • Singapore • Spain • United Kingdom • United States

BROOKS/COLE
CENGAGE Learning

Precalculus: Real Mathematics, Real People
Alternate Edition
Sixth Edition
Ron Larson

Acquiring Sponsoring Editor: Gary Whalen

Development Editor: Stacy Green

Assistant Editor: Cynthia Ashton

Editorial Assistant: Naomi Dreyer

Media Editor: Lynh Pham

Marketing Manager: Myriah FitzGibbon

Marketing Coordinator: Shannon Myers

Marketing Communications Manager: Darlene Macanan

Content Project Manager: Jessica Rasile

Senior Art Director: Jill Ort

Senior Print Buyer: Diane Gibbons

Senior Acquisitions Specialist, Images: Mandy Groszko

Senior Acquisitions Specialist, Text: Katie Huha

Text Designer: Larson Texts, Inc.

Cover Designer: Harold Burch

Compositor: Larson Texts, Inc.

For product information and technology assistance, contact us at
Cengage Learning Customer & Sales Support, 1-800-354-9706

For permission to use material from this text or product, submit all requests online at **www.cengage.com/permissions**.
Further permissions questions can be emailed to
permissionrequest@cengage.com

Library of Congress Control Number: 2010940163

ISBN-13: 978-1-111-42843-3
ISBN-10: 1-111-42843-3

Brooks/Cole
10 Davis Drive
Belmont, CA 94002-3098
USA

Cengage Learning is a leading provider of customized learning solutions with office locations around the globe, including Singapore, the United Kingdom, Australia, Mexico, Brazil, and Japan. Locate your local office at: **international.cengage.com/region**

Cengage Learning products are represented in Canada by Nelson Education, Ltd.

For your course and learning solutions, visit **www.cengage.com**

Purchase any of our products at your local college store or at our preferred online store **www.cengagebrain.com**

Printed in the United States of America
1 2 3 4 5 6 7 13 12 11 10

Rational Function (p. 298)

$$f(x) = \frac{1}{x}$$

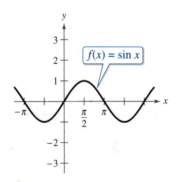

Domain: $(-\infty, 0) \cup (0, \infty)$
Range: $(-\infty, 0) \cup (0, \infty)$
No intercepts
Decreasing on $(-\infty, 0)$ and $(0, \infty)$
Odd function
Origin symmetry
Vertical asymptote: y-axis
Horizontal asymptote: x-axis

Exponential Function (p. 326)

$$f(x) = a^x, \ a > 0, \ a \neq 1$$

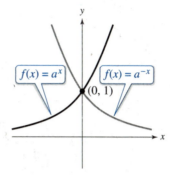

Domain: $(-\infty, \infty)$
Range: $(0, \infty)$
Intercept: $(0, 1)$
Increasing on $(-\infty, \infty)$
    for $f(x) = a^x$
Decreasing on $(-\infty, \infty)$
    for $f(x) = a^{-x}$
x-axis is a horizontal asymptote
Continuous

Logarithmic Function (p. 339)

$$f(x) = \log_a x, \ a > 0, \ a \neq 1$$

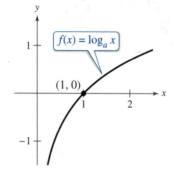

Domain: $(0, \infty)$
Range: $(-\infty, \infty)$
Intercept: $(1, 0)$
Increasing on $(0, \infty)$
y-axis is a vertical asymptote
Continuous
Reflection of graph of $f(x) = a^x$
    in the line $y = x$

Sine Function (p. 433)

$$f(x) = \sin x$$

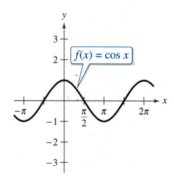

Domain: $(-\infty, \infty)$
Range: $[-1, 1]$
Period: 2π
x-intercepts: $(n\pi, 0)$
y-intercept: $(0, 0)$
Odd function
Origin symmetry

Cosine Function (p. 433)

$$f(x) = \cos x$$

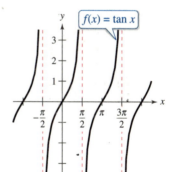

Domain: $(-\infty, \infty)$
Range: $[-1, 1]$
Period: 2π
x-intercepts: $\left(\frac{\pi}{2} + n\pi, 0\right)$
y-intercept: $(0, 1)$
Even function
y-axis symmetry

Tangent Function (p. 444)

$$f(x) = \tan x$$

Domain: $x \neq \frac{\pi}{2} + n\pi$
Range: $(-\infty, \infty)$
Period: π
x-intercepts: $(n\pi, 0)$
y-intercept: $(0, 0)$
Vertical asymptotes: $x = \frac{\pi}{2} + n\pi$
Odd function
Origin symmetry

Cosecant Function (*p. 447*)

$f(x) = \csc x$

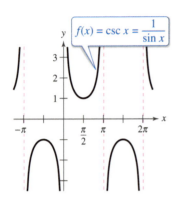

$f(x) = \csc x = \dfrac{1}{\sin x}$

Domain: $x \neq n\pi$
Range: $(-\infty, -1] \cup [1, \infty)$
Period: 2π
No intercepts
Vertical asymptotes: $x = n\pi$
Odd function
Origin symmetry

Secant Function (*p. 447*)

$f(x) = \sec x$

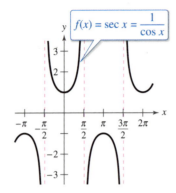

$f(x) = \sec x = \dfrac{1}{\cos x}$

Domain: $x \neq \dfrac{\pi}{2} + n\pi$

Range: $(-\infty, -1] \cup [1, \infty)$
Period: 2π
y-intercept: $(0, 1)$
Vertical asymptotes:
$$x = \dfrac{\pi}{2} + n\pi$$
Even function
y-axis symmetry

Cotangent Function (*p. 446*)

$f(x) = \cot x$

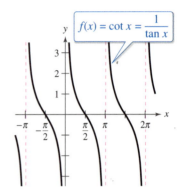

$f(x) = \cot x = \dfrac{1}{\tan x}$

Domain: $x \neq n\pi$
Range: $(-\infty, \infty)$
Period: π
x-intercepts: $\left(\dfrac{\pi}{2} + n\pi, 0\right)$

Vertical asymptotes: $x = n\pi$
Odd function
Origin symmetry

Inverse Sine Function (*p. 459*)

$f(x) = \arcsin x$

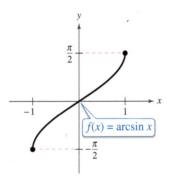

$f(x) = \arcsin x$

Domain: $[-1, 1]$
Range: $\left[-\dfrac{\pi}{2}, \dfrac{\pi}{2}\right]$
Intercept: $(0, 0)$
Odd function
Origin symmetry

Inverse Cosine Function (*p. 459*)

$f(x) = \arccos x$

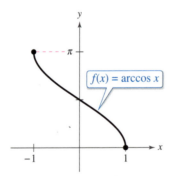

$f(x) = \arccos x$

Domain: $[-1, 1]$
Range: $[0, \pi]$
y-intercept: $\left(0, \dfrac{\pi}{2}\right)$

Inverse Tangent Function (*p. 459*)

$f(x) = \arctan x$

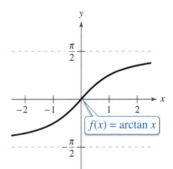

$f(x) = \arctan x$

Domain: $(-\infty, \infty)$
Range: $\left(-\dfrac{\pi}{2}, \dfrac{\pi}{2}\right)$
Intercept: $(0, 0)$
Horizontal asymptotes: $y = \pm\dfrac{\pi}{2}$

Odd function
Origin symmetry

Precalculus: Real Mathematics, Real People
Alternate Edition
Sixth Edition

Contents

$$f(x) = \frac{2x}{x - 3}$$

$$f(x) = \frac{2x}{x-3}$$

Preface

In the first five editions of this text, the subtitle was "*A Graphing Approach*." With the extensive revision to create the sixth edition, that subtitle no longer represents the text's essence. So, I have changed the title to "*Precalculus: Real Mathematics, Real People, Alternate Edition*." Both of the phrases "Real Mathematics" and "Real People" speak of a primary need in education today . . . the need for relevance. The mathematics in this text is both real and relevant. The people in this text are either already in or are preparing for real and relevant careers.

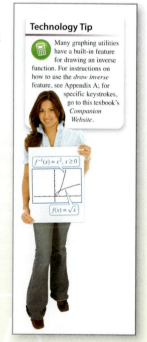

New to This Edition

NEW *Chapter Openers*
Each *Chapter Opener* highlights a real modeling data problem, showing a graph of the data, a section reference, and a short description of the data.

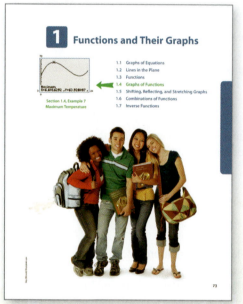

NEW *Explore the Concept*
Each *Explore the Concept* engages the student in active discovery of mathematical concepts, strengthens critical thinking skills, and helps build intuition.

NEW *What's Wrong?*
Each *What's Wrong?* points out common errors made using graphing utilities.

NEW *Vocabulary and Concept Check*
The *Vocabulary and Concept Check* appears at the beginning of the exercise set for each section. Each of these checks asks fill-in-the-blank, matching, and non-computational questions designed to help students learn mathematical terminology and to test basic understanding of that section's concepts.

NEW *Organization of Section Exercises*
The exercise sets are grouped into four categories: (1) *Vocabulary and Concept Check*, (2) *Procedures and Problem Solving*, (3) *Conclusions*, and (4) *Cumulative Mixed Review*. Many of the exercises are titled for easy reference.

Technology Tip

Many graphing utilities have a built-in feature for drawing an inverse function. For instructions on how to use the *draw inverse* feature, see Appendix A; for specific keystrokes, go to this texbook's *Companion Website*.

NEW *Algebraic-Graphical-Numerical Exercises*
These exercises allow students to solve a problem using multiple approaches—algebraic, graphical, and numerical. This helps students to see that a problem can be solved in more than one way and to see that different methods yield the same result.

NEW *Modeling Data Exercises*
These multi-part applications that involve real-life data offer students the opportunity to generate and analyze mathematical models.

NEW *Capstone Exercise*

The *Capstone* is a conceptual problem that synthesizes key topics and provides students with a better understanding of the concepts in a section. This exercise is excellent for classroom discussion or test preparation. I recommend it.

NEW *Why You Should Learn It Exercise*

An engaging real-life application of the concepts of the section. This application exercise is typically described in the section opener as a motivator for the section.

NEW *Error Analysis Exercise*

This exercise presents a sample solution that contains a common error which the students are asked to identify.

130. CAPSTONE Decide whether the two functions shown in each graph appear to be inverse functions of each other. Explain your reasoning.

REVISED *Library of Parent Functions*

To facilitate familiarity with the basic functions, I have compiled several elementary and nonelementary functions as a *Library of Parent Functions*. Each function is introduced at the first point of use in the text with a definition and description of basic characteristics. New to this edition are the *Library of Parent Functions Examples*, which are identified in the title of the example, and the *Review of Library of Parent Functions* after Chapter 5. A summary of the functions is presented on the inside front cover of this text.

REVISED *Side-by-Side Examples*

Many examples present side-by-side solutions with multiple approaches—algebraic, graphical, and numerical. I have revised the graphical solutions to be more visual. Often, the algebraic solution is formal, with step-by-step work. As a change in this edition, the graphical solution is often more of a visual check to ensure the reasonableness of the solution obtained algebraically.

REVISED *Exercise Sets*

The exercise sets have been carefully and extensively examined to ensure they are rigorous, relevant, and cover all topics suggested by our users. Many new skill building and challenging exercises have been added.

REVISED *Chapter Summaries*

The Chapter Summary now includes explanations and examples of the objectives taught in the chapter.

Calc Chat

For the past several years, we have maintained an independent website—**CalcChat.com**—that provides free solutions to all odd-numbered exercises in the text. Thousands of students using our textbooks have visited the site for practice and help with their homework. For the Sixth Edition, I was able to use information from *CalcChat.com*, including which solutions students accessed most often, to help guide the revision of the exercises.

Table of Contents Changes

In Chapter 9 (Sequences, Series, and Probability), old Section 9.4 (Mathematical Induction) is now Appendix F and has been moved to *CengageBrain.com*. We have moved Appendix B (Concepts in Statistics), Appendix C (Variation), Appendix D (Solving Linear Equations and Inequalities), and Appendix E (Systems of Inequalities) to *CengageBrain.com*.

$$f(x) = \frac{2x}{x-3}$$

Trusted Features

What you should learn/Why you should learn it

These summarize important topics in the section and why they are important in math and life.

Study Tip

These hints and tips can be used to reinforce concepts, help students learn how to study mathematics, caution students about common errors, or show alternative or additional steps to a solution of an example.

Technology Tip

Technology Tips provide graphing calculator tips or provide alternative methods of solving a problem using a graphing utility.

Checkpoint

Checkpoints guide students to an odd-numbered exercise that is similar to the example they just read.

Algebra of Calculus

Throughout the text, special emphasis is given to the algebraic techniques used in calculus. *Algebra of Calculus* examples and exercises are integrated throughout the text and are identified by the symbol $\int$.

What you should learn
- Find the slopes of lines.
- Write linear equations given points on lines and their slopes.
- Use slope-intercept forms of linear equations to sketch lines.
- Use slope to identify parallel and perpendicular lines.

Why you should learn it
The slope of a line can be used to solve real-life problems. For instance, in Exercise 97 on page 95, you will use a linear equation to model student enrollment at Penn State University.

Instructor Resources

Printed

Complete Solutions Manual
ISBN-10: 1-111-58217-3; ISBN-13: 978-1-111-58217-3
The CSM includes complete solutions for every exercise in the text.

Media

Enhanced WebAssign
ISBN-10: 0-538-73810-3; ISBN-13: 978-0-538-73810-1
Exclusively from Cengage Learning, Enhanced WebAssign® offers an extensive online program for Precalculus to encourage the practice that is so critical for concept mastery. The meticulously crafted pedagogy and exercises in our proven texts become even more effective in Enhanced WebAssign, supplemented by multimedia tutorial support and immediate feedback as students complete their assignments.
Key features include:
- Read It eBook pages, Watch It videos, Master It tutorials, and Chat About It links
- As many as 4000 homework problems that match your textbook's end-of-section exercises
- New! Premium eBook with highlighting, note-taking, and search features, as well as links to multimedia resources
- New! Personal Study Plans (based on diagnostic quizzing) that identify chapter topics that students still need to master
- Algorithmic problems, allowing you to assign unique versions to each student
- Practice Another Version feature (activated at the instructor's discretion) allows students to attempt the questions with new sets of values until they feel confident enough to work the original problem
- Graphing utility tool enables students to graph lines, segments, parabolas, and circles as they answer questions
- MathPad simplifies the input of mathematical symbols

ExamView Computerized Testing

ExamView testing software allows instructors to quickly create, deliver, and customize tests for class in print and online formats, and features automatic grading. This software includes a test bank with hundreds of questions customized directly to the text. ExamView is available within the PowerLecture CD-ROM.

Solution Builder

www.cengage.com/solutionbuilder

This online instructor database offers complete worked-out solutions to all exercises in the text, allowing you to create customized, secure solutions printouts (in PDF format) matched exactly to the problems you assign in class.

PowerLecture with ExamView

ISBN-10: 1-111-58223-8; ISBN-13: 978-1-111-58223-4

This CD-ROM provides the instructor with dynamic media tools for teaching. Create, deliver, and customize tests (both print and online) in minutes with ExamView® Computerized Testing Featuring Algorithmic Equations. Easily build solution sets for homework or exams using Solution Builder's online solutions manual. Microsoft® PowerPoint® lecture slides and figures from the book are also included on this CD-ROM.

CengageBrain.com

To access additional course materials and companion resources, please visit *www.CengageBrain.com*. At the CengageBrain.com home page, search for the ISBN of your title (from the back cover of your book) using the search box at the top of the page. This will take you to the product page where free companion resources can be found.

Student Resources

Printed

Student Solutions Manual

ISBN-10: 1-111-58227-0; ISBN-13: 978-1-111-58227-2

This manual offers step-by-step solutions for all odd-numbered text exercises as well as Chapter and Cumulative Tests. In addition to Chapter and Cumulative Tests, the manual also provides practice tests and practice test answers.

Media

Enhanced WebAssign

ISBN-10: 0-538-73810-3; ISBN-13: 978-0-538-73810-1

Enhanced WebAssign is designed for you to do your homework online. This proven and reliable system uses pedagogy and content found in this text, and then enhances it to help you learn Precalculus more effectively. Automatically graded homework allows you to focus on your learning and get interactive study assistance outside of class.

Text-Specific DVD

ISBN-10: 1-111-58226-2; ISBN-13: 978-1-111-58226-5

These text-specific DVDs cover all sections of the text and provide key explanations of key concepts, examples, exercises, and applications in a lecture-based format.

CengageBrain.com

To access additional course materials and companion resources, please visit *www.CengageBrain.com*. At the CengageBrain.com home page, search for the ISBN of your title (from the back cover of your book) using the search box at the top of the page. This will take you to the product page where free companion resources can be found.

Acknowledgments

I would like to thank my colleagues who have helped me develop this program. Their encouragement, criticisms, and suggestions have been invaluable to me.

Reviewers

Dave Bregenzer, *Utah State University*
Beth Burns, *Bowling Green State University*
Stephen Nicoloff, *Paradise Valley Community College*
Sandra Poinsett, *College of Southern Maryland*
Abdallah Shuaibi, *Truman College*
Diane Veneziale, *Burlington County College*
Ellen Vilas, *York Technical College*
Rich West, *Francis Marion University*
Vanessa White, *Southern University*
Paul Winterbottom, *Montgomery County Community College*
Cathleen Zucco-Teveloff, *Rowan University*

I would also like to thank the following reviewers, who have given me many useful insights to this and previous editions.

Tony Homayoon Akhlaghi, *Bellevue Community College;* Daniel D. Anderson, *University of Iowa;* Bruce Armbrust, *Lake Tahoe Community College;* Jamie Whitehead Ashby, *Texarkana College;* Teresa Barton, *Western New England College;* Kimberly Bennekin, *Georgia Perimeter College;* Charles M. Biles, *Humboldt State University;* Phyllis Barsch Bolin, *Oklahoma Christian University;* Khristo Boyadzhiev, *Ohio Northern University;* Dave Bregenzer, *Utah State University;* Anne E. Brown, *Indiana University-South Bend;* Diane Burleson, *Central Piedmont Community College;* Alexander Burstein, *University of Rhode Island;* Marilyn Carlson, *University of Kansas;* Victor M. Cornell, *Mesa Community College;* John Dersh, *Grand Rapids Community College;* Jennifer Dollar, *Grand Rapids Community College;* Marcia Drost, *Texas A & M University;* Cameron English, *Rio Hondo College;* Susan E. Enyart, *Otterbein College;* Patricia J. Ernst, *St. Cloud State University;* Eunice Everett, *Seminole Community College;* Kenny Fister, *Murray State University;* Susan C. Fleming, *Virginia Highlands Community College;* Jeff Frost, *Johnson County Community College;* James R. Fryxell, *College of Lake County;* Khadiga H. Gamgoum, *Northern Virginia Community College;* Nicholas E. Geller, *Collin County Community College;* Betty Givan, *Eastern Kentucky University;* Patricia K. Gramling, *Trident Technical College;* Michele Greenfield, *Middlesex County College;* Bernard Greenspan, *University of Akron;* Zenas Hartvigson, *University of Colorado at Denver;* Rodger Hergert, *Rock Valley College;* Allen Hesse, *Rochester Community College;* Rodney Holke-Farnam, *Hawkeye Community College;* Lynda Hollingsworth, *Northwest Missouri State University;* Jean M. Horn, *Northern Virginia Community College;* Spencer Hurd, *The Citadel;* Bill Huston, *Missouri Western State College;* Deborah Johnson, *Cambridge South Dorchester High School;* Francine Winston Johnson, *Howard Community College;* Luella Johnson, *State University of New York, College at Buffalo;* Susan Kellicut, *Seminole Community College;* John Kendall, *Shelby State Community College;* Donna M. Krawczyk, *University of Arizona;* Peter A. Lappan, *Michigan State University;* Charles G. Laws, *Cleveland State Community College;* JoAnn Lewin, *Edison Community College;* Richard J. Maher, *Loyola University;* Carl Main, *Florida College;* Marilyn McCollum, *North Carolina State University;* Judy McInerney, *Sandhills Community College;* David E. Meel, *Bowling Green University;* Beverly Michael, *University of Pittsburgh;* Roger B. Nelsen, *Lewis and Clark College;* Jon Odell, *Richland Community College;* Paul Oswood, *Ridgewater College;* Wing M. Park, *College of Lake County;* Rupa M. Patel, *University of Portland;* Robert Pearce, *South Plains College;* David R. Peterson, *University of Central Arkansas;* James Pommersheim, *Reed College;* Antonio Quesada, *University of Akron;* Laura Reger, *Milwaukee Area Technical College;* Jennifer Rhinehart, *Mars Hill College;* Lila F. Roberts, *Georgia Southern University;* Keith Schwingendorf, *Purdue University North Central;* George W. Shultz, *St. Petersburg Junior College;* Stephen Slack, *Kenyon College;* Judith Smalling, *St. Petersburg Junior College;* Pamela K. M. Smith, *Fort Lewis College;* Cathryn U. Stark, *Collin County Community College;* Craig M. Steenberg, *Lewis-Clark State College;* Mary Jane Sterling, *Bradley University;* G. Bryan Stewart, *Tarrant County Junior College;* Mahbobeh Vezvaei, *Kent State University;* Ellen Vilas, *York Technical College;* Hayat Weiss, *Middlesex Community College;* Howard L. Wilson, *Oregon State University;* Joel E. Wilson, *Eastern Kentucky University;* Michelle Wilson, *Franklin University;* Fred Worth, *Henderson State University;* Karl M. Zilm, *Lewis and Clark Community College*

I hope that you enjoy learning the mathematics presented in this text. More than that, I hope you gain a new appreciation for the relevance of mathematics to careers in science, technology, business, and medicine.

My thanks to Robert Hostetler, The Behrend College, The Pennsylvania State University, Bruce Edwards, University of Florida, and David Heyd, The Behrend College, The Pennsylvania State University, for their significant contributions to previous editions of this text.

I would also like to thank the staff of Larson Texts, Inc. who assisted in preparing the manuscript, rendering the art package, and typesetting and proofreading the pages and supplements.

On a personal level, I am grateful to my spouse, Deanna Gilbert Larson, for her love, patience, and support. Also, a special thanks goes to R. Scott O'Neil.

If you have suggestions for improving this text, please feel free to write me. Over the past two decades I have received many useful comments from both instructors and students, and I value these very much.

Ron Larson, Ph.D.
Professor of Mathematics
Penn State University
www.RonLarson.com

Prerequisites

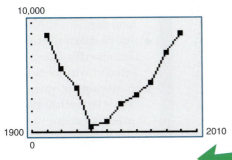

Section P.6, Example 5
U.S. Immigrants

P.1   Real Numbers

Real Numbers

Real numbers are used in everyday life to describe quantities such as age, miles per gallon, and population. Real numbers are represented by symbols such as

$$-5, \quad 9, \quad 0, \quad \tfrac{4}{3}, \quad 0.666\ldots, \quad 28.21, \quad \sqrt{2}, \quad \pi, \quad \text{and} \quad \sqrt[3]{-32}.$$

Here are some important **subsets** (each member of subset B is also a member of set A) of the set of real numbers.

$$\{1, 2, 3, 4, \ldots\} \qquad \text{Set of natural numbers}$$
$$\{0, 1, 2, 3, 4, \ldots\} \qquad \text{Set of whole numbers}$$
$$\{\ldots, -3, -2, -1, 0, 1, 2, 3, \ldots\} \qquad \text{Set of integers}$$

A real number is **rational** when it can be written as the ratio p/q of two integers, where $q \neq 0$. For instance, the numbers

$$\frac{1}{3} = 0.3333\ldots = 0.\overline{3}, \quad \frac{1}{8} = 0.125, \quad \text{and} \quad \frac{125}{111} = 1.126126\ldots = 1.\overline{126}$$

are rational. The decimal representation of a rational number either *repeats* (as in $\frac{173}{55} = 3.1\overline{45}$) or *terminates* (as in $\frac{1}{2} = 0.5$). A real number that cannot be written as the ratio of two integers is called **irrational.** Irrational numbers have infinite nonrepeating decimal representations. For instance, the numbers

$$\sqrt{2} = 1.4142135\ldots \approx 1.41$$

and

$$\pi = 3.1415926\ldots \approx 3.14$$

are irrational. (The symbol $\approx$ means "is approximately equal to.") Figure P.1 shows subsets of real numbers and their relationships to each other.

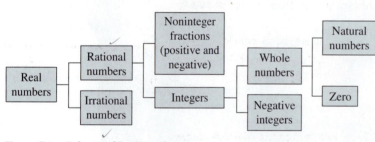

Figure P.1   Subsets of Real Numbers

Real numbers are represented graphically by a **real number line.** The point 0 on the real number line is the **origin.** Numbers to the right of 0 are positive and numbers to the left of 0 are negative, as shown in Figure P.2. The term **nonnegative** describes a number that is either positive or zero.

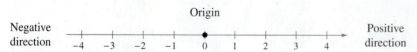

Figure P.2   The Real Number Line

There is a *one-to-one correspondence* between real numbers and points on the real number line. That is, every point on the real number line corresponds to exactly one real number, called its **coordinate,** and every real number corresponds to exactly one point on the real number line, as shown in Figure P.3.

What you should learn

- Represent and classify real numbers.
- Order real numbers and use inequalities.
- Find the absolute values of real numbers and the distance between two real numbers.
- Evaluate algebraic expressions and use the basic rules and properties of algebra.

Why you should learn it

Real numbers are used in every aspect of our lives, such as finding the surplus or deficit in the federal budget. See Exercises 91–96 on page 10.

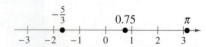

Every point on the real number line corresponds to exactly one real number.

Every real number corresponds to exactly one point on the real number line.

Figure P.3   One-to-One Correspondence

Ordering Real Numbers

One important property of real numbers is that they are **ordered.**

> **Definition of Order on the Real Number Line**
>
> If a and b are real numbers, then a is **less than** b when $b - a$ is positive. This order is denoted by the **inequality** $a < b$. This relationship can also be described by saying that b is **greater than** a and writing $b > a$. The inequality $a \leq b$ means that a is **less than or equal to** b, and the inequality $b \geq a$ means that b is **greater than or equal to** a. The symbols $<$, $>$, $\leq$, and $\geq$ are **inequality symbols.**

Geometrically, this definition implies that $a < b$ if and only if a lies to the *left* of b on the real number line, as shown in Figure P.4.

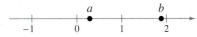

Figure P.4 $a < b$ *if and only if a lies to the left of b.*

Example 1   Interpreting Inequalities

Describe the subset of real numbers represented by each inequality.

a. $x \leq 2$

b. $x > -1$

c. $-2 \leq x < 3$

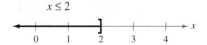

Figure P.5

Solution

a. The inequality $x \leq 2$ denotes all real numbers less than or equal to 2, as shown in Figure P.5.

b. The inequality $x > -1$ denotes all real numbers greater than -1, as shown in Figure P.6.

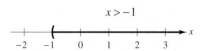

Figure P.6

c. The inequality $-2 \leq x < 3$ means that $x \geq -2$ *and* $x < 3$. The "double inequality" denotes all real numbers between -2 and 3, including -2 but not including 3, as shown in Figure P.7.

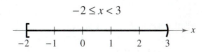

Figure P.7

✔CHECKPOINT   Now try Exercise 43(a).

Inequalities can be used to describe subsets of real numbers called **intervals.** In the bounded intervals below, the real numbers a and b are the **endpoints** of each interval.

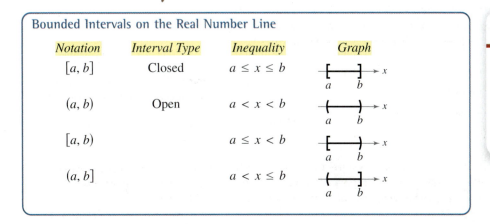

> **Bounded Intervals on the Real Number Line**
>
> | Notation | Interval Type | Inequality | Graph |
> |----------|---------------|------------|-------|
> | $[a, b]$ | Closed | $a \leq x \leq b$ | |
> | (a, b) | Open | $a < x < b$ | |
> | $[a, b)$ | | $a \leq x < b$ | |
> | $(a, b]$ | | $a < x \leq b$ | |

> **Study Tip**
>
> The endpoints of a closed interval are included in the interval. The endpoints of an open interval are *not* included in the interval.

The symbols ∞, **positive infinity,** and −∞, **negative infinity,** do not represent real numbers. They are simply convenient symbols used to describe the unboundedness of an interval such as

$$(1, \infty) \quad \text{or} \quad (-\infty, 3].$$

Unbounded Intervals on the Real Number Line

Notation	Interval Type	Inequality	Graph
$[a, \infty)$		$x \geq a$	
(a, ∞)	Open	$x > a$	
$(-\infty, b]$		$x \leq b$	
$(-\infty, b)$	Open	$x < b$	
$(-\infty, \infty)$	Entire real line	$-\infty < x < \infty$	

Study Tip

An interval is unbounded when it continues indefinitely in one or both directions.

Example 2   Using Inequalities to Represent Intervals

Use inequality notation to describe each of the following.

a. c is at most 2.

b. All x in the interval $(-3, 5]$

c. t is at least 4, but less than 11.

Solution

a. The statement "c is at most 2" can be represented by $c \leq 2$.

b. "All x in the interval $(-3, 5]$" can be represented by $-3 < x \leq 5$.

c. The statement "t is at least 4, but less than 11" can be represented by $4 \leq t < 11$.

✓**CHECKPOINT**   Now try Exercise 47.

Example 3   Interpreting Intervals

Give a verbal description of each interval.

a. $(-1, 0)$

b. $[2, \infty)$

c. $(-\infty, 0)$

Solution

a. This interval consists of all real numbers that are greater than -1 and less than 0.

b. This interval consists of all real numbers that are greater than or equal to 2.

c. This interval consists of all negative real numbers.

✓**CHECKPOINT**   Now try Exercise 61.

The **Law of Trichotomy** states that for any two real numbers a and b, precisely one of three relationships is possible:

$$a = b, \quad a < b, \quad \text{or} \quad a > b. \qquad \text{Law of Trichotomy}$$

Absolute Value and Distance

The **absolute value** of a real number is its *magnitude*, or the distance between the origin and the point representing the real number on the real number line.

> **Definition of Absolute Value**
>
> If a is a real number, then the **absolute value** of a is
>
> $$|a| = \begin{cases} a, & a \geq 0 \\ -a, & a < 0 \end{cases}.$$

Notice from this definition that the absolute value of a real number is never negative. For instance, if $a = -5$, then

$$|-5| = -(-5) = 5.$$

The absolute value of a real number is either positive or zero. Moreover, 0 is the only real number whose absolute value is 0. So, $|0| = 0$.

Example 4   Evaluating the Absolute Value of a Number

Evaluate $\dfrac{|x|}{x}$ for (a) $x > 0$ and (b) $x < 0$.

Solution

a. If $x > 0$, then $|x| = x$ and $\dfrac{|x|}{x} = \dfrac{x}{x} = 1$.

b. If $x < 0$, then $|x| = -x$ and $\dfrac{|x|}{x} = \dfrac{-x}{x} = -1$.

 CHECKPOINT   Now try Exercise 67.

> **Properties of Absolute Value**
>
> **1.** $|a| \geq 0$ **2.** $|-a| = |a|$
>
> **3.** $|ab| = |a||b|$ **4.** $\left|\dfrac{a}{b}\right| = \dfrac{|a|}{|b|}, \quad b \neq 0$

Absolute value can be used to define the distance between two points on the real number line. For instance, the distance between -3 and 4 is

$$|-3 - 4| = |-7| = 7$$

as shown in Figure P.8.

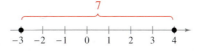

Figure P.8   *The distance between -3 and 4 is 7.*

> **Distance Between Two Points on the Real Number Line**
>
> Let a and b be real numbers. The **distance between a and b** is
>
> $$d(a, b) = |b - a| = |a - b|.$$

Algebraic Expressions and the Basic Rules of Algebra

One characteristic of algebra is the use of letters to represent numbers. The letters are **variables,** and combinations of letters and numbers are **algebraic expressions.** Here are a few examples of algebraic expressions.

$$5x, \quad 2x - 3, \quad \frac{4}{x^2 + 2}, \quad 7x + y$$

Definition of an Algebraic Expression

An **algebraic expression** is a combination of letters (**variables**) and real numbers (**constants**) combined using the operations of addition, subtraction, multiplication, division, and exponentiation.

The **terms** of an algebraic expression are those parts that are separated by *addition.* For example,

$$x^2 - 5x + 8 = x^2 + (-5x) + 8$$

has three terms: x^2 and $-5x$ are the **variable terms** and 8 is the **constant term.** The numerical factor of a term is called the **coefficient.** For instance, the coefficient of $-5x$ is -5, and the coefficient of x^2 is 1.

To **evaluate** an algebraic expression, substitute numerical values for each of the variables in the expression. Here are three examples.

Expression	Value of Variable	Substitute	Value of Expression
$-3x + 5$	$x = 3$	$-3(3) + 5$	$-9 + 5 = -4$
$3x^2 + 2x - 1$	$x = -1$	$3(-1)^2 + 2(-1) - 1$	$3 - 2 - 1 = 0$
$\dfrac{2x}{x + 1}$	$x = -3$	$\dfrac{2(-3)}{-3 + 1}$	$\dfrac{-6}{-2} = 3$

When an algebraic expression is evaluated, the Substitution Principle is used. It states, "If $a = b$, then a can be replaced by b in any expression involving a." In the first evaluation shown above, for instance, 3 is substituted for x in the expression $-3x + 5$.

There are four arithmetic operations with real numbers: addition, multiplication, subtraction, and division, denoted by the symbols

$$+, \quad \times \text{ or } \cdot, \quad -, \quad \text{and} \quad \div \text{ or } /.$$

Of these, addition and multiplication are the two primary operations. Subtraction and division are the inverse operations of addition and multiplication, respectively.

Subtraction: Add the opposite of b.

$$a - b = a + (-b)$$

Division: Multiply by the reciprocal of b.

$$\text{If } b \neq 0, \text{ then } a/b = a\left(\frac{1}{b}\right) = \frac{a}{b}.$$

In these definitions, $-b$ is the **additive inverse** (or opposite) of b, and

$$1/b$$

is the **multiplicative inverse** (or reciprocal) of b. In the fractional form

$$a/b$$

a is the **numerator** of the fraction and b is the **denominator.**

Because the properties of real numbers below are true for variables and algebraic expressions, as well as for real numbers, they are often called the **Basic Rules of Algebra.** Try to formulate a verbal description of each property. For instance, the Commutative Property of Addition states that *the order in which two real numbers are added does not affect their sum.*

Basic Rules of Algebra

Let a, b, and c be real numbers, variables, or algebraic expressions.

	Property	*Example*
Commutative Property of Addition:	$a + b = b + a$	$4x + x^2 = x^2 + 4x$
Commutative Property of Multiplication:	$ab = ba$	$(1 - x)x^2 = x^2(1 - x)$
Associative Property of Addition:	$(a + b) + c = a + (b + c)$	$(x + 5) + x^2 = x + (5 + x^2)$
Associative Property of Multiplication:	$(ab)c = a(bc)$	$(2x \cdot 3y)(8) = (2x)(3y \cdot 8)$
Distributive Properties:	$a(b + c) = ab + ac$	$3x(5 + 2x) = 3x \cdot 5 + 3x \cdot 2x$
	$(a + b)c = ac + bc$	$(y + 8)y = y \cdot y + 8 \cdot y$
Additive Identity Property:	$a + 0 = a$	$5y^2 + 0 = 5y^2$
Multiplicative Identity Property:	$a \cdot 1 = a$	$(4x^2)(1) = 4x^2$
Additive Inverse Property:	$a + (-a) = 0$	$6x^3 + (-6x^3) = 0$
Multiplicative Inverse Property:	$a \cdot \dfrac{1}{a} = 1, \quad a \neq 0$	$(x^2 + 4)\left(\dfrac{1}{x^2 + 4}\right) = 1$

Because subtraction is defined as "adding the opposite," the Distributive Properties are also true for subtraction. For instance, the "subtraction form" of $a(b + c) = ab + ac$ is written as

$$a(b - c) = ab - ac.$$

Properties of Negation and Equality

Let a, b, and c be real numbers, variables, or algebraic expressions.

Property	*Example*
1. $(-1)a = -a$	$(-1)7 = -7$
2. $-(-a) = a$	$-(-6) = 6$
3. $(-a)b = -(ab) = a(-b)$	$(-5)3 = -(5 \cdot 3) = 5(-3)$
4. $(-a)(-b) = ab$	$(-2)(-x) = 2x$
5. $-(a + b) = (-a) + (-b)$	$-(x + 8) = (-x) + (-8) = -x - 8$
6. If $a = b$, then $a + c = b + c$.	$\frac{1}{2} + 3 = 0.5 + 3$
7. If $a = b$, then $ac = bc$.	$4^2(2) = 16(2)$
8. If $a + c = b + c$, then $a = b$.	$1.4 - 1 = \frac{7}{5} - 1$
9. If $ac = bc$ and $c \neq 0$, then $a = b$.	$\dfrac{3}{4} = \dfrac{\sqrt{9}}{4}$

Study Tip

Be sure you see the difference between the *opposite of a number* and a *negative number.* If a is already negative, then its opposite, $-a$, is positive. For instance, if $a = -2$, then $-a = -(-2) = 2$.

Properties of Zero

Let a and b be real numbers, variables, or algebraic expressions.

1. $a + 0 = a$ and $a - 0 = a$ **2.** $a \cdot 0 = 0$

3. $\dfrac{0}{a} = 0$, $a \neq 0$ **4.** $\dfrac{a}{0}$ is undefined.

5. Zero-Factor Property: If $ab = 0$, then $a = 0$ or $b = 0$.

The "or" in the Zero-Factor Property includes the possibility that either or both factors may be zero. This is an **inclusive or,** and it is the way the word "or" is generally used in mathematics.

Properties and Operations of Fractions

Let a, b, c, and d be real numbers, variables, or algebraic expressions such that $b \neq 0$ and $d \neq 0$.

1. Equivalent Fractions: $\dfrac{a}{b} = \dfrac{c}{d}$ if and only if $ad = bc$.

2. Rules of Signs: $-\dfrac{a}{b} = \dfrac{-a}{b} = \dfrac{a}{-b}$ and $\dfrac{-a}{-b} = \dfrac{a}{b}$

3. Generate Equivalent Fractions: $\dfrac{a}{b} = \dfrac{ac}{bc}$, $c \neq 0$

4. Add or Subtract with Like Denominators: $\dfrac{a}{b} \pm \dfrac{c}{b} = \dfrac{a \pm c}{b}$

5. Add or Subtract with Unlike Denominators: $\dfrac{a}{b} \pm \dfrac{c}{d} = \dfrac{ad \pm bc}{bd}$

6. Multiply Fractions: $\dfrac{a}{b} \cdot \dfrac{c}{d} = \dfrac{ac}{bd}$

7. Divide Fractions: $\dfrac{a}{b} \div \dfrac{c}{d} = \dfrac{a}{b} \cdot \dfrac{d}{c} = \dfrac{ad}{bc}$, $c \neq 0$

Study Tip

In Property 1 of fractions, the phrase "if and only if" implies two statements. One statement is: If $a/b = c/d$, then $ad = bc$. The other statement is: If $ad = bc$, where $b \neq 0$ and $d \neq 0$, then $a/b = c/d$.

Example 5   Properties and Operations of Fractions

a. $\dfrac{x}{3} + \dfrac{2x}{5} = \dfrac{5 \cdot x + 3 \cdot 2x}{15} = \dfrac{11x}{15}$ Add fractions with unlike denominators.

b. $\dfrac{7}{x} \div \dfrac{3}{2} = \dfrac{7}{x} \cdot \dfrac{2}{3} = \dfrac{14}{3x}$ Divide fractions.

✓**CHECKPOINT**   Now try Exercise 125.

If a, b, and c are integers such that $ab = c$, then a and b are **factors** or **divisors** of c. A **prime number** is an integer that has exactly two positive factors: itself and 1. For example, 2, 3, 5, 7, and 11 are prime numbers. The numbers 4, 6, 8, 9, and 10 are **composite** because they can be written as the product of two or more prime numbers. The number 1 is neither prime nor composite. The **Fundamental Theorem of Arithmetic** states that every positive integer greater than 1 can be written as the product of prime numbers. For instance, the **prime factorization** of 24 is

$24 = 2 \cdot 2 \cdot 2 \cdot 3.$

P.1   Exercises

See www.CalcChat.com for worked-out solutions to odd-numbered exercises.
For instructions on how to use a graphing utility, see Appendix A.

Vocabulary and Concept Check

In Exercises 1–5, fill in the blank(s).

1. A real number is _rational_ when it can be written as the ratio $\dfrac{p}{q}$ of two integers, where $q \neq 0$.

2. _Irrational_ numbers have infinite nonrepeating decimal representations.

3. A _prime_ number is an integer with exactly two positive factors: itself and 1.

4. An algebraic expression is a combination of letters called _variables_ and real _algebraic expressions_ numbers called _constants_

5. The _term_ of an algebraic expression are those parts separated by addition.

6. Is $|5 - 2| = |2 - 5|$? _yes it is_

In Exercises 7–12, match each property with its name.

7. Commutative Property of Addition   c   (a) $a \cdot 1 = a$

8. Associative Property of Multiplication   d   (b) $a(b + c) = ab + ac$

9. Additive Inverse Property   c   (c) $a + b = b + a$

10. Distributive Property   b   (d) $(ab)c = a(bc)$

11. Associative Property of Addition   f   (e) $a + (-a) = 0$

12. Multiplicative Identity Property   a   (f) $(a + b) + c = a + (b + c)$

Procedures and Problem Solving

Identifying Subsets of Real Numbers   In Exercises 13–18, determine which numbers are (a) natural numbers, (b) whole numbers, (c) integers, (d) rational numbers, and (e) irrational numbers.

13. $\left\{ -9, -\frac{7}{2}, 5, \frac{2}{3}, \sqrt{2}, 0, 1, -4, -1 \right\}$

14. $\left\{ \sqrt{5}, -7, -\frac{7}{3}, 0, 3.12, \frac{5}{4}, -2, -8, 3 \right\}$ ✓

15. $\left\{ 2.01, 0.666\ldots, -13, 0.010110111\ldots, 1, -10, 20 \right\}$

16. $\left\{ 2.3030030003\ldots, 0.7575, -4.63, \sqrt{10}, -2, 0.3, 8 \right\}$

17. $\left\{ -\pi, -\frac{1}{3}, \frac{6}{3}, \frac{1}{2}\sqrt{2}, -7.5, -2, 3, -3 \right\}$

18. $\left\{ 25, -17, -\frac{12}{5}, \sqrt{9}, 3.12, \frac{1}{2}\pi, 6, -4, 18 \right\}$

Finding the Decimal Form of a Rational Number   In Exercises 19–24, use a calculator to find the decimal form of the rational number. If it is a nonterminating decimal, write the repeating pattern.

19. $\frac{5}{16}$

20. $\frac{17}{4}$

21. $\frac{41}{333}$

22. $\frac{3}{7}$

23. $-\frac{100}{11}$

24. $-\frac{218}{33}$

Writing a Decimal as a Fraction   In Exercises 25–28, use a calculator to rewrite the rational number as the ratio of two integers.

25. 4.6

26. 12.3

27. -6.5

28. -1.83

Writing an Inequality   In Exercises 29 and 30, approximate the numbers and place the correct inequality symbol (< or >) between them.

29.

30.

Plotting Real Numbers   In Exercises 31–36, plot the two real numbers on the real number line. Then place the correct inequality symbol (< or >) between them.

31. $-4, -8$

32. $-3.5, 1$

33. $\frac{3}{2}, 7$

34. $1, \frac{16}{3}$

35. $\frac{5}{6}, \frac{2}{3}$

36. $-\frac{8}{7}, -\frac{3}{7}$

Interpreting Inequalities   In Exercises 37–44, (a) verbally describe the subset of real numbers represented by the inequality, (b) sketch the subset on the real number line, and (c) state whether the interval is bounded or unbounded.

37. $x \leq 5$

38. $x > 3$

39. $x < 0$

40. $x \geq 4$

41. $-2 < x < 2$

42. $0 \leq x \leq 5$

✓ 43. $-1 \leq x < 0$

44. $0 < x \leq 6$

Using Inequalities to Represent Intervals　In Exercises 45–52, use inequality and interval notation to describe the set.

45. x is negative.　　**46.** z is at least 10.

✓ **47.** y is nonnegative.　　**48.** y is no more than 25.

49. t is at least 9 and at most 24.

50. k is less than 3 but no less than -1.

51. x is greater than -2 and at most 5.

52. The inflation rate r is at least 2.5% and at most 5%.

Using Interval Notation　In Exercises 53–58, use interval notation to describe the graph.

53.

![number line from -2 to 6, bracket at -1 to bracket at 5]

54.

![number line from -3 to 5, parenthesis at -2 to parenthesis at 4]

55.

![number line from -4 to 10, bracket at -4 to parenthesis at 8]

56.

![number line from -4 to 6, parenthesis at -4 to bracket at 4]

57.

![number line, bracket at -a to bracket at a+4]

58.

![number line, parenthesis at -c+2 to parenthesis at c+1]

Interpreting Intervals　In Exercises 59–62, give a verbal description of the interval.

59. $(-6, \infty)$　　**60.** $(-\infty, 4]$

✓ **61.** $(-\infty, 2]$　　**62.** $[1, \infty)$

Evaluating the Absolute Value of a Number　In Exercises 63–68, evaluate the expression.

63. $|-10|$　　**64.** $|0|$

65. $-3 - |-3|$　　**66.** $|-1| - |-2|$

✓ **67.** $\dfrac{|x + 2|}{x + 2}$　　**68.** $\dfrac{|x - 1|}{x - 1}$

Evaluating an Expression　In Exercises 69–74, evaluate the expression for the given values of x and y. Then use a graphing utility to verify your result.

69. $|2x + y|$ for $x = -2$ and $y = 3$

70. $|y - 4x|$ for $x = 2$ and $y = -3$

71. $|x| - 2|y|$ for $x = -2$ and $y = -1$

72. $|2x| - 3|y|$ for $x = -4$ and $y = 1$

73. $\left|\dfrac{3x + 2y}{|x|}\right|$ for $x = 4$ and $y = 1$

74. $\left|\dfrac{3|x| - 2y}{2x + y}\right|$ for $x = -2$ and $y = -4$

Writing an Inequality　In Exercises 75–80, place the correct symbol ($<$, $>$, or $=$) between the pair of real numbers.

75. $|-3|$ ▢ $-|-3|$　　**76.** $|-4|$ ▢ $|4|$

77. -5 ▢ $-|5|$　　**78.** $-|-6|$ ▢ $|-6|$

79. $-|-2|$ ▢ $-|2|$　　**80.** $-(-2)$ ▢ -2

Finding the Distance Between Two Real Numbers　In Exercises 81–86, find the distance between a and b.

81. $a = 126, b = 75$　　**82.** $a = -126, b = -75$

83. $a = -\frac{5}{2}, b = 0$　　**84.** $a = \frac{1}{4}, b = \frac{11}{4}$

85. $a = \frac{16}{5}, b = \frac{112}{75}$

86. $a = 9.34, b = -5.65$

Using Absolute Value Notation　In Exercises 87–90, use absolute value notation to describe the situation.

87. The distance between x and 5 is no more than 3.

88. The distance between x and -10 is at least 6.

89. y is at least six units from 0.

90. y is at most three units from a.

Why you should learn it　*(p. 2)*　In Exercises 91–96, use the bar graph, which shows the receipts of the federal government (in billions of dollars) for selected years from 1998 through 2008. In each exercise, you are given the expenditures of the federal government. Find the magnitude of the surplus or deficit for the year.　(Source: U.S. Office of Management and Budget)

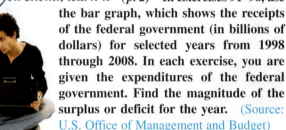

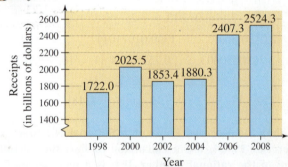

	Receipts	Expenditures	$\lvert$Receipts $-$ Expenditures$\rvert$
91. 1998	▢	$1652.7 billion	▢
92. 2000	▢	$1789.2 billion	▢
93. 2002	▢	$2011.2 billion	▢
94. 2004	▢	$2293.0 billion	▢
95. 2006	▢	$2655.4 billion	▢
96. 2008	▢	$2982.9 billion	▢

The properties of exponents listed on the preceding page apply to *all* integers m and n, not just positive integers. For instance, by Property 2, you can write

$$\frac{3^4}{3^{-5}} = 3^{4-(-5)} = 3^{4+5} = 3^9.$$

Example 1   Using Properties of Exponents

a. $(-3ab^4)(4ab^{-3}) = -12(a)(a)(b^4)(b^{-3}) = -12a^2b$

b. $(2xy^2)^3 = 2^3(x)^3(y^2)^3 = 8x^3y^6$

c. $3a(-4a^2)^0 = 3a(1) = 3a, \quad a \neq 0$

✔**CHECKPOINT**   Now try Exercise 27.

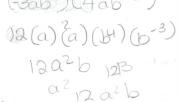

Example 2   Rewriting with Positive Exponents

a. $x^{-1} = \dfrac{1}{x}$ — Property 3

b. $\dfrac{1}{3x^{-2}} = \dfrac{1(x^2)}{3} = \dfrac{x^2}{3}$ — The exponent -2 does not apply to 3.

c. $\dfrac{1}{(3x)^{-2}} = (3x)^2 = 9x^2$ — The exponent -2 does apply to 3.

d. $\dfrac{12a^3b^{-4}}{4a^{-2}b} = \dfrac{12a^3 \cdot a^2}{4b \cdot b^4} = \dfrac{3a^5}{b^5}$ — Properties 3 and 1

e. $\left(\dfrac{3x^2}{y}\right)^{-2} = \dfrac{3^{-2}(x^2)^{-2}}{y^{-2}}$ — Properties 5 and 7

$\qquad = \dfrac{3^{-2}x^{-4}}{y^{-2}}$ — Property 6

$\qquad = \dfrac{y^2}{3^2x^4}$ — Property 3

$\qquad = \dfrac{y^2}{9x^4}$ — Simplify.

✔**CHECKPOINT**   Now try Exercise 31.

Example 3   Calculators and Exponents

Expression	Graphing Calculator Keystrokes	Display
a. $3^{-2} + 4^{-1}$	3 ⌃ (−) 2 + 4 ⌃ (−) 1 ⎡ENTER⎤	.3611111111
b. $\dfrac{3^5 + 1}{3^5 - 1}$	(3 ⌃ 5 + 1) ÷	
	(3 ⌃ 5 − 1) ⎡ENTER⎤	1.008264463

✔**CHECKPOINT**   Now try Exercise 35.

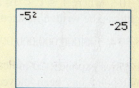
Study Tip

📖 Rarely in algebra is there only one way to solve a problem. Do not be concerned when the steps you use to solve a problem are not exactly the same as the steps presented in this text. The important thing is to use steps that you understand *and*, of course, that are justified by the rules of algebra. For instance, you might prefer the following steps for Example 2(e).

$$\left(\frac{3x^2}{y}\right)^{-2} = \left(\frac{y}{3x^2}\right)^2 = \frac{y^2}{9x^4}$$

Technology Tip

🖩 The graphing calculator keystrokes given in this text may not be the same as the keystrokes for your graphing calculator. Be sure you are familiar with the use of the keys on your calculator.

Scientific Notation

Exponents provide an efficient way of writing and computing with very large (or very small) numbers. For instance, there are about 359 billion billion gallons of water on Earth—that is, 359 followed by 18 zeros.

359,000,000,000,000,000,000

It is convenient to write such numbers in **scientific notation.** This notation has the form $\pm c \times 10^{n}$, where $1 \leq c < 10$ and n is an integer. So, the number of gallons of water on Earth can be written in scientific notation as

$$3.59 \times 100,000,000,000,000,000,000 = 3.59 \times 10^{20}.$$

The *positive* exponent 20 indicates that the number is *large* (10 or more) and that the decimal point has been moved 20 places. A *negative* exponent indicates that the number is *small* (less than 1). For instance, the mass (in grams) of one electron is approximately

$$9.0 \times 10^{-28} = 0.0000000000000000000000000009.$$

28 decimal places

Example 4   Scientific Notation

a. $1.345 \times 10^{2} = 134.5$

b. $0.0000782 = 7.82 \times 10^{-5}$

c. $-9.36 \times 10^{-6} = -0.00000936$

d. $836,100,000 = 8.361 \times 10^{8}$

✓CHECKPOINT   Now try Exercise 45.

Technology Tip

Most calculators automatically switch to scientific notation when they are showing large or small numbers that exceed the display range. Try evaluating

$86,500,000 \times 6000.$

A calculator that follows standard conventions should display

$\boxed{5.19 \ 11}$ or $\boxed{5.19 \ E \ 11}$

which is 5.19×10^{11}.

Example 5   Using Scientific Notation with a Calculator

Use a calculator to evaluate $65,000 \times 3,400,000,000.$

Solution

Because $65,000 = 6.5 \times 10^{4}$ and $3,400,000,000 = 3.4 \times 10^{9}$, you can multiply the two numbers using the following graphing calculator keystrokes.

$6.5 \ \boxed{EE} \ 4 \ \boxed{\times} \ 3.4 \ \boxed{EE} \ 9 \ \boxed{ENTER}$

After entering these keystrokes, the calculator should display $\boxed{2.21 \ E \ 14}$. So, the product of the two numbers is

$$(6.5 \times 10^{4})(3.4 \times 10^{9}) = 2.21 \times 10^{14} = 221,000,000,000,000.$$

✓CHECKPOINT   Now try Exercise 71.

Radicals and Their Properties

A **square root** of a number is one of its two equal factors. For instance, 5 is a square root of 25 because 5 is one of the two equal factors of

$$25 = 5 \cdot 5.$$

In a similar way, a **cube root** of a number is one of its three equal factors, as in

$$125 = 5^3.$$

Definition of the *n*th Root of a Number

Let a and b be real numbers and let $n \geq 2$ be a positive integer. If

$$a = b^n$$

then b is an ***n*th root of *a*.** If $n = 2$, then the root is a **square root.** If $n = 3$, then the root is a **cube root.**

Some numbers have more than one *n*th root. For example, both 5 and -5 are square roots of 25. The *principal square root* of 25, written as $\sqrt{25}$, is the positive root, 5. The **principal *n*th root** of a number is defined as follows.

Principal *n*th Root of a Number

Let a be a real number that has at least one *n*th root. The **principal *n*th root of *a*** is the *n*th root that has the same sign as a. It is denoted by a **radical symbol**

$$\sqrt[n]{a}. \qquad \text{Principal } n\text{th root}$$

The positive integer n is the **index** of the radical, and the number a is the **radicand**. When $n = 2$, omit the index and write $\sqrt{a}$ rather than $\sqrt[2]{a}$. (The plural of index is *indices*.)

A common misunderstanding when taking square roots of real numbers is that the square root sign implies both negative and positive roots. This is not correct. The square root sign implies only a positive root. When a negative root is needed, you must use the negative sign with the square root sign.

Incorrect: $\sqrt{4} = \pm 2$

Correct: $\sqrt{4} = 2$

Correct: $-\sqrt{4} = -2$

Example 6   Evaluating Expressions Involving Radicals

a. $\sqrt{36} = 6$ because $6^2 = 36$.

b. $-\sqrt{36} = -6$ because $-\left(\sqrt{36}\right) = -\left(\sqrt{6^2}\right) = -(6) = -6$.

c. $\sqrt[3]{\dfrac{125}{64}} = \dfrac{5}{4}$ because $\left(\dfrac{5}{4}\right)^3 = \dfrac{5^3}{4^3} = \dfrac{125}{64}$.

d. $\sqrt[5]{-32} = -2$ because $(-2)^5 = -32$.

e. $\sqrt[4]{-81}$ is not a real number because there is no real number that can be raised to the fourth power to produce -81.

✓**CHECKPOINT**   Now try Exercise 77.

Here are some generalizations about the *n*th roots of real numbers.

Generalizations About *n*th Roots of Real Numbers

Real number a	Integer n	Root(s) of a	Example
$a > 0$	$n > 0$, n is even.	$\sqrt[n]{a}, -\sqrt[n]{a}$	$\sqrt[4]{81} = 3, -\sqrt[4]{81} = -3$
$a > 0$ or $a < 0$	n is odd.	$\sqrt[n]{a}$	$\sqrt[3]{-8} = -2$
$a < 0$	n is even.	No real roots	$\sqrt{-4}$ is not a real number.
$a = 0$	n is even or odd.	$\sqrt[n]{0} = 0$	$\sqrt[5]{0} = 0$

Integers such as

$1,\quad 4,\quad 9,\quad 16,\quad 25,\quad$ and $\quad 36$ Perfect squares

are called **perfect squares** because they have integer square roots. Similarly, integers such as

$1,\quad 8,\quad 27,\quad 64,\quad$ and $\quad 125$ Perfect cubes

are called **perfect cubes** because they have integer cube roots.

Properties of Radicals

Let a and b be real numbers, variables, or algebraic expressions such that the indicated roots are real numbers, and let m and n be positive integers.

Property	*Example*
1. $\sqrt[n]{a^m} = \left(\sqrt[n]{a}\right)^m$	$\sqrt[3]{8^2} = \left(\sqrt[3]{8}\right)^2 = (2)^2 = 4$
2. $\sqrt[n]{a} \cdot \sqrt[n]{b} = \sqrt[n]{ab}$	$\sqrt{5} \cdot \sqrt{7} = \sqrt{5 \cdot 7} = \sqrt{35}$
3. $\dfrac{\sqrt[n]{a}}{\sqrt[n]{b}} = \sqrt[n]{\dfrac{a}{b}}$, $b \neq 0$	$\dfrac{\sqrt[4]{27}}{\sqrt[4]{9}} = \sqrt[4]{\dfrac{27}{9}} = \sqrt[4]{3}$
4. $\sqrt[m]{\sqrt[n]{a}} = \sqrt[mn]{a}$	$\sqrt[3]{\sqrt{10}} = \sqrt[6]{10}$
5. $\left(\sqrt[n]{a}\right)^n = a$	$\left(\sqrt{3}\right)^2 = 3$
6. For n even, $\sqrt[n]{a^n} = |a|$.	$\sqrt{(-12)^2} = |-12| = 12$
For n odd, $\sqrt[n]{a^n} = a$.	$\sqrt[3]{(-12)^3} = -12$

A common special case of Property 6 is

$$\sqrt{a^2} = |a|.$$

Example 7   Using Properties of Radicals

Use the properties of radicals to simplify each expression.

a. $\sqrt{8} \cdot \sqrt{2}$ **b.** $\left(\sqrt[3]{5}\right)^3$ **c.** $\sqrt[3]{x^3}$ **d.** $\sqrt[6]{y^6}$

Solution

a. $\sqrt{8} \cdot \sqrt{2} = \sqrt{8 \cdot 2} = \sqrt{16} = 4$

b. $\left(\sqrt[3]{5}\right)^3 = 5$

c. $\sqrt[3]{x^3} = x$

d. $\sqrt[6]{y^6} = |y|$

✓CHECKPOINT   Now try Exercise 93.

Simplifying Radicals

An expression involving radicals is in **simplest form** when the following conditions are satisfied.

1. All possible factors have been removed from the radical.

2. All fractions have radical-free denominators (accomplished by a process called *rationalizing the denominator*).

3. The index of the radical is reduced.

   To simplify a radical, factor the radicand into factors whose exponents are multiples of the index. The roots of these factors are written outside the radical, and the "leftover" factors make up the new radicand.

Example 8   Simplifying Even Roots

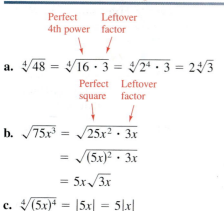

a. $\sqrt[4]{48} = \sqrt[4]{16 \cdot 3} = \sqrt[4]{2^4 \cdot 3} = 2\sqrt[4]{3}$

b. $\sqrt{75x^3} = \sqrt{25x^2 \cdot 3x}$ Find greatest square factor.

$\quad\quad\quad = \sqrt{(5x)^2 \cdot 3x}$

$\quad\quad\quad = 5x\sqrt{3x}$ Find root of perfect square.

c. $\sqrt[4]{(5x)^4} = |5x| = 5|x|$

✓**CHECKPOINT**   Now try Exercise 95(a).

Example 9   Simplifying Odd Roots

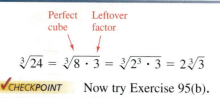

$\sqrt[3]{24} = \sqrt[3]{8 \cdot 3} = \sqrt[3]{2^3 \cdot 3} = 2\sqrt[3]{3}$

✓**CHECKPOINT**   Now try Exercise 95(b).

   Radical expressions can be combined (added or subtracted) when they are **like radicals**—that is, when they have the same index and radicand.

Example 10   Combining Radicals

a. $2\sqrt{48} - 3\sqrt{27} = 2\sqrt{16 \cdot 3} - 3\sqrt{9 \cdot 3}$ Find square factors.

$\quad\quad\quad\quad\quad\quad = 8\sqrt{3} - 9\sqrt{3}$ Find square roots and multiply by coefficients.

$\quad\quad\quad\quad\quad\quad = (8 - 9)\sqrt{3}$ Combine like terms.

$\quad\quad\quad\quad\quad\quad = -\sqrt{3}$ Simplify.

b. $\sqrt[3]{16x} - \sqrt[3]{54x^4} = \sqrt[3]{8 \cdot 2x} - \sqrt[3]{27 \cdot x^3 \cdot 2x}$ Find cube factors.

$\quad\quad\quad\quad\quad\quad\quad = 2\sqrt[3]{2x} - 3x\sqrt[3]{2x}$ Find cube roots.

$\quad\quad\quad\quad\quad\quad\quad = (2 - 3x)\sqrt[3]{2x}$ Combine like terms.

✓**CHECKPOINT**   Now try Exercise 99.

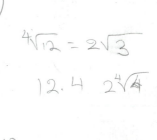

Rationalizing Denominators and Numerators

To rationalize a denominator or numerator of the form $a - b\sqrt{m}$ or $a + b\sqrt{m}$, multiply both numerator and denominator by a **conjugate:** $a + b\sqrt{m}$ and $a - b\sqrt{m}$ are conjugates of each other. If $a = 0$, then the rationalizing factor for $\sqrt{m}$ is itself, $\sqrt{m}$.

Example 11   Rationalizing Denominators

a. $\dfrac{5}{2\sqrt{3}} = \dfrac{5}{2\sqrt{3}} \cdot \dfrac{\sqrt{3}}{\sqrt{3}}$ $\sqrt{3}$ is rationalizing factor.

$= \dfrac{5\sqrt{3}}{2(3)}$ Multiply.

$= \dfrac{5\sqrt{3}}{6}$ Simplify.

b. $\dfrac{2}{\sqrt[3]{5}} = \dfrac{2}{\sqrt[3]{5}} \cdot \dfrac{\sqrt[3]{5^2}}{\sqrt[3]{5^2}}$ $\sqrt[3]{5^2}$ is rationalizing factor.

$= \dfrac{2\sqrt[3]{5^2}}{\sqrt[3]{5^3}}$ Multiply.

$= \dfrac{2\sqrt[3]{25}}{5}$ Simplify.

 CHECKPOINT   Now try Exercise 105.

> **Study Tip**
>
> Notice in Example 11(b) that the numerator and denominator are multiplied by $\sqrt[3]{5^2}$ to produce a perfect cube radicand.

Example 12   Rationalizing a Denominator with Two Terms

$\dfrac{2}{3 + \sqrt{7}} = \dfrac{2}{3 + \sqrt{7}} \cdot \dfrac{3 - \sqrt{7}}{3 - \sqrt{7}}$ Multiply numerator and denominator by conjugate of denominator.

$= \dfrac{2(3 - \sqrt{7})}{(3)^2 - (\sqrt{7})^2}$ Find products. In denominator, $(a + b)(a - b) = a^2 - ab + ab - b^2$
$\qquad\qquad\quad = a^2 - b^2$.

$= \dfrac{2(3 - \sqrt{7})}{2}$ Simplify.

$= 3 - \sqrt{7}$ Divide out common factors.

 CHECKPOINT   Now try Exercise 107.

In calculus, sometimes it is necessary to rationalize the numerator of an expression.

Example 13   Rationalizing a Numerator

$\dfrac{\sqrt{5} - \sqrt{7}}{2} = \dfrac{\sqrt{5} - \sqrt{7}}{2} \cdot \dfrac{\sqrt{5} + \sqrt{7}}{\sqrt{5} + \sqrt{7}}$ Multiply numerator and denominator by conjugate of numerator.

$= \dfrac{(\sqrt{5})^2 - (\sqrt{7})^2}{2(\sqrt{5} + \sqrt{7})}$ Find products. In numerator, $(a + b)(a - b) = a^2 - ab + ab - b^2$
$\qquad\qquad\quad = a^2 - b^2$.

$= \dfrac{-2}{2(\sqrt{5} + \sqrt{7})} = \dfrac{-1}{\sqrt{5} + \sqrt{7}}$ Simplify and divide out common factors.

 CHECKPOINT   Now try Exercise 111.

> **Study Tip**
>
> Do not confuse the expression $\sqrt{5} + \sqrt{7}$ with the expression $\sqrt{5 + 7}$. In general, $\sqrt{x + y}$ does not equal $\sqrt{x} + \sqrt{y}$. Similarly, $\sqrt{x^2 + y^2}$ does not equal $x + y$.

The symbol $\displaystyle\int$ indicates an example or exercise that highlights algebraic techniques specifically used in calculus.

Rational Exponents

> **Definition of Rational Exponents**
>
> If a is a real number and n is a positive integer such that the principal nth root of a exists, then $a^{1/n}$ is defined as
>
> $$a^{1/n} = \sqrt[n]{a}, \text{ where } 1/n \text{ is the } \textbf{rational exponent} \text{ of } a.$$
>
> Moreover, if m is a positive integer that has no common factor with n, then
>
> $$a^{m/n} = (a^{1/n})^m = \left(\sqrt[n]{a}\right)^m \quad \text{and} \quad a^{m/n} = (a^m)^{1/n} = \sqrt[n]{a^m}.$$

The numerator of a rational exponent denotes the *power* to which the base is raised, and the denominator denotes the *index* or the *root* to be taken.

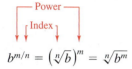

$$b^{m/n} = \left(\sqrt[n]{b}\right)^m = \sqrt[n]{b^m}$$

When you are working with rational exponents, the properties of integer exponents still apply. For instance,

$$2^{1/2}2^{1/3} = 2^{(1/2)+(1/3)} = 2^{5/6}.$$

Example 14   Changing from Radical to Exponential Form

a. $\sqrt{3} = 3^{1/2}$

b. $\sqrt{(3xy)^5} = \sqrt[2]{(3xy)^5} = (3xy)^{(5/2)}$

✓**CHECKPOINT**   Now try Exercise 115.

Example 15   Changing from Exponential to Radical Form

a. $(x^2 + y^2)^{3/2} = \left(\sqrt{x^2 + y^2}\right)^3$

$$= \sqrt{(x^2 + y^2)^3}$$

b. $2y^{3/4}z^{1/4} = 2(y^3z)^{1/4}$

$$= 2\sqrt[4]{y^3z}$$

c. $a^{-3/2} = \dfrac{1}{a^{3/2}} = \dfrac{1}{\sqrt{a^3}}$

✓**CHECKPOINT**   Now try Exercise 117.

Example 16   Simplifying with Rational Exponents

a. $(-32)^{-4/5} = \left(\sqrt[5]{-32}\right)^{-4} = (-2)^{-4} = \dfrac{1}{(-2)^4} = \dfrac{1}{16}$

b. $(-5x^{5/3})(3x^{-3/4}) = -15x^{(5/3)-(3/4)} = -15x^{11/12}, \quad x \neq 0$

c. $\sqrt[3]{\sqrt{125}} = \sqrt[6]{125} = \sqrt[6]{(5)^3} = 5^{3/6} = 5^{1/2} = \sqrt{5}$

d. $(2x - 1)^{4/3}(2x - 1)^{-1/3} = (2x - 1)^{(4/3)-(1/3)} = 2x - 1, \quad x \neq \dfrac{1}{2}$

✓**CHECKPOINT**   Now try Exercise 123.

P.2   Exercises

See www.CalcChat.com for worked-out solutions to odd-numbered exercises.
For instructions on how to use a graphing utility, see Appendix A.

Vocabulary and Concept Check

In Exercises 1–6, fill in the blank(s).

1. In the exponential form a^n, n is the _____ and a is the _____ .

2. One of the two equal factors of a number is called a _____ of the number.

3. The _____ of a number a is the nth root that has the same sign as a.

4. In the radical form $\sqrt[n]{a}$, the positive integer n is called the _____ of the radical and the number a is called the _____ .

5. Creating a radical-free denominator is known as _____ the denominator.

6. In the expression $b^{m/n}$, m denotes the _____ to which the base is raised and n denotes the _____ or root to be taken.

7. What is the conjugate of $2 + 3\sqrt{5}$?

8. When is an expression involving radicals in simplest form?

9. Is -10.678×10^3 written in scientific notation?

10. Is 64 a perfect square, a perfect cube, or both?

Procedures and Problem Solving

Evaluating an Expression   In Exercises 11–18, evaluate each expression.

11. (a) $4^2 \cdot 3$ (b) $3 \cdot 3^3$

12. (a) $\dfrac{5^5}{5^2}$ (b) $\dfrac{3^2}{3^4}$

13. (a) $(4^3)^0$ (b) -3^2

14. (a) $(2^3 \cdot 3^2)^2$ (b) $\left(-\dfrac{3}{5}\right)^3\left(\dfrac{5}{3}\right)^2$

15. (a) $\dfrac{3}{3^{-4}}$ (b) $24(-2)^{-5}$

16. (a) $\dfrac{4 \cdot 3^{-2}}{2^{-2} \cdot 3^{-1}}$ (b) $(-2)^0$

17. (a) $2^{-1} + 3^{-1}$ (b) $(3^{-1})^{-3}$

18. (a) $3^{-1} + 2^{-2}$ (b) $(3^{-2})^2$

Evaluating an Expression   In Exercises 19–24, evaluate the expression for the value of x.

19. $7x^{-2}$, $x = 2$

20. $6x^0 - (6x)^0$, $x = 7$

21. $2x^3$, $x = -3$

22. $-3x^4$, $x = 2$

23. $4x^2$, $x = -\dfrac{1}{2}$

24. $5(-x)^3$, $x = \dfrac{1}{3}$

Using Properties of Exponents   In Exercises 25–32, simplify each expression.

25. (a) $x^3(x^2)$ (b) $x^4(3x^5)$

26. (a) $4z^4(-2z^3)$ (b) $-5x^3(4x^2)$

27. (a) $(3x)^2$ (b) $(4x^3)^2$

28. (a) $6z^2(2z^5)^4$ (b) $(3x^5)^3(2x^7)^2$

29. (a) $\dfrac{7x^2}{x^3}$ (b) $\dfrac{12(x+y)^3}{9(x+y)}$

30. (a) $\dfrac{r^5}{r^9}$ (b) $\left(\dfrac{4}{y}\right)^3\left(\dfrac{3}{y}\right)^4$

31. (a) $[(x^2y^{-2})^{-1}]^{-1}$ (b) $\left(\dfrac{a^{-2}}{b^{-2}}\right)\left(\dfrac{b}{a}\right)^3$

32. (a) $(2x^5)^0$, $x \neq 0$ (b) $(5x^2z^6)^3(5x^2z^6)^{-3}$

Calculators and Exponents   In Exercises 33–38, use a calculator to evaluate the expression. (Round your answer to three decimal places.)

33. $(-4)^3(5^2)$

34. $(8^{-4})(10^3)$

35. $\dfrac{3^6}{7^3}$

36. $\dfrac{4^5}{9^3}$

37. $\dfrac{4^3 - 1}{3^{-4}}$

38. $\dfrac{3^2 - 2}{4^2 - 3}$

Scientific Notation   In Exercises 39–52, write the number in scientific notation.

39. 973.50

40. 28,022.2

41. 10,252.484

42. 525,252,118

43. -1110.25

44. $-5,222,145$

45. 0.0002485

46. 0.0000025

47. -0.0000025

48. -0.000125005

49. Land area of Earth: 57,300,000 square miles

50. Light year: 9,460,000,000,000 kilometers

51. Relative density of hydrogen: 0.0000899 gram per cubic centimeter

52. One micron (millionth of a meter): 0.00003937 inch

Writing a Number in Decimal Notation   In Exercises 53–64, write the number in decimal notation.

53. 1.25×10^5

54. 1.08×10^4

55. -4.816×10^8

56. -3.785×10^{10}

57. 7.65×10^7

58. 5.05×10^{-10}

59. -9.001×10^{-3}

60. -8.098×10^{-6}

61. Mean SAT Writing score of college-bound seniors in 2009: 4.93×10^2 (Source: The College Board)

62. Interior temperature of the Sun: 1.5×10^7 degrees Celsius

63. Charge of an electron: 1.6022×10^{-19} coulomb

64. Width of a human hair: 9.0×10^{-5} meter

Evaluating an Expression   In Exercises 65–70, evaluate the expression without using a calculator.

65. $(2.0 \times 10^7)(3.0 \times 10^{-3})$ **66.** $(1.4 \times 10^5)(5.2 \times 10^{-2})$

67. $\dfrac{7.0 \times 10^5}{4.0 \times 10^{-3}}$

68. $\dfrac{3.0 \times 10^{-3}}{6.0 \times 10^2}$

69. $\sqrt{25 \times 10^8}$

70. $\sqrt[3]{8 \times 10^{15}}$

Using Scientific Notation with a Calculator   In Exercises 71–74, use a calculator to evaluate each expression. (Round your answer to three decimal places.)

71. (a) $(9.3 \times 10^6)^3(6.1 \times 10^{-4})$

(b) $\dfrac{(2.414 \times 10^4)^6}{(1.68 \times 10^5)^5}$

72. (a) $750\left(1 + \dfrac{0.11}{365}\right)^{800}$

(b) $\dfrac{67{,}000{,}000 + 93{,}000{,}000}{0.0052}$

73. (a) $\sqrt{4.5 \times 10^9}$

(b) $\sqrt[3]{6.3 \times 10^4}$

74. (a) $(2.65 \times 10^{-4})^{1/3}$

(b) $\sqrt{9 \times 10^{-4}}$

$\sqrt[3]{2.65 \cdot 10^{-4}}$

Evaluating Expressions Involving Radicals   In Exercises 75–80, evaluate the expression without using a calculator.

75. $\sqrt{121}$

76. $\sqrt{49}$

77. $-\sqrt[3]{-64}$

78. $\dfrac{\sqrt[4]{81}}{3}$

79. $\left(\sqrt[3]{-125}\right)^3$

80. $\sqrt[4]{562^4}$

Approximating the Value of an Expression   In Exercises 81–92, use a calculator to approximate the value of the expression. (Round your answer to three decimal places.)

81. $\sqrt[5]{-27^3}$

82. $\sqrt[3]{45^2}$

83. $(3.4)^{2.5}$

84. $(6.1)^{-2.9}$

85. $(1.2^{-2})\sqrt{75} + 3\sqrt{8}$

86. $\dfrac{-5 + \sqrt{33}}{5}$

87. $\sqrt{\pi + 1}$

88. $\sqrt{10 - \pi}$

89. $\dfrac{3.14}{\pi} + \sqrt[3]{5}$

90. $\dfrac{\sqrt{10}}{2.5} - \pi^2$

91. $(2.8)^{-2} + 1.01 \times 10^6$

92. $2.12 \times 10^{-2} + \sqrt{15}$

Using Properties of Radicals   In Exercises 93 and 94, use the properties of radicals to simplify each expression.

93. (a) $\left(\sqrt[5]{3}\right)^5$

(b) $\sqrt[5]{96x^5}$

94. (a) $\sqrt{12} \cdot \sqrt{3}$

(b) $\sqrt[4]{x^4}$

Simplifying a Radical Expression   In Exercises 95–100, simplify each expression.

95. (a) $\sqrt{54xy^4}$

(b) $\sqrt[3]{\dfrac{32a^2}{b^2}}$

96. (a) $\sqrt[3]{54}$

(b) $\sqrt{32x^3y^4}$

97. (a) $2\sqrt{50} + 12\sqrt{8}$

(b) $10\sqrt{32} - 6\sqrt{18}$

98. (a) $5\sqrt{x} - 3\sqrt{x}$

(b) $-2\sqrt{9y} + 10\sqrt{y}$

99. (a) $3\sqrt{x+1} + 10\sqrt{x+1}$

(b) $7\sqrt{80x} - 2\sqrt{125x}$

100. (a) $5\sqrt{10x^2} - \sqrt{90x^2}$ (b) $8\sqrt[3]{27x} - \tfrac{1}{2}\sqrt[3]{64x}$

Evaluating Radical Expressions   In Exercises 101–104, complete the statement with <, =, or >.

101. $\sqrt{5} + \sqrt{3} \;\rule{1.2em}{0.8em}\; \sqrt{5 + 3}$

102. $\sqrt{\dfrac{3}{11}} \;\rule{1.2em}{0.8em}\; \dfrac{\sqrt{3}}{\sqrt{11}}$

103. $5 \;\rule{1.2em}{0.8em}\; \sqrt{3^2 + 2^2}$

104. $5 \;\rule{1.2em}{0.8em}\; \sqrt{3^2 + 4^2}$

Rationalizing a Denominator   In Exercises 105–108, rationalize the denominator of the expression. Then simplify your answer.

105. $\dfrac{1}{\sqrt{3}}$

106. $\dfrac{8}{\sqrt[3]{2}}$

107. $\dfrac{5}{\sqrt{14} - 2}$

108. $\dfrac{3}{\sqrt{5} + \sqrt{6}}$

Rationalizing a Numerator   In Exercises 109–112, rationalize the numerator of the expression. Then simplify your answer.

109. $\dfrac{\sqrt{12}}{2}$

110. $\dfrac{\sqrt{3}}{3}$

111. $\dfrac{\sqrt{5} + \sqrt{3}}{3}$

112. $\dfrac{\sqrt{7} - 3}{4}$

The symbol ∫ indicates an example or exercise that highlights algebraic techniques specifically used in calculus.

Reducing the Index of a Radical In Exercises 113 and 114, reduce the index of each radical and rewrite in radical form.

113. (a) $\sqrt[4]{3^2}$ (b) $\sqrt[6]{(x+1)^4}$
114. (a) $\sqrt[6]{x^3}$ (b) $\sqrt[4]{(3x^2)^4}$

Changing Forms In Exercises 115–122, fill in the missing form of the expression.

Radical Form	Rational Exponent Form
✓ 115. $\sqrt[3]{64}$	
116.	$-(144^{1/2})$
✓ 117.	$32^{1/5}$
118. $\sqrt[3]{614.125}$	
119. $\sqrt[3]{-216}$	
120.	$(-243)^{1/5}$
121. $\sqrt[4]{81^3}$	
122.	$16^{5/4}$

Simplifying with Rational Exponents In Exercises 123–130, simplify the expression.

✓ 123. $\dfrac{(2x^2)^{3/2}}{2^{1/2}x^4}$ 124. $\dfrac{x^{4/3}y^{2/3}}{(xy)^{1/3}}$

125. $\dfrac{x^{-3} \cdot x^{1/2}}{x^{3/2} \cdot x^{-1}}$ 126. $\dfrac{5^{-1/2} \cdot 5x^{5/2}}{(5x)^{3/2}}$

127. $32^{-3/5}$ 128. $\left(\dfrac{9}{4}\right)^{-1/2}$

129. $\left(-\dfrac{1}{27}\right)^{-1/3}$ 130. $-\left(\dfrac{1}{125}\right)^{-4/3}$

Rewriting a Radical Expression In Exercises 131 and 132, write each expression as a single radical. Then simplify your answer.

131. (a) $\sqrt{\sqrt{32}}$ (b) $\sqrt{\sqrt[4]{2x}}$
132. (a) $\sqrt{\sqrt{243(x+1)}}$ (b) $\sqrt{\sqrt[3]{10a^7b}}$

133. *Why you should learn it* (p. 12) A stream of water moving at the rate of v feet per second can carry particles of size $0.03\sqrt{v}$ inches. Find the size of the largest particle that can be carried by a stream flowing at the rate of $\frac{3}{4}$ foot per second.

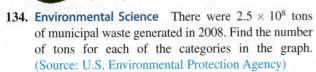

134. **Environmental Science** There were 2.5×10^8 tons of municipal waste generated in 2008. Find the number of tons for each of the categories in the graph. (Source: U.S. Environmental Protection Agency)

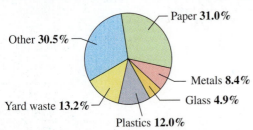

Paper **31.0%**
Other **30.5%**
Metals **8.4%**
Glass **4.9%**
Plastics **12.0%**
Yard waste **13.2%**

135. **Economics** The table shows the 2009 estimated populations and gross domestic products (GDP) for several countries, given in scientific notation. Find the per capita GDP for each country. (Sources: U.S. Census Bureau; Central Intelligence Agency)

Country	Population	GDP (in U.S. dollars)
Brazil	1.99×10^8	2.03×10^{12}
China	1.32×10^9	8.79×10^{12}
Germany	8.23×10^7	2.81×10^{12}
Iran	7.60×10^7	8.76×10^{11}
Ireland	4.58×10^6	1.77×10^{11}
India	1.16×10^9	3.56×10^{12}
Mexico	1.11×10^8	1.48×10^{12}

136. **Chemistry** A funnel is filled with hydrochloric acid to a height of h centimeters. The formula

$$t = 0.03[12^{5/2} - (12 - h)^{5/2}], \quad 0 \le h \le 12$$

represents the amount of time t (in seconds) it will take for the funnel to empty. Find t for $h = 7$ centimeters.

Conclusions

True or False? In Exercises 137 and 138, determine whether the statement is true or false. Justify your answer.

137. $\dfrac{x^{k+1}}{x} = x^k$

138. $(a^n)^k = a^{(n^k)}$

139. **Think About It** Verify that $a^0 = 1$, $a \ne 0$. (*Hint:* Use the property of exponents $a^m/a^n = a^{m-n}$.)

140. **CAPSTONE**
(a) Explain how to simplify the expression $(3x^3y^{-2})^{-2}$.
(b) Is the expression
$$\sqrt{\dfrac{4}{x^3}}$$
in simplest form? Why or why not?

141. **Exploration** List all possible digits that occur in the units place of the square of a positive integer. Use that list to determine whether $\sqrt{5233}$ is an integer.

142. **Think About It** Square the real number $5/\sqrt{3}$ and note that the radical is eliminated from the denominator. Is this equivalent to rationalizing the denominator? Why or why not?

P.3   Polynomials and Factoring

Polynomials

The most common type of algebraic expression is the **polynomial.** Some examples are

$$2x + 5, \quad 3x^4 - 7x^2 + 2x + 4, \quad \text{and} \quad 5x^2y^2 - xy + 3.$$

The first two are *polynomials in x* and the third is a *polynomial in x and y*. The terms of a polynomial in x have the form ax^k, where a is the **coefficient** and k is the **degree** of the term. For instance, the polynomial

$$2x^3 - 5x^2 + 1 = 2x^3 + (-5)x^2 + (0)x + 1$$

has coefficients $2, -5, 0,$ and 1.

Definition of a Polynomial in x

Let $a_0, a_1, a_2, \ldots, a_n$ be *real numbers* and let n be a *nonnegative integer*. A **polynomial in x** is an expression of the form

$$a_n x^n + a_{n-1} x^{n-1} + \cdots + a_1 x + a_0$$

where $a_n \neq 0$. The polynomial is of **degree** n, a_n is the **leading coefficient,** and a_0 is the **constant term.**

In **standard form,** a polynomial in x is written with descending powers of x. Polynomials with one, two, and three terms are called **monomials, binomials,** and **trinomials,** respectively.

Example 1   Writing Polynomials in Standard Form

Polynomial	Standard Form	Degree	Leading Coefficient
a. $4x^2 - 5x^7 - 2 + 3x$	$-5x^7 + 4x^2 + 3x - 2$	7	-5
b. $4 - 9x^2$	$-9x^2 + 4$	2	-9
c. 8	$8 \; (8 = 8x^0)$	0	8

✓**CHECKPOINT**   Now try Exercise 21.

A polynomial that has all zero coefficients is called the **zero polynomial,** denoted by 0. No degree is assigned to this particular polynomial. For polynomials in more than one variable, the degree of a *term* is the sum of the exponents of the variables in the term. The degree of the *polynomial* is the highest degree of its terms. For instance, the degree of the polynomial

$$-2x^3y^6 + 4xy - x^7y^4$$

is 11 because the sum of the exponents in the last term is the greatest. Expressions are not polynomials when a variable is underneath a radical or when a polynomial expression (with degree greater than 0) is in the denominator of a term. The following expressions are *not* polynomials.

$$x^3 - \sqrt{3x} = x^3 - (3x)^{1/2} \qquad \text{The exponent } 1/2 \text{ is not an integer.}$$

$$x^2 + \frac{5}{x} = x^2 + 5x^{-1} \qquad \text{The exponent } -1 \text{ is not a nonnegative integer.}$$

Operations with Polynomials and Special Products

You can add and subtract polynomials in much the same way you add and subtract real numbers. Simply add or subtract the *like terms* (terms having exactly the same variables to exactly the same powers) by adding their coefficients. For instance,

$$-3xy^2 \quad \text{and} \quad 5xy^2$$

are like terms and their sum is

$$-3xy^2 + 5xy^2 = (-3 + 5)xy^2$$
$$= 2xy^2.$$

Example 2 Sums and Differences of Polynomials

Perform the indicated operation.

a. $(5x^3 - 7x^2 - 3) + (x^3 + 2x^2 - x + 8)$

b. $(7x^4 - x^2 - 4x + 2) - (3x^4 - 4x^2 + 3x)$

Solution

a. $(5x^3 - 7x^2 - 3) + (x^3 + 2x^2 - x + 8)$

$$= (5x^3 + x^3) + (-7x^2 + 2x^2) - x + (-3 + 8) \qquad \text{Group like terms.}$$
$$= 6x^3 - 5x^2 - x + 5 \qquad \text{Combine like terms.}$$

b. $(7x^4 - x^2 - 4x + 2) - (3x^4 - 4x^2 + 3x)$

$$= 7x^4 - x^2 - 4x + 2 - 3x^4 + 4x^2 - 3x \qquad \text{Distributive Property}$$
$$= (7x^4 - 3x^4) + (-x^2 + 4x^2) + (-4x - 3x) + 2 \qquad \text{Group like terms.}$$
$$= 4x^4 + 3x^2 - 7x + 2 \qquad \text{Combine like terms.}$$

✓CHECKPOINT   Now try Exercise 27.

> **Study Tip**
>
> When a negative sign precedes an expression within parentheses, treat it like the coefficient (-1) and distribute the negative sign to each term inside the parentheses.
>
> $$-(x^2 - x + 3) = -x^2 + x - 3$$

To find the product of two polynomials, use the left and right Distributive Properties.

Example 3 Multiplying Polynomials: The FOIL Method

Find the product of $3x - 2$ and $5x + 7$.

Solution

Treating $5x + 7$ as a single quantity, you can use the Distributive Property to multiply $3x - 2$ by $5x + 7$ as follows.

$$(3x - 2)(5x + 7) = 3x(5x + 7) - 2(5x + 7)$$
$$= (3x)(5x) + (3x)(7) - (2)(5x) - (2)(7)$$
$$= 15x^2 + 21x - 10x - 14$$

Product of First terms	Product of Outer terms	Product of Inner terms	Product of Last terms

$$= 15x^2 + 11x - 14$$

✓CHECKPOINT   Now try Exercise 43.

Note that when using the **FOIL Method** (which can be used only to multiply two binomials), some of the terms in the product may be like terms that can be combined into one term.

Example 4   The Product of Two Trinomials

Find the product of $4x^2 + x - 2$ and $-x^2 + 3x + 5$.

Solution

When multiplying two polynomials, be sure to multiply *each* term of one polynomial by *each* term of the other. A vertical format is helpful.

$$
\begin{array}{r}
4x^2 + x - 2 \\
\times \quad -x^2 + 3x + 5 \\
\hline
20x^2 + 5x - 10 \\
12x^3 + 3x^2 - 6x \\
-4x^4 - x^3 + 2x^2 \\
\hline
-4x^4 + 11x^3 + 25x^2 - x - 10
\end{array}
$$

Write in standard form.

Write in standard form.

⟸ $5(4x^2 + x - 2)$

⟸ $3x(4x^2 + x - 2)$

⟸ $-x^2(4x^2 + x - 2)$

Combine like terms.

(handwritten: $-4x^2 + 12x^3$)

✔CHECKPOINT   Now try Exercise 61.

Some binomial products have special forms that occur frequently in algebra. You do not need to memorize these formulas because you can use the Distributive Property to multiply. Becoming familiar with these formulas, however, will enable you to manipulate the algebra more quickly.

Special Products

Let u and v be real numbers, variables, or algebraic expressions.

Special Product	*Example*

Sum and Difference of Same Terms

$$(u + v)(u - v) = u^2 - v^2$$ $(x + 4)(x - 4) = x^2 - 4^2 = x^2 - 16$

Square of a Binomial

$$(u + v)^2 = u^2 + 2uv + v^2$$ $(x + 3)^2 = x^2 + 2(x)(3) + 3^2 = x^2 + 6x + 9$

$$(u - v)^2 = u^2 - 2uv + v^2$$ $(3x - 2)^2 = (3x)^2 - 2(3x)(2) + 2^2 = 9x^2 - 12x + 4$

Cube of a Binomial

$$(u + v)^3 = u^3 + 3u^2v + 3uv^2 + v^3$$ $(x + 2)^3 = x^3 + 3x^2(2) + 3x(2^2) + 2^3 = x^3 + 6x^2 + 12x + 8$

$$(u - v)^3 = u^3 - 3u^2v + 3uv^2 - v^3$$ $(x - 1)^3 = x^3 - 3x^2(1) + 3x(1^2) - 1^3 = x^3 - 3x^2 + 3x - 1$

Example 5   The Product of Two Trinomials

To find the product of $x + y - 2$ and $x + y + 2$, begin by grouping $x + y$ in parentheses. Then write the product of the trinomials as a special product.

Difference   Sum

$$(x + y - 2)(x + y + 2) = [(x + y) - 2][(x + y) + 2]$$

$$= (x + y)^2 - 2^2$$ Sum and difference of same terms

$$= x^2 + 2xy + y^2 - 4$$

✔CHECKPOINT   Now try Exercise 63.

Factoring

The process of writing a polynomial as a product is called **factoring.** It is an important tool for solving equations and for simplifying rational expressions.

Unless noted otherwise, when you are asked to factor a polynomial, you can assume that you are looking for factors with integer coefficients. If a polynomial cannot be factored using integer coefficients, then it is **prime** or **irreducible over the integers.** For instance, the polynomial

$$x^2 - 3$$

is irreducible over the integers. Over the real numbers, this polynomial can be factored as

$$x^2 - 3 = (x + \sqrt{3})(x - \sqrt{3}).$$

A polynomial is **completely factored** when each of its factors is prime. So,

$$x^3 - x^2 + 4x - 4 = (x - 1)(x^2 + 4) \qquad \text{Completely factored}$$

is completely factored, but

$$x^3 - x^2 - 4x + 4 = (x - 1)(x^2 - 4) \qquad \text{Not completely factored}$$

is not completely factored. Its complete factorization is

$$x^3 - x^2 - 4x + 4 = (x - 1)(x + 2)(x - 2).$$

The simplest type of factoring involves a polynomial that can be written as the product of a monomial and another polynomial. The technique used here is the Distributive Property,

$$a(b + c) = ab + ac$$

in the *reverse* direction. For instance, the polynomial $5x^2 + 15x$ can be factored as follows.

$$5x^2 + 15x = 5x(x) + 5x(3) \qquad \text{$5x$ is a common factor.}$$
$$= 5x(x + 3)$$

The first step in completely factoring a polynomial is to remove (factor out) any common factors, as shown in the next example.

Example 6 Removing Common Factors

Factor each expression.

a. $6x^3 - 4x$ **b.** $-4x^2 + 12x - 16$

c. $3x^4 + 9x^3 + 6x^2$ **d.** $(x - 2)(2x) + (x - 2)(3)$

Solution

a. $6x^3 - 4x = 2x(3x^2) - 2x(2)$ $2x$ is a common factor.
$$= 2x(3x^2 - 2)$$

b. $-4x^2 + 12x - 16 = -4(x^2) + (-4)(-3x) + (-4)4$ -4 is a common factor.
$$= -4(x^2 - 3x + 4)$$

c. $3x^4 + 9x^3 + 6x^2 = 3x^2(x^2) + 3x^2(3x) + 3x^2(2)$ $3x^2$ is a common factor.
$$= 3x^2(x^2 + 3x + 2)$$
$$= 3x^2(x + 1)(x + 2)$$

d. $(x - 2)(2x) + (x - 2)(3) = (x - 2)(2x + 3)$ $x - 2$ is a common factor.

✓CHECKPOINT   Now try Exercise 73.

Factoring Special Polynomial Forms

Some polynomials have special forms that arise from the special product forms on page 25. You should learn to recognize these forms so that you can factor such polynomials easily.

Factoring Special Polynomial Forms

Factored Form	*Example*

Difference of Two Squares

$u^2 - v^2 = (u + v)(u - v)$ $\qquad 9x^2 - 4 = (3x)^2 - 2^2 = (3x + 2)(3x - 2)$

Perfect Square Trinomial

$u^2 + 2uv + v^2 = (u + v)^2$ $\qquad x^2 + 6x + 9 = x^2 + 2(x)(3) + 3^2 = (x + 3)^2$

$u^2 - 2uv + v^2 = (u - v)^2$ $\qquad x^2 - 6x + 9 = x^2 - 2(x)(3) + 3^2 = (x - 3)^2$

Sum or Difference of Two Cubes

$u^3 + v^3 = (u + v)(u^2 - uv + v^2)$ $\qquad x^3 + 8 = x^3 + 2^3 = (x + 2)(x^2 - 2x + 4)$

$u^3 - v^3 = (u - v)(u^2 + uv + v^2)$ $\qquad 27x^3 - 1 = (3x)^3 - 1^3 = (3x - 1)(9x^2 + 3x + 1)$

One of the easiest special polynomial forms to factor is the difference of two squares. Think of this form as follows.

$$u^2 - v^2 = (u + v)(u - v)$$

Difference   Opposite signs

To recognize perfect square terms, look for coefficients that are squares of integers and variables raised to *even powers*.

Example 7   Removing a Common Factor First

$3 - 12x^2 = 3(1 - 4x^2)$ 3 is a common factor.

$\qquad\quad = 3[1^2 - (2x)^2]$ Difference of two squares

$\qquad\quad = 3(1 + 2x)(1 - 2x)$ Factored form

✓CHECKPOINT   Now try Exercise 77.

Example 8   Factoring the Difference of Two Squares

Factor each expression.

a. $(x + 2)^2 - y^2$

b. $16x^4 - 81$

Solution

a. $(x + 2)^2 - y^2 = [(x + 2) + y][(x + 2) - y]$
$\qquad\qquad\qquad = (x + 2 + y)(x + 2 - y)$

b. $16x^4 - 81 = (4x^2)^2 - 9^2$ Difference of two squares

$\qquad\qquad = (4x^2 + 9)(4x^2 - 9)$

$\qquad\qquad = (4x^2 + 9)[(2x)^2 - 3^2]$ Difference of two squares

$\qquad\qquad = (4x^2 + 9)(2x + 3)(2x - 3)$ Factored form

✓CHECKPOINT   Now try Exercise 81.

Study Tip

In Example 7, note that the first step in factoring a polynomial is to check for a common factor. Once the common factor is removed, it is often possible to recognize patterns that were not immediately obvious.

A perfect square trinomial is the square of a binomial, as shown below.

$$u^2 + 2uv + v^2 = (u + v)^2 \quad \text{or} \quad u^2 - 2uv + v^2 = (u - v)^2$$

Like signs   Like signs

Note that the first and last terms are squares and the middle term is twice the product of u and v.

Example 9 Factoring Perfect Square Trinomials

Factor each trinomial.

a. $x^2 - 10x + 25$

b. $16x^2 + 8x + 1$

Solution

a. $x^2 - 10x + 25 = x^2 - 2(x)(5) + 5^2$ Rewrite in $u^2 - 2uv + v^2$ form.

$\qquad\qquad\qquad\quad = (x - 5)^2$

b. $16x^2 + 8x + 1 = (4x)^2 + 2(4x)(1) + 1^2$ Rewrite in $u^2 + 2uv + v^2$ form.

$\qquad\qquad\qquad\quad = (4x + 1)^2$

✓CHECKPOINT   Now try Exercise 87.

The next two formulas show the sums and differences of cubes. Pay special attention to the signs of the terms.

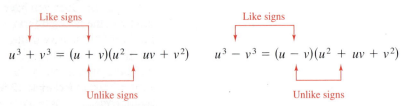

Like signs   Like signs

$$u^3 + v^3 = (u + v)(u^2 - uv + v^2) \qquad u^3 - v^3 = (u - v)(u^2 + uv + v^2)$$

Unlike signs   Unlike signs

Example 10 Factoring the Difference of Cubes

Factor $x^3 - 27$.

Solution

$x^3 - 27 = x^3 - 3^3$ Rewrite 27 as 3^3.

$\qquad\quad = (x - 3)(x^2 + 3x + 9)$ Factor.

✓CHECKPOINT   Now try Exercise 91.

Example 11 Factoring the Sum of Cubes

Factor $3x^3 + 192$.

Solution

$3x^3 + 192 = 3(x^3 + 64)$ 3 is a common factor.

$\qquad\qquad = 3(x^3 + 4^3)$ Rewrite 64 as 4^3.

$\qquad\qquad = 3(x + 4)(x^2 - 4x + 16)$ Factor.

✓CHECKPOINT   Now try Exercise 93.

Explore the Concept

Rewrite $u^6 - v^6$ as the difference of two squares. Then find a formula for completely factoring $u^6 - v^6$. Use your formula to factor completely $x^6 - 1$ and $x^6 - 64$.

Trinomials with Binomial Factors

To factor a trinomial of the form $ax^2 + bx + c$, use the following pattern.

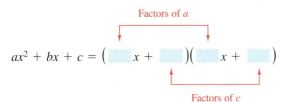

Factors of a

$$ax^2 + bx + c = (\quad x + \quad)(\quad x + \quad)$$

Factors of c

The goal is to find a combination of factors of a and c such that the outer and inner products add up to the middle term bx. For instance, in the trinomial

$$6x^2 + 17x + 5$$

you can write all possible factorizations and determine which one has outer and inner products that add up to $17x$.

$$(6x + 5)(x + 1), \quad (6x + 1)(x + 5), \quad (2x + 1)(3x + 5), \quad (2x + 5)(3x + 1)$$

You can see that $(2x + 5)(3x + 1)$ is the correct factorization because the outer (O) and inner (I) products add up to $17x$.

$$
\begin{array}{ccccc}
\text{F} & \text{O} & \text{I} & \text{L} & \text{O} + \text{I} \\
\end{array}
$$
$$(2x + 5)(3x + 1) = 6x^2 + 2x + 15x + 5 = 6x^2 + 17x + 5.$$

Example 12   Factoring a Trinomial: Leading Coefficient Is 1

Factor $x^2 - 7x + 12$.

Solution

The possible factorizations are as follows.

$$(x - 2)(x - 6)$$
$$(x - 1)(x - 12)$$
$$(x - 3)(x - 4)$$

Testing the middle term, you will find the correct factorization to be

$$x^2 - 7x + 12 = (x - 3)(x - 4). \qquad \text{O} + \text{I} = -4x + (-3x) = -7x$$

 CHECKPOINT   Now try Exercise 103.

Example 13   Factoring a Trinomial: Leading Coefficient Is Not 1

Factor $2x^2 + x - 15$.

Solution

The eight possible factorizations are as follows.

$$(2x - 1)(x + 15) \qquad (2x + 1)(x - 15)$$
$$(2x - 3)(x + 5) \qquad (2x + 3)(x - 5)$$
$$(2x - 5)(x + 3) \qquad (2x + 5)(x - 3)$$
$$(2x - 15)(x + 1) \qquad (2x + 15)(x - 1)$$

Testing the middle term, you will find the correct factorization to be

$$2x^2 + x - 15 = (2x - 5)(x + 3). \qquad \text{O} + \text{I} = 6x - 5x = x$$

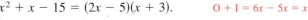 **CHECKPOINT**   Now try Exercise 111.

Factoring by Grouping

Sometimes polynomials with more than three terms can be factored by a method called **factoring by grouping.**

Example 14 Factoring by Grouping

Use factoring by grouping to factor $x^3 - 2x^2 - 3x + 6$.

Solution

$$
\begin{aligned}
x^3 - 2x^2 - 3x + 6 &= (x^3 - 2x^2) - (3x - 6) && \text{Group terms.} \\
&= x^2(x - 2) - 3(x - 2) && \text{Factor groups.} \\
&= (x - 2)(x^2 - 3) && (x - 2) \text{ is a common factor.}
\end{aligned}
$$

✓**CHECKPOINT** Now try Exercise 115.

Study Tip

When grouping terms, be sure to strategically group terms that have a common factor.

Factoring a trinomial can involve quite a bit of trial and error. Some of this trial and error can be lessened by using factoring by grouping. The key to this method of factoring is knowing how to rewrite the middle term. In general, to factor a trinomial

$$ax^2 + bx + c$$

by grouping, choose factors of the product ac that add up to b and use these factors to rewrite the middle term.

Example 15 Factoring a Trinomial by Grouping

Use factoring by grouping to factor $2x^2 + 5x - 3$.

Solution

In the trinomial $2x^2 + 5x - 3$, $a = 2$ and $c = -3$, which implies that the product ac is -6. Now, because -6 factors as $(6)(-1)$ and

$$6 - 1 = 5 = b$$

rewrite the middle term as

$$5x = 6x - x.$$

This produces the following.

$$
\begin{aligned}
2x^2 + 5x - 3 &= 2x^2 + 6x - x - 3 && \text{Rewrite middle term.} \\
&= (2x^2 + 6x) - (x + 3) && \text{Group terms.} \\
&= 2x(x + 3) - (x + 3) && \text{Factor groups.} \\
&= (x + 3)(2x - 1) && (x + 3) \text{ is a common factor.}
\end{aligned}
$$

So, the trinomial factors as $2x^2 + 5x - 3 = (x + 3)(2x - 1)$.

✓**CHECKPOINT** Now try Exercise 117.

Guidelines for Factoring Polynomials

1. Factor out any common factors using the Distributive Property.

2. Factor according to one of the special polynomial forms.

3. Factor as $ax^2 + bx + c = (mx + r)(nx + s)$.

4. Factor by grouping.

P.3   Exercises

Vocabulary and Concept Check

In Exercises 1–4, fill in the blank(s).

1. For the polynomial $a_n x^n + a_{n-1} x^{n-1} + \cdots + a_1 x + a_0$, the degree is _____ and the leading coefficient is _____ .

2. A polynomial with one term is called a _____ .

3. The letters in "FOIL" stand for the following.
   F _____   O _____   I _____   L _____

4. When a polynomial cannot be factored using integer coefficients, it is called _____ .

5. When is a polynomial completely factored?   *when each of his factor is form*

6. List four guidelines for factoring polynomials.

Procedures and Problem Solving

Identifying Polynomials   In Exercises 7–12, match the polynomial with its description. [The polynomials are labeled (a), (b), (c), (d), (e), and (f).]

(a) $6x$

(b) $1 - 4x^3$

(c) $x^3 + 2x^2 - 4x + 1$

(d) 7^0

(e) $-3x^5 + 2x^3 + x$

(f) $\frac{3}{4}x^4 + x^2 + 14$

7. A polynomial of degree zero

8. A trinomial of degree five

9. A binomial with leading coefficient -4

10. A monomial of positive degree

11. A trinomial with leading coefficient $\frac{3}{4}$

12. A third-degree polynomial with leading coefficient 1

Writing a Polynomial   In Exercises 13–16, write a polynomial that fits the description. (There are many correct answers.)

13. A third-degree polynomial with leading coefficient -2

14. A fifth-degree polynomial with leading coefficient 8

15. A fourth-degree polynomial with a negative leading coefficient

16. A third-degree trinomial with an even leading coefficient

Writing a Polynomial in Standard Form   In Exercises 17–22, write the polynomial in standard form. Then identify the degree and leading coefficient of the polynomial.

17. $3x + 4x^2 + 2$

18. $x^2 - 4 - 3x^4$

19. $-8 + x^7$

20. $23 - x^3$

21. $1 - x + 6x^4 - 2x^5$

22. $7 + 8x$

Classifying an Expression   In Exercises 23–26, determine whether the expression is a polynomial. If so, write the polynomial in standard form.

23. $3x + 4x^3 - 5$

24. $5x^4 - 2x^2 + x^{-2}$

25. $\sqrt{x^2 - x^4}$

26. $\dfrac{x^2 + 2x - 3}{6}$

Performing Operations with Polynomials   In Exercises 27–40, perform the operations and write the result in standard form.

27. $(t^2 - 3) + (6t^2 - 4t)$

28. $(4x + 1) + (-3x^2 - x + 9)$ *$2x^2 - x^2 - 5 - 3x + 1$*

29. $(8x + 5) - (6x - 12)$

30. $(x^2 - 5) - (2x^2 - 3x + 1)$

31. $(2x^3 - 9x^2 - 20) + (-2x^3 + 10x^2)$

32. $(y^3 - 6y + 3) + (5y^3 - 2y^2 + y - 10)$

33. $(15x^2 - 6) - (-8.1x^3 - 14.7x^2 - 17)$

34. $(15.6w - 14w - 17.4) - (16.9w^4 - 9.2w + 13)$

35. $-3z(5z - 1)$

36. $-7x(4 - x^3)$

37. $\left(5 - \frac{3}{2}y\right)(4y)$

38. $\left(\frac{1}{6}x + 1\right)(-2x^2)$

39. $3x(x^2 - 2x + 1)$

40. $y^2(4y^2 + 2y - 3)$

Multiplying Polynomials   In Exercises 41–68, multiply or find the special product.

41. $(x + 3)(x + 4)$

42. $(x - 5)(x + 10)$

43. $(3x - 5)(2x + 1)$

44. $(7x - 2)(4x - 3)$

45. $(2 - 5x)^2$

46. $(5x + 8y)^2$

47. $(x - 9)(x + 9)$

48. $(5x + 6)(5x - 6)$

49. $(x + 2y)(x - 2y)$

50. $(4a + 5b)(4a - 5b)$

51. $(2r^2 - 5)(2r^2 + 5)$

52. $(3a^3 - 4b^2)(3a^3 + 4b^2)$

53. $(x + 1)^3$

54. $(y - 4)^3$

55. $(2x - y)^3$

56. $(3x + 2y)^3$

57. $\left(\frac{1}{4}x - 3\right)\left(\frac{1}{4}x + 3\right)$

58. $\left(2x + \frac{1}{6}\right)\left(2x - \frac{1}{6}\right)$

59. $(2.4x + 3)^2$

60. $(1.8y - 5)^2$

61. $(-x^2 + x - 5)(3x^2 + 4x + 1)$

62. $(x^2 + 3x + 2)(2x^2 - x + 4)$

✓ **63.** $[(x + z) + 5][(x + z) - 5]$

64. $[(x - 3y) + z][(x - 3y) - z]$

65. $[(x - 3) + y]^2$ **66.** $[(x + 1) - y]^2$

67. $5x(x + 1) - 3x(x + 1)$

68. $(2x - 1)(x + 3) + 3(x + 3)$

Removing Common Factors In Exercises 69–74, factor out the common factor.

69. $5x - 40$ **70.** $4y + 20$

71. $2x^3 - 6x$ **72.** $3z^3 - 6z^2 + 9z$

✓ **73.** $3x(x - 5) + 8(x - 5)$ **74.** $(5x - 4)^2 + (5x - 4)$

Factoring the Difference of Two Squares In Exercises 75–82, factor the difference of two squares.

75. $x^2 - 64$ **76.** $x^2 - 81$

✓ **77.** $48y^2 - 27$ **78.** $50 - 98z^2$

79. $4x^2 - \frac{1}{9}$ **80.** $\frac{25}{36}y^2 - 49$

✓ **81.** $(x - 1)^2 - 4$ **82.** $25 - (z + 5)^2$

Factoring a Perfect Square Trinomial In Exercises 83–90, factor the perfect square trinomial.

83. $x^2 - 4x + 4$ **84.** $x^2 + 10x + 25$

85. $x^2 + x + \frac{1}{4}$ **86.** $x^2 - \frac{4}{3}x + \frac{4}{9}$

✓ **87.** $4x^2 - 12x + 9$ **88.** $25z^2 - 10z + 1$

89. $4x^2 - \frac{4}{3}x + \frac{1}{9}$ **90.** $9y^2 - \frac{3}{2}y + \frac{1}{16}$

Factoring the Sum or Difference of Cubes In Exercises 91–100, factor the sum or difference of cubes.

✓ **91.** $x^3 - 64$ **92.** $z^3 - 216$

✓ **93.** $x^3 + 1$ **94.** $y^3 + 125$

95. $x^3 - \frac{8}{27}$ **96.** $x^3 + \frac{8}{125}$

97. $8x^3 - 1$ **98.** $27x^3 + 64$

99. $(x + 2)^3 - y^3$ **100.** $(x - 3y)^3 - 8z^3$

Factoring a Trinomial In Exercises 101–114, factor the trinomial.

101. $x^2 + x - 2$ **102.** $x^2 + 6x + 8$

✓ **103.** $s^2 - 5s + 6$ **104.** $t^2 - t - 6$

105. $20 - y - y^2$ **106.** $24 + 5z - z^2$

107. $3x^2 - 5x + 2$ **108.** $3x^2 + 13x - 10$

109. $2x^2 - x - 1$ **110.** $2x^2 - x - 21$

✓ **111.** $5x^2 + 26x + 5$ **112.** $8x^2 - 45x - 18$

113. $-5u^2 - 13u + 6$ **114.** $-6x^2 + 23x + 4$

Factoring by Grouping In Exercises 115–120, factor by grouping.

✓ **115.** $x^3 - x^2 + 2x - 2$ **116.** $x^3 + 5x^2 - 5x - 25$

✓ **117.** $6x^2 + x - 2$ **118.** $3x^2 + 10x + 8$

119. $x^3 - 5x^2 + x - 5$ **120.** $x^3 - x^2 + 3x - 3$

Factoring an Expression Completely In Exercises 121–148, completely factor the expression.

121. $10x^2 - 40$ **122.** $7z^2 - 63$

123. $y^3 - y$ **124.** $x^3 - 9x^2$

125. $x^2 - 2x + 1$ **126.** $9x^2 - 6x + 1$

127. $1 - 4x + 4x^2$ **128.** $16 - 6x - x^2$

129. $2x^2 + 4x - 2x^3$ **130.** $7y^2 + 15y - 2y^3$

131. $9x^2 + 10x + 1$ **132.** $13x + 6 + 5x^2$

133. $\frac{1}{8}x^2 - \frac{1}{96}x - \frac{1}{16}$ **134.** $\frac{1}{81}x^2 + \frac{2}{9}x - 8$

135. $3x^3 + x^2 + 15x + 5$ **136.** $5 - x + 5x^2 - x^3$

137. $3u - 2u^2 + 6 - u^3$ **138.** $x^4 - 4x^3 + x^2 - 4x$

139. $2x^3 + x^2 - 8x - 4$ **140.** $3x^3 + x^2 - 27x - 9$

141. $(x^2 + 1)^2 - 4x^2$ **142.** $(x^2 + 8)^2 - 36x^2$

143. $3t^3 + 24$ **144.** $4x^3 - 32$

145. $4x(2x - 1) + 2(2x - 1)^2$

146. $5(3 - 4x)^2 - 8(3 - 4x)(5x - 1)$

147. $2(x + 1)(x - 3)^2 - 3(x + 1)^2(x - 3)$

148. $7(3x + 2)^2(1 - x)^2 + (3x + 2)(1 - x)^3$

149. MODELING DATA

After 2 years, an investment of \$1000 compounded annually at an interest rate r will yield an amount of $1000(1 + r)^2$.

(a) Write this polynomial in standard form.

(b) Use a calculator to evaluate the polynomial for the values of r shown in the table.

r	$2\frac{1}{2}\%$	3%	4%	$4\frac{1}{2}\%$	5%
$1000(1 + r)^2$					

(c) What conclusion can you make from the table?

150. Manufacturing A shipping company is constructing an open box made by cutting squares out of the corners of a piece of cardboard that is 18 centimeters by 26 centimeters (see figure). The edge of each cut-out square is x centimeters. Find the volume of the box in terms of x. Find the volume when $x = 1$, $x = 2$, and $x = 3$.

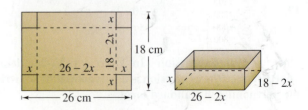

151. Public Safety The stopping distance of an automobile is the distance traveled during the driver's reaction time plus the distance traveled after the brakes are applied. In an experiment, these distances were measured (in feet) when the automobile was traveling at a speed of x miles per hour on dry, level pavement, as shown in the bar graph. The distance traveled during the reaction time R was $R = 1.1x$ and the braking distance B was $B = 0.0475x^2 - 0.001x + 0.23$.

(a) Determine the polynomial that represents the total stopping distance T.

(b) Use the result of part (a) to estimate T when $x = 30$, $x = 40$, and $x = 55$.

(c) Use the bar graph to make a statement about the total stopping distance required at increasing speeds.

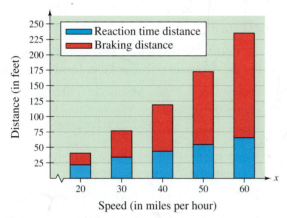

152. *Why you should learn it* (*p. 23*) A uniformly distributed load is placed on a one-inch-wide steel beam. When the span of the beam is x feet and its depth is 6 inches, the safe load S (in pounds) is approximated by

$$S_6 = (0.06x^2 - 2.42x + 38.71)^2.$$

When the depth is 8 inches, the safe load is approximated by

$$S_8 = (0.08x^2 - 3.30x + 51.93)^2.$$

(a) Use the bar graph to estimate the difference in the safe loads for these two beams when the span is 12 feet.

(b) How does the difference in safe load change as the span increases?

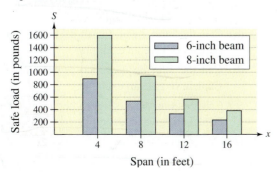

Geometric Modeling In Exercises 153 and 154, match the geometric factoring model with the correct factoring formula. For instance, the figure shown below is a factoring model for $2x^2 + 3x + 1 = (2x + 1)(x + 1)$.

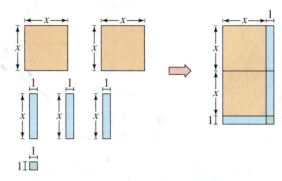

153.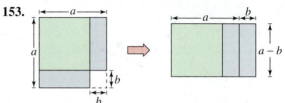

(a) $a^2 - b^2 = (a + b)(a - b)$

(b) $a^2 - 2ab + b^2 = (a - b)^2$

(c) $a^2 + 2ab + b^2 = (a + b)^2$

154.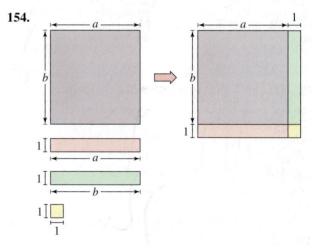

(a) $a^2 + 2a + 1 = (a + 1)^2$

(b) $ab + a + b + 1 = (a + 1)(b + 1)$

(c) $ab - a + b - 1 = (a + 1)(b - 1)$

Geometric Modeling In Exercises 155–158, draw a geometric factoring model to represent the factorization.

155. $3x^2 + 7x + 2 = (3x + 1)(x + 2)$

156. $x^2 + 4x + 3 = (x + 3)(x + 1)$

157. $2x^2 + 7x + 3 = (2x + 1)(x + 3)$

158. $x^2 + 3x + 2 = (x + 2)(x + 1)$

Geometry In Exercises 159–162, write an expression in factored form for the area of the shaded portion of the figure.

159.

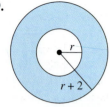

160.

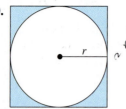

161.

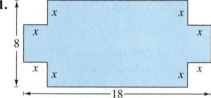

162.

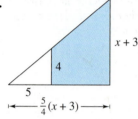

𝄐 **Factoring an Expression Completely** In Exercises 163–168, factor the expression completely.

163. $x^4(4)(2x + 1)^3(2x) + (2x + 1)^4(4x^3)$

164. $x^3(3)(x^2 + 1)^2(2x) + (x^2 + 1)^3(3x^2)$

165. $(2x - 5)^4(3)(5x - 4)^2(5) + (5x - 4)^3(4)(2x - 5)^3(2)$

166. $(x^2 - 5)^3(2)(4x + 3)(4) + (4x + 3)^2(3)(x^2 - 5)^2(x^2)$

167. $\dfrac{(5x - 1)(3) - (3x + 1)(5)}{(5x - 1)^2}$

168. $\dfrac{(2x + 3)(4) - (4x - 1)(2)}{(2x + 3)^2}$

Classifying an Expression In Exercises 169–172, find all values of b for which the trinomial can be factored with integer coefficients.

169. $x^2 + bx - 15$

170. $x^2 + bx - 12$

171. $x^2 + bx + 50$

172. $x^2 + bx + 24$

Classifying an Expression In Exercises 173–176, find two integer values of c such that the trinomial can be factored. (There are many correct answers.)

173. $2x^2 + 5x + c$ **174.** $3x^2 - x + c$

175. $3x^2 - 10x + c$ **176.** $2x^2 + 9x + c$

177. Geometry The cylindrical shell shown in the figure has a volume of

$$V = \pi R^2 h - \pi r^2 h.$$

(a) Factor the expression for the volume.

(b) From the result of part (a), show that the volume is

$2\pi(\text{average radius})(\text{thickness of the shell})h.$

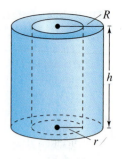

178. Chemistry The rate of change of an autocatalytic chemical reaction is $kQx - kx^2$, where Q is the amount of the original substance, x is the amount of substance formed, and k is a constant of proportionality. Factor the expression.

Conclusions

True or False? In Exercises 179 and 180, determine whether the statement is true or false. Justify your answer.

179. The product of two binomials is always a second-degree polynomial.

180. The product of two binomials is always a trinomial.

181. Exploration Find the degree of the product of two polynomials of degrees m and n.

182. Writing Write a paragraph explaining to a classmate why $(x + y)^2 \neq x^2 + y^2$.

183. Writing Write a paragraph explaining to a classmate a pattern that can be used to cube a binomial difference. Then use your pattern to cube the difference $(x - y)$.

184. Writing Explain what is meant when it is said that a polynomial is in factored form.

185. Think About It Is $(3x - 6)(x + 1)$ completely factored? Explain.

186. Think About It Must the sum of two second-degree polynomials be a second-degree polynomial? If not, give an example.

187. Error Analysis Describe the error.

$9x^2 - 9x - 54 = (3x + 6)(3x - 9)$
$= 3(x + 2)(x - 3)$

188. CAPSTONE A third-degree polynomial and a fourth-degree polynomial are added.

(a) Can the sum be a fourth-degree polynomial? Explain or give an example.

(b) Can the sum be a second-degree polynomial? Explain or give an example.

(c) Can the sum be a seventh-degree polynomial? Explain or give an example.

P.4 Rational Expressions

Domain of an Algebraic Expression

The set of real numbers for which an algebraic expression is defined is the **domain** of the expression. Two algebraic expressions are **equivalent** when they have the same domain and yield the same values for all numbers in their domain. For instance, the expressions

$$(x + 1) + (x + 2) \quad \text{and} \quad 2x + 3$$

are equivalent because

$$(x + 1) + (x + 2) = x + 1 + x + 2$$
$$= x + x + 1 + 2$$
$$= 2x + 3.$$

Example 1   Finding the Domain of an Algebraic Expression

Find the domain of the expression.

a. $2x^3 + 3x + 4$

b. $\sqrt{x - 2}$

c. $\dfrac{x + 2}{x - 3}$

Solution

a. The domain of the polynomial

$$2x^3 + 3x + 4$$

is the set of all real numbers. In fact, the domain of any polynomial is the set of all real numbers, unless the domain is specifically restricted.

b. The domain of the radical expression

$$\sqrt{x - 2}$$

is the set of real numbers greater than or equal to 2, because the square root of a negative number is not a real number.

c. The domain of the expression

$$\frac{x + 2}{x - 3}$$

is the set of all real numbers except

$$x = 3$$

which would result in division by zero, which is undefined.

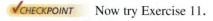 **CHECKPOINT**   Now try Exercise 11.

The quotient of two algebraic expressions is a **fractional expression.** Moreover, the quotient of two *polynomials* such as

$$\frac{1}{x}, \quad \frac{2x - 1}{x + 1}, \quad \text{or} \quad \frac{x^2 - 1}{x^2 + 1}$$

is a **rational expression.**

What you should learn

- Find domains of algebraic expressions.
- Simplify rational expressions.
- Add, subtract, multiply, and divide rational expressions.
- Simplify complex fractions.
- Simplify expressions from calculus.

Why you should learn it

Rational expressions are useful in estimating the temperature of food as it cools. For instance, a rational expression is used in Exercise 105 on page 44 to model the temperature of food as it cools in a refrigerator set at 40°F.

Simplifying Rational Expressions

Recall that a fraction is in simplest form when its numerator and denominator have no factors in common aside from ± 1. To write a fraction in simplest form, divide out common factors.

$$\frac{a \cdot \cancel{c}}{b \cdot \cancel{c}} = \frac{a}{b}, \quad c \neq 0$$

The key to success in simplifying rational expressions lies in your ability to *factor* polynomials. When simplifying rational expressions, be sure to factor each polynomial completely before concluding that the numerator and denominator have no factors in common.

Example 2 Simplifying a Rational Expression

Write $\dfrac{x^2 + 4x - 12}{3x - 6}$ in simplest form.

Solution

$$\frac{x^2 + 4x - 12}{3x - 6} = \frac{(x + 6)\cancel{(x - 2)}}{3\cancel{(x - 2)}} \qquad \text{Factor completely.}$$

$$= \frac{x + 6}{3}, \quad x \neq 2 \qquad \text{Divide out common factors.}$$

Note that the original expression is undefined when $x = 2$ (because division by zero is undefined). To make sure that the simplified expression is *equivalent* to the original expression, you must restrict the domain of the simplified expression by excluding the value $x = 2$.

✓**CHECKPOINT** Now try Exercise 33.

It may sometimes be necessary to change the sign of a factor by factoring out (-1) to simplify a rational expression, as shown in Example 3.

Example 3 Simplifying a Rational Expression

$$\frac{12 + x - x^2}{2x^2 - 9x + 4} = \frac{(4 - x)(3 + x)}{(2x - 1)(x - 4)} \qquad \text{Factor completely.}$$

$$= \frac{-\cancel{(x - 4)}(3 + x)}{(2x - 1)\cancel{(x - 4)}} \qquad (4 - x) = -(x - 4)$$

$$= -\frac{3 + x}{2x - 1}, \quad x \neq 4 \qquad \text{Divide out common factors.}$$

✓**CHECKPOINT** Now try Exercise 41.

In this text, when a rational expression is written, the domain is usually not listed with the expression. It is *implied* that the real numbers that make the denominator zero are excluded from the domain. Also, when performing operations with rational expressions, this text follows the convention of listing *by the simplified expression* all values of x that must be specifically excluded from the domain in order to make the domains of the simplified and original expressions agree. In Example 3, for instance, the restriction $x \neq 4$ is listed with the simplified expression to make the two domains agree. Note that the value $x = \frac{1}{2}$ is excluded from *both* domains, so it is not necessary to list this value.

Study Tip

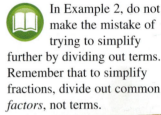

In Example 2, do not make the mistake of trying to simplify further by dividing out terms. Remember that to simplify fractions, divide out common *factors*, not terms.

Operations with Rational Expressions

To multiply or divide rational expressions, you can use the properties of fractions discussed in Section P.1. Recall that to divide fractions you invert the divisor and multiply.

Example 4   Multiplying Rational Expressions

$$\frac{2x^2 + x - 6}{x^2 + 4x - 5} \cdot \frac{x^3 - 3x^2 + 2x}{4x^2 - 6x} = \frac{(2x - 3)(x + 2)}{(x + 5)(x - 1)} \cdot \frac{x(x - 2)(x - 1)}{2x(2x - 3)}$$

$$= \frac{(x + 2)(x - 2)}{2(x + 5)}, \quad x \neq 0, x \neq 1, x \neq \frac{3}{2}$$

✔**CHECKPOINT**   Now try Exercise 57.

In Example 4, the restrictions

$$x \neq 0, \quad x \neq 1, \quad \text{and} \quad x \neq \frac{3}{2}$$

are listed with the simplified expression in order to make the two domains agree. Note that the value $x = -5$ is excluded from both domains, so it is not necessary to list this value.

Example 5   Dividing Rational Expressions

$$\frac{x^3 - 8}{x^2 - 4} \div \frac{x^2 + 2x + 4}{x^3 + 8} = \frac{x^3 - 8}{x^2 - 4} \cdot \frac{x^3 + 8}{x^2 + 2x + 4} \qquad \text{Invert and multiply.}$$

$$= \frac{(x - 2)(x^2 + 2x + 4)}{(x + 2)(x - 2)} \cdot \frac{(x + 2)(x^2 - 2x + 4)}{(x^2 + 2x + 4)}$$

$$= x^2 - 2x + 4, \quad x \neq \pm 2 \qquad \text{Divide out common factors.}$$

(handwritten annotations: $(x \cdot 2)(x^2 - 2x + 4)$, $x^3 - 8 \cdot x^3 + 8$, $x^2 - 4 \cdot x^2 + 2x + 4$, $(x+2)\ (x-2)$)

✔**CHECKPOINT**   Now try Exercise 59.

To add or subtract rational expressions, you can use the LCD (least common denominator) method or the basic definition

$$\frac{a}{b} \pm \frac{c}{d} = \frac{ad \pm bc}{bd}, \quad b \neq 0 \text{ and } d \neq 0. \qquad \text{Basic definition}$$

This definition provides an efficient way of adding or subtracting *two* fractions that have no common factors in their denominators.

Example 6   Subtracting Rational Expressions

Subtract $\dfrac{2}{3x + 4}$ from $\dfrac{x}{x - 3}$.

Solution

$$\frac{x}{x - 3} - \frac{2}{3x + 4} = \frac{x(3x + 4) - 2(x - 3)}{(x - 3)(3x + 4)} \qquad \text{Basic definition}$$

$$= \frac{3x^2 + 4x - 2x + 6}{(x - 3)(3x + 4)} \qquad \text{Distributive Property}$$

$$= \frac{3x^2 + 2x + 6}{(x - 3)(3x + 4)} \qquad \text{Combine like terms.}$$

✔**CHECKPOINT**   Now try Exercise 63.

Study Tip

 When subtracting rational expressions, remember to distribute the negative sign to *all* the terms in the quantity that is being subtracted.

For three or more fractions, or for fractions with a repeated factor in the denominators, the LCD method works well. Recall that the least common denominator of several fractions consists of the product of all prime factors in the denominators, with each factor given the highest power of its occurrence in any denominator. Here is a numerical example.

$$\frac{1}{6} + \frac{3}{4} - \frac{2}{3} = \frac{1 \cdot 2}{6 \cdot 2} + \frac{3 \cdot 3}{4 \cdot 3} - \frac{2 \cdot 4}{3 \cdot 4} \qquad \text{The LCD is 12.}$$

$$= \frac{2}{12} + \frac{9}{12} - \frac{8}{12}$$

$$= \frac{3}{12}$$

$$= \frac{1}{4}$$

Sometimes the numerator of the answer has a factor in common with the denominator. In such cases the answer should be simplified. For instance, in the example above, $\frac{3}{12}$ was simplified to $\frac{1}{4}$.

Example 7 Combining Rational Expressions: The LCD Method

Perform the operations and simplify.

$$\frac{3}{x - 1} - \frac{2}{x} + \frac{x + 3}{x^2 - 1}$$

Solution

Using the factored denominators

$$(x - 1), \quad x, \quad \text{and} \quad (x + 1)(x - 1)$$

you can see that the LCD is $x(x + 1)(x - 1)$.

$$\frac{3}{x - 1} - \frac{2}{x} + \frac{x + 3}{x^2 - 1}$$

$$= \frac{3}{x - 1} - \frac{2}{x} + \frac{x + 3}{(x + 1)(x - 1)}$$

$$= \frac{3(x)(x + 1)}{x(x + 1)(x - 1)} - \frac{2(x + 1)(x - 1)}{x(x + 1)(x - 1)} + \frac{(x + 3)(x)}{x(x + 1)(x - 1)}$$

$$= \frac{3(x)(x + 1) - 2(x + 1)(x - 1) + (x + 3)(x)}{x(x + 1)(x - 1)}$$

$$= \frac{3x^2 + 3x - 2x^2 + 2 + x^2 + 3x}{x(x + 1)(x - 1)} \qquad \text{Distributive Property}$$

$$= \frac{(3x^2 - 2x^2 + x^2) + (3x + 3x) + 2}{x(x + 1)(x - 1)} \qquad \text{Group like terms.}$$

$$= \frac{2x^2 + 6x + 2}{x(x + 1)(x - 1)} \qquad \text{Combine like terms.}$$

$$= \frac{2(x^2 + 3x + 1)}{x(x + 1)(x - 1)} \qquad \text{Factor.}$$

✔**CHECKPOINT**   Now try Exercise 69.

Complex Fractions

Fractional expressions with separate fractions in the numerator, denominator, or both are called **complex fractions.** For instance,

$$\dfrac{\left(\dfrac{1}{x}\right)}{x^2 + 1}$$

and

$$\dfrac{\left(\dfrac{1}{x}\right)}{\left(\dfrac{1}{x^2 + 1}\right)}$$

are complex fractions. A complex fraction can be simplified by combining the fractions in its numerator into a single fraction and then combining the fractions in its denominator into a single fraction. Then invert the denominator and multiply.

Example 8   Simplifying a Complex Fraction

$$\dfrac{\left(\dfrac{2}{x} - 3\right)}{\left(1 - \dfrac{1}{x - 1}\right)} = \dfrac{\left[\dfrac{2 - 3(x)}{x}\right]}{\left[\dfrac{1(x - 1) - 1}{x - 1}\right]} \qquad \text{Combine fractions.}$$

$$= \dfrac{\left(\dfrac{2 - 3x}{x}\right)}{\left(\dfrac{x - 2}{x - 1}\right)} \qquad \text{Simplify.}$$

$$= \dfrac{2 - 3x}{x} \cdot \dfrac{x - 1}{x - 2} \qquad \text{Invert and multiply.}$$

$$= \dfrac{(2 - 3x)(x - 1)}{x(x - 2)}, \quad x \neq 1$$

✓**CHECKPOINT**   Now try Exercise 77.

In Example 8, the restriction $x \neq 1$ is listed by the final expression to make its domain agree with the domain of the original expression.

Another way to simplify a complex fraction is to multiply each term in its numerator and denominator by the LCD of all fractions in its numerator and denominator. This method is applied to the fraction in Example 8 as follows.

$$\dfrac{\left(\dfrac{2}{x} - 3\right)}{\left(1 - \dfrac{1}{x - 1}\right)} = \dfrac{\left(\dfrac{2}{x} - 3\right)}{\left(1 - \dfrac{1}{x - 1}\right)} \cdot \dfrac{x(x - 1)}{x(x - 1)} \qquad \text{LCD is } x(x - 1).$$

$$= \dfrac{\left(\dfrac{2 - 3x}{x}\right) \cdot x(x - 1)}{\left(\dfrac{x - 2}{x - 1}\right) \cdot x(x - 1)} \qquad \text{Combine fractions.}$$

$$= \dfrac{(2 - 3x)(x - 1)}{x(x - 2)}, \quad x \neq 1 \qquad \text{Simplify.}$$

Simplifying Expressions from Calculus

The next four examples illustrate some methods for simplifying expressions involving negative exponents and radicals. These types of expressions occur frequently in calculus.

To simplify an expression with negative exponents, one method is to begin by factoring out the common factor with the lesser exponent. Remember that when factoring, you subtract exponents. For instance, in

$$3x^{-5/2} + 2x^{-3/2}$$

the lesser exponent is $-\frac{5}{2}$ and the common factor is $x^{-5/2}$.

$$3x^{-5/2} + 2x^{-3/2} = x^{-5/2}[3(1) + 2x^{-3/2-(-5/2)}]$$

$$= x^{-5/2}(3 + 2x^1)$$

$$= \frac{3 + 2x}{x^{5/2}}$$

Example 9   Simplifying an Expression with Negative Exponents

Simplify $x(1 - 2x)^{-3/2} + (1 - 2x)^{-1/2}$.

Solution

Begin by factoring out the common factor with the lesser exponent.

$$x(1 - 2x)^{-3/2} + (1 - 2x)^{-1/2} = (1 - 2x)^{-3/2}[x + (1 - 2x)^{(-1/2)-(-3/2)}]$$

$$= (1 - 2x)^{-3/2}[x + (1 - 2x)^1]$$

$$= \frac{1 - x}{(1 - 2x)^{3/2}}$$

✔CHECKPOINT   Now try Exercise 83.

A second method for simplifying an expression with negative exponents involves multiplying the numerator and denominator by another expression to eliminate the negative exponent.

Example 10   Simplifying an Expression with Negative Exponents

Simplify $\dfrac{(4 - x^2)^{1/2} + x^2(4 - x^2)^{-1/2}}{4 - x^2}$.

Solution

$$\frac{(4 - x^2)^{1/2} + x^2(4 - x^2)^{-1/2}}{4 - x^2} = \frac{(4 - x^2)^{1/2} + x^2(4 - x^2)^{-1/2}}{4 - x^2} \cdot \frac{(4 - x^2)^{1/2}}{(4 - x^2)^{1/2}}$$

$$= \frac{(4 - x^2)^1 + x^2(4 - x^2)^0}{(4 - x^2)^{3/2}}$$

$$= \frac{4 - x^2 + x^2}{(4 - x^2)^{3/2}}$$

$$= \frac{4}{(4 - x^2)^{3/2}}$$

✔CHECKPOINT   Now try Exercise 87.

Example 11   Rewriting a Difference Quotient

The following expression from calculus is an example of a *difference quotient*.

$$\frac{\sqrt{x + h} - \sqrt{x}}{h}$$

Rewrite this expression by rationalizing its numerator.

Solution

$$\frac{\sqrt{x + h} - \sqrt{x}}{h} = \frac{\sqrt{x + h} - \sqrt{x}}{h} \cdot \frac{\sqrt{x + h} + \sqrt{x}}{\sqrt{x + h} + \sqrt{x}}$$

$$= \frac{\left(\sqrt{x + h}\right)^2 - \left(\sqrt{x}\right)^2}{h\left(\sqrt{x + h} + \sqrt{x}\right)}$$

$$= \frac{x + h - x}{h\left(\sqrt{x + h} + \sqrt{x}\right)}$$

$$= \frac{h}{h\left(\sqrt{x + h} + \sqrt{x}\right)}$$

$$= \frac{1}{\sqrt{x + h} + \sqrt{x}}, \quad h \neq 0$$

Notice that the original expression is undefined when $h = 0$. So, you must exclude $h = 0$ from the domain of the simplified expression so that the expressions are equivalent.

 Now try Exercise 93.

Difference quotients, like that in Example 11, occur frequently in calculus. Often, they need to be rewritten in an equivalent form that can be evaluated when $h = 0$. Note that the equivalent form is not simpler than the original form, but it has the advantage that it is defined when $h = 0$.

Example 12   Rewriting a Difference Quotient

Rewrite the expression by rationalizing its numerator.

$$\frac{\sqrt{x - 4} - \sqrt{x}}{4}$$

Solution

$$\frac{\sqrt{x - 4} - \sqrt{x}}{4} = \frac{\sqrt{x - 4} - \sqrt{x}}{4} \cdot \frac{\sqrt{x - 4} + \sqrt{x}}{\sqrt{x - 4} + \sqrt{x}}$$

$$= \frac{\left(\sqrt{x - 4}\right)^2 - \left(\sqrt{x}\right)^2}{4\left(\sqrt{x - 4} + \sqrt{x}\right)}$$

$$= \frac{x - 4 - x}{4\left(\sqrt{x - 4} + \sqrt{x}\right)}$$

$$= \frac{-4}{4\left(\sqrt{x - 4} + \sqrt{x}\right)}$$

$$= -\frac{1}{\sqrt{x - 4} + \sqrt{x}}$$

CHECKPOINT   Now try Exercise 95.

P.4   Exercises

See www.CalcChat.com for worked-out solutions to odd-numbered exercises.
For instructions on how to use a graphing utility, see Appendix A.

Vocabulary and Concept Check

In Exercises 1–4, fill in the blank.

1. The set of real numbers for which an algebraic expression is defined is the _____ of the expression.

2. The quotient of two algebraic expressions is a fractional expression and the quotient of two polynomials is a _____ .

3. Fractional expressions with separate fractions in the numerator, denominator, or both are called _____ .

4. To simplify an expression with negative exponents, it is possible to begin by factoring out the common factor with the _____ exponent.

5. When is a rational expression in simplest form?

6. What values are excluded from the domain of a rational expression?

Procedures and Problem Solving

Finding the Domain of an Algebraic Expression In Exercises 7–22, find the domain of the expression.

7. $3x^2 - 4x + 7$

8. $2x^2 + 5x - 2$

9. $4x^3 + 3, \quad x \geq 0$

10. $6x^2 - 9, \quad x > 0$

11. $\dfrac{1}{3 - x}$

12. $\dfrac{x + 6}{3x + 2}$

13. $\dfrac{x^2 - 1}{x^2 - 2x + 1}$

14. $\dfrac{x^2 - 5x + 6}{x^2 - 4}$

15. $\dfrac{x^2 - 2x - 3}{x^2 - 6x + 9}$

16. $\dfrac{x^2 - x - 12}{x^2 - 8x + 16}$

17. $\sqrt{x + 7}$

18. $\sqrt{4 - x}$

19. $\sqrt{2x - 5}$

20. $\sqrt{4x + 5}$

21. $\dfrac{1}{\sqrt{x - 3}}$

22. $\dfrac{1}{\sqrt{x + 2}}$

Writing Equivalent Fractions In Exercises 23–28, find the missing factor in the numerator such that the two fractions are equivalent.

23. $\dfrac{5}{2x} = \dfrac{5(\quad)}{6x^2}$

24. $\dfrac{2}{3x^2} = \dfrac{2(\quad)}{3x^4}$

25. $\dfrac{3}{4} = \dfrac{3(\quad)}{4(x + 1)}$

26. $\dfrac{2}{5} = \dfrac{2(\quad)}{5(x - 3)}$

27. $\dfrac{x - 1}{4(x + 2)} = \dfrac{(x - 1)(\quad)}{4(x + 2)^2}$

28. $\dfrac{x + 3}{2(x - 1)} = \dfrac{(x + 3)(\quad)}{2(x - 1)^2}$

Simplifying a Rational Expression In Exercises 29–46, write the rational expression in simplest form.

29. $\dfrac{15x^2}{10x}$

30. $\dfrac{18y^2}{60y^5}$

31. $\dfrac{3xy}{xy + x}$

32. $\dfrac{2x^2y}{xy - y}$

33. $\dfrac{4y - 8y^2}{10y - 5}$

34. $\dfrac{9x^2 + 9x}{2x + 2}$

35. $\dfrac{x - 5}{10 - 2x}$

36. $\dfrac{12 - 4x}{x - 3}$

37. $\dfrac{y^2 - 16}{y + 4}$

38. $\dfrac{x^2 - 25}{5 - x}$

39. $\dfrac{x^3 + 5x^2 + 6x}{x^2 - 4}$

40. $\dfrac{x^2 + 8x - 20}{x^2 + 11x + 10}$

41. $\dfrac{y^2 - 7y + 12}{y^2 + 3y - 18}$

42. $\dfrac{-10 - x}{x^2 + 11x + 10}$

43. $\dfrac{2 - x + 2x^2 - x^3}{x - 2}$

44. $\dfrac{x^2 - 9}{x^3 + x^2 - 9x - 9}$

45. $\dfrac{z^3 - 8}{z^2 + 2z + 4}$

46. $\dfrac{y^3 - 2y^2 - 3y}{y^3 + 1}$

Comparing Rational Expressions In Exercises 47 and 48, complete the table. What can you conclude?

47.

x	0	1	2	3	4	5	6
$\dfrac{x^2 - 2x - 3}{x - 3}$							
$x + 1$							

48.

x	0	1	2	3	4	5	6
$\dfrac{x - 3}{x^2 - x - 6}$							
$\dfrac{1}{x + 2}$							

Error Analysis   **In Exercises 49 and 50, describe the error.**

49. $\dfrac{5x^3}{2x^3 + 4} = \dfrac{5x^3}{2x^3 + 4} = \dfrac{5}{2+4} = \dfrac{5}{6}$

50. $\dfrac{x^3 + 25x}{x^2 - 2x - 15} = \dfrac{x(x^2 + 25)}{(x - 5)(x + 3)}$

$= \dfrac{x(x - 5)(x + 5)}{(x - 5)(x + 3)} = \dfrac{x(x + 5)}{x + 3}$

Geometry   **In Exercises 51 and 52, find the ratio of the area of the shaded portion of the figure to the total area of the figure.**

51.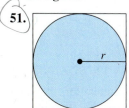

52.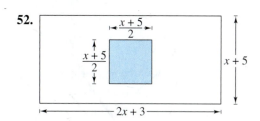

Multiplying or Dividing Rational Expressions **In Exercises 53–60, perform the multiplication or division and simplify.**

53. $\dfrac{5}{x - 1} \cdot \dfrac{x - 1}{25(x - 2)}$

54. $\dfrac{x + 13}{x^3(3 - x)} \cdot \dfrac{x(x - 3)}{5}$

55. $\dfrac{r}{r - 1} \div \dfrac{r^2}{r^2 - 1}$

56. $\dfrac{4y - 16}{5y + 15} \div \dfrac{4 - y}{2y + 6}$

57. $\dfrac{t^2 - t - 6}{t^2 + 6t + 9} \cdot \dfrac{t + 3}{t^2 - 4}$

58. $\dfrac{y^3 - 8}{2y^3} \cdot \dfrac{4y}{y^2 - 5y + 6}$

59. $\dfrac{3(x + y)}{4} \div \dfrac{x + y}{2}$

60. $\dfrac{x + 2}{5(x - 3)} \div \dfrac{x - 2}{5(x - 3)}$

Adding or Subtracting Rational Expressions **In Exercises 61–70, perform the addition or subtraction and simplify.**

61. $\dfrac{5}{x - 1} + \dfrac{x}{x - 1}$

62. $\dfrac{2x - 1}{x + 3} - \dfrac{1 - x}{x + 3}$

63. $\dfrac{6}{2x + 1} - \dfrac{x}{x + 3}$

64. $\dfrac{3}{x - 1} + \dfrac{5x}{3x + 4}$

65. $\dfrac{3}{x - 2} + \dfrac{5}{2 - x}$

66. $\dfrac{2x}{x - 5} - \dfrac{5}{5 - x}$

67. $\dfrac{1}{x^2 - x - 2} - \dfrac{x}{x^2 - 5x + 6}$

68. $\dfrac{2}{x^2 - x - 2} + \dfrac{10}{x^2 + 2x - 8}$

69. $-\dfrac{1}{x} + \dfrac{2}{x^2 + 1} - \dfrac{1}{x^3 + x}$

70. $\dfrac{2}{x + 1} + \dfrac{2}{x - 1} + \dfrac{1}{x^2 - 1}$

Error Analysis   **In Exercises 71 and 72, describe the error.**

71. $\dfrac{x - 4}{x} + \dfrac{2x - 3}{x + 1} = \dfrac{3x - 7}{2x + 1}$

72. $\dfrac{x + 4}{x + 2} - \dfrac{3x - 8}{x + 2} = \dfrac{x + 4 - 3x - 8}{x + 2}$

$= \dfrac{-2x - 4}{x + 2} = \dfrac{-2(x + 2)}{x + 2} = -2$

Simplifying a Complex Fraction **In Exercises 73–80, simplify the complex fraction.**

73. $\dfrac{\left(\dfrac{x}{2} - 1\right)}{(x - 2)}$

74. $\dfrac{(x - 4)}{\left(\dfrac{x}{4} - \dfrac{4}{x}\right)}$

75. $\dfrac{\left[\dfrac{x^2}{(x + 1)^2}\right]}{\left[\dfrac{x}{(x + 1)^3}\right]}$

76. $\dfrac{\left(\dfrac{x^2 - 1}{x}\right)}{\left[\dfrac{(x - 1)^2}{x}\right]}$

77. $\dfrac{\left[\dfrac{1}{(x + h)^2} - \dfrac{1}{x^2}\right]}{h}$

78. $\dfrac{\left(\dfrac{x + h}{x + h + 1} - \dfrac{x}{x + 1}\right)}{h}$

79. $\dfrac{\left(\sqrt{x} - \dfrac{1}{2\sqrt{x}}\right)}{\sqrt{x}}$

80. $\dfrac{\left(\dfrac{t^2}{\sqrt{t^2 + 1}} - \sqrt{t^2 + 1}\right)}{t^2}$

Simplifying an Expression with Negative Exponents **In Exercises 81–86, simplify the expression by removing the common factor with the lesser exponent.**

81. $x^5 - 2x^{-2}$

82. $x^5 - 5x^{-3}$

83. $x^2(x^2 + 1)^{-5} - (x^2 + 1)^{-4}$

84. $2x(x - 5)^{-3} - 4x^2(x - 5)^{-4}$

85. $2x^2(x - 1)^{1/2} - 5(x - 1)^{-1/2}$

86. $4x^3(2x - 1)^{3/2} - 2x(2x - 1)^{-1/2}$

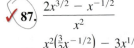 **Simplifying an Expression with Negative Exponents** In Exercises 87–92, simplify the expression.

✓ 87. $\dfrac{2x^{3/2} - x^{-1/2}}{x^2}$

88. $\dfrac{x^2\left(\frac{3}{2}x^{-1/2}\right) - 3x^{1/2}(x^2)}{x^4}$

89. $\dfrac{-x^2(x^2 + 1)^{-1/2} + 2x(x^2 + 1)^{-3/2}}{x^3}$

90. $\dfrac{x^3(4x^{-1/2}) - 3x^2\left(\frac{8}{3}x^{-3/2}\right)}{x^6}$

91. $\dfrac{(x^2 + 5)\left(\frac{1}{2}\right)(4x + 3)^{-1/2}(4) - (4x + 3)^{1/2}(2x)}{(x^2 + 5)^2}$

92. $\dfrac{(2x + 1)^{1/2}(3)(x - 5)^2 - (x - 5)^3\left(\frac{1}{2}\right)(2x + 1)^{-1/2}(2)}{2x + 1}$

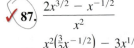 **Rewriting a Difference Quotient** In Exercises 93–98, rewrite the expression by rationalizing its numerator.

✓ 93. $\dfrac{\sqrt{x + 2} - \sqrt{x}}{2}$

94. $\dfrac{\sqrt{z - 3} - \sqrt{z}}{3}$

✓ 95. $\dfrac{\sqrt{x + 2} - \sqrt{2}}{x}$

96. $\dfrac{\sqrt{x + 5} - \sqrt{5}}{x}$

97. $\dfrac{\sqrt{x + 9} - 3}{x}$

98. $\dfrac{\sqrt{x + 4} - 2}{x}$

Probability In Exercises 99 and 100, consider an experiment in which a marble is tossed into a box whose base is shown in the figure. The probability that the marble will come to rest in the shaded portion of the base is equal to the ratio of the shaded area to the total area of the figure. Find the probability.

99. 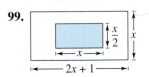   100.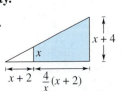

Finance In Exercises 101 and 102, the formula that approximates the annual interest rate r of a monthly installment loan is given by

$$r = \dfrac{\left[\dfrac{24(NM - P)}{N}\right]}{\left(P + \dfrac{NM}{12}\right)}$$

where N is the total number of payments, M is the monthly payment, and P is the amount financed.

101. (a) Approximate the annual interest rate for a four-year car loan of \$20,000 that has monthly payments of \$475.

    (b) Simplify the expression for the annual interest rate r, and then rework part (a).

102. (a) Approximate the annual interest rate for a five-year car loan of \$28,000 that has monthly payments of \$525.

    (b) Simplify the expression for the annual interest rate r, and then rework part (a).

103. **Using a Rate** A digital copier copies in color at a rate of 50 pages per minute.

    (a) Find the time required to copy one page.

    (b) Find the time required to copy x pages.

    (c) Find the time required to copy 120 pages.

104. **Electrical Engineering** The formula for the total resistance R_T (in ohms) of a parallel circuit is given by

$$R_T = \dfrac{1}{\dfrac{1}{R_1} + \dfrac{1}{R_2} + \dfrac{1}{R_3}}$$

    where R_1, R_2, and R_3 are the resistance values of the first, second, and third resistors, respectively.

    (a) Simplify the total resistance formula.

    (b) Find the total resistance in the parallel circuit when $R_1 = 6$ ohms, $R_2 = 4$ ohms, and $R_3 = 12$ ohms.

105. *Why you should learn it* (p. 35) When food (at room temperature) is placed in a refrigerator, the time required for the food to cool depends on the amount of food, the air circulation in the refrigerator, the original temperature of the food, and the temperature of the refrigerator. Consider the model that gives the temperature of food that is at 75°F and is placed in a 40°F refrigerator as

$$T = 10\left(\dfrac{4t^2 + 16t + 75}{t^2 + 4t + 10}\right)$$

    where T is the temperature (in degrees Fahrenheit) and t is the time (in hours).

    (a) Complete the table.

t	0	2	4	6	8	10
T						

t	12	14	16	18	20	22
T						

    (b) What value of T does the mathematical model appear to be approaching?

106. MODELING DATA

The table shows the rates (per 1000 of the total population) of marriages and divorces in the United States for the years 1994 through 2008. (Source: U.S. National Center for Health Statistics)

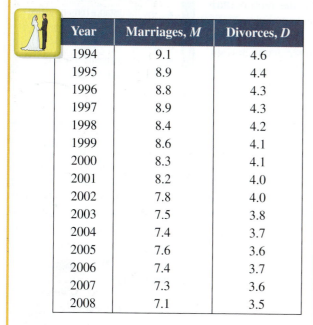

Year	Marriages, M	Divorces, D
1994	9.1	4.6
1995	8.9	4.4
1996	8.8	4.3
1997	8.9	4.3
1998	8.4	4.2
1999	8.6	4.1
2000	8.3	4.1
2001	8.2	4.0
2002	7.8	4.0
2003	7.5	3.8
2004	7.4	3.7
2005	7.6	3.6
2006	7.4	3.7
2007	7.3	3.6
2008	7.1	3.5

Mathematical models for the data are

Marriages: $M = \dfrac{0.04057t^2 - 0.9533t + 9.059}{0.00499t^2 - 0.1065t + 1}$

and

Divorces: $D = -0.07t + 4.8$

where t represents the year, with $t = 4$ corresponding to 1994.

(a) Using the models, create a table to estimate the number of marriages and the number of divorces for each of the given years. Compare these estimates with the actual data.

(b) Determine a model for the ratio of the number of marriages to the number of divorces. Use the model to find this ratio for each of the given years.

Simplifying a Difference Quotient In Exercises 107–110, simplify the expression.

107. $\dfrac{(x + h)^2 - x^2}{h}$, $h \neq 0$

108. $\dfrac{(x + h)^3 - x^3}{h}$, $h \neq 0$

109. $\dfrac{\dfrac{1}{(x + h)^2} - \dfrac{1}{x^2}}{h}$, $h \neq 0$

110. $\dfrac{\dfrac{1}{2(x + h)} - \dfrac{1}{2x}}{h}$, $h \neq 0$

Simplifying an Expression In Exercises 111 and 112, simplify the expression.

111. $\dfrac{4}{n}\left(\dfrac{n(n + 1)(2n + 1)}{6}\right) + 2n\left(\dfrac{4}{n}\right)$

112. $9\left(\dfrac{3}{n}\right)\left(\dfrac{n(n + 1)(2n + 1)}{6}\right) - n\left(\dfrac{3}{n}\right)$

Conclusions

True or False? In Exercises 113 and 114, determine whether the statement is true or false. Justify your answer.

113. $\dfrac{x^{2n} - 1^{2n}}{x^n - 1^n} = x^n + 1^n$

114. $\dfrac{x^2 - 3x + 2}{x - 1} = x - 2$

115. **Think About It** Explain why rewriting the difference quotient

$$\frac{\sqrt{x + h} - \sqrt{x}}{h}$$

by rationalizing its numerator does not result in an expression in simplest form.

116. **Think About It** Is the following statement true for all nonzero real numbers a and b? Explain.

$$\frac{ax - b}{b - ax} = -1$$

117. **Writing** Write a paragraph explaining to a classmate why $\sqrt{x + y} \neq \sqrt{x} + \sqrt{y}$.

118. **Writing** Write a paragraph explaining to a classmate why

$$\frac{1}{x + y} \neq \frac{1}{x} + \frac{1}{y}.$$

119. **Exploration** Simplify $2x^{-2} - x^{-4}$ by factoring out the common factor with the lesser exponent. Simplify

$$\frac{2x^{-2} - x^{-4}}{1}$$

by multiplying the numerator and denominator by an expression to eliminate the negative exponent. Discuss your results.

120. **CAPSTONE** In your own words, explain how to find the quotient of two rational expressions and express the result in simplest form. Discuss when it is necessary to indicate domain restrictions.

P.5   The Cartesian Plane

The Cartesian Plane

Just as you can represent real numbers by points on a real number line, you can represent ordered pairs of real numbers by points in a plane called the **rectangular coordinate system,** or the **Cartesian plane,** after the French mathematician René Descartes (1596–1650).

The Cartesian plane is formed by using two real number lines intersecting at right angles, as shown in Figure P.10. The horizontal real number line is usually called the **x-axis,** and the vertical real number line is usually called the **y-axis.** The point of intersection of these two axes is the **origin,** and the two axes divide the plane into four parts called **quadrants.**

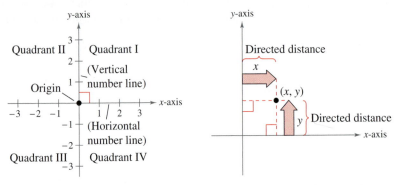

Figure P.10   *The Cartesian Plane*

Figure P.11   *Ordered Pair (x, y)*

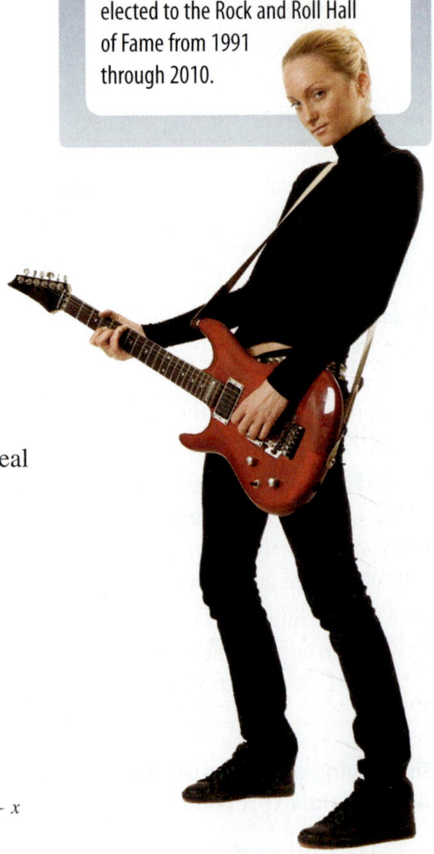

Each point in the plane corresponds to an **ordered pair** (x, y) of real numbers x and y, called **coordinates** of the point. The **x-coordinate** represents the directed distance from the y-axis to the point, and the **y-coordinate** represents the directed distance from the x-axis to the point, as shown in Figure P.11.

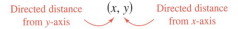

Directed distance from y-axis $\quad (x, y) \quad$ Directed distance from x-axis

The notation (x, y) denotes both a point in the plane and an open interval on the real number line. The context will tell you which meaning is intended.

Example 1   Plotting Points in the Cartesian Plane

Plot the points $(-1, 2)$, $(3, 4)$, $(0, 0)$, $(3, 0)$, and $(-2, -3)$ in the Cartesian plane.

Solution

To plot the point $(-1, 2)$, imagine a vertical line through -1 on the x-axis and a horizontal line through 2 on the y-axis. The intersection of these two lines is the point $(-1, 2)$. This point is one unit to the left of the y-axis and two units up from the x-axis. The other four points can be plotted in a similar way, as shown in Figure P.12.

✓CHECKPOINT   Now try Exercise 13.

Figure P.12

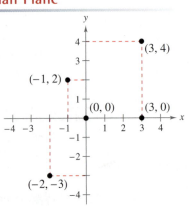

The beauty of a rectangular coordinate system is that it enables you to see relationships between two variables. It would be difficult to overestimate the importance of Descartes's introduction of coordinates in the plane. Today, his ideas are in common use in virtually every scientific and business-related field.

In the next example, data are represented graphically by points plotted in a rectangular coordinate system. This type of graph is called a **scatter plot.**

Example 2   Sketching a Scatter Plot

The numbers of employees E (in thousands) in dentist offices in the United States from 1999 through 2009 are shown in the table, where t represents the year. Sketch a scatter plot of the data by hand.   (Source: U.S. Bureau of Labor Statistics)

Dentist

Year, t	Employees, E
1999	667
2000	688
2001	705
2002	725
2003	744
2004	760
2005	774
2006	786
2007	808
2008	818
2009	818

Solution

Before you sketch the scatter plot, it is helpful to represent each pair of values by an ordered pair (t, E) as follows.

(1999, 667), (2000, 688), (2001, 705), (2002, 725), (2003, 744), (2004, 760), (2005, 774), (2006, 786), (2007, 808), (2008, 818), (2009, 818)

To sketch a scatter plot of the data shown in the table, first draw a vertical axis to represent the number of employees (in thousands) and a horizontal axis to represent the year. Then plot the resulting points, as shown in Figure P.13. Note that the break in the t-axis indicates that the numbers 0 through 1998 have been omitted.

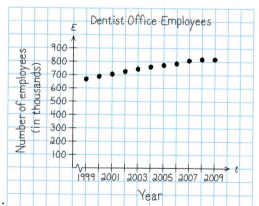

✔ CHECKPOINT   Now try Exercise 31.

Figure P.13

Technology Tip

You can use a graphing utility to graph the scatter plot in Example 2. For instructions on how to use the *list editor* and the *statistical plotting* feature, see Appendix A; for specific keystrokes, go to this textbook's *Companion Website.*

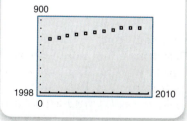

In Example 2, you could have let $t = 1$ represent the year 1999. In that case, the horizontal axis of the graph would not have been broken, and the tick marks would have been labeled 1 through 11 (instead of 1999 through 2009).

The Distance Formula

Recall from the Pythagorean Theorem that, for a right triangle with hypotenuse of length c and sides of lengths a and b, you have

$$a^2 + b^2 = c^2 \qquad \text{Pythagorean Theorem}$$

as shown in Figure P.14. (The converse is also true. That is, if $a^2 + b^2 = c^2$, then the triangle is a right triangle.)

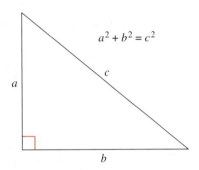

Figure P.14

Suppose you want to determine the distance d between two points (x_1, y_1) and (x_2, y_2) in the plane. With these two points, a right triangle can be formed, as shown in Figure P.15.

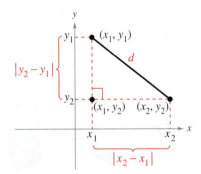

Figure P.15

The length of the vertical side of the triangle is

$$|y_2 - y_1| \qquad \text{Length of vertical side}$$

and the length of the horizontal side is

$$|x_2 - x_1|. \qquad \text{Length of horizontal side}$$

By the Pythagorean Theorem,

$$d^2 = |x_2 - x_1|^2 + |y_2 - y_1|^2$$
$$d = \sqrt{|x_2 - x_1|^2 + |y_2 - y_1|^2}$$
$$d = \sqrt{(x_2 - x_1)^2 + (y_2 - y_1)^2}.$$

This result is called the **Distance Formula.**

The Distance Formula

The distance d between the points (x_1, y_1) and (x_2, y_2) in the plane is

$$d = \sqrt{(x_2 - x_1)^2 + (y_2 - y_1)^2}.$$

Example 3   Finding a Distance

Find the distance between the points

$$\overset{x_1\; y_1}{(-2,\, 1)} \quad \text{and} \quad \overset{x_2\; y_2}{(3,\, 4)}.$$

Algebraic Solution

Let $(x_1, y_1) = (-2, 1)$ and $(x_2, y_2) = (3, 4)$. Then apply the Distance Formula as follows.

$$
\begin{aligned}
d &= \sqrt{(x_2 - x_1)^2 + (y_2 - y_1)^2} && \text{Distance Formula} \\
&= \sqrt{[3 - (-2)]^2 + (4 - 1)^2} && \text{Substitute for } x_1,\, y_1,\, x_2,\, \text{and } y_2. \\
&= \sqrt{(5)^2 + (3)^2} && \text{Simplify.} \\
&= \sqrt{34} && \text{Simplify.} \\
&\approx 5.83 && \text{Use a calculator.}
\end{aligned}
$$

So, the distance between the points is about 5.83 units.

You can use the Pythagorean Theorem to check that the distance is correct.

$$
\begin{aligned}
d^2 &\overset{?}{=} 3^2 + 5^2 && \text{Pythagorean Theorem} \\
\left(\sqrt{34}\right)^2 &\overset{?}{=} 3^2 + 5^2 && \text{Substitute for } d. \\
34 &= 34 && \text{Distance checks.} ✔
\end{aligned}
$$

Graphical Solution

Use centimeter graph paper to plot the points $A(-2, 1)$ and $B(3, 4)$. Carefully sketch the line segment from A to B. Then use a centimeter ruler to measure the length of the segment.

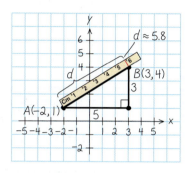

Figure P.16

The line segment measures about 5.8 centimeters, as shown in Figure P.16. So, the distance between the points is about 5.8 units.

✓CHECKPOINT   Now try Exercise 33.

When the Distance Formula is used, it does not matter which point is (x_1, y_1) and which is (x_2, y_2), because the result will be the same. For instance, in Example 3, let $(x_1, y_1) = (3, 4)$ and $(x_2, y_2) = (-2, 1)$. Then

$$d = \sqrt{(-2 - 3)^2 + (1 - 4)^2} = \sqrt{(-5)^2 + (-3)^2} = \sqrt{34} \approx 5.83.$$

Example 4   Verifying a Right Triangle

Show that the points

$$(2, 1), \quad (4, 0), \quad \text{and} \quad (5, 7)$$

are the vertices of a right triangle.

Solution

The three points are plotted in Figure P.17. Using the Distance Formula, you can find the lengths of the three sides as follows.

$$
\begin{aligned}
d_1 &= \sqrt{(5 - 2)^2 + (7 - 1)^2} = \sqrt{9 + 36} = \sqrt{45} \\
d_2 &= \sqrt{(4 - 2)^2 + (0 - 1)^2} = \sqrt{4 + 1} = \sqrt{5} \\
d_3 &= \sqrt{(5 - 4)^2 + (7 - 0)^2} = \sqrt{1 + 49} = \sqrt{50}
\end{aligned}
$$

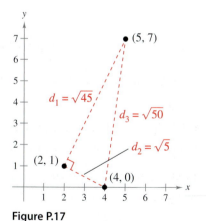

Figure P.17

Because

$$(d_1)^2 + (d_2)^2 = 45 + 5 = 50 = (d_3)^2$$

you can conclude that the triangle must be a right triangle.

✓CHECKPOINT   Now try Exercise 47.

The Midpoint Formula

To find the **midpoint** of the line segment that joins two points in a coordinate plane, find the average values of the respective coordinates of the two endpoints using the **Midpoint Formula.**

> **The Midpoint Formula** (See the proof on page 72.)
>
> The midpoint of the line segment joining the points (x_1, y_1) and (x_2, y_2) is given by the Midpoint Formula
>
> $$\text{Midpoint} = \left(\frac{x_1 + x_2}{2}, \frac{y_1 + y_2}{2}\right).$$

Example 5 Finding a Line Segment's Midpoint

Find the midpoint of the line segment joining the points $(-5, -3)$ and $(9, 3)$.

Solution

Let $(x_1, y_1) = (-5, -3)$ and $(x_2, y_2) = (9, 3)$.

$$\text{Midpoint} = \left(\frac{x_1 + x_2}{2}, \frac{y_1 + y_2}{2}\right) \qquad \text{Midpoint Formula}$$

$$= \left(\frac{-5 + 9}{2}, \frac{-3 + 3}{2}\right) \qquad \text{Substitute for } x_1, y_1, x_2, \text{ and } y_2.$$

$$= (2, 0) \qquad \text{Simplify.}$$

The midpoint of the line segment is $(2, 0)$, as shown in Figure P.18.

✓CHECKPOINT   Now try Exercise 59.

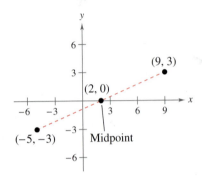

Figure P.18

Example 6 Estimating Annual Revenues

Verizon Communications had annual revenues of $88.1 billion in 2006 and $97.4 billion in 2008. Without knowing any additional information, what would you estimate the 2007 revenue to have been? (Source: Verizon Communications)

Solution

One solution to the problem is to assume that revenue followed a linear pattern. With this assumption, you can estimate the 2007 revenue by finding the midpoint of the line segment connecting the points $(2006, 88.1)$ and $(2008, 97.4)$.

$$\text{Midpoint} = \left(\frac{2006 + 2008}{2}, \frac{88.1 + 97.4}{2}\right)$$

$$\approx (2007, 92.8)$$

So, you would estimate the 2007 revenue to have been about $92.8 billion, as shown in Figure P.19. (The actual 2007 revenue was $93.5 billion.)

✓CHECKPOINT   Now try Exercise 65.

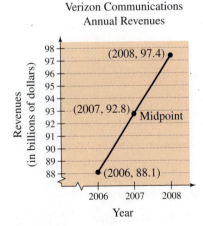

Figure P.19

The Equation of a Circle

The Distance Formula provides a convenient way to define circles. A **circle of radius** r with center at the point (h, k) is shown in Figure P.20. The point (x, y) is on this circle if and only if its distance from the center (h, k) is r. This means that a **circle** in the plane consists of all points (x, y) that are a given positive distance r from a fixed point (h, k). Using the Distance Formula, you can express this relationship by saying that the point (x, y) lies on the circle if and only if

$$\sqrt{(x - h)^2 + (y - k)^2} = r.$$

By squaring each side of this equation, you obtain the **standard form of the equation of a circle.**

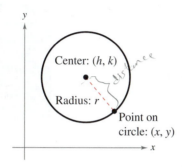

Center: (h, k)

Radius: r

Point on circle: (x, y)

Figure P.20

Standard Form of the Equation of a Circle

The **standard form of the equation of a circle** is

$$(x - h)^2 + (y - k)^2 = r^2.$$

The point (h, k) is the **center** of the circle, and the positive number r is the **radius** of the circle. The standard form of the equation of a circle whose center is the origin, $(h, k) = (0, 0)$, is $x^2 + y^2 = r^2$.

Example 7   Writing an Equation of a Circle

The point $(3, 4)$ lies on a circle whose center is at $(-1, 2)$, as shown in Figure P.21. Write the standard form of the equation of this circle.

Solution

The radius r of the circle is the distance between $(-1, 2)$ and $(3, 4)$.

$r = \sqrt{[3 - (-1)]^2 + (4 - 2)^2}$ Substitute for x, y, h, and k.

$ = \sqrt{4^2 + 2^2}$ Simplify.

$ = \sqrt{16 + 4}$ Simplify.

$ = \sqrt{20}$ Radius

Using $(h, k) = (-1, 2)$ and $r = \sqrt{20}$, the equation of the circle is

$(x - h)^2 + (y - k)^2 = r^2$ Equation of circle

$[x - (-1)]^2 + (y - 2)^2 = \left(\sqrt{20}\right)^2$ Substitute for h, k, and r.

$(x + 1)^2 + (y - 2)^2 = 20.$ Standard form

✔CHECKPOINT   Now try Exercise 69.

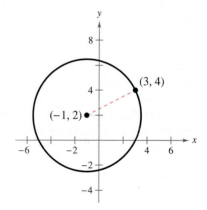

$(3, 4)$

$(-1, 2)$

Figure P.21

Application

Much of computer graphics, including the computer-generated goldfish tessellation shown at the right, consists of transformations of points in a coordinate plane. One type of transformation, a translation, is illustrated in Example 8. Other types of transformations include reflections, rotations, and stretches.

Example 8   Translating Points in the Plane

The triangle in Figure P.22 has vertices at the points

$$(-1, 2), \quad (1, -4), \quad \text{and} \quad (2, 3).$$

Shift the triangle three units to the right and two units upward and find the vertices of the shifted triangle, as shown in Figure P.23.

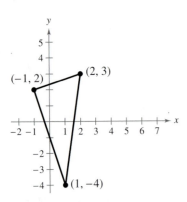

Figure P.22   **Figure P.23**

Solution

To shift the vertices three units to the right, add 3 to each of the x-coordinates. To shift the vertices two units upward, add 2 to each of the y-coordinates.

Original Point	Translated Point
$(-1, 2)$	$(-1 + 3, 2 + 2) = (2, 4)$
$(1, -4)$	$(1 + 3, -4 + 2) = (4, -2)$
$(2, 3)$	$(2 + 3, 3 + 2) = (5, 5)$

Plotting the translated points and sketching the line segments between them produces the shifted triangle shown in Figure P.23.

✓CHECKPOINT   Now try Exercise 87.

Example 8 shows how to translate points in a coordinate plane. The following transformed points are related to the original points as follows.

Original Point	Transformed Point	
(x, y)	$(-x, y)$	$(-x, y)$ is a reflection of the original point in the y-axis.
(x, y)	$(x, -y)$	$(x, -y)$ is a reflection of the original point in the x-axis.
(x, y)	$(-x, -y)$	$(-x, -y)$ is a reflection of the original point through the origin.

The figures provided with Example 8 were not really essential to the solution. Nevertheless, it is strongly recommended that you develop the habit of including sketches with your solutions because they serve as useful problem-solving tools.

P.5   Exercises

See www.CalcChat.com for worked-out solutions to odd-numbered exercises.
For instructions on how to use a graphing utility, see Appendix A.

Vocabulary and Concept Check

In Exercises 1–4, fill in the blank(s).

1. An ordered pair of real numbers can be represented in a plane called the rectangular coordinate system or the _____ plane.

2. The _____ is a result derived from the Pythagorean Theorem.

3. Finding the average values of the respective coordinates of the two endpoints of a line segment in a coordinate plane is also known as using the _____ .

4. The standard form of the equation of a circle is _____ , where the point (h, k) is the _____ of the circle and the positive number r is the _____ of the circle.

In Exercises 5–10, match each term with its definition.

5. x-axis c
6. y-axis f
7. origin a
8. quadrants d
9. x-coordinate e
10. y-coordinate b

(a) point of intersection of vertical axis and horizontal axis

(b) directed distance from the x-axis

(c) horizontal real number line

(d) four regions of the coordinate plane

(e) directed distance from the y-axis

(f) vertical real number line

Procedures and Problem Solving

Approximating Coordinates of Points   In Exercises 11 and 12, approximate the coordinates of the points.

11.

12.

Plotting Points in the Cartesian Plane   In Exercises 13–16, plot the points in the Cartesian plane.

13. $(-4, 2), (-3, -6), (0, 5), (1, -4)$
14. $(4, -2), (0, 0), (-4, 0), (-5, -5)$
15. $(3, 8), (0.5, -1), (5, -6), (-2, -2.5)$
16. $\left(1, -\frac{1}{2}\right), \left(-\frac{5}{2}, 2\right), (3, -3), \left(\frac{3}{2}, 1\right)$

Finding Coordinates of Points   In Exercises 17–20, find the coordinates of the point.

17. The point is located five units to the left of the y-axis and four units above the x-axis.

18. The point is located three units below the x-axis and two units to the right of the y-axis.

19. The point is on the y-axis and six units below the x-axis.

20. The point is on the x-axis and 11 units to the left of the y-axis.

Determining Quadrants   In Exercises 21–30, determine the quadrant(s) in which (x, y) is located so that the condition(s) is (are) satisfied.

21. $x > 0$ and $y < 0$
22. $x < 0$ and $y < 0$
23. $x = -4$ and $y > 0$
24. $x > 2$ and $y = 3$
25. $y < -5$
26. $x > 4$
27. $x < 0$ and $-y > 0$
28. $-x > 0$ and $y < 0$
29. $xy > 0$
30. $xy < 0$

Sketching a Scatter Plot   In Exercises 31 and 32, sketch a scatter plot of the data shown in the table.

31. **Accounting**   The table shows the sales y (in millions of dollars) for Apple for the years 2000 through 2009. (Source: Apple Inc.)

Year	Sales, y (in millions of dollars)
2000	7,983
2001	5,363
2002	5,742
2003	6,207
2004	8,279
2005	13,931
2006	19,315
2007	24,006
2008	32,479
2009	36,537

32. Meteorology The table shows the lowest temperature on record y (in degrees Fahrenheit) in Flagstaff, Arizona for each month x, where $x = 1$ represents January. (Source: U.S. National Oceanic and Atmospheric Administration)

Month, x	Temperature, y
1	-22
2	-23
3	-16
4	-2
5	14
6	22
7	32
8	24
9	23
10	-2
11	-13
12	-23

Finding a Distance In Exercises 33–42, find the distance between the points algebraically and confirm graphically by using centimeter graph paper and a centimeter ruler.

33. $(6, -3), (6, 5)$ **34.** $(-3, 0), (-3, -7)$

35. $(-3, -1), (2, -1)$ **36.** $(-11, 4), (-1, 4)$

37. $(-2, 6), (3, -6)$

38. $(8, 5), (0, 20)$

39. $\left(\frac{1}{2}, \frac{4}{3}\right), (2, -1)$

40. $\left(-\frac{2}{3}, 3\right), \left(-1, \frac{5}{4}\right)$

41. $(-4.2, 3.1), (-12.5, 4.8)$

42. $(9.5, -2.6), (-3.9, 8.2)$

Verifying a Right Triangle In Exercises 43–46, (a) find the length of each side of the right triangle and (b) show that these lengths satisfy the Pythagorean Theorem.

43.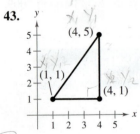

44.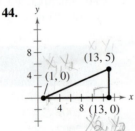

45.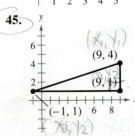

46.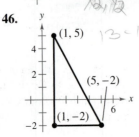

Verifying a Polygon In Exercises 47–54, show that the points form the vertices of the polygon.

47. Right triangle: $(4, 0), (2, 1), (-1, -5)$

48. Right triangle: $(-1, 3), (3, 5), (5, 1)$

49. Isosceles triangle: $(1, -3), (3, 2), (-2, 4)$

50. Isosceles triangle: $(2, 3), (4, 9), (-2, 7)$

51. Parallelogram: $(2, 5), (0, 9), (-2, 0), (0, -4)$

52. Parallelogram: $(0, 1), (3, 7), (4, 4), (1, -2)$

53. Rectangle: $(-5, 6), (0, 8), (-3, 1), (2, 3)$ (*Hint:* Show that the diagonals are of equal length.)

54. Rectangle: $(2, 4), (3, 1), (1, 2), (4, 3)$ (*Hint:* Show that the diagonals are of equal length.)

Finding a Line Segment's Midpoint In Exercises 55–64, (a) plot the points, (b) find the distance between the points, and (c) find the midpoint of the line segment joining the points.

55. $(0, 0), (8, 6)$

56. $(1, 12), (6, 0)$

57. $(-4, 10), (4, -5)$

58. $(-7, -4), (2, 8)$

59. $(-1, 2), (5, 4)$

60. $(2, 10), (10, 2)$

61. $\left(\frac{1}{2}, 1\right), \left(-\frac{5}{2}, \frac{4}{3}\right)$

62. $\left(-\frac{1}{3}, -\frac{1}{3}\right), \left(-\frac{1}{6}, -\frac{1}{2}\right)$

63. $(6.2, 5.4), (-3.7, 1.8)$

64. $(-16.8, 12.3), (5.6, 4.9)$

Estimating Annual Revenues In Exercises 65 and 66, use the Midpoint Formula to estimate the annual revenues (in millions of dollars) for Texas Roadhouse and Papa John's in 2006. The revenues for the two companies in 2003 and 2009 are shown in the tables. Assume that the revenues followed a linear pattern. (Sources: Texas Roadhouse, Inc.; Papa John's International)

65. Texas Roadhouse

Year	Annual revenue (in millions of dollars)
2003	287
2009	942

66. Papa John's Intl.

Year	Annual revenue (in millions of dollars)
2003	917
2009	1106

Writing an Equation of a Circle   In Exercises 67–80, write the standard form of the equation of the specified circle.

67. Center: $(0, 0)$; radius: 5
68. Center: $(0, 0)$; radius: 6
✓ 69. Center: $(2, -1)$; radius: 4
70. Center: $(-5, 3)$; radius: 2
71. Center: $(-1, 2)$; solution point: $(0, 0)$
72. Center: $(3, -2)$; solution point: $(-1, 1)$
73. Endpoints of a diameter: $(0, 0), (6, 8)$
74. Endpoints of a diameter: $(-4, -1), (4, 1)$
75. Center: $(-2, 1)$; tangent to the x-axis
76. Center: $(3, -2)$; tangent to the y-axis
77. The circle inscribed in the square with vertices $(7, -2)$, $(-1, -2), (-1, -10)$, and $(7, -10)$
78. The circle inscribed in the square with vertices $(-12, 10), (8, 10), (8, -10)$, and $(-12, -10)$

79. 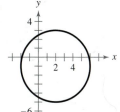   80.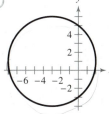

Sketching a Circle   In Exercises 81–86, find the center and radius, and sketch the circle.

81. $x^2 + y^2 = 9$
82. $x^2 + y^2 = 16$
83. $(x - 1)^2 + (y + 3)^2 = 4$
84. $x^2 + (y - 1)^2 = 49$
85. $\left(x - \frac{1}{2}\right)^2 + \left(y - \frac{1}{2}\right)^2 = \frac{9}{4}$
86. $\left(x - \frac{2}{3}\right)^2 + \left(y + \frac{1}{4}\right)^2 = \frac{25}{36}$

Translating Points in the Plane   In Exercises 87–90, the polygon is shifted to a new position in the plane. Find the coordinates of the vertices of the polygon in the new position.

✓ 87. 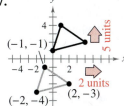   88.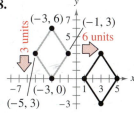

Viorel Sima 2010/used under license from Shutterstock.com

89. Original coordinates of vertices:
$(0, 2), (-3, 5), (-5, 2), (-2, -1)$
Shift: three units upward, one unit to the left

90. Original coordinates of vertices:
$(1, -1), (3, 2), (1, -2)$
Shift: two units downward, three units to the left

Education   In Exercises 91 and 92, refer to the scatter plot, which shows the mathematics entrance test scores x and the final examination scores y in an algebra course for a sample of 10 students.

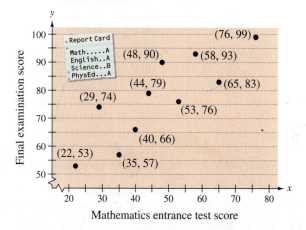

91. Find the entrance exam score of any student with a final exam score in the 80s.

92. Does a higher entrance exam score necessarily imply a higher final exam score? Explain.

93. *Why you should learn it* *(p. 46)*   The graph shows the numbers of performers who were elected to the Rock and Roll Hall of Fame from 1991 through 2010. Describe any trends in the data. From these trends, predict the number of performers elected in 2011. (Source: rockhall.com)

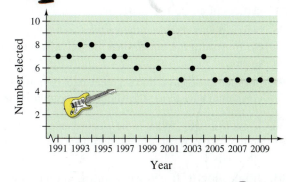

94. **Aviation**   A jet plane flies from Naples, Italy in a straight line to Rome, Italy, which is 120 kilometers north and 150 kilometers west of Naples. How far does the plane fly?

95. Physical Education In a football game, a quarterback throws a pass from the 15-yard line, 10 yards from the sideline, as shown in the figure. The pass is caught on the 40-yard line, 45 yards from the same sideline. How long is the pass?

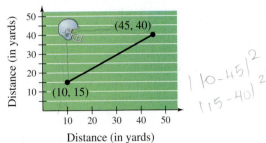

Distance (in yards)

$|10-45|^2$
$|15-40|^2$

96. Physical Education A major league baseball diamond is a square with 90-foot sides. In the figure, home plate is at the origin and the first base line lies on the positive x-axis. The right fielder fields the ball at the point $(300, 25)$. How far does the right fielder have to throw the ball to get a runner out at home plate? How far does the right fielder have to throw the ball to get a runner out at third base? (Round your answers to one decimal place.)

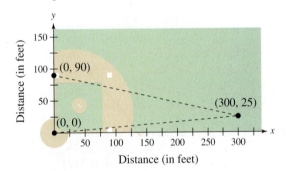

Distance (in feet)

Conclusions

True or False? In Exercises 97–99, determine whether the statement is true or false. Justify your answer.

97. The points $(-8, 4)$, $(2, 11)$, and $(-5, 1)$ represent the vertices of an isosceles triangle.

98. If four points represent the vertices of a polygon, and the four sides are equal, then the polygon must be a square.

99. In order to divide a line segment into 16 equal parts, you would have to use the Midpoint Formula 16 times.

100. Think About It What is the y-coordinate of any point on the x-axis? What is the x-coordinate of any point on the y-axis?

101. Think About It When plotting points on the rectangular coordinate system, is it true that the scales on the x- and y-axes must be the same? Explain.

102. Exploration A line segment has (x_1, y_1) as one endpoint and (x_m, y_m) as its midpoint. Find the other endpoint (x_2, y_2) of the line segment in terms of x_1, y_1, x_m, and y_m. Use the result to find the coordinates of the endpoint of a line segment when the coordinates of the other endpoint and midpoint are, respectively,

(a) $(1, -2)$, $(4, -1)$.   (b) $(-5, 11)$, $(2, 4)$.

103. Exploration Use the Midpoint Formula three times to find the three points that divide the line segment joining (x_1, y_1) and (x_2, y_2) into four parts. Use the result to find the points that divide the line segment joining the given points into four equal parts.

(a) $(1, -2)$, $(4, -1)$ (b) $(-2, -3)$, $(0, 0)$

104. CAPSTONE Use the plot of the point (x_0, y_0) in the figure. Match the transformation of the point with the correct plot. Explain your reasoning. [The plots are labeled (i), (ii), (iii), and (iv).]

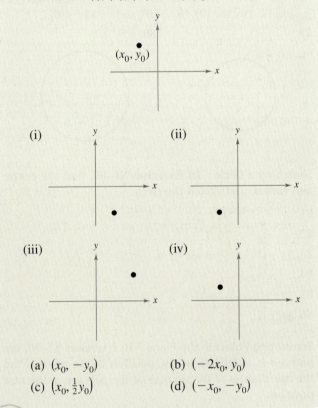

(a) $(x_0, -y_0)$ (b) $(-2x_0, y_0)$
(c) $(x_0, \frac{1}{2}y_0)$ (d) $(-x_0, -y_0)$

105. Proof Prove that the diagonals of the parallelogram in the figure intersect at their midpoints.

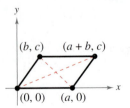

P.6   Representing Data Graphically

Line Plots

Statistics is the branch of mathematics that studies techniques for collecting, organizing, and interpreting data. In this section, you will study several ways to organize data. The first is a **line plot,** which uses a portion of a real number line to order numbers. Line plots are especially useful for ordering small sets of numbers (about 50 or less) by hand.

Many statistical measures can be obtained from a line plot. Two such measures are the *frequency* and *range* of the data. The **frequency** measures the number of times a value occurs in a data set. The **range** is the difference between the greatest and least data values. For example, consider the data values

20,   21,   21,   25,   32.

The frequency of 21 in the data set is 2 because 21 occurs twice. The range is 12 because the difference between the greatest and least data values is

$32 - 20 = 12.$

Example 1   Constructing a Line Plot

The scores from an economics class of 30 students are listed below. The scores are for a 100-point exam.

93, 70, 76, 67, 86, 93, 82, 78, 83, 86, 64, 78, 76, 66, 83
83, 96, 74, 69, 76, 64, 74, 79, 76, 88, 76, 81, 82, 74, 70

a. Use a line plot to organize the scores.

b. Which score occurs with the greatest frequency?

c. What is the range of scores?

Solution

a. Begin by scanning the data to find the least and greatest numbers. For the data, the least number is 64 and the greatest is 96. Next, draw a portion of a real number line that includes the interval

[64, 96].

To create the line plot, start with the first number, 93, and record an × above 93 on the number line. Continue recording an × for each number in the list until you obtain the line plot shown in Figure P.24.

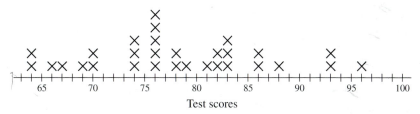

Figure P.24

b. From the line plot, you can see that 76 occurs with the greatest frequency.

c. Because the range is the difference between the greatest and least data values, the range of scores is

$96 - 64 = 32.$

 CHECKPOINT   Now try Exercise 9.

What you should learn

● Use line plots to order and analyze data.
● Use histograms to represent frequency distributions.
● Use bar graphs to represent and analyze data.
● Use line graphs to represent and analyze data.

Why you should learn it

Double bar graphs allow you to compare visually two sets of data over time. For example, in Exercises 17 and 18 on page 63, you are asked to estimate the difference in tuition between public and private institutions of higher education.

Histograms and Frequency Distributions

When you want to organize large sets of data, it is useful to group the data into intervals and plot the frequency of the data in each interval. A **frequency distribution** can be used to construct a **histogram.** A histogram uses a portion of a real number line as its horizontal axis. The bars of a histogram are not separated by spaces.

Example 2   Constructing a Histogram

The table at the right shows the percent of the resident population of each state and the District of Columbia that was at least 65 years old in 2008. Construct a frequency distribution and a histogram for the data.   (Source: U.S. Census Bureau)

Solution

To begin constructing a frequency distribution, you must first decide on the number of intervals. There are several ways to group the data. However, because the least number is 7.3 and the greatest is 17.4, it seems that six intervals would be appropriate. The first would be the interval

$[7, 9)$

the second would be

$[9, 11)$

and so on. By tallying the data into the six intervals, you obtain the frequency distribution shown below.

Interval	Tally
$[7, 9)$	|
$[9, 11)$	||||
$[11, 13)$	⊬⊬ ⊬⊬ ⊬⊬
$[13, 15)$	⊬⊬ ⊬⊬ ⊬⊬ ⊬⊬ ⊬⊬ ||
$[15, 17)$	|||
$[17, 19)$	|

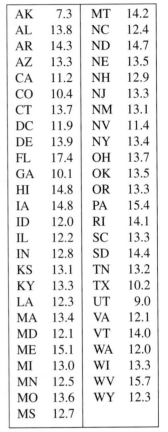

AK	7.3	MT	14.2
AL	13.8	NC	12.4
AR	14.3	ND	14.7
AZ	13.3	NE	13.5
CA	11.2	NH	12.9
CO	10.4	NJ	13.3
CT	13.7	NM	13.1
DC	11.9	NV	11.4
DE	13.9	NY	13.4
FL	17.4	OH	13.7
GA	10.1	OK	13.5
HI	14.8	OR	13.3
IA	14.8	PA	15.4
ID	12.0	RI	14.1
IL	12.2	SC	13.3
IN	12.8	SD	14.4
KS	13.1	TN	13.2
KY	13.3	TX	10.2
LA	12.3	UT	9.0
MA	13.4	VA	12.1
MD	12.1	VT	14.0
ME	15.1	WA	12.0
MI	13.0	WI	13.3
MN	12.5	WV	15.7
MO	13.6	WY	12.3
MS	12.7		

You can construct the histogram by drawing a vertical axis to represent the number of states and a horizontal axis to represent the percent of the population 65 and older. Then, for each interval, draw a vertical bar whose height is the total tally, as shown in Figure P.25. Note that the break in the horizontal axis indicates that the numbers 0 through 6 have been omitted.

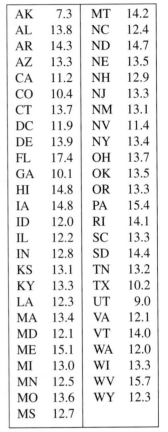

Figure P.25

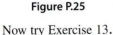   Now try Exercise 13.

Bar Graphs

A **bar graph** is similar to a histogram, except that the bars can be either horizontal or vertical and the labels of the bars are not necessarily numbers. For instance, the labels of the bars can be the months of a year, names of cities, or types of vehicles. Another difference between a bar graph and a histogram is that the bars in a bar graph are usually separated by spaces.

Example 3   Constructing a Bar Graph

The data below show the normal monthly precipitation (in inches) in Houston, Texas. Construct a bar graph for the data. What can you conclude?   (Source: U.S. National Oceanic and Atmospheric Administration)

Month	Precipitation
January	3.7
February	3.0
March	3.4
April	3.6
May	5.2
June	5.4
July	3.2
August	3.8
September	4.3
October	4.5
November	4.2
December	3.7

Solution

To create a bar graph, begin by drawing a vertical axis to represent the precipitation and a horizontal axis to represent the month. The bar graph is shown in Figure P.26.

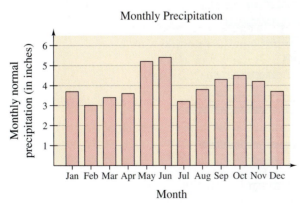

Figure P.26

From the graph, you can see that Houston receives a fairly consistent amount of rain throughout the year—the driest month tends to be February and the wettest month tends to be June.

✓CHECKPOINT   Now try Exercise 15.

Example 4   Constructing a Double Bar Graph

The table shows the percents of associate degrees awarded to males and females for selected fields of study in the United States in 2007. Construct a double bar graph for the data.   (Source: U.S. National Center for Education Statistics)

Field of Study	% Female	% Male
Agriculture and Natural Resources	37.9	62.1
Biological Sciences/Life Sciences	67.2	32.8
Accounting	77.5	22.5
Business	60.3	39.7
Communications	49.8	50.2
Education	86.4	13.6
Engineering	14.1	85.9
Family and Consumer Sciences	96.4	3.6
Emergency Medical Technician	32.4	67.6
Law and Legal Studies	89.9	10.1
Liberal/General Studies	62.5	37.5
Mathematics	33.7	66.3
Physical Sciences	43.0	57.0
Precision Production Trades	6.3	93.7
Psychology	76.6	23.4
Social Sciences	66.3	33.7

Solution

For the data, a horizontal bar graph seems to be appropriate. This makes it easier to label and read the bars. Such a graph is shown in Figure P.27.

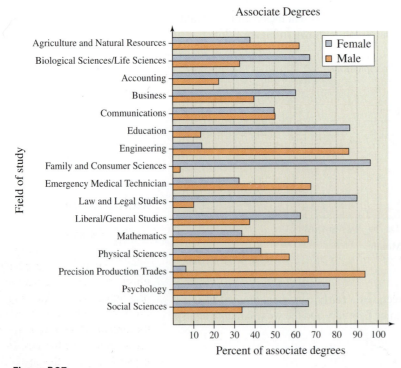

Figure P.27

✓CHECKPOINT   Now try Exercise 19.

Line Graphs

A **line graph** is similar to a standard coordinate graph. Line graphs are usually used to show trends over periods of time.

Example 5   Constructing a Line Graph

The table at the right shows the number of immigrants (in thousands) entering the United States for each decade from 1901 through 2000. Construct a line graph for the data. What can you conclude?   (Source: U.S. Immigration and Naturalization Service)

Decade	Number (in thousands)
1901–1910	8795
1911–1920	5736
1921–1930	4107
1931–1940	528
1941–1950	1035
1951–1960	2515
1961–1970	3322
1971–1980	4493
1981–1990	7338
1991–2000	9095

Solution

Begin by drawing a vertical axis to represent the number of immigrants in thousands. Then label the horizontal axis with decades and plot the points shown in the table. Finally, connect the points with line segments, as shown in Figure P.28. From the line graph, you can see that the number of immigrants hit a low point during the depression of the 1930s. Since then the number has steadily increased.

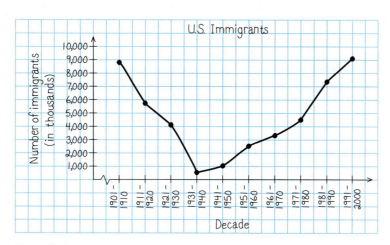

Figure P.28

✔CHECKPOINT   Now try Exercise 29.

You can use a graphing utility to check your sketch, as shown in Figure P.29.

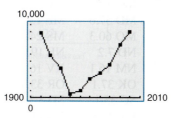

Figure P.29

Technology Tip

 You can use the *statistical plotting* feature of a graphing utility to create different types of graphs, such as line graphs. For instructions on how to use the *statistical plotting* feature, see Appendix A; for specific keystrokes, go to this textbook's *Companion Website*.

Economics In Exercises 21–24, use the line graph, which shows the average prices of a gallon of premium unleaded gasoline from 2000 through 2008. (Source: U.S. Energy Information Administration)

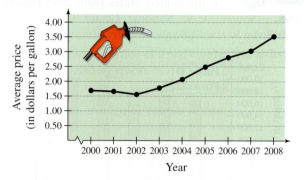

21. Describe the trend in the price of premium unleaded gasoline from 2000 to 2002.

22. Describe the trend in the price of premium unleaded gasoline from 2002 to 2008.

23. Approximate the percent increase in the price per gallon of premium unleaded gasoline from 2005 to 2008.

24. Approximate the percent increase in the price per gallon of premium unleaded gasoline from 2002 to 2008.

Agriculture In Exercises 25–28, use the line graph, which shows the average retail price (in dollars) of one dozen Grade A large eggs in the United States for each month in 2009. (Source: U.S. Bureau of Labor Statistics)

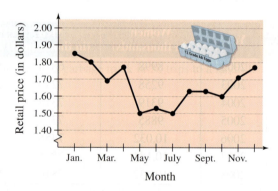

25. What is the highest price of one dozen Grade A large eggs shown in the graph? When did this price occur?

26. What was the difference between the highest price and the lowest price of one dozen Grade A large eggs in 2009?

27. Determine when the price of eggs showed the greatest rate of decrease from one month to the next.

28. Describe any trends shown by the line graph. Then predict the average price of eggs in February of 2010. Are you confident that your prediction is within $0.10 of the actual average retail price? Explain your reasoning.

29. Human Resources The table shows the total numbers of women in the workforce (in thousands) in the United States from 1998 through 2009. Construct a line graph for the data. Write a brief statement describing what the graph reveals. (Source: U.S. Bureau of Labor Statistics)

Year	Women in the workforce (in thousands)
1998	63,714
1999	64,855
2000	66,303
2001	66,848
2002	67,363
2003	68,272
2004	68,421
2005	69,288
2006	70,173
2007	70,988
2008	71,767
2009	72,019

30. Economics The table shows the total amounts (in trillions of dollars) of private fixed assets held by noncorporate entities in the United States from 2000 through 2008. Construct a line graph for the data. Write a brief statement describing what the graph reveals. (Source: U.S. Bureau of Economic Analysis)

Year	Fixed assets (in trillions of dollars)
2000	13.04
2001	13.95
2002	14.79
2003	15.94
2004	17.77
2005	19.74
2006	21.15
2007	21.56
2008	20.96

31. Human Resources The table shows the average hourly earnings (in dollars) of construction workers in the United States from 1994 through 2009. Use a graphing utility to construct a line graph for the data. (Source: U.S. Bureau of Labor Statistics)

Year	Hourly earnings (in dollars)
1994	14.38
1995	14.73
1996	15.11
1997	15.67
1998	16.23
1999	16.80
2000	17.48
2001	18.00
2002	18.52
2003	18.95
2004	19.23
2005	19.46
2006	20.02
2007	20.95
2008	21.87
2009	22.67

32. Networking The list shows the percent of households in each of the 50 states and the District of Columbia with Internet access in 2009. Use a graphing utility to organize the data in the graph of your choice. Explain your choice of graph. (Source: National Telecommunications and Information Administration)

AK 77.3	AL 56.1	AR 55.9
AZ 72.1	CA 73.1	CO 72.9
CT 74.8	DC 71.8	DE 71.1
FL 71.5	GA 67.6	HI 73.3
IA 67.8	ID 72.2	IL 68.3
IN 62.5	KS 70.5	KY 59.5
LA 60.6	MA 75.9	MD 73.4
ME 70.2	MI 68.0	MN 72.4
MO 63.7	MS 51.8	MT 64.1
NC 63.9	ND 67.1	NE 70.0
NH 78.7	NJ 76.8	NM 61.7
NV 72.4	NY 70.7	OH 66.9
OK 60.7	OR 76.0	PA 67.3
RI 71.6	SC 58.3	SD 65.5
TN 62.3	TX 63.9	UT 77.9
VA 71.0	VT 70.8	WA 77.9
WI 71.8	WV 59.7	WY 69.8

33. Athletics The table shows the numbers of participants (in thousands) in high school athletic programs in the United States from 1998 through 2009. Organize the data in an appropriate display. Explain your choice of graph. (Source: National Federation of State High School Associations)

Year	Female athletes (in thousands)	Male athletes (in thousands)
1998	2570	3763
1999	2653	3832
2000	2676	3862
2001	2784	3921
2002	2807	3961
2003	2856	3989
2004	2865	4038
2005	2908	4110
2006	2953	4207
2007	3022	4321
2008	3057	4372
2009	3114	4423

Conclusions

34. Writing Describe the differences between a bar graph and a histogram.

35. Writing Describe the differences between a line plot and a scatter plot.

36. CAPSTONE The graphs shown below represent the same data points.

(a) Which of the two graphs is misleading, and why?

(b) Discuss other ways in which graphs can be misleading.

(c) Why would it be beneficial for someone to use a misleading graph?

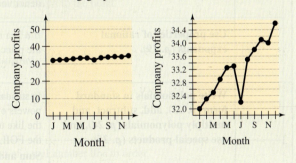

37. Think About It How can you decide which type of graph to use when you are organizing data?

P   Review Exercises

See www.CalcChat.com for worked-out solutions to odd-numbered exercises. For instructions on how to use a graphing utility, see Appendix A.

P.1

Identifying Subsets of Real Numbers   In Exercises 1 and 2, determine which numbers are (a) natural numbers, (b) whole numbers, (c) integers, (d) rational numbers, and (e) irrational numbers.

1. $\left\{11, -14, -\frac{8}{9}, \frac{5}{2}, \sqrt{6}, 0.4\right\}$
2. $\left\{\sqrt{15}, -22, -\frac{10}{3}, 0, 5.2, \frac{3}{7}\right\}$

Writing an Inequality   In Exercises 3 and 4, use a calculator to find the decimal form of each rational number. If it is a nonterminating decimal, write the repeating pattern. Then plot the numbers on the real number line and place the correct inequality symbol (< or >) between them.

3. (a) $\frac{5}{6}$ (b) $\frac{7}{8}$ 4. (a) $\frac{1}{3}$ (b) $\frac{9}{25}$

Interpreting Inequalities   In Exercises 5 and 6, (a) verbally describe the subset of real numbers represented by the inequality, (b) sketch the subset on the real number line, and (c) state whether the interval is bounded or unbounded.

5. $x \leq 7$ 6. $-3 < x \leq 6$

Using Interval Notation   In Exercises 7 and 8, use interval notation to describe the graph.

7.

8.

Finding the Distance Between Two Real Numbers   In Exercises 9 and 10, find the distance between a and b.

9. $a = -74$, $b = 48$ 10. $a = -123$, $b = -9$

Using Absolute Value Notation   In Exercises 11 and 12, use absolute value notation to describe the situation.

11. The distance between x and 7 is at least 6.
12. The distance between y and -30 is less than 5.

Evaluating an Expression   In Exercises 13–16, evaluate the expression for each value of x. (If not possible, state the reason.)

	Expression		*Values*	
13.	$9x - 2$	(a) $x = -1$	(b) $x = 3$	
14.	$x^2 - 11x + 24$	(a) $x = -2$	(b) $x = 2$	
15.	$-2x^2 - x + 3$	(a) $x = 3$	(b) $x = -3$	
16.	$\dfrac{4x}{x - 1}$	(a) $x = -1$	(b) $x = 1$	

Identifying Rules of Algebra   In Exercises 17 and 18, identify the rule of algebra illustrated by the statement.

17. $(t^2 + 1) + 3 = 3 + (t^2 + 1)$
18. $0 + (a - 5) = a - 5$

P.2

Using Properties of Exponents   In Exercises 19–22, simplify each expression.

19. (a) $(-2z)^3$ (b) $(a^2b^4)(3ab^{-2})$
20. (a) $\dfrac{(8y)^0}{y^2}$ (b) $\dfrac{40(b - 3)^5}{75(b - 3)^2}$
21. (a) $\dfrac{6^2u^3v^{-3}}{12u^{-2}v}$ (b) $\dfrac{3^{-4}m^{-1}n^{-3}}{9^{-2}mn^{-3}}$
22. (a) $(x^{-1} + y)^{-2}$ (b) $\left(\dfrac{y^{-2}}{x}\right)^{-1}\left(\dfrac{x^2}{y^{-2}}\right)$

Scientific Notation   In Exercises 23–26, write the number in scientific notation.

23. 2,585,000,000   24. $-3,250,000$
25. 0.000000125   26. -0.000002104

Writing a Number in Decimal Notation   In Exercises 27–30, write the number in decimal notation.

27. 1.28×10^5 28. -4.002×10^2
29. 1.80×10^{-5} 30. -4.02×10^{-2}

Using Properties of Radicals   In Exercises 31 and 32, use the properties of radicals to simplify the expression.

31. $\left(\sqrt[4]{78}\right)^4$
32. $\sqrt[5]{8} \cdot \sqrt[5]{4}$

Simplifying a Radical Expression   In Exercises 33–44, simplify the expression.

33. $\sqrt{25a^2}$ 34. $\sqrt[5]{64x^6}$
35. $\sqrt{\dfrac{81}{144}}$ 36. $\sqrt[3]{\dfrac{125}{216}}$
37. $\sqrt[3]{\dfrac{2x^3}{27}}$ 38. $\sqrt{\dfrac{75x^2}{y^4}}$
39. $\sqrt{48} - \sqrt{27}$ 40. $3\sqrt{32} + 4\sqrt{98}$
41. $8\sqrt{3x} - 5\sqrt{3x}$ 42. $-11\sqrt{36y} - 6\sqrt{y}$
43. $\sqrt{8x^3} + \sqrt{2x}$ 44. $3\sqrt{14x^2} - \sqrt{56x^2}$

Rationalizing a Denominator   In Exercises 45 and 46, rationalize the denominator of the expression. Then simplify your answer.

45. $\dfrac{1}{3 - \sqrt{5}}$ 46. $\dfrac{1}{\sqrt{x} - 1}$

Rationalizing a Numerator In Exercises 47 and 48, rationalize the numerator of the expression. Then simplify your answer.

47. $\dfrac{\sqrt{20}}{4}$

48. $\dfrac{\sqrt{2} - \sqrt{11}}{3}$

Simplifying with Rational Exponents In Exercises 49–52, simplify the expression.

49. $64^{5/2}$

50. $64^{-2/3}$

51. $(-3x^{2/5})(-2x^{1/2})$

52. $(x - 1)^{1/3}(x - 1)^{-1/4}$

P.3

Writing Polynomials in Standard Form In Exercises 53 and 54, write the polynomial in standard form. Then identify the degree and leading coefficient of the polynomial.

53. $15x^2 - 2x^5 + 3x^3 + 5 - x^4$

54. $-2x^4 + x^2 - 10 - x + x^3$

Performing Operations with Polynomials In Exercises 55–60, perform the operations and write the result in standard form.

55. $(3x^2 + 2x) - (1 - 5x)$

56. $(8y^2 + 2y) + (3y - 8)$

57. $(2x^3 - 5x^2 + 10x - 7) + (4x^2 - 7x - 2)$

58. $(6x^4 - 4x^3 - x + 3 - 20x^2) - (16 + 9x^4 - 11x^2)$

59. $-2a(a^2 + a - 3)$

60. $(y^2 - 4y)(y^3)$

Multiplying Polynomials In Exercises 61–68, multiply or find the special product.

61. $(x + 4)(x + 9)$

62. $(z + 1)(5z - 6)$

63. $(x + 8)(x - 8)$

64. $(7x + 4)(7x - 4)$

65. $(x - 4)^3$

66. $(2x - 1)^3$

67. $[(m - 7) + n][(m - 7) - n]$

68. $[(x - y) - 4][(x - y) + 4]$

69. Geometry Use the area model to write two different expressions for the area. Then equate the two expressions and name the algebraic property that is illustrated.

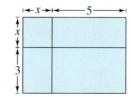

70. Compound Interest After 2 years, an investment of \$2500 compounded annually at an interest rate r will yield an amount of $2500(1 + r)^2$. Write this polynomial in standard form.

Removing Common Factors In Exercises 71–76, factor out the common factor.

71. $7x + 35$

72. $4b - 12$

73. $x^3 - x$

74. $x(x - 3) + 4(x - 3)$

75. $2x^3 + 18x^2 - 4x$

76. $-6x^4 - 3x^3 + 12x$

Factoring Polynomials In Exercises 77–86, factor the expression.

77. $x^2 - 169$

78. $9x^2 - \frac{1}{25}$

79. $x^2 + 6x + 9$

80. $4x^2 - 4x + 1$

81. $x^3 + 216$

82. $64x^3 - 27$

83. $x^2 - 6x - 27$

84. $x^2 - 9x + 14$

85. $2x^2 + 21x + 10$

86. $3x^2 + 14x + 8$

Factoring by Grouping In Exercises 87–90, factor by grouping.

87. $x^3 - 4x^2 - 3x + 12$

88. $x^3 - 6x^2 - x + 6$

89. $2x^2 - x - 15$

90. $6x^2 + x - 12$

P.4

Finding the Domain of an Algebraic Expression In Exercises 91–94, find the domain of the expression.

91. $-5x^2 - x - 1$

92. $9x^4 + 7, \quad x > 0$

93. $\dfrac{4}{2x - 3}$

94. $\sqrt{x + 12}$

Simplifying a Rational Expression In Exercises 95–98, write the rational expression in simplest form.

95. $\dfrac{4x^2}{4x^3 + 28x}$

96. $\dfrac{6xy}{xy + 2x}$

97. $\dfrac{x^2 - x - 30}{x^2 - 25}$

98. $\dfrac{x^2 - 9x + 18}{8x - 48}$

Performing Operations with Rational Expressions In Exercises 99–106, perform the operations and simplify your answer.

99. $\dfrac{x^2 - 4}{x^4 - 2x^2 - 8} \cdot \dfrac{x^2 + 2}{x^2}$

100. $\dfrac{2x - 1}{x + 1} \cdot \dfrac{x^2 - 1}{2x^2 - 7x + 3}$

101. $\dfrac{x^2(5x - 6)}{2x + 3} \div \dfrac{5x}{2x + 3}$

102. $\dfrac{4x - 6}{(x - 1)^2} \div \dfrac{2x^2 - 3x}{x^2 + 2x - 3}$

103. $x - 1 + \dfrac{1}{x + 2} + \dfrac{1}{x - 1}$

104. $2x + \dfrac{3}{2(x - 4)} - \dfrac{1}{2(x + 2)}$

105. $\dfrac{1}{x} - \dfrac{x - 1}{x^2 + 1}$

106. $\dfrac{1}{x - 1} + \dfrac{1 - x}{x^2 + x + 1}$

Simplifying a Complex Fraction In Exercises 107 and 108, simplify the complex fraction.

107. $\dfrac{\left(\dfrac{1}{x} - \dfrac{1}{y}\right)}{(x^2 - y^2)}$

108. $\dfrac{\left(\dfrac{1}{2x-3} - \dfrac{1}{2x+3}\right)}{\left(\dfrac{1}{2x} - \dfrac{1}{2x+3}\right)}$

P.5

Plotting Points in the Cartesian Plane In Exercises 109–112, plot the point in the Cartesian plane and determine the quadrant in which it is located.

109. $(8, -3)$

110. $(-4, -9)$

111. $\left(-\dfrac{5}{2}, 10\right)$

112. $(6.5, 0.5)$

Accounting In Exercises 113 and 114, use the table, which shows spectator sports revenues (in millions of dollars) for the years 2003 through 2008. (Source: U.S. Census Bureau)

Year	Revenue (in millions of dollars)
2003	22,445
2004	23,659
2005	24,471
2006	26,948
2007	29,277
2008	31,302

113. Sketch a scatter plot of the data.

114. Write a brief statement regarding spectator sports revenues over time.

Using the Distance and Midpoint Formulas In Exercises 115–118, (a) plot the points, (b) find the distance between the points, and (c) find the midpoint of the line segment joining the points.

115. $(-3, 8), (1, 5)$

116. $(-12, 5), (4, -7)$

117. $(5.6, 0), (0, 4.2)$

118. $(3.8, 2.6), (-1.2, -9.4)$

Writing an Equation of a Circle In Exercises 119 and 120, write the standard form of the equation of the specified circle.

119. Center: $(3, -1)$; solution point: $(-5, 1)$

120. Endpoints of a diameter: $(-4, 6), (10, -2)$

Translating Points in the Plane In Exercises 121 and 122, the polygon is shifted to a new position in the plane. Find the new coordinates of the vertices.

121. Original coordinates of vertices:

$(4, 8), (6, 8), (4, 3), (6, 3)$

Shift: three units downward, two units to the left

122. Original coordinates of vertices:

$(0, 1), (3, 3), (0, 5), (-3, 3)$

Shift: five units upward, four units to the right

P.6

123. **Merchandising** Use a line plot to organize the following sample of prices (in dollars) of running shoes. Which price occurred with the greatest frequency?

100, 65, 67, 88, 69, 60, 100, 100, 88, 79, 99, 75, 65, 89, 68, 74, 100, 66, 81, 95, 75, 69, 85, 91, 71

124. **Athletics** The list shows the average numbers of points per game for the top 20 NBA players for the 2009–2010 regular NBA season. Use a frequency distribution and a histogram to organize the data. (Source: National Basketball Association)

25.0, 20.6, 22.6, 23.1, 27.0, 20.8, 19.8, 24.0, 30.1, 25.5, 20.8, 26.5, 21.2, 20.6, 20.1, 28.2, 21.5, 20.2, 29.7, 24.1

125. **Meteorology** The normal daily maximum and minimum temperatures (in °F) for several months for the city of Nashville are shown in the table. Construct a double bar graph for the data. (Source: U.S. National Oceanic and Atmospheric Administration)

Month	Maximum	Minimum
Jan.	45.6	27.9
Feb.	51.4	31.2
Mar.	60.7	39.4
Apr.	69.8	47.1
May	77.5	56.7
Jun.	85.1	65.0

126. **Economics** The table shows the average retail prices (in dollars) of a pound of white all-purpose flour from 2000 through 2008. Construct a line graph for the data and state what information the graph reveals. (Source: U.S. Bureau of Labor Statistics)

Year	Retail price (in dollars)
2000	0.28
2001	0.28
2002	0.29
2003	0.29
2004	0.29
2005	0.30
2006	0.32
2007	0.40
2008	0.50

See www.CalcChat.com for worked-out solutions to odd-numbered exercises.
For instructions on how to use a graphing utility, see Appendix A.

P   Chapter Test

Take this test as you would take a test in class. After you are finished, check your work against the answers in the back of the book.

1. Use < or > to show the relationship between $-\frac{10}{3}$ and $-|-4|$.

2. Find the distance between the real numbers -16 and 38.

3. Identify the rule of algebra illustrated by $(5 - x) + 0 = 5 - x$.

In Exercises 4 and 5, evaluate each expression without using a calculator.

4. (a) $27\left(-\frac{2}{3}\right)$ (b) $\frac{5}{18} \div \frac{15}{8}$ (c) $\left(-\frac{2}{7}\right)^3$ (d) $\left(\frac{3^2}{2}\right)^{-3}$

5. (a) $\sqrt{5} \cdot \sqrt{125}$ (b) $\frac{\sqrt{72}}{\sqrt{2}}$ (c) $\frac{5.4 \times 10^8}{3 \times 10^3}$ (d) $(3 \times 10^4)^3$

In Exercises 6 and 7, simplify each expression.

6. (a) $3z^2(2z^3)^2$ (b) $(u - 2)^{-4}(u - 2)^{-3}$ (c) $\left(\frac{x^{-2}y^2}{3}\right)^{-1}$

7. (a) $9z\sqrt{8z} - 3\sqrt{2z^3}$ (b) $-5\sqrt{16y} + 10\sqrt{y}$ (c) $\sqrt[3]{\frac{16}{v^5}}$

8. Write the polynomial $3 - 2x^5 + 3x^3 - x^4$ in standard form. Identify the degree and leading coefficient.

In Exercises 9–12, perform the operations and simplify.

9. $(x^2 + 3) - [3x + (8 - x^2)]$ 10. $(2x - 5)(4x^2 + 6)$

11. $\dfrac{8x}{x - 3} + \dfrac{24}{3 - x}$ 12. $\dfrac{\left(\dfrac{2}{x} - \dfrac{2}{x + 1}\right)}{\left(\dfrac{4}{x^2 - 1}\right)}$

In Exercises 13–15, find the special product.

13. $\left(x + \sqrt{5}\right)\left(x - \sqrt{5}\right)$ 14. $(x - 2)^3$

15. $[(x + y) - z][(x + y) + z]$

In Exercises 16–18, factor the expression completely.

16. $2x^4 - 3x^3 - 2x^2$ 17. $x^3 + 2x^2 - 4x - 8$

18. $8x^3 - 64$

19. Rationalize each denominator: (a) $\dfrac{16}{\sqrt[3]{16}}$, (b) $\dfrac{6}{1 - \sqrt{3}}$, and (c) $\dfrac{1}{\sqrt{x + 2} - \sqrt{2}}$.

20. Write an expression for the area of the shaded region in the figure at the right and simplify the result.

21. Plot the points $(-2, 5)$ and $(6, 0)$. Find the coordinates of the midpoint of the line segment joining the points and the distance between the points.

22. The numbers (in millions) of votes cast for the Democratic candidates for president in 1984, 1988, 1992, 1996, 2000, 2004, and 2008 were 37.5, 41.7, 44.9, 47.4, 51.0, 58.9, and 69.5, respectively. Construct a bar graph for the data. (Source: Office of the Clerk, U.S. House of Representatives)

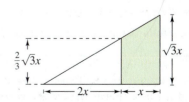

Figure for 20

Proofs in Mathematics

What does the word *proof* mean to you? In mathematics, the word *proof* is used to mean simply a valid argument. When you are proving a statement or theorem, you must use facts, definitions, and accepted properties in a logical order. You can also use previously proved theorems in your proof. For instance, the Distance Formula is used in the proof of the Midpoint Formula below. There are several different proof methods, which you will see in later chapters.

The Midpoint Formula   (p. 50)

The midpoint of the line segment joining the points (x_1, y_1) and (x_2, y_2) is given by the Midpoint Formula

$$\text{Midpoint} = \left(\frac{x_1 + x_2}{2}, \frac{y_1 + y_2}{2} \right).$$

Proof

Using the figure, you must show that $d_1 = d_2$ and $d_1 + d_2 = d_3$.

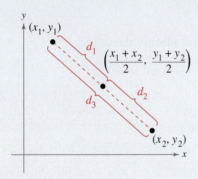

By the Distance Formula, you obtain

$$d_1 = \sqrt{\left(\frac{x_1 + x_2}{2} - x_1 \right)^2 + \left(\frac{y_1 + y_2}{2} - y_1 \right)^2}$$

$$= \frac{1}{2} \sqrt{(x_2 - x_1)^2 + (y_2 - y_1)^2}$$

$$d_2 = \sqrt{\left(x_2 - \frac{x_1 + x_2}{2} \right)^2 + \left(y_2 - \frac{y_1 + y_2}{2} \right)^2}$$

$$= \frac{1}{2} \sqrt{(x_2 - x_1)^2 + (y_2 - y_1)^2}$$

$$d_3 = \sqrt{(x_2 - x_1)^2 + (y_2 - y_1)^2}.$$

So, it follows that $d_1 = d_2$ and $d_1 + d_2 = d_3$.

The Cartesian Plane

The Cartesian plane was named after the French mathematician René Descartes (1596–1650). While Descartes was lying in bed, he noticed a fly buzzing around on the square ceiling tiles. He discovered that the position of the fly could be described by which ceiling tile the fly landed on. This led to the development of the Cartesian plane. Descartes felt that a coordinate plane could be used to facilitate description of the positions of objects.

1 Functions and Their Graphs

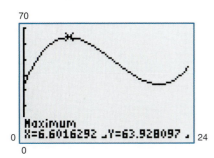

Section 1.4, Example 7
Maximum Temperature

Introduction to Library of Parent Functions

In Chapter 1, you will be introduced to the concept of a *function*. As you proceed through the text, you will see that functions play a primary role in modeling real-life situations.

There are three basic types of functions that have proven to be the most important in modeling real-life situations. These functions are algebraic functions, exponential and logarithmic functions, and trigonometric and inverse trigonometric functions. These three types of functions are referred to as the *elementary functions*, though they are often placed in the two categories of *algebraic functions* and *transcendental functions*. Each time a new type of function is studied in detail in this text, it will be highlighted in a box similar to those shown below. The graphs of these functions are shown on the inside covers of this text.

 ALGEBRAIC FUNCTIONS

These functions are formed by applying algebraic operations to the linear function $f(x) = x$.

Name	Function	Location
Linear	$f(x) = x$	Section 1.2
Quadratic	$f(x) = x^2$	Section 3.1
Cubic	$f(x) = x^3$	Section 3.2
Rational	$f(x) = \dfrac{1}{x}$	Section 3.6
Square root	$f(x) = \sqrt{x}$	Section 1.3

 TRANSCENDENTAL FUNCTIONS

These functions cannot be formed from the linear function by using algebraic operations.

Name	Function	Location
Exponential	$f(x) = a^x,\ a > 0,\ a \neq 1$	Section 4.1
Logarithmic	$f(x) = \log_a x,\ x > 0,\ a > 0,\ a \neq 1$	Section 4.2
Trigonometric	$f(x) = \sin x$	Section 5.4
	$f(x) = \cos x$	Section 5.4
	$f(x) = \tan x$	Section 5.5
	$f(x) = \csc x$	Section 5.5
	$f(x) = \sec x$	Section 5.5
	$f(x) = \cot x$	Section 5.5
Inverse trigonometric	$f(x) = \arcsin x$	Section 5.6
	$f(x) = \arccos x$	Section 5.6
	$f(x) = \arctan x$	Section 5.6

 NONELEMENTARY FUNCTIONS

Some useful nonelementary functions include the following.

Name	Function	Location
Absolute value	$f(x) = |x|$	Section 1.3
Greatest integer	$f(x) = [\![x]\!]$	Section 1.4

1.1   Graphs of Equations

The Graph of an Equation

News magazines often show graphs comparing the rate of inflation, the federal deficit, or the unemployment rate to the time of year. Businesses use graphs to report monthly sales statistics. Such graphs provide geometric pictures of the way one quantity changes with respect to another. Frequently, the relationship between two quantities is expressed as an **equation.** This section introduces the basic procedure for determining the geometric picture associated with an equation.

For an equation in the variables x and y, a point (a, b) is a **solution point** when substitution of a for x and b for y satisfies the equation. Most equations have *infinitely many* solution points. For example, the equation

$$3x + y = 5$$

has solution points

$$(0, 5), \quad (1, 2), \quad (2, -1), \quad (3, -4)$$

and so on. The set of all solution points of an equation is the **graph of the equation.**

Example 1   Determining Solution Points

Determine whether (a) $(2, 13)$ and (b) $(-1, -3)$ lie on the graph of

$$y = 10x - 7.$$

Solution

a.
$y = 10x - 7$ Write original equation.

$13 \stackrel{?}{=} 10(2) - 7$ Substitute 2 for x and 13 for y.

$13 = 13$ $(2, 13)$ is a solution. ✔

The point $(2, 13)$ *does* lie on the graph of $y = 10x - 7$ because it is a solution point of the equation.

b.
$y = 10x - 7$ Write original equation.

$-3 \stackrel{?}{=} 10(-1) - 7$ Substitute -1 for x and -3 for y.

$-3 \neq -17$ $(-1, -3)$ is not a solution.

The point $(-1, -3)$ *does not* lie on the graph of $y = 10x - 7$ because it is not a solution point of the equation.

✓CHECKPOINT   Now try Exercise 7.

The basic technique used for sketching the graph of an equation is the point-plotting method.

Sketching the Graph of an Equation by Point Plotting

1. If possible, rewrite the equation so that one of the variables is isolated on one side of the equation.

2. Make a table of values showing several solution points.

3. Plot these points on a rectangular coordinate system.

4. Connect the points with a smooth curve or line.

Example 2   Sketching a Graph by Point Plotting

Use point plotting and graph paper to sketch the graph of $3x + y = 6$.

Solution

In this case you can isolate the variable y.

$y = 6 - 3x$ Solve equation for y.

Using negative and positive values of x, and $x = 0$, you can obtain the following table of values (solution points).

x	-1	0	1	2	3
$y = 6 - 3x$	9	6	3	0	-3
Solution point	$(-1, 9)$	$(0, 6)$	$(1, 3)$	$(2, 0)$	$(3, -3)$

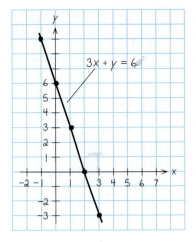

Figure 1.1

Next, plot the solution points and connect them, as shown in Figure 1.1. It appears that the graph is a straight line. You will study lines extensively in Section 1.2.

CHECKPOINT   Now try Exercise 11.

The points at which a graph touches or crosses an axis are called the **intercepts** of the graph. For instance, in Example 2 the point

$(0, 6)$ y-intercept

is the y-intercept of the graph because the graph crosses the y-axis at that point. The point

$(2, 0)$ x-intercept

is the x-intercept of the graph because the graph crosses the x-axis at that point.

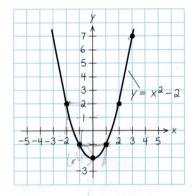

(a)

Example 3   Sketching a Graph by Point Plotting

Use point plotting and graph paper to sketch the graph of

$y = x^2 - 2.$

Solution

Because the equation is already solved for y, make a table of values by choosing several convenient values of x and calculating the corresponding values of y.

x	-2	-1	0	1	2	3
$y = x^2 - 2$	2	-1	-2	-1	2	7
Solution point	$(-2, 2)$	$(-1, -1)$	$(0, -2)$	$(1, -1)$	$(2, 2)$	$(3, 7)$

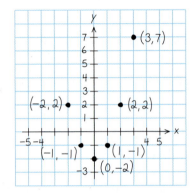

(b)
Figure 1.2

Next, plot the solution points, as shown in Figure 1.2(a). Finally, connect the points with a smooth curve, as shown in Figure 1.2(b). This graph is called a *parabola*. You will study parabolas in Section 3.1.

CHECKPOINT   Now try Exercise 13.

In this text, you will study two basic ways to create graphs: *by hand* and *using a graphing utility*. For instance, the graphs in Figures 1.1 and 1.2 were sketched by hand, and the graph in Figure 1.5 (on the next page) was created using a graphing utility.

Using a Graphing Utility

One of the disadvantages of the point-plotting method is that to get a good idea about the shape of a graph, you need to plot *many* points. With only a few points, you could misrepresent the graph of an equation. For instance, consider the equation

$$y = \frac{1}{30}x(x^4 - 10x^2 + 39).$$

When you plot the points $(-3, -3)$, $(-1, -1)$, $(0, 0)$, $(1, 1)$, and $(3, 3)$, as shown in Figure 1.3(a), you might think that the graph of the equation is a line. This is not correct. By plotting several more points and connecting the points with a smooth curve, you can see that the actual graph is not a line, as shown in Figure 1.3(b).

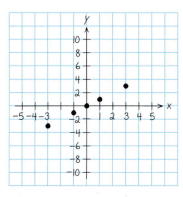

(a)

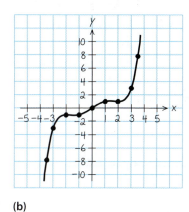

(b)

Figure 1.3

From this, you can see that the point-plotting method leaves you with a dilemma. This method can be very inaccurate when only a few points are plotted, and it is very time-consuming to plot a dozen (or more) points. Technology can help solve this dilemma. Plotting several (even several hundred) points on a rectangular coordinate system is something that a computer or calculator can do easily. For instance, you can enter the equation $y = \frac{1}{30}x(x^4 - 10x^2 + 39)$ in a graphing utility (see Figure 1.4) to obtain the graph shown in Figure 1.5.

Figure 1.4

$$y = \frac{1}{30}x(x^4 - 10x^2 + 39)$$

Figure 1.5

Using a Graphing Utility to Graph an Equation

To graph an equation involving x and y on a graphing utility, do the following.

   1. Rewrite the equation so that y is isolated on the left side.

   2. Enter the equation in the graphing utility.

   3. Determine a *viewing window* that shows all important features of the graph.

   4. Graph the equation.

Technology Tip

Many graphing utilities are capable of creating a table of values such as the following, which shows some points of the graph in Figure 1.3(b). For instructions on how to use the *table* feature, see Appendix A; for specific keystrokes, go to this textbook's *Companion Website*.

Technology Tip

By choosing different viewing windows for a graph, it is possible to obtain very different impressions of the graph's shape. For instance, Figure 1.6 shows a different viewing window for the graph of the equation in Figure 1.5. Note how Figure 1.6 does not show *all* of the important features of the graph as does Figure 1.5. For instructions on how to set up a viewing window, see Appendix A; for specific keystrokes, go to this textbook's *Companion Website*.

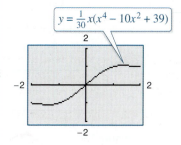

$$y = \frac{1}{30}x(x^4 - 10x^2 + 39)$$

Figure 1.6

Example 4 Using a Graphing Utility to Graph an Equation

To graph

$$y = -\frac{x^3}{2} + 2x$$

enter the equation in a graphing utility. Then use a standard viewing window (see Figure 1.7) to obtain the graph shown in Figure 1.8.

Figure 1.7

Figure 1.8

✔CHECKPOINT   Now try Exercise 43.

Example 5 Using a Graphing Utility to Graph a Circle

Use a graphing utility to graph $x^2 + y^2 = 9$.

Solution

The graph of $x^2 + y^2 = 9$ is a circle whose center is the origin and whose radius is 3. To graph the equation, begin by solving the equation for y.

$x^2 + y^2 = 9$	Write original equation.
$y^2 = 9 - x^2$	Subtract x^2 from each side.
$y = \pm\sqrt{9 - x^2}$	Take the square root of each side.

Remember that when you take the square root of a variable expression, you must account for both the positive and negative solutions. The graph of $y = \sqrt{9 - x^2}$ is the upper semicircle. The graph of $y = -\sqrt{9 - x^2}$ is the lower semicircle. Enter *both* equations in your graphing utility and generate the resulting graphs. In Figure 1.9, note that for a standard viewing window, the two graphs do not appear to form a circle. You can overcome this problem by using a *square setting*, in which the horizontal and vertical tick marks have equal spacing, as shown in Figure 1.10. On many graphing utilities, a square setting can be obtained by using a y to x ratio of 2 to 3. For instance, in Figure 1.10, the y to x ratio is

$$\frac{Y_{max} - Y_{min}}{X_{max} - X_{min}} = \frac{4 - (-4)}{6 - (-6)} = \frac{8}{12} = \frac{2}{3}.$$

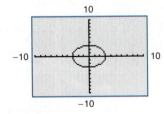

Figure 1.9

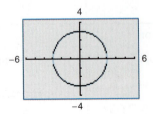

Figure 1.10

✔CHECKPOINT   Now try Exercise 57.

Technology Tip

The standard viewing window on many graphing utilities does not give a true geometric perspective because the screen is rectangular, which distorts the image. That is, perpendicular lines will not appear to be perpendicular and circles will not appear to be circular. To overcome this, you can use a *square setting*, as demonstrated in Example 5.

Technology Tip

Notice that when you graph a circle by graphing two separate equations for y, your graphing utility may not connect the two semicircles. This is because some graphing utilities are limited in their resolution. So, in this text, a blue curve is placed behind the graphing utility's display to indicate where the graph should appear.

Applications

Throughout this course, you will learn that there are many ways to approach a problem. Two of the three common approaches are illustrated in Example 6.

An Algebraic Approach: Use the rules of algebra.

A Graphical Approach: Draw and use a graph.

A Numerical Approach: Construct and use a table.

You should develop the habit of using at least two approaches to solve every problem in order to build your intuition and to check that your answer is reasonable.

The following two applications show how to develop mathematical models to represent real-world situations. You will see that both a graphing utility and algebra can be used to understand and solve the problems posed.

Example 6   Running a Marathon

A runner runs at a constant rate of 4.9 miles per hour. The verbal model and algebraic equation relating distance run and elapsed time are as follows.

Verbal Model: | Distance | = | Rate | · | Time |

Equation: $d = 4.9t$

a. Determine how far the runner can run in 3.1 hours.

b. Determine how long it will take the runner to run a 26.2-mile marathon.

Algebraic Solution

a. To begin, find how far the runner can run in 3.1 hours by substituting 3.1 for t in the equation.

$d = 4.9t$ Write original equation.

$\quad = 4.9(3.1)$ Substitute 3.1 for t.

$\quad \approx 15.2$ Use a calculator.

So, the runner can run about 15.2 miles in 3.1 hours. Use estimation to check your answer. Because 4.9 is about 5 and 3.1 is about 3, the distance is about

$\quad 5(3) = 15.$

So, 15.2 is reasonable.

b. You can find how long it will take to run a 26.2-mile marathon as follows. (For help with solving linear equations, see Appendix D at this textbook's *Companion Website*.)

$\quad d = 4.9t$ Write original equation.

$\quad 26.2 = 4.9t$ Substitute 26.2 for d.

$\quad \dfrac{26.2}{4.9} = t$ Divide each side by 4.9.

$\quad 5.3 \approx t$ Use a calculator.

So, it will take the runner about 5.3 hours to run 26.2 miles.

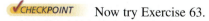 **CHECKPOINT**   Now try Exercise 63.

Graphical Solution

a. Use a graphing utility to graph $d = 4.9t$. (Represent d by y and t by x.) Choose a viewing window that shows the graph at $x = 3.1$.

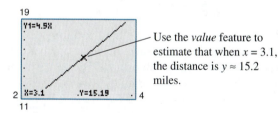

Use the *value* feature to estimate that when $x = 3.1$, the distance is $y \approx 15.2$ miles.

Figure 1.11

b. Adjust the viewing window so that it shows the graph at $y = 26.2$.

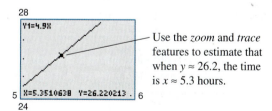

Use the *zoom* and *trace* features to estimate that when $y \approx 26.2$, the time is $x \approx 5.3$ hours.

Figure 1.12

Note that the viewing window on your graphing utility may differ slightly from those shown in Figures 1.11 and 1.12.

Example 7   Monthly Wage

You receive a monthly salary of $2000 plus a commission of 10% of sales. The verbal model and algebraic equation relating the wages, the salary, and the commission are as follows.

Verbal Model: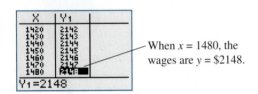

Equation: $y = 2000 + 0.1x$

a. Sales are $1480 in August. What are your wages for that month?

b. You receive $2225 for September. What are your sales for that month?

Numerical Solution

a. Enter $y = 2000 + 0.1x$ in a graphing utility. Then use the *table* feature of the graphing utility to create a table. Start the table at $x = 1400$ with a table step of 10.

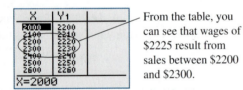

When $x = 1480$, the wages are $y = 2148.

b. Adjust the table to start at $x = 2000$ with a table step of 100.

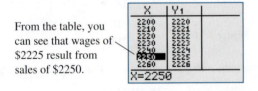

From the table, you can see that wages of $2225 result from sales between $2200 and $2300.

You can improve the estimate by starting the table at $x = 2200$ with a table step of 10.

From the table, you can see that wages of $2225 result from sales of $2250.

    Now try Exercise 65.

Graphical Solution

a. Use a graphing utility to graph $y = 2000 + 0.1x$. Choose a viewing window that shows the graph at $x = 1480$.

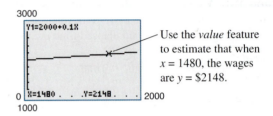

Use the *value* feature to estimate that when $x = 1480$, the wages are $y = 2148.

b. Use the graphing utility to find the value along the x-axis (sales) that corresponds to a y-value of 2225 (wages). Adjust the viewing window so that it shows the graph at $y = 2225$.

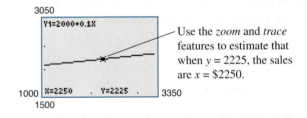

Use the *zoom* and *trace* features to estimate that when $y = 2225$, the sales are $x = 2250.

Remember to use a different approach to check that your answer is reasonable. For instance, to check the numerical solution to Example 7, use a graphical approach as shown above or use an algebraic approach as follows.

a. Substitute 1480 for x in the original equation and solve for y.

$$y = 2000 + 0.1(1480) = $2148$$

b. Substitute 2225 for y in the original equation and solve for x.

$$2225 = 2000 + 0.1x \quad \Longrightarrow \quad x = $2250$$

1.1 Exercises

See www.CalcChat.com for worked-out solutions to odd-numbered exercises.
For instructions on how to use a graphing utility, see Appendix A.

Vocabulary and Concept Check

In Exercises 1 and 2, fill in the blank.

1. For an equation in x and y, if substitution of a for x and b for y satisfies the equation, then the point (a, b) is a _____ .

2. The set of all solution points of an equation is the _____ of the equation.

3. Name three common approaches you can use to solve problems mathematically.

4. List the steps for sketching the graph of an equation by point plotting.

Procedures and Problem Solving

Determining Solution Points In Exercises 5–10, determine whether each point lies on the graph of the equation.

	Equation	Points	
5.	$y = \sqrt{x + 4}$	(a) $(0, 2)$	(b) $(12, 4)$
6.	$y = x^2 - 3x + 2$	(a) $(2, 0)$	(b) $(-2, 8)$
7.	$y = 4 - |x - 2|$	(a) $(1, 5)$	(b) $(1.2, 3.2)$
8.	$2x - y - 3 = 0$	(a) $(1, 2)$	(b) $(1, -1)$
9.	$x^2 + y^2 = 20$	(a) $(3, -2)$	(b) $(-4, 2)$
10.	$y = \frac{1}{3}x^3 - 2x^2$	(a) $\left(2, -\frac{16}{3}\right)$	(b) $(-3, 9)$

Sketching a Graph by Point Plotting In Exercises 11–14, complete the table. Use the resulting solution points to sketch the graph of the equation. Use a graphing utility to verify the graph.

11. $3x - 2y = 2$

x	-2	0	$\frac{2}{3}$	1	2
y					
Solution point					

12. $8x + 4y = 24$

x	-3	-1	0	2	3
y					
Solution point					

13. $2x + y = x^2$

x	-1	0	1	2	3
y					
Solution point					

14. $6x - 2y = -2x^2$

x	-4	-3	-2	0	1
y					
Solution point					

Matching an Equation with Its Graph In Exercises 15–18, match the equation with its graph. [The graphs are labeled (a), (b), (c), and (d).]

(a)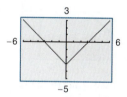

(b)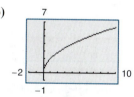

(c)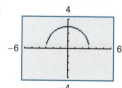

(d)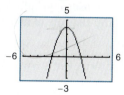

15. $y = 2\sqrt{x}$ 16. $y = 4 - x^2$

17. $y = \sqrt{9 - x^2}$ 18. $y = |x| - 3$

Sketching the Graph of an Equation In Exercises 19–32, sketch the graph of the equation.

19. $y = -4x + 1$ 20. $y = 2x - 3$

21. $y = 2 - x^2$ 22. $y = x^2 - 1$

23. $y = x^2 - 3x$ 24. $y = -x^2 - 4x$

25. $y = x^3 + 2$ 26. $y = x^3 - 3$

27. $y = \sqrt{x - 3}$ 28. $y = \sqrt{1 - x}$

29. $y = |x - 2|$ 30. $y = 4 - |x|$

31. $x = y^2 - 1$ 32. $x = y^2 + 4$

Using a Graphing Utility to Graph an Equation In Exercises 33–46, use a graphing utility to graph the equation. Use a standard viewing window. Approximate any x- or y-intercepts of the graph.

33. $y = x - 7$

34. $y = x + 1$

35. $y = 3 - \frac{1}{2}x$

36. $y = \frac{2}{3}x - 1$

37. $y = \dfrac{2x}{x - 1}$

38. $y = \dfrac{6}{x}$

39. $y = x\sqrt{x + 3}$

40. $y = (6 - x)\sqrt{x}$

41. $y = \sqrt[3]{x - 8}$

42. $y = \sqrt[3]{x + 1}$

✓ **43.** $x^2 - y = 4x - 3$

44. $2y - x^2 + 8 = 2x$

45. $y - 4x = x^2(x - 4)$

46. $x^3 + y = 1$

Describing the Viewing Window of a Graphing Utility In Exercises 47 and 48, describe the viewing window of the graph shown.

47. $y = -10x + 50$

48. $y = \sqrt{x + 2} - 1$

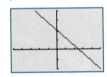

Verifying a Rule of Algebra In Exercises 49–52, explain how to use a graphing utility to verify that $y_1 = y_2$. Identify the rule of algebra that is illustrated.

49. $y_1 = \frac{1}{4}(x^2 - 8)$
$y_2 = \frac{1}{4}x^2 - 2$

50. $y_1 = \frac{1}{2}x + (x + 1)$
$y_2 = \frac{3}{2}x + 1$

51. $y_1 = \dfrac{1}{5}[10(x^2 - 1)]$
$y_2 = 2(x^2 - 1)$

52. $y_1 = (x - 3) \cdot \dfrac{1}{x - 3}$
$y_2 = 1$

Using a Graphing Utility to Graph an Equation In Exercises 53–56, use a graphing utility to graph the equation. Use the *trace* feature of the graphing utility to approximate the unknown coordinate of each solution point accurate to two decimal places. (*Hint:* You may need to use the *zoom* feature of the graphing utility to obtain the required accuracy.)

53. $y = \sqrt{5 - x}$

(a) $(3, y)$

(b) $(x, 3)$

54. $y = x^3(x - 3)$

(a) $(2.25, y)$

(b) $(x, 20)$

55. $y = x^5 - 5x$

(a) $(-0.5, y)$

(b) $(x, -4)$

56. $y = |x^2 - 6x + 5|$

(a) $(2, y)$

(b) $(x, 1.5)$

Using a Graphing Utility to Graph a Circle In Exercises 57–60, solve for y and use a graphing utility to graph each of the resulting equations in the same viewing window. (Adjust the viewing window so that the circle appears circular.)

✓ **57.** $x^2 + y^2 = 16$

58. $x^2 + y^2 = 36$

59. $(x - 1)^2 + (y - 2)^2 = 9$

60. $(x - 3)^2 + (y - 1)^2 = 25$

Determining Solution Points In Exercises 61 and 62, determine which point lies on the graph of the circle. (There may be more than one correct answer.)

61. $(x - 1)^2 + (y - 2)^2 = 25$

(a) $(1, 3)$

(b) $(-2, 6)$

(c) $(5, -1)$

(d) $\left(0, 2 + 2\sqrt{6}\right)$

62. $(x + 2)^2 + (y - 3)^2 = 25$

(a) $(-2, 3)$

(b) $(0, 0)$

(c) $(1, -1)$

(d) $\left(-1, 3 - 2\sqrt{6}\right)$

✓ **63. MODELING DATA**

A manufacturing plant purchases a new molding machine for \$225,000. The depreciated value (decreased value) y after t years is $y = 225{,}000 - 20{,}000t$, for $0 \le t \le 8$.

(a) Use the constraints of the model and a graphing utility to graph the equation using an appropriate viewing window.

(b) Use the *value* feature or the *zoom* and *trace* features of the graphing utility to determine the value of y when $t = 5.8$. Verify your answer algebraically.

(c) Use the *value* feature or the *zoom* and *trace* features of the graphing utility to determine the value of y when $t = 2.35$. Verify your answer algebraically.

64. MODELING DATA

You buy a personal watercraft for \$8100. The depreciated value y after t years is $y = 8100 - 929t$, for $0 \le t \le 6$.

(a) Use the constraints of the model and a graphing utility to graph the equation using an appropriate viewing window.

(b) Use the *zoom* and *trace* features of the graphing utility to determine the value of t when $y = 5545.25$. Verify your answer algebraically.

(c) Use the *value* feature or the *zoom* and *trace* features of the graphing utility to determine the value of y when $t = 5.5$. Verify your answer algebraically.

✓ 65. MODELING DATA

The table shows the median (middle) sales prices y (in thousands of dollars) of new one-family homes in the southern United States from 2000 through 2008. (Sources: U.S. Census Bureau; U.S. Department of Housing and Urban Development)

Year	Median sales price, y
2000	148.0
2001	155.4
2002	163.4
2003	168.1
2004	181.1
2005	197.3
2006	208.2
2007	217.7
2008	203.7

A model that represents the data is given by $y = -0.4221t^3 + 4.690t^2 - 3.47t + 150.9$, $0 \leq t \leq 8$, where t represents the year, with $t = 0$ corresponding to 2000.

(a) Use the model and the *table* feature of a graphing utility to find the median sales prices from 2000 through 2008. How well does the model fit the data? Explain.

(b) Use the graphing utility to graph the data from the table and the model in the same viewing window. How well does the model fit the data? Explain.

(c) Use the model to estimate the median sales prices in 2012 and 2014. Do the values seem reasonable? Explain.

(d) Use the *zoom* and *trace* features of the graphing utility to determine during which year(s) the median sales price was approximately $150,000.

66. *Why you should learn it* (p. 75)

The table shows the life expectancy y of a child (at birth) in the United States for each of the selected years from 1930 through 2000. (Source: U.S. National Center for Health Statistics)

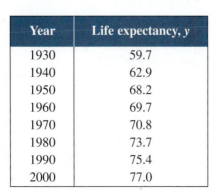

Year	Life expectancy, y
1930	59.7
1940	62.9
1950	68.2
1960	69.7
1970	70.8
1980	73.7
1990	75.4
2000	77.0

A model that represents the data is given by

$$y = \frac{59.617 + 1.18t}{1 + 0.012t}, \quad 0 \leq t \leq 70$$

where t is the time in years, with $t = 0$ corresponding to 1930.

(a) Use a graphing utility to graph the data from the table above and the model in the same viewing window. How well does the model fit the data? Explain.

(b) Find the y-intercept of the graph of the model. What does it represent in the context of the problem?

(c) Use the *zoom* and *trace* features of the graphing utility to determine the year when the life expectancy was 73.2. Verify your answer algebraically.

(d) Determine the life expectancy in 1948 both graphically and algebraically.

(e) Use the model to estimate the life expectancy of a child born in 2020.

Conclusions

True or False?   In Exercises 67 and 68, determine whether the statement is true or false. Justify your answer.

67. A parabola can have only one x-intercept.

68. The graph of a linear equation can have either no x-intercepts or only one x-intercept.

69. Writing   Your employer offers you a choice of wage scales: a monthly salary of $3000 plus commission of 7% of sales or a salary of $3400 plus a 5% commission. Write a short paragraph discussing how you would choose your option. At what sales level would the options yield the same salary?

70. CAPSTONE You open a savings account and deposit $200. Every week you withdraw $50. The account balance y after t weeks is $y = -50t + 200$, for $0 \leq t \leq 4$.

(a) Use point plotting and graph paper to sketch the graph of $y = -50t + 200$.

(b) Use a graphing utility to graph $y = -50t + 200$.

(c) Explain how to find an appropriate viewing window for the graph of the equation.

(d) Find the y-intercept of the graph of the model. What does it represent in the context of the problem?

Cumulative Mixed Review

Performing Operations with Polynomials   In Exercises 71 and 72, perform the operation and write the result in standard form.

71. $(9x - 4) + (2x^2 - x + 15)$

72. $(3x^2 - 5)(-x^2 + 1)$

1.2   Lines in the Plane

The Slope of a Line

In this section, you will study lines and their equations. The **slope** of a nonvertical line represents the number of units the line rises or falls vertically for each unit of horizontal change from left to right. For instance, consider the two points

$$(x_1, y_1) \quad \text{and} \quad (x_2, y_2)$$

on the line shown in Figure 1.13.

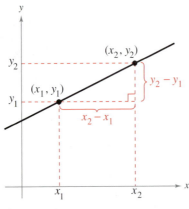

Figure 1.13

As you move from left to right along this line, a change of $(y_2 - y_1)$ units in the vertical direction corresponds to a change of $(x_2 - x_1)$ units in the horizontal direction. That is,

$$y_2 - y_1 = \text{the change in } y$$

and

$$x_2 - x_1 = \text{the change in } x.$$

The slope of the line is given by the ratio of these two changes.

> **Definition of the Slope of a Line**
>
> The **slope** m of the nonvertical line through (x_1, y_1) and (x_2, y_2) is
>
> $$m = \frac{y_2 - y_1}{x_2 - x_1} = \frac{\text{change in } y}{\text{change in } x}$$
>
> where $x_1 \neq x_2$.

When this formula for slope is used, the *order of subtraction* is important. Given two points on a line, you are free to label either one of them as (x_1, y_1) and the other as (x_2, y_2). Once you have done this, however, you must form the numerator and denominator using the same order of subtraction.

$$m = \frac{y_2 - y_1}{x_2 - x_1} \qquad m = \frac{y_1 - y_2}{x_1 - x_2} \qquad m = \frac{y_2 - y_1}{x_1 - x_2}$$

Correct   Correct   Incorrect

Throughout this text, the term *line* always means a *straight* line.

Example 1   Finding the Slope of a Line

Find the slope of the line passing through each pair of points.

a. $(-2, 0)$ and $(3, 1)$

b. $(-1, 2)$ and $(2, 2)$

c. $(0, 4)$ and $(1, -1)$

Solution

a. Difference in y-values

$$m = \frac{y_2 - y_1}{x_2 - x_1} = \frac{1 - 0}{3 - (-2)} = \frac{1}{3 + 2} = \frac{1}{5}$$

Difference in x-values

b. $m = \frac{2 - 2}{2 - (-1)} = \frac{0}{3} = 0$

c. $m = \frac{-1 - 4}{1 - 0} = \frac{-5}{1} = -5$

The graphs of the three lines are shown in Figure 1.14. Note that the *square setting* gives the correct "steepness" of the lines.

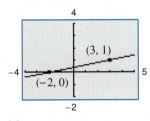
(a)

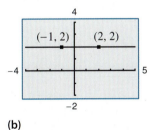
(b)

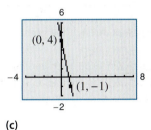
(c)

Figure 1.14

✔CHECKPOINT   Now try Exercise 15.

The definition of slope does not apply to vertical lines. For instance, consider the points $(3, 4)$ and $(3, 1)$ on the vertical line shown in Figure 1.15. Applying the formula for slope, you obtain

$$m = \frac{4 - 1}{3 - 3} = \frac{3}{0}.$$ Undefined

Because division by zero is undefined, the slope of a vertical line is undefined.

Figure 1.15

From the slopes of the lines shown in Figures 1.14 and 1.15, you can make the following generalizations about the slope of a line.

> ### The Slope of a Line
>
> **1.** A line with positive slope $(m > 0)$ *rises* from left to right.
>
> **2.** A line with negative slope $(m < 0)$ *falls* from left to right.
>
> **3.** A line with zero slope $(m = 0)$ is *horizontal*.
>
> **4.** A line with undefined slope is *vertical*.

The Point-Slope Form of the Equation of a Line

When you know the slope of a line *and* you also know the coordinates of one point on the line, you can find an equation of the line. For instance, in Figure 1.16, let (x_1, y_1) be a point on the line whose slope is m. When (x, y) is any *other* point on the line, it follows that

$$\frac{y - y_1}{x - x_1} = m.$$

This equation in the variables x and y can be rewritten in the **point-slope form** of the equation of a line.

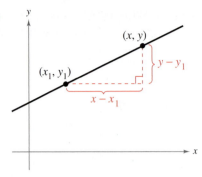

Figure 1.16

> **Point-Slope Form of the Equation of a Line**
>
> The **point-slope form** of the equation of the line that passes through the point (x_1, y_1) and has a slope of m is
>
> $$y - y_1 = m(x - x_1).$$

Example 2   The Point-Slope Form of the Equation of a Line

Find an equation of the line that passes through the point

$$(1, -2)$$

and has a slope of 3.

Solution

$y - y_1 = m(x - x_1)$	Point-slope form
$y - (-2) = 3(x - 1)$	Substitute for y_1, m, and x_1.
$y + 2 = 3x - 3$	Simplify.
$y = 3x - 5$	Solve for y.

The line is shown in Figure 1.17.

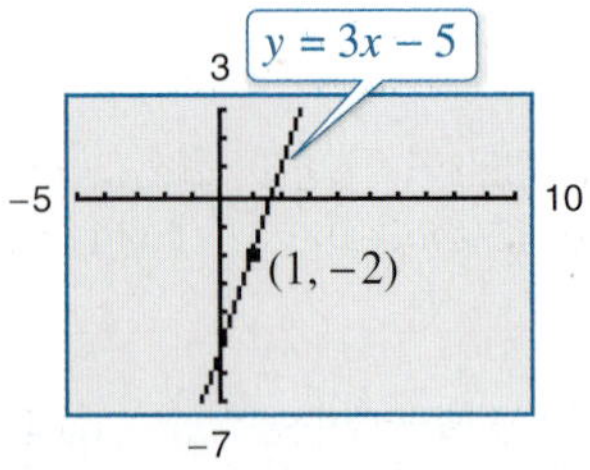

Figure 1.17

✓CHECKPOINT   Now try Exercise 25.

The point-slope form can be used to find an equation of a nonvertical line passing through two points

$$(x_1, y_1) \quad \text{and} \quad (x_2, y_2).$$

First, find the slope of the line.

$$m = \frac{y_2 - y_1}{x_2 - x_1}, \quad x_1 \neq x_2$$

Then use the point-slope form to obtain the equation

$$y - y_1 = \frac{y_2 - y_1}{x_2 - x_1}(x - x_1).$$

This is sometimes called the **two-point form** of the equation of a line.

Study Tip

When you find an equation of the line that passes through two given points, you need to substitute the coordinates of only one of the points into the point-slope form. It does not matter which point you choose because both points will yield the same result.

Example 3   A Linear Model for Profits Prediction

During 2006, Research In Motion's net profits were \$631.6 million, and in 2007 net profits were \$1293.9 million. Write a linear equation giving the net profits y in terms of the year x. Then use the equation to predict the net profits for 2008.   (Source: Research In Motion Limited)

Solution

Let $x = 0$ represent 2000. In Figure 1.18, let $(6, 631.6)$ and $(7, 1293.9)$ be two points on the line representing the net profits. The slope of this line is

$$m = \frac{1293.9 - 631.6}{7 - 6} = 662.3.$$

By the point-slope form, the equation of the line is as follows.

$$y - 631.6 = 662.3(x - 6)$$

$$y = 662.3x - 3342.2$$

Now, using this equation, you can predict the 2008 net profits ($x = 8$) to be

$$y = 662.3(8) - 3342.2 = 5298.4 - 3342.2 = \$1956.2 \text{ million.}$$

(In this case, the prediction is quite good—the actual net profits in 2008 were \$1968.8 million.)

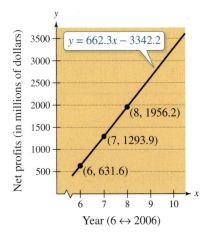

Figure 1.18

✔CHECKPOINT   Now try Exercise 33.

Library of Parent Functions: Linear Function

In the next section, you will be introduced to the precise meaning of the term *function*. The simplest type of function is the *parent linear function*

$$f(x) = x.$$

As its name implies, the graph of the parent linear function is a line. The basic characteristics of the parent linear function are summarized below and on the inside front cover of this text. (Note that some of the terms below will be defined later in the text.)

Graph of $f(x) = x$
Domain: $(-\infty, \infty)$
Range: $(-\infty, \infty)$
Intercept: $(0, 0)$
Increasing

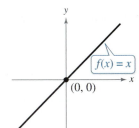

The function $f(x) = x$ is also referred to as the *identity function*. Later in this text, you will learn that the graph of the linear function $f(x) = mx + b$ is a line with slope m and y-intercept $(0, b)$. When $m = 0$, $f(x) = b$ is called a *constant function* and its graph is a horizontal line.

Sketching Graphs of Lines

Many problems in coordinate geometry can be classified as follows.

1. Given a graph (or parts of it), find its equation.

2. Given an equation, sketch its graph.

For lines, the first problem is solved easily by using the point-slope form. This formula, however, is not particularly useful for solving the second type of problem. The form that is better suited to graphing linear equations is the **slope-intercept form** of the equation of a line, $y = mx + b$.

> ### Slope-Intercept Form of the Equation of a Line
>
> The graph of the equation
>
> $$y = mx + b$$
>
> is a line whose slope is m and whose y-intercept is $(0, b)$.

Example 4   Using the Slope-Intercept Form

Determine the slope and y-intercept of each linear equation. Then describe its graph.

a. $x + y = 2$

b. $y = 2$

Algebraic Solution

a. Begin by writing the equation in slope-intercept form.

$x + y = 2$	Write original equation.
$y = 2 - x$	Subtract x from each side.
$y = -x + 2$	Write in slope-intercept form.

From the slope-intercept form of the equation, the slope is -1 and the y-intercept is

$(0, 2)$.

Because the slope is negative, you know that the graph of the equation is a line that falls one unit for every unit it moves to the right.

b. By writing the equation $y = 2$ in slope-intercept form

$$y = (0)x + 2$$

you can see that the slope is 0 and the y-intercept is

$(0, 2)$.

A zero slope implies that the line is horizontal.

Graphical Solution

a.

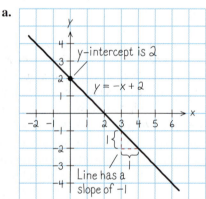

b.

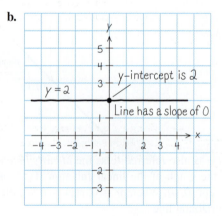

 CHECKPOINT   Now try Exercise 35.

From the slope-intercept form of the equation of a line, you can see that a horizontal line ($m = 0$) has an equation of the form

$y = b.$ Horizontal line

This is consistent with the fact that each point on a horizontal line through $(0, b)$ has a y-coordinate of b. Similarly, each point on a vertical line through $(a, 0)$ has an x-coordinate of a. So, a vertical line has an equation of the form

$x = a.$ Vertical line

This equation cannot be written in slope-intercept form because the slope of a vertical line is undefined. However, every line has an equation that can be written in the general form

$Ax + By + C = 0$ General form of the equation of a line

where A and B are not *both* zero.

Summary of Equations of Lines

1. General form: $Ax + By + C = 0$

2. Vertical line: $x = a$

3. Horizontal line: $y = b$

4. Slope-intercept form: $y = mx + b$

5. Point-slope form: $y - y_1 = m(x - x_1)$

Example 5   Different Viewing Windows

When a graphing utility is used to graph a line, it is important to realize that the graph of the line may not visually appear to have the slope indicated by its equation. This occurs because of the viewing window used for the graph. For instance, Figure 1.19 shows graphs of $y = 2x + 1$ produced on a graphing utility using three different viewing windows. Notice that the slopes in Figures 1.19(a) and (b) do not visually appear to be equal to 2. When you use a *square setting*, as in Figure 1.19(c), the slope visually appears to be 2.

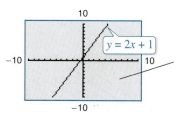

Using a *nonsquare setting*, you do *not* obtain a graph with a true geometric perspective. So, the slope does *not* visually appear to be 2.

(a) *Nonsquare setting*

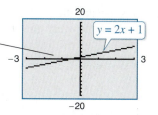

Using a *nonsquare setting*, you do *not* obtain a graph with a true geometric perspective. So, the slope does *not* visually appear to be 2.

(b) *Nonsquare setting*

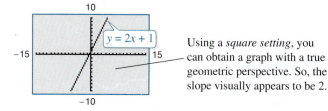

Using a *square setting*, you can obtain a graph with a true geometric perspective. So, the slope visually appears to be 2.

(c) *Square setting*

Figure 1.19

✓CHECKPOINT   Now try Exercise 61.

Parallel and Perpendicular Lines

The slope of a line is a convenient tool for determining whether two lines are parallel or perpendicular.

Explore the Concept

Graph the lines $y_1 = \frac{1}{2}x + 1$ and $y_2 = -2x + 1$ in the same viewing window. What do you observe?

Graph the lines $y_1 = 2x + 1$, $y_2 = 2x$, and $y_3 = 2x - 1$ in the same viewing window. What do you observe?

> **Parallel Lines**
>
> Two distinct nonvertical lines are **parallel** if and only if their slopes are equal. That is,
>
> $$m_1 = m_2.$$

Example 6 Equations of Parallel Lines

Find the slope-intercept form of the equation of the line that passes through the point $(2, -1)$ and is parallel to the line

$$2x - 3y = 5.$$

Solution

Begin by writing the equation of the given line in slope-intercept form.

$2x - 3y = 5$	Write original equation.
$-2x + 3y = -5$	Multiply by -1.
$3y = 2x - 5$	Add $2x$ to each side.
$y = \dfrac{2}{3}x - \dfrac{5}{3}$	Write in slope-intercept form.

Therefore, the given line has a slope of

$$m = \tfrac{2}{3}.$$

Any line parallel to the given line must also have a slope of $\frac{2}{3}$. So, the line through $(2, -1)$ has the following equation.

$y - y_1 = m(x - x_1)$	Point-slope form
$y - (-1) = \dfrac{2}{3}(x - 2)$	Substitute for y_1, m, and x_1.
$y + 1 = \dfrac{2}{3}x - \dfrac{4}{3}$	Simplify.
$y = \dfrac{2}{3}x - \dfrac{7}{3}$	Write in slope-intercept form.

Notice the similarity between the slope-intercept form of the original equation and the slope-intercept form of the parallel equation. The graphs of both equations are shown in Figure 1.20.

✔CHECKPOINT   Now try Exercise 67(a).

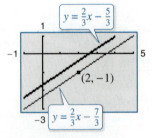

Figure 1.20

> **Perpendicular Lines**
>
> Two nonvertical lines are **perpendicular** if and only if their slopes are negative reciprocals of each other. That is,
>
> $$m_1 = -\frac{1}{m_2}.$$

Example 7   Equations of Perpendicular Lines

Find the slope-intercept form of the equation of the line that passes through the point $(2, -1)$ and is perpendicular to the line

$$2x - 3y = 5.$$

Solution

From Example 6, you know that the equation can be written in the slope-intercept form

$$y = \frac{2}{3}x - \frac{5}{3}.$$

You can see that the line has a slope of $\frac{2}{3}$. So, any line perpendicular to this line must have a slope of $-\frac{3}{2}$ $\left(\text{because } -\frac{3}{2} \text{ is the negative reciprocal of } \frac{2}{3}\right)$. So, the line through the point $(2, -1)$ has the following equation.

$$y - (-1) = -\frac{3}{2}(x - 2) \qquad \text{Write in point-slope form.}$$

$$y + 1 = -\frac{3}{2}x + 3 \qquad \text{Simplify.}$$

$$y = -\frac{3}{2}x + 2 \qquad \text{Write in slope-intercept form.}$$

The graphs of both equations are shown in Figure 1.21.

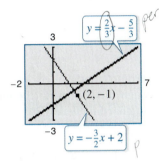

Figure 1.21

✓CHECKPOINT   Now try Exercise 67(b).

Example 8   Graphs of Perpendicular Lines

Use a graphing utility to graph the lines $y = x + 1$ and $y = -x + 3$ in the same viewing window. The lines are supposed to be perpendicular (they have slopes of $m_1 = 1$ and $m_2 = -1$). Do they appear to be perpendicular on the display?

Solution

When the viewing window is nonsquare, as in Figure 1.22, the two lines will not appear perpendicular. When, however, the viewing window is square, as in Figure 1.23, the lines will appear perpendicular.

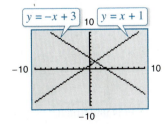

Figure 1.22

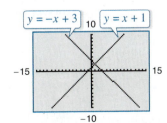

Figure 1.23

✓CHECKPOINT   Now try Exercise 81.

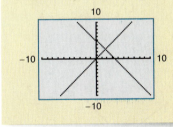

?   What's Wrong?

You use a graphing utility to graph $y_1 = 1.5x$ and $y_2 = -1.5x + 5$, as shown in the figure. You use the graph to conclude that the lines are perpendicular. What's wrong?

1.2   Exercises

See www.CalcChat.com for worked-out solutions to odd-numbered exercises.
For instructions on how to use a graphing utility, see Appendix A.

Vocabulary and Concept Check

1. Match each equation with its form.

    (a) $Ax + By + C = 0$ (i) vertical line

    (b) $x = a$ (ii) slope-intercept form

    (c) $y = b$ (iii) general form

    (d) $y = mx + b$ (iv) point-slope form

    (e) $y - y_1 = m(x - x_1)$ (v) horizontal line

In Exercises 2 and 3, fill in the blank.

2. For a line, the ratio of the change in y to the change in x is called the _____ of the line.

3. Two lines are _____ if and only if their slopes are equal.

4. What is the relationship between two lines whose slopes are -3 and $\frac{1}{3}$?

5. What is the slope of a line that is perpendicular to the line represented by $x = 3$?

6. Give the coordinates of a point on the line whose equation in point-slope form is $y - (-2) = \frac{1}{2}(x - 5)$.

Procedures and Problem Solving

Using Slope   In Exercises 7 and 8, identify the line that has the indicated slope.

7. (a) $m = \frac{2}{3}$ (b) m is undefined.   (c) $m = -2$

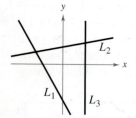

8. (a) $m = 0$ (b) $m = -\frac{3}{4}$ (c) $m = 1$

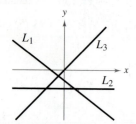

Estimating Slope   In Exercises 9 and 10, estimate the slope of the line.

9.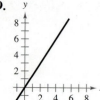

10.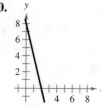

Sketching Lines   In Exercises 11 and 12, sketch the lines through the point with the indicated slopes on the same set of coordinate axes.

	Point			Slopes		
11.	$(2, 3)$	(a) 0	(b) 1	(c) 2	(d) -3
12.	$(-4, 1)$	(a) 3	(b) -3	(c) $\frac{1}{2}$	(d) Undefined

Finding the Slope of a Line   In Exercises 13–16, find the slope of the line passing through the pair of points. Then use a graphing utility to plot the points and use the *draw* feature to graph the line segment connecting the two points. (Use a *square setting*.)

13. $(0, -10), (-4, 0)$ 14. $(2, 4), (4, -4)$

✓ 15. $(-6, -1), (-6, 4)$ 16. $(-3, -2), (1, 6)$

Using Slope   In Exercises 17–24, use the point on the line and the slope of the line to find three additional points through which the line passes. (There are many correct answers.)

	Point	Slope
17.	$(2, 1)$	$m = 0$
18.	$(3, -2)$	$m = 0$
19.	$(1, 5)$	m is undefined.
20.	$(-4, 1)$	m is undefined.
21.	$(0, -9)$	$m = -2$
22.	$(-5, 4)$	$m = 2$
23.	$(7, -2)$	$m = \frac{1}{2}$
24.	$(-1, -6)$	$m = -\frac{1}{2}$

The Point-Slope Form of the Equation of a Line   In Exercises 25–32, find an equation of the line that passes through the given point and has the indicated slope. Sketch the line by hand. Use a graphing utility to verify your sketch, if possible.

✓ **25.** $(0, -2)$, $m = 3$ **26.** $(-3, 6)$, $m = -2$

27. $(2, -3)$, $m = -\frac{1}{2}$ **28.** $(-2, -5)$, $m = \frac{3}{4}$

29. $(6, -1)$, m is undefined

30. $(-10, 4)$, m is undefined

31. $\left(-\frac{1}{2}, \frac{3}{2}\right)$, $m = 0$ **32.** $(2.3, -8.5)$, $m = 0$

✓ **33. Finance**   The median player salary for the New York Yankees was $1.6 million in 2001 and $5.2 million in 2009. Write a linear equation giving the median salary y in terms of the year x. Then use the equation to predict the median salary in 2017.

34. Finance   The median player salary for the Dallas Cowboys was $441,300 in 2000 and $1,326,720 in 2008. Write a linear equation giving the median salary y in terms of the year x. Then use the equation to predict the median salary in 2016.

Using the Slope-Intercept Form   In Exercises 35–42, determine the slope and y-intercept (if possible) of the linear equation. Then describe its graph.

✓ **35.** $x - 2y = 4$ **36.** $3x + 4y = 1$

37. $2x - 5y + 10 = 0$ **38.** $4x - 3y - 9 = 0$

39. $x = -6$ **40.** $y = 12$

41. $3y + 2 = 0$ **42.** $2x - 5 = 0$

Using the Slope-Intercept Form   In Exercises 43–48, (a) find the slope and y-intercept (if possible) of the equation of the line algebraically, and (b) sketch the line by hand. Use a graphing utility to verify your answers to parts (a) and (b).

43. $5x - y + 3 = 0$ **44.** $2x + 3y - 9 = 0$

45. $5x - 2 = 0$ **46.** $3x + 7 = 0$

47. $3y + 5 = 0$ **48.** $-11 - 8y = 0$

Finding the Slope-Intercept Form   In Exercises 49 and 50, find the slope-intercept form of the equation of the line shown.

49.   **50.**

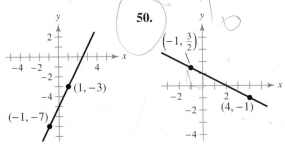

Finding the Slope-Intercept Form   In Exercises 51–60, write an equation of the line that passes through the points. Use the slope-intercept form (if possible). If not possible, explain why and use the general form. Use a graphing utility to graph the line (if possible).

51. $(5, -1), (-5, 5)$ **52.** $(4, 3), (-4, -4)$

53. $(-8, 1), (-8, 7)$ **54.** $(-1, 4), (6, 4)$

55. $\left(2, \frac{1}{2}\right), \left(\frac{1}{2}, \frac{5}{4}\right)$ **56.** $(1, 1), \left(6, -\frac{2}{3}\right)$

57. $\left(-\frac{1}{10}, -\frac{3}{5}\right), \left(\frac{9}{10}, -\frac{9}{5}\right)$ **58.** $\left(\frac{3}{4}, \frac{3}{2}\right), \left(-\frac{4}{3}, \frac{7}{4}\right)$

59. $(1, 0.6), (-2, -0.6)$ **60.** $(-8, 0.6), (2, -2.4)$

Different Viewing Windows   In Exercises 61 and 62, use a graphing utility to graph the equation using each viewing window. Describe the differences in the graphs.

✓ **61.** $y = 0.5x - 3$

Xmin = -5	Xmin = -2	Xmin = -5
Xmax = 10	Xmax = 10	Xmax = 10
Xscl = 1	Xscl = 1	Xscl = 1
Ymin = -1	Ymin = -4	Ymin = -7
Ymax = 10	Ymax = 1	Ymax = 3
Yscl = 1	Yscl = 1	Yscl = 1

62. $y = -8x + 5$

Xmin = -5	Xmin = -5	Xmin = -5
Xmax = 5	Xmax = 10	Xmax = 13
Xscl = 1	Xscl = 1	Xscl = 1
Ymin = -10	Ymin = -80	Ymin = -2
Ymax = 10	Ymax = 80	Ymax = 10
Yscl = 1	Yscl = 20	Yscl = 1

Parallel and Perpendicular Lines   In Exercises 63–66, determine whether the lines L_1 and L_2 passing through the pairs of points are parallel, perpendicular, or neither.

63. L_1: $(0, -1), (5, 9)$ **64.** L_1: $(-2, -1), (1, 5)$

 L_2: $(0, 3), (4, 1)$ L_2: $(1, 3), (5, -5)$

65. L_1: $(3, 6), (-6, 0)$ **66.** L_1: $(4, 8), (-4, 2)$

 L_2: $(0, -1), \left(5, \frac{7}{3}\right)$ L_2: $(3, -5), \left(-1, \frac{1}{3}\right)$

Equations of Parallel and Perpendicular Lines   In Exercises 67–76, write the slope-intercept forms of the equations of the lines through the given point (a) parallel to the given line and (b) perpendicular to the given line.

✓ **67.** $(2, 1)$, $4x - 2y = 3$ **68.** $(-3, 2)$, $x + y = 7$

69. $\left(-\frac{2}{3}, \frac{7}{8}\right)$, $3x + 4y = 7$ **70.** $\left(\frac{2}{5}, -1\right)$, $3x - 2y = 6$

71. $(-3.9, -1.4)$, $6x + 2y = 9$

72. $(-1.2, 2.4)$, $5x + 4y = 1$

73. $(3, -2)$, $x - 4 = 0$ **74.** $(3, -1)$, $y - 2 = 0$

75. $(-4, 1)$, $y + 2 = 0$ **76.** $(-2, 4)$, $x + 5 = 0$

Equations of Parallel Lines
In Exercises 77 and 78, the lines are parallel. Find the slope-intercept form of the equation of line y_2.

77.

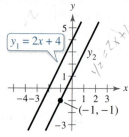

$y_1 = 2x + 4$

78.

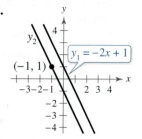

$y_1 = -2x + 1$

Equations of Perpendicular Lines
In Exercises 79 and 80, the lines are perpendicular. Find the slope-intercept form of the equation of line y_2.

79.

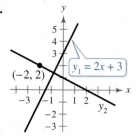

$y_1 = 2x + 3$

80.

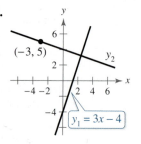

$y_1 = 3x - 4$

Graphs of Parallel and Perpendicular Lines
In Exercises 81–84, identify any relationships that exist among the lines, and then use a graphing utility to graph the three equations in the same viewing window. Adjust the viewing window so that each slope appears visually correct. Use the slopes of the lines to verify your results.

✓ **81.** (a) $y = 2x$ (b) $y = -2x$ (c) $y = \frac{1}{2}x$

82. (a) $y = \frac{2}{3}x$ (b) $y = -\frac{3}{2}x$ (c) $y = \frac{2}{3}x + 2$

83. (a) $y = -\frac{1}{2}x$ (b) $y = -\frac{1}{2}x + 3$ (c) $y = 2x - 4$

84. (a) $y = x - 8$ (b) $y = x + 1$ (c) $y = -x + 3$

85. Architectural Design The "rise to run" ratio of the roof of a house determines the steepness of the roof. The rise to run ratio of the roof in the figure is 3 to 4. Determine the maximum height in the attic of the house if the house is 32 feet wide.

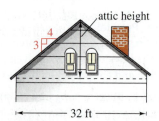

86. Highway Engineering When driving down a mountain road, you notice warning signs indicating that it is a "12% grade." This means that the slope of the road is $-\frac{12}{100}$. Approximate the amount of horizontal change in your position if you note from elevation markers that you have descended 2000 feet vertically.

Sean Locke/iStockphoto.com

87. MODELING DATA

The graph shows the sales y (in millions of dollars) of the Coca-Cola Bottling Company each year x from 2000 through 2008, where $x = 0$ represents 2000. (Source: Coca-Cola Bottling Company)

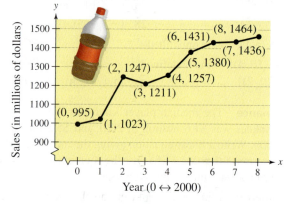

(a) Use the slopes to determine the years in which the sales showed the greatest increase and greatest decrease.

(b) Find the equation of the line between the years 2000 and 2008.

(c) Interpret the meaning of the slope of the line from part (b) in the context of the problem.

(d) Use the equation from part (b) to estimate the sales of the Coca-Cola Bottling Company in 2010. Do you think this is an accurate estimate? Explain.

88. MODELING DATA

The table shows the profits y (in millions of dollars) for Buffalo Wild Wings for each year x from 2002 through 2008, where $x = 2$ represents 2002. (Source: Buffalo Wild Wings Inc.)

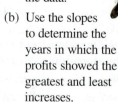

Year, x	Profits, y
2	3.1
3	3.9
4	7.2
5	8.9
6	16.3
7	19.7
8	24.4

(a) Sketch a graph of the data.

(b) Use the slopes to determine the years in which the profits showed the greatest and least increases.

(c) Find the equation of the line between the years 2002 and 2008.

(d) Interpret the meaning of the slope of the line from part (c) in the context of the problem.

(e) Use the equation from part (c) to estimate the profit for Buffalo Wild Wings in 2010. Do you think this is an accurate estimate? Explain.

Using a Rate of Change to Write an Equation   In Exercises 89–92, you are given the dollar value of a product in 2009 and the rate at which the value of the product is expected to change during the next 5 years. Write a linear equation that gives the dollar value V of the product in terms of the year t. (Let $t = 9$ represent 2009.)

	2009 Value	*Rate*
89.	$2540	$125 increase per year
90.	$156	$4.50 increase per year
91.	$20,400	$2000 decrease per year
92.	$245,000	$5600 decrease per year

93. Accounting   A school district purchases a high-volume printer, copier, and scanner for $25,000. After 10 years, the equipment will have to be replaced. Its value at that time is expected to be $2000.

(a) Write a linear equation giving the value V of the equipment for each year t during its 10 years of use.

(b) Use a graphing utility to graph the linear equation representing the depreciation of the equipment, and use the value or trace feature to complete the table. Verify your answers algebraically by using the equation you found in part (a).

t	0	1	2	3	4	5	6	7	8	9	10
V											

94. Meterology   Recall that water freezes at 0°C (32°F) and boils at 100°C (212°F).

(a) Find an equation of the line that shows the relationship between the temperature in degrees Celsius C and degrees Fahrenheit F.

(b) Use the result of part (a) to complete the table.

C		−10°	10°			177°
F	0°			68°	90°	

95. Business   A contractor purchases a bulldozer for $36,500. The bulldozer requires an average expenditure of $9.25 per hour for fuel and maintenance, and the operator is paid $18.50 per hour.

(a) Write a linear equation giving the total cost C of operating the bulldozer for t hours. (Include the purchase cost of the bulldozer.)

(b) Assuming that customers are charged $65 per hour of bulldozer use, write an equation for the revenue R derived from t hours of use.

(c) Use the profit formula ($P = R − C$) to write an equation for the profit gained from t hours of use.

(d) Use the result of part (c) to find the break-even point (the number of hours the bulldozer must be used to gain a profit of 0 dollars).

96. Real Estate   A real estate office handles an apartment complex with 50 units. When the rent per unit is $580 per month, all 50 units are occupied. However, when the rent is $625 per month, the average number of occupied units drops to 47. Assume that the relationship between the monthly rent p and the demand x is linear.

(a) Write the equation of the line giving the demand x in terms of the rent p.

(b) Use a graphing utility to graph the demand equation and use the *trace* feature to estimate the number of units occupied when the rent is $655. Verify your answer algebraically.

(c) Use the demand equation to predict the number of units occupied when the rent is lowered to $595. Verify your answer graphically.

97. *Why you should learn it* (p. 84)   In 1990, Penn State University had an enrollment of 75,365 students. By 2009, the enrollment had increased to 87,163. (Source: Penn State Fact Book)

(a) What was the average annual change in enrollment from 1990 to 2009?

(b) Use the average annual change in enrollment to estimate the enrollments in 1995, 2000, and 2005.

(c) Write the equation of a line that represents the given data. What is its slope? Interpret the slope in the context of the problem.

98. Writing   Using the results of Exercise 97, write a short paragraph discussing the concepts of *slope* and *average rate of change*.

Conclusions

True or False?   In Exercises 99 and 100, determine whether the statement is true or false. Justify your answer.

99. The line through $(−8, 2)$ and $(−1, 4)$ and the line through $(0, −4)$ and $(−7, 7)$ are parallel.

100. If the points $(10, −3)$ and $(2, −9)$ lie on the same line, then the point $\left(−12, −\frac{37}{2}\right)$ also lies on that line.

Exploration   In Exercises 101–104, use a graphing utility to graph the equation of the line in the form

$$\frac{x}{a} + \frac{y}{b} = 1, \quad a \neq 0, b \neq 0.$$

Use the graphs to make a conjecture about what a and b represent. Verify your conjecture.

101. $\dfrac{x}{5} + \dfrac{y}{−3} = 1$ **102.** $\dfrac{x}{−6} + \dfrac{y}{2} = 1$

103. $\dfrac{x}{4} + \dfrac{y}{−\frac{2}{3}} = 1$ **104.** $\dfrac{x}{\frac{1}{2}} + \dfrac{y}{5} = 1$

Using Intercepts In Exercises 105–108, use the results of Exercises 101–104 to write an equation of the line that passes through the points.

105. x-intercept: $(2, 0)$
y-intercept: $(0, 3)$

106. x-intercept: $(-5, 0)$
y-intercept: $(0, -4)$

107. x-intercept: $\left(-\frac{1}{6}, 0\right)$
y-intercept: $\left(0, -\frac{2}{3}\right)$

108. x-intercept: $\left(\frac{3}{4}, 0\right)$
y-intercept: $\left(0, \frac{4}{5}\right)$

Think About It In Exercises 109 and 110, determine which equation(s) may be represented by the graphs shown. (There may be more than one correct answer.)

109.

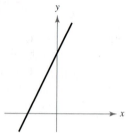

(a) $2x - y = -10$
(b) $2x + y = 10$
(c) $x - 2y = 10$
(d) $x + 2y = 10$

110.

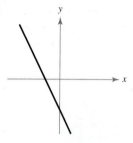

(a) $2x + y = 5$
(b) $2x + y = -5$
(c) $x - 2y = 5$
(d) $x - 2y = -5$

Think About It In Exercises 111 and 112, determine which pair of equations may be represented by the graphs shown.

111.

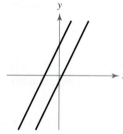

(a) $2x - y = 5$
$2x - y = 1$
(b) $2x + y = -5$
$2x + y = 1$
(c) $2x - y = -5$
$2x - y = 1$
(d) $x - 2y = -5$
$x - 2y = -1$

112.

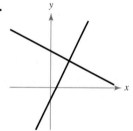

(a) $2x - y = 2$
$x + 2y = 12$
(b) $x - y = 1$
$x + y = 6$
(c) $2x + y = 2$
$x - 2y = 12$
(d) $x - 2y = 2$
$x + 2y = 12$

113. Think About It Does every line have both an x-intercept and a y-intercept? Explain.

114. Think About It Can every line be written in slope-intercept form? Explain.

115. Think About It Does every line have an infinite number of lines that are parallel to it? Explain.

116. CAPSTONE Match the description with its graph. Determine the slope of each graph and how it is interpreted in the given context. [The graphs are labeled (i), (ii), (iii), and (iv).]

(i)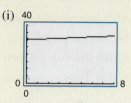

(ii)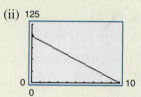

(iii)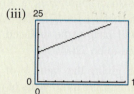

(iv)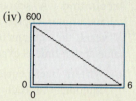

(a) You are paying $10 per week to repay a $100 loan.

(b) An employee is paid $12.50 per hour plus $1.50 for each unit produced per hour.

(c) A sales representative receives $30 per day for food plus $0.35 for each mile traveled.

(d) A computer that was purchased for $600 depreciates $100 per year.

Cumulative Mixed Review

Identifying Polynomials In Exercises 117–122, determine whether the expression is a polynomial. If it is, write the polynomial in standard form.

117. $x + 20$

118. $3x - 10x^2 + 1$

119. $4x^2 + x^{-1} - 3$

120. $2x^2 - 2x^4 - x^3 + 2$

121. $\dfrac{x^2 + 3x + 4}{x^2 - 9}$

122. $\sqrt{x^2 + 7x + 6}$

Factoring Trinomials In Exercises 123–126, factor the trinomial.

123. $x^2 - 6x - 27$

124. $x^2 - 11x + 28$

125. $2x^2 + 11x - 40$

126. $3x^2 - 16x + 5$

127. *Make a Decision* To work an extended application analyzing the numbers of bachelor's degrees earned by women in the United States from 1985 through 2007, visit this textbook's *Companion Website*. (Source: National Center for Education Statistics)

The *Make a Decision* exercise indicates a multipart exercise using large data sets. Go to this textbook's *Companion Website* to view these exercises.

1.3   Functions

Introduction to Functions

Many everyday phenomena involve two quantities that are related to each other by some rule of correspondence. The mathematical term for such a rule of correspondence is a **relation.** Here are two examples.

1. The simple interest I earned on an investment of $1000 for 1 year is related to the annual interest rate r by the formula $I = 1000r$.

2. The area A of a circle is related to its radius r by the formula $A = \pi r^2$.

Not all relations have simple mathematical formulas. For instance, people commonly match up NFL starting quarterbacks with touchdown passes, and time of day with temperature. In each of these cases, there is some relation that matches each item from one set with exactly one item from a different set. Such a relation is called a **function.**

Definition of a Function

A **function** f from a set A to a set B is a relation that assigns to each element x in the set A exactly one element y in the set B. The set A is the **domain** (or set of inputs) of the function f, and the set B contains the **range** (or set of outputs).

To help understand this definition, look at the function that relates the time of day to the temperature in Figure 1.24.

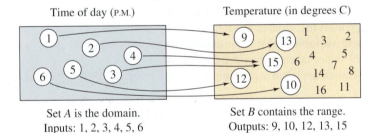

Time of day (P.M.)   Temperature (in degrees C)

Set A is the domain.   Set B contains the range.
Inputs: 1, 2, 3, 4, 5, 6   Outputs: 9, 10, 12, 13, 15

Figure 1.24

This function can be represented by the ordered pairs

$$\{(1, 9°), (2, 13°), (3, 15°), (4, 15°), (5, 12°), (6, 10°)\}.$$

In each ordered pair, the first coordinate (x-value) is the **input** and the second coordinate (y-value) is the **output.**

Characteristics of a Function from Set A to Set B

1. Each element of A must be matched with an element of B.

2. Some elements of B may not be matched with any element of A.

3. Two or more elements of A may be matched with the same element of B.

4. An element of A (the domain) cannot be matched with two different elements of B.

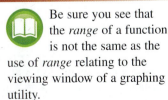

Study Tip

Be sure you see that the *range* of a function is not the same as the use of *range* relating to the viewing window of a graphing utility.

wolfram alpha co

To determine whether or not a relation is a function, you must decide whether each input value is matched with exactly one output value. When any input value is matched with two or more output values, the relation is not a function.

Example 1   Testing for Functions

Decide whether the relation represents y as a function of x.

a.

Input, x	2	2	3	4	5
Output, y	11	10	8	5	1

b.

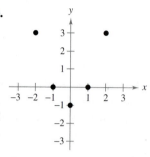

Figure 1.25

Solution

a. This table *does not* describe y as a function of x. The input value 2 is matched with two different y-values.

b. The graph in Figure 1.25 *does* describe y as a function of x. Each input value is matched with exactly one output value.

✓CHECKPOINT   Now try Exercise 11.

In algebra, it is common to represent functions by equations or formulas involving two variables. For instance, the equation $y = x^2$ represents the variable y as a function of the variable x. In this equation, x is the **independent variable** and y is the **dependent variable.** The domain of the function is the set of all values taken on by the independent variable x, and the range of the function is the set of all values taken on by the dependent variable y.

Example 2   Testing for Functions Represented Algebraically

Determine whether the equation represents y as a function of x.

a. $x^2 + y = 1$ **b.** $-x + y^2 = 1$

Solution

To determine whether y is a function of x, try to solve for y in terms of x.

a. $x^2 + y = 1$ Write original equation.

$\quad\quad y = 1 - x^2$ Solve for y.

Each value of x corresponds to exactly one value of y. So, y is a function of x.

b. $-x + y^2 = 1$ Write original equation.

$\quad\quad y^2 = 1 + x$ Add x to each side.

$\quad\quad y = \pm\sqrt{1 + x}$ Solve for y.

The $\pm$ indicates that for a given value of x there correspond two values of y. For instance, when $x = 3$, $y = 2$ or $y = -2$. So, y is not a function of x.

✓CHECKPOINT   Now try Exercise 23.

Explore the Concept

Use a graphing utility to graph $x^2 + y = 1$. Then use the graph to write a convincing argument that each x-value corresponds to at most one y-value.

Use a graphing utility to graph $-x + y^2 = 1$. (*Hint:* You will need to use two equations.) Does the graph represent y as a function of x? Explain.

Function Notation

When an equation is used to represent a function, it is convenient to name the function so that it can be referenced easily. For example, you know that the equation $y = 1 - x^2$ describes y as a function of x. Suppose you give this function the name "f." Then you can use the following **function notation.**

Input	*Output*	*Equation*
x	$f(x)$	$f(x) = 1 - x^2$

The symbol $f(x)$ is read as the *value of f at x* or simply *f of x.* The symbol $f(x)$ corresponds to the y-value for a given x. So, you can write $y = f(x)$. Keep in mind that f is the *name* of the function, whereas $f(x)$ is the *output value* of the function at the *input value x.* In function notation, the *input* is the independent variable and the *output* is the dependent variable. For instance, the function $f(x) = 3 - 2x$ has *function values* denoted by $f(-1), f(0)$, and so on. To find these values, substitute the specified input values into the given equation.

For $x = -1$, $f(-1) = 3 - 2(-1) = 3 + 2 = 5.$

For $x = 0$, $f(0) = 3 - 2(0) = 3 - 0 = 3.$

Although f is often used as a convenient function name and x is often used as the independent variable, you can use other letters. For instance,

$$f(x) = x^2 - 4x + 7, \quad f(t) = t^2 - 4t + 7, \quad \text{and} \quad g(s) = s^2 - 4s + 7$$

all define the same function. In fact, the role of the independent variable is that of a "placeholder." Consequently, the function could be written as

$$f(\quad) = (\quad)^2 - 4(\quad) + 7.$$

Example 3   Evaluating a Function

Let $g(x) = -x^2 + 4x + 1$. Find each value of the function.

a. $g(2)$

b. $g(t)$

c. $g(x + 2)$

Solution

a. Replacing x with 2 in $g(x) = -x^2 + 4x + 1$ yields the following.

$$g(2) = -(2)^2 + 4(2) + 1 = -4 + 8 + 1 = 5$$

b. Replacing x with t yields the following.

$$g(t) = -(t)^2 + 4(t) + 1 = -t^2 + 4t + 1$$

c. Replacing x with $x + 2$ yields the following.

$$g(x + 2) = -(x + 2)^2 + 4(x + 2) + 1 \qquad \text{Substitute } x + 2 \text{ for } x.$$
$$= -(x^2 + 4x + 4) + 4x + 8 + 1 \qquad \text{Multiply.}$$
$$= -x^2 - 4x - 4 + 4x + 8 + 1 \qquad \text{Distributive Property}$$
$$= -x^2 + 5 \qquad \text{Simplify.}$$

✓CHECKPOINT   Now try Exercise 31.

In Example 3, note that $g(x + 2)$ is not equal to $g(x) + g(2)$. In general,

$$g(u + v) \neq g(u) + g(v).$$

Library of Parent Functions: Absolute Value Function

The *parent absolute value function* given by $f(x) = |x|$ can be written as a piecewise-defined function. The basic characteristics of the parent absolute value function are summarized below and on the inside front cover of this text.

Graph of $f(x) = |x| = \begin{cases} x, & x \geq 0 \\ -x, & x < 0 \end{cases}$

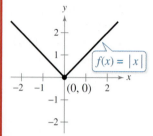

$f(x) = |x|$

$(0, 0)$

Domain: $(-\infty, \infty)$
Range: $[0, \infty)$
Intercept: $(0, 0)$
Decreasing on $(-\infty, 0)$
Increasing on $(0, \infty)$

A function defined by two or more equations over a specified domain is called a **piecewise-defined function.**

Example 4 A Piecewise–Defined Function

Evaluate the function when $x = -1$ and $x = 0$.

$$f(x) = \begin{cases} x^2 + 1, & x < 0 \\ x - 1, & x \geq 0 \end{cases}$$

Solution

Because $x = -1$ is less than 0, use $f(x) = x^2 + 1$ to obtain

$f(-1) = (-1)^2 + 1$ Substitute -1 for x.

$\quad\quad = 2.$ Simplify.

For $x = 0$, use $f(x) = x - 1$ to obtain

$f(0) = 0 - 1$ Substitute 0 for x.

$\quad\quad = -1.$ Simplify.

The graph of f is shown in Figure 1.26.

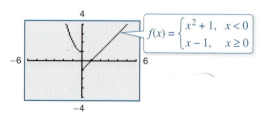

$f(x) = \begin{cases} x^2 + 1, & x < 0 \\ x - 1, & x \geq 0 \end{cases}$

Figure 1.26

✔CHECKPOINT   Now try Exercise 39.

Technology Tip

Most graphing utilities can graph piecewise-defined functions. For instructions on how to enter a piecewise-defined function into your graphing utility, consult your user's manual. You may find it helpful to set your graphing utility to *dot mode* before graphing such functions.

The Domain of a Function

The domain of a function can be described explicitly or it can be *implied* by the expression used to define the function. The **implied domain** is the set of all real numbers for which the expression is defined. For instance, the function

$$f(x) = \frac{1}{x^2 - 4}$$ Domain excludes *x*-values that result in division by zero.

has an implied domain that consists of all real numbers x other than $x = \pm 2$. These two values are excluded from the domain because division by zero is undefined. Another common type of implied domain is that used to avoid even roots of negative numbers. For example, the function

$$f(x) = \sqrt{x}$$ Domain excludes *x*-values that result in even roots of negative numbers.

is defined only for $x \geq 0$. So, its implied domain is the interval $[0, \infty)$. In general, the domain of a function *excludes* values that would cause division by zero *or* result in the even root of a negative number.

📚 Library of Parent Functions: Square Root Function

Radical functions arise from the use of rational exponents. The most common radical function is the *parent square root* function given by $f(x) = \sqrt{x}$. The basic characteristics of the parent square root function are summarized below and on the inside front cover of this text.

Graph of $f(x) = \sqrt{x}$

Domain: $[0, \infty)$
Range: $[0, \infty)$
Intercept: $(0, 0)$
Increasing on $(0, \infty)$

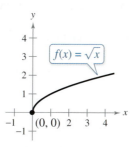

Study Tip

Because the square root function is not defined for $x < 0$, you must be careful when analyzing the domains of complicated functions involving the square root symbol.

Example 5   Finding the Domain of a Function

Find the domain of each function.

a. $f: \{(-3, 0), (-1, 4), (0, 2), (2, 2), (4, -1)\}$

b. $g(x) = -3x^2 + 4x + 5$

c. $h(x) = \dfrac{1}{x + 5}$

Solution

a. The domain of f consists of all first coordinates in the set of ordered pairs.

Domain $= \{-3, -1, 0, 2, 4\}$

b. The domain of g is the set of all *real* numbers.

c. Excluding x-values that yield zero in the denominator, the domain of h is the set of all real numbers x except $x = -5$.

 CHECKPOINT   Now try Exercise 55.

Example 6   Finding the Domain of a Function

Find the domain of each function.

a. Volume of a sphere: $V = \frac{4}{3}\pi r^3$ **b.** $k(x) = \sqrt{4 - 3x}$

Solution

a. Because this function represents the volume of a sphere, the values of the radius r must be positive (see Figure 1.27). So, the domain is the set of all real numbers r such that $r > 0$.

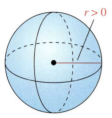

Figure 1.27

b. This function is defined only for x-values for which $4 - 3x \geq 0$. By solving this inequality, you will find that the domain of k is all real numbers that are less than or equal to $\frac{4}{3}$.

 CHECKPOINT   Now try Exercise 61.

In Example 6(a), note that the *domain of a function may be implied by the physical context*. For instance, from the equation $V = \frac{4}{3}\pi r^3$ you would have no reason to restrict r to positive values, but the physical context implies that a sphere cannot have a negative or zero radius.

For some functions, it may be easier to find the domain and range of the function by examining its graph.

Example 7   Finding the Domain and Range of a Function

Use a graphing utility to find the domain and range of the function $f(x) = \sqrt{9 - x^2}$.

Solution

Graph the function as $y = \sqrt{9 - x^2}$, as shown in Figure 1.28. Using the *trace* feature of a graphing utility, you can determine that the x-values extend from -3 to 3 and the y-values extend from 0 to 3. So, the domain of the function f is all real numbers such that

$-3 \leq x \leq 3$ Domain of f

and the range of f is all real numbers such that

$0 \leq y \leq 3$.   Range of f

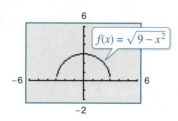

Figure 1.28

CHECKPOINT   Now try Exercise 65.

Applications

Example 8   Construction Employees

The number N (in millions) of employees in the construction industry in the United States increased in a linear pattern from 2003 through 2006 (see Figure 1.29). In 2007, the number dropped, then decreased through 2008 in a different linear pattern. These two patterns can be approximated by the function

$$N(t) = \begin{cases} 0.32t + 5.7, & 3 \le t \le 6 \\ -0.42t + 10.5, & 7 \le t \le 8 \end{cases}$$

where t represents the year, with $t = 3$ corresponding to 2003. Use this function to approximate the number of employees for each year from 2003 to 2008.
(Source: U.S. Bureau of Labor Statistics)

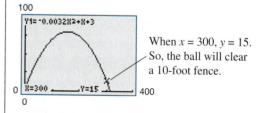

Figure 1.29

Solution

From 2003 to 2006, use $N(t) = 0.32t + 5.7$.

$$\underbrace{6.66,}_{2003} \quad \underbrace{6.98,}_{2004} \quad \underbrace{7.3,}_{2005} \quad \underbrace{7.62}_{2006}$$

From 2007 to 2008, use $N(t) = -0.42t + 10.5$.

$$\underbrace{7.56,}_{2007} \quad \underbrace{7.14}_{2008}$$

 CHECKPOINT   Now try Exercise 77.

Example 9   The Path of a Baseball

A baseball is hit at a point 3 feet above the ground at a velocity of 100 feet per second and an angle of 45°. The path of the baseball is given by the function

$$f(x) = -0.0032x^2 + x + 3$$

where $f(x)$ is the height of the baseball (in feet) and x is the horizontal distance from home plate (in feet). Will the baseball clear a 10-foot fence located 300 feet from home plate?

Algebraic Solution

The height of the baseball is a function of the horizontal distance from home plate. When $x = 300$, you can find the height of the baseball as follows.

$f(x) = -0.0032x^2 + x + 3$	Write original function.
$f(300) = -0.0032(300)^2 + 300 + 3$	Substitute 300 for x.
$= 15$	Simplify.

When $x = 300$, the height of the baseball is 15 feet. So, the baseball will clear a 10-foot fence.

 CHECKPOINT   Now try Exercise 79.

Graphical Solution

When $x = 300$, $y = 15$. So, the ball will clear a 10-foot fence.

Y1=-0.0032X2+X+3

X=300   Y=15

Difference Quotients

One of the basic definitions in calculus employs the ratio

$$\frac{f(x + h) - f(x)}{h}, \quad h \neq 0.$$

This ratio is called a **difference quotient,** as illustrated in Example 10.

Example 10   Evaluating a Difference Quotient

For $f(x) = x^2 - 4x + 7$, find $\dfrac{f(x + h) - f(x)}{h}$.

Solution

$$\frac{f(x + h) - f(x)}{h} = \frac{[(x + h)^2 - 4(x + h) + 7] - (x^2 - 4x + 7)}{h}$$

$$= \frac{x^2 + 2xh + h^2 - 4x - 4h + 7 - x^2 + 4x - 7}{h}$$

$$= \frac{2xh + h^2 - 4h}{h}$$

$$= \frac{h(2x + h - 4)}{h}$$

$$= 2x + h - 4, \ h \neq 0$$

✓CHECKPOINT   Now try Exercise 83.

> ### Study Tip
>
>   Notice in Example 10 that h cannot be zero in the original expression. Therefore, you must restrict the domain of the simplified expression by listing $h \neq 0$ so that the simplified expression is equivalent to the original expression.

Summary of Function Terminology

Function: A **function** is a relationship between two variables such that to each value of the independent variable there corresponds exactly one value of the dependent variable.

Function Notation: $y = f(x)$

  f is the *name* of the function.
  y is the **dependent variable,** or output value.
  x is the **independent variable,** or input value.
  $f(x)$ is the *value of the function at x.*

Domain: The **domain** of a function is the set of all values (inputs) of the independent variable for which the function is defined. If x is in the domain of f, then f is said to be *defined* at x. If x is not in the domain of f, then f is said to be *undefined* at x.

Range: The **range** of a function is the set of all values (outputs) assumed by the dependent variable (that is, the set of all function values).

Implied Domain: If f is defined by an algebraic expression and the domain is not specified, then the **implied domain** consists of all real numbers for which the expression is defined.

The symbol $\int$ indicates an example or exercise that highlights algebraic techniques specifically used in calculus.

1.3   Exercises

See www.CalcChat.com for worked-out solutions to odd-numbered exercises.
For instructions on how to use a graphing utility, see Appendix A.

Vocabulary and Concept Check

In Exercises 1 and 2, fill in the blanks.

1. A relation that assigns to each element x from a set of inputs, or _____ , exactly one element y in a set of outputs, or _____ , is called a _____ .

2. For an equation that represents y as a function of x, the _____ variable is the set of all x in the domain, and the _____ variable is the set of all y in the range.

3. Can the ordered pairs $(3, 0)$ and $(3, 5)$ represent a function?

4. To find $g(x + 1)$, what do you substitute for x in the function $g(x) = 3x - 2$?

5. Does the domain of the function $f(x) = \sqrt{1 + x}$ include $x = -2$?

6. Is the domain of a piecewise-defined function *implied* or *explicitly described*?

Procedures and Problem Solving

Testing for Functions   In Exercises 7–10, does the relation describe a function? Explain your reasoning.

7. *Domain   Range*

$$\begin{array}{cc} -2 & 5 \\ -1 & 6 \\ 0 & 7 \\ 1 & 8 \\ 2 & \end{array}$$

8. *Domain   Range*

$$\begin{array}{cc} -2 & 3 \\ -1 & 4 \\ 0 & 5 \\ 1 & \\ 2 & \end{array}$$

9. *Domain   Range*

10. *Domain   Range*
    (State)   (Electoral votes 2000–2010)

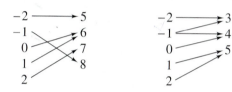

Testing for Functions   In Exercises 11 and 12, decide whether the relation represents y as a function of x. Explain your reasoning.

11.

Input, x	-3	-1	0	1	3
Output, y	-9	-1	0	1	9

12.

Input, x	0	1	2	1	0
Output, y	-4	-2	0	2	4

Testing for Functions   In Exercises 13 and 14, which sets of ordered pairs represent functions from A to B? Explain.

13. $A = \{0, 1, 2, 3\}$ and $B = \{-2, -1, 0, 1, 2\}$
    (a) $\{(0, 1), (1, -2), (2, 0), (3, 2)\}$
    (b) $\{(0, -1), (2, 2), (1, -2), (3, 0), (1, 1)\}$
    (c) $\{(0, 0), (1, 0), (2, 0), (3, 0)\}$

14. $A = \{a, b, c\}$ and $B = \{0, 1, 2, 3\}$
    (a) $\{(a, 1), (c, 2), (c, 3), (b, 3)\}$
    (b) $\{(a, 1), (b, 2), (c, 3)\}$
    (c) $\{(1, a), (0, a), (2, c), (3, b)\}$

Pharmacology   In Exercises 15 and 16, use the graph, which shows the average prices of name brand and generic drug prescriptions in the United States. (Source: National Association of Chain Drug Stores)

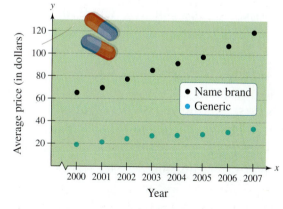

15. Is the average price of a name brand prescription a function of the year? Is the average price of a generic prescription a function of the year? Explain.

16. Let $b(t)$ and $g(t)$ represent the average prices of name brand and generic prescriptions, respectively, in year t. Find $b(2007)$ and $g(2000)$.

Testing for Functions Represented Algebraically In Exercises 17–28, determine whether the equation represents y as a function of x.

17. $x^2 + y^2 = 4$

18. $x = y^2 + 1$

19. $y = \sqrt{x^2 - 1}$

20. $y = \sqrt{x + 5}$

21. $2x + 3y = 4$

22. $x = -y + 5$

✓ **23.** $y^2 = x^2 - 1$

24. $x + y^2 = 3$

25. $y = |4 - x|$

26. $|y| = 4 - x$

27. $x = -7$

28. $y = 8$

Evaluating a Function In Exercises 29–44, evaluate the function at each specified value of the independent variable and simplify.

29. $f(t) = 3t + 1$

    (a) $f(2)$ (b) $f(-4)$ (c) $f(t + 2)$

30. $g(y) = 7 - 3y$

    (a) $g(0)$ (b) $g\left(\frac{7}{3}\right)$ (c) $g(s + 2)$

✓ **31.** $h(t) = t^2 - 2t$

    (a) $h(2)$ (b) $h(1.5)$ (c) $h(x + 2)$

32. $V(r) = \frac{4}{3}\pi r^3$

    (a) $V(3)$ (b) $V\left(\frac{3}{2}\right)$ (c) $V(2r)$

33. $f(y) = 3 - \sqrt{y}$

    (a) $f(4)$ (b) $f(0.25)$ (c) $f(4x^2)$

34. $f(x) = \sqrt{x + 8} + 2$

    (a) $f(-4)$ (b) $f(8)$ (c) $f(x - 8)$

35. $q(x) = \dfrac{1}{x^2 - 9}$

    (a) $q(-3)$ (b) $q(2)$ (c) $q(y + 3)$

36. $q(t) = \dfrac{2t^2 + 3}{t^2}$

    (a) $q(2)$ (b) $q(0)$ (c) $q(-x)$

37. $f(x) = \dfrac{|x|}{x}$

    (a) $f(9)$ (b) $f(-9)$ (c) $f(t)$

38. $f(x) = |x| + 4$

    (a) $f(5)$ (b) $f(-5)$ (c) $f(t)$

✓ **39.** $f(x) = \begin{cases} 2x + 1, & x < 0 \\ 2x + 2, & x \ge 0 \end{cases}$

    (a) $f(-1)$ (b) $f(0)$ (c) $f(2)$

40. $f(x) = \begin{cases} 2x + 5, & x \le 0 \\ 2 - x^2, & x > 0 \end{cases}$

    (a) $f(-2)$ (b) $f(0)$ (c) $f(1)$

41. $f(x) = \begin{cases} x^2 + 2, & x \le 1 \\ 2x^2 + 2, & x > 1 \end{cases}$

    (a) $f(-2)$ (b) $f(1)$ (c) $f(2)$

42. $f(x) = \begin{cases} x^2 - 4, & x \le 0 \\ 1 - 2x^2, & x > 0 \end{cases}$

    (a) $f(-2)$ (b) $f(0)$ (c) $f(1)$

43. $f(x) = \begin{cases} x + 2, & x < 0 \\ 4, & 0 \le x < 2 \\ x^2 + 1, & x \ge 2 \end{cases}$

    (a) $f(-2)$ (b) $f(1)$ (c) $f(4)$

44. $f(x) = \begin{cases} 5 - 2x, & x < 0 \\ 5, & 0 \le x < 1 \\ 4x + 1, & x \ge 1 \end{cases}$

    (a) $f(-2)$ (b) $f\left(\frac{1}{2}\right)$ (c) $f(1)$

Evaluating a Function In Exercises 45–48, assume that the domain of f is the set $A = \{-2, -1, 0, 1, 2\}$. Determine the set of ordered pairs representing the function f.

45. $f(x) = x^2$ **46.** $f(x) = x^2 - 3$

47. $f(x) = |x| + 2$ **48.** $f(x) = |x + 1|$

Evaluating a Function In Exercises 49 and 50, complete the table.

49. $h(t) = \frac{1}{2}|t + 3|$

t	-5	-4	-3	-2	-1
$h(t)$					

50. $f(s) = \dfrac{|s - 2|}{s - 2}$

s	0	1	$\frac{3}{2}$	$\frac{5}{2}$	4
$f(s)$					

Finding the Inputs That Have Outputs of Zero In Exercises 51–54, find all values of x such that $f(x) = 0$.

51. $f(x) = 15 - 3x$ **52.** $f(x) = 5x + 1$

53. $f(x) = \dfrac{3x - 4}{5}$ **54.** $f(x) = \dfrac{2x - 3}{7}$

Finding the Domain of a Function In Exercises 55–64, find the domain of the function.

✓ **55.** $f(x) = 5x^2 + 2x - 1$ **56.** $g(x) = 1 - 2x^2$

57. $h(t) = \dfrac{4}{t}$ **58.** $s(y) = \dfrac{3y}{y + 5}$

59. $f(x) = \sqrt[3]{x - 4}$ **60.** $f(x) = \sqrt[4]{x^2 + 3x}$

✓ **61.** $g(x) = \dfrac{1}{x} - \dfrac{3}{x + 2}$ **62.** $h(x) = \dfrac{10}{x^2 - 2x}$

63. $g(y) = \dfrac{y + 2}{\sqrt{y - 10}}$ **64.** $f(x) = \dfrac{\sqrt{x + 6}}{6 + x}$

Finding the Domain and Range of a Function   In Exercises 65–68, use a graphing utility to graph the function. Find the domain and range of the function.

✓ **65.** $f(x) = \sqrt{4 - x^2}$ **66.** $f(x) = \sqrt{x^2 + 1}$

67. $g(x) = |2x + 3|$ **68.** $g(x) = |x - 5|$

69. Geometry   Write the area A of a circle as a function of its circumference C.

70. Geometry   Write the area A of an equilateral triangle as a function of the length s of its sides.

71. Exploration   An open box of maximum volume is to be made from a square piece of material, 24 centimeters on a side, by cutting equal squares from the corners and turning up the sides (see figure).

(a) The table shows the volume V (in cubic centimeters) of the box for various heights x (in centimeters). Use the table to estimate the maximum volume.

Height, x	Volume, V
1	484
2	800
3	972
4	1024
5	980
6	864

(b) Plot the points (x, V) from the table in part (a). Does the relation defined by the ordered pairs represent V as a function of x?

(c) If V is a function of x, write the function and determine its domain.

(d) Use a graphing utility to plot the points from the table in part (a) with the function from part (c). How closely does the function represent the data? Explain.

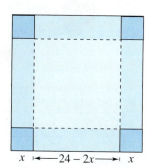

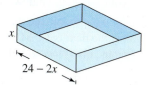

72. Geometry   A right triangle is formed in the first quadrant by the x- and y-axes and a line through the point $(2, 1)$ (see figure). Write the area A of the triangle as a function of x, and determine the domain of the function.

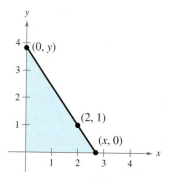

73. Geometry   A rectangle is bounded by the x-axis and the semicircle $y = \sqrt{36 - x^2}$ (see figure). Write the area A of the rectangle as a function of x, and determine the domain of the function.

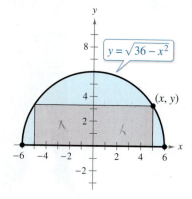

74. Geometry   A rectangular package to be sent by the U.S. Postal Service can have a maximum combined length and girth (perimeter of a cross section) of 108 inches (see figure).

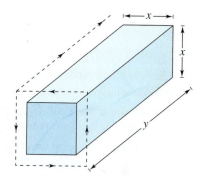

(a) Write the volume V of the package as a function of x. What is the domain of the function?

(b) Use a graphing utility to graph the function. Be sure to use an appropriate viewing window.

(c) What dimensions will maximize the volume of the package? Explain.

75. Business A company produces a handheld game console for which the variable cost is $68.20 per unit and the fixed costs are $248,000. The game console sells for $98.98. Let x be the number of units produced and sold.

(a) The total cost for a business is the sum of the variable cost and the fixed costs. Write the total cost C as a function of the number of units produced.

(b) Write the revenue R as a function of the number of units sold.

(c) Write the profit P as a function of the number of units sold. (*Note: $P = R - C$.*)

76. MODELING DATA

The table shows the revenue y (in thousands of dollars) of a landscaping business for each month of 2010, with $x = 1$ representing January.

Month, x	Revenue, y
1	5.2
2	5.6
3	6.6
4	8.3
5	11.5
6	15.8
7	12.8
8	10.1
9	8.6
10	6.9
11	4.5
12	2.7

The mathematical model below represents the data.

$$f(x) = \begin{cases} -1.97x + 26.3 \\ 0.505x^2 - 1.47x + 6.3 \end{cases}$$

(a) Identify the independent and dependent variables and explain what they represent in the context of the problem.

(b) What is the domain of each part of the piecewise-defined function? Explain your reasoning.

(c) Use the mathematical model to find $f(5)$. Interpret your result in the context of the problem.

(d) Use the mathematical model to find $f(11)$. Interpret your result in the context of the problem.

(e) How do the values obtained from the models in parts (c) and (d) compare with the actual data values?

✓ 77. Civil Engineering The numbers n (in billions) of miles traveled by vans, pickup trucks, and sport utility vehicles in the United States from 1990 through 2007 can be approximated by the model

$$n(t) = \begin{cases} -5.24t^2 + 69.5t + 581, & 0 \le t \le 6 \\ 25.7t + 664, & 6 < t \le 17 \end{cases}$$

where t represents the year, with $t = 0$ corresponding to 1990. The actual numbers are shown in the bar graph. (Source: U.S. Federal Highway Administration)

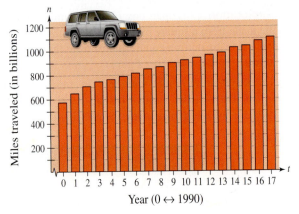

(a) Identify the independent and dependent variables and explain what they represent in the context of the problem.

(b) Use the *table* feature of a graphing utility to approximate the number of miles traveled by vans, pickup trucks, and sport utility vehicles each year from 1990 through 2007.

(c) Compare the values in part (b) with the actual values shown in the bar graph. How well does the model fit the data?

78. *Why you should learn it* (*p. 97*) The force F (in tons) of water against the face of a dam is estimated by the function

$$F(y) = 149.76\sqrt{10}y^{5/2}$$

where y is the depth of the water (in feet).

(a) Complete the table. What can you conclude from it?

y	5	10	20	30	40
$F(y)$					

(b) Use a graphing utility to graph the function. Describe your viewing window.

(c) Use the table to approximate the depth at which the force against the dam is 1,000,000 tons. Verify your answer graphically. How could you find a better estimate?

✓ 79. Projectile Motion A second baseman throws a baseball toward the first baseman 60 feet away. The path of the ball is given by

$$y = -0.004x^2 + 0.3x + 6$$

where y is the height (in feet) and x is the horizontal distance (in feet) from the second baseman. The first baseman can reach 8 feet high. Can the first baseman catch the ball without jumping? Explain.

80. Business The graph shows the sales (in millions of dollars) of Peet's Coffee & Tea from 2000 through 2008. Let $f(x)$ represent the sales in year x. (Source: Peet's Coffee & Tea, Inc.)

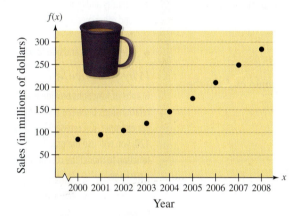

Year

(a) Find $\dfrac{f(2008) - f(2000)}{2008 - 2000}$ and interpret the result in the context of the problem.

(b) An approximate model for the function is

$$S(t) = 2.484t^2 + 5.71t + 84.0, \quad 0 \le t \le 8$$

where S is the sales (in millions of dollars) and $t = 0$ represents 2000. Complete the table and compare the results with the data in the graph.

t	0	1	2	3	4	5	6	7	8
$S(t)$									

∫ Evaluating a Difference Quotient In Exercises 81–86, find the difference quotient and simplify your answer.

81. $f(x) = 2x$, $\dfrac{f(x + c) - f(x)}{c}$, $c \neq 0$

82. $g(x) = 3x - 1$, $\dfrac{g(x + h) - g(x)}{h}$, $h \neq 0$

✓ 83. $f(x) = x^2 - x + 1$, $\dfrac{f(2 + h) - f(2)}{h}$, $h \neq 0$

84. $f(x) = x^3 + x$, $\dfrac{f(x + h) - f(x)}{h}$, $h \neq 0$

85. $f(t) = \dfrac{1}{t}$, $\dfrac{f(t) - f(1)}{t - 1}$, $t \neq 1$

86. $f(x) = \dfrac{4}{x + 1}$, $\dfrac{f(x) - f(7)}{x - 7}$, $x \neq 7$

Conclusions

True or False? In Exercises 87 and 88, determine whether the statement is true or false. Justify your answer.

87. The domain of the function $f(x) = x^4 - 1$ is $(-\infty, \infty)$, and the range of $f(x)$ is $(0, \infty)$.

88. The set of ordered pairs $\{(-8, -2), (-6, 0), (-4, 0), (-2, 2), (0, 4), (2, -2)\}$ represents a function.

Think About It In Exercises 89 and 90, write a square root function for the graph shown. Then, identify the domain and range of the function.

89.

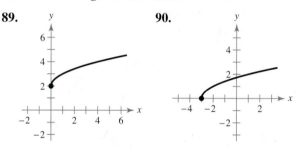

90.

91. Think About It Given $f(x) = x^2$, is f the independent variable? Why or why not?

92. CAPSTONE

(a) Describe any differences between a *relation* and a *function*.

(b) In your own words, explain the meanings of *domain* and *range*.

Cumulative Mixed Review

Operations with Rational Expressions In Exercises 93–96, perform the operation and simplify.

93. $12 - \dfrac{4}{x + 2}$

94. $\dfrac{3}{x^2 + x - 20} + \dfrac{x}{x^2 + 4x - 5}$

95. $\dfrac{2x^3 + 11x^2 - 6x}{5x} \cdot \dfrac{x + 10}{2x^2 + 5x - 3}$

96. $\dfrac{x + 7}{2(x - 9)} \div \dfrac{x - 7}{2(x - 9)}$

The symbol ∫ indicates an example or exercise that highlights algebraic techniques specifically used in calculus.

1.4   Graphs of Functions

The Graph of a Function

In Section 1.3, some functions were represented graphically by points on a graph in a coordinate plane in which the input values are represented by the horizontal axis and the output values are represented by the vertical axis. The **graph of a function** f is the collection of ordered pairs $(x, f(x))$ such that x is in the domain of f. As you study this section, remember the geometric interpretations of x and $f(x)$.

x = the directed distance from the y-axis

$f(x)$ = the directed distance from the x-axis

Example 1 shows how to use the graph of a function to find the domain and range of the function.

Example 1   Finding the Domain and Range of a Function

Use the graph of the function f shown in Figure 1.30 to find (a) the domain of f, (b) the function values $f(-1)$ and $f(2)$, and (c) the range of f.

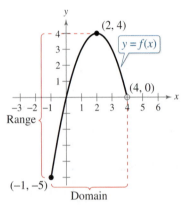

Figure 1.30

Solution

a. The closed dot at $(-1, -5)$ indicates that $x = -1$ is in the domain of f, whereas the open dot at $(4, 0)$ indicates that $x = 4$ is not in the domain. So, the domain of f is all x in the interval $[-1, 4)$.

b. Because $(-1, -5)$ is a point on the graph of f, it follows that

$f(-1) = -5.$

Similarly, because $(2, 4)$ is a point on the graph of f, it follows that

$f(2) = 4.$

c. Because the graph does not extend below $f(-1) = -5$ or above $f(2) = 4$, the range of f is the interval $[-5, 4]$.

✓**CHECKPOINT**   Now try Exercise 9.

The use of dots (open or closed) at the extreme left and right points of a graph indicates that the graph does not extend beyond these points. When no such dots are shown, assume that the graph extends beyond these points.

What you should learn

- Find the domains and ranges of functions and use the Vertical Line Test for functions.
- Determine intervals on which functions are increasing, decreasing, or constant.
- Determine relative maximum and relative minimum values of functions.
- Identify and graph step functions and other piecewise-defined functions.
- Identify even and odd functions.

Why you should learn it

Graphs of functions provide visual relationships between two variables. For example, in Exercise 92 on page 120, you will use the graph of a step function to model the cost of sending a package.

Example 2   **Finding the Domain and Range of a Function**

Find the domain and range of

$$f(x) = \sqrt{x - 4}.$$

Algebraic Solution

Because the expression under a radical cannot be negative, the domain of $f(x) = \sqrt{x - 4}$ is the set of all real numbers such that

$$x - 4 \geq 0.$$

Solve this linear inequality for x as follows. (For help with solving linear inequalities, see Appendix D at this textbook's *Companion Website*.)

$x - 4 \geq 0$	Write original inequality.
$x \geq 4$	Add 4 to each side.

So, the domain is the set of all real numbers greater than or equal to 4. Because the value of a radical expression is never negative, the range of $f(x) = \sqrt{x - 4}$ is the set of all nonnegative real numbers.

✔**CHECKPOINT**   Now try Exercise 13.

Graphical Solution

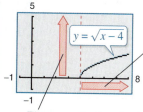

The x-coordinates of points on the graph extend from 4 to the right. So, the domain is the set of all real numbers greater than or equal to 4.

The y-coordinates of points on the graph extend from 0 upwards. So, the range is the set of all nonnegative real numbers.

By the definition of a function, at most one y-value corresponds to a given x-value. It follows, then, that a vertical line can intersect the graph of a function at most once. This leads to the **Vertical Line Test** for functions.

> **Vertical Line Test for Functions**
>
> A set of points in a coordinate plane is the graph of y as a function of x if and only if no vertical line intersects the graph at more than one point.

Example 3   **Vertical Line Test for Functions**

Use the Vertical Line Test to decide whether the graphs in Figure 1.31 represent y as a function of x.

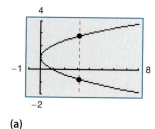
(a)

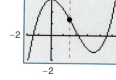
(b)

Figure 1.31

Solution

a. This is *not* a graph of y as a function of x because you can find a vertical line that intersects the graph twice.

b. This *is* a graph of y as a function of x because every vertical line intersects the graph at most once.

✔**CHECKPOINT**   Now try Exercise 21.

Technology Tip

 Most graphing utilities are designed to graph functions of x more easily than other types of equations. For instance, the graph shown in Figure 1.31(a) represents the equation $x - (y - 1)^2 = 0$. To use a graphing utility to duplicate this graph you must first solve the equation for y to obtain $y = 1 \pm \sqrt{x}$, and then graph the two equations $y_1 = 1 + \sqrt{x}$ and $y_2 = 1 - \sqrt{x}$ in the same viewing window.

Increasing and Decreasing Functions

The more you know about the graph of a function, the more you know about the function itself. Consider the graph shown in Figure 1.32. Moving from *left to right*, this graph falls from $x = -2$ to $x = 0$, is constant from $x = 0$ to $x = 2$, and rises from $x = 2$ to $x = 4$.

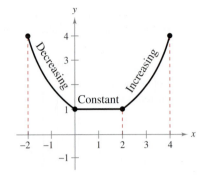

Figure 1.32

Increasing, Decreasing, and Constant Functions

A function f is **increasing** on an interval when, for any x_1 and x_2 in the interval,

$x_1 < x_2$ implies $f(x_1) < f(x_2)$.

A function f is **decreasing** on an interval when, for any x_1 and x_2 in the interval,

$x_1 < x_2$ implies $f(x_1) > f(x_2)$.

A function f is **constant** on an interval when, for any x_1 and x_2 in the interval,

$f(x_1) = f(x_2)$.

Example 4   Increasing and Decreasing Functions

In Figure 1.33, determine the open intervals on which each function is increasing, decreasing, or constant.

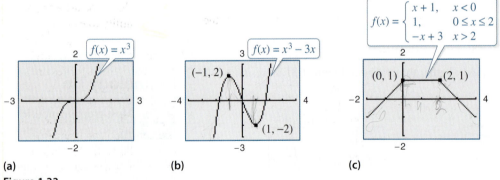

(a)   (b)   (c)

Figure 1.33

Solution

a. Although it might appear that there is an interval in which this function is constant, you can see that if $x_1 < x_2$, then $(x_1)^3 < (x_2)^3$, which implies that $f(x_1) < f(x_2)$. So, the function is increasing over the entire real line.

b. This function is increasing on the interval $(-\infty, -1)$, decreasing on the interval $(-1, 1)$, and increasing on the interval $(1, \infty)$.

c. This function is increasing on the interval $(-\infty, 0)$, constant on the interval $(0, 2)$, and decreasing on the interval $(2, \infty)$.

✓**CHECKPOINT**   Now try Exercise 25.

Relative Minimum and Maximum Values

The points at which a function changes its increasing, decreasing, or constant behavior are helpful in determining the relative maximum or relative minimum values of the function.

Definition of Relative Minimum and Relative Maximum

A function value $f(a)$ is called a **relative minimum** of f when there exists an interval (x_1, x_2) that contains a such that

$$x_1 < x < x_2 \quad \text{implies} \quad f(a) \leq f(x).$$

A function value $f(a)$ is called a **relative maximum** of f when there exists an interval (x_1, x_2) that contains a such that

$$x_1 < x < x_2 \quad \text{implies} \quad f(a) \geq f(x).$$

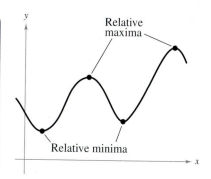

Figure 1.34

Figure 1.34 shows several different examples of relative minima and relative maxima. In Section 3.1, you will study a technique for finding the *exact points* at which a second-degree polynomial function has a relative minimum or relative maximum. For the time being, however, you can use a graphing utility to find reasonable approximations of these points.

Example 5   Approximating a Relative Minimum

Use a graphing utility to approximate the relative minimum of the function given by $f(x) = 3x^2 - 4x - 2$.

Solution

The graph of f is shown in Figure 1.35. By using the *zoom* and *trace* features of a graphing utility, you can estimate that the function has a relative minimum at the point

$$(0.67, -3.33). \qquad \text{See Figure 1.36.}$$

Later, in Section 3.1, you will be able to determine that the exact point at which the relative minimum occurs is $\left(\frac{2}{3}, -\frac{10}{3}\right)$.

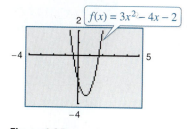

Figure 1.35

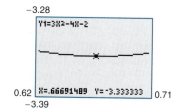

Figure 1.36

 CHECKPOINT   Now try Exercise 35.

Technology Tip

When you use a graphing utility to estimate the x- and y-values of a relative minimum or relative maximum, the *zoom* feature will often produce graphs that are nearly flat, as shown in Figure 1.36. To overcome this problem, you can manually change the vertical setting of the viewing window. The graph will stretch vertically when the values of Ymin and Ymax are closer together.

Technology Tip

Some graphing utilities have built-in programs that will find minimum or maximum values. These features are demonstrated in Example 6.

Example 6 Approximating Relative Minima and Maxima

Use a graphing utility to approximate the relative minimum and relative maximum of the function given by $f(x) = -x^3 + x$.

Solution

By using the *minimum* and *maximum* features of the graphing utility, you can estimate that the function has a relative minimum at the point

$(-0.58, -0.38)$ See Figure 1.37.

and a relative maximum at the point

$(0.58, 0.38)$.   See Figure 1.38.

If you take a course in calculus, you will learn a technique for finding the exact points at which this function has a relative minimum and a relative maximum.

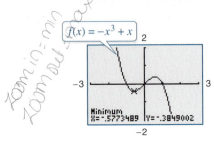

Figure 1.37

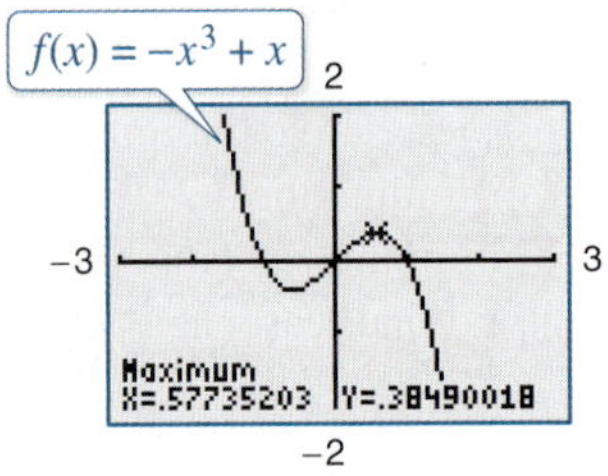

Figure 1.38

✔CHECKPOINT   Now try Exercise 37.

Example 7 Temperature

During a 24-hour period, the temperature y (in degrees Fahrenheit) of a certain city can be approximated by the model

$$y = 0.026x^3 - 1.03x^2 + 10.2x + 34, \quad 0 \le x \le 24$$

where x represents the time of day, with $x = 0$ corresponding to 6 A.M. Approximate the maximum and minimum temperatures during this 24-hour period.

Solution

Using the *maximum* feature of a graphing utility, you can determine that the maximum temperature during the 24-hour period was approximately 64°F. This temperature occurred at about 12:36 P.M. ($x \approx 6.6$), as shown in Figure 1.39. Using the *minimum* feature, you can determine that the minimum temperature during the 24-hour period was approximately 34°F, which occurred at about 1:48 A.M. ($x \approx 19.8$), as shown in Figure 1.40.

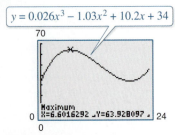

Figure 1.39

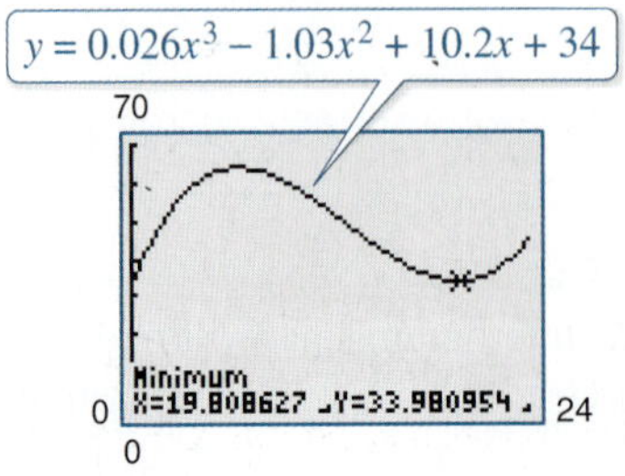

Figure 1.40

✔CHECKPOINT   Now try Exercise 95.

Step Functions and Piecewise-Defined Functions

 Library of Parent Functions: Greatest Integer Function

The *greatest integer function*, denoted by $[\![x]\!]$ and defined as the greatest integer less than or equal to x, has an infinite number of breaks or steps—one at each integer value in its domain. The basic characteristics of the greatest integer function are summarized below.

Graph of $f(x) = [\![x]\!]$

Domain: $(-\infty, \infty)$

Range: the set of integers

x-intercepts: in the interval $[0, 1)$

y-intercept: $(0, 0)$

Constant between each pair of consecutive integers

Jumps vertically one unit at each integer value

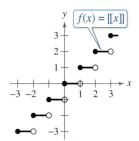

$f(x) = [\![x]\!]$

Technology Tip

 Most graphing utilities display graphs in *connected mode*, which works well for graphs that do not have breaks. For graphs that do have breaks, such as the greatest integer function, it is better to use *dot mode*. Graph the greatest integer function [often called Int (x)] in *connected* and *dot modes*, and compare the two results.

Because of the vertical jumps described above, the greatest integer function is an example of a **step function** whose graph resembles a set of stairsteps. Some values of the greatest integer function are as follows.

$$[\![-1]\!] = (\text{greatest integer} \le -1) = -1$$

$$\left[\!\!\left[-\tfrac{1}{2}\right]\!\!\right] = \left(\text{greatest integer} \le -\tfrac{1}{2}\right) = -1$$

$$\left[\!\!\left[\tfrac{1}{10}\right]\!\!\right] = \left(\text{greatest integer} \le \tfrac{1}{10}\right) = 0$$

$$[\![1.5]\!] = (\text{greatest integer} \le 1.5) = 1$$

In Section 1.3, you learned that a piecewise-defined function is a function that is defined by two or more equations over a specified domain. To sketch the graph of a piecewise-defined function, you need to sketch the graph of each equation on the appropriate portion of the domain.

Example 8   Sketching a Piecewise-Defined Function

Sketch the graph of

$$f(x) = \begin{cases} 2x + 3, & x \le 1 \\ -x + 4, & x > 1 \end{cases}$$

by hand.

Solution

This piecewise-defined function is composed of two linear functions. At and to the left of $x = 1$, the graph is the line given by

$$y = 2x + 3.$$

To the right of $x = 1$, the graph is the line given by

$$y = -x + 4$$

as shown in Figure 1.41. Notice that the point $(1, 5)$ is a solid dot and the point $(1, 3)$ is an open dot. This is because $f(1) = 5$.

CHECKPOINT   Now try Exercise 55.

Figure 1.41

$y = 2x + 3$

$y = -x + 4$

What's Wrong?

You use a graphing utility to graph

$$f(x) = \begin{cases} x^2 + 1, & x \le 0 \\ 4 - x, & x > 0 \end{cases}$$

by letting $y_1 = x^2 + 1$ and $y_2 = 4 - x$, as shown in the figure. You conclude that this is the graph of f. What's wrong?

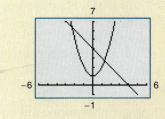

Even and Odd Functions

A graph has *symmetry with respect to the y-axis* if whenever (x, y) is on the graph, then so is the point $(-x, y)$. A graph has *symmetry with respect to the origin* if whenever (x, y) is on the graph, then so is the point $(-x, -y)$. A graph has *symmetry with respect to the x-axis* if whenever (x, y) is on the graph, then so is the point $(x, -y)$. A function whose graph is symmetric with respect to the y-axis is an **even function.** A function whose graph is symmetric with respect to the origin is an **odd function.** A graph that is symmetric with respect to the x-axis is not the graph of a function (except for the graph of $y = 0$). These three types of symmetry are illustrated in Figure 1.42.

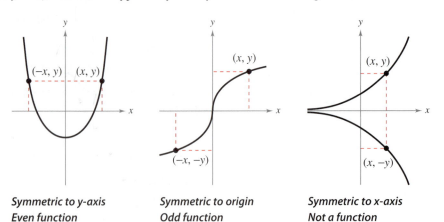

Symmetric to y-axis
Even function

Symmetric to origin
Odd function

Symmetric to x-axis
Not a function

Figure 1.42

Example 9   Even and Odd Functions

Use the figure to determine whether the function is even, odd, or neither.

a.

b.

c.

d.

Solution

a. The graph is symmetric with respect to the y-axis. So, the function is even.

b. The graph is symmetric with respect to the origin. So, the function is odd.

c. The graph is neither symmetric with respect to the origin nor with respect to the y-axis. So, the function is neither even nor odd.

d. The graph is symmetric with respect to the y-axis. So, the function is even.

✓**CHECKPOINT**   Now try Exercise 67.

Test for Even and Odd Functions

A function f is **even** when, for each x in the domain of f, $f(-x) = f(x)$.
A function f is **odd** when, for each x in the domain of f, $f(-x) = -f(x)$.

Example 10   Even and Odd Functions

Determine whether each function is even, odd, or neither.

a. $g(x) = x^3 - x$ it's neither even nor odd

b. $h(x) = x^2 + 1$ even

c. $f(x) = x^3 - 1$ it's

Algebraic Solution

a. This function is odd because

$$g(-x) = (-x)^3 - (-x)$$

$$= -x^3 + x$$

$$= -(x^3 - x)$$

$$= -g(x).$$

b. This function is even because

$$h(-x) = (-x)^2 + 1$$

$$= x^2 + 1$$

$$= h(x).$$

c. Substituting $-x$ for x produces

$$f(-x) = (-x)^3 - 1$$

$$= -x^3 - 1.$$

Because

$$f(x) = x^3 - 1$$

and

$$-f(x) = -x^3 + 1$$

you can conclude that

$$f(-x) \neq f(x)$$

and

$$f(-x) \neq -f(x).$$

So, the function is neither even nor odd.

✓CHECKPOINT   Now try Exercise 81.

Graphical Solution

a. In Figure 1.43, the graph is symmetric with respect to the origin. So, this function is odd.

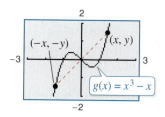

Figure 1.43

b. In Figure 1.44, the graph is symmetric with respect to the y-axis. So, this function is even.

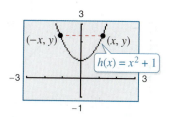

Figure 1.44

c. In Figure 1.45, the graph is neither symmetric with respect to the origin nor with respect to the y-axis. So, this function is neither even nor odd.

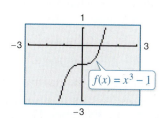

Figure 1.45

To help visualize symmetry with respect to the origin, place a pin at the origin of a graph and rotate the graph 180°. If the result after rotation coincides with the original graph, then the graph is symmetric with respect to the origin.

1.4   Exercises

See www.CalcChat.com for worked-out solutions to odd-numbered exercises.
For instructions on how to use a graphing utility, see Appendix A.

Vocabulary and Concept Check

In Exercises 1 and 2, fill in the blank.

1. A function f is _____ on an interval when, for any x_1 and x_2 in the interval, $x_1 < x_2$ implies $f(x_1) > f(x_2)$.

2. A function f is _____ when, for each x in the domain of f, $f(-x) = f(x)$.

3. The graph of a function f is the segment from $(1, 2)$ to $(4, 5)$, including the endpoints. What is the domain of f?

4. A vertical line intersects a graph twice. Does the graph represent a function?

5. Let f be a function such that $f(2) \geq f(x)$ for all values of x in the interval $(0, 3)$. Does $f(2)$ represent a relative minimum or a relative maximum?

6. Given $f(x) = [\![x]\!]$, in what interval does $f(x) = 5$?

Procedures and Problem Solving

Finding the Domain and Range of a Function In Exercises 7–10, use the graph of the function to find the domain and range of f. Then find $f(0)$.

7.

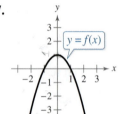

8.

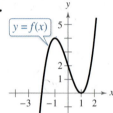

9. ✓

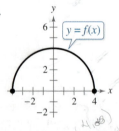

10.

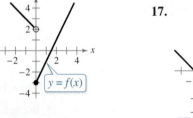

Finding the Domain and Range of a Function In Exercises 11–16, use a graphing utility to graph the function and estimate its domain and range. Then find the domain and range algebraically.

11. $f(x) = 2x^2 + 3$

12. $f(x) = -x^2 - 1$

✓ 13. $f(x) = \sqrt{x - 1}$

14. $h(t) = \sqrt{4 - t^2}$

15. $f(x) = |x + 3|$

16. $f(x) = -\frac{1}{4}|x - 5|$

Analyzing a Graph In Exercises 17 and 18, use the graph of the function to answer the questions.

(a) Determine the domain of the function.

(b) Determine the range of the function.

(c) Find the value(s) of x for which $f(x) = 0$.

(d) What are the values of x from part (c) referred to graphically?

(e) Find $f(0)$, if possible.

(f) What is the value from part (e) referred to graphically?

(g) What is the value of f at $x = 1$? What are the coordinates of the point?

(h) What is the value of f at $x = -1$? What are the coordinates of the point?

(i) The coordinates of the point on the graph of f at which $x = -3$ can be labeled $(-3, f(-3))$, or $(-3, \underline{\quad})$.

17.

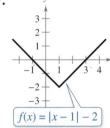

$$f(x) = |x - 1| - 2$$

18.

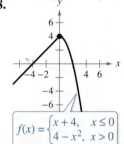

$$f(x) = \begin{cases} x + 4, & x \leq 0 \\ 4 - x^2, & x > 0 \end{cases}$$

Vertical Line Test for Functions In Exercises 19–22, use the Vertical Line Test to determine whether y is a function of x. Describe how you can use a graphing utility to produce the given graph.

19. $y = \frac{1}{2}x^2$

20. $x - y^2 = 1$

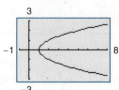

✓ **21.** $x^2 + y^2 = 25$ **22.** $x^2 = 2xy - 1$

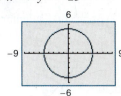   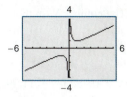

Increasing and Decreasing Functions In Exercises 23–26, determine the open intervals on which the function is increasing, decreasing, or constant.

23. $f(x) = \frac{3}{2}x$ **24.** $f(x) = x^2 - 4x$

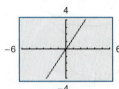   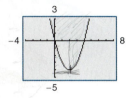

✓ **25.** $f(x) = x^3 - 3x^2 + 2$ **26.** $f(x) = \sqrt{x^2 - 1}$

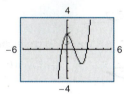   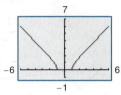

Increasing and Decreasing Functions In Exercises 27–34, (a) use a graphing utility to graph the function and (b) determine the open intervals on which the function is increasing, decreasing, or constant.

27. $f(x) = 3$ **28.** $f(x) = x$
29. $f(x) = x^{2/3}$ **30.** $f(x) = -x^{3/4}$
31. $f(x) = x\sqrt{x + 3}$ **32.** $f(x) = \sqrt{1 - x}$
33. $f(x) = |x + 1| + |x - 1|$
34. $f(x) = -|x + 4| - |x + 1|$

Approximating Relative Minima and Maxima In Exercises 35–46, use a graphing utility to graph the function and to approximate any relative minimum or relative maximum values of the function.

✓ **35.** $f(x) = x^2 - 6x$ **36.** $f(x) = 3x^2 - 2x - 5$
✓ **37.** $y = 2x^3 + 3x^2 - 12x$ **38.** $y = x^3 - 6x^2 + 15$
39. $h(x) = (x - 1)\sqrt{x}$ **40.** $g(x) = x\sqrt{4 - x}$
41. $f(x) = x^2 - 4x - 5$ **42.** $f(x) = 3x^2 - 12x$
43. $f(x) = x^3 - 3x$ **44.** $f(x) = -x^3 + 3x^2$
45. $f(x) = 3x^2 - 6x + 1$ **46.** $f(x) = 8x - 4x^2$

Library of Parent Functions In Exercises 47–52, sketch the graph of the function by hand. Then use a graphing utility to verify the graph.

47. $f(x) = [\![x]\!] + 2$ **48.** $f(x) = [\![x]\!] - 3$

49. $f(x) = [\![x - 1]\!] + 2$ **50.** $f(x) = [\![x - 2]\!] + 1$
51. $f(x) = [\![2x]\!]$ **52.** $f(x) = [\![4x]\!]$

Describing a Step Function In Exercises 53 and 54, use a graphing utility to graph the function. State the domain and range of the function. Describe the pattern of the graph.

53. $s(x) = 2\left(\frac{1}{4}x - [\![\frac{1}{4}x]\!]\right)$
54. $g(x) = 2\left(\frac{1}{4}x - [\![\frac{1}{4}x]\!]\right)^2$

Sketching a Piecewise-Defined Function In Exercises 55–62, sketch the graph of the piecewise-defined function by hand.

✓ **55.** $f(x) = \begin{cases} 2x + 3, & x < 0 \\ 3 - x, & x \geq 0 \end{cases}$

56. $f(x) = \begin{cases} x + 6, & x \leq -4 \\ 2x - 4, & x > -4 \end{cases}$

57. $f(x) = \begin{cases} \sqrt{4 + x}, & x < 0 \\ \sqrt{4 - x}, & x \geq 0 \end{cases}$

58. $f(x) = \begin{cases} 1 - (x - 1)^2, & x \leq 2 \\ \sqrt{x - 2}, & x > 2 \end{cases}$

59. $f(x) = \begin{cases} x + 3, & x \leq 0 \\ 3, & 0 < x \leq 2 \\ 2x - 1, & x > 2 \end{cases}$

60. $g(x) = \begin{cases} x + 5, & x \leq -3 \\ -2, & -3 < x < 1 \\ 5x - 4, & x \geq 1 \end{cases}$

61. $f(x) = \begin{cases} 2x + 1, & x \leq -1 \\ x^2 - 2, & x > -1 \end{cases}$

62. $h(x) = \begin{cases} 3 + x, & x < 0 \\ x^2 + 1, & x \geq 0 \end{cases}$

Even and Odd Functions In Exercises 63–72, use a graphing utility to graph the function and determine whether it is even, odd, or neither.

63. $f(x) = 5$ **64.** $f(x) = -9$
65. $f(x) = 3x - 2$ **66.** $f(x) = 5 - 3x$
✓ **67.** $h(x) = x^2 - 4$ **68.** $f(x) = -x^2 - 8$
69. $f(x) = \sqrt{1 - x}$ **70.** $g(t) = \sqrt[3]{t - 1}$
71. $f(x) = |x + 2|$ **72.** $f(x) = -|x - 5|$

Think About It In Exercises 73–78, find the coordinates of a second point on the graph of a function f if the given point is on the graph and the function is (a) even and (b) odd.

73. $\left(-\frac{3}{2}, 4\right)$ **74.** $\left(-\frac{5}{3}, -7\right)$
75. $(4, 9)$ **76.** $(5, -1)$
77. $(x, -y)$ **78.** $(2a, 2c)$

Algebraic-Graphical-Numerical In Exercises 79–86, determine whether the function is even, odd, or neither (a) algebraically, (b) graphically by using a graphing utility to graph the function, and (c) numerically by using the *table* feature of the graphing utility to compare $f(x)$ and $f(-x)$ for several values of x.

79. $f(t) = t^2 + 2t - 3$ **80.** $f(x) = x^6 - 2x^2 + 3$

✓ **81.** $g(x) = x^3 - 5x$ **82.** $h(x) = x^3 - 5$

83. $f(x) = x\sqrt{1 - x^2}$ **84.** $f(x) = x\sqrt{x + 5}$

85. $g(s) = 4s^{2/3}$ **86.** $f(s) = 4s^{3/2}$

Finding the Intervals Where a Function is Positive In Exercises 87–90, graph the function and determine the interval(s) (if any) on the real axis for which $f(x) \geq 0$. Use a graphing utility to verify your results.

87. $f(x) = 4 - x$ **88.** $f(x) = 4x + 2$

89. $f(x) = x^2 - 9$ **90.** $f(x) = x^2 - 4x$

91. Business The cost of using a telephone calling card is $1.05 for the first minute and $0.08 for each additional minute or portion of a minute.

(a) A customer needs a model for the cost C of using the calling card for a call lasting t minutes. Which of the following is the appropriate model?

$$C_1(t) = 1.05 + 0.08[\![t - 1]\!]$$

$$C_2(t) = 1.05 - 0.08[\![-(t - 1)]\!]$$

(b) Use a graphing utility to graph the appropriate model. Estimate the cost of a call lasting 18 minutes and 45 seconds.

92. *Why you should learn it* (p. 110) The cost of sending an overnight package from New York to Atlanta is $18.80 for a package weighing up to but not including 1 pound and $3.50 for each additional pound or portion of a pound. Use the greatest integer function to create a model for the cost C of overnight delivery of a package weighing x pounds, where $x > 0$. Sketch the graph of the function.

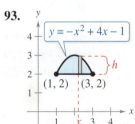

Using the Graph of a Function In Exercises 93 and 94, write the height h of the rectangle as a function of x.

93.

$y = -x^2 + 4x - 1$

$(1, 2)$ $(3, 2)$

94.

$(1, 3)$

$y = 4x - x^2$

95. MODELING DATA

The number N (in thousands) of existing condominiums and cooperative homes sold each year from 2000 through 2008 in the United States is approximated by the model

$$N = 0.4825t^4 - 11.293t^3 + 65.26t^2 - 48.8t + 578,$$
$$0 \leq t \leq 8$$

where t represents the year, with $t = 0$ corresponding to 2000. (Source: National Association of Realtors)

(a) Use a graphing utility to graph the model over the appropriate domain.

(b) Use the graph from part (a) to determine during which years the number of cooperative homes and condos was increasing. During which years was the number decreasing?

(c) Approximate the maximum number of cooperative homes and condos sold from 2000 through 2008.

96. Mechanical Engineering The intake pipe of a 100-gallon tank has a flow rate of 10 gallons per minute, and two drain pipes have a flow rate of 5 gallons per minute each. The graph shows the volume V of fluid in the tank as a function of time t. Determine in which pipes the fluid is flowing in specific subintervals of the one-hour interval of time shown on the graph. (There are many correct answers.)

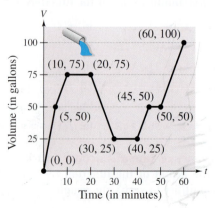

Conclusions

True or False?   **In Exercises 97 and 98, determine whether the statement is true or false. Justify your answer.**

97. A function with a square root cannot have a domain that is the set of all real numbers.

98. It is possible for an odd function to have the interval $[0, \infty)$ as its domain.

Think About It   In Exercises 99–104, match the graph of the function with the description that best fits the situation.

(a) The air temperature at a beach on a sunny day

(b) The height of a football kicked in a field goal attempt

(c) The number of children in a family over time

(d) The population of California as a function of time

(e) The depth of the tide at a beach over a 24-hour period

(f) The number of cupcakes on a tray at a party

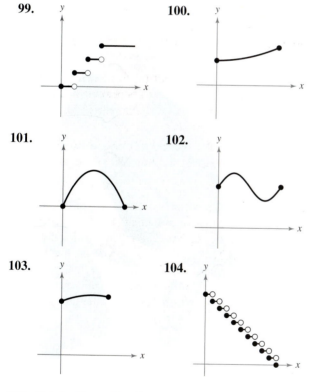

99.   **100.**

101.   **102.**

103.   **104.**

105. Think About It   Does the graph in Exercise 20 represent x as a function of y? Explain.

106. Think About It   Does the graph in Exercise 21 represent x as a function of y? Explain.

107. Think About It   Can you represent the greatest integer function using a piecewise-defined function?

108. Think About It   How does the graph of the greatest integer function differ from the graph of a line with a slope of zero?

109. Let f be an even function. Determine whether g is even, odd, or neither. Explain.

(a) $g(x) = -f(x)$ (b) $g(x) = f(-x)$

(c) $g(x) = f(x) - 2$ (d) $g(x) = -f(x - 2)$

110. CAPSTONE   Half of the graph of an odd function is shown.

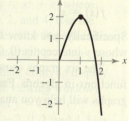

(a) Sketch a complete graph of the function.

(b) Find the domain and range of the function.

(c) Identify the open intervals on which the function is increasing, decreasing, or constant.

(d) Find any relative minimum and relative maximum values of the function.

111. Proof   Prove that a function of the following form is odd.

$$y = a_{2n+1}x^{2n+1} + a_{2n-1}x^{2n-1} + \cdots + a_3 x^3 + a_1 x$$

112. Proof   Prove that a function of the following form is even.

$$y = a_{2n}x^{2n} + a_{2n-2}x^{2n-2} + \cdots + a_2 x^2 + a_0$$

Cumulative Mixed Review

Identifying Terms and Coefficients   In Exercises 113–116, identify the terms. Then identify the coefficients of the variable terms of the expression.

113. $-2x^2 + 8x$ **114.** $10 + 3x$

115. $\dfrac{x}{3} - 5x^2 + x^3$ **116.** $7x^4 + \sqrt{2}x^2$

Evaluating a Function   In Exercises 117 and 118, evaluate the function at each specified value of the independent variable and simplify.

117. $f(x) = -x^2 - x + 3$

(a) $f(4)$ (b) $f(-2)$ (c) $f(x - 2)$

118. $f(x) = x\sqrt{x - 3}$

(a) $f(3)$ (b) $f(12)$ (c) $f(6)$

Evaluating a Difference Quotient   In Exercises 119 and 120, find the difference quotient and simplify your answer.

119. $f(x) = x^2 - 2x + 9$, $\dfrac{f(3 + h) - f(3)}{h}$, $h \neq 0$

120. $f(x) = 5 + 6x - x^2$, $\dfrac{f(6 + h) - f(6)}{h}$, $h \neq 0$

Example 4 Reflections and Shifts

Compare the graph of each function with the graph of

$$f(x) = \sqrt{x}.$$

a. $g(x) = -\sqrt{x}$
b. $h(x) = \sqrt{-x}$
c. $k(x) = -\sqrt{x + 2}$

Algebraic Solution

a. Relative to the graph of $f(x) = \sqrt{x}$, the graph of g is a reflection in the x-axis because

$$g(x) = -\sqrt{x}$$

$$= -f(x).$$

b. The graph of h is a reflection of the graph of $f(x) = \sqrt{x}$ in the y-axis because

$$h(x) = \sqrt{-x}$$

$$= f(-x).$$

c. From the equation

$$k(x) = -\sqrt{x + 2}$$

$$= -f(x + 2)$$

you can conclude that the graph of k is a left shift of two units, followed by a reflection in the x-axis, of the graph of $f(x) = \sqrt{x}$.

Graphical Solution

a. From the graph in Figure 1.54, you can see that the graph of g is a reflection of the graph of f in the x-axis. Note that the domain of g is $x \geq 0$.

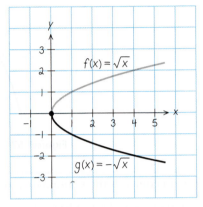

Figure 1.54

b. From the graph in Figure 1.55, you can see that the graph of h is a reflection of the graph of f in the y-axis. Note that the domain of h is $x \leq 0$.

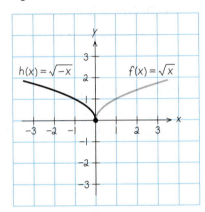

Figure 1.55

c. From the graph in Figure 1.56, you can see that the graph of k is a left shift of two units of the graph of f, followed by a reflection in the x-axis. Note that the domain of k is $x \geq -2$.

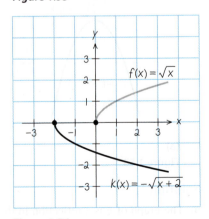

Figure 1.56

CHECKPOINT   Now try Exercise 35.

Nonrigid Transformations

Horizontal shifts, vertical shifts, and reflections are called **rigid transformations** because the basic shape of the graph is unchanged. These transformations change only the *position* of the graph in the coordinate plane. **Nonrigid transformations** are those that cause a *distortion*—a change in the shape of the original graph. For instance, a nonrigid transformation of the graph of $y = f(x)$ is represented by $g(x) = cf(x)$, where the transformation is a **vertical stretch** when $c > 1$ and a **vertical shrink** when $0 < c < 1$. Another nonrigid transformation of the graph of $y = f(x)$ is represented by $h(x) = f(cx)$, where the transformation is a **horizontal shrink** when $c > 1$ and a **horizontal stretch** when $0 < c < 1$.

Example 5   Nonrigid Transformations

Compare the graph of each function with the graph of $f(x) = |x|$.

a. $h(x) = 3|x|$
b. $g(x) = \frac{1}{3}|x|$

Solution

a. Relative to the graph of $f(x) = |x|$, the graph of

$$h(x) = 3|x| = 3f(x)$$

is a vertical stretch (each y-value is multiplied by 3) of the graph of f. (See Figure 1.57.)

b. Similarly, the graph of

$$g(x) = \frac{1}{3}|x| = \frac{1}{3}f(x)$$

is a vertical shrink $\left(\text{each } y\text{-value is multiplied by } \frac{1}{3}\right)$ of the graph of f. (See Figure 1.58.)

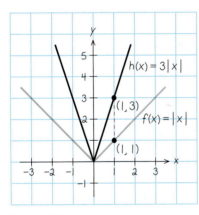

Figure 1.57

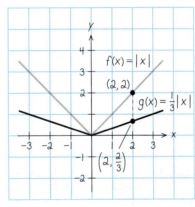

Figure 1.58

✓CHECKPOINT   Now try Exercise 41.

Example 6   Nonrigid Transformations

Compare the graph of $h(x) = f\left(\frac{1}{2}x\right)$ with the graph of $f(x) = 2 - x^3$.

Solution

Relative to the graph of $f(x) = 2 - x^3$, the graph of

$$h(x) = f\left(\tfrac{1}{2}x\right) = 2 - \left(\tfrac{1}{2}x\right)^3 = 2 - \tfrac{1}{8}x^3$$

is a horizontal stretch (each x-value is multiplied by 2) of the graph of f. (See Figure 1.59.)

✓CHECKPOINT   Now try Exercise 49.

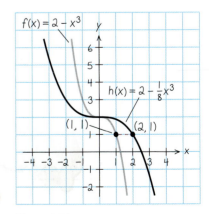

Figure 1.59

1.5   Exercises

See www.CalcChat.com for worked-out solutions to odd-numbered exercises.
For instructions on how to use a graphing utility, see Appendix A.

Vocabulary and Concept Check

1. Name three types of rigid transformations.

2. Match the rigid transformation of $y = f(x)$ with the correct representation, where $c > 0$.

   (a) $h(x) = f(x) + c$ (i) horizontal shift c units to the left
   (b) $h(x) = f(x) - c$ (ii) vertical shift c units upward
   (c) $h(x) = f(x - c)$ (iii) horizontal shift c units to the right
   (d) $h(x) = f(x + c)$ (iv) vertical shift c units downward

In Exercises 3 and 4, fill in the blanks.

3. A reflection in the x-axis of $y = f(x)$ is represented by $h(x) = $ _____ ,
   while a reflection in the y-axis of $y = f(x)$ is represented by $h(x) = $ _____ .

4. A nonrigid transformation of $y = f(x)$ represented by $cf(x)$ is a vertical stretch
   when _____ and a vertical shrink when _____ .

Procedures and Problem Solving

Sketching Transformations In Exercises 5–18, sketch
the graphs of the three functions by hand on the same
rectangular coordinate system. Verify your results with a
graphing utility.

5. $f(x) = x$

   $g(x) = x - 4$

   $h(x) = 3x$

6. $f(x) = \frac{1}{2}x$

   $g(x) = \frac{1}{2}x + 2$

   $h(x) = \frac{1}{2}(x - 2)$

7. $f(x) = x^2$

   $g(x) = x^2 + 2$

   $h(x) = (x - 2)^2$

8. $f(x) = x^2$

   $g(x) = x^2 - 4$

   $h(x) = (x + 2)^2 + 1$

9. $f(x) = -x^2$

   $g(x) = -x^2 + 1$

   $h(x) = -(x - 2)^2$

10. $f(x) = (x - 2)^2$

    $g(x) = (x + 2)^2 + 2$

    $h(x) = -(x - 2)^2 - 1$

11. $f(x) = x^2$

    $g(x) = \frac{1}{2}x^2$

    $h(x) = (2x)^2$

12. $f(x) = x^2$

    $g(x) = \frac{1}{4}x^2 + 2$

    $h(x) = -\frac{1}{4}x^2$

13. $f(x) = |x|$

    $g(x) = |x| - 1$

    $h(x) = |x - 3|$

14. $f(x) = |x|$

    $g(x) = |x + 3|$

    $h(x) = -2|x + 2| - 1$

15. $f(x) = \sqrt{x}$

    $g(x) = \sqrt{x + 1}$

    $h(x) = \sqrt{x - 2} + 1$

16. $f(x) = \sqrt{x}$

    $g(x) = \frac{1}{2}\sqrt{x}$

    $h(x) = -\sqrt{x + 4}$

17. $f(x) = \dfrac{1}{x}$

    $g(x) = \dfrac{1}{x} + 2$

    $h(x) = \dfrac{1}{x - 1} + 2$

18. $f(x) = \dfrac{1}{x}$

    $g(x) = \dfrac{1}{x} - 4$

    $h(x) = \dfrac{1}{x + 3} - 1$

Sketching Transformations In **Exercises 19 and 20**, use
the graph of f to sketch each graph. To print an enlarged
copy of the graph, go to the website *www.mathgraphs.com*.

19. (a) $y = f(x) + 2$

    (b) $y = -f(x)$

    (c) $y = f(x - 2)$

    (d) $y = f(x + 3)$

    (e) $y = 2f(x)$

    (f) $y = f(-x)$

    (g) $y = f\left(\frac{1}{2}x\right)$

20. (a) $y = f(x) - 1$

    (b) $y = f(x + 1)$

    (c) $y = f(x - 1)$

    (d) $y = -f(x - 2)$

    (e) $y = f(-x)$

    (f) $y = \frac{1}{2}f(x)$

    (g) $y = f(2x)$

Error Analysis In **Exercises 21 and 22**, describe the
error in graphing the function.

21. $f(x) = (x + 1)^2$

22. $f(x) = (x - 1)^2$

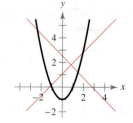

Library of Parent Functions   In Exercises 23–28, compare the graph of the function with the graph of its parent function.

✓ **23.** $y = \sqrt{x} + 2$

24. $y = \dfrac{1}{x} - 5$

25. $y = (x - 4)^3$

26. $y = |x + 5|$

27. $y = x^2 - 2$

28. $y = \sqrt{x - 2}$

Library of Parent Functions   In Exercises 29–34, identify the parent function and describe the transformation shown in the graph. Write an equation for the graphed function.

29.

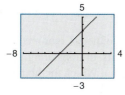

30.

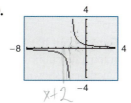

✓ **31.**

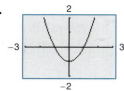

32.

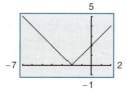

✓ **33.**

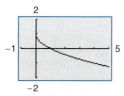

34.

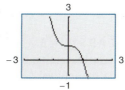

Rigid and Nonrigid Transformations   In Exercises 35–46, compare the graph of the function with the graph of its parent function.

✓ **35.** $y = -|x|$

36. $y = |-x|$

37. $y = (-x)^2$

38. $y = -x^3$

39. $y = \dfrac{1}{-x}$

40. $y = -\dfrac{1}{x}$

✓ **41.** $h(x) = 4|x|$

42. $p(x) = \frac{1}{2}x^2$

43. $g(x) = \frac{1}{4}x^3$

44. $y = 2\sqrt{x}$

45. $f(x) = \sqrt{4x}$

46. $y = \left|\frac{1}{2}x\right|$

Rigid and Nonrigid Transformations   In Exercises 47–50, use a graphing utility to graph the three functions in the same viewing window. Describe the graphs of g and h relative to the graph of f.

47. $f(x) = x^3 - 3x^2$
   $g(x) = f(x + 2)$
   $h(x) = \frac{1}{2}f(x)$

48. $f(x) = x^3 - 3x^2 + 2$
   $g(x) = f(x - 1)$
   $h(x) = f(3x)$

✓ **49.** $f(x) = x^3 - 3x^2$
   $g(x) = -\frac{1}{3}f(x)$
   $h(x) = f(-x)$

50. $f(x) = x^3 - 3x^2 + 2$
   $g(x) = -f(x)$
   $h(x) = f(2x)$

Describing Transformations   In Exercises 51–64, g is related to one of the six parent functions on page 122. (a) Identify the parent function f. (b) Describe the sequence of transformations from f to g. (c) Sketch the graph of g by hand. (d) Use function notation to write g in terms of the parent function f.

51. $g(x) = 2 - (x + 5)^2$

52. $g(x) = -(x + 10)^2 + 5$

53. $g(x) = 3 + 2(x - 4)^2$

54. $g(x) = -\frac{1}{4}(x + 2)^2 - 2$

55. $g(x) = 3(x - 2)^3$

56. $g(x) = -\frac{1}{2}(x + 1)^3$

57. $g(x) = (x - 1)^3 + 2$

58. $g(x) = -(x + 3)^3 - 10$

59. $g(x) = \dfrac{1}{x + 8} - 9$

60. $g(x) = \dfrac{1}{x - 7} + 4$

61. $g(x) = -2|x - 1| - 4$

62. $g(x) = \frac{1}{2}|x - 2| - 3$

63. $g(x) = -\frac{1}{2}\sqrt{x + 3} - 1$

64. $g(x) = -\sqrt{x + 1} - 6$

65. MODELING DATA

The amounts of fuel F (in billions of gallons) used by motor vehicles from 1991 through 2007 are given by the ordered pairs of the form $(t, F(t))$, where $t = 1$ represents 1991. A model for the data is

$$F(t) = -0.099(t - 24.7)^2 + 183.4.$$

(Source: U.S. Federal Highway Administration)

(1, 128.6)
(2, 132.9)
(3, 137.3)
(4, 140.8)
(5, 143.8)
(6, 147.4)
(7, 150.4)
(8, 155.4)
(9, 161.4)
(10, 162.5)   (14, 173.5)
(11, 163.5)   (15, 174.8)
(12, 168.7)   (16, 175.0)
(13, 170.0)   (17, 176.1)

(a) Describe the transformation of the parent function $f(t) = t^2$.

(b) Use a graphing utility to graph the model and the data in the same viewing window.

(c) Rewrite the function so that $t = 0$ represents 2000. Explain how you got your answer.

66. *Why you should learn it* (p. 122) The sales S (in millions of dollars) of the WD-40 Company from 2000 through 2008 can be approximated by the function

$$S(t) = 99\sqrt{t} + 2.37$$

where $t = 0$ represents 2000. (Source: WD-40 Company)

(a) Describe the transformation of the parent function $f(t) = \sqrt{t}$.

(b) Use a graphing utility to graph the model over the interval $0 \le t \le 8$.

(c) According to the model, in what year will the sales of WD-40 be approximately 400 million dollars?

(d) Rewrite the function so that $t = 0$ represents 2005. Explain how you got your answer.

Conclusions

True or False? In Exercises 67 and 68, determine whether the statement is true or false. Justify your answer.

67. The graph of $y = f(-x)$ is a reflection of the graph of $y = f(x)$ in the x-axis.

68. The graphs of $f(x) = |x| + 6$ and $f(x) = |-x| + 6$ are identical.

Exploration In Exercises 69–72, use the fact that the graph of $y = f(x)$ has x-intercepts at $x = 2$ and $x = -3$ to find the x-intercepts of the given graph. If not possible, state the reason.

69. $y = f(-x)$ **70.** $y = 2f(x)$

71. $y = f(x) + 2$ **72.** $y = f(x - 3)$

 Library of Parent Functions In Exercises 73–76, determine which equation(s) may be represented by the graph shown. There may be more than one correct answer.

73.   **74.**

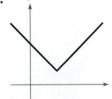

(a) $f(x) = |x + 2| + 1$ (a) $f(x) = -\sqrt{x} - 4$

(b) $f(x) = |x - 1| + 2$ (b) $f(x) = -4 - \sqrt{x}$

(c) $f(x) = |x - 2| + 1$ (c) $f(x) = -4 - \sqrt{-x}$

(d) $f(x) = 2 + |x - 2|$ (d) $f(x) = \sqrt{-x} - 4$

(e) $f(x) = |(x - 2) + 1|$ (e) $f(x) = \sqrt{-x} + 4$

(f) $f(x) = 1 - |x - 2|$ (f) $f(x) = \sqrt{x} - 4$

75.   **76.**

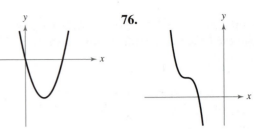

(a) $f(x) = (x - 2)^2 - 2$ (a) $f(x) = -(x - 4)^3 + 2$

(b) $f(x) = (x + 4)^2 - 4$ (b) $f(x) = -(x + 4)^3 + 2$

(c) $f(x) = (x - 2)^2 - 4$ (c) $f(x) = -(x - 2)^3 + 4$

(d) $f(x) = (x + 2)^2 - 4$ (d) $f(x) = (-x - 4)^3 + 2$

(e) $f(x) = 4 - (x - 2)^2$ (e) $f(x) = (x + 4)^3 + 2$

(f) $f(x) = 4 - (x + 2)^2$ (f) $f(x) = (-x + 4)^3 + 2$

77. Think About It You can use either of two methods to graph a function: plotting points, or translating a parent function as shown in this section. Which method do you prefer to use for each function? Explain.

(a) $f(x) = 3x^2 - 4x + 1$ (b) $f(x) = 2(x - 1)^2 - 6$

78. Think About It The graph of $y = f(x)$ passes through the points $(0, 1)$, $(1, 2)$, and $(2, 3)$. Find the corresponding points on the graph of $y = f(x + 2) - 1$.

79. Think About It Compare the graph of $g(x) = ax^2$ with the graph of $f(x) = x^2$ when (a) $0 < a < 1$ and (b) $a > 1$.

80. CAPSTONE Use the fact that the graph of $y = f(x)$ is increasing on the interval $(-\infty, 2)$ and decreasing on the interval $(2, \infty)$ to find the intervals on which the graph is increasing and decreasing. If not possible, state the reason.

(a) $y = f(-x)$ (b) $y = -f(x)$ (c) $y = 2f(x)$

(d) $y = f(x) - 3$ (e) $y = f(x + 1)$

Cumulative Mixed Review

Parallel and Perpendicular Lines In Exercises 81 and 82, determine whether the lines L_1 and L_2 passing through the pairs of points are parallel, perpendicular, or neither.

81. L_1: $(-2, -2)$, $(2, 10)$
    L_2: $(-1, 3)$, $(3, 9)$

82. L_1: $(-1, -7)$, $(4, 3)$
    L_2: $(1, 5)$, $(-2, -7)$

Finding the Domain of a Function In Exercises 83–86, find the domain of the function.

83. $f(x) = \dfrac{4}{9 - x}$ **84.** $f(x) = \dfrac{\sqrt{x - 5}}{x - 7}$

85. $f(x) = \sqrt{100 - x^2}$ **86.** $f(x) = \sqrt[3]{16 - x^2}$

1.6 Combinations of Functions

Arithmetic Combinations of Functions

Just as two real numbers can be combined by the operations of addition, subtraction, multiplication, and division to form other real numbers, two *functions* can be combined to create new functions. When

$$f(x) = 2x - 3 \quad \text{and} \quad g(x) = x^2 - 1$$

you can form the sum, difference, product, and quotient of f and g as follows.

$$f(x) + g(x) = (2x - 3) + (x^2 - 1)$$
$$= x^2 + 2x - 4 \qquad \text{Sum}$$
$$f(x) - g(x) = (2x - 3) - (x^2 - 1)$$
$$= -x^2 + 2x - 2 \qquad \text{Difference}$$
$$f(x) \cdot g(x) = (2x - 3)(x^2 - 1)$$
$$= 2x^3 - 3x^2 - 2x + 3 \qquad \text{Product}$$
$$\frac{f(x)}{g(x)} = \frac{2x - 3}{x^2 - 1}, \quad x \neq \pm 1 \qquad \text{Quotient}$$

The domain of an **arithmetic combination** of functions f and g consists of all real numbers that are common to the domains of f and g. In the case of the quotient

$$\frac{f(x)}{g(x)}$$

there is the further restriction that $g(x) \neq 0$.

> ### Sum, Difference, Product, and Quotient of Functions
>
> Let f and g be two functions with overlapping domains. Then, for all x common to both domains, the sum, difference, product, and quotient of f and g are defined as follows.
>
> **1.** Sum: $\quad (f + g)(x) = f(x) + g(x)$
>
> **2.** Difference: $(f - g)(x) = f(x) - g(x)$
>
> **3.** Product: $\quad (fg)(x) = f(x) \cdot g(x)$
>
> **4.** Quotient: $\left(\dfrac{f}{g}\right)(x) = \dfrac{f(x)}{g(x)}, \quad g(x) \neq 0$

What you should learn

- Add, subtract, multiply, and divide functions.
- Find compositions of one function with another function.
- Use combinations of functions to model and solve real-life problems.

Why you should learn it

You can model some situations by combining functions. For instance, in Exercise 79 on page 138, you will model the stopping distance of a car by combining the driver's reaction time with the car's braking distance.

Example 1 Finding the Sum of Two Functions

Given $f(x) = 2x + 1$ and $g(x) = x^2 + 2x - 1$, find $(f + g)(x)$. Then evaluate the sum when $x = 2$.

Solution

$$(f + g)(x) = f(x) + g(x)$$
$$= (2x + 1) + (x^2 + 2x - 1)$$
$$= x^2 + 4x$$

When $x = 2$, the value of this sum is $(f + g)(2) = 2^2 + 4(2) = 12$.

✓**CHECKPOINT** Now try Exercise 13(a).

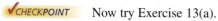

Example 2   Finding the Difference of Two Functions

Given $f(x) = 2x + 1$ and $g(x) = x^2 + 2x - 1$, find $(f - g)(x)$. Then evaluate the difference when $x = 2$.

Algebraic Solution

The difference of the functions f and g is

$$(f - g)(x) = f(x) - g(x)$$

$$= (2x + 1) - (x^2 + 2x - 1)$$

$$= -x^2 + 2.$$

When $x = 2$, the value of this difference is

$$(f - g)(2) = -(2)^2 + 2$$

$$= -2.$$

✔CHECKPOINT   Now try Exercise 13(b).

Graphical Solution

Enter the functions in a graphing utility (see Figure 1.60). Then graph the difference of the two functions, y_3, as shown in Figure 1.61.

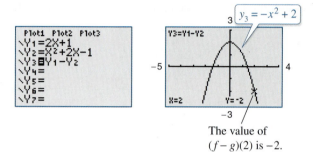

The value of $(f - g)(2)$ is -2.

Figure 1.60   **Figure 1.61**

Example 3   Finding the Product of Two Functions

Given $f(x) = x^2$ and $g(x) = x - 3$, find $(fg)(x)$. Then evaluate the product when $x = 4$.

Solution

$$(fg)(x) = f(x)g(x)$$

$$= (x^2)(x - 3)$$

$$= x^3 - 3x^2$$

When $x = 4$, the value of this product is

$$(fg)(4) = 4^3 - 3(4)^2 = 16.$$

✔CHECKPOINT   Now try Exercise 13(c).

In Examples 1–3, both f and g have domains that consist of all real numbers. So, the domain of both $(f + g)$ and $(f - g)$ is also the set of all real numbers. Remember that any restrictions on the domains of f or g must be considered when forming the sum, difference, product, or quotient of f and g. For instance, the domain of $f(x) = 1/x$ is all $x \neq 0$, and the domain of $g(x) = \sqrt{x}$ is $[0, \infty)$. This implies that the domain of $(f + g)$ is $(0, \infty)$.

Example 4   Finding the Quotient of Two Functions

Given $f(x) = \sqrt{x}$ and $g(x) = \sqrt{4 - x^2}$, find $(f/g)(x)$. Then find the domain of f/g.

Solution

$$\left(\frac{f}{g}\right)(x) = \frac{f(x)}{g(x)} = \frac{\sqrt{x}}{\sqrt{4 - x^2}}$$

The domain of f is $[0, \infty)$ and the domain of g is $[-2, 2]$. The intersection of these domains is $[0, 2]$. So, the domain of f/g is $[0, 2)$.

✔CHECKPOINT   Now try Exercise 13(d).

Compositions of Functions

Another way of combining two functions is to form the **composition** of one with the other. For instance, when $f(x) = x^2$ and $g(x) = x + 1$, the composition of f with g is

$$f(g(x)) = f(x + 1)$$

$$= (x + 1)^2.$$

This composition is denoted as $f \circ g$ and is read as "f composed with g."

Definition of Composition of Two Functions

The **composition** of the function f with the function g is

$$(f \circ g)(x) = f(g(x)).$$

The domain of $f \circ g$ is the set of all x in the domain of g such that $g(x)$ is in the domain of f. (See Figure 1.62.)

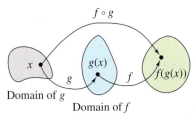

Figure 1.62

Example 5   Forming the Composition of f with g

Find $(f \circ g)(x)$ for

$$f(x) = \sqrt{x}, x \geq 0, \qquad \text{and} \qquad g(x) = x - 1, x \geq 1.$$

If possible, find $(f \circ g)(2)$ and $(f \circ g)(0)$.

Solution

The composition of f with g is

$$(f \circ g)(x) = f(g(x)) \qquad \text{Definition of } f \circ g$$

$$= f(x - 1) \qquad \text{Definition of } g(x)$$

$$= \sqrt{x - 1}, \quad x \geq 1. \qquad \text{Definition of } f(x)$$

The domain of $f \circ g$ is $[1, \infty)$. (See Figure 1.63.) So,

$$(f \circ g)(2) = \sqrt{2 - 1} = 1$$

is defined, but $(f \circ g)(0)$ is not defined because 0 is not in the domain of $f \circ g$.

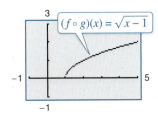

$(f \circ g)(x) = \sqrt{x - 1}$

Figure 1.63

✔CHECKPOINT   Now try Exercise 41.

Explore the Concept

Let $f(x) = x + 2$ and $g(x) = 4 - x^2$. Are the compositions $f \circ g$ and $g \circ f$ equal? You can use your graphing utility to answer this question by entering and graphing the following functions.

$$y_1 = (4 - x^2) + 2$$

$$y_2 = 4 - (x + 2)^2$$

What do you observe? Which function represents $f \circ g$ and which represents $g \circ f$?

The composition of f with g is generally not the same as the composition of g with f. This is illustrated in Example 6.

Example 6   Compositions of Functions

Given $f(x) = x + 2$ and $g(x) = 4 - x^2$, evaluate

(a) $(f \circ g)(x)$ and   (b) $(g \circ f)(x)$

when $x = 0$ and 1.

Algebraic Solution

a. $(f \circ g)(x) = f(g(x))$ Definition of $f \circ g$

$\qquad = f(4 - x^2)$ Definition of $g(x)$

$\qquad = (4 - x^2) + 2$ Definition of $f(x)$

$\qquad = -x^2 + 6$

$(f \circ g)(0) = -0^2 + 6 = 6$

$(f \circ g)(1) = -1^2 + 6 = 5$

b. $(g \circ f)(x) = g(f(x))$ Definition of $g \circ f$

$\qquad = g(x + 2)$ Definition of $f(x)$

$\qquad = 4 - (x + 2)^2$ Definition of $g(x)$

$\qquad = 4 - (x^2 + 4x + 4)$

$\qquad = -x^2 - 4x$

$(g \circ f)(0) = -0^2 - 4(0) = 0$

$(g \circ f)(1) = -1^2 - 4(1) = -5$

Note that $f \circ g \neq g \circ f$.

✓CHECKPOINT   Now try Exercise 43.

Graphical Solution

a. and b. Enter $y_1 = f(x)$, $y_2 = g(x)$, $y_3 = (f \circ g)(x)$, and $y_4 = (g \circ f)(x)$, as shown in Figure 1.64. Then use the *table* feature to find the desired function values (see Figure 1.65).

Figure 1.64

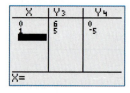

Figure 1.65

From the table you can see that $f \circ g \neq g \circ f$.

Example 7   Finding the Domain of a Composite Function

Find the domain of $f \circ g$ for the functions given by

$f(x) = x^2 - 9$ and $g(x) = \sqrt{9 - x^2}$.

Algebraic Solution

The composition of the functions is as follows.

$(f \circ g)(x) = f(g(x))$

$\qquad = f\left(\sqrt{9 - x^2}\right)$

$\qquad = \left(\sqrt{9 - x^2}\right)^2 - 9$

$\qquad = 9 - x^2 - 9$

$\qquad = -x^2$

From this, it might appear that the domain of the composition is the set of all real numbers. This, however, is not true. Because the domain of f is the set of all real numbers and the domain of g is $[-3, 3]$, the domain of $f \circ g$ is $[-3, 3]$.

✓CHECKPOINT   Now try Exercise 45.

Graphical Solution

The x-coordinates of points on the graph extend from -3 to 3. So, the domain of $f \circ g$ is $[-3, 3]$.

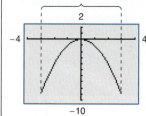

Example 8   A Case in Which $f \circ g = g \circ f$

Given

$$f(x) = 2x + 3 \text{ and } g(x) = \tfrac{1}{2}(x - 3)$$

find each composition.

a. $(f \circ g)(x)$

b. $(g \circ f)(x)$

Solution

a. $(f \circ g)(x) = f(g(x))$

$$= f\left(\frac{1}{2}(x - 3)\right)$$

$$= 2\left[\frac{1}{2}(x - 3)\right] + 3$$

$$= x - 3 + 3$$

$$= x$$

b. $(g \circ f)(x) = g(f(x))$

$$= g(2x + 3)$$

$$= \frac{1}{2}\Big[(2x + 3) - 3\Big]$$

$$= \frac{1}{2}(2x)$$

$$= x$$

✔CHECKPOINT   Now try Exercise 57.

In Examples 5–8, you formed the composition of two given functions. In calculus, it is also important to be able to identify two functions that make up a given composite function. Basically, to "decompose" a composite function, look for an "inner" and an "outer" function.

Example 9   Identifying a Composite Function

Write the function

$$h(x) = \frac{1}{(x - 2)^2}$$

as a composition of two functions.

Solution

One way to write h as a composition of two functions is to take the inner function to be $g(x) = x - 2$ and the outer function to be

$$f(x) = \frac{1}{x^2} = x^{-2}.$$

Then you can write

$$h(x) = \frac{1}{(x - 2)^2} = (x - 2)^{-2} = f(x - 2) = f(g(x)).$$

✔CHECKPOINT   Now try Exercise 75.

Study Tip

In Example 8, note that the two composite functions $f \circ g$ and $g \circ f$ are equal, and both represent the identity function. That is, $(f \circ g)(x) = x$ and $(g \circ f)(x) = x$. You will study this special case in the next section.

Explore the Concept

Write each function as a composition of two functions.

a. $h(x) = |x^3 - 2|$

b. $r(x) = |x^3| - 2$

What do you notice about the inner and outer functions?

Application

Example 10   Bacteria Count

The number N of bacteria in a refrigerated petri dish is given by

$$N(T) = 20T^2 - 80T + 500, \quad 2 \le T \le 14$$

where T is the temperature of the petri dish (in degrees Celsius). When the petri dish is removed from refrigeration, the temperature of the petri dish is given by

$$T(t) = 4t + 2, \quad 0 \le t \le 3$$

where t is the time (in hours).

a. Find the composition $N(T(t))$ and interpret its meaning in context.

b. Find the number of bacteria in the petri dish when $t = 2$ hours.

c. Find the time when the bacteria count reaches 2000.

Microbiologist

Solution

a. $N(T(t)) = 20(4t + 2)^2 - 80(4t + 2) + 500$

$$= 20(16t^2 + 16t + 4) - 320t - 160 + 500$$

$$= 320t^2 + 320t + 80 - 320t - 160 + 500$$

$$= 320t^2 + 420$$

The composite function $N(T(t))$ represents the number of bacteria as a function of the amount of time the petri dish has been out of refrigeration.

b. When $t = 2$, the number of bacteria is

$$N = 320(2)^2 + 420 = 1280 + 420 = 1700.$$

c. The bacteria count will reach $N = 2000$ when $320t^2 + 420 = 2000$. You can solve this equation for t algebraically as follows.

$$320t^2 + 420 = 2000$$

$$320t^2 = 1580$$

$$t^2 = \frac{79}{16}$$

$$t = \frac{\sqrt{79}}{4} \quad \Longrightarrow \quad t \approx 2.22 \text{ hours}$$

So, the count will reach 2000 when $t \approx 2.22$ hours. Note that the negative value is rejected because it is not in the domain of the composite function. To confirm your solution, graph the equation $N = 320t^2 + 420$, as shown in Figure 1.66. Then use the *zoom* and *trace* features to approximate $N = 2000$ when $t \approx 2.22$, as shown in Figure 1.67.

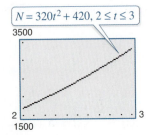

Figure 1.66

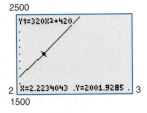

Figure 1.67

✓**CHECKPOINT**   Now try Exercise 85.

1.6   Exercises

See www.CalcChat.com for worked-out solutions to odd-numbered exercises.
For instructions on how to use a graphing utility, see Appendix A.

Vocabulary and Concept Check

In Exercises 1–4, fill in the blank(s).

1. Two functions f and g can be combined by the arithmetic operations of _____ , _____ , _____ , and _____ to create new functions.

2. The _____ of the function f with the function g is $(f \circ g)(x) = f(g(x))$.

3. The domain of $f \circ g$ is the set of all x in the domain of g such that _____ is in the domain of f.

4. To decompose a composite function, look for an _____ and an _____ function.

5. Given $f(x) = x^2 + 1$ and $(fg)(x) = 2x(x^2 + 1)$, what is $g(x)$?

6. Given $(f \circ g)(x) = f(x^2 + 1)$, what is $g(x)$?

Procedures and Problem Solving

Finding the Sum of Two Functions In Exercises 7–10, use the graphs of f and g to graph $h(x) = (f + g)(x)$. To print an enlarged copy of the graph, go to the website *www.mathgraphs.com*.

7.
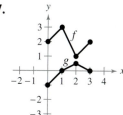

8.
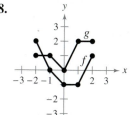

9.
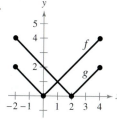

10.
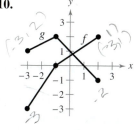

Finding Arithmetic Combinations of Functions In Exercises 11–18, find (a) $(f + g)(x)$, (b) $(f - g)(x)$, (c) $(fg)(x)$, and (d) $(f/g)(x)$. What is the domain of f/g?

11. $f(x) = x + 3$, $g(x) = x - 3$

12. $f(x) = 2x - 5$, $g(x) = 1 - x$

✓ 13. $f(x) = x^2$, $g(x) = 1 - x$

14. $f(x) = 2x - 5$, $g(x) = 5$

15. $f(x) = x^2 + 5$, $g(x) = \sqrt{1 - x}$

16. $f(x) = \sqrt{x^2 - 4}$, $g(x) = \dfrac{x^2}{x^2 + 1}$

17. $f(x) = \dfrac{1}{x}$, $g(x) = \dfrac{1}{x^2}$

18. $f(x) = \dfrac{x}{x + 1}$, $g(x) = x^3$

Evaluating an Arithmetic Combination of Functions In Exercises 19–32, evaluate the indicated function for $f(x) = x^2 - 1$ and $g(x) = x - 2$ algebraically. If possible, use a graphing utility to verify your answer.

19. $(f + g)(3)$
20. $(f - g)(-2)$
21. $(f - g)(0)$
22. $(f + g)(1)$
23. $(fg)(6)$
24. $(fg)(-4)$
25. $(f/g)(-5)$
26. $(f/g)(0)$
27. $(f - g)(2t)$
28. $(f + g)(t - 4)$
29. $(fg)(-5t)$
30. $(fg)(3t^2)$
31. $(f/g)(-t)$
32. $(f/g)(t + 2)$

Graphing an Arithmetic Combination of Functions In Exercises 33–36, use a graphing utility to graph the functions f, g, and h in the same viewing window.

33. $f(x) = \frac{1}{2}x$, $g(x) = x - 1$, $h(x) = f(x) + g(x)$

34. $f(x) = \frac{1}{3}x$, $g(x) = -x + 4$, $h(x) = f(x) - g(x)$

35. $f(x) = x^2$, $g(x) = -2x$, $h(x) = f(x) \cdot g(x)$

36. $f(x) = 4 - x^2$, $g(x) = x$, $h(x) = f(x)/g(x)$

Graphing a Sum of Functions In Exercises 37–40, use a graphing utility to graph f, g, and $f + g$ in the same viewing window. Which function contributes most to the magnitude of the sum when $0 \le x \le 2$? Which function contributes most to the magnitude of the sum when $x > 6$?

37. $f(x) = 3x$, $g(x) = -\dfrac{x^3}{10}$

38. $f(x) = \dfrac{x}{2}$, $g(x) = \sqrt{x}$

39. $f(x) = 3x + 2$, $g(x) = -\sqrt{x + 5}$

40. $f(x) = x^2 - \frac{1}{2}$, $g(x) = -3x^2 - 1$

Compositions of Functions   In Exercises 41–44, find (a) $f \circ g$, (b) $g \circ f$, and, if possible, (c) $(f \circ g)(0)$.

✓ **41.** $f(x) = x^2$, $g(x) = x - 1$

42. $f(x) = \sqrt[3]{x - 1}$, $g(x) = x^3 + 1$

✓ **43.** $f(x) = 3x + 5$, $g(x) = 5 - x$

44. $f(x) = x^3$, $g(x) = \dfrac{1}{x}$

Finding the Domain of a Composite Function   In Exercises 45–54, determine the domains of (a) f, (b) g, and (c) $f \circ g$. Use a graphing utility to verify your results.

✓ **45.** $f(x) = \sqrt{x + 4}$, $g(x) = x^2$

46. $f(x) = \sqrt{x + 3}$, $g(x) = \dfrac{x}{2}$

47. $f(x) = x^2 + 1$, $g(x) = \sqrt{x}$

48. $f(x) = x^{1/4}$, $g(x) = x^4$

49. $f(x) = \dfrac{1}{x}$, $g(x) = x + 3$

50. $f(x) = \dfrac{1}{x}$, $g(x) = \dfrac{1}{2x}$

51. $f(x) = |x - 4|$, $g(x) = 3 - x$

52. $f(x) = \dfrac{2}{|x|}$, $g(x) = x - 1$

53. $f(x) = x + 2$, $g(x) = \dfrac{1}{x^2 - 4}$

54. $f(x) = \dfrac{3}{x^2 - 1}$, $g(x) = x + 1$

Determining Whether $f \circ g = g \circ f$ In Exercises 55–60, (a) find $f \circ g$, $g \circ f$, and the domain of $f \circ g$. (b) Use a graphing utility to graph $f \circ g$ and $g \circ f$. Determine whether $f \circ g = g \circ f$.

55. $f(x) = \sqrt{x + 4}$, $g(x) = x^2$

56. $f(x) = \sqrt[3]{x + 1}$, $g(x) = x^3 - 1$

✓ **57.** $f(x) = \frac{1}{3}x - 3$, $g(x) = 3x + 9$

58. $f(x) = \sqrt{x}$, $g(x) = \sqrt{x}$

59. $f(x) = x^{2/3}$, $g(x) = x^6$

60. $f(x) = |x|$, $g(x) = -x^2 + 1$

Determining Whether $f \circ g = g \circ f$ In Exercises 61–66, (a) find $(f \circ g)(x)$ and $(g \circ f)(x)$, (b) determine algebraically whether $(f \circ g)(x) = (g \circ f)(x)$, and (c) use a graphing utility to complete a table of values for the two compositions to confirm your answer to part (b).

61. $f(x) = 5x + 4$, $g(x) = 4 - x$

62. $f(x) = \frac{1}{4}(x - 1)$, $g(x) = 4x + 1$

63. $f(x) = \sqrt{x + 6}$, $g(x) = x^2 - 5$

64. $f(x) = x^3 - 4$, $g(x) = \sqrt[3]{x + 10}$

65. $f(x) = |x|$, $g(x) = 2x^3$

66. $f(x) = \dfrac{6}{3x - 5}$, $g(x) = -x$

Evaluating Combinations of Functions   In Exercises 67–70, use the graphs of f and g to evaluate the functions.

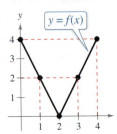
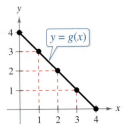

67. (a) $(f + g)(3)$ (b) $(f/g)(2)$

68. (a) $(f - g)(1)$ (b) $(fg)(4)$

69. (a) $(f \circ g)(2)$ (b) $(g \circ f)(2)$

70. (a) $(f \circ g)(1)$ (b) $(g \circ f)(3)$

𝑓 **Identifying a Composite Function**   In Exercises 71–78, find two functions f and g such that $(f \circ g)(x) = h(x)$. (There are many correct answers.)

71. $h(x) = (2x + 1)^2$ **72.** $h(x) = (1 - x)^3$

73. $h(x) = \sqrt[3]{x^2 - 4}$ **74.** $h(x) = \sqrt{9 - x}$

✓ **75.** $h(x) = \dfrac{1}{x + 2}$

76. $h(x) = \dfrac{4}{(5x + 2)^2}$

77. $h(x) = (x + 4)^2 + 2(x + 4)$

78. $h(x) = (x + 3)^{3/2} + 4(x + 3)^{1/2}$

79. *Why you should learn it* (p. 131)   The research and development department of an automobile manufacturer has determined that when required to stop quickly to avoid an accident, the distance (in feet) a car travels during the driver's reaction time is given by

$$R(x) = \tfrac{3}{4}x$$

where x is the speed of the car in miles per hour. The distance (in feet) traveled while the driver is braking is given by

$$B(x) = \tfrac{1}{15}x^2.$$

(a) Find the function that represents the total stopping distance T.

(b) Use a graphing utility to graph the functions R, B, and T in the same viewing window for $0 \le x \le 60$.

(c) Which function contributes most to the magnitude of the sum at higher speeds? Explain.

80. MODELING DATA

The table shows the total amounts (in billions of dollars) of private expenditures on health services and supplies in the United States (including Puerto Rico) for the years 1997 through 2007. The variables y_1, y_2, and y_3 represent out-of-pocket payments, insurance premiums, and other types of payments, respectively. (Source: U.S. Centers for Medicare and Medicaid Services)

Year	y_1	y_2	y_3
1997	162	359	52
1998	175	385	56
1999	184	417	59
2000	193	455	58
2001	200	498	58
2002	211	551	59
2003	225	604	65
2004	235	646	66
2005	247	690	70
2006	255	731	75
2007	269	775	80

The data are approximated by the following models, where t represents the year, with $t = 7$ corresponding to 1997.

$y_1 = 10.5t + 88$

$y_2 = 0.66t^2 + 27.6t + 123$

$y_3 = 0.23t^2 - 3.0t + 64$

(a) Use the models and the *table* feature of a graphing utility to create a table showing the values of y_1, y_2, and y_3 for each year from 1997 through 2007. Compare these models with the original data. Are the models a good fit? Explain.

(b) Use the graphing utility to graph y_1, y_2, y_3, and $y_T = y_1 + y_2 + y_3$ in the same viewing window. What does the function y_T represent?

81. Geometry A square concrete foundation was prepared as a base for a large cylindrical gasoline tank (see figure).

(a) Write the radius r of the tank as a function of the length x of the sides of the square.

(b) Write the area A of the circular base of the tank as a function of the radius r.

(c) Find and interpret $(A \circ r)(x)$.

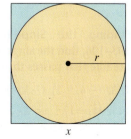

82. Geometry A pebble is dropped into a calm pond, causing ripples in the form of concentric circles. The radius (in feet) of the outermost ripple is given by $r(t) = 0.6t$, where t is the time (in seconds) after the pebble strikes the water. The area of the circle is given by $A(r) = \pi r^2$. Find and interpret $(A \circ r)(t)$.

83. Business A company owns two retail stores. The annual sales (in thousands of dollars) of the stores each year from 2004 through 2010 can be approximated by the models

$$S_1 = 830 + 1.2t^2 \quad \text{and} \quad S_2 = 390 + 75.4t$$

where t is the year, with $t = 4$ corresponding to 2004.

(a) Write a function T that represents the total annual sales of the two stores.

(b) Use a graphing utility to graph S_1, S_2, and T in the same viewing window.

84. Business The annual cost C (in thousands of dollars) and revenue R (in thousands of dollars) for a company each year from 2004 through 2010 can be approximated by the models

$$C = 260 - 8t + 1.6t^2 \quad \text{and} \quad R = 320 + 2.8t$$

where t is the year, with $t = 4$ corresponding to 2004.

(a) Write a function P that represents the annual profits of the company.

(b) Use a graphing utility to graph C, R, and P in the same viewing window.

✓ 85. Biology The number of bacteria in a refrigerated food product is given by

$$N(T) = 10T^2 - 20T + 600, \quad 1 \le T \le 20$$

where T is the temperature of the food in degrees Celsius. When the food is removed from the refrigerator, the temperature of the food is given by

$$T(t) = 2t + 1$$

where t is the time in hours.

(a) Find the composite function $N(T(t))$ or $(N \circ T)(t)$ and interpret its meaning in the context of the situation.

(b) Find $(N \circ T)(6)$ and interpret its meaning.

(c) Find the time when the bacteria count reaches 800.

86. Environmental Science The spread of a contaminant is increasing in a circular pattern on the surface of a lake. The radius of the contaminant can be modeled by $r(t) = 5.25\sqrt{t}$, where r is the radius in meters and t is time in hours since contamination.

(a) Find a function that gives the area A of the circular leak in terms of the time t since the spread began.

(b) Find the size of the contaminated area after 36 hours.

(c) Find when the size of the contaminated area is 6250 square meters.

87. Air Traffic Control   An air traffic controller spots two planes flying at the same altitude. Their flight paths form a right angle at point P. One plane is 150 miles from point P and is moving at 450 miles per hour. The other plane is 200 miles from point P and is moving at 450 miles per hour. Write the distance s between the planes as a function of time t.

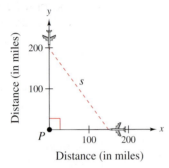

Distance (in miles)

88. Marketing   The suggested retail price of a new car is p dollars. The dealership advertised a factory rebate of $1200 and an 8% discount.

(a) Write a function R in terms of p giving the cost of the car after receiving the rebate from the factory.

(b) Write a function S in terms of p giving the cost of the car after receiving the dealership discount.

(c) Form the composite functions $(R \circ S)(p)$ and $(S \circ R)(p)$ and interpret each.

(d) Find $(R \circ S)(18{,}400)$ and $(S \circ R)(18{,}400)$. Which yields the lower cost for the car? Explain.

Conclusions

True or False?   **In Exercises 89 and 90, determine whether the statement is true or false. Justify your answer.**

89. A function that represents the graph of $f(x) = x^2$ shifted three units to the right is $f(g(x))$, where $g(x) = x + 3$.

90. Given two functions $f(x)$ and $g(x)$, you can calculate $(f \circ g)(x)$ if and only if the range of g is a subset of the domain of f.

Exploration   In Exercises 91 and 92, three siblings are of three different ages. The oldest is twice the age of the middle sibling, and the middle sibling is six years older than one-half the age of the youngest.

91. (a) Write a composite function that gives the oldest sibling's age in terms of the youngest. Explain how you arrived at your answer.

(b) The oldest sibling is 16 years old. Find the ages of the other two siblings.

92. (a) Write a composite function that gives the youngest sibling's age in terms of the oldest. Explain how you arrived at your answer.

(b) The youngest sibling is two years old. Find the ages of the other two siblings.

93. Proof   Prove that the product of two odd functions is an even function, and that the product of two even functions is an even function.

94. Proof   Use examples to hypothesize whether the product of an odd function and an even function is even or odd. Then prove your hypothesis.

95. Proof   Given a function f, prove that $g(x)$ is even and $h(x)$ is odd, where $g(x) = \frac{1}{2}[f(x) + f(-x)]$ and $h(x) = \frac{1}{2}[f(x) - f(-x)]$.

96. (a) Use the result of Exercise 95 to prove that any function can be written as a sum of even and odd functions. (*Hint:* Add the two equations in Exercise 95.)

(b) Use the result of part (a) to write each function as a sum of even and odd functions.

$$f(x) = x^2 - 2x + 1, \quad g(x) = \frac{1}{x+1}$$

97. Exploration   The function in Example 9 can be decomposed in other ways. For which of the following pairs of functions is $h(x) = \dfrac{1}{(x-2)^2}$ equal to $f(g(x))$?

(a) $g(x) = \dfrac{1}{x-2}$ and $f(x) = x^2$

(b) $g(x) = x^2$ and $f(x) = \dfrac{1}{x-2}$

(c) $g(x) = (x-2)^2$ and $f(x) = \dfrac{1}{x}$

98. CAPSTONE   Consider the functions $f(x) = x^2$ and $g(x) = \sqrt{x}$. Describe the restrictions that need to be made on the domains of f and g so that $f(g(x)) = g(f(x))$.

Cumulative Mixed Review

Evaluating an Equation   In Exercises 99–102, find three points that lie on the graph of the equation. (There are many correct answers.)

99. $y = -x^2 + x - 5$

100. $y = \frac{1}{5}x^3 - 4x^2 + 1$

101. $x^2 + y^2 = 24$

102. $y = \dfrac{x}{x^2 - 5}$

Finding the Slope-Intercept Form   In Exercises 103–106, find the slope-intercept form of the equation of the line that passes through the two points.

103. $(-4, -2), (-3, 8)$

104. $(1, 5), (-8, 2)$

105. $\left(\frac{3}{2}, -1\right), \left(-\frac{1}{3}, 4\right)$

106. $(0, 1.1), (-4, 3.1)$

1.7   Inverse Functions

Inverse Functions

What you should learn

- Find inverse functions informally and verify that two functions are inverse functions of each other.
- Use graphs of functions to decide whether functions have inverse functions.
- Determine whether functions are one-to-one.
- Find inverse functions algebraically.

Why you should learn it

Inverse functions can be helpful in further exploring how two variables relate to each other. For example, in Exercise 115 on page 150, you will use inverse functions to find the European shoe sizes from the corresponding U.S. shoe sizes.

Recall from Section 1.3 that a function can be represented by a set of ordered pairs. For instance, the function $f(x) = x + 4$ from the set $A = \{1, 2, 3, 4\}$ to the set $B = \{5, 6, 7, 8\}$ can be written as follows.

$$f(x) = x + 4: \ \{(1, 5), (2, 6), (3, 7), (4, 8)\}$$

In this case, by interchanging the first and second coordinates of each of these ordered pairs, you can form the **inverse function** of f, which is denoted by f^{-1}. It is a function from the set B to the set A, and can be written as follows.

$$f^{-1}(x) = x - 4: \ \{(5, 1), (6, 2), (7, 3), (8, 4)\}$$

Note that the domain of f is equal to the range of f^{-1}, and vice versa, as shown in Figure 1.68. Also note that the functions f and f^{-1} have the effect of "undoing" each other. In other words, when you form the composition of f with f^{-1} or the composition of f^{-1} with f, you obtain the identity function.

$$f(f^{-1}(x)) = f(x - 4) = (x - 4) + 4 = x$$

$$f^{-1}(f(x)) = f^{-1}(x + 4) = (x + 4) - 4 = x$$

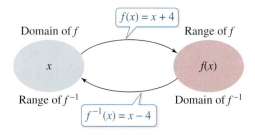

Domain of f $f(x) = x + 4$ Range of f

x $f(x)$

Range of f^{-1} $f^{-1}(x) = x - 4$ Domain of f^{-1}

Figure 1.68

Example 1   Finding Inverse Functions Informally

Find the inverse function of

$$f(x) = 4x.$$

Then verify that both $f(f^{-1}(x))$ and $f^{-1}(f(x))$ are equal to the identity function.

Solution

The function f *multiplies* each input by 4. To "undo" this function, you need to *divide* each input by 4. So, the inverse function of $f(x) = 4x$ is given by

$$f^{-1}(x) = \frac{x}{4}.$$

You can verify that both $f(f^{-1}(x))$ and $f^{-1}(f(x))$ are equal to the identity function as follows.

$$f(f^{-1}(x)) = f\left(\frac{x}{4}\right) = 4\left(\frac{x}{4}\right) = x$$

$$f^{-1}(f(x)) = f^{-1}(4x) = \frac{4x}{4} = x$$

 CHECKPOINT   Now try Exercise 7.

Don't be confused by the use of the exponent -1 to denote the inverse function f^{-1}. In this text, whenever f^{-1} is written, it always refers to the inverse function of the function f and not to the reciprocal of $f(x)$, which is given by

$$\frac{1}{f(x)}.$$

Example 2 Finding Inverse Functions Informally

Find the inverse function of $f(x) = x - 6$. Then verify that both $f(f^{-1}(x))$ and $f^{-1}(f(x))$ are equal to the identity function.

Solution

The function f subtracts 6 from each input. To "undo" this function, you need to *add* 6 to each input. So, the inverse function of $f(x) = x - 6$ is given by

$$f^{-1}(x) = x + 6.$$

You can verify that both $f(f^{-1}(x))$ and $f^{-1}(f(x))$ are equal to the identity function as follows.

$$f(f^{-1}(x)) = f(x + 6) = (x + 6) - 6 = x$$

$$f^{-1}(f(x)) = f^{-1}(x - 6) = (x - 6) + 6 = x$$

✓CHECKPOINT   Now try Exercise 9.

A table of values can help you understand inverse functions. For instance, the first table below shows several values of the function in Example 2. Interchange the rows of this table to obtain values of the inverse function.

x	-2	-1	0	1	2
$f(x)$	-8	-7	-6	-5	-4

x	-8	-7	-6	-5	-4
$f^{-1}(x)$	-2	-1	0	1	2

In the table at the left, each output is 6 less than the input, and in the table at the right, each output is 6 more than the input.

The formal definition of an inverse function is as follows.

Definition of Inverse Function

Let f and g be two functions such that

$$f(g(x)) = x \qquad \text{for every } x \text{ in the domain of } g$$

and

$$g(f(x)) = x \qquad \text{for every } x \text{ in the domain of } f.$$

Under these conditions, the function g is the **inverse function** of the function f. The function g is denoted by f^{-1} (read "f-inverse"). So,

$$f(f^{-1}(x)) = x \qquad \text{and} \qquad f^{-1}(f(x)) = x.$$

The domain of f must be equal to the range of f^{-1}, and the range of f must be equal to the domain of f^{-1}.

If the function g is the inverse function of the function f, then it must also be true that the function f is the inverse function of the function g. For this reason, you can say that the functions f and g are *inverse functions of each other*.

Example 3 Verifying Inverse Functions Algebraically

Show that the functions are inverse functions of each other.

$$f(x) = 2x^3 - 1 \quad \text{and} \quad g(x) = \sqrt[3]{\frac{x+1}{2}}$$

Solution

$$f(g(x)) = f\left(\sqrt[3]{\frac{x+1}{2}}\right)$$

$$= 2\left(\sqrt[3]{\frac{x+1}{2}}\right)^3 - 1$$

$$= 2\left(\frac{x+1}{2}\right) - 1$$

$$= x + 1 - 1$$

$$= x$$

$$g(f(x)) = g(2x^3 - 1)$$

$$= \sqrt[3]{\frac{(2x^3 - 1) + 1}{2}}$$

$$= \sqrt[3]{\frac{2x^3}{2}}$$

$$= \sqrt[3]{x^3}$$

$$= x$$

✔CHECKPOINT Now try Exercise 19.

Technology Tip

Most graphing utilities can graph $y = x^{1/3}$ in two ways:

$$y_1 = x \wedge (1/3) \quad \text{or}$$

$$y_1 = \sqrt[3]{x}.$$

On some graphing utilities, you may not be able to obtain the complete graph of $y = x^{2/3}$ by entering $y_1 = x \wedge (2/3)$. If not, you should use

$$y_1 = (x \wedge (1/3))^2 \quad \text{or}$$

$$y_1 = \sqrt[3]{x^2}.$$

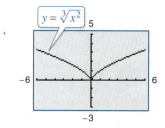

Example 4 Verifying Inverse Functions Algebraically

Which of the functions is the inverse function of $f(x) = \dfrac{5}{x-2}$?

$$g(x) = \frac{x-2}{5} \quad \text{or} \quad h(x) = \frac{5}{x} + 2$$

Solution

By forming the composition of f with g, you have

$$f(g(x)) = f\left(\frac{x-2}{5}\right) = \frac{5}{\left(\dfrac{x-2}{5}\right) - 2} = \frac{25}{x - 12} \neq x.$$

Because this composition is not equal to the identity function x, it follows that g *is not* the inverse function of f. By forming the composition of f with h, you have

$$f(h(x)) = f\left(\frac{5}{x} + 2\right) = \frac{5}{\left(\dfrac{5}{x} + 2\right) - 2} = \frac{5}{\left(\dfrac{5}{x}\right)} = x.$$

So, it appears that h is the inverse function of f. You can confirm this by showing that the composition of h with f is also equal to the identity function.

✔CHECKPOINT Now try Exercise 23.

The Graph of an Inverse Function

The graphs of a function f and its inverse function f^{-1} are related to each other in the following way. If the point

$$(a, b)$$

lies on the graph of f, then the point

$$(b, a)$$

must lie on the graph of f^{-1}, and vice versa. This means that the graph of f^{-1} is a reflection of the graph of f in the line $y = x$, as shown in Figure 1.69.

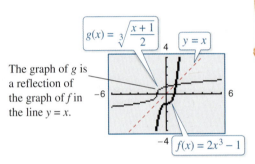

Figure 1.69

Technology Tip

Many graphing utilities have a built-in feature for drawing an inverse function. For instructions on how to use the *draw inverse* feature, see Appendix A; for specific keystrokes, go to this texbook's *Companion Website*.

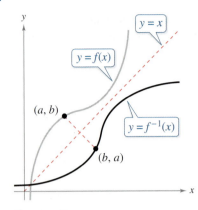

Example 5   Verifying Inverse Functions Graphically

Verify that the functions f and g from Example 3 are inverse functions of each other graphically.

Solution

From Figure 1.70, you can conclude that f and g are inverse functions of each other.

The graph of g is a reflection of the graph of f in the line $y = x$.

Figure 1.70

 Now try Exercise 33(b).

Example 6   Verifying Inverse Functions Numerically

Verify that the functions $f(x) = \dfrac{x - 5}{2}$ and $g(x) = 2x + 5$ are inverse functions of each other numerically.

Solution

You can verify that f and g are inverse functions of each other *numerically* by using a graphing utility. Enter $y_1 = f(x)$, $y_2 = g(x)$, $y_3 = f(g(x))$, and $y_4 = g(f(x))$, as shown in Figure 1.71. Then use the *table* feature to create a table (see Figure 1.72).

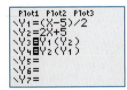

Figure 1.71   **Figure 1.72**

Note that the entries for x, y_3, and y_4 are the same. So, $f(g(x)) = x$ and $g(f(x)) = x$. You can conclude that f and g are inverse functions of each other.

 Now try Exercise 33(c).

The Existence of an Inverse Function

To have an inverse function, a function must be **one-to-one,** which means that no two elements in the domain of f correspond to the same element in the range of f.

> **Definition of a One-to-One Function**
>
> A function f is **one-to-one** when, for a and b in its domain, $f(a) = f(b)$ implies that $a = b$.

> **Existence of an Inverse Function**
>
> A function f has an inverse function f^{-1} if and only if f is one-to-one.

From its graph, it is easy to tell whether a function of x is one-to-one. Simply check to see that every horizontal line intersects the graph of the function at most once. This is called the **Horizontal Line Test.** For instance, Figure 1.73 shows the graph of $y = x^2$. On the graph, you can find a horizontal line that intersects the graph twice.

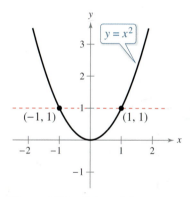

Figure 1.73 $f(x) = x^2$ is not one-to-one.

Two special types of functions that pass the Horizontal Line Test are those that are increasing or decreasing on their entire domains.

1. If f is *increasing* on its entire domain, then f is one-to-one.

2. If f is *decreasing* on its entire domain, then f is one-to-one.

Example 7   Testing for One-to-One Functions

Is the function $f(x) = \sqrt{x} + 1$ one-to-one?

Algebraic Solution

Let a and b be nonnegative real numbers with $f(a) = f(b)$.

$$\sqrt{a} + 1 = \sqrt{b} + 1 \qquad \text{Set } f(a) = f(b).$$
$$\sqrt{a} = \sqrt{b}$$
$$a = b$$

So, $f(a) = f(b)$ implies that $a = b$. You can conclude that f is one-to-one and *does* have an inverse function.

✔CHECKPOINT   Now try Exercise 67.

Graphical Solution

A horizontal line will intersect the graph at most once.

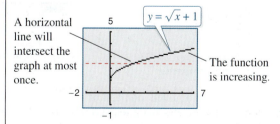

The function is increasing.

Figure 1.74

From Figure 1.74, you can conclude that f is one-to-one and *does* have an inverse function.

Finding Inverse Functions Algebraically

For simple functions, you can find inverse functions by inspection. For more complicated functions, however, it is best to use the following guidelines.

Finding an Inverse Function

1. Use the Horizontal Line Test to decide whether f has an inverse function.

2. In the equation for $f(x)$, replace $f(x)$ by y.

3. Interchange the roles of x and y, and solve for y.

4. Replace y by $f^{-1}(x)$ in the new equation.

5. Verify that f and f^{-1} are inverse functions of each other by showing that the domain of f is equal to the range of f^{-1}, the range of f is equal to the domain of f^{-1}, and $f(f^{-1}(x)) = x$ and $f^{-1}(f(x)) = x$.

What's Wrong?

You use a graphing utility to graph $y_1 = x^2$ and then use the *draw inverse* feature to conclude that $f(x) = x^2$ has an inverse function (see figure). What's wrong?

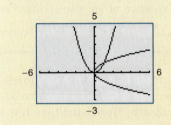

Example 8   Finding an Inverse Function Algebraically

Find the inverse function of

$$f(x) = \frac{5 - 3x}{2}.$$

Solution

The graph of f in Figure 1.75 passes the Horizontal Line Test. So, you know that f is one-to-one and has an inverse function.

$$f(x) = \frac{5 - 3x}{2} \qquad \text{Write original function.}$$

$$y = \frac{5 - 3x}{2} \qquad \text{Replace } f(x) \text{ by } y.$$

$$x = \frac{5 - 3y}{2} \qquad \text{Interchange } x \text{ and } y.$$

$$2x = 5 - 3y \qquad \text{Multiply each side by 2.}$$

$$3y = 5 - 2x \qquad \text{Isolate the } y\text{-term.}$$

$$y = \frac{5 - 2x}{3} \qquad \text{Solve for } y.$$

$$f^{-1}(x) = \frac{5 - 2x}{3} \qquad \text{Replace } y \text{ by } f^{-1}(x).$$

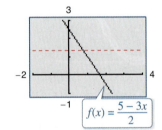

Figure 1.75

The domains and ranges of f and f^{-1} consist of all real numbers. Verify that $f(f^{-1}(x)) = x$ and $f^{-1}(f(x)) = x$.

✓CHECKPOINT   Now try Exercise 71.

A function f with an implied domain of all real numbers may not pass the Horizontal Line Test. In this case, the domain of f may be restricted so that f does have an inverse function. For instance, when the domain of $f(x) = x^2$ is restricted to the nonnegative real numbers, then f does have an inverse function.

Example 9   Finding an Inverse Function Algebraically

Find the inverse function of

$$f(x) = x^3 - 4.$$

Solution

The graph of f in Figure 1.76 passes the Horizontal Line Test. So, you know that f is one-to-one and has an inverse function.

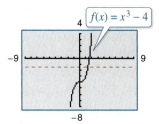

Figure 1.76

$$\begin{array}{ll} f(x) = x^3 - 4 & \text{Write original function.} \\ y = x^3 - 4 & \text{Replace } f(x) \text{ by } y. \\ x = y^3 - 4 & \text{Interchange } x \text{ and } y. \\ y^3 = x + 4 & \text{Isolate } y. \\ y = \sqrt[3]{x + 4} & \text{Solve for } y. \\ f^{-1}(x) = \sqrt[3]{x + 4} & \text{Replace } y \text{ by } f^{-1}(x). \end{array}$$

The domains and ranges of f and f^{-1} consist of all real numbers. You can verify that $f(f^{-1}(x)) = x$ and $f^{-1}(f(x)) = x$ as follows.

$$\begin{aligned} f(f^{-1}(x)) &= f\left(\sqrt[3]{x + 4}\right) & f^{-1}(f(x)) &= f^{-1}(x^3 - 4) \\ &= \left(\sqrt[3]{x + 4}\right)^3 - 4 & &= \sqrt[3]{(x^3 - 4) + 4} \\ &= x + 4 - 4 & &= \sqrt[3]{x^3} \\ &= x & &= x \end{aligned}$$

✔**CHECKPOINT**   Now try Exercise 73.

Example 10   Finding an Inverse Function Algebraically

Find the inverse function of

$$f(x) = \sqrt{2x - 3}.$$

Solution

The graph of f in Figure 1.77 passes the Horizontal Line Test. So, you know that f is one-to-one and has an inverse function.

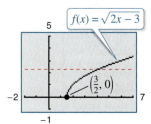

Figure 1.77

$$\begin{array}{ll} f(x) = \sqrt{2x - 3} & \text{Write original function.} \\ y = \sqrt{2x - 3} & \text{Replace } f(x) \text{ by } y. \\ x = \sqrt{2y - 3} & \text{Interchange } x \text{ and } y. \\ x^2 = 2y - 3 & \text{Square each side.} \\ 2y = x^2 + 3 & \text{Isolate } y. \\ y = \dfrac{x^2 + 3}{2} & \text{Solve for } y. \\ f^{-1}(x) = \dfrac{x^2 + 3}{2}, \quad x \geq 0 & \text{Replace } y \text{ by } f^{-1}(x). \end{array}$$

Note that the range of f is the interval $[0, \infty)$, which implies that the domain of f^{-1} is the interval $[0, \infty)$. Moreover, the domain of f is the interval $\left[\frac{3}{2}, \infty\right)$, which implies that the range of f^{-1} is the interval $\left[\frac{3}{2}, \infty\right)$. Verify that $f(f^{-1}(x)) = x$ and $f^{-1}(f(x)) = x$.

✔**CHECKPOINT**   Now try Exercise 77.

1.7 Exercises

See www.CalcChat.com for worked-out solutions to odd-numbered exercises.
For instructions on how to use a graphing utility, see Appendix A.

Vocabulary and Concept Check

In Exercises 1–4, fill in the blank(s).

1. If f and g are functions such that $f(g(x)) = x$ and $g(f(x)) = x$, then the function g is the _____ function of f, and is denoted by _____ .

2. The domain of f is the _____ of f^{-1}, and the _____ of f^{-1} is the range of f.

3. The graphs of f and f^{-1} are reflections of each other in the line _____ .

4. To have an inverse function, a function f must be _____ ; that is, $f(a) = f(b)$ implies $a = b$.

5. How many times can a horizontal line intersect the graph of a function that is one-to-one?

6. Can $(1, 4)$ and $(2, 4)$ be two ordered pairs of a one-to-one function?

Procedures and Problem Solving

Finding Inverse Functions Informally   In Exercises 7–14, find the inverse function of f informally. Verify that $f(f^{-1}(x)) = x$ and $f^{-1}(f(x)) = x$.

✓ 7. $f(x) = 6x$

8. $f(x) = \frac{1}{3}x$

✓ 9. $f(x) = x + 7$

10. $f(x) = x - 3$

11. $f(x) = 2x + 1$

12. $f(x) = (x - 1)/4$

13. $f(x) = \sqrt[3]{x}$

14. $f(x) = x^5$

Identifying Graphs of Inverse Functions   In Exercises 15–18, match the graph of the function with the graph of its inverse function. [The graphs of the inverse functions are labeled (a), (b), (c), and (d).]

(a)
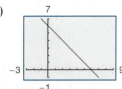

(b)
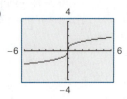

(c)
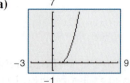

(d)
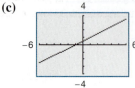

15.
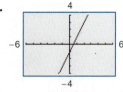

16.
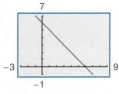

17.
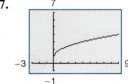

18.
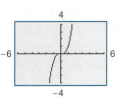

Verifying Inverse Functions Algebraically   In Exercises 19–24, show that f and g are inverse functions algebraically. Use a graphing utility to graph f and g in the same viewing window. Describe the relationship between the graphs.

✓ 19. $f(x) = x^3$, $g(x) = \sqrt[3]{x}$

20. $f(x) = \frac{1}{x}$, $g(x) = \frac{1}{x}$

21. $f(x) = \sqrt{x - 4}$; $g(x) = x^2 + 4$, $x \geq 0$

22. $f(x) = 9 - x^2$, $x \geq 0$; $g(x) = \sqrt{9 - x}$

✓ 23. $f(x) = 1 - x^3$, $g(x) = \sqrt[3]{1 - x}$

24. $f(x) = \frac{1}{1 + x}$, $x \geq 0$; $g(x) = \frac{1 - x}{x}$, $0 < x \leq 1$

Algebraic-Graphical-Numerical   In Exercises 25–34, show that f and g are inverse functions (a) algebraically, (b) graphically, and (c) numerically.

25. $f(x) = -\frac{7}{2}x - 3$, $g(x) = -\frac{2x + 6}{7}$

26. $f(x) = \frac{x - 9}{4}$, $g(x) = 4x + 9$

27. $f(x) = x^3 + 5$, $g(x) = \sqrt[3]{x - 5}$

28. $f(x) = \frac{x^3}{2}$, $g(x) = \sqrt[3]{2x}$

29. $f(x) = -\sqrt{x - 8}$, $g(x) = 8 + x^2$, $x \leq 0$

30. $f(x) = \sqrt[3]{3x - 10}$, $g(x) = \frac{x^3 + 10}{3}$

31. $f(x) = 2x$, $g(x) = \frac{x}{2}$

32. $f(x) = x - 5$, $g(x) = x + 5$

✓ 33. $f(x) = \frac{x - 1}{x + 5}$, $g(x) = -\frac{5x + 1}{x - 1}$

34. $f(x) = \frac{x + 3}{x - 2}$, $g(x) = \frac{2x + 3}{x - 1}$

Identifying Whether Functions Have Inverses   In Exercises 35–38, does the function have an inverse? Explain.

35. Domain   Range

1 can ⟶ $1
6 cans ⟶ $5
12 cans ⟶ $9
24 cans ⟶ $16

36. Domain   Range

1/2 hour ⟶ $40
1 hour ⟶ $70
2 hours ⟶ $120
4 hours

37. $\{(-3, 6), (-1, 5), (0, 6)\}$

38. $\{(2, 4), (3, 7), (7, 2)\}$

Recognizing One-to-One Functions   In Exercises 39–44, determine whether the graph is that of a function. If so, determine whether the function is one-to-one.

39.
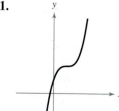

40.
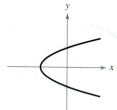

41.
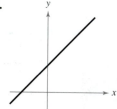

42.
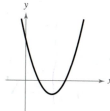

43.
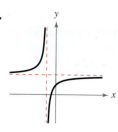

44.
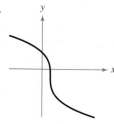

Using the Horizontal Line Test   In Exercises 45–56, use a graphing utility to graph the function and use the Horizontal Line Test to determine whether the function is one-to-one and thus has an inverse function.

45. $f(x) = 3 - \frac{1}{2}x$

46. $f(x) = \frac{1}{4}(x + 2)^2 - 1$

47. $h(x) = \dfrac{x^2}{x^2 + 1}$

48. $g(x) = \dfrac{4 - x}{6x^2}$

49. $h(x) = \sqrt{16 - x^2}$

50. $f(x) = -2x\sqrt{16 - x^2}$

51. $f(x) = 10$

52. $f(x) = -0.65$

53. $g(x) = (x + 5)^3$

54. $f(x) = x^5 - 7$

55. $h(x) = |x + 4| - |x - 4|$

56. $f(x) = -\dfrac{|x - 6|}{|x + 6|}$

Analyzing a Piecewise-Defined Function   In Exercises 57 and 58, sketch the graph of the piecewise-defined function by hand and use the graph to determine whether an inverse function exists.

57. $f(x) = \begin{cases} x^2, & 0 \le x \le 1 \\ x, & x > 1 \end{cases}$

58. $f(x) = \begin{cases} (x - 2)^3, & x < 3 \\ (x - 4)^2, & x \ge 3 \end{cases}$

Testing for One-to-One Functions   In Exercises 59–70, determine algebraically whether the function is one-to-one. Verify your answer graphically. If the function is one-to-one, find its inverse.

59. $f(x) = x^4$

60. $g(x) = x^2 - x^4$

61. $f(x) = \dfrac{3x + 4}{5}$

62. $f(x) = 3x + 5$

63. $f(x) = \dfrac{1}{x^2}$

64. $h(x) = \dfrac{4}{x^2}$

65. $f(x) = (x + 3)^2, \quad x \ge -3$

66. $q(x) = (x - 5)^2, \quad x \le 5$

✓ **67.** $f(x) = \sqrt{2x + 3}$

68. $f(x) = \sqrt{x - 2}$

69. $f(x) = |x - 2|, \quad x \le 2$

70. $f(x) = \dfrac{x^2}{x^2 + 1}$

Finding an Inverse Function Algebraically   In Exercises 71–80, find the inverse function of f algebraically. Use a graphing utility to graph both f and f^{-1} in the same viewing window. Describe the relationship between the graphs.

✓ **71.** $f(x) = 2x - 3$

72. $f(x) = 3x$

✓ **73.** $f(x) = x^5$

74. $f(x) = x^3 + 1$

75. $f(x) = x^{3/5}$

76. $f(x) = x^2, \quad x \ge 0$

✓ **77.** $f(x) = \sqrt{4 - x^2}, \quad 0 \le x \le 2$

78. $f(x) = \sqrt{16 - x^2}, \quad -4 \le x \le 0$

79. $f(x) = \dfrac{4}{x}$

80. $f(x) = \dfrac{6}{\sqrt{x}}$

Think About It   In Exercises 81–90, restrict the domain of the function f so that the function is one-to-one and has an inverse function. Then find the inverse function f^{-1}. State the domains and ranges of f and f^{-1}. Explain your results. (There are many correct answers.)

81. $f(x) = (x - 2)^2$ **82.** $f(x) = 1 - x^4$

83. $f(x) = |x + 2|$ **84.** $f(x) = |x - 2|$

85. $f(x) = (x + 3)^2$

86. $f(x) = (x - 4)^2$

87. $f(x) = -2x^2 + 5$

88. $f(x) = \frac{1}{2}x^2 - 1$

89. $f(x) = |x - 4| + 1$

90. $f(x) = -|x - 1| - 2$

Using the Properties of Inverse Functions   In Exercises 91 and 92, use the graph of the function f to complete the table and sketch the graph of f^{-1}.

91.

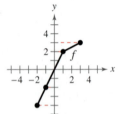

x	$f^{-1}(x)$
-4	
-2	
2	
3	

92.

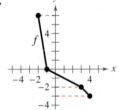

x	$f^{-1}(x)$
-3	
-2	
0	
6	

Using Graphs to Evaluate a Function   In Exercises 93–100, use the graphs of $y = f(x)$ and $y = g(x)$ to evaluate the function.

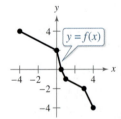

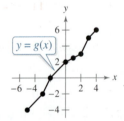

93. $f^{-1}(0)$ **94.** $g^{-1}(0)$

95. $(f \circ g)(2)$ **96.** $g(f(-4))$

97. $f^{-1}(g(0))$ **98.** $(g^{-1} \circ f)(3)$

99. $(g \circ f^{-1})(2)$ **100.** $(f^{-1} \circ g^{-1})(6)$

Using the *Draw Inverse* Feature   In Exercises 101–104, (a) use a graphing utility to graph the function f, (b) use the *draw inverse* feature of the graphing utility to draw the inverse relation of the function, and (c) determine whether the inverse relation is an inverse function. Explain your reasoning.

101. $f(x) = x^3 + x + 1$ **102.** $f(x) = x\sqrt{4 - x^2}$

103. $f(x) = \dfrac{3x^2}{x^2 + 1}$ **104.** $f(x) = \dfrac{4x}{\sqrt{x^2 + 15}}$

Evaluating a Composition of Functions   In Exercises 105–110, use the functions $f(x) = \frac{1}{8}x - 3$ and $g(x) = x^3$ to find the indicated value or function.

105. $(f^{-1} \circ g^{-1})(1)$ **106.** $(g^{-1} \circ f^{-1})(-3)$

107. $(f^{-1} \circ f^{-1})(6)$ **108.** $(g^{-1} \circ g^{-1})(-4)$

109. $(f \circ g)^{-1}$ **110.** $g^{-1} \circ f^{-1}$

Finding a Composition of Functions   In Exercises 111–114, use the functions $f(x) = x + 4$ and $g(x) = 2x - 5$ to find the specified function.

111. $g^{-1} \circ f^{-1}$ **112.** $f^{-1} \circ g^{-1}$

113. $(f \circ g)^{-1}$ **114.** $(g \circ f)^{-1}$

115. ***Why you should learn it*** *(p. 141)*   The table shows men's shoe sizes in the United States and the corresponding European shoe sizes. Let $y = f(x)$ represent the function that gives the men's European shoe size in terms of x, the men's U.S. size.

Men's U.S. shoe size	Men's European shoe size
8	41
9	42
10	43
11	45
12	46
13	47

(a) Is f one-to-one? Explain.

(b) Find $f(11)$.

(c) Find $f^{-1}(43)$, if possible.

(d) Find $f(f^{-1}(41))$.

(e) Find $f^{-1}(f(13))$.

116. Fashion Design   Let $y = g(x)$ represent the function that gives the women's European shoe size in terms of x, the women's U.S. size. A women's U.S. size 6 shoe corresponds to a European size 38. Find $g^{-1}(g(6))$.

117. Military Science   You can encode and decode messages using functions and their inverses. To code a message, first translate the letters to numbers using 1 for "A," 2 for "B," and so on. Use 0 for a space. So, "A ball" becomes

1 0 2 1 12 12.

Then, use a one-to-one function to convert to coded numbers. Using $f(x) = 2x - 1$, "A ball" becomes

1 −1 3 1 23 23.

(a) Encode "Call me later" using the function $f(x) = 5x + 4$.

(b) Find the inverse function of $f(x) = 5x + 4$ and use it to decode 119 44 9 104 4 104 49 69 29.

118. Production Management   Your wage is $10.00 per hour plus $0.75 for each unit produced per hour. So, your hourly wage y in terms of the number of units produced x is $y = 10 + 0.75x$.

(a) Find the inverse function. What does each variable in the inverse function represent?

(b) Use a graphing utility to graph the function and its inverse function.

(c) Use the *trace* feature of the graphing utility to find the hourly wage when 10 units are produced per hour.

(d) Use the *trace* feature of the graphing utility to find the number of units produced per hour when your hourly wage is $21.25.

Conclusions

True or False?   In Exercises 119 and 120, determine whether the statement is true or false. Justify your answer.

119. If f is an even function, then f^{-1} exists.

120. If the inverse function of f exists, and the graph of f has a y-intercept, then the y-intercept of f is an x-intercept of f^{-1}.

Think About It   In Exercises 121–124, determine whether the situation could be represented by a one-to-one function. If so, write a statement that describes the inverse function.

121. The number of miles n a marathon runner has completed in terms of the time t in hours

122. The population p of a town in terms of the year t from 1990 through 2010 given that the population was greatest in 2000

123. The depth of the tide d at a beach in terms of the time t over a 24-hour period

124. The height h in inches of a human child from age 2 to age 14 in terms of his or her age n in years

125. Writing   Describe the relationship between the graph of a function f and the graph of its inverse function f^{-1}.

126. Think About It   The domain of a one-to-one function f is $[0, 9]$ and the range is $[-3, 3]$. Find the domain and range of f^{-1}.

127. Think About It   The function $f(x) = \frac{9}{5}x + 32$ can be used to convert a temperature of x degrees Celsius to its corresponding temperature in degrees Fahrenheit.

(a) Using the expression for f, make a conceptual argument to show that f has an inverse function.

(b) What does $f^{-1}(50)$ represent?

128. Think About It   A function f is increasing over its entire domain. Does f have an inverse function? Explain.

129. Think About It   Describe a type of function that is *not* one-to-one on any interval of its domain.

130. CAPSTONE   Decide whether the two functions shown in each graph appear to be inverse functions of each other. Explain your reasoning.

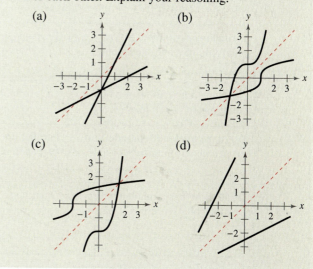

131. Proof   Prove that if f and g are one-to-one functions, then $(f \circ g)^{-1}(x) = (g^{-1} \circ f^{-1})(x)$.

132. Proof   Prove that if f is a one-to-one odd function, then f^{-1} is an odd function.

Cumulative Mixed Review

Simplifying a Rational Expression   In Exercises 133–136, write the rational expression in simplest form.

133. $\dfrac{27x^3}{3x^2}$

134. $\dfrac{5x^2y}{xy + 5x}$

135. $\dfrac{x^2 - 36}{6 - x}$

136. $\dfrac{x^2 + 3x - 40}{x^2 - 3x - 10}$

Testing for Functions   In Exercises 137–140, determine whether the equation represents y as a function of x.

137. $x = 5$

138. $y = \sqrt{x + 2}$

139. $x^2 + y^2 = 9$

140. $x - y^2 = 0$

1 Chapter Summary

What did you learn?	Explanation and Examples	Review Exercises
1.1 **Sketch graphs of equations by point plotting** *(p. 75)*.	**1.** If possible, rewrite the equation so that one of the variables is isolated on one side of the equation. **2.** Make a table of values showing several solution points. **3.** Plot these points on a rectangular coordinate system. **4.** Connect the points with a smooth curve or line.	1, 2
Graph equations using a graphing utility *(p. 77)*, **and use graphs of equations to solve real-life problems** *(p. 79)*.	**1.** Rewrite the equation so that y is isolated on the left side. **2.** Enter the equation in the graphing utility. **3.** Determine a *viewing window* that shows all important features of the graph. **4.** Graph the equation.	3–8
1.2 **Find the slopes of lines** *(p. 84)*.	**Slope m of nonvertical line:** $m = \dfrac{y_2 - y_1}{x_2 - x_1},\ x_1 \neq x_2$	9–16
Write linear equations given points on lines and their slopes *(p. 86)*.	The point-slope form of the equation of the line that passes through the point (x_1, y_1) and has a slope of m is $y - y_1 = m(x - x_1)$.	17–22
Use slope-intercept forms of linear equations to sketch lines *(p. 88)*.	The graph of the equation $y = mx + b$ is a line whose slope is m and whose y-intercept is $(0, b)$.	23–30
Use slope to identify parallel and perpendicular lines *(p. 90)*.	**Parallel lines:** Slopes are equal. **Perpendicular lines:** Slopes are negative reciprocals of each other.	31, 32
1.3 **Decide whether a relation between two variables represents a function** *(p. 97)*.	A function f from a set A to a set B is a relation that assigns to each element x in the set A exactly one element y in the set B. The set A is the domain (or set of inputs) of the function f, and the set B contains the range (or set of outputs).	33–40
Use function notation and evaluate functions *(p. 99)*, **and find the domains of functions** *(p. 101)*.	**Equation:** $f(x) = 5 - x^2$ **$f(2)$:** $f(2) = 5 - 2^2 = 1$ **Domain of $f(x) = 5 - x^2$:** All real numbers	41–46
Use functions to model and solve real-life problems *(p. 103)*.	A function can be used to model the number of construction employees in the United States. (See Example 8.)	47, 48
Evaluate difference quotients *(p. 104)*.	**Difference quotient:** $\dfrac{f(x + h) - f(x)}{h},\ h \neq 0$	49, 50
1.4 **Find the domains and ranges of functions** *(p. 110)*.		51–58
Use the Vertical Line Test for functions *(p. 111)*.	A set of points in a coordinate plane is the graph of y as a function of x if and only if no *vertical* line intersects the graph at more than one point.	59, 60

What did you learn?		Explanation and Examples	Review Exercises
1.4	**Determine intervals on which functions are increasing, decreasing, or constant** *(p. 112)*.	A function f is increasing on an interval when, for any x_1 and x_2 in the interval, $x_1 < x_2$ implies $f(x_1) < f(x_2)$. A function f is decreasing on an interval when, for any x_1 and x_2 in the interval, $x_1 < x_2$ implies $f(x_1) > f(x_2)$. A function f is constant on an interval when, for any x_1 and x_2 in the interval, $f(x_1) = f(x_2)$.	61–64
	Determine relative maximum and relative minimum values of functions *(p. 113)*.	A function value $f(a)$ is called a relative minimum of f when there exists an interval (x_1, x_2) that contains a such that $x_1 < x < x_2$ implies $f(a) \le f(x)$. A function value $f(a)$ is called a relative maximum of f when there exists an interval (x_1, x_2) that contains a such that $x_1 < x < x_2$ implies $f(a) \ge f(x)$.	65, 66
	Identify and graph step functions and other piecewise-defined functions *(p. 115)*, **and identify even and odd functions** *(p. 116)*.	**Greatest integer:** $f(x) = [\![x]\!]$ **Even:** For each x in the domain of f, $f(-x) = f(x)$. **Odd:** For each x in the domain of f, $f(-x) = -f(x)$.	67–78
1.5	**Recognize graphs of parent functions** *(p. 122)*.	**Linear:** $f(x) = x$; **Quadratic:** $f(x) = x^2$; **Cubic:** $f(x) = x^3$; **Absolute value:** $f(x) = |x|$; **Square root:** $f(x) = \sqrt{x}$; **Rational:** $f(x) = 1/x$ (See Figure 1.46, page 122.)	79–82
	Use vertical and horizontal shifts *(p. 123)*, **reflections** *(p. 125)*, **and nonrigid transformations** *(p. 127)* **to graph functions.**	**Vertical shifts:** $h(x) = f(x) + c$ or $h(x) = f(x) - c$ **Horizontal shifts:** $h(x) = f(x - c)$ or $h(x) = f(x + c)$ **Reflection in the x-axis:** $h(x) = -f(x)$ **Reflection in the y-axis:** $h(x) = f(-x)$ **Nonrigid transformations:** $h(x) = cf(x)$ or $h(x) = f(cx)$	83–94
1.6	**Add, subtract, multiply, and divide functions** *(p. 131)*, **find the compositions of functions** *(p. 133)*, **and write a function as a composition of two functions** *(p. 135)*.	$(f + g)(x) = f(x) + g(x) \qquad (f - g)(x) = f(x) - g(x)$ $(fg)(x) = f(x) \cdot g(x) \qquad (f/g)(x) = f(x)/g(x), g(x) \ne 0$ **Composition of Functions:** $(f \circ g)(x) = f(g(x))$	95–108
	Use combinations of functions to model and solve real-life problems *(p. 136)*.	A composite function can be used to represent the number of bacteria in a petri dish as a function of the amount of time the petri dish has been out of refrigeration. (See Example 10.)	109, 110
1.7	**Find inverse functions informally and verify that two functions are inverse functions of each other** *(p. 141)*.	Let f and g be two functions such that $f(g(x)) = x$ for every x in the domain of g and $g(f(x)) = x$ for every x in the domain of f. Under these conditions, the function g is the inverse function of the function f.	111–114
	Use graphs of functions to decide whether functions have inverse functions *(p. 144)*.	If the point (a, b) lies on the graph of f, then the point (b, a) must lie on the graph of f^{-1}, and vice versa. In short, f^{-1} is a reflection of f in the line $y = x$.	115, 116
	Determine whether functions are one-to-one *(p. 145)*.	A function f is one-to-one when, for a and b in its domain, $f(a) = f(b)$ implies $a = b$.	117–120
	Find inverse functions algebraically *(p. 146)*.	To find inverse functions, replace $f(x)$ by y, interchange the roles of x and y, and solve for y. Replace y by $f^{-1}(x)$.	121–128

1.1

Sketching a Graph by Point Plotting   In Exercises 1 and 2, complete the table. Use the resulting solution points to sketch the graph of the equation. Use a graphing utility to verify the graph.

1. $y = -\frac{1}{2}x + 2$

x	-2	0	2	3	4
y					
Solution point					

2. $y = x^2 - 3x$

x	-1	0	1	2	3
y					
Solution point					

Using a Graphing Utility to Graph an Equation   In Exercises 3–6, use a graphing utility to graph the equation. Approximate any x- or y-intercepts.

3. $y = \frac{1}{4}x^4 - 2x^2$ **4.** $y = \frac{1}{4}x^3 - 3x$

5. $y = x\sqrt{9 - x^2}$

6. $y = x\sqrt{x + 3}$

7. Economics   You purchase a compact car for $17,500. The depreciated value y after t years is

$$y = 17{,}500 - 1400t, \quad 0 \le t \le 6.$$

(a) Use the constraints of the model to determine an appropriate viewing window.

(b) Use a graphing utility to graph the equation.

(c) Use the *zoom* and *trace* features of the graphing utility to determine the value of t when $y = \$11{,}900$.

8. Business   The annual sales S (in billions of dollars) for Best Buy from 1995 through 2009 can be approximated by the model

$$S = 0.1473t^2 - 0.441t + 5.01, \quad 5 \le t \le 19$$

where t represents the year, with $t = 5$ corresponding to 1995.   (Source: Best Buy Company, Inc.)

(a) Use a graphing utility to graph the model.

(b) Use the *table* feature of the graphing utility to approximate the annual sales for Best Buy from 1995 through 2009.

(c) Use the *zoom* and *trace* features of the graphing utility to predict the first year that the sales for Best Buy exceed $60 billion. Confirm your result algebraically.

1.2

Finding the Slope of a Line   In Exercises 9–16, plot the two points and find the slope of the line passing through the points.

9. $(-3, 2), (8, 2)$ **10.** $(3, -1), (-3, -1)$

11. $(7, -1), (7, 12)$ **12.** $(8, -1), (8, 2)$

13. $\left(\frac{3}{2}, 1\right), \left(5, \frac{5}{2}\right)$ **14.** $\left(-\frac{3}{4}, \frac{5}{6}\right), \left(\frac{1}{2}, -\frac{5}{2}\right)$

15. $(-4.5, 6), (2.1, 3)$ **16.** $(-2.7, -6.3), (0, 1.8)$

The Point-Slope Form of the Equation of a Line   In Exercises 17–22, (a) use the point on the line and the slope of the line to find an equation of the line, and (b) find three additional points through which the line passes. (There are many correct answers.)

17. $(2, -1), m = \frac{1}{4}$ **18.** $(-3, 5), m = -\frac{3}{2}$

19. $(0, -5), m = \frac{3}{2}$ **20.** $\left(0, \frac{7}{8}\right), m = -\frac{4}{5}$

21. $(-2, 6), m = 0$ **22.** $(-8, 8), m = 0$

Finding the Slope-Intercept Form   In Exercises 23–30, write an equation of the line that passes through the points. Use the slope-intercept form, if possible. If not possible, explain why. Use a graphing utility to graph the line (if possible).

23. $(2, -1), (4, -1)$ **24.** $(0, 0), (0, 10)$

25. $\left(7, \frac{11}{3}\right), \left(9, \frac{11}{3}\right)$ **26.** $\left(\frac{5}{8}, 4\right), \left(\frac{5}{8}, -6\right)$

27. $(-1, 0), (6, 2)$ **28.** $(1, 6), (4, 2)$

29. $(3, -1), (-3, 2)$ **30.** $\left(-\frac{5}{2}, 1\right), \left(-4, \frac{2}{9}\right)$

Equations of Parallel and Perpendicular Lines   In Exercises 31 and 32, write the slope-intercept forms of the equations of the lines through the given point (a) parallel to the given line and (b) perpendicular to the given line. Verify your result with a graphing utility (use a *square setting*).

	Point	Line
31.	$(3, -2)$	$5x - 4y = 8$
32.	$(-8, 3)$	$2x + 3y = 5$

1.3

Testing for Functions   In Exercises 33 and 34, which set of ordered pairs represents a function from A to B? Explain.

33. $A = \{10, 20, 30, 40\}$ and $B = \{0, 2, 4, 6\}$

(a) $\{(20, 4), (40, 0), (20, 6), (30, 2)\}$

(b) $\{(10, 4), (20, 4), (30, 4), (40, 4)\}$

34. $A = \{u, v, w\}$ and $B = \{-2, -1, 0, 1, 2\}$

(a) $\{(u, -2), (v, 2), (w, 1)\}$

(b) $\{(w, -2), (v, 0), (w, 2)\}$

Testing for Functions Represented Algebraically In Exercises 35–40, determine whether the equation represents y as a function of x.

35. $16x^2 - y^2 = 0$

36. $x^3 + y^2 = 64$

37. $2x - y - 3 = 0$

38. $2x + y = 10$

39. $y = \sqrt{1 - x}$

40. $y = \sqrt{x^2 + 4}$

Evaluating a Function In Exercises 41 and 42, evaluate the function at each specified value of the independent variable, and simplify.

41. $f(x) = x^2 + 1$

(a) $f(1)$ (b) $f(-3)$
(c) $f(b^3)$ (d) $f(x - 1)$

42. $g(x) = \sqrt{x^2 + 1}$

(a) $g(-1)$ (b) $g(3)$
(c) $g(3x)$ (d) $g(x + 2)$

Finding the Domain of a Function In Exercises 43–46, find the domain of the function.

43. $f(x) = \dfrac{x - 1}{x + 2}$

44. $f(x) = \dfrac{x^2}{x^2 + 1}$

45. $f(x) = \sqrt{25 - x^2}$

46. $f(x) = \sqrt{x^2 - 16}$

47. Industrial Engineering A hand tool manufacturer produces a product for which the variable cost is $5.35 per unit and the fixed costs are $16,000. The company sells the product for $8.20 and can sell all that it produces.

(a) Write the total cost C as a function of x, the number of units produced.

(b) Write the profit P as a function of x.

48. Education The numbers n (in millions) of students enrolled in public schools in the United States from 2000 through 2008 can be approximated by

$$n(t) = \begin{cases} 0.76t + 61.4, & 0 \le t \le 4 \\ -0.3333t^3 + 6.6t^2 - 42.37t + 152.7, & 4 < t \le 8 \end{cases}$$

where t is the year, with $t = 0$ corresponding to 2000. (Source: U.S. Census Bureau)

(a) Use the *table* feature of a graphing utility to approximate the enrollments from 2000 through 2008.

(b) Use the graphing utility to graph the model and estimate the enrollments for the years 2009 through 2012. Do the values seem reasonable? Explain.

Evaluating a Difference Quotient In Exercises 49 and 50, find the difference quotient $\dfrac{f(x + h) - f(x)}{h}$ for the given function and simplify your answer.

49. $f(x) = 2x^2 + 3x - 1$ **50.** $f(x) = x^2 - 3x + 5$

1.4

Finding the Domain and Range of a Function In Exercises 51–58, use a graphing utility to graph the function and estimate its domain and range. Then find the domain and range algebraically.

51. $f(x) = 3 - 2x^2$

52. $f(x) = 2x^2 + 5$

53. $f(x) = \sqrt{x + 3} + 4$

54. $f(x) = 2 - \sqrt{x - 5}$

55. $h(x) = \sqrt{36 - x^2}$

56. $f(x) = \sqrt{x^2 - 9}$

57. $f(x) = |x + 5| + 2$

58. $f(x) = |x + 1| - 3$

Vertical Line Test for Functions In Exercises 59 and 60, use the Vertical Line Test to determine whether y is a function of x. Describe how to enter the equation into a graphing utility to produce the given graph.

59. $y - 4x = x^2$

60. $3x + y^2 - 2 = 0$

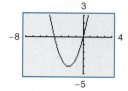

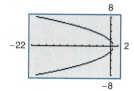

Increasing and Decreasing Functions In Exercises 61–64, (a) use a graphing utility to graph the function and (b) determine the open intervals on which the function is increasing, decreasing, or constant.

61. $f(x) = x^3 - 3x$

62. $f(x) = \sqrt{x^2 - 9}$

63. $f(x) = x\sqrt{x - 6}$

64. $f(x) = \dfrac{|x + 8|}{2}$

Approximating Relative Minima and Maxima In Exercises 65 and 66, use a graphing utility to approximate (to two decimal places) any relative minimum or relative maximum values of the function.

65. $f(x) = (x^2 - 4)^2$ **66.** $f(x) = x^3 - 4x^2 - 1$

Sketching Graphs In Exercises 67–70, sketch the graph of the function by hand.

67. $f(x) = \begin{cases} 3x + 5, & x < 0 \\ x - 4, & x \ge 0 \end{cases}$ **68.** $f(x) = \begin{cases} \frac{1}{2}x + 3, & x < 0 \\ 4 - x^2, & x \ge 0 \end{cases}$

69. $f(x) = [\![x]\!] + 3$ **70.** $f(x) = [\![x + 2]\!]$

Even and Odd Functions In Exercises 71–78, determine algebraically whether the function is even, odd, or neither. Verify your answer using a graphing utility.

71. $f(x) = x^2 + 6$

72. $f(x) = x^2 - x - 1$

73. $f(x) = (x^2 - 8)^2$

74. $f(x) = 2x^3 - x^2$

75. $f(x) = 3x^{5/2}$

76. $f(x) = 3x^{2/5}$

77. $f(x) = \sqrt{4 - x^2}$

78. $f(x) = x\sqrt{x^2 - 1}$

1.5

Library of Parent Functions In Exercises 79–82, identify the parent function and describe the transformation shown in the graph. Write an equation for the graphed function.

79.
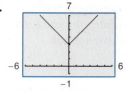

80.
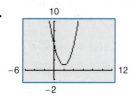

81.
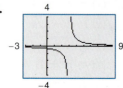

82.
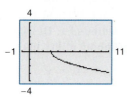

Sketching Transformations In Exercises 83–86, use the graph of $y = f(x)$ to graph the function.

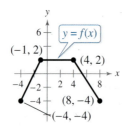

83. $y = f(-x)$

84. $y = -f(x)$

85. $y = f(x) - 2$

86. $y = f(x - 1)$

Describing Transformations In Exercises 87–94, h is related to one of the six parent functions on page 122. (a) Identify the parent function f. (b) Describe the sequence of transformations from f to h. (c) Sketch the graph of h by hand. (d) Use function notation to write h in terms of the parent function f.

87. $h(x) = \dfrac{1}{x} - 6$

88. $h(x) = -\dfrac{1}{x} - 3$

89. $h(x) = (x - 2)^3 + 5$

90. $h(x) = -(x - 2)^2 - 8$

91. $h(x) = -\sqrt{x} + 6$

92. $h(x) = \sqrt{x - 1} + 4$

93. $h(x) = |x| + 9$

94. $h(x) = |x + 8| - 1$

1.6

Evaluating a Combination of Functions In Exercises 95–102, let $f(x) = 3 - 2x$, $g(x) = \sqrt{x}$, and $h(x) = 3x^2 + 2$, and find the indicated values.

95. $(f - g)(4)$

96. $(f + h)(5)$

97. $(f + g)(25)$

98. $(g - h)(1)$

99. $(fh)(1)$

100. $(g/h)(1)$

101. $(h \circ g)(5)$

102. $(g \circ f)(-3)$

Identifying a Composite Function In Exercises 103–108, find two functions f and g such that $(f \circ g)(x) = h(x)$. (There are many correct answers.)

103. $h(x) = (x + 3)^2$

104. $h(x) = (1 - 2x)^3$

105. $h(x) = \sqrt{4x + 2}$

106. $h(x) = \sqrt[3]{(x + 2)^2}$

107. $h(x) = \dfrac{4}{x + 2}$

108. $h(x) = \dfrac{6}{(3x + 1)^3}$

Education In Exercises 109 and 110, the numbers (in thousands) of students taking the SAT (y_1) and the ACT (y_2) for the years 2000 through 2009 can be modeled by

$$y_1 = -2.61t^2 + 55.0t + 1244 \quad \text{and}$$

$$y_2 = 0.949t^3 - 8.02t^2 + 44.4t + 1056$$

where t represents the year, with $t = 0$ corresponding to 2000. (Source: College Entrance Examination Board and ACT, Inc.)

109. Use a graphing utility to graph y_1, y_2, and $y_1 + y_2$ in the same viewing window.

110. Use the model $y_1 + y_2$ to estimate the total number of students taking the SAT and ACT in 2010.

1.7

Finding Inverse Functions Informally In Exercises 111–114, find the inverse function of f informally. Verify that $f(f^{-1}(x)) = x$ and $f^{-1}(f(x)) = x$.

111. $f(x) = 6x$

112. $f(x) = x + 5$

113. $f(x) = \dfrac{1}{2}x + 3$

114. $f(x) = \dfrac{x - 4}{5}$

Algebraic-Graphical-Numerical In Exercises 115 and 116, show that f and g are inverse functions (a) algebraically, (b) graphically, and (c) numerically.

115. $f(x) = 3 - 4x$, $\quad g(x) = \dfrac{3 - x}{4}$

116. $f(x) = \sqrt{x + 1}$, $\quad g(x) = x^2 - 1$, $x \geq 0$

Using the Horizontal Line Test In Exercises 117–120, use a graphing utility to graph the function and use the Horizontal Line Test to determine whether the function is one-to-one and an inverse function exists.

117. $f(x) = \frac{1}{2}x - 3$

118. $f(x) = (x - 1)^2$

119. $h(t) = \dfrac{2}{t - 3}$

120. $g(x) = \sqrt{x + 6}$

Finding an Inverse Function Algebraically In Exercises 121–128, find the inverse function of f algebraically.

121. $f(x) = \dfrac{1}{2}x - 5$

122. $f(x) = \dfrac{7x + 3}{8}$

123. $f(x) = 4x^3 - 3$

124. $f(x) = 5x^3 + 2$

125. $f(x) = \sqrt{x + 10}$

126. $f(x) = 4\sqrt{6 - x}$

127. $f(x) = \frac{1}{4}x^2 + 1$, $x \geq 0$

128. $f(x) = 5 - \frac{1}{9}x^2$, $x \geq 0$

Take this test as you would take a test in class. After you are finished, check your work against the answers in the back of the book.

In Exercises 1–6, use the point-plotting method to graph the equation by hand and identify any x- and y-intercepts. Verify your results using a graphing utility.

1. $y = 2|x| - 1$

2. $y = 2x - \frac{8}{5}$

3. $y = 2x^2 - 4x$

4. $y = x^3 - x$

5. $y = -x^2 + 4$

6. $y = \sqrt{x - 2}$

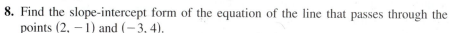

Figure for 9

7. Find equations of the lines that pass through the point $(0, 4)$ and are (a) parallel to and (b) perpendicular to the line $5x + 2y = 3$.

8. Find the slope-intercept form of the equation of the line that passes through the points $(2, -1)$ and $(-3, 4)$.

9. Does the graph at the right represent y as a function of x? Explain.

10. Evaluate $f(x) = |x + 2| - 15$ at each value of the independent variable and simplify.

    (a) $f(-8)$ (b) $f(14)$ (c) $f(t - 6)$

11. Find the domain of $f(x) = 10 - \sqrt{3 - x}$.

12. An electronics company produces a car stereo for which the variable cost is $25.60 and the fixed costs are $24,000. The product sells for $99.50. Write the total cost C as a function of the number of units produced and sold, x. Write the profit P as a function of the number of units produced and sold, x.

In Exercises 13 and 14, determine algebraically whether the function is even, odd, or neither.

13. $f(x) = 2x^3 - 3x$

14. $f(x) = 3x^4 + 5x^2$

In Exercises 15 and 16, determine the open intervals on which the function is increasing, decreasing, or constant.

15. $h(x) = \frac{1}{4}x^4 - 2x^2$

16. $g(t) = |t + 2| - |t - 2|$

In Exercises 17 and 18, use a graphing utility to approximate (to two decimal places) any relative minimum or relative maximum values of the function.

17. $f(x) = -x^3 - 5x^2 + 12$

18. $f(x) = x^5 - x^3 + 2$

In Exercises 19–21, (a) identify the parent function f, (b) describe the sequence of transformations from f to g, and (c) sketch the graph of g.

19. $g(x) = -2(x - 5)^3 + 3$

20. $g(x) = \sqrt{-x - 7}$

21. $g(x) = 4|-x| - 7$

22. Use the functions $f(x) = x^2$ and $g(x) = \sqrt{2 - x}$ to find the specified function and its domain.

    (a) $(f - g)(x)$ (b) $\left(\dfrac{f}{g}\right)(x)$ (c) $(f \circ g)(x)$ (d) $(g \circ f)(x)$

In Exercises 23–25, determine whether the function has an inverse function, and if so, find the inverse function.

23. $f(x) = x^3 + 8$

24. $f(x) = x^2 + 6$

25. $f(x) = \dfrac{3x\sqrt{x}}{8}$

2.1   Linear Equations and Problem Solving

Equations and Solutions of Equations

An **equation** in x is a statement that two algebraic expressions are equal. For example,

$$3x - 5 = 7, \quad x^2 - x - 6 = 0, \quad \text{and} \quad \sqrt{2x} = 4$$

are equations. To **solve** an equation in x means to find all values of x for which the equation is true. Such values are **solutions.** For instance, $x = 4$ is a solution of the equation $3x - 5 = 7$, because $3(4) - 5 = 7$ is a true statement.

The solutions of an equation depend on the kinds of numbers being considered. For instance, in the set of rational numbers, $x^2 = 10$ has no solution because there is no rational number whose square is 10. In the set of real numbers, however, the equation has two solutions: $x = \sqrt{10}$ and $x = -\sqrt{10}$.

An equation can be classified as an *identity*, a *conditional*, or a *contradiction*, as shown in the following table.

Equation	Definition	Example
Identity	An equation that is true for *every* real number in the domain of the variable	$\dfrac{x}{3x^2} = \dfrac{1}{3x}$ is a true statement for any nonzero real value of x.
Conditional	An equation that is true for just *some* (but not all) of the real numbers in the domain of the variable	$x^2 - 9 = 0$ is a true statement for $x = 3$ and $x = -3$, but not for any other real values.
Contradiction	An equation that is *false* for every real number in the domain of the variable	$2x + 1 = 2x - 3$ is a false statement for any real value of x.

A **linear equation in one variable** x is an equation that can be written in the standard form $ax + b = 0$, where a and b are real numbers, with $a \neq 0$.

To solve an equation involving fractional expressions, find the least common denominator (LCD) of all terms in the equation and multiply every term by this LCD. This procedure clears the equation of fractions.

Example 1   Solving an Equation Involving Fractions

Solve $\dfrac{x}{3} + \dfrac{3x}{4} = 2$.

Solution

$$\dfrac{x}{3} + \dfrac{3x}{4} = 2 \qquad \text{Write original equation.}$$

$$(12)\dfrac{x}{3} + (12)\dfrac{3x}{4} = (12)2 \qquad \text{Multiply each term by the LCD of 12.}$$

$$4x + 9x = 24 \qquad \text{Divide out and multiply.}$$

$$13x = 24 \qquad \text{Combine like terms.}$$

$$x = \dfrac{24}{13} \qquad \text{Divide each side by 13.}$$

✓CHECKPOINT   Now try Exercise 25.

What you should learn

- Solve equations involving fractional expressions.
- Write and use mathematical models to solve real-life problems.
- Use common formulas to solve real-life problems.

Why you should learn it

Linear equations are useful for modeling situations in which you need to find missing information. For instance, Exercise 58 on page 168 shows how to use a linear equation to determine the height of a flagpole by measuring its shadow.

Study Tip

For a review of solving one- and two-step linear equations, see Appendix D at this textbook's *Companion Website.*

After solving an equation, you should check each solution in the original equation. For instance, you can check the solution to Example 1 as follows.

$$\frac{x}{3} + \frac{3x}{4} = 2$$ Write original equation.

$$\frac{\frac{24}{13}}{3} + \frac{3\left(\frac{24}{13}\right)}{4} \stackrel{?}{=} 2$$ Substitute $\frac{24}{13}$ for x.

$$\frac{8}{13} + \frac{18}{13} \stackrel{?}{=} 2$$ Simplify.

$$2 = 2$$ Solution checks. ✓

When multiplying or dividing an equation by a *variable* expression, it is possible to introduce an **extraneous solution**—one that does not satisfy the original equation. The next example demonstrates the importance of checking your solution when you have multiplied or divided by a variable expression.

Example 2   An Equation with an Extraneous Solution

Solve $\dfrac{1}{x - 2} = \dfrac{3}{x + 2} - \dfrac{6x}{x^2 - 4}$.

Solution

The LCD is $x^2 - 4 = (x + 2)(x - 2)$. Multiplying each term by the LCD and simplifying produces the following.

$$\frac{1}{x - 2}(x + 2)(x - 2) = \frac{3}{x + 2}(x + 2)(x - 2) - \frac{6x}{x^2 - 4}(x + 2)(x - 2)$$

$$x + 2 = 3(x - 2) - 6x, \quad x \neq \pm 2$$

$$x + 2 = 3x - 6 - 6x$$

$$4x = -8$$

$$x = -2 \qquad \text{Extraneous solution}$$

In the original equation, $x = -2$ yields a denominator of zero. So, $x = -2$ is an extraneous solution, and the original equation has *no solution*. To check this result graphically, graph

$$y_1 = \frac{1}{x - 2} \quad \text{and} \quad y_2 = \frac{3}{x + 2} - \frac{6x}{x^2 - 4}$$

in the same viewing window, as shown in Figure 2.1.

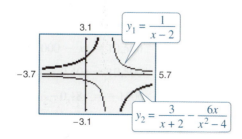

Figure 2.1

The graphs of the equations do not appear to intersect. This means that there is no point for which the left side of the equation y_1 is equal to the right side of the equation y_2. So, the equation appears to have *no solution*.

✓CHECKPOINT   Now try Exercise 39.

 What's Wrong?

To approximate the solution of $0.9x + 1 = 0.91x$, you use a graphing utility to graph $y_1 = 0.9x + 1$ and $y_2 = 0.91x$, as shown in the figure. You use the graph to conclude that the equation has no solution because the lines do not intersect. What's wrong?

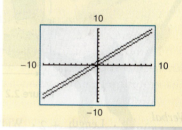

Common Formulas

Many common types of geometric, scientific, and investment problems use ready-made equations called **formulas.** Knowing these formulas will help you translate and solve a wide variety of real-life applications.

Common Formulas for Area A, Perimeter P, Circumference C, and Volume V

Square

$A = s^2$

$P = 4s$

Rectangle

$A = lw$

$P = 2l + 2w$

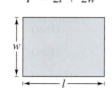

Circle

$A = \pi r^2$

$C = 2\pi r$

Triangle

$A = \dfrac{1}{2} bh$

$P = a + b + c$

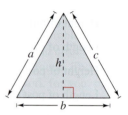

Cube

$V = s^3$

Rectangular Solid

$V = lwh$

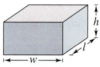

Circular Cylinder

$V = \pi r^2 h$

Sphere

$V = \dfrac{4}{3} \pi r^3$

Miscellaneous Common Formulas

Temperature: $F = \dfrac{9}{5} C + 32$ F = degrees Fahrenheit
C = degrees Celsius

Simple Interest: $I = Prt$ I = interest
P = principal (original deposit)
r = annual interest rate
t = time in years

Compound Interest: $A = P\left(1 + \dfrac{r}{n}\right)^{nt}$ A = balance
P = principal (original deposit)
r = annual interest rate
n = compoundings (number of times interest is calculated) per year
t = time in years

Distance: $d = rt$ d = distance traveled
r = rate
t = time

When working with applied problems, you may find it helpful to rewrite a common formula. For instance, the formula for the perimeter of a rectangle,

$$P = 2l + 2w$$

can be solved for w as

$$w = \tfrac{1}{2}(P - 2l).$$

Example 6   Using a Formula

A cylindrical can has a volume of 600 cubic centimeters and a radius of 4 centimeters, as shown in Figure 2.4. Find the height of the can.

Solution

The formula for the volume of a cylinder is

$$V = \pi r^2 h. \qquad \text{Volume of a cylinder}$$

To find the height of the can, solve for h.

$$h = \frac{V}{\pi r^2}$$

Then, using $V = 600$ and $r = 4$, find the height.

$$h = \frac{600}{\pi(4)^2} = \frac{600}{16\pi} \approx 11.94$$

The height of the can is about 11.94 centimeters. You can use unit analysis to check that your answer is reasonable.

$$\frac{600 \text{ cm}^3}{16\pi \text{ cm}^2} \approx 11.94 \text{ cm}$$

  Now try Exercise 65.

Figure 2.4

Example 7   Using a Formula

The average daily temperature in San Diego, California is 64.4°F. What is San Diego's average daily temperature in degrees Celsius?   (Source: U.S. National Oceanic and Atmospheric Administration)

Solution

First solve for C in the formula for temperature. Then use $F = 64.4$ to find the temperature in degrees Celsius.

$$F = \frac{9}{5}C + 32 \qquad \text{Formula for temperature}$$

$$F - 32 = \frac{9}{5}C \qquad \text{Subtract 32 from each side.}$$

$$\frac{5}{9}(F - 32) = C \qquad \text{Multiply each side by } \tfrac{5}{9}.$$

$$\frac{5}{9}(64.4 - 32) = C \qquad \text{Substitute 64.4 for } F.$$

$$18 = C \qquad \text{Simplify.}$$

The average daily temperature in San Diego is 18°C.

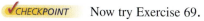 Now try Exercise 69.

> ### Study Tip
>
> Once you have rewritten the formula for temperature, you can easily find other Celsius values. Simply substitute other Fahrenheit values and evaluate.

See www.CalcChat.com for worked-out solutions to odd-numbered exercises. For instructions on how to use a graphing utility, see Appendix A.

2.1 Exercises

Vocabulary and Concept Check

In Exercises 1–4, fill in the blank.

1. A(n) _____ is a statement that two algebraic expressions are equal.

2. A linear equation in one variable is an equation that can be written in the standard form _____ .

3. When solving an equation, it is possible to introduce a(n) _____ solution, which is a value that does not satisfy the original equation.

4. Many real-life problems can be solved using ready-made equations called _____ .

5. Is the equation $x + 1 = 3$ an identity, a conditional equation, or a contradiction?

6. How can you clear the equation $\dfrac{x}{2} + 1 = \dfrac{1}{4}$ of fractions?

Procedures and Problem Solving

Checking Solutions of an Equation In Exercises 7–10, determine whether each value of x is a solution of the equation.

Equation	Values

7. $\dfrac{5}{2x} - \dfrac{4}{x} = 3$ (a) $x = -\dfrac{1}{2}$ (b) $x = 4$

            (c) $x = 0$ (d) $x = \dfrac{1}{4}$

8. $\dfrac{x}{2} + \dfrac{6x}{7} = \dfrac{19}{14}$ (a) $x = -2$ (b) $x = 1$

            (c) $x = \dfrac{1}{2}$ (d) $x = 7$

9. $\dfrac{\sqrt{x + 4}}{6} + 3 = 4$ (a) $x = -3$ (b) $x = 0$

            (c) $x = 21$ (d) $x = 32$

10. $\dfrac{\sqrt[3]{x - 8}}{3} = -\dfrac{2}{3}$ (a) $x = -16$ (b) $x = 0$

            (c) $x = 9$ (d) $x = 16$

Classifying Equations In Exercises 11–16, determine whether the equation is an identity, a conditional equation, or a contradiction.

11. $2(x - 1) = 2x - 2$

12. $x^2 - 8x + 5 = (x - 4)^2 - 11$

13. $-5(x - 1) = -5(x + 1)$

14. $(x + 3)(x - 5) = x^2 - 2(x + 7)$

15. $3 + \dfrac{1}{x + 1} = \dfrac{4x}{x + 1}$

16. $\dfrac{5}{x} + \dfrac{3}{x} = 24$

Solving an Equation Involving Fractions In Exercises 17–20, solve the equation using two methods. Then explain which method is easier.

17. $\dfrac{3x}{8} - \dfrac{4x}{3} = 4$

18. $\dfrac{3z}{8} - \dfrac{z}{10} = 6$

19. $\dfrac{2x}{5} + 5x = \dfrac{4}{3}$

20. $\dfrac{4y}{3} - 2y = \dfrac{16}{5}$

Solving Equations In Exercises 21–40, solve the equation (if possible).

21. $3x - 5 = 2x + 7$

22. $5x + 3 = 6 - 2x$

23. $3(y - 5) = 3 + 5y$

24. $5(z - 4) + 4z = 5 - 6z$

25. $\dfrac{x}{5} - \dfrac{x}{2} = 3$

26. $\dfrac{5x}{4} + \dfrac{1}{2} = x - \dfrac{1}{2}$

27. $\dfrac{2(z - 4)}{5} + 5 = 10z$

28. $\dfrac{3x}{2} + \dfrac{1}{4}(x - 2) = 10$

29. $\dfrac{100 - 4u}{3} = \dfrac{5u + 6}{4} + 6$

30. $\dfrac{17 + y}{y} + \dfrac{32 + y}{y} = 100$

31. $\dfrac{5x - 4}{5x + 4} = \dfrac{2}{3}$

32. $\dfrac{10x + 3}{5x + 6} = \dfrac{1}{2}$

33. $\dfrac{1}{x - 3} + \dfrac{1}{x + 3} = \dfrac{10}{x^2 - 9}$

34. $\dfrac{1}{x - 2} + \dfrac{3}{x + 3} = \dfrac{4}{x^2 + x - 6}$

35. $\dfrac{1}{x} + \dfrac{2}{x - 5} = 0$

36. $3 = 2 + \dfrac{2}{z + 2}$

37. $\dfrac{2}{(x - 4)(x - 2)} = \dfrac{1}{x - 4} + \dfrac{2}{x - 2}$

38. $\dfrac{2}{x(x-2)} + \dfrac{5}{x} = \dfrac{1}{x-2}$

✓ **39.** $\dfrac{3}{x^2 - 3x} + \dfrac{4}{x} = \dfrac{1}{x-3}$ **40.** $\dfrac{6}{x} - \dfrac{2}{x+3} = \dfrac{3(x+5)}{x(x+3)}$

Solving for a Variable   In Exercises 41–46, solve for the indicated variable.

41. *Area of a Triangle*

Solve for h: $A = \frac{1}{2}bh$

42. *Area of a Trapezoid*

Solve for b: $A = \frac{1}{2}(a+b)h$

43. *Investment at Compound Interest*

Solve for P: $A = P\left(1 + \dfrac{r}{n}\right)^{nt}$

44. *Investment at Simple Interest*

Solve for r: $A = P + Prt$

45. *Volume of a Right Circular Cylinder*

Solve for h: $V = \pi r^2 h$

46. *Volume of a Right Circular Cone*

Solve for h: $V = \frac{1}{3}\pi r^2 h$

MODELING DATA

In Exercises 47 and 48, use the following information. The relationship between the length of an adult's femur (thigh bone) and the height of the adult can be approximated by the linear equations

$y = 0.432x - 10.44$ **Female**

$y = 0.449x - 12.15$ **Male**

where y is the length of the femur in inches and x is the height of the adult in inches (see figure).

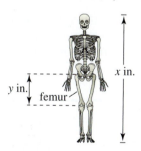

47. An anthropologist discovers a femur belonging to an adult human female. The bone is 16 inches long. Estimate the height of the female.

48. From the foot bones of an adult human male, an anthropologist estimates that the person's height was 69 inches. A few feet away from the site where the foot bones were discovered, the anthropologist discovers a male adult femur that is 19 inches long. Is it likely that both the foot bones and the thigh bone came from the same person?

✓ **49. Interior Design**   A room is 1.5 times as long as it is wide, and its perimeter is 25 meters.

(a) Draw a diagram that gives a visual representation of the problem. Identify the length as l and the width as w.

(b) Write l in terms of w and write an equation for the perimeter in terms of w.

(c) Find the dimensions of the room.

50. Woodworking   A picture frame has a total perimeter of 3 meters. The height of the frame is $\frac{2}{3}$ times its width.

(a) Draw a diagram that gives a visual representation of the problem. Identify the width as w and the height as h.

(b) Write h in terms of w and write an equation for the perimeter in terms of w.

(c) Find the dimensions of the picture frame.

51. Education   To get an A in a course, you must have an average of at least 90 on four tests of 100 points each. The scores on your first three tests were 87, 92, and 84.

(a) Write a verbal model for the test average for the course.

(b) What is the least you can score on the fourth test to get an A in the course?

52. Business   A store generates Monday through Thursday sales of $150, $125, $75, and $180. What sales on Friday would give a weekday average of $150?

53. Professional Sales   A salesperson is driving from the office to a client, a distance of about 250 kilometers. After 30 minutes, the salesperson passes a town that is 50 kilometers from the office. Assuming the salesperson continues at the same constant speed, how long will it take to drive from the office to the client?

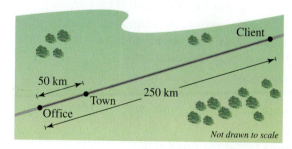

54. Professional Sales   On the first part of a 336-mile trip, a salesperson averaged 58 miles per hour. The salesperson averaged only 52 miles per hour on the last part of the trip because of an increased volume of traffic. The total time of the trip was 6 hours. Find the amount of time at each of the two speeds.

55. Professional Driving   A truck driver traveled at an average speed of 55 miles per hour on a 200-mile trip to pick up a load of freight. On the return trip (with the truck fully loaded), the average speed was 40 miles per hour. Find the average speed for the round trip.

56. Aviation An executive flew in the corporate jet to a meeting in a city 1500 kilometers away. After traveling the same amount of time on the return flight, the pilot mentioned that they still had 300 kilometers to go. The air speed of the plane was 600 kilometers per hour. How fast was the wind blowing? (Assume that the wind direction was parallel to the flight path and constant all day.)

57. Dendrology To determine the height of a pine tree, you measure the shadow cast by the tree and find it to be 20 feet long (see figure). Then you measure the shadow cast by a 36-inch tall oak sapling and find it to be 24 inches long. Estimate the pine tree's height.

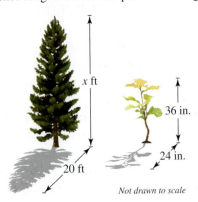

x ft

36 in.

24 in.

20 ft

Not drawn to scale

58. *Why you should learn it* *(p. 160)* A person who is 6 feet tall walks away from a flagpole toward the tip of the shadow of the flagpole. When the person is 30 feet from the flagpole, the tips of the person's shadow and the shadow cast by the flagpole coincide at a point 5 feet in front of the person.

(a) Draw a diagram that illustrates the problem. Let *h* represent the height of the flagpole.

(b) Find the height of the flagpole.

59. Simple Interest A certificate of deposit with an initial deposit of $8000 accumulates $400 interest in 2 years. Find the annual interest rate.

60. Finance You plan to invest $12,000 in two funds paying $4\frac{1}{2}\%$ and 5% simple interest. (There is more risk in the 5% fund.) Your goal is to obtain a total annual interest income of $560 from the investments. What is the least amount you can invest in the 5% fund in order to meet your objective?

61. Merchandising A grocer mixes peanuts that cost $2.69 per pound and walnuts that cost $4.29 per pound to make 100 pounds of a mixture that costs $3.49 per pound. How much of each kind of nut is put into the mixture?

62. Forestry A forester mixes gasoline and oil to make 2 gallons of mixture for a two-cycle chainsaw engine. This mixture is 32 parts gasoline and 1 part oil. How much gasoline must be added to bring the mixture to 40 parts gasoline and 1 part oil?

63. Retail Management A store has $40,000 of inventory in notebook computers and desktop computers. The profit on a notebook computer is 25% and the profit on a desktop computer is 20%. The profit for the entire stock is 23%. How much is invested in notebook computers and how much in desktop computers?

64. Retail Management A store has $4500 of inventory in 8 × 10 picture frames and 5 × 7 picture frames. The profit on an 8 × 10 frame is 25% and the profit on a 5 × 7 frame is 22%. The profit on the entire stock is 24%. How much is invested in the 8 × 10 picture frames and how much is invested in the 5 × 7 picture frames?

65. Sailing A triangular sail has an area of 182.25 square feet. The sail has a base of 13.5 feet. Find the height of the sail.

66. Geometry The figure shows three squares. The perimeter of square I is 20 inches and the perimeter of square II is 32 inches. Find the area of square III.

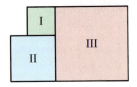

I

III

II

67. Geometry The volume of a rectangular package is 2304 cubic inches. The length of the package is 3 times its width, and the height is $1\frac{1}{2}$ times its width.

(a) Draw a diagram that illustrates the problem. Label the height, width, and length accordingly.

(b) Find the dimensions of the package.

68. Geometry The volume of a globe is about 47,712.94 cubic centimeters. Use a graphing utility to find the radius of the globe. Round your result to two decimal places.

69. Meteorology The line graph shows the temperatures (in degrees Fahrenheit) on a summer day in Buffalo, New York from 10:00 A.M. to 6:00 P.M. Create a new line graph showing the temperatures throughout the day in degrees Celsius.

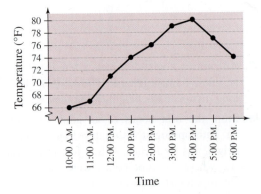

70. Meteorology   The average July 2009 temperature in the contiguous United States was 73.5°F. What was the average temperature in degrees Celsius? (U.S. National Oceanic and Atmospheric Administration)

71. Geography   You are driving on a Canadian freeway to a town that is 300 kilometers from your home. After 30 minutes you pass a freeway exit that you know is 50 kilometers from your home. Assuming that you continue at the same constant speed, how long will it take for the entire trip?

72. Parks and Recreation   A gondola tower in an amusement park casts a shadow that is 80 feet long while a sign that is 4 feet tall casts a shadow that is $3\frac{1}{2}$ feet long. Draw a diagram for the situation. Then find the height of the tower.

Physics   In Exercises 73 and 74, you have a uniform beam of length L with a fulcrum x feet from one end (see figure). Objects with weights W_1 and W_2 are placed at opposite ends of the beam. The beam will balance when

$$W_1 x = W_2(L - x).$$

Find x such that the beam will balance.

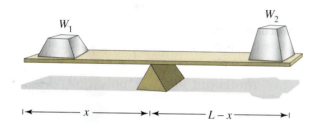

73. Two children weighing 50 pounds (W_1) and 75 pounds (W_2) are going to play on a seesaw that is 10 feet long.

74. A person weighing 200 pounds (W_1) is attempting to move a 550-pound rock (W_2) with a bar that is 5 feet long.

Conclusions

True or False?   In Exercises 75 and 76, determine whether the statement is true or false. Justify your answer.

75. The equation

$$x(3 - x) = 10$$

is a linear equation.

76. The volume of a cube with a side length of 9.5 inches is greater than the volume of a sphere with a radius of 5.9 inches.

Writing Equations   In Exercises 77–80, write a linear equation that has the given solution. (There are many correct answers.)

77. $x = -3$ **78.** $x = 0$

79. $x = \frac{1}{4}$ **80.** $x = -2.5$

81. Error Analysis   Describe the error in solving the equation.

$$\frac{1}{x + 1} + \frac{1}{x - 1} = \frac{2}{(x + 1)(x - 1)}$$

$$(x - 1) + (x + 1) = 2$$

$$2x = 2$$

$$x = 1$$

The solution is $x = 1$.

82. CAPSTONE   Consider the equation

$$\frac{6}{(x - 3)(x - 1)} = \frac{3}{x - 3} + \frac{4}{x - 1}.$$

Without performing any calculations, explain how to clear this equation of fractions. Is it possible that this process will introduce an extraneous solution? If so, describe two ways to determine whether a solution is extraneous.

83. Think About It   Find c such that $x = 3$ is a solution of the linear equation $2x - 5c = 10 + 3c - 3x$.

84. Think About It   Find c such that $x = 2$ is a solution of the linear equation $5x + 2c = 12 + 4x - 2c$.

Cumulative Mixed Review

Sketching a Graph   In Exercises 85–90, sketch the graph of the equation by hand. Verify using a graphing utility.

85. $y = \frac{5}{8}x - 2$

86. $y = \frac{3x - 5}{2} + 2$

87. $y = (x - 3)^2 + 7$

88. $y = \frac{1}{3}x^2 - 4$

89. $y = -\frac{1}{2}|x + 4| - 1$

90. $y = |x - 2| + 10$

Evaluating Combinations of Functions   In Exercises 91–96, evaluate the combination of functions for

$$f(x) = -x^2 + 4 \quad \text{and} \quad g(x) = 6x - 5.$$

91. $(f + g)(-3)$

92. $(g - f)(-1)$

93. $(fg)(8)$

94. $\left(\dfrac{f}{g}\right)\left(\dfrac{1}{2}\right)$

95. $(f \circ g)(4)$

96. $(g \circ f)(2)$

Finding Solutions Graphically

Polynomial equations of degree 1 or 2 can be solved in relatively straightforward ways. Solving polynomial equations of higher degrees can, however, be quite difficult, especially when you rely only on algebraic techniques. For such equations, a graphing utility can be very helpful.

Graphical Approximations of Solutions of an Equation

1. Write the equation in *general form*, $f(x) = 0$, with the nonzero terms on one side of the equation and zero on the other side.

2. Use a graphing utility to graph the function $y = f(x)$. Be sure the viewing window shows all the relevant features of the graph.

3. Use the *zero* or *root* feature or the *zoom* and *trace* features of the graphing utility to approximate the x-intercepts of the graph of f.

In Chapter 3 you will learn techniques for determining the number of solutions of a polynomial equation. For now, you should know that a polynomial equation of degree n cannot have more than n different solutions.

Example 3   Finding Solutions of an Equation Graphically

Use a graphing utility to approximate the solutions of

$$2x^3 - 3x + 2 = 0.$$

Solution

Graph the function $y = 2x^3 - 3x + 2$. You can see from the graph that there is one x-intercept. It lies between -2 and -1 and is approximately -1.5. By using the *zero* or *root* feature of a graphing utility, you can improve the approximation. Choose a left bound of $x = -2$ (see Figure 2.8) and a right bound of $x = -1$ (see Figure 2.9). To two-decimal-place accuracy, the solution is $x \approx -1.48$, as shown in Figure 2.10. Check this approximation on your calculator. You will find that the value of y is

$$y = 2(-1.48)^3 - 3(-1.48) + 2 \approx -0.04.$$

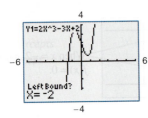

Figure 2.8   **Figure 2.9**

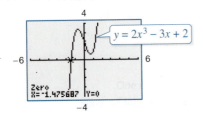

Figure 2.10

✔CHECKPOINT   Now try Exercise 41.

You can also use a graphing utility's *zoom* and *trace* features to approximate the solution of an equation. Here are some suggestions for using the *zoom-in* feature of a graphing utility.

1. With each successive zoom-in, adjust the *x*-scale (if necessary) so that the resulting viewing window shows at least one scale mark on each side of the solution.

2. The accuracy of the approximation will always be such that the error is less than the distance between two scale marks. For instance, to approximate the zero to the nearest hundredth, set the *x*-scale to 0.01. To approximate the zero to the nearest thousandth, set the *x*-scale to 0.001.

3. The graphing utility's *trace* feature can sometimes be used to add one more decimal place of accuracy without changing the viewing window.

Unless stated otherwise, all real solutions in this text will be approximated with an error of *at most* 0.01.

Example 4   Approximating Solutions of an Equation Graphically

Use a graphing utility to approximate the solutions of

$$x^2 + 3 = 5x.$$

Solution

In general form, this equation is

$$x^2 - 5x + 3 = 0. \qquad \text{Equation in general form}$$

So, you can begin by graphing

$$y = x^2 - 5x + 3 \qquad \text{Function to be graphed}$$

as shown in Figure 2.11. This graph has two *x*-intercepts, and by using the *zoom* and *trace* features you can approximate the corresponding solutions to be

$$x \approx 0.70 \quad \text{and} \quad x \approx 4.30$$

as shown in Figures 2.12 and 2.13.

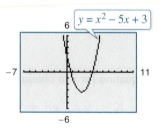

Figure 2.11

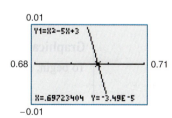

Figure 2.12

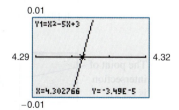

Figure 2.13

✓CHECKPOINT   Now try Exercise 47.

Technology Tip

Use the *zero* or *root* feature of a graphing utility to approximate the solutions of the equation in Example 4 to see that it yields a similar result.

Evaluating an Expression   In Exercises 77 and 78, evaluate the expression in two ways. (a) Calculate entirely on your calculator by storing intermediate results and then rounding the final answer to two decimal places. (b) Round both the numerator and denominator to two decimal places before dividing, and then round the final answer to two decimal places. Does the method in part (b) decrease the accuracy? Explain.

77. $\dfrac{1 + 0.73205}{1 - 0.73205}$ **78.** $\dfrac{1 + 0.86603}{1 - 0.86603}$

79. Professional Sales   On the first part of a 280-mile trip, a salesperson averaged 63 miles per hour. The salesperson averaged only 54 miles per hour on the last part of the trip because of an increased volume of traffic.

(a) Write the total time t for the trip as a function of the distance x traveled at an average speed of 63 miles per hour.

(b) Use a graphing utility to graph the time function. What is the domain of the function?

(c) Approximate the number of miles traveled at 63 miles per hour when the total time is 4 hours and 45 minutes.

80. Chemistry   A 55-gallon barrel contains a mixture with a concentration of 33% sodium chloride. You remove x gallons of this mixture and replace it with 100% sodium chloride.

(a) Write the amount A of sodium chloride in the final mixture as a function of x.

(b) Use a graphing utility to graph the concentration function. What is the domain of the function?

(c) Approximate (accurate to one decimal place) the value of x when the final mixture is 60% sodium chloride.

Geometry   In Exercises 81 and 82, (a) write a function for the area of the region, (b) use a graphing utility to graph the function, and (c) approximate the value of x when the area of the region is 180 square units.

81.   **82.**

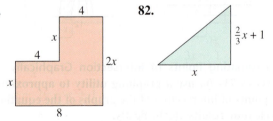

83. Accounting   The following information describes a possible negative income tax for a family consisting of two adults and two children. The plan would guarantee the poor a minimum income while encouraging a family to increase its private income ($0 \le x \le 20{,}000$). (A *subsidy* is a grant of money.)

Family's earned income: $I = x$

Subsidy: $S = 10{,}000 - \frac{1}{2}x$

Total income: $T = I + S$

(a) Write the total income T in terms of x.

(b) Use a graphing utility to find the earned income x when the subsidy is $6600. Verify your answer algebraically.

(c) Use the graphing utility to find the earned income x when the total income is $13,800. Verify your answer algebraically.

(d) Find the subsidy S graphically when the total income is $12,500.

84. Pharmacy   The numbers y (in millions) of drug prescriptions filled by mail order in the United States from 1995 through 2008 can be modeled by $y = 13.0t + 20$, $5 \le t \le 18$, where t is the year, with $t = 5$ corresponding to 1995. (Source: The National Association of Chain Drug Stores)

(a) According to the model, when did the number of prescriptions filled by mail order reach 100 million?

(b) What is the slope of the model and what does it tell you about the number of prescriptions filled by mail order in the United States?

(c) Do you think the model can be used to predict the numbers of prescriptions filled by mail order in the United States for years beyond 2008? If so, for what time period? Explain.

(d) Explain, both algebraically and graphically, how you could find when the number of prescriptions filled by mail order reaches 300 million.

85. *Why you should learn it*   (p. 170)   The populations (in thousands) of Maryland M and Arizona A from 1990 through 2008 can be modeled by

$M = 51.1t + 4785$, $0 \le t \le 18$

$A = 162.0t + 3522$, $0 \le t \le 18$

where t represents the year, with $t = 0$ corresponding to 1990. (Source: U.S. Census Bureau)

(a) Use a graphing utility to graph each model in the same viewing window over the appropriate domain. Approximate the point of intersection. Round your result to one decimal place. Explain the meaning of the coordinates of the point.

(b) Find the point of intersection algebraically. Round your result to one decimal place. What does the point of intersection represent?

(c) Explain the meaning of the slopes of both models and what they tell you about the population growth rates.

(d) Use the models to estimate the population of each state in 2014. Do the values seem reasonable? Explain.

86. Design and Construction Consider the swimming pool in the figure. (When finding its volume, use the fact that the volume is the area of the region on the vertical sidewall times the width of the pool.)

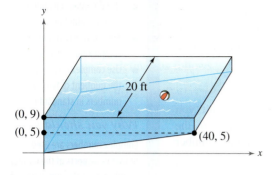

(a) Find the volume of the pool.

(b) Find an equation of the line representing the base of the pool.

(c) The depth of the water at the deep end of the pool is d feet. Show that the volume of water is

$$V(d) = \begin{cases} 80d^2, & 0 \le d \le 5 \\ 800d - 2000, & 5 < d \le 9 \end{cases}.$$

(d) Graph the volume function.

(e) Use a graphing utility to complete the table.

d	3	5	7	9
V				

(f) Approximate the depth of the water at the deep end when the volume is 4800 cubic feet.

(g) How many gallons of water are in the pool? (There are about 7.48 gallons of water in 1 cubic foot.)

Conclusions

True or False? In Exercises 87 and 88, determine whether the statement is true or false. Justify your answer.

87. To find the y-intercept of a graph, let $x = 0$ and solve the equation for y.

88. Every linear equation has at least one y-intercept or x-intercept.

89. Writing You graphically approximate the solution of the equation

$$\frac{x}{x-1} - \frac{99}{100} = 0$$

to be $x = -99.1$. Substituting this value for x produces

$$\frac{-99.1}{-99.1 - 1} - \frac{99}{100} = 0.00000999 = 9.99 \times 10^{-6} \approx 0.$$

Is -99.1 a good approximation of the solution? Write a short paragraph explaining why or why not.

90. CAPSTONE For each of the following, find the answer algebraically, numerically, and graphically.

(a) Find the x- and y-intercepts of the graph of

$$y = 2x + 2.$$

(b) Verify that the real numbers -1 and 1 are zero(s) of the function

$$f(x) = x^2 - 1.$$

(c) Find the points of intersection of the graphs of

$$y = 2x + 2$$

and

$$y = x^2 - 1.$$

Exploration In Exercises 91–94, use the table to solve each linear equation where

$$y_1 = f(x) \quad \text{and} \quad y_2 = g(x).$$

X	Y1	Y2
-2	-15	0
-1	-12	2
0	-9	4
1	-6	6
2	-3	8
3	0	10
4	3	12

X= -2

91. $f(x) = 0$

92. $g(x) = 0$

93. $g(x) = -f(x)$

94. $f(x) = -6g(x)$

Cumulative Mixed Review

Rationalizing a Denominator In Exercises 95–98, rationalize the denominator.

95. $\dfrac{12}{5\sqrt{3}}$

96. $\dfrac{4}{\sqrt{10} - 2}$

97. $\dfrac{3}{8 + \sqrt{11}}$

98. $\dfrac{14}{3\sqrt{10} - 1}$

Multiplying Polynomials In Exercises 99–102, find the product.

99. $(x + 6)(3x - 5)$

100. $(3x + 13)(4x - 7)$

101. $(2x - 9)(2x + 9)$

102. $(4x + 1)^2$

Many of the properties of real numbers are valid for complex numbers as well. Here are some examples.

    Associative Properties of Addition and Multiplication
    Commutative Properties of Addition and Multiplication
    Distributive Property of Multiplication over Addition

Notice how these properties are used when two complex numbers are multiplied.

$$(a + bi)(c + di) = a(c + di) + bi(c + di) \qquad \text{Distributive Property}$$
$$= ac + (ad)i + (bc)i + (bd)i^2 \qquad \text{Distributive Property}$$
$$= ac + (ad)i + (bc)i + (bd)(-1) \qquad i^2 = -1$$
$$= ac - bd + (ad)i + (bc)i \qquad \text{Commutative Property}$$
$$= (ac - bd) + (ad + bc)i \qquad \text{Associative Property}$$

The procedure above is similar to multiplying two polynomials and combining like terms, as in the FOIL Method discussed in Section P.3.

Example 2 Multiplying Complex Numbers

Perform the operation(s) and write the result in standard form.

a. $\sqrt{-4} \cdot \sqrt{-16}$ **b.** $(2 - i)(4 + 3i)$ **c.** $(3 + 2i)(3 - 2i)$

d. $4i(-1 + 5i)$ **e.** $(3 + 2i)^2$

Solution

a. $\sqrt{-4} \cdot \sqrt{-16} = (2i)(4i)$ Write each factor in *i*-form.
$$= 8i^2 \qquad \text{Multiply.}$$
$$= 8(-1) \qquad i^2 = -1$$
$$= -8 \qquad \text{Simplify.}$$

b. $(2 - i)(4 + 3i) = 8 + 6i - 4i - 3i^2$ Product of binomials
$$= 8 + 6i - 4i - 3(-1) \qquad i^2 = -1$$
$$= 8 + 3 + 6i - 4i \qquad \text{Group like terms.}$$
$$= 11 + 2i \qquad \text{Write in standard form.}$$

c. $(3 + 2i)(3 - 2i) = 9 - 6i + 6i - 4i^2$ Product of binomials
$$= 9 - 4(-1) \qquad i^2 = -1$$
$$= 9 + 4 \qquad \text{Simplify.}$$
$$= 13 \qquad \text{Write in standard form.}$$

d. $4i(-1 + 5i) = 4i(-1) + 4i(5i)$ Distributive Property
$$= -4i + 20i^2 \qquad \text{Simplify.}$$
$$= -4i + 20(-1) \qquad i^2 = -1$$
$$= -20 - 4i \qquad \text{Write in standard form.}$$

e. $(3 + 2i)^2 = 9 + 6i + 6i + 4i^2$ Product of binomials
$$= 9 + 12i + 4(-1) \qquad i^2 = -1$$
$$= 9 - 4 + 12i \qquad \text{Group like terms.}$$
$$= 5 + 12i \qquad \text{Write in standard form.}$$

✓**CHECKPOINT**   Now try Exercise 37.

Explore the Concept

Complete the following:

$i^1 = i$ $i^7 = $

$i^2 = -1$ $i^8 = $

$i^3 = -i$ $i^9 = $

$i^4 = 1$ $i^{10} = $

$i^5 = $ $i^{11} = $

$i^6 = $ $i^{12} = $

What pattern do you see? Write a brief description of how you would find *i* raised to any positive integer power.

Study Tip

Before you perform operations with complex numbers, be sure to rewrite the terms or factors in *i*-form first and then proceed with the operations, as shown in Example 2(a).

Complex Conjugates

Notice in Example 2(c) that the product of two complex numbers can be a real number. This occurs with pairs of complex numbers of the forms $a + bi$ and $a - bi$, called **complex conjugates.**

$$(a + bi)(a - bi) = a^2 - abi + abi - b^2i^2$$
$$= a^2 - b^2(-1)$$
$$= a^2 + b^2$$

Example 3   Multiplying Conjugates

Multiply each complex number by its complex conjugate.

a. $1 + i$ **b.** $3 - 5i$

Solution

a. The complex conjugate of $1 + i$ is $1 - i$.

$$(1 + i)(1 - i) = 1^2 - i^2 = 1 - (-1) = 2$$

b. The complex conjugate of $3 - 5i$ is $3 + 5i$.

$$(3 - 5i)(3 + 5i) = 3^2 - (5i)^2 = 9 - 25i^2 = 9 - 25(-1) = 34$$

✔CHECKPOINT   Now try Exercise 47.

To write the quotient of $a + bi$ and $c + di$ in standard form, where c and d are not both zero, multiply the numerator and denominator by the complex conjugate of the *denominator* to obtain

$$\frac{a + bi}{c + di} = \frac{a + bi}{c + di}\left(\frac{c - di}{c - di}\right)$$ Multiply numerator and denominator by complex conjugate of denominator.

$$= \frac{ac + bd}{c^2 + d^2} + \left(\frac{bc - ad}{c^2 + d^2}\right)i.$$ Standard form

Technology Tip

Some graphing utilities can perform operations with complex numbers. For instance, on some graphing utilities, to divide $2 + 3i$ by $4 - 2i$, use the following keystrokes.

(2 + 3 *i*)) ÷

(4 − 2 *i*)) ENTER

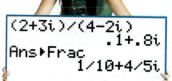

Example 4   Writing a Quotient of Complex Numbers in Standard Form

Write the quotient $\dfrac{2 + 3i}{4 - 2i}$ in standard form. $a + bi$

Solution

$$\frac{2 + 3i}{4 - 2i} = \frac{2 + 3i}{4 - 2i}\left(\frac{4 + 2i}{4 + 2i}\right)$$ Multiply numerator and denominator by complex conjugate of denominator.

$$= \frac{8 + 4i + 12i + 6i^2}{16 - 4i^2}$$ Expand.

$$= \frac{8 - 6 + 16i}{16 + 4}$$ $i^2 = -1$

$$= \frac{2 + 16i}{20}$$ Simplify.

$$= \frac{1}{10} + \frac{4}{5}i.$$ Write in standard form.

✔CHECKPOINT   Now try Exercise 59.

GLUE STOCK 2010/used under license from Shutterstock.com

2.3   Exercises

See www.CalcChat.com for worked-out solutions to odd-numbered exercises.
For instructions on how to use a graphing utility, see Appendix A.

Vocabulary and Concept Check

1. Match the type of complex number with its definition.
   - (a) real number
   - (b) imaginary number
   - (c) pure imaginary number
   - (i) $a + bi, a = 0, b \neq 0$
   - (ii) $a + bi, b = 0$
   - (iii) $a + bi, a \neq 0, b \neq 0$

In Exercises 2 and 3, fill in the blanks.

2. The imaginary unit i is defined as $i =$ _____ , where $i^2 =$ _____ .

3. The set of real multiples of the imaginary unit i combined with the set of real numbers is called the set of _____ numbers, which are written in the standard form _____ .

4. What method for multiplying two polynomials can you use when multiplying two complex numbers?

5. What is the additive inverse of the complex number $2 - 4i$?

6. What is the complex conjugate of the complex number $2 - 4i$?

Procedures and Problem Solving

Equality of Complex Numbers   In Exercises 7–10, find real numbers a and b such that the equation is true.

7. $a + bi = -9 + 4i$

8. $a + bi = 12 + 5i$

9. $(a - 1) + (b + 3)i = 5 + 8i$

10. $(a + 6) + 2bi = 6 - 5i$

Writing a Complex Number in Standard Form   In Exercises 11–20, write the complex number in standard form.

11. $5 + \sqrt{-16}$

12. $2 - \sqrt{-9}$

13. -6

14. 8

15. $-5i + i^2$

16. $-3i^2 + i$

17. $\left(\sqrt{-75}\right)^2$

18. $\left(\sqrt{-4}\right)^2 - 7$

19. $\sqrt{-0.09}$

20. $\sqrt{-0.0004}$

Adding and Subtracting Complex Numbers   In Exercises 21–30, perform the addition or subtraction and write the result in standard form.

21. $(4 + i) - (7 - 2i)$

22. $(11 - 2i) - (-3 + 6i)$

23. $\left(-1 + \sqrt{-8}\right) + \left(8 - \sqrt{-50}\right)$

24. $\left(7 + \sqrt{-18}\right) + \left(3 + \sqrt{-32}\right)$

✓ 25. $13i - (14 - 7i)$

26. $22 + (-5 + 8i) - 9i$

27. $\left(\frac{3}{2} + \frac{5}{2}i\right) + \left(\frac{5}{3} + \frac{11}{3}i\right)$

28. $\left(\frac{3}{4} + \frac{7}{5}i\right) - \left(\frac{5}{6} - \frac{1}{6}i\right)$

29. $(1.6 + 3.2i) + (-5.8 + 4.3i)$

30. $-(-3.7 - 12.8i) - (6.1 - 16.3i)$

Multiplying Complex Numbers   In Exercises 31–46, perform the operation and write the result in standard form.

31. $\sqrt{-6} \cdot \sqrt{-2}$

32. $\sqrt{-5} \cdot \sqrt{-10}$

33. $\left(\sqrt{-10}\right)^2$

34. $\left(\sqrt{-75}\right)^2$

35. $4(3 + 5i)$

36. $-6(5 - 3i)$

✓ 37. $(1 + i)(3 - 2i)$

38. $(6 - 2i)(2 - 3i)$

39. $4i(8 + 5i)$

40. $-3i(6 - i)$

41. $\left(\sqrt{14} + \sqrt{10}i\right)\left(\sqrt{14} - \sqrt{10}i\right)$

42. $\left(\sqrt{3} + \sqrt{15}i\right)\left(\sqrt{3} - \sqrt{15}i\right)$

43. $(6 + 7i)^2$

44. $(5 - 4i)^2$

45. $(4 + 5i)^2 - (4 - 5i)^2$

46. $(1 - 2i)^2 - (1 + 2i)^2$

Multiplying Conjugates   In Exercises 47–54, write the complex conjugate of the complex number. Then multiply the number by its complex conjugate.

✓ 47. $4 + 3i$

48. $7 - 5i$

49. $-6 - \sqrt{5}i$

50. $-3 + \sqrt{2}i$

51. $\sqrt{-20}$

52. $\sqrt{-13}$

53. $3 - \sqrt{-2}$

54. $1 + \sqrt{-8}$

Writing a Quotient of Complex Numbers in Standard Form   In Exercises 55–62, write the quotient in standard form.

55. $\dfrac{6}{i}$

56. $-\dfrac{5}{2i}$

57. $\dfrac{2}{4 - 5i}$ 58. $\dfrac{3}{1 - i}$

✓ 59. $\dfrac{2 + i}{2 - i}$ 60. $\dfrac{8 - 7i}{1 - 2i}$

61. $\dfrac{i}{(4 - 5i)^2}$ 62. $\dfrac{5i}{(2 + 3i)^2}$

Adding or Subtracting Quotients of Complex Numbers
In Exercises 63–66, perform the operation and write the result in standard form.

63. $\dfrac{2}{1 + i} - \dfrac{3}{1 - i}$ 64. $\dfrac{2i}{2 + i} + \dfrac{5}{2 - i}$

65. $\dfrac{i}{3 - 2i} + \dfrac{2i}{3 + 8i}$ 66. $\dfrac{1 + i}{i} - \dfrac{3}{4 - i}$

Expressions Involving Powers of i
In Exercises 67–72, simplify the complex number and write it in standard form.

67. $-6i^3 + i^2$ 68. $4i^2 - 2i^3$

69. $\left(\sqrt{-75}\right)^3$ 70. $\left(\sqrt{-2}\right)^6$

71. $\dfrac{1}{i^3}$ 72. $\dfrac{1}{(2i)^3}$

73. Cube each complex number. What do you notice?

    (a) 2   (b) $-1 + \sqrt{3}i$ (c) $-1 - \sqrt{3}i$

74. Raise each complex number to the fourth power and simplify.

    (a) 2   (b) -2 (c) $2i$ (d) $-2i$

75. Use the results of the Explore the Concept feature on page 182 to find each power of i.

    (a) i^{20} (b) i^{45} (c) i^{67} (d) i^{114}

76. **Why you should learn it** (p. 180) The opposition to current in an electrical circuit is called its impedance. The impedance z in a parallel circuit with two pathways satisfies the equation $1/z = 1/z_1 + 1/z_2$, where z_1 is the impedance (in ohms) of pathway 1, and z_2 is the impedance (in ohms) of pathway 2. Use the table to determine the impedance of each parallel circuit. (*Hint:* You can find the impedance of each pathway in a parallel circuit by adding the impedances of all components in the pathway.)

	Resistor	Inductor	Capacitor
Symbol	⟿	⟼	⊣⊢
	$a\ \Omega$	$b\ \Omega$	$c\ \Omega$
Impedance	a	bi	$-ci$

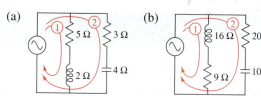

(a)   (b)

Conclusions

True or False? In Exercises 77–82, determine whether the statement is true or false. Justify your answer.

77. No complex number is equal to its complex conjugate.

78. $i^{44} + i^{150} - i^{74} - i^{109} + i^{61} = -1$

79. The sum of two imaginary numbers is always an imaginary number.

80. The product of two imaginary numbers is always an imaginary number.

81. The conjugate of the product of two complex numbers is equal to the product of the conjugates of the two complex numbers.

82. The conjugate of the sum of two complex numbers is equal to the sum of the conjugates of the two complex numbers.

83. **Error Analysis** Describe the error.

$$\sqrt{-6}\sqrt{-6} = \sqrt{(-6)(-6)} = \sqrt{36} = 6$$

84. **CAPSTONE** Consider the binomials $x + 5$ and $2x - 1$ and the complex numbers $1 + 5i$ and $2 - i$.

    (a) Find the sum of the binomials and the sum of the complex numbers.

    (b) Find the difference of the binomials and the difference of the complex numbers.

    (c) Describe the similarities and differences in your results for parts (a) and (b).

    (d) Find the product of the binomials and the product of the complex numbers.

    (e) Explain why the products you found in part (d) are not related in the same way as your results in parts (a) and (b).

    (f) Write a brief paragraph that compares operations with binomials and operations with complex numbers.

Cumulative Mixed Review

Multiplying Polynomials In Exercises 85–88, perform the operation and write the result in standard form.

85. $(4x - 5)(4x + 5)$ 86. $(x + 2)^3$

87. $\left(3x - \tfrac{1}{2}\right)(x + 4)$ 88. $(2x - 5)^2$

2.4   Solving Quadratic Equations Algebraically

Quadratic Equations

A **quadratic equation in x** is an equation that can be written in the general form

$$ax^2 + bx + c = 0$$

where a, b, and c are real numbers with $a \neq 0$. A quadratic equation in x is also known as a **second-degree polynomial equation in x.** You should be familiar with the following four methods for solving quadratic equations.

Solving a Quadratic Equation

Factoring: If $ab = 0$, then $a = 0$ or $b = 0$.   Zero-Factor Property

Example: $x^2 - x - 6 = 0$

$(x - 3)(x + 2) = 0$

$x - 3 = 0 \implies x = 3$

$x + 2 = 0 \implies x = -2$

Extracting Square Roots: If $u^2 = c$, then $u = \pm\sqrt{c}$.

Example: $(x + 3)^2 = 16$

$x + 3 = \pm 4$

$x = -3 \pm 4$

$x = 1 \quad \text{or} \quad x = -7$

Completing the Square: If $x^2 + bx = c$, then

$$x^2 + bx + \left(\frac{b}{2}\right)^2 = c + \left(\frac{b}{2}\right)^2$$

$$\left(x + \frac{b}{2}\right)^2 = c + \frac{b^2}{4}.$$

Example: $x^2 + 6x = 5$

$x^2 + 6x + 3^2 = 5 + 3^2$

$(x + 3)^2 = 14$

$x + 3 = \pm\sqrt{14}$

$x = -3 \pm \sqrt{14}$

Quadratic Formula: If $ax^2 + bx + c = 0$, then $x = \dfrac{-b \pm \sqrt{b^2 - 4ac}}{2a}$.

Example: $2x^2 + 3x - 1 = 0$

$$x = \frac{-3 \pm \sqrt{3^2 - 4(2)(-1)}}{2(2)}$$

$$x = \frac{-3 \pm \sqrt{17}}{4}$$

Note that you can solve every quadratic equation by completing the square or by using the Quadratic Formula.

What you should learn

- Solve quadratic equations by factoring.
- Solve quadratic equations by extracting square roots.
- Solve quadratic equations by completing the square.
- Use the Quadratic Formula to solve quadratic equations.
- Use quadratic equations to model and solve real-life problems.

Why you should learn it

Knowing how to solve quadratic equations algebraically can help you solve real-life problems, such as Exercise 92 on page 198, where you determine the annual per capita consumption of milk.

Technology Tip

Try programming the Quadratic Formula into a computer or graphing calculator. Programs for several graphing calculator models can be found at this textbook's *Companion Website*.

Example 1   Solving a Quadratic Equation by Factoring

Solve each quadratic equation by factoring.

a. $6x^2 = 3x$

b. $9x^2 - 6x = -1$

Solution

a.
$$6x^2 = 3x \qquad \text{Write original equation.}$$
$$6x^2 - 3x = 0 \qquad \text{Write in general form.}$$
$$3x(2x - 1) = 0 \qquad \text{Factor.}$$
$$3x = 0 \quad \Longrightarrow \quad x = 0 \qquad \text{Set 1st factor equal to 0.}$$
$$2x - 1 = 0 \quad \Longrightarrow \quad x = \tfrac{1}{2} \qquad \text{Set 2nd factor equal to 0.}$$

b.
$$9x^2 - 6x = -1 \qquad \text{Write original equation.}$$
$$9x^2 - 6x + 1 = 0 \qquad \text{Write in general form.}$$
$$(3x - 1)^2 = 0 \qquad \text{Factor.}$$
$$3x - 1 = 0 \quad \Longrightarrow \quad x = \tfrac{1}{3} \qquad \text{Set repeated factor equal to 0.}$$

Throughout the text, when solving equations, be sure to check your solutions either *algebraically* by substituting in the original equation or *graphically*.

Check

a.
$$6x^2 = 3x \qquad \text{Write original equation.}$$
$$6(0)^2 \stackrel{?}{=} 3(0) \qquad \text{Substitute 0 for } x.$$
$$0 = 0 \qquad \text{Solution checks. } \checkmark$$
$$6\left(\tfrac{1}{2}\right)^2 \stackrel{?}{=} 3\left(\tfrac{1}{2}\right) \qquad \text{Substitute } \tfrac{1}{2} \text{ for } x.$$
$$\tfrac{6}{4} = \tfrac{3}{2} \qquad \text{Solution checks. } \checkmark$$

b.
$$9x^2 - 6x = -1 \qquad \text{Write original equation.}$$
$$9\left(\tfrac{1}{3}\right)^2 - 6\left(\tfrac{1}{3}\right) \stackrel{?}{=} -1 \qquad \text{Substitute } \tfrac{1}{3} \text{ for } x.$$
$$1 - 2 \stackrel{?}{=} -1 \qquad \text{Simplify.}$$
$$-1 = -1 \qquad \text{Solution checks. } \checkmark$$

Similarly, you can check your solutions graphically using a graphing utility, as shown in Figure 2.19.

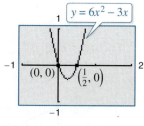

(a)   (b)

Figure 2.19

✔CHECKPOINT   Now try Exercise 11.

Extracting Square Roots

Example 2   Extracting Square Roots

Solve each quadratic equation by extracting square roots.

a. $4x^2 = 12$ **b.** $(x - 3)^2 = 7$ **c.** $(2x - 1)^2 = -9$

Solution

a. $4x^2 = 12$ Write original equation.

$\quad x^2 = 3$ Divide each side by 4.

$\quad x = \pm\sqrt{3}$ Take square root of each side.

This equation has two solutions: $x = \pm\sqrt{3} \approx \pm 1.73$. You can check your solutions graphically, as shown in Figure 2.20.

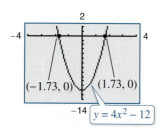

Figure 2.20

b. $(x - 3)^2 = 7$ Write original equation.

$\quad x - 3 = \pm\sqrt{7}$ Take square root of each side.

$\quad x = 3 \pm\sqrt{7}$ Add 3 to each side.

This equation has two solutions: $x = 3 + \sqrt{7} \approx 5.65$ and $x = 3 - \sqrt{7} \approx 0.35$. You can check your solutions graphically, as shown in Figure 2.21.

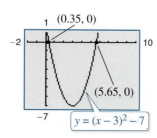

Figure 2.21

c. $(2x - 1)^2 = -9$ Write original equation.

$\quad 2x - 1 = \pm\sqrt{-9}$ Take square root of each side.

$\quad 2x - 1 = \pm 3i$ Write in i-form.

$\quad x = \frac{1}{2} \pm \frac{3}{2}i$ Solve for x.

This equation has two complex solutions: $x = \frac{1}{2} \pm \frac{3}{2}i$. Check these in the original equation. Notice from the graph shown in Figure 2.22 that you can conclude the equation has no real solution.

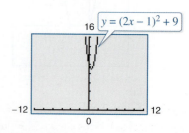

Figure 2.22

✔CHECKPOINT   Now try Exercise 25.

Study Tip

When you take the square root of a variable expression, you must account for both positive and negative solutions.

Completing the Square

Completing the square is best suited for quadratic equations in general form $ax^2 + bx + c = 0$ with $a = 1$ and b an even number (see page 186). When the leading coefficient of the quadratic is not 1, divide each side of the equation by this coefficient before completing the square (see Examples 4 and 5).

Example 3 Completing the Square: Leading Coefficient Is 1

Solve $x^2 + 2x - 6 = 0$ by completing the square.

Solution

$x^2 + 2x - 6 = 0$	Write original equation.
$x^2 + 2x = 6$	Add 6 to each side.
$x^2 + 2x + 1^2 = 6 + 1^2$	Add 1^2 to each side.

(Half of 2)^2

$(x + 1)^2 = 7$	Simplify.
$x + 1 = \pm\sqrt{7}$	Take square root of each side.
$x = -1 \pm \sqrt{7}$	Solutions

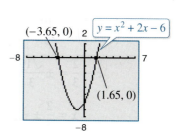

Using a calculator, the two solutions are $x \approx 1.65$ and $x \approx -3.65$, which agree with the graphical solutions shown in Figure 2.23.

Figure 2.23

✔**CHECKPOINT**   Now try Exercise 29.

Example 4 Completing the Square: Leading Coefficient Is Not 1

$2x^2 + 8x + 3 = 0$	Original equation
$2x^2 + 8x = -3$	Subtract 3 from each side.
$x^2 + 4x = -\dfrac{3}{2}$	Divide each side by 2.
$x^2 + 4x + 2^2 = -\dfrac{3}{2} + 2^2$	Add 2^2 to each side.

(Half of 4)^2

$(x + 2)^2 = \dfrac{5}{2}$	Simplify.
$x + 2 = \pm\sqrt{\dfrac{5}{2}}$	Take square root of each side.
$x + 2 = \pm\dfrac{\sqrt{10}}{2}$	Rationalize denominator.
$x = -2 \pm \dfrac{\sqrt{10}}{2}$	Solutions

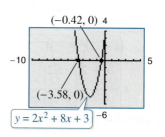

Using a calculator, the two solutions are $x \approx -0.42$ and $x \approx -3.58$, which agree with the graphical solutions shown in Figure 2.24.

Figure 2.24

✔**CHECKPOINT**   Now try Exercise 33.

Example 8 Quadratic Formula: One Repeated Solution

Solve the equation

$$8x^2 - 24x + 18 = 0$$

using the Quadratic Formula.

Algebraic Solution

This equation has a common factor of 2. You can simplify the equation by dividing each side of the equation by 2.

$8x^2 - 24x + 18 = 0$	Write original equation.
$4x^2 - 12x + 9 = 0$	Divide each side by 2.
$x = \dfrac{-b \pm \sqrt{b^2 - 4ac}}{2a}$	Quadratic Formula
$x = \dfrac{-(-12) \pm \sqrt{(-12)^2 - 4(4)(9)}}{2(4)}$	
$x = \dfrac{12 \pm \sqrt{0}}{8}$	Simplify.
$x = \dfrac{3}{2}$	Repeated solution

The equation has only one solution: $x = \frac{3}{2}$. Check this solution in the original equation.

✓CHECKPOINT   Now try Exercise 55.

Graphical Solution

Use a graphing utility to graph

$$y = 8x^2 - 24x + 18.$$

Use the *zero* feature of the graphing utility to approximate any values of x for which the function is equal to zero. From Figure 2.27, it appears that the function is equal to zero when $x = 1.5 = \frac{3}{2}$. This is the only solution of the equation $8x^2 - 24x + 18 = 0$.

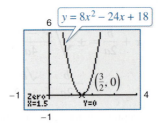

Figure 2.27

Example 9 Quadratic Formula: No Real Solution

Solve the equation

$$3x^2 - 2x + 5 = 0$$

using the Quadratic Formula.

Algebraic Solution

This equation is in general form, so use the Quadratic Formula with $a = 3$, $b = -2$, and $c = 5$.

$x = \dfrac{-b \pm \sqrt{b^2 - 4ac}}{2a}$	Quadratic Formula
$x = \dfrac{-(-2) \pm \sqrt{(-2)^2 - 4(3)(5)}}{2(3)}$	Substitute -2 for b, 3 for a, and 5 for c.
$x = \dfrac{2 \pm \sqrt{-56}}{6}$	Simplify.
$x = \dfrac{2 \pm 2\sqrt{14}i}{6}$	Simplify radical.
$x = \dfrac{1}{3} \pm \dfrac{\sqrt{14}}{3}i$	Solutions

The equation has no real solution, but it has two complex solutions:

$$x = \tfrac{1}{3}\left(1 + \sqrt{14}i\right) \quad \text{and} \quad x = \tfrac{1}{3}\left(1 - \sqrt{14}i\right).$$

✓CHECKPOINT   Now try Exercise 59.

Graphical Solution

Use a graphing utility to graph

$$y = 3x^2 - 2x + 5.$$

Note in Figure 2.28 that the graph of the function appears to have no x-intercept. From this you can conclude that the equation $3x^2 - 2x + 5 = 0$ has no real solution. You can solve the equation algebraically to find the complex solutions.

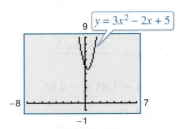

Figure 2.28

Applications

A common application of quadratic equations involves an object that is falling (or projected into the air). The general equation that gives the height of such an object is called a **position equation,** and on *Earth's* surface it has the form

$$s = -16t^2 + v_0 t + s_0.$$

In this equation, s represents the height of the object (in feet), v_0 represents the initial velocity of the object (in feet per second), s_0 represents the initial height of the object (in feet), and t represents the time (in seconds). Note that this position equation ignores air resistance.

Example 10   Falling Time

A construction worker on the 24th floor of a building project (see Figure 2.29) accidentally drops a wrench and yells, "Look out below!" Could a person at ground level hear this warning in time to get out of the way?

Solution

Assume that each floor of the building is 10 feet high, so that the wrench is dropped from a height of 235 feet (the construction worker's hand is 5 feet below the ceiling of the 24th floor). Because sound travels at about 1100 feet per second, it follows that a person at ground level hears the warning within 1 second of the time the wrench is dropped. To set up a mathematical model for the height of the wrench, use the position equation

$$s = -16t^2 + v_0 t + s_0. \qquad \text{\color{red}Position equation}$$

Because the object is dropped rather than thrown, the initial velocity is $v_0 = 0$ feet per second. So, with an initial height of $s_0 = 235$ feet, you have the model

$$s = -16t^2 + (0)t + 235 = -16t^2 + 235.$$

After falling for 1 second, the height of the wrench is

$$-16(1)^2 + 235 = 219 \text{ feet.}$$

After falling for 2 seconds, the height of the wrench is

$$-16(2)^2 + 235 = 171 \text{ feet.}$$

To find the number of seconds it takes the wrench to hit the ground, let the height s be zero and solve the equation for t.

$$s = -16t^2 + 235 \qquad \text{\color{red}Write position equation.}$$

$$0 = -16t^2 + 235 \qquad \text{\color{red}Substitute 0 for } s.$$

$$16t^2 = 235 \qquad \text{\color{red}Add } 16t^2 \text{ to each side.}$$

$$t^2 = \frac{235}{16} \qquad \text{\color{red}Divide each side by 16.}$$

$$t = \frac{\sqrt{235}}{4} \qquad \text{\color{red}Extract positive square root.}$$

$$t \approx 3.83 \qquad \text{\color{red}Use a calculator.}$$

The wrench will take about 3.83 seconds to hit the ground. If the person hears the warning 1 second after the wrench is dropped, the person still has almost 3 more seconds to get out of the way.

✓CHECKPOINT   Now try Exercise 87.

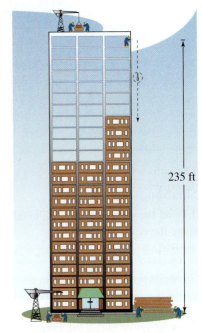

235 ft

Figure 2.29

Example 11 Quadratic Modeling: Internet Users

From 2000 through 2009, the estimated numbers of Internet users I (in millions) in the United States can be modeled by the quadratic equation

$$I = -1.21t^2 + 21.9t + 124, \quad 0 \le t \le 9$$

where t represents the year, with $t = 0$ corresponding to 2000 (see Figure 2.30). According to the model, in which year did the number of Internet users reach 212 million? (Sources: International Telecommunication Union, The Nielsen Company)

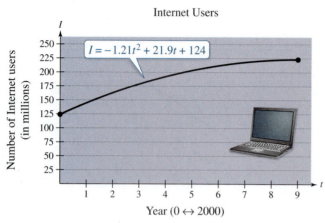

Internet Users

$I = -1.21t^2 + 21.9t + 124$

Year ($0 \leftrightarrow 2000$)

Figure 2.30

Solution

To find the year in which the number of Internet users reached 212 million, you can solve the equation

$$-1.21t^2 + 21.9t + 124 = 212.$$

To begin, write the equation in general form.

$$-1.21t^2 + 21.9t - 88 = 0$$

Then apply the Quadratic Formula.

$$t = \frac{-b \pm \sqrt{b^2 - 4ac}}{2a}$$

$$t = \frac{-21.9 \pm \sqrt{21.9^2 - 4(-1.21)(-88)}}{2(-1.21)}$$

$$t = \frac{-21.9 \pm \sqrt{53.69}}{-2.42} \qquad \Longrightarrow \qquad t \approx 6.02 \text{ or } 12.08$$

Choose the smaller value $t \approx 6.02$. Because $t = 0$ corresponds to 2000, it follows that $t \approx 6.02$ must correspond to 2006. So, the number of Internet users reached 212 million during 2006.

CHECKPOINT Now try Exercise 91.

Technology Tip

 You can solve Example 11 with your graphing utility by graphing $y_1 = -1.21x^2 + 21.9x + 124$ and $y_2 = 212$ in the same viewing window and finding their point of intersection. You should obtain $x \approx 6.02$.

Another type of application that often involves a quadratic equation is one dealing with the hypotenuse of a right triangle. These types of applications often use the Pythagorean Theorem, which states that $a^2 + b^2 = c^2$ where a and b are the legs of a right triangle and c is the hypotenuse, as indicated in Figure 2.31.

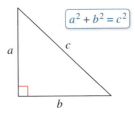

$a^2 + b^2 = c^2$

Figure 2.31

Example 12   An Application Involving the Pythagorean Theorem

An L-shaped sidewalk from the athletic center to the library on a college campus is shown in Figure 2.32. The sidewalk was constructed so that the length of one sidewalk forming the L is twice as long as the other. The length of the diagonal sidewalk that cuts across the grounds between the two buildings is 102 feet. How many feet does a person save by walking on the diagonal sidewalk?

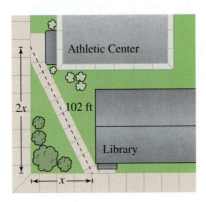

Figure 2.32

Solution

Using the Pythagorean Theorem, you have

$a^2 + b^2 = c^2$	Pythagorean Theorem
$x^2 + (2x)^2 = 102^2$	Substitute for a, b, and c.
$5x^2 = 10{,}404$	Combine like terms.
$x^2 = 2080.8$	Divide each side by 5.
$x = \pm\sqrt{2080.8}$	Take the square root of each side.
$x = \sqrt{2080.8}.$	Extract positive square root.

The total distance covered by walking on the L-shaped sidewalk is

$$x + 2x = 3x = 3\sqrt{2080.8} \approx 136.85 \text{ feet.}$$

Walking on the diagonal sidewalk saves a person about

$$136.85 - 102 = 34.85 \text{ feet.}$$

✓CHECKPOINT   Now try Exercise 95.

2.4 Exercises

See www.CalcChat.com for worked-out solutions to odd-numbered exercises.
For instructions on how to use a graphing utility, see Appendix A.

Vocabulary and Concept Check

In Exercises 1 and 2, fill in the blank.

1. An equation of the form $ax^2 + bx + c = 0$, where a, b, and c are real numbers and $a \neq 0$, is a _____ , or a second-degree polynomial equation in x.

2. The part of the Quadratic Formula $b^2 - 4ac$, known as the _____ , determines the type of solutions of a quadratic equation.

3. List four methods that can be used to solve a quadratic equation.

4. What does the equation $s = -16t^2 + v_0 t + s_0$ represent? What do v_0 and s_0 represent?

Procedures and Problem Solving

Writing a Quadratic Equation in General Form In Exercises 5–8, write the quadratic equation in general form. Do not solve the equation.

5. $2x^2 = 3 - 5x$

6. $x^2 = 25x + 26$

7. $\frac{1}{5}(3x^2 - 10) = 12x$

8. $x(x + 2) = 3x^2 + 1$

Solving a Quadratic Equation by Factoring In Exercises 9–20, solve the quadratic equation by factoring. Check your solutions in the original equation.

9. $6x^2 + 3x = 0$

10. $9x^2 - 1 = 0$

✓ 11. $x^2 - 2x - 8 = 0$

12. $x^2 - 10x + 9 = 0$

13. $x^2 + 10x + 25 = 0$

14. $4x^2 + 12x + 9 = 0$

15. $3 + 5x - 2x^2 = 0$

16. $2x^2 = 19x + 33$

17. $x^2 + 4x = 12$

18. $-x^2 + 8x = 12$

19. $(x + a)^2 - b^2 = 0$

20. $x^2 + 2ax + a^2 = 0$

Extracting Square Roots In Exercises 21–28, solve the equation by extracting square roots. List both the exact solutions and the decimal solutions rounded to the nearest hundredth.

21. $x^2 = 49$

22. $x^2 = 144$

23. $(x - 12)^2 = 16$

24. $(x - 5)^2 = 25$

✓ 25. $(3x - 1)^2 + 6 = 0$

26. $(2x + 3)^2 + 25 = 0$

27. $(x - 7)^2 = (x + 3)^2$

28. $(x + 5)^2 = (x + 4)^2$

Completing the Square In Exercises 29–42, solve the quadratic equation by completing the square. Verify your answer graphically.

✓ 29. $x^2 + 4x - 32 = 0$

30. $x^2 - 2x - 3 = 0$

31. $x^2 + 6x + 2 = 0$

32. $x^2 + 8x + 14 = 0$

✓ 33. $9x^2 - 18x + 3 = 0$

34. $4x^2 - 4x - 99 = 0$

35. $-6 + 2x - x^2 = 0$

36. $-x^2 + x - 1 = 0$

✓ 37. $2x^2 + 5x - 8 = 0$

38. $9x^2 - 12x - 14 = 0$

✓ 39. $x^2 - 4x + 13 = 0$

40. $x^2 - 6x + 34 = 0$

41. $x^2 + 2x + 17 = 0$

42. $x^2 + 18x + 117 = 0$

Graphing a Quadratic Equation In Exercises 43–46, (a) use a graphing utility to graph the equation, (b) use the graph to approximate any x-intercepts of the graph, and (c) verify your results algebraically.

43. $y = (x + 3)^2 - 4$

44. $y = 1 - (x - 2)^2$

45. $y = -4x^2 + 4x + 3$

46. $y = x^2 + 3x - 4$

Using a Graphing Utility to Find Solutions In Exercises 47–52, use a graphing utility to determine the number of real solutions of the quadratic equation.

47. $x^2 - 4x + 4 = 0$

48. $2x^2 - x - 1 = 0$

49. $\frac{4}{7}x^2 - 8x + 28 = 0$

50. $\frac{1}{3}x^2 - 5x + 25 = 0$

51. $-0.2x^2 + 1.2x - 8 = 0$

52. $9 + 2.4x - 8.3x^2 = 0$

Using the Quadratic Formula **In Exercises 53–60, use the Quadratic Formula to solve the equation. Use a graphing utility to verify your solutions graphically.**

✓ **53.** $2 + 2x - x^2 = 0$ **54.** $x^2 - 10x + 22 = 0$

✓ **55.** $-49x^2 + 28x - 4 = 0$ **56.** $9x^2 - 18x + 9 = 0$

57. $x^2 + 3x = -8$ **58.** $x^2 + 16 = -5x$

✓ **59.** $4x^2 + 16x + 17 = 0$ **60.** $9x^2 - 6x + 37 = 0$

Solving a Quadratic Equation **In Exercises 61–68, solve the equation using any convenient method.**

61. $x^2 - 2x - 1 = 0$ **62.** $11x^2 + 33x = 0$

63. $(x + 3)^2 = 81$ **64.** $(x - 1)^2 = -1$

65. $x^2 - 2x + \frac{13}{4} = 0$ **66.** $x^2 + 3x - \frac{3}{4} = 0$

67. $(x + 1)^2 = x^2$ **68.** $a^2x^2 - b^2 = 0, a \neq 0$

Finding x-Intercepts Algebraically **In Exercises 69–72, find algebraically the x-intercept(s), if any, of the graph of the equation.**

69.

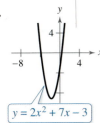

$y = 2x^2 + 7x - 3$

70.

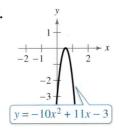

$y = -10x^2 + 11x - 3$

71.

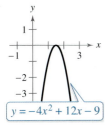

$y = -4x^2 + 12x - 9$

72.

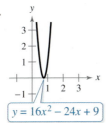

$y = 16x^2 - 24x + 9$

Think About It **In Exercises 73–82, find two quadratic equations having the given solutions. (There are many correct answers.)**

73. $-6, 5$ **74.** $-2, 1$

75. $-\frac{7}{3}, \frac{6}{7}$ **76.** $-\frac{2}{3}, \frac{4}{3}$

77. $5\sqrt{3}, -5\sqrt{3}$ **78.** $2\sqrt{5}, -2\sqrt{5}$

79. $1 + 2\sqrt{3}, 1 - 2\sqrt{3}$ **80.** $2 + 3\sqrt{5}, 2 - 3\sqrt{5}$

81. $2 + i, 2 - i$ **82.** $3 + 4i, 3 - 4i$

83. Architecture The floor of a one-story building is 14 feet longer than it is wide. The building has 1632 square feet of floor space.

  (a) Draw a diagram that gives a visual representation of the floor space. Represent the width as w and show the length in terms of w.

  (b) Write a quadratic equation for the area of the floor in terms of w.

  (c) Find the length and width of the building floor.

84. Design and Construction An above-ground swimming pool with a square base is to be constructed such that the surface area of the pool is 576 square feet. The height of the pool is to be 4 feet. (See figure.) What should the dimensions of the base be? (*Hint:* The surface area is $S = x^2 + 4xh$.)

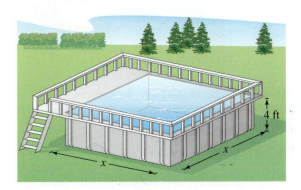

85. MODELING DATA

A gardener has 100 meters of fencing to enclose two adjacent rectangular gardens, as shown in the figure.

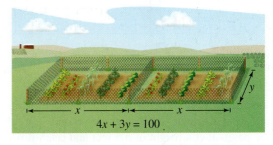

$4x + 3y = 100$

  (a) Write the area A of the enclosed region as a function of x.

  (b) Use a graphing utility to generate additional rows of the table. Use the table to estimate the dimensions that will produce a maximum area.

x	y	Area
2	$\frac{92}{3}$	$\frac{368}{3} \approx 123$
4	28	224

  (c) Use the graphing utility to graph the area function, and use the graph to estimate the dimensions that will produce a maximum area.

  (d) Use the graph to approximate the dimensions such that the enclosed area is 350 square meters.

  (e) Find the required dimensions of part (d) algebraically.

86. Geometry An open gift box is to be made from a square piece of material by cutting four-centimeter squares from each corner and turning up the sides (see figure). The volume of the finished gift box is to be 576 cubic centimeters. Find the size of the original piece of material.

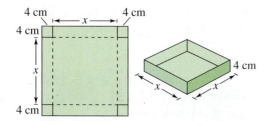

Using the Position Equation In Exercises 87–90, use the position equation given on page 193 as the model for the problem.

✓ **87. Falling Object** At 1815 feet tall, the CN Tower in Toronto, Ontario is the world's tallest self-supporting structure. An object is dropped from the top of the tower.

(a) Find the position equation $s = -16t^2 + v_0t + s_0$.

(b) Complete the table.

t	0	2	4	6	8	10	12
s							

(c) From the table in part (b), determine the time interval during which the object reaches the ground. Find the time algebraically.

88. Vertical Motion You throw a baseball straight up into the air at a velocity of 45 feet per second. You release the baseball at a height of 5.5 feet and catch it when it falls back to a height of 6 feet.

(a) Use the position equation to write a mathematical model for the height of the baseball.

(b) Find the height of the baseball after 0.5 second.

(c) How many seconds is the baseball in the air? (Use a graphing utility to verify your answer.)

89. Projectile Motion A cargo plane flying at 8000 feet over level terrain drops a 500-pound supply package.

(a) How long will it take the package to strike the ground?

(b) The plane is flying at 600 miles per hour. How far will the package travel horizontally during its descent?

90. Wildlife Management The game commission plane, flying at 1100 feet, drops oral rabies vaccination pellets into a game preserve.

(a) Use the position equation to write a mathematical model for the height of the vaccination pellets.

(b) The plane is flying at 95 miles per hour. How far will the pellets travel horizontally during its decent?

✓ **91. Education** The average salaries S (in thousands of dollars) of secondary classroom teachers in the United States from 2000 through 2007 can be approximated by the model $S = 0.010t^2 + 1.01t + 42.9$, $0 \le t \le 7$, where t represents the year, with $t = 0$ corresponding to 2000. (Source: National Education Association)

(a) Determine algebraically when the average salary of a secondary classroom teacher was $47,000.

(b) Verify your answer to part (a) by creating a table of values for the model.

(c) Use a graphing utility to graph the model.

(d) Use the model to predict when the average salary will reach $58,000.

(e) Do you believe the model could be used to predict the average salaries for years beyond 2007? Explain your reasoning.

92. *Why you should learn it* (p. 186) The annual per capita consumption M of milk (in gallons) in the United States from 1990 through 2008 can be approximated by the model

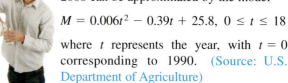

$$M = 0.006t^2 - 0.39t + 25.8, \quad 0 \le t \le 18$$

where t represents the year, with $t = 0$ corresponding to 1990. (Source: U.S. Department of Agriculture)

(a) Determine algebraically when the per capita consumption was 25 gallons per year and 22 gallons per year.

(b) Verify your answer to part (a) by creating a table of values for the model.

(c) Use a graphing utility to graph the model.

(d) Use the model to predict when the per capita consumption will reach 20 gallons per year.

(e) Do you believe the model could be used to predict the per capita consumption for years beyond 2008? Explain your reasoning.

93. Biology The metabolic rate of an ectothermic organism increases with increasing temperature within a certain range. Experimental data for oxygen consumption C (in microliters per gram per hour) of a beetle at certain temperatures yielded the model

$$C = 0.45x^2 - 1.65x + 50.75, \quad 10 \le x \le 25$$

where x is the air temperature (in degrees Celsius).

(a) Use a graphing utility to graph the consumption model over the specified domain.

(b) Use the graph to approximate the air temperature resulting in oxygen consumption of 150 microliters per gram per hour.

(c) When the temperature is increased from 10°C to 20°C, the oxygen consumption will be increased by approximately what factor?

94. Mechanical Engineering The distance d (in miles) a car can travel on one tank of fuel is approximated by $d = -0.024s^2 + 1.455s + 431.5, 0 < s \le 75$, where s is the average speed of the car (in miles per hour).

(a) Use a graphing utility to graph the function over the specified domain.

(b) Use the graph to determine the greatest distance that can be traveled on a tank of fuel. How long will the trip take?

(c) Determine the greatest distance that can be traveled in this car in 8 hours with no refueling. How fast should the car be driven? [*Hint:* The distance traveled in 8 hours is 8*s*. Graph this expression in the same viewing window as the graph in part (a) and approximate the point of intersection.]

✓ **95. Aviation** Two planes leave simultaneously from Chicago's O'Hare Airport, one flying due north and the other due east. The northbound plane is flying 50 miles per hour faster than the eastbound plane. After 3 hours the planes are 2440 miles apart. Find the speed of each plane. (*Hint:* Draw a diagram.)

96. Aviation A commercial jet flies to three cities whose locations form the vertices of a right triangle (see figure). The total flight distance (from Oklahoma City to Austin to New Orleans and back to Oklahoma City) is approximately 1348 miles. It is 560 miles between Oklahoma City and New Orleans. Approximate the other two distances.

Conclusions

True or False? In Exercises 97–100, determine whether the statement is true or false. Justify your answer.

97. The quadratic equation $-3x^2 - x = 10$ has two real solutions.

98. If $(2x - 3)(x + 5) = 8$, then $2x - 3 = 8$ or $x + 5 = 8$.

99. A quadratic equation with real coefficients can have one real solution and one imaginary solution.

100. A quadratic equation with real coefficients can have one repeated imaginary solution.

101. Exploration Given that a and b are nonzero real numbers, determine the solutions of the equations.

(a) $ax^2 + bx = 0$ (b) $ax^2 - ax = 0$

102. Proof Given that the solutions of a quadratic equation are $x = \left(-b \pm \sqrt{b^2 - 4ac}\right)/(2a)$, show that the sum of the solutions is $S = -b/a$.

103. Proof Given that the solutions of a quadratic equation are $x = \left(-b \pm \sqrt{b^2 - 4ac}\right)/(2a)$, show that the product of the solutions is $P = c/a$.

104. CAPSTONE Match the equation with a method you would use to solve it. Explain your reasoning. (Use each method once and do not solve the equations.)

(a) $3x^2 + 5x - 11 = 0$ (i) Factoring

(b) $x^2 + 10x = 3$ (ii) Extracting square roots

(c) $x^2 - 16x + 64 = 0$ (iii) Completing the square

(d) $x^2 - 15 = 0$ (iv) Quadratic Formula

Matching a Function with Its Graph In Exercises 105 and 106, determine which function the graph represents.

105. 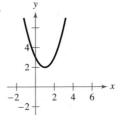   **106.**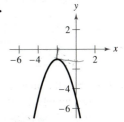

(a) $y = (x + 1)^2 - 2$ (a) $y = -(x - 2)^2 - 1$

(b) $y = (x - 2)^2 + 1$ (b) $y = -(x - 2)^2 + 1$

(c) $y = (x + 2)^2 + 1$ (c) $y = -x^2 - 2x + 1$

(d) $y = x^2 - x + 2$ (d) $y = -(x + 2)^2 - 1$

(e) $y = (x - 1)^2 + 2$ (e) $y = -(x - 1)^2 + 2$

Cumulative Mixed Review

Factoring an Expression In Exercises 107–110, completely factor the expression over the real numbers.

107. $x^5 - 27x^2$

108. $x^3 - 5x^2 - 14x$

109. $x^3 + 5x^2 - 2x - 10$

110. $5(x + 5)x^{1/3} + 4x^{4/3}$

111. *Make a Decision* To work an extended application analyzing the population of the United States, visit this textbook's *Companion Website*. (Data Source: U.S. Census Bureau)

2.5   Solving Other Types of Equations Algebraically

Polynomial Equations

In this section, the techniques for solving equations are extended to nonlinear and nonquadratic equations. At this point in the text, you have four basic algebraic methods for solving nonlinear equations—*factoring, extracting square roots, completing the square,* and the *Quadratic Formula.* The main goal of this section is to learn to *rewrite* nonlinear equations in a form to which you can apply one of these methods.

Example 1 shows how to use factoring to solve a **polynomial equation,** which is an equation that can be written in the general form

$$a_n x^n + a_{n-1} x^{n-1} + \cdots + a_2 x^2 + a_1 x + a_0 = 0.$$

Example 1   Solving a Polynomial Equation by Factoring

Solve $3x^4 = 48x^2$.

Solution

First write the polynomial equation in general form with zero on one side, factor the other side, and then set each factor equal to zero and solve.

$3x^4 = 48x^2$	Write original equation.
$3x^4 - 48x^2 = 0$	Write in general form.
$3x^2(x^2 - 16) = 0$	Factor out common factor.
$3x^2(x + 4)(x - 4) = 0$	Factor completely.
$3x^2 = 0 \implies x = 0$	Set 1st factor equal to 0.
$x + 4 = 0 \implies x = -4$	Set 2nd factor equal to 0.
$x - 4 = 0 \implies x = 4$	Set 3rd factor equal to 0.

You can check these solutions by substituting in the original equation, as follows.

Check

$3(0)^4 \stackrel{?}{=} 48(0)^2$	Substitute 0 for x.
$0 = 0$	0 checks. ✓
$3(-4)^4 \stackrel{?}{=} 48(-4)^2$	Substitute -4 for x.
$768 = 768$	-4 checks. ✓
$3(4)^4 \stackrel{?}{=} 48(4)^2$	Substitute 4 for x.
$768 = 768$	4 checks. ✓

So, you can conclude that the solutions are

$$x = 0, \quad x = -4, \quad \text{and} \quad x = 4.$$

✓**CHECKPOINT**   Now try Exercise 5.

A common mistake that is made in solving an equation such as that in Example 1 is to divide each side of the equation by the variable factor x^2. This loses the solution $x = 0$. When solving an equation, always write the equation in general form, then factor the equation and set each factor equal to zero. Do not divide each side of an equation by a variable factor in an attempt to simplify the equation.

What you should learn
- Solve polynomial equations of degree three or greater.
- Solve equations involving radicals.
- Solve equations involving fractions or absolute values.
- Use polynomial equations and equations involving radicals to model and to solve real-life problems.

Why you should learn it
Polynomial equations, radical equations, and absolute value equations can be used to model and solve real-life problems. For instance, in Exercise 82 on page 209, a radical equation can be used to model the total monthly cost of airplane flights between Chicago and Denver.

Example 2   Solving a Polynomial Equation by Factoring

Solve $2x^3 - 6x^2 + 6x - 18 = 0$.

Solution

This equation has a common factor of 2. You can simplify the equation by first dividing each side of the equation by 2.

$2x^3 - 6x^2 + 6x - 18 = 0$ Write original equation.

$x^3 - 3x^2 + 3x - 9 = 0$ Divide each side by 2.

$x^2(x - 3) + 3(x - 3) = 0$ Group terms.

$(x - 3)(x^2 + 3) = 0$ Factor by grouping.

$x - 3 = 0$ ➡ $x = 3$ Set 1st factor equal to 0.

$x^2 + 3 = 0$ ➡ $x = \pm\sqrt{3}\,i$ Set 2nd factor equal to 0.

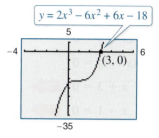

The equation has three solutions:

$x = 3, \quad x = \sqrt{3}\,i, \quad \text{and} \quad x = -\sqrt{3}\,i.$

Check these solutions in the original equation. Figure 2.33 verifies the real solution $x = 3$ graphically.

Figure 2.33

✓**CHECKPOINT**   Now try Exercise 9.

Occasionally, mathematical models involve equations that are of **quadratic type.** In general, an equation is of quadratic type when it can be written in the form

$$au^2 + bu + c = 0$$

where $a \ne 0$ and u is an algebraic expression.

Example 3   Solving an Equation of Quadratic Type

Solve $x^4 - 3x^2 + 2 = 0$.

Solution

This equation is of quadratic type with $u = x^2$.

$$(x^2)^2 - 3(x^2) + 2 = 0$$

To solve this equation, you can use the Quadratic Formula.

$x^4 - 3x^2 + 2 = 0$ Write original equation.

$\underbrace{(x^2)^2}_{u^2} - \underbrace{3(x^2)}_{3u} + 2 = 0$ Write in quadratic form.

$x^2 = \dfrac{-(-3) \pm \sqrt{(-3)^2 - 4(1)(2)}}{2(1)}$ Quadratic Formula

$x^2 = \dfrac{3 \pm 1}{2}$ Simplify.

$x^2 = 2$ ➡ $x = \pm\sqrt{2}$ Solutions

$x^2 = 1$ ➡ $x = \pm 1$ Solutions

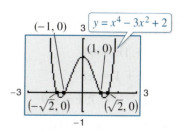

The equation has four solutions: $x = -1$, $x = 1$, $x = \sqrt{2}$, and $x = -\sqrt{2}$. Check these solutions in the original equation. Figure 2.34 verifies the solutions graphically.

Figure 2.34

✓**CHECKPOINT**   Now try Exercise 11.

Example 7   Solving an Equation Involving an Absolute Value

Solve $|x^2 - 3x| = -4x + 6$.

Solution

To solve an equation involving an absolute value algebraically, you must consider the fact that the expression inside the absolute value symbols can be positive or negative. This results in *two* separate equations, each of which must be solved.

First Equation

$$x^2 - 3x = -4x + 6 \qquad \text{Use positive expression.}$$

$$x^2 + x - 6 = 0 \qquad \text{Write in general form.}$$

$$(x + 3)(x - 2) = 0 \qquad \text{Factor.}$$

$$x + 3 = 0 \quad \Longrightarrow \quad x = -3 \qquad \text{Set 1st factor equal to 0.}$$

$$x - 2 = 0 \quad \Longrightarrow \quad x = 2 \qquad \text{Set 2nd factor equal to 0.}$$

Second Equation

$$-(x^2 - 3x) = -4x + 6 \qquad \text{Use negative expression.}$$

$$x^2 - 7x + 6 = 0 \qquad \text{Write in general form.}$$

$$(x - 1)(x - 6) = 0 \qquad \text{Factor.}$$

$$x - 1 = 0 \quad \Longrightarrow \quad x = 1 \qquad \text{Set 1st factor equal to 0.}$$

$$x - 6 = 0 \quad \Longrightarrow \quad x = 6 \qquad \text{Set 2nd factor equal to 0.}$$

Check

$$|(-3)^2 - 3(-3)| \overset{?}{=} -4(-3) + 6 \qquad \text{Substitute } -3 \text{ for } x.$$

$$18 = 18 \qquad -3 \text{ checks. } \checkmark$$

$$|2^2 - 3(2)| \overset{?}{=} -4(2) + 6 \qquad \text{Substitute 2 for } x.$$

$$2 \neq -2 \qquad 2 \text{ does not check.}$$

$$|1^2 - 3(1)| \overset{?}{=} -4(1) + 6 \qquad \text{Substitute 1 for } x.$$

$$2 = 2 \qquad 1 \text{ checks. } \checkmark$$

$$|6^2 - 3(6)| \overset{?}{=} -4(6) + 6 \qquad \text{Substitute 6 for } x.$$

$$18 \neq -18 \qquad 6 \text{ does not check.}$$

The solutions are

$$x = -3 \quad \text{and} \quad x = 1.$$

Figure 2.38 shows the solutions graphically.

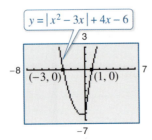

Figure 2.38

Explore the Concept

In Figure 2.38, the graph of

$$y = |x^2 - 3x| + 4x - 6$$

appears to be a straight line to the right of the y-axis. Is it? Explain your reasoning.

✓CHECKPOINT   Now try Exercise 69.

Applications

It would be impossible to categorize the many different types of applications that involve nonlinear and nonquadratic models. However, from the few examples and exercises that are given, you will gain some appreciation for the variety of applications that can occur.

Example 8   Reduced Rates

The athletic booster club of a university chartered a bus for an away football game at a cost of $960. In an attempt to lower the bus fare per person, the club invited nonmembers to go along. After five nonmembers joined the trip, the fare per person decreased by $9.60. How many club members are going on the trip?

Solution

Begin the solution by creating a verbal model and assigning labels.

Verbal Model: Cost per person · Number of people = Cost of trip

Labels: Cost of trip $= 960$ (dollars)

Number of athletic booster club members $= x$ (people)

Number of people $= x + 5$ (people)

Original cost per member $= \dfrac{960}{x}$ (dollars per person)

Cost per person $= \dfrac{960}{x} - 9.60$ (dollars per person)

Equation:

$$\left(\dfrac{960}{x} - 9.60\right)(x + 5) = 960$$

$$\left(\dfrac{960 - 9.6x}{x}\right)(x + 5) = 960 \qquad \text{Write } \left(\dfrac{960}{x} - 9.60\right) \text{ as a fraction.}$$

$$(960 - 9.6x)(x + 5) = 960x, \ x \neq 0 \qquad \text{Multiply each side by } x.$$

$$960x + 4800 - 9.6x^2 - 48x = 960x \qquad \text{Multiply.}$$

$$-9.6x^2 - 48x + 4800 = 0 \qquad \text{Subtract } 960x \text{ from each side.}$$

$$x^2 + 5x - 500 = 0 \qquad \text{Divide each side by } -9.6.$$

$$(x + 25)(x - 20) = 0 \qquad \text{Factor.}$$

$$x + 25 = 0 \implies x = -25$$

$$x - 20 = 0 \implies x = 20$$

Choosing the positive value of x, you can conclude that 20 athletic booster club members are going on the trip. Check this in the original statement of the problem, as follows.

$$\left(\dfrac{960}{20} - 9.60\right)(20 + 5) \overset{?}{=} 960 \qquad \text{Substitute 20 for}$$

$$(48 - 9.60)25 \overset{?}{=} 960 \qquad \text{Simplify.}$$

$$960 = 960 \qquad \text{20 checks.} \checkmark$$

     Now try Exercise 75.

Interest in a savings account is calculated by one of three basic methods: simple interest, interest compounded n times per year, and interest compounded continuously. The next example uses the formula for interest that is compounded n times per year.

$$A = P\left(1 + \frac{r}{n}\right)^{nt}$$

In this formula, A is the balance in the account, P is the principal (or original deposit), r is the annual interest rate (in decimal form), n is the number of compoundings per year, and t is the time in years. In Chapter 4, you will study a derivation of the formula above for interest compounded continuously.

Example 9   Compound Interest

When you were born, your grandparents deposited $5000 in a long-term investment in which the interest was compounded quarterly. Today, on your 25th birthday, the value of the investment is $25,062.59. What is the annual interest rate for this investment?

Solution

Formula: $A = P\left(1 + \dfrac{r}{n}\right)^{nt}$

Labels: Balance $= A = 25{,}062.59$ (dollars)

Principal $= P = 5000$ (dollars)

Time $= t = 25$ (years)

Compoundings per year $= n = 4$ (compoundings per year)

Annual interest rate $= r$ (percent in decimal form)

Equation: $25{,}062.59 = 5000\left(1 + \dfrac{r}{4}\right)^{4(25)}$

$\dfrac{25{,}062.59}{5000} = \left(1 + \dfrac{r}{4}\right)^{100}$ Divide each side by 5000.

$5.0125 \approx \left(1 + \dfrac{r}{4}\right)^{100}$ Use a calculator.

$(5.0125)^{1/100} \approx 1 + \dfrac{r}{4}$ Raise side to reciprocal power.

$1.01625 \approx 1 + \dfrac{r}{4}$ Use a calculator.

$0.01625 \approx \dfrac{r}{4}$ Subtract 1 from each side.

$0.065 \approx r$ Multiply each side by 4.

The annual interest rate is about 0.065, or 6.5%. Check this in the original statement of the problem. Note that you can use a graphing utility to check your answer, as shown in Figure 2.39.

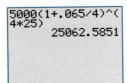

Figure 2.39

 CHECKPOINT   Now try Exercise 77.

2.5   Exercises

See www.CalcChat.com for worked-out solutions to odd-numbered exercises.
For instructions on how to use a graphing utility, see Appendix A.

Vocabulary and Concept Check

In Exercises 1 and 2, fill in the blank.

1. The general form of a _____ equation is
$a_n x^n + a_{n-1} x^{n-1} + \cdots + a_2 x^2 + a_1 x + a_0 = 0.$

2. To clear the equation $\dfrac{4}{x} + 5 = \dfrac{6}{x-3}$ of fractions, multiply each side of the equation by the least common denominator _____ .

3. Describe the step needed to remove the radical from the equation $\sqrt{x+2} = x.$

4. Is the equation $x^4 - 2x + 4 = 0$ of quadratic type?

Procedures and Problem Solving

Solving a Polynomial Equation by Factoring In Exercises 5–10, find all solutions of the equation algebraically. Use a graphing utility to verify the solutions graphically.

✓ 5. $4x^4 - 16x^2 = 0$ 6. $8x^4 - 18x^2 = 0$

7. $5x^3 + 30x^2 + 45x = 0$

8. $9x^4 - 24x^3 + 16x^2 = 0$

✓ 9. $x^3 + 5 = 5x^2 + x$

10. $x^4 - 2x^3 = 16 + 8x - 4x^3$

Solving an Equation of Quadratic Type In Exercises 11–14, find all solutions of the equation algebraically. Check your solutions.

✓ 11. $x^4 - 4x^2 + 3 = 0$ 12. $x^4 - 5x^2 - 36 = 0$

13. $4x^4 - 65x^2 + 16 = 0$ 14. $36t^4 + 29t^2 - 7 = 0$

Graphical Analysis In Exercises 15–18, (a) use a graphing utility to graph the equation, (b) use the graph to approximate any x-intercepts of the graph, (c) set $y = 0$ and solve the resulting equation, and (d) compare the result of part (c) with the x-intercepts of the graph.

15. $y = x^3 - 2x^2 - 3x$ 16. $y = 2x^4 - 15x^3 + 18x^2$

17. $y = x^4 - 10x^2 + 9$ 18. $y = x^4 - 29x^2 + 100$

Solving an Equation Involving Radicals In Exercises 19–38, find all solutions of the equation algebraically. Check your solutions.

19. $3\sqrt{x} - 10 = 0$ 20. $3\sqrt{x} - 6 = 0$

21. $\sqrt{x - 10} - 4 = 0$ 22. $\sqrt{2x + 5} + 3 = 0$

23. $\sqrt[3]{2x + 5} + 3 = 0$ 24. $\sqrt[3]{3x + 1} - 5 = 0$

25. $\sqrt[3]{2x + 1} + 8 = 0$ 26. $\sqrt[3]{4x - 3} + 2 = 0$

27. $\sqrt{5x - 26} + 4 = x$ 28. $x - \sqrt{8x - 31} = 5$

✓ 29. $\sqrt{x + 1} - 3x = 1$ 30. $\sqrt{x + 5} - 2x = 3$

31. $\sqrt{x + 1} = \sqrt{3x + 1}$ 32. $\sqrt{x + 5} = \sqrt{2x - 5}$

33. $2x + 9\sqrt{x} - 5 = 0$ 34. $6x - 7\sqrt{x} - 3 = 0$

35. $\sqrt{x} - \sqrt{x - 5} = 1$ 36. $\sqrt{x} + \sqrt{x - 20} = 10$

37. $3\sqrt{x - 5} - \sqrt{x - 1} = 0$

38. $4\sqrt{x - 3} - \sqrt{6x - 17} = 3$

Solving an Equation Involving Rational Exponents In Exercises 39–48, find all solutions of the equation algebraically. Check your solutions.

39. $3x^{1/3} + 2x^{2/3} = 5$ 40. $9t^{2/3} + 24t^{1/3} = -16$

✓ 41. $(x - 5)^{2/3} = 16$ 42. $(x - 1)^{3/2} = 8$

43. $(x - 9)^{2/3} = 25$ 44. $(x - 7)^{2/3} = 9$

45. $(x^2 - 5x - 2)^{1/3} = -2$ 46. $(x^2 - x - 22)^{4/3} = 16$

47. $3x(x - 1)^{1/2} + 2(x - 1)^{3/2} = 0$

48. $4x^2(x - 1)^{1/3} + 6x(x - 1)^{4/3} = 0$

Graphical Analysis In Exercises 49–52, (a) use a graphing utility to graph the equation, (b) use the graph to approximate any x-intercepts of the graph, (c) set $y = 0$ and solve the resulting equation, and (d) compare the result of part (c) with the x-intercepts of the graph.

49. $y = \sqrt{11x - 30} - x$

50. $y = 2x - \sqrt{15 - 4x}$

51. $y = \sqrt{7x + 36} - \sqrt{5x + 16} - 2$

52. $y = 3x - 3\sqrt{x} - 4$

Solving an Equation Involving Fractions In Exercises 53–64, find all solutions of the equation. Check your solutions.

✓ 53. $x = \dfrac{3}{x} + \dfrac{1}{2}$ 54. $\dfrac{4}{x} - \dfrac{5}{3} = \dfrac{x}{6}$

55. $\dfrac{1}{x} - \dfrac{1}{x + 1} = 3$ 56. $\dfrac{4}{x + 1} - \dfrac{3}{x + 2} = 1$

57. $\dfrac{20 - x}{x} = x$ 58. $4x + 1 = \dfrac{3}{x}$

59. $\dfrac{x}{x^2 - 4} + \dfrac{1}{x + 2} = 3$ **60.** $\dfrac{x - 2}{x} - \dfrac{1}{x + 2} = 0$

61. $\dfrac{1}{t^2} + \dfrac{8}{t} + 15 = 0$ **62.** $6 - \dfrac{1}{x} - \dfrac{1}{x^2} = 0$

63. $6\left(\dfrac{s}{s + 1}\right)^2 + 5\left(\dfrac{s}{s + 1}\right) - 6 = 0$

64. $8\left(\dfrac{t}{t - 1}\right)^2 - 2\left(\dfrac{t}{t - 1}\right) - 3 = 0$

Solving an Equation Involving an Absolute Value In Exercises 65–70, find all solutions of the equation algebraically. Check your solutions.

65. $|2x - 1| = 7$ **66.** $|3x + 2| = 8$

67. $|x| = x^2 + x - 3$ **68.** $|x - 10| = x^2 - 10x$

✓ **69.** $x - 10 = |x^2 - 10x|$ **70.** $|x + 1| = x^2 - 5$

Graphical Analysis In Exercises 71–74, (a) use a graphing utility to graph the equation, (b) use the graph to approximate any x-intercepts of the graph, (c) set $y = 0$ and solve the resulting equation, and (d) compare the result of part (c) with the x-intercepts of the graph.

71. $y = \dfrac{1}{x} - \dfrac{4}{x - 1} - 1$ **72.** $y = x + \dfrac{9}{x + 1} - 5$

73. $y = |x + 1| - 2$ **74.** $y = |x - 2| - 3$

✓ **75. Finance** Three students are planning to rent an apartment for a year and share equally in the cost. By adding a fourth person, each person saves $75 a month. How much is the monthly rent?

76. Finding a Speed A family drove 1080 miles to their vacation lodge. Because of traffic, their average speed on the return trip was decreased by 6 miles per hour and the trip took $2\frac{1}{2}$ hours longer. Determine their average speed on the way to the lodge.

✓ **77. Finance** A deposit of $2500 in a mutual fund reaches a balance of $3052.49 after 5 years. What annual interest rate on a certificate of deposit compounded monthly would yield an equivalent return?

78. Economics The demand equation for a high-definition television set is modeled by

$$p = 800 - \sqrt{0.2x + 1}$$

where x is the number of units demanded per month and p is the price per unit. Approximate the demand when the price is $750.

79. Physics The temperature T (in degrees Fahrenheit) of saturated steam increases as pressure increases. This relationship is approximated by the model

$$T = 75.82 - 2.11x + 43.51\sqrt{x}, \quad 5 \le x \le 40$$

where x is the absolute pressure (in pounds per square inch). Approximate the pressure for saturated steam at a temperature of 212°F.

80. Algebraic-Graphical-Numerical The average salaries S (in thousands of dollars) for football players in the National Football League from 2000 through 2008 can be approximated by the model

$$S = \sqrt{331,054.5t + 616,774}, \quad 0 \le t \le 8$$

where t is the year, with $t = 0$ corresponding to 2000. (Source: National Football League Players Association)

(a) Use the *table* feature of a graphing utility to estimate the average NFL player salary each year from 2000 through 2008.

(b) According to the table, when was the first year that the average salary exceeded $1.5 million?

(c) Find the answer to part (b) algebraically.

(d) Use the graphing utility to graph the model and find the answer to part (b).

81. MODELING DATA

The table shows the numbers N (in millions) of adults in the United States spending time on probation, on parole, or incarcerated from 1997 through 2007. (Source: U.S. Department of Justice)

Year	Number, N (in millions)
1997	5.726
1998	6.126
1999	6.331
2000	6.445
2001	6.582
2002	6.759
2003	6.925
2004	6.995
2005	7.052
2006	7.182
2007	7.328

(a) Use a graphing utility to create a scatter plot of the data, where t represents the year, with $t = 7$ corresponding to 1997.

(b) A model that approximates the data is

$$N = 0.0375t - \dfrac{13.3883}{t} + 7.431, \quad 7 \le t \le 17.$$

Graph the model and the data in the same viewing window. Is the model a good fit for the data?

(c) Use the model to predict the first year in which more than 7.5 million adults spend time on probation, on parole, or incarcerated.

82. *Why you should learn it* (p. 200)   An airline offers daily flights between Chicago and Denver. The total monthly cost C (in millions of dollars) of these flights is modeled by

$$C = \sqrt{0.2x + 1}$$

where x is the number of passengers (in thousands). The total cost of the flights for June is 2.5 million dollars. How many passengers flew in June?

83. Electrical Energy   A power station is on one side of a river that is $\frac{3}{4}$ mile wide, and a factory is 8 miles downstream on the other side of the river, as shown in the figure. It costs $24 per foot to run power lines over land and $30 per foot to run them under water.

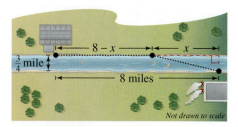

Not drawn to scale

(a) Write the total cost C of running power lines in terms of x (see figure).

(b) Find the total cost when $x = 3$.

(c) Find the length x when $C = \$1,098,662.40$.

(d) Use a graphing utility to graph the equation from part (a).

(e) Use your graph from part (d) to find the value of x that minimizes the cost.

84. Meteorology   A meteorologist is positioned 100 feet from the point at which a weather balloon is launched. When the balloon is at height h, the distance d (in feet) between the meteorologist and the balloon is given by

$$d = \sqrt{100^2 + h^2}.$$

(a) Use a graphing utility to graph the equation. Use the *trace* feature to approximate the value of h when $d = 200$.

(b) Complete the table. Use the table to approximate the value of h when $d = 200$.

h	160	165	170	175	180	185
d						

(c) Find h algebraically when $d = 200$.

(d) Compare the results of each method. In each case, what information did you gain that wasn't revealed by another solution method?

Conclusions

True or False?   In Exercises 85 and 86, determine whether the statement is true or false. Justify your answer.

85. An equation can never have more than one extraneous solution.

86. When solving an absolute value equation, you always have to check more than one solution.

Think About It   In Exercises 87–90, find an equation having the given solutions. (There are many correct answers.)

87. $\pm\sqrt{2}, 5$ **88.** $2, \pm\sqrt{3}$

89. $\pm 2, \pm i$ **90.** $\pm 3i, \pm 4$

Think About It   In Exercises 91 and 92, find x such that the distance between the points is 13.

91. $(1, 2), (x, -10)$ **92.** $(-8, 0), (x, 5)$

93. Think About It   Find a and b in the equation $x + |x - a| = b$ when the solution is $x = 9$. (There are many correct answers.)

94. Think About It   Find a and b in the equation $x + \sqrt{x - a} = b$ when the solution is $x = 20$. (There are many correct answers.)

95. Error Analysis   Describe the error.

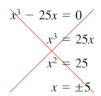

96. CAPSTONE   Make a study sheet of the techniques for solving nonlinear and nonquadratic equations. Illustrate each technique with an example. Discuss when extraneous solutions are possible.

Cumulative Mixed Review

Operations with Rational Expressions   In Exercises 97–100, simplify the expression.

97. $\dfrac{8}{3x} + \dfrac{3}{2x}$ **98.** $\dfrac{2}{x^2 - 4} - \dfrac{1}{x^2 - 3x + 2}$

99. $\dfrac{2}{z + 2} - \left(3 - \dfrac{2}{z}\right)$ **100.** $25y^2 \div \dfrac{xy}{5}$

Solving a Quadratic Equation   In Exercises 101 and 102, find all real solutions of the equation.

101. $x^2 - 22x + 121 = 0$

102. $x(x - 20) + 3(x - 20) = 0$

2.6   Solving Inequalities Algebraically and Graphically

Properties of Inequalities

Simple inequalities were reviewed in Section P.1. There, the inequality symbols

$$<, \quad \leq, \quad >, \quad \text{and} \quad \geq \qquad \textcolor{red}{\text{Inequality symbols}}$$

were used to compare two numbers and to denote subsets of real numbers. For instance, the simple inequality $x \geq 3$ denotes all real numbers x that are greater than or equal to 3.

In this section, you will study inequalities that contain more involved statements such as

$$5x - 7 > 3x + 9$$

and

$$-3 \leq 6x - 1 < 3.$$

As with an equation, you **solve an inequality** in the variable x by finding all values of x for which the inequality is true. These values are **solutions** of the inequality and are said to **satisfy** the inequality. For instance, the number 9 is a solution of the first inequality listed above because

$$5(9) - 7 > 3(9) + 9$$
$$45 - 7 > 27 + 9$$
$$38 > 36.$$

On the other hand, the number 7 is not a solution because

$$5(7) - 7 \not> 3(7) + 9$$
$$35 - 7 \not> 21 + 9$$
$$28 \not> 30.$$

The set of all real numbers that are solutions of an inequality is the **solution set** of the inequality.

The set of all points on the real number line that represent the solution set is the **graph of the inequality.** Graphs of many types of inequalities consist of intervals on the real number line.

The procedures for solving linear inequalities in one variable are much like those for solving linear equations. To isolate the variable, you can make use of the **properties of inequalities.** These properties are similar to the properties of equality, but there are two important exceptions. When each side of an inequality is multiplied or divided by a negative number, *the direction of the inequality symbol must be reversed* in order to maintain a true statement. Here is an example.

$$-2 < 5 \qquad \textcolor{red}{\text{Original inequality}}$$
$$(-3)(-2) > (-3)(5) \qquad \textcolor{red}{\text{Multiply each side by } -3 \text{ and reverse the inequality.}}$$
$$6 > -15 \qquad \textcolor{red}{\text{Simplify.}}$$

Two inequalities that have the same solution set are **equivalent inequalities.** For instance, the inequalities

$$x + 2 < 5 \quad \text{and} \quad x < 3$$

are equivalent. To obtain the second inequality from the first, you can subtract 2 from each side of the inequality. The properties listed at the top of the next page describe operations that can be used to create equivalent inequalities.

What you should learn

- Use properties of inequalities to solve linear inequalities.
- Solve inequalities involving absolute values.
- Solve polynomial inequalities.
- Solve rational inequalities.
- Use inequalities to model and solve real-life problems.

Why you should learn it

An inequality can be used to determine when a real-life quantity exceeds a given level. For instance, Exercises 99–102 on page 222 show how to use linear inequalities to determine when the number of Bed Bath & Beyond stores exceeds the number of Williams-Sonoma stores.

Properties of Inequalities

Let a, b, c, and d be real numbers.

1. *Transitive Property*

$a < b$ and $b < c$  $a < c$

2. *Addition of Inequalities*

$a < b$ and $c < d$  $a + c < b + d$

3. *Addition of a Constant*

$a < b$  $a + c < b + c$

4. *Multiplying by a Constant*

For $c > 0$, $a < b$  $ac < bc$

For $c < 0$, $a < b$  $ac > bc$

Each of the properties above is true when the symbol $<$ is replaced by $\leq$ and $>$ is replaced by $\geq$. For instance, another form of Property 3 is as follows.

$a \leq b$  $a + c \leq b + c$

The simplest type of inequality to solve is a **linear inequality** in one variable, such as $2x + 3 > 4$. (For help with solving one-step linear inequalities, see Appendix D at this textbook's *Companion Website*.)

Example 1   Solving a Linear Inequality

Solve $5x - 7 > 3x + 9$.

Solution

$5x - 7 > 3x + 9$	Original inequality
$2x - 7 > 9$	Subtract $3x$ from each side.
$2x > 16$	Add 7 to each side.
$x > 8$	Divide each side by 2.

So, the solution set is all real numbers that are greater than 8. The interval notation for this solution set is

$(8, \infty)$.

Figure 2.40   *Solution Interval:* $(8, \infty)$

The graph of this solution set is shown in Figure 2.40. Note that a parenthesis at 8 on the number line indicates that 8 *is not* part of the solution set.

✔**CHECKPOINT**   Now try Exercise 19.

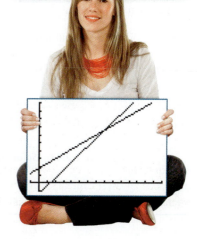

Note that the four inequalities forming the solution steps of Example 1 are all *equivalent* in the sense that each has the same solution set.

Checking the solution set of an inequality is not as simple as checking the solution(s) of an equation because there are simply too many x-values to substitute into the original inequality. However, you can get an indication of the validity of the solution set by substituting a few convenient values of x. For instance, in Example 1, try substituting $x = 6$ and $x = 10$ into the original inequality.

Example 2 Solving an Inequality

Solve $1 - \frac{3}{2}x \geq x - 4$.

Algebraic Solution

$1 - \frac{3}{2}x \geq x - 4$ Write original inequality.

$2 - 3x \geq 2x - 8$ Multiply each side by the LCD.

$2 - 5x \geq -8$ Subtract $2x$ from each side.

$-5x \geq -10$ Subtract 2 from each side.

$x \leq 2$ Divide each side by -5 and reverse the inequality.

The solution set is all real numbers that are less than or equal to 2. The interval notation for this solution set is $(-\infty, 2]$. The graph of this solution set is shown in Figure 2.41. Note that a bracket at 2 on the number line indicates that 2 *is* part of the solution set.

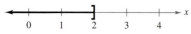

Figure 2.41   *Solution Interval:* $(-\infty, 2]$

✓CHECKPOINT   Now try Exercise 23.

Graphical Solution

Use a graphing utility to graph $y_1 = 1 - \frac{3}{2}x$ and $y_2 = x - 4$ in the same viewing window, as shown in Figure 2.42.

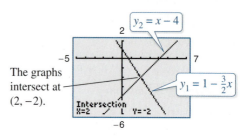

The graphs intersect at $(2, -2)$.

Figure 2.42

The graph of y_1 lies above the graph of y_2 to the left of their point of intersection, $(2, -2)$, which implies that $y_1 \geq y_2$ for all $x \leq 2$.

Sometimes it is possible to write two inequalities as a **double inequality,** as demonstrated in Example 3.

Example 3   Solving a Double Inequality

Solve $-3 \leq 6x - 1$ and $6x - 1 < 3$.

Algebraic Solution

$-3 \leq 6x - 1 < 3$ Write as a double inequality.

$-3 + 1 \leq 6x - 1 + 1 < 3 + 1$ Add 1 to each part.

$-2 \leq 6x < 4$ Simplify.

$\dfrac{-2}{6} \leq \dfrac{6x}{6} < \dfrac{4}{6}$ Divide each part by 6.

$-\dfrac{1}{3} \leq x < \dfrac{2}{3}$ Simplify.

The solution set is all real numbers that are greater than or equal to $-\frac{1}{3}$ *and* less than $\frac{2}{3}$. The interval notation for this solution set is $\left[-\frac{1}{3}, \frac{2}{3}\right)$. The graph of this solution set is shown in Figure 2.43.

Figure 2.43   *Solution Interval:* $\left[-\frac{1}{3}, \frac{2}{3}\right)$

✓CHECKPOINT   Now try Exercise 25.

Graphical Solution

Use a graphing utility to graph $y_1 = 6x - 1$, $y_2 = -3$, and $y_3 = 3$ in the same viewing window, as shown in Figure 2.44.

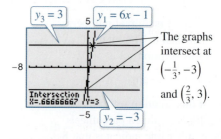

The graphs intersect at $\left(-\frac{1}{3}, -3\right)$ and $\left(\frac{2}{3}, 3\right)$.

Figure 2.44

The graph of y_1 lies above the graph of y_2 to the right of $\left(-\frac{1}{3}, -3\right)$ *and* the graph of y_1 lies below the graph of y_3 to the left of $\left(\frac{2}{3}, 3\right)$. This implies that $y_2 \leq y_1 < y_3$ when $-\frac{1}{3} \leq x < \frac{2}{3}$.

Inequalities Involving Absolute Values

> **Solving an Absolute Value Inequality**
>
> Let x be a variable or an algebraic expression and let a be a real number such that $a \geq 0$.
>
> **1.** The solutions of $|x| < a$ are all values of x that lie between $-a$ and a.
>
> $|x| < a$ if and only if $-a < x < a$.   Double inequality
>
> **2.** The solutions of $|x| > a$ are all values of x that are less than $-a$ or greater than a.
>
> $|x| > a$ if and only if $x < -a$ or $x > a$.   Compound inequality
>
> These rules are also valid when $<$ is replaced by $\leq$ and $>$ is replaced by $\geq$.

Example 4   Solving Absolute Value Inequalities

Solve each inequality.

a. $|x - 5| < 2$ **b.** $|x - 5| > 2$

Algebraic Solution

a. $|x - 5| < 2$ Write original inequality.

$\quad -2 < x - 5 < 2$ Write double inequality.

$\quad\quad 3 < x < 7$ Add 5 to each part.

The solution set is all real numbers that are greater than 3 *and* less than 7. The interval notation for this solution set is $(3, 7)$. The graph of this solution set is shown in Figure 2.45.

Figure 2.45

b. The absolute value inequality $|x - 5| > 2$ is equivalent to the following compound inequality: $x - 5 < -2$ or $x - 5 > 2$.

Solve first inequality: $x - 5 < -2$ Write first inequality.

$\quad\quad\quad\quad\quad\quad x < 3$ Add 5 to each side.

Solve second inequality: $x - 5 > 2$ Write second inequality.

$\quad\quad\quad\quad\quad\quad x > 7$ Add 5 to each side.

The solution set is all real numbers that are less than 3 *or* greater than 7. The interval notation for this solution set is $(-\infty, 3) \cup (7, \infty)$. The symbol $\cup$ is called a *union* symbol and is used to denote the combining of two sets. The graph of this solution set is shown in Figure 2.46.

Figure 2.46

✔CHECKPOINT   Now try Exercise 41.

Graphical Solution

Use a graphing utility to graph

$$y_1 = |x - 5| \quad \text{and} \quad y_2 = 2$$

in the same viewing window, as shown in Figure 2.47.

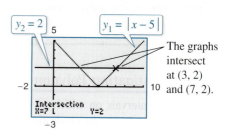

The graphs intersect at $(3, 2)$ and $(7, 2)$.

Figure 2.47

a. In Figure 2.47, you can see that the graph of y_1 lies below the graph of y_2 when

$$3 < x < 7.$$

This implies that the solution set is all real numbers greater than 3 *and* less than 7.

b. In Figure 2.47, you can see that the graph of y_1 lies above the graph of y_2 when

$$x < 3$$

or when

$$x > 7.$$

This implies that the solution set is all real numbers that are less than 3 *or* greater than 7.

Example 8   Unusual Solution Sets

a. The solution set of

$$x^2 + 2x + 4 > 0$$

consists of the entire set of real numbers, $(-\infty, \infty)$. In other words, the value of the quadratic $x^2 + 2x + 4$ is positive for every real value of x, as indicated in Figure 2.50. (Note that this quadratic inequality has *no* key numbers. In such a case, there is only one test interval— the entire real number line.)

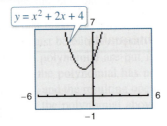

$y = x^2 + 2x + 4$

Figure 2.50

b. The solution set of

$$x^2 + 2x + 1 \le 0$$

consists of the single real number $\{-1\}$, because the quadratic

$$x^2 + 2x + 1$$

has one key number, $x = -1$, and it is the only value that satisfies the inequality, as indicated in Figure 2.51.

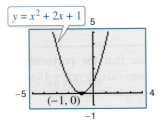

$y = x^2 + 2x + 1$

$(-1, 0)$

Figure 2.51

c. The solution set of

$$x^2 + 3x + 5 < 0$$

is empty. In other words, the quadratic

$$x^2 + 3x + 5$$

is not less than zero for any value of x, as indicated in Figure 2.52.

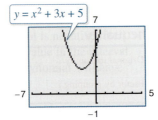

$y = x^2 + 3x + 5$

Figure 2.52

d. The solution set of

$$x^2 - 4x + 4 > 0$$

consists of all real numbers *except* the number 2. In interval notation, this solution set can be written as $(-\infty, 2) \cup (2, \infty)$. The graph of $y = x^2 - 4x + 4$ lies above the x-axis except at $x = 2$, where it touches it, as indicated in Figure 2.53.

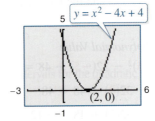

$y = x^2 - 4x + 4$

$(2, 0)$

Figure 2.53

 CHECKPOINT   Now try Exercise 73.

Technology Tip

One of the advantages of technology is that you can solve complicated polynomial inequalities that might be difficult, or even impossible, to factor. For instance, you could use a graphing utility to approximate the solution of the inequality

$$x^3 - 0.2x^2 - 3.16x + 1.4 < 0.$$

Rational Inequalities

The concepts of key numbers and test intervals can be extended to inequalities involving rational expressions. To do this, use the fact that the value of a rational expression can change sign only at its *zeros* (the *x*-values for which its numerator is zero) and its *undefined values* (the *x*-values for which its denominator is zero). These two types of numbers make up the *key numbers* of a rational inequality. When solving a rational inequality, begin by writing the inequality in general form with the rational expression on one side and zero on the other.

Example 9   Solving a Rational Inequality

Solve $\dfrac{2x - 7}{x - 5} \le 3$.

Algebraic Solution

$$\frac{2x - 7}{x - 5} - 3 \le 0 \qquad \text{Write in general form.}$$

$$\frac{2x - 7 - 3x + 15}{x - 5} \le 0 \qquad \text{Write as single fraction.}$$

$$\frac{-x + 8}{x - 5} \le 0 \qquad \text{Simplify.}$$

Now, in standard form you can see that the key numbers are $x = 5$ and $x = 8$, and you can proceed as follows.

Key Numbers: $x = 5$, $x = 8$

Test Intervals: $(-\infty, 5)$, $(5, 8)$, $(8, \infty)$

Test: Is $\dfrac{-x + 8}{x - 5} \le 0$?

Interval	x-Value	Polynomial Value	Conclusion
$(-\infty, 5)$	$x = 0$	$\dfrac{-0 + 8}{0 - 5} = -\dfrac{8}{5}$	Negative
$(5, 8)$	$x = 6$	$\dfrac{-6 + 8}{6 - 5} = 2$	Positive
$(8, \infty)$	$x = 9$	$\dfrac{-9 + 8}{9 - 5} = -\dfrac{1}{4}$	Negative

By testing these intervals, you can determine that the rational expression

$$\frac{-x + 8}{x - 5}$$

is negative in the open intervals $(-\infty, 5)$ and $(8, \infty)$. Moreover, because

$$\frac{-x + 8}{x - 5} = 0$$

when $x = 8$, you can conclude that the solution set of the inequality is $(-\infty, 5) \cup [8, \infty)$.

✓**CHECKPOINT**   Now try Exercise 83.

Graphical Solution

Use a graphing utility to graph

$$y_1 = \frac{2x - 7}{x - 5} \quad \text{and} \quad y_2 = 3$$

in the same viewing window, as shown in Figure 2.54.

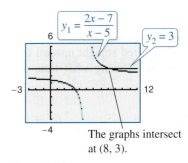

The graphs intersect at (8, 3).

Figure 2.54

The graph of y_1 lies below the graph of y_2 in the intervals $(-\infty, 5)$ and $[8, \infty)$. So, you can graphically estimate the solution set to be all real numbers less than 5 *or* greater than or equal to 8.

Note in Example 9 that $x = 5$ is not included in the solution set because the inequality is undefined when $x = 5$.

Solving an Absolute Value Inequality In Exercises 39–46, solve the inequality and sketch the solution on the real number line. Use a graphing utility to verify your solutions graphically.

39. $|5x| > 10$

40. $\left|\dfrac{x}{2}\right| \leq 1$

✓ **41.** $|x - 7| < 6$

42. $|x - 20| \geq 4$

43. $|x + 14| + 3 > 17$

44. $\left|\dfrac{x - 3}{2}\right| \geq 5$

45. $10|1 - x| < 5$

46. $3|4 - 5x| \leq 9$

Approximating Solutions In Exercises 47 and 48, use a graphing utility to graph the equation and graphically approximate the values of x that satisfy the specified inequalities. Then solve each inequality algebraically.

Equation	*Inequalities*
47. $y = |x - 3|$	(a) $y \leq 2$ (b) $y \geq 4$
48. $y = \left|\tfrac{1}{2}x + 1\right|$	(a) $y \leq 4$ (b) $y \geq 1$

Using Absolute Value Notation In Exercises 49–56, use absolute value notation to define the interval (or pair of intervals) on the real number line.

49.

50.

51.

52.

53. All real numbers within 10 units of 7

54. All real numbers no more than 8 units from -5

55. All real numbers at least 5 units from 3

56. All real numbers more than 3 units from -1

Investigating Polynomial Behavior In Exercises 57–62, determine the intervals on which the polynomial is entirely negative and those on which it is entirely positive.

57. $x^2 - 4x - 5$

58. $x^2 - 3x - 4$

✓ **59.** $2x^2 - 4x - 3$

60. $2x^2 - x - 5$

61. $x^2 - 4x + 5$

62. $-x^2 + 6x - 10$

Solving a Polynomial Inequality In Exercises 63–76, solve the inequality and graph the solution on the real number line. Use a graphing utility to verify your solution graphically.

63. $(x + 2)^2 < 25$

64. $(x - 3)^2 \geq 1$

✓ **65.** $x^2 + 4x + 4 \geq 9$

66. $x^2 - 6x + 9 < 16$

67. $x^3 - 4x \geq 0$

68. $x^4(x - 3) \leq 0$

✓ **69.** $2x^3 + 5x^2 > 6x + 9$

70. $2x^3 + 3x^2 < 11x + 6$

71. $x^3 - 3x^2 - x > -3$

72. $2x^3 + 13x^2 - 8x - 46 \geq 6$

✓ **73.** $3x^2 - 11x + 16 \leq 0$

74. $4x^2 + 12x + 9 \leq 0$

75. $x^2 + 3x + 8 > 0$

76. $4x^2 - 4x + 1 \leq 0$

Using Graphs to Find Solutions In Exercises 77 and 78, use the graph of the function to solve the equation or inequality.

(a) $f(x) = g(x)$ (b) $f(x) \geq g(x)$ (c) $f(x) > g(x)$

77.

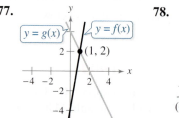

78.

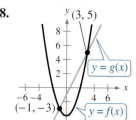

Approximating Solutions In Exercises 79 and 80, use a graphing utility to graph the equation and graphically approximate the values of x that satisfy the specified inequalities. Then solve each inequality algebraically.

Equation	*Inequalities*
79. $y = -x^2 + 2x + 3$	(a) $y \leq 0$ (b) $y \geq 3$
80. $y = x^3 - x^2 - 16x + 16$	(a) $y \leq 0$ (b) $y \geq 36$

Solving a Rational Inequality In Exercises 81–84, solve the inequality and graph the solution on the real number line. Use a graphing utility to verify your solution graphically.

81. $\dfrac{1}{x} - x > 0$

82. $\dfrac{1}{x} - 4 < 0$

✓ **83.** $\dfrac{x + 6}{x + 1} - 2 < 0$

84. $\dfrac{x + 12}{x + 2} - 3 \geq 0$

Approximating Solutions In Exercises 85 and 86, use a graphing utility to graph the equation and graphically approximate the values of x that satisfy the specified inequalities. Then solve each inequality algebraically.

Equation	*Inequalities*
85. $y = \dfrac{3x}{x - 2}$	(a) $y \leq 0$ (b) $y \geq 6$
86. $y = \dfrac{5x}{x^2 + 4}$	(a) $y \geq 1$ (b) $y \leq 0$

Finding the Domain of an Expression In Exercises 87–92, find the domain of x in the expression.

87. $\sqrt{x - 5}$

88. $\sqrt[4]{6x + 15}$

89. $\sqrt[3]{6 - x}$

90. $\sqrt[3]{2x^2 - 8}$

✓ **91.** $\sqrt{x^2 - 4}$

92. $\sqrt[4]{4 - x^2}$

93. MODELING DATA

The graph models the population P (in thousands) of Sacramento, California from 2000 through 2008, where t is the year, with $t = 0$ corresponding to 2000. Also shown is the line $y = 2000$. Use the graphs of the model and the horizontal line to write an equation or an inequality that could be solved to answer the question. Then answer the question. (Source: U.S. Census Bureau)

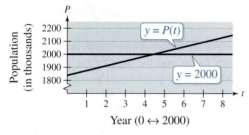

Year ($0 \leftrightarrow 2000$)

(a) In what year did the population of Sacramento reach two million?

(b) Over what time period is the population of Sacramento less than two million? greater than two million?

94. MODELING DATA

The graph models the population P (in thousands) of Pittsburgh, Pennsylvania from 2000 through 2008, where t is the year, with $t = 0$ corresponding to 2000. Also shown is the line $y = 2360$. Use the graphs of the model and the horizontal line to write an equation or an inequality that could be solved to answer the question. Then answer the question. (Source: U.S. Census Bureau)

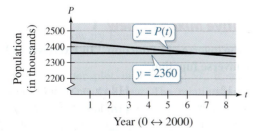

Year ($0 \leftrightarrow 2000$)

(a) In what year did the population of Pittsburgh reach 2.36 million?

(b) Over what time period is the population of Pittsburgh less than 2.36 million? greater than 2.36 million?

✓ **95. Height of a Projectile** A projectile is fired straight upward from ground level with an initial velocity of 160 feet per second.

(a) At what instant will it be back at ground level?

(b) When will the height exceed 384 feet?

96. Height of a Projectile A projectile is fired straight upward from ground level with an initial velocity of 128 feet per second.

(a) At what instant will it be back at ground level?

(b) When will the height be less than 128 feet?

97. MODELING DATA

The numbers D of doctorate degrees (in thousands) awarded to female students from 1990 through 2008 in the United States can be approximated by the model

$$D = 0.0510t^2 - 0.045t + 15.25, \quad 0 \leq t \leq 18,$$

where t is the year, with $t = 0$ corresponding to 1990. (Source: U.S. National Center for Education Statistics)

(a) Use a graphing utility to graph the model.

(b) Use the *zoom* and *trace* features to find when the number of degrees was between 20 and 25 thousand.

(c) Algebraically verify your results from part (b).

98. MODELING DATA

You want to determine whether there is a relationship between an athlete's weight x (in pounds) and the athlete's maximum bench-press weight y (in pounds). Sample data from 12 athletes are shown below.

(165, 170), (184, 185), (150, 200),

(210, 255), (196, 205), (240, 295),

(202, 190), (170, 175), (185, 195),

(190, 185), (230, 250), (160, 150)

(a) Use a graphing utility to plot the data.

(b) A model for the data is

$$y = 1.3x - 36.$$

Use the graphing utility to graph the equation in the same viewing window used in part (a).

(c) Use the graph to estimate the value of x that predicts a maximum bench-press weight of at least 200 pounds.

(d) Use the graph to write a statement about the accuracy of the model. If you think the graph indicates that an athlete's weight is not a good indicator of the athlete's maximum bench-press weight, list other factors that might influence an individual's maximum bench-press weight.

Consider a collection of ordered pairs of the form (x, y). If y tends to increase as x increases, then the collection is said to have a **positive correlation.** If y tends to decrease as x increases, then the collection is said to have a **negative correlation.** Figure 2.58 shows three examples: one with a positive correlation, one with a negative correlation, and one with no (discernible) correlation.

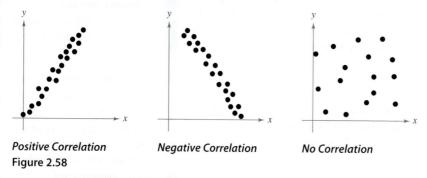

Positive Correlation
Figure 2.58

Negative Correlation

No Correlation

Example 2   Interpreting Correlation

On a Friday, 22 students in a class were asked to record the numbers of hours they spent studying for a test on Monday and the numbers of hours they spent watching television. The results are shown below. (The first coordinate is the number of hours and the second coordinate is the score obtained on the test.)

Study Hours: (0, 40), (1, 41), (2, 51), (3, 58), (3, 49), (4, 48), (4, 64),
(5, 55), (5, 69), (5, 58), (5, 75), (6, 68), (6, 63), (6, 93), (7, 84), (7, 67),
(8, 90), (8, 76), (9, 95), (9, 72), (9, 85), (10, 98)

TV Hours: (0, 98), (1, 85), (2, 72), (2, 90), (3, 67), (3, 93), (3, 95), (4, 68),
(4, 84), (5, 76), (7, 75), (7, 58), (9, 63), (9, 69), (11, 55), (12, 58), (14, 64),
(16, 48), (17, 51), (18, 41), (19, 49), (20, 40)

a. Construct a scatter plot for each set of data.

b. Determine whether the points are positively correlated, are negatively correlated, or have no discernible correlation. What can you conclude?

Solution

a. Scatter plots for the two sets of data are shown in Figure 2.59.

b. The scatter plot relating study hours and test scores has a positive correlation. This means that the more a student studied, the higher his or her score tended to be. The scatter plot relating television hours and test scores has a negative correlation. This means that the more time a student spent watching television, the lower his or her score tended to be.

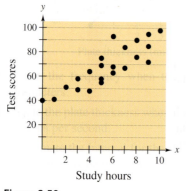

Study hours

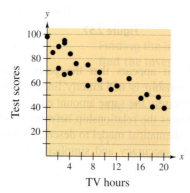

TV hours

Figure 2.59

✓CHECKPOINT   Now try Exercise 7.

Fitting a Line to Data

Finding a linear model to represent the relationship described by a scatter plot is called **fitting a line to data.** You can do this graphically by simply sketching the line that appears to fit the points, finding two points on the line, and then finding the equation of the line that passes through the two points.

Example 3   Fitting a Line to Data

Find a linear model that relates the year to the number of employees in the cellular telecommunications industry in the United States. (See Example 1.)

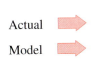

Year	Employees in the cellular telecommunications industry, E (in thousands)
2002	192
2003	206
2004	226
2005	233
2006	254
2007	267

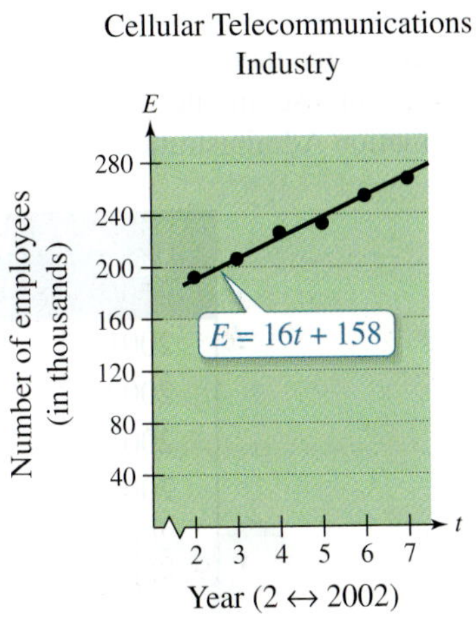

Cellular Telecommunications Industry

Figure 2.60

Solution

Let t represent the year, with $t = 2$ corresponding to 2002. After plotting the data in the table, draw the line that you think best represents the data, as shown in Figure 2.60. Two points that lie on this line are

$(3, 206)$ and $(6, 254)$.

Using the point-slope form, you can find the equation of the line to be

$E = 16(t - 3) + 206$

$= 16t + 158.$ Linear model

 CHECKPOINT   Now try Exercise 11.

Once you have found a model, you can measure how well the model fits the data by comparing the actual values with the values given by the model, as shown in the following table.

	t	2	3	4	5	6	7
Actual	E	192	206	226	233	254	267
Model	E	190	206	222	238	254	270

The sum of the squares of the differences between the actual values and the model values is called the **sum of the squared differences.** The model that has the least sum is called the **least squares regression line** for the data. For the model in Example 3, the sum of the squared differences is 54. The least squares regression line for the data is

$E = 15.0t + 162.$ Best-fitting linear model

Its sum of squared differences is 37. For more on the least squares regression line, see Appendix B.2 at this textbook's *Companion Website*.

Study Tip

The model in Example 3 is based on the two data points chosen. When different points are chosen, the model may change somewhat. For instance, when you choose $(5, 233)$ and $(7, 267)$, the new model is

$E = 17(t - 5) + 233$

$= 17t + 148.$

2.7   Exercises

See www.CalcChat.com for worked-out solutions to odd-numbered exercises.
For instructions on how to use a graphing utility, see Appendix A.

Vocabulary and Concept Check

In Exercises 1 and 2, fill in the blank.

1. Consider a collection of ordered pairs of the form (x, y). If y tends to increase as x increases, then the collection is said to have a _____ correlation.

2. To find the least squares regression line for data, you can use the _____ feature of a graphing utility.

3. In a collection of ordered pairs (x, y), y tends to decrease as x increases. Does the collection have a positive correlation or a negative correlation?

4. You find the least squares regression line for a set of data. The correlation coefficient is 0.114. Is the model a good fit?

Procedures and Problem Solving

✓ 5. **Constructing a Scatter Plot** The following ordered pairs give the years of experience x for 15 sales representatives and the monthly sales y (in thousands of dollars).

(1.5, 41.7), (1.0, 32.4), (0.3, 19.2), (3.0, 48.4),
(4.0, 51.2), (0.5, 28.5), (2.5, 50.4), (1.8, 35.5),
(2.0, 36.0), (1.5, 40.0), (3.5, 50.3), (4.0, 55.2),
(0.5, 29.1), (2.2, 43.2), (2.0, 41.6)

(a) Create a scatter plot of the data.

(b) Does the relationship between x and y appear to be approximately linear? Explain.

6. **Constructing a Scatter Plot** The following ordered pairs give the scores on two consecutive 15-point quizzes for a class of 18 students.

(7, 13), (9, 7), (14, 14), (15, 15), (10, 15), (9, 7),
(14, 11), (14, 15), (8, 10), (9, 10), (15, 9), (10, 11),
(11, 14), (7, 14), (11, 10), (14, 11), (10, 15), (9, 6)

(a) Create a scatter plot of the data.

(b) Does the relationship between consecutive quiz scores appear to be approximately linear? If not, give some possible explanations.

Interpreting Correlation In Exercises 7–10, the scatter plot of a set of data is shown. Determine whether the points are positively correlated, are negatively correlated, or have no discernible correlation.

✓ 7.
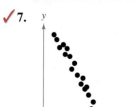

8.
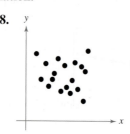

9.
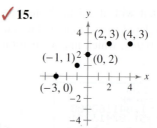

10.
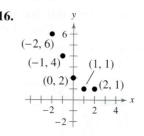

Fitting a Line to Data In Exercises 11–14, (a) create a scatter plot of the data, (b) draw a line of best fit that passes through two of the points, and (c) use the two points to find an equation of the line.

✓ 11. $(-3, -3), (3, 4), (1, 1), (3, 2), (4, 4), (-1, -1)$

12. $(-2, 3), (-2, 4), (-1, 2), (1, -2), (0, 0), (0, 1)$

13. $(0, 2), (-2, 1), (3, 3), (1, 3), (4, 4)$

14. $(3, 2), (2, 3), (1, 5), (4, 0), (5, 0)$

A Mathematical Model In Exercises 15 and 16, use the *regression* feature of a graphing utility to find a linear model for the data. Then use the graphing utility to decide how closely the model fits the data (a) graphically and (b) numerically. To print an enlarged copy of the graph, go to the website *www.mathgraphs.com*.

✓ 15.
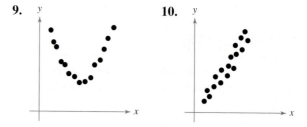

16.

17. MODELING DATA

Hooke's Law states that the force F required to compress or stretch a spring (within its elastic limits) is proportional to the distance d that the spring is compressed or stretched from its original length. That is, $F = kd$, where k is the measure of the stiffness of the spring and is called the *spring constant*. The table shows the elongation d in centimeters of a spring when a force of F kilograms is applied.

Force, F	Elongation, d
20	1.4
40	2.5
60	4.0
80	5.3
100	6.6

(a) Sketch a scatter plot of the data.

(b) Find the equation of the line that seems to best fit the data.

(c) Use the *regression* feature of a graphing utility to find a linear model for the data.

(d) Use the model from part (c) to estimate the elongation of the spring when a force of 55 kilograms is applied.

18. MODELING DATA

The numbers of subscribers S (in millions) to wireless networks from 2002 through 2008 are shown in the table. (Source: CTIA–The Wireless Association)

Year	Subscribers, S (in millions)
2002	140.8
2003	158.7
2004	182.1
2005	207.9
2006	233.0
2007	255.4
2008	270.3

(a) Use a graphing utility to create a scatter plot of the data, with $t = 2$ corresponding to 2002.

(b) Use the *regression* feature of the graphing utility to find a linear model for the data.

(c) Use the graphing utility to plot the data and graph the model in the same viewing window. Is the model a good fit? Explain.

(d) Use the model to predict the number of subscribers in 2015. Is your answer reasonable? Explain.

19. MODELING DATA

The total player payrolls T (in millions of dollars) for the Pittsburgh Steelers from 2004 through 2008 are shown in the table. (Source: USA Today)

Year	Total player payroll, T (in millions of dollars)
2004	78.0
2005	84.2
2006	94.0
2007	106.3
2008	128.8

(a) Use a graphing utility to create a scatter plot of the data, with $t = 4$ corresponding to 2004.

(b) Use the *regression* feature of the graphing utility to find a linear model for the data.

(c) Use the graphing utility to plot the data and graph the model in the same viewing window. Is the model a good fit? Explain.

(d) Use the model to predict the payrolls in 2010 and 2015. Do the results seem reasonable? Explain.

(e) What is the slope of your model? What does it tell you about the player payroll?

20. MODELING DATA

The mean salaries S (in thousands of dollars) of public school teachers in the United States from 2002 through 2008 are shown in the table. (Source: National Education Association)

Year	Mean salary, S (in thousands of dollars)
2002	44.7
2003	45.7
2004	46.6
2005	47.7
2006	49.0
2007	50.8
2008	52.3

(a) Use a graphing utility to create a scatter plot of the data, with $t = 2$ corresponding to 2002.

(b) Use the *regression* feature of the graphing utility to find a linear model for the data.

(c) Use the graphing utility to plot the data and graph the model in the same viewing window. Is the model a good fit? Explain.

(d) Use the model to predict the mean salaries in 2016 and 2018. Do the results seem reasonable? Explain.

2   Chapter Test

See www.CalcChat.com for worked-out solutions to odd-numbered exercises.
For instructions on how to use a graphing utility, see Appendix A.

Take this test as you would take a test in class. After you are finished, check your work against the answers in the back of the book.

In Exercises 1 and 2, solve the equation (if possible). Then use a graphing utility to verify your solution.

1. $\dfrac{12}{x} - 7 = -\dfrac{27}{x} + 6$

2. $\dfrac{4}{3x - 2} - \dfrac{9x}{3x + 2} = -3$

In Exercises 3–6, perform the operations and write the result in standard form.

3. $(-8 - 3i) + (-1 - 15i)$

4. $\left(10 + \sqrt{-20}\right) - \left(4 - \sqrt{-14}\right)$

5. $(2 + i)(6 - i)$

6. $(4 + 3i)^2 - (5 + i)^2$

In Exercises 7 and 8, write the quotient in standard form.

7. $\dfrac{8 + 5i}{6 - i}$

8. $(2i - 1) \div (3i + 2)$

In Exercises 9–12, use a graphing utility to approximate any solutions of the equation. [Remember to write the equation in the form $f(x) = 0$.]

9. $3x^2 - 6 = 0$

10. $8x^2 - 2 = 0$

11. $x^3 + 5x = 4x^2$

12. $x = x^3$

In Exercises 13–16, solve the equation using any convenient method. Use a graphing utility to verify the solutions graphically.

13. $x^2 - 10x + 9 = 0$

14. $x^2 + 12x - 2 = 0$

15. $4x^2 - 81 = 0$

16. $5x^2 + 14x - 3 = 0$

In Exercises 17–20, find all solutions of the equation algebraically. Use a graphing utility to verify the solutions graphically.

17. $3x^3 - 4x^2 - 12x + 16 = 0$

18. $x + \sqrt{22 - 3x} = 6$

19. $(x^2 + 6)^{2/3} = 16$

20. $|8x - 1| = 21$

In Exercises 21–24, solve the inequality and sketch the solution on the real number line. Use a graphing utility to verify your solution graphically.

21. $8x - 1 > 3x - 10$

22. $2|x - 8| < 10$

23. $6x^2 + 5x + 1 \geq 0$

24. $\dfrac{3 - 5x}{2 + 3x} < -2$

25. The table shows the numbers of cellular phone subscribers S (in millions) in the United States from 2003 through 2008, where t represents the year, with $t = 3$ corresponding to 2003. Use the *regression* feature of a graphing utility to find a linear model for the data and to identify the correlation coefficient. Use the model to find the year in which the number of subscribers exceeded 300 million. (Source: Cellular Telecommunications & Internet Association)

Year, t	Subscribers, S
3	158.7
4	182.1
5	207.9
6	233.0
7	255.4
8	270.3

Table for 25

See www.CalcChat.com for worked-out solutions to odd-numbered exercises.
For instructions on how to use a graphing utility, see Appendix A.

Take this test to review the material in Chapters P–2. After you are finished, check your work against the answers in the back of the book.

In Exercises 1–3, simplify the expression.

1. $\dfrac{14x^2y^{-3}}{32x^{-1}y^2}$

2. $8\sqrt{60} - 2\sqrt{135} - \sqrt{15}$

3. $\sqrt{28x^4y^3}$

In Exercises 4–6, perform the operation and simplify the result.

4. $4x - [2x + 5(2 - x)]$

5. $(x - 2)(x^2 + x - 3)$

6. $\dfrac{2}{x + 3} - \dfrac{1}{x + 1}$

In Exercises 7–9, factor the expression completely.

7. $25 - (x - 2)^2$

8. $x - 5x^2 - 6x^3$

9. $54 - 16x^3$

10. Find the midpoint of the line segment connecting the points $\left(-\frac{7}{2}, 4\right)$ and $\left(\frac{5}{2}, -8\right)$. Then find the distance between the points.

11. Write the standard form of the equation of a circle with center $\left(-\frac{1}{2}, -8\right)$ and a radius of 4.

In Exercises 12–14, use point plotting to sketch the graph of the equation.

12. $x - 3y + 12 = 0$

13. $y = x^2 - 9$

14. $y = \sqrt{4 - x}$

In Exercises 15–17, (a) write the general form of the equation of the line that satisfies the given conditions and (b) find three additional points through which the line passes.

15. The line contains the points $(-5, 8)$ and $(-1, 4)$.

16. The line contains the point $\left(-\frac{1}{2}, 1\right)$ and has a slope of -2.

17. The line has an undefined slope and contains the point $\left(-\frac{3}{7}, \frac{1}{8}\right)$.

18. Find the equation of the line that passes through the point $(2, 3)$ and is (a) parallel to and (b) perpendicular to the line $6x - y = 4$.

In Exercises 19 and 20, evaluate the function at each specified value of the independent variable and simplify.

19. $f(x) = \dfrac{x}{x - 2}$

   (a) $f(5)$ (b) $f(2)$ (c) $f(5 + 4s)$

20. $f(x) = \begin{cases} 3x - 8, & x < 0 \\ x^2 + 4, & x \geq 0 \end{cases}$

   (a) $f(-8)$ (b) $f(0)$ (c) $f(4)$

In Exercises 21–24, find the domain of the function.

21. $f(x) = (x + 2)(3x - 4)$

22. $f(t) = \sqrt{5 + 7t}$

23. $g(s) = \sqrt{9 - s^2}$

24. $h(x) = \dfrac{4}{5x + 2}$

25. Determine whether the function given by $g(x) = 3x - x^3$ is even, odd, or neither.

26. Does the graph at the right represent y as a function of x? Explain.

27. Use a graphing utility to graph the function $f(x) = 2|x - 5| - |x + 5|$. Then determine the open intervals over which the function is increasing, decreasing, or constant.

28. Compare the graph of each function with the graph of $f(x) = \sqrt[3]{x}$.

(a) $r(x) = \dfrac{1}{2}\sqrt[3]{x}$

(b) $h(x) = \sqrt[3]{x} + 2$

(c) $g(x) = -\sqrt[3]{x + 2}$

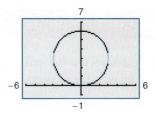

Figure for 26

In Exercises 29–32, evaluate the indicated function for

$$f(x) = x^2 + 2 \quad \text{and} \quad g(x) = 4x + 1.$$

29. $(f + g)(x)$

30. $(g - f)(x)$

31. $(g \circ f)(x)$

32. $(fg)(x)$

In Exercises 33–36, use a graphing utility to approximate the solutions of the equation. [Remember to write the equation in the form $f(x) = 0$.]

33. $4x^3 - 12x^2 + 8x = 0$

34. $\dfrac{5}{x} = \dfrac{10}{x - 3}$

35. $|3x + 4| - 2 = 0$

36. $\sqrt{x^2 + 1} + x - 9 = 0$

In Exercises 37–40, solve the inequality and sketch the solution on the real number line. Use a graphing utility to verify your solution graphically.

37. $\dfrac{x}{5} - 6 \leq -\dfrac{x}{2} + 6$

38. $2x^2 + x \geq 15$

39. $|7 + 8x| > 5$

40. $\dfrac{2(x - 2)}{x + 1} \leq 0$

41. A soccer ball has a volume of about 370.7 cubic inches. Find the radius of the soccer ball (accurate to three decimal places).

42. A rectangular plot of land with a perimeter of 546 feet has a width of x.

(a) Write the area A of the plot as a function of x.

(b) Use a graphing utility to graph the area function. What is the domain of the function?

(c) Approximate the dimensions of the plot when the area is 15,000 square feet.

43. The total revenues R (in millions of dollars) for Jack in the Box from 2001 through 2008 are shown in the table.   (Source: Jack in the Box, Inc.)

(a) Use the *regression* feature of a graphing utility to find a linear model for the data and to identify the correlation coefficient. Let t represent the year, with $t = 1$ corresponding to 2001.

(b) Use the graphing utility to plot the data and graph the model in the same viewing window.

(c) Use the model to predict the revenues for Jack in the Box in 2012 and 2014.

(d) In your opinion, is the model appropriate for predicting future revenue? Explain.

Year, t	Revenue, R
2001	1833.6
2002	1966.4
2003	2058.3
2004	2322.4
2005	2507.2
2006	2765.6
2007	2876.0
2008	2890.6

Table for 43

Proofs in Mathematics

Biconditional Statements

Recall from the Proofs in Mathematics in Chapter 1 that a conditional statement is a statement of the form "if p, then q." A statement of the form "p if and only if q" is called a **biconditional statement.** A biconditional statement, denoted by

$p \leftrightarrow q$ Biconditional statement

is the conjunction of the conditional statement $p \to q$ and its converse $q \to p$.

A biconditional statement can be either true or false. To be true, *both* the conditional statement and its converse must be true.

Example 1   Analyzing a Biconditional Statement

Consider the statement

$x = 3$ if and only if $x^2 = 9$.

a. Is the statement a biconditional statement?

b. Is the statement true?

Solution

a. The statement is a biconditional statement because it is of the form "p if and only if q."

b. The statement can be rewritten as the following conditional statement and its converse.

> *Conditional statement:* If $x = 3$, then $x^2 = 9$.

> *Converse:* If $x^2 = 9$, then $x = 3$.

The first of these statements is true, but the second is false because x could also equal -3. So, the biconditional statement is false.

Knowing how to use biconditional statements is an important tool for reasoning in mathematics.

Example 2   Analyzing a Biconditional Statement

Determine whether the biconditional statement is true or false. If it is false, provide a counterexample.

A number is divisible by 5 if and only if it ends in 0.

Solution

The biconditional statement can be rewritten as the following conditional statement and its converse.

> *Conditional statement:* If a number is divisible by 5, then it ends in 0.

> *Converse:* If a number ends in 0, then it is divisible by 5.

The conditional statement is false. A counterexample is the number 15, which is divisible by 5 but does not end in 0. So, the biconditional statement is false.

3.1   Quadratic Functions

The Graph of a Quadratic Function

In this and the next section, you will study the graphs of polynomial functions.

What you should learn
- Analyze graphs of quadratic functions.
- Write quadratic functions in standard form and use the results to sketch graphs of functions.
- Find minimum and maximum values of quadratic functions in real-life applications.

Why you should learn it
Quadratic functions can be used to model the design of a room. For instance, Exercise 63 on page 251 shows how the size of an indoor fitness room with a running track can be modeled.

Definition of Polynomial Function

Let n be a nonnegative integer and let $a_n, a_{n-1}, \ldots, a_2, a_1, a_0$ be real numbers with $a_n \neq 0$. The function given by

$$f(x) = a_n x^n + a_{n-1} x^{n-1} + \cdots + a_2 x^2 + a_1 x + a_0$$

is called a **polynomial function of x of degree n.**

Polynomial functions are classified by degree. For instance, the polynomial function

$$f(x) = a, \quad a \neq 0 \qquad \text{Constant function}$$

has degree 0 and is called a **constant function.** In Chapter 1, you learned that the graph of this type of function is a horizontal line. The polynomial function

$$f(x) = mx + b, \quad m \neq 0 \qquad \text{Linear function}$$

has degree 1 and is called a **linear function.** You learned in Chapter 1 that the graph of $f(x) = mx + b$ is a line whose slope is m and whose y-intercept is $(0, b)$. In this section, you will study second-degree polynomial functions, which are called **quadratic functions.**

Definition of Quadratic Function

Let a, b, and c be real numbers with $a \neq 0$. The function given by

$$f(x) = ax^2 + bx + c \qquad \text{Quadratic function}$$

is called a **quadratic function.**

Often real-life data can be modeled by quadratic functions. For instance, the table at the right shows the height h (in feet) of a projectile fired from a height of 6 feet with an initial velocity of 256 feet per second at any time t (in seconds). A quadratic model for the data in the table is

$$h(t) = -16t^2 + 256t + 6 \quad \text{for} \quad 0 \leq t \leq 16.$$

The graph of a quadratic function is a special type of U-shaped curve called a **parabola.** Parabolas occur in many real-life applications, especially those involving reflective properties, such as satellite dishes or flashlight reflectors. You will study these properties in a later chapter.

All parabolas are symmetric with respect to a line called the **axis of symmetry,** or simply the **axis** of the parabola. The point where the axis intersects the parabola is called the **vertex** of the parabola.

t	h
0	6
2	454
4	774
6	966
8	1030
10	966
12	774
14	454
16	6

Basic Characteristics of Quadratic Functions

Graph of $f(x) = ax^2$, $a > 0$	*Graph of* $f(x) = ax^2$, $a < 0$
Domain: $(-\infty, \infty)$	Domain: $(-\infty, \infty)$
Range: $[0, \infty)$	Range: $(-\infty, 0]$
Intercept: $(0, 0)$	Intercept: $(0, 0)$
Decreasing on $(-\infty, 0)$	Increasing on $(-\infty, 0)$
Increasing on $(0, \infty)$	Decreasing on $(0, \infty)$
Even function	Even function
Axis of symmetry: $x = 0$	Axis of symmetry: $x = 0$
Relative minimum or vertex: $(0, 0)$	Relative maximum or vertex: $(0, 0)$

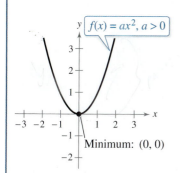
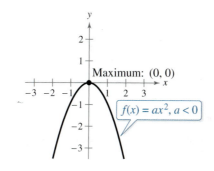

For the general quadratic form $f(x) = ax^2 + bx + c$, when the leading coefficient a is positive, the parabola opens upward; and when the leading coefficient a is negative, the parabola opens downward. Later in this section you will learn ways to find the coordinates of the vertex of a parabola.

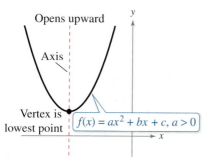
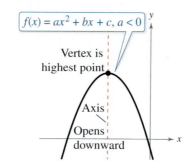

When sketching the graph of $f(x) = ax^2$, it is helpful to use the graph of $y = x^2$ as a reference, as discussed in Section 1.5. There you saw that when $a > 1$, the graph of $y = af(x)$ is a vertical stretch of the graph of $y = f(x)$. When $0 < a < 1$, the graph of $y = af(x)$ is a vertical shrink of the graph of $y = f(x)$. Notice in Figure 3.1 that the coefficient a determines how widely the parabola given by $f(x) = ax^2$ opens. When $|a|$ is small, the parabola opens more widely than when $|a|$ is large.

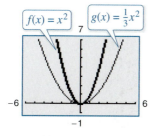
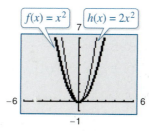

Vertical shrink

Figure 3.1

Vertical stretch

Library of Parent Functions: Quadratic Function

The *parent quadratic function* is $f(x) = x^2$, also known as the *squaring function*. The basic characteristics of the parent quadratic function are summarized below and on the inside front cover of this text.

Graph of $f(x) = x^2$
Domain: $(-\infty, \infty)$
Range: $[0, \infty)$
Intercept: $(0, 0)$
Decreasing on $(-\infty, 0)$
Increasing on $(0, \infty)$
Even function
Axis of symmetry: $x = 0$
Relative minimum or vertex: $(0, 0)$

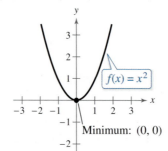

Example 1   Library of Parent Functions: $f(x) = x^2$

Sketch the graph of the function and describe how the graph is related to the graph of $f(x) = x^2$.

a. $g(x) = -x^2 + 1$

b. $h(x) = (x + 2)^2 - 3$

Solution

a. With respect to the graph of $f(x) = x^2$, the graph of g is obtained by a *reflection* in the x-axis and a vertical shift one unit *upward*, as shown in Figure 3.2. Confirm this with a graphing utility.

b. With respect to the graph of $f(x) = x^2$, the graph of h is obtained by a horizontal shift two units *to the left* and a vertical shift three units *downward*, as shown in Figure 3.3. Confirm this with a graphing utility.

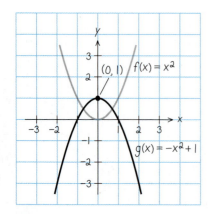

Figure 3.2

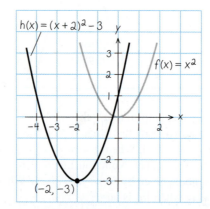

Figure 3.3

✔CHECKPOINT   Now try Exercise 11.

Recall from Section 1.5 that the graphs of $y = f(x \pm c)$, $y = f(x) \pm c$, $y = -f(x)$, and $y = f(-x)$ are rigid transformations of the graph of $y = f(x)$.

$y = f(x \pm c)$ Horizontal shift $y = -f(x)$ Reflection in x-axis

$y = f(x) \pm c$ Vertical shift $y = f(-x)$ Reflection in y-axis

The Standard Form of a Quadratic Function

The equation in Example 1(b) is written in the **standard form**

$$f(x) = a(x - h)^2 + k.$$

This form is especially convenient for sketching a parabola because it identifies the vertex of the parabola as (h, k).

Standard Form of a Quadratic Function

The quadratic function given by

$$f(x) = a(x - h)^2 + k, \quad a \neq 0$$

is in **standard form.** The graph of f is a parabola whose axis is the vertical line $x = h$ and whose vertex is the point (h, k). When $a > 0$, the parabola opens upward, and when $a < 0$, the parabola opens downward.

Example 2   Identifying the Vertex of a Quadratic Function

Describe the graph of

$$f(x) = 2x^2 + 8x + 7$$

and identify the vertex.

Solution

Write the quadratic function in standard form by completing the square. Recall that the first step is to factor out any coefficient of x^2 that is not 1.

$f(x) = 2x^2 + 8x + 7$	Write original function.
$= (2x^2 + 8x) + 7$	Group x-terms.
$= 2(x^2 + 4x) + 7$	Factor 2 out of x-terms.
$= 2(x^2 + 4x + 4 - 4) + 7$	Add and subtract $(4/2)^2 = 4$ within parentheses to complete the square.
$\left(\dfrac{4}{2}\right)^2$	
$= 2(x^2 + 4x + 4) - 2(4) + 7$	Regroup terms.
$= 2(x + 2)^2 - 1$	Write in standard form.

From the standard form, you can see that the graph of f is a parabola that opens upward with vertex

$$(-2, -1)$$

as shown in Figure 3.4. This corresponds to a left shift of two units and a downward shift of one unit relative to the graph of

$$y = 2x^2.$$

$f(x) = 2x^2 + 8x + 7$

(−2, −1)

Figure 3.4

✓**CHECKPOINT**   Now try Exercise 23.

Explore the Concept

Use a graphing utility to graph $y = ax^2$ with $a = -2, -1, -0.5, 0.5, 1,$ and 2. How does changing the value of a affect the graph?

Use a graphing utility to graph $y = (x - h)^2$ with $h = -4, -2, 2,$ and 4. How does changing the value of h affect the graph?

Use a graphing utility to graph $y = x^2 + k$ with $k = -4, -2, 2,$ and 4. How does changing the value of k affect the graph?

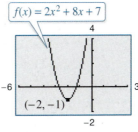

To find the x-intercepts of the graph of $f(x) = ax^2 + bx + c$, solve the equation $ax^2 + bx + c = 0$. When $ax^2 + bx + c$ does not factor, you can use the Quadratic Formula to find the x-intercepts, or a graphing utility to approximate the x-intercepts. Remember, however, that a parabola may not have x-intercepts.

Example 3   Identifying *x*-Intercepts of a Quadratic Function

Describe the graph of $f(x) = -x^2 + 6x - 8$ and identify any *x*-intercepts.

Solution

$$f(x) = -x^2 + 6x - 8$$ Write original function.

$$= -(x^2 - 6x) - 8$$ Factor -1 out of *x*-terms.

$$= -(x^2 - 6x + 9 - 9) - 8$$ Because $b = 6$, add and subtract $(6/2)^2 = 9$ within parentheses.

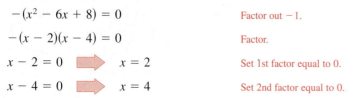

$$\left(\frac{6}{2}\right)^2$$

$$= -(x^2 - 6x + 9) - (-9) - 8$$ Regroup terms.

$$= -(x - 3)^2 + 1$$ Write in standard form.

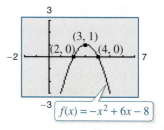

The graph of *f* is a parabola that opens downward with vertex $(3, 1)$, as shown in Figure 3.5. The *x*-intercepts are determined as follows.

Figure 3.5

$$-(x^2 - 6x + 8) = 0$$ Factor out -1.

$$-(x - 2)(x - 4) = 0$$ Factor.

$$x - 2 = 0 \implies x = 2$$ Set 1st factor equal to 0.

$$x - 4 = 0 \implies x = 4$$ Set 2nd factor equal to 0.

So, the *x*-intercepts are $(2, 0)$ and $(4, 0)$, as shown in Figure 3.5.

CHECKPOINT   Now try Exercise 31.

Example 4   Writing the Equation of a Parabola in Standard Form

Write the standard form of the equation of the parabola whose vertex is $(1, 2)$ and that passes through the point $(3, -6)$.

Solution

Because the vertex of the parabola is $(h, k) = (1, 2)$, the equation has the form

$$f(x) = a(x - 1)^2 + 2.$$ Substitute for *h* and *k* in standard form.

Because the parabola passes through the point $(3, -6)$, it follows that $f(3) = -6$. So, you obtain

$$f(x) = a(x - 1)^2 + 2$$ Write in standard form.

$$-6 = a(3 - 1)^2 + 2$$ Substitute -6 for $f(x)$ and 3 for *x*.

$$-6 = 4a + 2$$ Simplify.

$$-8 = 4a$$ Subtract 2 from each side.

$$-2 = a.$$ Divide each side by 4.

The equation in standard form is

$$f(x) = -2(x - 1)^2 + 2.$$

You can confirm this answer by graphing $f(x) = -2(x - 1)^2 + 2$ with a graphing utility, as shown in Figure 3.6. Use the *zoom* and *trace* features or the *maximum* and *value* features to confirm that its vertex is $(1, 2)$ and that it passes through the point $(3, -6)$.

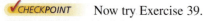CHECKPOINT   Now try Exercise 39.

 Study Tip

In Example 4, there are infinitely many different parabolas that have a vertex at $(1, 2)$. Of these, however, the only one that passes through the point $(3, -6)$ is the one given by

$$f(x) = -2(x - 1)^2 + 2.$$

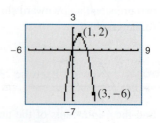

Figure 3.6

Finding Minimum and Maximum Values

Many applications involve finding the maximum or minimum value of a quadratic function. By completing the square of the quadratic function $f(x) = ax^2 + bx + c$, you can rewrite the function in standard form.

$$f(x) = a\left(x + \frac{b}{2a}\right)^2 + \left(c - \frac{b^2}{4a}\right) \qquad \text{Standard form}$$

So, the vertex of the graph of f is $\left(-\dfrac{b}{2a}, f\left(-\dfrac{b}{2a}\right)\right)$, which implies the following.

> ### Minimum and Maximum Values of Quadratic Functions
>
> Consider the function $f(x) = ax^2 + bx + c$ with vertex $\left(-\dfrac{b}{2a}, f\left(-\dfrac{b}{2a}\right)\right)$.
>
> **1.** If $a > 0$, then f has a *minimum* at $x = -\dfrac{b}{2a}$.
>
>   The minimum value is $f\left(-\dfrac{b}{2a}\right)$.
>
> **2.** If $a < 0$, then f has a *maximum* at $x = -\dfrac{b}{2a}$.
>
>   The maximum value is $f\left(-\dfrac{b}{2a}\right)$.

Example 5   The Maximum Height of a Projectile

A baseball is hit at a point 3 feet above the ground at a velocity of 100 feet per second and at an angle of 45° with respect to the ground. The path of the baseball is given by the function $f(x) = -0.0032x^2 + x + 3$, where $f(x)$ is the height of the baseball (in feet) and x is the horizontal distance from home plate (in feet). What is the maximum height reached by the baseball?

Algebraic Solution

For this quadratic function, you have

$$f(x) = ax^2 + bx + c = -0.0032x^2 + x + 3$$

which implies that $a = -0.0032$ and $b = 1$. Because the function has a maximum when $x = -b/(2a)$, you can conclude that the baseball reaches its maximum height when it is x feet from home plate, where x is

$$x = -\frac{b}{2a}$$

$$= -\frac{1}{2(-0.0032)}$$

$$= 156.25 \text{ feet.}$$

At this distance, the maximum height is

$$f(156.25) = -0.0032(156.25)^2 + 156.25 + 3$$

$$= 81.125 \text{ feet.}$$

✓ CHECKPOINT   Now try Exercise 65.

Graphical Solution

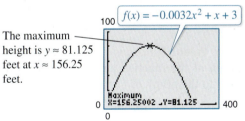

The maximum height is $y \approx 81.125$ feet at $x \approx 156.25$ feet.

3.1   Exercises

See www.CalcChat.com for worked-out solutions to odd-numbered exercises.
For instructions on how to use a graphing utility, see Appendix A.

Vocabulary and Concept Check

In Exercises 1 and 2, fill in the blanks.

1. A polynomial function of degree n and leading coefficient a_n is a function of the form $f(x) = a_n x^n + a_{n-1} x^{n-1} + \cdots + a_2 x^2 + a_1 x + a_0$, $a_n \neq 0$, where n is a _____ and $a_n, a_{n-1}, \ldots, a_2, a_1, a_0$ are _____ numbers.

2. A _____ function is a second-degree polynomial function, and its graph is called a _____ .

3. Is the quadratic function $f(x) = (x - 2)^2 + 3$ written in standard form? Identify the vertex of the graph of f.

4. Does the graph of the quadratic function $f(x) = -3x^2 + 5x + 2$ have a relative minimum value at its vertex?

Procedures and Problem Solving

Graphs of Quadratic Functions   In Exercises 5–8, match the quadratic function with its graph. [The graphs are labeled (a), (b), (c), and (d).]

(a)

(b)
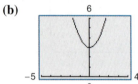

(c)
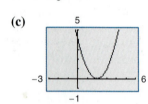

(d)
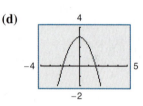

5. $f(x) = (x - 2)^2$

6. $f(x) = 3 - x^2$

7. $f(x) = x^2 + 3$

8. $f(x) = -(x - 4)^2$

Library of Parent Functions   In Exercises 9–16, sketch the graph of the function and describe how the graph is related to the graph of $y = x^2$.

9. $y = -x^2$

10. $y = x^2 - 1$

✓ 11. $y = (x + 3)^2$

12. $y = -(x + 3)^2 - 1$

13. $y = (x + 1)^2$

14. $y = -x^2 + 2$

15. $y = (x - 3)^2$

16. $y = -(x - 3)^2 + 1$

Identifying the Vertex of a Quadratic Function   In Exercises 17–30, describe the graph of the function and identify the vertex. Use a graphing utility to verify your results.

17. $f(x) = 25 - x^2$

18. $f(x) = x^2 - 7$

19. $f(x) = \frac{1}{2} x^2 - 4$

20. $f(x) = 16 - \frac{1}{4} x^2$

21. $f(x) = (x + 4)^2 - 3$

22. $f(x) = (x - 6)^2 + 3$

✓ 23. $h(x) = x^2 - 8x + 16$

24. $g(x) = x^2 + 2x + 1$

25. $f(x) = x^2 - x + \frac{5}{4}$

26. $f(x) = x^2 + 3x + \frac{1}{4}$

27. $f(x) = -x^2 + 2x + 5$

28. $f(x) = -x^2 - 4x + 1$

29. $h(x) = 4x^2 - 4x + 21$

30. $f(x) = 2x^2 - x + 1$

Identifying x-Intercepts of a Quadratic Function   In Exercises 31–36, describe the graph of the quadratic function. Identify the vertex and x-intercept(s). Use a graphing utility to verify your results.

✓ 31. $f(x) = -(x^2 + 2x - 3)$

32. $f(x) = -(x^2 + x - 30)$

33. $g(x) = x^2 + 8x + 11$

34. $f(x) = x^2 + 10x + 14$

35. $f(x) = -2x^2 + 16x - 31$

36. $f(x) = -4x^2 + 24x - 41$

Writing the Equation of a Parabola in Standard Form   In Exercises 37 and 38, write an equation of the parabola in standard form. Use a graphing utility to graph the equation and verify your result.

37.
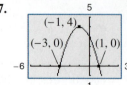

38.
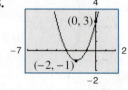

Writing the Equation of a Parabola in Standard Form
In Exercises 39–44, write the standard form of the quadratic function that has the indicated vertex and whose graph passes through the given point. Use a graphing utility to verify your result.

✓ **39.** Vertex: $(-2, 5)$;   Point: $(0, 9)$
40. Vertex: $(4, 1)$;   Point: $(6, -7)$
41. Vertex: $(1, -2)$;   Point: $(-1, 14)$
42. Vertex: $(-4, -1)$;   Point: $(-2, 4)$
43. Vertex: $\left(\frac{1}{2}, 1\right)$;   Point: $\left(-2, -\frac{21}{5}\right)$
44. Vertex: $\left(-\frac{1}{4}, -1\right)$;   Point: $\left(0, -\frac{17}{16}\right)$

Using a Graph to Identify x-Intercepts
In Exercises 45–48, determine the x-intercept(s) of the graph visually. Then find the x-intercept(s) algebraically to verify your answer.

45.

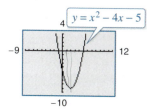

$y = x^2 - 4x - 5$

46.

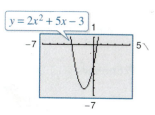

$y = 2x^2 + 5x - 3$

47.

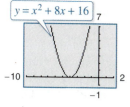

$y = x^2 + 8x + 16$

48.

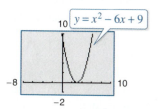

$y = x^2 - 6x + 9$

Graphing to Identify x-Intercepts
In Exercises 49–54, use a graphing utility to graph the quadratic function and find the x-intercepts of the graph. Then find the x-intercepts algebraically to verify your answer.

49. $y = x^2 - 4x$
50. $y = -2x^2 + 10x$
51. $y = 2x^2 - 7x - 30$
52. $y = 4x^2 + 25x - 21$
53. $y = -\frac{1}{2}(x^2 - 6x - 7)$
54. $y = \frac{7}{10}(x^2 + 12x - 45)$

Using the x-Intercepts to Write Equations
In Exercises 55–58, find two quadratic functions, one that opens upward and one that opens downward, whose graphs have the given x-intercepts. (There are many correct answers.)

55. $(-1, 0), (3, 0)$
56. $(0, 0), (10, 0)$
57. $(-3, 0), \left(-\frac{1}{2}, 0\right)$
58. $\left(-\frac{5}{2}, 0\right), (2, 0)$

Maximizing a Product of Two Numbers
In Exercises 59–62, find the two positive real numbers with the given sum whose product is a maximum.

59. The sum is 110.
60. The sum is 66.

61. The sum of the first and twice the second is 24.
62. The sum of the first and three times the second is 42.

63. *Why you should learn it* (p. 244) An indoor physical fitness room consists of a rectangular region with a semicircle on each end. The perimeter of the room is to be a 200-meter single-lane running track.

(a) Draw a diagram that illustrates the problem. Let x and y represent the length and width of the rectangular region, respectively.

(b) Determine the radius of the semicircular ends of the track. Determine the distance, in terms of y, around the inside edge of the two semicircular parts of the track.

(c) Use the result of part (b) to write an equation, in terms of x and y, for the distance traveled in one lap around the track. Solve for y.

(d) Use the result of part (c) to write the area A of the rectangular region as a function of x.

(e) Use a graphing utility to graph the area function from part (d). Use the graph to approximate the dimensions that will produce a rectangle of maximum area.

64. Algebraic-Graphical-Numerical A child care center has 200 feet of fencing to enclose two adjacent rectangular safe play areas (see figure). Use the following methods to determine the dimensions that will produce a maximum enclosed area.

(a) Write the total area A of the play areas as a function of x.

(b) Use the *table* feature of a graphing utility to create a table showing possible values of x and the corresponding total area A of the play areas. Use the table to estimate the dimensions that will produce the maximum enclosed area.

(c) Use the graphing utility to graph the area function. Use the graph to approximate the dimensions that will produce the maximum enclosed area.

(d) Write the area function in standard form to find algebraically the dimensions that will produce the maximum enclosed area.

(e) Compare your results from parts (b), (c), and (d).

✓ **65. Height of a Projectile**   The height y (in feet) of a punted football is approximated by

$$y = -\frac{16}{2025}x^2 + \frac{9}{5}x + \frac{3}{2}$$

where x is the horizontal distance (in feet) from where the football is punted.

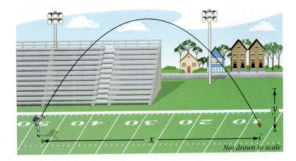

Not drawn to scale

(a) Use a graphing utility to graph the path of the football.

(b) How high is the football when it is punted? (*Hint:* Find y when $x = 0$.)

(c) What is the maximum height of the football?

(d) How far from the punter does the football strike the ground?

66. Physics   The path of a diver is approximated by

$$y = -\frac{4}{9}x^2 + \frac{24}{9}x + 12$$

where y is the height (in feet) and x is the horizontal distance (in feet) from the end of the diving board (see figure). What is the maximum height of the diver? Verify your answer using a graphing utility.

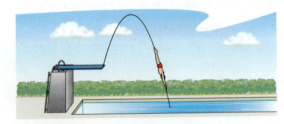

67. Geometry   To make a sign holder, you bend a 100-inch long steel wire x inches from each end to form two right angles. To use the sign holder, you insert each end 6 inches into the ground.

(a) Write a function for the rectangular area A enclosed by the sign holder in terms of x.

(b) Use the *table* feature of a graphing utility to determine the value of x that maximizes the rectangular area enclosed by the sign holder.

68. Aerodynamic Engineering   The number of horsepower H required to overcome wind drag on a certain automobile is approximated by

$$H(s) = 0.002s^2 + 0.05s - 0.029, \quad 0 \le s \le 100$$

where s is the speed of the car (in miles per hour).

(a) Use a graphing utility to graph the function.

(b) Graphically estimate the maximum speed of the car given that the power required to overcome wind drag is not to exceed 10 horsepower. Verify your result algebraically.

69. Economics   The monthly revenue R (in thousands of dollars) from the sales of a digital picture frame is approximated by $R(p) = -10p^2 + 1580p$, where p is the price per unit (in dollars).

(a) Find the monthly revenues for unit prices of $50, $70, and $90.

(b) Find the unit price that will yield a maximum monthly revenue.

(c) What is the maximum monthly revenue?

(d) Explain your results.

70. Economics   The weekly revenue R (in dollars) earned by a computer repair service is given by

$$R(p) = -12p^2 + 372p$$

where p is the price charged per service hour (in dollars).

(a) Find the weekly revenues for prices per service hour of $12, $16, and $20.

(b) Find the price that will yield a maximum weekly revenue.

(c) What is the maximum weekly revenue?

(d) Explain your results.

71. Public Health   From 1955 through 2000, the annual per capita consumption C of cigarettes by Americans (age 18 and older) can be modeled by

$$C(t) = -2.10t^2 + 70.9t + 3557, \quad 5 \le t \le 50$$

where t is the year, with $t = 5$ corresponding to 1955. (Source: U.S. Department of Agriculture)

(a) Use a graphing utility to graph the model.

(b) Use the graph of the model to approximate the year when the maximum annual consumption of cigarettes occurred. Approximate the maximum average annual consumption. Beginning in 1966, all cigarette packages were required by law to carry a health warning. Do you think the warning had any effect? Explain.

(c) In 2000, the U.S. population (age 18 and older) was 209,117,000. Of those, about 48,306,000 were smokers. What was the average annual cigarette consumption *per smoker* in 2000? What was the average daily cigarette consumption *per smoker*?

72. Demography The population P of Germany (in thousands) from 1999 through 2009 can be modeled by

$$P(t) = -8.87t^2 + 271.4t + 80{,}362, \quad 9 \le t \le 19$$

where t is the year, with $t = 9$ corresponding to 1999. (Source: U.S. Census Bureau)

(a) According to the model, in what year did Germany have its greatest population? What was the population?

(b) According to the model, what will Germany's population be in the year 2100? Is this result reasonable? Explain.

Conclusions

True or False? In Exercises 73 and 74, determine whether the statement is true or false. Justify your answer.

73. The function $f(x) = -12x^2 - 1$ has no x-intercepts.

74. The graphs of $f(x) = -4x^2 - 10x + 7$ and $g(x) = 12x^2 + 30x + 1$ have the same axis of symmetry.

Library of Parent Functions In Exercises 75 and 76, determine which equation(s) may be represented by the graph shown. (There may be more than one correct answer.)

75. (a) $f(x) = -(x - 4)^2 + 2$

(b) $f(x) = -(x + 2)^2 + 4$

(c) $f(x) = -(x + 2)^2 - 4$

(d) $f(x) = -x^2 - 4x - 8$

(e) $f(x) = -(x - 2)^2 - 4$

(f) $f(x) = -x^2 + 4x - 8$

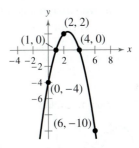

76. (a) $f(x) = (x - 1)^2 + 3$

(b) $f(x) = (x + 1)^2 + 3$

(c) $f(x) = (x - 3)^2 + 1$

(d) $f(x) = x^2 + 2x + 4$

(e) $f(x) = (x + 3)^2 + 1$

(f) $f(x) = x^2 + 6x + 10$

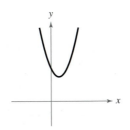

Describing Parabolas In Exercises 77–80, let z represent a positive real number. Describe how the family of parabolas represented by the given function compares with the graph of $g(x) = x^2$.

77. $f(x) = (x - z)^2$

78. $f(x) = x^2 - z$

79. $f(x) = z(x - 3)^2$

80. $f(x) = zx^2 + 4$

Think About It In Exercises 81–84, find the value of b such that the function has the given maximum or minimum value.

81. $f(x) = -x^2 + bx - 75$; Maximum value: 25

82. $f(x) = -x^2 + bx - 16$; Maximum value: 48

83. $f(x) = x^2 + bx + 26$; Minimum value: 10

84. $f(x) = x^2 + bx - 25$; Minimum value: -50

85. Proof Let x and y be two positive real numbers whose sum is S. Show that the maximum product of x and y occurs when x and y are both equal to $S/2$.

86. Proof Assume that the function given by $f(x) = ax^2 + bx + c$, $a \ne 0$, has two real zeros. Show that the x-coordinate of the vertex of the graph is the average of the zeros of f. (Hint: Use the Quadratic Formula.)

87. Writing The parabola in the figure below has an equation of the form $y = ax^2 + bx - 4$. Find the equation of this parabola in two different ways, by hand and with technology (graphing utility or computer software). Write a paragraph describing the methods you used and comparing the results of the two methods.

```
          y
           |     (2, 2)
   (1, 0)  |     (4, 0)
  ---+--+--+--+--+--+---- x
  -4 -2  |  2    6  8
        -2|
        -4|
          |(0, -4)
        -6|
          |(6, -10)
```

88. CAPSTONE The annual profit P (in dollars) of a company is modeled by a function of the form $P = at^2 + bt + c$, where t represents the year. Discuss which of the following models the company might prefer.

(a) a is positive and $t \ge -b/(2a)$.

(b) a is positive and $t \le -b/(2a)$.

(c) a is negative and $t \ge -b/(2a)$.

(d) a is negative and $t \le -b/(2a)$.

Cumulative Mixed Review

Finding Points of Intersection In Exercises 89–92, determine algebraically any point(s) of intersection of the graphs of the equations. Verify your results using the *intersect* feature of a graphing utility.

89. $x + y = 8$
$-\frac{2}{3}x + y = 6$

90. $y = 3x - 10$
$y = \frac{1}{4}x + 1$

91. $y = 9 - x^2$
$y = x + 3$

92. $y = x^3 + 2x - 1$
$y = -2x + 15$

93. *Make a Decision* To work an extended application analyzing the height of a basketball after it has been dropped, visit this textbook's *Companion Website*.

3.2 Polynomial Functions of Higher Degree

Graphs of Polynomial Functions

At this point, you should be able to sketch accurate graphs of polynomial functions of degrees 0, 1, and 2.

Function	*Graph*
$f(x) = a$	Horizontal line
$f(x) = ax + b$	Line of slope a
$f(x) = ax^2 + bx + c$	Parabola

The graphs of polynomial functions of degree greater than 2 are more difficult to sketch by hand. However, in this section you will learn how to recognize some of the basic features of the graphs of polynomial functions. Using these features along with point plotting, intercepts, and symmetry, you should be able to make reasonably accurate sketches *by hand*.

The graph of a polynomial function is **continuous.** Essentially, this means that the graph of a polynomial function has no breaks, holes, or gaps, as shown in Figure 3.7. Informally, you can say that a function is continuous when its graph can be drawn with a pencil without lifting the pencil from the paper.

What you should learn
- Use transformations to sketch graphs of polynomial functions.
- Use the Leading Coefficient Test to determine the end behavior of graphs of polynomial functions.
- Find and use zeros of polynomial functions as sketching aids.
- Use the Intermediate Value Theorem to help locate zeros of polynomial functions.

Why you should learn it
You can use polynomial functions to model various aspects of nature, such as the growth of a red oak tree, as shown in Exercise 112 on page 264.

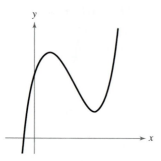

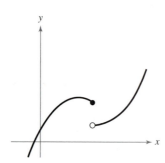

(a) **Polynomial functions have continuous graphs.**

(b) **Functions with graphs that are not continuous are not polynomial functions.**

Figure 3.7

Another feature of the graph of a polynomial function is that it has only smooth, rounded turns, as shown in Figure 3.8(a). It cannot have a sharp turn such as the one shown in Figure 3.8(b).

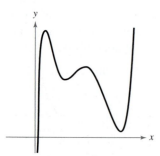

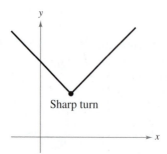

Sharp turn

(a) **Polynomial functions have graphs with smooth, rounded turns.**

(b) **Functions with graphs that have sharp turns are not polynomial functions.**

Figure 3.8

The graphs of polynomial functions of degree 1 are lines, and those of functions of degree 2 are parabolas. The graphs of all polynomial functions are smooth and continuous. A polynomial function of degree n has the form

$$f(x) = a_n x^n + a_{n-1} x^{n-1} + \cdots + a_2 x^2 + a_1 x + a_0$$

where n is a positive integer and $a_n \neq 0$.

The polynomial functions that have the simplest graphs are monomials of the form $f(x) = x^n$, where n is an integer greater than zero. The greater the value of n, the flatter the graph near the origin. When n is even, the graph is similar to the graph of $f(x) = x^2$ and touches the x-axis at the x-intercept. When n is odd, the graph is similar to the graph of $f(x) = x^3$ and crosses the x-axis at the x-intercept. Polynomial functions of the form $f(x) = x^n$ are often referred to as **power functions.**

Library of Parent Functions: Cubic Function

The basic characteristics of the *parent cubic function* $f(x) = x^3$ are summarized below and on the inside front cover of this text.

Graph of $f(x) = x^3$
Domain: $(-\infty, \infty)$
Range: $(-\infty, \infty)$
Intercept: $(0, 0)$
Increasing on $(-\infty, \infty)$
Odd function
Origin symmetry

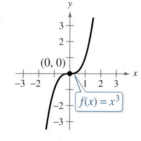

Example 1   Library of Parent Functions: $f(x) = x^3$

Sketch the graphs of (a) $g(x) = -x^3$, (b) $h(x) = x^3 + 1$, and (c) $k(x) = (x - 1)^3$.

Solution

a. With respect to the graph of $f(x) = x^3$, the graph of g is obtained by a *reflection* in the x-axis, as shown in Figure 3.9.

b. With respect to the graph of $f(x) = x^3$, the graph of h is obtained by a vertical shift one unit *upward*, as shown in Figure 3.10.

c. With respect to the graph of $f(x) = x^3$, the graph of k is obtained by a horizontal shift one unit *to the right*, as shown in Figure 3.11.

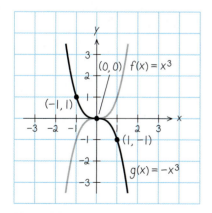

Figure 3.9

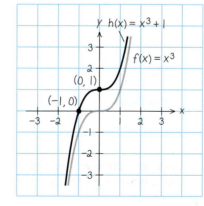

Figure 3.10

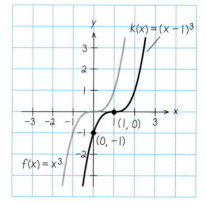

Figure 3.11

✓ CHECKPOINT   Now try Exercise 17.

The Leading Coefficient Test

In Example 1, note that all three graphs eventually rise or fall without bound as x moves to the right. Whether the graph of a polynomial eventually rises or falls can be determined by the polynomial function's degree (even or odd) and by its leading coefficient, as indicated in the **Leading Coefficient Test.**

Leading Coefficient Test

As x moves without bound to the left or to the right, the graph of the polynomial function

$$f(x) = a_n x^n + \cdots + a_1 x + a_0, \quad a_n \neq 0$$

eventually rises or falls in the following manner.

1. When n is odd:

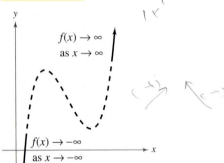

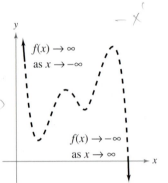

If the leading coefficient is positive ($a_n > 0$), then the graph falls to the left and rises to the right.

If the leading coefficient is negative ($a_n < 0$), then the graph rises to the left and falls to the right.

2. When n is even:

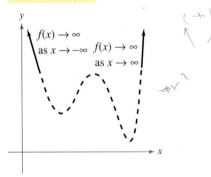

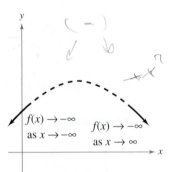

If the leading coefficient is positive ($a_n > 0$), then the graph rises to the left and right.

If the leading coefficient is negative ($a_n < 0$), then the graph falls to the left and right.

Note that the dashed portions of the graphs indicate that the test determines only the right-hand and left-hand behavior of the graph.

As you continue to study polynomial functions and their graphs, you will notice that the degree of a polynomial plays an important role in determining other characteristics of the polynomial function and its graph.

Explore the Concept

For each function, identify the degree of the function and whether the degree of the function is even or odd. Identify the leading coefficient and whether the leading coefficient is positive or negative. Use a graphing utility to graph each function. Describe the relationship between the degree and sign of the leading coefficient of the function, and the right- and left-hand behavior of the graph of the function.

a. $y = x^3 - 2x^2 - x + 1$

b. $y = 2x^5 + 2x^2 - 5x + 1$

c. $y = -2x^5 - x^2 + 5x + 3$

d. $y = -x^3 + 5x - 2$

e. $y = 2x^2 + 3x - 4$

f. $y = x^4 - 3x^2 + 2x - 1$

g. $y = -x^2 + 3x + 2$

h. $y = -x^6 - x^2 - 5x + 4$

Study Tip

The notation "$f(x) \to -\infty$ as $x \to -\infty$" indicates that the graph falls to the left. The notation "$f(x) \to \infty$ as $x \to \infty$" indicates that the graph rises to the right.

Example 2   Applying the Leading Coefficient Test

Use the Leading Coefficient Test to describe the right-hand and left-hand behavior of the graph of

$$f(x) = -x^3 + 4x.$$

Solution

Because the degree is odd and the leading coefficient is negative, the graph rises to the left and falls to the right, as shown in Figure 3.12.

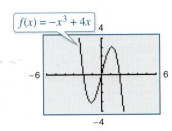

Figure 3.12

✔CHECKPOINT   Now try Exercise 29.

Example 3   Applying the Leading Coefficient Test

Use the Leading Coefficient Test to describe the right-hand and left-hand behavior of the graph of each polynomial function.

a. $f(x) = x^4 - 5x^2 + 4$

b. $f(x) = x^5 - x$

Solution

a. Because the degree is even and the leading coefficient is positive, the graph rises to the left and right, as shown in Figure 3.13.

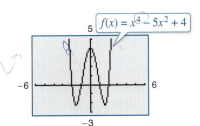

Figure 3.13

b. Because the degree is odd and the leading coefficient is positive, the graph falls to the left and rises to the right, as shown in Figure 3.14.

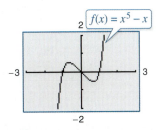

Figure 3.14

✔CHECKPOINT   Now try Exercise 31.

Explore the Concept

For each of the graphs in Examples 2 and 3, count the number of zeros of the polynomial function and the number of relative extrema, and compare these numbers with the degree of the polynomial. What do you observe?

In Examples 2 and 3, note that the Leading Coefficient Test tells you only whether the graph *eventually* rises or falls to the right or left. Other characteristics of the graph, such as intercepts and minimum and maximum points, must be determined by other tests.

Zeros of Polynomial Functions

It can be shown that for a polynomial function f of degree n, the following statements are true.

1. The function f has at most n real zeros. (You will study this result in detail in Section 3.4 on the Fundamental Theorem of Algebra.)

2. The graph of f has at most $n - 1$ relative **extrema** (relative minima or maxima).

Recall that a zero of a function f is a number x for which $f(x) = 0$. Finding the zeros of polynomial functions is one of the most important problems in algebra. You have already seen that there is a strong interplay between graphical and algebraic approaches to this problem. Sometimes you can use information about the graph of a function to help find its zeros. In other cases, you can use information about the zeros of a function to find a good viewing window.

Real Zeros of Polynomial Functions

If f is a polynomial function and a is a real number, then the following statements are equivalent.

    1. $x = a$ is a *zero* of the function f.

    2. $x = a$ is a *solution* of the polynomial equation $f(x) = 0$.

    3. $(x - a)$ is a *factor* of the polynomial $f(x)$.

    4. $(a, 0)$ is an *x-intercept* of the graph of f.

Finding zeros of polynomial functions is closely related to factoring and finding x-intercepts, as demonstrated in Examples 4, 5, and 6.

Example 4 Finding Zeros of a Polynomial Function

Find all real zeros of $f(x) = x^3 - x^2 - 2x$.

Algebraic Solution

$$f(x) = x^3 - x^2 - 2x \qquad \text{Write original function.}$$

$$0 = x^3 - x^2 - 2x \qquad \text{Substitute 0 for } f(x).$$

$$0 = x(x^2 - x - 2) \qquad \text{Remove common monomial factor.}$$

$$0 = x(x - 2)(x + 1) \qquad \text{Factor completely.}$$

So, the real zeros are

$$x = 0, \quad x = 2, \quad \text{and} \quad x = -1$$

and the corresponding x-intercepts are

$$(0, 0), \quad (2, 0), \quad \text{and} \quad (-1, 0).$$

Check

$$(0)^3 - (0)^2 - 2(0) = 0 \qquad x = 0 \text{ is a zero. } ✔$$

$$(2)^3 - (2)^2 - 2(2) = 0 \qquad x = 2 \text{ is a zero. } ✔$$

$$(-1)^3 - (-1)^2 - 2(-1) = 0 \qquad x = -1 \text{ is a zero. } ✔$$

✔**CHECKPOINT**   Now try Exercise 37.

Graphical Solution

The graph of f has the x-intercepts

$$(0, 0), \quad (2, 0), \quad \text{and} \quad (-1, 0)$$

as shown in Figure 3.15. So, the real zeros of f are

$$x = 0, \quad x = 2, \quad \text{and} \quad x = -1.$$

Use the *zero* or *root* feature of a graphing utility to verify these zeros.

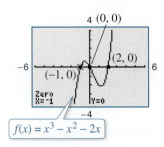

Figure 3.15

Example 5   Analyzing a Polynomial Function

Find all real zeros and relative extrema of $f(x) = -2x^4 + 2x^2$.

Solution

$$0 = -2x^4 + 2x^2 \qquad \text{Substitute 0 for } f(x).$$

$$0 = -2x^2(x^2 - 1) \qquad \text{Remove common monomial factor.}$$

$$0 = -2x^2(x - 1)(x + 1) \qquad \text{Factor completely.}$$

So, the real zeros are $x = 0$, $x = 1$, and $x = -1$, and the corresponding x-intercepts are $(0, 0)$, $(1, 0)$, and $(-1, 0)$, as shown in Figure 3.16. Using the *minimum* and *maximum* features of a graphing utility, you can approximate the three relative extrema to be $(-0.71, 0.5)$, $(0, 0)$, and $(0.71, 0.5)$.

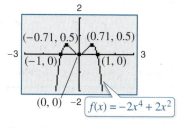

Figure 3.16

✓CHECKPOINT   Now try Exercise 59.

Repeated Zeros

For a polynomial function, a factor of $(x - a)^k$, $k > 1$, yields a **repeated zero** $x = a$ of **multiplicity** k.

   1. If k is odd, then the graph *crosses* the x-axis at $x = a$.

   2. If k is even, then the graph *touches* the x-axis (but does not cross the x-axis) at $x = a$.

Study Tip

In Example 5, note that because k is even, the factor $-2x^2$ yields the repeated zero $x = 0$. The graph touches (but does not cross) the x-axis at $x = 0$, as shown in Figure 3.16.

Example 6   Analyzing a Polynomial Function

Find all real zeros of $f(x) = x^5 - 3x^3 - x^2 - 4x - 1$.

Solution

From Figure 3.17, you can see that there are three zeros. Using the *zero* feature of a graphing utility, you can determine that the zeros are approximately $x \approx -1.86$, $x \approx -0.25$, and $x \approx 2.11$. It should be noted that this fifth-degree polynomial factors as

$$f(x) = x^5 - 3x^3 - x^2 - 4x - 1 = (x^2 + 1)(x^3 - 4x - 1).$$

The three zeros obtained above are the zeros of the cubic factor $x^3 - 4x - 1$. The quadratic factor $x^2 + 1$ has no real zeros, but does have two *complex* zeros. You will learn more about complex zeros in Section 3.4.

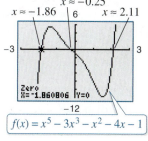

Figure 3.17

✓CHECKPOINT   Now try Exercise 61.

Example 7   Finding a Polynomial Function with Given Zeros

Find a polynomial function with zeros $-\frac{1}{2}$, 3, and 3. (There are many correct solutions.)

Solution

Note that the zero $x = -\frac{1}{2}$ corresponds to either $\left(x + \frac{1}{2}\right)$ or $(2x + 1)$. To avoid fractions, choose the second factor and write

$$f(x) = (2x + 1)(x - 3)^2$$

$$= (2x + 1)(x^2 - 6x + 9)$$

$$= 2x^3 - 11x^2 + 12x + 9.$$

✓CHECKPOINT   Now try Exercise 67.

Note in Example 7 that there are many polynomial functions with the indicated zeros. In fact, multiplying the function by any real number does not change the zeros of the function. For instance, multiply the function from Example 7 by $\frac{1}{2}$ to obtain

$$f(x) = x^3 - \tfrac{11}{2}x^2 + 6x + \tfrac{9}{2}.$$

Then find the zeros of the function. You will obtain the zeros $-\frac{1}{2}$, 3, and 3, as given in Example 7.

Example 8   Sketching the Graph of a Polynomial Function

Sketch the graph of

$$f(x) = -2x^3 + 6x^2 - \tfrac{9}{2}x.$$

Solution

1. *Apply the Leading Coefficient Test.*   Because the leading coefficient is negative and the degree is odd, you know that the graph eventually rises to the left and falls to the right (see Figure 3.18).

2. *Find the Real Zeros of the Polynomial.*   By factoring

$$
\begin{aligned}
f(x) &= -2x^3 + 6x^2 - \tfrac{9}{2}x \\
     &= -\tfrac{1}{2}x(4x^2 - 12x + 9) \\
     &= -\tfrac{1}{2}x(2x - 3)^2
\end{aligned}
$$

you can see that the real zeros of f are $x = 0$ (of odd multiplicity 1) and $x = \frac{3}{2}$ (of even multiplicity 2). So, the x-intercepts occur at $(0, 0)$ and $\left(\frac{3}{2}, 0\right)$. Add these points to your graph, as shown in Figure 3.18.

3. *Plot a Few Additional Points.*   To sketch the graph by hand, find a few additional points, as shown in the table. Then plot the points (see Figure 3.19).

x	-0.5	0.5	1	2
$f(x)$	4	-1	-0.5	-1

4. *Draw the Graph.*   Draw a continuous curve through the points, as shown in Figure 3.19. As indicated by the multiplicities of the zeros, the graph crosses the x-axis at $(0, 0)$ and touches (but does not cross) the x-axis at $\left(\frac{3}{2}, 0\right)$.

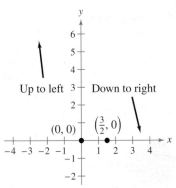

Figure 3.18

$$f(x) = -2x^3 + 6x^2 - \tfrac{9}{2}x$$

Figure 3.19

✓CHECKPOINT   Now try Exercise 87.

Study Tip

Observe in Example 8 that the sign of $f(x)$ is positive to the left of and negative to the right of the zero $x = 0$. Similarly, the sign of $f(x)$ is negative to the left and to the right of the zero $x = \frac{3}{2}$. This suggests that if a zero of a polynomial function is of *odd* multiplicity, then the sign of $f(x)$ changes from one side of the zero to the other side. If a zero is of *even* multiplicity, then the sign of $f(x)$ does not change from one side of the zero to the other side. The table below helps to illustrate this result. This sign analysis may be helpful in graphing polynomial functions.

x	-0.5	0	0.5
$f(x)$	4	0	-1
Sign	$+$		$-$

x	1	$\frac{3}{2}$	2
$f(x)$	-0.5	0	-1
Sign	$-$		$-$

The Intermediate Value Theorem

The **Intermediate Value Theorem** concerns the existence of real zeros of polynomial functions. The theorem states that if

$$(a, f(a)) \quad \text{and} \quad (b, f(b))$$

are two points on the graph of a polynomial function such that $f(a) \neq f(b)$, then for any number d between $f(a)$ and $f(b)$ there must be a number c between a and b such that $f(c) = d$. (See Figure 3.20.)

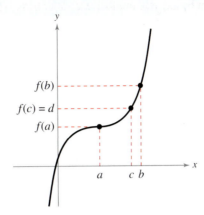

Figure 3.20

Intermediate Value Theorem

Let a and b be real numbers such that $a < b$. If f is a polynomial function such that $f(a) \neq f(b)$, then in the interval $[a, b]$, f takes on every value between $f(a)$ and $f(b)$.

This theorem helps you locate the real zeros of a polynomial function in the following way. If you can find a value $x = a$ at which a polynomial function is positive, and another value $x = b$ at which it is negative, then you can conclude that the function has at least one real zero between these two values. For example, the function $f(x) = x^3 + x^2 + 1$ is negative when $x = -2$ and positive when $x = -1$. Therefore, it follows from the Intermediate Value Theorem that f must have a real zero somewhere between -2 and -1.

Example 9   Approximating the Zeros of a Function

Find three intervals of length 1 in which the polynomial

$$f(x) = 12x^3 - 32x^2 + 3x + 5$$

is guaranteed to have a zero.

Graphical Solution

From Figure 3.21, you can see that the graph of f crosses the x-axis three times—between -1 and 0, between 0 and 1, and between 2 and 3. So, you can conclude that the function has zeros in the intervals $(-1, 0)$, $(0, 1)$, and $(2, 3)$.

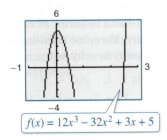

Figure 3.21

✓CHECKPOINT   Now try Exercise 95.

Numerical Solution

From the table in Figure 3.22, you can see that $f(-1)$ and $f(0)$ differ in sign. So, you can conclude from the Intermediate Value Theorem that the function has a zero between -1 and 0. Similarly, $f(0)$ and $f(1)$ differ in sign, so the function has a zero between 0 and 1. Likewise, $f(2)$ and $f(3)$ differ in sign, so the function has a zero between 2 and 3. So, you can conclude that the function has zeros in the intervals $(-1, 0)$, $(0, 1)$, and $(2, 3)$.

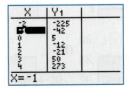

Figure 3.22

93. $g(t) = -\frac{1}{4}t^4 + 2t^2 - 4$

94. $g(x) = \frac{1}{10}(x^4 - 4x^3 - 2x^2 + 12x + 9)$

Approximating the Zeros of a Function   In Exercises **95–100**, (a) use the Intermediate Value Theorem and a graphing utility to find graphically any intervals of length 1 in which the polynomial function is guaranteed to have a zero, and (b) use the *zero* or *root* feature of the graphing utility to approximate the real zeros of the function. Verify your answers in part (a) by using the *table* feature of the graphing utility.

✓ **95.** $f(x) = x^3 - 3x^2 + 3$ **96.** $f(x) = -2x^3 - 6x^2 + 3$

97. $g(x) = 3x^4 + 4x^3 - 3$ **98.** $h(x) = x^4 - 10x^2 + 2$

99. $f(x) = x^4 - 3x^3 - 4x - 3$

100. $f(x) = x^3 - 4x^2 - 2x + 10$

Identifying Symmetry and x-Intercepts   In Exercises **101–108**, use a graphing utility to graph the function. Identify any symmetry with respect to the *x*-axis, *y*-axis, or origin. Determine the number of *x*-intercepts of the graph.

101. $f(x) = x^2(x + 6)$ **102.** $h(x) = x^3(x - 4)^2$

103. $g(t) = -\frac{1}{2}(t - 4)^2(t + 4)^2$

104. $g(x) = \frac{1}{8}(x + 1)^2(x - 3)^3$

105. $f(x) = x^3 - 4x$ **106.** $f(x) = x^4 - 2x^2$

107. $g(x) = \frac{1}{5}(x + 1)^2(x - 3)(2x - 9)$

108. $h(x) = \frac{1}{5}(x + 2)^2(3x - 5)^2$

109. Geometry   An open box is to be made from a square piece of material 36 centimeters on a side by cutting equal squares with sides of length x from the corners and turning up the sides (see figure).

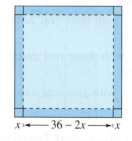
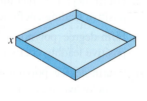

$x \longleftarrow 36 - 2x \longrightarrow x$

(a) Verify that the volume of the box is given by the function $V(x) = x(36 - 2x)^2$.

(b) Determine the domain of the function V.

(c) Use the *table* feature of a graphing utility to create a table that shows various box heights x and the corresponding volumes V. Use the table to estimate a range of dimensions within which the maximum volume is produced.

(d) Use the graphing utility to graph V and use the range of dimensions from part (c) to find the x-value for which $V(x)$ is maximum.

Kurhan 2010/used under license from Shutterstock.com

110. Geometry   An open box with locking tabs is to be made from a square piece of material 24 inches on a side. This is done by cutting equal squares from the corners and folding along the dashed lines, as shown in the figure.

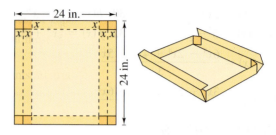

(a) Verify that the volume of the box is given by the function $V(x) = 8x(6 - x)(12 - x)$.

(b) Determine the domain of the function V.

(c) Sketch the graph of the function and estimate the value of x for which $V(x)$ is maximum.

111. Marketing   The total revenue R (in millions of dollars) for a company is related to its advertising expense by the function

$$R = 0.00001(-x^3 + 600x^2), \quad 0 \le x \le 400$$

where x is the amount spent on advertising (in tens of thousands of dollars). Use the graph of the function shown in the figure to estimate the point on the graph at which the function is increasing most rapidly. This point is called the **point of diminishing returns** because any expense above this amount will yield less return per dollar invested in advertising.

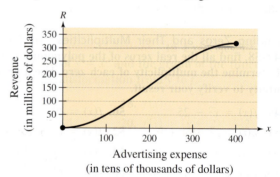

Advertising expense
(in tens of thousands of dollars)

112. *Why you should learn it* (*p. 254*)   The growth of a red oak tree is approximated by the function

$$G = -0.003t^3 + 0.137t^2 + 0.458t - 0.839$$

where G is the height of the tree (in feet) and t ($2 \le t \le 34$) is its age (in years). Use a graphing utility to graph the function and estimate the age of the tree when it is growing most rapidly. This point is called the **point of diminishing returns** because the increase in growth will be less with each additional year. (*Hint:* Use a viewing window in which $0 \le x \le 35$ and $0 \le y \le 60$.)

113. MODELING DATA

The U.S. production of crude oil y_1 (in quadrillions of British thermal units) and of solar and photovoltaic energy y_2 (in trillions of British thermal units) are shown in the table for the years 1999 through 2008, where t represents the year, with $t = 9$ corresponding to 1999. These data can be approximated by the models

$y_1 = 7.204t^3 - 301.60t^2 + 3854.2t - 3130$ and
$y_2 = 0.077t^3 - 2.31t^2 + 21.3t + 8.$

(Source: Energy Information Administration)

Year, t	y_1	y_2
9	12,451	69
10	12,358	66
11	12,282	65
12	12,163	64
13	12,026	64
14	11,503	65
15	10,963	66
16	10,801	72
17	10,721	81
18	10,519	91

(a) Use a graphing utility to plot the data and graph the model for y_1 in the same viewing window. How closely does the model represent the data?

(b) Extend the viewing window of the graphing utility to show the right-hand behavior of the model y_1. Would you use the model to estimate the production of crude oil in 2010? in 2020? Explain.

(c) Repeat parts (a) and (b) for y_2.

Conclusions

True or False?   In Exercises 114–118, determine whether the statement is true or false. Justify your answer.

114. It is possible for a sixth-degree polynomial to have only one zero.

115. The graph of the function

$f(x) = 2 + x - x^2 + x^3 - x^4 + x^5 + x^6 - x^7$

rises to the left and falls to the right.

116. The graph of the function $f(x) = 2x(x - 1)^2(x + 3)^3$ crosses the x-axis at $x = 1$.

117. The graph of the function $f(x) = 2x(x - 1)^2(x + 3)^3$ touches, but does not cross, the x-axis.

118. The graph of the function $f(x) = 2x(x - 1)^2(x + 3)^3$ rises to the left and falls to the right.

119. Exploration   Use a graphing utility to graph

$y_1 = x + 2$ and $y_2 = (x + 2)(x - 1)$.

Predict the shape of the graph of

$y_3 = (x + 2)(x - 1)(x - 3)$.

Use the graphing utility to verify your answer.

120. CAPSTONE   For each graph, describe a polynomial function that could represent the graph. (Indicate the degree of the function and the sign of its leading coefficient.)

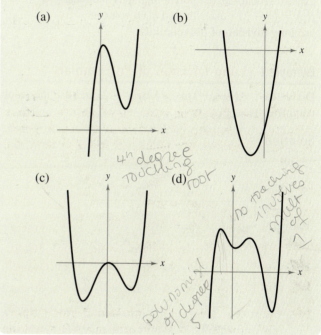

(a)

(b)

(c)

(d)

Cumulative Mixed Review

Evaluating Combinations of Functions   In Exercises 121–126, let $f(x) = 14x - 3$ and $g(x) = 8x^2$. Find the indicated value.

121. $(f + g)(-4)$

122. $(g - f)(3)$

123. $(fg)\left(-\dfrac{4}{7}\right)$

124. $\left(\dfrac{f}{g}\right)(-1.5)$

125. $(f \circ g)(-1)$

126. $(g \circ f)(0)$

Solving Inequalities   In Exercises 127–130, solve the inequality and sketch the solution on the real number line. Use a graphing utility to verify your solution graphically.

127. $3(x - 5) < 4x - 7$

128. $2x^2 - x \geq 1$

129. $\dfrac{5x - 2}{x - 7} \leq 4$

130. $|x + 8| - 1 \geq 15$

3.3 Real Zeros of Polynomial Functions

Long Division of Polynomials

Consider the graph of

$$f(x) = 6x^3 - 19x^2 + 16x - 4.$$

Notice in Figure 3.23 that $x = 2$ appears to be a zero of f. Because $f(2) = 0$, you know that $x = 2$ is a zero of the polynomial function f, and that $(x - 2)$ is a factor of $f(x)$. This means that there exists a second-degree polynomial $q(x)$ such that $f(x) = (x - 2) \cdot q(x)$. To find $q(x)$, you can use **long division of polynomials.**

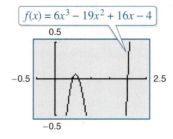

$f(x) = 6x^3 - 19x^2 + 16x - 4$

Figure 3.23

What you should learn

- Use long division to divide polynomials by other polynomials.
- Use synthetic division to divide polynomials by binomials of the form $(x - k)$.
- Use the Remainder and Factor Theorems.
- Use the Rational Zero Test to determine possible rational zeros of polynomial functions.
- Use Descartes's Rule of Signs and the Upper and Lower Bound Rules to find zeros of polynomials.

Why you should learn it

The Remainder Theorem can be used to determine the number of employees in education and health services in the United States in a given year based on a polynomial model, as shown in Exercise 104 on page 280.

Example 1 Long Division of Polynomials

Divide $6x^3 - 19x^2 + 16x - 4$ by $x - 2$, and use the result to factor the polynomial completely.

Solution

Partial quotients

$$
\begin{array}{r}
6x^2 - 7x + 2 \\
x - 2 \overline{\smash{)}\,6x^3 - 19x^2 + 16x - 4} \\
\underline{6x^3 - 12x^2} \\
-7x^2 + 16x \\
\underline{-7x^2 + 14x} \\
2x - 4 \\
\underline{2x - 4} \\
0
\end{array}
$$

Multiply: $6x^2(x - 2)$.
Subtract.
Multiply: $-7x(x - 2)$.
Subtract.
Multiply: $2(x - 2)$.
Subtract.

You can see that

$$6x^3 - 19x^2 + 16x - 4 = (x - 2)(6x^2 - 7x + 2)$$
$$= (x - 2)(2x - 1)(3x - 2).$$

Note that this factorization agrees with the graph of f (see Figure 3.23) in that the three x-intercepts occur at $x = 2$, $x = \frac{1}{2}$, and $x = \frac{2}{3}$.

✓**CHECKPOINT** Now try Exercise 9.

Note that in Example 1, the division process requires $-7x^2 + 14x$ to be subtracted from $-7x^2 + 16x$. Therefore, it is implied that

$$\frac{-7x^2 + 16x}{-(-7x^2 + 14x)} = \frac{-7x^2 + 16x}{7x^2 - 14x}$$

and instead is written simply as

$$
\begin{array}{r}
-7x^2 + 16x \\
\underline{-7x^2 + 14x} \\
2x
\end{array}.
$$

In Example 1, $x - 2$ is a factor of the polynomial

$$6x^3 - 19x^2 + 16x - 4$$

and the long division process produces a remainder of zero. Often, long division will produce a nonzero remainder. For instance, when you divide $x^2 + 3x + 5$ by $x + 1$, you obtain the following.

$$
\begin{array}{r}
x + 2 \quad \text{Quotient} \\
x + 1 \,\overline{)\, x^2 + 3x + 5} \quad \text{Dividend} \\
\underline{x^2 + x} \\
2x + 5 \\
\underline{2x + 2} \\
3 \quad \text{Remainder}
\end{array}
$$

Divisor $\Rightarrow$

In fractional form, you can write this result as follows.

$$
\underbrace{\frac{x^2 + 3x + 5}{x + 1}}_{\text{Divisor}} = \overbrace{x + 2}^{\text{Quotient}} + \frac{\overset{\text{Remainder}}{\downarrow}3}{\underbrace{x + 1}_{\text{Divisor}}}
$$

Dividend

This implies that

$$x^2 + 3x + 5 = (x + 1)(x + 2) + 3 \qquad \text{Multiply each side by } (x + 1).$$

which illustrates the following theorem, called the **Division Algorithm.**

The Division Algorithm

If $f(x)$ and $d(x)$ are polynomials such that $d(x) \neq 0$, and the degree of $d(x)$ is less than or equal to the degree of $f(x)$, then there exist unique polynomials $q(x)$ and $r(x)$ such that

$$f(x) = d(x)q(x) + r(x)$$

Dividend   Quotient
Divisor   Remainder

where $r(x) = 0$ *or* the degree of $r(x)$ is less than the degree of $d(x)$. If the remainder $r(x)$ is zero, then $d(x)$ *divides evenly* into $f(x)$.

The Division Algorithm can also be written as

$$\frac{f(x)}{d(x)} = q(x) + \frac{r(x)}{d(x)}.$$

In the Division Algorithm, the rational expression $f(x)/d(x)$ is **improper** because the degree of $f(x)$ is greater than or equal to the degree of $d(x)$. On the other hand, the rational expression $r(x)/d(x)$ is **proper** because the degree of $r(x)$ is less than the degree of $d(x)$.

Before you apply the Division Algorithm, follow these steps.

1. Write the dividend and divisor in descending powers of the variable.

2. Insert placeholders with zero coefficients for missing powers of the variable.

Note how these steps are applied in the next two examples.

Example 2 Long Division of Polynomials

Divide $8x^3 - 1$ by $2x - 1$.

Solution

Because there is no x^2-term or x-term in the dividend, you need to line up the subtraction by using zero coefficients (or leaving spaces) for the missing terms.

$$
\begin{array}{r}
4x^2 + 2x + 1 \\
2x - 1 \overline{\smash{)}\, 8x^3 + 0x^2 + 0x - 1} \\
\underline{8x^3 - 4x^2} \\
4x^2 + 0x \\
\underline{4x^2 - 2x} \\
2x - 1 \\
\underline{2x - 1} \\
0
\end{array}
$$

So, $2x - 1$ divides evenly into $8x^3 - 1$, and you can write

$$\frac{8x^3 - 1}{2x - 1} = 4x^2 + 2x + 1, \quad x \neq \frac{1}{2}.$$

✔**CHECKPOINT**   Now try Exercise 15.

You can check the result of Example 2 by multiplying.

$$(2x - 1)(4x^2 + 2x + 1) = 8x^3 + 4x^2 + 2x - 4x^2 - 2x - 1$$
$$= 8x^3 - 1$$

In each of the long division examples presented so far, the divisor has been a first-degree polynomial. The long division algorithm works just as well with polynomial divisors of degree two or more, as shown in Example 3.

Example 3 Long Division of Polynomials

Divide $-2 + 3x - 5x^2 + 4x^3 + 2x^4$ by $x^2 + 2x - 3$.

Solution

Begin by writing the dividend in descending powers of x.

$$
\begin{array}{r}
2x^2 + 1 \\
x^2 + 2x - 3 \overline{\smash{)}\, 2x^4 + 4x^3 - 5x^2 + 3x - 2} \\
\underline{2x^4 + 4x^3 - 6x^2} \\
x^2 + 3x - 2 \\
\underline{x^2 + 2x - 3} \\
x + 1
\end{array}
$$

Note that the first subtraction eliminated two terms from the dividend. When this happens, the quotient skips a term. You can write the result as

$$\frac{2x^4 + 4x^3 - 5x^2 + 3x - 2}{x^2 + 2x - 3} = 2x^2 + 1 + \frac{x + 1}{x^2 + 2x - 3}.$$

✔**CHECKPOINT**   Now try Exercise 17.

Synthetic Division

There is a nice shortcut for long division of polynomials when dividing by divisors of the form

$x - k$.

The shortcut is called **synthetic division.** The pattern for synthetic division of a cubic polynomial is summarized as follows. (The pattern for higher-degree polynomials is similar.)

Synthetic Division (of a Cubic Polynomial)

To divide $ax^3 + bx^2 + cx + d$ by $x - k$, use the following pattern.

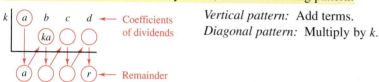

Vertical pattern: Add terms.
Diagonal pattern: Multiply by k.

This algorithm for synthetic division works *only* for divisors of the form $x - k$. Remember that

$x + k = x - (-k)$.

Example 4   Using Synthetic Division

Use synthetic division to divide

$x^4 - 10x^2 - 2x + 4$ by $x + 3$.

Solution

You should set up the array as follows. Note that a zero is included for each missing term in the dividend.

$$-3 \;\big|\; 1 \quad 0 \;-10 \;-2 \quad 4$$

Then, use the synthetic division pattern by adding terms in columns and multiplying the results by -3.

Divisor: $x + 3$ Dividend: $x^4 - 10x^2 - 2x + 4$

$$
\begin{array}{r|rrrrr}
-3 & 1 & 0 & -10 & -2 & 4 \\
   &   & -3 & 9 & 3 & -3 \\
\hline
   & 1 & -3 & -1 & 1 & 1 \\
\end{array}
$$
Remainder: 1

Quotient: $x^3 - 3x^2 - x + 1$

So, you have

$$\frac{x^4 - 10x^2 - 2x + 4}{x + 3} = x^3 - 3x^2 - x + 1 + \frac{1}{x + 3}.$$

CHECKPOINT   Now try Exercise 23.

Explore the Concept

Evaluate the polynomial $x^4 - 10x^2 - 2x + 4$ at $x = -3$. What do you observe?

The Remainder and Factor Theorems

The remainder obtained in the synthetic division process has an important interpretation, as described in the **Remainder Theorem.**

The Remainder Theorem   (See the proof on page 321.)

If a polynomial $f(x)$ is divided by $x - k$, then the remainder is

$r = f(k)$.

The Remainder Theorem tells you that synthetic division can be used to evaluate a polynomial function. That is, to evaluate a polynomial function $f(x)$ when $x = k$, divide $f(x)$ by $x - k$. The remainder will be $f(k)$.

Example 5   Using the Remainder Theorem

Use the Remainder Theorem to evaluate the following function at $x = -2$.

$$f(x) = 3x^3 + 8x^2 + 5x - 7$$

Solution

Using synthetic division, you obtain the following.

$$
\begin{array}{r|rrrr}
-2 & 3 & 8 & 5 & -7 \\
   &   & -6 & -4 & -2 \\
\hline
   & 3 & 2 & 1 & -9
\end{array}
$$

Because the remainder is $r = -9$, you can conclude that

$f(-2) = -9.$ $r = f(k)$

This means that $(-2, -9)$ is a point on the graph of f. You can check this by substituting $x = -2$ in the original function.

Check

$$
\begin{aligned}
f(-2) &= 3(-2)^3 + 8(-2)^2 + 5(-2) - 7 \\
&= 3(-8) + 8(4) - 10 - 7 \\
&= -24 + 32 - 10 - 7 \\
&= -9
\end{aligned}
$$

✓**CHECKPOINT**   Now try Exercise 43.

Another important theorem is the **Factor Theorem.** This theorem states that you can test whether a polynomial has $(x - k)$ as a factor by evaluating the polynomial at $x = k$. If the result is 0, then $(x - k)$ is a factor.

The Factor Theorem   (See the proof on page 321.)

A polynomial $f(x)$ has a factor

$(x - k)$

if and only if

$f(k) = 0$.

Example 6 Factoring a Polynomial: Repeated Division

Show that $(x - 2)$ and $(x + 3)$ are factors of

$$f(x) = 2x^4 + 7x^3 - 4x^2 - 27x - 18.$$

Then find the remaining factors of $f(x)$.

Algebraic Solution

Using synthetic division with the factor $(x - 2)$, you obtain the following.

$$
\begin{array}{r|rrrrr}
2 & 2 & 7 & -4 & -27 & -18 \\
  &   & 4 & 22 & 36 & 18 \\
\hline
  & 2 & 11 & 18 & 9 & 0
\end{array}
$$

0 remainder; $(x - 2)$ is a factor.

Take the result of this division and perform synthetic division again using the factor $(x + 3)$.

$$
\begin{array}{r|rrrr}
-3 & 2 & 11 & 18 & 9 \\
   &   & -6 & -15 & -9 \\
\hline
   & 2 & 5 & 3 & 0
\end{array}
$$

$2x^2 + 5x + 3$

0 remainder; $(x + 3)$ is a factor.

Because the resulting quadratic factors as

$$2x^2 + 5x + 3 = (2x + 3)(x + 1)$$

the complete factorization of $f(x)$ is

$$f(x) = (x - 2)(x + 3)(2x + 3)(x + 1).$$

✓**CHECKPOINT** Now try Exercise 53.

Graphical Solution

From the graph of

$$f(x) = 2x^4 + 7x^3 - 4x^2 - 27x - 18$$

you can see that there are four x-intercepts (see Figure 3.24). These occur at $x = -3$, $x = -\frac{3}{2}$, $x = -1$, and $x = 2$. (Check this algebraically.) This implies that $(x + 3)$, $\left(x + \frac{3}{2}\right)$, $(x + 1)$, and $(x - 2)$ are factors of $f(x)$. $\left[\text{Note that } \left(x + \frac{3}{2}\right) \text{ and } (2x + 3) \text{ are equivalent factors because they both yield the same zero, } x = -\frac{3}{2}.\right]$

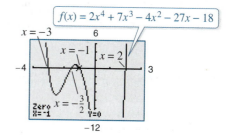

Figure 3.24

Note in Example 6 that the complete factorization of $f(x)$ implies that f has four real zeros:

$$x = 2, \quad x = -3, \quad x = -\frac{3}{2}, \quad \text{and} \quad x = -1.$$

This is confirmed by the graph of f, which is shown in Figure 3.24.

Using the Remainder in Synthetic Division

In summary, the remainder r, obtained in the synthetic division of $f(x)$ by $x - k$, provides the following information.

1. The remainder r gives the value of f at $x = k$. That is, $r = f(k)$.

2. If $r = 0$, then $(x - k)$ is a factor of $f(x)$.

3. If $r = 0$, then $(k, 0)$ is an x-intercept of the graph of f.

Throughout this text, the importance of developing several problem-solving strategies is emphasized. In the exercises for this section, try using more than one strategy to solve several of the exercises. For instance, when you find that $x - k$ divides evenly into $f(x)$, try sketching the graph of f. You should find that $(k, 0)$ is an x-intercept of the graph.

The Rational Zero Test

The **Rational Zero Test** relates the possible rational zeros of a polynomial (having integer coefficients) to the leading coefficient and to the constant term of the polynomial.

The Rational Zero Test

If the polynomial

$$f(x) = a_n x^n + a_{n-1} x^{n-1} + \cdots + a_2 x^2 + a_1 x + a_0$$

has integer coefficients, then every rational zero of f has the form

$$\text{Rational zero} = \frac{p}{q}$$

where p and q have no common factors other than 1, p is a factor of the constant term a_0, and q is a factor of the leading coefficient a_n.

To use the Rational Zero Test, first list all rational numbers whose numerators are factors of the constant term and whose denominators are factors of the leading coefficient.

$$\text{Possible rational zeros} = \frac{\text{factors of constant term}}{\text{factors of leading coefficient}}$$

Now that you have formed this list of *possible rational zeros*, use a trial-and-error method to determine which, if any, are actual zeros of the polynomial. Note that when the leading coefficient is 1, the possible rational zeros are simply the factors of the constant term. This case is illustrated in Example 7.

Study Tip

Use a graphing utility to graph the polynomial

$$y = x^3 - 53x^2 + 103x - 51$$

in a standard viewing window. From the graph alone, it appears that there is only one zero. From the Leading Coefficient Test, you know that because the degree of the polynomial is odd and the leading coefficient is positive, the graph falls to the left and rises to the right. So, the function must have another zero. From the Rational Zero Test, you know that ± 51 might be zeros of the function. When you zoom out several times, you will see a more complete picture of the graph. Your graph should confirm that $x = 51$ is a zero of f.

Example 7   Rational Zero Test with Leading Coefficient of 1

Find the rational zeros of $f(x) = x^3 + x + 1$.

Solution

Because the leading coefficient is 1, the possible rational zeros are simply the factors of the constant term.

Possible rational zeros: ± 1

By testing these possible zeros, you can see that neither works.

$$f(1) = (1)^3 + 1 + 1 = 3$$

$$f(-1) = (-1)^3 + (-1) + 1 = -1$$

So, you can conclude that the polynomial has *no* rational zeros. Note from the graph of f in Figure 3.25 that f does have one real zero between -1 and 0. However, by the Rational Zero Test, you know that this real zero is *not* a rational number.

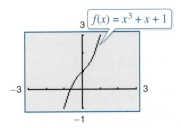

Figure 3.25

    Now try Exercise 57.

When the leading coefficient of a polynomial is not 1, the list of possible rational zeros can increase dramatically. In such cases, the search can be shortened in several ways.

1. A programmable calculator can be used to speed up the calculations.

2. A graphing utility can give a good estimate of the locations of the zeros.

3. The Intermediate Value Theorem, along with a table generated by a graphing utility, can give approximations of zeros.

4. The Factor Theorem and synthetic division can be used to test the possible rational zeros.

Finding the first zero is often the most difficult part. After that, the search is simplified by working with the lower-degree polynomial obtained in synthetic division, as shown in Example 8.

Example 8   Using the Rational Zero Test

Find the rational zeros of

$$f(x) = 2x^3 + 3x^2 - 8x + 3.$$

Solution

The leading coefficient is 2 and the constant term is 3.

Possible rational zeros:

$$\frac{\text{Factors of 3}}{\text{Factors of 2}} = \frac{\pm 1, \pm 3}{\pm 1, \pm 2} = \pm 1, \pm 3, \pm \frac{1}{2}, \pm \frac{3}{2}$$

By synthetic division, you can determine that $x = 1$ is a rational zero.

```
1 | 2   3   -8    3
  |     2    5   -3
  ------------------
    2   5   -3    0
```

So, $f(x)$ factors as

$$f(x) = (x - 1)(2x^2 + 5x - 3)$$

$$= (x - 1)(2x - 1)(x + 3)$$

and you can conclude that the rational zeros of f are $x = 1$, $x = \frac{1}{2}$, and $x = -3$, as shown in Figure 3.26.

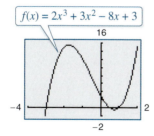

Figure 3.26

✓CHECKPOINT   Now try Exercise 59.

Remember that when you try to find the rational zeros of a polynomial function with many possible rational zeros, as in Example 8, you must use trial and error. There is no quick algebraic method to determine which of the possibilities is an actual zero; however, sketching a graph may be helpful.

Other Tests for Zeros of Polynomials

You know that an nth-degree polynomial function can have *at most* n real zeros. Of course, many nth-degree polynomials do not have that many real zeros. For instance, $f(x) = x^2 + 1$ has no real zeros, and $f(x) = x^3 + 1$ has only one real zero. The following theorem, called **Descartes's Rule of Signs,** sheds more light on the number of real zeros of a polynomial.

Decartes's Rule of Signs

Let $f(x) = a_n x^n + a_{n-1} x^{n-1} + \cdots + a_2 x^2 + a_1 x + a_0$ be a polynomial with real coefficients and $a_0 \neq 0$.

1. The number of *positive real zeros* of f is either equal to the number of variations in sign of $f(x)$ or less than that number by an even integer.

2. The number of *negative real zeros* of f is either equal to the number of variations in sign of $f(-x)$ or less than that number by an even integer.

A **variation in sign** means that two consecutive (nonzero) coefficients have opposite signs.

When using Descartes's Rule of Signs, a zero of multiplicity k should be counted as k zeros. For instance, the polynomial $x^3 - 3x + 2$ has two variations in sign, and so has either two positive or no positive real zeros. Because

$$x^3 - 3x + 2 = (x - 1)(x - 1)(x + 2)$$

you can see that the two positive real zeros are $x = 1$ of multiplicity 2.

Example 9   Using Descartes's Rule of Signs

Describe the possible real zeros of $f(x) = 3x^3 - 5x^2 + 6x - 4$.

Solution

The original polynomial has *three* variations in sign.

$$f(x) = 3x^3 - 5x^2 + 6x - 4$$

The polynomial

$$f(-x) = 3(-x)^3 - 5(-x)^2 + 6(-x) - 4$$
$$= -3x^3 - 5x^2 - 6x - 4$$

has no variations in sign. So, from Descartes's Rule of Signs, the polynomial $f(x) = 3x^3 - 5x^2 + 6x - 4$ has either three positive real zeros or one positive real zero, and has no negative real zeros. By using the *trace* feature of a graphing utility, you can see that the function has only one real zero (it is a positive number near $x = 1$), as shown in Figure 3.27.

Figure 3.27

✓CHECKPOINT   Now try Exercise 61.

Another test for zeros of a polynomial function is related to the sign pattern in the last row of the synthetic division array. This test can give you an upper or lower bound of the real zeros of f, which can help you eliminate possible real zeros. A real number c is an **upper bound** for the real zeros of f when no zeros are greater than c. Similarly, c is a **lower bound** when no real zeros of f are less than c.

Upper and Lower Bound Rules

Let $f(x)$ be a polynomial with real coefficients and a positive leading coefficient. Suppose $f(x)$ is divided by $x - c$, using synthetic division.

1. If $c > 0$ and each number in the last row is either positive or zero, then c is an **upper bound** for the real zeros of f.

2. If $c < 0$ and the numbers in the last row are alternately positive and negative (zero entries count as positive or negative), then c is a **lower bound** for the real zeros of f.

Example 10   Finding the Zeros of a Polynomial Function

Find the real zeros of $f(x) = 6x^3 - 4x^2 + 3x - 2$.

Solution

The possible real zeros are as follows.

$$\frac{\text{Factors of } 2}{\text{Factors of } 6} = \frac{\pm 1, \pm 2}{\pm 1, \pm 2, \pm 3, \pm 6} = \pm 1, \pm \frac{1}{2}, \pm \frac{1}{3}, \pm \frac{1}{6}, \pm \frac{2}{3}, \pm 2$$

The original polynomial $f(x)$ has three variations in sign. The polynomial

$$f(-x) = 6(-x)^3 - 4(-x)^2 + 3(-x) - 2$$

$$= -6x^3 - 4x^2 - 3x - 2$$

has no variations in sign. As a result of these two findings, you can apply Descartes's Rule of Signs to conclude that there are three positive real zeros or one positive real zero, and no negative real zeros. Trying $x = 1$ produces the following.

```
1 | 6   -4   3   -2
  |      6   2    5
  --------------------
    6    2   5    3
```

So, $x = 1$ is not a zero, but because the last row has all positive entries, you know that $x = 1$ is an upper bound for the real zeros. Therefore, you can restrict the search to zeros between 0 and 1. By trial and error, you can determine that $x = \frac{2}{3}$ is a zero. So,

$$f(x) = \left(x - \frac{2}{3}\right)(6x^2 + 3).$$

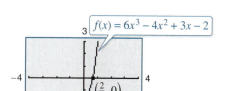

Figure 3.28

Because $6x^2 + 3$ has no real zeros, it follows that $x = \frac{2}{3}$ is the only real zero, as shown in Figure 3.28.

✔CHECKPOINT   Now try Exercise 71.

Explore the Concept

 Use a graphing utility to graph the polynomial

$$y_1 = 6x^3 - 4x^2 + 3x - 2.$$

Notice that the graph intersects the x-axis at the point $\left(\frac{2}{3}, 0\right)$. How does this information relate to the real zero found in Example 10? Use the graphing utility to graph

$$y_2 = x^4 - 5x^3 + 3x^2 + x.$$

How many times does the graph intersect the x-axis? How many real zeros does y_2 have?

Here are two additional hints that can help you find the real zeros of a polynomial.

1. When the terms of $f(x)$ have a common monomial factor, it should be factored out before applying the tests in this section. For instance, by writing

$$f(x) = x^4 - 5x^3 + 3x^2 + x = x(x^3 - 5x^2 + 3x + 1)$$

you can see that $x = 0$ is a zero of f and that the remaining zeros can be obtained by analyzing the cubic factor.

2. When you are able to find all but two zeros of $f(x)$, you can always use the Quadratic Formula on the remaining quadratic factor. For instance, after writing

$$f(x) = x^4 - 5x^3 + 3x^2 + x = x(x - 1)(x^2 - 4x - 1)$$

you can apply the Quadratic Formula to $x^2 - 4x - 1$ to conclude that the two remaining zeros are $x = 2 + \sqrt{5}$ and $x = 2 - \sqrt{5}$.

Note how these hints are applied in the next example.

Example 11 Finding the Zeros of a Polynomial Function

Find all the real zeros of $f(x) = 10x^4 - 15x^3 - 16x^2 + 12x$.

Solution

Remove the common monomial factor x to write

$$f(x) = 10x^4 - 15x^3 - 16x^2 + 12x = x(10x^3 - 15x^2 - 16x + 12).$$

So, $x = 0$ is a zero of f. You can find the remaining zeros of f by analyzing the cubic factor. Because the leading coefficient is 10 and the constant term is 12, there is a long list of possible rational zeros.

Possible rational zeros:

$$\frac{\text{Factors of 12}}{\text{Factors of 10}} = \frac{\pm 1, \pm 2, \pm 3, \pm 4, \pm 6, \pm 12}{\pm 1, \pm 2, \pm 5, \pm 10}$$

With so many possibilities (32, in fact), it is worth your time to use a graphing utility to focus on just a few. By using the *trace* feature of a graphing utility, it looks like three reasonable choices are $x = -\frac{6}{5}$, $x = \frac{1}{2}$, and $x = 2$ (see Figure 3.29). Synthetic division shows that only $x = 2$ works. (You could also use the Factor Theorem to test these choices.)

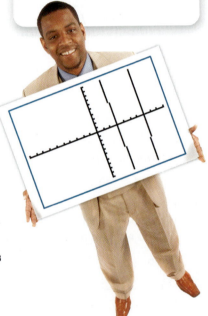

Figure 3.29

$$\begin{array}{r|rrrr} 2 & 10 & -15 & -16 & 12 \\ & & 20 & 10 & -12 \\ \hline & 10 & 5 & -6 & 0 \end{array}$$

So, $x = 2$ is one zero and you have

$$f(x) = x(x - 2)(10x^2 + 5x - 6).$$

Using the Quadratic Formula, you find that the two additional zeros are irrational numbers.

$$x = \frac{-5 + \sqrt{265}}{20} \approx 0.56 \quad \text{and} \quad x = \frac{-5 - \sqrt{265}}{20} \approx -1.06$$

✔**CHECKPOINT**   Now try Exercise 87.

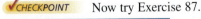

Explore the Concept

 Use a graphing utility to graph the polynomial

$$y = x^3 + 4.8x^2 - 127x + 309$$

in a standard viewing window. From the graph, what do the real zeros appear to be? Discuss how the mathematical tools of this section might help you realize that the graph does not show all the important features of the polynomial function. Now use the *zoom* feature to find all the zeros of this function.

3.3 Exercises

See www.CalcChat.com for worked-out solutions to odd-numbered exercises.
For instructions on how to use a graphing utility, see Appendix A.

Vocabulary and Concept Check

1. Two forms of the Division Algorithm are shown below. Identify and label each part.

$$f(x) = d(x)q(x) + r(x) \qquad \frac{f(x)}{d(x)} = q(x) + \frac{r(x)}{d(x)}$$

In Exercises 2–5, fill in the blank(s).

2. The rational expression $p(x)/q(x)$ is called _____ when the degree of the numerator is greater than or equal to that of the denominator.

3. Every rational zero of a polynomial function with integer coefficients has the form p/q, where p is a factor of the _____ and q is a factor of the _____.

4. The theorem that can be used to determine the possible numbers of positive real zeros and negative real zeros of a function is called _____ of _____ .

5. A real number c is a(n) _____ bound for the real zeros of f when no zeros are greater than c, and is a(n) _____ bound when no real zeros of f are less than c.

6. How many negative real zeros are possible for a polynomial function $f(x)$, given that $f(-x)$ has 5 variations in sign?

7. You divide the polynomial $f(x)$ by $(x - 4)$ and obtain a remainder of 7. What is $f(4)$?

8. What value should you write in the circle to check whether $(x - 3)$ is a factor of $f(x) = x^3 - 2x^2 + 3x + 4$?

$$\bigcirc \; \lfloor \underline{1 \quad -2 \quad 3 \quad 4}$$

Procedures and Problem Solving

Long Division of Polynomials In Exercises 9–22, use long division to divide.

✓ 9. Divide $2x^2 + 10x + 12$ by $x + 3$.

10. Divide $5x^2 - 17x - 12$ by $x - 4$.

11. Divide $x^4 + 5x^3 + 6x^2 - x - 2$ by $x + 2$.

12. Divide $x^3 - 4x^2 - 17x + 6$ by $x - 3$.

13. Divide $4x^3 - 7x^2 - 11x + 5$ by $4x + 5$.

14. Divide $2x^3 - 3x^2 - 50x + 75$ by $2x - 3$.

✓ 15. Divide $7x^3 + 3$ by $x + 2$.

16. Divide $8x^4 - 5$ by $2x + 1$.

✓ 17. $(x + 8 + 6x^3 + 10x^2) \div (2x^2 + 1)$

18. $(1 + 3x^2 + x^4) \div (3 - 2x + x^2)$

19. $(x^3 - 9) \div (x^2 + 1)$ 20. $(x^5 + 7) \div (x^3 - 1)$

21. $\dfrac{2x^3 - 4x^2 - 15x + 5}{(x - 1)^2}$ 22. $\dfrac{x^4}{(x - 1)^3}$

Using Synthetic Division In Exercises 23–32, use synthetic division to divide.

✓ 23. $(3x^3 - 17x^2 + 15x - 25) \div (x - 5)$

24. $(5x^3 + 18x^2 + 7x - 6) \div (x + 3)$

25. $(6x^3 + 7x^2 - x + 26) \div (x - 3)$

26. $(2x^3 + 14x^2 - 20x + 7) \div (x + 6)$

27. $(9x^3 - 18x^2 - 16x + 32) \div (x - 2)$

28. $(5x^3 + 6x + 8) \div (x + 2)$

29. $(x^3 + 512) \div (x + 8)$

30. $(x^3 - 729) \div (x - 9)$

31. $\dfrac{4x^3 + 16x^2 - 23x - 15}{x + \frac{1}{2}}$ 32. $\dfrac{3x^3 - 4x^2 + 5}{x - \frac{3}{2}}$

Verifying Quotients In Exercises 33–36, use a graphing utility to graph the two equations in the same viewing window. Use the graphs to verify that the expressions are equivalent. Verify the results algebraically.

33. $y_1 = \dfrac{x^2}{x + 2}, \quad y_2 = x - 2 + \dfrac{4}{x + 2}$

34. $y_1 = \dfrac{x^2 + 2x - 1}{x + 3}, \quad y_2 = x - 1 + \dfrac{2}{x + 3}$

35. $y_1 = \dfrac{x^4 - 3x^2 - 1}{x^2 + 5}, \quad y_2 = x^2 - 8 + \dfrac{39}{x^2 + 5}$

36. $y_1 = \dfrac{x^4 + x^2 - 1}{x^2 + 1}, \quad y_2 = x^2 - \dfrac{1}{x^2 + 1}$

Verifying the Remainder Theorem In Exercises 37–42, write the function in the form $f(x) = (x - k)q(x) + r(x)$ for the given value of k. Use a graphing utility to demonstrate that $f(k) = r$.

Function	Value of k
37. $f(x) = x^3 - x^2 - 14x + 11$	$k = 4$
38. $f(x) = 15x^4 + 10x^3 - 6x^2 + 14$	$k = -\frac{2}{3}$

Function   *Value of k*

39. $f(x) = x^3 + 3x^2 - 2x - 14$ $k = \sqrt{2}$

40. $f(x) = x^3 + 2x^2 - 5x - 4$ $k = -\sqrt{5}$

41. $f(x) = 4x^3 - 6x^2 - 12x - 4$ $k = 1 - \sqrt{3}$

42. $f(x) = -3x^3 + 8x^2 + 10x - 8$ $k = 2 + \sqrt{2}$

Using the Remainder Theorem   In Exercises 43–46, use the Remainder Theorem and synthetic division to evaluate the function at each given value. Use a graphing utility to verify your results.

✓ **43.** $f(x) = 2x^3 - 7x + 3$

  (a) $f(1)$ (b) $f(-2)$ (c) $f\left(\frac{1}{2}\right)$ (d) $f(2)$

44. $g(x) = 2x^6 + 3x^4 - x^2 + 3$

  (a) $g(2)$ (b) $g(1)$ (c) $g(3)$ (d) $g(-1)$

45. $h(x) = x^3 - 5x^2 - 7x + 4$

  (a) $h(3)$ (b) $h(2)$ (c) $h(-2)$ (d) $h(-5)$

46. $f(x) = 4x^4 - 16x^3 + 7x^2 + 20$

  (a) $f(1)$ (b) $f(-2)$ (c) $f(5)$ (d) $f(-10)$

Using the Factor Theorem   In Exercises 47–50, use synthetic division to show that x is a solution of the third-degree polynomial equation, and use the result to factor the polynomial completely. List all the real solutions of the equation.

Polynomial Equation   *Value of x*

47. $x^3 - 7x + 6 = 0$ $x = 2$

48. $x^3 - 28x - 48 = 0$ $x = -4$

49. $2x^3 - 15x^2 + 27x - 10 = 0$ $x = \frac{1}{2}$

50. $48x^3 - 80x^2 + 41x - 6 = 0$ $x = \frac{2}{3}$

Factoring a Polynomial   In Exercises 51–56, (a) verify the given factor(s) of the function f, (b) find the remaining factors of f, (c) use your results to write the complete factorization of f, and (d) list all real zeros of f. Confirm your results by using a graphing utility to graph the function.

Function   *Factor(s)*

51. $f(x) = 2x^3 + x^2 - 5x + 2$ $(x + 2)$

52. $f(x) = 3x^3 + 2x^2 - 19x + 6$ $(x + 3)$

✓ **53.** $f(x) = x^4 - 4x^3 - 15x^2$ $(x - 5), (x + 4)$
   $+ 58x - 40$

54. $f(x) = 8x^4 - 14x^3 - 71x^2$ $(x + 2), (x - 4)$
   $- 10x + 24$

55. $f(x) = 6x^3 + 41x^2 - 9x - 14$ $(2x + 1)$

56. $f(x) = 2x^3 - x^2 - 10x + 5$ $(2x - 1)$

Using the Rational Zero Test   In Exercises 57–60, use the Rational Zero Test to list all possible rational zeros of f. Then find the rational zeros.

✓ **57.** $f(x) = x^3 + 3x^2 - x - 3$

58. $f(x) = x^3 - 4x^2 - 4x + 16$

✓ **59.** $f(x) = 2x^4 - 17x^3 + 35x^2 + 9x - 45$

60. $f(x) = 4x^5 - 8x^4 - 5x^3 + 10x^2 + x - 2$

Using Descartes's Rule of Signs   In Exercises 61–64, use Descartes's Rule of Signs to determine the possible numbers of positive and negative real zeros of the function.

✓ **61.** $f(x) = 2x^4 - x^3 + 6x^2 - x + 5$

62. $f(x) = 3x^4 + 5x^3 - 6x^2 + 8x - 3$

63. $g(x) = 4x^3 - 5x + 8$

64. $g(x) = 2x^3 - 4x^2 - 5$

Finding the Zeros of a Polynomial Function   In Exercises 65–70, (a) use Descartes's Rule of Signs to determine the possible numbers of positive and negative real zeros of f, (b) list the possible rational zeros of f, (c) use a graphing utility to graph f so that some of the possible zeros in parts (a) and (b) can be disregarded, and (d) determine all the real zeros of f.

65. $f(x) = x^3 + x^2 - 4x - 4$

66. $f(x) = -3x^3 + 20x^2 - 36x + 16$

67. $f(x) = -2x^4 + 13x^3 - 21x^2 + 2x + 8$

68. $f(x) = 4x^4 - 17x^2 + 4$

69. $f(x) = 32x^3 - 52x^2 + 17x + 3$

70. $f(x) = x^4 - x^3 - 29x^2 - x - 30$

Finding the Zeros of a Polynomial Function   In Exercises 71–74, use synthetic division to verify the upper and lower bounds of the real zeros of f. Then find the real zeros of the function.

✓ **71.** $f(x) = x^4 - 4x^3 + 15$

  Upper bound: $x = 4$

  Lower bound: $x = -1$

72. $f(x) = 2x^3 - 3x^2 - 12x + 8$

  Upper bound: $x = 4$

  Lower bound: $x = -3$

73. $f(x) = x^4 - 4x^3 + 16x - 16$

  Upper bound: $x = 5$

  Lower bound: $x = -3$

74. $f(x) = 2x^4 - 8x + 3$

  Upper bound: $x = 3$

  Lower bound: $x = -4$

Occasionally, throughout this text, you will be asked to round to a place value rather than to a number of decimal places.

Rewriting to Use the Rational Zero Test **In Exercises 75–78, find the rational zeros of the polynomial function.**

75. $P(x) = x^4 - \frac{25}{4}x^2 + 9 = \frac{1}{4}(4x^4 - 25x^2 + 36)$

76. $f(x) = x^3 - \frac{3}{2}x^2 - \frac{23}{2}x + 6 = \frac{1}{2}(2x^3 - 3x^2 - 23x + 12)$

77. $f(x) = x^3 - \frac{1}{4}x^2 - x + \frac{1}{4} = \frac{1}{4}(4x^3 - x^2 - 4x + 1)$

78. $f(z) = z^3 + \frac{11}{6}z^2 - \frac{1}{2}z - \frac{1}{3} = \frac{1}{6}(6z^3 + 11z^2 - 3z - 2)$

A Cubic Polynomial with Two Terms **In Exercises 79–82, match the cubic function with the correct number of rational and irrational zeros.**

(a) Rational zeros: 0; Irrational zeros: 1

(b) Rational zeros: 3; Irrational zeros: 0

(c) Rational zeros: 1; Irrational zeros: 2

(d) Rational zeros: 1; Irrational zeros: 0

79. $f(x) = x^3 - 1$ **80.** $f(x) = x^3 - 2$

81. $f(x) = x^3 - x$ **82.** $f(x) = x^3 - 2x$

Using a Graph to Help Find Zeros **In Exercises 83–86, the graph of $y = f(x)$ is shown. Use the graph as an aid to find all the real zeros of the function.**

83. $y = 2x^4 - 9x^3 + 5x^2$
$+ 3x - 1$

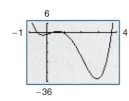

84. $y = x^4 - 5x^3 - 7x^2$
$+ 13x - 2$

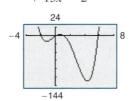

85. $y = -2x^4 + 17x^3$
$- 3x^2 - 25x - 3$

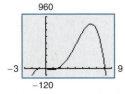

86. $y = -x^4 + 5x^3$
$- 10x - 4$

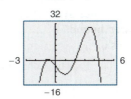

Finding the Zeros of a Polynomial Function **In Exercises 87–98, find all real zeros of the polynomial function.**

✓ **87.** $f(x) = 3x^4 - 14x^2 - 4x$

88. $g(x) = 4x^4 - 11x^3 - 22x^2 + 8x$

89. $f(z) = z^4 - z^3 - 2z - 4$

90. $f(x) = 4x^3 + 7x^2 - 11x - 18$

91. $g(y) = 2y^4 + 7y^3 - 26y^2 + 23y - 6$

92. $h(x) = x^5 - x^4 - 3x^3 + 5x^2 - 2x$

93. $f(x) = 4x^4 - 55x^2 - 45x + 36$

94. $z(x) = 4x^4 - 43x^2 - 9x + 90$

95. $g(x) = 8x^4 + 28x^3 + 9x^2 - 9x$

96. $h(x) = x^5 + 5x^4 - 5x^3 - 15x^2 - 6x$

97. $f(x) = 4x^5 + 12x^4 - 11x^3 - 42x^2 + 7x + 30$

98. $g(x) = 4x^5 + 8x^4 - 15x^3 - 23x^2 + 11x + 15$

Using a Rational Zero **In Exercises 99–102, (a) use the *zero* or *root* feature of a graphing utility to approximate (accurate to the nearest thousandth) the zeros of the function, (b) determine one of the exact zeros and use synthetic division to verify your result, and (c) factor the polynomial completely.**

99. $h(t) = t^3 - 2t^2 - 7t + 2$

100. $f(s) = s^3 - 12s^2 + 40s - 24$

101. $h(x) = x^5 - 7x^4 + 10x^3 + 14x^2 - 24x$

102. $g(x) = 6x^4 - 11x^3 - 51x^2 + 99x - 27$

103. MODELING DATA

The table shows the numbers S of cellular phone subscriptions per 100 people in the United States from 1991 through 2008. (Source: U.S. International Telecommunications Union)

Year	Subscriptions per 100 people, S
1991	3.0
1992	4.3
1993	6.2
1994	9.2
1995	12.7
1996	16.4
1997	20.3
1998	25.1
1999	30.8
2000	38.9
2001	45.1
2002	49.2
2003	55.2
2004	62.9
2005	71.5
2006	77.4
2007	85.2
2008	86.8

The data can be approximated by the model

$S = -0.0135t^3 + 0.545t^2 - 0.71t + 3.6, \quad 1 \le t \le 18$

where t represents the year, with $t = 1$ corresponding to 1991.

(a) Use a graphing utility to graph the data and the model in the same viewing window.

(b) How well does the model fit the data?

(c) Use the Remainder Theorem to evaluate the model for the year 2015. Is the value reasonable? Explain.

104. *Why you should learn it* (p. 266) The numbers of

employees E (in thousands) in education and health services in the United States from 1960 through 2008 are approximated by $E = -0.084t^3 + 10.32t^2 - 23.5t + 3167$, $0 \le t \le 48$, where t is the year, with $t = 0$ corresponding to 1960. (Source: U.S. Bureau of Labor Statistics)

(a) Use a graphing utility to graph the model over the domain.

(b) Estimate the number of employees in education and health services in 1960. Use the Remainder Theorem to estimate the number in 2000.

(c) Is this a good model for making predictions in future years? Explain.

105. Geometry A rectangular package sent by a delivery service can have a maximum combined length and girth (perimeter of a cross section) of 120 inches (see figure).

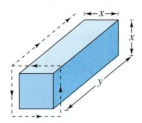

(a) Show that the volume of the package is given by the function $V(x) = 4x^2(30 - x)$.

(b) Use a graphing utility to graph the function and approximate the dimensions of the package that yield a maximum volume.

(c) Find values of x such that $V = 13,500$. Which of these values is a physical impossibility in the construction of the package? Explain.

106. Environmental Science The number of parts per million of nitric oxide emissions y from a car engine is approximated by $y = -5.05x^3 + 3857x - 38{,}411.25$, $13 \le x \le 18$, where x is the air-fuel ratio.

(a) Use a graphing utility to graph the model.

(b) There are two air-fuel ratios that produce 2400 parts per million of nitric oxide. One is $x = 15$. Use the graph to approximate the other.

(c) Find the second air-fuel ratio from part (b) algebraically. (*Hint:* Use the known value of $x = 15$ and synthetic division.)

Conclusions

True or False? **In Exercises 107 and 108, determine whether the statement is true or false. Justify your answer.**

107. If $(7x + 4)$ is a factor of some polynomial function f, then $\frac{4}{7}$ is a zero of f.

108. The value $x = \frac{1}{7}$ is a zero of the polynomial function
$$f(x) = 3x^5 - 2x^4 + x^3 - 16x^2 + 3x - 8.$$

Think About It **In Exercises 109 and 110, the graph of a cubic polynomial function $y = f(x)$ with integer zeros is shown. Find the factored form of f.**

109.
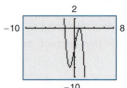

110.
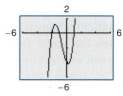

111. Think About It Let $y = f(x)$ be a fourth-degree polynomial with leading coefficient $a = -1$ and $f(\pm 1) = f(\pm 2) = 0$. Find the factored form of f.

112. Think About It Find the value of k such that $x - 3$ is a factor of $x^3 - kx^2 + 2kx - 12$.

113. Writing Complete each polynomial division. Write a brief description of the pattern that you obtain, and use your result to find a formula for the polynomial division $(x^n - 1)/(x - 1)$. Create a numerical example to test your formula.

(a) $\dfrac{x^2 - 1}{x - 1} = $

(b) $\dfrac{x^3 - 1}{x - 1} = $

(c) $\dfrac{x^4 - 1}{x - 1} = $

114. CAPSTONE A graph of $f(x)$ is shown, where $f(x) = 2x^5 - 3x^4 + x^3 - 8x^2 + 5x + 3$ and $f(-x) = -2x^5 - 3x^4 - x^3 - 8x^2 - 5x + 3$.

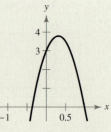

(a) How many negative real zeros does f have? Explain.

(b) How many positive real zeros are *possible* for f? Explain. What does this tell you about the eventual right-hand behavior of the graph?

(c) Is $x = -\frac{1}{3}$ a possible rational zero of f? Explain.

(d) Explain how to check whether $\left(x - \frac{3}{2}\right)$ is a factor of f and whether $x = \frac{3}{2}$ is an upper bound for the real zeros of f.

Cumulative Mixed Review

Solving a Quadratic Equation **In Exercises 115–118, use any convenient method to solve the quadratic equation.**

115. $9x^2 - 25 = 0$

116. $16x^2 - 21 = 0$

117. $2x^2 + 6x + 3 = 0$

118. $8x^2 - 22x + 15 = 0$

3.4   The Fundamental Theorem of Algebra

The Fundamental Theorem of Algebra

You know that an *n*th-degree polynomial can have at most *n* real zeros. In the complex number system, this statement can be improved. That is, in the complex number system, every *n*th-degree polynomial function has *precisely n zeros*. This important result is derived from the **Fundamental Theorem of Algebra,** first proved by the German mathematician Carl Friedrich Gauss (1777–1855).

> #### The Fundamental Theorem of Algebra
>
> If $f(x)$ is a polynomial of degree n, where $n > 0$, then f has at least one zero in the complex number system.

Using the Fundamental Theorem of Algebra and the equivalence of zeros and factors, you obtain the **Linear Factorization Theorem.**

> #### Linear Factorization Theorem   (See the proof on page 322.)
>
> If $f(x)$ is a polynomial of degree n, where $n > 0$, then f has precisely n linear factors
>
> $$f(x) = a_n(x - c_1)(x - c_2) \cdots (x - c_n)$$
>
> where $c_1, c_2, \ldots, c_n$ are complex numbers.

Note that neither the Fundamental Theorem of Algebra nor the Linear Factorization Theorem tells you *how* to find the zeros or factors of a polynomial. Such theorems are called *existence theorems*. To find the zeros of a polynomial function, you still must rely on other techniques.

Example 1   Zeros of Polynomial Functions

a. The first-degree polynomial $f(x) = x - 2$ has exactly *one* zero: $x = 2$.

b. Counting multiplicity, the second-degree polynomial function

$$f(x) = x^2 - 6x + 9$$
$$= (x - 3)(x - 3)$$

has exactly *two* zeros: $x = 3$ and $x = 3$. (This is called a *repeated zero*.)

c. The third-degree polynomial function

$$f(x) = x^3 + 4x = x(x^2 + 4) = x(x - 2i)(x + 2i)$$

has exactly *three* zeros: $x = 0$, $x = 2i$, and $x = -2i$.

d. The fourth-degree polynomial function

$$f(x) = x^4 - 1$$
$$= (x - 1)(x + 1)(x - i)(x + i)$$

has exactly *four* zeros: $x = 1$, $x = -1$, $x = i$, and $x = -i$.

✔**CHECKPOINT**   Now try Exercise 5.

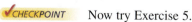

Finding Zeros of a Polynomial Function

Remember that the *n* zeros of a polynomial function can be real or complex, and they may be repeated. Examples 2 and 3 illustrate several cases.

Example 2   Real and Complex Zeros of a Polynomial Function

Confirm that the third-degree polynomial function

$$f(x) = x^3 + 4x$$

has exactly three zeros: $x = 0$, $x = 2i$, and $x = -2i$.

Solution

Factor the polynomial completely as $x(x - 2i)(x + 2i)$. So, the zeros are

$$x(x - 2i)(x + 2i) = 0$$

$$x = 0$$

$$x - 2i = 0 \implies x = 2i$$

$$x + 2i = 0 \implies x = -2i.$$

In the graph in Figure 3.30, only the real zero $x = 0$ appears as an x-intercept.

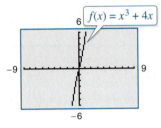

Figure 3.30

✔CHECKPOINT   Now try Exercise 11.

Example 3 shows how to use the methods described in Sections 3.2 and 3.3 (the Rational Zero Test, synthetic division, and factoring) to find all the zeros of a polynomial function, including complex zeros.

Example 3   Finding the Zeros of a Polynomial Function

Write $f(x) = x^5 + x^3 + 2x^2 - 12x + 8$ as the product of linear factors, and list all the zeros of f.

Solution

The possible rational zeros are $\pm 1, \pm 2, \pm 4$, and ± 8. The graph shown in Figure 3.31 indicates that 1 and -2 are likely zeros, and that 1 is possibly a repeated zero because it appears that the graph touches (but does not cross) the x-axis at this point. Using synthetic division, you can determine that -2 is a zero and 1 is a repeated zero of f. So, you have

$$f(x) = x^5 + x^3 + 2x^2 - 12x + 8$$

$$= (x - 1)(x - 1)(x + 2)(x^2 + 4).$$

By factoring $x^2 + 4$ as

$$x^2 - (-4) = \left(x - \sqrt{-4}\right)\left(x + \sqrt{-4}\right)$$

$$= (x - 2i)(x + 2i)$$

you obtain

$$f(x) = (x - 1)(x - 1)(x + 2)(x - 2i)(x + 2i)$$

which gives the following five zeros of f.

$$x = 1, x = 1, x = -2, x = 2i, \text{ and } x = -2i$$

Note from the graph of f shown in Figure 3.31 that the *real* zeros are the only ones that appear as x-intercepts.

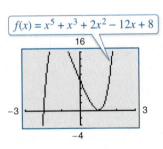

Figure 3.31

✔CHECKPOINT   Now try Exercise 35.

Conjugate Pairs

In Example 3, note that the two complex zeros are **conjugates.** That is, they are of the forms $a + bi$ and $a - bi$.

Complex Zeros Occur in Conjugate Pairs

Let $f(x)$ be a polynomial function that has *real coefficients*. If $a + bi$, where $b \neq 0$, is a zero of the function, then the conjugate $a - bi$ is also a zero of the function.

Be sure you see that this result is true only when the polynomial function has *real coefficients*. For instance, the result applies to the function $f(x) = x^2 + 1$, but not to the function $g(x) = x - i$.

Example 4   Finding a Polynomial with Given Zeros

Find a *fourth-degree* polynomial function with real coefficients that has -1, -1, and $3i$ as zeros.

Solution

Because $3i$ is a zero *and* the polynomial is stated to have real coefficients, you know that the conjugate $-3i$ must also be a zero. So, from the Linear Factorization Theorem, $f(x)$ can be written as

$$f(x) = a(x + 1)(x + 1)(x - 3i)(x + 3i).$$

For simplicity, let $a = 1$ to obtain

$$f(x) = (x^2 + 2x + 1)(x^2 + 9)$$

$$= x^4 + 2x^3 + 10x^2 + 18x + 9.$$

✓CHECKPOINT   Now try Exercise 47.

Example 5   Finding a Polynomial with Given Zeros

Find a *cubic* polynomial function f with real coefficients that has 2 and $1 - i$ as zeros, and $f(1) = 3$.

Solution

Because $1 - i$ is a zero of f, the conjugate $1 + i$ must also be a zero.

$$f(x) = a(x - 2)[x - (1 - i)][x - (1 + i)]$$

$$= a(x - 2)[x^2 - x(1 + i) - x(1 - i) + 1 - i^2]$$

$$= a(x - 2)(x^2 - 2x + 2)$$

$$= a(x^3 - 4x^2 + 6x - 4)$$

To find the value of a, use the fact that $f(1) = 3$ to obtain

$$a[(1)^3 - 4(1)^2 + 6(1) - 4)] = 3.$$

Thus, $a = -3$ and you can conclude that

$$f(x) = -3(x^3 - 4x^2 + 6x - 4)$$

$$= -3x^3 + 12x^2 - 18x + 12.$$

✓CHECKPOINT   Now try Exercise 51.

Factoring a Polynomial

The Linear Factorization Theorem states that you can write any nth-degree polynomial as the product of n linear factors.

$$f(x) = a_n(x - c_1)(x - c_2)(x - c_3) \cdots (x - c_n)$$

This result, however, includes the possibility that some of the values of c_i are complex. The following theorem states that even when you do not want to get involved with "complex factors," you can still write $f(x)$ as the product of linear and/or quadratic factors.

> **Factors of a Polynomial**   (See the proof on page 322.)
>
> Every polynomial of degree $n > 0$ with real coefficients can be written as the product of linear and quadratic factors with real coefficients, where the quadratic factors have no real zeros.

A quadratic factor with no real zeros is said to be **prime** or **irreducible over the reals.** Be sure you see that this is not the same as being *irreducible over the rationals.* For example, the quadratic

$$x^2 + 1 = (x - i)(x + i)$$

is irreducible over the reals (and therefore over the rationals). On the other hand, the quadratic

$$x^2 - 2 = \left(x - \sqrt{2}\right)\left(x + \sqrt{2}\right)$$

is irreducible over the rationals, but *reducible* over the reals.

Example 6   Factoring a Polynomial

Write the polynomial $f(x) = x^4 - x^2 - 20$

a. as the product of factors that are irreducible over the *rationals,*

b. as the product of linear factors and quadratic factors that are irreducible over the *reals,* and

c. in completely factored form.

Solution

a. Begin by factoring the polynomial into the product of two quadratic polynomials.

$$x^4 - x^2 - 20 = (x^2 - 5)(x^2 + 4)$$

Both of these factors are irreducible over the rationals.

b. By factoring over the reals, you have

$$x^4 - x^2 - 20 = \left(x + \sqrt{5}\right)\left(x - \sqrt{5}\right)(x^2 + 4)$$

where the quadratic factor is irreducible over the reals.

c. In completely factored form, you have

$$x^4 - x^2 - 20 = \left(x + \sqrt{5}\right)\left(x - \sqrt{5}\right)(x - 2i)(x + 2i).$$

✓CHECKPOINT   Now try Exercise 55.

In Example 6, notice from the completely factored form that the *fourth*-degree polynomial has *four* zeros.

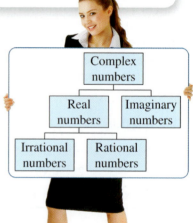

Study Tip

    Recall that irrational and rational numbers are subsets of the set of real numbers, and the real numbers are a subset of the set of complex numbers.

Throughout this chapter, the results and theorems have been stated in terms of zeros of polynomial functions. Be sure you see that the same results could have been stated in terms of solutions of polynomial equations. This is true because the zeros of the polynomial function

$$f(x) = a_n x^n + a_{n-1} x^{n-1} + \cdots + a_2 x^2 + a_1 x + a_0$$

are precisely the solutions of the polynomial equation

$$a_n x^n + a_{n-1} x^{n-1} + \cdots + a_2 x^2 + a_1 x + a_0 = 0.$$

Example 7   Finding the Zeros of a Polynomial Function

Find all the zeros of

$$f(x) = x^4 - 3x^3 + 6x^2 + 2x - 60$$

given that $1 + 3i$ is a zero of f.

Algebraic Solution

Because complex zeros occur in conjugate pairs, you know that $1 - 3i$ is also a zero of f. This means that both

$$x - (1 + 3i) \quad \text{and} \quad x - (1 - 3i)$$

are factors of f. Multiplying these two factors produces

$$[x - (1 + 3i)][x - (1 - 3i)] = [(x - 1) - 3i][(x - 1) + 3i]$$

$$= (x - 1)^2 - 9i^2$$

$$= x^2 - 2x + 10.$$

Using long division, you can divide $x^2 - 2x + 10$ into f to obtain the following.

$$
\begin{array}{r}
x^2 - x - 6 \\
x^2 - 2x + 10 \,\overline{)\, x^4 - 3x^3 + 6x^2 + 2x - 60} \\
\underline{x^4 - 2x^3 + 10x^2} \\
-x^3 - 4x^2 + 2x \\
\underline{-x^3 + 2x^2 - 10x} \\
-6x^2 + 12x - 60 \\
\underline{-6x^2 + 12x - 60} \\
0
\end{array}
$$

So, you have

$$f(x) = (x^2 - 2x + 10)(x^2 - x - 6)$$

$$= (x^2 - 2x + 10)(x - 3)(x + 2)$$

and you can conclude that the zeros of f are

$$x = 1 + 3i, \ x = 1 - 3i, \ x = 3, \text{ and } x = -2.$$

✔**CHECKPOINT**   Now try Exercise 61.

Graphical Solution

You can use a graphing utility to determine that $x = -2$ and $x = 3$ are x-intercepts of the graph of f (see Figure 3.32).

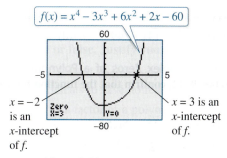

$x = -2$ is an x-intercept of f

$x = 3$ is an x-intercept of f

Figure 3.32

Because $1 + 3i$ is a zero of f, you know that the conjugate $1 - 3i$ must also be a zero. So, you can conclude that the zeros of f are

$$x = 1 + 3i, \ x = 1 - 3i, \ x = 3, \text{ and } x = -2.$$

In Example 7, if you were not told that $1 + 3i$ is a zero of f, you could still find all the zeros of the function by using synthetic division to find the real zeros -2 and 3. Then, you could factor the polynomial as $(x + 2)(x - 3)(x^2 - 2x + 10)$. Finally, by using the Quadratic Formula, you could determine that the zeros are $x = 1 + 3i$, $x = 1 - 3i$, $x = 3$, and $x = -2$.

3.5   Rational Functions and Asymptotes

Introduction to Rational Functions

A **rational function** can be written in the form

$$f(x) = \frac{N(x)}{D(x)}$$

where $N(x)$ and $D(x)$ are polynomials and $D(x)$ is not the zero polynomial.

In general, the *domain* of a rational function of x includes all real numbers except x-values that make the denominator zero. Much of the discussion of rational functions will focus on their graphical behavior near these x-values.

Example 1   Finding the Domain of a Rational Function

Find the domain of $f(x) = 1/x$ and discuss the behavior of f near any excluded x-values.

Solution

$(-\infty, 0) \cup (0, \infty)$

Because the denominator is zero when $x = 0$, the domain of f is all real numbers except $x = 0$. To determine the behavior of f near this excluded value, evaluate $f(x)$ to the left and right of $x = 0$, as indicated in the following tables.

x	-1	-0.5	-0.1	-0.01	-0.001	$\rightarrow 0$
$f(x)$	-1	-2	-10	-100	-1000	$\rightarrow -\infty$

x	$0 \leftarrow$	0.001	0.01	0.1	0.5	1
$f(x)$	$\infty \leftarrow$	1000	100	10	2	1

From the table, note that as x approaches 0 *from the left*, $f(x)$ decreases without bound. In contrast, as x approaches 0 *from the right*, $f(x)$ increases without bound. Because $f(x)$ decreases without bound from the left and increases without bound from the right, you can conclude that f is not continuous. The graph of f is shown in Figure 3.33.

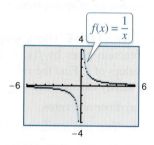

$$f(x) = \frac{1}{x}$$

Figure 3.33

 CHECKPOINT   Now try Exercise 5.

Technology Tip

The graphing utility graphs in this section and the next section were created using the *dot* mode. A blue curve is placed behind the graphing utility's display to indicate where the graph should appear. You will learn more about how graphing utilities graph rational functions in the next section.

What you should learn

- Find the domains of rational functions.
- Find vertical and horizontal asymptotes of graphs of rational functions.
- Use rational functions to model and solve real-life problems.

Why you should learn it

Rational functions are convenient in modeling a wide variety of real-life problems, such as environmental scenarios. For instance, Exercise 45 on page 294 shows how to determine the cost of removing pollutants from a river.

Explore the Concept

Use the *table* and *trace* features of a graphing utility to verify that the function $f(x) = 1/x$ in Example 1 is not continuous.

Vertical and Horizontal Asymptotes

In Example 1, the behavior of f near $x = 0$ is denoted as follows.

$$f(x) \to -\infty \text{ as } x \to 0^-$$

$f(x)$ decreases without bound as x approaches 0 from the left.

$$f(x) \to \infty \text{ as } x \to 0^+$$

$f(x)$ increases without bound as x approaches 0 from the right.

The line $x = 0$ is a **vertical asymptote** of the graph of f, as shown in Figure 3.34. From this figure you can see that the graph of f also has a **horizontal asymptote**—the line $y = 0$. This means the values of

$$f(x) = \frac{1}{x}$$

approach zero as x increases or decreases without bound.

$$f(x) \to 0 \text{ as } x \to -\infty$$

$f(x)$ approaches 0 as x decreases without bound.

$$f(x) \to 0 \text{ as } x \to \infty$$

$f(x)$ approaches 0 as x increases without bound.

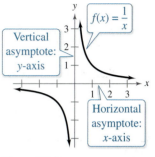

Figure 3.34

Definition of Vertical and Horizontal Asymptotes

1. The line $x = a$ is a **vertical asymptote** of the graph of f when

$$f(x) \to \infty \quad \text{or} \quad f(x) \to -\infty$$

as $x \to a$, either from the right or from the left.

2. The line $y = b$ is a **horizontal asymptote** of the graph of f when

$$f(x) \to b$$

as $x \to \infty$ or $x \to -\infty$.

Explore the Concept

Use a table of values to determine whether the functions in Figure 3.35 are continuous. When the graph of a function has an asymptote, can you conclude that the function is not continuous? Explain.

Figure 3.35 shows the vertical and horizontal asymptotes of the graphs of three rational functions. Note in Figure 3.35 that eventually (as $x \to \infty$ or $x \to -\infty$) the distance between the horizontal asymptote and the points on the graph must approach zero.

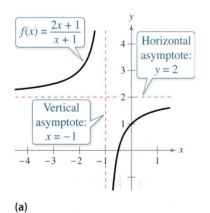

(a)

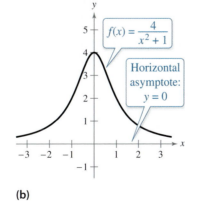

(b)

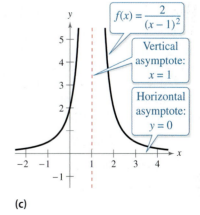

(c)

Figure 3.35

The graphs of $f(x) = 1/x$ in Figure 3.34 and $f(x) = (2x + 1)/(x + 1)$ in Figure 3.35(a) are **hyperbolas.** You will study hyperbolas later in this text.

Vertical and Horizontal Asymptotes of a Rational Function

Let f be the rational function

$$f(x) = \frac{N(x)}{D(x)} = \frac{a_n x^n + a_{n-1} x^{n-1} + \cdots + a_1 x + a_0}{b_m x^m + b_{m-1} x^{m-1} + \cdots + b_1 x + b_0}$$

where $N(x)$ and $D(x)$ have no common factors.

1. The graph of f has *vertical* asymptotes at the zeros of $D(x)$.

2. The graph of f has at most one *horizontal* asymptote determined by comparing the degrees of $N(x)$ and $D(x)$.

   a. If $n < m$, then the graph of f has the line $y = 0$ (the x-axis) as a horizontal asymptote.

   b. If $n = m$, then the graph of f has the line

$$y = \frac{a_n}{b_m}$$

   as a horizontal asymptote, where a_n is the leading coefficient of the numerator and b_m is the leading coefficient of the denominator.

   c. If $n > m$, then the graph of f has no horizontal asymptote.

What's Wrong?

You use a graphing utility to graph

$$y_1 = \frac{2x^3 + 1000x^2 + x}{x^3 + 1000x^2 + x + 1000}$$

as shown in the figure. You use the graph to conclude that the graph of y_1 has the line $y = 1$ as a horizontal asymptote. What's wrong?

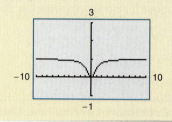

Example 2   Finding Vertical and Horizontal Asymptotes

Find all asymptotes of the graph of each rational function.

a. $f(x) = \dfrac{2x}{3x^2 + 1}$ **b.** $f(x) = \dfrac{2x^2}{x^2 - 1}$

Solution

a. For this rational function, the degree of the numerator is *less than* the degree of the denominator, so the graph has the line $y = 0$ as a horizontal asymptote. To find any vertical asymptotes, set the denominator equal to zero and solve the resulting equation for x.

$$3x^2 + 1 = 0 \qquad \qquad \text{Set denominator equal to zero.}$$

Because this equation has no real solutions, you can conclude that the graph has no vertical asymptote. The graph of the function is shown in Figure 3.36.

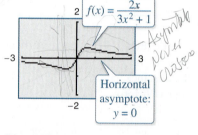

Figure 3.36

b. For this rational function, the degree of the numerator is *equal* to the degree of the denominator. The leading coefficient of the numerator is 2 and the leading coefficient of the denominator is 1, so the graph has the line $y = 2$ as a horizontal asymptote. To find any vertical asymptotes, set the denominator equal to zero and solve the resulting equation for x.

$$x^2 - 1 = 0 \qquad \qquad \text{Set denominator equal to zero.}$$
$$(x + 1)(x - 1) = 0 \qquad \qquad \text{Factor.}$$
$$x + 1 = 0 \implies x = -1 \qquad \text{Set 1st factor equal to 0.}$$
$$x - 1 = 0 \implies x = 1 \qquad \text{Set 2nd factor equal to 0.}$$

This equation has two real solutions, $x = -1$ and $x = 1$, so the graph has the lines $x = -1$ and $x = 1$ as vertical asymptotes, as shown in Figure 3.37.

✓**CHECKPOINT**   Now try Exercise 17.

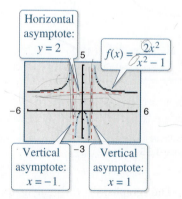

Figure 3.37

Values for which a rational function is undefined (the denominator is zero) result in a vertical asymptote or a hole in the graph, as shown in Example 3.

Example 3   Finding Asymptotes and Holes

Find all asymptotes and holes in the graph of

$$f(x) = \frac{x^2 + x - 2}{x^2 - x - 6}.$$

Solution

For this rational function the degree of the numerator is *equal to* the degree of the denominator. The leading coefficients of the numerator and denominator are both 1, so the graph has the line $y = 1$ as a horizontal asymptote. To find any vertical asymptotes, first factor the numerator and denominator as follows.

$$f(x) = \frac{x^2 + x - 2}{x^2 - x - 6} = \frac{(x - 1)(x + 2)}{(x + 2)(x - 3)} = \frac{x - 1}{x - 3}, \quad x \neq -2$$

By setting the denominator $x - 3$ (of the simplified function) equal to zero, you can determine that the graph has the line $x = 3$ as a vertical asymptote, as shown in Figure 3.38. To find any holes in the graph, note that the function is undefined at $x = -2$ and $x = 3$. Because $x = -2$ is not a vertical asymptote of the function, there is a hole in the graph at $x = -2$. To find the y-coordinate of the hole, substitute $x = -2$ into the simplified form of the function.

$$y = \frac{x - 1}{x - 3} = \frac{-2 - 1}{-2 - 3} = \frac{3}{5}$$

So, the graph of the rational function has a hole at $\left(-2, \frac{3}{5}\right)$.

✔CHECKPOINT   Now try Exercise 23.

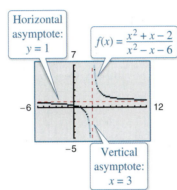

Horizontal asymptote: $y = 1$

$f(x) = \dfrac{x^2 + x - 2}{x^2 - x - 6}$

Vertical asymptote: $x = 3$

Figure 3.38

Example 4   Finding a Function's Domain and Asymptotes

For the function f, find (a) the domain of f, (b) the vertical asymptote of f, and (c) the horizontal asymptote of f.

$$f(x) = \frac{3x^3 + 7x^2 + 2}{-2x^3 + 16}$$

Solution

a. Because the denominator is zero when $-2x^3 + 16 = 0$, solve this equation to determine that the domain of f is all real numbers except $x = 2$.

b. Because the denominator of f has a zero at $x = 2$, and 2 is not a zero of the numerator, the graph of f has the vertical asymptote $x = 2$.

c. Because the degrees of the numerator and denominator are the same, and the leading coefficient of the numerator is 3 and the leading coefficient of the denominator is -2, the horizontal asymptote of f is $y = -\frac{3}{2}$.

✔CHECKPOINT   Now try Exercise 25.

Application

There are many examples of asymptotic behavior in real life. For instance, Example 5 shows how a vertical asymptote can be used to analyze the cost of removing pollutants from smokestack emissions.

Example 5   Cost-Benefit Model

A utility company burns coal to generate electricity. The cost C (in dollars) of removing $p\%$ of the smokestack pollutants is given by

$$C = \frac{80,000p}{100 - p}$$

for $0 \le p < 100$. Use a graphing utility to graph this function. You are a member of a state legislature that is considering a law that would require utility companies to remove 90% of the pollutants from their smokestack emissions. The current law requires 85% removal. How much additional cost would the utility company incur as a result of the new law?

Solution

The graph of this function is shown in Figure 3.39. Note that the graph has a vertical asymptote at

$$p = 100.$$

Because the current law requires 85% removal, the current cost to the utility company is

$$C = \frac{80,000p}{100 - p} \qquad \text{Write original function.}$$

$$C = \frac{80,000(85)}{100 - 85} \qquad \text{Substitute 85 for } p.$$

$$\approx \$453,333. \qquad \text{Simplify.}$$

If the new law increases the percent removal to 90%, the cost will be

$$C = \frac{80,000p}{100 - p} \qquad \text{Write original function.}$$

$$C = \frac{80,000(90)}{100 - 90} \qquad \text{Substitute 90 for } p.$$

$$= \$720,000. \qquad \text{Simplify.}$$

So, the new law would require the utility company to spend an additional

$$720,000 - 453,333 = \$266,667. \qquad \text{Subtract 85\% removal cost from 90\% removal cost.}$$

Explore the Concept

The *table* feature of a graphing utility can be used to estimate vertical and horizontal asymptotes of rational functions. Use the *table* feature to find any vertical or horizontal asymptotes of

$$f(x) = \frac{2x}{x + 1}.$$

Write a statement explaining how you found the asymptote(s) using the table.

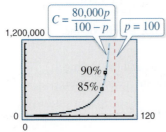

Figure 3.39

Now try Exercise 45.

3.5   Exercises

See www.CalcChat.com for worked-out solutions to odd-numbered exercises.
For instructions on how to use a graphing utility, see Appendix A.

Vocabulary and Concept Check

In Exercises 1 and 2, fill in the blank.

1. Functions of the form $f(x) = N(x)/D(x)$, where $N(x)$ and $D(x)$ are polynomials and $D(x)$ is not the zero polynomial, are called _____ .

2. If $f(x) \to \pm\infty$ as $x \to a$ from the left (or right), then $x = a$ is a _____ of the graph of f.

3. What feature of the graph of $y = \dfrac{9}{x - 3}$ can you find by solving $x - 3 = 0$?

4. Is $y = \dfrac{2}{3}$ a horizontal asymptote of the function $y = \dfrac{2x}{3x^2 - 5}$?

Procedures and Problem Solving

Finding the Domain of a Rational Function   In Exercises 5–10, (a) find the domain of the function, (b) complete each table, and (c) discuss the behavior of f near any excluded x-values.

x	$f(x)$
0.5	
0.9	
0.99	
0.999	

x	$f(x)$
1.5	
1.1	
1.01	
1.001	

x	$f(x)$
5	
10	
100	
1000	

x	$f(x)$
-5	
-10	
-100	
-1000	

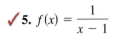 **5.** $f(x) = \dfrac{1}{x - 1}$

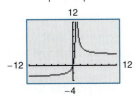

6. $f(x) = \dfrac{5x}{x - 1}$

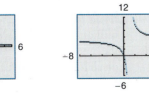

7. $f(x) = \dfrac{3x}{|x - 1|}$

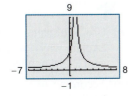

8. $f(x) = \dfrac{3}{|x - 1|}$

9. $f(x) = \dfrac{3x^2}{x^2 - 1}$

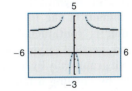

10. $f(x) = \dfrac{4x}{x^2 - 1}$

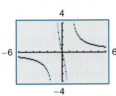

Identifying Graphs of Rational Functions   In Exercises 11–16, match the function with its graph. [The graphs are labeled (a), (b), (c), (d), (e), and (f).]

(a)

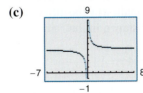

(b)

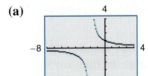

(c)

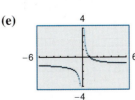

(d)

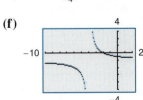

(e)

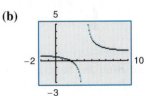

(f)

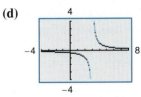

11. $f(x) = \dfrac{2}{x + 2}$

12. $f(x) = \dfrac{1}{x - 3}$

13. $f(x) = \dfrac{4x + 1}{x}$

14. $f(x) = \dfrac{1 - x}{x}$

15. $f(x) = \dfrac{x - 2}{x - 4}$

16. $f(x) = -\dfrac{x + 2}{x + 4}$

Finding Vertical and Horizontal Asymptotes
In Exercises 17–20, find any asymptotes of the graph of the rational function. Verify your answers by using a graphing utility to graph the function.

17. $f(x) = \dfrac{1}{x^2}$

18. $f(x) = \dfrac{3}{(x-2)^3}$

19. $f(x) = \dfrac{2x^2}{x^2 + x - 6}$

20. $f(x) = \dfrac{x^2 - 4x}{x^2 - 4}$

Finding Asymptotes and Holes
In Exercises 21–24, find any asymptotes and holes in the graph of the rational function. Verify your answers by using a graphing utility.

21. $f(x) = \dfrac{x(2 + x)}{2x - x^2}$

22. $f(x) = \dfrac{x^2 + 2x + 1}{2x^2 - x - 3}$

23. $f(x) = \dfrac{x^2 - 25}{x^2 + 5x}$

24. $f(x) = \dfrac{3 - 14x - 5x^2}{3 + 7x + 2x^2}$

Finding a Function's Domain and Asymptotes
In Exercises 25–28, (a) find the domain of the function, (b) decide whether the function is continuous, and (c) identify any horizontal and vertical asymptotes. Verify your answer to part (a) both graphically by using a graphing utility and numerically by creating a table of values.

25. $f(x) = \dfrac{3x^2 + x - 5}{x^2 + 1}$

26. $f(x) = \dfrac{3x^2 + 1}{x^2 + x + 9}$

27. $f(x) = \dfrac{x^2 + 3x - 4}{-x^3 + 27}$

28. $f(x) = \dfrac{4x^3 - x^2 + 3}{3x^3 + 24}$

Algebraic-Graphical-Numerical
In Exercises 29–32, (a) determine the domains of f and g, (b) find any vertical asymptotes and holes in the graphs of f and g, (c) compare f and g by completing the table, (d) use a graphing utility to graph f and g, and (e) explain why the differences in the domains of f and g are not shown in their graphs.

29. $f(x) = \dfrac{x^2 - 16}{x - 4}$, $g(x) = x + 4$

x	1	2	3	4	5	6	7
$f(x)$							
$g(x)$							

30. $f(x) = \dfrac{x^2 - 9}{x - 3}$, $g(x) = x + 3$

x	0	1	2	3	4	5	6
$f(x)$							
$g(x)$							

31. $f(x) = \dfrac{x^2 - 1}{x^2 - 2x - 3}$, $g(x) = \dfrac{x - 1}{x - 3}$

x	−2	−1	0	1	2	3	4
$f(x)$							
$g(x)$							

32. $f(x) = \dfrac{x^2 - 4}{x^2 - 3x + 2}$, $g(x) = \dfrac{x + 2}{x - 1}$

x	−3	−2	−1	0	1	2	3
$f(x)$							
$g(x)$							

Exploration   In Exercises 33–36, determine the value that the function f approaches as the magnitude of x increases. Is $f(x)$ greater than or less than this value when x is positive and large in magnitude? What about when x is negative and large in magnitude?

33. $f(x) = 4 - \dfrac{1}{x}$

34. $f(x) = 2 + \dfrac{1}{x - 3}$

35. $f(x) = \dfrac{2x - 1}{x - 3}$

36. $f(x) = \dfrac{2x - 1}{x^2 + 1}$

Finding the Zeros of a Rational Function
In Exercises 37–44, find the zeros (if any) of the rational function. Use a graphing utility to verify your answer.

37. $g(x) = \dfrac{x^2 - 4}{x + 3}$

38. $g(x) = \dfrac{x^3 - 8}{x^2 + 4}$

39. $f(x) = 1 - \dfrac{2}{x - 5}$

40. $h(x) = 5 + \dfrac{3}{x^2 + 1}$

41. $g(x) = \dfrac{x^2 - 2x - 3}{x^2 + 1}$

42. $g(x) = \dfrac{x^2 - 5x + 6}{x^2 + 4}$

43. $f(x) = \dfrac{2x^2 - 5x + 2}{2x^2 - 7x + 3}$

44. $f(x) = \dfrac{2x^2 + 3x - 2}{x^2 + x - 2}$

45. *Why you should learn it* (p. 288)   The cost C (in millions of dollars) of removing $p\%$ of the industrial and municipal pollutants discharged into a river is given by

$$C = \dfrac{255p}{100 - p}, \quad 0 \le p < 100.$$

(a) Find the costs of removing 10%, 40%, and 75% of the pollutants.

(b) Use a graphing utility to graph the cost function. Explain why you chose the values that you used in your viewing window.

(c) According to this model, is it possible to remove 100% of the pollutants? Explain.

46. MODELING DATA

The endpoints of the interval over which distinct vision is possible are called the *near point* and *far point* of the eye (see figure). With increasing age these points normally change. The table shows the approximate near points y (in inches) for various ages x (in years).

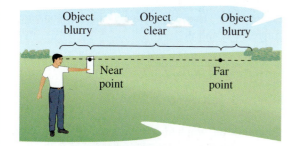

Object blurry — Object clear — Object blurry

Near point   Far point

Age, x	Near point, y
16	3.0
32	4.7
44	9.8
50	19.7
60	39.4

(a) Find a rational model for the data. Take the reciprocals of the near points to generate the points $(x, 1/y)$. Use the *regression* feature of a graphing utility to find a linear model for the data. The resulting line has the form

$$\frac{1}{y} = ax + b.$$

Solve for y.

(b) Use the *table* feature of the graphing utility to create a table showing the predicted near point based on the model for each of the ages in the original table.

(c) Do you think the model can be used to predict the near point for a person who is 70 years old? Explain.

47. Physics

Consider a physics laboratory experiment designed to determine an unknown mass. A flexible metal meter stick is clamped to a table with 50 centimeters overhanging the edge (see figure). Known masses M ranging from 200 grams to 2000 grams are attached to the end of the meter stick. For each mass, the meter stick is displaced vertically and then allowed to oscillate. The average time t (in seconds) of one oscillation for each mass is recorded in the table.

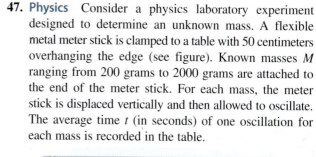

|←50 cm→|

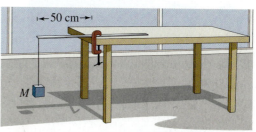

Mass, M	Time, t
200	0.450
400	0.597
600	0.712
800	0.831
1000	0.906
1200	1.003
1400	1.088
1600	1.126
1800	1.218
2000	1.338

A model for the data is given by

$$t = \frac{38M + 16{,}965}{10(M + 5000)}.$$

(a) Use the *table* feature of a graphing utility to create a table showing the estimated time based on the model for each of the masses shown in the table. What can you conclude?

(b) Use the model to approximate the mass of an object when the average time for one oscillation is 1.056 seconds.

48. Biology

The game commission introduces 100 deer into newly acquired state game lands. The population N of the herd is given by

$$N = \frac{20(5 + 3t)}{1 + 0.04t}, \quad t \geq 0$$

where t is the time in years.

(a) Use a graphing utility to graph the model.

(b) Find the populations when $t = 5$, $t = 10$, and $t = 25$.

(c) What is the limiting size of the herd as time increases? Explain.

49. Military Science The table shows the numbers N (in thousands) of Department of Defense personnel from 1990 through 2008. The data can be modeled by

$$N = \frac{77.095t^2 - 216.04t + 2050}{0.052t^2 - 0.08t + 1}, \quad 0 \le t \le 18$$

where t is the year, with $t = 0$ corresponding to 1990. (Source: U.S. Department of Defense)

Year	Number, N (in thousands)
1990	2044
1991	1986
1992	1807
1993	1705
1994	1610
1995	1518
1996	1472
1997	1439
1998	1407
1999	1386
2000	1384
2001	1385
2002	1414
2003	1434
2004	1427
2005	1389
2006	1385
2007	1380
2008	1402

(a) Use a graphing utility to plot the data and graph the model in the same viewing window. How well does the model represent the data?

(b) Use the model to estimate the numbers of Department of Defense personnel in the years 2009, 2010, and 2011. Are the estimates reasonable?

(c) Find the horizontal asymptote of the graph of the model. Does it seem realistic that the numbers of Department of Defense personnel will keep getting closer to the value of the asymptote? Explain.

Conclusions

True or False? In Exercises 50 and 51, determine whether the statement is true or false. Justify your answer.

50. A rational function can have infinitely many vertical asymptotes.

51. A rational function must have at least one vertical asymptote.

52. Think About It Describe the possible features of the graph of a rational function f at $x = c$, when c is not in the domain of f.

53. Think About It A real zero of the numerator of a rational function f is $x = c$. Must $x = c$ also be a zero of f? Explain.

54. Think About It When the graph of a rational function f has a vertical asymptote at $x = 4$, can f have a common factor of $(x - 4)$ in the numerator and denominator? Explain.

55. Exploration Use a graphing utility to compare the graphs of y_1 and y_2.

$$y_1 = \frac{3x^3 - 5x^2 + 4x - 5}{2x^2 - 6x + 7}, \quad y_2 = \frac{3x^3}{2x^2}$$

Start with a viewing window of $-5 \le x \le 5$ and $-10 \le y \le 10$, and then zoom out. Make a conjecture about how the graph of a rational function f is related to the graph of $y = a_n x^n / b_m x^m$, where $a_n x^n$ is the leading term of the numerator of f and $b_m x^m$ is the leading term of the denominator of f.

56. CAPSTONE Write a rational function f that has the specified characteristics. (There are many correct answers.)

(a) Vertical asymptote: $x = 2$

    Horizontal asymptote: $y = 0$

    Zero: $x = 1$

(b) Vertical asymptote: $x = -1$

    Horizontal asymptote: $y = 0$

    Zero: $x = 2$

(c) Vertical asymptotes: $x = -2, x = 1$

    Horizontal asymptote: $y = 2$

    Zeros: $x = 3, x = -3$

(d) Vertical asymptotes: $x = -1, x = 2$

    Horizontal asymptote: $y = -2$

    Zeros: $x = -2, x = 3$

Cumulative Mixed Review

Finding the Equation of a Line In Exercises 57–60, write the general form of the equation of the line that passes through the points.

57. $(3, 2), (0, -1)$ **58.** $(-6, 1), (4, -5)$

59. $(2, 7), (3, 10)$ **60.** $(0, 0), (-9, 4)$

Long Division of Polynomials In Exercises 61–64, divide using long division.

61. $(x^2 + 5x + 6) \div (x - 4)$

62. $(x^2 - 10x + 15) \div (x - 3)$

63. $(2x^4 + x^2 - 11) \div (x^2 + 5)$

64. $(4x^5 + 3x^3 - 10) \div (2x + 3)$

3.6   Graphs of Rational Functions

The Graph of a Rational Function

To sketch the graph of a rational function, use the following guidelines.

> **Guidelines for Graphing Rational Functions**
>
> Let
>
> $$f(x) = N(x)/D(x)$$
>
> where $N(x)$ and $D(x)$ are polynomials.
>
> 1. **Simplify f, if possible.** Any restrictions on the domain of f not in the simplified function should be listed.
>
> 2. **Find and plot the y-intercept** (if any) by evaluating $f(0)$.
>
> 3. **Find the zeros of the numerator** (if any) by setting the numerator equal to zero. Then plot the corresponding x-intercepts.
>
> 4. **Find the zeros of the denominator** (if any) by setting the denominator equal to zero. Then sketch the corresponding vertical asymptotes using dashed vertical lines and plot the corresponding holes using open circles.
>
> 5. **Find and sketch any other asymptotes** of the graph using dashed lines.
>
> 6. **Plot at least one point *between* and one point *beyond* each x-intercept and vertical asymptote.**
>
> 7. **Use smooth curves to complete the graph** between and beyond the vertical asymptotes, excluding any points where f is not defined.

What you should learn
- Analyze and sketch graphs of rational functions.
- Sketch graphs of rational functions that have slant asymptotes.
- Use graphs of rational functions to model and solve real-life problems.

Why you should learn it
The graph of a rational function provides a good indication of the behavior of a mathematical model. Exercise 89 on page 305 models the concentration of a chemical in the bloodstream after injection.

Technology Tip

Some graphing utilities have difficulty graphing rational functions that have vertical asymptotes. Often, the utility will connect parts of the graph that are not supposed to be connected. Notice that the graph in Figure 3.40(a) should consist of two *unconnected* portions—one to the left of $x = 2$ and the other to the right of $x = 2$. To eliminate this problem, you can try changing the *mode* of the graphing utility to *dot mode*. The problem with this mode is that the graph is then represented as a collection of dots rather than as a smooth curve, as shown in Figure 3.40(b). In this text, a blue curve is placed behind the graphing utility's display to indicate where the graph should appear. [See Figure 3.40(b).]

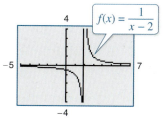

(a) Connected mode
Figure 3.40

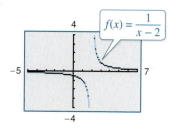

(b) Dot mode

Library of Parent Functions: Rational Function

The simplest type of rational function is the *parent rational function* $f(x) = 1/x$, also known as the *reciprocal function*. The basic characteristics of the parent rational function are summarized below and on the inside front cover of this text.

Graph of $f(x) = \dfrac{1}{x}$

Domain: $(-\infty, 0) \cup (0, \infty)$

Range: $(-\infty, 0) \cup (0, \infty)$

No intercepts

Decreasing on $(-\infty, 0)$ and $(0, \infty)$

Odd function

Origin symmetry

Vertical asymptote: *y*-axis

Horizontal asymptote: *x*-axis

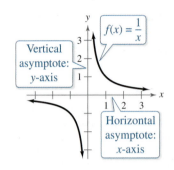

Explore the Concept

 Use a graphing utility to graph

$$f(x) = 1 + \dfrac{1}{x - \dfrac{1}{x}}.$$

Set the graphing utility to *dot* mode and use a decimal viewing window. Use the *trace* feature to find three "holes" or "breaks" in the graph. Do all three holes or breaks represent zeros of the denominator

$$x - \dfrac{1}{x}?$$

Explain.

Example 1 Library of Parent Functions: $f(x) = 1/x$

Sketch the graph of the function and describe how the graph is related to the graph of $f(x) = 1/x$.

a. $g(x) = \dfrac{-1}{x + 2}$

b. $h(x) = \dfrac{1}{x - 1} + 3$

Solution

a. With respect to the graph of $f(x) = 1/x$, the graph of g is obtained by a *reflection* in the *y*-axis and a horizontal shift two units *to the left*, as shown in Figure 3.41. Confirm this with a graphing utility.

b. With respect to the graph of $f(x) = 1/x$, the graph of h is obtained by a horizontal shift one unit *to the right* and a vertical shift three units *upward*, as shown in Figure 3.42. Confirm this with a graphing utility.

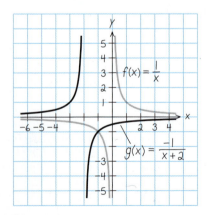

Figure 3.41

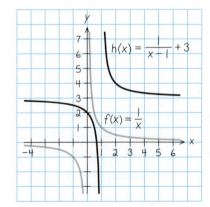

Figure 3.42

✔CHECKPOINT Now try Exercise 5.

Example 2   Sketching the Graph of a Rational Function

Sketch the graph of

$$g(x) = \frac{3}{x - 2}$$

by hand.

Solution

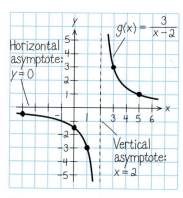

y-intercept:	$\left(0, -\frac{3}{2}\right)$, because $g(0) = -\frac{3}{2}$
x-intercept:	None, because $3 \neq 0$
Vertical asymptote:	$x = 2$, zero of denominator
Horizontal asymptote:	$y = 0$, because degree of $N(x) <$ degree of $D(x)$

Additional points:

x	-4	1	2	3	5
$g(x)$	-0.5	-3	Undefined	3	1

By plotting the intercept, asymptotes, and a few additional points, you can obtain the graph shown in Figure 3.43. Confirm this with a graphing utility.

Figure 3.43

✔CHECKPOINT   Now try Exercise 17.

Note that the graph of g in Example 2 is a vertical stretch and a right shift of the graph of

$$f(x) = \frac{1}{x}$$

because

$$g(x) = \frac{3}{x - 2} = 3\left(\frac{1}{x - 2}\right) = 3f(x - 2).$$

Study Tip

 Note in Examples 2–6 that the vertical asymptotes are included in the tables of additional points. This is done to emphasize numerically the behavior of the graph of the function.

Example 3   Sketching the Graph of a Rational Function

Sketch the graph of

$$f(x) = \frac{2x - 1}{x}$$

by hand.

Solution

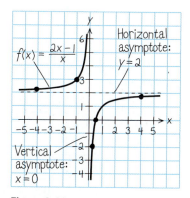

y-intercept:	None, because $x = 0$ is not in the domain
x-intercept:	$\left(\frac{1}{2}, 0\right)$, because $2x - 1 = 0$ when $x = \frac{1}{2}$
Vertical asymptote:	$x = 0$, zero of denominator
Horizontal asymptote:	$y = 2$, because degree of $N(x) =$ degree of $D(x)$

Additional points:

x	-4	-1	0	$\frac{1}{4}$	4
$f(x)$	2.25	3	Undefined	-2	1.75

By plotting the intercept, asymptotes, and a few additional points, you can obtain the graph shown in Figure 3.44. Confirm this with a graphing utility.

Figure 3.44

✔CHECKPOINT   Now try Exercise 21.

Example 4 Sketching the Graph of a Rational Function

Sketch the graph of $f(x) = \dfrac{x}{x^2 - x - 2}$.

Solution

Factor the denominator to determine more easily the zeros of the denominator.

$$f(x) = \frac{x}{x^2 - x - 2}$$

$$= \frac{x}{(x + 1)(x - 2)}$$

y-intercept:	$(0, 0)$, because $f(0) = 0$
x-intercept:	$(0, 0)$
Vertical asymptotes:	$x = -1$, $x = 2$, zeros of denominator
Horizontal asymptote:	$y = 0$, because degree of $N(x) <$ degree of $D(x)$
Additional points:	

x	-3	-1	-0.5	1	2	3
$f(x)$	-0.3	Undefined	0.4	-0.5	Undefined	0.75

The graph is shown in Figure 3.45.

 CHECKPOINT Now try Exercise 27.

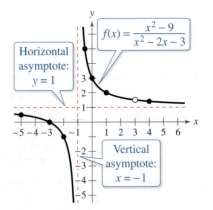

Figure 3.45

Example 5 Sketching the Graph of a Rational Function

Sketch the graph of $f(x) = \dfrac{x^2 - 9}{x^2 - 2x - 3}$.

Solution

By factoring the numerator and denominator, you have

$$f(x) = \frac{x^2 - 9}{x^2 - 2x - 3}$$

$$= \frac{(x - 3)(x + 3)}{(x - 3)(x + 1)}$$

$$= \frac{x + 3}{x + 1}, \quad x \neq 3.$$

y-intercept:	$(0, 3)$, because $f(0) = 3$
x-intercept:	$(-3, 0)$, because $x + 3 = 0$ when $x = -3$
Vertical asymptote:	$x = -1$, zero of (simplified) denominator
Hole:	$\left(3, \frac{3}{2}\right)$, f is not defined at $x = 3$
Horizontal asymptote:	$y = 1$, because degree of $N(x) =$ degree of $D(x)$
Additional points:	

x	-5	-2	-1	-0.5	1	3	4
$f(x)$	0.5	-1	Undefined	5	2	Undefined	1.4

The graph is shown in Figure 3.46.

 CHECKPOINT Now try Exercise 29.

Figure 3.46 *Hole at $x = 3$*

Slant Asymptotes

Consider a rational function whose denominator is of degree 1 or greater. If the degree of the numerator is exactly *one more* than the degree of the denominator, then the graph of the function has a **slant** (or **oblique**) **asymptote.** For example, the graph of

$$f(x) = \frac{x^2 - x}{x + 1}$$

has a slant asymptote, as shown in Figure 3.47. To find the equation of a slant asymptote, use long division. For instance, by dividing $x + 1$ into $x^2 - x$, you have

$$f(x) = \frac{x^2 - x}{x + 1} = \underbrace{x - 2}_{\substack{\text{Slant asymptote} \\ (y = x - 2)}} + \frac{2}{x + 1}.$$

As x increases or decreases without bound, the remainder term

$$\frac{2}{x + 1}$$

approaches 0, so the graph of f approaches the line $y = x - 2$, as shown in Figure 3.47.

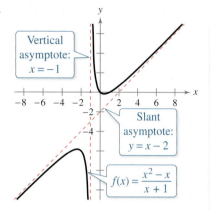

Vertical asymptote: $x = -1$

Slant asymptote: $y = x - 2$

$f(x) = \dfrac{x^2 - x}{x + 1}$

Figure 3.47

Explore the Concept

Do you think it is possible for the graph of a rational function to cross its horizontal asymptote or its slant asymptote? Use the graphs of the following functions to investigate this question. Write a summary of your conclusion. Explain your reasoning.

$$f(x) = \frac{x}{x^2 + 1}$$

$$g(x) = \frac{2x}{3x^2 - 2x + 1}$$

$$h(x) = \frac{x^3}{x^2 + 1}$$

Example 6   A Rational Function with a Slant Asymptote

Sketch the graph of $f(x) = \dfrac{x^2 - x - 2}{x - 1}$.

Solution

First write $f(x)$ in two different ways. Factoring the numerator

$$f(x) = \frac{x^2 - x - 2}{x - 1} = \frac{(x - 2)(x + 1)}{x - 1}$$

enables you to recognize the *x*-intercepts. Long division

$$f(x) = \frac{x^2 - x - 2}{x - 1} = x - \frac{2}{x - 1}$$

enables you to recognize that the line $y = x$ is a slant asymptote of the graph.

y-intercept:	$(0, 2)$, because $f(0) = 2$
x-intercepts:	$(-1, 0)$ and $(2, 0)$
Vertical asymptote:	$x = 1$, zero of denominator
Horizontal asymptote:	None, because degree of $N(x) >$ degree of $D(x)$
Slant asymptote:	$y = x$
Additional points:	

x	-2	0.5	1	1.5	3
$f(x)$	-1.33	4.5	Undefined	-2.5	2

The graph is shown in Figure 3.48.

✔CHECKPOINT   Now try Exercise 51.

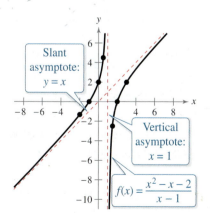

Slant asymptote: $y = x$

Vertical asymptote: $x = 1$

$f(x) = \dfrac{x^2 - x - 2}{x - 1}$

Figure 3.48

Application

Example 7 Publishing

A rectangular page is designed to contain 48 square inches of print. The margins on each side of the page are $1\frac{1}{2}$ inches wide. The margins at the top and bottom are each 1 inch deep. What should the dimensions of the page be so that the minimum amount of paper is used?

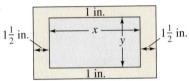

Figure 3.49

Graphical Solution

Let A be the area to be minimized. From Figure 3.49, you can write

$$A = (x + 3)(y + 2).$$

The printed area inside the margins is modeled by $48 = xy$ or $y = 48/x$. To find the minimum area, rewrite the equation for A in terms of just one variable by substituting $48/x$ for y.

$$A = (x + 3)\left(\frac{48}{x} + 2\right) = \frac{(x + 3)(48 + 2x)}{x}, \quad x > 0$$

The graph of this rational function is shown in Figure 3.50. Because x represents the width of the printed area, you need consider only the portion of the graph for which x is positive. Using the *minimum* feature of a graphing utility, you can approximate the minimum value of A to occur when $x \approx 8.5$ inches. The corresponding value of y is $48/8.5 \approx 5.6$ inches. So, the dimensions should be

$$x + 3 \approx 11.5 \text{ inches by } y + 2 \approx 7.6 \text{ inches.}$$

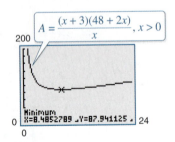

Figure 3.50

✓CHECKPOINT   Now try Exercise 85.

Numerical Solution

Let A be the area to be minimized. From Figure 3.49, you can write

$$A = (x + 3)(y + 2).$$

The printed area inside the margins is modeled by $48 = xy$ or $y = 48/x$. To find the minimum area, rewrite the equation for A in terms of just one variable by substituting $48/x$ for y.

$$A = (x + 3)\left(\frac{48}{x} + 2\right) = \frac{(x + 3)(48 + 2x)}{x}, \quad x > 0$$

Use the *table* feature of a graphing utility to create a table of values for the function

$$y_1 = \frac{(x + 3)(48 + 2x)}{x}$$

beginning at $x = 1$. From the table, you can see that the minimum value of y_1 occurs when x is somewhere between 8 and 9, as shown in Figure 3.51. To approximate the minimum value of y_1 to one decimal place, change the table to begin at $x = 8$ and set the table step to 0.1. The minimum value of y_1 occurs when $x \approx 8.5$, as shown in Figure 3.52. The corresponding value of y is $48/8.5 \approx 5.6$ inches. So, the dimensions should be

$$x + 3 \approx 11.5 \text{ inches by } y + 2 \approx 7.6 \text{ inches.}$$

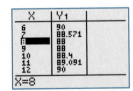

Figure 3.51

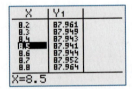

Figure 3.52

If you go on to take a course in calculus, you will learn an analytic technique for finding the exact value of x that produces a minimum area in Example 7. In this case, that value is

$$x = 6\sqrt{2} \approx 8.485.$$

3.6 Exercises

See www.CalcChat.com for worked-out solutions to odd-numbered exercises.
For instructions on how to use a graphing utility, see Appendix A.

Vocabulary and Concept Check

In Exercises 1 and 2, fill in the blank(s).

1. For the rational function $f(x) = N(x)/D(x)$, if the degree of $N(x)$ is exactly one more than the degree of $D(x)$, then the graph of f has a _____ (or oblique) _____ .

2. The graph of $f(x) = 1/x$ has a _____ asymptote at $x = 0$.

3. Does the graph of $f(x) = \dfrac{x^3 - 1}{x^2 + 2}$ have a slant asymptote?

4. Using long division, you find that $f(x) = \dfrac{x^2 + 1}{x + 1} = x - 1 + \dfrac{2}{x + 1}$. What is the slant asymptote of the graph of f?

Procedures and Problem Solving

Library of Parent Functions: $f(x) = 1/x$ In Exercises 5–8, sketch the graph of the function g and describe how the graph is related to the graph of $f(x) = 1/x$.

✓ 5. $g(x) = \dfrac{1}{x - 4}$

6. $g(x) = \dfrac{-1}{x} - 5$

7. $g(x) = \dfrac{-1}{x + 3}$

8. $g(x) = \dfrac{1}{x + 6} - 2$

Describing a Transformation of $f(x) = 2/x$ In Exercises 9–12, use a graphing utility to graph $f(x) = 2/x$ and the function g in the same viewing window. Describe the relationship between the two graphs.

9. $g(x) = f(x) + 1$

10. $g(x) = f(x - 1)$

11. $g(x) = -f(x)$

12. $g(x) = \frac{1}{2}f(x + 2)$

Describing a Transformation of $f(x) = 2/x^2$ In Exercises 13–16, use a graphing utility to graph $f(x) = 2/x^2$ and the function g in the same viewing window. Describe the relationship between the two graphs.

13. $g(x) = f(x) - 2$

14. $g(x) = -f(x)$

15. $g(x) = f(x - 2)$

16. $g(x) = \frac{1}{4}f(x)$

Sketching the Graph of a Rational Function In Exercises 17–32, sketch the graph of the rational function by hand. As sketching aids, check for intercepts, vertical asymptotes, horizontal asymptotes, and holes. Use a graphing utility to verify your graph.

✓ 17. $f(x) = \dfrac{1}{x + 2}$

18. $f(x) = \dfrac{1}{x - 6}$

19. $C(x) = \dfrac{5 + 2x}{1 + x}$

20. $P(x) = \dfrac{1 - 3x}{1 - x}$

✓ 21. $f(t) = \dfrac{1 - 2t}{t}$

22. $g(x) = \dfrac{1}{x + 2} + 2$

23. $f(x) = \dfrac{x^2}{x^2 - 4}$

24. $g(x) = \dfrac{x}{x^2 - 9}$

25. $g(x) = \dfrac{4(x + 1)}{x(x - 4)}$

26. $h(x) = \dfrac{2}{x^2(x - 3)}$

✓ 27. $f(x) = \dfrac{3x}{x^2 - x - 2}$

28. $f(x) = \dfrac{2x}{x^2 + x - 2}$

✓ 29. $f(x) = \dfrac{x^2 + 3x}{x^2 + x - 6}$

30. $g(x) = \dfrac{5(x + 4)}{x^2 + x - 12}$

31. $f(x) = \dfrac{x^2 - 1}{x + 1}$

32. $f(x) = \dfrac{x^2 - 16}{x - 4}$

Finding the Domain and Asymptotes In Exercises 33–42, use a graphing utility to graph the function. Determine its domain and identify any vertical or horizontal asymptotes.

33. $f(x) = \dfrac{2 + x}{1 - x}$

34. $f(x) = \dfrac{3 - x}{2 - x}$

35. $f(t) = \dfrac{3t + 1}{t}$

36. $h(x) = \dfrac{x - 2}{x - 3}$

37. $h(t) = \dfrac{4}{t^2 + 1}$

38. $g(x) = -\dfrac{x}{(x - 2)^2}$

39. $f(x) = \dfrac{x + 1}{x^2 - x - 6}$

40. $f(x) = \dfrac{x + 4}{x^2 + x - 6}$

41. $f(x) = \dfrac{20x}{x^2 + 1} - \dfrac{1}{x}$

42. $f(x) = 5\left(\dfrac{1}{x - 4} - \dfrac{1}{x + 2}\right)$

Exploration In Exercises 43–48, use a graphing utility to graph the function. What do you observe about its asymptotes?

43. $h(x) = \dfrac{6x}{\sqrt{x^2 + 1}}$

44. $f(x) = -\dfrac{x}{\sqrt{9 + x^2}}$

45. $g(x) = \dfrac{4|x - 2|}{x + 1}$

46. $f(x) = -\dfrac{8|3 + x|}{x - 2}$

47. $f(x) = \dfrac{4(x - 1)^2}{x^2 - 4x + 5}$

48. $g(x) = \dfrac{3x^4 - 5x + 3}{x^4 + 1}$

A Rational Function with a Slant Asymptote In Exercises 49–56, sketch the graph of the rational function by hand. As sketching aids, check for intercepts, vertical asymptotes, and slant asymptotes.

49. $f(x) = \dfrac{2x^2 + 1}{x}$

50. $g(x) = \dfrac{1 - x^2}{x}$

✓ **51.** $h(x) = \dfrac{x^2}{x - 1}$

52. $f(x) = \dfrac{x^3}{x^2 - 1}$

53. $g(x) = \dfrac{x^3}{2x^2 - 8}$

54. $f(x) = \dfrac{x^3}{x^2 + 4}$

55. $f(x) = \dfrac{x^3 + 2x^2 + 4}{2x^2 + 1}$

56. $f(x) = \dfrac{2x^2 - 5x + 5}{x - 2}$

Finding the x-Intercepts In Exercises 57–60, use the graph to estimate any x-intercepts of the rational function. Set $y = 0$ and solve the resulting equation to confirm your result.

57. $y = \dfrac{x + 1}{x - 3}$

58. $y = \dfrac{2x}{x - 3}$

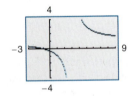

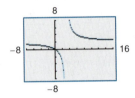

59. $y = \dfrac{1}{x} - x$

60. $y = x - 3 + \dfrac{2}{x}$

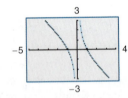

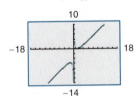

Finding the Domain and Asymptotes In Exercises 61–64, use a graphing utility to graph the rational function. Determine the domain of the function and identify any asymptotes.

61. $y = \dfrac{2x^2 + x}{x + 1}$

62. $y = \dfrac{x^2 + 5x + 8}{x + 3}$

63. $y = \dfrac{1 + 3x^2 - x^3}{x^2}$

64. $y = \dfrac{12 - 2x - x^2}{2(4 + x)}$

Finding Asymptotes and Holes In Exercises 65–70, find all vertical asymptotes, horizontal asymptotes, slant asymptotes, and holes in the graph of the function. Then use a graphing utility to verify your result.

65. $f(x) = \dfrac{x^2 - 5x + 4}{x^2 - 4}$

66. $f(x) = \dfrac{x^2 - 2x - 8}{x^2 - 9}$

67. $f(x) = \dfrac{2x^2 - 5x + 2}{2x^2 - x - 6}$

68. $f(x) = \dfrac{3x^2 - 8x + 4}{2x^2 - 3x - 2}$

69. $f(x) = \dfrac{2x^3 - x^2 - 2x + 1}{x^2 + 3x + 2}$

70. $f(x) = \dfrac{2x^3 + x^2 - 8x - 4}{x^2 - 3x + 2}$

Finding x-Intercepts Graphically In Exercises 71–82, use a graphing utility to graph the function and determine any x-intercepts. Set $y = 0$ and solve the resulting equation to confirm your result.

71. $y = \dfrac{1}{x + 5} + \dfrac{4}{x}$

72. $y = \dfrac{2}{x + 1} - \dfrac{3}{x}$

73. $y = \dfrac{1}{x + 2} + \dfrac{2}{x + 4}$

74. $y = \dfrac{2}{x + 2} - \dfrac{3}{x - 1}$

75. $y = x - \dfrac{6}{x - 1}$

76. $y = x - \dfrac{9}{x}$

77. $y = x + 2 - \dfrac{1}{x + 1}$

78. $y = 2x - 1 + \dfrac{1}{x - 2}$

79. $y = x + 1 + \dfrac{2}{x - 1}$

80. $y = x + 2 + \dfrac{2}{x + 2}$

81. $y = x + 3 - \dfrac{2}{2x - 1}$

82. $y = x - 1 - \dfrac{2}{2x - 3}$

83. Chemistry A 1000-liter tank contains 50 liters of a 25% brine solution. You add x liters of a 75% brine solution to the tank.

(a) Show that the concentration C (the ratio of brine to the total solution) of the final mixture is given by

$$C = \dfrac{3x + 50}{4(x + 50)}.$$

(b) Determine the domain of the function based on the physical constraints of the problem.

(c) Use a graphing utility to graph the function. As the tank is filled, what happens to the rate at which the concentration of brine increases? What percent does the concentration of brine appear to approach?

84. Geometry A rectangular region of length x and width y has an area of 500 square meters.

(a) Write the width y as a function of x.

(b) Determine the domain of the function based on the physical constraints of the problem.

(c) Sketch a graph of the function and determine the width of the rectangle when $x = 30$ meters.

85. Publishing A page that is x inches wide and y inches high contains 30 square inches of print. The margins at the top and bottom are 2 inches deep and the margins on each side are 1 inch wide (see figure).

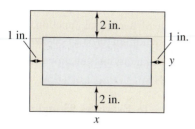

(a) Show that the total area A of the page is given by
$$A = \frac{2x(2x + 11)}{x - 2}.$$

(b) Determine the domain of the function based on the physical constraints of the problem.

(c) Use a graphing utility to graph the area function and approximate the page size such that the minimum amount of paper will be used. Verify your answer numerically using the *table* feature of the graphing utility.

86. Geometry A right triangle is formed in the first quadrant by the x-axis, the y-axis, and a line segment through the point $(3, 2)$ (see figure).

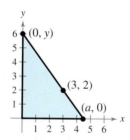

(a) Show that an equation of the line segment is given by
$$y = \frac{2(a - x)}{a - 3}, \quad 0 \le x \le a.$$

(b) Show that the area of the triangle is given by
$$A = \frac{a^2}{a - 3}.$$

(c) Use a graphing utility to graph the area function and estimate the value of a that yields a minimum area. Estimate the minimum area. Verify your answer numerically using the *table* feature of the graphing utility.

87. Cost Management The ordering and transportation cost C (in thousands of dollars) for the components used in manufacturing a product is given by
$$C = 100\left(\frac{200}{x^2} + \frac{x}{x + 30}\right), \quad x \ge 1$$

where x is the order size (in hundreds). Use a graphing utility to graph the cost function. From the graph, estimate the order size that minimizes cost.

88. Cost Management The cost C of producing x units of a product is given by $C = 0.2x^2 + 10x + 5$, and the average cost per unit is given by
$$\bar{C} = \frac{C}{x} = \frac{0.2x^2 + 10x + 5}{x}, \quad x > 0.$$

Sketch the graph of the average cost function, and estimate the number of units that should be produced to minimize the average cost per unit.

89. *Why you should learn it* (p. 297) The concentration C of a chemical in the bloodstream t hours after injection into muscle tissue is given by
$$C = \frac{3t^2 + t}{t^3 + 50}, \quad t \ge 0.$$

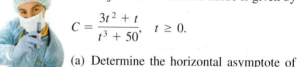

(a) Determine the horizontal asymptote of the function and interpret its meaning in the context of the problem.

(b) Use a graphing utility to graph the function and approximate the time when the bloodstream concentration is greatest.

(c) Use the graphing utility to determine when the concentration is less than 0.345.

90. Algebraic-Graphical-Numerical A driver averaged 50 miles per hour on the round trip between Baltimore, Maryland and Philadelphia, Pennsylvania, 100 miles away. The average speeds for going and returning were x and y miles per hour, respectively.

(a) Show that $y = \frac{25x}{x - 25}$.

(b) Determine the vertical and horizontal asymptotes of the function.

(c) Use a graphing utility to complete the table. What do you observe?

x	30	35	40	45	50	55	60
y							

(d) Use the graphing utility to graph the function.

(e) Is it possible to average 20 miles per hour in one direction and still average 50 miles per hour on the round trip? Explain.

91. MODELING DATA

Data are recorded at 124 monitoring sites throughout the United States to study national trends in air quality. The table shows the mean amount A of carbon monoxide (in parts per million) recorded at these sites in each year from 1999 through 2008. (Source: Environmental Protection Agency)

Year	Amount, A (in parts per million)
1999	3.9
2000	3.5
2001	3.3
2002	2.9
2003	2.7
2004	2.5
2005	2.3
2006	2.2
2007	2.0
2008	1.9

(a) Use the *regression* feature of a graphing utility to find a linear model for the data. Let $t = 9$ represent 1999. Use the graphing utility to plot the data and graph the model in the same viewing window.

(b) Find a rational model for the data. Take the reciprocal of A to generate the points $(t, 1/A)$. Use the *regression* feature of the graphing utility to find a linear model for these data. The resulting line has the form $1/A = at + b$. Solve for A. Use the graphing utility to plot the data and graph the rational model in the same viewing window.

(c) Use the *table* feature of the graphing utility to create a table showing the mean amounts of carbon monoxide generated by each model for the years in the original table. Which model do you prefer? Why?

92. **Biology** A herd of elk is released onto state game lands. The expected population P of the herd can be modeled by the equation $P = (10 + 2.7t)/(1 + 0.1t)$, where t is the time in years since the initial number of elk were released.

(a) State the domain of the model. Explain your answer.

(b) Find the initial number of elk in the herd.

(c) Find the populations of elk after 25, 50, and 100 years.

(d) Is there a limit to the size of the herd? If so, what is the expected population?

Use a graphing utility to confirm your results for parts (a) through (d).

Alex Staroseltsev 2010/used under license from Shutterstock.com

Conclusions

True or False? In Exercises 93 and 94, determine whether the statement is true or false. Justify your answer.

93. The graph of a rational function is continuous only when the denominator is a constant polynomial.

94. The graph of a rational function can never cross one of its asymptotes.

Think About It In Exercises 95 and 96, use a graphing utility to graph the function. Explain why there is no vertical asymptote when a superficial examination of the function might indicate that there should be one.

95. $h(x) = \dfrac{6 - 2x}{3 - x}$

96. $g(x) = \dfrac{x^2 + x - 2}{x - 1}$

97. **Writing** Write a set of guidelines for finding all the asymptotes of a rational function given that the degree of the numerator is not more than 1 greater than the degree of the denominator.

98. **CAPSTONE** Write a rational function that has the specified characteristics. (There are many correct answers.)

(a) Vertical asymptote: $x = -2$

Slant asymptote: $y = x + 1$

Zero of the function: $x = 2$

(b) Vertical asymptote: $x = -4$

Slant asymptote: $y = x - 2$

Zero of the function: $x = 3$

Cumulative Mixed Review

Simplifying Exponential Expressions In Exercises 99–102, simplify the expression.

99. $\left(\dfrac{x}{8}\right)^{-3}$

100. $(4x^2)^{-2}$

101. $\dfrac{3^{7/6}}{3^{1/6}}$

102. $\dfrac{(x^{-2})(x^{1/2})}{(x^{-1})(x^{5/2})}$

Finding the Domain and Range of a Function In Exercises 103–106, use a graphing utility to graph the function and find its domain and range.

103. $f(x) = \sqrt{6 + x^2}$

104. $f(x) = \sqrt{121 - x^2}$

105. $f(x) = -|x + 9|$

106. $f(x) = -x^2 + 9$

107. **Make a Decision** To work an extended application analyzing the median sales prices of existing one-family homes, visit this textbook's *Companion Website*. (Data Source: National Association of Realtors)

3.7   Quadratic Models

Classifying Scatter Plots

In real life, many relationships between two variables are parabolic, as in Section 3.1, Example 5. A scatter plot can be used to give you an idea of which type of model will best fit a set of data.

Example 1   Classifying Scatter Plots

Decide whether each set of data could be better modeled by a linear model,

$$y = ax + b$$

a quadratic model,

$$y = ax^2 + bx + c$$

or neither.

a. (0.9, 1.7), (1.2, 2.0), (1.3, 1.9), (1.4, 2.1), (1.6, 2.5), (1.8, 2.8), (2.1, 3.0), (2.5, 3.4), (2.9, 3.7), (3.2, 3.9), (3.3, 4.1), (3.6, 4.4), (4.0, 4.7), (4.2, 4.8), (4.3, 5.0)

b. (0.9, 3.2), (1.2, 4.0), (1.3, 4.1), (1.4, 4.4), (1.6, 5.1), (1.8, 6.0), (2.1, 7.6), (2.5, 9.8), (2.9, 12.4), (3.2, 14.3), (3.3, 15.2), (3.6, 18.1), (4.0, 22.7), (4.2, 24.9), (4.3, 27.2)

c. (0.9, 1.2), (1.2, 6.5), (1.3, 9.3), (1.4, 11.6), (1.6, 15.2), (1.8, 16.9), (2.1, 14.7), (2.5, 8.1), (2.9, 3.7), (3.2, 5.8), (3.3, 7.1), (3.6, 11.5), (4.0, 20.2), (4.2, 23.7), (4.3, 26.9)

Solution

a. Begin by entering the data into a graphing utility. Then display the scatter plot, as shown in Figure 3.53. From the scatter plot, it appears the data follow a linear pattern. So, the data can be better modeled by a linear function.

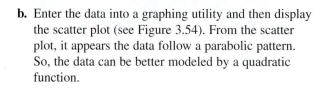

Figure 3.53

b. Enter the data into a graphing utility and then display the scatter plot (see Figure 3.54). From the scatter plot, it appears the data follow a parabolic pattern. So, the data can be better modeled by a quadratic function.

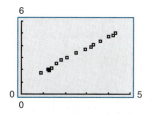

Figure 3.54

c. Enter the data into a graphing utility and then display the scatter plot (see Figure 3.55). From the scatter plot, it appears the data do not follow either a linear or a parabolic pattern. So, the data cannot be modeled by either a linear function or a quadratic function.

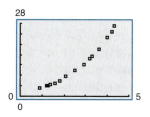

Figure 3.55

✓**CHECKPOINT**   Now try Exercise 5.

LeggNet/iStockphoto.com

What you should learn

- Classify scatter plots.
- Use scatter plots and a graphing utility to find quadratic models for data.
- Choose a model that best fits a set of data.

Why you should learn it

Many real-life situations can be modeled by quadratic equations. For instance, in Exercise 17 on page 311, a quadratic equation is used to model the monthly precipitation for San Francisco, California.

Fitting a Quadratic Model to Data

In Section 2.7, you created scatter plots of data and used a graphing utility to find the least squares regression lines for the data. You can use a similar procedure to find a model for nonlinear data. Once you have used a scatter plot to determine the type of model that would best fit a set of data, there are several ways that you can actually find the model. Each method is best used with a computer or calculator, rather than with hand calculations.

Example 2   Fitting a Quadratic Model to Data

A study was done to compare the speed x (in miles per hour) with the mileage y (in miles per gallon) of an automobile. The results are shown in the table.

a. Use a graphing utility to create a scatter plot of the data.

b. Use the *regression* feature of the graphing utility to find a model that best fits the data.

c. Approximate the speed at which the mileage is the greatest.

Speed, x	Mileage, y
15	22.3
20	25.5
25	27.5
30	29.0
35	28.7
40	29.9
45	30.4
50	30.2
55	30.0
60	28.8
65	27.4
70	25.3
75	23.3

Solution

a. Begin by entering the data into a graphing utility and displaying the scatter plot, as shown in Figure 3.56. From the scatter plot, you can see that the data appear to follow a parabolic pattern.

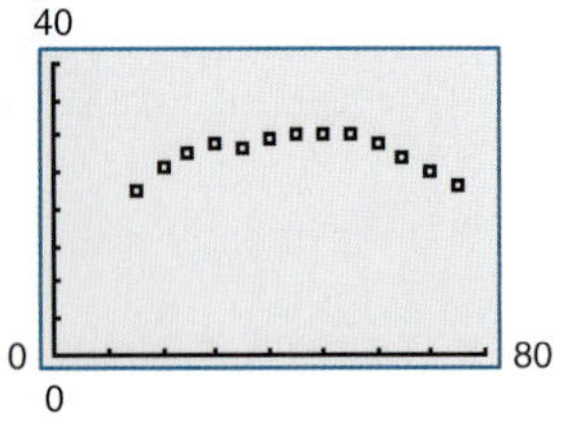

Figure 3.56

b. Using the *regression* feature of the graphing utility, you can find the quadratic model, as shown in Figure 3.57. So, the quadratic equation that best fits the data is given by

$$y = -0.0082x^2 + 0.75x + 13.5. \qquad \text{Quadratic model}$$

```
QuadReg
 y=ax²+bx+c
 a=-.0082137862
 b=.7468231768
 c=13.46223776
```

Figure 3.57

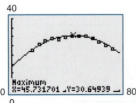

c. Graph the data and the model in the same viewing window, as shown in Figure 3.58. Use the *maximum* feature or the *zoom* and *trace* features of the graphing utility to approximate the speed at which the mileage is greatest. You should obtain a maximum of approximately $(46, 31)$, as shown in Figure 3.58. So, the speed at which the mileage is greatest is about 46 miles per hour.

Figure 3.58

CHECKPOINT   Now try Exercise 17.

Example 3   Fitting a Quadratic Model to Data

A basketball is dropped from a height of about 5.25 feet. The height of the basketball is recorded 23 times at intervals of about 0.02 second. The results are shown in the table. Use a graphing utility to find a model that best fits the data. Then use the model to predict the time when the basketball will hit the ground.

Time, x	Height, y
0.0	5.23594
0.02	5.20353
0.04	5.16031
0.06	5.09910
0.08	5.02707
0.099996	4.95146
0.119996	4.85062
0.139992	4.74979
0.159988	4.63096
0.179988	4.50132
0.199984	4.35728
0.219984	4.19523
0.23998	4.02958
0.25993	3.84593
0.27998	3.65507
0.299976	3.44981
0.319972	3.23375
0.339961	3.01048
0.359961	2.76921
0.379951	2.52074
0.399941	2.25786
0.419941	1.98058
0.439941	1.63488

Solution

Begin by entering the data into a graphing utility and displaying the scatter plot, as shown in Figure 3.59.

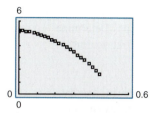

Figure 3.59

From the scatter plot, you can see that the data show a parabolic trend. So, using the *regression* feature of the graphing utility, you can find the quadratic model, as shown in Figure 3.60. The quadratic model that best fits the data is given by $y = -15.449x^2 - 1.30x + 5.2$.

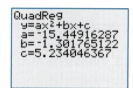

Figure 3.60

You can graph the data and the model in the same viewing window to see that the model fits the data well, as shown in Figure 3.61.

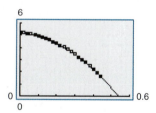

Figure 3.61

Using this model, you can predict the time when the basketball will hit the ground by substituting 0 for y and solving the resulting equation for x.

$y = -15.449x^2 - 1.30x + 5.2$ Write original model.

$0 = -15.449x^2 - 1.30x + 5.2$ Substitute 0 for y.

$x = \dfrac{-b \pm \sqrt{b^2 - 4ac}}{2a}$ Quadratic Formula

$ = \dfrac{-(-1.30) \pm \sqrt{(-1.30)^2 - 4(-15.449)(5.2)}}{2(-15.449)}$ Substitute for a, b, and c.

$ \approx 0.54$ Choose positive solution.

So, the solution is about 0.54 second. In other words, the basketball will continue to fall for about $0.54 - 0.44 = 0.1$ second more before hitting the ground.

 CHECKPOINT   Now try Exercise 19.

Choosing a Model

Sometimes it is not easy to distinguish from a scatter plot which type of model will best fit the data. You should first find several models for the data, using the *Library of Parent Functions*, and then choose the model that best fits the data by comparing the *y*-values of each model with the actual *y*-values.

Example 4   Choosing a Model

The table shows the amounts *y* (in gallons per person) of regular soft drinks consumed in the United States in the years 2000 through 2007. Use the *regression* feature of a graphing utility to find a linear model and a quadratic model for the data. Determine which model better fits the data.   (Source: United States Department of Agriculture)

Year	Amounts, y
2000	39.4
2001	39.0
2002	38.5
2003	37.5
2004	37.0
2005	36.3
2006	35.4
2007	33.9

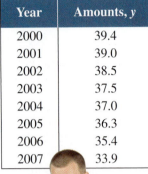

Solution

Let *x* represent the year, with $x = 0$ corresponding to 2000. Begin by entering the data into a graphing utility. Using the *regression* feature of the graphing utility, a linear model for the data is

$$y = -0.76x + 39.8 \qquad \text{Linear model}$$

and a quadratic model for the data is

$$y = -0.056x^2 - 0.37x + 39.4. \qquad \text{Quadratic model}$$

Plot the data and the linear model in the same viewing window, as shown in Figure 3.62. Then plot the data and the quadratic model in the same viewing window, as shown in Figure 3.63. To determine which model fits the data better, compare the *y*-values given by each model with the actual *y*-values. The model whose *y*-values are closest to the actual values is the better fit. In this case, the better-fitting model is the quadratic model.

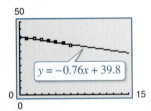

Figure 3.62

Figure 3.63

 CHECKPOINT   Now try Exercise 21.

Technology Tip

When you use the regression feature of a graphing utility, the program may output an "r^2-value." This r^2-value is the **coefficient of determination** of the data and gives a measure of how well the model fits the data. The coefficient of determination for the linear model in Example 4 is

$$r^2 \approx 0.97105$$

and the coefficient of determination for the quadratic model is

$$r^2 \approx 0.99226.$$

Because the coefficient of determination for the quadratic model is closer to 1, the quadratic model better fits the data.

3.7   Exercises

See www.CalcChat.com for worked-out solutions to odd-numbered exercises.
For instructions on how to use a graphing utility, see Appendix A.

Vocabulary and Concept Check

1. What type of model best represents data that follow a parabolic pattern?

2. Which coefficient of determination indicates a better model for a set of data, $r^2 = 0.0365$, or $r^2 = 0.9688$?

Procedures and Problem Solving

Classifying Scatter Plots   In Exercises 3–8, determine whether the scatter plot could best be modeled by a linear model, a quadratic model, or neither.

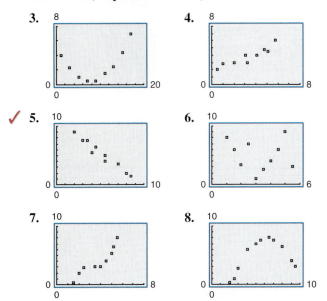

3.

4.

✓ **5.**

6.

7.

8.

Using a Scatter Plot   In Exercises 9–16, (a) use a graphing utility to create a scatter plot of the data, (b) determine whether the data could be better modeled by a linear model or a quadratic model, (c) use the *regression* feature of the graphing utility to find a model for the data, (d) use the graphing utility to graph the model with the scatter plot from part (a), and (e) create a table comparing the original data with the data given by the model.

9. (0, 2.1), (1, 2.4), (2, 2.5), (3, 2.8), (4, 2.9), (5, 3.0), (6, 3.0), (7, 3.2), (8, 3.4), (9, 3.5), (10, 3.6)

10. (−2, 11.0), (−1, 10.7), (0, 10.4), (1, 10.3), (2, 10.1), (3, 9.9), (4, 9.6), (5, 9.4), (6, 9.4), (7, 9.2), (8, 9.0)

11. (0, 3480), (5, 2235), (10, 1250), (15, 565), (20, 150), (25, 12), (30, 145), (35, 575), (40, 1275), (45, 2225), (50, 3500), (55, 5010)

12. (0, 6140), (2, 6815), (4, 7335), (6, 7710), (8, 7915), (10, 7590), (12, 7975), (14, 7700), (16, 7325), (18, 6820), (20, 6125), (22, 5325)

13. (1, 4.0), (2, 6.5), (3, 8.8), (4, 10.6), (5, 13.9), (6, 15.0), (7, 17.5), (8, 20.1), (9, 24.0), (10, 27.1)

14. (−6, 10.7), (−4, 9.0), (−2, 7.0), (0, 5.4), (2, 3.5), (4, 1.7), (6, −0.1), (8, −1.8), (10, −3.6), (12, −5.3)

15. (0, 587), (5, 551), (10, 512), (15, 478), (20, 436), (25, 430) (30, 424), (35, 420), (40, 423), (45, 429), (50, 444)

16. (2, 34.3), (3, 33.8), (4, 32.6), (5, 30.1), (6, 27.8), (7, 22.5), (8, 19.1), (9, 14.8), (10, 9.4). (11, 3.7), (12, −1.6)

✓ **17.** *Why you should learn it* (p. 307)   The table shows the monthly normal precipitation P (in inches) for San Francisco, California. (Source: U.S. National Oceanic and Atmospheric Administration)

Month	Precipitation, P
January	4.45
February	4.01
March	3.26
April	1.17
May	0.38
June	0.11
July	0.03
August	0.07
September	0.20
October	1.40
November	2.49
December	2.89

(a) Use a graphing utility to create a scatter plot of the data. Let t represent the month, with $t = 1$ corresponding to January.

(b) Use the *regression* feature of the graphing utility to find a quadratic model for the data.

(c) Use the graphing utility to graph the model with the scatter plot from part (a).

(d) Use the graph from part (c) to determine in which month the normal precipitation in San Francisco is the least.

18. MODELING DATA

The table shows the annual sales S (in billions of dollars) of department stores in the United States from 2003 through 2008. (Source: U.S. Census Bureau)

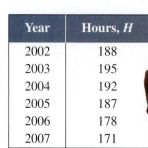

Year	Sales, S (in billions of dollars)
2003	221.0
2004	222.0
2005	220.7
2006	219.0
2007	215.9
2008	206.1

(a) Use a graphing utility to create a scatter plot of the data. Let t represent the year, with $t = 3$ corresponding to 2003.

(b) Use the *regression* feature of the graphing utility to find a quadratic model for the data.

(c) Use the graphing utility to graph the model with the scatter plot from part (a).

(d) Use the model to estimate the first year when the annual sales of department stores will be below $180 billion. Is this a good model for predicting future sales? Explain.

✓ 19. MODELING DATA

The table shows the percents P of the U.S. population who used the Internet from 2003 through 2008. (Source: U.S. Census Bureau)

Year	Percent, P
2003	55.58
2004	63.00
2005	66.33
2006	69.83
2007	71.83
2008	74.00

(a) Use a graphing utility to create a scatter plot of the data. Let t represent the year, with $t = 3$ corresponding to 2003.

(b) Use the *regression* feature of the graphing utility to find a quadratic model for the data.

(c) Use the graphing utility to graph the model with the scatter plot from part (a).

(d) According to the model, in what year will the percent of the U.S. population who use the Internet fall below 60%? Is this a good model for making future predictions? Explain.

20. MODELING DATA

The table shows the average numbers of hours H that adults in the United States spent reading newspapers each year from 2002 through 2007. (Source: Veronis Suhler Stevenson)

Year	Hours, H
2002	188
2003	195
2004	192
2005	187
2006	178
2007	171

(a) Use a graphing utility to create a scatter plot of the data. Let t represent the year, with $t = 2$ corresponding to 2002.

(b) A cubic model for the data is

$$H = 0.500t^3 - 8.43t^2 + 38.9t + 140$$

which has an r^2-value of 0.9965. Use the graphing utility to graph the model with the scatter plot from part (a). Is the cubic model a good fit for the data? Explain.

(c) Use the *regression* feature of the graphing utility to find a quadratic model for the data and identify the coefficient of determination.

(d) Use the graphing utility to graph the quadratic model with the scatter plot from part (a). Is the quadratic model a good fit for the data? Explain.

(e) Which model is a better fit for the data? Explain.

(f) The projected average numbers of hours H^* that adults spent reading newspapers each year from 2008 through 2010 are shown in the table. Use the models from parts (b) and (c) to *predict* the average numbers of hours for 2008 through 2010. Explain why your values may differ from those in the table.

Year	2008	2009	2010
H^*	164	159	155

✓ 21. MODELING DATA

The table shows the numbers of televisions T (in millions) in homes in the United States from 1997 through 2008. (Source: The Nielsen Company)

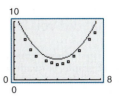

Year	Televisions, T (in millions)
1997	229
1998	235
1999	240
2000	245
2001	248
2002	254
2003	260
2004	268
2005	287
2006	301
2007	311
2008	310

(a) Use a graphing utility to create a scatter plot of the data. Let t represent the year, with $t = 7$ corresponding to 1997.

(b) Use the *regression* feature of the graphing utility to find a linear model for the data and identify the coefficient of determination.

(c) Use the graphing utility to graph the linear model with the scatter plot from part (a).

(d) Use the *regression* feature of the graphing utility to find a quadratic model for the data and identify the coefficient of determination.

(e) Use the graphing utility to graph the quadratic model with the scatter plot from part (a).

(f) Which model is a better fit for the data? Explain.

(g) Use each model to approximate the year when the number of televisions in homes will reach 350 million.

Conclusions

True or False? In Exercises 22–24, determine whether the statement is true or false. Justify your answer.

22. The graph of a quadratic model with a negative leading coefficient will have a maximum value at its vertex.

23. The graph of a quadratic model with a positive leading coefficient will have a minimum value at its vertex.

24. Data that are positively correlated are always better modeled by a linear equation than by a quadratic equation.

25. Writing Explain why the parabola shown in the figure is not a good fit for the data.

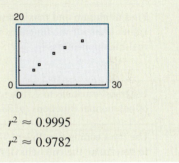

26. CAPSTONE The r^2-values representing the coefficients of determination for the least squares linear model and the least squares quadratic model for the data shown are given below. Which is which? Explain your reasoning.

$$r^2 \approx 0.9995$$
$$r^2 \approx 0.9782$$

Cumulative Mixed Review

Compositions of Functions In Exercises 27–30, find (a) $f \circ g$ and (b) $g \circ f$.

27. $f(x) = 2x - 1$, $g(x) = x^2 + 3$

28. $f(x) = 5x + 8$, $g(x) = 2x^2 - 1$

29. $f(x) = x^3 - 1$, $g(x) = \sqrt[3]{x + 1}$

30. $f(x) = \sqrt[3]{x + 5}$, $g(x) = x^3 - 5$

Testing for One-to-One Functions In Exercises 31–34, determine algebraically whether the function is one-to-one. If it is, find its inverse function. Verify your answer graphically.

31. $f(x) = 2x + 5$

32. $f(x) = \dfrac{x - 4}{5}$

33. $f(x) = x^2 + 5, x \geq 0$

34. $f(x) = 2x^2 - 3, x \geq 0$

Multiplying Complex Conjugates In Exercises 35–38, write the complex conjugate of the complex number. Then multiply the number by its complex conjugate.

35. $1 - 3i$

36. $-2 + 4i$

37. $-5i$

38. $8i$

3   Chapter Summary

	What did you learn?	Explanation and Examples	Review Exercises
3.1	Analyze graphs of quadratic functions (*p. 244*).	Let a, b, and c be real numbers with $a \neq 0$. The function $f(x) = ax^2 + bx + c$ is called a quadratic function. Its graph is a "U-shaped" curve called a parabola.	1–6
	Write quadratic functions in standard form and use the results to sketch graphs of functions (*p. 247*).	The quadratic function $f(x) = a(x - h)^2 + k$, $a \neq 0$, is in standard form. The graph of f is a parabola whose axis is the vertical line $x = h$ and whose vertex is the point (h, k). The parabola opens upward when $a > 0$ and opens downward when $a < 0$.	7–12
	Find minimum and maximum values of quadratic functions in real-life applications (*p. 249*).	Consider $f(x) = ax^2 + bx + c$ with vertex $\left(-\dfrac{b}{2a}, f\left(-\dfrac{b}{2a}\right)\right)$. If $a > 0$, then f has a *minimum* at $x = -b/(2a)$. If $a < 0$, then f has a *maximum* at $x = -b/(2a)$.	13, 14
3.2	Use transformations to sketch graphs of polynomial functions (*p. 254*).	The graph of a polynomial function is continuous (no breaks, holes, or gaps) and has only smooth, rounded turns.	15–18
	Use the Leading Coefficient Test to determine the end behavior of graphs of polynomial functions (*p. 256*).	Consider $f(x) = a_n x^n + \cdots + a_1 x + a_0$, $a_n \neq 0$. ***n* is odd:** If $a_n > 0$, then the graph falls to the left and rises to the right. If $a_n < 0$, then the graph rises to the left and falls to the right. ***n* is even:** If $a_n > 0$, then the graph rises to the left and right. If $a_n < 0$, then the graph falls to the left and right.	19–22
	Find and use zeros of polynomial functions as sketching aids (*p. 258*).	If f is a polynomial function and a is a real number, the following are equivalent: (1) $x = a$ is a *zero* of the function f, (2) $x = a$ is a *solution* of the polynomial equation $f(x) = 0$, (3) $(x - a)$ is a *factor* of the polynomial $f(x)$, and (4) $(a, 0)$ is an *x-intercept* of the graph of f.	23–34
	Use the Intermediate Value Theorem to help locate zeros of polynomial functions (*p. 261*).	Let a and b be real numbers such that $a < b$. If f is a polynomial function such that $f(a) \neq f(b)$, then, in $[a, b]$, f takes on every value between $f(a)$ and $f(b)$.	35–38
3.3	Use long division to divide polynomials by other polynomials (*p. 266*).	Dividend $\quad$ Quotient $\quad$ Remainder $$\frac{x^2 + 3x + 5}{x + 1} = x + 2 + \frac{3}{x + 1}$$ Divisor $\longrightarrow x + 1 \qquad\qquad x + 1 \longleftarrow$ Divisor	39–44
	Use synthetic division to divide polynomials by binomials of the form $(x - k)$ (*p. 269*).	Divisor: $x + 3$ $\quad$ Dividend: $x^4 - 10x^2 - 2x + 4$ $\begin{array}{r|rrrrr} -3 & 1 & 0 & -10 & -2 & 4 \\ & & -3 & 9 & 3 & -3 \\ \hline & 1 & -3 & -1 & 1 & \boxed{1} \end{array}$ $\leftarrow$ Remainder: 1 Quotient: $x^3 - 3x^2 - x + 1$	45–50
	Use the Remainder Theorem and the Factor Theorem (*p. 270*).	**The Remainder Theorem:** If a polynomial $f(x)$ is divided by $x - k$, then the remainder is $r = f(k)$. **The Factor Theorem:** A polynomial $f(x)$ has a factor $(x - k)$ if and only if $f(k) = 0$.	51–54
	Use the Rational Zero Test to determine possible rational zeros of polynomial functions (*p. 272*).	The Rational Zero Test relates the possible rational zeros of a polynomial to the leading coefficient and to the constant term of the polynomial.	55, 56

	What did you learn?	Explanation and Examples	Review Exercises
3.3	**Use Descartes's Rule of Signs** *(p. 274)* **and the Upper and Lower Bound Rules** *(p. 275)* **to find zeros of polynomials.**	**Descartes's Rule of Signs** Let $f(x) = a_n x^n + a_{n-1}x^{n-1} + \cdots + a_2 x^2 + a_1 x + a_0$ be a polynomial with real coefficients and $a_0 \neq 0$. **1.** The number of *positive real zeros* of f is either equal to the number of variations in sign of $f(x)$ or less than that number by an even integer. **2.** The number of *negative real zeros* of f is either equal to the number of variations in sign of $f(-x)$ or less than that number by an even integer.	57–64
3.4	**Use the Fundamental Theorem of Algebra to determine the number of zeros of a polynomial function** *(p. 281)*.	**The Fundamental Theorem of Algebra** If $f(x)$ is a polynomial of degree n, where $n > 0$, then f has at least one zero in the complex number system. **Linear Factorization Theorem** If $f(x)$ is a polynomial of degree n, where $n > 0$, then f has precisely n linear factors $$f(x) = a_n(x - c_1)(x - c_2) \cdots (x - c_n)$$ where $c_1, c_2, \ldots, c_n$ are complex numbers.	65–68
	Find all zeros of polynomial functions, including complex zeros *(p. 282)*, **and find conjugate pairs of complex zeros** *(p. 283)*.	**Complex Zeros Occur in Conjugate Pairs** Let $f(x)$ be a polynomial function that has real coefficients. If $a + bi$ $(b \neq 0)$ is a zero of the function, then the conjugate $a - bi$ is also a zero of the function.	69–86
	Find zeros of polynomials by factoring *(p. 284)*.	Every polynomial of degree $n > 0$ with real coefficients can be written as the product of linear and quadratic factors with real coefficients, where the quadratic factors have no real zeros.	87–90
3.5	**Find the domains** *(p. 288)* **and vertical and horizontal asymptotes** *(p. 289)* **of rational functions.**	The domain of a rational function of x includes all real numbers except x-values that make the denominator zero.	91–102
	Use rational functions to model and solve real-life problems *(p. 292)*.	A rational function can be used to model the cost of removing a given percent of smokestack pollutants at a utility company that burns coal. (See Example 5.)	103, 104
3.6	**Analyze and sketch graphs of rational functions** *(p. 297)*, **including functions with slant asymptotes** *(p. 301)*.	Consider a rational function whose denominator is of degree 1 or greater. If the degree of the numerator is exactly *one more* than the degree of the denominator, then the graph of the function has a slant asymptote.	105–118
	Use rational functions to model and solve real-life problems *(p. 302)*.	A rational function can be used to model the area of a page. The model can be used to determine the dimensions of the page that use the minimum amount of paper. (See Example 7.)	119, 120
3.7	**Classify scatter plots** *(p. 307)*, **find quadratic models for data** *(p. 308)*, **and choose a model that best fits a set of data** *(p. 310)*.	Sometimes it is not easy to distinguish from a scatter plot which type of model will best fit the data. You should first find several models for the data and then choose the model that best fits the data by comparing the y-values of each model with the actual y-values.	121–123

3   Review Exercises

See www.CalcChat.com for worked-out solutions to odd-numbered exercises.
For instructions on how to use a graphing utility, see Appendix A.

3.1

Library of Parent Functions   In Exercises 1–6, sketch the graph of each function and describe how the graph is related to the graph of $y = x^2$.

1. $y = x^2 - 2$
2. $y = x^2 + 4$
3. $y = (x - 2)^2$
4. $y = -(x + 4)^2$
5. $y = (x + 5)^2 - 2$
6. $y = -(x - 4)^2 + 1$

Identifying the Vertex of a Quadratic Function   In Exercises 7–10, describe the graph of the function and identify the vertex. Then, sketch the graph of the function. Identify any x-intercepts.

7. $f(x) = \left(x + \frac{3}{2}\right)^2 + 1$
8. $f(x) = (x - 4)^2 - 4$
9. $f(x) = \frac{1}{3}(x^2 + 5x - 4)$
10. $f(x) = 3x^2 - 12x + 11$

Writing the Equation of a Parabola in Standard Form   In Exercises 11 and 12, write the standard form of the quadratic function that has the indicated vertex and whose graph passes through the given point. Use a graphing utility to verify your result.

11. Vertex: $(1, -4)$;   Point: $(2, -3)$
12. Vertex: $(2, 3)$;   Point: $(0, 2)$

13. **Geometry**   A rectangle is inscribed in the region bounded by the x-axis, the y-axis, and the graph of $x + 2y - 8 = 0$, as shown in the figure.

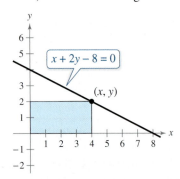

(a) Write the area A of the rectangle as a function of x. Determine the domain of the function in the context of the problem.

(b) Use a graphing utility to graph the area function. Use the graph to approximate the dimensions that will produce a maximum area.

(c) Write the area function in standard form to find algebraically the dimensions that will produce a maximum area. Compare your results with your answer from part (b).

14. **Physical Education**   A college has 1500 feet of portable rink boards to form three adjacent outdoor ice rinks, as shown in the figure. Determine the dimensions that will produce the maximum total area of ice surface.

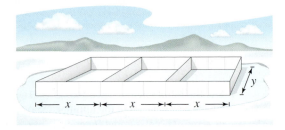

3.2

Library of Parent Functions   In Exercises 15–18, sketch the graph of $y = x^3$ and the graph of the function f. Describe the transformation from y to f.

15. $f(x) = (x + 4)^3$
16. $f(x) = x^3 - 4$
17. $f(x) = -x^3 + 2$
18. $f(x) = (x + 3)^3 - 2$

Comparing End Behavior   In Exercises 19 and 20, use a graphing utility to graph the functions f and g in the same viewing window. Zoom out far enough to see the right-hand and left-hand behavior of each graph. Do the graphs of f and g have the same right-hand and left-hand behavior? Explain why or why not.

19. $f(x) = \frac{1}{2}x^3 - 2x + 1$, $g(x) = \frac{1}{2}x^3$
20. $f(x) = -x^4 + 2x^3$, $g(x) = -x^4$

Applying the Leading Coefficient Test   In Exercises 21 and 22, use the Leading Coefficient Test to describe the right-hand and left-hand behavior of the graph of the polynomial function.

21. $f(x) = -x^2 + 6x + 9$
22. $f(x) = \frac{1}{2}x^3 + 2x$

Finding Zeros of a Polynomial Function   In Exercises 23–28, (a) find the zeros algebraically, (b) use a graphing utility to graph the function, and (c) use the graph to approximate any zeros and compare them with those in part (a).

23. $g(x) = x^4 - x^3 - 2x^2$
24. $h(x) = -2x^3 - x^2 + x$
25. $f(t) = t^3 - 3t$
26. $f(x) = -(x + 6)^3 - 8$
27. $f(x) = x(x + 3)^2$
28. $f(t) = t^4 - 4t^2$

Finding a Polynomial Function with Given Zeros   In Exercises 29–32, find a polynomial function that has the given zeros. (There are many correct answers.)

29. $-2, 1, 1, 5$
30. $-3, 0, 1, 4$
31. $3, 2 - \sqrt{3}, 2 + \sqrt{3}$
32. $-7, 4 - \sqrt{6}, 4 + \sqrt{6}$

Sketching the Graph of a Polynomial Function In Exercises 33 and 34, sketch the graph of the function by (a) applying the Leading Coefficient Test, (b) finding the zeros of the polynomial, (c) plotting sufficient solution points, and (d) drawing a continuous curve through the points.

33. $f(x) = x^4 - 2x^3 - 12x^2 + 18x + 27$

34. $f(x) = 18 + 27x - 2x^2 - 3x^3$

Approximating the Zeros of a Function In Exercises 35–38, (a) use the Intermediate Value Theorem and a graphing utility to find graphically any intervals of length 1 in which the polynomial function is guaranteed to have a zero and (b) use the *zero* or *root* feature of the graphing utility to approximate the real zeros of the function. Verify your results in part (a) by using the *table* feature of the graphing utility.

35. $f(x) = x^3 + 2x^2 - x - 1$

36. $f(x) = 0.24x^3 - 2.6x - 1.4$

37. $f(x) = x^4 - 6x^2 - 4$

38. $f(x) = 2x^4 + \frac{7}{2}x^3 - 2$

 3.3

Long Division of Polynomials In Exercises 39–44, use long division to divide.

39. $\dfrac{24x^2 - x - 8}{3x - 2}$

40. $\dfrac{4x^2 + 7}{3x - 2}$

41. $\dfrac{x^4 - 3x^2 + 2}{x^2 - 1}$

42. $\dfrac{3x^4 + x^2 - 1}{x^2 - 1}$

43. $(5x^3 - 13x^2 - x + 2) \div (x^2 - 3x + 1)$

44. $\dfrac{6x^4 + 10x^3 + 13x^2 - 5x + 2}{2x^2 - 1}$

Using Synthetic Division In Exercises 45–50, use synthetic division to divide.

45. $(0.25x^4 - 4x^3) \div (x + 2)$

46. $(0.1x^3 + 0.3x^2 - 0.5) \div (x - 5)$

47. $(6x^4 - 4x^3 - 27x^2 + 18x) \div \left(x - \frac{2}{3}\right)$

48. $(2x^3 + 2x^2 - x + 2) \div \left(x - \frac{1}{2}\right)$

49. $(3x^3 - 10x^2 + 12x - 22) \div (x - 4)$

50. $(2x^3 + 6x^2 - 14x + 9) \div (x - 1)$

Using the Remainder Theorem In Exercises 51 and 52, use the Remainder Theorem and synthetic division to evaluate the function at each given value. Use a graphing utility to verify your results.

51. $f(x) = x^4 + 10x^3 - 24x^2 + 20x + 44$

    (a) $f(-3)$ (b) $f(-2)$

52. $g(t) = 2t^5 - 5t^4 - 8t + 20$

    (a) $g(-4)$ (b) $g(\sqrt{2})$

Factoring a Polynomial In Exercises 53 and 54, (a) verify the given factor(s) of the function f, (b) find the remaining factors of f, (c) use your results to write the complete factorization of f, and (d) list all real zeros of f. Confirm your results by using a graphing utility to graph the function.

	Function	Factor(s)
53.	$f(x) = x^3 + 4x^2 - 25x - 28$	$(x - 4)$
54.	$f(x) = x^4 - 4x^3 - 7x^2 + 22x + 24$	$(x + 2), (x - 3)$

Using the Rational Zero Test In Exercises 55 and 56, use the Rational Zero Test to list all possible rational zeros of f. Use a graphing utility to verify that all the zeros of f are contained in the list.

55. $f(x) = 4x^3 - 11x^2 + 10x - 3$

56. $f(x) = 10x^3 + 21x^2 - x - 6$

Using Descartes's Rule of Signs In Exercises 57 and 58, use Descartes's Rule of Signs to determine the possible numbers of positive and negative real zeros of the function.

57. $g(x) = 5x^3 - 6x + 9$ 58. $f(x) = 2x^5 - 3x^2 + 2x - 1$

Finding the Zeros of a Polynomial Function In Exercises 59 and 60, use synthetic division to verify the upper and lower bounds of the real zeros of f. Then find the real zeros of the function.

59. $f(x) = 4x^3 - 3x^2 + 4x - 3$

    Upper bound: $x = 1$; Lower bound: $x = -\frac{1}{4}$

60. $f(x) = 2x^3 - 5x^2 - 14x + 8$

    Upper bound: $x = 8$; Lower bound: $x = -4$

Finding the Zeros of a Polynomial Function In Exercises 61–64, find all the real zeros of the polynomial function.

61. $f(x) = 6x^3 + 31x^2 - 18x - 10$

62. $f(x) = x^3 - 1.3x^2 - 1.7x + 0.6$

63. $f(x) = 6x^4 - 25x^3 + 14x^2 + 27x - 18$

64. $f(x) = 5x^4 + 126x^2 + 25$

3.4

Zeros of a Polynomial Function In Exercises 65–68, confirm that the function has the indicated zero(s).

65. $f(x) = x^2 + 6x + 9$;   Repeated zero: -3

66. $f(x) = x^2 - 10x + 25$;   Repeated zero: 5

67. $f(x) = x^3 + 16x$; $0, -4i, 4i$

68. $f(x) = x^3 + 144x$; $0, -12i, 12i$

Using the Factored Form of a Function In Exercises 69 and 70, find all the zeros of the function.

69. $f(x) = 3x(x - 2)^2$ 70. $f(x) = (x - 4)(x + 9)^2$

Finding the Zeros of a Polynomial Function In Exercises 71–76, find all the zeros of the function and write the polynomial as a product of linear factors. Verify your results by using a graphing utility to graph the function.

71. $h(x) = x^3 - 7x^2 + 18x - 24$

72. $f(x) = 2x^3 - 5x^2 - 9x + 40$

73. $f(x) = 2x^4 - 5x^3 + 10x - 12$

74. $g(x) = 3x^4 - 4x^3 + 7x^2 + 10x - 4$

75. $f(x) = x^5 + x^4 + 5x^3 + 5x^2$

76. $f(x) = x^5 - 5x^3 + 4x$

Using the Zeros to Find the x-Intercepts In Exercises 77–82, (a) find all the zeros of the function, (b) write the polynomial as a product of linear factors, and (c) use your factorization to determine the x-intercepts of the graph of the function. Use a graphing utility to verify that the real zeros are the only x-intercepts.

77. $f(x) = x^3 - 4x^2 + 6x - 4$

78. $f(x) = x^3 - 5x^2 - 7x + 51$

79. $f(x) = -3x^3 - 19x^2 - 4x + 12$

80. $f(x) = 2x^3 - 9x^2 + 22x - 30$

81. $f(x) = x^4 + 34x^2 + 225$

82. $f(x) = x^4 + 10x^3 + 26x^2 + 10x + 25$

Finding a Polynomial with Given Zeros In Exercises 83–86, find a polynomial function with real coefficients that has the given zeros. (There are many correct answers.)

83. $4, -2, 5i$

84. $2, -2, 2i$

85. $1, -4, -3 + 5i$

86. $-4, -4, 1 + \sqrt{3}i$

Factoring a Polynomial In Exercises 87 and 88, write the polynomial (a) as the product of factors that are irreducible over the *rationals*, (b) as the product of linear and quadratic factors that are irreducible over the *reals*, and (c) in completely factored form.

87. $f(x) = x^4 - 2x^3 + 8x^2 - 18x - 9$

   (*Hint:* One factor is $x^2 + 9$.)

88. $f(x) = x^4 - 4x^3 + 3x^2 + 8x - 16$

   (*Hint:* One factor is $x^2 - x - 4$.)

Finding the Zeros of a Polynomial Function In Exercises 89 and 90, use the given zero to find all the zeros of the function.

Function	Zero
89. $f(x) = x^3 + 3x^2 + 4x + 12$	$-2i$
90. $f(x) = 2x^3 - 7x^2 + 14x + 9$	$2 + \sqrt{5}i$

3.5

Finding a Function's Domain and Asymptotes In Exercises 91–102, (a) find the domain of the function, (b) decide whether the function is continuous, and (c) identify any horizontal and vertical asymptotes.

91. $f(x) = \dfrac{2 - x}{x + 3}$

92. $f(x) = \dfrac{4x}{x - 8}$

93. $f(x) = \dfrac{2}{x^2 - 3x - 18}$

94. $f(x) = \dfrac{2x^2 + 3}{x^2 + x + 3}$

95. $f(x) = \dfrac{7 + x}{7 - x}$

96. $f(x) = \dfrac{6x}{x^2 - 1}$

97. $f(x) = \dfrac{4x^2}{2x^2 - 3}$

98. $f(x) = \dfrac{3x^2 - 11x - 4}{x^2 + 2}$

99. $f(x) = \dfrac{2x - 10}{x^2 - 2x - 15}$

100. $f(x) = \dfrac{x^3 - 4x^2}{x^2 + 3x + 2}$

101. $f(x) = \dfrac{x - 2}{|x| + 2}$

102. $f(x) = \dfrac{2x}{|2x - 1|}$

103. Criminology The cost C (in millions of dollars) for the U.S. government to seize $p\%$ of an illegal drug as it enters the country is given by

$$C = \frac{528p}{100 - p}, \quad 0 \le p < 100.$$

(a) Find the costs of seizing 25%, 50%, and 75% of the illegal drug.

(b) Use a graphing utility to graph the function. Be sure to choose an appropriate viewing window. Explain why you chose the values you used in your viewing window.

(c) According to this model, would it be possible to seize 100% of the drug? Explain.

104. Biology A biology class performs an experiment comparing the quantity of food consumed by a certain kind of moth with the quantity supplied. The model for the experimental data is given by

$$y = \frac{1.568x - 0.001}{6.360x + 1}, \quad x > 0$$

where x is the quantity (in milligrams) of food supplied and y is the quantity (in milligrams) eaten (see figure). At what level of consumption will the moth become satiated?

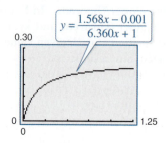

3.6

Finding Asymptotes and Holes In Exercises 105–108, find all of the vertical, horizontal, and slant asymptotes, and any holes in the graph of the function. Then use a graphing utility to verify your result.

105. $f(x) = \dfrac{x^2 - 5x + 4}{x^2 - 1}$ **106.** $f(x) = \dfrac{2x^2 - 7x + 3}{2x^2 - 3x - 9}$

107. $f(x) = \dfrac{3x^2 + 5x - 2}{x + 1}$ **108.** $f(x) = \dfrac{2x^2 + 5x + 3}{x - 2}$

Sketching the Graph of a Rational Function In Exercises 109–118, sketch the graph of the rational function by hand. As sketching aids, check for intercepts, vertical asymptotes, horizontal asymptotes, slant asymptotes, and holes.

109. $f(x) = \dfrac{2x - 1}{x - 5}$ **110.** $f(x) = \dfrac{x - 3}{x - 2}$

111. $f(x) = \dfrac{2x^2}{x^2 - 4}$ **112.** $f(x) = \dfrac{5x}{x^2 + 1}$

113. $f(x) = \dfrac{2}{(x + 1)^2}$ **114.** $f(x) = \dfrac{4}{(x - 1)^2}$

115. $f(x) = \dfrac{2x^3}{x^2 + 1}$ **116.** $f(x) = \dfrac{x^3}{3x^2 - 6}$

117. $f(x) = \dfrac{x^2 - x + 1}{x - 3}$ **118.** $f(x) = \dfrac{2x^2 + 7x + 3}{x + 1}$

119. Biology A Parks and Wildlife Commission releases 80,000 fish into a lake. After t years, the population N of the fish (in thousands) is given by

$$N = \dfrac{20(4 + 3t)}{1 + 0.05t}, \quad t \geq 0.$$

(a) Use a graphing utility to graph the function and find the populations when $t = 5$, $t = 10$, and $t = 25$.

(b) What is the maximum number of fish in the lake as time passes? Explain your reasoning.

∫ **120. Publishing** A page that is x inches wide and y inches high contains 30 square inches of print. The top and bottom margins are 2 inches deep and the margins on each side are 2 inches wide.

(a) Draw a diagram that illustrates the problem.

(b) Show that the total area A of the page is given by

$$A = \dfrac{2x(2x + 7)}{x - 4}.$$

(c) Determine the domain of the function based on the physical constraints of the problem.

(d) Use a graphing utility to graph the area function and approximate the page size such that the minimum amount of paper will be used.

3.7

Classifying Scatter Plots In Exercises 121 and 122, determine whether the scatter plot could best be modeled by a linear model, a quadratic model, or neither.

121.   **122.**

123. MODELING DATA

The table shows the numbers of FM radio stations S in the United States from 2000 through 2009. (Source: Federal Communication Commission)

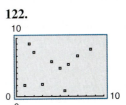

Year	FM stations, S
2000	5892
2001	6051
2002	6161
2003	6207
2004	6217
2005	6215
2006	6252
2007	6290
2008	6309
2009	6427

(a) Use a graphing utility to create a scatter plot of the data. Let t represent the year, with $t = 0$ corresponding to 2000.

(b) A cubic model for the data is

$S = 2.520t^3 - 37.51t^2 + 192.4t + 5895.$

Use the graphing utility to graph this model with the scatter plot from part (a).

(c) Use the *regression* feature of the graphing utility to find a quadratic model for the data. Then graph the model with the scatter plot from part (a).

(d) Which model is a better fit for the data? Explain.

(e) Use the model you chose in part (d) to predict the number of FM radio stations in 2012.

Conclusions

True or False? In Exercises 124 and 125, determine whether the statement is true or false. Justify your answer.

124. A fourth-degree polynomial with real coefficients can have -5, $-8i$, $4i$, and 5 as its zeros.

125. The sum of two complex numbers cannot be a real number.

3   Chapter Test

See www.CalcChat.com for worked-out solutions to odd-numbered exercises.
For instructions on how to use a graphing utility, see Appendix A.

Take this test as you would take a test in class. After you are finished, check your work against the answers given in the back of the book.

1. Identify the vertex and intercepts of the graph of $y = x^2 + 4x + 3$.

2. Write an equation of the parabola shown at the right.

3. The path of a ball is given by $y = -\frac{1}{20}x^2 + 3x + 5$, where y is the height (in feet) and x is the horizontal distance (in feet).

    (a) Find the maximum height of the ball.

    (b) Which term represents the height at which the ball was thrown? Does changing this term change the maximum height of the ball? Explain.

4. Find all the real zeros of $f(x) = 4x^3 + 4x^2 + x$. Determine the multiplicity of each zero.

5. Sketch the graph of the function $f(x) = -x^3 + 7x + 6$.

6. Divide using long division: $(3x^3 + 4x - 1) \div (x^2 + 1)$.

7. Divide using synthetic division: $(2x^4 - 5x^2 - 3) \div (x - 2)$.

8. Use synthetic division to evaluate $f(-2)$ for $f(x) = 3x^4 - 6x^2 + 5x - 1$.

In Exercises 9 and 10, list all the possible rational zeros of the function. Use a graphing utility to graph the function and find all the rational zeros.

9. $g(t) = 2t^4 - 3t^3 + 16t - 24$

10. $h(x) = 3x^5 + 2x^4 - 3x - 2$

11. Find all the zeros of the function $f(x) = x^3 - 7x^2 + 11x + 19$ and write the polynomial as a product of linear factors.

In Exercises 12–14, find a polynomial with real coefficients that has the given zeros. (There are many correct answers.)

12. $0, 2, 2 + i$

13. $1 - \sqrt{3}i, 2, 2$

14. $0, 1 + i$

In Exercises 15–17, sketch the graph of the rational function. As sketching aids, check for intercepts, vertical asymptotes, horizontal asymptotes, and slant asymptotes.

15. $h(x) = \dfrac{4}{x^2} - 1$

16. $g(x) = \dfrac{x^2 + 2}{x - 1}$

17. $f(x) = \dfrac{2x^2 + 9}{5x^2 + 2}$

18. The table shows the amounts A (in billions of dollars) spent on military procurement by the Department of Defense for the years 2002 through 2008. (Source: U.S. Office of Management and Budget)

    (a) Use a graphing utility to create a scatter plot of the data. Let t represent the year, with $t = 2$ corresponding to 2002.

    (b) Use the *regression* feature of the graphing utility to find a quadratic model for the data.

    (c) Use the graphing utility to graph the quadratic model with the scatter plot from part (a). Is the quadratic model a good fit for the data?

    (d) Use the model to estimate the amounts spent on military procurement in 2010 and 2012.

    (e) Do you believe the model is useful for predicting the amounts spent on military procurement for years beyond 2008? Explain.

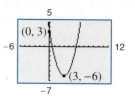

Figure for 2

Year	Military procurement, A (in billions of dollars)
2002	62.5
2003	67.9
2004	76.2
2005	82.3
2006	89.8
2007	99.6
2008	117.4

Table for 18

Proofs in Mathematics

These two pages contain proofs of four important theorems about polynomial functions. The first two theorems are from Section 3.3, and the second two theorems are from Section 3.4.

The Remainder Theorem   (p. 270)

If a polynomial $f(x)$ is divided by $x - k$, then the remainder is

$r = f(k)$.

Proof

From the Division Algorithm, you have

$$f(x) = (x - k)q(x) + r(x)$$

and because either $r(x) = 0$ or the degree of $r(x)$ is less than the degree of $x - k$, you know that $r(x)$ must be a constant. That is, $r(x) = r$. Now, by evaluating $f(x)$ at $x = k$, you have

$$f(k) = (k - k)q(k) + r$$

$$= (0)q(k) + r$$

$$= r.$$

To be successful in algebra, it is important that you understand the connection among the *factors* of a polynomial, the *zeros* of a polynomial function, and the *solutions* or *roots* of a polynomial equation. The Factor Theorem is the basis for this connection.

The Factor Theorem   (p. 270)

A polynomial $f(x)$ has a factor $(x - k)$ if and only if $f(k) = 0$.

Proof

Using the Division Algorithm with the factor $(x - k)$, you have

$$f(x) = (x - k)q(x) + r(x).$$

By the Remainder Theorem, $r(x) = r = f(k)$, and you have

$$f(x) = (x - k)q(x) + f(k)$$

where $q(x)$ is a polynomial of lesser degree than $f(x)$. If $f(k) = 0$, then

$$f(x) = (x - k)q(x)$$

and you see that $(x - k)$ is a factor of $f(x)$. Conversely, if $(x - k)$ is a factor of $f(x)$, then division of $f(x)$ by $(x - k)$ yields a remainder of 0. So, by the Remainder Theorem, you have $f(k) = 0$.

> ### Linear Factorization Theorem   (p. 281)
>
> If $f(x)$ is a polynomial of degree n, where $n > 0$, then f has precisely n linear factors
>
> $$f(x) = a_n(x - c_1)(x - c_2) \cdots (x - c_n)$$
>
> where $c_1, c_2, \ldots, c_n$ are complex numbers.

Proof

Using the Fundamental Theorem of Algebra, you know that f must have at least one zero, c_1. Consequently, $(x - c_1)$ is a factor of $f(x)$, and you have

$$f(x) = (x - c_1)f_1(x).$$

If the degree of $f_1(x)$ is greater than zero, then apply the Fundamental Theorem again to conclude that f_1 must have a zero c_2, which implies that

$$f(x) = (x - c_1)(x - c_2)f_2(x).$$

It is clear that the degree of $f_1(x)$ is $n - 1$, that the degree of $f_2(x)$ is $n - 2$, and that you can repeatedly apply the Fundamental Theorem n times until you obtain

$$f(x) = a_n(x - c_1)(x - c_2) \cdots (x - c_n)$$

where a_n is the leading coefficient of the polynomial $f(x)$.

> ### Factors of a Polynomial   (p. 284)
>
> Every polynomial of degree $n > 0$ with real coefficients can be written as the product of linear and quadratic factors with real coefficients, where the quadratic factors have no real zeros.

Proof

To begin, you use the Linear Factorization Theorem to conclude that $f(x)$ can be *completely* factored in the form

$$f(x) = d(x - c_1)(x - c_2)(x - c_3) \cdots (x - c_n).$$

If each c_i is real, then there is nothing more to prove. If any c_i is complex ($c_i = a + bi$, $b \neq 0$), then, because the coefficients of $f(x)$ are real, you know that the conjugate $c_j = a - bi$ is also a zero. By multiplying the corresponding factors, you obtain

$$(x - c_i)(x - c_j) = [x - (a + bi)][x - (a - bi)]$$
$$= x^2 - 2ax + (a^2 + b^2)$$

where each coefficient is real.

The Fundamental Theorem of Algebra

The Linear Factorization Theorem is closely related to the Fundamental Theorem of Algebra. The Fundamental Theorem of Algebra has a long and interesting history. In the early work with polynomial equations, the Fundamental Theorem of Algebra was thought to have been not true, because imaginary solutions were not considered. In fact, in the very early work by mathematicians such as Abu al-Khwarizmi (c. 800 A.D.), negative solutions were also not considered.

Once imaginary numbers were accepted, several mathematicians attempted to give a general proof of the Fundamental Theorem of Algebra. These included Gottfried von Leibniz (1702), Jean d'Alembert (1746), Leonhard Euler (1749), Joseph-Louis Lagrange (1772), and Pierre Simon Laplace (1795). The mathematician usually credited with the first correct proof of the Fundamental Theorem of Algebra is Carl Friedrich Gauss, who published the proof in his doctoral thesis in 1799.

4 Exponential and Logarithmic Functions

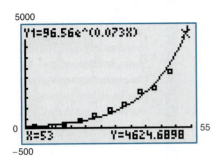

Section 4.6, Example 4
Internal Revenue Service

4.1 Exponential Functions and Their Graphs

Exponential Functions

So far, this text has dealt mainly with **algebraic functions,** which include polynomial functions and rational functions. In this chapter you will study two types of nonalgebraic functions—*exponential functions* and *logarithmic functions*. These functions are examples of **transcendental functions.**

> #### Definition of Exponential Function
>
> The **exponential function *f* with base *a*** is denoted by
>
> $$f(x) = a^x$$
>
> where $a > 0$, $a \neq 1$, and x is any real number.

Note that in the definition of an exponential function, the base $a = 1$ is excluded because it yields

$$f(x) = 1^x = 1. \qquad \text{Constant function}$$

This is a constant function, not an exponential function.

You have already evaluated a^x for integer and rational values of x. For example, you know that

$$4^3 = 64 \quad \text{and} \quad 4^{1/2} = 2.$$

However, to evaluate 4^x for any real number x, you need to interpret forms with *irrational* exponents. For the purposes of this text, it is sufficient to think of

$$a^{\sqrt{2}} \left(\text{where } \sqrt{2} \approx 1.41421356 \right)$$

as the number that has the successively closer approximations

$$a^{1.4}, a^{1.41}, a^{1.414}, a^{1.4142}, a^{1.41421}, \ldots .$$

Example 1 shows how to use a calculator to evaluate exponential functions.

Example 1 Evaluating Exponential Functions

Use a calculator to evaluate each function at the indicated value of x.

Function	Value
a. $f(x) = 2^x$	$x = -3.1$
b. $f(x) = 2^{-x}$	$x = \pi$
c. $f(x) = 0.6^x$	$x = \frac{3}{2}$
d. $f(x) = 1.05^{2x}$	$x = 12$

Solution

Function Value	Graphing Calculator Keystrokes	Display
a. $f(-3.1) = 2^{-3.1}$	2 ⌃ (−) 3.1 ENTER	0.1166291
b. $f(\pi) = 2^{-\pi}$	2 ⌃ (−) π ENTER	0.1133147
c. $f\left(\frac{3}{2}\right) = (0.6)^{3/2}$	.6 ⌃ (3 ÷ 2) ENTER	0.4647580
d. $f(12) = (1.05)^{2(12)}$	1.05 ⌃ (2 × 12) ENTER	3.2250999

✔**CHECKPOINT** Now try Exercise 7.

Graphs of Exponential Functions

The graphs of all exponential functions have similar characteristics, as shown in Examples 2, 3, and 4.

Example 2   Graphs of $y = a^x$

In the same coordinate plane, sketch the graph of each function by hand.

a. $f(x) = 2^x$

b. $g(x) = 4^x$

Solution

The table below lists some values for each function. By plotting these points and connecting them with smooth curves, you obtain the graphs shown in Figure 4.1. Note that both graphs are increasing. Moreover, the graph of $g(x) = 4^x$ is increasing more rapidly than the graph of $f(x) = 2^x$.

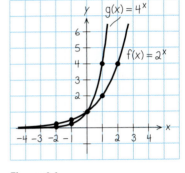

Figure 4.1

x	-2	-1	0	1	2	3
2^x	$\frac{1}{4}$	$\frac{1}{2}$	1	2	4	8
4^x	$\frac{1}{16}$	$\frac{1}{4}$	1	4	16	64

✔CHECKPOINT   Now try Exercise 9.

Example 3   Graphs of $y = a^{-x}$

In the same coordinate plane, sketch the graph of each function by hand.

a. $F(x) = 2^{-x}$

b. $G(x) = 4^{-x}$

Solution

The table below lists some values for each function. By plotting these points and connecting them with smooth curves, you obtain the graphs shown in Figure 4.2. Note that both graphs are decreasing. Moreover, the graph of $G(x) = 4^{-x}$ is decreasing more rapidly than the graph of $F(x) = 2^{-x}$.

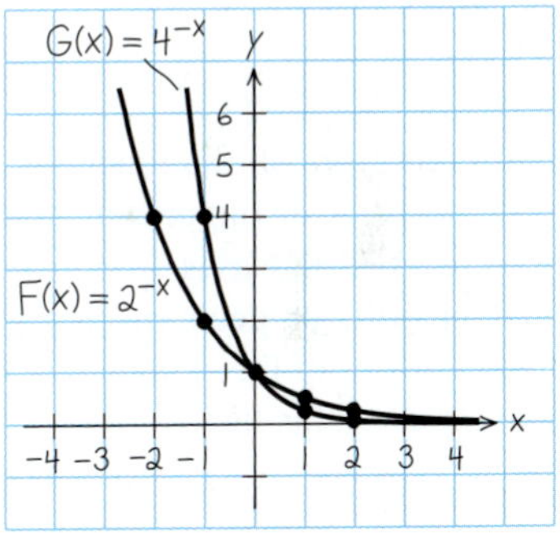

Figure 4.2

x	-3	-2	-1	0	1	2
2^{-x}	8	4	2	1	$\frac{1}{2}$	$\frac{1}{4}$
4^{-x}	64	16	4	1	$\frac{1}{4}$	$\frac{1}{16}$

✔CHECKPOINT   Now try Exercise 11.

The properties of exponents can also be applied to real-number exponents. For review, these properties are listed below.

1. $a^x a^y = a^{x+y}$ **2.** $\dfrac{a^x}{a^y} = a^{x-y}$ **3.** $a^{-x} = \dfrac{1}{a^x} = \left(\dfrac{1}{a}\right)^x$ **4.** $a^0 = 1$

5. $(ab)^x = a^x b^x$ **6.** $(a^x)^y = a^{xy}$ **7.** $\left(\dfrac{a}{b}\right)^x = \dfrac{a^x}{b^x}$ **8.** $|a^2| = |a|^2 = a^2$

Study Tip

In Example 3, note that the functions $F(x) = 2^{-x}$ and $G(x) = 4^{-x}$ can be rewritten with positive exponents.

$$F(x) = 2^{-x} = \left(\frac{1}{2}\right)^x \quad \text{and}$$

$$G(x) = 4^{-x} = \left(\frac{1}{4}\right)^x$$

Comparing the functions in Examples 2 and 3, observe that

$$F(x) = 2^{-x} = f(-x) \qquad \text{and} \qquad G(x) = 4^{-x} = g(-x).$$

Consequently, the graph of F is a reflection (in the y-axis) of the graph of f, as shown in Figure 4.3. The graphs of G and g have the same relationship, as shown in Figure 4.4.

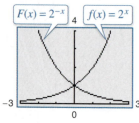

Figure 4.3

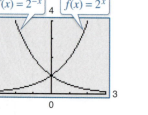

Figure 4.4

The graphs in Figures 4.3 and 4.4 are typical of the graphs of the exponential functions

$$f(x) = a^x \qquad \text{and} \qquad f(x) = a^{-x}.$$

They have one y-intercept and one horizontal asymptote (the x-axis), and they are continuous. The basic characteristics of these exponential functions are summarized below.

 ## Library of Parent Functions: Exponential Function

The *parent exponential function*

$$f(x) = a^x, \ a > 0, \ a \neq 1$$

is different from all the functions you have studied so far because the variable x is an *exponent*. A distinguishing characteristic of an exponential function is its rapid increase as x increases (for $a > 1$). Many real-life phenomena with patterns of rapid growth (or decline) can be modeled by exponential functions. The basic characteristics of the exponential function are summarized below and on the inside front cover of this text.

Graph of $f(x) = a^x, \ a > 1$

Domain: $(-\infty, \infty)$

Range: $(0, \infty)$

Intercept: $(0, 1)$

Increasing on $(-\infty, \infty)$

x-axis is a horizontal asymptote

$(a^x \to 0 \text{ as } x \to -\infty)$

Continuous

Graph of $f(x) = a^{-x}, \ a > 1$

Domain: $(-\infty, \infty)$

Range: $(0, \infty)$

Intercept: $(0, 1)$

Decreasing on $(-\infty, \infty)$

x-axis is a horizontal asymptote

$(a^{-x} \to 0 \text{ as } x \to \infty)$

Continuous

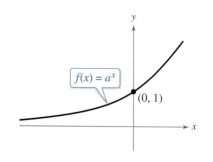

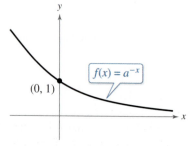

Explore the Concept

Use a graphing utility to graph $y = a^x$ for $a = 3, 5,$ and 7 in the same viewing window. (Use a viewing window in which $-2 \leq x \leq 1$ and $0 \leq y \leq 2$.) How do the graphs compare with each other? Which graph is on the top in the interval $(-\infty, 0)$? Which is on the bottom? Which graph is on the top in the interval $(0, \infty)$? Which is on the bottom? Repeat this experiment with the graphs of $y = b^x$ for $b = \frac{1}{3}, \frac{1}{5},$ and $\frac{1}{7}$. (Use a viewing window in which $-1 \leq x \leq 2$ and $0 \leq y \leq 2$.) What can you conclude about the shape of the graph of $y = b^x$ and the value of b?

In the following example, the graph of

$$y = a^x$$

is used to graph functions of the form

$$f(x) = b \pm a^{x+c}$$

where b and c are any real numbers.

Example 4   Library of Parent Functions: $f(x) = a^x$

Each of the following graphs is a transformation of the graph of $f(x) = 3^x$.

a. Because $g(x) = 3^{x+1} = f(x + 1)$, the graph of g can be obtained by shifting the graph of f one unit to the *left*, as shown in Figure 4.5.

b. Because $h(x) = 3^x - 2 = f(x) - 2$, the graph of h can be obtained by shifting the graph of f *downward* two units, as shown in Figure 4.6.

c. Because $k(x) = -3^x = -f(x)$, the graph of k can be obtained by *reflecting* the graph of f in the x-axis, as shown in Figure 4.7.

d. Because $j(x) = 3^{-x} = f(-x)$, the graph of j can be obtained by *reflecting* the graph of f in the y-axis, as shown in Figure 4.8.

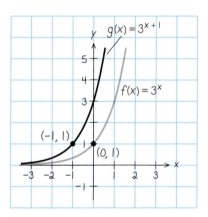

Figure 4.5

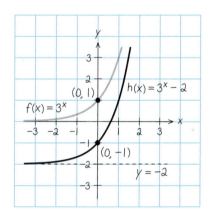

Figure 4.6

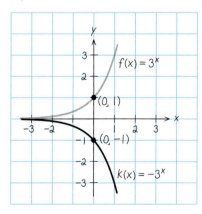

Figure 4.7

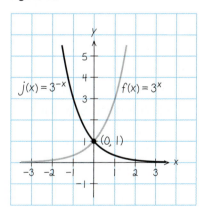

Figure 4.8

 CHECKPOINT   Now try Exercise 21.

Notice that the transformations in Figures 4.5, 4.7, and 4.8 keep the x-axis ($y = 0$) as a horizontal asymptote, but the transformation in Figure 4.6 yields a new horizontal asymptote of $y = -2$. Also, be sure to note how the y-intercept is affected by each transformation.

What's Wrong?

You use a graphing utility to graph $f(x) = 3^x$ and $g(x) = 3^{x+2}$, as shown in the figure. You use the graph to conclude that the graph of g can be obtained by shifting the graph of f upward two units. What's wrong?

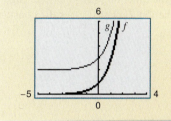

Explore the Concept

 The following table shows some points on the graphs in Figure 4.5. The functions $f(x)$ and $g(x)$ are represented by Y₁ and Y₂, respectively. Explain how you can use the table to describe the transformation.

X	Y₁	Y₂
-3	.03704	.11111
-2	.11111	.33333
-1	.33333	1
0	1	3
1	3	9
2	9	27
3	27	81

X = -3

The Natural Base e

For many applications, the convenient choice for a base is the irrational number

$$e = 2.718281828 \ldots .$$

This number is called the **natural base.** The function

$$f(x) = e^x$$

is called the **natural exponential function** and its graph is shown in Figure 4.9. The graph of the natural exponential function has the same basic characteristics as the graph of the function $f(x) = a^x$ (see page 326). Be sure you see that for the natural exponential function $f(x) = e^x$, e is the constant $2.718281828 \ldots$, whereas x is the variable.

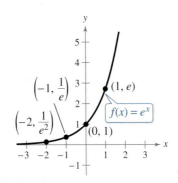

Figure 4.9 *The Natural Exponential Function*

Explore the Concept

Use your graphing utility to graph the functions

$$y_1 = 2^x$$
$$y_2 = e^x$$
$$y_3 = 3^x$$

in the same viewing window. From the relative positions of these graphs, make a guess as to the value of the real number e. Then try to find a number a such that the graphs of $y_2 = e^x$ and $y_4 = a^x$ are as close to each other as possible.

In Example 5, you will see that the number e can be approximated by the expression

$$\left(1 + \frac{1}{x}\right)^x \text{ for large values of } x.$$

Example 5 Approximation of the Number e

Evaluate the expression

$$\left(1 + \frac{1}{x}\right)^x$$

for several large values of x to see that the values approach

$$e \approx 2.718281828$$

as x increases without bound.

Graphical Solution

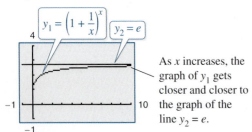

$y_1 = \left(1 + \frac{1}{x}\right)^x$ $y_2 = e$

As x increases, the graph of y_1 gets closer and closer to the graph of the line $y_2 = e$.

Figure 4.10

✔CHECKPOINT Now try Exercise 27.

Numerical Solution

Enter $y_1 = [1 + (1/x)]^x$.

Use the *table* feature (in *ask* mode) to evaluate y_1 for increasing values of x.

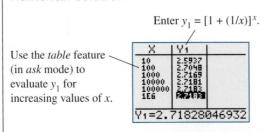

Figure 4.11

From Figure 4.11, it seems reasonable to conclude that

$$\left(1 + \frac{1}{x}\right)^x \to e \text{ as } x \to \infty.$$

Example 6   Evaluating the Natural Exponential Function

Use a calculator to evaluate the function

$$f(x) = e^x$$

at each indicated value of x.

a. $x = -2$

b. $x = 0.25$

c. $x = -0.4$

d. $x = \frac{2}{3}$

Solution

Function Value	Graphing Calculator Keystrokes	Display
a. $f(-2) = e^{-2}$	$\boxed{e^x}$ $\boxed{(-)}$ 2 $\boxed{\text{ENTER}}$	0.1353353
b. $f(0.25) = e^{0.25}$	$\boxed{e^x}$ .25 $\boxed{\text{ENTER}}$	1.2840254
c. $f(-0.4) = e^{-0.4}$	$\boxed{e^x}$ $\boxed{(-)}$ .4 $\boxed{\text{ENTER}}$	0.6703200
d. $f\left(\frac{2}{3}\right) = e^{2/3}$	$\boxed{e^x}$ $\boxed{(}$ 2 $\boxed{\div}$ 3 $\boxed{)}$ $\boxed{\text{ENTER}}$	1.9477340

 CHECKPOINT   Now try Exercise 29.

Example 7   Graphing Natural Exponential Functions

Sketch the graph of each natural exponential function.

a. $f(x) = 2e^{0.24x}$

b. $g(x) = \frac{1}{2}e^{-0.58x}$

Solution

To sketch these two graphs, you can use a calculator to construct a table of values, as shown below.

x	-3	-2	-1	0	1	2	3
$f(x)$	0.974	1.238	1.573	2.000	2.542	3.232	4.109
$g(x)$	2.849	1.595	0.893	0.500	0.280	0.157	0.088

After constructing the table, plot the points and connect them with smooth curves. Note that the graph in Figure 4.12 is increasing, whereas the graph in Figure 4.13 is decreasing. Use a graphing calculator to verify these graphs.

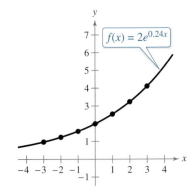

Figure 4.12

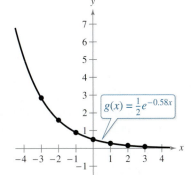

Figure 4.13

 CHECKPOINT   Now try Exercise 47.

Explore the Concept

 Use a graphing utility to graph $y = (1 + x)^{1/x}$. Describe the behavior of the graph near $x = 0$. Is there a y-intercept? How does the behavior of the graph near $x = 0$ relate to the result of Example 5? Use the *table* feature of the graphing utility to create a table that shows values of y for values of x near $x = 0$ to help you describe the behavior of the graph near this point.

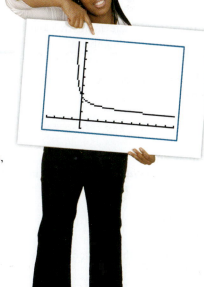

Applications

One of the most familiar examples of exponential growth is an investment earning *continuously compounded interest*. Suppose a principal P is invested at an annual interest rate r, compounded once a year. If the interest is added to the principal at the end of the year, then the new balance P_1 is

$$P_1 = P + Pr = P(1 + r).$$

This pattern of multiplying the previous principal by $1 + r$ is then repeated each successive year, as shown in the table.

Time in years	Balance after each compounding
0	$P = P$
1	$P_1 = P(1 + r)$
2	$P_2 = P_1(1 + r) = P(1 + r)(1 + r) = P(1 + r)^2$
⋮	⋮
t	$P_t = P(1 + r)^t$

To accommodate more frequent (quarterly, monthly, or daily) compounding of interest, let n be the number of compoundings per year and let t be the number of years. (The product nt represents the total number of times the interest will be compounded.) Then the interest rate per compounding period is r/n, and the account balance after t years is

$$A = P\left(1 + \frac{r}{n}\right)^{nt}. \qquad \text{Amount (balance) with } n \text{ compoundings per year}$$

When the number of compoundings n increases without bound, the process approaches what is called **continuous compounding.** In the formula for n compoundings per year, let $m = n/r$. This produces

$$A = P\left(1 + \frac{r}{n}\right)^{nt} = P\left(1 + \frac{1}{m}\right)^{mrt} = P\left[\left(1 + \frac{1}{m}\right)^{m}\right]^{rt}.$$

As m increases without bound, you know from Example 5 that

$$\left(1 + \frac{1}{m}\right)^{m}$$

approaches e. So, for continuous compounding, it follows that

$$P\left[\left(1 + \frac{1}{m}\right)^{m}\right]^{rt} \quad \Longrightarrow \quad P[e]^{rt}$$

and you can write $A = Pe^{rt}$. This result is part of the reason that e is the "natural" choice for a base of an exponential function.

Formulas for Compound Interest

After t years, the balance A in an account with principal P and annual interest rate r (in decimal form) is given by the following formulas.

1. For n compoundings per year: $A = P\left(1 + \dfrac{r}{n}\right)^{nt}$

2. For continuous compounding: $A = Pe^{rt}$

Example 8   Finding the Balance for Compound Interest

A total of $9000 is invested at an annual interest rate of 2.5%, compounded annually. Find the balance in the account after 5 years.

Algebraic Solution

In this case,

$$P = 9000, r = 2.5\% = 0.025, n = 1, t = 5.$$

Using the formula for compound interest with n compoundings per year, you have

$$A = P\left(1 + \frac{r}{n}\right)^{nt}$$ Formula for compound interest

$$= 9000\left(1 + \frac{0.025}{1}\right)^{1(5)}$$ Substitute for P, r, n, and t.

$$= 9000(1.025)^5$$ Simplify.

$$\approx \$10,182.67.$$ Use a calculator.

So, the balance in the account after 5 years will be about $10,182.67.

    Now try Exercise 57.

Graphical Solution

Substitute the values for P, r, and n into the formula for compound interest with n compoundings per year and simplify to obtain

$$A = 9000(1.025)^t.$$

Use a graphing utility to graph $A = 9000(1.025)^t$. Then use the *value* feature to approximate the value of A when $t = 5$, as shown in Figure 4.14.

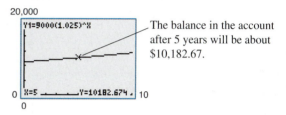

The balance in the account after 5 years will be about $10,182.67.

Figure 4.14

Example 9   Finding Compound Interest

A total of $12,000 is invested at an annual interest rate of 3%. Find the balance after 4 years for each type of compounding.

a. Quarterly

b. Continuous

Solution

a. For quarterly compoundings, $n = 4$. So, after 4 years at 3%, the balance is

$$A = P\left(1 + \frac{r}{n}\right)^{nt}$$ Formula for compound interest

$$= 12{,}000\left(1 + \frac{0.03}{4}\right)^{4(4)}$$ Substitute for P, r, n, and t.

$$\approx \$13{,}523.91.$$ Use a calculator.

b. For continuous compounding, the balance is

$$A = Pe^{rt}$$ Formula for continuous compounding

$$= 12{,}000e^{0.03(4)}$$ Substitute for P, r, and t.

$$\approx \$13{,}529.96.$$ Use a calculator.

Note that a continuous-compounding account yields more than a quarterly-compounding account.

    Now try Exercise 59.

Financial Analyst

Example 9 illustrates the following general rule. For a given principal, interest rate, and time, the more often the interest is compounded per year, the greater the balance will be. Moreover, the balance obtained by continuous compounding is greater than the balance obtained by compounding n times per year.

Example 10 Radioactive Decay

Let y represent a mass, in grams, of radioactive strontium (^{90}Sr), whose half-life is 29 years. The quantity of strontium present after t years is

$$y = 10\left(\tfrac{1}{2}\right)^{t/29}.$$

a. What is the initial mass (when $t = 0$)?

b. How much of the initial mass is present after 80 years?

Algebraic Solution

a. $y = 10\left(\dfrac{1}{2}\right)^{t/29}$ Write original equation.

$= 10\left(\dfrac{1}{2}\right)^{0/29}$ Substitute 0 for t.

$= 10$ Simplify.

So, the initial mass is 10 grams.

b. $y = 10\left(\dfrac{1}{2}\right)^{t/29}$ Write original equation.

$= 10\left(\dfrac{1}{2}\right)^{80/29}$ Substitute 80 for t.

$\approx 10\left(\dfrac{1}{2}\right)^{2.759}$ Simplify.

≈ 1.48 Use a calculator.

So, about 1.48 grams are present after 80 years.

 Now try Exercise 71.

Graphical Solution

a. When $t = 0$, $y = 10$. So, the initial mass is 10 grams.

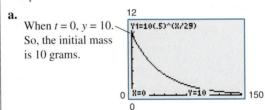

b. When $t = 80$, $y \approx 1.48$. So, about 1.48 grams are present after 80 years.

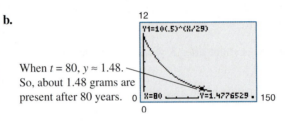

Example 11 Population Growth

The approximate number of fruit flies in an experimental population after t hours is given by

$$Q(t) = 20e^{0.03t}$$

where $t \geq 0$.

a. Find the initial number of fruit flies in the population.

b. How large is the population of fruit flies after 72 hours?

c. Graph Q.

Solution

a. To find the initial population, evaluate $Q(t)$ when $t = 0$.

$$Q(0) = 20e^{0.03(0)} = 20e^0 = 20(1) = 20 \text{ flies}$$

b. After 72 hours, the population size is

$$Q(72) = 20e^{0.03(72)} = 20e^{2.16} \approx 173 \text{ flies}.$$

c. The graph of Q is shown in Figure 4.15.

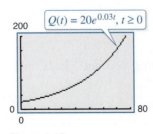

Figure 4.15

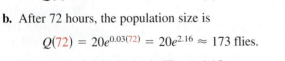 Now try Exercise 73.

4.1   Exercises

See www.CalcChat.com for worked-out solutions to odd-numbered exercises.
For instructions on how to use a graphing utility, see Appendix A.

Vocabulary and Concept Check

In Exercises 1 and 2, fill in the blank(s).

1. Exponential and logarithmic functions are examples of nonalgebraic functions, also called _____ functions.

2. The exponential function $f(x) = e^x$ is called the _____ function, and the base e is called the _____ base.

3. What type of transformation of the graph of $f(x) = 5^x$ is the graph of $f(x + 1)$?

4. The formula $A = Pe^{rt}$ gives the balance A of an account earning what type of interest?

Procedures and Problem Solving

Evaluating Exponential Functions   In Exercises 5–8, use a calculator to evaluate the function at the indicated value of x. Round your result to three decimal places.

Function	Value
5. $f(x) = 3.4^x$	$x = 6.8$
6. $f(x) = 1.2^x$	$x = \frac{1}{3}$
✓ 7. $g(x) = 5^x$	$x = -\pi$
8. $h(x) = 8.6^{-3x}$	$x = -\sqrt{2}$

Graphs of $y = a^x$ and $y = a^{-x}$ In Exercises 9–16, graph the exponential function by hand. Identify any asymptotes and intercepts and determine whether the graph of the function is increasing or decreasing.

✓ 9. $g(x) = 5^x$ 10. $f(x) = \left(\frac{3}{2}\right)^x$

✓ 11. $f(x) = 5^{-x}$ 12. $h(x) = \left(\frac{3}{2}\right)^{-x}$

13. $h(x) = 3^x$ 14. $g(x) = 10^x$

15. $g(x) = 3^{-x}$ 16. $f(x) = 10^{-x}$

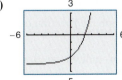 **Library of Parent Functions   In Exercises 17–20, use the graph of $y = 2^x$ to match the function with its graph. [The graphs are labeled (a), (b), (c), and (d).]**

(a)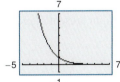

(b)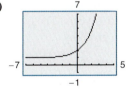

(c)

(d)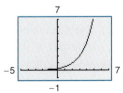

17. $f(x) = 2^{x-2}$ 18. $f(x) = 2^{-x}$

19. $f(x) = 2^x - 4$ 20. $f(x) = 2^x + 1$

 Library of Parent Functions   In Exercises 21–26, use the graph of f to describe the transformation that yields the graph of g. Then sketch the graphs of f and g by hand.

✓ 21. $f(x) = 3^x$, $g(x) = 3^{x-5}$

22. $f(x) = -2^x$, $g(x) = 5 - 2^x$

23. $f(x) = \left(\frac{3}{5}\right)^x$, $g(x) = -\left(\frac{3}{5}\right)^{x+4}$

24. $f(x) = 0.3^x$, $g(x) = -0.3^x + 5$

25. $f(x) = 4^x$, $g(x) = 4^{x-2} - 3$

26. $f(x) = \left(\frac{1}{2}\right)^x$, $g(x) = \left(\frac{1}{2}\right)^{-(x+4)}$

Approximation of a Power with Base e In Exercises 27 and 28, show that the value of $f(x)$ approaches the value of $g(x)$ as x increases without bound (a) graphically and (b) numerically.

✓ 27. $f(x) = [1 + (2/x)]^x$, $g(x) = e^2$

28. $f(x) = [1 + (3/x)]^x$, $g(x) = e^3$

Evaluating the Natural Exponential Function   In Exercises 29–32, use a calculator to evaluate the function at the indicated value of x. Round your result to the nearest thousandth.

Function	Value
✓ 29. $f(x) = e^x$	$x = 9.2$
30. $f(x) = e^{-x}$	$x = -\frac{3}{4}$
31. $g(x) = 50e^{4x}$	$x = 0.02$
32. $h(x) = -5.5e^{-x}$	$x = 200$

Graphing an Exponential Function   In Exercises 33–48, use a graphing utility to construct a table of values for the function. Then sketch the graph of the function. Identify any asymptotes of the graph.

33. $f(x) = \left(\frac{5}{2}\right)^x$ 34. $f(x) = \left(\frac{5}{2}\right)^{-x}$

35. $f(x) = 6^x$ 36. $f(x) = 2^{x-1}$

37. $f(x) = 3^{x+2}$ **38.** $y = 2^{-x^2}$

39. $y = 3^{x-2} + 1$ **40.** $y = 4^{x+1} - 2$

41. $f(x) = e^{-x}$ **42.** $s(t) = 3e^{-0.2t}$

43. $f(x) = 3e^{x+4}$ **44.** $f(x) = 2e^{-0.5x}$

45. $f(x) = 2 + e^{x-5}$ **46.** $g(x) = 2 - e^{-x}$

✓ **47.** $s(t) = 2e^{0.12t}$ **48.** $g(x) = 1 + e^{-x}$

Finding Asymptotes   In Exercises 49–52, use a graphing utility to (a) graph the function and (b) find any asymptotes numerically by creating a table of values for the function.

49. $f(x) = \dfrac{8}{1 + e^{-0.5x}}$ **50.** $g(x) = \dfrac{8}{1 + e^{-0.5/x}}$

51. $f(x) = -\dfrac{6}{2 - e^{0.2x}}$ **52.** $f(x) = \dfrac{6}{2 - e^{0.2/x}}$

Finding Points of Intersection   In Exercises 53 and 54, use a graphing utility to find the point(s) of intersection, if any, of the graphs of the functions. Round your result to three decimal places.

53. $y = 20e^{0.05x}$ **54.** $y = 100e^{0.01x}$

     $y = 1500$ $y = 12{,}500$

Approximating Relative Extrema   In Exercises 55 and 56, (a) use a graphing utility to graph the function, (b) use the graph to find the open intervals on which the function is increasing and decreasing, and (c) approximate any relative maximum or minimum values.

55. $f(x) = x^2e^{-x}$ **56.** $f(x) = 2x^2e^{x+1}$

Finding the Balance for Compound Interest   In Exercises 57–60, complete the table to determine the balance A for \$2500 invested at rate r for t years and compounded n times per year.

n	1	2	4	12	365	Continuous
A						

✓ **57.** $r = 2\%$, $t = 10$ years   **58.** $r = 6\%$, $t = 10$ years

✓ **59.** $r = 4\%$, $t = 20$ years   **60.** $r = 3\%$, $t = 40$ years

Finding the Balance for Compound Interest   In Exercises 61–64, complete the table to determine the balance A for \$12,000 invested at rate r for t years, compounded continuously.

t	1	10	20	30	40	50
A						

61. $r = 4\%$ **62.** $r = 6\%$

63. $r = 3.5\%$ **64.** $r = 2.5\%$

Finding the Amount of an Annuity   In Exercises 65–68, you build an annuity by investing P dollars every month at interest rate r, compounded monthly. Find the amount A accrued after n months using the formula

$$A = P\left[\frac{(1 + r/12)^n - 1}{r/12}\right]$$

where r is in decimal form.

65. $P = \$25$, $r = 0.12$, $n = 48$ months

66. $P = \$100$, $r = 0.09$, $n = 60$ months

67. $P = \$200$, $r = 0.06$, $n = 72$ months

68. $P = \$75$, $r = 0.03$, $n = 24$ months

69. MODELING DATA

There are three options for investing \$500. The first earns 7% compounded annually, the second earns 7% compounded quarterly, and the third earns 7% compounded continuously.

(a) Find equations that model the growth of each investment and use a graphing utility to graph each model in the same viewing window over a 20-year period.

(b) Use the graph from part (a) to determine which investment yields the highest return after 20 years. What are the differences in earnings among the three investments?

70. Radioactive Decay   Let Q represent a mass, in grams, of radioactive radium (^{226}Ra), whose half-life is 1599 years. The quantity of radium present after t years is given by

$$Q = 25\left(\tfrac{1}{2}\right)^{t/1599}.$$

(a) Determine the initial quantity (when $t = 0$).

(b) Determine the quantity present after 1000 years.

(c) Use a graphing utility to graph the function over the interval $t = 0$ to $t = 5000$.

(d) When will the quantity of radium be 0 grams? Explain.

71. Radioactive Decay   Let Q represent a mass, in grams, of carbon 14 (^{14}C), whose half-life is 5715 years. The quantity present after t years is given by $Q = 10\left(\frac{1}{2}\right)^{t/5715}$.

(a) Determine the initial quantity (when $t = 0$).

(b) Determine the quantity present after 2000 years.

(c) Sketch the graph of the function over the interval $t = 0$ to $t = 10{,}000$.

72. Algebraic-Graphical-Numerical   Suppose the annual rate of inflation is 4% for the next 10 years. The approximate cost C of goods or services during these years is $C(t) = P(1.04)^t$, where t is the time (in years) and P is the present cost. An oil change for your car presently costs $23.95. Use the following methods to approximate the cost 10 years from now.

(a) Use a graphing utility to graph the function and then use the *value* feature.

(b) Use the *table* feature of the graphing utility to find a numerical approximation.

(c) Use a calculator to evaluate the cost function algebraically.

73. Population Growth   The projected populations of California for the years 2015 through 2030 can be modeled by $P = 34.706e^{0.0097t}$, where P is the population (in millions) and t is the time (in years), with $t = 15$ corresponding to 2015.   (Source: U.S. Census Bureau)

(a) Use a graphing utility to graph the function for the years 2015 through 2030.

(b) Use the *table* feature of the graphing utility to create a table of values for the same time period as in part (a).

(c) According to the model, in what year will the population of California exceed 50 million?

74. *Why you should learn it* (p. 324)   In early 2010, a new sedan had a manufacturer's suggested retail price of $31,915. After t years, the sedan's value is given by

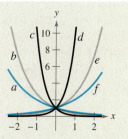

$$V(t) = 31{,}915\left(\frac{4}{5}\right)^t.$$

(a) Use a graphing utility to graph the function.

(b) Use the graphing utility to create a table of values that shows the value V for $t = 1$ to $t = 10$ years.

(c) According to the model, when will the sedan have no value?

Conclusions

True or False?   In Exercises 75 and 76, determine whether the statement is true or false. Justify your answer.

75. $f(x) = 1^x$ is not an exponential function.

76. $e = \dfrac{271{,}801}{99{,}990}$

77.   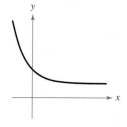 **Library of Parent Functions**   Determine which equation(s) may be represented by the graph shown. (There may be more than one correct answer.)

(a) $y = e^x + 1$

(b) $y = -e^{-x} + 1$

(c) $y = e^{-x} - 1$

(d) $y = e^{-x} + 1$

78. Exploration   Use a graphing utility to graph $y_1 = e^x$ and each of the functions $y_2 = x^2$, $y_3 = x^3$, $y_4 = \sqrt{x}$, and $y_5 = |x|$ in the same viewing window.

(a) Which function increases at the fastest rate for "large" values of x?

(b) Use the result of part (a) to make a conjecture about the rates of growth of $y_1 = e^x$ and $y = x^n$, where n is a natural number and x is "large."

(c) Use the results of parts (a) and (b) to describe what is implied when it is stated that a quantity is growing exponentially.

79. Think About It   Graph $y = 3^x$ and $y = 4^x$. Use the graph to solve the inequality $3^x < 4^x$.

80. CAPSTONE   The figure shows the graphs of $y = 2^x$, $y = e^x$, $y = 10^x$, $y = 2^{-x}$, $y = e^{-x}$, and $y = 10^{-x}$. Match each function with its graph. [The graphs are labeled (a) through (f).] Explain your reasoning.

Think About It   In Exercises 81–84, place the correct symbol (< or >) between the two of numbers.

81. e^π ▢ π^e **82.** 2^{10} ▢ 10^2

83. 5^{-3} ▢ 3^{-5} **84.** $4^{1/2}$ ▢ $\left(\frac{1}{2}\right)^4$

Cumulative Mixed Review

Inverse Functions   In Exercises 85–88, determine whether the function has an inverse function. If it does, find f^{-1}.

85. $f(x) = 5x - 7$ **86.** $f(x) = -\frac{2}{3}x + \frac{5}{2}$

87. $f(x) = \sqrt[3]{x + 8}$ **88.** $f(x) = \sqrt{x^2 + 6}$

89. *Make a Decision*   To work an extended application analyzing the population per square mile in the United States, visit this textbook's *Companion Website*.   (Data Source: U.S. Census Bureau)

4.2   Logarithmic Functions and Their Graphs

Logarithmic Functions

In Section 1.7, you studied the concept of an inverse function. There, you learned that when a function is one-to-one—that is, when the function has the property that no horizontal line intersects its graph more than once—the function must have an inverse function. By looking back at the graphs of the exponential functions introduced in Section 4.1, you will see that every function of the form

$$f(x) = a^x, \quad a > 0, a \neq 1$$

passes the Horizontal Line Test and therefore must have an inverse function. This inverse function is called the **logarithmic function with base a.**

> **Definition of Logarithmic Function**
>
> For $x > 0$, $a > 0$, and $a \neq 1$,
>
> $$y = \log_a x \quad \text{if and only if} \quad x = a^y.$$
>
> The function given by
>
> $$f(x) = \log_a x \qquad \text{Read as "log base } a \text{ of } x."$$
>
> is called the **logarithmic function with base a.**

From the definition above, you can see that every logarithmic equation can be written in an equivalent exponential form and every exponential equation can be written in logarithmic form. The equations

$$y = \log_a x \quad \text{and} \quad x = a^y$$

are equivalent.

When evaluating logarithms, remember that *a logarithm is an exponent.* This means that $\log_a x$ is the exponent to which a must be raised to obtain x. For instance, $\log_2 8 = 3$ because 2 must be raised to the third power to get 8.

Example 1   Evaluating Logarithms

Use the definition of logarithmic function to evaluate each logarithm at the indicated value of x.

Function	*Value*
a. $f(x) = \log_2 x$	$x = 32$
b. $f(x) = \log_3 x$	$x = 1$
c. $f(x) = \log_4 x$	$x = 2$
d. $f(x) = \log_{10} x$	$x = \frac{1}{100}$

Solution

a. $f(32) = \log_2 32 = 5$ because $2^5 = 32$.

b. $f(1) = \log_3 1 = 0$ because $3^0 = 1$.

c. $f(2) = \log_4 2 = \frac{1}{2}$ because $4^{1/2} = \sqrt{4} = 2$.

d. $f\left(\frac{1}{100}\right) = \log_{10} \frac{1}{100} = -2$ because $10^{-2} = \frac{1}{10^2} = \frac{1}{100}$.

 CHECKPOINT   Now try Exercise 23.

What you should learn

- Recognize and evaluate logarithmic functions with base a.
- Graph logarithmic functions with base a.
- Recognize, evaluate, and graph natural logarithmic functions.
- Use logarithmic functions to model and solve real-life problems.

Why you should learn it

Logarithmic functions are useful in modeling data that represent quantities that increase or decrease slowly. For instance, Exercise 114 on page 345 shows how to use a logarithmic function to model the minimum required ventilation rates in public school classrooms.

Study Tip

 In this text, the parentheses in $\log_a(u)$ are sometimes omitted when u is an expression involving exponents, radicals, products, or quotients. For instance, $\log_{10}(2x)$ can be written as $\log_{10} 2x$. To evaluate $\log_{10} 2x$, find the logarithm of the product $2x$.

The logarithmic function with base 10 is called the **common logarithmic function.** On most calculators, this function is denoted by (LOG). Example 2 shows how to use a calculator to evaluate common logarithmic functions. You will learn how to use a calculator to calculate logarithms to any base in the next section.

Example 2 Evaluating Common Logarithms on a Calculator

Use a calculator to evaluate the function

$$f(x) = \log_{10} x$$

at each value of x.

a. $x = 10$ **b.** $x = 2.5$
c. $x = -2$ **d.** $x = \frac{1}{4}$

Solution

Function Value	Graphing Calculator Keystrokes	Display
a. $f(10) = \log_{10} 10$	(LOG) 10 (ENTER)	1
b. $f(2.5) = \log_{10} 2.5$	(LOG) 2.5 (ENTER)	0.3979400
c. $f(-2) = \log_{10}(-2)$	(LOG) ((-)) 2 (ENTER)	ERROR
d. $f\left(\frac{1}{4}\right) = \log_{10} \frac{1}{4}$	(LOG) (() 1 (÷) 4 ()) (ENTER)	-0.6020600

Note that the calculator displays an error message when you try to evaluate $\log_{10}(-2)$. In this case, there is no *real* power to which 10 can be raised to obtain -2.

✓CHECKPOINT   Now try Exercise 27.

The following properties follow directly from the definition of the logarithmic function with base a.

Technology Tip

 Some graphing utilities do not give an error message for $\log_{10}(-2)$. Instead, the graphing utility will display a complex number. For the purpose of this text, however, it will be said that the domain of a logarithmic function is the set of positive *real* numbers.

Properties of Logarithms

1. $\log_a 1 = 0$ because $a^0 = 1$.

2. $\log_a a = 1$ because $a^1 = a$.

3. $\log_a a^x = x$ and $a^{\log_a x} = x$.   Inverse Properties

4. If $\log_a x = \log_a y$, then $x = y$.   One-to-One Property

---

Example 3 Using Properties of Logarithms

a. Solve for x: $\log_2 x = \log_2 3$
b. Solve for x: $\log_4 4 = x$
c. Simplify: $\log_5 5^x$
d. Simplify: $7^{\log_7 14}$

Solution

a. Using the One-to-One Property (Property 4), you can conclude that $x = 3$.
b. Using Property 2, you can conclude that $x = 1$.
c. Using the Inverse Property (Property 3), it follows that $\log_5 5^x = x$.
d. Using the Inverse Property (Property 3), it follows that $7^{\log_7 14} = 14$.

✓CHECKPOINT   Now try Exercise 31.

Graphs of Logarithmic Functions

To sketch the graph of

$$y = \log_a x$$

you can use the fact that the graphs of inverse functions are reflections of each other in the line $y = x$.

Example 4   Graphs of Exponential and Logarithmic Functions

In the same coordinate plane, sketch the graph of each function by hand.

a. $f(x) = 2^x$

b. $g(x) = \log_2 x$

Solution

a. For $f(x) = 2^x$, construct a table of values. By plotting these points and connecting them with a smooth curve, you obtain the graph of f shown in Figure 4.16.

x	-2	-1	0	1	2	3
$f(x) = 2^x$	$\frac{1}{4}$	$\frac{1}{2}$	1	2	4	8

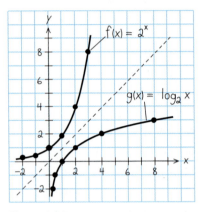

b. Because $g(x) = \log_2 x$ is the inverse function of $f(x) = 2^x$, the graph of g is obtained by plotting the points $(f(x), x)$ and connecting them with a smooth curve. The graph of g is a reflection of the graph of f in the line $y = x$, as shown in Figure 4.16.

Figure 4.16

✔CHECKPOINT   Now try Exercise 41.

Before you can confirm the result of Example 4 using a graphing utility, you need to know how to enter $\log_2 x$. You will learn how to do this using the *change-of-base formula* discussed in Section 4.3.

Example 5   Sketching the Graph of a Logarithmic Function

Sketch the graph of the common logarithmic function $f(x) = \log_{10} x$ by hand.

Solution

Begin by constructing a table of values. Note that some of the values can be obtained without a calculator by using the Inverse Property of Logarithms. Others require a calculator. Next, plot the points and connect them with a smooth curve, as shown in Figure 4.17.

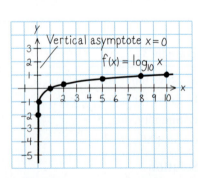

	Without calculator				With calculator		
x	$\frac{1}{100}$	$\frac{1}{10}$	1	10	2	5	8
$f(x) = \log_{10}x$	-2	-1	0	1	0.301	0.699	0.903

Figure 4.17

✔CHECKPOINT   Now try Exercise 45.

The nature of the graph in Figure 4.17 is typical of functions of the form $f(x) = \log_a x$, $a > 1$. They have one x-intercept and one vertical asymptote. Notice how slowly the graph rises for $x > 1$.

Library of Parent Functions: Logarithmic Function

The *parent logarithmic function*

$$f(x) = \log_a x, \quad a > 0, \ a \neq 1$$

is the inverse function of the exponential function. Its domain is the set of positive real numbers and its range is the set of all real numbers. This is the opposite of the exponential function. Moreover, the logarithmic function has the *y*-axis as a vertical asymptote, whereas the exponential function has the *x*-axis as a horizontal asymptote. Many real-life phenomena with slow rates of growth can be modeled by logarithmic functions. The basic characteristics of the logarithmic function are summarized below and on the inside front cover of this text.

Graph of $f(x) = \log_a x$, $a > 1$

Domain: $(0, \infty)$

Range: $(-\infty, \infty)$

Intercept: $(1, 0)$

Increasing on $(0, \infty)$

y-axis is a vertical asymptote
$(\log_a x \to -\infty$ as $x \to 0^+)$

Continuous

Reflection of graph of $f(x) = a^x$
in the line $y = x$

Explore the Concept

 Use a graphing utility to graph $y = \log_{10} x$ and $y = 8$ in the same viewing window. Find a viewing window that shows the point of intersection. What is the point of intersection? Use the point of intersection to complete the equation $\log_{10} \boxed{} = 8$.

Example 6   Library of Parent Functions $f(x) = \log_a x$

Each of the following functions is a transformation of the graph of

$$f(x) = \log_{10} x.$$

a. Because $g(x) = \log_{10}(x - 1) = f(x - 1)$, the graph of *g* can be obtained by shifting the graph of *f* one unit to the *right*, as shown in Figure 4.18.

b. Because $h(x) = 2 + \log_{10} x = 2 + f(x)$, the graph of *h* can be obtained by shifting the graph of *f* two units *upward*, as shown in Figure 4.19.

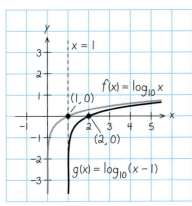

Figure 4.18

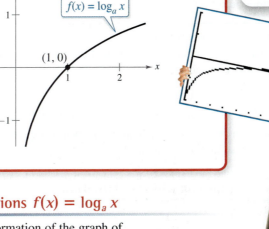

Figure 4.19

Notice that the transformation in Figure 4.19 keeps the *y*-axis as a vertical asymptote, but the transformation in Figure 4.18 yields the new vertical asymptote $x = 1$.

 CHECKPOINT   Now try Exercise 55.

The Natural Logarithmic Function

By looking back at the graph of the natural exponential function introduced in Section 4.1, you will see that $f(x) = e^x$ is one-to-one and so has an inverse function. This inverse function is called the **natural logarithmic function** and is denoted by the special symbol ln x, read as "the natural log of x" or "el en of x."

> **The Natural Logarithmic Function**
>
> For $x > 0$,
>
> $y = \ln x$ if and only if $x = e^y$.
>
> The function given by
>
> $f(x) = \log_e x = \ln x$
>
> is called the **natural logarithmic function.**

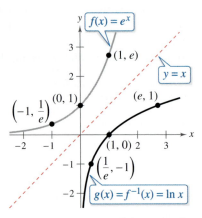

Reflection of graph of $f(x) = e^x$ *in the line y = x*

Figure 4.20

The equations $y = \ln x$ and $x = e^y$ are equivalent. Note that the natural logarithm ln x is written without a base. The base is understood to be e.

Because the functions

$$f(x) = e^x \quad \text{and} \quad g(x) = \ln x$$

are inverse functions of each other, their graphs are reflections of each other in the line $y = x$. This reflective property is illustrated in Figure 4.20.

Example 7 Evaluating the Natural Logarithmic Function

Use a calculator to evaluate the function

$$f(x) = \ln x$$

at each indicated value of x.

a. $x = 2$

b. $x = 0.3$

c. $x = -1$

Technology Tip

On most calculators, the natural logarithm is denoted by (LN), as illustrated in Example 7.

Solution

Function Value	Graphing Calculator Keystrokes	Display
a. $f(2) = \ln 2$	(LN) 2 (ENTER)	0.6931472
b. $f(0.3) = \ln 0.3$	(LN) .3 (ENTER)	-1.2039728
c. $f(-1) = \ln(-1)$	(LN) (−) 1 (ENTER)	ERROR

✓CHECKPOINT Now try Exercise 77.

The four properties of logarithms listed on page 337 are also valid for natural logarithms.

> **Properties of Natural Logarithms**
>
> **1.** $\ln 1 = 0$ because $e^0 = 1$.
>
> **2.** $\ln e = 1$ because $e^1 = e$.
>
> **3.** $\ln e^x = x$ and $e^{\ln x} = x$.   Inverse Properties
>
> **4.** If $\ln x = \ln y$, then $x = y$.   One-to-One Property

Study Tip

In Example 7(c), be sure you see that $\ln(-1)$ gives an error message on most calculators. This occurs because the domain of ln x is the set of *positive* real numbers (see Figure 4.20). So, $\ln(-1)$ is undefined.

Example 8   Using Properties of Natural Logarithms

Use the properties of natural logarithms to rewrite each expression.

a. $\ln \dfrac{1}{e}$ **b.** $e^{\ln 5}$ **c.** $4 \ln 1$ **d.** $2 \ln e$

Solution

a. $\ln \dfrac{1}{e} = \ln e^{-1} = -1$ *Inverse Property*   **b.** $e^{\ln 5} = 5$ *Inverse Property*

c. $4 \ln 1 = 4(0) = 0$ *Property 1*   **d.** $2 \ln e = 2(1) = 2$ *Property 2*

✓**CHECKPOINT**   Now try Exercise 81.

Example 9   Finding the Domains of Logarithmic Functions

Find the domain of each function.

a. $f(x) = \ln(x - 2)$ **b.** $g(x) = \ln(2 - x)$ **c.** $h(x) = \ln x^2$

Algebraic Solution

a. Because $\ln(x - 2)$ is defined only when

$$x - 2 > 0$$

it follows that the domain of f is $(2, \infty)$.

b. Because $\ln(2 - x)$ is defined only when

$$2 - x > 0$$

it follows that the domain of g is $(-\infty, 2)$.

c. Because $\ln x^2$ is defined only when

$$x^2 > 0$$

it follows that the domain of h is all real numbers except $x = 0$.

✓**CHECKPOINT**   Now try Exercise 89.

Graphical Solution

a.

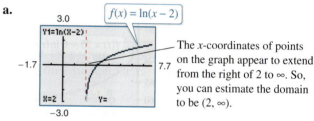

The x-coordinates of points on the graph appear to extend from the right of 2 to ∞. So, you can estimate the domain to be $(2, \infty)$.

b.

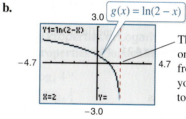

The x-coordinates of points on the graph appear to extend from $-\infty$ to the left of 2. So, you can estimate the domain to be $(-\infty, 2)$.

c.

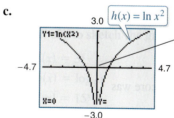

The x-coordinates of points on the graph appear to include all real numbers except 0. So, you can estimate the domain to be all real numbers except $x = 0$.

In Example 9, suppose you had been asked to analyze the function $h(x) = \ln|x - 2|$. How would the domain of this function compare with the domains of the functions given in parts (a) and (b) of the example?

Technology Tip

When a graphing utility graphs a logarithmic function, it may appear that the graph has an endpoint. This is because some graphing utilities have a limited resolution. So, in this text, a blue curve is placed behind the graphing utility's display to indicate where the graph should appear.

Application

Example 7 Finding a Mathematical Model

The table shows the mean distance x from the sun and the period y (the time it takes a planet to orbit the sun) for each of the six planets that are closest to the sun. In the table, the mean distance is given in astronomical units (where the Earth's mean distance is defined as 1.0), and the period is given in years. The points in the table are plotted in Figure 4.22. Find an equation that relates y and x.

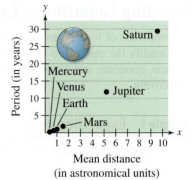

Figure 4.22

Planet	Mercury	Venus	Earth	Mars	Jupiter	Saturn
Mean distance, x	0.387	0.723	1.000	1.524	5.203	9.555
Period, y	0.241	0.615	1.000	1.881	11.860	29.420

Solution

From Figure 4.22, it is not clear how to find an equation that relates y and x. To solve this problem, take the natural log of each of the x- and y-values in the table. This produces the following results.

Planet	Mercury	Venus	Earth	Mars	Jupiter	Saturn
$\ln x = X$	−0.949	−0.324	0.000	0.421	1.649	2.257
$\ln y = Y$	−1.423	−0.486	0.000	0.632	2.473	3.382

Now, by plotting the points in the table, you can see that all six of the points appear to lie in a line, as shown in Figure 4.23. To find an equation of the line through these points, you can use one of the following methods.

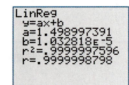

Figure 4.23

Method 1: Algebraic

Choose any two points to determine the slope of the line. Using the two points $(0.421, 0.632)$ and $(0, 0)$, you can determine that the slope of the line is

$$m = \frac{0.632 - 0}{0.421 - 0} \approx 1.5 = \frac{3}{2}.$$

By the point-slope form, the equation of the line is

$$Y = \tfrac{3}{2}X$$

where $Y = \ln y$ and $X = \ln x$. You can therefore conclude that

$$\ln y = \frac{3}{2} \ln x.$$

✓**CHECKPOINT** Now try Exercise 109.

Method 2: Graphical

Using the *linear regression* feature of a graphing utility, you can find a linear model for the data, as shown in Figure 4.24. You can approximate this model to be $Y = 1.5X$, where $Y = \ln y$ and $X = \ln x$. From the model, you can see that the slope of the line is $\frac{3}{2}$. So, you can conclude that

$$\ln y = \tfrac{3}{2} \ln x.$$

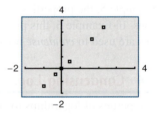

Figure 4.24

In Example 7, try to convert the final equation to $y = f(x)$ form. You will get a function of the form $y = ax^b$, which is called a *power model*.

4.3   Exercises

See www.CalcChat.com for worked-out solutions to odd-numbered exercises.
For instructions on how to use a graphing utility, see Appendix A.

Vocabulary and Concept Check

In Exercises 1 and 2, fill in the blank(s).

1. You can evaluate logarithms to any base using the _____ formula.

2. Two properties of logarithms are _____ $= n \log_a u$ and $\ln(uv) = $ _____.

3. Is $\log_3 24 = \dfrac{\ln 3}{\ln 24}$ or $\log_3 24 = \dfrac{\ln 24}{\ln 3}$ correct?

4. Which property of logarithms can you use to condense the expression $\ln x - \ln 2$?

Procedures and Problem Solving

Changing the Base   In Exercises 5–12, rewrite the logarithm as a ratio of (a) common logarithms and (b) natural logarithms.

5. $\log_5 x$
6. $\log_3 x$
7. $\log_{1/5} x$
8. $\log_{1/3} x$
9. $\log_a \frac{3}{10}$
10. $\log_a \frac{4}{5}$
11. $\log_{2.6} x$
12. $\log_{7.1} x$

Changing the Base   In Exercises 13–20, evaluate the logarithm using the change-of-base formula. Round your result to three decimal places.

✓ 13. $\log_3 7$
14. $\log_7 4$
15. $\log_{1/2} 4$
16. $\log_{1/8} 64$
17. $\log_6 0.9$
18. $\log_4 0.045$
✓ 19. $\log_{15} 1460$
20. $\log_{20} 175$

Using Properties of Logarithms   In Exercises 21–24, rewrite the expression in terms of $\ln 4$ and $\ln 5$.

✓ 21. $\ln 20$
22. $\ln 500$
23. $\ln \frac{25}{4}$
24. $\ln \frac{5}{2}$

Using Properties to Evaluate Logarithms   In Exercises 25–28, approximate the logarithm using the properties of logarithms, given the values $\log_b 2 \approx 0.3562$, $\log_b 3 \approx 0.5646$, and $\log_b 5 \approx 0.8271$. Round your result to four decimal places.

25. $\log_b 25$
26. $\log_b 30$
27. $\log_b \sqrt{3}$
28. $\log_b \frac{16}{25}$

Graphing a Logarithm   In Exercises 29–36, use the change-of-base formula $\log_a x = (\ln x)/(\ln a)$ and a graphing utility to graph the function.

29. $f(x) = \log_3(x + 2)$
30. $f(x) = \log_2(x - 1)$
31. $f(x) = \log_{1/2}(x - 2)$
32. $f(x) = \log_{1/3}(x + 1)$
33. $f(x) = \log_{1/4} x^2$
34. $f(x) = \log_3 \sqrt{x}$
35. $f(x) = \log_{1/2}\left(\dfrac{x}{2}\right)$
36. $f(x) = \log_5\left(\dfrac{x}{3}\right)$

Simplifying a Logarithm   In Exercises 37–44, use the properties of logarithms to rewrite and simplify the logarithmic expression.

37. $\log_4 8$
38. $\log_9 243$
39. $\log_2 4^2 \cdot 3^4$
40. $\log_3 9^2 \cdot 2^4$
41. $\ln 5e^6$
42. $\ln 8e^3$
43. $\ln \dfrac{6}{e^2}$
44. $\ln \dfrac{e^5}{7}$

Using Properties of Logarithms   In Exercises 45 and 46, use the properties of logarithms to verify the equation.

✓ 45. $\log_5 \frac{1}{250} = -3 - \log_5 2$
46. $-\ln 24 = -(3 \ln 2 + \ln 3)$

Expanding Logarithmic Expressions   In Exercises 47–64, use the properties of logarithms to expand the expression as a sum, difference, and/or constant multiple of logarithms. (Assume all variables are positive.)

47. $\log_{10} 5x$
48. $\log_{10} 10z$
49. $\log_{10} \dfrac{t}{8}$
50. $\log_{10} \dfrac{7}{z}$
51. $\log_8 x^4$
52. $\log_6 z^{-3}$
53. $\ln \sqrt{z}$
54. $\ln \sqrt[3]{t}$
55. $\ln xyz$
56. $\ln \dfrac{xy}{z}$
57. $\log_6 ab^3c^2$
58. $\log_4 xy^6 z^4$
59. $\ln \sqrt[3]{\dfrac{x}{y}}$
60. $\ln \sqrt{\dfrac{x^2}{y^3}}$
61. $\ln \dfrac{x^2 - 1}{x^3}, \quad x > 1$
62. $\ln \dfrac{x}{\sqrt{x^2 + 1}}$
✓ 63. $\ln \dfrac{x^4\sqrt{y}}{z^5}$
64. $\log_b \dfrac{\sqrt{x}y^4}{z^4}$

Algebraic-Graphical-Numerical In Exercises 65–68, (a) use a graphing utility to graph the two equations in the same viewing window and (b) use the *table* feature of the graphing utility to create a table of values for each equation. (c) What do the graphs and tables suggest? Verify your conclusion algebraically.

65. $y_1 = \ln[x^3(x + 4)]$, $y_2 = 3 \ln x + \ln(x + 4)$

66. $y_1 = \ln\left(\dfrac{\sqrt{x}}{x - 2}\right)$, $y_2 = \dfrac{1}{2} \ln x - \ln(x - 2)$

67. $y_1 = \ln\left(\dfrac{x^4}{x - 2}\right)$, $y_2 = 4 \ln x - \ln(x - 2)$

68. $y_1 = \ln 4x^3$, $y_2 = \ln 4 + 3 \ln x$

Condensing Logarithmic Expressions In Exercises 69–84, condense the expression to the logarithm of a single quantity.

69. $\ln x + \ln 4$ **70.** $\ln y + \ln z$

71. $\log_4 z - \log_4 y$ **72.** $\log_5 8 - \log_5 t$

73. $2 \log_2(x + 3)$ **74.** $\frac{5}{2} \log_7(z - 4)$

75. $\frac{1}{2} \ln(x^2 + 4)$ **76.** $2 \ln x + \ln(x + 1)$

77. $\ln x - 3 \ln(x + 1)$ **78.** $\ln x - 2 \ln(x + 2)$

79. $\ln(x - 2) - \ln(x + 2)$ **80.** $3 \ln x + 2 \ln y - 4 \ln z$

✓ **81.** $\ln x - 2[\ln(x + 2) + \ln(x - 2)]$

82. $4[\ln z + \ln(z + 5)] - 2 \ln(z - 5)$

83. $\frac{1}{3}[2 \ln(x + 3) + \ln x - \ln(x^2 - 1)]$

84. $2[\ln x - \ln(x + 1) - \ln(x - 1)]$

Algebraic-Graphical-Numerical In Exercises 85–88, (a) use a graphing utility to graph the two equations in the same viewing window and (b) use the *table* feature of the graphing utility to create a table of values for each equation. (c) What do the graphs and tables suggest? Verify your conclusion algebraically.

85. $y_1 = 2[\ln 8 - \ln(x^2 + 1)]$, $y_2 = \ln\left[\dfrac{64}{(x^2 + 1)^2}\right]$

86. $y_1 = 2[\ln 6 + \ln(x^2 + 1)]$, $y_2 = \ln[36(x^2 + 1)^2]$

87. $y_1 = \ln x + \frac{1}{2} \ln(x + 1)$, $y_2 = \ln\left(x\sqrt{x + 1}\right)$

88. $y_1 = \dfrac{1}{2} \ln x - \ln(x + 2)$, $y_2 = \ln\left(\dfrac{\sqrt{x}}{x + 2}\right)$

Algebraic-Graphical-Numerical In Exercises 89–92, (a) use a graphing utility to graph the two equations in the same viewing window and (b) use the *table* feature of the graphing utility to create a table of values for each equation. (c) Are the expressions equivalent? Explain. Verify your conclusion algebraically.

89. $y_1 = \ln x^2$, $y_2 = 2 \ln x$

90. $y_1 = 2(\ln 2 + \ln x)$, $y_2 = \ln 4x^2$

91. $y_1 = \ln(x - 2) + \ln(x + 2)$, $y_2 = \ln(x^2 - 4)$

92. $y_1 = \frac{1}{4} \ln[x^4(x^2 + 1)]$, $y_2 = \ln x + \frac{1}{4} \ln(x^2 + 1)$

Using Properties to Evaluate Logarithms In Exercises 93–106, find the exact value of the logarithm without using a calculator. If this is not possible, state the reason.

93. $\log_3 9$ **94.** $\log_6 \sqrt[3]{6}$

95. $\log_4 16^{3.4}$ **96.** $\log_5\left(\frac{1}{125}\right)$

97. $\log_2(-4)$ **98.** $\log_4(-16)$

99. $\log_5 75 - \log_5 3$ **100.** $\log_4 2 + \log_4 32$

101. $\ln e^3 - \ln e^7$

102. $\ln e^6 - 2 \ln e^5$

103. $2 \ln e^4$

104. $\ln e^{4.5}$

105. $\ln \dfrac{1}{\sqrt{e}}$

106. $\ln \sqrt[5]{e^3}$

107. *Why you should learn it* (p. 347) The relationship between the number of decibels β and the intensity of a sound I in watts per square meter is given by

$$\beta = 10 \log_{10}\left(\dfrac{I}{10^{-12}}\right).$$

(a) Use the properties of logarithms to write the formula in a simpler form.

(b) Use a graphing utility to complete the table. Verify your answers algebraically.

I	10^{-4}	10^{-6}	10^{-8}	10^{-10}	10^{-12}	10^{-14}
β						

108. Psychology Students participating in a psychology experiment attended several lectures and were given an exam. Every month for the next year, the students were retested to see how much of the material they remembered. The average scores for the group are given by the human memory model

$$f(t) = 90 - 15 \log_{10}(t + 1), \quad 0 \le t \le 12$$

where t is the time (in months).

(a) Use a graphing utility to graph the function over the specified domain.

(b) What was the average score on the original exam $(t = 0)$?

(c) What was the average score after 6 months?

(d) What was the average score after 12 months?

(e) When did the average score decrease to 75?

109. MODELING DATA

A beaker of liquid at an initial temperature of 78°C is placed in a room at a constant temperature of 21°C. The temperature of the liquid is measured every 5 minutes during a half-hour period. The results are recorded as ordered pairs of the form (t, T), where t is the time (in minutes) and T is the temperature (in degrees Celsius).

(0, 78.0°), (5, 66.0°), (10, 57.5°), (15, 51.2°), (20, 46.3°), (25, 42.5°), (30, 39.6°)

(a) The graph of the temperature of the room should be an asymptote of the graph of the model for the data. Subtract the room temperature from each of the temperatures in the ordered pairs. Use a graphing utility to plot the data points (t, T) and $(t, T - 21)$.

(b) An exponential model for the data $(t, T - 21)$ is given by

$$T - 21 = 54.4(0.964)^t.$$

Solve for T and graph the model. Compare the result with the plot of the original data.

(c) Take the natural logarithms of the revised temperatures. Use the graphing utility to plot the points $(t, \ln(T - 21))$ and observe that the points appear linear. Use the *regression* feature of the graphing utility to fit a line to the data. The resulting line has the form

$$\ln(T - 21) = at + b.$$

Use the properties of logarithms to solve for T. Verify that the result is equivalent to the model in part (b).

(d) Fit a rational model to the data. Take the reciprocals of the y-coordinates of the revised data points to generate the points

$$\left(t, \frac{1}{T - 21}\right).$$

Use the graphing utility to plot these points and observe that they appear linear. Use the *regression* feature of the graphing utility to fit a line to the data. The resulting line has the form

$$\frac{1}{T - 21} = at + b.$$

Solve for T, and use the graphing utility to graph the rational function and the original data points.

110. **Writing** Write a short paragraph explaining why the transformations of the data in Exercise 109 were necessary to obtain the models. Why did taking the logarithms of the temperatures lead to a linear scatter plot? Why did taking the reciprocals of the temperatures lead to a linear scatter plot?

Conclusions

True or False? In Exercises 111–116, determine whether the statement is true or false given that $f(x) = \ln x$, where $x > 0$. Justify your answer.

111. $f(ax) = f(a) + f(x), \ a > 0$

112. $f(x - a) = f(x) - f(a), \ x > a$

113. $\sqrt{f(x)} = \frac{1}{2}f(x)$ 114. $[f(x)]^n = nf(x)$

115. If $f(x) < 0$, then $0 < x < e$.

116. If $f(x) > 0$, then $x > e$.

117. **Error Analysis** Describe the error.

$$\ln\left(\frac{x^2}{\sqrt{x^2 + 4}}\right) \ne \frac{\ln x^2}{\ln\sqrt{x^2 + 4}}$$

118. **Think About It** Consider the functions below.

$$f(x) = \ln\frac{x}{2}, \quad g(x) = \frac{\ln x}{\ln 2}, \quad h(x) = \ln x - \ln 2$$

Which two functions have identical graphs? Verify your answer by using a graphing utility to graph all three functions in the same viewing window.

119. **Exploration** For how many integers between 1 and 20 can the natural logarithms be approximated given that $\ln 2 \approx 0.6931$, $\ln 3 \approx 1.0986$, and $\ln 5 \approx 1.6094$? Approximate these logarithms. (Do not use a calculator.)

120. **CAPSTONE** Show that each expression is equivalent to $\ln 8$. Then write three more expressions that are equivalent to $\ln 8$.

(a) $3 \ln 2$ (b) $-\ln\frac{1}{8}$ (c) $-3\ln\frac{1}{2}$

(d) $\ln 16 - \ln 2$ (e) $(\log_{10} 2 + \log_{10} 4) \div \log_{10} e$

121. **Think About It** Does $y_1 = \ln[x(x - 2)]$ have the same domain as $y_2 = \ln x + \ln(x - 2)$? Explain.

122. **Proof** Prove that $\dfrac{\log_a x}{\log_{a/b} x} = 1 + \log_a \dfrac{1}{b}$.

Cumulative Mixed Review

Using Rules of Exponents In Exercises 123–126, simplify the expression.

123. $\dfrac{24xy^{-2}}{16x^{-3}y}$ 124. $\left(\dfrac{2x^3}{3y}\right)^{-3}$

125. $(18x^3y^4)^{-3}(18x^3y^4)^4$ 126. $xy(x^{-1} + y^{-1})^{-1}$

Solving Polynomial Equations In Exercises 127–130, find all solutions of the equation.

127. $x^2 - 6x + 2 = 0$ 128. $2x^3 + 20x^2 + 50x = 0$

129. $x^4 - 19x^2 + 48 = 0$ 130. $9x^4 - 37x^2 + 4 = 0$

4.4   Solving Exponential and Logarithmic Equations

Introduction

So far in this chapter, you have studied the definitions, graphs, and properties of exponential and logarithmic functions. In this section, you will study procedures for *solving equations* involving exponential and logarithmic functions.

There are two basic strategies for solving exponential or logarithmic equations. The first is based on the One-to-One Properties and the second is based on the Inverse Properties. For $a > 0$ and $a \neq 1$, the following properties are true for all x and y for which

$$\log_a x \quad \text{and} \quad \log_a y$$

are defined.

One-to-One Properties

$a^x = a^y$ if and only if $x = y$.

$\log_a x = \log_a y$ if and only if $x = y$.

Inverse Properties

$a^{\log_a x} = x$

$\log_a a^x = x$

Example 1   Solving Simple Exponential and Logarithmic Equations

	Original Equation	Rewritten Equation	Solution	Property
a.	$2^x = 32$	$2^x = 2^5$	$x = 5$	One-to-One
b.	$\log_4 x - \log_4 8 = 0$	$\log_4 x = \log_4 8$	$x = 8$	One-to-One
c.	$\ln x - \ln 3 = 0$	$\ln x = \ln 3$	$x = 3$	One-to-One
d.	$\left(\frac{1}{3}\right)^x = 9$	$3^{-x} = 3^2$	$x = -2$	One-to-One
e.	$e^x = 7$	$\ln e^x = \ln 7$	$x = \ln 7$	Inverse
f.	$\ln x = -3$	$e^{\ln x} = e^{-3}$	$x = e^{-3}$	Inverse
g.	$\log_{10} x = -1$	$10^{\log_{10} x} = 10^{-1}$	$x = 10^{-1} = \frac{1}{10}$	Inverse
h.	$\log_3 x = 4$	$3^{\log_3 x} = 3^4$	$x = 81$	Inverse

 CHECKPOINT   Now try Exercise 27.

The strategies used in Example 1 are summarized as follows.

Strategies for Solving Exponential and Logarithmic Equations

1. Rewrite the original equation in a form that allows the use of the One-to-One Properties of exponential or logarithmic functions.

2. Rewrite an *exponential* equation in logarithmic form and apply the Inverse Property of logarithmic functions.

3. Rewrite a *logarithmic* equation in exponential form and apply the Inverse Property of exponential functions.

Solving Exponential Equations

Example 2   Solving Exponential Equations

Solve each equation.

a. $e^x = 72$

b. $3(2^x) = 42$

Algebraic Solution	Graphical Solution

Algebraic Solution

a.

$e^x = 72$	Write original equation.
$\ln e^x = \ln 72$	Take natural log of each side.
$x = \ln 72$	Inverse Property
$x \approx 4.28$	Use a calculator.

The solution is $x = \ln 72 \approx 4.28$. Check this in the original equation.

b.

$3(2^x) = 42$	Write original equation.
$2^x = 14$	Divide each side by 3.
$\log_2 2^x = \log_2 14$	Take log (base 2) of each side.
$x = \log_2 14$	Inverse Property
$x = \dfrac{\ln 14}{\ln 2}$	Change-of-base formula
$x \approx 3.81$	Use a calculator.

The solution is $x = \log_2 14 \approx 3.81$. Check this in the original equation.

Graphical Solution

To solve an equation using a graphing utility, you can graph the left- and right-hand sides of the equation and use the *intersect* feature.

a.

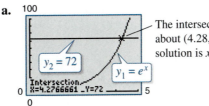

The intersection point is about (4.28, 72). So, the solution is $x \approx 4.28$.

b.

The intersection point is about (3.81, 42). So, the solution is $x \approx 3.81$.

CHECKPOINT   Now try Exercise 33.

Example 3   Solving an Exponential Equation

Solve $4e^{2x} - 3 = 2$.

Algebraic Solution

$4e^{2x} - 3 = 2$	Write original equation.
$4e^{2x} = 5$	Add 3 to each side.
$e^{2x} = \dfrac{5}{4}$	Divide each side by 4.
$\ln e^{2x} = \ln \dfrac{5}{4}$	Take natural log of each side.
$2x = \ln \dfrac{5}{4}$	Inverse Property
$x = \dfrac{1}{2} \ln \dfrac{5}{4}$	Divide each side by 2.
$x \approx 0.11$	Use a calculator.

The solution is

$$x = \tfrac{1}{2} \ln \tfrac{5}{4} \approx 0.11.$$

Check this in the original equation.

CHECKPOINT   Now try Exercise 61.

Graphical Solution

Rather than using the procedure in Example 2, another way to solve the equation graphically is first to rewrite the equation as

$$4e^{2x} - 5 = 0$$

and then use a graphing utility to graph

$$y = 4e^{2x} - 5.$$

Use the *zero* or *root* feature of the graphing utility to approximate the value of x for which $y = 0$, as shown in Figure 4.25.

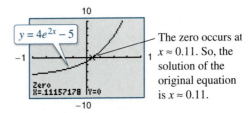

The zero occurs at $x \approx 0.11$. So, the solution of the original equation is $x \approx 0.11$.

Figure 4.25

Example 4 Solving an Exponential Equation

Solve $2(3^{2t-5}) - 4 = 11$.

Solution

$$2(3^{2t-5}) - 4 = 11$$ Write original equation.

$$2(3^{2t-5}) = 15$$ Add 4 to each side.

$$3^{2t-5} = \frac{15}{2}$$ Divide each side by 2.

$$\log_3 3^{2t-5} = \log_3 \frac{15}{2}$$ Take log (base 3) of each side.

$$2t - 5 = \log_3 \frac{15}{2}$$ Inverse Property

$$2t = 5 + \log_3 7.5$$ Add 5 to each side.

$$t = \frac{5}{2} + \frac{1}{2}\log_3 7.5$$ Divide each side by 2.

$$t \approx 3.42$$ Use a calculator.

The solution is $t = \frac{5}{2} + \frac{1}{2}\log_3 7.5 \approx 3.42$. Check this in the original equation.

 ✓CHECKPOINT   Now try Exercise 65.

Study Tip

 Remember that to evaluate a logarithm such as $\log_3 7.5$, you need to use the change-of-base formula.

$$\log_3 7.5 = \frac{\ln 7.5}{\ln 3} \approx 1.834$$

When an equation involves two or more exponential expressions, you can still use a procedure similar to that demonstrated in the previous three examples. However, the algebra is a bit more complicated.

Example 5 Solving an Exponential Equation in Quadratic Form

Solve $e^{2x} - 3e^x + 2 = 0$.

Algebraic Solution

$$e^{2x} - 3e^x + 2 = 0$$ Write original equation.

$$(e^x)^2 - 3e^x + 2 = 0$$ Write in quadratic form.

$$(e^x - 2)(e^x - 1) = 0$$ Factor.

$$e^x - 2 = 0$$ Set 1st factor equal to 0.

$$e^x = 2$$ Add 2 to each side.

$$x = \ln 2$$ Solution

$$e^x - 1 = 0$$ Set 2nd factor equal to 0.

$$e^x = 1$$ Add 1 to each side.

$$x = \ln 1$$ Inverse Property

$$x = 0$$ Solution

The solutions are

$$x = \ln 2 \approx 0.69 \text{ and } x = 0.$$

Check these in the original equation.

 ✓CHECKPOINT   Now try Exercise 71.

Graphical Solution

Use a graphing utility to graph $y = e^{2x} - 3e^x + 2$ and then find the zeros.

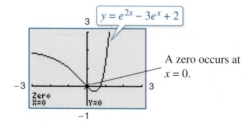

A zero occurs at $x = 0$.

Figure 4.26

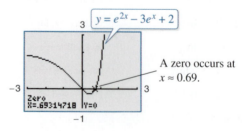

A zero occurs at $x \approx 0.69$.

Figure 4.27

From Figures 4.26 and 4.27, you can conclude that the solutions are $x = 0$ and $x \approx 0.69$.

Solving Logarithmic Equations

To solve a logarithmic equation, you can write it in exponential form.

$$\ln x = 3 \qquad \text{Logarithmic form}$$

$$e^{\ln x} = e^3 \qquad \text{Exponentiate each side.}$$

$$x = e^3 \qquad \text{Exponential form}$$

This procedure is called *exponentiating* each side of an equation. It is applied after the logarithmic expression has been isolated.

Example 6   Solving Logarithmic Equations

Solve each logarithmic equation.

a. $\ln 3x = 2$

b. $\log_3(5x - 1) = \log_3(x + 7)$

Solution

a.

$$\ln 3x = 2 \qquad \text{Write original equation.}$$

$$e^{\ln 3x} = e^2 \qquad \text{Exponentiate each side.}$$

$$3x = e^2 \qquad \text{Inverse Property}$$

$$x = \tfrac{1}{3}e^2 \qquad \text{Multiply each side by } \tfrac{1}{3}.$$

$$x \approx 2.46 \qquad \text{Use a calculator.}$$

The solution is $x = \tfrac{1}{3}e^2 \approx 2.46$. Check this in the original equation.

b.

$$\log_3(5x - 1) = \log_3(x + 7) \qquad \text{Write original equation.}$$

$$5x - 1 = x + 7 \qquad \text{One-to-One Property}$$

$$x = 2 \qquad \text{Solve for } x.$$

The solution is $x = 2$. Check this in the original equation.

✓**CHECKPOINT**   Now try Exercise 93.

Example 7   Solving a Logarithmic Equation

Solve $5 + 2 \ln x = 4$.

Algebraic Solution

$$5 + 2 \ln x = 4 \qquad \text{Write original equation.}$$

$$2 \ln x = -1 \qquad \text{Subtract 5 from each side.}$$

$$\ln x = -\tfrac{1}{2} \qquad \text{Divide each side by 2.}$$

$$e^{\ln x} = e^{-1/2} \qquad \text{Exponentiate each side.}$$

$$x = e^{-1/2} \qquad \text{Inverse Property}$$

$$x \approx 0.61 \qquad \text{Use a calculator.}$$

The solution is $x = e^{-1/2} \approx 0.61$. Check this in the original equation.

✓**CHECKPOINT**   Now try Exercise 97.

Graphical Solution

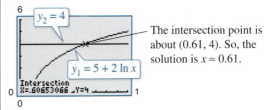

The intersection point is about $(0.61, 4)$. So, the solution is $x \approx 0.61$.

Example 8 Solving a Logarithmic Equation

Solve $2 \log_5 3x = 4$.

Solution

$2 \log_5 3x = 4$	Write original equation.
$\log_5 3x = 2$	Divide each side by 2.
$5^{\log_5 3x} = 5^2$	Exponentiate each side (base 5).
$3x = 25$	Inverse Property
$x = \frac{25}{3}$	Divide each side by 3.

The solution is $x = \frac{25}{3}$. Check this in the original equation. Or, perform a graphical check by graphing

$$y_1 = 2 \log_5 3x = 2\left(\frac{\log_{10} 3x}{\log_{10} 5}\right) \quad \text{and} \quad y_2 = 4$$

in the same viewing window. The two graphs should intersect at

$$x = \frac{25}{3} \approx 8.33$$

and

$$y = 4$$

as shown in Figure 4.28.

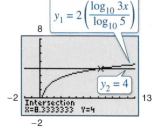

Figure 4.28

✔CHECKPOINT   Now try Exercise 99.

Because the domain of a logarithmic function generally does not include all real numbers, you should be sure to check for extraneous solutions of logarithmic equations, as shown in the next example.

Example 9 Checking for Extraneous Solutions

Solve $\ln(x - 2) + \ln(2x - 3) = 2 \ln x$.

Algebraic Solution

$\ln(x - 2) + \ln(2x - 3) = 2 \ln x$	Write original equation.
$\ln[(x - 2)(2x - 3)] = \ln x^2$	Use properties of logarithms.
$\ln(2x^2 - 7x + 6) = \ln x^2$	Multiply binomials.
$2x^2 - 7x + 6 = x^2$	One-to-One Property
$x^2 - 7x + 6 = 0$	Write in general form.
$(x - 6)(x - 1) = 0$	Factor.
$x - 6 = 0 \implies x = 6$	Set 1st factor equal to 0.
$x - 1 = 0 \implies x = 1$	Set 2nd factor equal to 0.

Finally, by checking these two "solutions" in the original equation, you can conclude that $x = 1$ is not valid. This is because when $x = 1$,

$$\ln(x - 2) + \ln(2x - 3) = \ln(-1) + \ln(-1)$$

which is invalid because -1 is not in the domain of the natural logarithmic function. So, the only solution is $x = 6$.

✔CHECKPOINT   Now try Exercise 109.

Graphical Solution

First rewrite the original equation as

$$\ln(x - 2) + \ln(2x - 3) - 2 \ln x = 0.$$

Then use a graphing utility to graph the equation

$$y = \ln(x - 2) + \ln(2x - 3) - 2 \ln x$$

and find the zeros (see Figure 4.29).

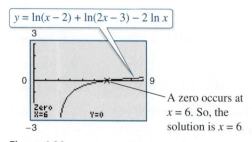

A zero occurs at $x = 6$. So, the solution is $x = 6$.

Figure 4.29

Example 10   The Change-of-Base Formula

Prove the change-of-base formula: $\log_a x = \dfrac{\log_b x}{\log_b a}$.

Solution

Begin by letting

$$y = \log_a x$$

and writing the equivalent exponential form

$$a^y = x.$$

Now, taking the logarithms *with base b* of each side produces the following.

$$\log_b a^y = \log_b x$$

$$y \log_b a = \log_b x \qquad \text{Power Property}$$

$$y = \frac{\log_b x}{\log_b a} \qquad \text{Divide each side by } \log_b a.$$

$$\log_a x = \frac{\log_b x}{\log_b a} \qquad \text{Replace } y \text{ with } \log_a x.$$

✓**CHECKPOINT**   Now try Exercise 113.

Equations that involve combinations of algebraic functions, exponential functions, and/or logarithmic functions can be very difficult to solve by algebraic procedures. Here again, you can take advantage of a graphing utility.

Example 11   Approximating the Solution of an Equation

Approximate (to three decimal places) the solution of $\ln x = x^2 - 2$.

Solution

First, rewrite the equation as

$$\ln x - x^2 + 2 = 0.$$

Then use a graphing utility to graph

$$y = -x^2 + 2 + \ln x$$

as shown in Figure 4.30. From this graph, you can see that the equation has two solutions. Next, using the *zero* or *root* feature, you can approximate the two solutions to be $x \approx 0.138$ and $x \approx 1.564$.

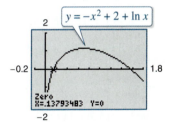

Figure 4.30

Check

$$\ln x = x^2 - 2 \qquad \text{Write original equation.}$$

$$\ln(0.138) \overset{?}{\approx} (0.138)^2 - 2 \qquad \text{Substitute 0.138 for } x.$$

$$-1.9805 \approx -1.9810 \qquad \text{Solution checks. } ✓$$

$$\ln(1.564) \overset{?}{\approx} (1.564)^2 - 2 \qquad \text{Substitute 1.564 for } x.$$

$$0.4472 \approx 0.4461 \qquad \text{Solution checks. } ✓$$

So, the two solutions $x \approx 0.138$ and $x \approx 1.564$ seem reasonable.

✓**CHECKPOINT**   Now try Exercise 119.

Applications

Example 12   Doubling an Investment

You have deposited $500 in an account that pays 6.75% interest, compounded continuously. How long will it take your money to double?

Solution

Using the formula for continuous compounding, you can find that the balance in the account is

$A = Pe^{rt} = 500e^{0.0675t}.$

To find the time required for the balance to double, let $A = 1000$, and solve the resulting equation for t.

$500e^{0.0675t} = 1000$ Substitute 1000 for A.

$e^{0.0675t} = 2$ Divide each side by 500.

$\ln e^{0.0675t} = \ln 2$ Take natural log of each side.

$0.0675t = \ln 2$ Inverse Property

$t = \dfrac{\ln 2}{0.0675}$ Divide each side by 0.0675.

$t \approx 10.27$ Use a calculator.

The balance in the account will double after approximately 10.27 years. This result is demonstrated graphically in Figure 4.31.

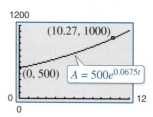

Figure 4.31

 CHECKPOINT   Now try Exercise 143.

Example 13   Average Salary for Public School Teachers

From 1985 through 2007, the average salary y (in thousands of dollars) for public school teachers for the year t can be modeled by the equation

$y = -2.983 + 15.206 \ln t, \quad 5 \le t \le 27$

where $t = 5$ represents 1985. During which year did the average salary for public school teachers reach $45,000?   (Source: National Education Association)

Solution

$-2.983 + 15.206 \ln t = y$ Write original equation.

$-2.983 + 15.206 \ln t = 45$ Substitute 45 for y.

$15.206 \ln t = 47.983$ Add 2.983 to each side.

$\ln t = \dfrac{47.983}{15.206}$ Divide each side by 15.206.

$e^{\ln t} = e^{47.983/15.206}$ Exponentiate each side.

$t = e^{47.983/15.206}$ Inverse Property

$t \approx 23.47$ Use a calculator.

The solution is $t \approx 23.47$ years. Because $t = 5$ represents 1985, it follows that the average salary for public school teachers reached $45,000 in 2003.

 CHECKPOINT   Now try Exercise 149.

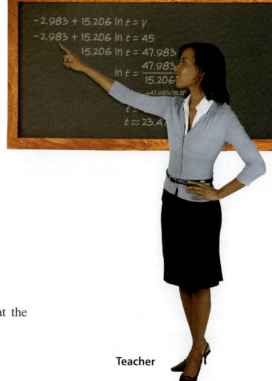

Teacher

4.4 Exercises

See www.CalcChat.com for worked-out solutions to odd-numbered exercises.
For instructions on how to use a graphing utility, see Appendix A.

Vocabulary and Concept Check

In Exercise 1 and 2, fill in the blank.

1. To solve exponential and logarithmic equations, you can use the following One-to-One and Inverse Properties.

    (a) $a^x = a^y$ if and only if _____ .   (b) $\log_a x = \log_a y$ if and only if _____ .

    (c) $a^{\log_a x} = $ _____   (d) $\log_a a^x = $ _____

2. An _____ solution does not satisfy the original equation.

3. What is the value of $\ln e^7$?

4. Can you solve $5^x = 125$ using a One-to-One Property?

5. What is the first step in solving the equation $3 + \ln x = 10$?

6. Do you solve $\log_4 x = 2$ by using a One-to-One Property or an Inverse Property?

Procedures and Problem Solving

Checking Solutions In Exercises 7–14, determine whether each x-value is a solution of the equation.

7. $4^{2x-7} = 64$

    (a) $x = 5$

    (b) $x = 2$

8. $2^{3x+1} = 32$

    (a) $x = -1$

    (b) $x = 2$

9. $3e^{x+2} = 75$

    (a) $x = -2 + e^{25}$

    (b) $x = -2 + \ln 25$

    (c) $x \approx 1.2189$

10. $4e^{x-1} = 60$

    (a) $x = 1 + \ln 15$

    (b) $x \approx 3.7081$

    (c) $x = \ln 16$

11. $\log_4(3x) = 3$

    (a) $x \approx 21.3560$

    (b) $x = -4$

    (c) $x = \frac{64}{3}$

12. $\log_6\left(\frac{5}{3}x\right) = 2$

    (a) $x \approx 20.2882$

    (b) $x = \frac{108}{5}$

    (c) $x = 7.2$

13. $\ln(x - 1) = 3.8$

    (a) $x = 1 + e^{3.8}$

    (b) $x \approx 45.7012$

    (c) $x = 1 + \ln 3.8$

14. $\ln(2 + x) = 2.5$

    (a) $x = e^{2.5} - 2$

    (b) $x \approx \frac{4073}{400}$

    (c) $x = \frac{1}{2}$

Solving Equations Graphically In Exercises 15–22, use a graphing utility to graph f and g in the same viewing window. Approximate the point of intersection of the graphs of f and g. Then solve the equation $f(x) = g(x)$ algebraically.

15. $f(x) = 2^x$

    $g(x) = 8$

16. $f(x) = 27^x$

    $g(x) = 9$

17. $f(x) = 5^{x-2} - 15$

    $g(x) = 10$

18. $f(x) = 2^{-x+1} - 3$

    $g(x) = 13$

19. $f(x) = 4 \log_3 x$

    $g(x) = 20$

20. $f(x) = 3 \log_5 x$

    $g(x) = 6$

21. $f(x) = \ln e^{x+1}$

    $g(x) = 2x + 5$

22. $f(x) = \ln e^{x-2}$

    $g(x) = 3x + 2$

Solving an Exponential Equation In Exercises 23–36, solve the exponential equation.

23. $4^x = 16$

24. $3^x = 243$

25. $5^x = \frac{1}{625}$

26. $7^x = \frac{1}{49}$

✓ 27. $\left(\frac{1}{8}\right)^x = 64$

28. $\left(\frac{1}{2}\right)^x = 32$

29. $\left(\frac{2}{3}\right)^x = \frac{81}{16}$

30. $\left(\frac{3}{4}\right)^x = \frac{27}{64}$

31. $e^x = 14$

32. $e^x = 66$

✓ 33. $6(10^x) = 216$

34. $5(8^x) = 325$

35. $2^{x+3} = 256$

36. $3^{x-1} = \frac{1}{81}$

Solving a Logarithmic Equation In Exercises 37–46, solve the logarithmic equation.

37. $\ln x - \ln 5 = 0$

38. $\ln x - \ln 2 = 0$

39. $\ln x = -9$

40. $\ln x = -14$

41. $\log_x 625 = 4$

42. $\log_x 25 = 2$

43. $\log_{10} x = -1$

44. $\log_{10} x = -\frac{1}{2}$

45. $\ln(2x - 1) = 5$

46. $\ln(3x + 5) = 8$

Using Inverse Properties In Exercises 47–54, simplify the expression.

47. $\ln e^{x^2}$

48. $\ln e^{2x-1}$

49. $e^{\ln x^2}$

50. $e^{\ln(x^2+2)}$

51. $-1 + \ln e^{2x}$

52. $-4 + e^{\ln x^4}$

53. $5 + e^{\ln(x^2+1)}$

54. $3 - \ln(e^{x^2+2})$

Solving an Exponential Equation In Exercises 55–80, solve the exponential equation algebraically. Round your result to three decimal places. Use a graphing utility to verify your answer.

55. $8^{3x} = 360$

56. $6^{5x} = 3000$

57. $5^{-t/2} = 0.20$

58. $4^{-3t} = 0.10$

59. $250e^{0.02x} = 10,000$

60. $100e^{0.005x} = 125,000$

✓ **61.** $500e^{-x} = 300$

62. $1000e^{-4x} = 75$

63. $7 - 2e^x = 5$

64. $-14 + 3e^x = 11$

✓ **65.** $5(2^{3-x}) - 13 = 100$

66. $6(8^{-2-x}) + 15 = 2601$

67. $\left(1 + \dfrac{0.10}{12}\right)^{12t} = 2$

68. $\left(16 + \dfrac{0.878}{26}\right)^{3t} = 30$

69. $5000\left[\dfrac{(1 + 0.005)^x}{0.005}\right] = 250,000$

70. $250\left[\dfrac{(1 + 0.01)^x}{0.01}\right] = 150,000$

✓ **71.** $e^{2x} - 4e^x - 5 = 0$

72. $e^{2x} - 5e^x + 6 = 0$

73. $e^x = e^{x^2-2}$

74. $e^{2x} = e^{x^2-8}$

75. $e^{x^2-3x} = e^{x-2}$

76. $e^{-x^2} = e^{x^2-2x}$

77. $\dfrac{400}{1 + e^{-x}} = 350$

78. $\dfrac{525}{1 + e^{-x}} = 275$

79. $\dfrac{40}{1 - 5e^{-0.01x}} = 200$

80. $\dfrac{50}{1 - 2e^{-0.001x}} = 1000$

Algebraic-Graphical-Numerical In Exercises 81 and 82, (a) complete the table to find an interval containing the solution of the equation, (b) use a graphing utility to graph both sides of the equation to estimate the solution, and (c) solve the equation algebraically. Round your results to three decimal places.

81. $e^{3x} = 12$

x	0.6	0.7	0.8	0.9	1.0
e^{3x}					

82. $20(100 - e^{x/2}) = 500$

x	5	6	7	8	9
$20(100 - e^{x/2})$					

Solving an Exponential Equation Graphically In Exercises 83–86, use the *zero* or *root* feature or the *zoom* and *trace* features of a graphing utility to approximate the solution of the exponential equation accurate to three decimal places.

83. $\left(1 + \dfrac{0.065}{365}\right)^{365t} = 4$

84. $\left(4 - \dfrac{2.471}{40}\right)^{9t} = 21$

85. $\dfrac{3000}{2 + e^{2x}} = 2$

86. $\dfrac{119}{e^{6x} - 14} = 7$

Finding the Zero of a Function In Exercises 87–90, use a graphing utility to graph the function and approximate its zero accurate to three decimal places.

87. $g(x) = 6e^{1-x} - 25$

88. $f(x) = 3e^{3x/2} - 962$

89. $g(t) = e^{0.09t} - 3$

90. $h(t) = e^{0.125t} - 8$

Solving a Logarithmic Equation In Exercises 91–112, solve the logarithmic equation algebraically. Round the result to three decimal places. Verify your answer using a graphing utility.

91. $\ln x = -3$

92. $\ln x = -4$

✓ **93.** $\ln 4x = 2.1$

94. $\ln 2x = 1.5$

95. $\log_5(3x + 2) = \log_5(6 - x)$

96. $\log_9(4 + x) = \log_9(2x - 1)$

✓ **97.** $-2 + 2\ln 3x = 17$

98. $3 + 2\ln x = 10$

✓ **99.** $7\log_4(0.6x) = 12$

100. $4\log_{10}(x - 6) = 11$

101. $\log_{10}(z - 3) = 2$

102. $\log_{10} x^2 = 6$

103. $\ln \sqrt{x + 2} = 1$

104. $\ln \sqrt{x - 8} = 5$

105. $\ln(x + 1)^2 = 2$

106. $\ln(x^2 + 1) = 8$

107. $\log_4 x - \log_4(x - 1) = \frac{1}{2}$

108. $\log_3 x + \log_3(x - 8) = 2$

✓ **109.** $\ln(x + 5) = \ln(x - 1) - \ln(x + 1)$

110. $\ln(x + 1) - \ln(x - 2) = \ln x$

111. $\log_{10} 8x - \log_{10}(1 + \sqrt{x}) = 2$

112. $\log_{10} 4x - \log_{10}(12 + \sqrt{x}) = 2$

The Change-of-Base Formula In Exercises 113 and 114, use the method of Example 10 to prove the change-of-base formula for the indicated base.

✓ **113.** $\log_a x = \dfrac{\log_e x}{\log_e a}$

114. $\log_a x = \dfrac{\log_{10} x}{\log_{10} a}$

Algebraic-Graphical-Numerical In Exercises 115–118, (a) complete the table to find an interval containing the solution of the equation, (b) use a graphing utility to graph both sides of the equation to estimate the solution, and (c) solve the equation algebraically. Round your results to three decimal places.

115. $\ln 2x = 2.4$

x	2	3	4	5	6
$\ln 2x$					

116. $3\ln 5x = 10$

x	4	5	6	7	8
$3\ln 5x$					

117. $6 \log_3(0.5x) = 11$

x	12	13	14	15	16
$6 \log_3(0.5x)$					

118. $5 \log_{10}(x - 2) = 11$

x	150	155	160	165	170
$5 \log_{10}(x - 2)$					

Approximating the Solution of an Equation   In Exercises 119–124, use the *zero* or *root* feature of a graphing utility to approximate the solution of the logarithmic equation.

✓ **119.** $\log_{10} x = x^3 - 3$ **120.** $\log_{10} x^2 = 4$
121. $\ln x + \ln(x - 2) = 1$ **122.** $\ln x + \ln(x + 1) = 2$
123. $\ln(x - 3) + \ln(x + 3) = 1$
124. $\ln x + \ln(x^2 + 4) = 10$

Finding the Point of Intersection   In Exercises 125–130, use a graphing utility to approximate the point of intersection of the graphs. Round your result to three decimal places.

125. $y_1 = 7$
$\quad y_2 = 2^{x-1} - 5$

126. $y_1 = 4$
$\quad y_2 = 3^{x+1} - 2$

127. $y_1 = 80$
$\quad y_2 = 4e^{-0.2x}$

128. $y_1 = 500$
$\quad y_2 = 1500e^{-x/2}$

129. $y_1 = 3.25$
$\quad y_2 = \frac{1}{2}\ln(x + 2)$

130. $y_1 = 1.05$
$\quad y_2 = \ln \sqrt{x - 2}$

Solving Exponential and Logarithmic Equations   In Exercises 131–138, solve the equation algebraically. Round the result to three decimal places. Verify your answer using a graphing utility.

131. $2x^2 e^{2x} + 2x e^{2x} = 0$ **132.** $-x^2 e^{-x} + 2x e^{-x} = 0$
133. $-x e^{-x} + e^{-x} = 0$ **134.** $e^{-2x} - 2x e^{-2x} = 0$
135. $2x \ln x + x = 0$ **136.** $\dfrac{1 - \ln x}{x^2} = 0$
137. $\dfrac{1 + \ln x}{2} = 0$ **138.** $2x \ln\left(\dfrac{1}{x}\right) - x = 0$

Solving a Model for x   In Exercises 139–142, the equation represents the given type of model, which you will use in Section 4.5. Solve the equation for x.

Model type	*Equation*
139. Exponential growth	$y = ae^{bx}$
140. Exponential decay	$y = ae^{-bx}$
141. Gaussian	$y = ae^{-(x-b)^2/c}$
142. Logarithmic	$y = a + b \ln x$

Doubling and Tripling an Investment   In Exercises 143–146, find the time required for a $1000 investment to (a) double at interest rate r, compounded continuously, and (b) triple at interest rate r, compounded continuously. Round your results to two decimal places.

✓ **143.** $r = 7.5\%$ **144.** $r = 6\%$
145. $r = 2.5\%$ **146.** $r = 3.75\%$

147. Economics   The demand x for a handheld electronic organizer is given by

$$p = 5000\left(1 - \frac{4}{4 + e^{-0.002x}}\right)$$

where p is the price in dollars. Find the demands x for prices of (a) $p = \$300$ and (b) $p = \$250$.

148. *Why you should learn it* (p. 354)   The percent m of American males between the ages of 18 and 24 who are no more than x inches tall is modeled by

$$m(x) = \frac{100}{1 + e^{-0.6114(x - 69.71)}}$$

and the percent f of American females between the ages of 18 and 24 who are no more than x inches tall is modeled by

$$f(x) = \frac{100}{1 + e^{-0.66607(x - 64.51)}}.$$

(Source: U.S. National Center for Health Statistics)

(a) Use a graphing utility to graph the two functions in the same viewing window.

(b) Use the graphs in part (a) to determine the horizontal asymptotes of the functions. Interpret their meanings in the context of the problem.

(c) What is the average height for each sex?

✓ **149. Finance**   The numbers y of commercial banks in the United States from 1999 through 2009 can be modeled by

$$y = 13{,}107 - 2077.6 \ln t, \quad 9 \le t \le 19$$

where t represents the year, with $t = 9$ corresponding to 1999. In what year were there about 7100 commercial banks?   (Source: Federal Deposit Insurance Corp.)

150. Forestry   The yield V (in millions of cubic feet per acre) for a forest at age t years is given by $V = 6.7e^{-48.1/t}$.

(a) Use a graphing utility to graph the function.

(b) Determine the horizontal asymptote of the function. Interpret its meaning in the context of the problem.

(c) Find the time necessary to obtain a yield of 1.3 million cubic feet.

151. Science An object at a temperature of 160°C was removed from a furnace and placed in a room at 20°C. The temperature T of the object was measured after each hour h and recorded in the table. A model for the data is given by $T = 20[1 + 7(2^{-h})]$.

Hour, h	Temperature
0	160°
1	90°
2	56°
3	38°
4	29°
5	24°

(a) Use a graphing utility to plot the data and graph the model in the same viewing window.

(b) Identify the horizontal asymptote of the graph. Interpret its meaning in the context of the problem.

(c) Approximate the time when the temperature of the object is 100°C.

152. MODELING DATA

The table shows the numbers N of college-bound seniors intending to major in computer or information sciences who took the SAT exam from 2001 through 2009. The data can be modeled by the logarithmic function $N = 77,010 - 21,554.3 \ln t$, where t represents the year, with $t = 1$ corresponding to 2001. (Source: The College Board)

Year	Number, N
2001	73,466
2002	68,051
2003	53,449
2004	45,879
2005	42,890
2006	37,943
2007	33,965
2008	30,495
2009	31,022

(a) According to the model, in what year would 25,325 seniors intending to major in computer or information sciences take the SAT exam?

(b) Use a graphing utility to graph the model with the data, and use the graph to verify your answer in part (a).

(c) Do you think this is a good model for predicting future values? Explain.

Conclusions

True or False? In Exercises 153 and 154, determine whether the statement is true or false. Justify your answer.

153. An exponential equation must have at least one solution.

154. A logarithmic equation can have at most one extraneous solution.

155. Error Analysis Describe the error.

$$2e^x = 10$$
$$\ln(2e^x) = \ln 10$$
$$2x = \ln 10$$
$$x = \tfrac{1}{2} \ln 10$$

156. CAPSTONE Write two or three sentences stating the general guidelines that you follow when you solve (a) exponential equations and (b) logarithmic equations.

157. Think About It Would you use a One-to-One Property or an Inverse Property to solve $5^x = 34$? Explain.

158. Exploration Let $f(x) = \log_a x$ and $g(x) = a^x$, where $a > 1$.

(a) Let $a = 1.2$ and use a graphing utility to graph the two functions in the same viewing window. What do you observe? Approximate any points of intersection of the two graphs.

(b) Determine the value(s) of a for which the two graphs have one point of intersection.

(c) Determine the value(s) of a for which the two graphs have two points of intersection.

159. Think About It Is the time required for a continuously compounded investment to quadruple twice as long as the time required for it to double? Give a reason for your answer and verify your answer algebraically.

160. Writing Write a paragraph explaining whether or not the time required for a continuously compounded investment to double is dependent on the size of the investment.

Cumulative Mixed Review

Sketching Graphs In Exercises 161–166, sketch the graph of the function.

161. $f(x) = 3x^3 - 4$

162. $f(x) = -(x + 1)^3 + 2$

163. $f(x) = |x| + 9$

164. $f(x) = |x + 2| - 8$

165. $f(x) = \begin{cases} 2x, & x < 0 \\ -x^2 + 4, & x \geq 0 \end{cases}$

166. $f(x) = \begin{cases} x - 9, & x \leq -1 \\ x^2 + 1, & x > -1 \end{cases}$

4.5   Exponential and Logarithmic Models

Introduction

There are many examples of exponential and logarithmic models in real life. In Section 4.1, you used the formula

$$A = Pe^{rt} \qquad \text{Exponential model}$$

to find the balance in an account when the interest was compounded continuously. In Section 4.2, Example 10, you used the human memory model

$$f(t) = 75 - 6\ln(t + 1). \qquad \text{Logarithmic model}$$

The five most common types of mathematical models involving exponential functions or logarithmic functions are as follows.

1. **Exponential growth model:** $y = ae^{bx}, \quad b > 0$

2. **Exponential decay model:** $y = ae^{-bx}, \quad b > 0$

3. **Gaussian model:** $y = ae^{-(x-b)^2/c}$

4. **Logistic growth model:** $y = \dfrac{a}{1 + be^{-rx}}$

5. **Logarithmic models:** $y = a + b\ln x, \quad y = a + b\log_{10} x$

The basic shapes of these graphs are shown in Figure 4.32.

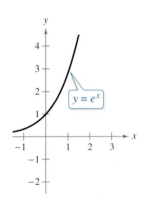

Exponential Growth Model

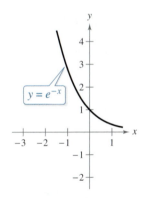

Exponential Decay Model

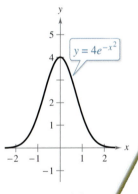

Gaussian Model

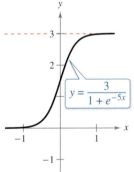

Logistic Growth Model
Figure 4.32

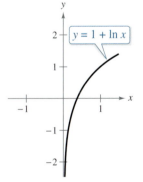

Natural Logarithmic Model

Common Logarithmic Model

You can often gain quite a bit of insight into a situation modeled by an exponential or logarithmic function by identifying and interpreting the function's asymptotes.

What you should learn

- Recognize the five most common types of models involving exponential or logarithmic functions.
- Use exponential growth and decay functions to model and solve real-life problems.
- Use Gaussian functions to model and solve real-life problems.
- Use logistic growth functions to model and solve real-life problems.
- Use logarithmic functions to model and solve real-life problems.

Why you should learn it

Exponential decay models are used in carbon dating. For instance, in Exercise 37 on page 374, you will use an exponential decay model to estimate the age of a piece of ancient charcoal.

Archaeologist

Explore the Concept

Use a graphing utility to graph each model shown in Figure 4.32. Use the *table* and *trace* features of the graphing utility to identify the asymptotes of the graph of each function.

Exponential Growth and Decay

Example 1 Demography

Estimates of the world population (in millions) from 2003 through 2009 are shown in the table. A scatter plot of the data is shown in Figure 4.33. (Source: U.S. Census Bureau)

Year	Population, P
2003	6313
2004	6387
2005	6462
2006	6538
2007	6615
2008	6691
2009	6768

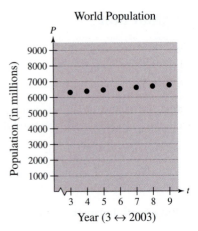

World Population

Figure 4.33

An exponential growth model that approximates these data is given by

$$P = 6097e^{0.0116t}, \quad 3 \le t \le 9$$

where P is the population (in millions) and $t = 3$ represents 2003. Compare the values given by the model with the estimates shown in the table. According to this model, when will the world population reach 7.1 billion?

Algebraic Solution

The following table compares the two sets of population figures.

Year	2003	2004	2005	2006	2007	2008	2009
Population	6313	6387	6462	6538	6615	6691	6768
Model	6313	6387	6461	6536	6613	6690	6768

From the table, it appears that the model is a good fit for the data. To find when the world population will reach 7.1 billion, let

$$P = 7100$$

in the model and solve for t.

$6097e^{0.0116t} = P$	Write original equation.
$6097e^{0.0116t} = 7100$	Substitute 7100 for P.
$e^{0.0116t} \approx 1.16451$	Divide each side by 6097.
$\ln e^{0.0116t} \approx \ln 1.16451$	Take natural log of each side.
$0.0116t \approx 0.15230$	Inverse Property
$t \approx 13.1$	Divide each side by 0.0116.

According to the model, the world population will reach 7.1 billion in 2013.

 Now try Exercise 33.

Graphical Solution

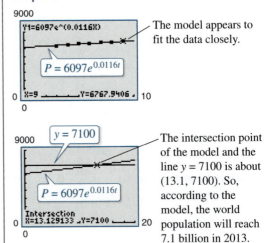

The model appears to fit the data closely.

The intersection point of the model and the line $y = 7100$ is about $(13.1, 7100)$. So, according to the model, the world population will reach 7.1 billion in 2013.

An exponential model increases (or decreases) by the same percent each year. What is the annual percent increase for the model in Example 1?

In Example 1, you were given the exponential growth model. Sometimes you must find such a model. One technique for doing this is shown in Example 2.

Example 2 Modeling Population Growth

In a research experiment, a population of fruit flies is increasing according to the law of exponential growth. After 2 days there are 100 flies, and after 4 days there are 300 flies. How many flies will there be after 5 days?

Solution

Let y be the number of flies at time t (in days). From the given information, you know that $y = 100$ when $t = 2$ and $y = 300$ when $t = 4$. Substituting this information into the model $y = ae^{bt}$ produces

$$100 = ae^{2b} \quad \text{and} \quad 300 = ae^{4b}.$$

To solve for b, solve for a in the first equation.

$$100 = ae^{2b} \quad \Longrightarrow \quad a = \frac{100}{e^{2b}} \qquad \text{Solve for } a \text{ in the first equation.}$$

Then substitute the result into the second equation.

$$300 = ae^{4b} \qquad\qquad \text{Write second equation.}$$

$$300 = \left(\frac{100}{e^{2b}}\right)e^{4b} \qquad \text{Substitute } \tfrac{100}{e^{2b}} \text{ for } a.$$

$$300 = 100e^{2b} \qquad\qquad \text{Simplify.}$$

$$3 = e^{2b} \qquad\qquad \text{Divide each side by 100.}$$

$$\ln 3 = \ln e^{2b} \qquad\qquad \text{Take natural log of each side.}$$

$$\ln 3 = 2b \qquad\qquad \text{Inverse Property}$$

$$\frac{1}{2}\ln 3 = b \qquad\qquad \text{Solve for } b.$$

Using $b = \tfrac{1}{2}\ln 3$ and the equation you found for a, you can determine that

$$a = \frac{100}{e^{2[(1/2)\ln 3]}} \qquad \text{Substitute } \tfrac{1}{2}\ln 3 \text{ for } b.$$

$$= \frac{100}{e^{\ln 3}} \qquad\qquad \text{Simplify.}$$

$$= \frac{100}{3} \qquad\qquad \text{Inverse Property}$$

$$\approx 33.33. \qquad\qquad \text{Simplify.}$$

So, with $a \approx 33.33$ and

$$b = \tfrac{1}{2}\ln 3 \approx 0.5493$$

the exponential growth model is

$$y = 33.33e^{0.5493t},$$

as shown in Figure 4.34. This implies that after 5 days, the population will be

$$y = 33.33e^{0.5493(5)} \approx 520 \text{ flies.}$$

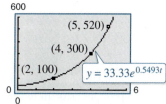

Figure 4.34

 CHECKPOINT   Now try Exercise 35.

In living organic material, the ratio of the content of radioactive carbon isotopes (carbon 14) to the content of nonradioactive carbon isotopes (carbon 12) is about 1 to 10^{12}. When organic material dies, its carbon 12 content remains fixed, whereas its radioactive carbon 14 begins to decay with a half-life of 5700 years. To estimate the age of dead organic material, scientists use the following formula, which denotes the ratio of carbon 14 to carbon 12 present at any time t (in years).

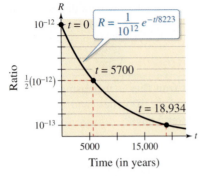

Figure 4.35

$$R = \frac{1}{10^{12}} e^{-t/8223} \qquad \text{Carbon dating model}$$

The graph of R is shown in Figure 4.35. Note that R decreases as t increases.

Example 3 Carbon Dating

The ratio of carbon 14 to carbon 12 in a newly discovered fossil is

$$R = \frac{1}{10^{13}}.$$

Estimate the age of the fossil.

Algebraic Solution

In the carbon dating model, substitute the given value of R to obtain the following.

$$\frac{1}{10^{12}} e^{-t/8223} = R \qquad \text{Write original model.}$$

$$\frac{e^{-t/8223}}{10^{12}} = \frac{1}{10^{13}} \qquad \text{Substitute } \tfrac{1}{10^{13}} \text{ for } R.$$

$$e^{-t/8223} = \frac{1}{10} \qquad \text{Multiply each side by } 10^{12}.$$

$$\ln e^{-t/8223} = \ln \frac{1}{10} \qquad \text{Take natural log of each side.}$$

$$-\frac{t}{8223} \approx -2.3026 \qquad \text{Inverse Property}$$

$$t \approx 18{,}934 \qquad \text{Multiply each side by } -8223.$$

So, to the nearest thousand years, you can estimate the age of the fossil to be 19,000 years.

  Now try Exercise 37.

Graphical Solution

Use a graphing utility to graph the formula for the ratio of carbon 14 to carbon 12 at any time t as

$$y_1 = \frac{1}{10^{12}} e^{-x/8223}.$$

In the same viewing window, graph $y_2 = 1/(10^{13})$, as shown in Figure 4.36.

Use the *intersect* feature to estimate that $x \approx 18{,}934$ when $y = 1/(10^{13})$.

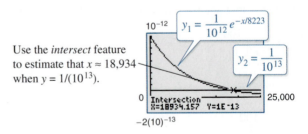

Figure 4.36

So, to the nearest thousand years, you can estimate the age of the fossil to be 19,000 years.

The carbon dating model in Example 3 assumed that the carbon 14 to carbon 12 ratio was one part in 10,000,000,000,000. Suppose an error in measurement occurred and the actual ratio was only one part in 8,000,000,000,000. The fossil age corresponding to the actual ratio would then be approximately 17,000 years. Try checking this result.

Gaussian Models

As mentioned at the beginning of this section, Gaussian models are of the form

$$y = ae^{-(x-b)^2/c}.$$

This type of model is commonly used in probability and statistics to represent populations that are **normally distributed.** For *standard* normal distributions, the model takes the form

$$y = \frac{1}{\sqrt{2\pi}}e^{-x^2/2}.$$

The graph of a Gaussian model is called a **bell-shaped curve.** Try graphing the normal distribution curve with a graphing utility. Can you see why it is called a bell-shaped curve?

The average value for a population can be found from the bell-shaped curve by observing where the maximum *y*-value of the function occurs. The *x*-value corresponding to the maximum *y*-value of the function represents the average value of the independent variable—in this case, *x*.

Example 4   SAT Scores

In 2009, the Scholastic Aptitude Test (SAT) mathematics scores for college-bound seniors roughly followed the normal distribution

$$y = 0.0034e^{-(x-515)^2/26,912}, \quad 200 \le x \le 800$$

where *x* is the SAT score for mathematics. Use a graphing utility to graph this function and estimate the average SAT score.   (Source: College Board)

Solution

The graph of the function is shown in Figure 4.37. On this bell-shaped curve, the maximum value of the curve represents the average score. Using the *maximum* feature of the graphing utility, you can see that the average mathematics score for college-bound seniors in 2009 was 515.

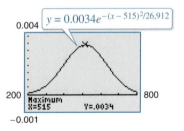

✓CHECKPOINT   Now try Exercise 41.

Figure 4.37

In Example 4, note that 50% of the seniors who took the test received scores lower than 515 (see Figure 4.38).

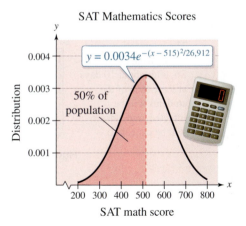

Figure 4.38

Logistic Growth Models

Some populations initially have rapid growth, followed by a declining rate of growth, as indicated by the graph in Figure 4.39. One model for describing this type of growth pattern is the **logistic curve** given by the function

$$y = \frac{a}{1 + be^{-rx}}$$

where y is the population size and x is the time. An example is a bacteria culture that is initially allowed to grow under ideal conditions, and then under less favorable conditions that inhibit growth. A logistic growth curve is also called a **sigmoidal curve.**

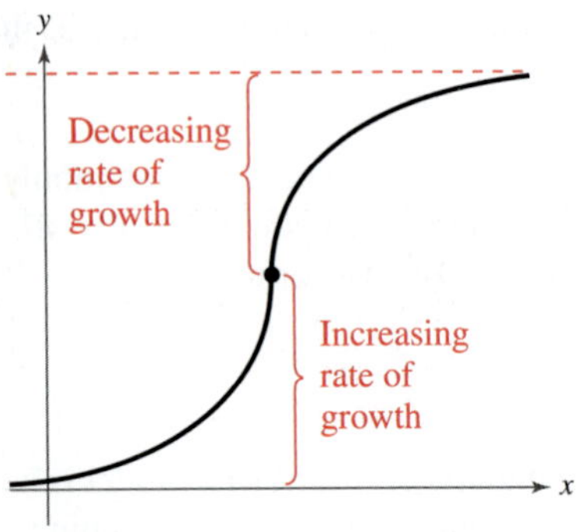

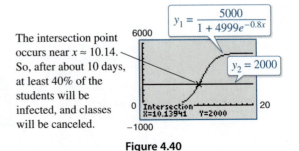

Decreasing rate of growth

Increasing rate of growth

Figure 4.39 *Logistic Curve*

Example 5 Spread of a Virus

On a college campus of 5000 students, one student returns from vacation with a contagious flu virus. The spread of the virus is modeled by

$$y = \frac{5000}{1 + 4999e^{-0.8t}}, \quad t \geq 0$$

where y is the total number of students infected after t days. The college will cancel classes when 40% or more of the students are infected.

a. How many students are infected after 5 days?

b. After how many days will the college cancel classes?

Algebraic Solution

a. After 5 days, the number of students infected is

$$y = \frac{5000}{1 + 4999e^{-0.8(5)}}$$

$$= \frac{5000}{1 + 4999e^{-4}}$$

$$\approx 54.$$

b. Classes are canceled when the number of infected students is $(0.40)(5000) = 2000$.

$$2000 = \frac{5000}{1 + 4999e^{-0.8t}}$$

$$1 + 4999e^{-0.8t} = 2.5$$

$$e^{-0.8t} = \frac{1.5}{4999}$$

$$\ln e^{-0.8t} = \ln \frac{1.5}{4999}$$

$$-0.8t = \ln \frac{1.5}{4999}$$

$$t = -\frac{1}{0.8} \ln \frac{1.5}{4999}$$

$$t \approx 10.14$$

So, after about 10 days, at least 40% of the students will be infected, and classes will be canceled.

✓CHECKPOINT Now try Exercise 43.

Graphical Solution

a.

6000

Use the *value* feature to estimate that $y \approx 54$ when $x = 5$. So, after 5 days, about 54 students will be infected.

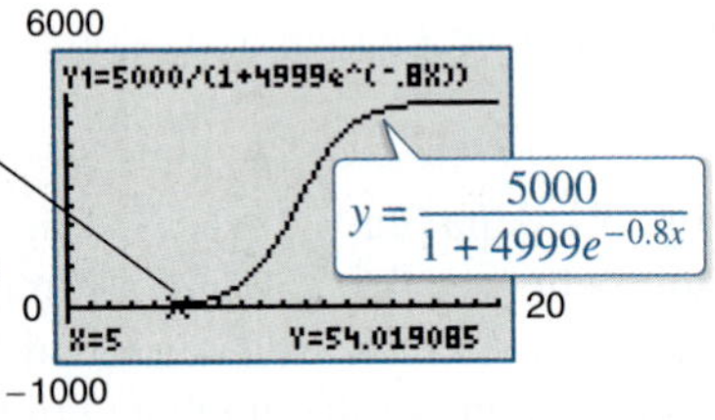

Y1=5000/(1+4999e^(-.8X))

$$y = \frac{5000}{1 + 4999e^{-0.8x}}$$

X=5   Y=54.019085

0   20

−1000

b. Classes are canceled when the number of infected students is $(0.40)(5000) = 2000$. Use a graphing utility to graph

$$y_1 = \frac{5000}{1 + 4999e^{-0.8x}} \quad \text{and} \quad y_2 = 2000$$

in the same viewing window. Use the *intersect* feature of the graphing utility to find the point of intersection of the graphs, as shown in Figure 4.40.

The intersection point occurs near $x \approx 10.14$. So, after about 10 days, at least 40% of the students will be infected, and classes will be canceled.

6000 $$y_1 = \frac{5000}{1 + 4999e^{-0.8x}}$$

$$y_2 = 2000$$

0   20

Intersection
X=10.13941   Y=2000

−1000

Figure 4.40

Logarithmic Models

On the Richter scale, the magnitude R of an earthquake of intensity I is given by

$$R = \log_{10} \frac{I}{I_0}$$

where $I_0 = 1$ is the minimum intensity used for comparison. Intensity is a measure of the wave energy of an earthquake.

Example 6   Magnitudes of Earthquakes

In 2009, Crete, Greece experienced an earthquake that measured 6.4 on the Richter scale. Also in 2009, the north coast of Indonesia experienced an earthquake that measured 7.6 on the Richter scale. Find the intensity of each earthquake and compare the two intensities.

Solution

Because $I_0 = 1$ and $R = 6.4$, you have

$$6.4 = \log_{10} \frac{I}{1}$$

$$10^{6.4} = 10^{\log_{10} I} \qquad \Longrightarrow \qquad 10^{6.4} = I.$$

For $R = 7.6$, you have

$$7.6 = \log_{10} \frac{I}{1}$$

$$10^{7.6} = 10^{\log_{10} I} \qquad \Longrightarrow \qquad 10^{7.6} = I.$$

Note that an increase of 1.2 units on the Richter scale (from 6.4 to 7.6) represents an increase in intensity by a factor of

$$\frac{10^{7.6}}{10^{6.4}} = 10^{1.2} \approx 16.$$

In other words, the intensity of the earthquake near the north coast of Indonesia was about 16 times as great as the intensity of the earthquake in Greece.

 CHECKPOINT   Now try Exercise 45.

Earthquake Relief Worker

Example 7   pH Levels

Acidity, or pH level, is a measure of the hydrogen ion concentration $[H^+]$ (measured in moles of hydrogen per liter) of a solution. Use the model given by $pH = -\log_{10}[H^+]$ to determine the hydrogen ion concentration of milk of magnesia, which has a pH of 10.5.

Solution

$pH = -\log_{10}[H^+]$	Write original model.
$10.5 = -\log_{10}[H^+]$	Substitute 10.5 for pH.
$-10.5 = \log_{10}[H^+]$	Multiply each side by -1.
$10^{-10.5} = 10^{\log_{10}[H^+]}$	Exponentiate each side (base 10).
$3.16 \times 10^{-11} = [H^+]$	Simplify.

So, the hydrogen ion concentration of milk of magnesia is 3.16×10^{-11} mole of hydrogen per liter.

 CHECKPOINT   Now try Exercise 51.

4.5 Exercises

Vocabulary and Concept Check

1. Match the equation with its model.

   (a) Exponential growth model   (i) $y = ae^{-bx},\ b > 0$

   (b) Exponential decay model   (ii) $y = a + b \ln x$

   (c) Logistic growth model   (iii) $y = \dfrac{a}{1 + be^{-rx}}$

   (d) Gaussian model   (iv) $y = ae^{bx},\ b > 0$

   (e) Natural logarithmic model   (v) $y = a + b \log_{10} x$

   (f) Common logarithmic model   (vi) $y = ae^{-(x-b)^2/c}$

In Exercises 2 and 3, fill in the blank.

2. Gaussian models are commonly used in probability and statistics to represent populations that are _____ distributed.

3. Logistic growth curves are also called _____ curves.

4. Which model in Exercise 1 has a graph called a bell-shaped curve?

5. Does the model $y = 120e^{-0.25x}$ represent exponential growth or exponential decay?

6. Which model in Exercise 1 has a graph with two horizontal asymptotes?

Procedures and Problem Solving

Identifying Graphs of Models   In Exercises 7–12, match the function with its graph. [The graphs are labeled (a), (b), (c), (d), (e), and (f).]

7. $y = 2e^{x/4}$ 8. $y = 6e^{-x/4}$

9. $y = 6 + \log_{10}(x + 2)$ 10. $y = 3e^{-(x-2)^2/5}$

11. $y = \ln(x + 1)$ 12. $y = \dfrac{4}{1 + e^{-2x}}$

(a)
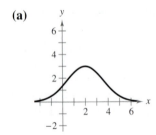

(b)
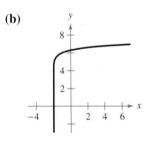

(c)
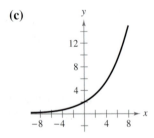

(d)
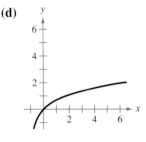

(e)
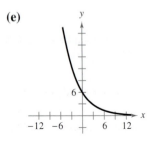

(f)
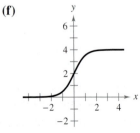

Using a Compound Interest Formula   In Exercises 13–20, complete the table for a savings account in which interest is compounded continuously.

	Initial Investment	Annual % Rate	Time to Double	Amount After 10 Years
13.	$10,000	3.5%		
14.	$2000	1.5%		
15.	$7500		21 years	
16.	$1000		12 years	
17.	$5000			$5665.74
18.	$300			$385.21
19.		4.5%		$100,000.00
20.		2%		$2500.00

21. **Tripling an Investment**   Complete the table for the time t (in years) necessary for P dollars to triple when interest is compounded continuously at rate r. Create a scatter plot of the data.

r	2%	4%	6%	8%	10%	12%
t						

22. Tripling an Investment Complete the table for the time t (in years) necessary for P dollars to triple when interest is compounded annually at rate r. Create a scatter plot of the data.

r	2%	4%	6%	8%	10%	12%
t						

23. Finance When $1 is invested in an account over a 10-year period, the amount A in the account after t years is given by

$$A = 1 + 0.075[\![t]\!] \quad \text{or} \quad A = e^{0.07t}$$

depending on whether the account pays simple interest at $7\frac{1}{2}\%$ or continuous compound interest at 7%. Use a graphing utility to graph each function in the same viewing window. Which grows at a greater rate? (Remember that $[\![t]\!]$ is the greatest integer function discussed in Section 1.4.)

24. Finance When $1 is invested in an account over a 10-year period, the amount A in the account after t years is given by

$$A = 1 + 0.06[\![t]\!] \quad \text{or} \quad A = \left(1 + \frac{0.055}{365}\right)^{[\![365t]\!]}$$

depending on whether the account pays simple interest at 6% or compound interest at $5\frac{1}{2}\%$ compounded daily. Use a graphing utility to graph each function in the same viewing window. Which grows at a greater rate?

Radioactive Decay In Exercises 25–28, complete the table for the radioactive isotope.

	Isotope	Half-Life (years)	Initial Quantity	Amount After 1000 Years
25.	^{226}Ra	1599	10 g	
26.	^{226}Ra	1599		1.5 g
27.	^{14}C	5700	3 g	
28.	^{239}Pu	24,100		0.4 g

Identifying a Model In Exercises 29–32, find the exponential model $y = ae^{bx}$ that fits the points shown in the graph or table.

29.
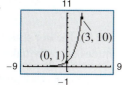

30.
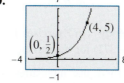

31.

x	0	5
y	4	1

32.

x	0	3
y	1	$\frac{1}{4}$

✓ **33. Demography** The populations P (in thousands) of Pittsburgh, Pennsylvania from 2000 through 2008 can be modeled by $P = 333.68e^{-0.0099t}$, where t is the year, with $t = 0$ corresponding to 2000. (Source: U.S. Census Bureau)

(a) According to the model, was the population of Pittsburgh increasing or decreasing from 2000 through 2008? Explain your reasoning.

(b) What were the populations of Pittsburgh in 2000, 2005, and 2008?

(c) According to the model, when will the population of Pittsburgh be approximately 290,000?

34. MODELING DATA

The table shows the populations (in millions) of five countries in 2005 and the projected populations (in millions) for 2015. (Source: U.S. Census Bureau)

Country	2005	2015
Australia	20.2	22.8
Canada	32.4	35.1
Hungary	10.0	9.7
Philippines	90.4	109.6
Turkey	72.7	82.5

(a) Find the exponential growth or decay model, $y = ae^{bt}$ or $y = ae^{-bt}$, for the population of each country, where t is the year, with $t = 5$ corresponding to 2005. Use the model to predict the population of each country in 2030.

(b) You can see that the populations of Canada and the Philippines are growing at different rates. What constant in the equation $y = ae^{bt}$ is determined by these different growth rates? Discuss the relationship between the different growth rates and the magnitude of the constant.

(c) The population of Turkey is increasing while the population of Hungary is decreasing. What constant in the equation $y = ae^{bt}$ reflects this difference? Explain.

✓ **35. Demography** The populations P (in thousands) of San Antonio, Texas from 2000 through 2008 can be modeled by

$$P = 1155.4e^{kt}$$

where t is the year, with $t = 0$ corresponding to 2000. In 2002, the population was 1,200,000. (Source: U.S. Census Bureau)

(a) Find the value of k for the model. Round your result to four decimal places.

(b) Use your model to predict the population in 2015.

36. Demography The populations P (in thousands) of Raleigh, North Carolina from 2000 through 2008 can be modeled by $P = 289.81e^{kt}$, where t is the year, with $t = 0$ corresponding to 2000. In 2006, the population was 363,000. (Source: U.S. Census Bureau)

(a) Find the value of k for the model. Round your result to four decimal places.

(b) Use your model to predict the population in 2015.

37. *Why you should learn it* (p. 365) Carbon 14 (^{14}C) dating assumes that the carbon dioxide on Earth today has the same radioactive content as it did centuries ago. If this is true, then the amount of ^{14}C absorbed by a tree that grew several centuries ago should be the same as the amount of ^{14}C absorbed by a tree growing today. A piece of ancient charcoal contains only 15% as much radioactive carbon as a piece of modern charcoal. How long ago was the tree burned to make the ancient charcoal given that the half-life of ^{14}C is 5700 years?

38. Radioactive Decay The half-life of radioactive radium (^{226}Ra) is 1599 years. What percent of a present amount of radioactive radium will remain after 100 years?

---

39. MODELING DATA

A new 2009 luxury sedan that sold for $49,200 has a book value V of $32,590 after 2 years.

(a) Find a linear model for the value V of the sedan.

(b) Find an exponential model for the value V of the sedan. Round the numbers in the model to four decimal places.

(c) Use a graphing utility to graph the two models in the same viewing window.

(d) Which model represents a greater depreciation rate in the first year?

(e) For what years is the value of the sedan greater using the linear model? the exponential model?

---

40. MODELING DATA

A new laptop computer that sold for $935 in 2009 has a book value V of $385 after 2 years.

(a) Find a linear model for the value V of the laptop.

(b) Find an exponential model for the value V of the laptop. Round the numbers in the model to four decimal places.

(c) Use a graphing utility to graph the two models in the same viewing window.

(d) Which model represents a greater depreciation rate in the first year?

(e) For what years is the value of the laptop greater using the linear model? the exponential model?

---

41. Psychology The IQ scores for adults roughly follow the normal distribution $y = 0.0266e^{-(x-100)^2/450}$, $70 \le x \le 115$, where x is the IQ score.

(a) Use a graphing utility to graph the function.

(b) Use the graph in part (a) to estimate the average IQ score.

42. Marketing The sales S (in thousands of units) of a cleaning solution after x hundred dollars is spent on advertising are given by $S = 10(1 - e^{kx})$. When $500 is spent on advertising, 2500 units are sold.

(a) Complete the model by solving for k.

(b) Estimate the number of units that will be sold when advertising expenditures are raised to $700.

43. Forestry A conservation organization releases 100 animals of an endangered species into a game preserve. The organization believes that the preserve has a carrying capacity of 1000 animals and that the growth of the herd will follow the logistic curve

$$p(t) = \frac{1000}{1 + 9e^{-0.1656t}}$$

where t is measured in months.

(a) What is the population after 5 months?

(b) After how many months will the population reach 500?

(c) Use a graphing utility to graph the function. Use the graph to determine the values of p at which the horizontal asymptotes occur. Identify the asymptote that is most relevant in the context of the problem and interpret its meaning.

44. Biology The number Y of yeast organisms in a culture is given by the model

$$Y = \frac{663}{1 + 72e^{-0.547t}}, \quad 0 \le t \le 18$$

where t represents the time (in hours).

(a) Use a graphing utility to graph the model.

(b) Use the model to predict the populations for the 19th hour and the 30th hour.

(c) According to this model, what is the limiting value of the population?

(d) Why do you think this population of yeast follows a logistic growth model instead of an exponential growth model?

Geology In Exercises 45 and 46, use the Richter scale (see page 371) for measuring the magnitudes of earthquakes.

45. Find the intensities I of the following earthquakes measuring R on the Richter scale (let $I_0 = 1$). (Source: U.S. Geological Survey)

(a) Haiti in 2010, $R = 7.0$

(b) Samoa Islands in 2009, $R = 8.1$

(c) Virgin Islands in 2008, $R = 6.1$

46. Find the magnitudes R of the following earthquakes of intensity I (let $I_0 = 1$).

(a) $I = 39,811,000$

(b) $I = 12,589,000$

(c) $I = 251,200$

Audiology   In Exercises 47–50, use the following information for determining sound intensity. The level of sound β (in decibels) with an intensity I is

$$\beta = 10 \log_{10} \frac{I}{I_0}$$

where I_0 is an intensity of 10^{-12} watt per square meter, corresponding roughly to the faintest sound that can be heard by the human ear. In Exercises 47 and 48, find the level of each sound β.

47. (a) $I = 10^{-10}$ watt per m^2 (quiet room)

(b) $I = 10^{-5}$ watt per m^2 (busy street corner)

(c) $I \approx 10^0$ watt per m^2 (threshold of pain)

48. (a) $I = 10^{-4}$ watt per m^2 (door slamming)

(b) $I = 10^{-3}$ watt per m^2 (loud car horn)

(c) $I = 10^{-2}$ watt per m^2 (siren at 30 meters)

49. As a result of the installation of a muffler, the noise level of an engine was reduced from 88 to 72 decibels. Find the percent decrease in the intensity level of the noise due to the installation of the muffler.

50. As a result of the installation of noise suppression materials, the noise level in an auditorium was reduced from 93 to 80 decibels. Find the percent decrease in the intensity level of the noise due to the installation of these materials.

Chemistry   In Exercises 51–54, use the acidity model given in Example 7.

✓ **51.** Find the pH when $[\text{H}^+] = 2.3 \times 10^{-5}$.

52. Compute $[\text{H}^+]$ for a solution for which pH $= 5.8$.

53. A grape has a pH of 3.5, and baking soda has a pH of 8.0. The hydrogen ion concentration of the grape is how many times that of the baking soda?

54. The pH of a solution is decreased by one unit. The hydrogen ion concentration is increased by what factor?

55. Finance   The total interest u paid on a home mortgage of P dollars at interest rate r for t years is given by

$$u = P \left[\frac{rt}{1 - \left(\dfrac{1}{1 + r/12} \right)^{12t}} - 1 \right].$$

Consider a $120,000 home mortgage at $7\frac{1}{2}\%$.

(a) Use a graphing utility to graph the total interest function.

(b) Approximate the length of the mortgage when the total interest paid is the same as the amount of the mortgage. Is it possible that a person could pay twice as much in interest charges as the amount of his or her mortgage?

56. Finance   A $120,000 home mortgage for 30 years at $7\frac{1}{2}\%$ has a monthly payment of $839.06. Part of the monthly payment goes toward the interest charge on the unpaid balance, and the remainder of the payment is used to reduce the principal. The amount that goes toward the interest is given by

$$u = M - \left(M - \frac{Pr}{12} \right) \left(1 + \frac{r}{12} \right)^{12t}$$

and the amount that goes toward reduction of the principal is given by

$$v = \left(M - \frac{Pr}{12} \right) \left(1 + \frac{r}{12} \right)^{12t}.$$

In these formulas, P is the size of the mortgage, r is the interest rate, M is the monthly payment, and t is the time (in years).

(a) Use a graphing utility to graph each function in the same viewing window. (The viewing window should show all 30 years of mortgage payments.)

(b) In the early years of the mortgage, the larger part of the monthly payment goes for what purpose? Approximate the time when the monthly payment is evenly divided between interest and principal reduction.

(c) Repeat parts (a) and (b) for a repayment period of 20 years ($M = \$966.71$). What can you conclude?

57. Forensics   At 8:30 A.M., a coroner was called to the home of a person who had died during the night. In order to estimate the time of death, the coroner took the person's temperature twice. At 9:00 A.M. the temperature was 85.7°F, and at 11:00 A.M. the temperature was 82.8°F. From these two temperatures the coroner was able to determine that the time elapsed since death and the body temperature were related by the formula

$$t = -10 \ln \frac{T - 70}{98.6 - 70}$$

where t is the time (in hours elapsed since the person died) and T is the temperature (in degrees Fahrenheit) of the person's body. Assume that the person had a normal body temperature of 98.6°F at death and that the room temperature was a constant 70°F. Use the formula to estimate the time of death of the person. (This formula is derived from a general cooling principle called Newton's Law of Cooling.)

58. Culinary Arts You take a five-pound package of steaks out of a freezer at 11 A.M. and place it in a refrigerator. Will the steaks be thawed in time to be grilled at 6 P.M.? Assume that the refrigerator temperature is 40°F and the freezer temperature is 0°F. Use the formula for Newton's Law of Cooling

$$t = -5.05 \ln \frac{T - 40}{0 - 40}$$

where t is the time in hours (with $t = 0$ corresponding to 11 A.M.) and T is the temperature of the package of steaks (in degrees Fahrenheit).

Conclusions

True or False? In Exercises 59 and 60, determine whether the statement is true or false. Justify your answer.

59. The domain of a logistic growth function cannot be the set of real numbers.

60. The graph of a logistic growth function will always have an x-intercept.

61. Think About It Can the graph of a Gaussian model ever have an x-intercept? Explain.

62. CAPSTONE For each graph, state whether an exponential, Gaussian, logarithmic, logistic, or quadratic model will fit the data best. Explain your reasoning. Then describe a real-life situation that could be represented by the data.

(a) 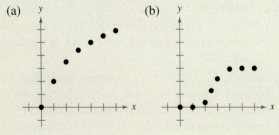 (b)

(c) 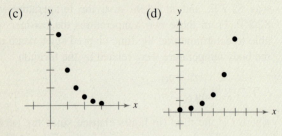 (d)

(e) 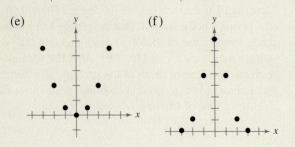 (f)

Identifying Models In Exercises 63–66, identify the type of model you studied in this section that has the given characteristic.

63. The maximum value of the function occurs at the average value of the independent variable.

64. A horizontal asymptote of its graph represents the limiting value of a population.

65. Its graph shows a steadily increasing rate of growth.

66. The only asymptote of its graph is a vertical asymptote.

Cumulative Mixed Review

Identifying Graphs of Linear Equations In Exercises 67–70, match the equation with its graph, and identify any intercepts. [The graphs are labeled (a), (b), (c), and (d).]

(a) 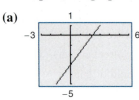 (b)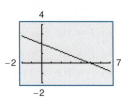

(c) 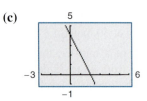 (d)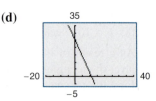

67. $4x - 3y - 9 = 0$

68. $2x + 5y - 10 = 0$

69. $y = 25 - 2.25x$

70. $\dfrac{x}{2} + \dfrac{y}{4} = 1$

Applying the Leading Coefficient Test In Exercises 71–74, use the Leading Coefficient Test to determine the right-hand and left-hand behavior of the graph of the polynomial function.

71. $f(x) = 2x^3 - 3x^2 + x - 1$

72. $f(x) = 5 - x^2 - 4x^4$

73. $g(x) = -1.6x^5 + 4x^2 - 2$

74. $g(x) = 7x^6 + 9.1x^5 - 3.2x^4 + 25x^3$

Using Synthetic Division In Exercises 75 and 76, divide using synthetic division.

75. $(2x^3 - 8x^2 + 3x - 9) \div (x - 4)$

76. $(x^4 - 3x + 1) \div (x + 5)$

77. *Make a Decision* To work an extended application analyzing the net sales for Kohl's Corporation from 1992 through 2008, visit this textbook's *Companion Website*. (Data Source: Kohl's Illinois, Inc.)

4.6   Nonlinear Models

Classifying Scatter Plots

In Section 2.7, you saw how to fit linear models to data, and in Section 3.7, you saw how to fit quadratic models to data. In real life, many relationships between two variables are represented by different types of growth patterns. A scatter plot can be used to give you an idea of which type of model will best fit a set of data.

Example 1   Classifying Scatter Plots

Decide whether each set of data could best be modeled by a linear model, $y = ax + b$, an exponential model, $y = ab^x$, or a logarithmic model, $y = a + b \ln x$.

a. (2, 1), (2.5, 1.2), (3, 1.3), (3.5, 1.5), (4, 1.8), (4.5, 2), (5, 2.4), (5.5, 2.5), (6, 3.1), (6.5, 3.8), (7, 4.5), (7.5, 5), (8, 6.5), (8.5, 7.8), (9, 9), (9.5, 10)

b. (2, 2), (2.5, 3.1), (3, 3.8), (3.5, 4.3), (4, 4.6), (4.5, 5.3), (5, 5.6), (5.5, 5.9), (6, 6.2), (6.5, 6.4), (7, 6.9), (7.5, 7.2), (8, 7.6), (8.5, 7.9), (9, 8), (9.5, 8.2)

c. (2, 1.9), (2.5, 2.5), (3, 3.2), (3.5, 3.6), (4, 4.3), (4.5, 4.7), (5, 5.2), (5.5, 5.7), (6, 6.4), (6.5, 6.8), (7, 7.2), (7.5, 7.9), (8, 8.6), (8.5, 8.9), (9, 9.5), (9.5, 9.9)

Solution

a. From Figure 4.41, it appears that the data can best be modeled by an exponential function.

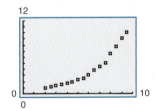

Figure 4.41

b. From Figure 4.42, it appears that the data can best be modeled by a logarithmic function.

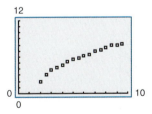

Figure 4.42

c. From Figure 4.43, it appears that the data can best be modeled by a linear function.

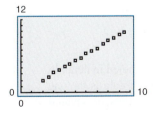

Figure 4.43

✓CHECKPOINT   Now try Exercise 13.

Fitting Nonlinear Models to Data

Once you have used a scatter plot to determine the type of model that would best fit a set of data, there are several ways that you can actually find the model. Each method is best used with a computer or calculator, rather than with hand calculations.

Example 2   Fitting a Model to Data

Fit the following data from Example 1(a) to an exponential model and a power model. Identify the coefficient of determination and determine which model fits the data better.

(2, 1), (2.5, 1.2), (3, 1.3), (3.5, 1.5),

(4, 1.8), (4.5, 2), (5, 2.4), (5.5, 2.5),

(6, 3.1), (6.5, 3.8), (7, 4.5), (7.5, 5),

(8, 6.5), (8.5, 7.8), (9, 9), (9.5, 10)

Solution

Begin by entering the data into a graphing utility. Then use the *regression* feature of the graphing utility to find exponential and power models for the data, as shown in Figure 4.44.

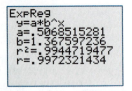
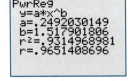

Exponential Model
Figure 4.44

Power Model

So, an exponential model for the data is $y = 0.507(1.368)^x$, and a power model for the data is $y = 0.249x^{1.518}$. Plot the data and each model in the same viewing window, as shown in Figure 4.45. To determine which model fits the data better, compare the coefficients of determination for each model. The model whose r^2-value is closest to 1 is the model that better fits the data. In this case, the better-fitting model is the exponential model.

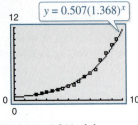

$y = 0.507(1.368)^x$

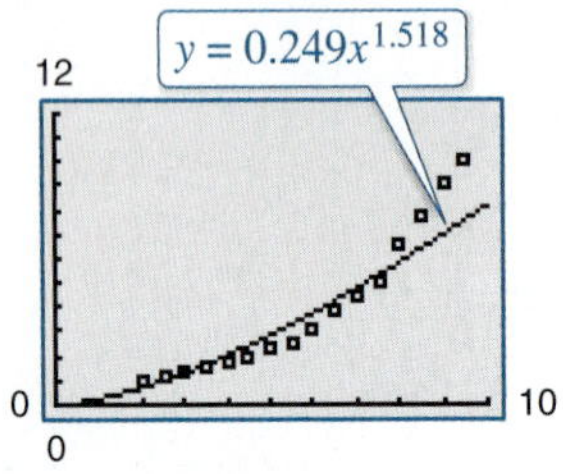

$y = 0.249x^{1.518}$

Exponential Model
Figure 4.45

Power Model

✔CHECKPOINT   Now try Exercise 31.

Deciding which model best fits a set of data is a question that is studied in detail in statistics. Recall from Section 2.7 that the model that best fits a set of data is the one whose *sum of squared differences* is the least. In Example 2, the sums of squared differences are 0.90 for the exponential model and 14.30 for the power model.

Example 3   Fitting a Model to Data

The table shows the yield y (in milligrams) of a chemical reaction after x minutes. Use a graphing utility to find a logarithmic model and a linear model for the data and identify the coefficient of determination for each model. Determine which model fits the data better.

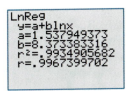

Minutes, x	Yield, y
1	1.5
2	7.4
3	10.2
4	13.4
5	15.8
6	16.3
7	18.2
8	18.3

Solution

Begin by entering the data into a graphing utility. Then use the *regression* feature of the graphing utility to find logarithmic and linear models for the data, as shown in Figure 4.46.

```
LnReg
 y=a+blnx
 a=1.537949373
 b=8.373383316
 r²=.9934905682
 r=.9967399702
```

```
LinReg
 y=ax+b
 a=2.289285714
 b=2.335714286
 r²=.9005643856
 r=.9489807088
```

Logarithmic Model
Figure 4.46

Linear Model

So, a logarithmic model for the data is $y = 1.538 + 8.373 \ln x$ and a linear model for the data is $y = 2.29x + 2.3$. Plot the data and each model in the same viewing window, as shown in Figure 4.47. To determine which model fits the data better, compare the coefficients of determination for each model. The model whose coefficient of determination is closer to 1 is the model that better fits the data. In this case, the better-fitting model is the logarithmic model.

Explore the Concept

Use a graphing utility to find a quadratic model for the data in Example 3. Do you think this model fits the data better than the logarithmic model in Example 3? Explain your reasoning.

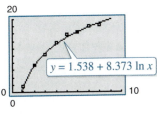

$y = 1.538 + 8.373 \ln x$

Logarithmic Model
Figure 4.47

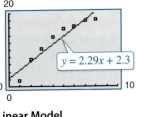

$y = 2.29x + 2.3$

Linear Model

✓CHECKPOINT   Now try Exercise 33.

In Example 3, the sum of the squared differences for the logarithmic model is 1.59 and the sum of the squared differences for the linear model is 24.31.

Modeling With Exponential and Logistic Functions

Example 4 Fitting an Exponential Model to Data

The table at the right shows the amounts of revenue R (in billions of dollars) collected by the Internal Revenue Service (IRS) for selected years from 1963 through 2008. Use a graphing utility to find a model for the data. Then use the model to estimate the revenue collected in 2013.   (Source: IRS Data Book)

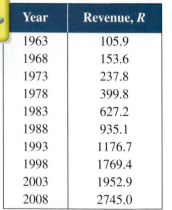

Year	Revenue, R
1963	105.9
1968	153.6
1973	237.8
1978	399.8
1983	627.2
1988	935.1
1993	1176.7
1998	1769.4
2003	1952.9
2008	2745.0

Solution

Let x represent the year, with $x = 3$ corresponding to 1963. Begin by entering the data into a graphing utility and displaying the scatter plot, as shown in Figure 4.48.

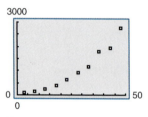

Figure 4.48

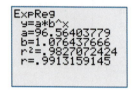

Figure 4.49

From the scatter plot, it appears that an exponential model is a good fit. Use the *regression* feature of the graphing utility to find the exponential model, as shown in Figure 4.49. Change the model to a natural exponential model, as follows.

$R = 96.56(1.076)^x$ Write original model.

$= 96.56e^{(\ln 1.076)x}$ $b = e^{\ln b}$

$\approx 96.56e^{0.073x}$ Simplify.

Graph the data and the natural exponential model

$R = 96.56e^{0.073x}$

in the same viewing window, as shown in Figure 4.50. From the model, you can see that the revenue collected by the IRS from 1963 through 2008 had an average annual increase of about 7%. From this model, you can estimate the 2013 revenue to be

$R = 96.56e^{0.073x}$ Write natural exponential model.

$= 96.56e^{0.073(53)}$ Substitute 53 for x.

$\approx \$4624.7$ billion   Use a calculator.

which is more than twice the amount collected in 2003. You can also use the *value* feature of the graphing utility to approximate the revenue in 2013 to be \$4624.7 billion, as shown in Figure 4.50.

Study Tip

You can change an exponential model of the form

$y = ab^x$

to one of the form

$y = ae^{cx}$

by rewriting b in the form

$b = e^{\ln b}$.

For instance,

$y = 3(2^x)$

can be written as

$y = 3(2^x) = 3e^{(\ln 2)x} \approx 3e^{0.693x}$.

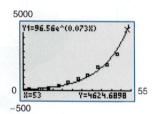

Figure 4.50

 Now try Exercise 35.

The next example demonstrates how to use a graphing utility to fit a logistic model to data.

Example 5   Fitting a Logistic Model to Data

To estimate the amount of defoliation caused by the gypsy moth during a given year, a forester counts the number x of egg masses on $\frac{1}{40}$ of an acre (circle of radius 18.6 feet) in the fall. The percent of defoliation y the next spring is shown in the table.   (Source: USDA, Forest Service)

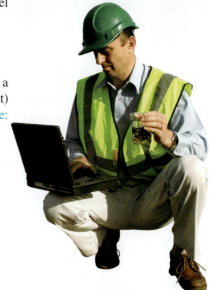

Forester

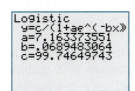

Egg masses, x	Percent of defoliation, y
0	12
25	44
50	81
75	96
100	99

a. Use the *regression* feature of a graphing utility to find a logistic model for the data.

b. How closely does the model represent the data?

Graphical Solution

a. Enter the data into a graphing utility. Using the *regression* feature of the graphing utility, you can find the logistic model, as shown in Figure 4.51.

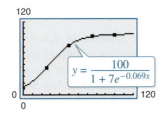

```
Logistic
y=c/(1+ae^(-bx))
a=7.163373551
b=.0689483064
c=99.74649743
```

Figure 4.51

You can approximate this model to be

$$y = \frac{100}{1 + 7e^{-0.069x}}.$$

b. You can use the graphing utility to graph the actual data and the model in the same viewing window. In Figure 4.52, it appears that the model is a good fit for the actual data.

120

$$y = \frac{100}{1 + 7e^{-0.069x}}$$

0 ⌞_____⌟ 120
0

Figure 4.52

✓**CHECKPOINT**   Now try Exercise 37.

Numerical Solution

a. Enter the data into a graphing utility. Using the *regression* feature of the graphing utility, you can approximate the logistic model to be

$$y = \frac{100}{1 + 7e^{-0.069x}}.$$

b. You can see how well the model fits the data by comparing the actual values of y with the values of y given by the model, which are labeled y^* in the table below.

x	0	25	50	75	100
y	12	44	81	96	99
y^*	12.5	44.5	81.8	96.2	99.3

In the table, you can see that the model appears to be a good fit for the actual data.

4.6   Exercises

See www.CalcChat.com for worked-out solutions to odd-numbered exercises. For instructions on how to use a graphing utility, see Appendix A.

Vocabulary and Concept Check

In Exercises 1 and 2, fill in the blank.

1. A power model has the form _____ .

2. An exponential model of the form $y = ab^x$ can be rewritten as a natural exponential model of the form _____.

3. What type of visual display can you create to get an idea of which type of model will best fit the data set?

4. A power model for a set of data has a coefficient of determination of $r^2 \approx 0.901$ and an exponential model for the data has a coefficient of determination of $r^2 \approx 0.967$. Which model fits the data better?

Procedures and Problem Solving

Classifying Scatter Plots   In Exercises 5–12, determine whether the scatter plot could best be modeled by a linear model, a quadratic model, an exponential model, a logarithmic model, or a logistic model.

5.
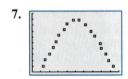

6.
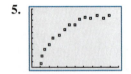

7.

8.
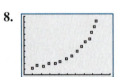

9.
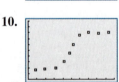

10.
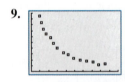

11.
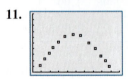

12.
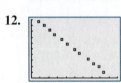

Classifying Scatter Plots   In Exercises 13–18, use a graphing utility to create a scatter plot of the data. Decide whether the data could best be modeled by a linear model, an exponential model, or a logarithmic model.

✓ 13. (1, 2.0), (1.5, 3.5), (2, 4.0), (4, 5.8), (6, 7.0), (8, 7.8)

14. (1, 5.8), (1.5, 6.0), (2, 6.5), (4, 7.6), (6, 8.9), (8, 10.0)

15. (1, 4.4), (1.5, 4.7), (2, 5.5), (4, 9.9), (6, 18.1), (8, 33.0)

16. (1, 11.0), (1.5, 9.6), (2, 8.2), (4, 4.5), (6, 2.5), (8, 1.4)

17. (1, 7.5), (1.5, 7.0), (2, 6.8), (4, 5.0), (6, 3.5), (8, 2.0)

18. (1, 5.0), (1.5, 6.0), (2, 6.4), (4, 7.8), (6, 8.6), (8, 9.0)

Finding an Exponential Model   In Exercises 19–22, use the *regression* feature of a graphing utility to find an exponential model $y = ab^x$ for the data and identify the coefficient of determination. Use the graphing utility to plot the data and graph the model in the same viewing window.

19. (0, 5), (1, 6), (2, 7), (3, 9), (4, 13)

20. (0, 4.0), (2, 6.9), (4, 18.0), (6, 32.3), (8, 59.1), (10, 118.5)

21. (0, 10.0), (1, 6.1), (2, 4.2), (3, 3.8), (4, 3.6)

22. (−3, 120.2), (0, 80.5), (3, 64.8), (6, 58.2), (10, 55.0)

Finding a Logarithmic Model   In Exercises 23–26, use the *regression* feature of a graphing utility to find a logarithmic model $y = a + b \ln x$ for the data and identify the coefficient of determination. Use the graphing utility to plot the data and graph the model in the same viewing window.

23. (1, 2.0), (2, 3.0), (3, 3.5), (4, 4.0), (5, 4.1), (6, 4.2), (7, 4.5)

24. (1, 8.5), (2, 11.4), (4, 12.8), (6, 13.6), (8, 14.2), (10, 14.6)

25. (1, 10), (2, 6), (3, 6), (4, 5), (5, 3), (6, 2)

26. (3, 14.6), (6, 11.0), (9, 9.0), (12, 7.6), (15, 6.5)

Finding a Power Model   In Exercises 27–30, use the *regression* feature of a graphing utility to find a power model $y = ax^b$ for the data and identify the coefficient of determination. Use the graphing utility to plot the data and graph the model in the same viewing window.

27. (1, 2.0), (2, 3.4), (5, 6.7), (6, 7.3), (10, 12.0)

28. (0.5, 1.0), (2, 12.5), (4, 33.2), (6, 65.7), (8, 98.5), (10, 150.0)

29. (1, 10.0), (2, 4.0), (3, 0.7), (4, 0.1)

30. (2, 450), (4, 385), (6, 345), (8, 332), (10, 312)

✓ 31. MODELING DATA

The table shows the yearly sales S (in millions of dollars) of Whole Foods Market for the years 2001 through 2008. (Source: Whole Foods Market)

Year	Sales, S
2001	2272.2
2002	2690.5
2003	3148.6
2004	3865.0
2005	4701.3
2006	5607.4
2007	6591.8
2008	7953.9

(a) Use the *regression* feature of a graphing utility to find an exponential model and a power model for the data and identify the coefficient of determination for each model. Let t represent the year, with $t = 1$ corresponding to 2001.

(b) Use the graphing utility to graph each model with the data.

(c) Use the coefficients of determination to determine which model best fits the data.

32. MODELING DATA

The table shows the annual amounts A (in billions of dollars) spent in the U.S. by the cruise lines and passengers of the North American cruise industry from 2003 through 2008. (Source: Cruise Lines International Association)

Year	Amount, A
2003	12.92
2004	14.70
2005	16.18
2006	17.64
2007	18.70
2008	19.07

(a) Use the *regression* feature of a graphing utility to find a linear model, an exponential model, and a logarithmic model for the data. Let t represent the year, with $t = 3$ corresponding to 2003.

(b) Use the graphing utility to graph each model with the data. Use the graphs to determine which model best fits the data.

(c) Use the model you chose in part (b) to predict the amount spent in 2009. Is the amount reasonable?

✓ 33. MODELING DATA

The populations P (in millions) of the United States for the years 1995 through 2008 are shown in the table, where t represents the year, with $t = 5$ corresponding to 1995. (Source: U.S. Census Bureau)

Year	Population, P
1995	266.6
1996	269.7
1997	272.9
1998	276.1
1999	279.3
2000	282.4
2001	285.3
2002	288.0
2003	290.7
2004	293.3
2005	296.0
2006	298.8
2007	301.7
2008	304.5

(a) Use the *regression* feature of a graphing utility to find a linear model for the data and to identify the coefficient of determination. Plot the model and the data in the same viewing window.

(b) Use the *regression* feature of the graphing utility to find a power model for the data and to identify the coefficient of determination. Plot the model and the data in the same viewing window.

(c) Use the *regression* feature of the graphing utility to find an exponential model for the data and to identify the coefficient of determination. Plot the model and the data in the same viewing window.

(d) Use the *regression* feature of the graphing utility to find a logarithmic model for the data and to identify the coefficient of determination. Plot the model and the data in the same viewing window.

(e) Which model is the best fit for the data? Explain.

(f) Use each model to predict the populations of the United States for the years 2009 through 2014.

(g) Which model is the best choice for predicting the future population of the United States? Explain.

(h) Were your choices of models the same for parts (e) and (g)? If not, explain why your choices were different.

34. *Why you should learn it* (p. 377)   The atmospheric

pressure decreases with increasing altitude. At sea level, the average air pressure is approximately 1.03323 kilograms per square centimeter, and this pressure is called one atmosphere. Variations in weather conditions cause changes in the atmospheric pressure of up to ±5 percent. The ordered pairs (h, p) give the pressures p (in atmospheres) for various altitudes h (in kilometers).

$$(0, 1), (10, 0.25), (20, 0.06), (5, 0.55),$$
$$(15, 0.12), (25, 0.02)$$

(a) Use the *regression* feature of a graphing utility to attempt to find the logarithmic model $p = a + b \ln h$ for the data. Explain why the result is an error message.

(b) Use the *regression* feature of the graphing utility to find the logarithmic model $h = a + b \ln p$ for the data.

(c) Use the graphing utility to plot the data and graph the logarithmic model in the same viewing window.

(d) Use the model to estimate the altitude at which the pressure is 0.75 atmosphere.

(e) Use the graph in part (c) to estimate the pressure at an altitude of 13 kilometers.

✓ 35. MODELING DATA

The table shows the numbers N of office supply stores operated by Staples from 2001 through 2008. (Source: Staples, Inc.)

Year	Number, N
2001	1436
2002	1488
2003	1559
2004	1680
2005	1780
2006	1884
2007	2038
2008	2218

(a) Use the *regression* feature of a graphing utility to find an exponential model for the data. Let t represent the year, with $t = 1$ corresponding to 2001.

(b) Rewrite the model as a natural exponential model.

(c) Use the natural exponential model to predict the number of Staples stores in 2009. Is the number reasonable?

36. MODELING DATA

A beaker of liquid at an initial temperature of 78°C is placed in a room at a constant temperature of 21°C. The temperature of the liquid is measured every 5 minutes for a period of $\frac{1}{2}$ hour. The results are recorded in the table, where t is the time (in minutes) and T is the temperature (in degrees Celsius).

Time, t	Temperature, T
0	78.0°
5	66.0°
10	57.5°
15	51.2°
20	46.3°
25	42.5°
30	39.6°

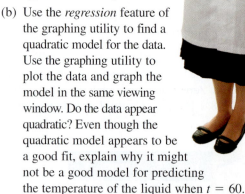

(a) Use the *regression* feature of a graphing utility to find a linear model for the data. Use the graphing utility to plot the data and graph the model in the same viewing window. Do the data appear linear? Explain.

(b) Use the *regression* feature of the graphing utility to find a quadratic model for the data. Use the graphing utility to plot the data and graph the model in the same viewing window. Do the data appear quadratic? Even though the quadratic model appears to be a good fit, explain why it might not be a good model for predicting the temperature of the liquid when $t = 60$.

(c) The graph of the temperature of the room should be an asymptote of the graph of the model. Subtract the room temperature from each of the temperatures in the table. Use the *regression* feature of the graphing utility to find an exponential model for the revised data. Add the room temperature to this model. Use the graphing utility to plot the original data and graph the model in the same viewing window.

(d) Explain why the procedure in part (c) was necessary for finding the exponential model.

37. MODELING DATA

The table shows the percents P of women in different age groups (in years) who have been married at least once.   (Source: U.S. Census Bureau)

Age group	Percent, P
20–24	21.1
25–29	54.5
30–34	73.9
35–39	84.0
40–44	86.5
45–54	89.7
55–64	93.1
65–74	95.8

(a) Use the *regression* feature of a graphing utility to find a logistic model for the data. Let x represent the midpoint of the age group.

(b) Use the graphing utility to graph the model with the original data. How closely does the model represent the data?

38. MODELING DATA

The table shows the annual sales S (in millions of dollars) of AutoZone for the years from 2002 through 2009.   (Source: AutoZone, Inc.)

Year	Sales, S
2002	5325.5
2003	5457.1
2004	5637.0
2005	5710.9
2006	5948.4
2007	6169.8
2008	6522.7
2009	6816.8

(a) Use the *regression* feature of a graphing utility to find a logarithmic model, an exponential model, and a power model for the data. Let t represent the year, with $t = 2$ corresponding to 2002.

(b) Use each of the following methods to choose the model that best fits the data. Compare your results.
 (i) Create a table of values for each model.
 (ii) Use the graphing utility to graph each model with the data.
 (iii) Find and compare the coefficients of determination for the models.

Conclusions

True or False? In Exercises 39 and 40, determine whether the statement is true or false. Justify your answer.

39. The exponential model $y = ae^{bx}$ represents a growth model when $b > 0$.

40. To change an exponential model of the form $y = ab^x$ to one of the form $y = ae^{cx}$, rewrite b as $b = \ln e^b$.

41. Writing In your own words, explain how to fit a model to a set of data using a graphing utility.

42. CAPSTONE You use a graphing utility to create the scatter plot of a set of data.

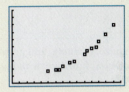

(a) What types of models are likely to fit the data well? Explain.

(b) Discuss the methods you can use to find the model of best fit for the data. Which method would you prefer? Explain.

Cumulative Mixed Review

Using the Slope-Intercept Form In Exercises 43–46, find the slope and y-intercept of the equation of the line. Then sketch the line by hand.

43. $2x + 5y = 10$

44. $3x - 2y = 9$

45. $1.2x + 3.5y = 10.5$

46. $0.4x - 2.5y = 12.0$

Writing the Equation of a Parabola in Standard Form In Exercises 47–50, write an equation of the parabola in standard form.

47.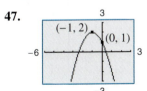

48.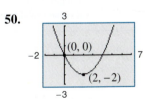

49.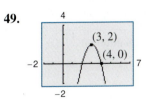

50.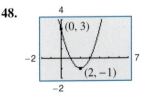

4   Chapter Summary

	What did you learn?	Explanation and Examples	Review Exercises
4.1	**Recognize and evaluate exponential functions with base a (p. 324).**	The exponential function f with base a is denoted by $f(x) = a^x$, where $a > 0$, $a \neq 1$, and x is any real number.	1–4
	Graph exponential functions with base a (p. 325).		5–12
	Recognize, evaluate, and graph exponential functions with base e (p. 328).	The function $f(x) = e^x$ is called the natural exponential function.	13–18
	Use exponential functions to model and solve real-life problems (p. 330).	Exponential functions are used in compound interest formulas (see Example 8) and in radioactive decay models (see Example 10).	19–22
4.2	**Recognize and evaluate logarithmic functions with base a (p. 336).**	For $x > 0$, $a > 0$, and $a \neq 1$, $y = \log_a x$ if and only if $x = a^y$. The function $f(x) = \log_a x$ is called the logarithmic function with base a.	23–36
	Graph logarithmic functions with base a (p. 338), and recognize, evaluate, and graph natural logarithmic functions (p. 340).	The graphs of $g(x) = \log_a x$ and $f(x) = a^x$ are reflections of each other in the line $y = x$. The function defined by $g(x) = \ln x$, $x > 0$, is called the natural logarithmic function.	37–48
	Use logarithmic functions to model and solve real-life problems (p. 342).	A logarithmic function is used in the human memory model. (See Example 10.)	49, 50
4.3	**Rewrite logarithms with different bases (p. 347).**	Let a, b, and x be positive real numbers such that $a \neq 1$ and $b \neq 1$. Then $\log_a x$ can be converted to a different base as follows. *Base b* *Base 10* *Base e* $\log_a x = \dfrac{\log_b x}{\log_b a}$ $\log_a x = \dfrac{\log_{10} x}{\log_{10} a}$ $\log_a x = \dfrac{\ln x}{\ln a}$	51–58

What did you learn?	Explanation and Examples	Review Exercises
4.3 **Use properties of logarithms to evaluate, rewrite, expand, or condense logarithmic expressions** *(p. 348)*.	Let a be a positive real number ($a \neq 1$), let n be a real number, and let u and v be positive real numbers. 1. **Product Property:** $\log_a(uv) = \log_a u + \log_a v$ $\ln(uv) = \ln u + \ln v$ 2. **Quotient Property:** $\log_a(u/v) = \log_a u - \log_a v$ $\ln(u/v) = \ln u - \ln v$ 3. **Power Property:** $\log_a u^n = n \log_a u, \quad \ln u^n = n \ln u$	59–78
Use logarithmic functions to model and solve real-life problems *(p. 350)*.	A logarithmic function can be used to find an equation that relates the periods of several planets and their distances from the sun. (See Example 7.)	79, 80
4.4 **Solve simple exponential and logarithmic equations** *(p. 354)*.	Solve simple exponential or logarithmic equations using the One-to-One Properties and Inverse Properties of exponential and logarithmic functions.	81–94
Solve more complicated exponential *(p. 355)* **and logarithmic** *(p. 357)* **equations.**	To solve more complicated equations, rewrite the equations so that the One-to-One Properties and Inverse Properties of exponential and logarithmic functions can be used. (See Examples 2–8.)	95–118
Use exponential and logarithmic equations to model and solve real-life problems *(p. 360)*.	Exponential and logarithmic equations can be used to find how long it will take to double an investment (see Example 12) and to find the year in which the average salary for public school teachers reached $45,000 (see Example 13).	119, 120
4.5 **Recognize the five most common types of models involving exponential or logarithmic functions** *(p. 365)*.	1. **Exponential growth model:** $y = ae^{bx}, \quad b > 0$ 2. **Exponential decay model:** $y = ae^{-bx}, \quad b > 0$ 3. **Gaussian model:** $y = ae^{-(x-b)^2/c}$ 4. **Logistic growth model:** $y = \dfrac{a}{1 + be^{-rx}}$ 5. **Logarithmic models:** $y = a + b \ln x$ $y = a + b \log_{10} x$	121–126
Use exponential growth and decay functions to model and solve real-life problems *(p. 366)*.	An exponential growth function can be used to model the world population (see Example 1), and an exponential decay function can be used to estimate the age of a fossil (see Example 3).	127
Use Gaussian functions *(p. 369)*, **logistic growth functions** *(p. 370)*, **and logarithmic functions** *(p. 371)* **to model and solve real-life problems.**	A Gaussian function can be used to model SAT math scores for college-bound seniors (see Example 4). A logistic growth function can be used to model the spread of a flu virus (see Example 5). A logarithmic function can be used to find the intensity of an earthquake using its magnitude (see Example 6).	128–130
4.6 **Classify scatter plots** *(p. 377)*, **and use scatter plots and a graphing utility to find models for data and choose the model that best fits a set of data** *(p. 378)*.	You can use a scatter plot and a graphing utility to choose a model that best fits a set of data that represents the yield of a chemical reaction. (See Example 3.)	131, 132
Use a graphing utility to find exponential and logistic models for data *(p. 380)*.	An exponential model can be used to estimate the amount of revenue collected by the Internal Revenue Service for a given year (see Example 4), and a logistic model can be used to estimate the percent of defoliation caused by the gypsy moth (see Example 5).	133, 134

4   Review Exercises

See www.CalcChat.com for worked-out solutions to odd-numbered exercises. For instructions on how to use a graphing utility, see Appendix A.

4.1

Evaluating Exponential Functions In Exercises 1–4, use a calculator to evaluate the function at the indicated value of x. Round your result to four decimal places.

1. $f(x) = 1.45^x$, $x = 2\pi$ **2.** $f(x) = 7^x$, $x = -\sqrt{11}$
3. $g(x) = 60^{2x}$, $x = -1.1$ **4.** $g(x) = 25^{-3x}$, $x = \frac{3}{2}$

  Library of Parent Functions In Exercises 5–8, match the function with its graph. [The graphs are labeled (a), (b), (c), and (d).]

(a)
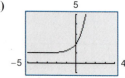

(b)
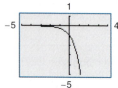

(c)
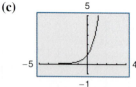

(d)
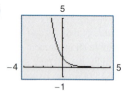

5. $f(x) = 4^x$ **6.** $f(x) = 4^{-x}$
7. $f(x) = -4^x$ **8.** $f(x) = 4^x + 1$

Graphs of $y = a^x$ and $y = a^{-x}$ In Exercises 9–12, graph the exponential function by hand. Identify any asymptotes and intercepts and determine whether the graph of the function is increasing or decreasing.

9. $f(x) = 6^x$ **10.** $f(x) = 0.3^x$
11. $g(x) = 6^{-x}$ **12.** $g(x) = 0.3^{-x}$

Graphing an Exponential Function In Exercises 13–18, use a graphing utility to construct a table of values for the function. Then sketch the graph of the function. Identify any asymptotes of the graph.

13. $h(x) = e^{x-1}$ **14.** $f(x) = e^{x+2}$
15. $h(x) = -e^x$ **16.** $f(x) = 3 - e^{-x}$
17. $f(x) = 4e^{-0.5x}$ **18.** $f(x) = 2 + e^{x+3}$

Finding the Balance for Compound Interest In Exercises 19 and 20, complete the table to determine the balance A for $10,000 invested at rate r for t years, compounded continuously.

t	1	10	20	30	40	50
A						

19. $r = 8\%$ **20.** $r = 3\%$

21. Economics A new SUV costs $32,000. The value V of the SUV after t years is modeled by $V(t) = 32,000\left(\frac{3}{4}\right)^t$.

(a) Use a graphing utility to graph the function.
(b) Find the value of the SUV after 2 years.
(c) According to the model, when does the SUV depreciate most rapidly? Is this realistic? Explain.

22. Radioactive Decay Let Q represent the mass, in grams, of a quantity of plutonium 241 (^{241}Pu), whose half-life is 14 years. The quantity of plutonium present after t years is given by $Q = 50\left(\frac{1}{2}\right)^{t/14}$.

(a) Determine the initial quantity (when $t = 0$).
(b) Determine the quantity present after 10 years.
(c) Use a graphing utility to graph the function over the interval $t = 0$ to $t = 50$.

4.2

Rewriting Equations In Exercises 23–32, write the logarithmic equation in exponential form or write the exponential equation in logarithmic form.

23. $\log_5 125 = 3$ **24.** $\log_9 81 = 2$
25. $\log_{64} 2 = \frac{1}{6}$ **26.** $\log_{10}\left(\frac{1}{100}\right) = -2$
27. $4^3 = 64$ **28.** $3^5 = 243$
29. $125^{2/3} = 25$ **30.** $12^{-1} = \frac{1}{12}$
31. $\left(\frac{1}{2}\right)^{-3} = 8$ **32.** $\left(\frac{2}{3}\right)^{-2} = \frac{9}{4}$

Evaluating Logarithms In Exercises 33–36, use the definition of logarithmic function to evaluate the function at the indicated value of x without using a calculator.

Function	Value
33. $f(x) = \log_6 x$	$x = 216$
34. $f(x) = \log_7 x$	$x = 1$
35. $f(x) = \log_4 x$	$x = \frac{1}{4}$
36. $f(x) = \log_{10} x$	$x = 0.00001$

Sketching the Graph of a Logarithmic Function In Exercises 37–40, find the domain, vertical asymptote, and x-intercept of the logarithmic function, and sketch its graph by hand.

37. $g(x) = -\log_2 x + 5$ **38.** $g(x) = \log_5(x - 3)$
39. $f(x) = \log_2(x - 1) + 6$ **40.** $f(x) = \log_5(x + 2) - 3$

Evaluating the Natural Logarithmic Function In Exercises 41–44, use a calculator to evaluate the function $f(x) = \ln x$ at the indicated value of x. Round your result to three decimal places, if necessary.

41. $x = 21.5$ **42.** $x = 0.46$
43. $x = \sqrt{6}$ **44.** $x = \frac{5}{6}$

Analyzing Graphs of Functions In Exercises 45–48, use a graphing utility to graph the logarithmic function. Determine the domain and identify any vertical asymptote and x-intercept.

45. $f(x) = \ln x + 3$ **46.** $f(x) = \ln(x - 3)$

47. $h(x) = \frac{1}{2} \ln x$ **48.** $f(x) = \frac{1}{4} \ln x$

49. Aeronautics The time t (in minutes) for a small plane to climb to an altitude of h feet is given by

$$t = 50 \log_{10}[18{,}000/(18{,}000 - h)]$$

where 18,000 feet is the plane's absolute ceiling.

(a) Determine the domain of the function appropriate for the context of the problem.

(b) Use a graphing utility to graph the function and identify any asymptotes.

(c) As the plane approaches its absolute ceiling, what can be said about the time required to further increase its altitude?

(d) Find the amount of time it will take for the plane to climb to an altitude of 4000 feet.

50. Real Estate The model

$$t = 12.542 \ln[x/(x - 1000)], \quad x > 1000$$

approximates the length of a home mortgage of $150,000 at 8% in terms of the monthly payment. In the model, t is the length of the mortgage in years and x is the monthly payment in dollars.

(a) Use the model to approximate the length of a $150,000 mortgage at 8% when the monthly payment is $1254.68.

(b) Approximate the total amount paid over the term of the mortgage with a monthly payment of $1254.68. What amount of the total is interest costs?

4.3

Changing the Base In Exercises 51–54, evaluate the logarithm using the change-of-base formula. Do each problem twice, once with common logarithms and once with natural logarithms. Round your results to three decimal places.

51. $\log_4 9$ **52.** $\log_{1/2} 9$

53. $\log_{14} 364$ **54.** $\log_3 0.28$

Graphing a Logarithm with Any Base In Exercises 55–58, use the change-of-base formula and a graphing utility to graph the function.

55. $f(x) = \log_2(x - 1)$

56. $f(x) = 2 - \log_3 x$

57. $f(x) = -\log_{1/2}(x + 2)$

58. $f(x) = \log_{1/3}(x - 1) + 1$

Using Properties to Evaluate Logarithms In Exercises 59–62, approximate the logarithm using the properties of logarithms, given the values $\log_b 2 \approx 0.3562$, $\log_b 3 \approx 0.5646$, and $\log_b 5 \approx 0.8271$.

59. $\log_b 9$ **60.** $\log_b \frac{4}{9}$

61. $\log_b \sqrt{5}$ **62.** $\log_b 50$

Simplifying a Logarithm In Exercises 63–66, use the properties of logarithms to rewrite and simplify the logarithmic expression.

63. $\ln(5e^{-2})$

64. $\ln \sqrt{e^5}$

65. $\log_{10} 200$

66. $\log_{10} 0.002$

Expanding Logarithmic Expressions In Exercises 67–72, use the properties of logarithms to expand the expression as a sum, difference, and/or constant multiple of logarithms. (Assume all variables are positive.)

67. $\log_5 5x^2$ **68.** $\log_4 16xy^2$

69. $\log_{10} \dfrac{5\sqrt{y}}{x^2}$ **70.** $\ln \dfrac{\sqrt{x}}{4}$

71. $\ln \dfrac{x + 3}{xy}$ **72.** $\ln \dfrac{xy^5}{\sqrt{z}}$

Condensing Logarithmic Expressions In Exercises 73–78, condense the expression to the logarithm of a single quantity.

73. $\log_2 9 + \log_2 x$

74. $\log_6 y - 2 \log_6 z$

75. $\frac{1}{2} \ln(2x - 1) - 2 \ln(x + 1)$

76. $5 \ln(x - 2) - \ln(x + 2) - 3 \ln x$

77. $\ln 3 + \frac{1}{3} \ln(4 - x^2) - \ln x$

78. $3[\ln x - 2 \ln(x^2 + 1)] + 2 \ln 5$

79. Public Service The number of miles s of roads cleared of snow in 1 hour is approximated by the model

$$s = 25 - \frac{13 \ln(h/12)}{\ln 3}, \quad 2 \le h \le 15$$

where h is the depth of the snow (in inches).

(a) Use a graphing utility to graph the function.

(b) Complete the table.

h	4	6	8	10	12	14
s						

(c) Using the graph of the function and the table, what conclusion can you make about the number of miles of roads cleared as the depth of the snow increases?

80. Psychology Students in a sociology class were given an exam and then retested monthly with an equivalent exam. The average scores for the class are given by the human memory model $f(t) = 85 - 17 \log_{10}(t + 1)$, where t is the time in months and $0 \le t \le 10$. When will the average score decrease to 68?

4.4

Solving an Exponential or Logarithmic Equation In Exercises 81–94, solve the equation for x without using a calculator.

81. $10^x = 10,000$
82. $7^x = 343$

83. $6^x = \frac{1}{216}$
84. $6^{x-2} = 1296$

85. $2^{x+1} = \frac{1}{16}$
86. $4^{x/2} = 64$

87. $\log_8 x = 4$
88. $\log_x 729 = 6$

89. $\log_2(x - 1) = 3$
90. $\log_5(2x + 1) = 2$

91. $\ln x = 4$
92. $\ln x = -3$

93. $\ln(x - 1) = 2$
94. $\ln(2x + 1) = -4$

Solving an Exponential Equation In Exercises 95–104, solve the exponential equation algebraically. Round your result to three decimal places.

95. $3e^{-5x} = 132$
96. $14e^{3x+2} = 560$

97. $2^x + 13 = 35$
98. $6^x - 28 = -8$

99. $-4(5^x) = -68$
100. $2(12^x) = 190$

101. $2e^{x-3} - 1 = 4$
102. $-e^{x/2} + 1 = \frac{1}{2}$

103. $e^{2x} - 7e^x + 10 = 0$
104. $e^{2x} - 6e^x + 8 = 0$

Solving a Logarithmic Equation In Exercises 105–114, solve the logarithmic equation algebraically. Round your result to three decimal places.

105. $\ln 3x = 6.4$
106. $\ln 5x = 4.5$

107. $\ln x - \ln 5 = 2$
108. $\ln x - \ln 3 = 4$

109. $\ln \sqrt{x + 1} = 2$
110. $\ln \sqrt{x + 40} = 3$

111. $\log_4(x - 1) = \log_4(x - 2) - \log_4(x + 2)$

112. $\log_5(x + 2) - \log_5 x = \log_5(x + 5)$

113. $\log_{10}(1 - x) = -1$

114. $\log_{10}(-x - 4) = 2$

∫ **Solving an Exponential or Logarithmic Equation** In Exercises 115–118, solve the equation algebraically. Round your result to three decimal places.

115. $xe^x + e^x = 0$
116. $2xe^{2x} + e^{2x} = 0$

117. $x \ln x + x = 0$
118. $\dfrac{1 - \ln x}{x^2} = 0$

119. Finance You deposit $7550 in an account that pays 6.9% interest, compounded continuously. How long will it take for the money to double?

120. Economics The demand x for a 32-inch plasma television is modeled by

$$p = 5000\left(1 - \frac{4}{4 + e^{-0.0005x}}\right).$$

Find the demands x for prices of (a) $p = 450 and (b) $p = 400.

4.5

Identifying Graphs of Models In Exercises 121–126, match the function with its graph. [The graphs are labeled (a), (b), (c), (d), (e), and (f).]

(a)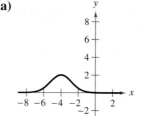

(b)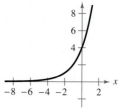

(c)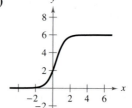

(d)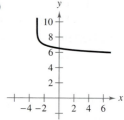

(e)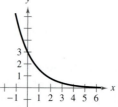

(f)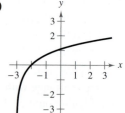

121. $y = 3e^{-2x/3}$
122. $y = 4e^{2x/3}$

123. $y = \ln(x + 3)$
124. $y = 7 - \log_{10}(x + 3)$

125. $y = 2e^{-(x+4)^2/3}$
126. $y = \dfrac{6}{1 + 2e^{-2x}}$

127. Demography The populations P (in thousands) of North Carolina from 1990 through 2008 can be modeled by $P = 6707.7e^{kt}$, where t is the year, with $t = 0$ corresponding to 1990. In 2008, the population was about 9,222,000. Find the value of k and use the result to predict the population in the year 2020. (Source: U.S. Census Bureau)

128. Education The scores for a biology test follow a normal distribution modeled by $y = 0.0499e^{-(x-74)^2/128}$, where x is the test score and $40 \le x \le 100$.

(a) Use a graphing utility to graph the function.

(b) Use the graph to estimate the average test score.

129. Education   The average number N of words per minute that the students in a first grade class could read orally after t weeks of school is modeled by

$$N = \frac{62}{1 + 5.4e^{-0.24t}}.$$

Find the numbers of weeks it took the class to read at average rates of (a) 40 words per minute and (b) 60 words per minute.

130. Geology   On the Richter scale, the magnitude R of an earthquake of intensity I is modeled by

$$R = \log_{10} \frac{I}{I_0}$$

where $I_0 = 1$ is the minimum intensity used for comparison. Find the intensities I of the following earthquakes measuring R on the Richter scale.

(a) $R = 7.1$ (b) $R = 8.4$ (c) $R = 5.5$

4.6

Classifying Scatter Plots   In Exercises 131 and 132, determine whether the scatter plot could best be modeled by a linear model, an exponential model, a logarithmic model, or a logistic model.

131.

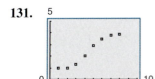

132.

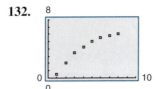

133. MODELING DATA

Each ordered pair (t, N) represents the year t and the number N (in thousands) of female participants in high school athletic programs during nine school years, with $t = 1$ corresponding to the 2000–2001 school year. (Source: National Federation of State High School Associations)

(1, 2784), (2, 2807), (3, 2856), (4, 2865), (5, 2908), (6, 2953), (7, 3022), (8, 3057), (9, 3114)

(a) Use the *regression* feature of a graphing utility to find a linear model, an exponential model, and a power model for the data and identify the coefficient of determination for each model.

(b) Use the graphing utility to graph each model with the original data.

(c) Determine which model best fits the data. Explain.

(d) Use the model you chose in part (c) to predict the number of participants during the 2009–2010 school year.

(e) Use the model you chose in part (c) to predict the school year in which about 3,720,000 girls will participate.

134. MODELING DATA

You plant a tree when it is 1 meter tall and check its height h (in meters) every 10 years, as shown in the table.

Year	Height, h
0	1
10	3
20	7.5
30	14.5
40	19
50	20.5
60	21

(a) Use the *regression* feature of a graphing utility to find a logistic model for the data. Let x represent the year.

(b) Use the graphing utility to graph the model with the original data.

(c) How closely does the model represent the data?

(d) What is the limiting height of the tree?

Conclusions

True or False?   In Exercises 135–138, determine whether the equation or statement is true or false. Justify your answer.

135. $e^{x-1} = \dfrac{e^x}{e}$

136. $\ln(x + y) = \ln(xy)$

137. The domain of the function $f(x) = \ln x$ is the set of all real numbers.

138. The logarithm of the quotient of two numbers is equal to the difference of the logarithms of the numbers.

139. Think About It   Without using a calculator, explain why you know that $2^{\sqrt{2}}$ is greater than 2, but less than 4.

140. Exploration

(a) Use a graphing utility to compare the graph of the function $y = e^x$ with the graph of each function below. [$n!$ (read as "n factorial") is defined as $n! = 1 \cdot 2 \cdot 3 \cdots (n - 1) \cdot n$.]

$$y_1 = 1 + \frac{x}{1!}, \quad y_2 = 1 + \frac{x}{1!} + \frac{x^2}{2!},$$

$$y_3 = 1 + \frac{x}{1!} + \frac{x^2}{2!} + \frac{x^3}{3!}$$

(b) Identify the pattern of successive polynomials given in part (a). Extend the pattern one more term and compare the graph of the resulting polynomial function with the graph of $y = e^x$. What do you think this pattern implies?

4 Chapter Test

See www.CalcChat.com for worked-out solutions to odd-numbered exercises.
For instructions on how to use a graphing utility, see Appendix A.

Take this test as you would take a test in class. After you are finished, check your work against the answers given in the back of the book.

In Exercises 1–3, use a graphing utility to construct a table of values for the function. Then sketch a graph of the function. Identify any asymptotes and intercepts.

1. $f(x) = 10^{-x}$
2. $f(x) = -6^{x-2}$
3. $f(x) = 1 - e^{2x}$

In Exercises 4–6, evaluate the expression.

4. $\log_7 7^{-0.89}$
5. $4.6 \ln e^2$
6. $5 - \log_{10} 1000$

In Exercises 7–9, find the domain, vertical asymptote, and x-intercept of the logarithmic function, and sketch its graph by hand.

7. $f(x) = -\log_{10} x - 6$
8. $f(x) = \ln(x - 4)$
9. $f(x) = 1 + \ln(x + 6)$

In Exercises 10–12, evaluate the logarithm using the change-of-base formula. Round your result to three decimal places.

10. $\log_7 44$
11. $\log_{2/5} 0.9$
12. $\log_{12} 64$

In Exercises 13–15, use the properties of logarithms to expand the expression as a sum, difference, and/or multiple of logarithms.

13. $\log_2 3a^4$
14. $\ln \dfrac{5\sqrt{x}}{6}$
15. $\ln \dfrac{x\sqrt{x+1}}{2e^4}$

In Exercises 16–18, condense the expression to the logarithm of a single quantity.

16. $\log_3 13 + \log_3 y$
17. $4 \ln x - 4 \ln y$
18. $\ln x - \ln(x + 2) + \ln(2x - 3)$

In Exercises 19–22, solve the equation for x.

19. $3^x = 81$
20. $5^{2x} = 2500$
21. $\log_7 x = 3$
22. $\log_{10}(x - 4) = 5$

In Exercises 23–26, solve the equation algebraically. Round your result to three decimal places.

23. $\dfrac{1025}{8 + e^{4x}} = 5$
24. $-xe^{-x} + e^{-x} = 0$
25. $\log_{10} x - \log_{10}(8 - 5x) = 2$
26. $2x \ln x - x = 0$

27. The half-life of radioactive actinium (^{227}Ac) is 22 years. What percent of a present amount of radioactive actinium will remain after 19 years?

28. The table shows the annual revenues R (in millions of dollars) for Daktronics from 2001 through 2008.   (Source: Daktronics, Inc.)

(a) Use the *regression* feature of a graphing utility to find a logarithmic model, an exponential model, and a power model for the data. Let t represent the year, with $t = 1$ corresponding to 2001.

(b) Use the graphing utility to graph each model with the original data.

(c) Determine which model best fits the data. Use the model to predict the revenue of Daktronics in 2015.

Year	Revenue, R
2001	148.8
2002	177.8
2003	209.9
2004	230.3
2005	309.4
2006	433.2
2007	499.7
2008	581.9

Table for 28

3–4 Cumulative Test

See www.CalcChat.com for worked-out solutions to odd-numbered exercises.
For instructions on how to use a graphing utility, see Appendix A.

Take this test to review the material in Chapters 3 and 4. After you are finished, check your work against the answers in the back of the book.

In Exercises 1–3, sketch the graph of the function. Use a graphing utility to verify the graph.

1. $f(x) = -(x - 2)^2 + 5$

2. $f(x) = x^2 - 6x + 5$

3. $f(x) = x^3 + 2x^2 - 9x - 18$

4. Find all the zeros of $f(x) = x^3 + 2x^2 + 4x + 8$.

5. Use a graphing utility to approximate any real zeros of $g(x) = x^3 + 4x^2 - 11$ accurate to three decimal places.

6. Divide $(4x^2 + 14x - 9)$ by $(x + 3)$ using long division.

7. Divide $(2x^3 - 5x^2 + 6x - 20)$ by $(x - 6)$ using synthetic division.

8. Find a polynomial function with real coefficients that has the zeros $0, -3$, and $1 + \sqrt{5}i$.

In Exercises 9–11, sketch the graph of the rational function. Identify any asymptotes. Use a graphing utility to verify the graph.

9. $f(x) = \dfrac{2x}{x - 3}$

10. $f(x) = \dfrac{5x}{x^2 + x - 6}$

11. $f(x) = \dfrac{x^2 - 3x + 8}{x - 2}$

12. Write a rational function whose graph has no vertical asymptotes and has a horizontal asymptote at $y = 4$.

In Exercises 13 and 14, use the graph of f to describe the transformation that yields the graph of g.

13. $f(x) = 3^x$
    $g(x) = 3^{x+1} - 5$

14. $f(x) = \log_{10} x$
    $g(x) = -\log_{10}(x + 3)$

In Exercises 15–18, (a) determine the domain of the function, (b) find all intercepts, if any, of the graph of the function, and (c) identify any asymptotes of the graph of the function.

15. $f(x) = 8^{-x+1} + 2$

16. $f(x) = 2e^{0.01x}$

17. $f(x) = \dfrac{\ln x}{x}$

18. $f(x) = \log_6(2x - 1) + 1$

In Exercises 19–22, evaluate the expression without using a calculator.

19. $\log_2 64$

20. $\log_2\left(\dfrac{1}{16}\right)$

21. $\ln e^{10}$

22. $\ln \dfrac{1}{e^3}$

In Exercises 23–26, evaluate the logarithm using the change-of-base formula. Round your result to three decimal places.

23. $\log_5 16$

24. $\log_9 6.8$

25. $\log_{3/4}(8.61)$

26. $\log_{7/8}\left(\dfrac{3}{2}\right)$

27. Use the properties of logarithms to expand $\ln\left(\dfrac{x^2 - 4}{x^2 + 1}\right)$, where $x \neq 2$.

28. Write $2 \ln x - \ln(x - 1) + \ln(x + 1)$ as a logarithm of a single quantity.

In Exercises 29–34, solve the equation algebraically. Round your result to three decimal places and verify your result graphically.

29. $6e^{2x} = 72$

30. $4^{x-5} + 21 = 30$

31. $\log_2 x + \log_2 5 = 6$

32. $250e^{0.05x} = 500,000$

33. $2x^2e^{2x} - 2xe^{2x} = 0$

34. $\ln(2x - 5) - \ln x = 1$

35. The average cost $\overline{C}$ (in dollars) of recycling a waste product x (in pounds) is given by

$$\overline{C}(x) = \frac{450,000 + 5x}{x}, \quad x > 0.$$

Find the average costs $\overline{C}$ of recycling $x = 10,000$ pounds, $x = 100,000$ pounds, and $x = 1,000,000$ pounds. According to this model, what is the limiting average cost as the number of pounds increases?

36. Find the amounts in an account when $200 is deposited earning 6% interest for 30 years, if the interested is compounded (a) monthly and (b) continuously. Recall that

$$A = P\left(1 + \frac{r}{n}\right)^{nt} \text{ and } A = Pe^{rt}.$$

37. If the inflation rate averages 4.5% over the next 10 years, the approximate cost C of goods or services t years from now is given by

$$C(t) = P(1.045)^t$$

where P is the present cost. If the price of a tire is presently $69.95, estimate the price 10 years from now.

38. The population p of a species t years after it is introduced into a new habitat is given by $p(t) = 1200/(1 + 3e^{-t/5})$.

   (a) Determine the population size that was introduced into the habitat.

   (b) Determine the population after 5 years.

   (c) After how many years will the population be 800?

39. A rectangular plot of land with a perimeter of 546 feet has a width of x.

   (a) Write the area A of the plot as a function of x.

   (b) Use a graphing utility to graph the area function. What is the domain of the function?

   (c) Approximate the dimensions of the plot when the area is 15,000 square feet.

40. The table shows the average prices y (in dollars) received by commercial trout producers per pound of trout in the United States from 2001 to 2008.   (Source: U.S. Department of Agriculture)

   (a) Use the *regression* feature of a graphing utility to find a quadratic model, an exponential model, and a power model for the data and identify the coefficient of determination for each model. Let t represent the year, with $t = 1$ corresponding to 2001.

   (b) Use the graphing utility to graph each model with the original data.

   (c) Determine which model best fits the data. Explain.

   (d) Use the model you chose in part (c) to predict the average price of one pound of trout in 2010. Is your answer reasonable? Explain.

Year	Average price, y (in dollars)
2001	1.13
2002	1.08
2003	1.04
2004	1.03
2005	1.05
2006	1.11
2007	1.19
2008	1.38

Table for 40

Proofs in Mathematics

Each of the following three properties of logarithms can be proved by using properties of exponential functions.

> ## Properties of Logarithms (p. 348)
>
> Let a be a positive real number such that $a \neq 1$, and let n be a real number. If u and v are positive real numbers, then the following properties are true.
>
> | | Logarithm with Base a | Natural Logarithm |
> |---|---|---|
> | **1. Product Property:** | $\log_a(uv) = \log_a u + \log_a v$ | $\ln(uv) = \ln u + \ln v$ |
> | **2. Quotient Property:** | $\log_a \dfrac{u}{v} = \log_a u - \log_a v$ | $\ln \dfrac{u}{v} = \ln u - \ln v$ |
> | **3. Power Property:** | $\log_a u^n = n \log_a u$ | $\ln u^n = n \ln u$ |

Proof

Let

$$x = \log_a u \quad \text{and} \quad y = \log_a v.$$

The corresponding exponential forms of these two equations are

$$a^x = u \quad \text{and} \quad a^y = v.$$

To prove the Product Property, multiply u and v to obtain

$$uv = a^x a^y = a^{x+y}.$$

The corresponding logarithmic form of $uv = a^{x+y}$ is

$$\log_a(uv) = x + y.$$

So,

$$\log_a(uv) = \log_a u + \log_a v.$$

To prove the Quotient Property, divide u by v to obtain

$$\frac{u}{v} = \frac{a^x}{a^y} = a^{x-y}.$$

The corresponding logarithmic form of $u/v = a^{x-y}$ is

$$\log_a \frac{u}{v} = x - y.$$

So,

$$\log_a \frac{u}{v} = \log_a u - \log_a v.$$

To prove the Power Property, substitute a^x for u in the expression $\log_a u^n$, as follows.

$$\log_a u^n = \log_a (a^x)^n \qquad \text{Substitute } a^x \text{ for } u.$$
$$= \log_a a^{nx} \qquad \text{Property of exponents}$$
$$= nx \qquad \text{Inverse Property of logarithms}$$
$$= n \log_a u \qquad \text{Substitute } \log_a u \text{ for } x.$$

So, $\log_a u^n = n \log_a u.$

Slide Rules

The slide rule was invented by William Oughtred (1574–1660) in 1625. The slide rule is a computational device with a sliding portion and a fixed portion. A slide rule enables you to perform multiplication by using the Product Property of logarithms. There are other slide rules that allow for the calculation of roots and trigonometric functions. Slide rules were used by mathematicians and engineers until the invention of the handheld calculator in 1972.

Progressive Summary (Chapters P–4)

This chart outlines the topics that have been covered so far in this text. Progressive Summary charts appear after Chapters 2, 4, 7, and 10. In each Progressive Summary, new topics encountered for the first time appear in red.

ALGEBRAIC FUNCTIONS

Polynomial, Rational, Radical

■ Rewriting

Polynomial form ↔ Factored form
Operations with polynomials
Rationalize denominators
Simplify rational expressions
Exponent form ↔ Radical form
Operations with complex numbers

■ Solving

Equation	Strategy
Linear	Isolate variable
Quadratic	Factor, set to zero
	Extract square roots
	Complete the square
	Quadratic Formula
Polynomial	Factor, set to zero
	Rational Zero Test
Rational	Multiply by LCD
Radical	Isolate, raise to power
Absolute value	Isolate, form two equations

■ Analyzing

Graphically	Algebraically
Intercepts	Domain, Range
Symmetry	Transformations
Slope	Composition
Asymptotes	Standard forms
End behavior	of equations
Minimum values	Leading Coefficient
Maximum values	Test
	Synthetic division
	Descartes's Rule of Signs

Numerically

Table of values

TRANSCENDENTAL FUNCTIONS

Exponential, Logarithmic

■ Rewriting

Exponential form ↔ Logarithmic form
Condense/expand logarithmic expressions

■ Solving

Equation	Strategy
Exponential	Take logarithm of each side
Logarithmic	Exponentiate each side

■ Analyzing

Graphically	Algebraically
Intercepts	Domain, Range
Asymptotes	Transformations
	Composition
	Inverse Properties

Numerically

Table of values

OTHER TOPICS

■ Rewriting

■ Solving

■ Analyzing

5 Trigonometric Functions

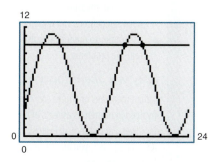

Section 5.4, Example 8
A Depth of at Least 10 Feet

Chris Schmidt/iStockphoto.com

397

5.1   Angles and Their Measure

Angles

As derived from the Greek language, the word **trigonometry** means "measurement of triangles." Initially, trigonometry dealt with relationships among the sides and angles of triangles and was used in the development of astronomy, navigation, and surveying. With the development of calculus and the physical sciences in the 17th century, a different perspective arose—one that viewed the classic trigonometric relationships as *functions* having the set of real numbers as their domains. Consequently, the applications of trigonometry expanded to include a vast number of physical phenomena involving rotations and vibrations, including the following.

- sound waves
- light rays
- planetary orbits
- vibrating strings
- pendulums
- orbits of atomic particles

This text incorporates *both* perspectives, starting with angles and their measure.

What you should learn

- Describe angles.
- Use degree measure.
- Use radian measure and convert between degrees and radians.
- Use angles to model and solve real-life problems.

Why you should learn it

Radian measures of angles are involved in numerous aspects of our daily lives. For instance, in Exercise 106 on page 407, you are asked to determine the measure of the angle generated as a skater performs an axel jump.

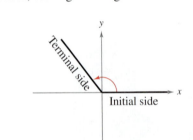

Figure 5.1   **Figure 5.2**

An **angle** is determined by rotating a ray (half-line) about its endpoint. The starting position of the ray is the **initial side** of the angle, and the position after rotation is the **terminal side,** as shown in Figure 5.1. The endpoint of the ray is the **vertex** of the angle. This perception of an angle fits a coordinate system in which the origin is the vertex and the initial side coincides with the positive *x*-axis. Such an angle is in **standard position,** as shown in Figure 5.2. **Positive angles** are generated by counterclockwise rotation, and **negative angles** by clockwise rotation, as shown in Figure 5.3. Angles are labeled with Greek letters such as α (alpha), β (beta), and θ (theta), as well as uppercase letters such as *A*, *B*, and *C*. In Figure 5.4, note that angles α and β have the same initial and terminal sides. Such angles are **coterminal.**

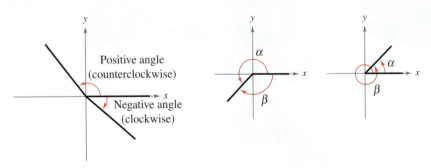

Figure 5.3   **Figure 5.4**

Degree Measure

The **measure of an angle** is determined by the amount of rotation from the initial side to the terminal side. The most common unit of angle measure is the **degree,** denoted by the symbol °. A measure of one degree (1°) is equivalent to a rotation of $\frac{1}{360}$ of a complete revolution about the vertex. To measure angles, it is convenient to mark degrees on the circumference of a circle, as shown in Figure 5.5. So, a full revolution (counterclockwise) corresponds to 360°, a half revolution to 180°, a quarter revolution to 90°, and so on.

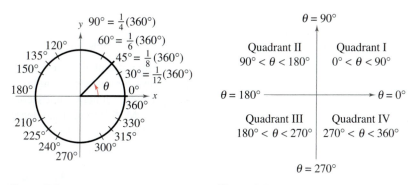

Figure 5.5   **Figure 5.6**

Recall that the four quadrants in a coordinate system are numbered I, II, III, and IV. Figure 5.6 shows which angles between 0° and 360° lie in each of the four quadrants. Figure 5.7 shows several common angles with their degree measures. Note that angles between 0° and 90° are **acute** and angles between 90° and 180° are **obtuse.**

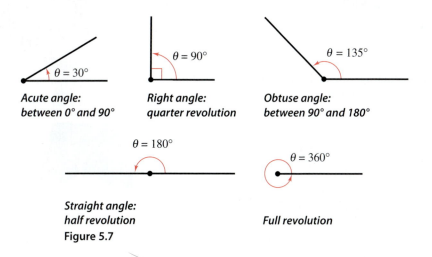

Acute angle:
between 0° and 90°   *Right angle:*
quarter revolution   *Obtuse angle:*
between 90° and 180°

Straight angle:
half revolution
Figure 5.7   *Full revolution*

> ## Study Tip
>
> The phrase "the terminal side of θ lies in a quadrant" is often abbreviated by simply saying that "θ lies in a quadrant." The terminal sides of the "quadrant angles" 0°, 90°, 180°, and 270° do not lie within quadrants.

Two angles are coterminal when they have the same initial and terminal sides. For instance, the angles

0°   and   360°

are coterminal, as are the angles

30°   and   390°.

You can find an angle that is coterminal to a given angle θ by adding or subtracting 360° (one revolution), as demonstrated in Example 1 on the next page. A given angle θ has infinitely many coterminal angles. For instance, $\theta = 30°$ is coterminal with

$$30° + n(360°)$$

where n is an integer.

Example 1 Finding Coterminal Angles

Find two coterminal angles (one positive and one negative) for (a) $\theta = 390°$ and (b) $\theta = -120°$.

Solution

a. For the positive angle $\theta = 390°$, subtract $360°$ to obtain a positive coterminal angle.

$$390° - 360° = 30° \qquad \text{See Figure 5.8.}$$

Subtract $2(360°) = 720°$ to obtain a negative coterminal angle.

$$390° - 720° = -330°$$

b. For the negative angle $\theta = -120°$, add $360°$ to obtain a positive coterminal angle.

$$-120° + 360° = 240° \qquad \text{See Figure 5.9.}$$

Subtract $360°$ to obtain a negative coterminal angle.

$$-120° - 360° = -480°$$

Technology Tip

Historically, fractional parts of degrees were expressed in *minutes* and *seconds*, using the prime ($'$) and double prime ($''$) notations, respectively. That is,

$$1' = \text{one minute} = \tfrac{1}{60}(1°)$$

$$1'' = \text{one second} = \tfrac{1}{3600}(1°).$$

Many calculators have special keys for converting angles in degrees, minutes, and seconds ($\text{D}° \, \text{M}' \, \text{S}''$) to decimal degree form, and vice versa.

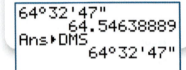

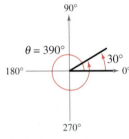

Figure 5.8

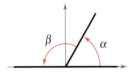

Figure 5.9

 CHECKPOINT Now try Exercise 27.

Two positive angles α and β are **complementary** (complements of each other) when their sum is $90°$. Two positive angles are **supplementary** (supplements of each other) when their sum is $180°$. (See Figure 5.10.)

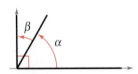

Complementary angles
Figure 5.10

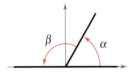

Supplementary angles

Example 2 Complementary and Supplementary Angles

a. The complement of $72°$ is

$$90° - 72° = 18°.$$

The supplement of $72°$ is

$$180° - 72° = 108°.$$

b. Because $148°$ is greater than $90°$, $148°$ has no complement. (Remember that complements are *positive* angles.) The supplement is

$$180° - 148° = 32°.$$

 CHECKPOINT Now try Exercise 41.

Radian Measure

A second way to measure angles is in *radians*. This type of measure is especially useful in calculus. To define a radian, you can use a **central angle** of a circle, one whose vertex is the center of the circle, as shown in Figure 5.11.

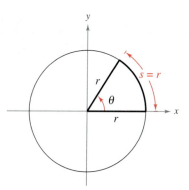

Definition of Radian

One **radian** (rad) is the measure of a central angle θ that intercepts an arc s equal in length to the radius r of the circle. (See Figure 5.11.) Algebraically this means that

$$\theta = \frac{s}{r}$$

where θ is measured in radians.

Arc length = radius when $\theta = 1$ radian
Figure 5.11

Because the circumference of a circle is $2\pi r$ units, it follows that a central angle of one full revolution (counterclockwise) corresponds to an arc length of $s = 2\pi r$. Moreover, because $2\pi \approx 6.28$, there are just over six radius lengths in a full circle, as shown in Figure 5.12. Because the units of measure for s and r are the same, the ratio s/r has no units—it is simply a real number.

Because 2π radians corresponds to one complete revolution, degrees and radians are related by the equations

$$360° = 2\pi \text{ rad} \quad \text{and} \quad 180° = \pi \text{ rad}.$$

From the second equation, you obtain

$$1° = \frac{\pi}{180} \text{ rad} \quad \text{and} \quad 1 \text{ rad} = \frac{180°}{\pi}$$

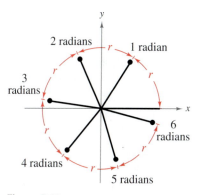

Figure 5.12

which lead to the following conversion rules.

Conversions Between Degrees and Radians

1. To convert degrees to radians, multiply degrees by $\dfrac{\pi \text{ rad}}{180°}$.

2. To convert radians to degrees, multiply radians by $\dfrac{180°}{\pi \text{ rad}}$.

To apply these two conversion rules, use the basic relationship π rad $=180°$. (See Figure 5.13.)

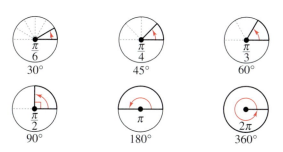

Figure 5.13

When no units of angle measure are specified, *radian measure is implied*. For instance, $\theta = \pi$ or $\theta = 2$ implies that $\theta = \pi$ radians or $\theta = 2$ radians.

Example 3   Converting from Degrees to Radians

a. $135° = (135 \text{ deg})\left(\dfrac{\pi \text{ rad}}{180 \text{ deg}}\right) = \dfrac{3\pi}{4}$ radians   Multiply by $\dfrac{\pi}{180}$.

b. $-270° = (-270 \text{ deg})\left(\dfrac{\pi \text{ rad}}{180 \text{ deg}}\right) = -\dfrac{3\pi}{2}$ radians   Multiply by $\dfrac{\pi}{180}$.

✓**CHECKPOINT**   Now try Exercise 59.

Example 4   Converting from Radians to Degrees

a. $-\dfrac{\pi}{2} \text{ rad} = \left(-\dfrac{\pi}{2} \text{ rad}\right)\left(\dfrac{180 \text{ deg}}{\pi \text{ rad}}\right) = -90°$ Multiply by $\dfrac{180}{\pi}$.

b. $2 \text{ rad} = (2 \text{ rad})\left(\dfrac{180 \text{ deg}}{\pi \text{ rad}}\right) = \dfrac{360}{\pi} \approx 114.59°$ Multiply by $\dfrac{180}{\pi}$.

✓**CHECKPOINT**   Now try Exercise 63.

Use a calculator with a "radian-to-degree" conversion key to verify the result shown in part (b) of Example 4.

Example 5   Finding Angles

Find each angle.

a. The complement of $\theta = \pi/12$ **b.** The supplement of $\theta = 5\pi/6$

c. A coterminal angle to $\theta = 17\pi/6$

Solution

a. In radian measure, the complement of an angle is found by subtracting the angle from $\pi/2$ ($\pi/2 = 90°$). So, the complement of $\theta = \pi/12$ is $\pi/2 - \theta$, which is

$$\frac{\pi}{2} - \frac{\pi}{12} = \frac{6\pi}{12} - \frac{\pi}{12} = \frac{5\pi}{12}.$$ See Figure 5.14.

b. In radian measure, the supplement of an angle is found by subtracting the angle from π ($\pi = 180°$). So, the supplement of $\theta = 5\pi/6$ is $\pi - \theta$, which is

$$\pi - \frac{5\pi}{6} = \frac{6\pi}{6} - \frac{5\pi}{6} = \frac{\pi}{6}.$$ See Figure 5.15.

c. In radian measure, a coterminal angle is found by adding or subtracting 2π ($2\pi = 360°$). For $\theta = 17\pi/6$, subtract 2π to obtain a coterminal angle.

$$\frac{17\pi}{6} - 2\pi = \frac{17\pi}{6} - \frac{12\pi}{6} = \frac{5\pi}{6}.$$ See Figure 5.16.

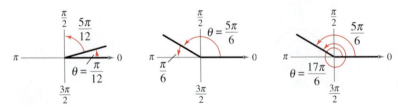

Figure 5.14   Figure 5.15   Figure 5.16

✓**CHECKPOINT**   Now try Exercise 83.

Linear and Angular Speed

The *radian measure* formula $\theta = s/r$ can be used to measure **arc length** along a circle.

Arc Length

For a circle of radius r, a central angle θ (in radian measure) intercepts an arc of length s given by

$$s = r\theta.$$ Length of circular arc

Example 6   Finding Arc Length

A circle has a radius of 4 inches. Find the length of the arc intercepted by a central angle of 240°, as shown in Figure 5.17.

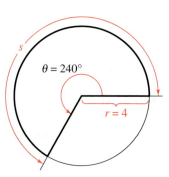

Figure 5.17

Solution

To use the formula $s = r\theta$, first convert 240° to radian measure.

$$240° = (240 \text{ deg})\left(\frac{\pi \text{ rad}}{180 \text{ deg}}\right)$$ Convert from degrees to radians.

$$= \frac{4\pi}{3} \text{ radians}$$ Simplify.

Then, using a radius of $r = 4$ inches, you can find the arc length to be

$$s = r\theta$$ Length of circular arc

$$= 4\left(\frac{4\pi}{3}\right)$$ Substitute for r and θ.

$$= \frac{16\pi}{3}$$ Simplify.

$$\approx 16.76 \text{ inches.}$$ Use a calculator.

Note that the units for $r\theta$ are determined by the units for r because θ is given in radian measure and therefore has no units.

 CHECKPOINT   Now try Exercise 93.

The formula for the length of a circular arc can be used to analyze the motion of a particle moving at a constant speed along a circular path.

Linear and Angular Speed

Consider a particle moving at a constant speed along a circular arc of radius r. If s is the length of the arc traveled in time t, then the **linear speed** of the particle is

$$\text{Linear speed} = \frac{\text{arc length}}{\text{time}} = \frac{s}{t}.$$

Moreover, if θ is the angle (in radian measure) corresponding to the arc length s, then the **angular speed** of the particle is

$$\text{Angular speed} = \frac{\text{central angle}}{\text{time}} = \frac{\theta}{t}.$$

Linear speed measures how fast the particle moves, and angular speed measures how fast the angle changes.

5.3 Trigonometric Functions of Any Angle

Introduction

In Section 5.2, the definitions of trigonometric functions were restricted to acute angles. In this section, the definitions are extended to cover *any* angle. When θ is an *acute* angle, the definitions here coincide with those given in the preceding section.

Definition of Trigonometric Functions of Any Angle

Let θ be an angle in standard position with (x, y) a point on the terminal side of θ and $r = \sqrt{x^2 + y^2} \neq 0$.

$$\sin \theta = \frac{y}{r} \qquad \cos \theta = \frac{x}{r}$$

$$\tan \theta = \frac{y}{x}, \quad x \neq 0 \qquad \cot \theta = \frac{x}{y}, \quad y \neq 0$$

$$\sec \theta = \frac{r}{x}, \quad x \neq 0 \qquad \csc \theta = \frac{r}{y}, \quad y \neq 0$$

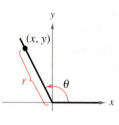

Because $r = \sqrt{x^2 + y^2}$ *cannot* be zero, it follows that the sine and cosine functions are defined for any real value of θ. When $x = 0$, however, the tangent and secant of θ are undefined. For example, the tangent of $90°$ is undefined. Similarly, when $y = 0$, the cotangent and cosecant of θ are undefined.

Example 1 Evaluating Trigonometric Functions

Let $(-3, 4)$ be a point on the terminal side of θ (see Figure 5.30). Find the sine, cosine, and tangent of θ.

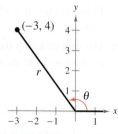

Figure 5.30

Solution

Referring to Figure 5.30, you can see that $x = -3$, $y = 4$, and

$$r = \sqrt{x^2 + y^2}$$
$$= \sqrt{(-3)^2 + 4^2}$$
$$= \sqrt{25}$$
$$= 5.$$

So, you have $\sin \theta = \dfrac{y}{r} = \dfrac{4}{5}$, $\cos \theta = \dfrac{x}{r} = -\dfrac{3}{5}$, and $\tan \theta = \dfrac{y}{x} = -\dfrac{4}{3}$.

✔CHECKPOINT   Now try Exercise 13.

What you should learn

● Evaluate trigonometric functions of any angle.
● Use reference angles to evaluate trigonometric functions.
● Evaluate trigonometric functions of real numbers.

Why you should learn it

You can use trigonometric functions to model and solve real-life problems. For instance, Exercise 136 on page 430 shows you how trigonometric functions can be used to model the monthly sales of a seasonal product, such as wakeboards.

The *signs* of the trigonometric functions in the four quadrants can be determined easily from the definitions of the functions. For instance, because

$$\cos \theta = \frac{x}{r}$$

it follows that $\cos \theta$ is positive wherever $x > 0$, which is in Quadrants I and IV. (Remember, r is always positive.) In a similar manner, you can verify the results shown in Figure 5.31.

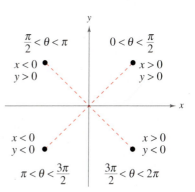

Quadrant II	Quadrant I
$\sin \theta$: +	$\sin \theta$: +
$\cos \theta$: −	$\cos \theta$: +
$\tan \theta$: −	$\tan \theta$: +
Quadrant III	Quadrant IV
$\sin \theta$: −	$\sin \theta$: −
$\cos \theta$: −	$\cos \theta$: +
$\tan \theta$: +	$\tan \theta$: −

Example 2   Evaluating Trigonometric Functions

Given $\sin \theta = -\frac{2}{3}$ and $\tan \theta > 0$, find $\cos \theta$ and $\cot \theta$.

Solution

Note that θ lies in Quadrant III because that is the only quadrant in which the sine is negative and the tangent is positive. Moreover, using

$$\sin \theta = \frac{y}{r} = -\frac{2}{3}$$

and the fact that y is negative in Quadrant III, you can let

$$y = -2 \quad \text{and} \quad r = 3.$$

Because x is negative in Quadrant III,

$$x = -\sqrt{9 - 4} = -\sqrt{5}$$

and you have the following.

$$\cos \theta = \frac{x}{r} = \frac{-\sqrt{5}}{3} \qquad \text{Exact value}$$

$$\approx -0.75 \qquad \text{Approximate value}$$

$$\cot \theta = \frac{x}{y} = \frac{-\sqrt{5}}{-2} \qquad \text{Exact value}$$

$$\approx 1.12 \qquad \text{Approximate value}$$

✓**CHECKPOINT**   Now try Exercise 33.

$\dfrac{\pi}{2} < \theta < \pi$ $\quad$ $0 < \theta < \dfrac{\pi}{2}$

$x < 0$ $\quad$ $x > 0$
$y > 0$ $\quad$ $y > 0$

$x < 0$ $\quad$ $x > 0$
$y < 0$ $\quad$ $y < 0$

$\pi < \theta < \dfrac{3\pi}{2}$ $\quad$ $\dfrac{3\pi}{2} < \theta < 2\pi$

Figure 5.31

Example 3   Trigonometric Functions of Quadrant Angles

Evaluate the sine and cosine functions at the angles 0, $\dfrac{\pi}{2}$, π, and $\dfrac{3\pi}{2}$.

Solution

To begin, choose a point on the terminal side of each angle, as shown in Figure 5.32. For each of the four given points, $r = 1$, and you have the following.

$$\sin 0 = \frac{y}{r} = \frac{0}{1} = 0 \qquad\qquad \cos 0 = \frac{x}{r} = \frac{1}{1} = 1 \qquad (x, y) = (1, 0)$$

$$\sin \frac{\pi}{2} = \frac{y}{r} = \frac{1}{1} = 1 \qquad\qquad \cos \frac{\pi}{2} = \frac{x}{r} = \frac{0}{1} = 0 \qquad (x, y) = (0, 1)$$

$$\sin \pi = \frac{y}{r} = \frac{0}{1} = 0 \qquad\qquad \cos \pi = \frac{x}{r} = \frac{-1}{1} = -1 \qquad (x, y) = (-1, 0)$$

$$\sin \frac{3\pi}{2} = \frac{y}{r} = \frac{-1}{1} = -1 \qquad \cos \frac{3\pi}{2} = \frac{x}{r} = \frac{0}{1} = 0 \qquad (x, y) = (0, -1)$$

✓**CHECKPOINT**   Now try Exercise 43.

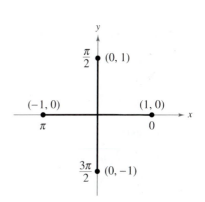

Figure 5.32

5.3   Exercises

Vocabulary and Concept Check

In Exercises 1–6, let θ be an angle in standard position with (x, y) a point on the terminal side of θ and $r = \sqrt{x^2 + y^2} \neq 0$. Fill in the blank.

1. $\sin \theta = $ _____

2. $\dfrac{r}{y} = $ _____

3. $\tan \theta = $ _____

4. $\sec \theta = $ _____

5. $\dfrac{x}{r} = $ _____

6. $\dfrac{x}{y} = $ _____

In Exercises 7 and 8, fill in the blank.

7. A function f is _____ if $f(-t) = -f(t)$.

8. A function f is _____ if $f(-t) = f(t)$.

9. What do you call the acute angle formed by the terminal side of an angle θ in standard position and the horizontal axis?

10. In which quadrants is $\cos \theta$ positive?

11. For which of the quadrant angles 0, $\pi/2$, π, and $3\pi/2$ is the sine function equal to 0?

12. Is the value of $\cos 170°$ equal to the value of $\cos 10°$?

Procedures and Problem Solving

Evaluating Trigonometric Functions   In Exercises 13–18, determine the exact values of the six trigonometric functions of the angle θ.

✓ 13.

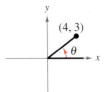

14.

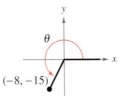

15.

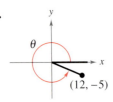

16.

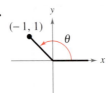

17.

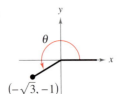

18.

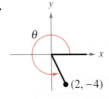

Evaluating Trigonometric Functions   In Exercises 19–26, the point is on the terminal side of an angle in standard position. Determine the exact values of the six trigonometric functions of the angle.

19. $(7, 24)$

20. $(8, 15)$

21. $(5, -12)$

22. $(-24, 10)$

23. $(-4, 10)$

24. $(-5, -6)$

25. $(-10, 8)$

26. $(3, -9)$

Determining a Quadrant   In Exercises 27–30, state the quadrant in which θ lies.

27. $\sin \theta < 0$ and $\cos \theta < 0$

28. $\sec \theta > 0$ and $\cot \theta < 0$

29. $\cot \theta > 0$ and $\cos \theta > 0$

30. $\tan \theta > 0$ and $\csc \theta < 0$

Evaluating Trigonometric Functions   In Exercises 31–38, find the values of the six trigonometric functions of θ.

Function Value	Constraint
31. $\sin \theta = \frac{3}{5}$	θ lies in Quadrant II.
32. $\cos \theta = -\frac{4}{5}$	θ lies in Quadrant III.
✓ 33. $\tan \theta = -\frac{15}{8}$	$\sin \theta < 0$
34. $\csc \theta = 4$	$\cot \theta < 0$
35. $\sec \theta = -2$	$0 \leq \theta \leq \pi$
36. $\sin \theta = 0$	$\dfrac{\pi}{2} \leq \theta \leq \dfrac{3\pi}{2}$
37. $\cot \theta$ is undefined.	$\dfrac{\pi}{2} \leq \theta \leq \dfrac{3\pi}{2}$
38. $\tan \theta$ is undefined.	$\pi \leq \theta \leq 2\pi$

An Angle Formed by a Line Through the Origin In Exercises 39–42, the terminal side of θ lies on the given line in the specified quadrant. Find the values of the six trigonometric functions of θ by finding a point on the line.

Line	Quadrant
39. $y = -x$	II
40. $y = \frac{1}{3}x$	III
41. $2x - y = 0$	III
42. $4x + 3y = 0$	IV

Trigonometric Function of a Quadrant Angle In Exercises 43–50, evaluate the trigonometric function of the quadrant angle, if possible.

✓ **43.** $\sec \pi$ **44.** $\tan \dfrac{\pi}{2}$

45. $\cot \dfrac{3\pi}{2}$ **46.** $\csc 0$

47. $\sec 0$ **48.** $\csc \dfrac{3\pi}{2}$

49. $\cot \pi$ **50.** $\csc \dfrac{\pi}{2}$

Finding a Reference Angle In Exercises 51–62, find the reference angle θ' for the special angle θ. Sketch θ in standard position and label θ'.

51. $\theta = 120°$ **52.** $\theta = 225°$

53. $\theta = 150°$ **54.** $\theta = 315°$

55. $\theta = -45°$ **56.** $\theta = -330°$

57. $\theta = \dfrac{5\pi}{3}$ **58.** $\theta = \dfrac{3\pi}{4}$

59. $\theta = -\dfrac{5\pi}{6}$ **60.** $\theta = -\dfrac{2\pi}{3}$

61. $\theta = \dfrac{11\pi}{6}$ **62.** $\theta = -\dfrac{5\pi}{3}$

Finding a Reference Angle In Exercises 63–70, find the reference angle θ'. Sketch θ in standard position and label θ'.

63. $\theta = 208°$

64. $\theta = 322°$

65. $\theta = -292°$

66. $\theta = -165°$

67. $\theta = \dfrac{11\pi}{5}$

68. $\theta = \dfrac{17\pi}{7}$

✓ **69.** $\theta = -1.8$

70. $\theta = 4.5$

Trigonometric Functions of an Angle In Exercises 71–82, evaluate the sine, cosine, and tangent of the angle without using a calculator.

71. $225°$ **72.** $300°$

73. $-750°$ **74.** $-495°$

75. $5\pi/3$ **76.** $3\pi/4$

77. $-\pi/6$ **78.** $-4\pi/3$

79. $11\pi/4$ **80.** $10\pi/3$

✓ **81.** $-17\pi/6$ **82.** $-20\pi/3$

Using Trigonometric Identities In Exercises 83–88, find the indicated trigonometric value in the specified quadrant.

	Function	Quadrant	Trigonometric Value
✓ **83.**	$\sin \theta = -\frac{3}{5}$	IV	$\cos \theta$
84.	$\cot \theta = -3$	II	$\sin \theta$
85.	$\csc \theta = -2$	IV	$\cot \theta$
86.	$\cos \theta = \frac{5}{8}$	I	$\sec \theta$
87.	$\sec \theta = -\frac{9}{4}$	III	$\tan \theta$
88.	$\tan \theta = -\frac{5}{4}$	IV	$\csc \theta$

Using Trigonometric Identities In Exerciss 89–92, use the given values to find the remaining trigonometric functions of the angle without using the trigonometric keys of your calculator. Round your answer to three decimal places.

89. $\sin 110° \approx 0.9397$ **90.** $\sin 250° \approx -0.9397$

 $\cos 110° \approx -0.3420$ $\cos 250° \approx -0.3420$

91. $\sin 350° \approx -0.1736$ **92.** $\sin 280° \approx -0.9848$

 $\cos 350° \approx 0.9848$ $\cos 280° \approx 0.1736$

Using Trigonometric Identities In Exercises 93–98, use the given value and the trigonometric identities to find the remaining trigonometric functions of the angle.

93. $\sin \theta = \frac{2}{5}$, $\cos \theta < 0$

94. $\cos \theta = -\frac{3}{7}$, $\sin \theta < 0$

95. $\tan \theta = -4$, $\cos \theta < 0$

96. $\cot \theta = -5$, $\sin \theta > 0$

97. $\csc \theta = -\frac{3}{2}$, $\tan \theta < 0$

98. $\sec \theta = -\frac{4}{3}$, $\cot \theta > 0$

Using a Calculator In Exercises 99–106, use a calculator to evaluate the trigonometric function. Round your answer to four decimal places. (Be sure the calculator is set in the correct angle mode.)

99. $\sin 10°$ **100.** $\sec 235°$

✓ **101.** $\cot(-220°)$ **102.** $\csc 320°$

103. $\tan 2\pi/9$ **104.** $\tan 11\pi/9$

105. $\csc(-8\pi/9)$ **106.** $\cos(-15\pi/14)$

Solving for θ In Exercises 107–112, find two solutions of each equation. Give your solutions in both degrees $(0° \leq \theta < 360°)$ and radians $(0 \leq \theta < 2\pi)$. Do not use a calculator.

107. (a) $\sin \theta = \frac{1}{2}$ (b) $\sin \theta = -\frac{1}{2}$

108. (a) $\cos \theta = \dfrac{\sqrt{2}}{2}$ (b) $\cos \theta = -\dfrac{\sqrt{2}}{2}$

109. (a) $\csc \theta = \dfrac{2\sqrt{3}}{3}$ (b) $\cot \theta = -1$

110. (a) $\csc \theta = -\sqrt{2}$ (b) $\csc \theta = 2$

111. (a) $\sec \theta = -\dfrac{2\sqrt{3}}{3}$ (b) $\cos \theta = -\dfrac{1}{2}$

112. (a) $\cot \theta = -\sqrt{3}$ (b) $\sec \theta = \sqrt{2}$

Estimation In Exercises 113 and 114, use the figure and a straightedge to approximate the value of each trigonometric function. Check your approximation using a graphing utility. To print an enlarged copy of the graph, go to the website *www.mathgraphs.com*.

✓ 113. (a) $\sin 5$ 114. (a) $\sin 0.75$
     (b) $\cos 2$ (b) $\cos 2.5$

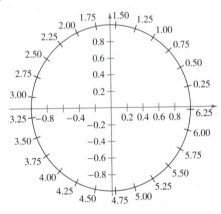

Estimation In Exercises 115 and 116, use the figure above and a straightedge to approximate the solution of each equation, where $0 \leq t < 2\pi$. Check your approximation using a graphing utility. To print an enlarged copy of the graph, go to the website *www.mathgraphs.com*.

115. (a) $\sin t = 0.25$ 116. (a) $\sin t = -0.75$
     (b) $\cos t = -0.25$ (b) $\cos t = 0.75$

Using the Value of a Trigonometric Function In Exercises 117–120, use the value of the trigonometric function to evaluate the indicated functions.

117. $\sin t = \frac{1}{3}$ 118. $\cos t = -\frac{3}{4}$
    (a) $\sin(-t)$ (a) $\cos(-t)$
    (b) $\csc(-t)$ (b) $\sec(-t)$

119. $\cos(-t) = -\frac{1}{5}$ 120. $\sin(-t) = \frac{3}{8}$
    (a) $\cos t$ (a) $\sin t$
    (b) $\sec(-t)$ (b) $\csc t$

Evaluating Trigonometric Functions In Exercises 121–134, find the exact value of each function for the given angle for $f(\theta) = \sin \theta$ and $g(\theta) = \cos \theta$. Do not use a calculator.

(a) $(f + g)(\theta)$ (b) $(g - f)(\theta)$ (c) $[g(\theta)]^2$

(d) $(fg)(\theta)$ (e) $f(2\theta)$ (f) $g(-\theta)$

121. $\theta = 30°$ 122. $\theta = 60°$

123. $\theta = 315°$ 124. $\theta = 225°$

125. $\theta = 150°$ 126. $\theta = 300°$

127. $\theta = 7\pi/6$ 128. $\theta = 5\pi/6$

129. $\theta = 4\pi/3$ 130. $\theta = 5\pi/3$

131. $\theta = 270°$ 132. $\theta = 180°$

133. $\theta = 7\pi/2$ 134. $\theta = 5\pi/2$

135. **Meteorology** The normal daily high temperature T (in degrees Fahrenheit) in Savannah, Georgia can be approximated by

$$T = 76.35 + 15.95 \cos\left(\frac{\pi t}{6} - \frac{7\pi}{6}\right)$$

where t is the time (in months), with $t = 1$ corresponding to January. Find the normal daily high temperature for each month. (Source: National Climatic Data Center)

(a) January   (b) July   (c) October

136. **Why you should learn it** *(p. 420)* A company that produces wakeboards forecasts monthly sales S over a two-year period to be

$$S = 2.7 + 0.142t + 2.2 \sin\left(\frac{\pi t}{6} - \frac{\pi}{2}\right)$$

where S is measured in hundreds of units and t is the time (in months), with $t = 1$ corresponding to January 2010. Estimate sales for each month.

(a) January 2010   (b) February 2011

(c) May 2010   (d) June 2011

137. **Aeronautics** An airplane flying at an altitude of 6 miles is on a flight path that passes directly over an observer (see figure). Let θ be the angle of elevation from the observer to the plane. Find the distance from the observer to the plane when (a) $\theta = 30°$, (b) $\theta = 90°$, and (c) $\theta = 120°$.

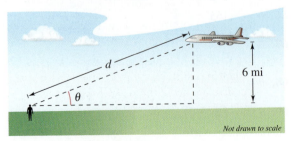

Not drawn to scale

138. Physics The displacement from equilibrium of an oscillating weight suspended by a spring is given by

$$y(t) = \frac{1}{4} \cos 6t$$

where y is the displacement (in feet) and t is the time (in seconds) (see figure). Find the displacement when (a) $t = 0$, (b) $t = \frac{1}{4}$, and (c) $t = \frac{1}{2}$.

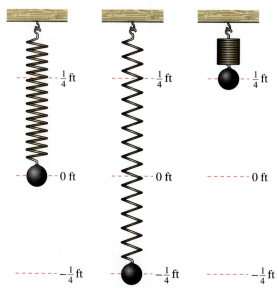

Equilibrium   Maximum negative   Maximum positive
displacement   displacement
Simple Harmonic Motion

Conclusions

True or False? **In Exercises 139–142, determine whether the statement is true or false. Justify your answer.**

139. $\sin \theta < \tan \theta$ in Quadrant I

140. $\sin \theta < \cos \theta$ for $0° < \theta < 45°$

141. $\sin \theta = -\sqrt{1 - \cos^2 \theta}$ for $90° < \theta < 180°$

142. $\cos \theta = -\sqrt{1 - \sin^2 \theta}$ for $90° < \theta < 180°$

143. Exploration

(a) Use a graphing utility to complete the table.

θ	0°	20°	40°	60°	80°
$\sin \theta$					
$\sin(180° - \theta)$					

(b) Make a conjecture about the relationship between $\sin \theta$ and $\sin(180° - \theta)$.

144. Exploration Use the procedure in Exercise 143 and a graphing utility to create a table of values and make a conjecture about the relationship between $\cos \theta$ and $\cos(180° - \theta)$ for an acute angle θ.

145. Exploration

(a) Use a graphing utility to complete the table.

θ	0	0.3	0.6	0.9	1.2	1.5
$\cos\left(\dfrac{3\pi}{2} - \theta\right)$						
$-\sin \theta$						

(b) Make a conjecture about the relationship between $\cos\left(\dfrac{3\pi}{2} - \theta\right)$ and $-\sin \theta$.

146. Exploration Use a graphing utility to create a table of values to compare $\tan t$ with $\tan(t + 2\pi)$, $\tan(t + \pi)$, and $\tan(t + \pi/2)$ for $t = 0, 0.3, 0.6, 0.9, 1.2$, and 1.5. Use your results to make a conjecture about the period of the tangent function. Explain your reasoning.

147. Writing Consider an angle in standard position with $r = 12$ centimeters, as shown in the figure. Write a short paragraph describing the changes in the magnitudes of x, y, $\sin \theta$, $\cos \theta$, and $\tan \theta$ as θ increases continually from $0°$ to $90°$.

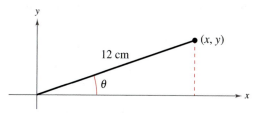

148. Think About It Because $f(t) = \sin t$ is an odd function and $g(t) = \cos t$ is an even function, what can be said about the function $h(t) = f(t)g(t)$?

149. Error Analysis Your classmate uses a calculator to evaluate $\tan(\pi/2)$ and gets a result of 0.0274224385. Describe the error.

150. CAPSTONE Write a study sheet that will help you remember how to evaluate the six trigonometric functions of any angle θ in standard position. Include figures and diagrams as needed.

Cumulative Mixed Review

Solving Equations **In Exercises 151–156, solve the equation. Round your answer to three decimal places, if necessary.**

151. $3x - 7 = 14$

152. $44 - 9x = 61$

153. $4^{3-x} = 726$

154. $\ln x = -6$

155. $\dfrac{3}{x - 1} = \dfrac{x + 2}{9}$

156. $\dfrac{5}{x} = \dfrac{x + 4}{2x}$

Basic Sine and Cosine Curves

In this section, you will study techniques for sketching the graphs of the sine and cosine functions. The graph of the sine function is a **sine curve.** In Figure 5.42, the black portion of the graph represents one period of the function and is called **one cycle** of the sine curve. The gray portion of the graph indicates that the basic sine wave repeats indefinitely to the right and left. The graph of the cosine function is shown in Figure 5.43. To produce these graphs with a graphing utility, make sure you set the graphing utility to *radian* mode.

Recall from Section 5.3 that the domain of the sine and cosine functions is the set of all real numbers. Moreover, the range of each function is the interval

$$[-1, 1]$$

and each function has a period of 2π. Do you see how this information is consistent with the basic graphs shown in Figures 5.42 and 5.43?

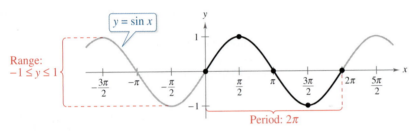

Figure 5.42

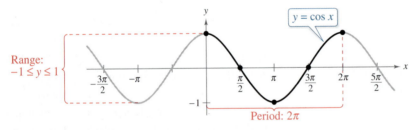

Figure 5.43

To sketch the graphs of the basic sine and cosine functions by hand, it helps to note five *key points* in one period of each graph: the *intercepts*, the *maximum points*, and the *minimum points*. The table below lists the five key points on the graphs of

$$y = \sin x \quad \text{and} \quad y = \cos x.$$

x	0	$\dfrac{\pi}{2}$	π	$\dfrac{3\pi}{2}$	2π
$\sin x$	0	1	0	-1	0
$\cos x$	1	0	-1	0	1

Note in Figures 5.42 and 5.43 that the sine curve is symmetric with respect to the *origin*, whereas the cosine curve is symmetric with respect to the *y-axis*. These properties of symmetry follow from the fact that the sine function is odd whereas the cosine function is even.

What you should learn

- Sketch the graphs of basic sine and cosine functions.
- Use amplitude and period to help sketch the graphs of sine and cosine functions.
- Sketch translations of graphs of sine and cosine functions.
- Use sine and cosine functions to model real-life data.

Why you should learn it

Sine and cosine functions are often used in scientific calculations. For instance, in Exercise 87 on page 441, you can use a trigonometric function to model the percent of the moon's face that is illuminated for any given day in 2012.

📚 Library of Parent Functions: Sine and Cosine Functions

The basic characteristics of the parent sine function and parent cosine function are listed below and summarized on the inside front cover of this text.

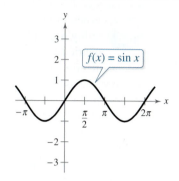

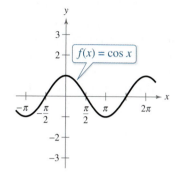

Domain: $(-\infty, \infty)$
Range: $[-1, 1]$
Period: 2π
x-intercepts: $(n\pi, 0)$
y-intercept: $(0, 0)$
Odd function
Origin symmetry

Domain: $(-\infty, \infty)$
Range: $[-1, 1]$
Period: 2π
x-intercepts: $\left(\dfrac{\pi}{2} + n\pi, 0\right)$
y-intercept: $(0, 1)$
Even function
y-axis symmetry

Example 1   Library of Parent Functions: $f(x) = \sin x$

Sketch the graph of $g(x) = 2 \sin x$ by hand on the interval $[-\pi, 4\pi]$.

Solution

Note that $g(x) = 2 \sin x = 2(\sin x)$ indicates that the y-values of the key points will have twice the magnitude of those on the graph of $f(x) = \sin x$. Divide the period 2π into four equal parts to get the key points

Intercept	Maximum	Intercept	Minimum	Intercept
$(0, 0)$,	$\left(\dfrac{\pi}{2}, 2\right)$,	$(\pi, 0)$,	$\left(\dfrac{3\pi}{2}, -2\right)$, and	$(2\pi, 0)$.

By connecting these key points with a smooth curve and extending the curve in both directions over the interval $[-\pi, 4\pi]$, you obtain the graph shown in Figure 5.44. Use a graphing utility to confirm this graph. Be sure to set the graphing utility to *radian* mode.

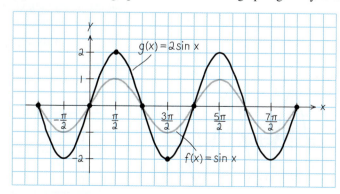

Figure 5.44

✓CHECKPOINT   Now try Exercise 43.

Explore the Concept

    Enter the Graphing a Sine Function Program, found at this textbook's *Companion Website*, into your graphing utility. This program simultaneously draws the unit circle and the corresponding points on the sine curve, as shown below. After the circle and sine curve are drawn, you can connect the points on the unit circle with their corresponding points on the sine curve by pressing ENTER. Discuss the relationship that is illustrated.

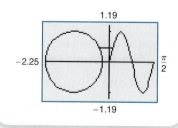

Amplitude and Period of Sine and Cosine Curves

In the rest of this section, you will study the graphic effect of each of the constants a, b, c, and d in equations of the forms

$$y = d + a \sin(bx - c) \quad \text{and} \quad y = d + a \cos(bx - c).$$

The constant factor a in $y = a \sin x$ acts as a *scaling factor*—a *vertical stretch* or *vertical shrink* of the basic sine curve. When $|a| > 1$, the basic sine curve is stretched, and when $|a| < 1$, the basic sine curve is shrunk. The result is that the graph of $y = a \sin x$ ranges between $-a$ and a instead of between -1 and 1. The absolute value of a is the **amplitude** of the function $y = a \sin x$. The range of the function $y = a \sin x$ for $a > 0$ is $-a \le y \le a$.

> **Definition of Amplitude of Sine and Cosine Curves**
>
> The **amplitude** of
>
> $$y = a \sin x \text{ and } y = a \cos x$$
>
> represents half the distance between the maximum and minimum values of the function and is given by
>
> $$\text{Amplitude} = |a|.$$

Example 2 Scaling: Vertical Shrinking and Stretching

On the same set of coordinate axes, sketch the graph of each function by hand.

a. $y = \frac{1}{2} \cos x$

b. $y = 3 \cos x$

Solution

a. Because the amplitude of $y = \frac{1}{2} \cos x$ is $\frac{1}{2}$, the maximum value is $\frac{1}{2}$ and the minimum value is $-\frac{1}{2}$. Divide one cycle, $0 \le x \le 2\pi$, into four equal parts to get the key points

Maximum	Intercept	Minimum	Intercept	Maximum
$\left(0, \frac{1}{2}\right),$	$\left(\frac{\pi}{2}, 0\right),$	$\left(\pi, -\frac{1}{2}\right),$	$\left(\frac{3\pi}{2}, 0\right),$ and	$\left(2\pi, \frac{1}{2}\right).$

b. A similar analysis shows that the amplitude of $y = 3 \cos x$ is 3, and the key points are

Maximum	Intercept	Minimum	Intercept	Maximum
$(0, 3),$	$\left(\frac{\pi}{2}, 0\right),$	$(\pi, -3),$	$\left(\frac{3\pi}{2}, 0\right),$ and	$(2\pi, 3).$

The graphs of these two functions are shown in Figure 5.45. Notice that the graph of

$$y = \frac{1}{2} \cos x$$

is a vertical shrink of the graph of $y = \cos x$ and the graph of

$$y = 3 \cos x$$

is a vertical stretch of the graph of $y = \cos x$. Use a graphing utility to confirm these graphs.

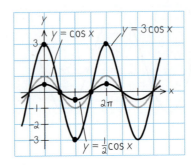

Figure 5.45

✓ **CHECKPOINT**   Now try Exercise 45.

Technology Tip

When using a graphing utility to graph trigonometric functions, pay special attention to the viewing window you use. For instance, try graphing $y = [\sin(10x)]/10$ in the standard viewing window in *radian* mode. What do you observe? Use the *zoom* feature to find a viewing window that displays a good view of the graph.

? What's Wrong?

You use a graphing utility to confirm the graph of $y = \frac{1}{2} \cos x$ in Example 2 and obtain the screen shown below. What's wrong?

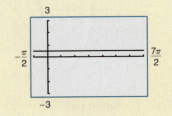

You know from Section 1.5 that the graph of $y = -f(x)$ is a *reflection* in the x-axis of the graph of $y = f(x)$. For instance, the graph of $y = -3 \cos x$ is a reflection of the graph of $y = 3 \cos x$, as shown in Figure 5.46.

Next, consider the effect of the *positive* real number b on the graphs of $y = a \sin bx$ and $y = a \cos bx$. Because $y = a \sin x$ completes one cycle from $x = 0$ to $x = 2\pi$, it follows that $y = a \sin bx$ completes one cycle from $x = 0$ to $x = 2\pi/b$.

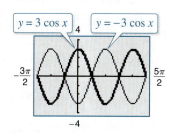

Figure 5.46

Period of Sine and Cosine Functions

Let b be a positive real number. The **period** of

$$y = a \sin bx \quad \text{and} \quad y = a \cos bx$$

is given by

$$\text{Period} = \frac{2\pi}{b}.$$

Note that when $0 < b < 1$, the period of $y = a \sin bx$ is greater than 2π and represents a *horizontal stretching* of the graph of $y = a \sin x$. Similarly, when $b > 1$, the period of $y = a \sin bx$ is less than 2π and represents a *horizontal shrinking* of the graph of $y = a \sin x$. When b is negative, the identities

$$\sin(-x) = -\sin x \quad \text{and} \quad \cos(-x) = \cos x$$

are used to rewrite the function.

Example 3 Scaling: Horizontal Stretching

Sketch the graph of $y = \sin \dfrac{x}{2}$ by hand.

Solution

The amplitude is 1. Moreover, because $b = \frac{1}{2}$, the period is

$$\frac{2\pi}{b} = \frac{2\pi}{\frac{1}{2}} = 4\pi. \qquad \textcolor{red}{\text{Substitute for } b.}$$

Now, divide the period-interval $[0, 4\pi]$ into four equal parts using the values $\pi, 2\pi,$ and 3π to obtain the key points on the graph

Intercept	*Maximum*	*Intercept*	*Minimum*	*Intercept*
$(0, 0),$	$(\pi, 1),$	$(2\pi, 0),$	$(3\pi, -1),$ and	$(4\pi, 0).$

The graph is shown in Figure 5.47. Use a graphing utility to confirm this graph.

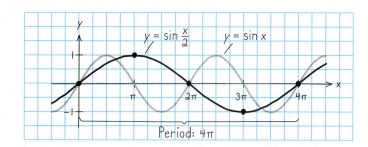

Figure 5.47

 CHECKPOINT   Now try Exercise 47.

Study Tip

In general, to divide a period-interval into four equal parts, successively add "period/4," starting with the left endpoint of the interval. For instance, for the period-interval $[-\pi/6, \pi/2]$ of length $2\pi/3$, you would successively add

$$\frac{2\pi/3}{4} = \frac{\pi}{6}$$

to get $-\pi/6, 0, \pi/6, \pi/3,$ and $\pi/2$ as the x-values of the key points on the graph.

Translations of Sine and Cosine Curves

The constant c in the general equations

$$y = a \sin(bx - c) \qquad \text{and} \qquad y = a \cos(bx - c)$$

creates *horizontal translations* (shifts) of the basic sine and cosine curves. Comparing $y = a \sin bx$ with $y = a \sin(bx - c)$, you find that the graph of $y = a \sin(bx - c)$ completes one cycle from $bx - c = 0$ to $bx - c = 2\pi$. By solving for x, you can find the interval for one cycle to be

Left endpoint   Right endpoint

$$\frac{c}{b} \le x \le \frac{c}{b} + \frac{2\pi}{b}.$$

Period

This implies that the period of $y = a \sin(bx - c)$ is $2\pi/b$, and the graph of $y = a \sin bx$ is shifted by an amount c/b. The number c/b is the **phase shift.**

Graphs of Sine and Cosine Functions

The graphs of $y = a \sin(bx - c)$ and $y = a \cos(bx - c)$ have the following characteristics. (Assume $b > 0$.)

$$\text{Amplitude} = |a| \qquad \text{Period} = \frac{2\pi}{b}$$

The left and right endpoints of a one-cycle interval can be determined by solving the equations

$$bx - c = 0 \qquad \text{and} \qquad bx - c = 2\pi.$$

Example 4   Horizontal Translation

Analyze the graph of $y = \dfrac{1}{2} \sin\left(x - \dfrac{\pi}{3}\right)$.

Algebraic Solution

The amplitude is $\frac{1}{2}$ and the period is 2π. By solving the equations

$$x - \frac{\pi}{3} = 0 \qquad \text{and} \qquad x - \frac{\pi}{3} = 2\pi$$

$$x = \frac{\pi}{3} \qquad\qquad\qquad x = \frac{7\pi}{3}$$

you see that the interval

$$\left[\frac{\pi}{3}, \frac{7\pi}{3}\right]$$

corresponds to one cycle of the graph. Dividing this interval into four equal parts produces the following key points.

Intercept	Maximum	Intercept	Minimum	Intercept
$\left(\dfrac{\pi}{3}, 0\right)$	$\left(\dfrac{5\pi}{6}, \dfrac{1}{2}\right)$	$\left(\dfrac{4\pi}{3}, 0\right)$	$\left(\dfrac{11\pi}{6}, -\dfrac{1}{2}\right)$	$\left(\dfrac{7\pi}{3}, 0\right)$

✓CHECKPOINT   Now try Exercise 49.

Graphical Solution

Use a graphing utility set in *radian* mode to graph

$$y = \left(\frac{1}{2}\right) \sin\left(x - \frac{\pi}{3}\right)$$

as shown in Figure 5.48. Use the *minimum*, *maximum*, and *zero* or *root* features of the graphing utility to approximate the key points $(1.05, 0)$, $(2.62, 0.5)$, $(4.19, 0)$, $(5.76, -0.5)$, and $(7.33, 0)$.

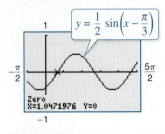

$$y = \frac{1}{2} \sin\left(x - \frac{\pi}{3}\right)$$

Zero
X=1.0471976 Y=0

Figure 5.48

Example 5   Horizontal Translation

Analyze the graph of $y = -3 \cos(2\pi x + 4\pi)$.

Algebraic Solution

The amplitude is 3 and the period is

$$\frac{2\pi}{2\pi} = 1.$$

By solving the equations

$$2\pi x + 4\pi = 0 \qquad \text{and} \qquad 2\pi x + 4\pi = 2\pi$$

$$2\pi x = -4\pi \qquad\qquad\qquad 2\pi x = -2\pi$$

$$x = -2 \qquad\qquad\qquad\qquad x = -1$$

you see that the interval $[-2, -1]$ corresponds to one cycle of the graph. Dividing this interval into four equal parts produces the key points

Minimum	Intercept	Maximum	Intercept	Minimum

$(-2, -3)$, $(-7/4, 0)$, $(-3/2, 3)$, $(-5/4, 0)$, and $(-1, -3)$.

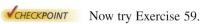 **CHECKPOINT**   Now try Exercise 51.

Graphical Solution

Use a graphing utility set in *radian* mode to graph $y = -3 \cos(2\pi x + 4\pi)$, as shown in Figure 5.49. Use the *minimum*, *maximum*, and *zero* or *root* features of the graphing utility to approximate the key points $(-2, -3), (-1.75, 0), (-1.5, 3), (-1.25, 0),$ and $(-1, -3)$.

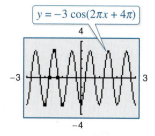

Figure 5.49

The final type of transformation is the *vertical translation* caused by the constant d in the equations

$$y = d + a \sin(bx - c) \qquad \text{and} \qquad y = d + a \cos(bx - c).$$

The shift is d units upward for $d > 0$ and d units downward for $d < 0$. In other words, the graph oscillates about the horizontal line $y = d$ instead of about the x-axis.

Example 6   Vertical Translation

Use a graphing utility to analyze the graph of $y = 2 + 3 \cos 2x$.

Solution

The amplitude is 3 and the period is π. The key points over the interval $[0, \pi]$ are

$$(0, 5), \qquad (\pi/4, 2), \qquad (\pi/2, -1), \qquad (3\pi/4, 2), \qquad \text{and} \qquad (\pi, 5).$$

The graph is shown in Figure 5.50. Compared with the graph of $f(x) = 3 \cos 2x$, the graph of $y = 2 + 3 \cos 2x$ is shifted upward two units.

CHECKPOINT   Now try Exercise 59.

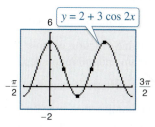

Figure 5.50

Example 7   Finding an Equation of a Graph

Find the amplitude, period, and phase shift of the sine function whose graph is shown in Figure 5.51. Write an equation of this graph.

Solution

The amplitude of this sine curve is 2. The period is 2π, and there is a right phase shift of $\pi/2$. So, you can write

$$y = 2 \sin\left(x - \frac{\pi}{2}\right).$$

CHECKPOINT   Now try Exercise 75.

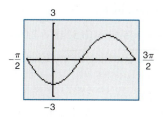

Figure 5.51

Mathematical Modeling

Sine and cosine functions can be used to model many real-life situations, including electric currents, musical tones, radio waves, tides, and weather patterns.

Example 8   Finding a Trigonometric Model

Throughout the day, the depth of the water at the end of a dock varies with the tides. The table shows the depths y (in feet) at various times during the morning.

Time	Depth, y
Midnight	3.4
2 A.M.	8.7
4 A.M.	11.3
6 A.M.	9.1
8 A.M.	3.8
10 A.M.	0.1
Noon	1.2

a. Use a trigonometric function to model the data. Let t be the time, with $t = 0$ corresponding to midnight.

b. A boat needs at least 10 feet of water to moor at the dock. During what times in the evening can it safely dock?

Solution

a. Begin by graphing the data, as shown in Figure 5.52. You can use either a sine or cosine model. Suppose you use a cosine model of the form

$$y = a \cos(bt - c) + d.$$

The difference between the maximum height and minimum height of the graph is twice the amplitude of the function. So, the amplitude is

$$a = \tfrac{1}{2}[(\text{maximum depth}) - (\text{minimum depth})] = \tfrac{1}{2}(11.3 - 0.1) = 5.6.$$

The cosine function completes one half of a cycle between the times at which the maximum and minimum depths occur. So, the period p is

$$p = 2[(\text{time of min. depth}) - (\text{time of max. depth})] = 2(10 - 4) = 12$$

which implies that $b = 2\pi/p \approx 0.524$. Because high tide occurs 4 hours after midnight, consider the left endpoint to be $c/b = 4$, so $c \approx 2.094$. Moreover, because the average depth is

$$\tfrac{1}{2}(11.3 + 0.1) = 5.7$$

it follows that $d = 5.7$. So, you can model the depth with the function

$$y = 5.6 \cos(0.524t - 2.094) + 5.7.$$

b. Using a graphing utility, graph the model with the line

$$y = 10.$$

Using the *intersect* feature, you can determine that the depth is at least 10 feet between 2:42 P.M. ($t \approx 14.7$) and 5:18 P.M. ($t \approx 17.3$), as shown in Figure 5.53.

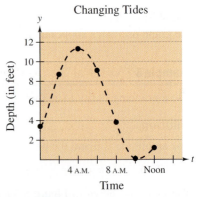

Figure 5.52

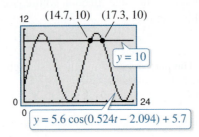

Figure 5.53

✓CHECKPOINT   Now try Exercise 87.

5.4   Exercises

Vocabulary and Concept Check

In Exercises 1–4, fill in the blank.

1. The _____ of a sine or cosine curve represents half the distance between the maximum and minimum values of the function.

2. One period of a sine function is called _____ of the sine curve.

3. The period of a sine or cosine function is given by _____ .

4. For the equation $y = a \sin(bx - c)$, $\dfrac{c}{b}$ is the _____ of the graph of the equation.

5. What is the period of the sine function $y = \sin x$?

6. How do you find the period of a cosine function of the form $y = \cos bx$?

7. Describe the effect of the constant d on the graph of $y = \sin x + d$.

8. What is the amplitude of $y = 6 \sin x$?

Procedures and Problem Solving

 Library of Parent Functions   In Exercises 9 and 10, use the graph of the function to answer the following.

(a) Find the x-intercepts of the graph of $y = f(x)$.

(b) Find the y-intercepts of the graph of $y = f(x)$.

(c) Find the intervals on which the graph of $y = f(x)$ is increasing and the intervals on which the graph of $y = f(x)$ is decreasing.

(d) Find the relative extrema of the graph of $y = f(x)$.

9. $f(x) = \sin x$ **10.** $f(x) = \cos x$

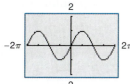

Finding the Period and Amplitude   In Exercises 11–20, find the period and amplitude.

11. $y = 3 \sin 2x$ **12.** $y = 2 \cos 3x$

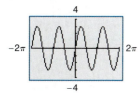

13. $y = \dfrac{5}{2} \cos \dfrac{x}{2}$ **14.** $y = -3 \sin \dfrac{x}{3}$

15. $y = \dfrac{2}{3} \sin \pi x$ **16.** $y = \dfrac{3}{2} \cos \dfrac{\pi x}{2}$

17. $y = -2 \sin x$ **18.** $y = -\cos \dfrac{2x}{5}$

19. $y = \dfrac{1}{4} \cos \dfrac{2x}{3}$ **20.** $y = \dfrac{5}{2} \cos \dfrac{x}{4}$

Describing the Relationship Between Graphs   In Exercises 21–28, describe the relationship between the graphs of f and g. Consider amplitudes, periods, and shifts.

21. $f(x) = \sin x$ **22.** $f(x) = \cos x$
    $g(x) = \sin(x - \pi)$ $g(x) = \cos(x + \pi)$

23. $f(x) = \cos 2x$ **24.** $f(x) = \sin 3x$
    $g(x) = -\cos 2x$ $g(x) = \sin(-3x)$

25. $f(x) = \cos x$ **26.** $f(x) = \sin x$
    $g(x) = -5 \cos x$ $g(x) = -\dfrac{1}{2} \sin x$

27. $f(x) = \sin 2x$ **28.** $f(x) = \cos 4x$
    $g(x) = 3 + \sin 2x$ $g(x) = -2 + \cos 4x$

Describing the Relationship Between Graphs   In Exercises 29–32, describe the relationship between the graphs of f and g. Consider amplitudes, periods, and shifts.

29. 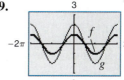   **30.**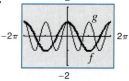

31. 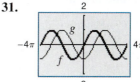   **32.**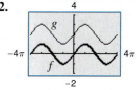

Sketching Graphs of Sine or Cosine Functions　In Exercises 33–38, sketch the graphs of f and g in the same coordinate plane. (Include two full periods.)

33. $f(x) = \sin x$

$g(x) = -4 \sin x$

34. $f(x) = \sin x$

$g(x) = \sin \dfrac{x}{3}$

35. $f(x) = \cos x$

$g(x) = 1 + \cos x$

36. $f(x) = 2 \cos 2x$

$g(x) = -\cos 4x$

37. $f(x) = -\dfrac{1}{2} \sin \dfrac{x}{2}$

$g(x) = 3 - \dfrac{1}{2} \sin \dfrac{x}{2}$

38. $f(x) = 4 \sin \pi x$

$g(x) = 4 \sin \pi x - 3$

Graphing Sine and Cosine Functions　In Exercises 39–42, use a graphing utility to graph f and g in the same viewing window. (Include two full periods.) Make a conjecture about the functions.

39. $f(x) = \sin x$

$g(x) = \cos\left(x - \dfrac{\pi}{2}\right)$

40. $f(x) = \sin x$

$g(x) = -\cos\left(x + \dfrac{\pi}{2}\right)$

41. $f(x) = \cos x$

$g(x) = -\sin\left(x - \dfrac{\pi}{2}\right)$

42. $f(x) = \cos x$

$g(x) = -\cos(x - \pi)$

Graphing a Sine or Cosine Function　In Exercises 43–56, sketch the graph of the function. Use a graphing utility to verify your sketch. (Include two full periods.)

✓ **43.** $y = 3 \sin x$

44. $y = 5 \sin x$

✓ **45.** $y = \dfrac{1}{4} \cos x$

46. $y = \dfrac{3}{4} \cos x$

✓ **47.** $y = \cos \dfrac{x}{2}$

48. $y = \sin 4x$

✓ **49.** $y = \sin\left(x - \dfrac{\pi}{4}\right)$

50. $y = \sin(x - \pi)$

✓ **51.** $y = -8 \cos(x + \pi)$

52. $y = 3 \cos\left(x + \dfrac{\pi}{2}\right)$

53. $y = 1 - \sin \dfrac{2\pi x}{3}$

54. $y = 2 \cos x - 3$

55. $y = \dfrac{2}{3} \cos\left(\dfrac{x}{2} - \dfrac{\pi}{4}\right)$

56. $y = -2 \cos(4x + \pi)$

Identifying Amplitude and Period　In Exercises 57–70, use a graphing utility to graph the function. (Include two full periods.) Identify the amplitude and period of the graph.

57. $y = -2 \sin \dfrac{2\pi x}{3}$

58. $y = -10 \cos \dfrac{\pi x}{6}$

✓ **59.** $y = -4 + 5 \cos \dfrac{\pi t}{12}$

60. $y = 2 - 2 \sin \dfrac{2\pi x}{3}$

61. $y = \dfrac{2}{3} \cos\left(\dfrac{x}{2} - \dfrac{\pi}{4}\right)$

62. $y = -3 \cos(6x + \pi)$

63. $y = -2 \sin(4x + \pi)$

64. $y = -4 \sin\left(\dfrac{2}{3}x - \dfrac{\pi}{3}\right)$

65. $y = \cos\left(2\pi x - \dfrac{\pi}{2}\right) + 1$

66. $y = 3 \cos\left(\dfrac{\pi x}{2} + \dfrac{\pi}{2}\right) - 2$

67. $y = 5 \sin(\pi - 2x) + 10$

68. $y = 5 \cos(\pi - 2x) + 6$

69. $y = \dfrac{1}{100} \sin 120\pi t$

70. $y = -\dfrac{1}{100} \cos 50\pi t$

Finding an Equation of a Graph　In Exercises 71–74, find a and d for the function $f(x) = a \cos x + d$ such that the graph of f matches the figure.

71.

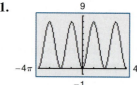

72.

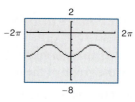

73.

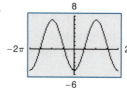

74.

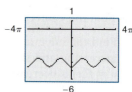

Finding an Equation of a Graph　In Exercises 75–78, find a, b, and c for the function $f(x) = a \sin(bx - c)$ such that the graph of f matches the graph shown.

✓ **75.**

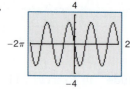

76.

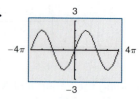

77.

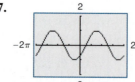

78.

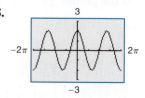

Graphing Sine and Cosine Functions　In Exercises 79 and 80, use a graphing utility to graph y_1 and y_2 for all real numbers x in the interval $[-2\pi, 2\pi]$. Use the graphs to find the real numbers x such that $y_1 = y_2$.

79. $y_1 = \sin x$

$y_2 = -\dfrac{1}{2}$

80. $y_1 = \cos x$

$y_2 = -1$

81. Health   For a person at rest, the velocity v (in liters per second) of air flow during a respiratory cycle (the time from the beginning of one breath to the beginning of the next) is given by $v = 0.85 \sin(\pi t/3)$, where t is the time (in seconds). (Inhalation occurs when $v > 0$, and exhalation occurs when $v < 0$.)

(a) Use a graphing utility to graph v.

(b) Find the time for one full respiratory cycle.

(c) Find the number of cycles per minute.

(d) The model is for a person at rest. How might the model change for a person who is exercising? Explain.

82. Economics   A company that produces snowboards, which are seasonal products, forecasts monthly sales for one year to be

$$S = 74.50 + 43.75 \cos \frac{\pi t}{6}$$

where S is the sales in thousands of units and t is the time in months, with $t = 1$ corresponding to January.

(a) Use a graphing utility to graph the sales function over the one-year period.

(b) Use the graph in part (a) to determine the months of maximum and minimum sales.

83. Physics   You are riding a Ferris wheel. Your height h (in feet) above the ground at any time t (in seconds) can be modeled by

$$h = 25 \sin \frac{\pi}{15}(t - 75) + 30.$$

The Ferris wheel turns for 135 seconds before it stops to let the first passengers off.

(a) Use a graphing utility to graph the model.

(b) What are the minimum and maximum heights above the ground?

84. Health   The pressure P (in millimeters of mercury) against the walls of the blood vessels of a person is modeled by

$$P = 100 - 20 \cos \frac{8\pi}{3}t$$

where t is the time (in seconds). Use a graphing utility to graph the model. One cycle is equivalent to one heartbeat. What is the person's pulse rate in heartbeats per minute?

85. Agriculture   The daily consumption C (in gallons) of diesel fuel on a farm is modeled by

$$C = 30.3 + 21.6 \sin\left(\frac{2\pi t}{365} + 10.9\right)$$

where t is the time in days, with $t = 1$ corresponding to January 1.

(a) What is the period of the model? Is it what you expected? Explain.

(b) What is the average daily fuel consumption? Which term of the model did you use? Explain.

(c) Use a graphing utility to graph the model. Use the graph to approximate the time of the year when consumption exceeds 40 gallons per day.

86. Physics   The motion of an oscillating weight suspended from a spring was measured by a motion detector. The data were collected, and the approximate maximum displacements from equilibrium ($y = 2$) are labeled in the figure. The distance y from the motion detector is measured in centimeters, and the time t is measured in seconds.

(a) Is y a function of t? Explain.

(b) Approximate the amplitude and period.

(c) Find a model for the data.

(d) Use a graphing utility to graph the model in part (c). Compare the result with the data in the figure.

✓ **87. Why you should learn it**   (p. 432)   The percent y (in decimal form) of the moon's face that is illuminated on day x of the year 2012, where $x = 1$ represents January 1, is shown in the table. (Source: U.S. Naval Observatory)

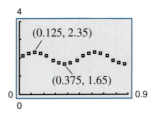

Day, x	Percent, y
23	0.0
31	0.5
38	1.0
45	0.5
52	0.0
61	0.5

(a) Create a scatter plot of the data.

(b) Find a trigonometric model for the data.

(c) Add the graph of your model in part (b) to the scatter plot. How well does the model fit the data?

(d) What is the period of the model?

(e) Estimate the percent illumination of the moon on June 21, 2013. (Assume there are 366 days in 2012.)

88. MODELING DATA

The table shows the average daily high temperatures for Quillayute, Washington Q and Chicago, Illinois C (in degrees Fahrenheit) for month t, with $t = 1$ corresponding to January. (Source: U.S. Weather Bureau and the National Weather Service)

Month, t	Quillayute, Q	Chicago, C
1	46.6	29.6
2	49.2	34.7
3	51.8	46.1
4	55.7	58.0
5	60.4	69.9
6	63.8	79.2
7	68.2	83.5
8	69.3	81.2
9	67.3	73.9
10	59.2	62.1
11	50.8	47.1
12	46.5	34.4

(a) A model for the temperature in Quillayute is given by

$$Q(t) = 57.9 + 11.1 \sin(0.549t - 2.438).$$

Find a trigonometric model for Chicago.

(b) Use a graphing utility to graph the data and the model for the temperatures in Quillayute in the same viewing window. How well does the model fit the data?

(c) Use the graphing utility to graph the data and the model for the temperatures in Chicago in the same viewing window. How well does the model fit the data?

(d) Use the models to estimate the average daily high temperature in each city. Which term of the models did you use? Explain.

(e) What is the period of each model? Are the periods what you expected? Explain.

(f) Which city has the greater variability in temperature throughout the year? Which factor of the models determines this variability? Explain.

Conclusions

True or False? In Exercises 89–92, determine whether the statement is true or false. Justify your answer.

89. The graph of the function given by $g(x) = \sin(x + 2\pi)$ translates the graph of $f(x) = \sin x$ one period to the right.

90. The graph of $y = 6 - \dfrac{3}{4}\sin\dfrac{3x}{10}$ has a period of $\dfrac{20\pi}{3}$.

91. The function $y = \frac{1}{2}\cos 2x$ has an amplitude that is twice that of the function $y = \cos x$.

92. The graph of $y = -\cos x$ is a reflection of the graph of $y = \sin\left(x + \dfrac{\pi}{2}\right)$ in the x-axis.

93. Writing Sketch the graph of $y = \cos bx$ for $b = \frac{1}{2}$, 2, and 3. How does the value of b affect the graph? How many complete cycles of the graph of y occur between 0 and 2π for each value of b?

94. Writing Sketch the graph of $y = \sin(x - c)$ for $c = -\dfrac{\pi}{4}$, 0, and $\dfrac{\pi}{4}$. How does the value of c affect the graph?

Library of Parent Functions In Exercises 95–98, determine which function is represented by the graph. Do not use a calculator.

95.

(a) $f(x) = 2\sin 2x$

(b) $f(x) = -2\sin\dfrac{x}{2}$

(c) $f(x) = -2\cos 2x$

(d) $f(x) = 2\cos\dfrac{x}{2}$

(e) $f(x) = -2\sin 2x$

96.

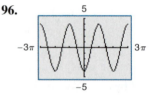

(a) $f(x) = 4\cos(x + \pi)$

(b) $f(x) = 4\cos 4x$

(c) $f(x) = 4\sin(x - \pi)$

(d) $f(x) = -4\cos(x + \pi)$

(e) $f(x) = 1 - \sin\dfrac{x}{2}$

97.

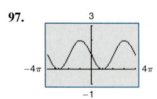

(a) $f(x) = 1 + \sin\dfrac{x}{2}$

(b) $f(x) = 1 + \cos\dfrac{x}{2}$

(c) $f(x) = 1 - \sin\dfrac{x}{2}$

(d) $f(x) = 1 - \cos 2x$

(e) $f(x) = 1 - \sin 2x$

98.

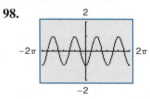

(a) $f(x) = \cos 2x$

(b) $f(x) = \sin\left(\dfrac{x}{2} - \pi\right)$

(c) $f(x) = \sin(2x + \pi)$

(d) $f(x) = \cos(2x - \pi)$

(e) $f(x) = \sin\dfrac{x}{2}$

99. Exploration   In Section 5.3, it was shown that $f(x) = \cos x$ is an even function and $g(x) = \sin x$ is an odd function. Use a graphing utility to graph h and use the graph to determine whether h is even, odd, or neither.

(a) $h(x) = \cos^2 x$

(b) $h(x) = \sin^2 x$

(c) $h(x) = \sin x \cos x$

100. Conjecture   If f is an even function and g is an odd function, use the results of Exercise 99 to make a conjecture about each of the following.

(a) $h(x) = [f(x)]^2$

(b) $h(x) = [g(x)]^2$

(c) $h(x) = f(x)g(x)$

101. Exploration   Use a graphing utility to explore the ratio $(\sin x)/x$, which appears in calculus.

(a) Complete the table. Round your results to four decimal places.

x	-1	-0.1	-0.01	-0.001
$\dfrac{\sin x}{x}$				

x	0	0.001	0.01	0.1	1
$\dfrac{\sin x}{x}$					

(b) Use the graphing utility to graph the function
$$f(x) = \frac{\sin x}{x}.$$

Use the *zoom* and *trace* features to describe the behavior of the graph as x approaches 0.

(c) Write a brief statement regarding the value of the ratio based on your results in parts (a) and (b).

102. Exploration   Use a graphing utility to explore the ratio $(1 - \cos x)/x$, which appears in calculus.

(a) Complete the table. Round your results to four decimal places.

x	-1	-0.1	-0.01	-0.001
$\dfrac{1 - \cos x}{x}$				

x	0	0.001	0.01	0.1	1
$\dfrac{1 - \cos x}{x}$					

(b) Use the graphing utility to graph the function
$$f(x) = \frac{1 - \cos x}{x}.$$

Use the *zoom* and *trace* features to describe the behavior of the graph as x approaches 0.

(c) Write a brief statement regarding the value of the ratio based on your results in parts (a) and (b).

103. Exploration   Using calculus, it can be shown that the sine and cosine functions can be approximated by the polynomials

$$\sin x \approx x - \frac{x^3}{3!} + \frac{x^5}{5!} \quad \text{and} \quad \cos x \approx 1 - \frac{x^2}{2!} + \frac{x^4}{4!}$$

where x is in radians.

(a) Use a graphing utility to graph the sine function and its polynomial approximation in the same viewing window. How do the graphs compare?

(b) Use the graphing utility to graph the cosine function and its polynomial approximation in the same viewing window. How do the graphs compare?

(c) Study the patterns in the polynomial approximations of the sine and cosine functions and predict the next term in each. Then repeat parts (a) and (b). How did the accuracy of the approximations change when an additional term was added?

104. CAPSTONE   Use a graphing utility to graph the function given by $y = d + a \sin(bx - c)$ for several different values of a, b, c, and d. Write a paragraph describing how the values of a, b, c, and d affect the graph.

Cumulative Mixed Review

Finding the Slope of a Line   In Exercises 105 and 106, plot the points and find the slope of the line passing through the points.

105. $(0, 1), (2, 7)$ **106.** $(-1, 4), (3, -2)$

Converting from Radians to Degrees   In Exercises 107 and 108, convert the angle measure from radians to degrees. Round your answer to three decimal places.

107. 8.5 **108.** -0.48

109. *Make a Decision*   To work an extended application analyzing the mean monthly high temperature and normal precipitation in Honolulu, Hawaii, visit this textbook's *Companion Website*. (Data Source: NOAA)

5.5   Graphs of Other Trigonometric Functions

Graph of the Tangent Function

Recall that the tangent function is odd. That is, $\tan(-x) = -\tan x$. Consequently, the graph of $y = \tan x$ is symmetric with respect to the origin. You also know from the identity $\tan x = \sin x / \cos x$ that the tangent function is undefined when $\cos x = 0$. Two such values are $x = \pm \pi/2 \approx \pm 1.5708$.

x	$-\dfrac{\pi}{2}$	-1.57	-1.5	$-\dfrac{\pi}{4}$	0	$\dfrac{\pi}{4}$	1.5	1.57	$\dfrac{\pi}{2}$
$\tan x$	Undef.	-1255.8	-14.1	-1	0	1	14.1	1255.8	Undef.

> tan x approaches $-\infty$ as x approaches $-\pi/2$ from the right.

> tan x approaches ∞ as x approaches $\pi/2$ from the left.

As indicated in the table, $\tan x$ increases without bound as x approaches $\pi/2$ from the left, and it decreases without bound as x approaches $-\pi/2$ from the right. So, the graph of $y = \tan x$ has *vertical asymptotes* at $x = \pi/2$ and $x = -\pi/2$, as shown in Figure 5.54. Moreover, because the period of the tangent function is π, vertical asymptotes also occur at $x = \pi/2 + n\pi$, where n is an integer. The domain of the tangent function is the set of all real numbers other than $x = \pi/2 + n\pi$, and the range is the set of all real numbers.

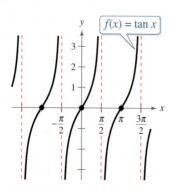
   Sketching the graph of $y = a \tan(bx - c)$ is similar to sketching the graph of $y = a \sin(bx - c)$ in that you locate key points that identify the intercepts and asymptotes. Two consecutive asymptotes can be found by solving the equations $bx - c = -\pi/2$ and $bx - c = \pi/2$. The midpoint between two consecutive asymptotes is an x-intercept of the graph. The period of the function $y = a \tan(bx - c)$ is the distance between two consecutive asymptotes. The amplitude of a tangent function is not defined. After plotting the asymptotes and the x-intercept, plot a few additional points between the two asymptotes and sketch one cycle. Finally, sketch one or two additional cycles to the left and right.

Example 1   Library of Parent Functions: $f(x) = \tan x$

Sketch the graph of $y = \tan \dfrac{x}{2}$ by hand.

Solution

By solving the equations $x/2 = -\pi/2$ and $x/2 = \pi/2$, you can see that two consecutive asymptotes occur at $x = -\pi$ and $x = \pi$. Between these two asymptotes, plot a few points, including the x-intercept, as shown in the table. Three cycles of the graph are shown in Figure 5.55. Use a graphing utility to confirm this graph.

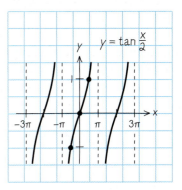

Figure 5.55

x	$-\pi$	$-\dfrac{\pi}{2}$	0	$\dfrac{\pi}{2}$	π
$\tan \dfrac{x}{2}$	Undef.	-1	0	1	Undef.

✔CHECKPOINT   Now try Exercise 9.

Example 2   Library of Parent Functions: $f(x) = \tan x$

Sketch the graph of $y = -3 \tan 2x$ by hand.

Solution

By solving the equations $2x = -\pi/2$ and $2x = \pi/2$, you can see that two consecutive asymptotes occur at $x = -\pi/4$ and $x = \pi/4$. Between these two asymptotes, plot a few points, including the x-intercept, as shown in the table. Three complete cycles of the graph are shown in Figure 5.56. You can use a graphing utility to confirm this graph, as shown in Figure 5.57.

x	$-\dfrac{\pi}{4}$	$-\dfrac{\pi}{8}$	0	$\dfrac{\pi}{8}$	$\dfrac{\pi}{4}$
$-3 \tan 2x$	Undef.	3	0	-3	Undef.

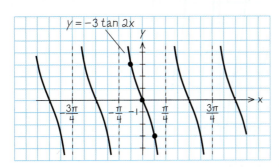

Figure 5.56

✔CHECKPOINT   Now try Exercise 11.

Technology Tip

 Your graphing utility may connect parts of the graphs of tangent, cotangent, secant, and cosecant functions that are not supposed to be connected. So, in this text, these functions are graphed on a graphing utility using the *dot* mode. A blue curve is placed behind the graphing utility's display to indicate where the graph should appear. (See Figure 5.57.)

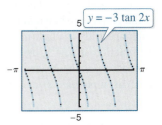

Figure 5.57

By comparing the graphs in Examples 1 and 2, you can see that the graph of

$$y = a \tan(bx - c)$$

increases between consecutive vertical asymptotes when $a > 0$ and decreases between consecutive vertical asymptotes when $a < 0$. In other words, the graph for $a < 0$ is a reflection in the x-axis of the graph for $a > 0$.

Graph of the Cotangent Function

 Library of Parent Functions: Cotangent Function

The graph of the parent cotangent function is similar to the graph of the parent tangent function. It also has a period of π. However, from the identity

$$f(x) = \cot x = \frac{\cos x}{\sin x}$$

you can see that the cotangent function has vertical asymptotes when $\sin x$ is zero, which occurs at $x = n\pi$, where n is an integer. The basic characteristics of the parent cotangent function are summarized below and on the inside front cover of this text.

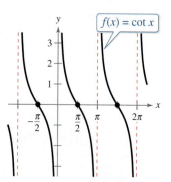

Domain: all real numbers $x, x \neq n\pi$

Range: $(-\infty, \infty)$

Period: π

x-intercepts: $\left(\dfrac{\pi}{2} + n\pi, 0\right)$

Vertical asymptotes: $x = n\pi$

Odd function

Origin symmetry

Example 3 Library of Parent Functions: $f(x) = \cot x$

Sketch the graph of $y = 2 \cot \dfrac{x}{3}$ by hand.

Solution

To locate two consecutive vertical asymptotes of the graph, solve the equations $x/3 = 0$ and $x/3 = \pi$ to see that two consecutive asymptotes occur at $x = 0$ and $x = 3\pi$. Then, between these two asymptotes, plot a few points, including the x-intercept, as shown in the table. Three cycles of the graph are shown in Figure 5.58. Use a graphing utility to confirm this graph. [Enter the function as $y = 2/\tan(x/3)$.] Note that the period is 3π, the distance between consecutive asymptotes.

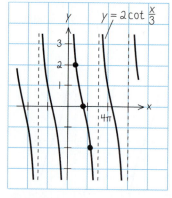

Figure 5.58

x	0	$\dfrac{3\pi}{4}$	$\dfrac{3\pi}{2}$	$\dfrac{9\pi}{4}$	3π
$2 \cot \dfrac{x}{3}$	Undef.	2	0	-2	Undef.

 CHECKPOINT   Now try Exercise 13.

Graphs of the Reciprocal Functions

The graphs of the two remaining trigonometric functions can be obtained from the graphs of the sine and cosine functions using the reciprocal identities

$$\csc x = \frac{1}{\sin x} \quad \text{and} \quad \sec x = \frac{1}{\cos x}.$$

For instance, at a given value of x, the y-coordinate for $\sec x$ is the reciprocal of the y-coordinate for $\cos x$. Of course, when $\cos x = 0$, the reciprocal does not exist. Near such values of x, the behavior of the secant function is similar to that of the tangent function. In other words, the graphs of

$$\tan x = \frac{\sin x}{\cos x} \quad \text{and} \quad \sec x = \frac{1}{\cos x}$$

have vertical asymptotes at $x = \pi/2 + n\pi$, where n is an integer (i.e., the values at which the cosine is zero). Similarly,

$$\cot x = \frac{\cos x}{\sin x} \quad \text{and} \quad \csc x = \frac{1}{\sin x}$$

have vertical asymptotes where $\sin x = 0$—that is, at $x = n\pi$.

To sketch the graph of a secant or cosecant function, you should first make a sketch of its reciprocal function. For instance, to sketch the graph of $y = \csc x$, first sketch the graph of $y = \sin x$. Then take the reciprocals of the y-coordinates to obtain points on the graph of $y = \csc x$. You can use this procedure to obtain the graphs shown in Figure 5.59.

Explore the Concept

Use a graphing utility to graph the functions $y_1 = \cos x$ and $y_2 = \sec x = 1/\cos x$ in the same viewing window. How are the graphs related? What happens to the graph of the secant function as x approaches the zeros of the cosine function?

Library of Parent Functions: Cosecant and Secant Functions

The basic characteristics of the parent cosecant and secant functions are summarized below and on the inside front cover of this text.

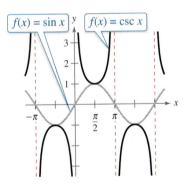

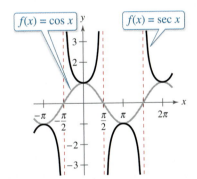

Domain: *all real numbers* $x, x \neq n\pi$

Range: $(-\infty, -1] \cup [1, \infty)$
Period: 2π
No intercepts

Vertical asymptotes: $x = n\pi$

Odd function
Origin symmetry

Figure 5.59

Domain: *all real numbers* $x, x \neq \dfrac{\pi}{2} + n\pi$

Range: $(-\infty, -1] \cup [1, \infty)$
Period: 2π
y-intercept: $(0, 1)$

Vertical asymptotes: $x = \dfrac{\pi}{2} + n\pi$

Even function
y-axis symmetry

In comparing the graphs of the cosecant and secant functions with those of the sine and cosine functions, note that the "hills" and "valleys" are interchanged. For example, a hill (or maximum point) on the sine curve corresponds to a valley (a local minimum) on the cosecant curve, and a valley (or minimum point) on the sine curve corresponds to a hill (a local maximum) on the cosecant curve, as shown in Figure 5.60. Additionally, x-intercepts of the sine and cosine functions become vertical asymptotes of the cosecant and secant functions, respectively (see Figure 5.60).

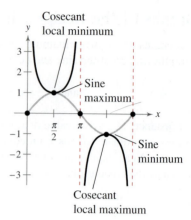

Figure 5.60

Example 4   Library of Parent Functions: $f(x) = \csc x$

Sketch the graph of $y = 2 \csc\left(x + \dfrac{\pi}{4}\right)$ by hand.

Solution

Begin by sketching the graph of $y = 2 \sin\left(x + \dfrac{\pi}{4}\right)$. For this function, the amplitude is 2 and the period is 2π. By solving the equations

$$x + \frac{\pi}{4} = 0 \quad \text{and} \quad x + \frac{\pi}{4} = 2\pi$$

you can see that one cycle of the sine function corresponds to the interval from

$$x = -\frac{\pi}{4} \quad \text{to} \quad x = \frac{7\pi}{4}.$$

The graph of this sine function is represented by the gray curve in Figure 5.61. Because the sine function is zero at the endpoints of this interval, the corresponding cosecant function

$$y = 2 \csc\left(x + \frac{\pi}{4}\right) = 2\left(\frac{1}{\sin[x + (\pi/4)]}\right)$$

has vertical asymptotes at $x = -\dfrac{\pi}{4}$, $x = \dfrac{3\pi}{4}$, $x = \dfrac{7\pi}{4}$, and so on. The graph of the cosecant function is represented by the black curve in Figure 5.61.

   Now try Exercise 15.

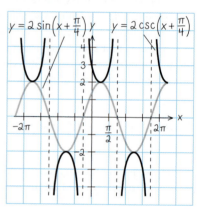

Figure 5.61

Example 5   Library of Parent Functions: $f(x) = \sec x$

Sketch the graph of $y = \sec 2x$ by hand.

Solution

Begin by sketching the graph of $y = \cos 2x$, as indicated by the gray curve in Figure 5.62. Then, form the graph of $y = \sec 2x$ as the black curve in the figure. Note that the x-intercepts of $y = \cos 2x$

$$\left(-\frac{\pi}{4}, 0\right), \quad \left(\frac{\pi}{4}, 0\right), \quad \left(\frac{3\pi}{4}, 0\right), \dots$$

correspond to the vertical asymptotes

$$x = -\frac{\pi}{4}, \quad x = \frac{\pi}{4}, \quad x = \frac{3\pi}{4}, \dots$$

of the graph of $y = \sec 2x$. Moreover, notice that the period of $y = \cos 2x$ and $y = \sec 2x$ is π.

   Now try Exercise 19.

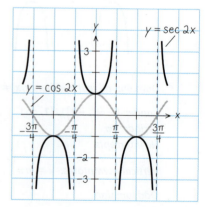

Figure 5.62

Damped Trigonometric Graphs

A *product* of two functions can be graphed using properties of the individual functions. For instance, consider the function

$$f(x) = x \sin x$$

as the product of the functions $y = x$ and $y = \sin x$. Using properties of absolute value and the fact that $|\sin x| \le 1$, you have $0 \le |x| |\sin x| \le |x|$. Consequently,

$$-|x| \le x \sin x \le |x|$$

which means that the graph of $f(x) = x \sin x$ lies between the lines $y = -x$ and $y = x$. Furthermore, because

$$f(x) = x \sin x = \pm x \qquad \text{at} \qquad x = \frac{\pi}{2} + n\pi$$

and

$$f(x) = x \sin x = 0 \qquad \text{at} \qquad x = n\pi$$

the graph of f touches the line $y = -x$ or the line $y = x$ at $x = \pi/2 + n\pi$ and has x-intercepts at $x = n\pi$. A sketch of f is shown in Figure 5.63. In the function $f(x) = x \sin x$, the factor x is called the **damping factor.**

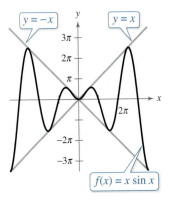

Figure 5.63

Example 6   Analyzing a Damped Sine Curve

Analyze the graph of $f(x) = e^{-x} \sin 3x$.

Solution

Consider $f(x)$ as the product of the two functions

$$y = e^{-x} \qquad \text{and} \qquad y = \sin 3x$$

each of which has the set of real numbers as its domain. For any real number x, you know that $e^{-x} \ge 0$ and $|\sin 3x| \le 1$. So, $|e^{-x}| |\sin 3x| \le e^{-x}$, which means that

$$-e^{-x} \le e^{-x} \sin 3x \le e^{-x}.$$

Furthermore, because

$$f(x) = e^{-x} \sin 3x = \pm e^{-x} \qquad \text{at} \qquad x = \frac{\pi}{6} + \frac{n\pi}{3}$$

and

$$f(x) = e^{-x} \sin 3x = 0 \qquad \text{at} \qquad x = \frac{n\pi}{3}$$

the graph of f touches the curves $y = -e^{-x}$ and $y = e^{-x}$ at $x = \pi/6 + n\pi/3$ and has intercepts at $x = n\pi/3$. The graph is shown in Figure 5.64.

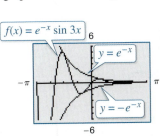

Figure 5.64

✔CHECKPOINT   Now try Exercise 55.

Study Tip

Do you see why the graph of $f(x) = x \sin x$ touches the lines $y = \pm x$ at $x = \pi/2 + n\pi$ and why the graph has x-intercepts at $x = n\pi$? Recall that the sine function is equal to ± 1 at $\pi/2, 3\pi/2, 5\pi/2, \ldots$ (odd multiples of $\pi/2$) and is equal to 0 at $\pi, 2\pi, 3\pi, \ldots$ (multiples of π).

Library of Parent Functions: Trigonometric Functions

Figure 5.65 summarizes the six basic trigonometric functions.

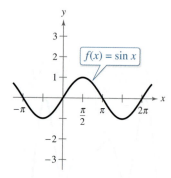

Domain: all real numbers x
Range: [−1, 1]
Period: 2π

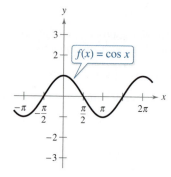

Domain: all real numbers x
Range: [−1, 1]
Period: 2π

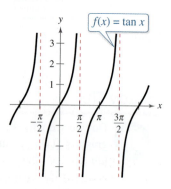

Domain: all real numbers x,

$$x \neq \frac{\pi}{2} + n\pi$$

Range: (−∞, ∞)
Period: π

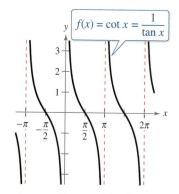

Domain: all real numbers x,

$$x \neq n\pi$$

Range: (−∞, ∞)
Period: π

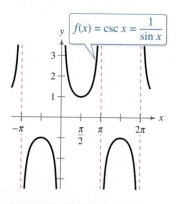

Domain: all real numbers x,
$$x \neq n\pi$$
Range: (−∞, −1] ∪ [1, ∞)
Period: 2π

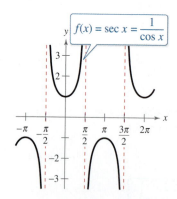

Domain: all real numbers x,
$$x \neq \frac{\pi}{2} + n\pi$$
Range: (−∞, −1] ∪ [1, ∞)
Period: 2π

Figure 5.65

5.5   Exercises

See www.CalcChat.com for worked-out solutions to odd-numbered exercises.
For instructions on how to use a graphing utility, see Appendix A.

Vocabulary and Concept Check

In Exercises 1 and 2, fill in the blank.

1. The graphs of the tangent, cotangent, secant, and cosecant functions have _____ asymptotes.

2. To sketch the graph of a secant or cosecant function, first make a sketch of its _____ function.

3. Which two parent trigonometric functions have a period of π and a range that consists of the set of all real numbers?

4. What is the damping factor of the function $f(x) = e^{2x} \sin x$?

Procedures and Problem Solving

Library of Parent Functions   In Exercises 5–8, use the graph of the function to answer the following.

(a) Find all x-intercepts of the graph of $y = f(x)$.

(b) Find all y-intercepts of the graph of $y = f(x)$.

(c) Find the intervals on which the graph of $y = f(x)$ is increasing and the intervals on which the graph of $y = f(x)$ is decreasing.

(d) Find all relative extrema, if any, of the graph of $y = f(x)$.

(e) Find all vertical asymptotes, if any, of the graph of $y = f(x)$.

5. $f(x) = \tan x$

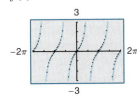

6. $f(x) = \cot x$

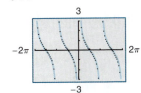

7. $f(x) = \sec x$

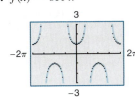

8. $f(x) = \csc x$

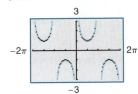

Library of Parent Functions   In Exercises 9–28, sketch the graph of the function. (Include two full periods.) Use a graphing utility to verify your result.

✓ 9. $y = \frac{1}{2} \tan x$

10. $y = \frac{1}{4} \tan x$

✓ 11. $y = -2 \tan 2x$

12. $y = -3 \tan 4x$

✓ 13. $y = \frac{1}{2} \cot \frac{x}{2}$

14. $y = 3 \cot \pi x$

✓ 15. $y = 3 \csc \frac{x}{2}$

16. $y = -\csc \frac{x}{3}$

17. $y = -\frac{1}{2} \sec x$

18. $y = \frac{1}{4} \sec x$

✓ 19. $y = \sec \pi x - 3$

20. $y = -2 \sec 4x + 2$

21. $y = 2 \tan \frac{\pi x}{4}$

22. $y = -\frac{1}{2} \tan \pi x$

23. $y = \frac{1}{2} \sec (2x - \pi)$

24. $y = -\sec(x + \pi)$

25. $y = \csc(\pi - x)$

26. $y = \csc(2x - \pi)$

27. $y = 2 \cot\left(x - \frac{\pi}{2}\right)$

28. $y = \frac{1}{4} \cot(x + \pi)$

Comparing Trigonometric Graphs   In Exercises 29–34, use a graphing utility to graph the function (include two full periods). Graph the corresponding reciprocal function in same viewing window. Describe and compare the graphs.

29. $y = 2 \csc 3x$

30. $y = -\csc(4x - \pi)$

31. $y = -2 \sec 4x$

32. $y = \frac{1}{4} \sec \pi x$

33. $y = \frac{1}{3} \sec\left(\frac{\pi x}{2} + \frac{\pi}{2}\right)$

34. $y = \frac{1}{2} \csc(2x - \pi)$

Solving a Trigonometric Equation Graphically   In Exercises 35–40, use a graph of the function to approximate the solution of the equation on the interval $[-2\pi, 2\pi]$.

35. $\tan x = 1$

36. $\cot x = -\sqrt{3}$

37. $\sec x = -2$

38. $\csc x = \sqrt{2}$

39. $\tan x = \sqrt{3}$

40. $\sec x = -\sqrt{2}$

Even and Odd Trigonometric Functions   In Exercises 41–46, use the graph of the function to determine whether the function is even, odd, or neither.

41. $f(x) = \sec x$

42. $f(x) = \tan x$

43. $f(x) = \csc 2x$

44. $f(x) = \cot 2x$

45. $f(x) = \tan\left(x - \frac{\pi}{2}\right)$

46. $f(x) = \sec(x + \pi)$

Using Graphs to Compare Functions In Exercises 47–50, use a graphing utility to graph the two equations in the same viewing window. Use the graphs to determine whether the expressions are equivalent. Verify the results algebraically.

47. $y_1 = \sin x \csc x$, $y_2 = 1$

48. $y_1 = \sin x \sec x$, $y_2 = \tan x$

49. $y_1 = \dfrac{\cos x}{\sin x}$, $y_2 = \cot x$

50. $y_1 = \sec^2 x - 1$, $y_2 = \tan^2 x$

Identifying Damped Trigonometric Graphs In Exercises 51–54, match the function with its graph. Describe the behavior of the function as x approaches zero. [The graphs are labeled (a), (b), (c), and (d).]

(a)
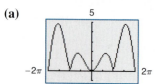

(b)
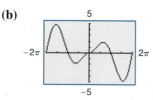

(c)
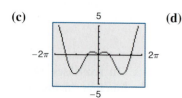

(d)
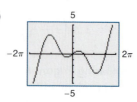

51. $f(x) = x \cos x$

52. $f(x) = |x \sin x|$

53. $g(x) = |x| \sin x$

54. $g(x) = |x| \cos x$

Analyzing a Damped Trigonometric Graph In Exercises 55–58, use a graphing utility to graph the function and the damping factor of the function in the same viewing window. Then analyze the graph of the function using the method in Example 6.

✓ **55.** $f(x) = e^{-x} \cos x$

56. $f(x) = e^{-2x} \sin x$

57. $h(x) = e^{-x^2/4} \cos x$

58. $g(x) = e^{-x^2/2} \sin x$

Exploration In Exercises 59 and 60, use a graphing utility to graph the function. Use the graph to determine the behavior of the function as $x \to c$.

(a) $x \to \dfrac{\pi^+}{2}$ $\left(\text{as } x \text{ approaches } \dfrac{\pi}{2} \text{ from the right}\right)$

(b) $x \to \dfrac{\pi^-}{2}$ $\left(\text{as } x \text{ approaches } \dfrac{\pi}{2} \text{ from the left}\right)$

(c) $x \to -\dfrac{\pi^+}{2}$ $\left(\text{as } x \text{ approaches } -\dfrac{\pi}{2} \text{ from the right}\right)$

(d) $x \to -\dfrac{\pi^-}{2}$ $\left(\text{as } x \text{ approaches } -\dfrac{\pi}{2} \text{ from the left}\right)$

59. $f(x) = \tan x$ **60.** $f(x) = \sec x$

Exploration In Exercises 61 and 62, use a graphing utility to graph the function. Use the graph to determine the behavior of the function as $x \to c$.

(a) As $x \to 0^+$, the value of $f(x) \to$ ▢.

(b) As $x \to 0^-$, the value of $f(x) \to$ ▢.

(c) As $x \to \pi^+$, the value of $f(x) \to$ ▢.

(d) As $x \to \pi^-$, the value of $f(x) \to$ ▢.

61. $f(x) = \cot x$

62. $f(x) = \csc x$

63. Aviation A plane flying at an altitude of 5 miles over level ground will pass directly over a radar antenna (see figure). Let d be the ground distance from the antenna to the point directly under the plane and let x be the angle of elevation to the plane from the antenna. (d is positive as the plane approaches the antenna.) Write d as a function of x and graph the function over the interval $0 < x < \pi$.

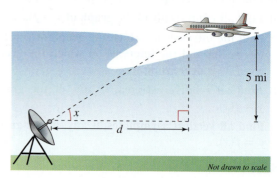

5 mi

Not drawn to scale

64. Why you should learn it (p. 444) A television camera is on a reviewing platform 36 meters from the street on which a parade will be passing from left to right (see figure). Write the distance d from the camera to a particular unit in the parade as a function of the angle x, and graph the function over the interval $-\pi/2 < x < \pi/2$. (Consider x as negative when a unit in the parade approaches from the left.)

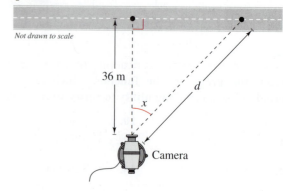

Not drawn to scale

36 m

d

x

Camera

65. Harmonic Motion An object weighing W pounds is suspended from a ceiling by a steel spring (see figure). The weight is pulled downward (positive direction) from its equilibrium position and released. The resulting motion of the weight is described by the function $y = \frac{1}{2}e^{-t/4} \cos 4t$, where y is the distance in feet and t is the time in seconds $(t > 0)$.

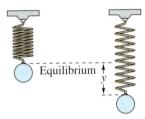

Equilibrium

(a) Use a graphing utility to graph the function.

(b) Describe the behavior of the displacement function for increasing values of time t.

66. Mechanical Engineering A crossed belt connects a 10-centimeter pulley on an electric motor with a 20-centimeter pulley on a saw arbor (see figure). The electric motor runs at 1700 revolutions per minute.

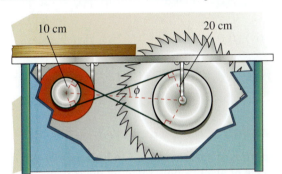

10 cm   20 cm

(a) Determine the number of revolutions per minute of the saw.

(b) How does crossing the belt affect the saw in relation to the motor?

(c) Let L be the total length of the belt. Write L as a function of ϕ, where ϕ is measured in radians. What is the domain of the function? (*Hint:* Add the lengths of the straight sections of the belt and the length of belt around each pulley.)

(d) Use a graphing utility to complete the table.

ϕ	0.3	0.6	0.9	1.2	1.5
L					

(e) As ϕ increases, do the lengths of the straight sections of the belt change faster or slower than the lengths of the belts around each pulley?

(f) Use the graphing utility to graph the function over the appropriate domain.

67. MODELING DATA

The motion of an oscillating weight suspended by a spring was measured by a motion detector. The data were collected, and the approximate maximum (positive and negative) displacements from equilibrium are shown in the graph. The displacement y is measured in centimeters and the time t is measured in seconds.

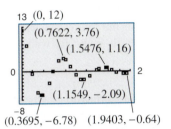

13 (0, 12)
(0.7622, 3.76)
(1.5476, 1.16)
0   2
(1.1549, −2.09)
−8
(0.3695, −6.78)   (1.9403, −0.64)

(a) Is y a function of t? Explain.

(b) Approximate the frequency of the oscillations.

(c) Fit a model of the form $y = ab^t \cos ct$ to the data. Use the result of part (b) to approximate c. Use the *regression* feature of a graphing utility to fit an exponential model to the positive maximum displacements of the weight.

(d) Rewrite the model in the form $y = ae^{kt} \cos ct$.

(e) Use the graphing utility to graph the model. Compare the result with the data in the graph above.

Conclusions

True or False? In Exercises 68–71, determine whether the statement is true or false. Justify your answer.

68. The graph of $y = -\frac{1}{8} \tan\left(\frac{x}{2} + \pi\right)$ has an asymptote at $x = -3\pi$.

69. For the graph of $y = 2^x \sin x$, as x approaches $-\infty$, y approaches 0.

70. The graph of $y = \csc x$ can be obtained on a calculator by graphing the reciprocal of $y = \sin x$.

71. The graph of $y = \sec x$ can be obtained on a calculator by graphing a translation of the reciprocal of $y = \sin x$.

72. Exploration Consider the functions

$$f(x) = 2 \sin x \quad \text{and} \quad g(x) = \frac{1}{2} \csc x$$

on the interval $(0, \pi)$.

(a) Use a graphing utility to graph f and g in the same viewing window.

(b) Approximate the interval in which $f > g$.

(c) Describe the behavior of each of the functions as x approaches π. How is the behavior of g related to the behavior of f as x approaches π?

73. Exploration Consider the functions given by

$$f(x) = \tan\frac{\pi x}{2} \quad \text{and} \quad g(x) = \frac{1}{2}\sec\frac{\pi x}{2}$$

on the interval $(-1, 1)$,

(a) Use a graphing utility to graph f and g in the same viewing window.

(b) Approximate the interval in which $f < g$.

(c) Approximate the interval in which $2f < 2g$. How does the result compare with that of part (b)? Explain.

74. Exploration

(a) Use a graphing utility to graph each function.

$$y_1 = \frac{4}{\pi}\left(\sin \pi x + \frac{1}{3}\sin 3\pi x\right)$$

$$y_2 = \frac{4}{\pi}\left(\sin \pi x + \frac{1}{3}\sin 3\pi x + \frac{1}{5}\sin 5\pi x\right)$$

(b) Identify the pattern in part (a) and find a function y_3 that continues the pattern one more term. Use the graphing utility to graph y_3.

(c) The graphs in parts (a) and (b) approximate the periodic function in the figure. Find a function y_4 that is a better approximation.

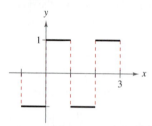

Exploration In Exercises 75 and 76, use a graphing utility to explore the ratio $f(x)$, which appears in calculus.

(a) **Complete the table. Round your results to four decimal places.**

x	-1	-0.1	-0.01	-0.001
$f(x)$				

x	0	0.001	0.01	0.1	1
$f(x)$					

(b) **Use the graphing utility to graph the function $f(x)$. Use the *zoom* and *trace* features to describe the behavior of the graph as x approaches 0.**

(c) **Write a brief statement regarding the value of the ratio based on your results in parts (a) and (b).**

75. $f(x) = \dfrac{\tan x}{x}$ **76.** $f(x) = \dfrac{\tan 3x}{3x}$

77. Exploration Using calculus, it can be shown that the tangent function can be approximated by the polynomial

$$\tan x \approx x + \frac{2x^3}{3!} + \frac{16x^5}{5!}$$

where x is in radians. Use a graphing utility to graph the tangent function and its polynomial approximation in the same viewing window. How do the graphs compare?

78. CAPSTONE Determine which function is represented by each graph. Do not use a calculator.

(a) 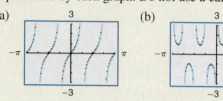 (b)

(i) $f(x) = \tan 2x$ (i) $f(x) = \sec 4x$

(ii) $f(x) = \tan\dfrac{x}{2}$ (ii) $f(x) = \csc 4x$

(iii) $f(x) = 2\tan x$ (iii) $f(x) = \csc\dfrac{x}{4}$

(iv) $f(x) = -\tan 2x$ (iv) $f(x) = \sec\dfrac{x}{4}$

(v) $f(x) = -\tan\dfrac{x}{2}$ (v) $f(x) = \csc(4x - \pi)$

Cumulative Mixed Review

Properties of Real Numbers In Exercises 79–82, identify the rule of algebra illustrated by the statement.

79. $5(a - 9) = 5a - 45$

80. $7(\frac{1}{7}) = 1$

81. $(3 + x) + 0 = 3 + x$

82. $(a + b) + 10 = a + (b + 10)$

Finding an Inverse Function In Exercises 83–86, determine whether the function is one-to-one. If it is, find its inverse function.

83. $f(x) = -10$

84. $f(x) = (x - 7)^2 + 3$

85. $f(x) = \sqrt{3x - 14}$

86. $f(x) = \sqrt[3]{x - 5}$

Finding the Domain, Intercepts, and Asymptotes of a Function In Exercises 87–90, identify the domain, any intercepts, and any asymptotes of the function.

87. $y = x^2 + 3x - 4$ **88.** $y = \ln x^4$

89. $f(x) = 3^{x+1} + 2$ **90.** $f(x) = \dfrac{x - 7}{x^2 + 4x + 4}$

5.6   Inverse Trigonometric Functions

Inverse Sine Function

Recall from Section 1.7 that for a function to have an inverse function, it must be one-to-one—that is, it must pass the Horizontal Line Test. In Figure 5.66 it is obvious that $y = \sin x$ does not pass the test because different values of x yield the same y-value.

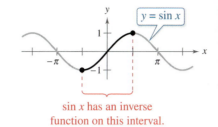

sin x has an inverse function on this interval.

Figure 5.66

However, when you restrict the domain to the interval $-\pi/2 \leq x \leq \pi/2$ (corresponding to the black portion of the graph in Figure 5.66), the following properties hold.

1. On the interval $[-\pi/2, \pi/2]$, the function $y = \sin x$ is increasing.

2. On the interval $[-\pi/2, \pi/2]$, $y = \sin x$ takes on its full range of values, $-1 \leq \sin x \leq 1$.

3. On the interval $[-\pi/2, \pi/2]$, $y = \sin x$ is one-to-one.

So, on the restricted domain $-\pi/2 \leq x \leq \pi/2$, $y = \sin x$ has a unique inverse function called the **inverse sine function.** It is denoted by

$$y = \arcsin x \qquad \text{or} \qquad y = \sin^{-1} x.$$

The notation $\sin^{-1} x$ is consistent with the inverse function notation $f^{-1}(x)$. The arcsin x notation (read as "the arcsine of x") comes from the association of a central angle with its intercepted *arc length* on a unit circle. So, arcsin x means the angle (or arc) whose sine is x. Both notations, arcsin x and $\sin^{-1} x$, are commonly used in mathematics, so remember that $\sin^{-1} x$ denotes the *inverse* sine function rather than $1/\sin x$. The values of arcsin x lie in the interval $-\pi/2 \leq \arcsin x \leq \pi/2$. The graph of $y = \arcsin x$ is shown in Example 2.

Definition of Inverse Sine Function

The **inverse sine function** is defined by

$$y = \arcsin x \qquad \text{if and only if} \qquad \sin y = x$$

where $-1 \leq x \leq 1$ and $-\pi/2 \leq y \leq \pi/2$. The domain of $y = \arcsin x$ is $[-1, 1]$ and the range is $[-\pi/2, \pi/2]$.

When evaluating the inverse sine function, it helps to remember the phrase "the arcsine of x is the angle (or number) whose sine is x."

What you should learn

- Evaluate and graph inverse sine functions.
- Evaluate and graph other inverse trigonometric functions.
- Evaluate compositions of trigonometric functions.

Why you should learn it

Inverse trigonometric functions can be useful in exploring how aspects of a real-life problem relate to each other. Exercise 99 on page 464 investigates the relationship between the height of a cone-shaped pile of rock salt, the angle of the cone shape, and the diameter of its base.

Study Tip

In this text, the parentheses in arcsin(u) are sometimes omitted when u is an expression involving exponents, radicals, products, or quotients. For instance, arcsin($2x$) can be written as arcsin $2x$. To evaluate arcsin $2x$, find the arcsine of the product $2x$. The other inverse trigonometric functions can be written and evaluated in the same manner.

Example 1   Evaluating the Inverse Sine Function

If possible, find the exact value.

a. $\arcsin\left(-\dfrac{1}{2}\right)$ **b.** $\sin^{-1}\dfrac{\sqrt{3}}{2}$ **c.** $\sin^{-1}2$

Solution

a. Because $\sin\left(-\dfrac{\pi}{6}\right) = -\dfrac{1}{2}$, and $-\dfrac{\pi}{6}$ lies in $\left[-\dfrac{\pi}{2},\dfrac{\pi}{2}\right]$, it follows that

$$\arcsin\left(-\dfrac{1}{2}\right) = -\dfrac{\pi}{6}. \qquad \text{Angle whose sine is } -\tfrac{1}{2}$$

b. Because $\sin\dfrac{\pi}{3} = \dfrac{\sqrt{3}}{2}$, and $\dfrac{\pi}{3}$ lies in $\left[-\dfrac{\pi}{2},\dfrac{\pi}{2}\right]$, it follows that

$$\sin^{-1}\dfrac{\sqrt{3}}{2} = \dfrac{\pi}{3}. \qquad \text{Angle whose sine is } \sqrt{3}/2$$

c. It is not possible to evaluate $y = \sin^{-1}x$ at $x = 2$ because there is no angle whose sine is 2. Remember that the domain of the inverse sine function is $[-1, 1]$.

 CHECKPOINT   Now try Exercise 5.

Example 2   Graphing the Arcsine Function

Sketch a graph of $y = \arcsin x$ by hand.

Solution

By definition, the equations

$$y = \arcsin x \qquad \text{and} \qquad \sin y = x$$

are equivalent for $-\pi/2 \le y \le \pi/2$. So, their graphs are the same. For the interval $[-\pi/2, \pi/2]$, you can assign values to y in the second equation to make a table of values.

y	$-\dfrac{\pi}{2}$	$-\dfrac{\pi}{4}$	$-\dfrac{\pi}{6}$	0	$\dfrac{\pi}{6}$	$\dfrac{\pi}{4}$	$\dfrac{\pi}{2}$
$x = \sin y$	-1	$-\dfrac{\sqrt{2}}{2}$	$-\dfrac{1}{2}$	0	$\dfrac{1}{2}$	$\dfrac{\sqrt{2}}{2}$	1

Then plot the points and connect them with a smooth curve. The resulting graph of $y = \arcsin x$ is shown in Figure 5.67. Note that it is the reflection (in the line $y = x$) of the black portion of the graph in Figure 5.66. Use a graphing utility to confirm this graph. Be sure you see that Figure 5.67 shows the *entire* graph of the inverse sine function. Remember that the domain of $y = \arcsin x$ is the closed interval $[-1, 1]$ and the range is the closed interval $[-\pi/2, \pi/2]$.

 CHECKPOINT   Now try Exercise 15.

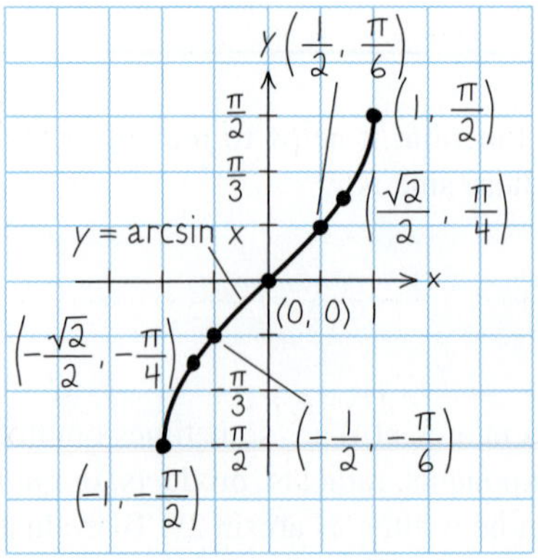

Figure 5.67

Study Tip

As with the trigonometric functions, much of the work with the inverse trigonometric functions can be done by *exact* calculations rather than by calculator approximations. Exact calculations help to increase your understanding of the inverse functions by relating them to the triangle definitions of the trigonometric functions.

Other Inverse Trigonometric Functions

The cosine function is decreasing and one-to-one on the interval $0 \leq x \leq \pi$, as shown in Figure 5.68.

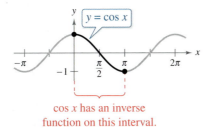

cos x has an inverse function on this interval.

Figure 5.68

Consequently, on this interval the cosine function has an inverse function—the **inverse cosine function**—denoted by

$$y = \arccos x \qquad \text{or} \qquad y = \cos^{-1} x.$$

Because $y = \arccos x$ and $x = \cos y$ are equivalent for $0 \leq y \leq \pi$, their graphs are the same, and can be confirmed by the following table of values.

y	0	$\dfrac{\pi}{6}$	$\dfrac{\pi}{3}$	$\dfrac{\pi}{2}$	$\dfrac{2\pi}{3}$	$\dfrac{5\pi}{6}$	π
$x = \cos y$	1	$\dfrac{\sqrt{3}}{2}$	$\dfrac{1}{2}$	0	$-\dfrac{1}{2}$	$-\dfrac{\sqrt{3}}{2}$	-1

Similarly, you can define an **inverse tangent function** by restricting the domain of $y = \tan x$ to the interval $(-\pi/2, \pi/2)$. The inverse tangent function is denoted by

$$y = \arctan x \qquad \text{or} \qquad y = \tan^{-1} x.$$

Because $y = \arctan x$ and $x = \tan y$ are equivalent for $-\pi/2 < y < \pi/2$, their graphs are the same, and can be confirmed by the following table of values.

y	$-\dfrac{\pi}{4}$	$-\dfrac{\pi}{6}$	0	$\dfrac{\pi}{6}$	$\dfrac{\pi}{4}$
$x = \tan y$	-1	$-\dfrac{\sqrt{3}}{3}$	0	$\dfrac{\sqrt{3}}{3}$	1

The following list summarizes the definitions of the three most common inverse trigonometric functions. (Their graphs are shown on page 459.) The remaining three are defined in Exercises 107–109.

> **Definition of the Inverse Trigonometric Functions**
>
> | *Function* | *Domain* | *Range* |
> |---|---|---|
> | $y = \arcsin x$ if and only if $\sin y = x$ | $-1 \leq x \leq 1$ | $-\dfrac{\pi}{2} \leq y \leq \dfrac{\pi}{2}$ |
> | $y = \arccos x$ if and only if $\cos y = x$ | $-1 \leq x \leq 1$ | $0 \leq y \leq \pi$ |
> | $y = \arctan x$ if and only if $\tan y = x$ | $-\infty < x < \infty$ | $-\dfrac{\pi}{2} < y < \dfrac{\pi}{2}$ |

Example 3   Evaluating Inverse Trigonometric Functions

Find the exact value.

a. $\arccos \dfrac{\sqrt{2}}{2}$

b. $\cos^{-1}(-1)$

c. $\arctan 0$

d. $\tan^{-1}(-1)$

Solution

a. Because $\cos(\pi/4) = \sqrt{2}/2$, and $\pi/4$ lies in $[0, \pi]$, it follows that

$$\arccos \dfrac{\sqrt{2}}{2} = \dfrac{\pi}{4}. \qquad \text{\color{red}Angle whose cosine is } \dfrac{\sqrt{2}}{2}$$

b. Because $\cos \pi = -1$, and π lies in $[0, \pi]$, it follows that

$$\cos^{-1}(-1) = \pi. \qquad \text{\color{red}Angle whose cosine is } -1$$

c. Because $\tan 0 = 0$, and 0 lies in $(-\pi/2, \pi/2)$, it follows that

$$\arctan 0 = 0. \qquad \text{\color{red}Angle whose tangent is } 0$$

d. Because $\tan(-\pi/4) = -1$, and $-\pi/4$ lies in $(-\pi/2, \pi/2)$, it follows that

$$\tan^{-1}(-1) = -\dfrac{\pi}{4}. \qquad \text{\color{red}Angle whose tangent is } -1$$

✔CHECKPOINT   Now try Exercise 9.

Example 4   Calculators and Inverse Trigonometric Functions

Use a calculator to approximate the value (if possible).

a. $\arctan(-8.45)$ **b.** $\sin^{-1} 0.2447$ **c.** $\arccos 2$

Solution

Function	*Mode*	*Graphing Calculator Keystrokes*
a. $\arctan(-8.45)$	Radian	(TAN⁻¹) (((−) 8.45) (ENTER)

From the display, it follows that $\arctan(-8.45) \approx -1.4530$.

b. $\sin^{-1} 0.2447$	Radian	(SIN⁻¹) ((0.2447) (ENTER)

From the display, it follows that $\sin^{-1} 0.2447 \approx 0.2472$.

c. $\arccos 2$	Radian	(COS⁻¹) ((2) (ENTER)

The calculator should display an *error message*, because the domain of the inverse cosine function is $[-1, 1]$.

✔CHECKPOINT   Now try Exercise 23.

Technology Tip

You can use the (SIN⁻¹), (COS⁻¹), and (TAN⁻¹) keys on your calculator to approximate values of other inverse trigonometric functions. To evaluate the inverse cosecant function, the inverse secant function, or the inverse cotangent function, you can use the inverse sine, inverse cosine, and inverse tangent functions, respectively. For instance, to evaluate $\sec^{-1} 3.4$, enter the expression as shown below.

$$\cos^{-1}(1/3.4)$$
$$1.272264126$$

Study Tip

Remember that the domain of the inverse sine function and the inverse cosine function is $[-1, 1]$, as indicated in Example 4(c).

📚 Library of Parent Functions: Inverse Trigonometric Functions

The parent inverse sine function, parent inverse cosine function, and parent inverse tangent function are summarized below and on the inside front cover of the text.

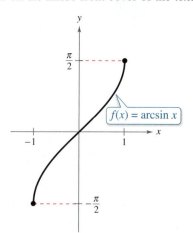

$f(x) = \arcsin x$

Domain: $[-1, 1]$

Range: $\left[-\dfrac{\pi}{2}, \dfrac{\pi}{2}\right]$

Intercept: $(0, 0)$

Odd function

Origin symmetry

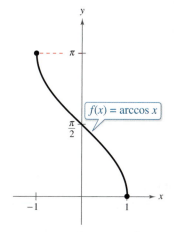

$f(x) = \arccos x$

Domain: $[-1, 1]$

Range: $[0, \pi]$

y-intercept: $\left(0, \dfrac{\pi}{2}\right)$

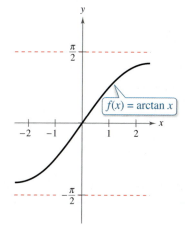

$f(x) = \arctan x$

Domain: $(-\infty, \infty)$

Range: $\left(-\dfrac{\pi}{2}, \dfrac{\pi}{2}\right)$

Intercept: $(0, 0)$

Horizontal asymptotes: $y = \pm\dfrac{\pi}{2}$

Odd function

Origin symmetry

Example 5   Library of Parent Functions: $f(x) = \arccos x$

Compare the graph of each function with the graph of $f(x) = \arccos x$.

a. $g(x) = \arccos(x - 2)$ **b.** $h(x) = \arccos(-x)$

Solution

a. Because $g(x) = \arccos(x - 2) = f(x - 2)$, the graph of g can be obtained by shifting the graph of f two units to the *right*, as shown in Figure 5.69.

b. Because $h(x) = \arccos(-x) = f(-x)$, the graph of h can be obtained by *reflecting* the graph of f in the *y*-axis, as shown in Figure 5.70.

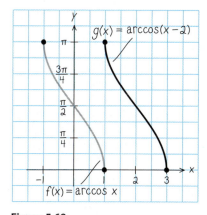

Figure 5.69

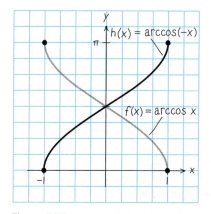

Figure 5.70

 CHECKPOINT   Now try Exercise 37.

Compositions of Functions

Recall from Section 1.7 that for all x in the domains of f and f^{-1}, inverse functions have the properties

$$f(f^{-1}(x)) = x \quad \text{and} \quad f^{-1}(f(x)) = x.$$

Inverse Properties

If $-1 \le x \le 1$ and $-\pi/2 \le y \le \pi/2$, then

$$\sin(\arcsin x) = x \quad \text{and} \quad \arcsin(\sin y) = y.$$

If $-1 \le x \le 1$ and $0 \le y \le \pi$, then

$$\cos(\arccos x) = x \quad \text{and} \quad \arccos(\cos y) = y.$$

If x is a real number and $-\pi/2 < y < \pi/2$, then

$$\tan(\arctan x) = x \quad \text{and} \quad \arctan(\tan y) = y.$$

Keep in mind that these inverse properties do not apply for arbitrary values of x and y. For instance,

$$\arcsin\left(\sin\frac{3\pi}{2}\right) = \arcsin(-1) = -\frac{\pi}{2} \ne \frac{3\pi}{2}.$$

In other words, the property

$$\arcsin(\sin y) = y$$

is not valid for values of y outside the interval $[-\pi/2, \pi/2]$.

Example 6 Using Inverse Properties

If possible, find the exact value.

a. $\tan[\arctan(-5)]$

b. $\arcsin\left(\sin\dfrac{5\pi}{3}\right)$

c. $\cos(\cos^{-1}\pi)$

Solution

a. Because -5 lies in the domain of the arctangent function, the inverse property applies, and you have

$$\tan[\arctan(-5)] = -5.$$

b. In this case, $5\pi/3$ does not lie within the range of the arcsine function, $-\pi/2 \le y \le \pi/2$. However, $5\pi/3$ is coterminal with

$$\frac{5\pi}{3} - 2\pi = -\frac{\pi}{3}$$

which does lie in the range of the arcsine function, and you have

$$\arcsin\left(\sin\frac{5\pi}{3}\right) = \arcsin\left[\sin\left(-\frac{\pi}{3}\right)\right] = -\frac{\pi}{3}.$$

c. The expression $\cos(\cos^{-1}\pi)$ is not defined because $\cos^{-1}\pi$ is not defined. Remember that the domain of the inverse cosine function is $[-1, 1]$.

✓**CHECKPOINT** Now try Exercise 55.

Explore the Concept

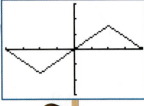 Use a graphing utility to graph $y = \arcsin(\sin x)$. What are the domain and range of this function? Explain why $\arcsin(\sin 4)$ does not equal 4. Now graph $y = \sin(\arcsin x)$ and determine the domain and range. Explain why $\sin(\arcsin 4)$ is not defined.

Example 7 shows how to use right triangles to find exact values of compositions of inverse functions.

Example 7   Evaluating Compositions of Functions

Find the exact value.

a. $\tan\left(\arccos \dfrac{2}{3}\right)$ **b.** $\cos\left[\arcsin\left(-\dfrac{3}{5}\right)\right]$

Algebraic Solution

a. If you let $u = \arccos \dfrac{2}{3}$, then $\cos u = \dfrac{2}{3}$. Because $\cos u$ is positive, u is a first-quadrant angle. You can sketch and label angle u as shown in Figure 5.71. Consequently,

$$\tan\left(\arccos \dfrac{2}{3}\right) = \tan u = \dfrac{\text{opp}}{\text{adj}} = \dfrac{\sqrt{5}}{2}.$$

b. If you let $u = \arcsin\left(-\dfrac{3}{5}\right)$, then $\sin u = -\dfrac{3}{5}$. Because $\sin u$ is negative, u is a fourth-quadrant angle. You can sketch and label angle u as shown in Figure 5.72. Consequently,

$$\cos\left[\arcsin\left(-\dfrac{3}{5}\right)\right] = \cos u = \dfrac{\text{adj}}{\text{hyp}} = \dfrac{4}{5}.$$

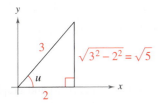

Figure 5.71   **Figure 5.72**

Graphical Solution

a. Use a graphing utility set in *radian* mode to graph $y = \tan(\arccos x)$, as shown in Figure 5.73.

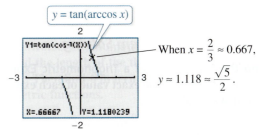

When $x = \dfrac{2}{3} \approx 0.667$, $y \approx 1.118 \approx \dfrac{\sqrt{5}}{2}$.

Figure 5.73

b. Use the graphing utility set in *radian* mode to graph $y = \cos(\arcsin x)$, as shown in Figure 5.74.

When $x = -\dfrac{3}{5} = -0.6$, $y = 0.8 = \dfrac{4}{5}$.

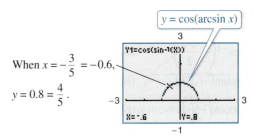

Figure 5.74

✔CHECKPOINT   Now try Exercise 67.

Example 8   Some Problems from Calculus   algebra of calculus

Write each of the following as an algebraic expression in x.

a. $\sin(\arccos 3x)$, $0 \le x \le \dfrac{1}{3}$ **b.** $\cot(\arccos 3x)$, $0 \le x < \dfrac{1}{3}$

Solution

If you let $u = \arccos 3x$, then $\cos u = 3x$, where $-1 \le 3x \le 1$. Because

$$\cos u = \text{adj/hyp} = (3x)/1$$

you can sketch a right triangle with acute angle u, as shown in Figure 5.75. From this triangle, you can easily convert each expression to algebraic form.

a. $\sin(\arccos 3x) = \sin u = \dfrac{\text{opp}}{\text{hyp}} = \sqrt{1 - 9x^2}$, $0 \le x \le \dfrac{1}{3}$

b. $\cot(\arccos 3x) = \cot u = \dfrac{\text{adj}}{\text{opp}} = \dfrac{3x}{\sqrt{1 - 9x^2}}$, $0 \le x < \dfrac{1}{3}$

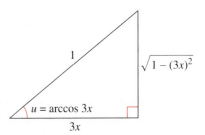

Figure 5.75

✔CHECKPOINT   Now try Exercise 73.

∫ **Writing an Expression** In Exercises 73–80, write an algebraic expression that is equivalent to the expression. (*Hint:* Sketch a right triangle, as demonstrated in Example 7.)

✓ **73.** $\cot(\arctan x)$ **74.** $\sin(\arctan x)$

75. $\sin[\arccos(x + 2)]$ **76.** $\sec[\arcsin(x - 1)]$

77. $\tan\left(\arccos \dfrac{x}{5}\right)$ **78.** $\cot\left(\arctan \dfrac{4}{x}\right)$

79. $\csc\left(\arctan \dfrac{x}{\sqrt{7}}\right)$ **80.** $\cos\left(\arcsin \dfrac{x - h}{r}\right)$

Completing an Equation In Exercises 81–84, complete the equation.

81. $\arctan \dfrac{14}{x} = \arcsin(\quad)$, $x > 0$

82. $\arcsin \dfrac{\sqrt{36 - x^2}}{6} = \arccos(\quad)$, $0 \le x \le 6$

83. $\arccos \dfrac{3}{\sqrt{x^2 - 2x + 10}} = \arcsin(\quad)$

84. $\arccos \dfrac{x - 2}{2} = \arctan(\quad)$, $2 < x < 4$

Graphing an Inverse Trigonometric Function In Exercises 85–90, use a graphing utility to graph the function.

85. $y = 2 \arccos x$ **86.** $y = \arcsin \dfrac{x}{2}$

87. $f(x) = \arcsin(x - 2)$ **88.** $g(t) = \arccos(t + 2)$

89. $f(x) = \arctan 2x$ **90.** $f(x) = \arccos \dfrac{x}{4}$

Using a Trigonometric Identity In Exercises 91 and 92, write the function in terms of the sine function by using the identity

$$A \cos \omega t + B \sin \omega t = \sqrt{A^2 + B^2} \sin\left(\omega t + \arctan \dfrac{A}{B}\right).$$

Use a graphing utility to graph both forms of the function. What does the graph imply?

91. $f(t) = 3 \cos 2t + 3 \sin 2t$

92. $f(t) = 4 \cos \pi t + 3 \sin \pi t$

Finding an Inverse Trigonometric Function In Exercises 93–98, find the value. If not possible, state the reason.

93. As $x \to 1^-$, the value of $\arcsin x \to$ ▢.

94. As $x \to 1^-$, the value of $\arccos x \to$ ▢.

95. As $x \to \infty$, the value of $\arctan x \to$ ▢.

96. As $x \to -1^+$, the value of $\arcsin x \to$ ▢.

97. As $x \to -1^+$, the value of $\arccos x \to$ ▢.

98. As $x \to -\infty$, the value of $\arctan x \to$ ▢.

99. *Why you should learn it* (p. 455) Different types of granular substances naturally settle at different angles when stored in cone-shaped piles. This angle θ is called the *angle of repose*. When rock salt is stored in a cone-shaped pile 11 feet high, the diameter of the pile's base is about 34 feet. (Source: Bulk-Store Structures, Inc.)

(a) Draw a diagram that gives a visual representation of the problem. Label all quantities.

(b) Find the angle of repose for rock salt.

(c) How tall is a pile of rock salt that has a base diameter of 40 feet?

100. Photography A television camera at ground level is filming the lift-off of a space shuttle at a point 750 meters from the launch pad (see figure). Let θ be the angle of elevation to the shuttle and let s be the height of the shuttle.

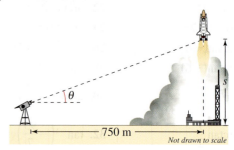

Not drawn to scale

(a) Write θ as a function of s.

(b) Find θ when $s = 400$ meters and $s = 1600$ meters.

101. MODELING DATA

A photographer takes a picture of a three-foot painting hanging in an art gallery. The camera lens is 1 foot below the lower edge of the painting (see figure). The angle β subtended by the camera lens x feet from the painting is $\beta = \arctan[3x/(x^2 + 4)]$, $x > 0$.

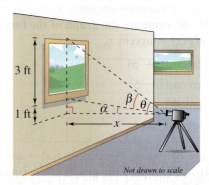

Not drawn to scale

(a) Use a graphing utility to graph β as a function of x.

(b) Move the cursor along the graph to approximate the distance from the picture when β is maximum.

(c) Identify the asymptote of the graph and discuss its meaning in the context of the problem.

102. Angle of Elevation   An airplane flies at an altitude of 6 miles toward a point directly over an observer. Consider θ and x as shown in the figure.

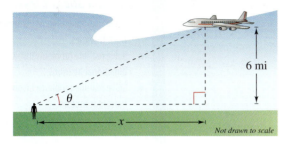

6 mi

Not drawn to scale

(a) Write θ as a function of x.

(b) Find θ when $x = 10$ miles and $x = 3$ miles.

103. Criminal Justice   A police car with its spotlight on is parked 20 meters from a warehouse. Consider θ and x as shown in the figure.

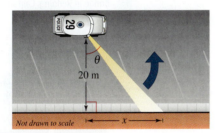

20 m

Not drawn to scale x

(a) Write θ as a function of x.

(b) Find θ when $x = 5$ meters and $x = 12$ meters.

Conclusions

True or False?   In Exercises 104–106, determine whether the statement is true or false. Justify your answer.

104. $\sin \dfrac{5\pi}{6} = \dfrac{1}{2}$ ⟹ $\arcsin \dfrac{1}{2} = \dfrac{5\pi}{6}$

105. $\tan \dfrac{5\pi}{4} = 1$ ⟹ $\arctan 1 = \dfrac{5\pi}{4}$

106. $\arctan x = \dfrac{\arcsin x}{\arccos x}$

107. Define the inverse cotangent function by restricting the domain of the cotangent function to the interval $(0, \pi)$, and sketch the graph of the inverse function.

108. Define the inverse secant function by restricting the domain of the secant function to the intervals $[0, \pi/2)$ and $(\pi/2, \pi]$, and sketch the graph of the inverse function.

109. Define the inverse cosecant function by restricting the domain of the cosecant function to the intervals $[-\pi/2, 0)$ and $(0, \pi/2]$, and sketch the graph of the inverse function.

110. CAPSTONE   Use the results of Exercises 107–109 to explain how to graph (a) the inverse cotangent function, (b) the inverse secant function, and (c) the inverse cosecant function on a graphing utility.

Evaluating a Trigonometric Expression   In Exercises 111–114, use the results of Exercises 107–109 to evaluate the expression without using a calculator.

111. $\operatorname{arcsec} \sqrt{2}$ **112.** $\operatorname{arcsec} 1$

113. $\operatorname{arccot}\left(-\sqrt{3}\right)$ **114.** $\operatorname{arccsc} 2$

Proof   In Exercises 115–117, prove the identity.

115. $\arcsin(-x) = -\arcsin x$

116. $\arctan(-x) = -\arctan x$

117. $\arcsin x + \arccos x = \dfrac{\pi}{2}$

∫ **118. Finding the Area of a Plane Region**   In calculus, it is shown that the area of the region bounded by the graphs of $y = 0$, $y = 1/(x^2 + 1)$, $x = a$, and $x = b$ is given by

$$\text{Area} = \arctan b - \arctan a$$

(see figure). Find the area for each value of a and b.

(a) $a = 0, b = 1$

(b) $a = -1, b = 1$

(c) $a = 0, b = 3$

(d) $a = -1, b = 3$

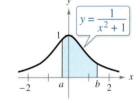

Cumulative Mixed Review

Simplifying a Radical Expression   In Exercises 119–122, simplify the radical expression.

119. $\dfrac{4}{4\sqrt{2}}$ **120.** $\dfrac{2}{\sqrt{3}}$

121. $\dfrac{2\sqrt{3}}{6}$ **122.** $\dfrac{5\sqrt{5}}{2\sqrt{10}}$

Evaluating Trigonometric Functions   In Exercises 123–126, sketch a right triangle corresponding to the trigonometric function of the acute angle θ. Use the Pythagorean Theorem to determine the third side and then find the other five trigonometric functions of θ.

123. $\sin \theta = \dfrac{5}{6}$ **124.** $\tan \theta = 2$

125. $\sin \theta = \dfrac{3}{4}$ **126.** $\sec \theta = 3$

5.7 Applications and Models

Applications Involving Right Triangles

In this section, the three angles of a right triangle are denoted by the letters A, B, and C (where C is the right angle), and the lengths of the sides opposite these angles by the letters a, b, and c (where c is the hypotenuse).

Example 1   Solving a Right Triangle

Solve the right triangle shown in Figure 5.76 for all unknown sides and angles.

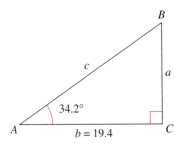

Figure 5.76

Solution

Because $C = 90°$, it follows that

$$A + B = 90°$$

and

$$B = 90° - 34.2° = 55.8°.$$

To solve for a, use the fact that

$$\tan A = \frac{\text{opp}}{\text{adj}} = \frac{a}{b} \implies a = b \tan A.$$

So, $a = 19.4 \tan 34.2° \approx 13.18$. Similarly, to solve for c, use the fact that

$$\cos A = \frac{\text{adj}}{\text{hyp}} = \frac{b}{c} \implies c = \frac{b}{\cos A}.$$

So, $c = \dfrac{19.4}{\cos 34.2°} \approx 23.46$.

✓**CHECKPOINT**   Now try Exercise 5.

Example 2   Finding a Side of a Right Triangle

A safety regulation states that the maximum angle of elevation for a rescue ladder is 72°. A fire department's longest ladder is 110 feet. What is the maximum safe rescue height?

Solution

A sketch is shown in Figure 5.77. From the equation $\sin A = a/c$, it follows that

$$a = c \sin A = 110 \sin 72° \approx 104.62.$$

So, the maximum safe rescue height is about 104.62 feet above the height of the fire truck.

✓**CHECKPOINT**   Now try Exercise 19.

What you should learn

- Solve real-life problems involving right triangles.
- Solve real-life problems involving directional bearings.
- Solve real-life problems involving harmonic motion.

Why you should learn it

You can use trigonometric functions to model and solve real-life problems. For instance, Exercise 24 on page 472 shows you how a trigonometric function can be used to model the length of the shadow of the Sundial Bridge in Redding, California.

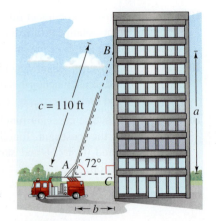

Figure 5.77

Example 3   Finding a Side of a Right Triangle

At a point 200 feet from the base of a building, the angle of elevation to the *bottom* of a smokestack is 35°, and the angle of elevation to the *top* is 53°, as shown in Figure 5.78. Find the height *s* of the smokestack alone.

Solution

This problem involves two right triangles. For the smaller right triangle, use the fact that

$$\tan 35° = \frac{a}{200}$$

to conclude that the height of the building is

$$a = 200 \tan 35°.$$

Now, for the larger right triangle, use the equation

$$\tan 53° = \frac{a + s}{200}$$

to conclude that

$$s = 200 \tan 53° - a.$$

So, the height of the smokestack is

$$s = 200 \tan 53° - a$$

$$= 200 \tan 53° - 200 \tan 35°$$

$$\approx 125.37 \text{ feet.}$$

✔CHECKPOINT   Now try Exercise 25.

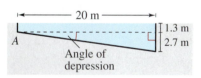

Figure 5.78

Example 4   Finding an Angle of Depression

A swimming pool is 20 meters long and 12 meters wide. The bottom of the pool is slanted such that the water depth is 1.3 meters at the shallow end and 4 meters at the deep end, as shown in Figure 5.79. Find the angle of depression of the bottom of the pool.

Figure 5.79

Solution

Using the tangent function, you see that

$$\tan A = \frac{\text{opp}}{\text{adj}}$$

$$= \frac{2.7}{20}$$

$$= 0.135.$$

So, the angle of depression is

$$A = \arctan 0.135 \approx 0.1342 \text{ radian} \approx 7.69°.$$

✔CHECKPOINT   Now try Exercise 31.

Trigonometry and Bearings

In surveying and navigation, directions are generally given in terms of **bearings.** A bearing measures the acute angle a path or line of sight makes with a fixed north-south line, as shown in Figure 5.80. For instance, the bearing of S 35° E in Figure 5.80(a) means 35 degrees east of south.

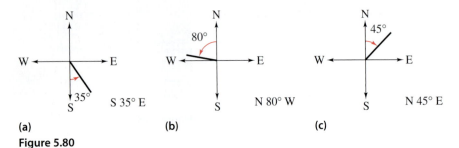

(a)   (b)   (c)

Figure 5.80

Example 5   Finding Directions in Terms of Bearings

A ship leaves port at noon and heads due west at 20 knots, or 20 nautical miles (nm) per hour. At 2 P.M. the ship changes course to N 54° W, as shown in Figure 5.81. Find the ship's bearing and distance from the port of departure at 3 P.M.

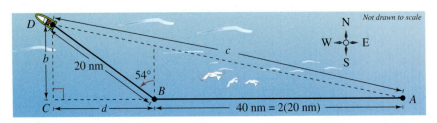

Figure 5.81

Solution

For triangle BCD, you have

$$B = 90° - 54° = 36°.$$

The two sides of this triangle can be determined to be

$$b = 20 \sin 36° \quad \text{and} \quad d = 20 \cos 36°.$$

In triangle ACD, you can find angle A as follows.

$$\tan A = \frac{b}{d + 40} = \frac{20 \sin 36°}{20 \cos 36° + 40} \approx 0.2092494$$

$$A \approx \arctan 0.2092494 \approx 0.2062732 \text{ radian} \approx 11.82°$$

The angle with the north-south line is

$$90° - 11.82° = 78.18°.$$

So, the bearing of the ship is N 78.18° W. Finally, from triangle ACD, you have

$$\sin A = \frac{b}{c}$$

which yields

$$c = \frac{b}{\sin A} = \frac{20 \sin 36°}{\sin 11.82°} \approx 57.39 \text{ nautical miles.} \qquad \text{Distance from port}$$

✓CHECKPOINT   Now try Exercise 37.

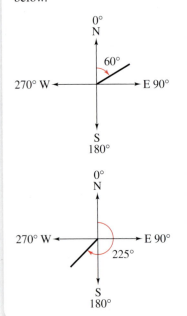

Harmonic Motion

The periodic nature of the trigonometric functions is useful for describing the motion of a point on an object that vibrates, oscillates, rotates, or is moved by wave motion.

For example, consider a ball that is bobbing up and down on the end of a spring, as shown in Figure 5.82. Suppose that 10 centimeters is the maximum distance the ball moves vertically upward or downward from its equilibrium (at-rest) position. Suppose further that the time it takes for the ball to move from its maximum displacement above zero to its maximum displacement below zero and back again is

$t = 4$ seconds.

Assuming the ideal conditions of perfect elasticity and no friction or air resistance, the ball would continue to move up and down in a uniform and regular manner.

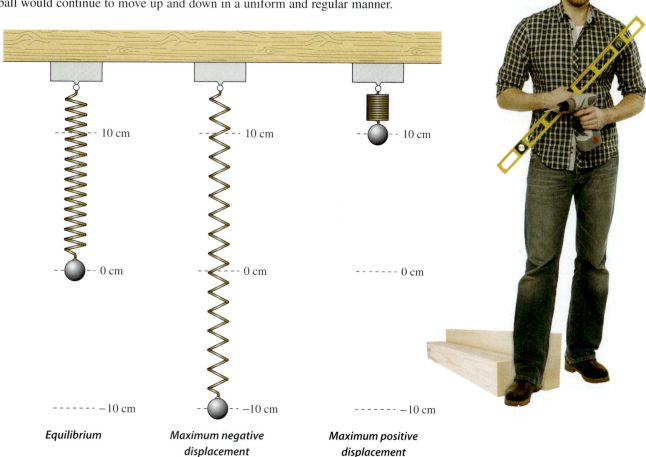

| Equilibrium | Maximum negative displacement | Maximum positive displacement |

Figure 5.82

From this spring you can conclude that the period (time for one complete cycle) of the motion is

Period = 4 seconds

its amplitude (maximum displacement from equilibrium) is

Amplitude = 10 centimeters

and its **frequency** (number of cycles per second) is

Frequency = $\dfrac{1}{4}$ cycle per second.

Motion of this nature can be described by a sine or cosine function, and is called **simple harmonic motion.**

> **Definition of Simple Harmonic Motion**
>
> A point that moves on a coordinate line is said to be in **simple harmonic motion** when its distance d from the origin at time t is given by either
>
> $$d = a \sin \omega t \quad \text{or} \quad d = a \cos \omega t$$
>
> where a and ω are real numbers such that $\omega > 0$. The motion has amplitude $|a|$, period $2\pi/\omega$, and frequency $\omega/(2\pi)$.

Example 6 Simple Harmonic Motion

Write the equation for the simple harmonic motion of the ball illustrated in Figure 5.82, where the period is 4 seconds. What is the frequency of this motion?

Solution

Because the spring is at equilibrium ($d = 0$) when $t = 0$, you use the equation

$$d = a \sin \omega t.$$

Moreover, because the maximum displacement from zero is 10 and the period is 4, you have the following.

$$\text{Amplitude} = |a| = 10$$

$$\text{Period} = \frac{2\pi}{\omega} = 4 \quad \Longrightarrow \quad \omega = \frac{\pi}{2}$$

Consequently, the equation of motion is

$$d = 10 \sin \frac{\pi}{2} t.$$

Note that the choice of $a = 10$ or $a = -10$ depends on whether the ball initially moves up or down. The frequency is

$$\text{Frequency} = \frac{\omega}{2\pi} = \frac{\pi/2}{2\pi} = \frac{1}{4} \text{ cycle per second.}$$

✔**CHECKPOINT**   Now try Exercise 55.

One illustration of the relationship between sine waves and harmonic motion is the wave motion that results when a stone is dropped into a calm pool of water. The waves move outward in roughly the shape of sine (or cosine) waves, as shown in Figure 5.83. As an example, suppose you are fishing and your fishing bob is attached so that it does not move horizontally. As the waves move outward from the dropped stone, your fishing bob will move up and down in simple harmonic motion, as shown in Figure 5.84.

Figure 5.83

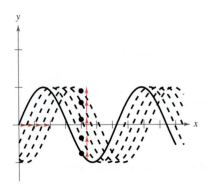

Figure 5.84

Example 7   Simple Harmonic Motion

Given the equation for simple harmonic motion

$$d = 6 \cos \frac{3\pi}{4} t$$

find (a) the maximum displacement, (b) the frequency, (c) the value of d when $t = 4$, and (d) the least positive value of t for which $d = 0$.

Algebraic Solution

The given equation has the form $d = a \cos \omega t$, with

$$a = 6 \quad \text{and} \quad \omega = \frac{3\pi}{4}.$$

a. The maximum displacement (from the point of equilibrium) is given by the amplitude. So, the maximum displacement is 6.

b. Frequency $= \dfrac{\omega}{2\pi}$

$$= \frac{3\pi/4}{2\pi}$$

$$= \frac{3}{8} \text{ cycle per unit of time}$$

c. $d = 6 \cos \left[\dfrac{3\pi}{4}(4) \right]$

$$= 6 \cos 3\pi$$

$$= 6(-1)$$

$$= -6$$

d. To find the least positive value of t for which $d = 0$, solve the equation

$$d = 6 \cos \frac{3\pi}{4} t = 0.$$

First divide each side by 6 to obtain

$$\cos \frac{3\pi}{4} t = 0.$$

You know that $\cos t = 0$ when

$$t = \frac{\pi}{2}, \frac{3\pi}{2}, \frac{5\pi}{2}, \dots \dots$$

Multiply these values by

$$\frac{4}{3\pi}$$

to obtain

$$t = \frac{2}{3}, 2, \frac{10}{3}, \dots \dots$$

So, the least positive value of t is $t = \frac{2}{3}$.

✓**CHECKPOINT**   Now try Exercise 59.

Graphical Solution

Use a graphing utility in radian mode to graph

$$d = 6 \cos \frac{3\pi}{4} t.$$

a.

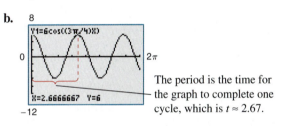

$d = 6 \cos \dfrac{3\pi}{4} t$

Maximum
X=2.6666688 Y=6

The maximum displacement from the point of equilibrium ($d = 0$) is 6.

b.

Y1=6cos((3π/4)X)

X=2.6666667 Y=6

The period is the time for the graph to complete one cycle, which is $t \approx 2.67$.

You can estimate the frequency as follows.

$$\text{Frequency} \approx \frac{1}{2.67} \approx 0.37 \text{ cycle per unit of time.}$$

c.

Y1=6cos((3π/4)X)

X=4   Y=-6

The value of d when $t = 4$ is $d = -6$.

d.

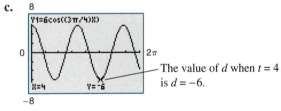

The least positive value of t for which $d = 0$ is $t \approx 0.67$.

Zero
X=.66666667 Y=0

5.7   Exercises

See www.CalcChat.com for worked-out solutions to odd-numbered exercises.
For instructions on how to use a graphing utility, see Appendix A.

Vocabulary and Concept Check

In Exercises 1 and 2, fill in the blank.

1. A point that moves on a coordinate line is said to be in simple _____ if its distance from the origin at time t is given by either $d = a \sin \omega t$ or $d = a \cos \omega t$.

2. A _____ measures the acute angle a path or line of sight makes with a fixed north-south line.

3. Does the bearing of N 20° E mean 20 degrees north of east?

4. What is the amplitude of the simple harmonic motion described by $d = 3 \sin \frac{\pi}{2}t$?

Procedures and Problem Solving

Solving a Right Triangle   In Exercises 5–14, solve the right triangle shown in the figure.

✓ **5.** $A = 30°$, $b = 10$ **6.** $B = 60°$, $c = 15$
7. $B = 71°$, $b = 14$ **8.** $A = 7.4°$, $a = 20.5$
9. $a = 6$, $b = 12$ **10.** $a = 25$, $c = 45$
11. $b = 16$, $c = 54$ **12.** $b = 1.32$, $c = 18.9$
13. $A = 12° 15'$, $c = 430.5$ **14.** $B = 65° 12'$, $a = 145.5$

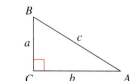

Figure for 5–14

Figure for 15–18

Finding an Altitude   In Exercises 15–18, find the altitude of the isosceles triangle shown in the figure.

15. $\theta = 52°$, $b = 8$ inches
16. $\theta = 18°$, $b = 12$ meters
17. $\theta = 41.6°$, $b = 18.5$ feet
18. $\theta = 72.94°$, $b = 3.26$ centimeters

✓ **19. Home Maintenance**   A ladder that is 20 feet long leans against the side of a house. The angle of elevation of the ladder is 80°. Find the height from the top of the ladder to the ground.

20. Electrical Maintenance   An electrician is running wire from the electric box on a house to a utility pole 75 feet away. The angle of elevation to the connection on the pole is 16°. How much wire does the electrician need?

21. ROTC   A cadet rappelling down a cliff on a rope needs help. A cadet on the ground pulls tight on the end of the rope that hangs down from the rappelling cadet to lock the cadet in place. The length of the rope between the two cadets is 120 feet, and the angle of elevation of the rope is 66°. How high above the ground is the cadet on the rope?

22. Home Maintenance   Snow is thrown from a snow blower with a 66° angle of elevation from a chute that is 2 feet off the ground. Is it possible for the snow to clear an 8-foot high fence that is 4 feet from the snow blower? Explain.

23. Angle of Depression   The sonar of a navy cruiser detects a submarine that is 4000 feet from the cruiser. The angle between the water level and the submarine is 31.5°. How deep is the submarine?

31.5°

4000 ft

Not drawn to scale

24. *Why you should learn it*   (p. 466)   The Sundial Bridge in Redding, California is supported by cables attached to a 217-foot sundial that leans backward at a 42° angle. (See figure.)

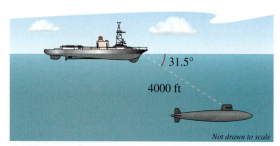

217 ft

h

42°

d

(a) Determine the height h of the sundial.

(b) The distance d shows how far the shadow extends horizontally from the tip of the sundial. Draw the right triangle formed by d and h. Label the angle of elevation of the sun as θ.

(c) Write d as a function of θ.

(d) Use a graphing utility to complete the table.

θ	10°	20°	30°	40°	50°
d					

(e) The angle measure increases in equal increments in the table. Does the distance also increase in equal increments? Explain.

✓ **25. Architecture** From a point 50 feet in front of a church, the angles of elevation to the base of the steeple and the top of the steeple are 35° and 47° 40′, respectively.

(a) Draw right triangles that give a visual representation of the problem. Label the known and unknown quantities.

(b) Use a trigonometric function to write an equation involving the unknown quantity.

(c) Find the height of the steeple.

26. Finding a Height From a point 100 feet in front of a public library, the angles of elevation to the base of the flagpole and the top of the flagpole are 28° and 39° 45′, respectively. (See figure.) Find the height of the flagpole.

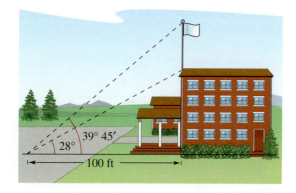

27. Finding a Height Before a parade you are holding one of the tethers attached to the top of a character balloon that will be in the parade. The balloon is upright and the bottom is floating approximately 20 feet above ground level. You are standing approximately 100 feet from the balloon (see figure).

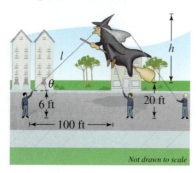

Not drawn to scale

(a) Find an expression for the length l of the tether in terms of h, the height of the balloon from top to bottom.

(b) Find an expression for the angle of elevation θ from you to the top of the balloon.

(c) The angle of elevation to the top of the balloon is 35°. Find the height h.

28. Parks and Recreation The designers of a park are creating a water slide and have sketched a preliminary drawing (see figure). The length of the stairs is 30 feet, and its angle of elevation is 45°.

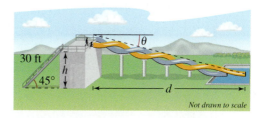

Not drawn to scale

(a) Find the height h of the slide.

(b) Find the angle of depression θ from the top of the slide to the end of the slide at the ground in terms of the horizontal distance d a rider travels.

(c) The designers want the angle of depression of the slide to be at least 25° and at most 30°. Find an interval for how far a rider travels horizontally.

29. Architecture The front of an A-frame cottage has the shape of an isosceles triangle. It stands 28 feet high and is 20 feet wide at its base. What is the angle of elevation of its roof?

30. Parks and Recreation The height of an outdoor basketball backboard is $12\frac{1}{2}$ feet, and the backboard casts a shadow $17\frac{1}{3}$ feet long.

(a) Draw a right triangle that gives a visual representation of the problem. Label the known and unknown quantities.

(b) Use a trigonometric function to write an equation involving the unknown angle of elevation.

(c) Find the angle of elevation of the sun.

✓ **31. Communications** A Global Positioning System satellite orbits 12,500 miles above Earth's surface (see figure). Find the angle of depression from the satellite to the horizon. Assume the radius of Earth is 4000 miles.

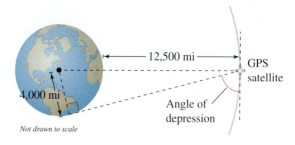

Not drawn to scale

32. Marine Transportation Find the angle of depression from the top of a lighthouse 250 feet above water level to the water line of a ship $2\frac{1}{2}$ miles offshore.

33. Aviation When an airplane leaves the runway, its angle of climb is 18° and its speed is 275 feet per second. Find the plane's altitude after 1 minute.

34. Aviation How long will it take the plane in Exercise 33 to climb to an altitude of 10,000 feet? 16,000 feet?

35. Topography A sign on the roadway at the top of a mountain indicates that for the next 4 miles the grade is 9.5° (see figure). Find the change in elevation for a car descending the 4-mile stretch.

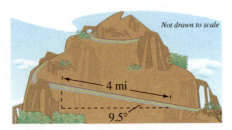

36. Topography A ski slope on a mountain has an angle of elevation of 25.2°. The vertical height of the slope is 1808 feet. How long is the slope?

✓ **37. Marine Transportation** A ship leaves port at noon and has a bearing of S 29° W. The ship sails at 20 knots. How many nautical miles south and how many nautical miles west does the ship travel by 6:00 P.M.?

38. Aviation An airplane flying at 600 miles per hour has a bearing of 52°. After flying for 1.5 hours, how far north and how far east has the plane traveled from its point of departure?

39. Geography A surveyor wants to find the distance across a pond (see figure). The bearing from A to B is N 32° W. The surveyor walks 50 meters from A to C, and at C the bearing to B is N 68° W.

(a) Find the bearing from A to C.

(b) Find the distance from A to B.

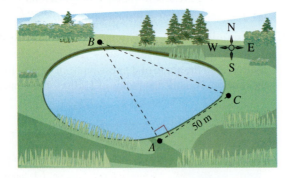

40. Forestry Fire tower A is 30 kilometers due west of fire tower B. A fire is spotted from the towers, and the bearings to the fire from A and B are E 14° N and W 34° N, respectively (see figure). Find the distance d of the fire from the line segment AB.

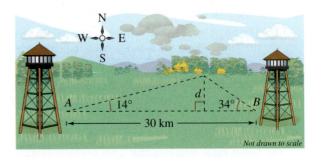

Not drawn to scale

41. Marine Transportation A ship is 45 miles east and 30 miles south of port. The captain wants to sail directly to port. What bearing should the captain take?

42. Aviation A plane is 160 miles north and 85 miles east of an airport. The pilot wants to fly directly to the airport. What bearing should the pilot take?

43. Marine Transportation An observer in a lighthouse 350 feet above sea level observes two ships in the same vertical plane as the lighthouse. The angles of depression to the ships are 4° and 6.5° (see figure). How far apart are the ships?

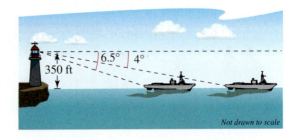

Not drawn to scale

44. Aviation A passenger in an airplane flying at an altitude of 10 kilometers sees two towns due east of the plane. The angles of depression to the towns are 28° and 55° (see figure). How far apart are the towns?

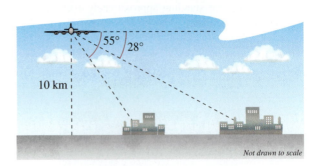

Not drawn to scale

45. Aviation The angle of elevation to a plane approaching your home is 16°. One minute later, it is 57°. You assume that the speed of the plane is 550 miles per hour. Approximate the altitude of the plane.

46. Topography While traveling across flat land, you notice a mountain directly in front of you. The angle of elevation to the peak is 2.5°. After you drive 18 miles closer to the mountain, the angle of elevation is 10°. Approximate the height of the mountain.

47. Cinema A park is showing a movie on the lawn. The base of the screen is 6 feet off the ground and the screen is 22 feet high (see figure).

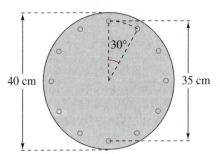

(a) Find the angles of elevation to the top of the screen from distances of 15 feet and 100 feet.

(b) You are lying on the ground and the angle of elevation to the top of the screen is 42°. How far are you from the screen?

48. Geometry You want to move a set of box springs through two hallways that meet at right angles. Each hallway has a width of 3 feet (see figure).

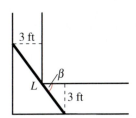

(a) Show that the length L for a given angle β can be written as $L(\beta) = 3 \csc \beta + 3 \sec \beta$.

(b) Graph the function in part (a) for the interval

$$0 < \beta < \frac{\pi}{2}.$$

(c) For what value(s) of β is the value of L the least? Interpret the meaning of your answer in the context of the problem.

Geometry **In Exercises 49 and 50, find the angle α between the two nonvertical lines L_1 and L_2 (assume L_1 and L_2 are not perpendicular). The angle α satisfies the equation**

$$\tan \alpha = \left| \frac{m_2 - m_1}{1 + m_2 m_1} \right|$$

where m_1 and m_2 are the slopes of L_1 and L_2, respectively.

49. L_1: $3x - 2y = 5$

 L_2: $x + y = 1$

50. L_1: $2x + y = 8$

 L_2: $x - 5y = -4$

51. Geometry Determine the angle between the diagonal of a cube and the diagonal of its base, as shown in the figure.

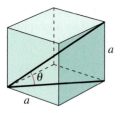 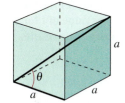

Figure for 51 Figure for 52

52. Geometry Determine the angle between the diagonal of a cube and its edge, as shown in the figure.

53. Mechanical Engineering Write the distance y from one side of a hexagonal nut to the opposite side as a function of r, as shown in the figure.

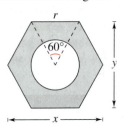

54. Mechanical Engineering The figure shows a circular piece of sheet metal of diameter 40 centimeters. The piece contains 12 equally spaced bolt holes. Determine the straight-line distance between the centers of two consecutive bolt holes.

Harmonic Motion In Exercises 55–58, find a model for simple harmonic motion satisfying the specified conditions.

	Displacement ($t = 0$)	Amplitude	Period
✓ **55.**	0	8 centimeters	2 seconds
56.	0	3 meters	6 seconds
57.	3 inches	3 inches	1.5 seconds
58.	2 feet	2 feet	10 seconds

Harmonic Motion In Exercises 59–62, for the simple harmonic motion described by the trigonometric function, find (a) the maximum displacement, (b) the frequency, (c) the value of d when $t = 5$, and (d) the least positive value of t for which $d = 0$. Use a graphing utility to verify your results.

✓ **59.** $d = 4 \cos 8\pi t$ **60.** $d = \frac{1}{2} \cos 20\pi t$

61. $d = \frac{1}{16} \sin 140\pi t$ **62.** $d = \frac{1}{64} \sin 792\pi t$

63. Music A point on the end of a tuning fork moves in the simple harmonic motion described by $d = a \sin \omega t$. A tuning fork for middle C has a frequency of 264 vibrations per second. Find ω.

64. Harmonic Motion A buoy oscillates in simple harmonic motion as waves go past. The buoy moves a total of 3.5 feet from its high point to its low point (see figure), and it returns to its high point every 10 seconds. Write an equation that describes the motion of the buoy, where the high point corresponds to the time $t = 0$.

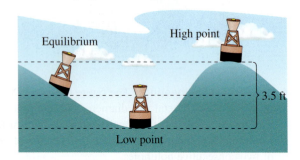

65. Harmonic Motion A ball that is bobbing up and down on the end of a spring has a maximum displacement of 3 inches. Its motion (in ideal conditions) is modeled by

$$y = \frac{1}{4} \cos 16t, \quad t > 0$$

where y is measured in feet and t is the time in seconds.

(a) Use a graphing utility to graph the function.

(b) What is the period of the oscillations?

(c) Determine the first time the ball passes the point of equilibrium ($y = 0$).

66. Home Maintenance You are washing your house. Use the following steps to find the shortest ladder that will reach over the greenhouse to the side of the house (see figure).

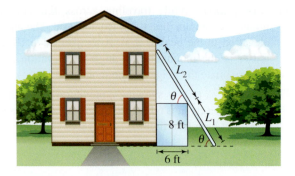

(a) Complete four rows of the table.

θ	L_1	L_2	$L_1 + L_2$
0.1	$\dfrac{8}{\sin 0.1}$	$\dfrac{6}{\cos 0.1}$	86.16
0.2	$\dfrac{8}{\sin 0.2}$	$\dfrac{6}{\cos 0.2}$	46.39

(b) Use the *table* feature of a graphing utility to generate additional rows of the table. Use the table to estimate the minimum length of the ladder.

(c) Write the length $L_1 + L_2$ as a function of θ.

(d) Use the graphing utility to graph the function. Use the graph to estimate the minimum length. How does your estimate compare with that in part (b)?

67. Irrigation Engineering The cross sections of an irrigation canal are isosceles trapezoids, where the lengths of three of the sides are 8 feet (see figure). The objective is to find the angle θ that maximizes the area of the cross sections. [*Hint:* The area of a trapezoid is given by $(h/2)(b_1 + b_2)$.]

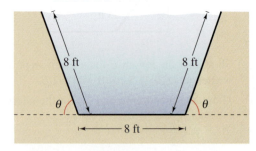

(a) Complete seven rows of the table.

Base 1	Base 2	Altitude	Area
8	$8 + 16 \cos 10°$	$8 \sin 10°$	22.06
8	$8 + 16 \cos 20°$	$8 \sin 20°$	42.46

(b) Use the *table* feature of a graphing utility to generate additional rows of the table. Use the table to estimate the maximum cross-sectional area.

(c) Write the area A as a function of θ.

(d) Use the graphing utility to graph the function. Use the graph to estimate the maximum cross-sectional area. How does your estimate compare with that in part (b)?

68. MODELING DATA

The table shows the average sales S (in millions of dollars) of an outerwear manufacturer for each month t, where $t = 1$ represents January.

Month, t	Sales, S
1	6.73
2	5.58
3	4.00
4	2.43
5	1.27
6	0.85
7	1.27
8	2.43
9	4.00
10	5.58
11	6.73
12	7.15

(a) Create a scatter plot of the data.

(b) Find a trigonometric model that fits the data. Graph the model on your scatter plot. How well does the model fit the data?

(c) What is the period of the model? Do you think it is reasonable given the context? Explain your reasoning.

(d) Interpret the meaning of the model's amplitude in the context of the problem.

69. **Climatology** The numbers of hours H of daylight in Denver, Colorado on the 15th of each month are: 1(9.67), 2(10.72), 3(11.92), 4(13.25), 5(14.37), 6(14.97), 7(14.72), 8(13.77), 9(12.48), 10(11.18), 11(10.00), 12(9.38). The month is represented by t, with $t = 1$ corresponding to January. A model for the data is given by

$$H(t) = 12.13 + 2.77 \sin[(\pi t/6) - 1.60].$$

(a) Use a graphing utility to graph the data points and the model in the same viewing window.

(b) What is the period of the model? Is it what you expected? Explain.

(c) What is the amplitude of the model? What does it represent in the context of the problem? Explain.

Dimiter Petrov 2010/used under license from Shutterstock.com

70. **Writing** Find two bearings perpendicular to N 32° E and explain how you found them.

Conclusions

True or False? In Exercises 71–74, determine whether the statement is true or false. Justify your answer.

71. Simple harmonic motion does not involve a damping factor.

72. The tangent function can be used to model harmonic motion.

73. In a function representing simple harmonic motion, the amplitude is equal to the period divided by 2π.

74. An example of a bearing used in aviation is S 25° W.

75. **Think About It** Draw a right triangle with one leg longer than the other and label its sides and angles. Describe the different combinations of known side lengths and angle measures that are sufficient to solve the right triangle.

76. **CAPSTONE** You stand near a monument with a laser measuring tool that you can aim at any part of the monument to find a distance d. Describe a procedure you can use to find the height of the monument without moving from your position.

Not drawn to scale

Cumulative Mixed Review

Equation of a Line in Standard Form In Exercises 77–80, write the standard form of the equation of the line that has the specified characteristics.

77. $m = 4$, passes through $(-1, 2)$

78. $m = -\frac{1}{2}$, passes through $(\frac{1}{3}, 0)$

79. Passes through $(-2, 6)$ and $(3, 2)$

80. Passes through $(\frac{1}{4}, -\frac{2}{3})$ and $(-\frac{1}{2}, \frac{1}{3})$

Finding the Domain of a Function In Exercises 81–84, find the domain of the function.

81. $f(x) = 3x + 8$

82. $f(x) = -x^2 - 1$

83. $g(x) = \sqrt[3]{x + 2}$

84. $g(x) = \sqrt{7 - x}$

5   Review Exercises

See www.CalcChat.com for worked-out solutions to odd-numbered exercises.
For instructions on how to use a graphing utility, see Appendix A.

5.1

Estimating an Angle   In Exercises 1 and 2, estimate the number of degrees in the angle.

1.

2.

Using Degree Measure   In Exercises 3–6, (a) sketch the angle in standard position, (b) determine the quadrant in which the angle lies, and (c) list one positive and one negative coterminal angle.

3. $45°$
4. $210°$
5. $-135°$
6. $-405°$

Complementary and Supplementary Angles   In Exercises 7–10, find (if possible) the complement and supplement of the angle.

7. $5°$
8. $84°$
9. $157°$
10. $108°$

Converting to Decimal Degree Form   In Exercises 11–14, use the angle-conversion capabilities of a graphing utility to convert the angle measure to decimal degree form. Round your answer to three decimal places.

11. $135°\,16'\,45''$
12. $-234°\,40''$
13. $6°\,34'\,19''$
14. $242°\,24'\,9''$

Converting to D°M′S″ Form   In Exercises 15–18, use the angle-conversion capabilities of a graphing utility to convert the angle measure to D°M′S″ form.

15. $135.29°$
16. $25.8°$
17. $-85.36°$
18. $-327.93°$

Using Radian Measure   In Exercises 19–22, (a) sketch the angle in standard position, (b) determine the quadrant in which the angle lies, and (c) list one positive and one negative coterminal angle.

19. $\dfrac{4\pi}{3}$
20. $\dfrac{11\pi}{6}$
21. $-\dfrac{5\pi}{6}$
22. $-\dfrac{7\pi}{4}$

Complementary and Supplementary Angles   In Exercises 23–26, find (if possible) the complement and supplement of the angle.

23. $\dfrac{\pi}{8}$
24. $\dfrac{\pi}{12}$
25. $\dfrac{3\pi}{10}$
26. $\dfrac{2\pi}{21}$

Converting from Degrees to Radians   In Exercises 27–30, convert the angle measure from degrees to radians. Round your answer to three decimal places.

27. $94°$
28. $-72°$
29. $415°$
30. $-355°$

Converting from Radians to Degrees   In Exercises 31–34, convert the angle measure from radians to degrees. Round your answer to three decimal places.

31. $\dfrac{5\pi}{7}$
32. $-\dfrac{3\pi}{5}$
33. -3.5
34. 1.55

35. **Geometry**   Find the radian measure of the central angle of a circle with a radius of 12 feet that intercepts an arc of length 25 feet.

36. **Geometry**   Find the radian measure of the central angle of a circle with a radius of 60 inches that intercepts an arc of length 245 inches.

37. **Geometry**   Find the length of the arc on a circle with a radius of 20 meters intercepted by a central angle of $138°$.

38. **Geometry**   Find the length of the arc on a circle with a radius of 15 centimeters intercepted by a central angle of $60°$.

39. **Finding Linear Speed**   The radius of a compact disc is 6 centimeters. Find the linear speed of a point on the circumference of the disc if it is rotating at a speed of 500 revolutions per minute.

40. **Finding Angular Speed**   A car is moving at a rate of 28 miles per hour, and the diameter of its wheels is about $2\frac{1}{3}$ feet.

(a) Find the number of revolutions per minute the wheels are rotating.

(b) Find the angular speed of the wheels in radians per minute.

5.2

Evaluating Trigonometric Functions   In Exercises 41–48, find the exact values of the six trigonometric functions of the angle θ.

41.

42.

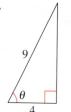

43.

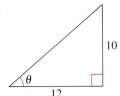

12

44.

10

2

θ

45. $\sin \theta = \frac{7}{24}$

46. $\cos \theta = \frac{2}{3}$

47. $\tan \theta = \frac{1}{4}$

48. $\cot \theta = \frac{9}{40}$

Using a Calculator In Exercises 49–52, use a calculator to evaluate each function. Round your answers to four decimal places.

49. (a) $\cos 84°$ (b) $\sin 6°$

50. (a) $\csc 52° \, 12'$ (b) $\sec 54° \, 7'$

51. (a) $\cos \dfrac{\pi}{4}$ (b) $\sec \dfrac{\pi}{4}$

52. (a) $\tan \dfrac{3\pi}{20}$ (b) $\cot \dfrac{3\pi}{20}$

Using Trigonometric Identities In Exercises 53 and 54, use trigonometric identities to transform one side of the equation into the other.

53. $\csc \theta \tan \theta = \sec \theta$ **54.** $\dfrac{\cot \theta + \tan \theta}{\cot \theta} = \sec^2 \theta$

55. Surveying A surveyor is trying to determine the width of a river (see figure). From point P, the engineer walks downstream 125 feet and sights to point Q. From this sighting, it is determined that $\theta = 62°$. How wide is the river?

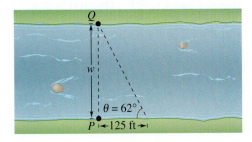

56. Architecture An escalator 152 feet in length rises to a platform and makes a 30° angle with the ground.

(a) Draw a right triangle that gives a visual representation of the problem. Show the known quantities and use a variable to indicate the height of the platform above the ground.

(b) Use a trigonometric function to write an equation involving the unknown quantity.

(c) Find the height of the platform above the ground.

5.3

Evaluating Trigonometric Functions In Exercises 57–62, the point is on the terminal side of an angle in standard position. Determine the exact values of the six trigonometric functions of the angle.

57. $(12, 16)$ **58.** $(2, 10)$

59. $(-7, 2)$ **60.** $(3, -4)$

61. $\left(\frac{2}{3}, \frac{5}{8}\right)$ **62.** $\left(-\frac{10}{3}, -\frac{2}{3}\right)$

Evaluating Trigonometric Functions In Exercises 63–66, find the values of the other five trigonometric functions of θ satisfying the given conditions.

63. $\sec \theta = \frac{6}{5}, \quad \tan \theta < 0$

64. $\tan \theta = -\frac{12}{5}, \quad \sin \theta > 0$

65. $\sin \theta = \frac{3}{8}, \quad \cos \theta < 0$

66. $\cos \theta = -\frac{2}{5}, \quad \sin \theta > 0$

Finding a Reference Angle In Exercises 67–74, find the reference angle θ'. Sketch θ in standard position and label θ'.

67. $\theta = 330°$

68. $\theta = -240°$

69. $\theta = \dfrac{5\pi}{4}$

70. $\theta = -\dfrac{9\pi}{4}$

71. $\theta = 264°$

72. $\theta = 635°$

73. $\theta = -\dfrac{6\pi}{5}$

74. $\theta = \dfrac{17\pi}{3}$

Trigonometric Functions of a Nonacute Angle In Exercises 75–82, evaluate the sine, cosine, and tangent of the angle without using a calculator.

75. $240°$

76. $315°$

77. $-210°$

78. $-315°$

79. $-\dfrac{9\pi}{4}$

80. $\dfrac{11\pi}{6}$

81. 4π

82. $\dfrac{7\pi}{3}$

Using a Calculator   In Exercises 83–86, use a calculator to evaluate the trigonometric function. Round your answer to four decimal places.

83. $\tan 33°$

84. $\csc 105°$

85. $\sec \dfrac{12\pi}{5}$

86. $\sin\left(-\dfrac{\pi}{9}\right)$

5.4

Sketching Graphs   In Exercises 87–90, sketch the graph of the function.

87. $f(x) = 3 \sin x$

88. $f(x) = 2 \cos x$

89. $f(x) = \frac{1}{4} \cos x$

90. $f(x) = \frac{7}{2} \sin x$

Finding the Period and Amplitude   In Exercises 91–94, find the period and amplitude.

91.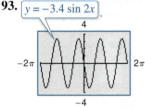

92.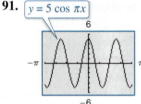

93.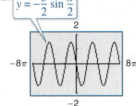

94.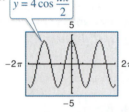

Graphing Sine and Cosine Functions   In Exercises 95–106, sketch the graph of the function. (Include two full periods.)

95. $f(x) = 3 \cos 2\pi x$

96. $f(x) = -2 \sin \pi x$

97. $f(x) = 5 \sin \dfrac{2x}{5}$

98. $f(x) = 8 \cos\left(-\dfrac{x}{4}\right)$

99. $f(x) = -\dfrac{5}{2} \cos \dfrac{x}{4}$

100. $f(x) = -\dfrac{1}{2} \sin \dfrac{\pi x}{4}$

101. $f(x) = \frac{5}{2} \sin(x - \pi)$

102. $f(x) = 3 \cos(x + \pi)$

103. $f(x) = 2 - \cos \dfrac{\pi x}{2}$

104. $f(x) = \frac{1}{2} \sin \pi x - 3$

105. $f(x) = -3 \cos\left(\dfrac{x}{2} - \dfrac{\pi}{4}\right)$

106. $f(x) = 4 - 2 \cos(4x + \pi)$

Finding an Equation of a Graph   In Exercises 107–110, find a, b, and c for the function $f(x) = a \cos(bx - c)$ such that the graph of f matches the graph shown.

107.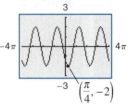

108.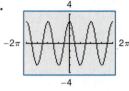

109.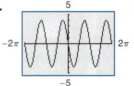

110.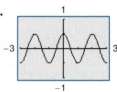

Economics   In Exercises 111 and 112, use a graphing utility to graph the sales function over 1 year, where S is the sales (in thousands of units) of a seasonal product, and t is the time (in months), with $t = 1$ corresponding to January. Determine the months of maximum and minimum sales.

111. $S = 48.4 - 6.1 \cos \dfrac{\pi t}{6}$

112. $S = 56.25 + 9.50 \sin \dfrac{\pi t}{6}$

5.5

 Library of Parent Functions   In Exercises 113–126, sketch the graph of the function. (Include two full periods.)

113. $f(x) = -\tan \dfrac{\pi x}{4}$

114. $f(x) = 4 \tan \pi x$

115. $f(x) = \dfrac{1}{4} \tan\left(x - \dfrac{\pi}{2}\right)$

116. $f(x) = 2 + 2 \tan \dfrac{x}{3}$

117. $f(x) = 3 \cot \dfrac{x}{2}$

118. $f(x) = \dfrac{1}{2} \cot \dfrac{\pi x}{2}$

119. $f(x) = \dfrac{1}{2} \cot\left(x - \dfrac{\pi}{2}\right)$

120. $f(x) = 4 \cot\left(x + \dfrac{\pi}{4}\right)$

121. $f(x) = \frac{1}{4} \sec x$

122. $f(x) = \frac{1}{2} \csc x$

123. $f(x) = \frac{1}{4} \csc 2x$

124. $f(x) = \frac{1}{2} \sec 2\pi x$

125. $f(x) = \sec\left(x - \dfrac{\pi}{4}\right)$

126. $f(x) = \dfrac{1}{2} \csc(2x + \pi)$

Comparing Trigonometric Graphs In Exercises 127–134, use a graphing utility to graph the function. (Include two full periods.) Graph the corresponding reciprocal function in the same viewing window. Compare the graphs.

127. $f(x) = \dfrac{1}{4} \tan \dfrac{\pi x}{2}$ **128.** $f(x) = \tan\left(x + \dfrac{\pi}{4}\right)$

129. $f(x) = 4 \cot(2x - \pi)$ **130.** $f(x) = -2 \cot(4x + \pi)$

131. $f(x) = 2 \sec(x - \pi)$ **132.** $f(x) = -2 \csc(x - \pi)$

133. $f(x) = \csc\left(3x - \dfrac{\pi}{2}\right)$

134. $f(x) = 3 \csc\left(2x + \dfrac{\pi}{4}\right)$

Analyzing a Damped Trigonometric Graph In Exercises 135–138, use a graphing utility to graph the function and the damping factor of the function in the same viewing window. Then, analyze the graph of the function using the method in Example 6 in Section 5.5.

135. $f(x) = e^x \sin 2x$ **136.** $f(x) = e^x \cos x$

137. $f(x) = 2x \cos x$ **138.** $f(x) = x \sin \pi x$

5.6

Finding the Exact Value of a Trigonometric Expression In Exercises 139–142, find the exact value of each expression without using a calculator.

139. (a) $\arcsin(-1)$ (b) $\arcsin 0$

140. (a) $\arcsin\left(-\dfrac{1}{2}\right)$ (b) $\arcsin\left(-\dfrac{\sqrt{3}}{2}\right)$

141. (a) $\cos^{-1}\dfrac{\sqrt{2}}{2}$ (b) $\cos^{-1}\left(-\dfrac{\sqrt{3}}{2}\right)$

142. (a) $\tan^{-1}\left(-\sqrt{3}\right)$ (b) $\tan^{-1} 1$

Calculators and Inverse Trigonometric Functions In Exercises 143–150, use a calculator to approximate the value of the expression. Round your answer to the nearest hundredth.

143. $\arccos 0.42$ **144.** $\arcsin 0.63$

145. $\sin^{-1}(-0.94)$ **146.** $\cos^{-1}(-0.12)$

147. $\arctan(-12)$ **148.** $\arctan 21$

149. $\tan^{-1} 0.81$ **150.** $\tan^{-1} 6.4$

Using an Inverse Trigonometric Function In Exercises 151 and 152, use an inverse trigonometric function to write θ as a function of x.

151.

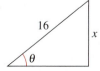

152.

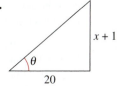

Writing an Expression In Exercises 153–156, write an algebraic expression that is equivalent to the expression.

153. $\sec[\arcsin(x - 1)]$

154. $\tan\left(\arccos \dfrac{x}{2}\right)$

155. $\sin\left(\arccos \dfrac{x^2}{4 - x^2}\right)$

156. $\csc(\arcsin 10x)$

5.7

157. Angle of Elevation The height of a radio transmission tower is 70 meters, and it casts a shadow of length 45 meters (see figure). Find the angle of elevation of the sun.

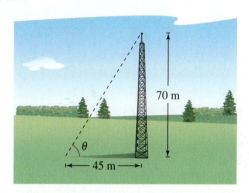

158. Finding a Height An observer 2.5 miles from the launch pad of a space shuttle launch measures the angle of elevation to the base of the shuttle to be 25° soon after lift-off (see figure). How high is the shuttle at that instant? (Assume the shuttle is still moving vertically.)

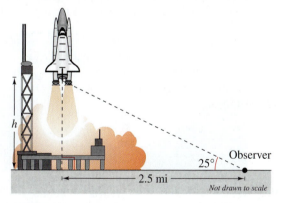

159. Topography A train travels 3.5 kilometers on a straight track with a grade of 1° 10′ (see figure). What is the vertical rise of the train in that distance?

160. Topography A road sign at the top of a mountain indicates that for the next 4 miles the grade is 12%. Find the angle of the grade and the change in elevation for a car descending the 4-mile stretch.

161. Aviation A passenger in an airplane flying at an altitude of 37,000 feet sees two towns due west of the airplane. The angles of depression to the towns are 32° and 76° (see figure). How far apart are the towns?

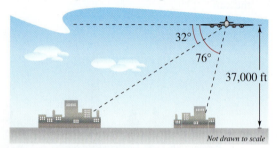

Not drawn to scale

162. Aviation From city *A* to city *B*, a plane flies 650 miles at a bearing of 48°. From city *B* to city *C*, the plane flies 810 miles at a bearing of 115°. Find the distance and bearing from *A* to *C*.

163. Harmonic Motion A buoy oscillates in simple harmonic motion as waves go past. The buoy moves a total of 6 feet from its high point to its low point, returning to its high point every 15 seconds (see figure). Write an equation that describes the motion of the buoy, where the high point corresponds to the time $t = 0$.

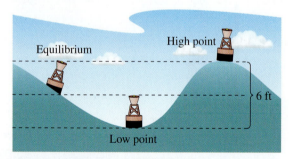

164. Harmonic Motion Your fishing bobber is oscillating in simple harmonic motion caused by waves. Your bobber moves a total of 7 inches from its high point to its low point and returns to its high point every 2 seconds. Write an equation modeling the motion of your bobber, where the high point corresponds to the time $t = 0$.

Conclusions

True or False? In Exercises 165–168, determine whether the statement is true or false. Justify your answer.

165. $y = \sin \theta$ is not a function because $\sin 30° = \sin 150°$.

166. The equation $y = \cos x$ does not have an inverse function on the interval $-\dfrac{\pi}{2} \le x \le \dfrac{\pi}{2}$.

167. You can use the cotangent function to model simple harmonic motion.

168. The sine of any nonacute angle θ is equal to the sine of the reference angle for θ.

169. Numerical Analysis A 3000-pound automobile is negotiating a circular interchange of radius 300 feet at a speed of *s* miles per hour (see figure). The relationship between the speed and the angle θ (in degrees) at which the roadway should be banked so that no lateral frictional force is exerted on the tires is $\tan \theta = 0.672s^2/3000$.

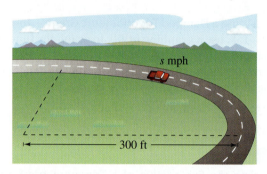

(a) Use a graphing utility to complete the table.

s	10	20	30	40	50	60
θ						

(b) In the table, *s* is incremented by 10, but θ does not increase by equal increments. Explain.

∫ **170. Exploration** Using calculus, it can be shown that the secant function can be approximated by the polynomial

$$\sec x \approx 1 + \frac{x^2}{2!} + \frac{5x^4}{4!}$$

where *x* is in radians. Use a graphing utility to graph the secant function and its polynomial approximation in the same viewing window. How do the graphs compare?

∫ **171. Exploration** Using calculus, it can be shown that the arctangent function can be approximated by the polynomial

$$\arctan x \approx x - \frac{x^3}{3} + \frac{x^5}{5} - \frac{x^7}{7}$$

where *x* is in radians.

(a) Use a graphing utility to graph the arctangent function and its polynomial approximation in the same viewing window. How do the graphs compare?

(b) Study the pattern in the polynomial approximation of the arctangent function and guess the next term. Then repeat part (a). How does the accuracy of the approximation change when an additional term is added?

**Take this test as you would take a test in class. After you are finished, check your
work against the answers given in the back of the book.**

1. Consider an angle that measures $\dfrac{5\pi}{4}$ radians.

   (a) Sketch the angle in standard position.

   (b) Determine two coterminal angles (one positive and one negative).

   (c) Convert the angle to degree measure.

2. A truck is moving at a rate of 100 kilometers per hour, and the diameter of its
   wheels is 1.25 meters. Find the angular speed of the wheels in radians per second.

3. Find the exact values of the six trigonometric functions of the angle θ shown in the
   figure.

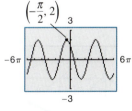

Figure for 3

4. Given that $\tan \theta = \frac{7}{2}$ and θ is an acute angle, find the other five trigonometric
   functions of θ.

5. Determine the reference angle θ' of the angle $\theta = 255°$. Sketch θ in standard
   position and label θ'.

6. Determine the quadrant in which θ lies when $\sec \theta < 0$ and $\tan \theta > 0$.

7. Find two exact values of θ in degrees $(0 \le \theta < 360°)$ when $\cos \theta = -\sqrt{2}/2$.

8. Use a calculator to approximate two values of θ in radians $(0 \le \theta < 2\pi)$ for which
   $\csc \theta = 1.030$. Round your answer to two decimal places.

9. Find the five remaining trigonometric functions of θ, given that $\cos \theta = -\frac{3}{5}$ and
   $\sin \theta > 0$.

In Exercises 10–15, sketch the graph of the function. (Include two full periods.)

10. $g(x) = -2 \sin\left(x - \dfrac{\pi}{4}\right)$

11. $f(x) = \dfrac{1}{2} \tan 4x$

12. $f(x) = \frac{1}{2} \sec(x - \pi)$

13. $f(x) = 2 \cos(\pi - 2x) + 3$

14. $f(x) = 2 \csc\left(x + \dfrac{\pi}{2}\right)$

15. $f(x) = 2 \cot\left(x - \dfrac{\pi}{2}\right)$

**In Exercises 16 and 17, use a graphing utility to graph the function. If the function
is periodic, find its period.**

16. $y = \sin 2\pi x + 2 \cos \pi x$

17. $y = 6e^{-0.12t} \cos(0.25t), \quad 0 \le t \le 32$

18. Find a, b, and c for the function $f(x) = a \sin(bx + c)$ such that the graph of f
    matches the graph at the right.

Figure for 18

19. Find the exact value of $\tan\left(\arccos \frac{2}{3}\right)$ without using a calculator.

In Exercises 20–22, use a graphing utility to graph the function.

20. $f(x) = 2 \arcsin\left(\dfrac{1}{2}x\right)$

21. $f(x) = 2 \arccos x$

22. $f(x) = \arctan \dfrac{x}{2}$

23. A plane is 64 miles north and 80 miles east of an airport. What bearing should the
    pilot take to fly directly to the airport?

Proofs in Mathematics

The Pythagorean Theorem

The Pythagorean Theorem is one of the most famous theorems in mathematics. More than 100 different proofs now exist. James A. Garfield, the twentieth president of the United States, developed a proof of the Pythagorean Theorem in 1876. His proof, shown below, involved the fact that a trapezoid can be formed from two congruent right triangles and an isosceles right triangle.

The Pythagorean Theorem

In a right triangle, the sum of the squares of the lengths of the legs is equal to the square of the length of the hypotenuse, where a and b are the legs and c is the hypotenuse.

$$a^2 + b^2 = c^2$$

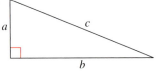

Proof

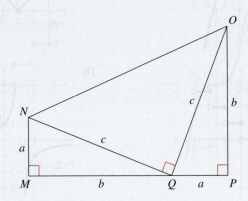

Area of   Area of   Area of   Area of
trapezoid $MNOP$ = $\triangle MNQ$ + $\triangle PQO$ + $\triangle NOQ$

$$\frac{1}{2}(a + b)(a + b) = \frac{1}{2}ab + \frac{1}{2}ab + \frac{1}{2}c^2$$

$$\frac{1}{2}(a + b)(a + b) = ab + \frac{1}{2}c^2$$

$$(a + b)(a + b) = 2ab + c^2$$

$$a^2 + 2ab + b^2 = 2ab + c^2$$

$$a^2 + b^2 = c^2$$

6 Analytic Trigonometry

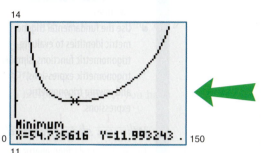

Section 6.3, Example 12
Minimum Surface Area
of a Honeycomb

When factoring trigonometric expressions, it is helpful to find a polynomial form that fits the expression, as shown in Example 3.

Example 3 Factoring Trigonometric Expressions

Factor each expression.

a. $\sec^2 \theta - 1$

b. $4 \tan^2 \theta + \tan \theta - 3$

Solution

a. Here the expression is a difference of two squares, which factors as

$$\sec^2 \theta - 1 = (\sec \theta - 1)(\sec \theta + 1).$$

b. This expression has the polynomial form $ax^2 + bx + c$ and it factors as

$$4 \tan^2 \theta + \tan \theta - 3 = (4 \tan \theta - 3)(\tan \theta + 1).$$

 CHECKPOINT   Now try Exercise 45.

On occasion, factoring or simplifying can best be done by first rewriting the expression in terms of just *one* trigonometric function or in terms of *sine or cosine alone*. These strategies are illustrated in Examples 4 and 5.

Example 4 Factoring a Trigonometric Expression

Factor $\csc^2 x - \cot x - 3$.

Solution

Use the identity

$$\csc^2 x = 1 + \cot^2 x$$

to rewrite the expression in terms of the cotangent.

$$\csc^2 x - \cot x - 3 = (1 + \cot^2 x) - \cot x - 3 \qquad \text{Pythagorean identity}$$

$$= \cot^2 x - \cot x - 2 \qquad \text{Combine like terms.}$$

$$= (\cot x - 2)(\cot x + 1) \qquad \text{Factor.}$$

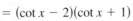 **CHECKPOINT**   Now try Exercise 51.

Example 5 Simplifying a Trigonometric Expression

Simplify $\sin t + \cot t \cos t$.

Solution

Begin by rewriting $\cot t$ in terms of sine and cosine.

$$\sin t + \cot t \cos t = \sin t + \left(\frac{\cos t}{\sin t} \right) \cos t \qquad \text{Quotient identity}$$

$$= \frac{\sin^2 t + \cos^2 t}{\sin t} \qquad \text{Add fractions.}$$

$$= \frac{1}{\sin t} \qquad \text{Pythagorean identity}$$

$$= \csc t \qquad \text{Reciprocal identity}$$

 CHECKPOINT   Now try Exercise 61.

GalaxyPhoto 2010/used under license from Shutterstock.com

Technology Tip

 You can use the *table* feature of a graphing utility to check the result of Example 5. To do this, enter

$$y_1 = \sin x + \left(\frac{\cos x}{\sin x} \right) \cos x$$

and

$$y_2 = \frac{1}{\sin x}.$$

Now, create a table that shows the values of y_1 and y_2 for different values of x. The values of y_1 and y_2 *appear* to be identical, so the expressions *appear* to be equivalent. To show that the expressions are equivalent, you need to show their equivalence algebraically, as in Example 5.

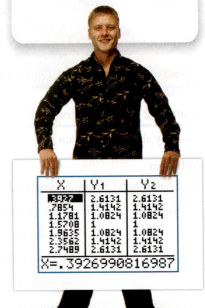

The next two examples involve techniques for rewriting expressions in forms that are used in calculus.

Example 6   Rewriting a Trigonometric Expression

Rewrite $\dfrac{1}{1 + \sin x}$ so that it is *not* in fractional form.

Solution

From the Pythagorean identity

$$\cos^2 x = 1 - \sin^2 x = (1 - \sin x)(1 + \sin x)$$

you can see that multiplying both the numerator and the denominator by $(1 - \sin x)$ will produce a monomial denominator.

$$\dfrac{1}{1 + \sin x} = \dfrac{1}{1 + \sin x} \cdot \dfrac{1 - \sin x}{1 - \sin x} \qquad \text{Multiply numerator and denominator by } (1 - \sin x).$$

$$= \dfrac{1 - \sin x}{1 - \sin^2 x} \qquad \text{Multiply.}$$

$$= \dfrac{1 - \sin x}{\cos^2 x} \qquad \text{Pythagorean identity}$$

$$= \dfrac{1}{\cos^2 x} - \dfrac{\sin x}{\cos^2 x} \qquad \text{Write as separate fractions.}$$

$$= \dfrac{1}{\cos^2 x} - \dfrac{\sin x}{\cos x} \cdot \dfrac{1}{\cos x} \qquad \text{Write as separate fractions.}$$

$$= \sec^2 x - \tan x \sec x \qquad \text{Reciprocal and quotient identities}$$

✔CHECKPOINT   Now try Exercise 65.

Example 7   Trigonometric Substitution

Use the substitution $x = 2 \tan \theta$, $0 < \theta < \pi/2$, to write $\sqrt{4 + x^2}$ as a trigonometric function of θ.

Solution

Begin by letting $x = 2 \tan \theta$. Then you can obtain

$$\sqrt{4 + x^2} = \sqrt{4 + (2 \tan \theta)^2} \qquad \text{Substitute } 2 \tan \theta \text{ for } x.$$

$$= \sqrt{4(1 + \tan^2 \theta)} \qquad \text{Distributive Property}$$

$$= \sqrt{4 \sec^2 \theta} \qquad \text{Pythagorean identity}$$

$$= 2 \sec \theta. \qquad \sec \theta > 0 \text{ for } 0 < \theta < \frac{\pi}{2}$$

✔CHECKPOINT   Now try Exercise 79.

Figure 6.1 shows the right triangle illustration of the substitution in Example 7. For $0 < \theta < \pi/2$, you have

$$\text{opp} = x, \ \text{adj} = 2, \ \text{and hyp} = \sqrt{4 + x^2}.$$

Try using these expressions to obtain the result shown in Example 7.

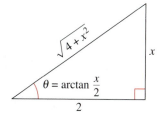

Figure 6.1

6.1   Exercises

See www.CalcChat.com for worked-out solutions to odd-numbered exercises.
For instructions on how to use a graphing utility, see Appendix A.

Vocabulary and Concept Check

1. Match each function with an equivalent expression.

(a) $\sin u$ (i) $\dfrac{1}{\sec u}$

(b) $\cos u$ (ii) $\dfrac{1}{\cot u}$

(c) $\tan u$ (iii) $\dfrac{1}{\csc u}$

2. Match each expression with an equivalent expression.

(a) $\sin^2 u$ (i) $1 + \cot^2 u$

(b) $\sec^2 u$ (ii) $1 - \cos^2 u$

(c) $\csc^2 u$ (iii) $1 + \tan^2 u$

In Exercises 3–6, fill in the blank to complete the trigonometric identity.

3. $\cos\left(\dfrac{\pi}{2} - u\right) = $ _____

4. $\csc\left(\dfrac{\pi}{2} - u\right) = $ _____

5. $\cos(-u) = $ _____

6. $\tan(-u) = $ _____

Procedures and Problem Solving

Using Identities to Evaluate a Function In Exercises 7–20, use the given values to evaluate (if possible) all six trigonometric functions.

7. $\sin x = \dfrac{1}{2}, \quad \cos x = \dfrac{\sqrt{3}}{2}$

8. $\csc \theta = 2, \quad \tan \theta = \dfrac{\sqrt{3}}{3}$

9. $\sec \theta = \sqrt{2}, \quad \sin \theta = -\dfrac{\sqrt{2}}{2}$

10. $\tan x = \dfrac{\sqrt{3}}{3}, \quad \cos x = -\dfrac{\sqrt{3}}{2}$

11. $\tan x = \dfrac{7}{24}, \quad \sec x = -\dfrac{25}{24}$

12. $\cot \phi = -5, \quad \sin \phi = \dfrac{\sqrt{26}}{26}$

✓ **13.** $\sec \phi = -\dfrac{17}{15}, \quad \sin \phi = \dfrac{8}{17}$

14. $\cos\left(\dfrac{\pi}{2} - x\right) = \dfrac{3}{5}, \quad \cos x = \dfrac{4}{5}$

15. $\sin(-x) = -\dfrac{2}{3}, \quad \tan x = -\dfrac{2\sqrt{5}}{5}$

16. $\csc(-x) = -5, \quad \cos x = \dfrac{\sqrt{24}}{5}$

17. $\tan \theta = 2, \quad \sin \theta < 0$

18. $\sec \theta = -3, \quad \tan \theta < 0$

19. $\csc \theta$ is undefined, $\quad \cos \theta < 0$

20. $\tan \theta$ is undefined, $\quad \sin \theta > 0$

Matching Trigonometric Expressions In Exercises 21–26, match the trigonometric expression with one of the following.

(a) $\sec x$ (b) -1 (c) $\cot x$

(d) 1 (e) $-\tan x$ (f) $\sin x$

21. $\sec x \cos x$ **22.** $\tan x \csc x$

23. $\cot^2 x - \csc^2 x$ **24.** $(1 - \cos^2 x)(\csc x)$

25. $\dfrac{\sin(-x)}{\cos(-x)}$ **26.** $\dfrac{\sin[(\pi/2) - x]}{\cos[(\pi/2) - x]}$

Matching Trigonometric Expressions In Exercises 27–32, match the trigonometric expression with one of the following.

(a) $\csc x$ (b) $\tan x$ (c) $\sin^2 x$

(d) $\sin x \tan x$ (e) $\sec^2 x$ (f) $\sec^2 x + \tan^2 x$

27. $\sin x \sec x$ **28.** $\cos^2 x(\sec^2 x - 1)$

29. $\sec^4 x - \tan^4 x$ **30.** $\cot x \sec x$

31. $\dfrac{\sec^2 x - 1}{\sin^2 x}$ **32.** $\dfrac{\cos^2[(\pi/2) - x]}{\cos x}$

Simplifying a Trigonometric Expression In Exercises 33–44, use the fundamental identities to simplify the expression. Use the *table* feature of a graphing utility to check your result numerically.

33. $\cot x \sin x$ **34.** $\cos \beta \tan \beta$

✓ **35.** $\sin \phi(\csc \phi - \sin \phi)$ **36.** $\sec^2 x(1 - \sin^2 x)$

37. $\dfrac{\csc x}{\cot x}$

38. $\dfrac{\sec \theta}{\csc \theta}$

39. $\sec \alpha \cdot \dfrac{\sin \alpha}{\tan \alpha}$

40. $\dfrac{\tan^2 \theta}{\sec^2 \theta}$

41. $\sin\left(\dfrac{\pi}{2} - x\right) \csc x$

42. $\cot\left(\dfrac{\pi}{2} - x\right) \cos x$

43. $\dfrac{\cos^2 y}{1 - \sin y}$

44. $\dfrac{1}{\cot^2 x + 1}$

Factoring Trigonometric Expressions   In Exercises 45–54, factor the expression and use the fundamental identities to simplify. Use a graphing utility to check your result graphically.

✓ **45.** $\cot^2 x - \cot^2 x \cos^2 x$

46. $\sec^2 x \tan^2 x + \sec^2 x$

47. $\dfrac{\cos^2 x - 4}{\cos x - 2}$

48. $\dfrac{\csc^2 x - 1}{\csc x - 1}$

49. $\tan^4 x + 2 \tan^2 x + 1$

50. $1 - 2 \sin^2 x + \sin^4 x$

✓ **51.** $\sin^4 x - \cos^4 x$

52. $\sec^4 x - \tan^4 x$

53. $\csc^3 x - \csc^2 x - \csc x + 1$

54. $\sec^3 x - \sec^2 x - \sec x + 1$

Simplifying a Trigonometric Expression   In Exercises 55–58, perform the multiplication and use the fundamental identities to simplify.

55. $(\sin x + \cos x)^2$

56. $(\tan x + \sec x)(\tan x - \sec x)$

57. $(\csc x + 1)(\csc x - 1)$

58. $(5 - 5 \sin x)(5 + 5 \sin x)$

Simplifying a Trigonometric Expression   In Exercises 59–64, perform the addition or subtraction and use the fundamental identities to simplify.

59. $\dfrac{1}{1 + \cos x} + \dfrac{1}{1 - \cos x}$

60. $\dfrac{1}{\sec x + 1} - \dfrac{1}{\sec x - 1}$

✓ **61.** $\tan x - \dfrac{\sec^2 x}{\tan x}$

62. $\dfrac{\cos x}{1 + \sin x} + \dfrac{1 + \sin x}{\cos x}$

63. $\tan x + \dfrac{\cos x}{1 + \sin x}$

64. $\dfrac{\tan x}{1 + \sec x} + \dfrac{1 + \sec x}{\tan x}$

∫ **Rewriting a Trigonometric Expression**   In Exercises 65–70, rewrite the expression so that it is *not* in fractional form.

✓ **65.** $\dfrac{\sin^2 y}{1 - \cos y}$

66. $\dfrac{5}{\tan x + \sec x}$

67. $\dfrac{\sin x}{\tan x}$

68. $\dfrac{\tan y}{\sin^2 y + \cos^2 y}$

69. $\dfrac{3}{\sec x - \tan x}$

70. $\dfrac{\tan^2 x}{\csc x + 1}$

Graphing Trigonometric Functions   In Exercises 71–74, use a graphing utility to complete the table and graph the functions in the same viewing window. Make a conjecture about y_1 and y_2.

x	0.2	0.4	0.6	0.8	1.0	1.2	1.4
y_1							
y_2							

71. $y_1 = \cos\left(\dfrac{\pi}{2} - x\right), \quad y_2 = \sin x$

72. $y_1 = \cos x + \sin x \tan x, \quad y_2 = \sec x$

73. $y_1 = \dfrac{\cos x}{1 - \sin x}, \quad y_2 = \dfrac{1 + \sin x}{\cos x}$

74. $y_1 = \sec^4 x - \sec^2 x, \quad y_2 = \tan^2 x + \tan^4 x$

Graphing Trigonometric Functions   In Exercises 75–78, use a graphing utility to determine which of the six trigonometric functions is equal to the expression.

75. $\cos x \cot x + \sin x$

76. $\sin x(\cot x + \tan x)$

77. $\sec x - \dfrac{\cos x}{1 + \sin x}$

78. $\dfrac{1}{2}\left(\dfrac{1 + \sin \theta}{\cos \theta} + \dfrac{\cos \theta}{1 + \sin \theta}\right)$

∫ **Trigonometric Substitution**   In Exercises 79–90, use the trigonometric substitution to write the algebraic expression as a trigonometric function of θ, where $0 < \theta < \pi/2$.

✓ **79.** $\sqrt{25 - x^2}, \quad x = 5 \sin \theta$

80. $\sqrt{64 - 16x^2}, \quad x = 2 \cos \theta$

81. $\sqrt{x^2 - 9}, \quad x = 3 \sec \theta$

82. $\sqrt{x^2 + 100}, \quad x = 10 \tan \theta$

83. $\sqrt{9 - x^2}, \quad x = 3 \sin \theta$

84. $\sqrt{4 - x^2}, \quad x = 2 \cos \theta$

85. $\sqrt{4x^2 + 9}, \quad 2x = 3 \tan \theta$

86. $\sqrt{9x^2 + 4}, \quad 3x = 2 \tan \theta$

87. $\sqrt{16x^2 - 9}, \quad 4x = 3 \sec \theta$

88. $\sqrt{9x^2 - 25}, \quad 3x = 5 \sec \theta$

89. $\sqrt{2 - x^2}, \quad x = \sqrt{2} \sin \theta$

90. $\sqrt{5 - x^2}, \quad x = \sqrt{5} \cos \theta$

Solving a Trigonometric Equation   In Exercises 91–94, use a graphing utility to solve the equation for θ, where $0 \le \theta < 2\pi$.

91. $\sin \theta = \sqrt{1 - \cos^2 \theta}$

92. $\cos \theta = -\sqrt{1 - \sin^2 \theta}$

93. $\sec \theta = \sqrt{1 + \tan^2 \theta}$

94. $\tan \theta = \sqrt{\sec^2 \theta - 1}$

Rewriting Expressions In Exercises 95–100, rewrite the expression as a single logarithm and simplify the result. (*Hint:* Begin by using the properties of logarithms.)

95. $\ln|\cos\theta| - \ln|\sin\theta|$

96. $\ln|\csc\theta| + \ln|\tan\theta|$

97. $\ln(1 + \sin x) - \ln|\sec x|$

98. $\ln|\cot t| + \ln(1 + \tan^2 t)$

99. $\ln|\sec x| + \ln|\sin x|$

100. $\ln|\cot x| + \ln|\sin x|$

Using Identities In Exercises 101–106, show that the identity is *not* true for all values of θ. (There are many correct answers.)

101. $\cos\theta = \sqrt{1 - \sin^2\theta}$

102. $\tan\theta = \sqrt{\sec^2\theta - 1}$

103. $\sin\theta = \sqrt{1 - \cos^2\theta}$

104. $\sec\theta = \sqrt{1 + \tan^2\theta}$

105. $\csc\theta = \sqrt{1 + \cot^2\theta}$

106. $\cot\theta = \sqrt{\csc^2\theta - 1}$

Using Identities In Exercises 107–110, use the *table* feature of a graphing utility to demonstrate the identity for each value of θ.

107. $\csc^2\theta - \cot^2\theta = 1$, (a) $\theta = 132°$ (b) $\theta = \dfrac{2\pi}{7}$

108. $\tan^2\theta + 1 = \sec^2\theta$, (a) $\theta = 346°$ (b) $\theta = 3.1$

109. $\cos\left(\dfrac{\pi}{2} - \theta\right) = \sin\theta$, (a) $\theta = 80°$ (b) $\theta = 0.8$

110. $\sin(-\theta) = -\sin\theta$, (a) $\theta = 250°$ (b) $\theta = \frac{1}{2}$

111. *Why you should learn it* (p. 490)   The forces acting on an object weighing W units on an inclined plane positioned at an angle of θ with the horizontal are modeled by

$$\mu W \cos\theta = W \sin\theta$$

where μ is the coefficient of friction (see figure). Solve the equation for μ and simplify the result.

112. Rate of Change The rate of change of the function $f(x) = -\csc x - \sin x$ is given by the expression $\csc x \cot x - \cos x$. Show that this expression can also be written as $\cos x \cot^2 x$.

113. Rate of Change The rate of change of the function $f(x) = \sec x + \cos x$ is given by the expression $\sec x \tan x - \sin x$. Show that this expression can also be written as $\sin x \tan^2 x$.

Conclusions

True or False? In Exercises 114 and 115, determine whether the statement is true or false. Justify your answer.

114. $\sin\theta \csc\theta = 1$

115. $\cos\theta \sec\phi = 1$

116. CAPSTONE

(a) Use the definitions of sine and cosine to derive the Pythagorean identity $\sin^2\theta + \cos^2\theta = 1$.

(b) Use the Pythagorean identity $\sin^2\theta + \cos^2\theta = 1$ to derive the other Pythagorean identities, $1 + \tan^2\theta = \sec^2\theta$ and $1 + \cot^2\theta = \csc^2\theta$. Discuss how to remember these identities and other fundamental identities.

Evaluating Trigonometric Functions In Exercises 117–120, fill in the blanks. (*Note:* $x \to c^+$ indicates that x approaches c from the right, and $x \to c^-$ indicates that x approaches c from the left.)

117. As $x \to \dfrac{\pi}{2}^-$, $\sin x \to$ ____ and $\csc x \to$ ____ .

118. As $x \to 0^+$, $\cos x \to$ ____ and $\sec x \to$ ____ .

119. As $x \to \dfrac{\pi}{2}^-$, $\tan x \to$ ____ and $\cot x \to$ ____ .

120. As $x \to \pi^+$, $\sin x \to$ ____ and $\csc x \to$ ____ .

121. Write each of the other trigonometric functions of θ in terms of $\sin\theta$.

122. Write each of the other trigonometric functions of θ in terms of $\cos\theta$.

Cumulative Mixed Review

Adding or Subtracting Rational Expressions In Exercises 123–126, perform the addition or subtraction and simplify.

123. $\dfrac{1}{x + 5} + \dfrac{x}{x - 8}$

124. $\dfrac{6x}{x - 4} - \dfrac{3}{4 - x}$

125. $\dfrac{2x}{x^2 - 4} - \dfrac{7}{x + 4}$

126. $\dfrac{x}{x^2 - 25} + \dfrac{x^2}{x - 5}$

Graphing Trigonometric Functions In Exercises 127–130, sketch the graph of the function. (Include two full periods.)

127. $f(x) = \dfrac{1}{2}\sin\pi x$

128. $f(x) = -2\tan\dfrac{\pi x}{2}$

129. $f(x) = \dfrac{1}{2}\cot\left(x + \dfrac{\pi}{4}\right)$

130. $f(x) = \dfrac{3}{2}\cos(x - \pi) + 3$

6.2   Verifying Trigonometric Identities

Verifying Trigonometric Identities

In this section, you will study techniques for verifying trigonometric identities. In the next section, you will study techniques for solving trigonometric equations. The key to both verifying identities *and* solving equations is your ability to use the fundamental identities and the rules of algebra to rewrite trigonometric expressions.

Remember that a *conditional equation* is an equation that is true for only some of the values in its domain. For example, the conditional equation

$$\sin x = 0 \qquad \text{Conditional equation}$$

is true only for

$$x = n\pi$$

where *n* is an integer. When you find these values, you are *solving* the equation.

On the other hand, an equation that is true for all real values in the domain of the variable is an *identity*. For example, the familiar equation

$$\sin^2 x = 1 - \cos^2 x \qquad \text{Identity}$$

is true for all real numbers *x*. So, it is an identity.

Verifying that a trigonometric equation is an identity is quite different from solving an equation. There is no well-defined set of rules to follow in verifying trigonometric identities, and the process is best learned by practice.

Guidelines for Verifying Trigonometric Identities

1. **Work with one side of the equation at a time.** It is often better to work with the more complicated side first.

2. **Look for opportunities to factor an expression, add fractions, square a binomial, or create a monomial denominator.**

3. **Look for opportunities to use the fundamental identities.** Note which functions are in the final expression you want. Sines and cosines pair up well, as do secants and tangents, and cosecants and cotangents.

4. **When the preceding guidelines do not help, try converting all terms to sines and cosines.**

5. **Always try *something*.** Even making an attempt that leads to a dead end provides insight.

Verifying trigonometric identities is a useful process when you need to convert a trigonometric expression into a form that is more useful algebraically. When you verify an identity, you cannot assume that the two sides of the equation are equal because you are trying to verify that they are equal. As a result, when verifying identities, you cannot use operations such as adding the same quantity to each side of the equation or cross multiplication.

Example 1 Verifying a Trigonometric Identity

Verify the identity.

$$\frac{\sec^2 \theta - 1}{\sec^2 \theta} = \sin^2 \theta$$

Solution

Because the left side is more complicated, start with it.

$$\frac{\sec^2 \theta - 1}{\sec^2 \theta} = \frac{(\tan^2 \theta + 1) - 1}{\sec^2 \theta} \qquad \text{\color{red}Pythagorean identity}$$

$$= \frac{\tan^2 \theta}{\sec^2 \theta} \qquad \text{\color{red}Simplify.}$$

$$= \tan^2 \theta (\cos^2 \theta) \qquad \text{\color{red}Reciprocal identity}$$

$$= \frac{\sin^2 \theta}{\cos^2 \theta}(\cos^2 \theta) \qquad \text{\color{red}Quotient identity}$$

$$= \sin^2 \theta \qquad \text{\color{red}Simplify.}$$

✓**CHECKPOINT** Now try Exercise 15.

There can be more than one way to verify an identity. Here is another way to verify the identity in Example 1.

$$\frac{\sec^2 \theta - 1}{\sec^2 \theta} = \frac{\sec^2 \theta}{\sec^2 \theta} - \frac{1}{\sec^2 \theta} \qquad \text{\color{red}Rewrite as the difference of fractions.}$$

$$= 1 - \cos^2 \theta \qquad \text{\color{red}Reciprocal identity}$$

$$= \sin^2 \theta \qquad \text{\color{red}Pythagorean identity}$$

Remember that an identity is true only for all real values in the domain of the variable. For instance, in Example 1 the identity is not true when $\theta = \pi/2$ because $\sec^2 \theta$ is not defined when $\theta = \pi/2$.

Example 2 Combining Fractions Before Using Identities

Verify the identity.

$$2 \sec^2 \alpha = \frac{1}{1 - \sin \alpha} + \frac{1}{1 + \sin \alpha}$$

Algebraic Solution

The right side is more complicated, so start with it.

$$\frac{1}{1 - \sin \alpha} + \frac{1}{1 + \sin \alpha} = \frac{1 + \sin \alpha + 1 - \sin \alpha}{(1 - \sin \alpha)(1 + \sin \alpha)} \qquad \text{\color{red}Add fractions.}$$

$$= \frac{2}{1 - \sin^2 \alpha} \qquad \text{\color{red}Simplify.}$$

$$= \frac{2}{\cos^2 \alpha} \qquad \text{\color{red}Pythagorean identity}$$

$$= 2 \sec^2 \alpha \qquad \text{\color{red}Reciprocal identity}$$

✓**CHECKPOINT** Now try Exercise 39.

Numerical Solution

Use a graphing utility to create a table that shows the values of $y_1 = 2/\cos^2 x$ and $y_2 = 1/(1 - \sin x) + 1/(1 + \sin x)$ for different values of x, as shown in Figure 6.2.

X	Y1	Y2
-.5	2.5969	2.5969
-.25	2.1304	2.1304
0	2	2
.25	2.1304	2.1304
.5	2.5969	2.5969
.75	3.7357	3.7357
1	6.851	6.851

X=-.5

The values appear to be identical, so the equation appears to be an identity.

Figure 6.2

Technology Tip

Although a graphing utility can be useful in helping to verify an identity, you must use algebraic techniques to produce a valid proof. For example, graph the two functions

$$y_1 = \sin 50x$$

$$y_2 = \sin 2x$$

in a trigonometric viewing window. On some graphing utilities the graphs appear to be identical. However, $\sin 50x \neq \sin 2x$.

In Example 2, you needed to write the Pythagorean identity $\sin^2 u + \cos^2 u = 1$ in the equivalent form

$$\cos^2 u = 1 - \sin^2 u.$$

When verifying identities, you may find it useful to write Pythagorean identities in one of these equivalent forms.

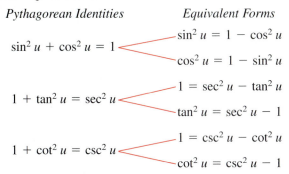

Pythagorean Identities   *Equivalent Forms*

$\sin^2 u + \cos^2 u = 1$ $\sin^2 u = 1 - \cos^2 u$
                                $\cos^2 u = 1 - \sin^2 u$

$1 + \tan^2 u = \sec^2 u$ $1 = \sec^2 u - \tan^2 u$
                                $\tan^2 u = \sec^2 u - 1$

$1 + \cot^2 u = \csc^2 u$ $1 = \csc^2 u - \cot^2 u$
                                $\cot^2 u = \csc^2 u - 1$

Example 3   Verifying a Trigonometric Identity

Verify the identity $(\tan^2 x + 1)(\cos^2 x - 1) = -\tan^2 x$.

Algebraic Solution

By applying identities before multiplying, you obtain the following.

$(\tan^2 x + 1)(\cos^2 x - 1) = (\sec^2 x)(-\sin^2 x)$ Pythagorean identities

$\qquad\qquad = -\dfrac{\sin^2 x}{\cos^2 x}$ Reciprocal identity

$\qquad\qquad = -\left(\dfrac{\sin x}{\cos x}\right)^2$ Property of exponents

$\qquad\qquad = -\tan^2 x$ Quotient identity

✔CHECKPOINT   Now try Exercise 47.

Graphical Solution

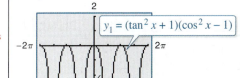

$y_1 = (\tan^2 x + 1)(\cos^2 x - 1)$

$y_2 = -\tan^2 x$

Because the graphs appear to coincide, the given equation appears to be an identity.

Example 4   Converting to Sines and Cosines

Verify the identity $\tan x + \cot x = \sec x \csc x$.

Solution

In this case there appear to be no fractions to add, no products to find, and no opportunities to use the Pythagorean identities. So, try converting the left side to sines and cosines.

$\tan x + \cot x = \dfrac{\sin x}{\cos x} + \dfrac{\cos x}{\sin x}$ Quotient identities

$\qquad = \dfrac{\sin^2 x + \cos^2 x}{\cos x \sin x}$ Add fractions.

$\qquad = \dfrac{1}{\cos x \sin x}$ Pythagorean identity

$\qquad = \dfrac{1}{\cos x} \cdot \dfrac{1}{\sin x}$ Product of fractions

$\qquad = \sec x \csc x$ Reciprocal identities

✔CHECKPOINT   Now try Exercise 49.

Recall from algebra that *rationalizing the denominator* using conjugates is, on occasion, a powerful simplification technique. A related form of this technique works for simplifying trigonometric expressions as well. For instance, to simplify

$$\frac{1}{1 - \cos x}$$

multiply the numerator and the denominator by $(1 + \cos x)$.

$$\frac{1}{1 - \cos x} = \frac{1}{1 - \cos x}\left(\frac{1 + \cos x}{1 + \cos x}\right)$$

$$= \frac{1 + \cos x}{1 - \cos^2 x}$$

$$= \frac{1 + \cos x}{\sin^2 x}$$

$$= \csc^2 x(1 + \cos x)$$

As shown above, $\csc^2 x(1 + \cos x)$ is considered a simplified form of

$$\frac{1}{1 - \cos x}$$

because the expression does not contain any fractions.

Example 5   Verifying a Trigonometric Identity

Verify the identity.

$$\sec x + \tan x = \frac{\cos x}{1 - \sin x}$$

Algebraic Solution

Begin with the *right* side because you can create a monomial denominator by multiplying the numerator and denominator by $(1 + \sin x)$.

$$\frac{\cos x}{1 - \sin x} = \frac{\cos x}{1 - \sin x}\left(\frac{1 + \sin x}{1 + \sin x}\right) \qquad \text{Multiply numerator and denominator by } (1 + \sin x).$$

$$= \frac{\cos x + \cos x \sin x}{1 - \sin^2 x} \qquad \text{Multiply.}$$

$$= \frac{\cos x + \cos x \sin x}{\cos^2 x} \qquad \text{Pythagorean identity}$$

$$= \frac{\cos x}{\cos^2 x} + \frac{\cos x \sin x}{\cos^2 x} \qquad \text{Write as separate fractions.}$$

$$= \frac{1}{\cos x} + \frac{\sin x}{\cos x} \qquad \text{Simplify.}$$

$$= \sec x + \tan x \qquad \text{Identities}$$

Graphical Solution

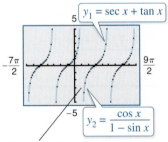

$y_1 = \sec x + \tan x$

$y_2 = \dfrac{\cos x}{1 - \sin x}$

Because the graphs appear to coincide, the given equation appears to be an identity.

✔**CHECKPOINT**   Now try Exercise 55.

In Examples 1 through 5, you have been verifying trigonometric identities by working with one side of the equation and converting it to the form given on the other side. On occasion it is practical to work with each side *separately* to obtain one common form equivalent to both sides. This is illustrated in Example 6.

Example 6 Working with Each Side Separately

Verify the identity.

$$\frac{\cot^2 \theta}{1 + \csc \theta} = \frac{1 - \sin \theta}{\sin \theta}$$

Algebraic Solution

Working with the left side, you have

$$\frac{\cot^2 \theta}{1 + \csc \theta} = \frac{\csc^2 \theta - 1}{1 + \csc \theta}$$ Pythagorean identity

$$= \frac{(\csc \theta - 1)(\csc \theta + 1)}{1 + \csc \theta}$$ Factor.

$$= \csc \theta - 1.$$ Simplify.

Now, simplifying the right side, you have

$$\frac{1 - \sin \theta}{\sin \theta} = \frac{1}{\sin \theta} - \frac{\sin \theta}{\sin \theta}$$ Write as separate fractions.

$$= \csc \theta - 1.$$ Reciprocal identity

The identity is verified because both sides are equal to $\csc \theta - 1$.

 CHECKPOINT   Now try Exercise 57.

Numerical Solution

Use a graphing utility to create a table that shows the values of

$$y_1 = \frac{\cot^2 x}{1 + \csc x}$$

and

$$y_2 = \frac{1 - \sin x}{\sin x}$$

for different values of x, as shown in Figure 6.3.

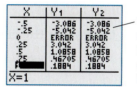

The values for y_1 and y_2 appear to be identical, so the equation appears to be an identity.

Figure 6.3

In Example 7, powers of trigonometric functions are rewritten as more complicated sums of products of trigonometric functions. This is a common procedure used in calculus.

Example 7 Examples from Calculus

Verify each identity.

a. $\tan^4 x = \tan^2 x \sec^2 x - \tan^2 x$

b. $\sin^3 x \cos^4 x = (\cos^4 x - \cos^6 x)\sin x$

c. $\csc^4 x \cot x = \csc^2 x(\cot x + \cot^3 x)$

Solution

a. $\tan^4 x = (\tan^2 x)(\tan^2 x)$ Write as separate factors.

$$= \tan^2 x(\sec^2 x - 1)$$ Pythagorean identity

$$= \tan^2 x \sec^2 x - \tan^2 x$$ Multiply.

b. $\sin^3 x \cos^4 x = \sin^2 x \cos^4 x \sin x$ Write as separate factors.

$$= (1 - \cos^2 x)\cos^4 x \sin x$$ Pythagorean identity

$$= (\cos^4 x - \cos^6 x)\sin x$$ Multiply.

c. $\csc^4 x \cot x = \csc^2 x \csc^2 x \cot x$ Write as separate factors.

$$= \csc^2 x(1 + \cot^2 x) \cot x$$ Pythagorean identity

$$= \csc^2 x(\cot x + \cot^3 x)$$ Multiply.

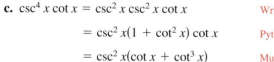 **CHECKPOINT**   Now try Exercise 69.

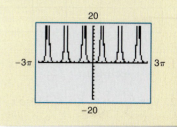

? What's Wrong?

To determine the validity of the statement

$$\tan^2 x \sin^2 x \overset{?}{=} \frac{5}{6} \tan^2 x,$$ you use

a graphing utility to graph $y_1 = \tan^2 x \sin^2 x$ and

$y_2 = \frac{5}{6} \tan^2 x$, as shown in the

figure. You use the graph to conclude that the statement is an identity. What's wrong?

6.2 **Exercises**

See www.CalcChat.com for worked-out solutions to odd-numbered exercises.
For instructions on how to use a graphing utility, see Appendix A.

Vocabulary and Concept Check

In Exercises 1–8, fill in the blank to complete the trigonometric identity.

1. $\dfrac{1}{\tan u} = $ _____

2. $\dfrac{1}{\csc u} = $ _____

3. $\dfrac{\sin u}{\cos u} = $ _____

4. $\dfrac{1}{\sec u} = $ _____

5. $\sin^2 u + $ _____ $= 1$

6. $\tan\left(\dfrac{\pi}{2} - u\right) = $ _____

7. $\sin(-u) = $ _____

8. $\sec(-u) = $ _____

9. Is a graphical solution sufficient to verify a trigonometric identity?

10. Is a conditional equation true for all real values in its domain?

Procedures and Problem Solving

Verifying a Trigonometric Identity In Exercises 11–20, verify the identity.

11. $\sin t \csc t = 1$

12. $\sec y \cos y = 1$

13. $\dfrac{\csc^2 x}{\cot x} = \csc x \sec x$

14. $\dfrac{\sin^2 t}{\tan^2 t} = \cos^2 t$

✓ **15.** $\cos^2 \beta - \sin^2 \beta = 1 - 2\sin^2 \beta$

16. $\cos^2 \beta - \sin^2 \beta = 2\cos^2 \beta - 1$

17. $\tan^2 \theta + 6 = \sec^2 \theta + 5$

18. $2 - \csc^2 z = 1 - \cot^2 z$

19. $(1 + \sin x)(1 - \sin x) = \cos^2 x$

20. $\tan^2 y(\csc^2 y - 1) = 1$

Algebraic-Graphical-Numerical In Exercises 21–28, use a graphing utility to complete the table and graph the functions in the same viewing window. Use both the table and the graph as evidence that $y_1 = y_2$. Then verify the identity algebraically.

x	0.2	0.4	0.6	0.8	1.0	1.2	1.4
y_1							
y_2							

21. $y_1 = \dfrac{1}{\sec x \tan x}$, $\quad y_2 = \csc x - \sin x$

22. $y_1 = \dfrac{\csc x - 1}{1 - \sin x}$, $\quad y_2 = \csc x$

23. $y_1 = \csc x - \sin x$, $\quad y_2 = \cos x \cot x$

24. $y_1 = \sec x - \cos x$, $\quad y_2 = \sin x \tan x$

25. $y_1 = \sin x + \cos x \cot x$, $\quad y_2 = \csc x$

26. $y_1 = \cos x + \sin x \tan x$, $\quad y_2 = \sec x$

27. $y_1 = \dfrac{1}{\tan x} + \dfrac{1}{\cot x}$, $\quad y_2 = \tan x + \cot x$

28. $y_1 = \dfrac{1}{\sin x} - \dfrac{1}{\csc x}$, $\quad y_2 = \csc x - \sin x$

Error Analysis In Exercises 29 and 30, describe the error.

29. $(1 + \tan x)[1 + \cot(-x)]$
$= (1 + \tan x)(1 + \cot x)$
$= 1 + \cot x + \tan x + \tan x \cot x$
$= 1 + \cot x + \tan x + 1$
$= 2 + \cot x + \tan x$

30. $\dfrac{1 + \sec(-\theta)}{\sin(-\theta) + \tan(-\theta)} = \dfrac{1 - \sec \theta}{\sin \theta - \tan \theta}$

$= \dfrac{1 - \sec \theta}{(\sin \theta)\left[1 - \left(\dfrac{1}{\cos \theta}\right)\right]}$

$= \dfrac{1 - \sec \theta}{\sin \theta(1 - \sec \theta)}$

$= \dfrac{1}{\sin \theta} = \csc \theta$

Verifying a Trigonometric Identity In Exercises 31 and 32, fill in the missing step(s).

31. $\sec^4 x - 2\sec^2 x + 1 = (\sec^2 x - 1)^2$

$$= \underline{}$$

$$= \tan^4 x$$

32. $\dfrac{\tan x - \cot x}{\tan x + \cot x} = \dfrac{\dfrac{\sin x}{\cos x} - \dfrac{\cos x}{\sin x}}{\dfrac{\sin x}{\cos x} + \dfrac{\cos x}{\sin x}}$

$$= \underline{}$$

$$= \dfrac{\sin^2 x - \cos^2 x}{1}$$

$$= \sin^2 x - \cos^2 x$$

$$= \underline{}$$

$$= 1 - 2\cos^2 x$$

Verifying a Trigonometric Identity In Exercises 33–38, verify the identity.

33. $\sin^{1/2} x \cos x - \sin^{5/2} x \cos x = \cos^3 x \sqrt{\sin x}$

34. $\sec^6 x(\sec x \tan x) - \sec^4 x(\sec x \tan x) = \sec^5 x \tan^3 x$

35. $\cot\left(\dfrac{\pi}{2} - x\right)\csc x = \sec x$

36. $\dfrac{\sec[(\pi/2) - x]}{\tan[(\pi/2) - x]} = \sec x$

37. $\dfrac{\csc(-x)}{\sec(-x)} = -\cot x$

38. $(1 + \sin y)[1 + \sin(-y)] = \cos^2 y$

Verifying a Trigonometric Identity In Exercises 39–46, verify the identity algebraically. Use the *table* feature of a graphing utility to check your result numerically.

✓ **39.** $\dfrac{\cos x - \cos y}{\sin x + \sin y} + \dfrac{\sin x - \sin y}{\cos x + \cos y} = 0$

40. $\dfrac{\tan x + \cot y}{\tan x \cot y} = \tan y + \cot x$

41. $\dfrac{\cos \theta}{1 - \sin \theta} = \sec \theta + \tan \theta$

42. $(\sec \theta - \tan \theta)(\csc \theta + 1) = \cot \theta$

43. $\sin^2\left(\dfrac{\pi}{2} - x\right) + \sin^2 x = 1$

44. $\sec^2 y - \cot^2\left(\dfrac{\pi}{2} - y\right) = 1$

45. $\sin x \csc\left(\dfrac{\pi}{2} - x\right) = \tan x$

46. $\sec^2\left(\dfrac{\pi}{2} - x\right) - 1 = \cot^2 x$

Verifying a Trigonometric Identity In Exercises 47–58, verify the identity algebraically. Use a graphing utility to check your result graphically.

✓ **47.** $2\sec^2 x - 2\sec^2 x \sin^2 x - \sin^2 x - \cos^2 x = 1$

48. $\csc x(\csc x - \sin x) + \dfrac{\sin x - \cos x}{\sin x} + \cot x = \csc^2 x$

✓ **49.** $\dfrac{\cot x \tan x}{\sin x} = \csc x$

50. $\dfrac{1 + \csc \theta}{\sec \theta} - \cot \theta = \cos \theta$

51. $\csc \theta \tan \theta = \sec \theta$

52. $\sin \theta \csc \theta - \sin^2 \theta = \cos^2 \theta$

53. $1 - \dfrac{\sin^2 \theta}{1 - \cos \theta} = -\cos \theta$

54. $\dfrac{\tan \theta}{1 + \sec \theta} + \dfrac{1 + \sec \theta}{\tan \theta} = 2\csc \theta$

✓ **55.** $\dfrac{\sin \beta}{1 - \cos \beta} = \dfrac{1 + \cos \beta}{\sin \beta}$

56. $\dfrac{\cot \alpha}{\csc \alpha - 1} = \dfrac{\csc \alpha + 1}{\cot \alpha}$

✓ **57.** $\dfrac{\tan^3 \alpha - 1}{\tan \alpha - 1} = \tan^2 \alpha + \tan \alpha + 1$

58. $\dfrac{\sin^3 \beta + \cos^3 \beta}{\sin \beta + \cos \beta} = 1 - \sin \beta \cos \beta$

Graphing a Trigonometric Function In Exercises 59–62, use a graphing utility to graph the trigonometric function. Use the graph to make a conjecture about a simplification of the expression. Verify the resulting identity algebraically.

59. $y = \dfrac{1}{\cot x + 1} + \dfrac{1}{\tan x + 1}$

60. $y = \dfrac{\cos x}{1 - \tan x} + \dfrac{\sin x \cos x}{\sin x - \cos x}$

61. $y = \dfrac{1}{\sin x} - \dfrac{\cos^2 x}{\sin x}$ **62.** $y = \sin t + \dfrac{\cot^2 t}{\csc t}$

Verifying an Identity Involving Logarithms In Exercises 63 and 64, use the properties of logarithms and trigonometric identities to verify the identity.

63. $\ln|\cot \theta| = \ln|\cos \theta| - \ln|\sin \theta|$

64. $\ln|\sec \theta| = -\ln|\cos \theta|$

Using Cofunction Identities In Exercises 65–68, use the cofunction identities to evaluate the expression without using a calculator.

65. $\sin^2 35° + \sin^2 55°$ **66.** $\cos^2 14° + \cos^2 76°$

67. $\cos^2 20° + \cos^2 52° + \cos^2 38° + \cos^2 70°$

68. $\sin^2 18° + \sin^2 40° + \sin^2 50° + \sin^2 72°$

∫ **Examples from Calculus**   In Exercises 69–72, powers of trigonometric functions are rewritten to be useful in calculus. Verify the identity.

✓ **69.** $\tan^5 x = \tan^3 x \sec^2 x - \tan^3 x$

70. $\sec^4 x \tan^2 x = (\tan^2 x + \tan^4 x)\sec^2 x$

71. $\cos^3 x \sin^2 x = (\sin^2 x - \sin^4 x)\cos x$

72. $\sin^4 x + \cos^4 x = 1 - 2\cos^2 x + 2\cos^4 x$

Verifying a Trigonometric Identity   In Exercises 73–76, verify the identity.

73. $\tan(\sin^{-1} x) = \dfrac{x}{\sqrt{1 - x^2}}$

74. $\cos(\sin^{-1} x) = \sqrt{1 - x^2}$

75. $\tan\left(\sin^{-1}\dfrac{x - 1}{4}\right) = \dfrac{x - 1}{\sqrt{16 - (x - 1)^2}}$

76. $\tan\left(\cos^{-1}\dfrac{x + 1}{2}\right) = \dfrac{\sqrt{4 - (x + 1)^2}}{x + 1}$

77. *Why you should learn it* (p. 497)   The length s of a shadow cast by a vertical gnomon (a device used to tell time) of height h when the angle of the sun above the horizon is θ (see figure) can be modeled by the equation

$$s = \frac{h\sin(90° - \theta)}{\sin\theta}.$$

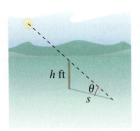

(a) Verify that the equation for s is equal to $h\cot\theta$.

(b) Use a graphing utility to complete the table. Let $h = 5$ feet.

θ	15°	30°	45°	60°	75°	90°
s						

(c) Use your table from part (b) to determine the angles of the sun that result in the maximum and minimum lengths of the shadow.

(d) Based on your results from part (c), what time of day do you think it is when the angle of the sun above the horizon is 90°?

78. Rate of Change   The rate of change of the function $f(x) = \sin x + \csc x$ is given by $\cos x - \csc x \cot x$. Show that the expression for the rate of change can also be given by $-\cos x \cot^2 x$.

Conclusions

True or False?   In Exercises 79–82, determine whether the statement is true or false. Justify your answer.

79. There can be more than one way to verify a trigonometric identity.

80. Of the six trigonometric functions, two are even.

81. The equation $\sin^2\theta + \cos^2\theta = 1 + \tan^2\theta$ is an identity, because $\sin^2(0) + \cos^2(0) = 1$ and $1 + \tan^2(0) = 1$.

82. $\sin(x^2) = \sin^2(x)$

Verifying a Trigonometric Identity   In Exercises 83 and 84, (a) verify the identity and (b) determine whether the identity is true for the given value of x. Explain.

83. $\dfrac{\sin x}{1 + \cos x} = \dfrac{1 - \cos x}{\sin x}$, $x = 0$

84. $\dfrac{\sec x}{\tan x} = \dfrac{\tan x}{\sec x - \cos x}$, $x = \pi$

∫ **Using Trigonometric Substitution**   In Exercises 85–88, use the trigonometric substitution to write the algebraic expression as a trigonometric function of θ, where $0 < \theta < \pi/2$. Assume $a > 0$.

85. $\sqrt{a^2 - u^2}$, $u = a\sin\theta$

86. $\sqrt{a^2 - u^2}$, $u = a\cos\theta$

87. $\sqrt{a^2 + u^2}$, $u = a\tan\theta$

88. $\sqrt{u^2 - a^2}$, $u = a\sec\theta$

Think About It   In Exercises 89–92, explain why the equation is not an identity and find one value of the variable for which the equation is not true.

89. $\sqrt{\tan^2 x} = \tan x$ **90.** $\sin\theta = \sqrt{1 - \cos^2\theta}$

91. $\tan\theta = \sqrt{\sec^2\theta - 1}$

92. $\sqrt{\sin^2 x + \cos^2 x} = \sin x + \cos x$

93. Verify that for all integers n, $\cos\left[\dfrac{(2n + 1)\pi}{2}\right] = 0.$

94. CAPSTONE   Write a study sheet explaining the difference between a trigonometric identity and a conditional equation. Include suggestions on how to verify a trigonometric identity.

Cumulative Mixed Review

Graphing an Exponential Function   In Exercises 95–98, use a graphing utility to construct a table of values for the function. Then sketch the graph of the function. Identify any asymptotes of the graph.

95. $f(x) = 2^x + 3$ **96.** $f(x) = -2^{x-3}$

97. $f(x) = 2^{-x} + 1$ **98.** $f(x) = 2^{x-1} + 3$

6.3   Solving Trigonometric Equations

Introduction

To solve a trigonometric equation, use standard algebraic techniques such as collecting like terms and factoring. Your preliminary goal is to isolate the trigonometric function involved in the equation.

Example 1   Solving a Trigonometric Equation

Solve $2 \sin x - 1 = 0$.

Solution

$2 \sin x - 1 = 0$	Write original equation.
$2 \sin x = 1$	Add 1 to each side.
$\sin x = \frac{1}{2}$	Divide each side by 2.

To solve for x, note in Figure 6.4 that the equation $\sin x = \frac{1}{2}$ has solutions $x = \pi/6$ and $x = 5\pi/6$ in the interval $[0, 2\pi)$. Moreover, because $\sin x$ has a period of 2π, there are infinitely many other solutions, which can be written as

$$x = \frac{\pi}{6} + 2n\pi \qquad \text{and} \qquad x = \frac{5\pi}{6} + 2n\pi \qquad \text{General solution}$$

where n is an integer, as shown in Figure 6.4.

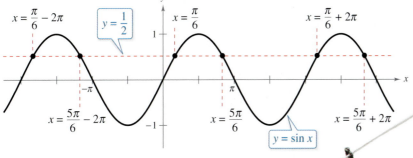

Figure 6.4

✓**CHECKPOINT**   Now try Exercise 29.

Another way to show that the equation $\sin x = \frac{1}{2}$ has infinitely many solutions is indicated in Figure 6.5. Any angles that are coterminal with $\pi/6$ or $5\pi/6$ are also solutions of the equation.

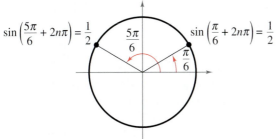

Figure 6.5

Example 2 Collecting Like Terms

Find all solutions of

$$\sin x + \sqrt{2} = -\sin x$$

in the interval $[0, 2\pi)$.

Algebraic Solution

Rewrite the equation so that $\sin x$ is isolated on one side of the equation.

$$\sin x + \sqrt{2} = -\sin x \qquad \text{Write original equation.}$$

$$\sin x + \sin x = -\sqrt{2} \qquad \text{Add } \sin x \text{ to and subtract } \sqrt{2} \text{ from each side.}$$

$$2 \sin x = -\sqrt{2} \qquad \text{Combine like terms.}$$

$$\sin x = -\frac{\sqrt{2}}{2} \qquad \text{Divide each side by 2.}$$

The solutions in the interval $[0, 2\pi)$ are

$$x = \frac{5\pi}{4} \qquad \text{and} \qquad x = \frac{7\pi}{4}.$$

✓**CHECKPOINT** Now try Exercise 37.

Numerical Solution

Use a graphing utility set in *radian* mode to create a table that shows the values of $y_1 = \sin x + \sqrt{2}$ and $y_2 = -\sin x$ for different values of x. Your table should go from $x = 0$ to $x = 2\pi$ using increments of $\pi/8$, as shown in Figure 6.6.

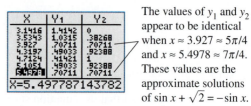

The values of y_1 and y_2 appear to be identical when $x \approx 3.927 \approx 5\pi/4$ and $x \approx 5.4978 \approx 7\pi/4$. These values are the approximate solutions of $\sin x + \sqrt{2} = -\sin x$.

Figure 6.6

Example 3 Extracting Square Roots

Solve $3 \tan^2 x - 1 = 0$.

Solution

Rewrite the equation so that $\tan x$ is isolated on one side of the equation.

$$3 \tan^2 x - 1 = 0 \qquad \text{Write original equation.}$$

$$3 \tan^2 x = 1 \qquad \text{Add 1 to each side.}$$

$$\tan^2 x = \frac{1}{3} \qquad \text{Divide each side by 3.}$$

$$\tan x = \pm\frac{1}{\sqrt{3}} \qquad \text{Extract square roots.}$$

$$\tan x = \pm\frac{\sqrt{3}}{3} \qquad \text{Rationalize the denominator.}$$

Because $\tan x$ has a period of π, first find all solutions in the interval $[0, \pi)$. These are

$$x = \frac{\pi}{6} \qquad \text{and} \qquad x = \frac{5\pi}{6}.$$

Finally, add multiples of π to each of these solutions to get the general form

$$x = \frac{\pi}{6} + n\pi \qquad \text{and} \qquad x = \frac{5\pi}{6} + n\pi \qquad \text{General solution}$$

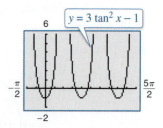

Figure 6.7

where n is an integer. You can confirm this answer by graphing $y = 3 \tan^2 x - 1$ with a graphing utility, as shown in Figure 6.7. The graph has x-intercepts at $\pi/6$, $5\pi/6$, $7\pi/6$, and so on. These x-intercepts correspond to the solutions of $3 \tan^2 x - 1 = 0$.

✓**CHECKPOINT** Now try Exercise 39.

The equations in Examples 1–3 involved only one trigonometric function. When two or more functions occur in the same equation, collect all terms on one side and try to separate the functions by factoring or by using appropriate identities. This may produce factors that yield no solutions, as illustrated in Example 4.

Example 4   Factoring

Solve $\cot x \cos^2 x = 2 \cot x$.

Solution

Begin by rewriting the equation so that all terms are collected on one side of the equation.

$$\cot x \cos^2 x = 2 \cot x \qquad \text{Write original equation.}$$

$$\cot x \cos^2 x - 2 \cot x = 0 \qquad \text{Subtract 2 cot } x \text{ from each side.}$$

$$\cot x (\cos^2 x - 2) = 0 \qquad \text{Factor.}$$

By setting each of these factors equal to zero, you obtain the following.

$$\cot x = 0 \qquad \text{and} \qquad \cos^2 x - 2 = 0$$

$$\cos^2 x = 2$$

$$\cos x = \pm \sqrt{2}$$

In the interval $(0, \pi)$, the equation $\cot x = 0$ has the solution

$$x = \frac{\pi}{2}.$$

No solution is obtained for

$$\cos x = \pm \sqrt{2}$$

because $\pm \sqrt{2}$ are outside the range of the cosine function. Because $\cot x$ has a period of π, the general form of the solution is obtained by adding multiples of π to $x = \pi/2$, to get

$$x = \frac{\pi}{2} + n\pi \qquad \text{General solution}$$

where n is an integer. The graph of $y = \cot x \cos^2 x - 2 \cot x$ (in *dot* mode), shown in Figure 6.8, confirms this result. From the graph you can see that the x-intercepts occur at

$$-\frac{\pi}{2}, \quad \frac{\pi}{2}, \quad \frac{3\pi}{2}, \quad \frac{5\pi}{2}$$

and so on. These x-intercepts correspond to the solutions of

$$\cot x \cos^2 x = 2 \cot x.$$

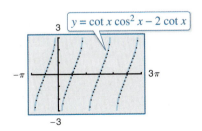

Figure 6.8

 CHECKPOINT   Now try Exercise 41.

Explore the Concept

 Using the equation in Example 4, explain what happens when each side of the equation is divided by $\cot x$. Why is this an incorrect method to use when solving an equation?

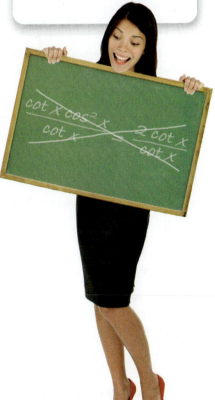

Equations of Quadratic Type

Many trigonometric equations are of quadratic type $ax^2 + bx + c = 0$, as shown below. To solve equations of this type, factor the quadratic or, when factoring is not possible, use the Quadratic Formula.

Quadratic in sin x

$2 \sin^2 x - \sin x - 1 = 0$

$2(\sin x)^2 - \sin x - 1 = 0$

Quadratic in sec x

$\sec^2 x - 3 \sec x - 2 = 0$

$(\sec x)^2 - 3(\sec x) - 2 = 0$

Example 5 Factoring an Equation of Quadratic Type

Find all solutions of $2 \sin^2 x - \sin x - 1 = 0$ in the interval $[0, 2\pi)$.

Algebraic Solution

Treating the equation as a quadratic in $\sin x$ and factoring produces the following.

$2 \sin^2 x - \sin x - 1 = 0$ Write original equation.

$(2 \sin x + 1)(\sin x - 1) = 0$ Factor.

Setting each factor equal to zero, you obtain the following solutions in the interval $[0, 2\pi)$.

$2 \sin x + 1 = 0$ and $\sin x - 1 = 0$

$\sin x = -\dfrac{1}{2}$ $\sin x = 1$

$x = \dfrac{7\pi}{6}, \dfrac{11\pi}{6}$ $x = \dfrac{\pi}{2}$

✓**CHECKPOINT**   Now try Exercise 49.

Graphical Solution

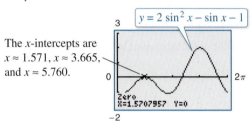

The x-intercepts are $x \approx 1.571$, $x \approx 3.665$, and $x \approx 5.760$.

Figure 6.9

From Figure 6.9, you can conclude that the approximate solutions of $2 \sin^2 x - \sin x - 1 = 0$ in the interval $[0, 2\pi)$ are

$x \approx 1.571 \approx \dfrac{\pi}{2}$, $x \approx 3.665 \approx \dfrac{7\pi}{6}$, and $x \approx 5.760 \approx \dfrac{11\pi}{6}$.

When working with an equation of quadratic type, be sure that the equation involves a *single* trigonometric function, as shown in the next example.

Example 6 Rewriting with a Single Trigonometric Function

Solve $2 \sin^2 x + 3 \cos x - 3 = 0$.

Solution

Begin by rewriting the equation so that it has only cosine functions.

$2 \sin^2 x + 3 \cos x - 3 = 0$ Write original equation.

$2(1 - \cos^2 x) + 3 \cos x - 3 = 0$ Pythagorean identity

$2 \cos^2 x - 3 \cos x + 1 = 0$ Combine like terms and multiply each side by -1.

$(2 \cos x - 1)(\cos x - 1) = 0$ Factor.

By setting each factor equal to zero, you can find the solutions in the interval $[0, 2\pi)$ to be $x = 0$, $x = \pi/3$, and $x = 5\pi/3$. Because $\cos x$ has a period of 2π, the general solution is

$x = 2n\pi$, $x = \dfrac{\pi}{3} + 2n\pi$, $x = \dfrac{5\pi}{3} + 2n\pi$ General solution

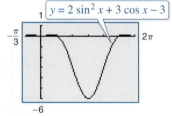

where n is an integer. The graph of $y = 2 \sin^2 x + 3 \cos x - 3$, shown in Figure 6.10, confirms this result.

Figure 6.10

✓**CHECKPOINT**   Now try Exercise 51.

Sometimes you must square each side of an equation to obtain a quadratic. Because this procedure can introduce extraneous solutions, you should check any solutions in the original equation to see whether they are valid or extraneous.

Example 7 Squaring and Converting to Quadratic Type

Find all solutions of $\cos x + 1 = \sin x$ in the interval $[0, 2\pi)$.

Solution

It is not clear how to rewrite this equation in terms of a single trigonometric function. Notice what happens when you square each side of the equation.

$\cos x + 1 = \sin x$	Write original equation.
$\cos^2 x + 2\cos x + 1 = \sin^2 x$	Square each side.
$\cos^2 x + 2\cos x + 1 = 1 - \cos^2 x$	Pythagorean identity
$2\cos^2 x + 2\cos x = 0$	Combine like terms.
$2\cos x(\cos x + 1) = 0$	Factor.

Setting each factor equal to zero produces the following.

$$2\cos x = 0 \quad \text{and} \quad \cos x + 1 = 0$$

$$\cos x = 0 \qquad\qquad \cos x = -1$$

$$x = \frac{\pi}{2}, \frac{3\pi}{2} \qquad\qquad x = \pi$$

Because you squared the original equation, check for extraneous solutions.

Check

$\cos \dfrac{\pi}{2} + 1 \overset{?}{=} \sin \dfrac{\pi}{2}$	Substitute $\pi/2$ for x.
$0 + 1 = 1$	Solution checks. ✓
$\cos \dfrac{3\pi}{2} + 1 \overset{?}{=} \sin \dfrac{3\pi}{2}$	Substitute $3\pi/2$ for x.
$0 + 1 \neq -1$	Solution does not check.
$\cos \pi + 1 \overset{?}{=} \sin \pi$	Substitute π for x.
$-1 + 1 = 0$	Solution checks. ✓

Of the three possible solutions, $x = 3\pi/2$ is extraneous. So, in the interval $[0, 2\pi)$, the only solutions are $x = \pi/2$ and $x = \pi$. The graph of $y = \cos x + 1 - \sin x$, shown in Figure 6.11, confirms this result because the graph has two x-intercepts (at $x = \pi/2$ and $x = \pi$) in the interval $[0, 2\pi)$.

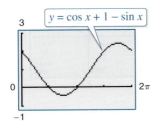

Figure 6.11

✓CHECKPOINT Now try Exercise 53.

Explore the Concept

Use a graphing utility to confirm the solutions found in Example 7 in two different ways. Do both methods produce the same x-values? Which method do you prefer? Why?

1. Graph both sides of the equation and find the x-coordinates of the points at which the graphs intersect.

   Left side: $y = \cos x + 1$

   Right side: $y = \sin x$

2. Graph the equation $y = \cos x + 1 - \sin x$ and find the x-intercepts of the graph.

103. Geometry The area of a rectangle inscribed in one arc of the graph of $y = \cos x$ (see figure) is given by

$$A = 2x \cos x, \quad 0 \le x \le \frac{\pi}{2}.$$

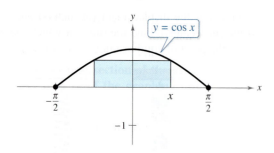

(a) Use a graphing utility to graph the area function, and approximate the area of the largest inscribed rectangle.

(b) Determine the values of x for which $A \ge 1$.

104. MODELING DATA

The unemployment rates r in the United States from 2000 through 2008 are shown in the table, where t represents the year, with $t = 0$ corresponding to 2000. (Source: U.S. Bureau of Labor Statistics)

Year, t	Rate, r
0	4.0
1	4.7
2	5.8
3	6.0
4	5.5
5	5.1
6	4.6
7	4.6
8	5.8

(a) Use a graphing utility to create a scatter plot of the data.

(b) A model for the data is given by

$$r = 0.90 \sin(1.04t - 1.62) + 5.23.$$

Graph the model with the scatter plot from part (a). Is the model a good fit for the data? Explain.

(c) What term in the model gives the average unemployment rate? What is the rate?

(d) Economists study the lengths of business cycles, such as unemployment rates. Based on this short span of time, use the model to determine the length of this cycle.

(e) Use the model to estimate the next time the unemployment rate will be 5% or less.

Approximating the Number of Intersections In Exercises 105 and 106, use the graph to approximate the number of points of intersection of the graphs of y_1 and y_2.

105. $y_1 = 2 \sin x$
$y_2 = 3x + 1$

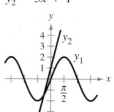

106. $y_1 = 2 \sin x$
$y_2 = \frac{1}{2}x + 1$

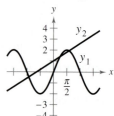

Conclusions

True or False? In Exercises 107–110, determine whether the statement is true or false. Justify your answer.

107. All trigonometric equations have either an infinite number of solutions or no solution.

108. The solutions of any trigonometric equation can always be found from its solutions in the interval $[0, 2\pi)$.

109. If you correctly solve a trigonometric equation down to the statement $\sin x = 3.4$, then you can finish solving the equation by using an inverse trigonometric function.

110. The equation $2 \sin 3t - 1 = 0$ has three times the number of solutions in the interval $[0, 2\pi)$ as the equation $2 \sin t - 1 = 0$.

111. Writing Describe the difference between verifying an identity and solving an equation.

112. CAPSTONE Consider the equation $2 \sin x = 1$. Explain the similarities and differences among finding all solutions in the interval $[0, \pi/2)$, finding all solutions in the interval $[0, 2\pi)$, and finding the general solution.

Cumulative Mixed Review

Converting from Degrees to Radians In Exercises 113–116, convert the angle measure from degrees to radians. Round your answer to three decimal places.

113. $124°$ **114.** $486°$

115. $-0.41°$ **116.** $-210.55°$

117. Make a Decision To work an extended application analyzing the normal daily high temperatures in Phoenix and in Seattle, visit this textbook's *Companion Website*. (Data Source: NOAA)

6.4   Sum and Difference Formulas

Using Sum and Difference Formulas

In this section and the following section, you will study the uses of several trigonometric identities and formulas.

Sum and Difference Formulas   (See proofs on page 540.)

$$\sin(u + v) = \sin u \cos v + \cos u \sin v$$

$$\sin(u - v) = \sin u \cos v - \cos u \sin v$$

$$\cos(u + v) = \cos u \cos v - \sin u \sin v$$

$$\cos(u - v) = \cos u \cos v + \sin u \sin v$$

$$\tan(u + v) = \frac{\tan u + \tan v}{1 - \tan u \tan v}$$

$$\tan(u - v) = \frac{\tan u - \tan v}{1 + \tan u \tan v}$$

Example 1 shows how **sum and difference formulas** can be used to find exact values of trigonometric functions involving sums or differences of special angles.

What you should learn

● Use sum and difference formulas to evaluate trigonometric functions, verify trigonometric identities, and solve trigonometric equations.

Why you should learn it

You can use sum and difference formulas to rewrite trigonometric expressions. For instance, Exercise 89 on page 522 shows how to use sum and difference formulas to rewrite a trigonometric expression in a form that helps you find the equation of a standing wave.

Example 1   Evaluating a Trigonometric Function

Find the exact value of (a) $\cos 75°$ and (b) $\sin \dfrac{\pi}{12}$.

Solution

a. Using the fact that $75° = 30° + 45°$ with the formula for $\cos(u + v)$ yields

$$\cos 75° = \cos(30° + 45°)$$

$$= \cos 30° \cos 45° - \sin 30° \sin 45°$$

$$= \frac{\sqrt{3}}{2}\left(\frac{\sqrt{2}}{2}\right) - \frac{1}{2}\left(\frac{\sqrt{2}}{2}\right)$$

$$= \frac{\sqrt{6} - \sqrt{2}}{4}.$$

Try checking this result on your calculator. You will find that $\cos 75° \approx 0.259$.

b. Using the fact that

$$\frac{\pi}{12} = \frac{\pi}{3} - \frac{\pi}{4}$$

with the formula for $\sin(u - v)$ yields

$$\sin \frac{\pi}{12} = \sin\left(\frac{\pi}{3} - \frac{\pi}{4}\right)$$

$$= \sin \frac{\pi}{3} \cos \frac{\pi}{4} - \cos \frac{\pi}{3} \sin \frac{\pi}{4}$$

$$= \frac{\sqrt{3}}{2}\left(\frac{\sqrt{2}}{2}\right) - \frac{1}{2}\left(\frac{\sqrt{2}}{2}\right)$$

$$= \frac{\sqrt{6} - \sqrt{2}}{4}.$$

 CHECKPOINT   Now try Exercise 9.

6.5 Multiple-Angle and Product-to-Sum Formulas

Multiple-Angle Formulas

In this section, you will study four additional categories of trigonometric identities.

1. The first category involves *functions of multiple angles* such as

$$\sin ku \quad \text{and} \quad \cos ku.$$

2. The second category involves *squares of trigonometric functions* such as

$$\sin^2 u.$$

3. The third category involves *functions of half-angles* such as

$$\sin \frac{u}{2}.$$

4. The fourth category involves *products of trigonometric functions* such as

$$\sin u \cos v.$$

You should learn the **double-angle formulas** below because they are used often in trigonometry and calculus.

> **Double-Angle Formulas** (See the proofs on page 541.)
>
> $$\sin 2u = 2 \sin u \cos u \qquad \tan 2u = \frac{2 \tan u}{1 - \tan^2 u}$$
>
> $$\cos 2u = \cos^2 u - \sin^2 u$$
> $$= 2 \cos^2 u - 1$$
> $$= 1 - 2 \sin^2 u$$

Example 1 Solving a Multiple-Angle Equation

Solve $2 \cos x + \sin 2x = 0$.

Solution

Begin by rewriting the equation so that it involves functions of x (rather than $2x$). Then factor and solve as usual.

$2 \cos x + \sin 2x = 0$	Write original equation.
$2 \cos x + 2 \sin x \cos x = 0$	Double-angle formula
$2 \cos x(1 + \sin x) = 0$	Factor.
$2 \cos x = 0 \qquad 1 + \sin x = 0$	Set factors equal to zero.
$\cos x = 0 \qquad \sin x = -1$	Isolate trigonometric functions.
$x = \dfrac{\pi}{2}, \dfrac{3\pi}{2} \qquad x = \dfrac{3\pi}{2}$	Solutions in $[0, 2\pi)$

So, the general solution is

$$x = \frac{\pi}{2} + 2n\pi \quad \text{and} \quad x = \frac{3\pi}{2} + 2n\pi \qquad \text{General solution}$$

where n is an integer. Try verifying this solution graphically.

✓CHECKPOINT Now try Exercise 11.

What you should learn

- Use multiple-angle formulas to rewrite and evaluate trigonometric functions.
- Use power-reducing formulas to rewrite and evaluate trigonometric functions.
- Use half-angle formulas to rewrite and evaluate trigonometric functions.
- Use product-to-sum and sum-to-product formulas to rewrite and evaluate trigonometric functions.

Why you should learn it

You can use a variety of trigonometric formulas to rewrite trigonometric functions in more convenient forms. For instance, Exercise 126 on page 533 shows you how to use a half-angle formula to determine the apex angle of a sound wave cone caused by the speed of an airplane.

Example 2   Evaluating Functions Involving Double Angles

Use the following to find $\sin 2\theta$, $\cos 2\theta$, and $\tan 2\theta$.

$$\cos \theta = \frac{5}{13}, \qquad \frac{3\pi}{2} < \theta < 2\pi$$

Solution

In Figure 6.24, you can see that

$$\sin \theta = \frac{y}{r} = -\frac{12}{13}$$

and

$$\tan \theta = -\frac{12}{5}.$$

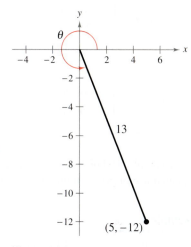

Figure 6.24

Consequently, using each of the double-angle formulas, you can write the double angles as follows.

$$\sin 2\theta = 2 \sin \theta \cos \theta \qquad\qquad \cos 2\theta = 2 \cos^2 \theta - 1$$

$$= 2\left(-\frac{12}{13}\right)\left(\frac{5}{13}\right) \qquad\qquad = 2\left(\frac{25}{169}\right) - 1$$

$$= -\frac{120}{169} \qquad\qquad\qquad = -\frac{119}{169}$$

$$\tan 2\theta = \frac{2 \tan \theta}{1 - \tan^2 \theta} = \frac{2(-12/5)}{1 - (-12/5)^2} = \frac{120}{119}$$

✓**CHECKPOINT**   Now try Exercise 21.

The double-angle formulas are not restricted to the angles 2θ and θ. Other *double* combinations, such as 4θ and 2θ or 6θ and 3θ, are also valid. Here are two examples.

$$\sin 4\theta = 2 \sin 2\theta \cos 2\theta \qquad \text{and} \qquad \cos 6\theta = \cos^2 3\theta - \sin^2 3\theta$$

By using double-angle formulas together with the sum formulas derived in the preceding section, you can form other multiple-angle formulas.

Example 3   Deriving a Triple-Angle Formula

Rewrite $\sin 3x$ in terms of $\sin x$.

Solution

$\sin 3x = \sin(2x + x)$	Rewrite as a sum.
$= \sin 2x \cos x + \cos 2x \sin x$	Sum formula
$= 2 \sin x \cos x \cos x + (1 - 2 \sin^2 x) \sin x$	Double-angle formula
$= 2 \sin x \cos^2 x + \sin x - 2 \sin^3 x$	Multiply.
$= 2 \sin x(1 - \sin^2 x) + \sin x - 2 \sin^3 x$	Pythagorean identity
$= 2 \sin x - 2 \sin^3 x + \sin x - 2 \sin^3 x$	Multiply.
$= 3 \sin x - 4 \sin^3 x$	Simplify.

✓**CHECKPOINT**   Now try Exercise 27.

6 Chapter Summary

	What did you learn?	**Explanation and Examples**	**Review Exercises**
6.1	Recognize and write the fundamental trigonometric identities (*p. 490*).	**Reciprocal Identities** $\sin u = 1/\csc u \qquad \cos u = 1/\sec u \qquad \tan u = 1/\cot u$ $\csc u = 1/\sin u \qquad \sec u = 1/\cos u \qquad \cot u = 1/\tan u$ **Quotient Identities** $\tan u = \dfrac{\sin u}{\cos u} \qquad \cot u = \dfrac{\cos u}{\sin u}$ **Pythagorean Identities** $\sin^2 u + \cos^2 u = 1 \qquad\qquad 1 + \tan^2 u = \sec^2 u$ $1 + \cot^2 u = \csc^2 u$ **Cofunction Identities** $\sin[(\pi/2) - u] = \cos u \qquad \cos[(\pi/2) - u] = \sin u$ $\tan[(\pi/2) - u] = \cot u \qquad \cot[(\pi/2) - u] = \tan u$ $\sec[(\pi/2) - u] = \csc u \qquad \csc[(\pi/2) - u] = \sec u$ **Even/Odd Identities** $\sin(-u) = -\sin u \quad \cos(-u) = \cos u \quad \tan(-u) = -\tan u$ $\csc(-u) = -\csc u \quad \sec(-u) = \sec u \quad \cot(-u) = -\cot u$	1–10
	Use the fundamental trigonometric identities to evaluate trigonometric functions, and simplify and rewrite trigonometric expressions (*p. 491*).	On occasion, factoring or simplifying trigonometric expressions can best be done by first rewriting the expression in terms of just *one* trigonometric function or in terms of *sine or cosine alone*.	11–26
6.2	Verify trigonometric identities (*p. 497*).	**Guidelines for Verifying Trigonometric Identities** 1. Work with one side of the equation at a time. 2. Look to factor an expression, add fractions, square a binomial, or create a monomial denominator. 3. Look to use the fundamental identities. Note which functions are in the final expression you want. Sines and cosines pair up well, as do secants and tangents, and cosecants and cotangents. 4. When the preceding guidelines do not help, try converting all terms to sines and cosines. 5. Always try *something*.	27–38
6.3	Use standard algebraic techniques to solve trigonometric equations (*p. 505*).	Use standard algebraic techniques such as collecting like terms, extracting square roots, and factoring to solve trigonometric equations.	39–50
	Solve trigonometric equations of quadratic type (*p. 508*).	To solve trigonometric equations of quadratic type $ax^2 + bx + c = 0$, factor the quadratic or, when this is not possible, use the Quadratic Formula.	51–54
	Solve trigonometric equations involving multiple angles (*p. 510*).	To solve equations that contain forms such as $\sin ku$ or $\cos ku$, first solve the equation for ku, then divide your result by k.	55–62
	Use inverse trigonometric functions to solve trigonometric equations (*p. 511*).	After factoring an equation and setting the factors equal to 0, you may get an equation such as $\tan x - 3 = 0$. In this case, use inverse trigonometric functions to solve. (See Example 10.)	63–66

What did you learn?	Explanation and Examples	Review Exercises
6.4 Use sum and difference formulas to evaluate trigonometric functions, verify trigonometric identities, and solve trigonometric equations (*p. 517*).	**Sum and Difference Formulas** $\sin(u + v) = \sin u \cos v + \cos u \sin v$ $\sin(u - v) = \sin u \cos v - \cos u \sin v$ $\cos(u + v) = \cos u \cos v - \sin u \sin v$ $\cos(u - v) = \cos u \cos v + \sin u \sin v$ $\tan(u + v) = \dfrac{\tan u + \tan v}{1 - \tan u \tan v}$ $\tan(u - v) = \dfrac{\tan u - \tan v}{1 + \tan u \tan v}$	67–88
6.5 Use multiple-angle formulas to rewrite and evaluate trigonometric functions (*p. 524*).	**Double-Angle Formulas** $\sin 2u = 2 \sin u \cos u \qquad \cos 2u = \cos^2 u - \sin^2 u$ $\qquad\qquad\qquad\qquad\qquad = 2 \cos^2 u - 1$ $\tan 2u = \dfrac{2 \tan u}{1 - \tan^2 u} \qquad = 1 - 2 \sin^2 u$	89–98
Use power-reducing formulas to rewrite and evaluate trigonometric functions (*p. 526*).	**Power-Reducing Formulas** $\sin^2 u = \dfrac{1 - \cos 2u}{2}$ $\cos^2 u = \dfrac{1 + \cos 2u}{2}$ $\tan^2 u = \dfrac{1 - \cos 2u}{1 + \cos 2u}$	99–104
Use half-angle formulas to rewrite and evaluate trigonometric functions (*p. 527*).	**Half-Angle Formulas** $\sin \dfrac{u}{2} = \pm \sqrt{\dfrac{1 - \cos u}{2}} \qquad \cos \dfrac{u}{2} = \pm \sqrt{\dfrac{1 + \cos u}{2}}$ $\tan \dfrac{u}{2} = \dfrac{1 - \cos u}{\sin u} = \dfrac{\sin u}{1 + \cos u}$ The signs of $\sin \dfrac{u}{2}$ and $\cos \dfrac{u}{2}$ depend on the quadrant in which $\dfrac{u}{2}$ lies.	105–118
Use product-to-sum formulas and sum-to-product formulas to rewrite and evaluate trigonometric functions (*p. 528*).	**Product-to-Sum Formulas** $\sin u \sin v = (1/2)[\cos(u - v) - \cos(u + v)]$ $\cos u \cos v = (1/2)[\cos(u - v) + \cos(u + v)]$ $\sin u \cos v = (1/2)[\sin(u + v) + \sin(u - v)]$ $\cos u \sin v = (1/2)[\sin(u + v) - \sin(u - v)]$ **Sum-to-Product Formulas** $\sin u + \sin v = 2 \sin\left(\dfrac{u + v}{2}\right) \cos\left(\dfrac{u - v}{2}\right)$ $\sin u - \sin v = 2 \cos\left(\dfrac{u + v}{2}\right) \sin\left(\dfrac{u - v}{2}\right)$ $\cos u + \cos v = 2 \cos\left(\dfrac{u + v}{2}\right) \cos\left(\dfrac{u - v}{2}\right)$ $\cos u - \cos v = -2 \sin\left(\dfrac{u + v}{2}\right) \sin\left(\dfrac{u - v}{2}\right)$	119–130

6   Review Exercises

See www.CalcChat.com for worked-out solutions to odd-numbered exercises.
For instructions on how to use a graphing utility, see Appendix A.

6.1

Recognizing Fundamental Trigonometric Identities   In Exercises 1–10, name the trigonometric function that is equivalent to the expression.

1. $\dfrac{1}{\cos x}$

2. $\dfrac{1}{\sin x}$

3. $\dfrac{1}{\sec x}$

4. $\dfrac{1}{\tan x}$

5. $\sqrt{1 - \cos^2 x}$

6. $\sqrt{1 + \tan^2 x}$

7. $\csc\left(\dfrac{\pi}{2} - x\right)$

8. $\cot\left(\dfrac{\pi}{2} - x\right)$

9. $\sec(-x)$

10. $\tan(-x)$

Using Identities to Evaluate a Function   In Exercises 11–14, use the given values to evaluate (if possible) the remaining trigonometric functions of the angle.

11. $\sin x = \dfrac{4}{5}, \quad \cos x = \dfrac{3}{5}$

12. $\tan \theta = \dfrac{2}{3}, \quad \sec \theta = \dfrac{\sqrt{13}}{3}$

13. $\sin\left(\dfrac{\pi}{2} - x\right) = \dfrac{1}{\sqrt{2}}, \quad \sin x = -\dfrac{1}{\sqrt{2}}$

14. $\csc\left(\dfrac{\pi}{2} - \theta\right) = 3, \quad \sin \theta = \dfrac{2\sqrt{2}}{3}$

Simplifying a Trigonometric Expression   In Exercises 15–24, use the fundamental identities to simplify the expression. Use the *table* feature of a graphing utility to check your result numerically.

15. $\dfrac{1}{\tan^2 x + 1}$

16. $\dfrac{\sec^2 x - 1}{\sec x - 1}$

17. $\dfrac{\sin^2 \theta + \cos^2 \theta}{\sin \theta}$

18. $\dfrac{\sec^2(-\theta)}{\csc^2 \theta}$

19. $\tan^2 \theta(\csc^2 \theta - 1)$

20. $\csc^2 x(1 - \cos^2 x)$

21. $\tan\left(\dfrac{\pi}{2} - x\right) \sec x$

22. $\dfrac{\sin(-x)\cot x}{\sin\left(\dfrac{\pi}{2} - x\right)}$

23. $\dfrac{\sin^2 \alpha - \cos^2 \alpha}{\sin^2 \alpha - \sin \alpha \cos \alpha}$

24. $\dfrac{\sin^3 \beta + \cos^3 \beta}{\sin \beta + \cos \beta}$

∫ 25. **Rate of Change**   The rate of change of the function $f(x) = 2\sqrt{\sin x}$ is given by $\sin^{-1/2} x \cos x$. Show that the rate of change can also be written as $\cot x \sqrt{\sin x}$.

∫ 26. **Rate of Change**   The rate of change of the function $f(x) = \csc x - \cot x$ is given by $\csc^2 x - \csc x \cot x$. Show that the rate of change can also be written as $(1 - \cos x)/\sin^2 x$.

6.2

Verifying a Trigonometric Identity   In Exercises 27–38, verify the identity.

27. $\cos x(\tan^2 x + 1) = \sec x$

28. $\sec^2 x \cot x - \cot x = \tan x$

29. $\sin^3 \theta + \sin \theta \cos^2 \theta = \sin \theta$

30. $\cot^2 x - \cos^2 x = \cot^2 x \cos^2 x$

31. $\sin^5 x \cos^2 x = (\cos^2 x - 2\cos^4 x + \cos^6 x)\sin x$

32. $\cos^3 x \sin^2 x = (\sin^2 x - \sin^4 x)\cos x$

33. $\sqrt{\dfrac{1 - \sin \theta}{1 + \sin \theta}} = \dfrac{1 - \sin \theta}{|\cos \theta|}$

34. $\sqrt{1 - \cos x} = \dfrac{|\sin x|}{\sqrt{1 + \cos x}}$

35. $\dfrac{\csc(-x)}{\sec(-x)} = -\cot x$

36. $\dfrac{1 + \sec(-x)}{\sin(-x) + \tan(-x)} = -\csc x$

37. $\csc^2\left(\dfrac{\pi}{2} - x\right) - 1 = \tan^2 x$

38. $\tan\left(\dfrac{\pi}{2} - x\right) \sec x = \csc x$

6.3

Solving a Trigonometric Equation   In Exercises 39–50, solve the equation.

39. $2 \sin x - 1 = 0$

40. $\tan x + 1 = 0$

41. $\sin x = \sqrt{3} - \sin x$

42. $4 \cos x = 1 + 2 \cos x$

43. $3\sqrt{3} \tan x = 3$

44. $\frac{1}{2} \sec x - 1 = 0$

45. $3 \csc^2 x = 4$

46. $4 \tan^2 x - 1 = \tan^2 x$

47. $4 \cos^2 x - 3 = 0$

48. $\sin x(\sin x + 1) = 0$

49. $\sin x - \tan x = 0$

50. $\csc x - 2 \cot x = 0$

Solving a Trigonometric Equation   In Exercises 51–54, find all solutions of the equation in the interval $[0, 2\pi)$. Use a graphing utility to check your answers.

51. $2 \cos^2 x - \cos x = 1$

52. $2 \sin^2 x - 3 \sin x = -1$

53. $\cos^2 x + \sin x = 1$

54. $\sin^2 x + 2 \cos x = 2$

Functions of Multiple Angles   In Exercises 55–58, find all solutions of the multiple-angle equation in the interval $[0, 2\pi)$.

55. $2 \sin 2x = \sqrt{2}$

56. $\sqrt{3} \tan 3x = 0$

57. $\cos 4x(\cos x - 1) = 0$

58. $3 \csc^2 5x = -4$

Functions of Multiple Angles In Exercises 59–62, solve the multiple-angle equation.

59. $2 \sin 2x - 1 = 0$

60. $2 \cos 4x + \sqrt{3} = 0$

61. $2 \sin^2 3x - 1 = 0$

62. $4 \cos^2 2x - 3 = 0$

Using Inverse Functions In Exercises 63–66, use inverse functions where necessary to find all solutions of the equation in the interval $[0, 2\pi)$.

63. $\sin^2 x - 2 \sin x = 0$

64. $3 \cos^2 x + 5 \cos x = 0$

65. $\tan^2 x + \tan x - 12 = 0$

66. $\sec^2 x + 6 \tan x + 4 = 0$

6.4

Evaluating Trigonometric Functions In Exercises 67–70, find the exact values of the sine, cosine, and tangent of the angle.

67. $285° = 315° - 30°$

68. $345° = 300° + 45°$

69. $\dfrac{31\pi}{12} = \dfrac{11\pi}{6} + \dfrac{3\pi}{4}$

70. $\dfrac{23\pi}{12} = \dfrac{7\pi}{6} + \dfrac{3\pi}{4}$

Rewriting a Trigonometric Expression In Exercises 71–74, write the expression as the sine, cosine, or tangent of an angle.

71. $\sin 130° \cos 50° + \cos 130° \sin 50°$

72. $\cos 45° \cos 120° - \sin 45° \sin 120°$

73. $\dfrac{\tan 25° + \tan 50°}{1 - \tan 25° \tan 50°}$

74. $\dfrac{\tan 63° - \tan 112°}{1 + \tan 63° \tan 112°}$

Evaluating a Trigonometric Expression In Exercises 75–80, find the exact value of the trigonometric function given that $\sin u = \frac{4}{5}$ and $\cos v = -\frac{7}{25}$. (Both u and v are in Quadrant II.)

75. $\sin(u + v)$

76. $\tan(u + v)$

77. $\tan(u - v)$

78. $\sin(u - v)$

79. $\cos(u + v)$

80. $\cos(u - v)$

Verifying a Trigonometric Identity In Exercises 81–86, verify the identity.

81. $\cos\left(x + \dfrac{\pi}{2}\right) = -\sin x$

82. $\sin\left(x - \dfrac{3\pi}{2}\right) = \cos x$

83. $\cot\left(\dfrac{\pi}{2} - x\right) = \tan x$

84. $\sin(\pi - x) = \sin x$

85. $\cos 3x = 4 \cos^3 x - 3 \cos x$

86. $\dfrac{\sin(\alpha + \beta)}{\cos \alpha \cos \beta} = \tan \alpha + \tan \beta$

Solving a Trigonometric Equation In Exercises 87 and 88, find the solutions of the equation in the interval $[0, 2\pi)$.

87. $\sin\left(x + \dfrac{\pi}{2}\right) - \sin\left(x - \dfrac{\pi}{2}\right) = \sqrt{2}$

88. $\cos\left(x + \dfrac{\pi}{4}\right) - \cos\left(x - \dfrac{\pi}{4}\right) = 1$

6.5

Evaluating Functions Involving Double Angles In Exercises 89–92, find the exact values of $\sin 2u$, $\cos 2u$, and $\tan 2u$ using the double-angle formulas.

89. $\sin u = \dfrac{5}{7}, \quad 0 < u < \dfrac{\pi}{2}$

90. $\cos u = \dfrac{4}{5}, \quad \dfrac{3\pi}{2} < u < 2\pi$

91. $\tan u = -\dfrac{2}{9}, \quad \dfrac{\pi}{2} < u < \pi$

92. $\cos u = -\dfrac{2}{\sqrt{5}}, \quad \dfrac{\pi}{2} < u < \pi$

Verifying Trigonometric Identities In Exercises 93–96, use double-angle formulas to verify the identity algebraically. Use a graphing utility to check your result graphically.

93. $6 \sin x \cos x = 3 \sin 2x$

94. $4 \sin x \cos x + 2 = 2 \sin 2x + 2$

95. $1 - 4 \sin^2 x \cos^2 x = \cos^2 2x$

96. $\sin 4x = 8 \cos^3 x \sin x - 4 \cos x \sin x$

97. Projectile Motion A baseball leaves the hand of the first baseman at an angle of θ with the horizontal and with an initial velocity of $v_0 = 80$ feet per second. The ball is caught by the second baseman 100 feet away. Find θ where the range r of a projectile is given by $r = \frac{1}{32} v_0^2 \sin 2\theta$.

98. Projectile Motion Use the equation in Exercise 97 to find θ when a golf ball is hit with an initial velocity of $v_0 = 50$ feet per second and lands 77 feet away.

∫ **Reducing a Power** In Exercises 99–104, use the power-reducing formulas to rewrite the expression in terms of the first power of the cosine.

99. $\sin^6 x$

100. $\cos^4 x \sin^4 x$

101. $\cos^4 2x$

102. $\sin^4 2x$

103. $\tan^2 4x$

104. $\sin^2 2x \tan^2 2x$

Using a Half-Angle Formula In Exercises 105–108, use the half-angle formulas to determine the exact values of the sine, cosine, and tangent of the angle.

105. $15°$

106. $112° \, 30'$

107. $\dfrac{7\pi}{8}$

108. $\dfrac{11\pi}{12}$

Using a Half-Angle Formula In Exercises 109–112, find the exact values of $\sin(u/2)$, $\cos(u/2)$, and $\tan(u/2)$ using the half-angle formulas.

109. $\sin u = \dfrac{3}{5}$, $\quad 0 < u < \dfrac{\pi}{2}$

110. $\tan u = \dfrac{4}{3}$, $\quad \pi < u < \dfrac{3\pi}{2}$

111. $\cos u = -\dfrac{2}{7}$, $\quad \dfrac{\pi}{2} < u < \pi$

112. $\sec u = -6$, $\quad \dfrac{\pi}{2} < u < \pi$

Using a Half-Angle Formula In Exercises 113–116, use the half-angle formulas to simplify the expression.

113. $-\sqrt{\dfrac{1 + \cos 8x}{2}}$

114. $\sqrt{\dfrac{1 - \cos 6x}{2}}$

115. $\dfrac{\sin 10x}{1 + \cos 10x}$

116. $\dfrac{1 - \cos 12x}{\sin 12x}$

Agriculture In Exercises 117 and 118, a trough for feeding cattle is 4 meters long and its cross sections are isosceles triangles with two equal sides of $\frac{1}{2}$ meter (see figure). The angle between the equal sides is θ.

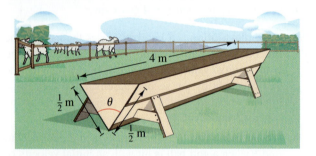

117. Write the trough's volume as a function of $\theta/2$.

118. Write the volume of the trough as a function of θ and determine the value of θ such that the volume is maximum.

Writing Products as Sums In Exercises 119–122, use the product-to-sum formulas to write the product as a sum or difference.

119. $6 \sin \dfrac{\pi}{4} \cos \dfrac{\pi}{4}$

120. $4 \sin 15° \sin 45°$

121. $\sin 5\alpha \sin 4\alpha$

122. $\cos 6\theta \sin 8\theta$

Writing Sums as Products In Exercises 123–126, use the sum-to-product formulas to write the sum or difference as a product.

123. $\cos 5\theta + \cos 4\theta$

124. $\sin 3\theta + \sin 2\theta$

125. $\sin\left(x + \dfrac{\pi}{4}\right) - \sin\left(x - \dfrac{\pi}{4}\right)$

126. $\cos\left(x + \dfrac{\pi}{6}\right) - \cos\left(x - \dfrac{\pi}{6}\right)$

Harmonic Motion In Exercises 127–130, a weight is attached to a spring suspended vertically from a ceiling. When a driving force is applied to the system, the weight moves vertically from its equilibrium position. This motion is described by the model

$$y = 1.5 \sin 8t - 0.5 \cos 8t$$

where y is the distance from equilibrium in feet and t is the time in seconds.

127. Write the model in the form

$$y = \sqrt{a^2 + b^2}\, \sin(Bt + C).$$

128. Use a graphing utility to graph the model.

129. Find the amplitude of the oscillations of the weight.

130. Find the frequency of the oscillations of the weight.

Conclusions

True or False? In Exercises 131–134, determine whether the statement is true or false. Justify your answer.

131. If $\dfrac{\pi}{2} < \theta < \pi$, then $\cos \dfrac{\theta}{2} < 0$.

132. $\sin(x + y) = \sin x + \sin y$

133. $4 \sin(-x) \cos(-x) = -2 \sin 2x$

134. $4 \sin 45° \cos 15° = 1 + \sqrt{3}$

135. **Think About It** List the reciprocal identities, quotient identities, and Pythagorean identities from memory.

136. **Think About It** Is $\cos \theta = \sqrt{1 - \sin^2 \theta}$ an identity? Explain.

137. **Think About It** Is any trigonometric equation with an infinite number of solutions an identity? Explain.

138. **Think About It** Does the equation $a \sin x - b = 0$ have a solution when $|a| < |b|$? Explain.

Think About It In Exercises 139 and 140, use the graphs of y_1 and y_2 to determine how to change y_2 to a new function y_3 such that $y_1 = y_3$.

139. $y_1 = \sec^2\left(\dfrac{\pi}{2} - x\right)$

$y_2 = \cot^2 x$

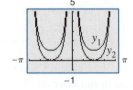

140. $y_1 = \dfrac{\cos 3x}{\cos x}$

$y_2 = (2 \sin x)^2$

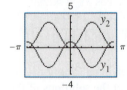

See www.CalcChat.com for worked-out solutions to odd-numbered exercises.
For instructions on how to use a graphing utility, see Appendix A.

6 Chapter Test

Take this test as you would take a test in class. After you are finished, check your work against the answers in the back of the book.

1. Given $\tan \theta = \frac{3}{2}$ and $\cos \theta < 0$, use the fundamental identities to evaluate the other five trigonometric functions of θ.

2. Use the fundamental identities to simplify $\csc^2 \beta (1 - \cos^2 \beta)$.

3. Factor and simplify $\dfrac{\sec^4 x - \tan^4 x}{\sec^2 x + \tan^2 x}$.

4. Add and simplify $\dfrac{\cos \theta}{\sin \theta} + \dfrac{\sin \theta}{\cos \theta}$.

5. Determine the values of θ, $0 \le \theta < 2\pi$, for which $\tan \theta = -\sqrt{\sec^2 \theta - 1}$ is true.

6. Use a graphing utility to graph the functions $y_1 = \sin x + \cos x \cot x$ and $y_2 = \csc x$. Make a conjecture about y_1 and y_2. Verify your result algebraically.

In Exercises 7–12, verify the identity.

7. $\sin \theta \sec \theta = \tan \theta$

8. $\sec^2 x \tan^2 x + \sec^2 x = \sec^4 x$

9. $\dfrac{\csc \alpha + \sec \alpha}{\sin \alpha + \cos \alpha} = \cot \alpha + \tan \alpha$

10. $\cos\left(x + \dfrac{\pi}{2}\right) = -\sin x$

11. $\sin(n\pi + \theta) = (-1)^n \sin \theta$, n is an integer.

12. $(\sin x + \cos x)^2 = 1 + \sin 2x$

13. Find the exact value of $\tan 255°$.

14. Rewrite $\sin^4 x \tan^2 x$ in terms of the first power of the cosine.

15. Use a half-angle formula to simplify the expression $\dfrac{\sin 4\theta}{1 + \cos 4\theta}$.

16. Write $4 \cos 2\theta \sin 4\theta$ as a sum or difference.

17. Write $\sin 3\theta - \sin 4\theta$ as a product.

In Exercises 18–21, find all solutions of the equation in the interval $[0, 2\pi)$.

18. $\tan^2 x + \tan x = 0$

19. $\sin 2\alpha - \cos \alpha = 0$

20. $4 \cos^2 x - 3 = 0$

21. $\csc^2 x - \csc x - 2 = 0$

22. Use a graphing utility to approximate the solutions of the equation $3 \cos x - x = 0$ accurate to three decimal places.

23. Use the figure to find the exact values of $\sin 2u$, $\cos 2u$, and $\tan 2u$.

24. The *index of refraction n* of a transparent material is the ratio of the speed of light in a vacuum to the speed of light in the material. For the triangular glass prism in the figure, $n = 1.5$ and $\alpha = 60°$. Find the angle θ for the glass prism given that

$$n = \dfrac{\sin\left(\dfrac{\theta}{2} + \dfrac{\alpha}{2}\right)}{\sin \dfrac{\theta}{2}}.$$

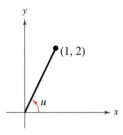

Figure for 23

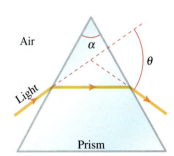

Figure for 24

Proofs in Mathematics

Sum and Difference Formulas   (p. 517)

$$\sin(u + v) = \sin u \cos v + \cos u \sin v$$

$$\sin(u - v) = \sin u \cos v - \cos u \sin v$$

$$\cos(u + v) = \cos u \cos v - \sin u \sin v$$

$$\cos(u - v) = \cos u \cos v + \sin u \sin v$$

$$\tan(u + v) = \frac{\tan u + \tan v}{1 - \tan u \tan v}$$

$$\tan(u - v) = \frac{\tan u - \tan v}{1 + \tan u \tan v}$$

Proof

You can use the figures at the right for the proofs of the formulas for $\cos(u \pm v)$. In the top figure, let A be the point $(1, 0)$ and then use u and v to locate the points $B(x_1, y_1)$, $C(x_2, y_2)$, and $D(x_3, y_3)$ on the unit circle. So, $x_i^2 + y_i^2 = 1$ for $i = 1, 2,$ and 3. For convenience, assume that $0 < v < u < 2\pi$. In the bottom figure, note that arcs AC and BD have the same length. So, line segments AC and BD are also equal in length, which implies that

$$\sqrt{(x_2 - 1)^2 + (y_2 - 0)^2} = \sqrt{(x_3 - x_1)^2 + (y_3 - y_1)^2}$$

$$x_2^2 - 2x_2 + 1 + y_2^2 = x_3^2 - 2x_1x_3 + x_1^2 + y_3^2 - 2y_1y_3 + y_1^2$$

$$(x_2^2 + y_2^2) + 1 - 2x_2 = (x_3^2 + y_3^2) + (x_1^2 + y_1^2) - 2x_1x_3 - 2y_1y_3$$

$$1 + 1 - 2x_2 = 1 + 1 - 2x_1x_3 - 2y_1y_3$$

$$x_2 = x_3x_1 + y_3y_1.$$

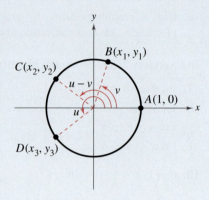

Finally, by substituting the values $x_2 = \cos(u - v)$, $x_3 = \cos u$, $x_1 = \cos v$, $y_3 = \sin u$, and $y_1 = \sin v$, you obtain $\cos(u - v) = \cos u \cos v + \sin u \sin v$. The formula for $\cos(u + v)$ can be established by considering $u + v = u - (-v)$ and using the formula just derived to obtain

$$\cos(u + v) = \cos[u - (-v)] = \cos u \cos(-v) + \sin u \sin(-v)$$

$$= \cos u \cos v - \sin u \sin v.$$

You can use the sum and difference formulas for sine and cosine to prove the formulas for $\tan(u \pm v)$.

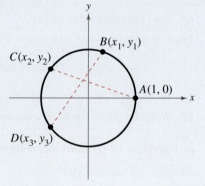

$$\tan(u \pm v) = \frac{\sin(u \pm v)}{\cos(u \pm v)} \qquad \text{Quotient identity}$$

$$= \frac{\sin u \cos v \pm \cos u \sin v}{\cos u \cos v \mp \sin u \sin v} \qquad \text{Sum and difference formulas}$$

$$= \frac{\dfrac{\sin u \cos v \pm \cos u \sin v}{\cos u \cos v}}{\dfrac{\cos u \cos v \mp \sin u \sin v}{\cos u \cos v}} \qquad \text{Divide numerator and denominator by } \cos u \cos v.$$

$$= \dfrac{\dfrac{\sin u \cos v}{\cos u \cos v} \pm \dfrac{\cos u \sin v}{\cos u \cos v}}{\dfrac{\cos u \cos v}{\cos u \cos v} \mp \dfrac{\sin u \sin v}{\cos u \cos v}}$$ Write as separate fractions.

$$= \dfrac{\dfrac{\sin u}{\cos u} \pm \dfrac{\sin v}{\cos v}}{1 \mp \dfrac{\sin u}{\cos u} \cdot \dfrac{\sin v}{\cos v}}$$ Product of fractions

$$= \dfrac{\tan u \pm \tan v}{1 \mp \tan u \tan v}$$ Quotient identity

---

Double-Angle Formulas   (p. 524)

$$\sin 2u = 2 \sin u \cos u$$

$$\cos 2u = \cos^2 u - \sin^2 u$$
$$= 2 \cos^2 u - 1$$
$$= 1 - 2 \sin^2 u$$

$$\tan 2u = \dfrac{2 \tan u}{1 - \tan^2 u}$$

Proof

To prove all three formulas, let $v = u$ in the corresponding sum formulas.

$$\sin 2u = \sin(u + u) = \sin u \cos u + \cos u \sin u = 2 \sin u \cos u$$

$$\cos 2u = \cos(u + u) = \cos u \cos u - \sin u \sin u = \cos^2 u - \sin^2 u$$

$$\tan 2u = \tan(u + u) = \dfrac{\tan u + \tan u}{1 - \tan u \tan u} = \dfrac{2 \tan u}{1 - \tan^2 u}$$

Power-Reducing Formulas   (p. 526)

$$\sin^2 u = \dfrac{1 - \cos 2u}{2} \qquad \cos^2 u = \dfrac{1 + \cos 2u}{2} \qquad \tan^2 u = \dfrac{1 - \cos 2u}{1 + \cos 2u}$$

Proof

To prove the first formula, solve for $\sin^2 u$ in the double-angle formula $\cos 2u = 1 - 2 \sin^2 u$, as follows.

$$\cos 2u = 1 - 2 \sin^2 u$$ Write double-angle formula.

$$2 \sin^2 u = 1 - \cos 2u$$ Subtract $\cos 2u$ from and add $2 \sin^2 x$ to each side.

$$\sin^2 u = \dfrac{1 - \cos 2u}{2}$$ Divide each side by 2.

Trigonometry and Astronomy

Trigonometry was used by early astronomers to calculate measurements in the universe. Trigonometry was used to calculate the circumference of Earth and the distance from Earth to the moon. Another major accomplishment in astronomy using trigonometry was computing distances to stars.

In a similar way, you can prove the second formula by solving for $\cos^2 u$ in the double-angle formula

$$\cos 2u = 2 \cos^2 u - 1.$$

To prove the third formula, use a quotient identity, as follows.

$$\tan^2 u = \frac{\sin^2 u}{\cos^2 u}$$

$$= \frac{\dfrac{1 - \cos 2u}{2}}{\dfrac{1 + \cos 2u}{2}}$$

$$= \frac{1 - \cos 2u}{1 + \cos 2u}$$

Sum-to-Product Formulas   (p. 528)

$$\sin u + \sin v = 2 \sin\left(\frac{u + v}{2}\right) \cos\left(\frac{u - v}{2}\right)$$

$$\sin u - \sin v = 2 \cos\left(\frac{u + v}{2}\right) \sin\left(\frac{u - v}{2}\right)$$

$$\cos u + \cos v = 2 \cos\left(\frac{u + v}{2}\right) \cos\left(\frac{u - v}{2}\right)$$

$$\cos u - \cos v = -2 \sin\left(\frac{u + v}{2}\right) \sin\left(\frac{u - v}{2}\right)$$

Proof

To prove the first formula, let $x = u + v$ and $y = u - v$. Then substitute $u = (x + y)/2$ and $v = (x - y)/2$ in the product-to-sum formula.

$$\sin u \cos v = \frac{1}{2}[\sin(u + v) + \sin(u - v)]$$

$$\sin\left(\frac{x + y}{2}\right) \cos\left(\frac{x - y}{2}\right) = \frac{1}{2}(\sin x + \sin y)$$

$$2 \sin\left(\frac{x + y}{2}\right) \cos\left(\frac{x - y}{2}\right) = \sin x + \sin y$$

The other sum-to-product formulas can be proved in a similar manner.

---

7 Additional Topics in Trigonometry

```
180+tan⁻¹(-2.4065
)
        112.5648997
```

Section 7.3, Example 10
Direction of an Airplane

7.1   Law of Sines

Introduction

In Chapter 5, you looked at techniques for solving right triangles. In this section and the next, you will solve **oblique triangles**—triangles that have no right angles. As standard notation, the angles of a triangle are labeled

A, B,   and C

and their opposite sides are labeled

a, b,   and c

as shown in Figure 7.1.

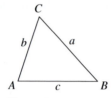

Figure 7.1

To solve an oblique triangle, you need to know the measure of at least one side and the measures of any two other parts of the triangle—two sides, two angles, or one angle and one side. This breaks down into the following four cases.

1. Two angles and any side (AAS or ASA)

2. Two sides and an angle opposite one of them (SSA)

3. Three sides (SSS)

4. Two sides and their included angle (SAS)

The first two cases can be solved using the **Law of Sines,** whereas the last two cases can be solved using the Law of Cosines (see Section 7.2).

You can use the Law of Sines to solve real-life problems involving oblique triangles. For instance, Exercise 46 on page 551 shows how the Law of Sines can be used to help determine the distance from a boat to the shoreline.

What you should learn
- Use the Law of Sines to solve oblique triangles (AAS or ASA).
- Use the Law of Sines to solve oblique triangles (SSA).
- Find areas of oblique triangles and use the Law of Sines to model and solve real-life problems.

Why you should learn it

Law of Sines   (See the proof on page 604.)

If ABC is a triangle with sides a, b, and c, then

$$\frac{a}{\sin A} = \frac{b}{\sin B} = \frac{c}{\sin C}.$$

Oblique Triangles

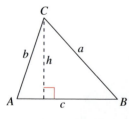

A is acute.
Figure 7.2

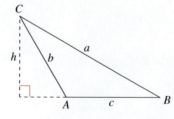

A is obtuse.

Study Tip

Notice in Figure 7.2 that the height h of each triangle can be found using the formula

$$\frac{h}{b} = \sin A$$

or

$$h = b \sin A.$$

The Law of Sines can also be written in the reciprocal form

$$\frac{\sin A}{a} = \frac{\sin B}{b} = \frac{\sin C}{c}.$$

Example 1   Given Two Angles and One Side—AAS

For the triangle in Figure 7.3,

$$C = 102.3°, B = 28.7°, \quad \text{and} \quad b = 27.4 \text{ feet.}$$

Find the remaining angle and sides.

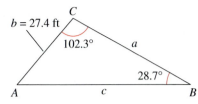

$b = 27.4$ ft

102.3°

a

28.7°

A c B

Figure 7.3

Solution

The third angle of the triangle is

$$A = 180° - B - C$$
$$= 180° - 28.7° - 102.3°$$
$$= 49.0°.$$

By the Law of Sines, you have

$$\frac{a}{\sin A} = \frac{b}{\sin B} = \frac{c}{\sin C}.$$

Using $b = 27.4$ produces

$$a = \frac{b}{\sin B}(\sin A) = \frac{27.4}{\sin 28.7°}(\sin 49.0°) \approx 43.06 \text{ feet}$$

and

$$c = \frac{b}{\sin B}(\sin C) = \frac{27.4}{\sin 28.7}(\sin 102.3°) \approx 55.75 \text{ feet}.$$

 CHECKPOINT   Now try Exercise 9.

Study Tip

When you are solving a triangle, a careful sketch is useful as a quick test for the feasibility of an answer. Remember that the longest side lies opposite the largest angle, and the shortest side lies opposite the smallest angle.

Example 2   Given Two Angles and One Side—ASA

A pole tilts *toward* the sun at an 8° angle from the vertical, and it casts a 22-foot shadow. The angle of elevation from the tip of the shadow to the top of the pole is 43°. How tall is the pole?

Solution

In Figure 7.4, $A = 43°$ and

$$B = 90° + 8° = 98°.$$

So, the third angle is

$$C = 180° - A - B$$
$$= 180° - 43° - 98°$$
$$= 39°.$$

By the Law of Sines, you have

$$\frac{a}{\sin A} = \frac{c}{\sin C}.$$

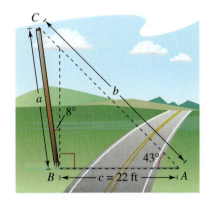

Figure 7.4

Because $c = 22$ feet, the length of the pole is

$$a = \frac{c}{\sin C}(\sin A) = \frac{22}{\sin 39°}(\sin 43°) \approx 23.84 \text{ feet}.$$

 CHECKPOINT   Now try Exercise 41.

For practice, try reworking Example 2 for a pole that tilts *away from* the sun under the same conditions.

The Ambiguous Case (SSA)

In Examples 1 and 2, you saw that two angles and one side determine a unique triangle. However, if two sides and one opposite angle are given, then three possible situations can occur: (1) no such triangle exists, (2) one such triangle exists, or (3) two distinct triangles satisfy the conditions.

The Ambiguous Case (SSA)

Consider a triangle in which you are given a, b, and A. (Notice that $h = b \sin A$.)

	A is acute.	*A* is acute.	*A* is acute.	*A* is acute.	*A* is obtuse.	*A* is obtuse.
Sketch						
Necessary condition	$a < h$	$a = h$	$a \geq b$	$h < a < b$	$a \leq b$	$a > b$
Possible triangles	None	One	One	Two	None	One

Example 3   Single-Solution Case–SSA

For the triangle in Figure 7.5,

$a = 22$ inches, $b = 12$ inches, and $A = 42°$.

Find the remaining side and angles.

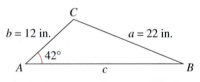

Figure 7.5   *One solution: $a \geq b$*

Solution

By the Law of Sines, you have

$$\frac{\sin B}{b} = \frac{\sin A}{a} \qquad \text{Reciprocal form}$$

$$\sin B = b\left(\frac{\sin A}{a}\right) \qquad \text{Multiply each side by } b.$$

$$\sin B = 12\left(\frac{\sin 42°}{22}\right) \qquad \text{Substitute for } A, a, \text{ and } b.$$

$$B \approx 21.41°. \qquad \text{\textit{B} is acute.}$$

Now you can determine that

$$C \approx 180° - 42° - 21.41° = 116.59°.$$

Then the remaining side is given by

$$\frac{c}{\sin C} = \frac{a}{\sin A} \qquad \text{Law of Sines}$$

$$c = \frac{a}{\sin A}(\sin C) \qquad \text{Multiply each side by } \sin C.$$

$$c = \frac{22}{\sin 42°}(\sin 116.59°) \qquad \text{Substitute for } a, A, \text{ and } C.$$

$$c \approx 29.40 \text{ inches.} \qquad \text{Simplify.}$$

✓**CHECKPOINT**   Now try Exercise 19.

Example 4   No-Solution Case—SSA

Show that there is no triangle for which $a = 15$, $b = 25$, and $A = 85°$.

Solution

Begin by making the sketch shown in Figure 7.6. From this figure it appears that no triangle is formed. You can verify this by using the Law of Sines.

$$\frac{\sin B}{b} = \frac{\sin A}{a} \qquad \text{Reciprocal form}$$

$$\sin B = b\left(\frac{\sin A}{a}\right) \qquad \text{Multiply each side by}$$

$$\sin B = 25\left(\frac{\sin 85°}{15}\right) \approx 1.6603 > 1$$

This contradicts the fact that $|\sin B| \le 1$. So, no triangle can be formed having sides $a = 15$ and $b = 25$ and an angle of $A = 85°$.

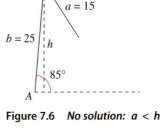

Figure 7.6   *No solution: $a < h$*

✓**CHECKPOINT**   Now try Exercise 27.

Example 5   Two-Solution Case—SSA

Find two triangles for which $a = 12$ meters, $b = 31$ meters, and $A = 20.5°$.

Solution

Because $h = b \sin A = 31(\sin 20.5°) \approx 10.86$ meters, you can conclude that there are two possible triangles (because $h < a < b$). By the Law of Sines, you have

$$\frac{\sin B}{b} = \frac{\sin A}{a} \qquad \text{Reciprocal form}$$

$$\sin B = b\left(\frac{\sin A}{a}\right) = 31\left(\frac{\sin 20.5°}{12}\right) \approx 0.9047.$$

There are two angles

$$B_1 \approx 64.8° \text{ and } B_2 \approx 180° - 64.8° = 115.2°$$

between $0°$ and $180°$ whose sine is 0.9047. For $B_1 \approx 64.8°$, you obtain

$$C \approx 180° - 20.5° - 64.8° = 94.7°$$

$$c = \frac{a}{\sin A}(\sin C) = \frac{12}{\sin 20.5°}(\sin 94.7°) \approx 34.15 \text{ meters.}$$

For $B_2 \approx 115.2°$, you obtain

$$C \approx 180° - 20.5° - 115.2° = 44.3°$$

$$c = \frac{a}{\sin A}(\sin C) = \frac{12}{\sin 20.5°}(\sin 44.3°) \approx 23.93 \text{ meters.}$$

The resulting triangles are shown in Figure 7.7.

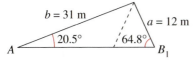

Figure 7.7   *Two solutions: $h < a < b$*

✓**CHECKPOINT**   Now try Exercise 29.

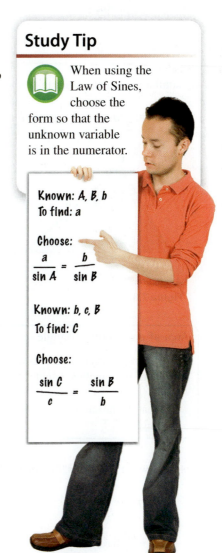

Study Tip

When using the Law of Sines, choose the form so that the unknown variable is in the numerator.

Known: A, B, b
To find: a

Choose:
$$\frac{a}{\sin A} = \frac{b}{\sin B}$$

Known: b, c, B
To find: C

Choose:
$$\frac{\sin C}{c} = \frac{\sin B}{b}$$

Area of an Oblique Triangle

The procedure used to prove the Law of Sines leads to a simple formula for the area of an oblique triangle. Referring to Figure 7.8, note that each triangle has a height of

$$h = b \sin A.$$

To see this when A is obtuse, substitute the reference angle $180° - A$ for A. Now the height of the triangle is given by

$$h = b \sin(180° - A).$$

Using the difference formula for sine, the height is given by

$$h = b(\sin 180° \cos A - \cos 180° \sin A) \qquad \sin(u - v) = \sin u \cos v - \cos u \sin v$$

$$= b[0 \cdot \cos A - (-1) \cdot \sin A]$$

$$= b \sin A.$$

Consequently, the area of each triangle is given by

$$\text{Area} = \frac{1}{2}(\text{base})(\text{height})$$

$$= \frac{1}{2}(c)(b \sin A)$$

$$= \frac{1}{2}bc \sin A.$$

By similar arguments, you can develop the formulas

$$\text{Area} = \frac{1}{2}ab \sin C = \frac{1}{2}ac \sin B.$$

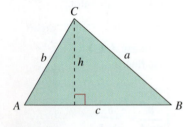

A is acute.

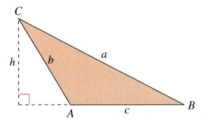

A is obtuse.

Figure 7.8

Area of an Oblique Triangle

The area of any triangle is one-half the product of the lengths of two sides times the sine of their included angle. That is,

$$\text{Area} = \frac{1}{2}bc \sin A = \frac{1}{2}ab \sin C = \frac{1}{2}ac \sin B.$$

Note that when angle A is $90°$, the formula gives the area of a right triangle as

$$\text{Area} = \frac{1}{2}bc$$

$$= \frac{1}{2}(\text{base})(\text{height}).$$

Similar results are obtained for angles C and B equal to $90°$.

Example 6   Finding the Area of an Oblique Triangle

Find the area of a triangular lot having two sides of lengths 90 meters and 52 meters and an included angle of 102°.

Solution

Consider $a = 90$ meters, $b = 52$ meters, and $C = 102°$, as shown in Figure 7.9. Then the area of the triangle is

Area $= \dfrac{1}{2} ab \sin C$ Formula for area

$= \dfrac{1}{2}(90)(52)(\sin 102°)$ Substitute for a, b, and C.

≈ 2288.87 square meters.   Simplify.

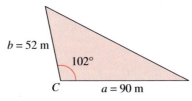

$b = 52$ m

$102°$

C $a = 90$ m

Figure 7.9

  Now try Exercise 35.

Example 7   An Application of the Law of Sines

The course for a boat race starts at point A and proceeds in the direction S 52° W to point B, then in the direction S 40° E to point C, and finally back to point A, as shown in Figure 7.10. Point C lies 8 kilometers directly south of point A. Approximate the total distance of the race course.

Solution

Because lines BD and AC are parallel, it follows that

$\angle BCA \cong \angle DBC$.

Consequently, triangle ABC has the measures shown in Figure 7.11. For angle B, you have

$B = 180° - 52° - 40° = 88°$.

Using the Law of Sines

$\dfrac{a}{\sin 52°} = \dfrac{b}{\sin 88°} = \dfrac{c}{\sin 40°}$

you can let $b = 8$ and obtain

$a = \dfrac{8}{\sin 88°}(\sin 52°) \approx 6.31$

and

$c = \dfrac{8}{\sin 88°}(\sin 40°) \approx 5.15$.

The total length of the course is approximately

Length $\approx 8 + 6.31 + 5.15 = 19.46$ kilometers.

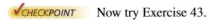 Now try Exercise 43.

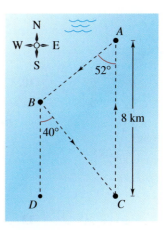

Figure 7.10

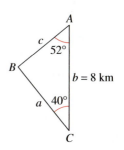

Figure 7.11

See www.CalcChat.com for worked-out solutions to odd-numbered exercises. For instructions on how to use a graphing utility, see Appendix A.

7.1 Exercises

Vocabulary and Concept Check

In Exercises 1–4, fill in the blank(s).

1. An _____ triangle is one that has no right angles.

2. Law of Sines: $\dfrac{a}{\sin A} =$ _____ $= \dfrac{c}{\sin C}$

3. To find the area of any triangle, use one of the following three formulas:
   Area = _____ , _____ , or _____ .

4. Two _____ and one _____ determine a unique triangle.

5. Which two cases can be solved using the Law of Sines?

6. Is the longest side of an oblique triangle always opposite the largest angle of the triangle?

Procedures and Problem Solving

Using the Law of Sines In Exercises 7–26, use the Law of Sines to solve the triangle.

7.
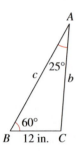

8.
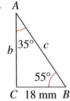

9.
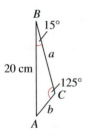

10.
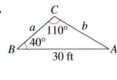

11.
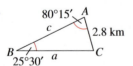

12.
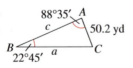

13. $A = 36°$, $a = 8$, $b = 5$
14. $A = 76°$, $a = 34$, $b = 21$
15. $A = 102.4°$, $C = 16.7°$, $a = 21.6$
16. $A = 24.3°$, $C = 54.6°$, $c = 2.68$
17. $A = 110° \, 15'$, $a = 48$, $b = 16$
18. $B = 2° \, 45'$, $b = 6.2$, $c = 5.8$
19. $A = 110°$, $a = 125$, $b = 100$
20. $A = 55°$, $B = 42°$, $c = \frac{3}{4}$
21. $B = 28°$, $C = 104°$, $a = 3\frac{5}{8}$

22. $B = 40°$, $C = 105°$, $c = 20$
23. $B = 10°$, $C = 135°$, $c = 45$
24. $A = 5°40'$, $B = 8°15'$, $b = 4.8$
25. $C = 85°20'$, $a = 35$, $c = 50$
26. $B = 2°45'$, $b = 6.2$, $c = 5.8$

Using the Law of Sines In Exercises 27–30, use the Law of Sines to solve the triangle. If two solutions exist, find both.

27. $A = 76°$, $a = 18$, $b = 20$
28. $A = 110°$, $a = 125$, $b = 200$
29. $A = 58°$, $a = 11.4$, $b = 12.8$
30. $A = 58°$, $a = 4.5$, $b = 12.8$

Using the Law of Sines In Exercises 31–34, find the value(s) of b such that the triangle has (a) one solution, (b) two solutions, and (c) no solution.

31. $A = 36°$, $a = 5$
32. $A = 60°$, $a = 10$
33. $A = 10°$, $a = 10.8$
34. $A = 88°$, $a = 315.6$

Finding the Area of a Triangle In Exercises 35–40, find the area of the triangle having the indicated angle and sides.

35. $C = 110°$, $a = 6$, $b = 10$
36. $B = 130°$, $a = 92$, $c = 30$
37. $A = 38° \, 45'$, $b = 67$, $c = 85$
38. $A = 5° \, 15'$, $b = 4.5$, $c = 22$
39. $B = 75° \, 15'$, $a = 103$, $c = 58$
40. $C = 85° \, 45'$, $a = 16$, $b = 20$

✓ **41. Physics**   A flagpole at a right angle to the horizontal is located on a slope that makes an angle of 14° with the horizontal. The flagpole casts a 16-meter shadow up the slope when the angle of elevation from the tip of the shadow to the sun is 20°.

(a) Draw a triangle that represents the problem. Show the known quantities on the triangle and use a variable to indicate the height of the flagpole.

(b) Write an equation involving the unknown quantity.

(c) Find the height of the flagpole.

42. Architecture   A bridge is to be built across a small lake from a gazebo to a dock (see figure). The bearing from the gazebo to the dock is S 41° W. From a tree 100 meters from the gazebo, the bearings to the gazebo and the dock are S 74° E and S 28° E, respectively. Find the distance from the gazebo to the dock.

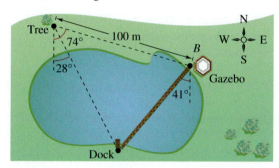

✓ **43. Aerodynamics**   A plane flies 500 kilometers with a bearing of 316° (clockwise from north) from Naples to Elgin (see figure). The plane then flies 720 kilometers from Elgin to Canton. (Canton is due west of Naples.) Find the bearing of the flight from Elgin to Canton.

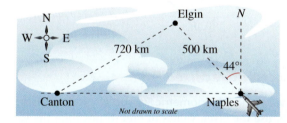

Not drawn to scale

44. Mechanical Engineering   The circular arc of a railroad curve has a chord of length 3000 feet and a central angle of 40°.

(a) Draw a diagram that visually represents the problem. Show the known quantities on the diagram and use the variables r and s to represent the radius of the arc and the length of the arc, respectively.

(b) Find the radius r of the circular arc.

(c) Find the length s of the circular arc.

45. Environmental Science   The bearing from the Pine Knob fire tower to the Colt Station fire tower is N 65° E, and the two towers are 30 kilometers apart. A fire spotted by rangers in each tower has a bearing of N 80° E from Pine Knob and S 70° E from Colt Station. Find the distance of the fire from each tower.

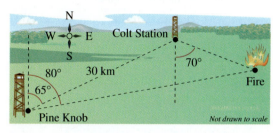

Not drawn to scale

46. Why you should learn it (p. 544)   A boat is sailing due east parallel to the shoreline at a speed of 10 miles per hour. At a given time the bearing to a lighthouse is S 70° E, and 15 minutes later the bearing is S 63° E (see figure). The lighthouse is located at the shoreline. Find the distance d from the boat to the shoreline.

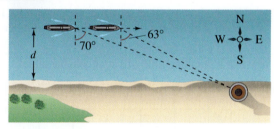

47. Angle of Elevation   A 10-meter telephone pole casts a 17-meter shadow directly down a slope when the angle of elevation of the sun is 42° (see figure). Find θ, the angle of elevation of the ground.

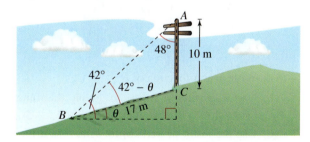

48. Aviation   The angles of elevation θ and ϕ to an airplane are being continuously monitored at two observation points A and B, respectively, which are 2 miles apart, and the airplane is east of both points in the same vertical plane.

(a) Draw a diagram that illustrates the problem.

(b) Write an equation giving the distance d between the plane and point B in terms of θ and ϕ.

49. MODELING DATA

The Leaning Tower of Pisa in Italy leans because it was built on unstable soil—a mixture of clay, sand, and water. The tower is approximately 58.36 meters tall from its foundation (see figure). The top of the tower leans about 5.45 meters off center.

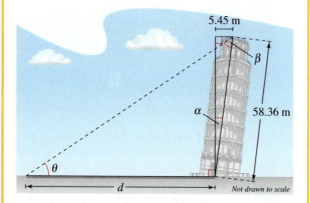

Not drawn to scale

(a) Find the angle of lean α of the tower.

(b) Write β as a function of d and θ, where θ is the angle of elevation to the sun.

(c) Use the Law of Sines to write an equation for the length d of the shadow cast by the tower in terms of θ.

(d) Use a graphing utility to complete the table.

θ	10°	20°	30°	40°	50°	60°
d						

50. Exploration In the figure, α and β are positive angles.

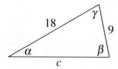

(a) Write α as a function of β.

(b) Use a graphing utility to graph the function. Determine its domain and range.

(c) Use the result of part (b) to write c as a function of β.

(d) Use the graphing utility to graph the function in part (c). Determine its domain and range.

(e) Use the graphing utility to complete the table. What can you conclude?

β	0.4	0.8	1.2	1.6	2.0	2.4	2.8
α							
c							

Conclusions

True or False? In Exercises 51–53, determine whether the statement is true or false. Justify your answer.

51. If any three sides or angles of an oblique triangle are known, then the triangle can be solved.

52. If a triangle contains an obtuse angle, then it must be oblique.

53. Two angles and one side of a triangle do not necessarily determine a unique triangle.

54. Writing Can the Law of Sines be used to solve a right triangle? If so, write a short paragraph explaining how to use the Law of Sines to solve the following triangle. Is there an easier way to solve the triangle? Explain.

$$B = 50°, C = 90°, a = 10$$

55. Think About It Given $A = 36°$ and $a = 5$, find values of b such that the triangle has (a) one solution, (b) two solutions, and (c) no solution.

56. CAPSTONE In the figure, a triangle is to be formed by drawing a line segment of length a from $(4, 3)$ to the positive x-axis. For what value(s) of a can you form (a) one triangle, (b) two triangles, and (c) no triangles? Explain your reasoning.

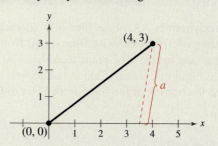

Cumulative Mixed Review

Evaluating Trigonometric Functions In Exercises 57 and 58, use the given values to find (if possible) the values of the remaining four trigonometric functions of θ.

57. $\cos \theta = \frac{5}{13}, \ \sin \theta = -\frac{12}{13}$

58. $\tan \theta = \frac{2}{9}, \ \csc \theta = -\frac{\sqrt{85}}{2}$

Writing Products as Sums or Differences In Exercises 59–62, write the product as a sum or difference.

59. $6 \sin 8\theta \cos 3\theta$

60. $2 \cos 2\theta \cos 5\theta$

61. $3 \cos \frac{\pi}{6} \sin \frac{5\pi}{3}$

62. $\frac{5}{2} \sin \frac{3\pi}{4} \sin \frac{5\pi}{6}$

7.2 Law of Cosines

Introduction

Two cases remain in the list of conditions needed to solve an oblique triangle—SSS and SAS. To use the Law of Sines, you must know at least one side and its opposite angle. When you are given three sides (SSS), or two sides and their included angle (SAS), none of the ratios in the Law of Sines would be complete. In such cases you can use the **Law of Cosines.**

> **Law of Cosines** (See the proof on page 605.)
>
> | Standard Form | Alternative Form |
> |---|---|
> | $a^2 = b^2 + c^2 - 2bc \cos A$ | $\cos A = \dfrac{b^2 + c^2 - a^2}{2bc}$ |
> | $b^2 = a^2 + c^2 - 2ac \cos B$ | $\cos B = \dfrac{a^2 + c^2 - b^2}{2ac}$ |
> | $c^2 = a^2 + b^2 - 2ab \cos C$ | $\cos C = \dfrac{a^2 + b^2 - c^2}{2ab}$ |

What you should learn

- Use the Law of Cosines to solve oblique triangles (SSS or SAS).
- Use the Law of Cosines to model and solve real-life problems.
- Use Heron's Area Formula to find areas of triangles.

Why you should learn it

You can use the Law of Cosines to solve real-life problems involving oblique triangles. For instance, Exercise 52 on page 558 shows you how the Law of Cosines can be used to determine the lengths of the guy wires that anchor a tower.

Example 1 Three Sides of a Triangle—SSS

Find the three angles of the triangle shown in Figure 7.12.

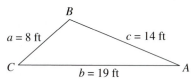

Figure 7.12

Solution

It is a good idea first to find the angle opposite the longest side—side b in this case. Using the alternative form of the Law of Cosines, you find that

$$\cos B = \frac{a^2 + c^2 - b^2}{2ac} \qquad \textcolor{red}{\text{Alternative form}}$$

$$= \frac{8^2 + 14^2 - 19^2}{2(8)(14)} \qquad \textcolor{red}{\text{Substitute for } a, b, \text{ and } c.}$$

$$\approx -0.45089. \qquad \textcolor{red}{\text{Simplify.}}$$

Because $\cos B$ is negative, you know that B is an *obtuse* angle given by $B \approx 116.80°$. At this point it is simpler to use the Law of Sines to determine A.

$$\sin A = a\left(\frac{\sin B}{b}\right) \approx 8\left(\frac{\sin 116.80°}{19}\right) \approx 0.37583$$

You know that A must be acute, because B is obtuse, and a triangle can have, at most, one obtuse angle. So, $A \approx 22.08°$ and

$$C \approx 180° - 22.08° - 116.80° = 41.12°$$

 ✓CHECKPOINT Now try Exercise 7.

Explore the Concept

 What familiar formula do you obtain when you use the third form of the Law of Cosines

$$c^2 = a^2 + b^2 - 2ab \cos C$$

and you let $C = 90°$? What is the relationship between the Law of Cosines and this formula?

Do you see why it was wise to find the largest angle *first* in Example 1? Knowing the cosine of an angle, you can determine whether the angle is acute or obtuse. That is,

$\cos \theta > 0$ for $0° < \theta < 90°$ Acute

$\cos \theta < 0$ for $90° < \theta < 180°$.   Obtuse

So, in Example 1, once you found that angle B was obtuse, you knew that angles A and C were both acute. Furthermore, if the largest angle is acute, then the remaining two angles are also acute.

Example 2   Two Sides and the Included Angle—SAS

Find the remaining angles and side of the triangle shown in Figure 7.13.

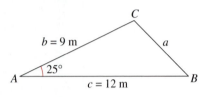

Figure 7.13

Solution

Use the Law of Cosines to find the unknown side a in the figure.

$a^2 = b^2 + c^2 - 2bc \cos A$

$a^2 = 9^2 + 12^2 - 2(9)(12) \cos 25°$

$a^2 \approx 29.2375$

$a \approx 5.4072$

Because $a \approx 5.4072$ meters, you now know the ratio

$$\frac{\sin A}{a}$$

and you can use the reciprocal form of the Law of Sines to solve for B.

$\dfrac{\sin B}{b} = \dfrac{\sin A}{a}$

$\sin B = b\left(\dfrac{\sin A}{a}\right)$

$\sin B \approx 9\left(\dfrac{\sin 25°}{5.4072}\right)$

$\sin B \approx 0.7034$

There are two angles between $0°$ and $180°$ whose sine is 0.7034,

$B_1 \approx 44.7°$ and $B_2 \approx 180° - 44.7° = 135.3°$.

For $B_1 \approx 44.7°$,

$C_1 \approx 180° - 25° - 44.7° = 110.3°$.

For $B_2 \approx 135.3°$,

$C_2 \approx 180° - 25° - 135.3° = 19.7°$.

Because side c is the longest side of the triangle, C must be the largest angle of the triangle. So, $B \approx 44.7°$ and $C \approx 110.3°$.

✔CHECKPOINT   Now try Exercise 11.

Study Tip

When solving an oblique triangle given three sides, you use the alternative form of the Law of Cosines to solve for an angle. When solving an oblique triangle given two sides and their included angle, you use the standard form of the Law of Cosines to solve for an unknown side.

Explore the Concept

In Example 2, suppose $A = 115°$. After solving for a, which angle would you solve for next, B or C? Are there two possible solutions for that angle? If so, how can you determine which angle is the correct measure?

Applications

Example 3   An Application of the Law of Cosines

The pitcher's mound on a women's softball field is 43 feet from home plate and the distance between the bases is 60 feet, as shown in Figure 7.14. (The pitcher's mound is *not* halfway between home plate and second base.) How far is the pitcher's mound from first base?

Solution

In triangle *HPF*, $H = 45°$ (line *HP* bisects the right angle at *H*), $f = 43$, and $p = 60$. Using the Law of Cosines for this SAS case, you have

$$h^2 = f^2 + p^2 - 2fp \cos H \qquad \text{Law of Cosines}$$

$$= 43^2 + 60^2 - 2(43)(60) \cos 45° \qquad \text{Substitute for } H, f, \text{ and } p.$$

$$\approx 1800.33. \qquad \text{Simplify.}$$

So, the approximate distance from the pitcher's mound to first base is

$$h \approx \sqrt{1800.33}$$

$$\approx 42.43 \text{ feet.}$$

✓**CHECKPOINT**   Now try Exercise 47.

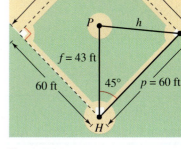

Figure 7.14

Example 4   An Application of the Law of Cosines

A ship travels 60 miles due east, then adjusts its course northward, as shown in Figure 7.15. After traveling 80 miles in the new direction, the ship is 139 miles from its point of departure. Describe the bearing from point *B* to point *C*.

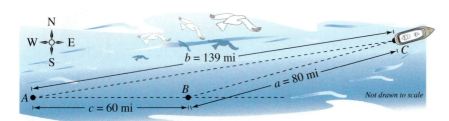

Figure 7.15

Solution

You have $a = 80$, $b = 139$, and $c = 60$; so, using the alternative form of the Law of Cosines, you have

$$\cos B = \frac{a^2 + c^2 - b^2}{2ac} \qquad \text{Alternative form}$$

$$= \frac{80^2 + 60^2 - 139^2}{2(80)(60)} \qquad \text{Substitute for } a, b, \text{ and } c.$$

$$\approx -0.97094. \qquad \text{Simplify.}$$

So, $B \approx \arccos(-0.97094) \approx 166.15°$. Therefore, the bearing measured from due north from point *B* to point *C* is

$$166.15° - 90° = 76.15°$$

or N 76.15° E.

✓**CHECKPOINT**   Now try Exercise 49.

Heron's Area Formula

The Law of Cosines can be used to establish the following formula for the area of a triangle. This formula is called **Heron's Area Formula** after the Greek mathematician Heron (ca. 100 B.C.).

Heron's Area Formula (See the proof on page 606.)

Given any triangle with sides of lengths a, b, and c, the area of the triangle is given by

$$\text{Area} = \sqrt{s(s-a)(s-b)(s-c)}$$

where $s = \dfrac{a+b+c}{2}$.

Example 5 Using Heron's Area Formula

Find the area of the triangle shown in Figure 7.16.

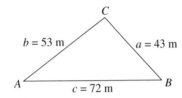

Figure 7.16

Solution

Because

$$s = \frac{a+b+c}{2}$$

$$= \frac{168}{2}$$

$$= 84$$

Heron's Area Formula yields

$$\text{Area} = \sqrt{s(s-a)(s-b)(s-c)}$$

$$= \sqrt{84(84-43)(84-53)(84-72)}$$

$$= \sqrt{84(41)(31)(12)}$$

$$\approx 1131.89 \text{ square meters.}$$

✓**CHECKPOINT**   Now try Exercise 55.

You have now studied three different formulas for the area of a triangle.

Formulas for Area of a Triangle

1. Standard Formula: Area $= \frac{1}{2}bh$

2. Oblique Triangle: Area $= \frac{1}{2}bc \sin A = \frac{1}{2}ab \sin C = \frac{1}{2}ac \sin B$

3. Heron's Area Formula: Area $= \sqrt{s(s-a)(s-b)(s-c)}$

Explore the Concept

Can the formulas at the bottom of the page be used to find the area of any type of triangle? Explain the advantages and disadvantages of using one formula over another.

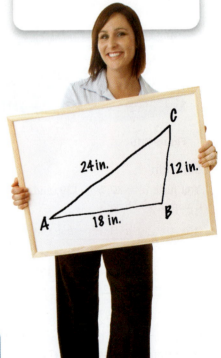

7.2   Exercises

See www.CalcChat.com for worked-out solutions to odd-numbered exercises.
For instructions on how to use a graphing utility, see Appendix A.

Vocabulary and Concept Check

In Exercises 1 and 2, fill in the blank(s).

1. The standard form of the Law of Cosines for $\cos C = \dfrac{a^2 + b^2 - c^2}{2ab}$ is _____ .

2. Three different formulas for the area of a triangle are given by Area = _____ , Area $= \frac{1}{2}bc \sin A = \frac{1}{2}ab \sin C = \frac{1}{2}ac \sin B$, and Area = _____ .

In Exercises 3–6, one of the cases for the known measures of an oblique triangle is given. State whether the Law of Cosines can be used to solve the triangle.

3. ASA   4. SAS   5. SSS   6. AAS

Procedures and Problem Solving

Using the Law of Cosines   In Exercises 7–24, use the Law of Cosines to solve the triangle.

✓ 7.
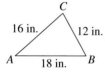

8.
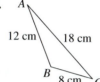

9.
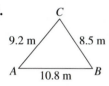

10.
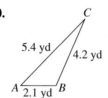

✓ 11.
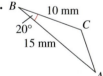

12.
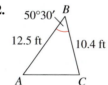

13. $a = 6, \quad b = 8, \quad c = 12$

14. $a = 9, \quad b = 3, \quad c = 11$

15. $A = 50°, \quad b = 15, \quad c = 30$

16. $C = 108°, \quad a = 10, \quad b = 7$

17. $a = 9, \quad b = 12, \quad c = 15$

18. $a = 45, \quad b = 30, \quad c = 72$

19. $a = 75.4, \quad b = 48, \quad c = 48$

20. $a = 1.42, \quad b = 0.75, \quad c = 1.25$

21. $B = 8° \, 15', \quad a = 26, \quad c = 18$

22. $B = 10° \, 35', \quad a = 40, \quad c = 30$

23. $B = 75° \, 20', \quad a = 6.2, \quad c = 9.5$

24. $C = 15° \, 15', \quad a = 6.25, \quad b = 2.15$

Finding Measures in a Parallelogram   In Exercises 25–30, complete the table by solving the parallelogram shown in the figure. (The lengths of the diagonals are given by c and d.)

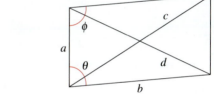

	a	b	c	d	θ	ϕ
25.	4	8			30°	
26.	25	35				120°
27.	10	14	20			
28.	40	60		80		
29.	15		25	20		
30.		25	50	35		

Solving a Triangle   In Exercises 31–36, determine whether the Law of Sines or the Law of Cosines can be used to find another measure of the triangle. Then solve the triangle.

31. $a = 8, \quad c = 5, \quad B = 40°$

32. $a = 10, \quad b = 12, \quad C = 70°$

33. $A = 24°, \quad a = 4, \quad b = 18$

34. $a = 11, \quad b = 13, \quad c = 7$

35. $A = 42°, \quad B = 35°, \quad c = 1.2$

36. $a = 160, \quad B = 12°, \quad C = 7°$

Using Heron's Area Formula   In Exercises 37–46, use Heron's Area Formula to find the area of the triangle.

37. $a = 12, \quad b = 24, \quad c = 18$

38. $a = 25, \quad b = 35, \quad c = 32$

39. $a = 5$, $b = 8$, $c = 10$

40. $a = 13$, $b = 17$, $c = 8$

41.

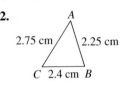

42.

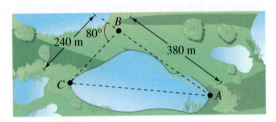

43. $a = 3.5$, $b = 10.2$, $c = 9$

44. $a = 75.4$, $b = 52$, $c = 52$

45. $a = 10.59$, $b = 6.65$, $c = 12.31$

46. $a = 4.45$, $b = 1.85$, $c = 3$

✓ **47. Surveying** To approximate the length of a marsh, a surveyor walks 380 meters from point A to point B. Then the surveyor turns 80° and walks 240 meters to point C (see figure). Approximate the length AC of the marsh.

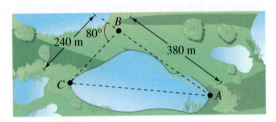

48. Geometry Determine the angle θ in the design of the streetlight shown in the figure.

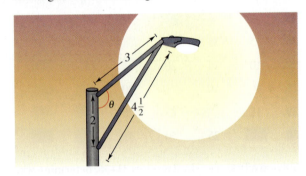

✓ **49. Geography** On a map, Minneapolis is 165 millimeters due west of Albany, Phoenix is 216 millimeters from Minneapolis, and Phoenix is 368 millimeters from Albany (see figure).

(a) Find the bearing of Minneapolis from Phoenix.

(b) Find the bearing of Albany from Phoenix.

50. Marine Transportation Two ships leave a port at 9 A.M. One travels at a bearing of N 53° W at 12 miles per hour, and the other travels at a bearing of S 67° W at 16 miles per hour. Approximate how far apart the ships are at noon.

51. Surveying A triangular parcel of ground has sides of lengths 725 feet, 650 feet, and 575 feet. Find the measure of the largest angle.

52. *Why you should learn it* (p. 553) A 100-foot vertical tower is to be erected on the side of a hill that makes a 6° angle with the horizontal (see figure). Find the length of each of the two guy wires that will be anchored 75 feet uphill and downhill from the base of the tower.

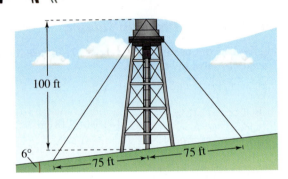

53. Structural Engineering Q is the midpoint of the line segment $\overline{PR}$ in the truss rafter shown in the figure. What are the lengths of the line segments $\overline{PQ}$, $\overline{QS}$, and $\overline{RS}$?

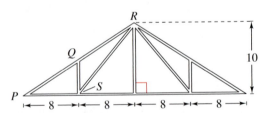

54. Architecture A retractable awning above a patio lowers at an angle of 50° from the exterior wall at a height of 10 feet above the ground (see figure). No direct sunlight is to enter the door when the angle of elevation of the sun is greater than 70°. What is the length x of the awning?

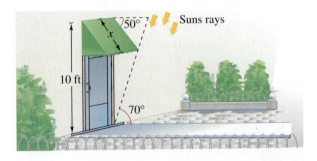

✓ **55. Architecture** The Landau Building in Cambridge, Massachusetts has a triangular-shaped base. The lengths of the sides of the triangular base are 145 feet, 257 feet, and 290 feet. Find the area of the base of the building.

56. Geometry A parking lot has the shape of a parallelogram (see figure). The lengths of two adjacent sides are 70 meters and 100 meters. The angle between the two sides is 70°. What is the area of the parking lot?

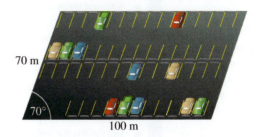

70 m

70°

100 m

57. Mechanical Engineering An engine has a seven-inch connecting rod fastened to a crank (see figure).

(a) Use the Law of Cosines to write an equation giving the relationship between x and θ.

(b) Write x as a function of θ. (Select the sign that yields positive values of x.)

(c) Use a graphing utility to graph the function in part (b).

(d) Use the graph in part (c) to determine the total distance the piston moves in one cycle.

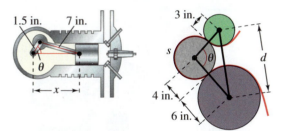

1.5 in.   7 in.

θ

x

3 in.

s

θ

d

4 in.

6 in.

Figure for 57   Figure for 58

58. MODELING DATA

In a process with continuous paper, the paper passes across three rollers of radii 3 inches, 4 inches, and 6 inches (see figure). The centers of the three-inch and six-inch rollers are d inches apart, and the length of the arc in contact with the paper on the four-inch roller is s inches.

(a) Use the Law of Cosines to write an equation giving the relationship between d and θ.

(b) Write θ as a function of d.

(c) Write s as a function of θ.

(d) Complete the table.

d (inches)	9	10	12	13	14	15	16
θ (degrees)							
s (inches)							

Conclusions

True or False? In Exercises 59 and 60, determine whether the statement is true or false. Justify your answer.

59. Two sides and their included angle determine a unique triangle.

60. In Heron's Area Formula, s is the average of the lengths of the three sides of the triangle.

Proof In Exercises 61 and 62, use the Law of Cosines to prove the identity.

61. $\dfrac{1}{2}bc(1 + \cos A) = \left(\dfrac{a+b+c}{2}\right)\left(\dfrac{-a+b+c}{2}\right)$

62. $\dfrac{1}{2}bc(1 - \cos A) = \left(\dfrac{a-b+c}{2}\right)\left(\dfrac{a+b-c}{2}\right)$

63. Writing Describe how the Law of Cosines can be used to solve the ambiguous case of the oblique triangle ABC, where $a = 12$ feet, $b = 30$ feet, and $A = 20°$. Is the result the same as when the Law of Sines is used to solve the triangle? Describe the advantages and the disadvantages of each method.

64. CAPSTONE Consider the cases SSS, AAS, ASA, SAS, and SSA.

(a) For which of these cases are you unable to solve the triangle using only the Law of Sines?

(b) For each case described in part (a), which form of the Law of Cosines is most convenient to use?

65. Proof Use a half-angle formula and the Law of Cosines to show that, for any triangle,

$$\cos\left(\frac{C}{2}\right) = \sqrt{\frac{s(s-c)}{ab}}$$

where $s = \frac{1}{2}(a + b + c)$.

66. Proof Use a half-angle formula and the Law of Cosines to show that, for any triangle,

$$\sin\left(\frac{C}{2}\right) = \sqrt{\frac{(s-a)(s-b)}{ab}}$$

where $s = \frac{1}{2}(a + b + c)$.

Cumulative Mixed Review

Evaluating an Inverse Trigonometric Function In Exercises 67–70, evaluate the expression without using a calculator.

67. $\arcsin(-1)$

68. $\cos^{-1} 0$

69. $\tan^{-1} \sqrt{3}$

70. $\arcsin\left(-\dfrac{\sqrt{3}}{2}\right)$

7.3   Vectors in the Plane

Introduction

Many quantities in geometry and physics, such as area, time, and temperature, can be represented by a single real number. Other quantities, such as force and velocity, involve both *magnitude* and *direction* and cannot be completely characterized by a single real number. To represent such a quantity, you can use a **directed line segment,** as shown in Figure 7.17. The directed line segment $\overrightarrow{PQ}$ has **initial point** P and **terminal point** Q. Its **magnitude,** or **length,** is denoted by $\|\overrightarrow{PQ}\|$ and can be found by using the Distance Formula.

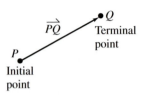

Figure 7.17

Figure 7.18

Two directed line segments that have the same magnitude and direction are *equivalent.* For example, the directed line segments in Figure 7.18 are all equivalent. The set of all directed line segments that are equivalent to a given directed line segment $\overrightarrow{PQ}$ is a **vector v in the plane,** written

$$\mathbf{v} = \overrightarrow{PQ}.$$

Vectors are denoted by lowercase, boldface letters such as **u**, **v**, and **w**.

Example 1   Equivalent Directed Line Segments

Let **u** be represented by the directed line segment from

$$P(0, 0) \quad \text{to} \quad Q(3, 2)$$

and let **v** be represented by the directed line segment from

$$R(1, 2) \quad \text{to} \quad S(4, 4)$$

as shown in Figure 7.19. Show that $\mathbf{u} = \mathbf{v}$.

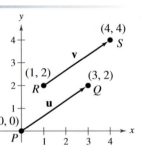

Figure 7.19

Solution

From the Distance Formula, it follows that $\overrightarrow{PQ}$ and $\overrightarrow{RS}$ have the *same magnitude.*

$$\|\overrightarrow{PQ}\| = \sqrt{(3 - 0)^2 + (2 - 0)^2} = \sqrt{13}$$

$$\|\overrightarrow{RS}\| = \sqrt{(4 - 1)^2 + (4 - 2)^2} = \sqrt{13}$$

Moreover, both line segments have the *same direction,* because they are both directed toward the upper right on lines having the same slope.

$$\text{Slope of } \overrightarrow{PQ} = \frac{2 - 0}{3 - 0} = \frac{2}{3}$$

$$\text{Slope of } \overrightarrow{RS} = \frac{4 - 2}{4 - 1} = \frac{2}{3}$$

So, $\overrightarrow{PQ}$ and $\overrightarrow{RS}$ have the same magnitude and direction, and it follows that $\mathbf{u} = \mathbf{v}$.

✓**CHECKPOINT**   Now try Exercise 11.

Patrick Hermans 2010/used under license from Shutterstock.com

Component Form of a Vector

The directed line segment whose initial point is the origin is often the most convenient representative of a set of equivalent directed line segments. This representative of the vector **v** is in **standard position.**

A vector whose initial point is at the origin $(0, 0)$ can be uniquely represented by the coordinates of its terminal point (v_1, v_2). This is the **component form of a vector v,** written as

$$\mathbf{v} = \langle v_1, v_2 \rangle.$$

The coordinates v_1 and v_2 are the *components* of **v**. If both the initial point and the terminal point lie at the origin, then **v** is the **zero vector** and is denoted by $\mathbf{0} = \langle 0, 0 \rangle$.

Component Form of a Vector

The component form of the vector with initial point $P(p_1, p_2)$ and terminal point $Q(q_1, q_2)$ is given by

$$\overrightarrow{PQ} = \langle q_1 - p_1, q_2 - p_2 \rangle = \langle v_1, v_2 \rangle = \mathbf{v}.$$

The **magnitude** (or length) of **v** is given by

$$\|\mathbf{v}\| = \sqrt{(q_1 - p_1)^2 + (q_2 - p_2)^2} = \sqrt{v_1^2 + v_2^2}.$$

If $\|\mathbf{v}\| = 1$, then **v** is a **unit vector.** Moreover, $\|\mathbf{v}\| = 0$ if and only if **v** is the zero vector **0**.

Two vectors $\mathbf{u} = \langle u_1, u_2 \rangle$ and $\mathbf{v} = \langle v_1, v_2 \rangle$ are *equal* if and only if $u_1 = v_1$ and $u_2 = v_2$. For instance, in Example 1, the vector **u** from $P(0, 0)$ to $Q(3, 2)$ is

$$\mathbf{u} = \overrightarrow{PQ} = \langle 3 - 0, 2 - 0 \rangle = \langle 3, 2 \rangle$$

and the vector **v** from $R(1, 2)$ to $S(4, 4)$ is

$$\mathbf{v} = \overrightarrow{RS} = \langle 4 - 1, 4 - 2 \rangle = \langle 3, 2 \rangle.$$

Technology Tip

You can graph vectors with a graphing utility by graphing directed line segments. Consult the user's guide for your graphing utility for specific instructions.

Example 2   Finding the Component Form of a Vector

Find the component form and magnitude of the vector **v** that has initial point $(4, -7)$ and terminal point $(-1, 5)$.

Solution

Let

$$P(4, -7) = (p_1, p_2)$$

and

$$Q(-1, 5) = (q_1, q_2)$$

as shown in Figure 7.20. Then, the components of $\mathbf{v} = \langle v_1, v_2 \rangle$ are

$$v_1 = q_1 - p_1 = -1 - 4 = -5$$

$$v_2 = q_2 - p_2 = 5 - (-7) = 12.$$

So, $\mathbf{v} = \langle -5, 12 \rangle$ and the magnitude of **v** is

$$\|\mathbf{v}\| = \sqrt{(-5)^2 + 12^2} = \sqrt{169} = 13.$$

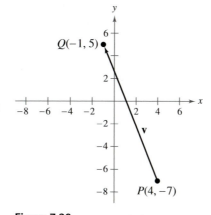

Figure 7.20

✓**CHECKPOINT**   Now try Exercise 15.

Vector Operations

The two basic vector operations are **scalar multiplication** and **vector addition.** Geometrically, the product of a vector **v** and a scalar k is the vector that is $|k|$ times as long as **v**. If k is positive, then $k\mathbf{v}$ has the same direction as **v**, and if k is negative, then $k\mathbf{v}$ has the opposite direction of **v**, as shown in Figure 7.21.

To add two vectors **u** and **v** geometrically, first position them (without changing their lengths or directions) so that the initial point of the second vector **v** coincides with the terminal point of the first vector **u**. The sum

u + **v**

is the vector formed by joining the initial point of the first vector **u** with the terminal point of the second vector **v**, as shown in Figure 7.22. This technique is called the **parallelogram law** for vector addition because the vector **u** + **v**, often called the **resultant** of vector addition, is the diagonal of a parallelogram having **u** and **v** as its adjacent sides.

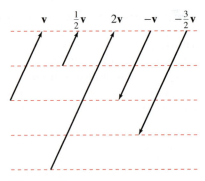

Figure 7.21

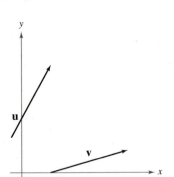

Figure 7.22

Definition of Vector Addition and Scalar Multiplication

Let $\mathbf{u} = \langle u_1, u_2 \rangle$ and $\mathbf{v} = \langle v_1, v_2 \rangle$ be vectors and let k be a scalar (a real number). Then the **sum** of **u** and **v** is the vector

$$\mathbf{u} + \mathbf{v} = \langle u_1 + v_1, u_2 + v_2 \rangle \qquad \text{Sum}$$

and the **scalar multiple** of k times **u** is the vector

$$k\mathbf{u} = k\langle u_1, u_2 \rangle = \langle ku_1, ku_2 \rangle. \qquad \text{Scalar multiple}$$

The **negative** of $\mathbf{v} = \langle v_1, v_2 \rangle$ is

$$\begin{aligned} -\mathbf{v} &= (-1)\mathbf{v} \\ &= \langle -v_1, -v_2 \rangle \qquad \text{Negative} \end{aligned}$$

and the **difference** of **u** and **v** is

$$\begin{aligned} \mathbf{u} - \mathbf{v} &= \mathbf{u} + (-\mathbf{v}) \qquad \text{Add } (-\mathbf{v}). \text{ See Figure 7.23.} \\ &= \langle u_1 - v_1, u_2 - v_2 \rangle. \qquad \text{Difference} \end{aligned}$$

To represent **u** − **v** geometrically, you can use directed line segments with the *same* initial point. The difference

u − **v**

is the vector from the terminal point of **v** to the terminal point of **u**, which is equal to

u + (−**v**)

as shown in Figure 7.23.

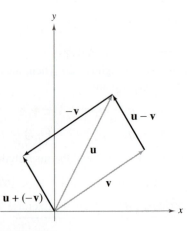

Figure 7.23

The component definitions of vector addition and scalar multiplication are illustrated in Example 3. In this example, notice that each of the vector operations can be interpreted geometrically.

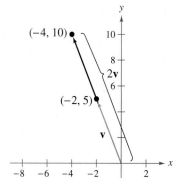

Figure 7.24

Example 3 Vector Operations

Let $\mathbf{v} = \langle -2, 5 \rangle$ and $\mathbf{w} = \langle 3, 4 \rangle$, and find each of the following vectors.

a. $2\mathbf{v}$ **b.** $\mathbf{w} - \mathbf{v}$ **c.** $\mathbf{v} + 2\mathbf{w}$

Solution

a. Because $\mathbf{v} = \langle -2, 5 \rangle$, you have

$$2\mathbf{v} = 2\langle -2, 5 \rangle$$
$$= \langle 2(-2), 2(5) \rangle$$
$$= \langle -4, 10 \rangle.$$

A sketch of $2\mathbf{v}$ is shown in Figure 7.24.

b. The difference of $\mathbf{w}$ and $\mathbf{v}$ is

$$\mathbf{w} - \mathbf{v} = \langle 3, 4 \rangle - \langle -2, 5 \rangle$$
$$= \langle 3 - (-2), 4 - 5 \rangle$$
$$= \langle 5, -1 \rangle.$$

A sketch of $\mathbf{w} - \mathbf{v}$ is shown in Figure 7.25. Note that the figure shows the vector difference $\mathbf{w} - \mathbf{v}$ as the sum $\mathbf{w} + (-\mathbf{v})$.

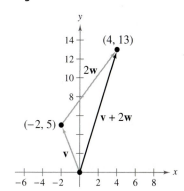

Figure 7.25

c. The sum of $\mathbf{v}$ and $2\mathbf{w}$ is

$$\mathbf{v} + 2\mathbf{w} = \langle -2, 5 \rangle + 2\langle 3, 4 \rangle$$
$$= \langle -2, 5 \rangle + \langle 2(3), 2(4) \rangle$$
$$= \langle -2, 5 \rangle + \langle 6, 8 \rangle$$
$$= \langle -2 + 6, 5 + 8 \rangle$$
$$= \langle 4, 13 \rangle.$$

A sketch of $\mathbf{v} + 2\mathbf{w}$ is shown in Figure 7.26.

 CHECKPOINT   Now try Exercise 37.

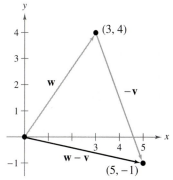

Figure 7.26

Vector addition and scalar multiplication share many of the properties of ordinary arithmetic.

Properties of Vector Addition and Scalar Multiplication

Let $\mathbf{u}$, $\mathbf{v}$, and $\mathbf{w}$ be vectors and let c and d be scalars. Then the following properties are true.

1. $\mathbf{u} + \mathbf{v} = \mathbf{v} + \mathbf{u}$ 2. $(\mathbf{u} + \mathbf{v}) + \mathbf{w} = \mathbf{u} + (\mathbf{v} + \mathbf{w})$

3. $\mathbf{u} + \mathbf{0} = \mathbf{u}$ 4. $\mathbf{u} + (-\mathbf{u}) = \mathbf{0}$

5. $c(d\mathbf{u}) = (cd)\mathbf{u}$ 6. $(c + d)\mathbf{u} = c\mathbf{u} + d\mathbf{u}$

7. $c(\mathbf{u} + \mathbf{v}) = c\mathbf{u} + c\mathbf{v}$ 8. $1(\mathbf{u}) = \mathbf{u},\ 0(\mathbf{u}) = \mathbf{0}$

9. $\|c\mathbf{v}\| = |c|\,\|\mathbf{v}\|$

Study Tip

 Property 9 can be stated as follows: The magnitude of the vector $c\mathbf{v}$ is the absolute value of c times the magnitude of $\mathbf{v}$.

Example 10   Using Vectors to Find Speed and Direction

An airplane is traveling at a speed of 500 miles per hour with a bearing of 330° at a fixed altitude with a negligible wind velocity, as shown in Figure 7.34(a). As the airplane reaches a certain point, it encounters a wind blowing with a velocity of 70 miles per hour in the direction N 45° E, as shown in Figure 7.34(b). What are the resultant speed and direction of the airplane?

Study Tip

Recall from Section 5.7 that in air navigation, bearings can be measured in degrees clockwise from north. In Figure 7.34, north is in the positive y-direction.

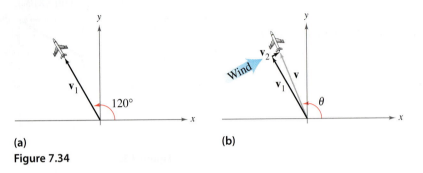

(a)

(b)

Figure 7.34

Solution

Using Figure 7.34, the velocity of the airplane (alone) is

$$\mathbf{v}_1 = 500\langle \cos 120°, \sin 120° \rangle$$
$$= \langle -250, 250\sqrt{3} \rangle$$

and the velocity of the wind is

$$\mathbf{v}_2 = 70\langle \cos 45°, \sin 45° \rangle$$
$$= \langle 35\sqrt{2}, 35\sqrt{2} \rangle.$$

So, the velocity of the airplane (in the wind) is

$$\mathbf{v} = \mathbf{v}_1 + \mathbf{v}_2$$
$$= \langle -250 + 35\sqrt{2}, 250\sqrt{3} + 35\sqrt{2} \rangle$$
$$\approx \langle -200.5, 482.5 \rangle$$

and the resultant speed of the airplane is

$$\|\mathbf{v}\| \approx \sqrt{(-200.5)^2 + (482.5)^2} \approx 522.5 \text{ miles per hour.}$$

Finally, given that θ is the direction angle of the flight path and

$$\tan \theta \approx \frac{482.5}{-200.5} \approx -2.4065$$

you have

$$\theta \approx 180° + \arctan(-2.4065) \approx 180° - 67.4° = 112.6°.$$

You can use a graphing utility in *degree* mode to check this calculation, as shown in Figure 7.35. So, the true direction of the airplane is approximately 337.4°.

```
180+tan⁻¹(-2.4065
)
        112.5648997
```

Figure 7.35

Airplane Pilot

✓CHECKPOINT   Now try Exercise 105.

7.3   Exercises

See www.CalcChat.com for worked-out solutions to odd-numbered exercises.
For instructions on how to use a graphing utility, see Appendix A.

Vocabulary and Concept Check

In Exercises 1–8, fill in the blank(s).

1. A _____ can be used to represent a quantity that involves both magnitude and direction.

2. The directed line segment $\overrightarrow{PQ}$ has _____ point P and _____ point Q.

3. The _____ of the directed line segment $\overrightarrow{PQ}$ is denoted by $\|\overrightarrow{PQ}\|$.

4. The set of all directed line segments that are equivalent to a given directed line segment $\overrightarrow{PQ}$ is a _____ **v** in the plane.

5. The directed line segment whose initial point is the origin is said to be in _____ .

6. The two basic vector operations are scalar _____ and vector _____ .

7. The vector **u** + **v** is called the _____ of vector addition.

8. The vector sum $v_1\mathbf{i} + v_2\mathbf{j}$ is called a _____ of the vectors **i** and **j**, and the scalars v_1 and v_2 are called the _____ and _____ components of **v**, respectively.

9. What two characteristics determine whether two directed line segments are equivalent?

10. What do you call a vector that has a magnitude of 1?

Procedures and Problem Solving

Equivalent Directed Line Segments   In Exercises 11 and 12, show that u = v.

✓ **11.**

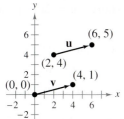

12.

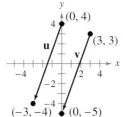

Finding the Component Form of a Vector   In Exercises 13–24, find the component form and the magnitude of the vector v.

13.

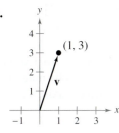

14.

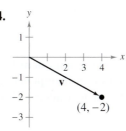

✓ **15.**

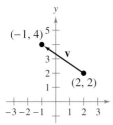

16.

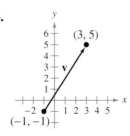

17.

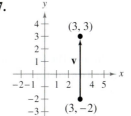

18.

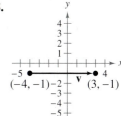

Initial Point	Terminal Point
19. $(-3, -5)$	$(5, 1)$
20. $(-3, 11)$	$(9, 40)$
21. $\left(\frac{2}{5}, 1\right)$	$\left(1, \frac{2}{5}\right)$
22. $\left(\frac{7}{2}, 0\right)$	$\left(0, -\frac{7}{2}\right)$
23. $\left(-\frac{2}{3}, -1\right)$	$\left(\frac{1}{2}, \frac{4}{5}\right)$
24. $\left(\frac{5}{2}, -2\right)$	$\left(1, \frac{2}{5}\right)$

Sketching the Graph of a Vector   In Exercises 25–30, use the figure to sketch a graph of the specified vector. To print an enlarged copy of the graph, go to the website www.mathgraphs.com.

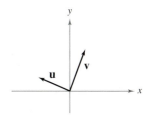

25. $-\mathbf{v}$

26. $3\mathbf{u}$

27. $\mathbf{u} + \mathbf{v}$

28. $\mathbf{u} - \mathbf{v}$

29. $\mathbf{u} + 2\mathbf{v}$

30. $\mathbf{v} - \frac{1}{2}\mathbf{u}$

102. MODELING DATA

To carry a 100-pound cylindrical weight, two people lift on the ends of short ropes that are tied to an eyelet on the top center of the cylinder. Each rope makes an angle of θ degrees with the vertical (see figure).

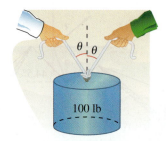

100 lb

(a) Write the tension T of each rope as a function of θ. Determine the domain of the function.

(b) Use a graphing utility to complete the table.

θ	10°	20°	30°	40°	50°	60°
T						

(c) Use the graphing utility to graph the tension function.

(d) Explain why the tension increases as θ increases.

103. MODELING DATA

Forces with magnitudes of 150 newtons and 220 newtons act on a hook (see figure).

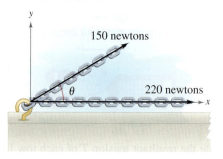

150 newtons

220 newtons

(a) Find the direction and magnitude of the resultant of the forces when $\theta = 30°$.

(b) Write the magnitude M of the resultant and the direction α of the resultant as functions of θ, where $0° \le \theta \le 180°$.

(c) Use a graphing utility to complete the table.

θ	0°	30°	60°	90°	120°	150°	180°
M							
α							

(d) Use the graphing utility to graph the two functions.

(e) Explain why one function decreases for increasing θ, whereas the other does not.

104. MODELING DATA

A commercial jet is flying from Miami to Seattle. The jet's velocity with respect to the air is 580 miles per hour, and its bearing is 332°. The wind, at the altitude of the plane, is blowing from the southwest with a velocity of 60 miles per hour.

(a) Draw a figure that gives a visual representation of the problem.

(b) Write the velocity of the wind as a vector in component form.

(c) Write the velocity of the jet relative to the air as a vector in component form.

(d) What is the speed of the jet with respect to the ground?

(e) What is the true direction of the jet?

✓ **105. Aviation** An airplane is flying in the direction 148° with an airspeed of 860 kilometers per hour. Because of the wind, its groundspeed and direction are 800 kilometers per hour and 140°, respectively. Find the direction and speed of the wind.

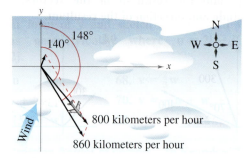

800 kilometers per hour

860 kilometers per hour

106. MODELING DATA

A tetherball weighing 1 pound is pulled outward from the pole by a horizontal force **u** until the rope makes an angle of θ degrees with the pole (see figure).

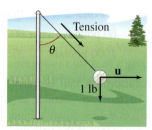

Tension

1 lb

(a) Write the tension T in the rope and the magnitude of **u** as functions of θ. Determine the domains of the functions.

(b) Use a graphing utility to graph the two functions for $0° \le \theta \le 60°$.

(c) Compare T and $\|\mathbf{u}\|$ as θ increases.

Conclusions

True or False?   In Exercises 107–110, determine whether the statement is true or false. Justify your answer.

107. If **u** and **v** have the same magnitude and direction, then **u** = **v**.

108. If **u** is a unit vector in the direction of **v**, then **v** = $\|\mathbf{v}\|\mathbf{u}$.

109. If **v** = $a\mathbf{i} + b\mathbf{j} = \mathbf{0}$, then $a = -b$.

110. If **u** = $a\mathbf{i} + b\mathbf{j}$ is a unit vector, then $a^2 + b^2 = 1$.

True or False?   In Exercises 111–118, use the figure to determine whether the statement is true or false. Justify your answer.

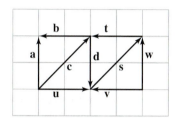

111. $\mathbf{a} = -\mathbf{d}$

112. $\mathbf{c} = \mathbf{s}$

113. $\mathbf{a} + \mathbf{u} = \mathbf{c}$

114. $\mathbf{v} + \mathbf{w} = -\mathbf{s}$

115. $\mathbf{a} + \mathbf{w} = -2\mathbf{d}$

116. $\mathbf{a} + \mathbf{d} = \mathbf{0}$

117. $\mathbf{u} - \mathbf{v} = -2(\mathbf{b} + \mathbf{t})$

118. $\mathbf{t} - \mathbf{w} = \mathbf{b} - \mathbf{a}$

119. Think About It   Consider two forces of equal magnitude acting on a point.

(a) If the magnitude of the resultant is the sum of the magnitudes of the two forces, make a conjecture about the angle between the forces.

(b) If the resultant of the forces is **0**, make a conjecture about the angle between the forces.

(c) Can the magnitude of the resultant be greater than the sum of the magnitudes of the two forces? Explain.

120. Exploration   Consider two forces

$$\mathbf{F}_1 = \langle 10, 0 \rangle \quad \text{and} \quad \mathbf{F}_2 = 5\langle \cos \theta, \sin \theta \rangle.$$

(a) Find $\|\mathbf{F}_1 + \mathbf{F}_2\|$ as a function of θ.

(b) Use a graphing utility to graph the function for $0 \le \theta < 2\pi$.

(c) Use the graph in part (b) to determine the range of the function. What is its maximum, and for what value of θ does it occur? What is its minimum, and for what value of θ does it occur?

(d) Explain why the magnitude of the resultant is never 0.

121. Proof   Prove that $(\cos \theta)\mathbf{i} + (\sin \theta)\mathbf{j}$ is a unit vector for any value of θ.

122. Writing   Write a program for your graphing utility that graphs two vectors and their difference given the vectors in component form.

123. Writing   Give geometric descriptions of (a) vector addition and (b) scalar multiplication.

124. CAPSTONE   The initial and terminal points of vector **v** are $(3, -4)$ and $(9, 1)$, respectively.

(a) Write **v** in component form.

(b) Write **v** as the linear combination of the standard unit vectors **i** and **j**.

(c) Sketch **v** with its initial point at the origin.

(d) Find the magnitude of **v**.

Finding the Difference of Two Vectors   In Exercises 125 and 126, use the program in Exercise 122 to find the difference of the vectors shown in the graph.

125. **126.**

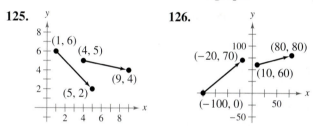

Cumulative Mixed Review

Simplifying an Expression   In Exercises 127–132, simplify the expression.

127. $\left(\dfrac{6x^4}{7y^{-2}}\right)(14x^{-1}y^5)$

128. $(5s^5 t^{-5})\left(\dfrac{3s^{-2}}{50t^{-1}}\right)$

129. $(18x)^0(4xy)^2(3x^{-1})$

130. $(5ab^2)(a^{-3}b^0)(2a^0b)^{-2}$

131. $(2.1 \times 10^9)(3.4 \times 10^{-4})$

132. $(6.5 \times 10^6)(3.8 \times 10^4)$

Solving an Equation   In Exercises 133–136, solve the equation.

133. $\cos x(\cos x + 1) = 0$

134. $\sin x(2 \sin x + \sqrt{2}) = 0$

135. $3 \sec x + 4 = 10$

136. $\cos x \cot x - \cos x = 0$

Rewriting the expression for the angle between two vectors in the form

$$\mathbf{u} \cdot \mathbf{v} = \|\mathbf{u}\| \|\mathbf{v}\| \cos \theta \qquad \text{Alternative form of dot product}$$

produces an alternative way to calculate the dot product. From this form, you can see that because $\|\mathbf{u}\|$ and $\|\mathbf{v}\|$ are always positive,

$$\mathbf{u} \cdot \mathbf{v} \text{ and } \cos \theta$$

will always have the same sign. Figure 7.38 shows the five possible orientations of two vectors.

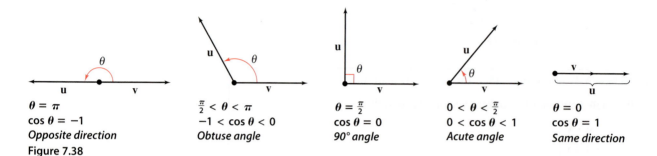

$\theta = \pi$	$\frac{\pi}{2} < \theta < \pi$	$\theta = \frac{\pi}{2}$	$0 < \theta < \frac{\pi}{2}$	$\theta = 0$
$\cos \theta = -1$	$-1 < \cos \theta < 0$	$\cos \theta = 0$	$0 < \cos \theta < 1$	$\cos \theta = 1$
Opposite direction	*Obtuse angle*	*90° angle*	*Acute angle*	*Same direction*

Figure 7.38

Definition of Orthogonal Vectors

The vectors $\mathbf{u}$ and $\mathbf{v}$ are **orthogonal** when $\mathbf{u} \cdot \mathbf{v} = 0$.

The terms *orthogonal* and *perpendicular* mean essentially the same thing—meeting at right angles. Even though the angle between the zero vector and another vector is not defined, it is convenient to extend the definition of orthogonality to include the zero vector. In other words, the zero vector is orthogonal to every vector $\mathbf{u}$ because $\mathbf{0} \cdot \mathbf{u} = 0$.

Example 4 Determining Orthogonal Vectors

Are the vectors

$$\mathbf{u} = \langle 2, -3 \rangle \text{ and } \mathbf{v} = \langle 6, 4 \rangle$$

orthogonal?

Solution

Begin by finding the dot product of the two vectors.

$$\mathbf{u} \cdot \mathbf{v} = \langle 2, -3 \rangle \cdot \langle 6, 4 \rangle = 2(6) + (-3)(4) = 0$$

Because the dot product is 0, the two vectors are orthogonal, as shown in Figure 7.39.

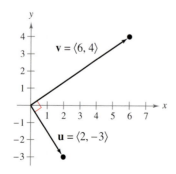

Figure 7.39

✔**CHECKPOINT** Now try Exercise 43.

Technology Tip

The graphing utility program Finding the Angle Between Two Vectors, found at this textbook's *Companion Website,* graphs two vectors $\mathbf{u} = \langle a, b \rangle$ and $\mathbf{v} = \langle c, d \rangle$ in standard position and finds the measure of the angle between them. Use the program to verify Examples 3 and 4.

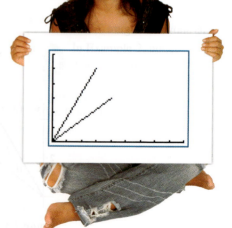

Finding Vector Components

You have already seen applications in which two vectors are added to produce a resultant vector. Many applications in physics and engineering pose the reverse problem—decomposing a given vector into the sum of two **vector components.**

Consider a boat on an inclined ramp, as shown in Figure 7.40. The force **F** due to gravity pulls the boat *down* the ramp and *against* the ramp. These two orthogonal forces, $\mathbf{w}_1$ and $\mathbf{w}_2$, are vector components of **F**. That is,

$\qquad \mathbf{F} = \mathbf{w}_1 + \mathbf{w}_2.$ Vector components of **F**

The negative of component $\mathbf{w}_1$ represents the force needed to keep the boat from rolling down the ramp, and $\mathbf{w}_2$ represents the force that the tires must withstand against the ramp. A procedure for finding $\mathbf{w}_1$ and $\mathbf{w}_2$ is shown below.

Figure 7.40

Definition of Vector Components

Let **u** and **v** be nonzero vectors such that

$\qquad \mathbf{u} = \mathbf{w}_1 + \mathbf{w}_2$

where $\mathbf{w}_1$ and $\mathbf{w}_2$ are orthogonal and $\mathbf{w}_1$ is parallel to (or a scalar multiple of) **v**, as shown in Figure 7.41. The vectors $\mathbf{w}_1$ and $\mathbf{w}_2$ are called **vector components** of **u**. The vector $\mathbf{w}_1$ is the **projection** of **u** onto **v** and is denoted by

$\qquad \mathbf{w}_1 = \text{proj}_{\mathbf{v}}\mathbf{u}.$

The vector $\mathbf{w}_2$ is given by

$\qquad \mathbf{w}_2 = \mathbf{u} - \mathbf{w}_1.$

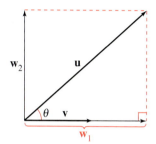

θ is acute.

θ is obtuse.
Figure 7.41

From the definition of vector components, you can see that it is easy to find the component $\mathbf{w}_2$ once you have found the projection of **u** onto **v**. To find the projection, you can use the dot product, as follows.

$$\mathbf{u} = \mathbf{w}_1 + \mathbf{w}_2$$
$$\mathbf{u} = c\mathbf{v} + \mathbf{w}_2 \qquad \text{\color{red}$\mathbf{w}_1$ is a scalar multiple of \textbf{v}.}$$
$$\mathbf{u} \cdot \mathbf{v} = (c\mathbf{v} + \mathbf{w}_2) \cdot \mathbf{v} \qquad \text{\color{red}Take dot product of each side with \textbf{v}.}$$
$$\mathbf{u} \cdot \mathbf{v} = c\mathbf{v} \cdot \mathbf{v} + \mathbf{w}_2 \cdot \mathbf{v}$$
$$\mathbf{u} \cdot \mathbf{v} = c\|\mathbf{v}\|^2 + 0 \qquad \text{\color{red}$\mathbf{w}_2$ and \textbf{v} are orthogonal.}$$

So,

$$c = \frac{\mathbf{u} \cdot \mathbf{v}}{\|\mathbf{v}\|^2}$$

and

$$\mathbf{w}_1 = \text{proj}_{\mathbf{v}}\mathbf{u} = c\mathbf{v} = \frac{\mathbf{u} \cdot \mathbf{v}}{\|\mathbf{v}\|^2}\mathbf{v}.$$

Projection of u onto v

Let **u** and **v** be nonzero vectors. The projection of **u** onto **v** is given by

$$\text{proj}_{\mathbf{v}}\mathbf{u} = \left(\frac{\mathbf{u} \cdot \mathbf{v}}{\|\mathbf{v}\|^2}\right)\mathbf{v}.$$

7.4 Exercises

See www.CalcChat.com for worked-out solutions to odd-numbered exercises.
For instructions on how to use a graphing utility, see Appendix A.

Vocabulary and Concept Check

1. For two vectors **u** and **v**, does $\mathbf{u} \cdot \mathbf{v} = \mathbf{v} \cdot \mathbf{u}$?

2. What is the dot product of two orthogonal vectors?

3. Is the dot product of two vectors an angle, a vector, or a scalar?

In Exercises 4–6, fill in the blank(s).

4. If θ is the angle between two nonzero vectors **u** and **v**, then $\cos\theta = $ _____ .

5. The projection of **u** onto **v** is given by $\text{proj}_\mathbf{v}\mathbf{u} = $ _____ .

6. The work W done by a constant force **F** as its point of application moves along the vector $\overrightarrow{PQ}$ is given by either $W = $ _____ or $W = $ _____ .

Procedures and Problem Solving

Finding a Dot Product In Exercises 7–10, find the dot product of u and v.

✓ 7. $\mathbf{u} = \langle 6, 3 \rangle$
   $\mathbf{v} = \langle 2, -4 \rangle$

8. $\mathbf{u} = \langle -4, 1 \rangle$
   $\mathbf{v} = \langle 2, -3 \rangle$

9. $\mathbf{u} = 5\mathbf{i} + \mathbf{j}$
   $\mathbf{v} = 3\mathbf{i} - \mathbf{j}$

10. $\mathbf{u} = 3\mathbf{i} + 2\mathbf{j}$
    $\mathbf{v} = -2\mathbf{i} + \mathbf{j}$

Using Properties of Dot Products In Exercises 11–16, use the vectors $\mathbf{u} = \langle 2, 2 \rangle$, $\mathbf{v} = \langle -3, 4 \rangle$, and $\mathbf{w} = \langle 1, -4 \rangle$ to find the indicated quantity. State whether the result is a vector or a scalar.

11. $\mathbf{u} \cdot \mathbf{u}$

12. $\mathbf{v} \cdot \mathbf{w}$

13. $\mathbf{u} \cdot 2\mathbf{v}$

14. $4\mathbf{u} \cdot \mathbf{v}$

✓ 15. $(3\mathbf{w} \cdot \mathbf{v})\mathbf{u}$

16. $(\mathbf{u} \cdot 2\mathbf{v})\mathbf{w}$

Finding the Magnitude of a Vector In Exercises 17–22, use the dot product to find the magnitude of u.

17. $\mathbf{u} = \langle -5, 12 \rangle$

18. $\mathbf{u} = \langle 2, -4 \rangle$

19. $\mathbf{u} = 20\mathbf{i} + 25\mathbf{j}$

20. $\mathbf{u} = 6\mathbf{i} - 10\mathbf{j}$

21. $\mathbf{u} = -4\mathbf{j}$

22. $\mathbf{u} = 9\mathbf{i}$

Finding the Angle Between Two Vectors In Exercises 23–30, find the angle θ between the vectors.

✓ 23. $\mathbf{u} = \langle -1, 0 \rangle$
   $\mathbf{v} = \langle 0, 2 \rangle$

24. $\mathbf{u} = \langle 4, 4 \rangle$
   $\mathbf{v} = \langle -2, 0 \rangle$

25. $\mathbf{u} = 3\mathbf{i} + 4\mathbf{j}$
   $\mathbf{v} = -2\mathbf{i} + 3\mathbf{j}$

26. $\mathbf{u} = 2\mathbf{i} - 3\mathbf{j}$
    $\mathbf{v} = \mathbf{i} - 2\mathbf{j}$

27. $\mathbf{u} = 2\mathbf{i}$
    $\mathbf{v} = -3\mathbf{j}$

28. $\mathbf{u} = 4\mathbf{j}$
    $\mathbf{v} = -3\mathbf{i}$

29. $\mathbf{u} = \cos\left(\dfrac{\pi}{3}\right)\mathbf{i} + \sin\left(\dfrac{\pi}{3}\right)\mathbf{j}$

    $\mathbf{v} = \cos\left(\dfrac{3\pi}{4}\right)\mathbf{i} + \sin\left(\dfrac{3\pi}{4}\right)\mathbf{j}$

30. $\mathbf{u} = \cos\left(\dfrac{\pi}{4}\right)\mathbf{i} + \sin\left(\dfrac{\pi}{4}\right)\mathbf{j}$

    $\mathbf{v} = \cos\left(\dfrac{2\pi}{3}\right)\mathbf{i} + \sin\left(\dfrac{2\pi}{3}\right)\mathbf{j}$

Finding the Angle Between Two Vectors In Exercises 31–34, graph the vectors and find the degree measure of the angle between the vectors.

31. $\mathbf{u} = 2\mathbf{i} - 4\mathbf{j}$
    $\mathbf{v} = 3\mathbf{i} - 5\mathbf{j}$

32. $\mathbf{u} = -6\mathbf{i} - 3\mathbf{j}$
    $\mathbf{v} = -8\mathbf{i} + 4\mathbf{j}$

33. $\mathbf{u} = 6\mathbf{i} - 2\mathbf{j}$
    $\mathbf{v} = 8\mathbf{i} - 5\mathbf{j}$

34. $\mathbf{u} = 2\mathbf{i} - 3\mathbf{j}$
    $\mathbf{v} = 4\mathbf{i} + 3\mathbf{j}$

Finding the Angles in a Triangle In Exercises 35–38, use vectors to find the interior angles of the triangle with the given vertices.

35. $(1, 2), (3, 4), (2, 5)$

36. $(-3, -4), (1, 7), (8, 2)$

37. $(-3, 0), (2, 2), (0, 6)$

38. $(-3, 5), (-1, 9), (7, 9)$

Using the Angle Between Two Vectors In Exercises 39–42, find $\mathbf{u} \cdot \mathbf{v}$, where θ is the angle between u and v.

39. $\|\mathbf{u}\| = 9, \|\mathbf{v}\| = 36, \theta = \dfrac{3\pi}{4}$

40. $\|\mathbf{u}\| = 4, \|\mathbf{v}\| = 12, \theta = \dfrac{\pi}{3}$

41. $\|\mathbf{u}\| = 4, \|\mathbf{v}\| = 10, \theta = \dfrac{2\pi}{3}$

42. $\|\mathbf{u}\| = 100, \|\mathbf{v}\| = 250, \theta = \dfrac{\pi}{6}$

Determining Orthogonal Vectors In Exercises 43–46, determine whether u and v are orthogonal.

✓ 43. $\mathbf{u} = \langle 10, -6 \rangle$
   $\mathbf{v} = \langle 9, 15 \rangle$

44. $\mathbf{u} = \langle 12, 4 \rangle$
    $\mathbf{v} = \left\langle \tfrac{1}{4}, -\tfrac{1}{3} \right\rangle$

45. $\mathbf{u} = \mathbf{j}$
  $\mathbf{v} = \mathbf{i} - \mathbf{j}$

46. $\mathbf{u} = 2\mathbf{i} - 2\mathbf{j}$
  $\mathbf{v} = -\mathbf{i} - \mathbf{j}$

A Relationship of Two Vectors   **In Exercises 47–50, determine whether u and v are orthogonal, parallel, or neither.**

47. $\mathbf{u} = \langle 10, 20 \rangle$
  $\mathbf{v} = \langle -5, 10 \rangle$

48. $\mathbf{u} = \langle 15, 9 \rangle$
  $\mathbf{v} = \langle -5, -3 \rangle$

49. $\mathbf{u} = -\frac{3}{5}\mathbf{i} + \frac{7}{10}\mathbf{j}$
  $\mathbf{v} = 12\mathbf{i} - 14\mathbf{j}$

50. $\mathbf{u} = -\frac{9}{10}\mathbf{i} + 3\mathbf{j}$
  $\mathbf{v} = -5\mathbf{i} - \frac{3}{2}\mathbf{j}$

Finding an Unknown Vector Component   **In Exercises 51–56, find the value of k such that the vectors u and v are orthogonal.**

51. $\mathbf{u} = 2\mathbf{i} - k\mathbf{j}$
  $\mathbf{v} = 3\mathbf{i} + 2\mathbf{j}$

52. $\mathbf{u} = 3\mathbf{i} + 2\mathbf{j}$
  $\mathbf{v} = 2\mathbf{i} - k\mathbf{j}$

53. $\mathbf{u} = \mathbf{i} + 4\mathbf{j}$
  $\mathbf{v} = 2k\mathbf{i} - 5\mathbf{j}$

54. $\mathbf{u} = -3k\mathbf{i} + 5\mathbf{j}$
  $\mathbf{v} = 2\mathbf{i} - 4\mathbf{j}$

55. $\mathbf{u} = -3k\mathbf{i} + 2\mathbf{j}$
  $\mathbf{v} = -6\mathbf{i}$

56. $\mathbf{u} = 4\mathbf{i} - 4k\mathbf{j}$
  $\mathbf{v} = 3\mathbf{j}$

Decomposing a Vector into Components   **In Exercises 57–60, find the projection of u onto v. Then write u as the sum of two orthogonal vectors, one of which is proj$_\mathbf{v}$ u.**

✓ **57.** $\mathbf{u} = \langle 3, 4 \rangle$
  $\mathbf{v} = \langle 8, 2 \rangle$

58. $\mathbf{u} = \langle 4, 2 \rangle$
  $\mathbf{v} = \langle 1, -2 \rangle$

59. $\mathbf{u} = \langle 0, 3 \rangle$
  $\mathbf{v} = \langle 2, 15 \rangle$

60. $\mathbf{u} = \langle -5, -1 \rangle$
  $\mathbf{v} = \langle -1, 1 \rangle$

Finding the Projection of u onto v Mentally   **In Exercises 61–64, use the graph to determine mentally the projection of u onto v. (The coordinates of the terminal points of the vectors in standard position are given.) Use the formula for the projection of u onto v to verify your result.**

61.
62.

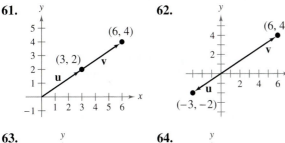

63.
64.

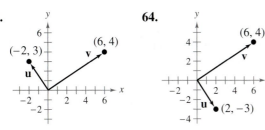

Finding Orthogonal Vectors   **In Exercises 65–68, find two vectors in opposite directions that are orthogonal to the vector u. (There are many correct answers.)**

65. $\mathbf{u} = \langle 2, 6 \rangle$

66. $\mathbf{u} = \langle -7, 5 \rangle$

67. $\mathbf{u} = \frac{1}{2}\mathbf{i} - \frac{3}{4}\mathbf{j}$

68. $\mathbf{u} = -\frac{5}{2}\mathbf{i} - 3\mathbf{j}$

Finding Work   **In Exercises 69 and 70, find the work done in moving a particle from P to Q if the magnitude and direction of the force are given by v.**

69. $P = (0, 0)$, $Q = (4, 7)$, $\mathbf{v} = \langle 1, 4 \rangle$

70. $P = (1, 3)$, $Q = (-3, 5)$, $\mathbf{v} = -2\mathbf{i} + 3\mathbf{j}$

71. Business   The vector $\mathbf{u} = \langle 1225, 2445 \rangle$ gives the numbers of hours worked by employees of a temp agency at two pay levels. The vector $\mathbf{v} = \langle 12.20, 8.50 \rangle$ gives the hourly wage (in dollars) paid at each level, respectively. (a) Find the dot product $\mathbf{u} \cdot \mathbf{v}$ and explain its meaning in the context of the problem. (b) Identify the vector operation used to increase wages by 2 percent.

72. Business   The vector $\mathbf{u} = \langle 3240, 2450 \rangle$ gives the numbers of hamburgers and hot dogs, respectively, sold at a fast food stand in one week. The vector $\mathbf{v} = \langle 1.75, 1.25 \rangle$ gives the prices in dollars of the food items. (a) Find the dot product $\mathbf{u} \cdot \mathbf{v}$ and explain its meaning in the context of the problem. (b) Identify the vector operation used to increase prices by $2\frac{1}{2}$ percent.

✓ **73.** *Why you should learn it* (p. 574)   A truck with a gross weight of 30,000 pounds is parked on a slope of $d°$ (see figure). Assume that the only force to overcome is the force of gravity.

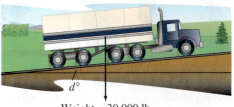

Weight = 30,000 lb

(a) Find the force required to keep the truck from rolling down the hill in terms of the slope d.

(b) Use a graphing utility to complete the table.

d	0°	1°	2°	3°	4°	5°
Force						

d	6°	7°	8°	9°	10°
Force					

(c) Find the force perpendicular to the hill when $d = 5°$.

Trigonometric Form of a Complex Number

In Section 2.3, you learned how to add, subtract, multiply, and divide complex numbers. To work effectively with *powers* and *roots* of complex numbers, it is helpful to write complex numbers in trigonometric form. In Figure 7.49, consider the nonzero complex number $a + bi$. By letting θ be the angle from the positive real axis (measured counterclockwise) to the line segment connecting the origin and the point (a, b), you can write

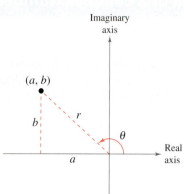

$$a = r \cos \theta \quad \text{and} \quad b = r \sin \theta$$

where

Figure 7.49

$$r = \sqrt{a^2 + b^2}.$$

Consequently, you have

$$a + bi = (r \cos \theta) + (r \sin \theta)i$$

from which you can obtain the **trigonometric form of a complex number.**

Trigonometric Form of a Complex Number

The **trigonometric form** of the complex number $z = a + bi$ is given by

$$z = r(\cos \theta + i \sin \theta)$$

where $a = r \cos \theta$, $b = r \sin \theta$, $r = \sqrt{a^2 + b^2}$, and $\tan \theta = b/a$. The number r is the **modulus** of z, and θ is called an **argument** of z.

The trigonometric form of a complex number is also called the *polar form*. Because there are infinitely many choices for θ, the trigonometric form of a complex number is not unique. Normally, θ is restricted to the interval $0 \le \theta < 2\pi$, although on occasion it is convenient to use $\theta < 0$.

Example 2   Writing a Complex Number in Trigonometric Form

Write the complex number

$$z = -2i$$

in trigonometric form.

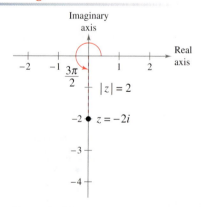

Solution

The absolute value of z is

$$r = |-2i| = \sqrt{0^2 + (-2)^2} = \sqrt{4} = 2.$$

With $a = 0$, you cannot use $\tan \theta = b/a$ to find θ. Because $z = -2i$ lies on the negative imaginary axis (see Figure 7.50), choose $\theta = 3\pi/2$. So, the trigonometric form is

Figure 7.50

$$z = r(\cos \theta + i \sin \theta)$$

$$= 2\left(\cos \frac{3\pi}{2} + i \sin \frac{3\pi}{2}\right).$$

✓**CHECKPOINT**   Now try Exercise 23.

Example 3   Writing a Complex Number in Trigonometric Form

Write the complex number $z = -2 - 2\sqrt{3}i$ in trigonometric form.

Solution

The absolute value of z is

$$r = \left| -2 - 2\sqrt{3}i \right| = \sqrt{(-2)^2 + \left(-2\sqrt{3}\right)^2} = \sqrt{16} = 4$$

and the angle θ is given by

$$\tan \theta = \frac{b}{a} = \frac{-2\sqrt{3}}{-2} = \sqrt{3}.$$

Because $\tan(\pi/3) = \sqrt{3}$ and $z = -2 - 2\sqrt{3}i$ lies in Quadrant III, choose θ to be $\theta = \pi + \pi/3 = 4\pi/3$. So, the trigonometric form is

$$z = r(\cos \theta + i \sin \theta)$$

$$= 4\left(\cos \frac{4\pi}{3} + i \sin \frac{4\pi}{3} \right).$$

See Figure 7.51.

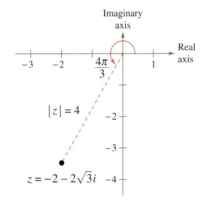

Figure 7.51

✔CHECKPOINT   Now try Exercise 29.

Example 4   Writing a Complex Number in Standard Form

Write the complex number in standard form $a + bi$.

$$z = \sqrt{8}\left[\cos\left(-\frac{\pi}{3}\right) + i \sin\left(-\frac{\pi}{3}\right) \right]$$

Solution

Because $\cos(-\pi/3) = 1/2$ and $\sin(-\pi/3) = -\sqrt{3}/2$, you can write

$$z = \sqrt{8}\left[\cos\left(-\frac{\pi}{3}\right) + i \sin\left(-\frac{\pi}{3}\right) \right]$$

$$= \sqrt{8}\left[\frac{1}{2} - \frac{\sqrt{3}}{2}i \right]$$

$$= 2\sqrt{2}\left[\frac{1}{2} - \frac{\sqrt{3}}{2}i \right]$$

$$= \sqrt{2} - \sqrt{6}i.$$

✔CHECKPOINT   Now try Exercise 47.

? What's Wrong?

You use a graphing utility to check the answer to Example 3, as shown in the figure. You determine that $r = 4$ and $\theta \approx 1.047 \approx \pi/3$. Your value for θ does not agree with the value found in the example. What's wrong?

Technology Tip

A graphing utility can be used to convert a complex number in trigonometric form to standard form. For instance, enter the complex number $\sqrt{2}(\cos \pi/4 + i \sin \pi/4)$ in your graphing utility and press ENTER. You should obtain the standard form $1 + i$, as shown below.

Multiplication and Division of Complex Numbers

The trigonometric form adapts nicely to multiplication and division of complex numbers. Suppose you are given two complex numbers

$$z_1 = r_1(\cos \theta_1 + i \sin \theta_1)$$

and

$$z_2 = r_2(\cos \theta_2 + i \sin \theta_2).$$

The product of z_1 and z_2 is

$$
\begin{aligned}
z_1 z_2 &= r_1 r_2 (\cos \theta_1 + i \sin \theta_1)(\cos \theta_2 + i \sin \theta_2) \\
&= r_1 r_2 [(\cos \theta_1 \cos \theta_2 - \sin \theta_1 \sin \theta_2) + i(\sin \theta_1 \cos \theta_2 + \cos \theta_1 \sin \theta_2)] \\
&= r_1 r_2 [\cos(\theta_1 + \theta_2) + i \sin(\theta_1 + \theta_2)]. \qquad \text{\color{red}Sum and difference formulas}
\end{aligned}
$$

This establishes the first part of the following rule. The second part is left for you to verify (see Exercise 171).

Product and Quotient of Two Complex Numbers

Let $z_1 = r_1(\cos \theta_1 + i \sin \theta_1)$ and $z_2 = r_2(\cos \theta_2 + i \sin \theta_2)$ be complex numbers.

$$z_1 z_2 = r_1 r_2 [\cos(\theta_1 + \theta_2) + i \sin(\theta_1 + \theta_2)] \qquad \text{\color{red}Product}$$

$$\frac{z_1}{z_2} = \frac{r_1}{r_2}[\cos(\theta_1 - \theta_2) + i \sin(\theta_1 - \theta_2)], \quad z_2 \neq 0 \qquad \text{\color{red}Quotient}$$

Note that this rule says that to *multiply* two complex numbers you multiply moduli and add arguments, whereas to *divide* two complex numbers you divide moduli and subtract arguments.

Example 5   Multiplying Complex Numbers in Trigonometric Form

Find the product $z_1 z_2$ of the complex numbers.

$$z_1 = 3\left(\cos \frac{\pi}{4} + i \sin \frac{\pi}{4}\right)$$

$$z_2 = 2\left(\cos \frac{3\pi}{4} + i \sin \frac{3\pi}{4}\right)$$

Solution

$$
\begin{aligned}
z_1 z_2 &= 3\left(\cos \frac{\pi}{4} + i \sin \frac{\pi}{4}\right) \cdot 2\left(\cos \frac{3\pi}{4} + i \sin \frac{3\pi}{4}\right) \\
&= 6\left[\cos\left(\frac{\pi}{4} + \frac{3\pi}{4}\right) + i \sin\left(\frac{\pi}{4} + \frac{3\pi}{4}\right)\right] \\
&= 6(\cos \pi + i \sin \pi) \\
&= 6[-1 + i(0)] \\
&= -6
\end{aligned}
$$

The numbers z_1, z_2, and $z_1 z_2$ are plotted in Figure 7.52.

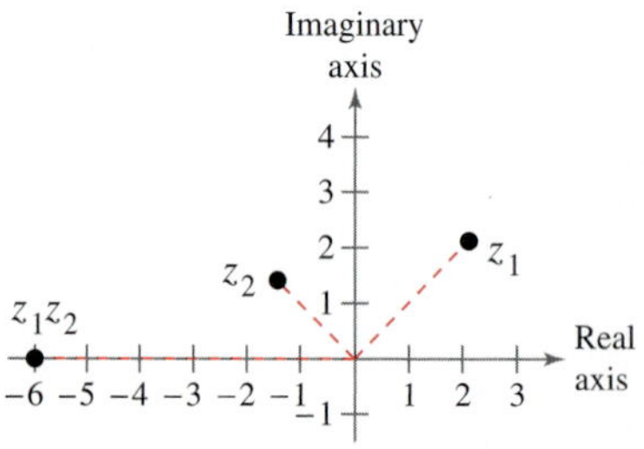

Figure 7.52

✓CHECKPOINT   Now try Exercise 65.

Example 6   Multiplying Complex Numbers in Trigonometric Form

Find the product z_1z_2 of the complex numbers.

$$z_1 = 2\left(\cos\frac{2\pi}{3} + i\sin\frac{2\pi}{3}\right) \qquad z_2 = 8\left(\cos\frac{11\pi}{6} + i\sin\frac{11\pi}{6}\right)$$

Solution

$$z_1z_2 = 2\left(\cos\frac{2\pi}{3} + i\sin\frac{2\pi}{3}\right) \cdot 8\left(\cos\frac{11\pi}{6} + i\sin\frac{11\pi}{6}\right)$$

$$= 16\left[\cos\left(\frac{2\pi}{3} + \frac{11\pi}{6}\right) + i\sin\left(\frac{2\pi}{3} + \frac{11\pi}{6}\right)\right]$$

$$= 16\left(\cos\frac{5\pi}{2} + i\sin\frac{5\pi}{2}\right)$$

$$= 16\left(\cos\frac{\pi}{2} + i\sin\frac{\pi}{2}\right)$$

$$= 16[0 + i(1)]$$

$$= 16i$$

You can check this result by first converting to the standard forms

$$z_1 = -1 + \sqrt{3}i \quad \text{and} \quad z_2 = 4\sqrt{3} - 4i$$

and then multiplying algebraically, as in Section 2.3.

$$z_1z_2 = (-1 + \sqrt{3}i)(4\sqrt{3} - 4i)$$

$$= -4\sqrt{3} + 4i + 12i + 4\sqrt{3}$$

$$= 16i$$

 CHECKPOINT   Now try Exercise 67.

Example 7   Dividing Complex Numbers in Trigonometric Form

Find the quotient

$$\frac{z_1}{z_2}$$

of the complex numbers.

$$z_1 = 24(\cos 300° + i\sin 300°) \qquad z_2 = 8(\cos 75° + i\sin 75°)$$

Solution

$$\frac{z_1}{z_2} = \frac{24(\cos 300° + i\sin 300°)}{8(\cos 75° + i\sin 75°)}$$

$$= \frac{24}{8}[\cos(300° - 75°) + i\sin(300° - 75°)]$$

$$= 3(\cos 225° + i\sin 225°)$$

$$= 3\left[\left(-\frac{\sqrt{2}}{2}\right) + i\left(-\frac{\sqrt{2}}{2}\right)\right]$$

$$= -\frac{3\sqrt{2}}{2} - \frac{3\sqrt{2}}{2}i$$

CHECKPOINT   Now try Exercise 73.

Technology Tip

Some graphing utilities can multiply and divide complex numbers in trigonometric form. If you have access to such a graphing utility, use it to find z_1z_2 and z_1/z_2 in Examples 6 and 7.

```
2(cos(2π/3)+isin
(2π/3))*8(cos(11
π/6)+isin(11π/6)
)
                 16i
```

Powers of Complex Numbers

The trigonometric form of a complex number is used to raise a complex number to a power. To accomplish this, consider repeated use of the multiplication rule.

$$z = r(\cos \theta + i \sin \theta)$$

$$z^2 = r(\cos \theta + i \sin \theta)r(\cos \theta + i \sin \theta) = r^2(\cos 2\theta + i \sin 2\theta)$$

$$z^3 = r^2(\cos 2\theta + i \sin 2\theta)r(\cos \theta + i \sin \theta) = r^3(\cos 3\theta + i \sin 3\theta)$$

$$z^4 = r^4(\cos 4\theta + i \sin 4\theta)$$

$$z^5 = r^5(\cos 5\theta + i \sin 5\theta)$$

$$\vdots$$

This pattern leads to **DeMoivre's Theorem,** which is named after the French mathematician Abraham DeMoivre (1667–1754).

DeMoivre's Theorem

If $z = r(\cos \theta + i \sin \theta)$ is a complex number and n is a positive integer, then

$$z^n = [r(\cos \theta + i \sin \theta)]^n$$

$$= r^n(\cos n\theta + i \sin n\theta).$$

Explore the Concept

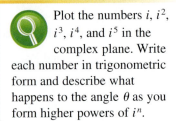

Plot the numbers i, i^2, i^3, i^4, and i^5 in the complex plane. Write each number in trigonometric form and describe what happens to the angle θ as you form higher powers of i^n.

Example 8   Finding a Power of a Complex Number

Use DeMoivre's Theorem to find

$$\left(1 + \sqrt{3}i\right)^{12}.$$

Solution

First convert the complex number to trigonometric form using

$$r = \sqrt{(1)^2 + \left(\sqrt{3}\right)^2} = 2$$

and

$$\theta = \arctan \frac{\sqrt{3}}{1} = \frac{\pi}{3}.$$

So, the trigonometric form is

$$1 + \sqrt{3}i = 2\left(\cos \frac{\pi}{3} + i \sin \frac{\pi}{3}\right).$$

Then, by DeMoivre's Theorem, you have

$$\left(1 + \sqrt{3}i\right)^{12} = \left[2\left(\cos \frac{\pi}{3} + i \sin \frac{\pi}{3}\right)\right]^{12}$$

$$= 2^{12}\left(\cos \frac{12\pi}{3} + i \sin \frac{12\pi}{3}\right)$$

$$= 4096(\cos 4\pi + i \sin 4\pi)$$

$$= 4096(1 + 0)$$

$$= 4096.$$

✓*CHECKPOINT*   Now try Exercise 107.

Roots of Complex Numbers

Recall that a consequence of the Fundamental Theorem of Algebra is that a polynomial equation of degree n has n solutions in the complex number system. So, an equation such as $x^6 = 1$ has six solutions, and in this particular case you can find the six solutions by factoring and using the Quadratic Formula.

$$x^6 - 1 = 0$$

$$(x^3 - 1)(x^3 + 1) = 0$$

$$(x - 1)(x^2 + x + 1)(x + 1)(x^2 - x + 1) = 0$$

Consequently, the solutions are

$$x = \pm 1, \qquad x = \frac{-1 \pm \sqrt{3}i}{2}, \qquad \text{and} \qquad x = \frac{1 \pm \sqrt{3}i}{2}.$$

Each of these numbers is a sixth root of 1. In general, the **nth root of a complex number** is defined as follows.

Definition of an nth Root of a Complex Number

The complex number $u = a + bi$ is an **nth root** of the complex number z when

$$z = u^n = (a + bi)^n.$$

To find a formula for an nth root of a complex number, let u be an nth root of z, where $u = s(\cos \beta + i \sin \beta)$ and $z = r(\cos \theta + i \sin \theta)$. By DeMoivre's Theorem and the fact that $u^n = z$, you have

$$s^n (\cos n\beta + i \sin n\beta) = r(\cos \theta + i \sin \theta).$$

Taking the absolute value of each side of this equation, it follows that $s^n = r$. Substituting back into the previous equation and dividing by r, you get

$$\cos n\beta + i \sin n\beta = \cos \theta + i \sin \theta.$$

So, it follows that

$$\cos n\beta = \cos \theta \qquad \text{and} \qquad \sin n\beta = \sin \theta.$$

Because both sine and cosine have a period of 2π, these last two equations have solutions if and only if the angles differ by a multiple of 2π. Consequently, there must exist an integer k such that

$$n\beta = \theta + 2\pi k$$

$$\beta = \frac{\theta + 2\pi k}{n}.$$

By substituting this value of β into the trigonometric form of u, you get the result stated in the theorem on the next page.

Explore the Concept

The nth roots of a complex number are useful for solving some polynomial equations. For instance, explain how you can use DeMoivre's Theorem to solve the polynomial equation $x^4 + 16 = 0$. [*Hint:* Write -16 as $16(\cos \pi + i \sin \pi)$.]

> ### nth Roots of a Complex Number
>
> For a positive integer n, the complex number $z = r(\cos \theta + i \sin \theta)$ has exactly n distinct nth roots given by
>
> $$\sqrt[n]{r}\left(\cos \frac{\theta + 2\pi k}{n} + i \sin \frac{\theta + 2\pi k}{n}\right)$$
>
> where $k = 0, 1, 2, \ldots, n - 1$.

When $k > n - 1$, the roots begin to repeat. For instance, when $k = n$, the angle

$$\frac{\theta + 2\pi n}{n} = \frac{\theta}{n} + 2\pi$$

is coterminal with θ/n, which is also obtained when $k = 0$.

The formula for the nth roots of a complex number z has a nice geometrical interpretation, as shown in Figure 7.53. Note that because the nth roots of z all have the same magnitude $\sqrt[n]{r}$, they all lie on a circle of radius $\sqrt[n]{r}$ with center at the origin. Furthermore, because successive nth roots have arguments that differ by

$$\frac{2\pi}{n}$$

the nth roots are equally spaced around the circle.

You have already found the sixth roots of 1 by factoring and by using the Quadratic Formula. Example 9 shows how you can solve the same problem with the formula for nth roots.

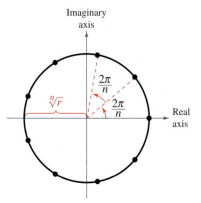

Figure 7.53

Example 9 Finding the nth Roots of a Real Number

Find all the sixth roots of 1.

Solution

First write 1 in the trigonometric form

$$1 = 1(\cos 0 + i \sin 0).$$

Then, by the nth root formula with $n = 6$ and $r = 1$, the roots have the form

$$\sqrt[6]{1}\left(\cos \frac{0 + 2\pi k}{6} + i \sin \frac{0 + 2\pi k}{6}\right) = \cos \frac{\pi k}{3} + i \sin \frac{\pi k}{3}.$$

So, for $k = 0, 1, 2, 3, 4,$ and 5, the sixth roots are as follows. (See Figure 7.54.)

$$\cos 0 + i \sin 0 = 1$$

$$\cos \frac{\pi}{3} + i \sin \frac{\pi}{3} = \frac{1}{2} + \frac{\sqrt{3}}{2}i \qquad \text{\color{red} Incremented by } \frac{2\pi}{n} = \frac{2\pi}{6} = \frac{\pi}{3}$$

$$\cos \frac{2\pi}{3} + i \sin \frac{2\pi}{3} = -\frac{1}{2} + \frac{\sqrt{3}}{2}i$$

$$\cos \pi + i \sin \pi = -1$$

$$\cos \frac{4\pi}{3} + i \sin \frac{4\pi}{3} = -\frac{1}{2} - \frac{\sqrt{3}}{2}i$$

$$\cos \frac{5\pi}{3} + i \sin \frac{5\pi}{3} = \frac{1}{2} - \frac{\sqrt{3}}{2}i$$

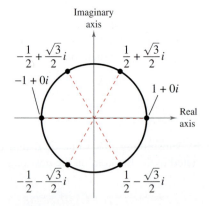

Figure 7.54

✔ **CHECKPOINT**   Now try Exercise 147.

In Figure 7.54, notice that the roots obtained in Example 9 all have a magnitude of 1 and are equally spaced around the unit circle. Also notice that the complex roots occur in conjugate pairs, as discussed in Section 3.4. The n distinct nth roots of 1 are called the **nth roots of unity.**

Example 10   Finding the nth Roots of a Complex Number

Find the three cube roots of $z = -2 + 2i$.

Solution

The absolute value of z is

$$r = |-2 + 2i| = \sqrt{(-2)^2 + 2^2} = \sqrt{8}$$

and the angle θ is given by

$$\tan \theta = \frac{b}{a} = \frac{2}{-2} = -1.$$

Because z lies in Quadrant II, the trigonometric form of z is

$$z = -2 + 2i = \sqrt{8}(\cos 135° + i \sin 135°).$$

By the formula for nth roots, the cube roots have the form

$$\sqrt[6]{8}\left(\cos \frac{135° + 360°k}{3} + i \sin \frac{135° + 360°k}{3}\right).$$

Finally, for $k = 0$, 1, and 2, you obtain the roots

$$\sqrt[6]{8}\left(\cos \frac{135° + 360°(0)}{3} + i \sin \frac{135° + 360°(0)}{3}\right)$$

$$= \sqrt{2}(\cos 45° + i \sin 45°)$$

$$= 1 + i$$

$$\sqrt[6]{8}\left(\cos \frac{135° + 360°(1)}{3} + i \sin \frac{135° + 360°(1)}{3}\right)$$

$$= \sqrt{2}(\cos 165° + i \sin 165°)$$

$$\approx -1.3660 + 0.3660i$$

$$\sqrt[6]{8}\left(\cos \frac{135° + 360°(2)}{3} + i \sin \frac{135° + 360°(2)}{3}\right)$$

$$= \sqrt{2}(\cos 285° + i \sin 285°)$$

$$\approx 0.3660 - 1.3660i.$$

See Figure 7.55.

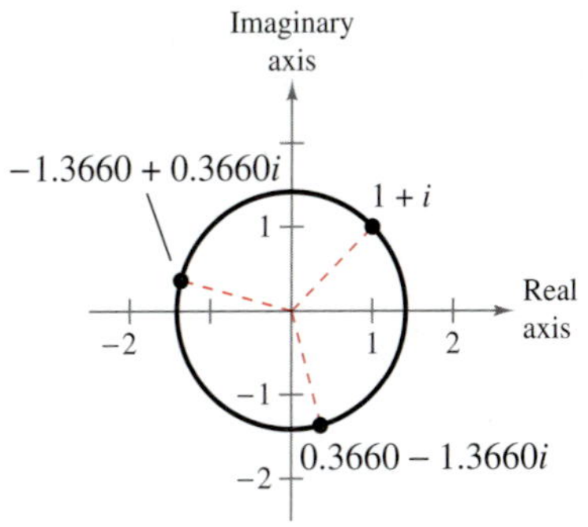

Figure 7.55

 CHECKPOINT   Now try Exercise 151.

Explore the Concept

 Use a graphing utility set in *parametric* and *radian* modes to display the graphs of X1T = cos T and Y1T = sin T. Set the viewing window so that $-1.5 \leq X \leq 1.5$ and $-1 \leq Y \leq 1$. Then, using $0 \leq T \leq 2\pi$, set the "Tstep" to $2\pi/n$ for various values of n. Explain how the graphing utility can be used to obtain the nth roots of unity.

7.5 Exercises

See www.CalcChat.com for worked-out solutions to odd-numbered exercises.
For instructions on how to use a graphing utility, see Appendix A.

Vocabulary and Concept Check

In Exercises 1–3, fill in the blank.

1. The _____ of a complex number $a + bi$ is the distance between the origin $(0, 0)$ and the point (a, b).

2. _____ Theorem states that if $z = r(\cos \theta + i \sin \theta)$ is a complex number and n is a positive integer, then $z^n = r^n(\cos n\theta + i \sin n\theta)$.

3. The complex number $u = a + bi$ is an _____ of the complex number z when $z = u^n = (a + bi)^n$.

4. What is the trigonometric form of the complex number $z = a + bi$?

5. When a complex number is written in trigonometric form, what does r represent?

6. When a complex number is written in trigonometric form, what does θ represent?

Procedures and Problem Solving

Finding the Absolute Value of a Complex Number In Exercises 7–14, plot the complex number and find its absolute value.

7. $4i$

8. $-2i$

9. -5

10. 8

✓ 11. $-4 + 4i$

12. $-5 - 12i$

13. $9 + 7i$

14. $10 - 3i$

Writing a Complex Number in Trigonometric Form In Exercises 15–22, write the complex number in trigonometric form without using a calculator.

15.
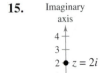

16.

17.

18.
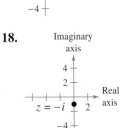

19.
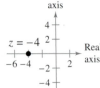

20.

21.
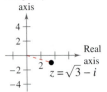

22.
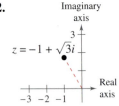

Writing a Complex Number in Trigonometric Form In Exercises 23–46, represent the complex number graphically, and find the trigonometric form of the number.

✓ 23. $-8i$

24. $4i$

25. $-5i$

26. $12i$

27. $5 - 5i$

28. $2 + 2i$

✓ 29. $\sqrt{3} + i$

30. $-1 - \sqrt{3}i$

31. $1 + i$

32. $4 - 4i$

33. $-2(1 + \sqrt{3}i)$

34. $\frac{5}{2}(\sqrt{3} - i)$

35. $-7 + 4i$

36. $5 - i$

37. 3

38. 6

39. $3 + \sqrt{3}i$

40. $2\sqrt{2} - i$

41. $-1 - 2i$

42. $1 + 3i$

43. $5 + 2i$

44. $-3 + i$

45. $3\sqrt{2} - 7i$

46. $-8 - 5\sqrt{3}i$

Writing a Complex Number in Standard Form In Exercises 47–58, represent the complex number graphically, and find the standard form of the number.

✓ **47.** $2(\cos 120° + i \sin 120°)$ **48.** $5(\cos 135° + i \sin 135°)$

49. $\frac{3}{2}(\cos 330° + i \sin 330°)$ **50.** $\frac{3}{4}(\cos 315° + i \sin 315°)$

51. $3.75\left(\cos\dfrac{3\pi}{4} + i \sin\dfrac{3\pi}{4}\right)$

52. $1.5\left(\cos\dfrac{\pi}{2} + i \sin\dfrac{\pi}{2}\right)$

53. $6\left(\cos\dfrac{\pi}{3} + i \sin\dfrac{\pi}{3}\right)$

54. $8\left(\cos\dfrac{5\pi}{6} + i \sin\dfrac{5\pi}{6}\right)$

55. $4\left(\cos\dfrac{3\pi}{2} + i \sin\dfrac{3\pi}{2}\right)$

56. $7(\cos 0 + i \sin 0)$

57. $3[\cos(18° \, 45') + i \sin(18° \, 45')]$

58. $6[\cos(230° \, 30') + i \sin(230° \, 30')]$

Writing a Complex Number in Standard Form In Exercises 59–62, use a graphing utility to represent the complex number in standard form.

59. $5\left(\cos\dfrac{\pi}{9} + i \sin\dfrac{\pi}{9}\right)$

60. $12\left(\cos\dfrac{3\pi}{5} + i \sin\dfrac{3\pi}{5}\right)$

61. $9(\cos 58° + i \sin 58°)$

62. $2(\cos 155° + i \sin 155°)$

Representing a Power In Exercises 63 and 64, represent the powers z, z^2, z^3, and z^4 graphically. Describe the pattern.

63. $z = \dfrac{\sqrt{2}}{2}(1 + i)$

64. $z = \dfrac{1}{2}\left(1 + \sqrt{3}i\right)$

Multiplying or Dividing Complex Numbers In Exercises 65–78, perform the operation and leave the result in trigonometric form.

✓ **65.** $\left[2\left(\cos\dfrac{\pi}{2} + i \sin\dfrac{\pi}{2}\right)\right]\left[5\left(\cos\dfrac{3\pi}{2} + i \sin\dfrac{3\pi}{2}\right)\right]$

66. $\left[3\left(\cos\dfrac{\pi}{3} + i \sin\dfrac{\pi}{3}\right)\right]\left[4\left(\cos\dfrac{\pi}{6} + i \sin\dfrac{\pi}{6}\right)\right]$

✓ **67.** $\left[\dfrac{2}{3}\left(\cos\dfrac{4\pi}{3} + i \sin\dfrac{4\pi}{3}\right)\right]\left[9\left(\cos\dfrac{5\pi}{3} + i \sin\dfrac{5\pi}{3}\right)\right]$

68. $\left[\dfrac{3}{2}\left(\cos\dfrac{\pi}{6} + i \sin\dfrac{\pi}{6}\right)\right]\left[6\left(\cos\dfrac{\pi}{4} + i \sin\dfrac{\pi}{4}\right)\right]$

69. $\left[\frac{5}{3}(\cos 140° + i \sin 140°)\right]\left[\frac{2}{3}(\cos 60° + i \sin 60°)\right]$

70. $\left[\frac{1}{2}(\cos 115° + i \sin 115°)\right]\left[\frac{4}{5}(\cos 300° + i \sin 300°)\right]$

71. $\left[\frac{11}{20}(\cos 290° + i \sin 290°)\right]\left[\frac{2}{5}(\cos 200° + i \sin 200°)\right]$

72. $(\cos 5° + i \sin 5°)(\cos 20° + i \sin 20°)$

✓ **73.** $\dfrac{\cos 50° + i \sin 50°}{\cos 20° + i \sin 20°}$

74. $\dfrac{5(\cos 4.3 + i \sin 4.3)}{4(\cos 2.1 + i \sin 2.1)}$

75. $\dfrac{2(\cos 120° + i \sin 120°)}{4(\cos 40° + i \sin 40°)}$

76. $\dfrac{\cos\left(\dfrac{7\pi}{4}\right) + i \sin\left(\dfrac{7\pi}{4}\right)}{\cos \pi + i \sin \pi}$

77. $\dfrac{18(\cos 54° + i \sin 54°)}{3(\cos 102° + i \sin 102°)}$

78. $\dfrac{9(\cos 20° + i \sin 20°)}{5(\cos 75° + i \sin 75°)}$

Operations with Complex Numbers in Trigonometric Form In Exercises 79–94, (a) write the trigonometric forms of the complex numbers, (b) perform the indicated operation using the trigonometric forms, and (c) perform the indicated operation using the standard forms and check your result with that of part (b).

79. $(2 - 2i)(1 + i)$

80. $(3 - 3i)(1 - i)$

81. $(2 + 2i)(1 - i)$

82. $\left(\sqrt{3} + i\right)(1 + i)$

83. $-2i(1 + i)$

84. $3i\left(1 + \sqrt{2}i\right)$

85. $-2i\left(\sqrt{3} - i\right)$

86. $-i\left(1 + \sqrt{3}i\right)$

87. $2(1 - i)$

88. $-4(1 + i)$

89. $\dfrac{3 + 4i}{1 - \sqrt{3}i}$

90. $\dfrac{2 + 2i}{1 + \sqrt{3}i}$

91. $\dfrac{5}{2 + 2i}$

92. $\dfrac{2}{\sqrt{3} - i}$

93. $\dfrac{4i}{-1 + i}$

94. $\dfrac{2i}{1 - \sqrt{3}i}$

Sketching the Graph of Complex Numbers **In Exercises 95–106, sketch the graph of all complex numbers z satisfying the given condition.**

95. $|z| = 2$

96. $|z| = 5$

97. $|z| = 4$

98. $|z| = 6$

99. $|z| = 7$

100. $|z| = 8$

101. $\theta = \dfrac{\pi}{6}$

102. $\theta = \dfrac{\pi}{4}$

103. $\theta = \dfrac{\pi}{3}$

104. $\theta = \dfrac{5\pi}{6}$

105. $\theta = \dfrac{2\pi}{3}$

106. $\theta = \dfrac{3\pi}{4}$

Finding a Power of a Complex Number **In Exercises 107–126, use DeMoivre's Theorem to find the indicated power of the complex number. Write the result in standard form.**

✓ 107. $(1 + i)^3$

108. $(2 + 2i)^6$

109. $(-1 + i)^6$

110. $(3 - 2i)^8$

111. $2(\sqrt{3} + i)^5$

112. $4(1 - \sqrt{3}i)^3$

113. $[5(\cos 20° + i \sin 20°)]^3$

114. $[3(\cos 150° + i \sin 150°)]^4$

115. $\left(\cos \dfrac{5\pi}{4} + i \sin \dfrac{5\pi}{4}\right)^{10}$

116. $\left[2\left(\cos \dfrac{\pi}{2} + i \sin \dfrac{\pi}{2}\right)\right]^{12}$

117. $[2(\cos 1.25 + i \sin 1.25)]^4$

118. $[4(\cos 2.8 + i \sin 2.8)]^5$

119. $[2(\cos \pi + i \sin \pi)]^8$

120. $(\cos 0 + i \sin 0)^{20}$

121. $(3 - 2i)^5$

122. $(\sqrt{5} - 4i)^4$

123. $[4(\cos 10° + i \sin 10°)]^6$

124. $[3(\cos 15° + i \sin 15°)]^4$

125. $\left[3\left(\cos \dfrac{\pi}{8} + i \sin \dfrac{\pi}{8}\right)\right]^2$

126. $\left[2\left(\cos \dfrac{\pi}{10} + i \sin \dfrac{\pi}{10}\right)\right]^5$

127. Show that $-\frac{1}{2}(1 + \sqrt{3}i)$ is a sixth root of 1.

128. Show that $2^{-1/4}(1 - i)$ is a fourth root of -2.

Finding Square Roots of a Complex Number **In Exercises 129–136, find the square roots of the complex number.**

129. $2i$

130. $5i$

131. $-3i$

132. $-6i$

133. $2 - 2i$

134. $2 + 2i$

135. $1 + \sqrt{3}i$

136. $1 - \sqrt{3}i$

Finding the nth Roots of a Complex Number **In Exercises 137–152, (a) use the theorem on page 590 to find the indicated roots of the complex number, (b) represent each of the roots graphically, and (c) write each of the roots in standard form.**

137. Square roots of $5(\cos 120° + i \sin 120°)$

138. Square roots of $16(\cos 60° + i \sin 60°)$

139. Fourth roots of $8\left(\cos \dfrac{2\pi}{3} + i \sin \dfrac{2\pi}{3}\right)$

140. Fifth roots of $32\left(\cos \dfrac{5\pi}{6} + i \sin \dfrac{5\pi}{6}\right)$

141. Cube roots of $-25i$

142. Fourth roots of $625i$

143. Cube roots of $-\frac{125}{2}(1 + \sqrt{3}i)$

144. Cube roots of $-4\sqrt{2}(1 - i)$

145. Cube roots of $64i$

146. Fourth roots of i

✓ 147. Fifth roots of 1

148. Cube roots of 1000

149. Cube roots of -125

150. Fourth roots of -4

✓ 151. Fifth roots of $128(-1 + i)$

152. Sixth roots of $729i$

Why you should learn it (*p. 583*) **In Exercises 153–160, use the theorem on page 590 to find all the solutions of the equation, and represent the solutions graphically.**

153. $x^4 - i = 0$ 154. $x^3 + 1 = 0$

155. $x^5 + 243 = 0$

156. $x^3 - 27 = 0$

157. $x^4 + 16i = 0$

158. $x^6 - 64i = 0$

159. $x^3 - (1 - i) = 0$

160. $x^4 + (1 + i) = 0$

Electrical Engineering   In Exercises 161–166, use the formula to find the missing quantity for the given conditions. The formula

$$E = I \cdot Z$$

where E represents voltage, I represents current, and Z represents impedance (a measure of opposition to a sinusoidal electric current), is used in electrical engineering. Each variable is a complex number.

161. $I = 10 + 2i$

$Z = 4 + 3i$

162. $I = 12 + 2i$

$Z = 3 + 5i$

163. $I = 2 + 4i$

$E = 5 + 5i$

164. $I = 10 + 2i$

$E = 4 + 5i$

165. $E = 12 + 24i$

$Z = 12 + 20i$

166. $E = 15 + 12i$

$Z = 25 + 24i$

Conclusions

True or False?   In Exercises 167–170, determine whether the statement is true or false. Justify your answer.

167. $\frac{1}{2}(1 - \sqrt{3}i)$ is a ninth root of -1.

168. $\sqrt{3} + i$ is a solution of the equation $x^2 - 8i = 0$.

169. The product of two complex numbers is 0 only when the modulus of one (or both) of the complex numbers is 0.

170. Geometrically, the nth roots of any complex number z are all equally spaced around the unit circle centered at the origin.

171. Given two complex numbers $z_1 = r_1(\cos \theta_1 + i \sin \theta_1)$ and $z_2 = r_2(\cos \theta_2 + i \sin \theta_2)$, $z_2 \neq 0$, show that

$$\frac{z_1}{z_2} = \frac{r_1}{r_2}[\cos(\theta_1 - \theta_2) + i \sin(\theta_1 - \theta_2)].$$

172. Show that $\bar{z} = r[\cos(-\theta) + i \sin(-\theta)]$ is the complex conjugate of $z = r(\cos \theta + i \sin \theta)$.

173. Use trigonometric forms of z and $\bar{z}$ in Exercise 172 to find the following.

(a) $z\bar{z}$

(b) $\dfrac{z}{\bar{z}}$, $\bar{z} \neq 0$

174. Show that the negative of $z = r(\cos \theta + i \sin \theta)$ is

$$-z = r[\cos(\theta + \pi) + i \sin(\theta + \pi)].$$

175. Writing   The famous formula

$$e^{a+bi} = e^a(\cos b + i \sin b)$$

is called Euler's Formula, after the Swiss mathematician Leonhard Euler (1707–1783). This formula gives rise to the equation

$$e^{\pi i} + 1 = 0.$$

This equation relates the five most famous numbers in mathematics—0, 1, π, e, and i—in a single equation. Show how Euler's Formula can be used to derive this equation. Write a short paragraph summarizing your work.

176. CAPSTONE   Use the graph of the roots of a complex number.

(a) Write each of the roots in trigonometric form.

(b) Identify the complex number whose roots are given. Use a graphing utility to verify your results.

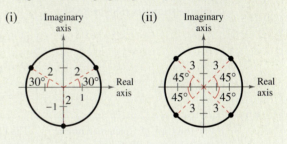

Cumulative Mixed Review

Harmonic Motion   In Exercises 177–180, for the simple harmonic motion described by the trigonometric function, find the maximum displacement from equilibrium and the lowest possible positive value of t for which $d = 0$.

177. $d = 16 \cos \dfrac{\pi}{4} t$

178. $d = \dfrac{1}{16} \sin \dfrac{5\pi}{4} t$

179. $d = \dfrac{1}{8} \cos 12\pi t$

180. $d = \dfrac{1}{12} \sin 60\pi t$

7 Chapter Summary

	What did you learn?	Explanation and Examples	Review Exercises
7.1	Use the Law of Sines to solve oblique triangles (AAS or ASA) (*p. 544*).	**Law of Sines** If ABC is a triangle with sides a, b, and c, then $$\frac{a}{\sin A} = \frac{b}{\sin B} = \frac{c}{\sin C}.$$ A is acute. A is obtuse.	1–6
	Use the Law of Sines to solve oblique triangles (SSA) (*p. 546*).	If two sides and one opposite angle are given, then three possible situations can occur: (1) no such triangle exists (see Example 4), (2) one such triangle exists (see Example 3), or (3) two distinct triangles satisfy the conditions (see Example 5).	7–10
	Find areas of oblique triangles (*p. 548*), and use the Law of Sines to model and solve real-life problems (*p. 549*).	The area of any triangle is one-half the product of the lengths of two sides times the sine of their included angle. That is, Area $= \frac{1}{2}bc \sin A = \frac{1}{2}ab \sin C = \frac{1}{2}ac \sin B$. The Law of Sines can be used to approximate the total distance of a boat race course. (See Example 7.)	11–16
7.2	Use the Law of Cosines to solve oblique triangles (SSS or SAS) (*p. 553*).	**Law of Cosines** *Standard Form* *Alternative Form* $a^2 = b^2 + c^2 - 2bc \cos A$ $\cos A = \dfrac{b^2 + c^2 - a^2}{2bc}$ $b^2 = a^2 + c^2 - 2ac \cos B$ $\cos B = \dfrac{a^2 + c^2 - b^2}{2ac}$ $c^2 = a^2 + b^2 - 2ab \cos C$ $\cos C = \dfrac{a^2 + b^2 - c^2}{2ab}$	17–24
	Use the Law of Cosines to model and solve real-life problems (*p. 555*).	The Law of Cosines can be used to find the distance between the pitcher's mound and first base on a women's softball field. (See Example 3.)	25, 26
	Use Heron's Area Formula to find areas of triangles (*p. 556*).	**Heron's Area Formula:** Given any triangle with sides of lengths a, b, and c, the area of the triangle is Area $= \sqrt{s(s - a)(s - b)(s - c)}$, where $s = (a + b + c)/2$.	27–30
7.3	Represent vectors as directed line segments (*p. 560*).		31, 32
	Write the component forms of vectors (*p. 561*).	The component form of the vector with initial point $P(p_1, p_2)$ and terminal point $Q(q_1, q_2)$ is given by $\overrightarrow{PQ} = \langle q_1 - p_1, q_2 - p_2 \rangle = \langle v_1, v_2 \rangle = \mathbf{v}$.	31–34
	Perform basic vector operations and represent vectors graphically (*p. 562*).	Let $\mathbf{u} = \langle u_1, u_2 \rangle$ and $\mathbf{v} = \langle v_1, v_2 \rangle$ be vectors and let k be a scalar (a real number). $\mathbf{u} + \mathbf{v} = \langle u_1 + v_1, u_2 + v_2 \rangle$ $k\mathbf{u} = \langle ku_1, ku_2 \rangle$ $-\mathbf{v} = \langle -v_1, -v_2 \rangle$ $\mathbf{u} - \mathbf{v} = \langle u_1 - v_1, u_2 - v_2 \rangle$	35–44

What did you learn?	Explanation and Examples	Review Exercises
7.3 Write vectors as linear combinations of unit vectors (*p. 564*).	$\mathbf{v} = \langle v_1, v_2 \rangle = v_1\langle 1, 0 \rangle + v_2\langle 0, 1 \rangle = v_1\mathbf{i} + v_2\mathbf{j}$ The scalars v_1 and v_2 are the horizontal and vertical components of $\mathbf{v}$, respectively. The vector sum $v_1\mathbf{i} + v_2\mathbf{j}$ is the linear combination of the vectors $\mathbf{i}$ and $\mathbf{j}$.	45–50
Find the direction angles of vectors (*p. 566*).	If $\mathbf{u} = 2\mathbf{i} + 2\mathbf{j}$, then the direction angle is $\tan \theta = 2/2 = 1.$ So, $\theta = 45°$.	51–56
Use vectors to model and solve real-life problems (*p. 567*).	Vectors can be used to find the resultant speed and direction of an airplane. (See Example 10.)	57–60
Find the dot product of two vectors and use the properties of the dot product (*p. 574*).	The dot product of $\mathbf{u} = \langle u_1, u_2 \rangle$ and $\mathbf{v} = \langle v_1, v_2 \rangle$ is $\mathbf{u} \cdot \mathbf{v} = u_1v_1 + u_2v_2$.	61–68
7.4 Find the angle between two vectors and determine whether two vectors are orthogonal (*p. 575*).	If θ is the angle between two nonzero vectors $\mathbf{u}$ and $\mathbf{v}$, then $\cos \theta = \dfrac{\mathbf{u} \cdot \mathbf{v}}{\|\mathbf{u}\| \|\mathbf{v}\|}.$ Vectors $\mathbf{u}$ and $\mathbf{v}$ are orthogonal when $\mathbf{u} \cdot \mathbf{v} = 0$.	69–84
Write vectors as the sums of two vector components (*p. 577*).	Many applications in physics and engineering require the decomposition of a given vector into the sum of two vector components. (See Example 6.)	85–88
Use vectors to find the work done by a force (*p. 579*).	The work W done by a constant force $\mathbf{F}$ as its point of application moves along the vector $\overrightarrow{PQ}$ is given by either of the following. **1.** $W = \|\text{proj}_{\overrightarrow{PQ}}\mathbf{F}\| \|\overrightarrow{PQ}\|$ Projection form **2.** $W = \mathbf{F} \cdot \overrightarrow{PQ}$ Dot product form	89, 90
Plot complex numbers in the complex plane and find absolute values of complex numbers (*p. 583*).	A complex number $z = a + bi$ can be represented as the point (a, b) in the complex plane. The horizontal axis is the real axis and the vertical axis is the imaginary axis. The absolute value of $z = a + bi$ is $|a + bi| = \sqrt{a^2 + b^2}.$	91–94
Write trigonometric forms of complex numbers (*p. 584*).	The trigonometric form of the complex number $z = a + bi$ is $z = r(\cos \theta + i \sin \theta)$ where $a = r\cos \theta$, $b = r\sin \theta$, $r = \sqrt{a^2 + b^2}$, and $\tan \theta = b/a$.	95–98
7.5 Multiply and divide complex numbers written in trigonometric form (*p. 586*).	Let $z_1 = r_1(\cos \theta_1 + i \sin \theta_1)$ and $z_2 = r_2(\cos \theta_2 + i \sin \theta_2)$. $z_1z_2 = r_1r_2[\cos(\theta_1 + \theta_2) + i \sin(\theta_1 + \theta_2)]$ $\dfrac{z_1}{z_2} = \dfrac{r_1}{r_2}[\cos(\theta_1 - \theta_2) + i \sin(\theta_1 - \theta_2)], \quad z_2 \neq 0$	99–106
Use DeMoivre's Theorem to find powers of complex numbers (*p. 588*).	**DeMoivre's Theorem:** If $z = r(\cos \theta + i \sin \theta)$ is a complex number and n is a positive integer, then $z^n = [r(\cos \theta + i \sin \theta)]^n = r^n(\cos n\theta + i \sin n\theta)$.	107–110
Find nth roots of complex numbers (*p. 589*).	The complex number $u = a + bi$ is an nth root of the complex number z when $z = u^n = (a + bi)^n$.	111–122

7   Review Exercises

See www.CalcChat.com for worked-out solutions to odd-numbered exercises.
For instructions on how to use a graphing utility, see Appendix A.

7.1

Using the Law of Sines   In Exercises 1–10, use the Law of Sines to solve the triangle. If two solutions exist, find both.

1. $A = 32°$, $B = 50°$, $a = 16$
2. $A = 38°$, $B = 58°$, $a = 12$
3. $B = 25°$, $C = 105°$, $c = 25$
4. $B = 20°$, $C = 115°$, $c = 30$
5. $A = 60° \ 15'$, $B = 45° \ 30'$, $b = 4.8$
6. $A = 82° \ 45'$, $B = 28° \ 45'$, $b = 40.2$
7. $A = 75°$, $a = 2.5$, $b = 16.5$
8. $A = 15°$, $a = 5$, $b = 10$
9. $B = 115°$, $a = 9$, $b = 14.5$
10. $B = 150°$, $a = 10$, $b = 3$

Finding the Area of a Triangle   In Exercises 11–14, find the area of the triangle having the indicated angle and sides.

11. $A = 33°$, $b = 7$, $c = 10$
12. $B = 80°$, $a = 4$, $c = 8$
13. $C = 122°$, $b = 18$, $a = 29$
14. $C = 100°$, $a = 120$, $b = 74$

15. **Landscaping**   A tree stands on a hillside of slope 28° from the horizontal. From a point 75 feet down the hill, the angle of elevation to the top of the tree is 45° (see figure). Find the height of the tree.

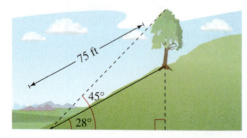

16. **Surveying**   A surveyor finds that a tree on the opposite bank of a river has a bearing of N 22° 30′ E from a certain point and a bearing of N 15° W from a point 400 feet downstream. Find the width of the river.

7.2

Using the Law of Cosines   In Exercises 17–24, use the Law of Cosines to solve the triangle.

17.

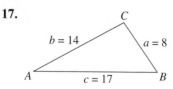

18.

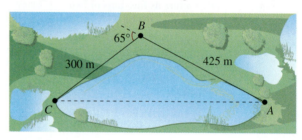

19. $a = 9$, $b = 12$, $c = 20$
20. $a = 7$, $b = 15$, $c = 19$
21. $a = 6.5$, $b = 10.2$, $c = 16$
22. $a = 6.2$, $b = 6.4$, $c = 2.1$
23. $C = 65°$, $a = 25$, $b = 12$
24. $B = 48°$, $a = 18$, $c = 12$

25. **Geometry**   The lengths of the diagonals of a parallelogram are 10 feet and 16 feet. Find the lengths of the sides of the parallelogram if the diagonals intersect at an angle of 28°.

26. **Surveying**   To approximate the length of a marsh, a surveyor walks 425 meters from point A to point B. The surveyor then turns 65° and walks 300 meters to point C (see figure). Approximate the length AC of the marsh.

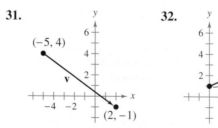

Using Heron's Area Theorem   In Exercises 27–30, use Heron's Area Formula to find the area of the triangle.

27. $a = 3$, $b = 6$, $c = 8$
28. $a = 15$, $b = 8$, $c = 10$
29. $a = 64.8$, $b = 49.2$, $c = 24.1$
30. $a = 8.55$, $b = 5.14$, $c = 12.73$

7.3

Finding the Component Form of a Vector   In Exercises 31–34, find the component form and the magnitude of the vector v.

31.

32.

33. Initial point: $(0, 10)$; terminal point: $(7, 3)$
34. Initial point: $(1, 5)$; terminal point: $(15, 9)$

Vector Operations In Exercises 35–40, find (a) **u** + **v**, (b) **u** − **v**, (c) 3**u**, and (d) 2**v** + 5**u**. Then sketch each resultant vector.

35. $\mathbf{u} = \langle -1, -3 \rangle, \mathbf{v} = \langle -3, 6 \rangle$
36. $\mathbf{u} = \langle 4, 5 \rangle, \mathbf{v} = \langle 0, -1 \rangle$
37. $\mathbf{u} = \langle -5, 2 \rangle, \mathbf{v} = \langle 4, 4 \rangle$
38. $\mathbf{u} = \langle 1, -8 \rangle, \mathbf{v} = \langle 3, -2 \rangle$
39. $\mathbf{u} = 2\mathbf{i} - \mathbf{j}, \mathbf{v} = 5\mathbf{i} + 3\mathbf{j}$
40. $\mathbf{u} = -6\mathbf{j}, \mathbf{v} = \mathbf{i} + \mathbf{j}$

Vector Operations In Exercises 41–44, find the component form of **w** and sketch the specified vector operations geometrically, where **u** = 6**i** − 5**j** and **v** = 10**i** + 3**j**.

41. $\mathbf{w} = 3\mathbf{v}$
42. $\mathbf{w} = \frac{1}{2}\mathbf{v}$
43. $\mathbf{w} = 4\mathbf{u} + 5\mathbf{v}$
44. $\mathbf{w} = 3\mathbf{v} - 2\mathbf{u}$

Finding a Unit Vector In Exercises 45–48, find a unit vector in the direction of the given vector. Verify that the result has a magnitude of 1.

45. $\mathbf{u} = \langle 0, -6 \rangle$
46. $\mathbf{v} = \langle -12, -5 \rangle$
47. $\mathbf{v} = 5\mathbf{i} - 2\mathbf{j}$
48. $\mathbf{w} = -7\mathbf{i}$

Writing a Linear Combination In Exercises 49 and 50, the initial and terminal points of a vector are given. Write the vector as a linear combination of the standard unit vectors **i** and **j**.

49. Initial point: $(-8, 3)$
    Terminal point: $(1, -5)$
50. Initial point: $(2, -3.2)$
    Terminal point: $(-6.4, 10.8)$

Finding the Magnitude and Direction Angle of a Vector In Exercises 51–56, find the magnitude and the direction angle of the vector **v**.

51. $\mathbf{v} = 7(\cos 60°\mathbf{i} + \sin 60°\mathbf{j})$
52. $\mathbf{v} = 3(\cos 150°\mathbf{i} + \sin 150°\mathbf{j})$
53. $\mathbf{v} = 5\mathbf{i} + 4\mathbf{j}$
54. $\mathbf{v} = -4\mathbf{i} + 7\mathbf{j}$
55. $\mathbf{v} = -3\mathbf{i} - 3\mathbf{j}$
56. $\mathbf{v} = 8\mathbf{i} - \mathbf{j}$

57. **Resultant Force** Forces with magnitudes of 85 pounds and 50 pounds act on a single point. The angle between the forces is 15°. Describe the resultant force.

58. **Physics** A 180-pound weight is supported by two ropes, as shown in the figure. Find the tension in each rope.

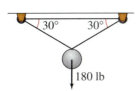

59. **Physics** In a manufacturing process, an electric hoist lifts 200-pound ingots. Find the tension in the supporting cables (see figure).

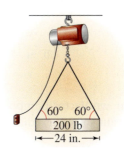

60. **Aviation** An airplane has an airspeed of 430 miles per hour at a bearing of 135°. The wind velocity is 35 miles per hour in the direction N 30° E. Find the resultant speed and direction of the plane.

7.4

Finding a Dot Product In Exercises 61–64, find the dot product of **u** and **v**.

61. $\mathbf{u} = \langle 0, -2 \rangle$
    $\mathbf{v} = \langle 1, 10 \rangle$
62. $\mathbf{u} = \langle -7, 12 \rangle$
    $\mathbf{v} = \langle -4, -14 \rangle$
63. $\mathbf{u} = 6\mathbf{i} - \mathbf{j}$
    $\mathbf{v} = 2\mathbf{i} + 5\mathbf{j}$
64. $\mathbf{u} = 8\mathbf{i} - 7\mathbf{j}$
    $\mathbf{v} = 3\mathbf{i} - 4\mathbf{j}$

Using Properties of Dot Products In Exercises 65–68, use the vectors $\mathbf{u} = \langle -3, -4 \rangle$ and $\mathbf{v} = \langle 2, 1 \rangle$ to find the indicated quantity.

65. $\mathbf{u} \cdot \mathbf{u}$
66. $\|\mathbf{v}\| - 3$
67. $4\mathbf{u} \cdot \mathbf{v}$
68. $(\mathbf{u} \cdot \mathbf{v})\mathbf{u}$

Finding the Angle Between Two Vectors In Exercises 69–72, find the angle θ between the vectors.

69. $\mathbf{u} = \langle 2\sqrt{2}, -4 \rangle, \quad \mathbf{v} = \langle -\sqrt{2}, 1 \rangle$
70. $\mathbf{u} = \langle 3, 1 \rangle, \quad \mathbf{v} = \langle 4, 5 \rangle$
71. $\mathbf{u} = \cos \frac{7\pi}{4}\mathbf{i} + \sin \frac{7\pi}{4}\mathbf{j}, \mathbf{v} = \cos \frac{5\pi}{6}\mathbf{i} + \sin \frac{5\pi}{6}\mathbf{j}$
72. $\mathbf{u} = \cos 45°\mathbf{i} + \sin 45°\mathbf{j}$
    $\mathbf{v} = \cos 300°\mathbf{i} + \sin 300°\mathbf{j}$

Finding the Angle Between Two Vectors In Exercises 73–76, graph the vectors and find the degree measure of the angle between the vectors.

73. $\mathbf{u} = 4\mathbf{i} + \mathbf{j}$
    $\mathbf{v} = \mathbf{i} - 4\mathbf{j}$
74. $\mathbf{u} = 6\mathbf{i} + 2\mathbf{j}$
    $\mathbf{v} = -3\mathbf{i} - \mathbf{j}$
75. $\mathbf{u} = 7\mathbf{i} - 5\mathbf{j}$
    $\mathbf{v} = 10\mathbf{i} + 3\mathbf{j}$
76. $\mathbf{u} = -5.3\mathbf{i} + 2.8\mathbf{j}$
    $\mathbf{v} = -8.1\mathbf{i} - 4\mathbf{j}$

A Relationship of Two Vectors In Exercises 77–80, determine whether u and v are orthogonal, parallel, or neither.

77. $u = \langle 39, -12 \rangle$
    $v = \langle -26, 8 \rangle$

78. $u = \langle 8, -4 \rangle$
    $v = \langle 5, 10 \rangle$

79. $u = \langle 8, 5 \rangle$
    $v = \langle -2, 4 \rangle$

80. $u = \langle -15, 51 \rangle$
    $v = \langle 20, -68 \rangle$

Finding an Unknown Vector Component In Exercises 81–84, find the value of k such that the vectors u and v are orthogonal.

81. $u = i - kj$
    $v = i + 2j$

82. $u = 2i + j$
    $v = -i - kj$

83. $u = ki - j$
    $v = 2i - 2j$

84. $u = ki - 2j$
    $v = i + 4j$

Decomposing a Vector into Components In Exercises 85–88, find the projection of u onto v. Then write u as the sum of two orthogonal vectors, one of which is $\text{proj}_v \, u$.

85. $u = \langle -4, 3 \rangle$, $v = \langle -8, -2 \rangle$
86. $u = \langle 5, 6 \rangle$, $v = \langle 10, 0 \rangle$
87. $u = \langle 2, 7 \rangle$, $v = \langle 1, -1 \rangle$
88. $u = \langle -3, 5 \rangle$, $v = \langle -5, 2 \rangle$

89. **Finding Work** Determine the work done by a crane lifting an 18,000-pound truck 48 inches.

90. **Physics** A 500-pound motorcycle is stopped on a hill inclined at $12°$. What force is required to keep the motorcycle from rolling back down the hill?

7.5

Finding the Absolute Value of a Complex Number In Exercises 91–94, plot the complex number and find its absolute value.

91. $7i$
92. $-6i$
93. $5 + 3i$
94. $-10 - 4i$

Writing a Complex Number in Trigonometric Form In Exercises 95–98, write the complex number in trigonometric form without using a calculator.

95. $2 - 2i$
96. $-2 + 2i$
97. $-\sqrt{3} - i$
98. $-\sqrt{3} + i$

Multiplying or Dividing Complex Numbers In Exercises 99–102, perform the operation and leave the result in trigonometric form.

99. $\left[\dfrac{5}{2} \left(\cos \dfrac{\pi}{2} + i \sin \dfrac{\pi}{2} \right) \right] \left[4 \left(\cos \dfrac{\pi}{4} + i \sin \dfrac{\pi}{4} \right) \right]$

100. $\left[2 \left(\cos \dfrac{2\pi}{3} + i \sin \dfrac{2\pi}{3} \right) \right] \left[3 \left(\cos \dfrac{\pi}{6} + i \sin \dfrac{\pi}{6} \right) \right]$

101. $\dfrac{20(\cos 320° + i \sin 320°)}{5(\cos 80° + i \sin 80°)}$

102. $\dfrac{3(\cos 230° + i \sin 230°)}{9(\cos 95° + i \sin 95°)}$

Operations with Complex Numbers in Trigonometric Form In Exercises 103–106, (a) write the trigonometric forms of the complex numbers, (b) perform the indicated operation using the trigonometric forms, and (c) perform the indicated operation using the standard forms and check your result with that of part (b).

103. $(2 - 2i)(3 + 3i)$
104. $(4 + 4i)(-1 - i)$
105. $\dfrac{3 - 3i}{2 + 2i}$
106. $\dfrac{-1 - i}{-2 - 2i}$

Finding a Power In Exercises 107–110, use DeMoivre's Theorem to find the indicated power of the complex number. Write the result in standard form.

107. $\left[5 \left(\cos \dfrac{\pi}{12} + i \sin \dfrac{\pi}{12} \right) \right]^4$

108. $\left[2 \left(\cos \dfrac{4\pi}{15} + i \sin \dfrac{4\pi}{15} \right) \right]^5$

109. $(2 + 3i)^6$
110. $(1 - i)^8$

Finding Square Roots In Exercises 111–114, find the square roots of the complex number.

111. $-\sqrt{3} + i$
112. $\sqrt{3} - i$
113. $-2i$
114. $-5i$

Finding Roots In Exercises 115–118, (a) use the theorem on page 590 to find the indicated roots of the complex number, (b) represent each of the roots graphically, and (c) write each of the roots in standard form.

115. Sixth roots of $-729i$
116. Fourth roots of $256i$
117. Cube roots of 8
118. Fifth roots of -1024

Solving an Equation In Exercises 119–122, use the theorem on page 590 to find all solutions of the equation, and represent the solutions graphically.

119. $x^4 + 256 = 0$
120. $x^5 - 32i = 0$
121. $x^3 + 8i = 0$
122. $x^4 + 81 = 0$

Conclusions

True or False? In Exercises 123 and 124, determine whether the statement is true or false. Justify your answer.

123. The Law of Sines is true if one of the angles in the triangle is a right angle.

124. When the Law of Sines is used, the solution is always unique.

7 Chapter Test

See www.CalcChat.com for worked-out solutions to odd-numbered exercises.
For instructions on how to use a graphing utility, see Appendix A.

Take this test as you would take a test in class. After you are finished, check your work against the answers in the back of the book.

In Exercises 1–6, use the given information to solve the triangle. If two solutions exist, find both solutions.

1. $A = 36°$, $B = 98°$, $c = 16$

2. $a = 2$, $b = 4$, $c = 5$

3. $A = 35°$, $b = 8$, $c = 12$

4. $A = 25°$, $b = 28$, $a = 18$

5. $B = 130°$, $c = 10.1$, $b = 5.2$

6. $A = 150°$, $b = 4.8$, $a = 9.4$

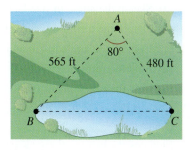

7. Find the length of the pond shown at the right.

Figure for 7

8. A triangular parcel of land has borders of lengths 55 meters, 85 meters, and 100 meters. Find the area of the parcel of land.

9. Find the component form and magnitude of the vector **w** that has initial point $(-8, -12)$ and terminal point $(4, 1)$.

In Exercises 10–13, find (a) $2\mathbf{v} + \mathbf{u}$, (b) $\mathbf{u} - 3\mathbf{v}$, and (c) $5\mathbf{u} - \mathbf{v}$.

10. $\mathbf{u} = \langle 0, -4 \rangle$, $\mathbf{v} = \langle 4, 6 \rangle$

11. $\mathbf{u} = \langle -5, 2 \rangle$, $\mathbf{v} = \langle -1, -10 \rangle$

12. $\mathbf{u} = \mathbf{i} - \mathbf{j}$, $\mathbf{v} = 6\mathbf{i} + 9\mathbf{j}$

13. $\mathbf{u} = 2\mathbf{i} + 3\mathbf{j}$, $\mathbf{v} = -\mathbf{i} - 2\mathbf{j}$

14. Find a unit vector in the direction of $\mathbf{v} = 6\mathbf{i} - 4\mathbf{j}$.

15. Find the component form of the vector **v** with $\|\mathbf{v}\| = 12$, in the same direction as $\mathbf{u} = \langle 3, -5 \rangle$.

16. Forces with magnitudes of 250 pounds and 130 pounds act on an object at angles of $45°$ and $-60°$, respectively, with the positive x-axis. Find the direction and magnitude of the resultant of these forces.

17. Find the dot product of $\mathbf{u} = \langle -9, 4 \rangle$ and $\mathbf{v} = \langle 1, 2 \rangle$.

18. Find the angle between the vectors $\mathbf{u} = 7\mathbf{i} + 2\mathbf{j}$ and $\mathbf{v} = -4\mathbf{j}$.

19. Are the vectors $\mathbf{u} = \langle 9, -12 \rangle$ and $\mathbf{v} = \langle -4, -3 \rangle$ orthogonal? Explain.

20. Find the projection of $\mathbf{u} = \langle 6, 7 \rangle$ onto $\mathbf{v} = \langle -5, -1 \rangle$. Then write **u** as the sum of two orthogonal vectors, one of which is $\text{proj}_{\mathbf{v}}\, \mathbf{u}$.

21. Write the complex number $z = -6 + 6i$ in trigonometric form.

22. Write the complex number $100(\cos 240° + i \sin 240°)$ in standard form.

In Exercises 23 and 24, use DeMoivre's Theorem to find the indicated power of the complex number. Write the result in standard form.

23. $\left[3\left(\cos \dfrac{5\pi}{6} + i \sin \dfrac{5\pi}{6} \right) \right]^8$

24. $(3 - 3i)^6$

25. Find the fourth roots of $128\left(1 + \sqrt{3}\,i\right)$.

26. Find all solutions of the equation $x^4 - 625i = 0$ and represent the solutions graphically.

5-7   Cumulative Test

See www.CalcChat.com for worked-out solutions to odd-numbered exercises.
For instructions on how to use a graphing utility, see Appendix A.

Take this test to review the material in Chapters 5–7. After you are finished, check your work against the answers in the back of the book.

1. Consider the angle $\theta = -150°$.

   (a) Sketch the angle in standard position.

   (b) Determine a coterminal angle in the interval $[0°, 360°)$.

   (c) Convert the angle to radian measure.

   (d) Find the reference angle θ'.

   (e) Find the exact values of the six trigonometric functions of θ.

2. Convert the angle $\theta = 2.55$ radians to degrees. Round your answer to one decimal place.

3. Find $\cos \theta$ when $\tan \theta = -\frac{12}{5}$ and $\sin \theta > 0$.

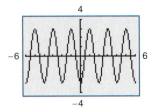

In Exercises 4–6, sketch the graph of the function by hand. (Include two full periods.) Use a graphing utility to verify your graph.

4. $f(x) = 3 - 2 \sin \pi x$ 5. $f(x) = \tan 3x$ 6. $f(x) = \frac{1}{2} \sec(x + \pi)$

7. Find positive values of a, b, and c such that the graph of the function $h(x) = a \cos(bx + c)$ matches the graph in the figure at the right.

Figure for 7

In Exercises 8 and 9, find the exact value of the expression without using a calculator.

8. $\sin\left(\arctan \frac{3}{4}\right)$

9. $\tan\left[\arcsin\left(-\frac{1}{2}\right)\right]$

10. Write an algebraic expression equivalent to $\sin(\arctan 2x)$.

11. Subtract and simplify: $\dfrac{\sin \theta - 1}{\cos \theta} - \dfrac{\cos \theta}{\sin \theta - 1}$.

In Exercises 12–14, verify the identity.

12. $\cot^2 \alpha(\sec^2 \alpha - 1) = 1$

13. $\sin(x + y) \sin(x - y) = \sin^2 x - \sin^2 y$

14. $\sin^2 x \cos^2 x = \frac{1}{8}(1 - \cos 4x)$

In Exercises 15 and 16, solve the equation.

15. $\sin^2 x + 2 \sin x + 1 = 0$

16. $3 \tan \theta - \cot \theta = 0$

17. Approximate the solutions to the equation $\cos^2 x - 5 \cos x - 1 = 0$ in the interval $[0, 2\pi)$.

In Exercises 18 and 19, use a graphing utility to graph the function and approximate its zeros in the interval $[0, 2\pi)$. If possible, find the exact values of the zeros algebraically.

18. $y = \dfrac{1 + \sin x}{\cos x} + \dfrac{\cos x}{1 + \sin x} - 4$

19. $y = \tan^3 x - \tan^2 x + 3 \tan x - 3$

20. Given that $\sin u = \frac{12}{13}$, $\cos v = \frac{3}{5}$, and angles u and v are both in Quadrant I, find $\tan(u - v)$.

21. If $\tan \theta = \frac{1}{2}$, find the exact value of $\tan 2\theta$, $0 < \theta < \frac{\pi}{2}$.

22. If $\tan \theta = \frac{4}{3}$, find the exact value of $\sin \frac{\theta}{2}$, $\pi < \theta < \frac{3\pi}{2}$.

23. Write $\cos 8x + \cos 4x$ as a product.

In Exercises 24–27, verify the identity.

24. $\tan x(1 - \sin^2 x) = \frac{1}{2} \sin 2x$

25. $\sin 3\theta \sin \theta = \frac{1}{2}(\cos 2\theta - \cos 4\theta)$

26. $\sin 3x \cos 2x = \frac{1}{2}(\sin 5x + \sin x)$

27. $\dfrac{2 \cos 3x}{\sin 4x - \sin 2x} = \csc x$

In Exercises 28–31, use the information to solve the triangle shown at the right.

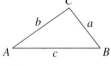

28. $A = 46°$, $a = 14$, $b = 5$ **29.** $A = 32°$, $b = 8$, $c = 10$

30. $A = 24°$, $C = 101°$, $a = 10$ **31.** $a = 24$, $b = 30$, $c = 47$

Figure for 28–31

32. Two sides of a triangle have lengths 14 inches and 19 inches. Their included angle measures 82°. Find the area of the triangle.

33. Find the area of a triangle with sides of lengths 12 inches, 16 inches, and 18 inches.

34. Write the vector $\mathbf{u} = \langle 3, 5 \rangle$ as a linear combination of the standard unit vectors $\mathbf{i}$ and $\mathbf{j}$.

35. Find a unit vector in the direction of $\mathbf{v} = \mathbf{i} - 2\mathbf{j}$.

36. Find $\mathbf{u} \cdot \mathbf{v}$ for $\mathbf{u} = 3\mathbf{i} + 4\mathbf{j}$ and $\mathbf{v} = \mathbf{i} - 2\mathbf{j}$.

37. Find k such that $\mathbf{u} = \mathbf{i} + 2k\mathbf{j}$ and $\mathbf{v} = 2\mathbf{i} - \mathbf{j}$ are orthogonal.

38. Find the projection of $\mathbf{u} = \langle 8, -2 \rangle$ onto $\mathbf{v} = \langle 1, 5 \rangle$. Then write $\mathbf{u}$ as the sum of two orthogonal vectors, one of which is $\text{proj}_{\mathbf{v}} \mathbf{u}$.

39. Find the trigonometric form of the complex number plotted at the right.

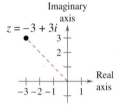

Figure for 39

40. Write the complex number $6\sqrt{3}\left(\cos \dfrac{5\pi}{6} + i \sin \dfrac{5\pi}{6}\right)$ in standard form.

41. Find the product $[4(\cos 30° + i \sin 30°)][6(\cos 120° + i \sin 120°)]$. Write the answer in standard form.

42. Find the square roots of $2 + i$.

43. Find the three cube roots of 1.

44. Write all the solutions of the equation $x^4 + 625 = 0$.

45. From a point 200 feet from a flagpole, the angles of elevation to the bottom and top of the flag are 16° 45′ and 18°, respectively. Approximate the height of the flag to the nearest foot.

46. Write a model for a particle in simple harmonic motion with a maximum displacement of 7 inches and a period of 8 seconds.

47. An airplane's velocity with respect to the air is 500 kilometers per hour, with a bearing of 30°. The airplane is in a steady wind blowing from the northwest with a velocity of 50 kilometers per hour. What is the true direction of the airplane? What is its speed relative to the ground?

48. Forces of 60 pounds and 100 pounds have a resultant force of 125 pounds. Find the angle between the two forces.

Proofs in Mathematics

Law of Sines   (p. 544)

If ABC is a triangle with sides a, b, and c, then

$$\frac{a}{\sin A} = \frac{b}{\sin B} = \frac{c}{\sin C}.$$

Oblique Triangles

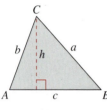

A is acute.   A is obtuse.

Proof

Let h be the altitude of either triangle shown in the figure above. Then you have

$$\sin A = \frac{h}{b} \qquad \text{or} \qquad h = b \sin A$$

$$\sin B = \frac{h}{a} \qquad \text{or} \qquad h = a \sin B.$$

Equating these two values of h, you have

$$a \sin B = b \sin A \qquad \text{or} \qquad \frac{a}{\sin A} = \frac{b}{\sin B}.$$

Note that $\sin A \neq 0$ and $\sin B \neq 0$ because no angle of a triangle can have a measure of $0°$ or $180°$. In a similar manner, construct an altitude from vertex B to side AC (extended in the obtuse triangle), as shown at the right. Then you have

$$\sin A = \frac{h}{c} \qquad \text{or} \qquad h = c \sin A$$

$$\sin C = \frac{h}{a} \qquad \text{or} \qquad h = a \sin C.$$

Equating these two values of h, you have

$$a \sin C = c \sin A \qquad \text{or} \qquad \frac{a}{\sin A} = \frac{c}{\sin C}.$$

By the Transitive Property of Equality, you know that

$$\frac{a}{\sin A} = \frac{b}{\sin B} = \frac{c}{\sin C}.$$

So, the Law of Sines is established.

Law of Tangents

Besides the Law of Sines and the Law of Cosines, there is also a Law of Tangents, which was developed by Francois Viète (1540–1603). The Law of Tangents follows from the Law of Sines and the sum-to-product formulas for sine and is defined as follows.

$$\frac{a + b}{a - b} = \frac{\tan[(A + B)/2]}{\tan[(A - B)/2]}$$

The Law of Tangents can be used to solve a triangle when two sides and the included angle are given (SAS). Before calculators were invented, the Law of Tangents was used to solve the SAS case instead of the Law of Cosines, because computation with a table of tangent values was easier.

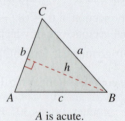

A is acute.

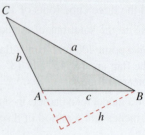

A is obtuse.

Law of Cosines (p. 553)

Standard Form

$$a^2 = b^2 + c^2 - 2bc \cos A$$

$$b^2 = a^2 + c^2 - 2ac \cos B$$

$$c^2 = a^2 + b^2 - 2ab \cos C$$

Alternative Form

$$\cos A = \frac{b^2 + c^2 - a^2}{2bc}$$

$$\cos B = \frac{a^2 + c^2 - b^2}{2ac}$$

$$\cos C = \frac{a^2 + b^2 - c^2}{2ab}$$

Proof

To prove the first formula, consider the top triangle at the right, which has three acute angles. Note that vertex B has coordinates $(c, 0)$. Furthermore, C has coordinates (x, y), where

$$x = b \cos A \quad \text{and} \quad y = b \sin A.$$

Because a is the distance from vertex C to vertex B, it follows that

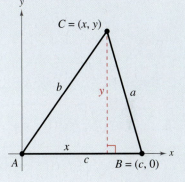

$a = \sqrt{(x - c)^2 + (y - 0)^2}$	Distance Formula
$a^2 = (x - c)^2 + (y - 0)^2$	Square each side.
$a^2 = (b \cos A - c)^2 + (b \sin A)^2$	Substitute for x and y.
$a^2 = b^2 \cos^2 A - 2bc \cos A + c^2 + b^2 \sin^2 A$	Expand.
$a^2 = b^2(\sin^2 A + \cos^2 A) + c^2 - 2bc \cos A$	Factor out b^2.
$a^2 = b^2 + c^2 - 2bc \cos A.$	$\sin^2 A + \cos^2 A = 1$

To prove the second formula, consider the bottom triangle at the right, which also has three acute angles. Note that vertex A has coordinates $(c, 0)$. Furthermore, C has coordinates (x, y), where

$$x = a \cos B \quad \text{and} \quad y = a \sin B.$$

Because b is the distance from vertex C to vertex A, it follows that

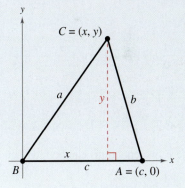

$b = \sqrt{(x - c)^2 + (y - 0)^2}$	Distance Formula
$b^2 = (x - c)^2 + (y - 0)^2$	Square each side.
$b^2 = (a \cos B - c)^2 + (a \sin B)^2$	Substitute for x and y.
$b^2 = a^2 \cos^2 B - 2ac \cos B + c^2 + a^2 \sin^2 B$	Expand.
$b^2 = a^2(\sin^2 B + \cos^2 B) + c^2 - 2ac \cos B$	Factor out a^2.
$b^2 = a^2 + c^2 - 2ac \cos B.$	$\sin^2 B + \cos^2 B = 1$

A similar argument is used to establish the third formula.

Heron's Area Formula (p. 556)

Given any triangle with sides of lengths a, b, and c, the area of the triangle is given by

$$\text{Area} = \sqrt{s(s-a)(s-b)(s-c)}$$

where $s = \dfrac{a+b+c}{2}$.

Proof

From Section 7.1, you know that

$$\text{Area} = \frac{1}{2}bc\sin A \qquad\qquad \text{Formula for the area of an oblique triangle}$$

$$(\text{Area})^2 = \frac{1}{4}b^2c^2\sin^2 A \qquad\qquad \text{Square each side.}$$

$$\text{Area} = \sqrt{\frac{1}{4}b^2c^2\sin^2 A} \qquad\qquad \text{Take the square root of each side.}$$

$$= \sqrt{\frac{1}{4}b^2c^2(1-\cos^2 A)} \qquad\qquad \text{Pythagorean Identity}$$

$$= \sqrt{\left[\frac{1}{2}bc(1+\cos A)\right]\left[\frac{1}{2}bc(1-\cos A)\right]}. \qquad \text{Factor.}$$

Using the Law of Cosines, you can show that

$$\frac{1}{2}bc(1+\cos A) = \frac{a+b+c}{2}\cdot\frac{-a+b+c}{2}$$

and

$$\frac{1}{2}bc(1-\cos A) = \frac{a-b+c}{2}\cdot\frac{a+b-c}{2}.$$

Letting

$$s = \frac{a+b+c}{2}$$

these two equations can be rewritten as

$$\frac{1}{2}bc(1+\cos A) = s(s-a)$$

and

$$\frac{1}{2}bc(1-\cos A) = (s-b)(s-c).$$

By substituting into the last formula for area, you can conclude that

$$\text{Area} = \sqrt{s(s-a)(s-b)(s-c)}.$$

Properties of the Dot Product (p. 574)

Let **u**, **v**, and **w** be vectors in the plane or in space and let c be a scalar.

1. $\mathbf{u} \cdot \mathbf{v} = \mathbf{v} \cdot \mathbf{u}$
2. $\mathbf{0} \cdot \mathbf{v} = 0$
3. $\mathbf{u} \cdot (\mathbf{v} + \mathbf{w}) = \mathbf{u} \cdot \mathbf{v} + \mathbf{u} \cdot \mathbf{w}$
4. $\mathbf{v} \cdot \mathbf{v} = \|\mathbf{v}\|^2$
5. $c(\mathbf{u} \cdot \mathbf{v}) = c\mathbf{u} \cdot \mathbf{v} = \mathbf{u} \cdot c\mathbf{v}$

Proof

Let $\mathbf{u} = \langle u_1, u_2 \rangle$, $\mathbf{v} = \langle v_1, v_2 \rangle$, $\mathbf{w} = \langle w_1, w_2 \rangle$, $\mathbf{0} = \langle 0, 0 \rangle$, and let c be a scalar.

1. $\mathbf{u} \cdot \mathbf{v} = u_1 v_1 + u_2 v_2 = v_1 u_1 + v_2 u_2 = \mathbf{v} \cdot \mathbf{u}$

2. $\mathbf{0} \cdot \mathbf{v} = 0 \cdot v_1 + 0 \cdot v_2 = 0$

3. $\mathbf{u} \cdot (\mathbf{v} + \mathbf{w}) = \mathbf{u} \cdot \langle v_1 + w_1, v_2 + w_2 \rangle$

$$= u_1(v_1 + w_1) + u_2(v_2 + w_2)$$

$$= u_1 v_1 + u_1 w_1 + u_2 v_2 + u_2 w_2$$

$$= (u_1 v_1 + u_2 v_2) + (u_1 w_1 + u_2 w_2) = \mathbf{u} \cdot \mathbf{v} + \mathbf{u} \cdot \mathbf{w}$$

4. $\mathbf{v} \cdot \mathbf{v} = v_1^2 + v_2^2 = \left(\sqrt{v_1^2 + v_2^2} \right)^2 = \|\mathbf{v}\|^2$

5. $c(\mathbf{u} \cdot \mathbf{v}) = c(\langle u_1, u_2 \rangle \cdot \langle v_1, v_2 \rangle)$

$$= c(u_1 v_1 + u_2 v_2)$$

$$= (cu_1)v_1 + (cu_2)v_2$$

$$= \langle cu_1, cu_2 \rangle \cdot \langle v_1, v_2 \rangle$$

$$= c\mathbf{u} \cdot \mathbf{v}$$

Angle Between Two Vectors (p. 575)

If θ is the angle between two nonzero vectors **u** and **v**, then

$$\cos \theta = \frac{\mathbf{u} \cdot \mathbf{v}}{\|\mathbf{u}\| \|\mathbf{v}\|}.$$

Proof

Consider the triangle determined by vectors **u**, **v**, and $\mathbf{v} - \mathbf{u}$, as shown in the figure. By the Law of Cosines, you can write

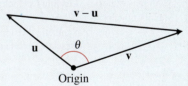

$$\|\mathbf{v} - \mathbf{u}\|^2 = \|\mathbf{u}\|^2 + \|\mathbf{v}\|^2 - 2\|\mathbf{u}\| \|\mathbf{v}\| \cos \theta$$

$$(\mathbf{v} - \mathbf{u}) \cdot (\mathbf{v} - \mathbf{u}) = \|\mathbf{u}\|^2 + \|\mathbf{v}\|^2 - 2\|\mathbf{u}\| \|\mathbf{v}\| \cos \theta$$

$$(\mathbf{v} - \mathbf{u}) \cdot \mathbf{v} - (\mathbf{v} - \mathbf{u}) \cdot \mathbf{u} = \|\mathbf{u}\|^2 + \|\mathbf{v}\|^2 - 2\|\mathbf{u}\| \|\mathbf{v}\| \cos \theta$$

$$\mathbf{v} \cdot \mathbf{v} - \mathbf{u} \cdot \mathbf{v} - \mathbf{v} \cdot \mathbf{u} + \mathbf{u} \cdot \mathbf{u} = \|\mathbf{u}\|^2 + \|\mathbf{v}\|^2 - 2\|\mathbf{u}\| \|\mathbf{v}\| \cos \theta$$

$$\|\mathbf{v}\|^2 - 2\mathbf{u} \cdot \mathbf{v} + \|\mathbf{u}\|^2 = \|\mathbf{u}\|^2 + \|\mathbf{v}\|^2 - 2\|\mathbf{u}\| \|\mathbf{v}\| \cos \theta$$

$$\cos \theta = \frac{\mathbf{u} \cdot \mathbf{v}}{\|\mathbf{u}\| \|\mathbf{v}\|}.$$

Progressive Summary (Chapters P–7)

This chart outlines the topics that have been covered so far in this text. Progressive Summary charts appear after Chapters 2, 4, 7, and 10. In each Progressive Summary, new topics encountered for the first time appear in red.

ALGEBRAIC FUNCTIONS

Polynomial, Rational, Radical

■ **Rewriting**

Polynomial form ↔ Factored form
Operations with polynomials
Rationalize denominators
Simplify rational expressions
Operations with complex numbers

■ **Solving**

Equation	*Strategy*
Linear	Isolate variable
Quadratic	Factor, set to zero
	Extract square roots
	Complete the square
	Quadratic Formula
Polynomial	Factor, set to zero
	Rational Zero Test
Rational	Multiply by LCD
Radical	Isolate, raise to power
Absolute value	Isolate, form two equations

■ **Analyzing**

Graphically	*Algebraically*
Intercepts	Domain, Range
Symmetry	Transformations
Slope	Composition
Asymptotes	Standard forms
End behavior	of equations
Minimum values	Leading Coefficient
Maximum values	Test
	Synthetic division
	Descartes's Rule of Signs

Numerically

Table of values

TRANSCENDENTAL FUNCTIONS

Exponential, Logarithmic
Trigonometric, Inverse Trigonometric

■ **Rewriting**

Exponential form ↔ Logarithmic form
Condense/expand logarithmic expressions
Simplify trigonometric expressions
Prove trigonometric identities
Use conversion formulas
Operations with vectors
Powers and roots of complex numbers

■ **Solving**

Equation	*Strategy*
Exponential	Take logarithm of each side
Logarithmic	Exponentiate each side
Trigonometric	Isolate function
. . . .	Factor, use inverse function
Multiple angle . . . . . or high powers	Use trigonometric identities

■ **Analyzing**

Graphically	*Algebraically*
Intercepts	Domain, Range
Asymptotes	Transformations
Minimum values	Composition
Maximum values	Inverse Properties
	Amplitude, Period
	Reference angles

Numerically

Table of values

OTHER TOPICS

■ **Rewriting**

■ **Solving**

■ **Analyzing**

8 Linear Systems and Matrices

[C] [E]
　　[[2250 2598]]

Section 8.5, Example 12
Softball Team Expenses

8.1 Solving Systems of Equations

The Methods of Substitution and Graphing

So far in this text, most problems have involved either a function of one variable or a single equation in two variables. However, many problems in science, business, and engineering involve two or more equations in two or more variables. To solve such problems, you need to find solutions of **systems of equations.** Here is an example of a system of two equations in two unknowns, x and y.

$$\begin{cases} 2x + y = 5 & \text{Equation 1} \\ 3x - 2y = 4 & \text{Equation 2} \end{cases}$$

A **solution** of this system is an ordered pair that satisfies each equation in the system. Finding the set of all such solutions is called **solving the system of equations.** For instance, the ordered pair $(2, 1)$ is a solution of this system. To check this, you can substitute 2 for x and 1 for y in *each* equation.

Check $(2, 1)$ *in Equation 1:*

$$2x + y = 5$$
$$2(2) + 1 \overset{?}{=} 5$$
$$5 = 5 \checkmark$$

Check $(2, 1)$ *in Equation 2:*

$$3x - 2y = 4$$
$$3(2) - 2(1) \overset{?}{=} 4$$
$$4 = 4 \checkmark$$

In this section, you will study two ways to solve systems of equations, beginning with the **method of substitution.**

The Method of Substitution

To use the **method of substitution** to solve a system of two equations in x and y, perform the following steps.

1. Solve one of the equations for one variable in terms of the other.

2. Substitute the expression found in Step 1 into the other equation to obtain an equation in one variable.

3. Solve the equation obtained in Step 2.

4. Back-substitute the value(s) obtained in Step 3 into the expression obtained in Step 1 to find the value(s) of the other variable.

5. Check that each solution satisfies *both* of the original equations.

When using the **method of graphing,** note that the solution of the system corresponds to the **point(s) of intersection** of the graphs.

The Method of Graphing

To use the **method of graphing** to solve a system of two equations in x and y, perform the following steps.

1. Solve both equations for y in terms of x.

2. Use a graphing utility to graph both equations in the same viewing window.

3. Use the *intersect* feature or the *zoom* and *trace* features of the graphing utility to approximate the point(s) of intersection of the graphs.

4. Check that each solution satisfies *both* of the original equations.

What you should learn

- Use the methods of substitution and graphing to solve systems of equations in two variables.
- Use systems of equations to model and solve real-life problems.

Why you should learn it

You can use systems of equations in situations in which the variables must satisfy two or more conditions. For instance, Exercise 88 on page 618 shows how to use a system of equations to compare two models for estimating the number of board feet in a 16-foot log.

Study Tip

When using the method of substitution, it does not matter which variable you choose to solve for first. Whether you solve for y first or x first, you will obtain the same solution. When making your choice, you should choose the variable and equation that are easier to work with.

Example 1   Solving a System of Equations

Solve the system of equations.

$$\begin{cases} x + y = 4 & \text{Equation 1} \\ x - y = 2 & \text{Equation 2} \end{cases}$$

Algebraic Solution

Begin by solving for y in Equation 1.

$$y = 4 - x \qquad \text{Solve for } y \text{ in Equation 1.}$$

Next, substitute this expression for y into Equation 2 and solve the resulting single-variable equation for x.

$x - y = 2$	Write Equation 2.
$x - (4 - x) = 2$	Substitute $4 - x$ for y.
$x - 4 + x = 2$	Distributive Property
$2x - 4 = 2$	Combine like terms.
$2x = 6$	Add 4 to each side.
$x = 3$	Divide each side by 2.

Finally, you can solve for y by *back-substituting* $x = 3$ into the equation $y = 4 - x$ to obtain

$y = 4 - x$	Write revised Equation 1.
$y = 4 - 3$	Substitute 3 for x.
$y = 1.$	Solve for y.

The solution is the ordered pair

$$(3, 1).$$

Check this as follows.

Check $(3, 1)$ *in Equation 1:*

$x + y = 4$	Write Equation 1.
$3 + 1 \overset{?}{=} 4$	Substitute for x and y.
$4 = 4$	Solution checks in Equation 1. ✔

Check $(3, 1)$ *in Equation 2:*

$x - y = 2$	Write Equation 2.
$3 - 1 \overset{?}{=} 2$	Substitute for x and y.
$2 = 2$	Solution checks in Equation 2. ✔

Because $(3, 1)$ satisfies both equations in the system, it is a solution of the system of equations.

 **CHECKPOINT**   Now try Exercise 19.

Graphical Solution

Begin by solving both equations for y. Then use a graphing utility to graph the equations

$$y_1 = 4 - x$$

and

$$y_2 = x - 2$$

in the same viewing window. Use the *intersect* feature (see Figure 8.1) to approximate the point of intersection of the graphs.

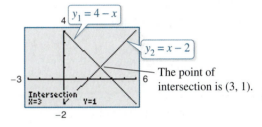

Figure 8.1

Check that $(3, 1)$ is the exact solution as follows.

Check $(3, 1)$ *in Equation 1:*

$x + y = 4$	Write Equation 1.
$3 + 1 \overset{?}{=} 4$	Substitute for x and y.
$4 = 4$	Solution checks in Equation 1. ✔

Check $(3, 1)$ *in Equation 2:*

$x - y = 2$	Write Equation 2.
$3 - 1 \overset{?}{=} 2$	Substitute for x and y.
$2 = 2$	Solution checks in Equation 2. ✔

Because $(3, 1)$ satisfies both equations in the system, it is a solution of the system of equations.

In the algebraic solution of Example 1, note that the term *back-substitution* implies that you work *backwards*. First you solve for one of the variables, and then you substitute that value *back* into one of the equations in the system to find the value of the other variable.

Example 2 Solving a System by Substitution

A total of $12,000 is invested in two funds paying 9% and 11% simple interest. The yearly interest is $1180. How much is invested at each rate?

Solution

Verbal Model:

| 9% fund | + | 11% fund | = | Total investment |

| 9% fund | + | 11% fund | = | Total interest |

Labels:

Amount in 9% fund $= x$	(dollars)
Interest for 9% fund $= 0.09x$	(dollars)
Amount in 11% fund $= y$	(dollars)
Interest for 11% fund $= 0.11y$	(dollars)
Total investment $= \$12,000$	(dollars)
Total interest $= \$1180$	(dollars)

System:
$$\begin{cases} x + y = 12{,}000 & \text{Equation 1} \\ 0.09x + 0.11y = 1180 & \text{Equation 2} \end{cases}$$

To begin, it is convenient to multiply each side of Equation 2 by 100. This eliminates the need to work with decimals.

$$9x + 11y = 118{,}000 \qquad \text{Revised Equation 2}$$

To solve this system, you can solve for x in Equation 1.

$$x = 12{,}000 - y \qquad \text{Revised Equation 1}$$

Next, substitute this expression for x into revised Equation 2 and solve the resulting equation for y.

$9x + 11y = 118{,}000$	Write revised Equation 2.
$9(12{,}000 - y) + 11y = 118{,}000$	Substitute $12{,}000 - y$ for x.
$108{,}000 - 9y + 11y = 118{,}000$	Distributive Property
$2y = 10{,}000$	Combine like terms.
$y = 5000$	Divide each side by 2.

Finally, back-substitute the value $y = 5000$ to solve for x.

$x = 12{,}000 - y$	Write revised Equation 1.
$x = 12{,}000 - 5000$	Substitute 5000 for y.
$x = 7000$	Simplify.

The solution is

$$(7000, 5000).$$

So, $7000 is invested at 9% and $5000 is invested at 11% to yield yearly interest of $1180. Check this in the original system.

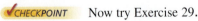 Now try Exercise 29.

The equations in Examples 1 and 2 are linear. Substitution and graphing can also be used to solve systems in which one or both of the equations are nonlinear.

Technology Tip

 Remember that a good way to check the answers you obtain in this section is to use a graphing utility. For instance, enter the two equations in Example 2

$$y_1 = 12{,}000 - x$$

$$y_2 = \frac{1180 - 0.09x}{0.11}$$

and find an appropriate viewing window that shows where the lines intersect. Then use the *intersect* feature or the *zoom* and *trace* features to find the point of intersection.

Example 3   Substitution: Two-Solution Case

Solve the system of equations.

$$\begin{cases} x^2 + 4x - y = 7 & \text{Equation 1} \\ 2x - y = -1 & \text{Equation 2} \end{cases}$$

Algebraic Solution

Begin by solving for y in Equation 2 to obtain

$$y = 2x + 1. \qquad \text{Solve for } y \text{ in Equation 2.}$$

Next, substitute this expression for y into Equation 1 and solve for x.

$x^2 + 4x - y = 7$	Write Equation 1.
$x^2 + 4x - (2x + 1) = 7$	Substitute $2x + 1$ for y.
$x^2 + 4x - 2x - 1 = 7$	Distributive Property
$x^2 + 2x - 8 = 0$	Write in general form.
$(x + 4)(x - 2) = 0$	Factor.
$x + 4 = 0 \quad \Longrightarrow \quad x = -4$	Set 1st factor equal to 0.
$x - 2 = 0 \quad \Longrightarrow \quad x = 2$	Set 2nd factor equal to 0.

Back-substituting these values of x into revised Equation 2 produces

$$y = 2(-4) + 1 = -7 \quad \text{and} \quad y = 2(2) + 1 = 5.$$

So, the solutions are

$$(-4, -7) \quad \text{and} \quad (2, 5).$$

Check these in the original system.

 **CHECKPOINT**   Now try Exercise 33.

Graphical Solution

Solve each equation for y and use a graphing utility to graph the equations in the same viewing window.

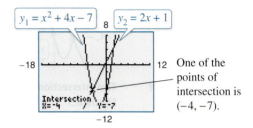

One of the points of intersection is $(-4, -7)$.

Figure 8.2

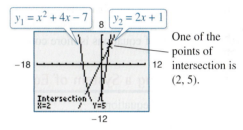

One of the points of intersection is $(2, 5)$.

Figure 8.3

From Figures 8.2 and 8.3, the solutions are $(-4, -7)$ and $(2, 5)$. Check these in the original system.

Example 4   Substitution: No-Solution Case

Solve the system of equations.

$$\begin{cases} -x + y = 4 & \text{Equation 1} \\ x^2 + y = 3 & \text{Equation 2} \end{cases}$$

Solution

Begin by solving for y in Equation 1 to obtain $y = x + 4$. Next, substitute this expression for y into Equation 2 and solve for x.

$x^2 + y = 3$	Write Equation 2.
$x^2 + (x + 4) = 3$	Substitute $x + 4$ for y.
$x^2 + x + 1 = 0$	Simplify.
$x = \dfrac{-1 \pm \sqrt{3}i}{2}$	Quadratic Formula

Because this yields two complex values, the equation $x^2 + x + 1 = 0$ has no *real* solution. So, the original system of equations has no *real* solution.

CHECKPOINT   Now try Exercise 35.

Explore the Concept

 Graph the system of equations in Example 4. Do the graphs of the equations intersect? Why or why not?

81. Marketing Research The daily DVD rentals of a newly released animated film and a newly released horror film from a movie rental store can be modeled by the equations

$$\begin{cases} N = 360 - 24x & \text{Animated film} \\ N = 24 + 18x & \text{Horror film} \end{cases}$$

where N is the number of DVDs rented and x represents the week, with $x = 1$ corresponding to the first week of release.

(a) Use the *table* feature of a graphing utility to find the numbers of rentals of each movie for each of the first 12 weeks of release.

(b) Use the results of part (a) to determine the solution to the system of equations.

(c) Solve the system of equations algebraically.

(d) Compare your results from parts (b) and (c).

(e) Interpret the results in the context of the situation.

82. Economics You want to buy either a wood pellet stove or an electric furnace. The pellet stove costs $2160 and produces heat at a cost of $15.15 per 1 million Btu (British thermal units). The electric furnace costs $1250 and produces heat at a cost of $33.25 per 1 million Btu.

(a) Write a function for the total cost y of buying the pellet stove and producing x million Btu of heat.

(b) Write a function for the total cost y of buying the electric furnace and producing x million Btu of heat.

(c) Use a graphing utility to graph and solve the system of equations formed by the two cost functions.

(d) Solve the system of equations algebraically.

(e) Interpret the results in the context of the situation.

✓ **83. Break-Even Analysis** A small software company invests $16,000 to produce a software package that will sell for $55.95. Each unit can be produced for $9.45.

(a) Write the cost and revenue functions for x units produced and sold.

(b) Use a graphing utility to graph the cost and revenue functions in the same viewing window. Use the graph to approximate the number of units that must be sold to break even, and verify the result algebraically.

84. Professional Sales You are offered two jobs selling college textbooks. One company offers an annual salary of $30,000 plus a year-end bonus of 1% of your total sales. The other company offers an annual salary of $25,000 plus a year-end bonus of 2% of your total sales. How much would you have to sell in a year to make the second offer the better offer?

85. Geometry What are the dimensions of a rectangular tract of land with a perimeter of 40 miles and an area of 96 square miles?

86. Geometry What are the dimensions of an isosceles right triangle with a two-inch hypotenuse and an area of 1 square inch?

87. Finance You are deciding how to invest a total of $20,000 in two funds paying 5.5% and 7.5% simple interest. You want to earn a total of $1300 in interest from the investments each year.

(a) Write a system of equations in which one equation represents the total amount invested and the other equation represents the $1300 yearly interest. Let x and y represent the amounts invested at 5.5% and 7.5%, respectively.

(b) Use a graphing utility to graph the two equations in the same viewing window.

(c) How much of the $20,000 should you invest at 5.5% to earn $1300 in interest per year? Explain your reasoning.

88. *Why you should learn it* (p. 610) You are offered two different rules for estimating the number of board feet in a 16-foot log. (A board foot is a unit of measure for lumber equal to a board 1 foot square and 1 inch thick.) One rule is the Doyle Log Rule modeled by

$$V = (D - 4)^2, \quad 5 \le D \le 40$$

where D is the diameter (in inches) of the log and V is its volume in (board feet). The other rule is the Scribner Log Rule modeled by

$$V = 0.79D^2 - 2D - 4, \quad 5 \le D \le 40.$$

(a) Use a graphing utility to graph the two log rules in the same viewing window.

(b) For what diameter do the two rules agree?

(c) You are selling large logs by the board foot. Which rule would you use? Explain your reasoning.

89. Algebraic-Graphical-Numerical The populations (in thousands) of Arizona A and Indiana I from 2000 through 2008 can by modeled by the system

$$\begin{cases} A = 171.9t + 5118 & \text{Arizona} \\ I = 35.7t + 6080 & \text{Indiana} \end{cases}$$

where t is the year, with $t = 0$ corresponding to 2000. (Source: U.S. Census Bureau)

(a) Record in a table the populations given by the models for the two states in the years 2000 through 2008.

(b) According to the table, in what year(s) was the population of Arizona greater than that of Indiana?

(c) Use a graphing utility to graph the models in the same viewing window. Estimate the point of intersection of the models.

(d) Find the point of intersection algebraically.

(e) Summarize your findings of parts (b) through (d).

90. MODELING DATA

The table shows the yearly revenues (in millions of dollars) of the online travel companies Expedia and Priceline.com from 2004 through 2008.   (Sources: Expedia; Priceline.com)

Year	Expedia	Priceline.com
2004	1843	914
2005	2120	963
2006	2238	1123
2007	2665	1391
2008	2937	1885

(a) Use the *regression* feature of a graphing utility to find a linear model for the yearly revenue E of Expedia and a quadratic model for the yearly revenue P of Priceline.com. Let x represent the year, with $x = 4$ corresponding to 2004.

(b) Use the graphing utility to graph the models with the original data in the same viewing window.

(c) Use the graph in part (b) to approximate the first year when the revenues of Priceline.com will be greater than the revenues of Expedia.

(d) Algebraically approximate the first year when the revenues of Priceline.com will be greater than the revenues of Expedia.

(e) Compare your results from parts (c) and (d).

Conclusions

True or False?   In Exercises 91 and 92, determine whether the statement is true or false. Justify your answer.

91. In order to solve a system of equations by substitution, you must always solve for y in one of the two equations and then back-substitute.

92. If a system consists of a parabola and a circle, then it can have at most two solutions.

93. **Think About It**   When solving a system of equations by substitution, how do you recognize that the system has no solution?

94. **Exploration**   Find the equations of lines whose graphs intersect the graph of the parabola $y = x^2$ at (a) two points, (b) one point, and (c) no points. (There are many correct answers.)

95. **Exploration**   Create systems of two linear equations in two variables that have (a) no solution, (b) one distinct solution, and (c) infinitely many solutions. (There are many correct answers.)

96. **Exploration**   Create a system of linear equations in two variables that has the solution $(2, -1)$ as its only solution. (There are many correct answers.)

97. **Exploration**   Consider the system of equations.

$$\begin{cases} y = b^x \\ y = x^b \end{cases}$$

(a) Use a graphing utility to graph the system of equations for $b = 2$ and $b = 4$.

(b) For a fixed value of $b > 1$, make a conjecture about the number of points of intersection of the graphs in part (a).

98. **CAPSTONE**   Consider the system of equations

$$\begin{cases} ax + by = c \\ dx + cy = f \end{cases}.$$

(a) Find values of a, b, c, d, e, and f such that the system has one distinct solution. (There is more than one correct answer.)

(b) Explain how to solve the system in part (a) by the method of substitution and graphically.

(c) Write a brief paragraph describing any advantages of the method of substitution over the graphical method of solving a system of equations.

Cumulative Mixed Review

Finding the Slope-Intercept Form   In Exercises 99–104, write an equation of the line passing through the two points. Use the slope-intercept form, if possible. If not possible, explain why.

99. $(-2, 7), (5, 5)$

100. $(3, 4), (10, 6)$

101. $(6, 3), (10, 3)$

102. $(4, -2), (4, 5)$

103. $\left(\frac{3}{5}, 0\right), (4, 6)$

104. $\left(-\frac{7}{3}, 8\right), \left(\frac{5}{2}, \frac{1}{2}\right)$

Finding the Domain and Asymptotes of a Function   In Exercises 105–110, find the domain of the function and identify any horizontal or vertical asymptotes.

105. $f(x) = \dfrac{5}{x - 6}$

106. $f(x) = \dfrac{2x - 7}{3x + 2}$

107. $f(x) = \dfrac{x^2 + 2}{x^2 - 16}$

108. $f(x) = 3 - \dfrac{2}{x^2}$

109. $f(x) = \dfrac{x + 1}{x^2 + 1}$

110. $f(x) = \dfrac{x - 4}{x^2 + 16}$

8.2   Systems of Linear Equations in Two Variables

The Method of Elimination

In Section 8.1, you studied two methods for solving a system of equations: substitution and graphing. Now you will study the **method of elimination** to solve a system of linear equations in two variables. The key step in this method is to obtain, for one of the variables, coefficients that differ only in sign so that *adding* the equations eliminates the variable.

$$
\begin{array}{ll}
3x + 5y = 7 & \text{Equation 1} \\
\underline{-3x - 2y = -1} & \text{Equation 2} \\
3y = 6 & \text{Add equations.}
\end{array}
$$

Note that by adding the two equations, you eliminate the x-terms and obtain a single equation in y. Solving this equation for y produces

$$y = 2$$

which you can then back-substitute into one of the original equations to solve for x.

Example 1   Solving a System by Elimination

Solve the system of linear equations.

$$
\begin{cases}
3x + 2y = 4 & \text{Equation 1} \\
5x - 2y = 8 & \text{Equation 2}
\end{cases}
$$

Solution

Because the coefficients of y differ only in sign, you can eliminate the y-terms by adding the two equations.

$$
\begin{array}{ll}
3x + 2y = 4 & \text{Write Equation 1.} \\
\underline{5x - 2y = 8} & \text{Write Equation 2.} \\
8x = 12 & \text{Add equations.} \\
x = \tfrac{3}{2} & \text{Solve for } x.
\end{array}
$$

So, $x = \tfrac{3}{2}$. By back-substituting into Equation 1, you can solve for y.

$$
\begin{array}{ll}
3x + 2y = 4 & \text{Write Equation 1.} \\
3\left(\tfrac{3}{2}\right) + 2y = 4 & \text{Substitute } \tfrac{3}{2} \text{ for } x. \\
y = -\tfrac{1}{4} & \text{Solve for } y.
\end{array}
$$

The solution is

$$\left(\tfrac{3}{2}, -\tfrac{1}{4}\right).$$

You can check the solution *algebraically* by substituting into the original system, or graphically as shown in Section 8.1.

Check

$$
\begin{array}{ll}
3\left(\tfrac{3}{2}\right) + 2\left(-\tfrac{1}{4}\right) \overset{?}{=} 4 & \text{Substitute into Equation 1.} \\
\tfrac{9}{2} - \tfrac{1}{2} = 4 & \text{Equation 1 checks. } \checkmark \\
5\left(\tfrac{3}{2}\right) - 2\left(-\tfrac{1}{4}\right) \overset{?}{=} 8 & \text{Substitute into Equation 2.} \\
\tfrac{15}{2} + \tfrac{1}{2} = 8 & \text{Equation 2 checks. } \checkmark
\end{array}
$$

✓**CHECKPOINT**   Now try Exercise 13.

$89.95 $79.50

What you should learn

- Use the method of elimination to solve systems of linear equations in two variables.
- Graphically interpret the number of solutions of a system of linear equations in two variables.
- Use systems of linear equations in two variables to model and solve real-life problems.

Why you should learn it

You can use systems of linear equations to model many business applications. For instance, Exercise 84 on page 627 shows how to use a system of linear equations to determine number of running shoes sold.

Explore the Concept

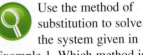 Use the method of substitution to solve the system given in Example 1. Which method is easier?

> ### The Method of Elimination
>
> To use the **method of elimination** to solve a system of two linear equations in x and y, perform the following steps.
>
> **1.** Obtain coefficients for x (or y) that differ only in sign by multiplying all terms of one or both equations by suitably chosen constants.
>
> **2.** Add the equations to eliminate one variable; solve the resulting equation.
>
> **3.** Back-substitute the value obtained in Step 2 into either of the original equations and solve for the other variable.
>
> **4.** Check your solution in both of the original equations.

Example 2   Solving a System by Elimination

Solve the system of linear equations.

$$\begin{cases} 5x + 3y = 9 & \text{Equation 1} \\ 2x - 4y = 14 & \text{Equation 2} \end{cases}$$

Algebraic Solution

You can obtain coefficients of y that differ only in sign by multiplying Equation 1 by 4 and multiplying Equation 2 by 3.

$5x + 3y = 9$ ⟹ $20x + 12y = 36$ Multiply Equation 1 by 4.

$\underline{2x - 4y = 14}$ ⟹ $\underline{6x - 12y = 42}$ Multiply Equation 2 by 3.

$\qquad\qquad\qquad\qquad 26x \qquad\; = 78$ Add equations.

From this equation, you can see that $x = 3$. By back-substituting this value of x into Equation 2, you can solve for y.

$2x - 4y = 14$ Write Equation 2.

$2(3) - 4y = 14$ Substitute 3 for x.

$-4y = 8$ Combine like terms.

$y = -2$ Solve for y.

The solution is $(3, -2)$. You can check the solution algebraically by substituting into the original system.

✓**CHECKPOINT**   Now try Exercise 15.

Graphical Solution

Solve each equation for y and use a graphing utility to graph the equations in the same viewing window.

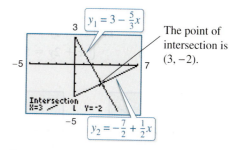

The point of intersection is $(3, -2)$.

Figure 8.9

From Figure 8.9, the solution is

$$(3, -2).$$

Check this in the original system.

In Example 2, the two systems of linear equations (the original system and the system obtained by multiplying by constants)

$$\begin{cases} 5x + 3y = 9 \\ 2x - 4y = 14 \end{cases} \quad \text{and} \quad \begin{cases} 20x + 12y = 36 \\ 6x - 12y = 42 \end{cases}$$

are called **equivalent systems** because they have precisely the same solution set. Each of the following operations can be performed on a system of linear equations to produce an equivalent system.

1. Interchange two equations.

2. Multiply one of the equations by a nonzero constant.

3. Add a multiple of one of the equations to any other equation in the system.

✓ **29.** $\begin{cases} \frac{2}{5}x - \frac{3}{2}y = 4 \\ \frac{1}{5}x - \frac{3}{4}y = -2 \end{cases}$ **30.** $\begin{cases} \frac{2}{3}x + \frac{1}{6}y = \frac{2}{3} \\ 4x + y = 4 \end{cases}$

✓ **31.** $\begin{cases} \frac{3}{4}x + y = \frac{1}{8} \\ \frac{9}{4}x + 3y = \frac{3}{8} \end{cases}$ **32.** $\begin{cases} \frac{1}{4}x + \frac{1}{6}y = 1 \\ -3x - 2y = 0 \end{cases}$

33. $\begin{cases} \dfrac{x+3}{4} + \dfrac{y-1}{3} = 1 \\ 2x - y = 12 \end{cases}$ **34.** $\begin{cases} \dfrac{x+2}{4} + \dfrac{y-1}{4} = 1 \\ x - y = 4 \end{cases}$

35. $\begin{cases} -5x + 6y = -3 \\ 20x - 24y = 12 \end{cases}$ **36.** $\begin{cases} 7x + 8y = 6 \\ -14x - 16y = -12 \end{cases}$

37. $\begin{cases} 2.5x - 3y = 1.5 \\ 2x - 2.4y = 1.2 \end{cases}$ **38.** $\begin{cases} 6.3x + 7.2y = 5.4 \\ 5.6x + 6.4y = 4.8 \end{cases}$

39. $\begin{cases} 0.2x - 0.5y = -27.8 \\ 0.3x + 0.4y = 68.7 \end{cases}$ **40.** $\begin{cases} 0.2x + 0.6y = -1 \\ x - 0.5y = 2 \end{cases}$

41. $\begin{cases} 0.05x - 0.03y = 0.21 \\ 0.07x + 0.02y = 0.16 \end{cases}$ **42.** $\begin{cases} 0.2x + 0.4y = -0.2 \\ x + 0.5y = -2.5 \end{cases}$

43. $\begin{cases} \dfrac{1}{x} + \dfrac{3}{y} = 2 \\ \dfrac{4}{x} - \dfrac{1}{y} = -5 \end{cases}$ **44.** $\begin{cases} \dfrac{2}{x} - \dfrac{1}{y} = 0 \\ \dfrac{4}{x} - \dfrac{3}{y} = -1 \end{cases}$

45. $\begin{cases} \dfrac{1}{x} + \dfrac{2}{y} = 5 \\ \dfrac{3}{x} - \dfrac{4}{y} = -5 \end{cases}$ **46.** $\begin{cases} \dfrac{2}{x} - \dfrac{1}{y} = 5 \\ \dfrac{6}{x} + \dfrac{1}{y} = 11 \end{cases}$

Solving a System Graphically   In Exercises 47–52, use a graphing utility to graph the lines in the system. Use the graphs to determine whether the system is consistent or inconsistent. If the system is consistent, determine the solution. Verify your results algebraically.

47. $\begin{cases} 2x - 5y = 0 \\ x - y = 3 \end{cases}$ **48.** $\begin{cases} 2x + y = 5 \\ x - 2y = -1 \end{cases}$

49. $\begin{cases} \frac{3}{5}x - y = 3 \\ -3x + 5y = 9 \end{cases}$ **50.** $\begin{cases} 4x - 6y = 9 \\ \frac{16}{3}x - 8y = 12 \end{cases}$

51. $\begin{cases} 8x - 14y = 5 \\ 2x - 3.5y = 1.25 \end{cases}$ **52.** $\begin{cases} -x + 7y = 3 \\ -\frac{1}{7}x + y = 5 \end{cases}$

Solving a System Graphically   In Exercises 53–60, use a graphing utility to graph the two equations. Use the graphs to approximate the solution of the system. Round your results to three decimal places.

53. $\begin{cases} 6y = 42 \\ 6x - y = 16 \end{cases}$ **54.** $\begin{cases} 4y = -8 \\ 7x - 2y = 25 \end{cases}$

55. $\begin{cases} \frac{3}{2}x - \frac{1}{5}y = 8 \\ -2x + 3y = 3 \end{cases}$ **56.** $\begin{cases} \frac{3}{4}x - \frac{5}{2}y = -9 \\ -x + 6y = 28 \end{cases}$

57. $\begin{cases} \frac{1}{3}x + y = -\frac{1}{3} \\ 5x - 3y = 7 \end{cases}$ **58.** $\begin{cases} 5x - y = -4 \\ 2x + \frac{3}{5}y = \frac{2}{5} \end{cases}$

59. $\begin{cases} 0.5x + 2.2y = 9 \\ 6x + 0.4y = -22 \end{cases}$ **60.** $\begin{cases} 2.4x + 3.8y = -17.6 \\ 4x - 0.2y = -3.2 \end{cases}$

Solving a System   In Exercises 61–68, use any method to solve the system.

61. $\begin{cases} 3x - 5y = 7 \\ 2x + y = 9 \end{cases}$ **62.** $\begin{cases} -x + 3y = 17 \\ 4x + 3y = 7 \end{cases}$

63. $\begin{cases} y = 2x - 5 \\ y = 5x - 11 \end{cases}$ **64.** $\begin{cases} 7x + 3y = 16 \\ y = x + 2 \end{cases}$

65. $\begin{cases} x - 5y = 21 \\ 6x + 5y = 21 \end{cases}$ **66.** $\begin{cases} y = -2x - 17 \\ y = 2 - 3x \end{cases}$

67. $\begin{cases} -5x + 9y = 13 \\ y = x - 4 \end{cases}$ **68.** $\begin{cases} 4x - 3y = 6 \\ -5x + 7y = -1 \end{cases}$

Exploration   In Exercises 69–72, find a system of linear equations that has the given solution. (There are many correct answers.)

69. $(0, 8)$ **70.** $(3, -4)$

71. $\left(3, \frac{5}{2}\right)$ **72.** $\left(-\frac{2}{3}, -10\right)$

Economics   In Exercises 73–76, find the *point of equilibrium* of the demand and supply equations. The point of equilibrium is the price p and the number of units x that satisfy both the demand and supply equations.

	Demand	*Supply*
73.	$p = 500 - 0.4x$	$p = 380 + 0.1x$
74.	$p = 100 - 0.05x$	$p = 25 + 0.1x$
75.	$p = 140 - 0.00002x$	$p = 80 + 0.00001x$
76.	$p = 400 - 0.0002x$	$p = 225 + 0.0005x$

✓ **77. Aviation**   An airplane flying into a headwind travels the 1800-mile flying distance between New York City and Albuquerque, New Mexico in 3 hours and 36 minutes. On the return flight, the same distance is traveled in 3 hours. Find the airspeed of the plane and the speed of the wind, assuming that both remain constant.

78. Nutrition   Two cheeseburgers and one small order of French fries from a fast-food restaurant contain a total of 830 calories. Three cheeseburgers and two small orders of French fries contain a total of 1360 calories. Find the number of calories in each item.

79. Business   A minor league baseball team had a total attendance one evening of 1175. The tickets for adults and children sold for $5.00 and $3.50, respectively. The ticket revenue that night was $5087.50.

(a) Create a system of linear equations to find the numbers of adults A and children C at the game.

(b) Solve your system of equations by elimination or by substitution. Explain your choice.

(c) Use the *intersect* feature or the *zoom* and *trace* features of a graphing utility to solve your system.

80. Chemistry Thirty liters of a 40% acid solution is obtained by mixing a 25% solution with a 50% solution.

(a) Write a system of equations in which one equation represents the amount of final mixture required and the other represents the percent of acid in the final mixture. Let x and y represent the amounts of the 25% and 50% solutions, respectively.

(b) Use a graphing utility to graph the two equations in part (a) in the same viewing window. As the amount of the 25% solution increases, how does the amount of the 50% solution change?

(c) How much of each solution is required to obtain the specified concentration of the final mixture?

81. Business A grocer sells oranges for $0.95 each and grapefruits for $1.05 each. You purchased a mix of 16 oranges and grapefruits and paid $15.90. How many of each type of fruit did you buy?

82. Sales The sales S (in millions of dollars) of Fossil and Aeropostale clothing stores from 2000 to 2008 can be modeled by

$$\begin{cases} S - 138.98t = 413.5 \\ S - 212.35t = 135.3 \end{cases}$$

where t is the year, with $t = 0$ corresponding to 2000. (Sources: Fossil, Inc.; Aeropostale, Inc.)

(a) Solve the system of equations using the method of your choice. Explain why you chose that method.

(b) Interpret the meaning of the solution in the context of the problem.

83. Economics Revenues for a movie rental store on a particular Friday evening are $867.50 for 310 rentals. The rental fee for movies is $3.00 each and the rental fee for video games is $2.50 each. Determine the number of each type that are rented that evening.

84. *Why you should learn it* (p.620) On a Saturday

night, the manager of a shoe store evaluates the receipts of the previous week's sales. Two hundred fifty pairs of two different styles of running shoes were sold. One style sold for $75.50 and the other sold for $89.95. The receipts totaled $20,031. The cash register that was supposed to record the number of each type of shoe sold malfunctioned. Can you recover the information? If so, how many shoes of each type were sold?

Fitting a Line to Data To find the least squares regression line $y = ax + b$ for a set of points

$(x_1, y_1), (x_2, y_2), \ldots, (x_n, y_n)$

you can solve the following system for a and b.

$$\begin{cases} nb + \left(\sum_{i=1}^{n} x_i\right)a = \left(\sum_{i=1}^{n} y_i\right) \\ \left(\sum_{i=1}^{n} x_i\right)b + \left(\sum_{i=1}^{n} x_i^2\right)a = \left(\sum_{i=1}^{n} x_i y_i\right) \end{cases}$$

In Exercises 85–88, the sums have been evaluated. Solve the given system for a and b to find the least squares regression line for the points. Use a graphing utility to confirm the result.

85. $\begin{cases} 5b + 10a = 20.2 \\ 10b + 30a = 50.1 \end{cases}$

86. $\begin{cases} 5b + 10a = 11.7 \\ 10b + 30a = 25.6 \end{cases}$

87. $\begin{cases} 5b + 10a = 2.7 \\ 10b + 30a = -19.6 \end{cases}$

88. $\begin{cases} 5b + 10a = -2.3 \\ 10b + 30a = -19.4 \end{cases}$

89. MODELING DATA

Four test plots were used to explore the relationship between wheat yield (in bushels per acre) and the amount of fertilizer applied (in hundreds of pounds per acre). The results are shown in the table.

Fertilizer, x	Yield, y
1.0	32
1.5	41
2.0	48
2.5	53

(a) Find the least squares regression line $y = ax + b$ for the data by solving the system for a and b.

$$\begin{cases} 4b + 7.0a = 174 \\ 7b + 13.5a = 322 \end{cases}$$

(b) Use the *regression* feature of a graphing utility to confirm the result in part (a).

(c) Use the graphing utility to plot the data and graph the linear model from part (a) in the same viewing window.

(d) Use the linear model from part (a) to predict the yield for a fertilizer application of 160 pounds per acre.

90. MODELING DATA

A candy store manager wants to know the demand for a candy bar as a function of the price. The daily sales for different prices of the product are shown in the table.

Price, x	Demand, y
$1.00	45
$1.20	37
$1.50	23

(a) Find the least squares regression line $y = ax + b$ for the data by solving the system for a and b.

$$\begin{cases} 3.00b + 3.70a = 105.00 \\ 3.70b + 4.69a = 123.90 \end{cases}$$

(b) Use the *regression* feature of a graphing utility to confirm the result in part (a).

(c) Use the graphing utility to plot the data and graph the linear model from part (a) in the same viewing window.

(d) Use the linear model from part (a) to predict the demand when the price is $1.75.

Conclusions

True or False? In Exercises 91–93, determine whether the statement is true or false. Justify your answer.

91. If a system of linear equations has two distinct solutions, then it has an infinite number of solutions.

92. If a system of linear equations has no solution, then the lines must be parallel.

93. Solving a system of equations graphically using a graphing utility always yields an exact solution.

94. Writing Briefly explain whether or not it is possible for a consistent system of linear equations to have exactly two solutions.

95. Think About It Give examples of (a) a system of linear equations that has no solution and (b) a system of linear equations that has an infinite number of solutions. (There are many correct answers.)

96. CAPSTONE Find all value(s) of k for which the system of linear equations

$$\begin{cases} x + 3y = 9 \\ 2x + 6y = k \end{cases}$$

has (a) infinitely many solutions and (b) no solution.

Solving a System In Exercises 97 and 98, solve the system of equations for u and v. While solving for these variables, consider the transcendental functions as constants. (Systems of this type are found in a course in differential equations.)

97. $\begin{cases} u \sin x + v \cos x = 0 \\ u \cos x - v \sin x = \sec x \end{cases}$

98. $\begin{cases} u \cos 2x + v \sin 2x = 0 \\ u(-2 \sin 2x) + v(2 \cos 2x) = \csc 2x \end{cases}$

Cumulative Mixed Review

Solving an Inequality In Exercises 99–104, solve the inequality and graph the solution on a real number line.

99. $-11 - 6x \geq 33$

100. $-6 \leq 3x - 10 < 6$

101. $|x - 8| < 10$

102. $|x + 10| \geq -3$

103. $2x^2 + 3x - 35 < 0$

104. $3x^2 + 12x > 0$

Rewriting a Logarithmic Expression In Exercises 105–110, write the expression as the logarithm of a single quantity.

105. $\ln x + \ln 6$

106. $\ln x - 5 \ln(x + 3)$

107. $\log_9 12 - \log_9 x$

108. $\frac{1}{4} \log_6 3 + \frac{1}{4} \log_6 x$

109. $2 \ln x - \ln(x + 2)$

110. $\frac{1}{2} \ln(x^2 + 4) - \ln x$

111. Make a Decision To work an extended application analyzing the average undergraduate tuition, room, and board charges at private colleges in the United States from 1985 through 2008, visit this textbook's *Companion Website.* (Data Source: National Center for Education Statistics)

8.3   Multivariable Linear Systems

Row-Echelon Form and Back-Substitution

The method of elimination can be applied to a system of linear equations in more than two variables. When elimination is used to solve a system of linear equations, the goal is to rewrite the system in a form to which back-substitution can be applied. To see how this works, consider the following two systems of linear equations.

System of Three Linear Equations in Three Variables (See Example 2):

$$\begin{cases} x - 2y + 3z = 9 \\ -x + 3y + z = -2 \\ 2x - 5y + 5z = 17 \end{cases}$$

Equivalent System in Row-Echelon Form (See Example 1):

$$\begin{cases} x - 2y + 3z = 9 \\ y + 4z = 7 \\ z = 2 \end{cases}$$

The second system is said to be in **row-echelon form,** which means that it has a "stair-step" pattern with leading coefficients of 1. After comparing the two systems, it should be clear that it is easier to solve the system in row-echelon form, using back-substitution.

Example 1   Using Back-Substitution in Row-Echelon Form

Solve the system of linear equations.

$$\begin{cases} x - 2y + 3z = 9 & \text{Equation 1} \\ y + 4z = 7 & \text{Equation 2} \\ z = 2 & \text{Equation 3} \end{cases}$$

Solution

From Equation 3, you know the value of z. To solve for y, substitute $z = 2$ into Equation 2 to obtain

$$y + 4(2) = 7 \qquad \text{Substitute 2 for } z.$$
$$y = -1. \qquad \text{Solve for } y.$$

Next, substitute $y = -1$ and $z = 2$ into Equation 1 to obtain

$$x - 2(-1) + 3(2) = 9 \qquad \text{Substitute } -1 \text{ for } y \text{ and 2 for } z.$$
$$x = 1. \qquad \text{Solve for } x.$$

The solution is

$$x = 1, \quad y = -1, \quad \text{and} \quad z = 2$$

which can be written as the **ordered triple**

$$(1, -1, 2).$$

Check this in the original system of equations.

 **Now try Exercise 13.**

What you should learn

- Use back-substitution to solve linear systems in row-echelon form.
- Use Gaussian elimination to solve systems of linear equations.
- Solve nonsquare systems of linear equations.
- Graphically interpret three-variable linear systems.
- Use systems of linear equations to write partial fraction decompositions of rational expressions.
- Use systems of linear equations in three or more variables to model and solve real-life problems

Why you should learn it

Systems of linear equations in three or more variables can be used to model and solve real-life problems. For instance, Exercise 90 on page 641 shows how to use a system of linear equations to analyze the numbers of par-3, par-4, and par-5 holes on a golf course.

Gaussian Elimination

Two systems of equations are *equivalent* when they have the same solution set. To solve a system that is not in row-echelon form, first convert it to an *equivalent* system that *is* in row-echelon form by using one or more of the elementary row operations shown below. This process is called **Gaussian elimination**, after the German mathematician Carl Friedrich Gauss (1777–1855).

> **Elementary Row Operations for Systems of Equations**
>
> **1.** Interchange two equations.
>
> **2.** Multiply one of the equations by a nonzero constant.
>
> **3.** Add a multiple of one equation to another equation.

Example 2   Using Gaussian Elimination to Solve a System

Solve the system of linear equations.

$$\begin{cases} x - 2y + 3z = 9 & \text{Equation 1} \\ -x + 3y + z = -2 & \text{Equation 2} \\ 2x - 5y + 5z = 17 & \text{Equation 3} \end{cases}$$

Solution

Because the leading coefficient of the first equation is 1, you can begin by saving the x at the upper left and eliminating the other x-terms from the first column.

$$\begin{cases} x - 2y + 3z = 9 \\ y + 4z = 7 \\ 2x - 5y + 5z = 17 \end{cases}$$

Adding the first equation to the second equation produces a new second equation.

$$\begin{cases} x - 2y + 3z = 9 \\ y + 4z = 7 \\ -y - z = -1 \end{cases}$$

Adding -2 times the first equation to the third equation produces a new third equation.

Now that all but the first x have been eliminated from the first column, go to work on the second column. (You need to eliminate y from the third equation.)

$$\begin{cases} x - 2y + 3z = 9 \\ y + 4z = 7 \\ 3z = 6 \end{cases}$$

Adding the second equation to the third equation produces a new third equation.

Finally, you need a coefficient of 1 for z in the third equation.

$$\begin{cases} x - 2y + 3z = 9 \\ y + 4z = 7 \\ z = 2 \end{cases}$$

Multiplying the third equation by $\frac{1}{3}$ produces a new third equation.

This is the same system that was solved in Example 1. As in that example, you can conclude that the solution is

$$x = 1, \quad y = -1, \quad \text{and} \quad z = 2$$

written as $(1, -1, 2)$.

✔**CHECKPOINT**   Now try Exercise 21.

Study Tip

Arithmetic errors are often made when elementary row operations are performed. You should note the operation performed in each step so that you can go back and check your work.

Nons

So far,
means t
system
system
many e

Exampl

Solve t

$$\begin{cases} \cdot \\ 2. \end{cases}$$

Soluti

Begin t

$$\begin{cases} x \\ \end{cases}$$

$$\begin{cases} x \\ \end{cases}$$

Solve f

y

By bac

x

$)$

Finally

x

So, ev

$(a$

is a so

✓CHE(

In
of the
in the

Chec

$1 - 2$

$2(1) -$

The goal of Gaussian elimination is to use elementary row operations on a system in order to isolate one variable. You can then solve for the value of the variable and use back-substitution to find the values of the remaining variables.

The next example involves an inconsistent system—one that has no solution. The key to recognizing an inconsistent system is that at some stage in the elimination process, you obtain a false statement such as

$$0 = -2. \qquad \text{False statement}$$

Example 3   An Inconsistent System

Solve the system of linear equations.

$$\begin{cases} x - 3y + z = 1 & \text{Equation 1} \\ 2x - y - 2z = 2 & \text{Equation 2} \\ x + 2y - 3z = -1 & \text{Equation 3} \end{cases}$$

Solution

$$\begin{cases} x - 3y + z = 1 \\ 5y - 4z = 0 \\ x + 2y - 3z = -1 \end{cases}$$

> Adding -2 times the first equation to the second equation produces a new second equation.

$$\begin{cases} x - 3y + z = 1 \\ 5y - 4z = 0 \\ 5y - 4z = -2 \end{cases}$$

> Adding -1 times the first equation to the third equation produces a new third equation.

$$\begin{cases} x - 3y + z = 1 \\ 5y - 4z = 0 \\ 0 = -2 \end{cases}$$

> Adding -1 times the second equation to the third equation produces a new third equation.

Because $0 = -2$ is a false statement, you can conclude that this system is inconsistent and so has no solution. Moreover, because this system is equivalent to the original system, you can conclude that the original system also has no solution.

✓**CHECKPOINT**   Now try Exercise 27.

As with a system of linear equations in two variables, the number of solutions of a system of linear equations in more than two variables must fall into one of three categories.

The Number of Solutions of a Linear System

For a system of linear equations, exactly one of the following is true.

1. There is exactly one solution.

2. There are infinitely many solutions.

3. There is no solution.

A system of linear equations is called *consistent* when it has at least one solution. A consistent system with exactly one solution is **independent.** A consistent system with infinitely many solutions is **dependent.** A system of linear equations is called *inconsistent* when it has no solution.

Example

Solve the

$\begin{cases} & x \\ & \\ & -\end{cases}$

Solutio

$\begin{cases} x \\ \\ \end{cases}$

$\begin{cases} x \\ \\ \end{cases}$

This resu
no addit

$\begin{cases} x \\ \end{cases}$

In the la
the prev

$x =$

Finally,
all of th

$x =$

So, eve

$(2a$

is a solu

 CHECK

In
instance

$(b,$

This de

Example 6 Partial Fraction Decomposition: Distinct Linear Factors

Write the partial fraction decomposition of

$$\frac{x + 7}{x^2 - x - 6}.$$

Solution

The expression is proper, so factor the denominator. Because

$$x^2 - x - 6 = (x - 3)(x + 2)$$

you should include one partial fraction with a constant numerator for each linear factor of the denominator and write

$$\frac{x + 7}{x^2 - x - 6} = \frac{A}{x - 3} + \frac{B}{x + 2}.$$

Multiplying each side of this equation by the least common denominator

$$(x - 3)(x + 2)$$

leads to the basic equation

$$\begin{aligned} x + 7 &= A(x + 2) + B(x - 3) & \text{Basic equation} \\ &= Ax + 2A + Bx - 3B & \text{Distributive Property} \\ &= (A + B)x + 2A - 3B. & \text{Write in polynomial form.} \end{aligned}$$

Because two polynomials are equal if and only if the coefficients of like terms are equal, you can equate the coefficients of like terms to opposite sides of the equation.

$$x + 7 = (A + B)x + (2A - 3B) \qquad \text{Equate coefficients of like terms.}$$

You can now write the following system of linear equations.

$$\begin{cases} A + B = 1 & \text{Equation 1} \\ 2A - 3B = 7 & \text{Equation 2} \end{cases}$$

You can solve the system of linear equations as follows.

$$\begin{array}{lll} A + B = 1 & \Rightarrow & 3A + 3B = 3 & \text{Multiply Equation 1 by 3.} \\ \underline{2A - 3B = 7} & \Rightarrow & \underline{2A - 3B = 7} & \text{Write Equation 2.} \\ & & 5A \quad\;\; = 10 & \text{Add equations.} \end{array}$$

From this equation, you can see that

$$A = 2.$$

By back-substituting this value of A into Equation 1, you can solve for B as follows.

$$\begin{aligned} A + B &= 1 & \text{Write Equation 1.} \\ 2 + B &= 1 & \text{Substitute 2 for} \\ B &= -1 & \text{Solve for } B. \end{aligned}$$

So, the partial fraction decomposition is

$$\frac{x + 7}{x^2 - x - 6} = \frac{2}{x - 3} - \frac{1}{x + 2}.$$

Check this result by combining the two partial fractions on the right side of the equation, or by using a graphing utility.

 CHECKPOINT   Now try Exercise 61.

Technology Tip

 You can graphically check the decomposition found in Example 6. To do this, use a graphing utility to graph

$$y_1 = \frac{x + 7}{x^2 - x - 6} \quad \text{and}$$

$$y_2 = \frac{2}{x - 3} - \frac{1}{x + 2}$$

in the same viewing window. The graphs should be identical.

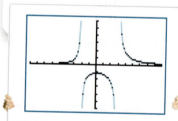

The next example shows how to find the partial fraction decomposition for a rational function whose denominator has a repeated linear factor.

Example 7 Partial Fraction Decomposition: Repeated Linear Factors

Write the partial fraction decomposition of

$$\frac{5x^2 + 20x + 6}{x^3 + 2x^2 + x}.$$

Solution

The expression is proper, so factor the denominator. Because the denominator factors as

$$x^3 + 2x^2 + x = x(x^2 + 2x + 1)$$

$$= x(x + 1)^2$$

you should include one partial fraction with a constant numerator for each power of x and $(x + 1)$ and write

$$\frac{5x^2 + 20x + 6}{x^3 + 2x^2 + x} = \frac{A}{x} + \frac{B}{x + 1} + \frac{C}{(x + 1)^2}.$$

Multiplying by the LCD

$$x(x + 1)^2$$

leads to the basic equation

$$5x^2 + 20x + 6 = A(x + 1)^2 + Bx(x + 1) + Cx \qquad \text{Basic equation}$$

$$= Ax^2 + 2Ax + A + Bx^2 + Bx + Cx \qquad \text{Expand.}$$

$$= (A + B)x^2 + (2A + B + C)x + A. \qquad \text{Polynomial form}$$

Equating coefficients of like terms on opposite sides of the equation

$$5x^2 + 20x + 6 = (A + B)x^2 + (2A + B + C)x + A$$

produces the following system of linear equations.

$$\begin{cases} A + B & = 5 & \text{Equation 1} \\ 2A + B + C = 20 & \text{Equation 2} \\ A & = 6 & \text{Equation 3} \end{cases}$$

Substituting 6 for A in Equation 1 yields

$$6 + B = 5$$

$$B = -1.$$

Substituting 6 for A and -1 for B in Equation 2 yields

$$2(6) + (-1) + C = 20$$

$$C = 9.$$

So, the partial fraction decomposition is

$$\frac{5x^2 + 20x + 6}{x^3 + 2x^2 + x} = \frac{6}{x} - \frac{1}{x + 1} + \frac{9}{(x + 1)^2}.$$

Check this result by combining the three partial fractions on the right side of the equation, or by using a graphing utility.

✓CHECKPOINT   Now try Exercise 65.

Explore the Concept

Partial fraction decomposition is practical only for rational functions whose denominators factor "nicely." For example, the factorization of the expression $x^2 - x - 5$ is

$$\left(x - \frac{1 - \sqrt{21}}{2}\right)\left(x - \frac{1 + \sqrt{21}}{2}\right).$$

Write the basic equation and try to complete the decomposition for

$$\frac{x + 7}{x^2 - x - 5}.$$

What problems do you encounter?

Applications

Example 8   Vertical Motion

The height at time t of an object that is moving in a (vertical) line with constant acceleration a is given by the *position equation*

$$s = \tfrac{1}{2}at^2 + v_0t + s_0.$$

The height s is measured in feet, the acceleration a is measured in feet per second squared, t is measured in seconds, v_0 is the initial velocity (in feet per second) at $t = 0$, and s_0 is the initial height (in feet). Find the values of a, v_0, and s_0 when

$$s = 52 \text{ at } t = 1, \quad s = 52 \text{ at } t = 2, \quad \text{and } s = 20 \text{ at } t = 3$$

and interpret the result. (See Figure 8.21.)

Solution

You can obtain three linear equations in a, v_0, and s_0 as follows.

When $t = 1$: $\tfrac{1}{2}a(1)^2 + v_0(1) + s_0 = 52$ ⟹ $a + 2v_0 + 2s_0 = 104$

When $t = 2$: $\tfrac{1}{2}a(2)^2 + v_0(2) + s_0 = 52$ ⟹ $2a + 2v_0 + s_0 = 52$

When $t = 3$: $\tfrac{1}{2}a(3)^2 + v_0(3) + s_0 = 20$ ⟹ $9a + 6v_0 + 2s_0 = 40$

Solving this system yields $a = -32$, $v_0 = 48$, and $s_0 = 20$. This solution results in a position equation of

$$s = -16t^2 + 48t + 20$$

and implies that the object was thrown upward at a velocity of 48 feet per second from a height of 20 feet.

✔CHECKPOINT   Now try Exercise 73.

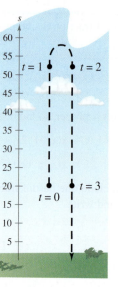

Figure 8.21

Example 9   Data Analysis: Curve-Fitting

Find a quadratic equation

$$y = ax^2 + bx + c$$

whose graph passes through the points $(-1, 3)$, $(1, 1)$, and $(2, 6)$.

Solution

Because the graph of $y = ax^2 + bx + c$ passes through the points $(-1, 3)$, $(1, 1)$, and $(2, 6)$, you can write the following.

When $x = -1$, $y = 3$: $a(-1)^2 + b(-1) + c = 3$

When $x = 1$, $y = 1$: $a(1)^2 + b(1) + c = 1$

When $x = 2$, $y = 6$: $a(2)^2 + b(2) + c = 6$

This produces the following system of linear equations.

$$\begin{cases} a - b + c = 3 & \text{Equation 1} \\ a + b + c = 1 & \text{Equation 2} \\ 4a + 2b + c = 6 & \text{Equation 3} \end{cases}$$

The solution of this system is $a = 2$, $b = -1$, and $c = 0$. So, the equation of the parabola is $y = 2x^2 - x$, and its graph is shown in Figure 8.22.

✔CHECKPOINT   Now try Exercise 77.

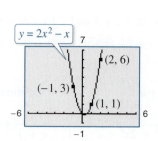

Figure 8.22

8.3   Exercises

See www.CalcChat.com for worked-out solutions to odd-numbered exercises.
For instructions on how to use a graphing utility, see Appendix A.

Vocabulary and Concept Check

In Exercises 1–6, fill in the blank.

1. A system of equations that is in _____ form has a "stair-step" pattern with leading coefficients of 1.

2. A solution of a system of three linear equations in three unknowns can be written as an _____ , which has the form (x, y, z).

3. The process used to write a system of equations in row-echelon form is called _____ elimination.

4. A system of equations is called _____ when the number of equations differs from the number of variables in the system.

5. Solutions of equations in three variables can be pictured using a _____ coordinate system.

6. The process of writing a rational expression as the sum of two or more simpler rational expressions is called _____ .

7. Is a consistent system with exactly one solution *independent* or *dependent*?

8. Is a consistent system with infinitely many solutions *independent* or *dependent*?

Procedures and Problem Solving

Checking Solutions   In Exercises 9–12, determine whether each ordered triple is a solution of the system of equations.

9. $\begin{cases} 3x - y + z = 1 \\ 2x \quad\;\; - 3z = -14 \\ \quad\;\; 5y + 2z = 8 \end{cases}$
   (a) $(3, 5, -3)$ (b) $(-1, 0, 4)$
   (c) $(0, -1, 3)$ (d) $(1, 0, 4)$

10. $\begin{cases} 3x + 4y - z = 17 \\ 5x - y + 2z = -2 \\ 2x - 3y + 7z = -21 \end{cases}$
    (a) $(1, 5, 6)$ (b) $(-2, -4, 2)$
    (c) $(1, 3, -2)$ (d) $(0, 7, 0)$

11. $\begin{cases} 4x + y - z = 0 \\ -8x - 6y + z = -\frac{7}{4} \\ 3x - y \quad\;\; = -\frac{9}{4} \end{cases}$
    (a) $(0, 1, 1)$ (b) $\left(-\frac{3}{2}, \frac{5}{4}, -\frac{5}{4}\right)$
    (c) $\left(-\frac{1}{2}, \frac{3}{4}, -\frac{5}{4}\right)$ (d) $\left(-\frac{1}{2}, \frac{1}{6}, -\frac{3}{4}\right)$

12. $\begin{cases} -4x - y - 8z = -6 \\ \quad\;\; y + z = 0 \\ 4x - 7y \quad\;\; = 6 \end{cases}$
    (a) $(-2, -2, 2)$ (b) $\left(-\frac{33}{2}, -10, 10\right)$
    (c) $\left(\frac{1}{8}, -\frac{1}{2}, \frac{1}{2}\right)$ (d) $\left(-\frac{11}{2}, -4, 4\right)$

Using Back-Substitution   In Exercises 13–18, use back-substitution to solve the system of linear equations.

✓ 13. $\begin{cases} 2x - y + 5z = 16 \\ \quad\;\; y + 2z = 2 \\ \quad\quad\;\; z = 2 \end{cases}$

14. $\begin{cases} 4x - 3y - 2z = 21 \\ \quad\;\; 6y - 5z = -8 \\ \quad\quad\;\; z = -2 \end{cases}$

15. $\begin{cases} 2x + y - 3z = 10 \\ \quad\;\; y + z = 12 \\ \quad\quad\;\; z = 2 \end{cases}$

16. $\begin{cases} x - y + 2z = 22 \\ \quad 3y - 8z = -9 \\ \quad\quad\;\; z = -3 \end{cases}$

17. $\begin{cases} 4x - 2y + z = 8 \\ \quad - y + z = 4 \\ \quad\quad\;\; z = 11 \end{cases}$

18. $\begin{cases} 5x \quad - 8z = 22 \\ 3y - 5z = 10 \\ \quad\quad\;\; z = -4 \end{cases}$

Performing Row Operations   In Exercises 19 and 20, perform the row operation and write the equivalent system. What did the operation accomplish?

19. Add Equation 1 to Equation 2.
    $\begin{cases} x - 2y + 3z = 5 & \text{Equation 1} \\ -x + 3y - 5z = 4 & \text{Equation 2} \\ 2x \quad\;\; - 3z = 0 & \text{Equation 3} \end{cases}$

20. Add -2 times Equation 1 to Equation 3.
    $\begin{cases} x - 2y + 3z = 5 & \text{Equation 1} \\ -x + 3y - 5z = 4 & \text{Equation 2} \\ 2x \quad\;\; - 3z = 0 & \text{Equation 3} \end{cases}$

Solving a System of Linear Equations In Exercises 21–42, solve the system of linear equations and check any solution algebraically.

✓ 21. $\begin{cases} x + y + z = 6 \\ 2x - y + z = 3 \\ 3x \qquad - z = 0 \end{cases}$ 22. $\begin{cases} x + y + z = 3 \\ x - 2y + 4z = 5 \\ 3y + 4z = 5 \end{cases}$

23. $\begin{cases} 2x \qquad + 2z = 2 \\ 5x + 3y \qquad = 4 \\ 3y - 4z = 4 \end{cases}$ 24. $\begin{cases} 2x + 4y + z = 1 \\ x - 2y - 3z = 2 \\ x + y - z = -1 \end{cases}$

25. $\begin{cases} 4x + y - 3z = 11 \\ 2x - 3y + 2z = 9 \\ x + y + z = -3 \end{cases}$ 26. $\begin{cases} 2x + 4y + z = -4 \\ 2x - 4y + 6z = 13 \\ 4x - 2y + z = 6 \end{cases}$

✓ 27. $\begin{cases} 3x - 2y + 4z = 1 \\ x + y - 2z = 3 \\ 2x - 3y + 6z = 8 \end{cases}$ 28. $\begin{cases} 5x - 3y + 2z = 3 \\ 2x + 4y - z = 7 \\ x - 11y + 4z = 3 \end{cases}$

29. $\begin{cases} 3x + 3y + 5z = 1 \\ 3x + 5y + 9z = 0 \\ 5x + 9y + 17z = 0 \end{cases}$ 30. $\begin{cases} 2x + y + 3z = 1 \\ 2x + 6y + 8z = 3 \\ 6x + 8y + 18z = 5 \end{cases}$

✓ 31. $\begin{cases} 3x - 3y + 6z = 6 \\ x + 2y - z = 5 \\ 5x - 8y + 13z = 7 \end{cases}$ 32. $\begin{cases} x + 4z = 13 \\ 4x - 2y + z = 7 \\ 2x - 2y - 7z = -19 \end{cases}$

33. $\begin{cases} x - 2y + 3z = 4 \\ 3x - y + 2z = 0 \\ x + 3y - 4z = -2 \end{cases}$ 34. $\begin{cases} -x + 3y + z = 4 \\ 4x - 2y - 5z = -7 \\ 2x + 4y - 3z = 12 \end{cases}$

35. $\begin{cases} x + 4z = 1 \\ x + y + 10z = 10 \\ 2x - y + 2z = -5 \end{cases}$ 36. $\begin{cases} 3x - 2y - 6z = -4 \\ -3x + 2y + 6z = 1 \\ x - y - 5z = -3 \end{cases}$

37. $\begin{cases} x + 2y + z = 1 \\ x - 2y + 3z = -3 \\ 2x + y + z = -1 \end{cases}$ 38. $\begin{cases} x - 2y + z = 2 \\ 2x + 2y - 3z = -4 \\ 5x + z = 1 \end{cases}$

✓ 39. $\begin{cases} x - 2y + 5z = 2 \\ 4x - z = 0 \end{cases}$ 40. $\begin{cases} 12x + 5y + z = 0 \\ 23x + 4y - z = 0 \end{cases}$

41. $\begin{cases} 2x - 3y + z = -2 \\ -4x + 9y = 7 \end{cases}$ 42. $\begin{cases} 10x - 3y + 2z = 0 \\ 19x - 5y - z = 0 \end{cases}$

Exploration In Exercises 43–46, find a system of linear equations that has the given solution. (There are many correct answers.)

43. $(3, -4, 2)$ 44. $(-5, -2, 1)$
45. $\left(-6, -\frac{1}{2}, -\frac{7}{4}\right)$ 46. $\left(-\frac{3}{2}, 4, -7\right)$

Sketching a Plane In Exercises 47–50, sketch the plane represented by the linear equation. Then list four points that lie in the plane.

47. $2x + 3y + 4z = 12$ 48. $x + y + z = 6$
49. $2x + y + z = 4$ 50. $x + 2y + 2z = 6$

Writing the Partial Fraction Decomposition In Exercises 51–56, write the form of the partial fraction decomposition of the rational expression. Do not solve for the constants.

51. $\dfrac{7}{x^2 - 14x}$ 52. $\dfrac{x - 2}{x^2 + 4x + 3}$

53. $\dfrac{12}{x^3 - 10x^2}$ 54. $\dfrac{x^2 - 3x + 2}{4x^3 + 11x^2}$

55. $\dfrac{4x^2 + 3}{(x - 5)^3}$ 56. $\dfrac{6x + 5}{(x + 2)^4}$

Partial Fraction Decomposition In Exercises 57–70, write the partial fraction decomposition for the rational expression. Check your result algebraically by combining fractions, and check your result graphically by using a graphing utility to graph the rational expression and the partial fractions in the same viewing window.

57. $\dfrac{1}{x^2 - 1}$ 58. $\dfrac{1}{4x^2 - 9}$

59. $\dfrac{1}{x^2 + x}$ 60. $\dfrac{3}{x^2 - 3x}$

✓ 61. $\dfrac{5 - x}{2x^2 + x - 1}$ 62. $\dfrac{x - 2}{x^2 + 4x + 3}$

63. $\dfrac{x^2 + 12x + 12}{x^3 - 4x}$ 64. $\dfrac{x^2 + 12x - 9}{x^3 - 9x}$

✓ 65. $\dfrac{4x^2 + 2x - 1}{x^2(x + 1)}$ 66. $\dfrac{2x - 3}{(x - 1)^2}$

67. $\dfrac{2x^3 - x^2 + x + 5}{x^2 + 3x + 2}$ 68. $\dfrac{x^3 + 2x^2 - x + 1}{x^2 + 3x - 4}$

69. $\dfrac{x^4}{(x - 1)^3}$ 70. $\dfrac{4x^4}{(2x - 1)^3}$

Writing the Partial Fraction Decomposition In Exercises 71 and 72, write the partial fraction decomposition for the rational function. Identify the graph of the rational function and the graph of each term of its decomposition. State any relationship between the vertical asymptotes of the rational function and the vertical asymptotes of the terms of the decomposition.

71. $y = \dfrac{x - 12}{x(x - 4)}$ 72. $y = \dfrac{2(4x - 3)}{x^2 - 9}$

Vertical Motion   In Exercises 73–76, an object moving vertically is at the given heights at the specified times. Find the position equation $s = \frac{1}{2}at^2 + v_0 t + s_0$ for the object.

✓ **73.** At $t = 1$ second, $s = 128$ feet.
     At $t = 2$ seconds, $s = 80$ feet.
     At $t = 3$ seconds, $s = 0$ feet.

74. At $t = 1$ second, $s = 32$ feet.
     At $t = 2$ seconds, $s = 32$ feet.
     At $t = 3$ seconds, $s = 0$ feet.

75. At $t = 1$ second, $s = 352$ feet.
     At $t = 2$ seconds, $s = 272$ feet.
     At $t = 3$ seconds, $s = 160$ feet.

76. At $t = 1$ second, $s = 132$ feet.
     At $t = 2$ seconds, $s = 100$ feet.
     At $t = 3$ seconds, $s = 36$ feet.

Data Analysis: Curve-Fitting   In Exercises 77–80, find the equation of the parabola

$$y = ax^2 + bx + c$$

that passes through the points. To verify your result, use a graphing utility to plot the points and graph the parabola.

✓ **77.** $(0, 0), (2, -2), (4, 0)$ **78.** $(0, 3), (1, 4), (2, 3)$
79. $(2, 0), (3, -1), (4, 0)$
80. $(-2, -3), (-1, 0), \left(\frac{1}{2}, -3\right)$

Finding the Equation of a Circle   In Exercises 81–84, find the equation of the circle

$$x^2 + y^2 + Dx + Ey + F = 0$$

that passes through the points. To verify your result, use a graphing utility to plot the points and graph the circle.

81. $(0, 0), (5, 5), (10, 0)$ **82.** $(0, 0), (0, 6), (3, 3)$
83. $(-3, -1), (2, 4), (-6, 8)$
84. $(0, 0), (0, -2), (3, 0)$

85. Finance   A small corporation borrowed $775,000 to expand its software line. Some of the money was borrowed at 8%, some at 9%, and some at 10%. How much was borrowed at each rate given that the annual interest was $67,500 and the amount borrowed at 8% was four times the amount borrowed at 10%?

86. Finance   A small corporation borrowed $800,000 to expand its line of toys. Some of the money was borrowed at 8%, some at 9%, and some at 10%. How much was borrowed at each rate given that the annual interest was $67,000 and the amount borrowed at 8% was five times the amount borrowed at 10%?

Investment Portfolio   In Exercises 87 and 88, consider an investor with a portfolio totaling $500,000 that is invested in certificates of deposit, municipal bonds, blue-chip stocks, and growth or speculative stocks. How much is invested in each type of investment?

87. The certificates of deposit pay 3% annually, and the municipal bonds pay 5% annually. Over a five-year period, the investor expects the blue-chip stocks to return 8% annually and the growth stocks to return 10% annually. The investor wants a combined annual return of 5% and also wants to have only one-fourth of the portfolio invested in stocks.

88. The certificates of deposit pay 2% annually, and the municipal bonds pay 4% annually. Over a five-year period, the investor expects the blue-chip stocks to return 10% annually and the growth stocks to return 14% annually. The investor wants a combined annual return of 6% and also wants to have only one-fourth of the portfolio invested in stocks.

89. Physical Education   In the 2010 Women's NCAA Championship basketball game, the University of Connecticut defeated Stanford University by a score of 53 to 47. Connecticut won by scoring a combination of two-point baskets, three-point baskets, and one-point free throws. The number of two-point baskets was four more than the number of free throws. The number of free throws was three more than the number of three-point baskets. What combination of scoring accounted for Connecticut's 53 points?   (Source: NCAA)

90. *Why you should learn it*   *(p. 629)*   The Augusta National Golf Club in Augusta, Georgia is an 18-hole course that consists of par-3 holes, par-4 holes, and par-5 holes. A golfer who shoots par has a total of 72 strokes for the entire course. There are two more par-4 holes than twice the number of par-5 holes, and the number of par-3 holes is equal to the number of par-5 holes. Find the numbers of par-3, par-4, and par-5 holes on the course.   (Source: Augusta National, Inc.)

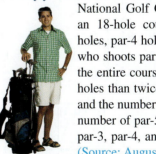

91. Electrical Engineering   When Kirchhoff's Laws are applied to the electrical network in the figure, the currents I_1, I_2, and I_3 are the solution of the system

$$\begin{cases} I_1 - I_2 + I_3 = 0 \\ 3I_1 + 2I_2 \quad\quad = 7. \\ \quad\quad 2I_2 + 4I_3 = 8 \end{cases}$$

Find the currents.

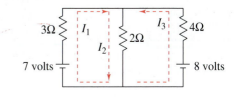

92. Physics A system of pulleys is loaded with 128-pound and 32-pound weights (see figure). The tensions t_1 and t_2 in the ropes and the acceleration a of the 32-pound weight are modeled by the following system, where t_1 and t_2 are measured in pounds and a is in feet per second squared. Solve the system.

$$\begin{cases} t_1 - 2t_2 & = 0 \\ t_1 & - 2a = 128 \\ t_2 + & a = 32 \end{cases}$$

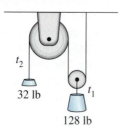

32 lb

128 lb

t_2

t_1

Fitting a Parabola **To find the least squares regression parabola $y = ax^2 + bx + c$ for a set of points**

$$(x_1, y_1), (x_2, y_2), \ldots, (x_n, y_n)$$

you can solve the following system of linear equations for a, b, and c.

$$\begin{cases} nc + \left(\sum_{i=1}^{n} x_i\right)b + \left(\sum_{i=1}^{n} x_i^2\right)a = \sum_{i=1}^{n} y_i \\ \left(\sum_{i=1}^{n} x_i\right)c + \left(\sum_{i=1}^{n} x_i^2\right)b + \left(\sum_{i=1}^{n} x_i^3\right)a = \sum_{i=1}^{n} x_i y_i \\ \left(\sum_{i=1}^{n} x_i^2\right)c + \left(\sum_{i=1}^{n} x_i^3\right)b + \left(\sum_{i=1}^{n} x_i^4\right)a = \sum_{i=1}^{n} x_i^2 y_i \end{cases}$$

In Exercises 93–96, the sums have been evaluated. Solve the given system for a and b to find the least squares regression parabola for the points. Use a graphing utility to confirm the result.

93. $\begin{cases} 4c & + 40a = 19 \\ 40b & = -12 \\ 40c & + 544a = 160 \end{cases}$

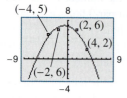

94. $\begin{cases} 5c & + 10a = 8 \\ 10b & = 12 \\ 10c & + 34a = 22 \end{cases}$

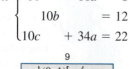

95. $\begin{cases} 4c + 9b + 29a = 20 \\ 9c + 29b + 99a = 70 \\ 29c + 99b + 353a = 254 \end{cases}$

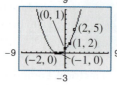

96. $\begin{cases} 4c + 6b + 14a = 25 \\ 6c + 14b + 36a = 21 \\ 14c + 36b + 98a = 33 \end{cases}$

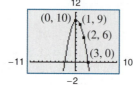

97. MODELING DATA

During the testing of a new automobile braking system, the speeds x (in miles per hour) and the stopping distances y (in feet) were recorded in the table.

Speed, x	Stopping distance, y
30	55
40	105
50	188

(a) Use the data to create a system of linear equations. Then find the least squares regression parabola for the data by solving the system.

(b) Use a graphing utility to graph the parabola and the data in the same viewing window.

(c) Use the model to estimate the stopping distance for a speed of 70 miles per hour.

98. MODELING DATA

A wildlife management team studied the reproduction rates of deer in three five-acre tracts of a wildlife preserve. In each tract, the number of females x and the percent of females y that had offspring the following year were recorded. The results are shown in the table.

Number, x	Percent, y
120	68
140	55
160	30

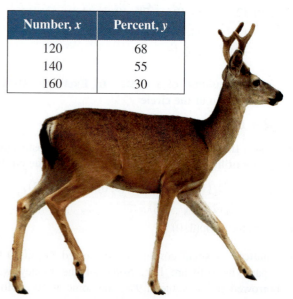

(a) Use the data to create a system of linear equations. Then find the least squares regression parabola for the data by solving the system.

(b) Use a graphing utility to graph the parabola and the data in the same viewing window.

(c) Use the model to predict the percent of females that had offspring when there were 170 females.

99. Thermodynamics The magnitude of the range R of exhaust temperatures (in degrees Fahrenheit) in an experimental diesel engine is approximated by the model

$$R = \frac{2000(4 - 3x)}{(11 - 7x)(7 - 4x)}, \quad 0 \le x \le 1$$

where x is the relative load (in foot-pounds).

(a) Write the partial fraction decomposition of the rational function.

(b) The decomposition in part (a) is the difference of two fractions. The absolute values of the terms give the expected maximum and minimum temperatures of the exhaust gases. Use a graphing utility to graph each term.

100. Environment The predicted cost C (in thousands of dollars) for a company to remove $p\%$ of a chemical from its waste water is given by the model

$$C = \frac{120p}{10{,}000 - p^2}, \quad 0 \le p < 100.$$

Write the partial fraction decomposition of the rational function. Verify your result by using the *table* feature of a graphing utility to create a table comparing the original function with the partial fractions.

Conclusions

True or False? In Exercises 101 and 102, determine whether the statement is true or false. Justify your answer.

101. The system

$$\begin{cases} x + 4y - 5z = 8 \\ \quad\;\; 2y + \;\; z = 5 \\ \quad\qquad\quad\; z = 1 \end{cases}$$

is in row-echelon form.

102. If a system of three linear equations is inconsistent, then its graph has no points common to all three equations.

103. Error Analysis You are tutoring a student in algebra. In trying to find a partial fraction decomposition, your student writes the following.

$$\frac{x^2 + 1}{x(x - 1)} = \frac{A}{x} + \frac{B}{x - 1}$$

$$x^2 + 1 = A(x - 1) + Bx \qquad \text{\color{red}Basic equation}$$

$$x^2 + 1 = (A + B)x - A$$

Your student then forms the following system of linear equations.

$$\begin{cases} A + B = 0 \\ -A \qquad = 1 \end{cases}$$

Solve the system and check the partial fraction decomposition it yields. Has your student worked the problem correctly? If not, what went wrong?

104. CAPSTONE Find values of a, b, and c (if possible) such that the system of linear equations has (a) a unique solution, (b) no solution, and (c) an infinite number of solutions.

$$\begin{cases} x + y \qquad\;\; = 2 \\ \quad\; y + z = 2 \\ x \quad\;\; + z = 2 \\ ax + by + cz = 0 \end{cases}$$

105. Think About It Are the two systems of equations equivalent? Give reasons for your answer.

$$\begin{cases} x + 3y - \;\; z = 6 \\ 2x - \;\; y + 2z = 1 \\ 3x + 2y - \;\; z = 2 \end{cases} \qquad \begin{cases} x + 3y - \;\; z = \;\;\;\; 6 \\ \quad\quad -7y + 4z = \;\;\;\; 1 \\ \quad\quad -7y - 4z = -16 \end{cases}$$

106. Writing When using Gaussian elimination to solve a system of linear equations, explain how you can recognize that the system has no solution. Give an example that illustrates your answer.

ƒ Lagrange Multiplier In Exercises 107 and 108, find values of x, y, and λ that satisfy the system. These systems arise in certain optimization problems in calculus. (λ is called a *Lagrange multiplier*.)

107. $$\begin{cases} y + \lambda = 0 \\ x + \lambda = 0 \\ x + y - 10 = 0 \end{cases}$$

108. $$\begin{cases} 2x + \lambda = 0 \\ 2y + \lambda = 0 \\ x + y - 4 = 0 \end{cases}$$

Cumulative Mixed Review

Finding Zeros In Exercises 109–112, (a) determine the real zeros of f and (b) sketch the graph of f.

109. $f(x) = x^3 + x^2 - 12x$

110. $f(x) = -8x^4 + 32x^2$

111. $f(x) = 2x^3 + 5x^2 - 21x - 36$

112. $f(x) = 6x^3 - 29x^2 - 6x + 5$

Solving a Trigonometric Equation In Exercises 113 and 114, solve the equation.

113. $4\sqrt{3} \tan \theta - 3 = 1$

114. $6 \cos x - 2 = 1$

115. *Make a Decision* To work an extended application analyzing the earnings per share for Wal-Mart Stores, Inc. from 1988 through 2008, visit this textbook's *Companion Website*. (Data Source: Wal-Mart Stores, Inc.)

8.4   Matrices and Systems of Equations

Matrices

In this section, you will study a streamlined technique for solving systems of linear equations. This technique involves the use of a rectangular array of real numbers called a **matrix.** The plural of matrix is *matrices.*

> **Definition of Matrix**
>
> If m and n are positive integers, then an $m \times n$ (read "m by n") matrix is a rectangular array
>
> $$\begin{array}{c} \text{Column 1} \quad \text{Column 2} \quad \text{Column 3} \quad \ldots \quad \text{Column } n \\ \begin{array}{c} \text{Row 1} \\ \text{Row 2} \\ \text{Row 3} \\ \vdots \\ \text{Row } m \end{array} \begin{bmatrix} a_{11} & a_{12} & a_{13} & \cdots & a_{1n} \\ a_{21} & a_{22} & a_{23} & \cdots & a_{2n} \\ a_{31} & a_{32} & a_{33} & \cdots & a_{3n} \\ \vdots & \vdots & \vdots & \vdots & \vdots \\ a_{m1} & a_{m2} & a_{m3} & \cdots & a_{mn} \end{bmatrix} \end{array}$$
>
> in which each **entry** a_{ij} of the matrix is a real number. An $m \times n$ matrix has m rows and n columns.

The entry in the ith row and jth column is denoted by the *double subscript* notation a_{ij}. For instance, the entry a_{23} is the entry in the second row and third column. A matrix having m rows and n columns is said to be of **dimension** $m \times n$. If $m = n$, then the matrix is **square** of dimension $m \times m$ (or $n \times n$). For a square matrix, the entries

$$a_{11}, a_{22}, a_{33}, \ldots$$

are the **main diagonal** entries.

Example 1   Dimension of a Matrix

Determine the dimension of each matrix.

a. $\begin{bmatrix} 2 \end{bmatrix}$ **b.** $\begin{bmatrix} 1 & -3 & 0 & \frac{1}{2} \end{bmatrix}$ **c.** $\begin{bmatrix} 0 & 0 \\ 0 & 0 \end{bmatrix}$

d. $\begin{bmatrix} 5 & 0 \\ 2 & -2 \\ -7 & 4 \end{bmatrix}$ **e.** $\begin{bmatrix} -2 \\ 0 \\ 1 \end{bmatrix}$

Pharmacist

Solution

a. This matrix has *one* row and *one* column. The dimension of the matrix is 1×1.

b. This matrix has *one* row and *four* columns. The dimension of the matrix is 1×4.

c. This matrix has *two* rows and *two* columns. The dimension of the matrix is 2×2.

d. This matrix has *three* rows and *two* columns. The dimension of the matrix is 3×2.

e. This matrix has *three* rows and *one* column. The dimension of the matrix is 3×1.

✓**CHECKPOINT**   Now try Exercise 9.

A matrix that has only one row [such as the matrix in Example 1(b)] is called a **row matrix,** and a matrix that has only one column [such as the matrix in Example 1(e)] is called a **column matrix.**

A matrix derived from a system of linear equations (each written in standard form with the constant term on the right) is the **augmented matrix** of the system. Moreover, the matrix derived from the coefficients of the system (but not including the constant terms) is the **coefficient matrix** of the system. The matrix derived from the constant terms of the system is the **constant matrix** of the system.

System:
$$\begin{cases} x - 4y + 3z = 5 \\ -x + 3y - z = -3 \\ 2x - 4z = 6 \end{cases}$$

Augmented Matrix:
$$\left[\begin{array}{ccc:c} 1 & -4 & 3 & 5 \\ -1 & 3 & -1 & -3 \\ 2 & 0 & -4 & 6 \end{array}\right]$$

Coefficient Matrix:
$$\begin{bmatrix} 1 & -4 & 3 \\ -1 & 3 & -1 \\ 2 & 0 & -4 \end{bmatrix}$$

Constant Matrix:
$$\begin{bmatrix} 5 \\ -3 \\ 6 \end{bmatrix}$$

Note the use of 0 for the missing coefficient of the y-variable in the third equation, and also note the fourth column (of constant terms) in the augmented matrix. The optional dotted line in the augmented matrix helps to separate the coefficients of the linear system from the constant terms.

When forming either the coefficient matrix or the augmented matrix of a system, you should begin by vertically aligning the variables in the equations and using 0's for any missing coefficients of variables.

Example 2   Writing an Augmented Matrix

Write the augmented matrix for the system of linear equations.

$$\begin{cases} x + 3y = 9 \\ -y + 4z = -2 \\ x - 5z = 0 \end{cases}$$

What is the dimension of the augmented matrix?

Solution

Begin by writing the linear system and aligning the variables.

$$\begin{cases} x + 3y = 9 \\ -y + 4z = -2 \\ x - 5z = 0 \end{cases}$$

Next, use the coefficients and constant terms as the matrix entries. Include zeros for the coefficients of the missing variables.

$$\begin{matrix} R_1 \\ R_2 \\ R_3 \end{matrix} \left[\begin{array}{ccc:c} 1 & 3 & 0 & 9 \\ 0 & -1 & 4 & -2 \\ 1 & 0 & -5 & 0 \end{array}\right]$$

The augmented matrix has three rows and four columns, so it is a 3×4 matrix. The notation R_n is used to designate each row in the matrix. For instance, Row 1 is represented by R_1.

✓CHECKPOINT   Now try Exercise 15.

Elementary Row Operations

In Section 8.3, you studied three operations that can be used on a system of linear equations to produce an equivalent system.

1. Interchange two equations.

2. Multiply one of the equations by a nonzero constant.

3. Add a multiple of one equation to another equation.

In matrix terminology, these three operations correspond to **elementary row operations.** An elementary row operation on an augmented matrix of a given system of linear equations produces a new augmented matrix corresponding to a new (but equivalent) system of linear equations. Two matrices are **row-equivalent** when one can be obtained from the other by a sequence of elementary row operations.

> ### Elementary Row Operations for Matrices
>
> **1.** Interchange two rows.
>
> **2.** Multiply one of the rows by a nonzero constant.
>
> **3.** Add a multiple of one row to another row.

Although elementary row operations are simple to perform, they involve a lot of arithmetic. Because it is easy to make a mistake, you should get in the habit of noting the elementary row operations performed in each step so that you can go back and check your work.

Example 3 demonstrates the elementary row operations described above.

Example 3 Elementary Row Operations

a. Interchange the first and second rows of the original matrix.

Original Matrix

$$\begin{bmatrix} 0 & 1 & 3 & 4 \\ -1 & 2 & 0 & 3 \\ 2 & -3 & 4 & 1 \end{bmatrix}$$

New Row-Equivalent Matrix

$$\begin{matrix} R_2 \\ R_1 \end{matrix} \begin{bmatrix} -1 & 2 & 0 & 3 \\ 0 & 1 & 3 & 4 \\ 2 & -3 & 4 & 1 \end{bmatrix}$$

b. Multiply the first row of the original matrix by $\frac{1}{2}$.

Original Matrix

$$\begin{bmatrix} 2 & -4 & 6 & -2 \\ 1 & 3 & -3 & 0 \\ 5 & -2 & 1 & 2 \end{bmatrix}$$

New Row-Equivalent Matrix

$$\tfrac{1}{2}R_1 \rightarrow \begin{bmatrix} 1 & -2 & 3 & -1 \\ 1 & 3 & -3 & 0 \\ 5 & -2 & 1 & 2 \end{bmatrix}$$

c. Add -2 times the first row of the original matrix to the third row.

Original Matrix

$$\begin{bmatrix} 1 & 2 & -4 & 3 \\ 0 & 3 & -2 & -1 \\ 2 & 1 & 5 & -2 \end{bmatrix}$$

New Row-Equivalent Matrix

$$\begin{matrix} \\ \\ -2R_1 + R_3 \rightarrow \end{matrix} \begin{bmatrix} 1 & 2 & -4 & 3 \\ 0 & 3 & -2 & -1 \\ 0 & -3 & 13 & -8 \end{bmatrix}$$

Note that the elementary row operation is written beside the row that is *changed*.

✓CHECKPOINT Now try Exercise 25.

Technology Tip

 Most graphing utilities can perform elementary row operations on matrices. For instructions on how to use the *matrix* feature and the elementary row operations features of a graphing utility, see Appendix A; for specific keystrokes, go to this textbook's *Companion Website*.

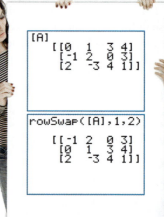

```
[A]
  [[0   1   3   4]
   [-1  2   0   3]
   [2  -3   4   1]]

rowSwap([A],1,2)
  [[-1  2   0   3]
   [0   1   3   4]
   [2  -3   4   1]]
```

Gaussian Elimination with Back-Substitution

In Example 2 of Section 8.3, you used Gaussian elimination with back-substitution to solve a system of linear equations. The next example demonstrates the matrix version of Gaussian elimination. The basic difference between the two methods is that with matrices you do not need to keep writing the variables.

Example 4 Comparing Linear Systems and Matrix Operations

Linear System

$$\begin{cases} x - 2y + 3z = 9 \\ -x + 3y + z = -2 \\ 2x - 5y + 5z = 17 \end{cases}$$

Associated Augmented Matrix

$$\begin{bmatrix} 1 & -2 & 3 & \vdots & 9 \\ -1 & 3 & 1 & \vdots & -2 \\ 2 & -5 & 5 & \vdots & 17 \end{bmatrix}$$

Add the first equation to the second equation.

$$\begin{cases} x - 2y + 3z = 9 \\ y + 4z = 7 \\ 2x - 5y + 5z = 17 \end{cases}$$

Add the first row to the second row: $R_1 + R_2$.

$$R_1 + R_2 \rightarrow \begin{bmatrix} 1 & -2 & 3 & \vdots & 9 \\ 0 & 1 & 4 & \vdots & 7 \\ 2 & -5 & 5 & \vdots & 17 \end{bmatrix}$$

Add -2 times the first equation to the third equation.

$$\begin{cases} x - 2y + 3z = 9 \\ y + 4z = 7 \\ -y - z = -1 \end{cases}$$

Add -2 times the first row to the third row: $-2R_1 + R_3$.

$$-2R_1 + R_3 \rightarrow \begin{bmatrix} 1 & -2 & 3 & \vdots & 9 \\ 0 & 1 & 4 & \vdots & 7 \\ 0 & -1 & -1 & \vdots & -1 \end{bmatrix}$$

Add the second equation to the third equation.

$$\begin{cases} x - 2y + 3z = 9 \\ y + 4z = 7 \\ 3z = 6 \end{cases}$$

Add the second row to the third row: $R_2 + R_3$.

$$R_2 + R_3 \rightarrow \begin{bmatrix} 1 & -2 & 3 & \vdots & 9 \\ 0 & 1 & 4 & \vdots & 7 \\ 0 & 0 & 3 & \vdots & 6 \end{bmatrix}$$

Multiply the third equation by $\frac{1}{3}$.

$$\begin{cases} x - 2y + 3z = 9 \\ y + 4z = 7 \\ z = 2 \end{cases}$$

Multiply the third row by $\frac{1}{3}$: $\frac{1}{3}R_3$.

$$\frac{1}{3}R_3 \rightarrow \begin{bmatrix} 1 & -2 & 3 & \vdots & 9 \\ 0 & 1 & 4 & \vdots & 7 \\ 0 & 0 & 1 & \vdots & 2 \end{bmatrix}$$

At this point, you can use back-substitution to find that the solution is

$$x = 1, y = -1, \text{ and } z = 2$$

as was done in Example 2 of Section 8.3.

✓CHECKPOINT   Now try Exercise 31.

Remember that you should check a solution by substituting the values of x, y, and z into each equation in the original system. For instance, you can check the solution to Example 4 as follows.

Equation 1	*Equation 2*	*Equation 3*
$x - 2y + 3z = 9$	$-x + 3y + z = -2$	$2x - 5y + 5z = 17$
$1 - 2(-1) + 3(2) \stackrel{?}{=} 9$	$-1 + 3(-1) + 2 \stackrel{?}{=} -2$	$2(1) - 5(-1) + 5(2) \stackrel{?}{=} 17$
$9 = 9$ ✓	$-2 = -2$ ✓	$17 = 17$ ✓

The last matrix in Example 4 is in **row-echelon form.** The term *echelon* refers to the stair-step pattern formed by the nonzero elements of the matrix. To be in this form, a matrix must have the following properties.

Row–Echelon Form and Reduced Row–Echelon Form

A matrix in **row-echelon form** has the following properties.

   1. Any rows consisting entirely of zeros occur at the bottom of the matrix.

   2. For each row that does not consist entirely of zeros, the first nonzero entry is 1 (called a **leading 1**).

   3. For two successive (nonzero) rows, the leading 1 in the higher row is farther to the left than the leading 1 in the lower row.

A matrix in *row-echelon form* is in **reduced row-echelon form** when every column that has a leading 1 has zeros in every position above and below its leading 1.

Technology Tip

Some graphing utilities can automatically transform a matrix to row-echelon form and reduced row-echelon form. For instructions on how to use the *row-echelon form* feature and the *reduced row-echelon form* feature of a graphing utility, see Appendix A; for specific keystrokes, go to this textbook's *Companion Website.*

It is worth mentioning that the row-echelon form of a matrix is not unique. That is, two different sequences of elementary row operations may yield different row-echelon forms. The *reduced* row-echelon form of a given matrix, however, is unique.

Example 5   Row–Echelon Form

Determine whether each matrix is in row-echelon form. If it is, determine whether the matrix is in reduced row-echelon form.

a. $\begin{bmatrix} 1 & 2 & -1 & 4 \\ 0 & 1 & 0 & 3 \\ 0 & 0 & 1 & -2 \end{bmatrix}$
b. $\begin{bmatrix} 1 & 2 & -1 & 2 \\ 0 & 0 & 0 & 0 \\ 0 & 1 & 2 & -4 \end{bmatrix}$

c. $\begin{bmatrix} 1 & -5 & 2 & -1 & 3 \\ 0 & 0 & 1 & 3 & -2 \\ 0 & 0 & 0 & 1 & 4 \\ 0 & 0 & 0 & 0 & 1 \end{bmatrix}$
d. $\begin{bmatrix} 1 & 0 & 0 & -1 \\ 0 & 1 & 0 & 2 \\ 0 & 0 & 1 & 3 \\ 0 & 0 & 0 & 0 \end{bmatrix}$

e. $\begin{bmatrix} 1 & 2 & -3 & 4 \\ 0 & 2 & 1 & -1 \\ 0 & 0 & 1 & -3 \end{bmatrix}$
f. $\begin{bmatrix} 0 & 1 & 0 & 5 \\ 0 & 0 & 1 & 3 \\ 0 & 0 & 0 & 0 \end{bmatrix}$

Solution

The matrices in (a), (c), (d), and (f) are in row-echelon form. The matrices in (d) and (f) are in *reduced* row-echelon form because every column that has a leading 1 has zeros in every position above and below its leading 1. The matrix in (b) is not in row-echelon form because the row of all zeros does not occur at the bottom of the matrix. The matrix in (e) is not in row-echelon form because the first nonzero entry in Row 2 is not a leading 1.

✓**CHECKPOINT**   Now try Exercise 35.

Every matrix is row-equivalent to a matrix in row-echelon form. For instance, in Example 5, you can change the matrix in part (e) to row-echelon form by multiplying its second row by $\frac{1}{2}$. What elementary row operation could you perform on the matrix in part (b) so that it would be in row-echelon form?

Gaussian elimination with back-substitution works well for solving systems of linear equations by hand or with a computer. For this algorithm, the order in which the elementary row operations are performed is important. You should operate *from left to right by columns*, using elementary row operations to obtain zeros in all entries directly below the leading 1's.

Example 6   Gaussian Elimination with Back-Substitution

Solve the system of equations.

$$\begin{cases} y + z - 2w = -3 \\ x + 2y - z = 2 \\ 2x + 4y + z - 3w = -2 \\ x - 4y - 7z - w = -19 \end{cases}$$

Solution

$$\begin{bmatrix} 0 & 1 & 1 & -2 & \vdots & -3 \\ 1 & 2 & -1 & 0 & \vdots & 2 \\ 2 & 4 & 1 & -3 & \vdots & -2 \\ 1 & -4 & -7 & -1 & \vdots & -19 \end{bmatrix}$$

Write augmented matrix.

$$\begin{array}{c} R_2 \\ R_1 \end{array} \begin{bmatrix} 1 & 2 & -1 & 0 & \vdots & 2 \\ 0 & 1 & 1 & -2 & \vdots & -3 \\ 2 & 4 & 1 & -3 & \vdots & -2 \\ 1 & -4 & -7 & -1 & \vdots & -19 \end{bmatrix}$$

Interchange R_1 and R_2 so first column has leading 1 in upper left corner.

$$\begin{array}{c} \\ \\ -2R_1 + R_3 \rightarrow \\ -R_1 + R_4 \rightarrow \end{array} \begin{bmatrix} 1 & 2 & -1 & 0 & \vdots & 2 \\ 0 & 1 & 1 & -2 & \vdots & -3 \\ 0 & 0 & 3 & -3 & \vdots & -6 \\ 0 & -6 & -6 & -1 & \vdots & -21 \end{bmatrix}$$

Perform operations on R_3 and R_4 so first column has zeros below its leading 1.

$$\begin{array}{c} \\ \\ \\ 6R_2 + R_4 \rightarrow \end{array} \begin{bmatrix} 1 & 2 & -1 & 0 & \vdots & 2 \\ 0 & 1 & 1 & -2 & \vdots & -3 \\ 0 & 0 & 3 & -3 & \vdots & -6 \\ 0 & 0 & 0 & -13 & \vdots & -39 \end{bmatrix}$$

Perform operations on R_4 so second column has zeros below its leading 1.

$$\begin{array}{c} \\ \\ \frac{1}{3}R_3 \rightarrow \\ -\frac{1}{13}R_4 \rightarrow \end{array} \begin{bmatrix} 1 & 2 & -1 & 0 & \vdots & 2 \\ 0 & 1 & 1 & -2 & \vdots & -3 \\ 0 & 0 & 1 & -1 & \vdots & -2 \\ 0 & 0 & 0 & 1 & \vdots & 3 \end{bmatrix}$$

Perform operations on R_3 and R_4 so third and fourth columns have leading 1's.

The matrix is now in row-echelon form, and the corresponding system is

$$\begin{cases} x + 2y - z = 2 \\ y + z - 2w = -3 \\ z - w = -2 \\ w = 3 \end{cases}$$

Using back-substitution, you can determine that the solution is

$$x = -1, \quad y = 2, \quad z = 1, \quad \text{and} \quad w = 3.$$

Check this in the original system of equations.

✓**CHECKPOINT**   Now try Exercise 61.

The following steps summarize the procedure used in Example 6.

Gaussian Elimination with Back-Substitution

1. Write the augmented matrix of the system of linear equations.

2. Use elementary row operations to rewrite the augmented matrix in row-echelon form.

3. Write the system of linear equations corresponding to the matrix in row-echelon form and use back-substitution to find the solution.

Remember that it is possible for a system to have no solution. If, in the elimination process, you obtain a row with zeros except for the last entry, then you can conclude that the system is inconsistent.

Example 7   A System with No Solution

Solve the system of equations.

$$\begin{cases} x - y + 2z = 4 \\ x + z = 6 \\ 2x - 3y + 5z = 4 \\ 3x + 2y - z = 1 \end{cases}$$

Solution

$$\begin{bmatrix} 1 & -1 & 2 & \vdots & 4 \\ 1 & 0 & 1 & \vdots & 6 \\ 2 & -3 & 5 & \vdots & 4 \\ 3 & 2 & -1 & \vdots & 1 \end{bmatrix}$$ Write augmented matrix.

$$\begin{matrix} \\ -R_1 + R_2 \rightarrow \\ -2R_1 + R_3 \rightarrow \\ -3R_1 + R_4 \rightarrow \end{matrix} \begin{bmatrix} 1 & -1 & 2 & \vdots & 4 \\ 0 & 1 & -1 & \vdots & 2 \\ 0 & -1 & 1 & \vdots & -4 \\ 0 & 5 & -7 & \vdots & -11 \end{bmatrix}$$ Perform row operations.

$$\begin{matrix} \\ \\ R_2 + R_3 \rightarrow \\ \\ \end{matrix} \begin{bmatrix} 1 & -1 & 2 & \vdots & 4 \\ 0 & 1 & -1 & \vdots & 2 \\ 0 & 0 & 0 & \vdots & -2 \\ 0 & 5 & -7 & \vdots & -11 \end{bmatrix}$$ Perform row operations.

Note that the third row of this matrix consists of zeros except for the last entry. This means that the original system of linear equations is *inconsistent*. You can see why this is true by converting back to a system of linear equations.

$$\begin{cases} x - y + 2z = 4 \\ y - z = 2 \\ 0 = -2 \\ 5y - 7z = -11 \end{cases}$$

Because the third equation

$$0 = -2$$

is not possible, the system has no solution.

✓CHECKPOINT   Now try Exercise 59.

Gauss–Jordan Elimination

With Gaussian elimination, elementary row operations are applied to a matrix to obtain a (row-equivalent) row-echelon form of the matrix. A second method of elimination, called **Gauss-Jordan elimination** after Carl Friedrich Gauss (1777–1855) and Wilhelm Jordan (1842–1899), continues the reduction process until a *reduced* row-echelon form is obtained. This procedure is demonstrated in Example 8.

Example 8   Gauss–Jordan Elimination

Use Gauss-Jordan elimination to solve the system.

$$\begin{cases} x - 2y + 3z = 9 \\ -x + 3y + z = -2 \\ 2x - 5y + 5z = 17 \end{cases}$$

Solution

In Example 4, Gaussian elimination was used to obtain the row-echelon form

$$\begin{bmatrix} 1 & -2 & 3 & \vdots & 9 \\ 0 & 1 & 4 & \vdots & 7 \\ 0 & 0 & 1 & \vdots & 2 \end{bmatrix}.$$

Now, rather than using back-substitution, apply additional elementary row operations until you obtain a matrix in *reduced* row-echelon form. To do this, you must produce zeros above each of the leading 1's, as follows.

$$\begin{array}{c} 2R_2 + R_1 \rightarrow \end{array} \begin{bmatrix} 1 & 0 & 11 & \vdots & 23 \\ 0 & 1 & 4 & \vdots & 7 \\ 0 & 0 & 1 & \vdots & 2 \end{bmatrix}$$

Perform operations on R_1 so second column has a zero above its leading 1.

$$\begin{array}{c} -11R_3 + R_1 \rightarrow \\ -4R_3 + R_2 \rightarrow \end{array} \begin{bmatrix} 1 & 0 & 0 & \vdots & 1 \\ 0 & 1 & 0 & \vdots & -1 \\ 0 & 0 & 1 & \vdots & 2 \end{bmatrix}$$

Perform operations on R_1 and R_2 so third column has zeros above its leading 1.

The matrix is now in reduced row-echelon form.
Converting back to a system of linear equations, you have

$$\begin{cases} x = 1 \\ y = -1. \\ z = 2 \end{cases}$$

Now you can simply read the solution,

$$x = 1, \quad y = -1, \quad \text{and} \quad z = 2$$

which can be written as the ordered triple

$$(1, -1, 2).$$

You can check this result using the *reduced row-echelon form* feature of a graphing utility, as shown in Figure 8.23.

✔CHECKPOINT   Now try Exercise 63.

In Example 8, note that the solution is the same as the one obtained using Gaussian elimination in Example 4. The advantage of Gauss-Jordan elimination is that, from the reduced row-echelon form, you can simply read the solution without the need for back-substitution.

Technology Tip

 For a demonstration of a graphical approach to Gauss-Jordan elimination on a 2×3 matrix, see the Visualizing Row Operations Program, available for several models of graphing calculators at this textbook's *Companion Website*.

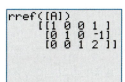

Figure 8.23

The elimination procedures described in this section employ an algorithmic approach that is easily adapted to computer programs. However, the procedure makes no effort to avoid fractional coefficients. For instance, in the elimination procedure for the system

$$\begin{cases} 2x - 5y + 5z = 17 \\ 3x - 2y + 3z = 11 \\ -3x + 3y = -6 \end{cases}$$

you may be inclined to multiply the first row by $\frac{1}{2}$ to produce a leading 1, which will result in working with fractional coefficients. For hand computations, you can sometimes avoid fractions by judiciously choosing the order in which you apply elementary row operations.

Example 9   A System with an Infinite Number of Solutions

Solve the system.

$$\begin{cases} 2x + 4y - 2z = 0 \\ 3x + 5y = 1 \end{cases}$$

Solution

$$\begin{bmatrix} 2 & 4 & -2 & \vdots & 0 \\ 3 & 5 & 0 & \vdots & 1 \end{bmatrix}$$

$$\frac{1}{2}R_1 \rightarrow \begin{bmatrix} 1 & 2 & -1 & \vdots & 0 \\ 3 & 5 & 0 & \vdots & 1 \end{bmatrix}$$

$$-3R_1 + R_2 \rightarrow \begin{bmatrix} 1 & 2 & -1 & \vdots & 0 \\ 0 & -1 & 3 & \vdots & 1 \end{bmatrix}$$

$$-R_2 \rightarrow \begin{bmatrix} 1 & 2 & -1 & \vdots & 0 \\ 0 & 1 & -3 & \vdots & -1 \end{bmatrix}$$

$$-2R_2 + R_1 \rightarrow \begin{bmatrix} 1 & 0 & 5 & \vdots & 2 \\ 0 & 1 & -3 & \vdots & -1 \end{bmatrix}$$

The corresponding system of equations is

$$\begin{cases} x + 5z = 2 \\ y - 3z = -1 \end{cases}.$$

Solving for x and y in terms of z, you have

$$x = -5z + 2 \quad \text{and} \quad y = 3z - 1.$$

To write a solution of the system that does not use any of the three variables of the system, let a represent any real number and let $z = a$. Now substitute a for z in the equations for x and y.

$$x = -5z + 2 = -5a + 2$$

$$y = 3z - 1 = 3a - 1$$

So, the solution set has the form

$$(-5a + 2, 3a - 1, a).$$

Recall from Section 8.3 that a solution set of this form represents an infinite number of solutions. Try substituting values for a to obtain a few solutions. Then check each solution in the original system of equations.

✓CHECKPOINT   Now try Exercise 65.

8.4   Exercises

See www.CalcChat.com for worked-out solutions to odd-numbered exercises.
For instructions on how to use a graphing utility, see Appendix A.

Vocabulary and Concept Check

In Exercises 1–3, fill in the blank.

1. A rectangular array of real numbers that can be used to solve a system of linear equations is called a _____ .

2. A matrix in row-echelon form is in _____ when every column that has a leading 1 has zeros in every position above and below its leading 1.

3. The process of using row operations to write a matrix in reduced row-echelon form is called _____ .

In Exercises 4–6, refer to the system of linear equations $\begin{cases} -2x + 3y = 5 \\ 6x + 7y = 4 \end{cases}$.

4. Is the coefficient matrix for the system a *square* matrix?

5. Is the augmented matrix for the system of dimension 3×2?

6. Is the augmented matrix row-equivalent to its reduced row-echelon form?

Procedures and Problem Solving

Dimension of a Matrix   In Exercises 7–12, determine the dimension of the matrix.

7. $\begin{bmatrix} 7 & 0 \end{bmatrix}$

8. $\begin{bmatrix} 3 & -1 & 2 & 6 \end{bmatrix}$

✓ **9.** $\begin{bmatrix} 4 \\ 32 \\ 3 \end{bmatrix}$

10. $\begin{bmatrix} 5 & 4 & 2 \\ 3 & -5 & 1 \\ 7 & -2 & 9 \end{bmatrix}$

11. $\begin{bmatrix} 33 & 45 \\ -9 & 20 \end{bmatrix}$

12. $\begin{bmatrix} 3 & -1 & 6 & 4 \\ -2 & 5 & 7 & 7 \end{bmatrix}$

Writing an Augmented Matrix   In Exercises 13–18, write the augmented matrix for the system of linear equations. What is the dimension of the augmented matrix?

13. $\begin{cases} 4x - 3y = -5 \\ -x + 3y = 12 \end{cases}$

14. $\begin{cases} 7x + 4y = 22 \\ 5x - 9y = 15 \end{cases}$

✓ **15.** $\begin{cases} x + 10y - 2z = 2 \\ 5x - 3y + 4z = 0 \\ 2x + y = 6 \end{cases}$

16. $\begin{cases} x - 3y + z = 1 \\ 4y = 0 \\ 7z = -5 \end{cases}$

17. $\begin{cases} 7x - 5y + z = 13 \\ 19x - 8z = 10 \end{cases}$

18. $\begin{cases} 9x + 2y - 3z = 20 \\ -25y + 11z = -5 \end{cases}$

Writing a System of Equations   In Exercises 19–22, write the system of linear equations represented by the augmented matrix. (Use the variables x, y, z, and w, if applicable.)

19. $\begin{bmatrix} 3 & 4 & \vdots & 9 \\ 1 & -1 & \vdots & -3 \end{bmatrix}$

20. $\begin{bmatrix} 7 & -5 & \vdots & 0 \\ 8 & 3 & \vdots & -2 \end{bmatrix}$

21. $\begin{bmatrix} 9 & 12 & 3 & \vdots & 0 \\ -2 & 18 & 5 & \vdots & 10 \\ 1 & 7 & -8 & \vdots & -4 \end{bmatrix}$

22. $\begin{bmatrix} 6 & 2 & -1 & -5 & \vdots & -25 \\ -1 & 0 & 7 & 3 & \vdots & 7 \\ 4 & -1 & -10 & 6 & \vdots & 23 \\ 0 & 8 & 1 & -11 & \vdots & -21 \end{bmatrix}$

Identifying an Elementary Row Operation   In Exercises 23–26, identify the elementary row operation performed to obtain the new row-equivalent matrix.

Original Matrix	New Row-Equivalent Matrix

23. $\begin{bmatrix} -3 & 6 & 0 \\ 5 & 2 & -2 \end{bmatrix}$ $\begin{bmatrix} -18 & 0 & 6 \\ 5 & 2 & -2 \end{bmatrix}$

24. $\begin{bmatrix} 3 & -1 & -4 \\ -4 & 3 & 7 \end{bmatrix}$ $\begin{bmatrix} 3 & -1 & -4 \\ 5 & 0 & -5 \end{bmatrix}$

✓ **25.** $\begin{bmatrix} 0 & -1 & -5 & 5 \\ -1 & 3 & -7 & 6 \\ 4 & -5 & 1 & 3 \end{bmatrix}$ $\begin{bmatrix} -1 & 3 & -7 & 6 \\ 0 & -1 & -5 & 5 \\ 4 & -5 & 1 & 3 \end{bmatrix}$

26. $\begin{bmatrix} -1 & -2 & 3 & -2 \\ 2 & -5 & 1 & -7 \\ 5 & 4 & -7 & 6 \end{bmatrix}$ $\begin{bmatrix} -1 & -2 & 3 & -2 \\ 2 & -5 & 1 & -7 \\ 0 & -6 & 8 & -4 \end{bmatrix}$

Elementary Row Operations   In Exercises 27–30, fill in the blanks using elementary row operations to form a row-equivalent matrix.

27. $\begin{bmatrix} 1 & 4 & 3 \\ 2 & 10 & 5 \end{bmatrix}$
$\begin{bmatrix} 1 & 4 & 3 \\ 0 & \boxed{} & -1 \end{bmatrix}$

28. $\begin{bmatrix} 3 & 6 & 8 \\ 4 & -3 & 6 \end{bmatrix}$
$\begin{bmatrix} 1 & \boxed{} & \frac{8}{3} \\ 4 & -3 & 6 \end{bmatrix}$

29. $\begin{bmatrix} 1 & 1 & 4 & -1 \\ 3 & 8 & 10 & 3 \\ -2 & 1 & 12 & 6 \end{bmatrix}$ **30.** $\begin{bmatrix} 2 & 4 & 8 & 3 \\ 1 & -1 & -3 & 2 \\ 2 & 6 & 4 & 9 \end{bmatrix}$

$\begin{bmatrix} 1 & 1 & 4 & -1 \\ 0 & 5 & & \\ 0 & 3 & & \end{bmatrix}$ $\begin{bmatrix} 1 & & & \\ 1 & -1 & -3 & 2 \\ 2 & 6 & 4 & 9 \end{bmatrix}$

$\begin{bmatrix} 1 & 1 & 4 & -1 \\ 0 & 1 & -\frac{2}{5} & \frac{6}{5} \\ 0 & 3 & & \end{bmatrix}$ $\begin{bmatrix} 1 & 2 & 4 & \frac{3}{2} \\ 0 & & -7 & \frac{1}{2} \\ 0 & 2 & & \end{bmatrix}$

Comparing Linear Systems and Matrix Operations In Exercises 31 and 32, (a) perform the row operations to solve the augmented matrix, (b) write and solve the system of linear equations represented by the augmented matrix, and (c) compare the two solution methods. Which do you prefer?

✓ **31.** $\begin{bmatrix} -3 & 4 & \vdots & 22 \\ 6 & -4 & \vdots & -28 \end{bmatrix}$

(i) Add R_2 to R_1.

(ii) Add -2 times R_1 to R_2.

(iii) Multiply R_2 by $-\frac{1}{4}$.

(iv) Multiply R_1 by $\frac{1}{3}$.

32. $\begin{bmatrix} 7 & 13 & 1 & \vdots & -4 \\ -3 & -5 & -1 & \vdots & -4 \\ 3 & 6 & 1 & \vdots & -2 \end{bmatrix}$

(i) Add R_2 to R_1.

(ii) Multiply R_1 and $\frac{1}{4}$.

(iii) Add R_3 to R_2.

(iv) Add -3 times R_1 to R_3.

(v) Add -2 times R_2 to R_1.

33. Repeat steps (i) through (iv) for the matrix in Exercise 31 using a graphing utility.

34. Repeat steps (i) through (v) for the matrix in Exercise 32 using a graphing utility.

Row-Echelon Form In Exercises 35–40, determine whether the matrix is in row-echelon form. If it is, determine if it is also in reduced row-echelon form.

✓ **35.** $\begin{bmatrix} 1 & 0 & 0 & 0 \\ 0 & 1 & 1 & 5 \\ 0 & 0 & 0 & 0 \end{bmatrix}$ **36.** $\begin{bmatrix} 1 & 3 & 0 & 0 \\ 0 & 0 & 1 & 8 \\ 0 & 0 & 0 & 0 \end{bmatrix}$

37. $\begin{bmatrix} 3 & 0 & 3 & 7 \\ 0 & -2 & 0 & 4 \\ 0 & 0 & 1 & 5 \end{bmatrix}$ **38.** $\begin{bmatrix} 1 & 0 & 2 & 1 \\ 0 & 1 & -3 & 10 \\ 0 & 0 & 1 & 0 \end{bmatrix}$

39. $\begin{bmatrix} 1 & 0 & 0 & 1 \\ 0 & 1 & 0 & -1 \\ 0 & 0 & 0 & 2 \end{bmatrix}$ **40.** $\begin{bmatrix} 1 & 0 & 1 & 0 \\ 0 & 1 & 0 & 2 \\ 0 & 0 & 1 & 0 \end{bmatrix}$

Using Gaussian Elimination In Exercises 41–44, write the matrix in row-echelon form. Remember that the row-echelon form of a matrix is not unique.

41. $\begin{bmatrix} 1 & 2 & 3 & 0 \\ -1 & 4 & 0 & -5 \\ 2 & 6 & 3 & 10 \end{bmatrix}$

42. $\begin{bmatrix} 1 & 2 & -1 & 3 \\ 3 & 7 & -5 & 14 \\ -2 & -1 & -3 & 8 \end{bmatrix}$

43. $\begin{bmatrix} 1 & -1 & -1 & 1 \\ 5 & -4 & 1 & 8 \\ -6 & 8 & 18 & 0 \end{bmatrix}$

44. $\begin{bmatrix} 1 & -3 & 0 & -7 \\ -3 & 10 & 1 & 23 \\ 4 & -10 & 2 & -24 \end{bmatrix}$

Using a Graphing Utility In Exercises 45–48, use the matrix capabilities of a graphing utility to write the matrix in reduced row-echelon form.

45. $\begin{bmatrix} 3 & 3 & 3 \\ -1 & 0 & -4 \\ 2 & 4 & -2 \end{bmatrix}$ **46.** $\begin{bmatrix} 1 & 3 & 2 \\ 5 & 15 & 9 \\ 2 & 6 & 10 \end{bmatrix}$

47. $\begin{bmatrix} -4 & 1 & 0 & 6 \\ 1 & -2 & 3 & -4 \end{bmatrix}$

48. $\begin{bmatrix} 5 & 1 & 2 & 4 \\ -1 & 5 & 10 & -32 \end{bmatrix}$

Using Back-Substitution In Exercises 49–52, write the system of linear equations represented by the augmented matrix. Then use back-substitution to find the solution. (Use the variables x, y, and z, if applicable.)

49. $\begin{bmatrix} 1 & -2 & \vdots & 4 \\ 0 & 1 & \vdots & -3 \end{bmatrix}$

50. $\begin{bmatrix} 1 & 8 & \vdots & 12 \\ 0 & 1 & \vdots & 3 \end{bmatrix}$

51. $\begin{bmatrix} 1 & -1 & 2 & \vdots & 4 \\ 0 & 1 & -1 & \vdots & 2 \\ 0 & 0 & 1 & \vdots & -2 \end{bmatrix}$

52. $\begin{bmatrix} 1 & 2 & -2 & \vdots & -1 \\ 0 & 1 & 1 & \vdots & 9 \\ 0 & 0 & 1 & \vdots & -3 \end{bmatrix}$

Interpreting Reduced Row-Echelon Form In Exercises 53–56, an augmented matrix that represents a system of linear equations (in the variables x and y or x, y, and z) has been reduced using Gauss-Jordan elimination. Write the solution represented by the augmented matrix.

53. $\begin{bmatrix} 1 & 0 & \vdots & 7 \\ 0 & 1 & \vdots & -5 \end{bmatrix}$ **54.** $\begin{bmatrix} 1 & 0 & \vdots & -2 \\ 0 & 1 & \vdots & 4 \end{bmatrix}$

55. $\begin{bmatrix} 1 & 0 & 0 & \vdots & -4 \\ 0 & 1 & 0 & \vdots & -8 \\ 0 & 0 & 1 & \vdots & 2 \end{bmatrix}$

56. $\begin{bmatrix} 1 & 0 & 0 & \vdots & 3 \\ 0 & 1 & 0 & \vdots & -1 \\ 0 & 0 & 1 & \vdots & 0 \end{bmatrix}$

Gaussian Elimination with Back-Substitution In Exercises 57–62, use matrices to solve the system of equations, if possible. Use Gaussian elimination with back-substitution.

57. $\begin{cases} x + 2y = 7 \\ 2x + y = 8 \end{cases}$

58. $\begin{cases} 2x + 6y = 16 \\ 2x + 3y = 7 \end{cases}$

✓ **59.** $\begin{cases} -x + y = -22 \\ 3x + 4y = 4 \\ 4x - 8y = 32 \end{cases}$

60. $\begin{cases} x + 2y = 0 \\ x + y = 6 \\ 3x - 2y = 8 \end{cases}$

✓ **61.** $\begin{cases} 3x + 2y - z + w = 0 \\ x - y + 4z + 2w = 25 \\ -2x + y + 2z - w = 2 \\ x + y + z + w = 6 \end{cases}$

62. $\begin{cases} x - 4y + 3z - 2w = 9 \\ 3x - 2y + z - 4w = -13 \\ -4x + 3y - 2z + w = -4 \\ -2x + y - 4z + 3w = -10 \end{cases}$

Gauss-Jordan Elimination In Exercises 63–68, use matrices to solve the system of equations, if possible. Use Gauss-Jordan elimination.

✓ **63.** $\begin{cases} x - 3z = -2 \\ 3x + y - 2z = 5 \\ 2x + 2y + z = 4 \end{cases}$

64. $\begin{cases} 2x - y + 3z = 24 \\ 2y - z = 14 \\ 7x - 5y = 6 \end{cases}$

✓ **65.** $\begin{cases} x + y - 5z = 3 \\ x - 2z = 1 \\ 2x - y - z = 0 \end{cases}$

66. $\begin{cases} 2x + 3z = 3 \\ 4x - 3y + 7z = 5 \\ 8x - 9y + 15z = 9 \end{cases}$

67. $\begin{cases} -x + y - z = -14 \\ 2x - y + z = 21 \\ 3x + 2y + z = 19 \end{cases}$

68. $\begin{cases} 2x + 2y - z = 2 \\ x - 3y + z = 28 \\ -x + y = 14 \end{cases}$

Using a Graphing Utility In Exercises 69–72, use the matrix capabilities of a graphing utility to reduce the augmented matrix corresponding to the system of equations, and solve the system.

69. $\begin{cases} 3x + 3y + 12z = 6 \\ x + y + 4z = 2 \\ 2x + 5y + 20z = 10 \\ -x + 2y + 8z = 4 \end{cases}$

70. $\begin{cases} x + y + z = 0 \\ 2x + 3y + z = 0 \\ 3x + 5y + z = 0 \end{cases}$

71. $\begin{cases} 2x + 10y + 2z = 6 \\ x + 5y + 2z = 6 \\ x + 5y + z = 3 \\ -3x - 15y - 3z = -9 \end{cases}$

72. $\begin{cases} 2x + y - z + 2w = -6 \\ 3x + 4y + w = 1 \\ x + 5y + 2z + 6w = -3 \\ 5x + 2y - z - w = 3 \end{cases}$

Comparing Solutions of Two Systems In Exercises 73–76, determine whether the two systems of linear equations yield the same solution. If so, find the solution.

73. (a) $\begin{cases} x - 2y + z = -6 \\ y - 5z = 16 \\ z = -3 \end{cases}$
(b) $\begin{cases} x + y - 2z = 6 \\ y + 3z = -8 \\ z = -3 \end{cases}$

74. (a) $\begin{cases} x - 3y + 4z = -11 \\ y - z = -4 \\ z = 2 \end{cases}$
(b) $\begin{cases} x + 4y = -11 \\ y + 3z = 4 \\ z = 2 \end{cases}$

75. (a) $\begin{cases} x - 4y + 5z = 27 \\ y - 7z = -54 \\ z = 8 \end{cases}$
(b) $\begin{cases} x - 6y + z = 15 \\ y + 5z = 42 \\ z = 8 \end{cases}$

76. (a) $\begin{cases} x + 3y - z = 19 \\ y + 6z = -18 \\ z = -4 \end{cases}$
(b) $\begin{cases} x - y + 3z = -15 \\ y - 2z = 14 \\ z = -4 \end{cases}$

Data Analysis: Curve Fitting In Exercises 77–80, use a system of equations to find the equation of the parabola $y = ax^2 + bx + c$ that passes through the points. Solve the system using matrices. Use a graphing utility to verify your result.

77.

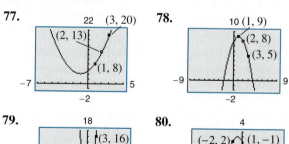

78.

79.

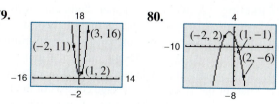

80.

Curve Fitting   In Exercises 81 and 82, use a system of equations to find the quadratic function $f(x) = ax^2 + bx + c$ that satisfies the equations. Solve the system using matrices.

81. $f(-2) = -15$
$\quad f(-1) = 7$
$\quad f(1) = -3$

82. $f(-2) = -3$
$\quad f(1) = -3$
$\quad f(2) = -11$

Curve Fitting   In Exercises 83 and 84, use a system of equations to find the cubic function $f(x) = ax^3 + bx^2 + cx + d$ that satisfies the equations. Solve the system using matrices.

83. $f(-2) = -7$
$\quad f(-1) = 2$
$\quad f(1) = -4$
$\quad f(2) = -7$

84. $f(-2) = -17$
$\quad f(-1) = -5$
$\quad f(1) = 1$
$\quad f(2) = 7$

85. Electrical Engineering   The currents in an electrical network are given by the solution of the system

$$\begin{cases} I_1 - I_2 + I_3 = 0 \\ 2I_1 + 2I_2 \quad\quad = 7 \\ \quad\quad 2I_2 + 4I_3 = 8 \end{cases}$$

where I_1, I_2, and I_3 are measured in amperes. Solve the system of equations using matrices.

86. Finance   A corporation borrowed $1,500,000 to expand its line of shoes. Some of the money was borrowed at 3%, some at 4%, and some at 6%. Use a system of equations to determine how much was borrowed at each rate given that the annual interest was $74,000 and the amount borrowed at 6% was four times the amount borrowed at 3%. Solve the system using matrices.

87. Using Matrices   A food server examines the amount of money earned in tips after working an 8-hour shift. The server has a total of $95 in denominations of $1, $5, $10, and $20 bills. The total number of paper bills is 26. The number of $5 bills is 4 times the number of $10 bills, and the number of $1 bills is 1 less than twice the number of $5 bills. Write a system of linear equations to represent the situation. Then use matrices to find the number of each denomination.

88. Marketing   A wholesale paper company sells a 100-pound package of computer paper that consists of three grades, glossy, semi-gloss, and matte, for printing photographs. Glossy costs $5.50 per pound, semi-gloss costs $4.25 per pound, and matte costs $3.75 per pound. One half of the 100-pound package consists of the two less expensive grades. The cost of the 100-pound package is $480. Set up and solve a system of equations, using matrices, to find the number of pounds of each grade of paper in a 100-pound package.

89. Partial Fractions   Use a system of equations to write the partial fraction decomposition of the rational expression. Solve the system using matrices.

$$\frac{8x^2}{(x-1)^2(x+1)} = \frac{A}{x+1} + \frac{B}{x-1} + \frac{C}{(x-1)^2}$$

90. MODELING DATA

A video of the path of a ball thrown by a baseball player was analyzed with a grid covering the TV screen. The video was paused three times, and the position of the ball was measured each time. The coordinates obtained are shown in the table (x and y are measured in feet).

Horizontal distance, x	Height, y
0	5.0
15	9.6
30	12.4

(a) Use a system of equations to find the equation of the parabola $y = ax^2 + bx + c$ that passes through the points. Solve the system using matrices.

(b) Use a graphing utility to graph the parabola.

(c) Graphically approximate the maximum height of the ball and the point at which the ball strikes the ground.

(d) Algebraically approximate the maximum height of the ball and the point at which the ball strikes the ground.

91. Why you should learn it   (p. 644)   The table shows the average retail prices y (in dollars) of prescriptions from 2006 through 2008. (Source: National Association of Chain Drug Stores)

Year	Price, y (in dollars)
2006	65.82
2007	68.77
2008	71.70

(a) Use a system of equations to find the equation of the parabola $y = at^2 + bt + c$ that passes through the points. Let t represent the year, with $t = 6$ corresponding to 2006. Solve the system using matrices.

(b) Use a graphing utility to graph the parabola and plot the data points.

(c) Use the equation in part (a) to estimate the average retail prices in 2010, 2015, and 2020.

(d) Are your estimates in part (c) reasonable? Explain.

92. MODELING DATA

The table shows the average annual salaries y (in thousands of dollars) for public school classroom teachers in the United States from 2006 through 2008. (Source: Educational Research Service)

Year	Annual salary, y (in thousands of dollars)
2006	48.2
2007	49.3
2008	51.3

(a) Use a system of equations to find the equation of the parabola $y = at^2 + bt + c$ that passes through the points. Let t represent the year, with $t = 6$ corresponding to 2006. Solve the system using matrices.

(b) Use a graphing utility to graph the parabola and plot the data points.

(c) Use the equation in part (a) to estimate the average annual salaries in 2010, 2015, and 2020.

(d) Are your estimates in part (c) reasonable? Explain.

93. **Network Analysis** Water flowing through a network of pipes (in thousands of cubic meters per hour) is shown below.

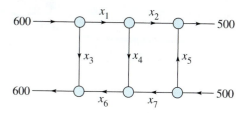

(a) Use matrices to solve this system for the water flow represented by x_i, $i = 1, 2, 3, 4, 5, 6$, and 7.

(b) Find the network flow pattern when $x_6 = 0$ and $x_7 = 0$.

(c) Find the network flow pattern when $x_5 = 400$ and $x_6 = 500$.

94. **Network Analysis** The flow of water (in thousands of cubic meters per hour) into and out of the right side of the network of pipes in Exercise 93 is increased from 500 to 700.

(a) Draw a diagram of the new network.

(b) Use matrices to solve this system for the water flow represented by x_i, $i = 1, 2, 3, 4, 5, 6$, and 7.

(c) Find the network flow pattern when $x_1 = 600$ and $x_7 = 200$.

(d) Find the network flow pattern when $x_4 = 150$ and $x_5 = 350$.

Conclusions

True or False? In Exercises 95 and 96, determine whether the statement is true or false. Justify your answer.

95. When using Gaussian elimination to solve a system of linear equations, you may conclude that the system is inconsistent before you complete the process of rewriting the augmented matrix in row-echelon form.

96. You cannot write an augmented matrix for a dependent system of linear equations in reduced row-echelon form.

97. **Think About It** The augmented matrix represents a system of linear equations (in the variables x, y, and z) that has been reduced using Gauss-Jordan elimination. Write a system of three equations in three variables with *nonzero* coefficients that is represented by the reduced matrix. (There are many correct answers.)

$$\begin{bmatrix} 1 & 0 & 3 & \vdots & -2 \\ 0 & 1 & 4 & \vdots & 1 \\ 0 & 0 & 0 & \vdots & 0 \end{bmatrix}$$

98. **Error Analysis** Describe the errors.

$$\begin{bmatrix} 1 & 1 & \vdots & 4 \\ 2 & 3 & \vdots & 5 \end{bmatrix}$$

$$-2R_1 + R_2 \rightarrow \begin{bmatrix} 1 & 1 & \vdots & 4 \\ 0 & 1 & \vdots & 5 \end{bmatrix}$$

$$-R_2 + R_1 \rightarrow \begin{bmatrix} 1 & 0 & \vdots & 4 \\ 0 & 1 & \vdots & 5 \end{bmatrix}$$

99. **Think About It** Can a 2×4 augmented matrix whose entries are all nonzero real numbers represent an independent system of linear equations? Explain.

100. **CAPSTONE** Determine all values of a and b for which the augmented matrix has each given number of solutions.

$$\begin{bmatrix} 1 & 2 & \vdots & -4 \\ 0 & a & \vdots & b \end{bmatrix}$$

(a) Exactly one solution

(b) Infinitely many solutions

(c) No solution

Cumulative Mixed Review

Graphing a Rational Function In Exercises 101–104, sketch the graph of the function. Identify any asymptotes.

101. $f(x) = \dfrac{7}{-x - 1}$

102. $f(x) = \dfrac{4x}{5x^2 + 2}$

103. $f(x) = \dfrac{x^2 - 2x - 3}{x - 4}$

104. $f(x) = \dfrac{x^2 - 36}{x + 1}$

8.5   Operations with Matrices

Equality of Matrices

In Section 8.4, you used matrices to solve systems of linear equations. There is a rich mathematical theory of matrices, and its applications are numerous. This section and the next two introduce some fundamentals of matrix theory. It is standard mathematical convention to represent matrices in any of the following three ways.

Representation of Matrices

1. A matrix can be denoted by an uppercase letter such as

A, B, or C.

2. A matrix can be denoted by a representative element enclosed in brackets, such as

$[a_{ij}]$, $[b_{ij}]$, or $[c_{ij}]$.

3. A matrix can be denoted by a rectangular array of numbers such as

$$A = [a_{ij}] = \begin{bmatrix} a_{11} & a_{12} & a_{13} & \cdots & a_{1n} \\ a_{21} & a_{22} & a_{23} & \cdots & a_{2n} \\ a_{31} & a_{32} & a_{33} & \cdots & a_{3n} \\ \vdots & \vdots & \vdots & & \vdots \\ a_{m1} & a_{m2} & a_{m3} & \cdots & a_{mn} \end{bmatrix}.$$

Two matrices

$A = [a_{ij}]$ and $B = [b_{ij}]$

are **equal** when they have the same dimension $(m \times n)$ and all of their corresponding entries are equal.

What you should learn

- Decide whether two matrices are equal.
- Add and subtract matrices and multiply matrices by scalars.
- Multiply two matrices.
- Use matrix operations to model and solve real-life problems.

Why you should learn it

Matrix algebra provides a systematic way of performing mathematical operations on large arrays of numbers. In Exercise 89 on page 670, you will use matrix multiplication to help analyze the labor and wage requirements for a boat manufacturer.

Example 1   Equality of Matrices

Solve for $a_{11}, a_{12}, a_{21},$ and a_{22} in the following matrix equation.

$$\begin{bmatrix} a_{11} & a_{12} \\ a_{21} & a_{22} \end{bmatrix} = \begin{bmatrix} 2 & -1 \\ -3 & 0 \end{bmatrix}$$

Solution

Because two matrices are equal only when their corresponding entries are equal, you can conclude that

$a_{11} = 2,$ $a_{12} = -1,$ $a_{21} = -3,$ and $a_{22} = 0.$

✔CHECKPOINT   Now try Exercise 9.

Be sure you see that for two matrices to be equal, they must have the same dimension *and* their corresponding entries must be equal. For instance,

$$\begin{bmatrix} 2 & -1 \\ \sqrt{4} & \frac{1}{2} \end{bmatrix} = \begin{bmatrix} 2 & -1 \\ 2 & 0.5 \end{bmatrix} \quad \text{but} \quad \begin{bmatrix} 2 & -1 \\ 3 & 4 \\ 0 & 0 \end{bmatrix} \neq \begin{bmatrix} 2 & -1 \\ 3 & 4 \end{bmatrix}.$$

Matrix Addition and Scalar Multiplication

You can add two matrices (of the same dimension) by adding their corresponding entries.

Definition of Matrix Addition

If $A = [a_{ij}]$ and $B = [b_{ij}]$ are matrices of dimension $m \times n$, then their sum is the $m \times n$ matrix given by

$$A + B = [a_{ij} + b_{ij}].$$

The sum of two matrices of different dimensions is undefined.

Example 2   **Addition of Matrices**

a. $\begin{bmatrix} -1 & 2 \\ 0 & 1 \end{bmatrix} + \begin{bmatrix} 1 & 3 \\ -1 & 2 \end{bmatrix} = \begin{bmatrix} -1+1 & 2+3 \\ 0+(-1) & 1+2 \end{bmatrix} = \begin{bmatrix} 0 & 5 \\ -1 & 3 \end{bmatrix}$

b. $\begin{bmatrix} 1 \\ -3 \\ -2 \end{bmatrix} + \begin{bmatrix} -1 \\ 3 \\ 2 \end{bmatrix} = \begin{bmatrix} 0 \\ 0 \\ 0 \end{bmatrix}$

c. The sum of

$$A = \begin{bmatrix} 2 & 1 & 0 \\ 4 & 0 & -1 \end{bmatrix} \quad \text{and} \quad B = \begin{bmatrix} 0 & 1 \\ -1 & 3 \end{bmatrix}$$

is undefined because A is of dimension 2×3 and B is of dimension 2×2.

✓**CHECKPOINT**   Now try Exercise 15(a).

In operations with matrices, numbers are usually referred to as **scalars.** In this text, scalars will always be real numbers. You can multiply a matrix A by a scalar c by multiplying each entry in A by c.

Definition of Scalar Multiplication

If $A = [a_{ij}]$ is an $m \times n$ matrix and c is a scalar, then the **scalar multiple** of A by c is the $m \times n$ matrix given by

$$cA = [ca_{ij}].$$

Technology Tip

Try using a graphing utility to find the sum of the two matrices in Example 2(c). Your graphing utility should display an error message similar to the one shown below.

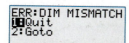

Example 3   **Scalar Multiplication**

For the following matrix, find $3A$.

$$A = \begin{bmatrix} 2 & 2 & 4 \\ -3 & 0 & -1 \\ 2 & 1 & 2 \end{bmatrix}$$

Solution

$$3A = 3\begin{bmatrix} 2 & 2 & 4 \\ -3 & 0 & -1 \\ 2 & 1 & 2 \end{bmatrix} = \begin{bmatrix} 3(2) & 3(2) & 3(4) \\ 3(-3) & 3(0) & 3(-1) \\ 3(2) & 3(1) & 3(2) \end{bmatrix} = \begin{bmatrix} 6 & 6 & 12 \\ -9 & 0 & -3 \\ 6 & 3 & 6 \end{bmatrix}$$

✓**CHECKPOINT**   Now try Exercise 15(c).

The symbol $-A$ represents the negation of A, which is the scalar product $(-1)A$. Moreover, if A and B are of the same dimension, then $A - B$ represents the sum of A and $(-1)B$. That is,

$$A - B = A + (-1)B.$$ Subtraction of matrices

The order of operations for matrix expressions is similar to that for real numbers. In particular, you perform scalar multiplication before matrix addition and subtraction, as shown in Example 4.

Example 4   Scalar Multiplication and Matrix Subtraction

For the following matrices, find $3A - B$.

$$A = \begin{bmatrix} 2 & 2 & 4 \\ -3 & 0 & -1 \\ 2 & 1 & 2 \end{bmatrix} \quad \text{and} \quad B = \begin{bmatrix} 2 & 0 & 0 \\ 1 & -4 & 3 \\ -1 & 3 & 2 \end{bmatrix}$$

Solution
Note that A is the same matrix from Example 3, where you found $3A$.

$$3A - B = \begin{bmatrix} 6 & 6 & 12 \\ -9 & 0 & -3 \\ 6 & 3 & 6 \end{bmatrix} - \begin{bmatrix} 2 & 0 & 0 \\ 1 & -4 & 3 \\ -1 & 3 & 2 \end{bmatrix} \qquad \text{Perform scalar multiplication first.}$$

$$= \begin{bmatrix} 4 & 6 & 12 \\ -10 & 4 & -6 \\ 7 & 0 & 4 \end{bmatrix} \qquad \text{Subtract corresponding entries.}$$

✓**CHECKPOINT**   Now try Exercise 15(d).

The properties of matrix addition and scalar multiplication are similar to those of addition and multiplication of real numbers. One important property of addition of real numbers is that the number 0 is the additive identity. That is, $c + 0 = c$ for any real number c. For matrices, a similar property holds. That is, if A is an $m \times n$ matrix and O is the $m \times n$ **zero matrix** consisting entirely of zeros, then $A + O = A$.

In other words, O is the **additive identity** for the set of all $m \times n$ matrices. For example, the following matrices are the additive identities for the sets of all 2×3 and 2×2 matrices.

$$O = \begin{bmatrix} 0 & 0 & 0 \\ 0 & 0 & 0 \end{bmatrix} \qquad \text{and} \qquad O = \begin{bmatrix} 0 & 0 \\ 0 & 0 \end{bmatrix}$$

2×3 zero matrix 2×2 zero matrix

Properties of Matrix Addition and Scalar Multiplication

Let A, B, and C be $m \times n$ matrices and let c and d be scalars.

1. $A + B = B + A$ Commutative Property of Matrix Addition
2. $A + (B + C) = (A + B) + C$ Associative Property of Matrix Addition
3. $(cd)A = c(dA)$ Associative Property of Scalar Multiplication
4. $1A = A$ Scalar Identity
5. $A + O = A$ Additive Identity
6. $c(A + B) = cA + cB$ Distributive Property
7. $(c + d)A = cA + dA$ Distributive Property

Explore the Concept

What do you observe about the relationship between the corresponding entries of A and B below? Use a graphing utility to find $A + B$. What conclusion can you make about the entries of A and B and the sum $A + B$?

$$A = \begin{bmatrix} -1 & 5 \\ 2 & -6 \end{bmatrix}$$

$$B = \begin{bmatrix} 1 & -5 \\ -2 & 6 \end{bmatrix}$$

Study Tip

Note that the Associative Property of Matrix Addition allows you to write expressions such as $A + B + C$ without ambiguity because the same sum occurs no matter how the matrices are grouped. This same reasoning applies to sums of four or more matrices.

Example 5   Using the Distributive Property

$$3\left(\begin{bmatrix} -2 & 0 \\ 4 & 1 \end{bmatrix} + \begin{bmatrix} 4 & -2 \\ 3 & 7 \end{bmatrix}\right) = 3\begin{bmatrix} -2 & 0 \\ 4 & 1 \end{bmatrix} + 3\begin{bmatrix} 4 & -2 \\ 3 & 7 \end{bmatrix}$$

$$= \begin{bmatrix} -6 & 0 \\ 12 & 3 \end{bmatrix} + \begin{bmatrix} 12 & -6 \\ 9 & 21 \end{bmatrix}$$

$$= \begin{bmatrix} 6 & -6 \\ 21 & 24 \end{bmatrix}$$

✔**CHECKPOINT**   Now try Exercise 23.

In Example 5, you could add the two matrices first and then multiply the resulting matrix by 3. The result would be the same.

The algebra of real numbers and the algebra of matrices have many similarities. For example, compare the following solutions.

Real Numbers	*m × n Matrices*
(Solve for x.)	*(Solve for X.)*
$x + a = b$	$X + A = B$
$x + a + (-a) = b + (-a)$	$X + A + (-A) = B + (-A)$
$x + 0 = b - a$	$X + O = B - A$
$x = b - a$	$X = B - A$

The algebra of real numbers and the algebra of matrices also have important differences, which will be discussed later.

Example 6   Solving a Matrix Equation

Solve for X in the equation

$$3X + A = B$$

where

$$A = \begin{bmatrix} 1 & -2 \\ 0 & 3 \end{bmatrix} \quad \text{and} \quad B = \begin{bmatrix} -3 & 4 \\ 2 & 1 \end{bmatrix}.$$

Solution

Begin by solving the equation for X to obtain

$$3X = B - A$$

$$X = \frac{1}{3}(B - A).$$

Now, using the matrices A and B, you have

$$X = \frac{1}{3}\left(\begin{bmatrix} -3 & 4 \\ 2 & 1 \end{bmatrix} - \begin{bmatrix} 1 & -2 \\ 0 & 3 \end{bmatrix}\right) \qquad \text{Substitute the matrices.}$$

$$= \frac{1}{3}\begin{bmatrix} -4 & 6 \\ 2 & -2 \end{bmatrix} \qquad \text{Subtract matrix } A \text{ from matrix } B.$$

$$= \begin{bmatrix} -\frac{4}{3} & 2 \\ \frac{2}{3} & -\frac{2}{3} \end{bmatrix}. \qquad \text{Multiply the resulting matrix by } \frac{1}{3}.$$

✔**CHECKPOINT**   Now try Exercise 31.

Matrix Multiplication

Another basic matrix operation is **matrix multiplication.** At first glance, the following definition may seem unusual. You will see later, however, that this definition of the product of two matrices has many practical applications.

Definition of Matrix Multiplication

If $A = [a_{ij}]$ is an $m \times n$ matrix and $B = [b_{ij}]$ is an $n \times p$ matrix, then the product AB is an $m \times p$ matrix given by

$$AB = [c_{ij}]$$

where $c_{ij} = a_{i1}b_{1j} + a_{i2}b_{2j} + a_{i3}b_{3j} + \cdots + a_{in}b_{nj}$.

The definition of matrix multiplication indicates a *row-by-column* multiplication, where the entry in the ith row and jth column of the product AB is obtained by multiplying the entries in the ith row of A by the corresponding entries in the jth column of B and then adding the results. The general pattern for matrix multiplication is as follows.

$$\begin{bmatrix} a_{11} & a_{12} & a_{13} & \cdots & a_{1n} \\ a_{21} & a_{22} & a_{23} & \cdots & a_{2n} \\ a_{31} & a_{32} & a_{33} & \cdots & a_{3n} \\ \vdots & \vdots & \vdots & & \vdots \\ a_{i1} & a_{i2} & a_{i3} & \cdots & a_{in} \\ \vdots & \vdots & \vdots & & \vdots \\ a_{m1} & a_{m2} & a_{m3} & \cdots & a_{mn} \end{bmatrix} \begin{bmatrix} b_{11} & b_{12} & \cdots & b_{1j} & \cdots & b_{1p} \\ b_{21} & b_{22} & \cdots & b_{2j} & \cdots & b_{2p} \\ b_{31} & b_{32} & \cdots & b_{3j} & \cdots & b_{3p} \\ \vdots & \vdots & & \vdots & & \vdots \\ b_{n1} & b_{n2} & \cdots & b_{nj} & \cdots & b_{np} \end{bmatrix} = \begin{bmatrix} c_{11} & c_{12} & \cdots & c_{1j} & \cdots & c_{1p} \\ c_{21} & c_{22} & \cdots & c_{2j} & \cdots & c_{2p} \\ \vdots & \vdots & & \vdots & & \vdots \\ c_{i1} & c_{i2} & \cdots & c_{ij} & \cdots & c_{ip} \\ \vdots & \vdots & & \vdots & & \vdots \\ c_{m1} & c_{m2} & \cdots & c_{mj} & \cdots & c_{mp} \end{bmatrix}$$

$$a_{i1}b_{1j} + a_{i2}b_{2j} + a_{i3}b_{3j} + \cdots + a_{in}b_{nj} = c_{ij}$$

Example 7 Finding the Product of Two Matrices

Find the product AB using $A = \begin{bmatrix} -1 & 3 \\ 4 & -2 \\ 5 & 0 \end{bmatrix}$ and $B = \begin{bmatrix} -3 & 2 \\ -4 & 1 \end{bmatrix}$.

Solution

First, note that the product AB is defined because the number of columns of A is equal to the number of rows of B. Moreover, the product AB has dimension 3×2. To find the entries of the product, multiply each row of A by each column of B.

$$AB = \begin{bmatrix} -1 & 3 \\ 4 & -2 \\ 5 & 0 \end{bmatrix} \begin{bmatrix} -3 & 2 \\ -4 & 1 \end{bmatrix}$$

$$= \begin{bmatrix} (-1)(-3) + (3)(-4) & (-1)(2) + (3)(1) \\ (4)(-3) + (-2)(-4) & (4)(2) + (-2)(1) \\ (5)(-3) + (0)(-4) & (5)(2) + (0)(1) \end{bmatrix}$$

$$= \begin{bmatrix} -9 & 1 \\ -4 & 6 \\ -15 & 10 \end{bmatrix}$$

✓**CHECKPOINT** Now try Exercise 35.

Be sure you understand that for the product of two matrices to be defined, the number of *columns* of the first matrix must equal the number of *rows* of the second matrix. That is, the middle two indices must be the same. The outside two indices give the dimension of the product, as shown in the following diagram.

$$\underset{m \times n}{A} \times \underset{n \times p}{B} = \underset{m \times p}{AB}$$

Equal

Dimension of *AB*

Example 8 Matrix Multiplication

a. $\begin{bmatrix} 1 & 0 & 3 \\ 2 & -1 & -2 \end{bmatrix} \begin{bmatrix} -2 & 4 & 2 \\ 1 & 0 & 0 \\ -1 & 1 & -1 \end{bmatrix} = \begin{bmatrix} -5 & 7 & -1 \\ -3 & 6 & 6 \end{bmatrix}$

      2×3 3×3 2×3

b. $\begin{bmatrix} 3 & 4 \\ -2 & 5 \end{bmatrix} \begin{bmatrix} 1 & 0 \\ 0 & 1 \end{bmatrix} = \begin{bmatrix} 3 & 4 \\ -2 & 5 \end{bmatrix}$

      2×2 2×2 2×2

c. $\begin{bmatrix} 1 & 2 \\ 1 & 1 \end{bmatrix} \begin{bmatrix} -1 & 2 \\ 1 & -1 \end{bmatrix} = \begin{bmatrix} 1 & 0 \\ 0 & 1 \end{bmatrix}$

      2×2 2×2 2×2

d. $\begin{bmatrix} 6 & 2 & 0 \\ 3 & -1 & 2 \\ 1 & 4 & 6 \end{bmatrix} \begin{bmatrix} 1 \\ 2 \\ -3 \end{bmatrix} = \begin{bmatrix} 10 \\ -5 \\ -9 \end{bmatrix}$

      3×3 3×1 3×1

e. The product *AB* for the following matrices is not defined.

$$A = \begin{bmatrix} -2 & 1 \\ 1 & -3 \\ 1 & 4 \end{bmatrix} \quad \text{and} \quad B = \begin{bmatrix} -2 & 3 & 1 & 4 \\ 0 & 1 & -1 & 2 \\ 2 & -1 & 0 & 1 \end{bmatrix}$$

        3×2 3×4

✔**CHECKPOINT** Now try Exercise 37.

Example 9 Matrix Multiplication

a. $\begin{bmatrix} 1 & -2 & -3 \end{bmatrix} \begin{bmatrix} 2 \\ -1 \\ 1 \end{bmatrix} = \begin{bmatrix} 1 \end{bmatrix}$ **b.** $\begin{bmatrix} 2 \\ -1 \\ 1 \end{bmatrix} \begin{bmatrix} 1 & -2 & -3 \end{bmatrix} = \begin{bmatrix} 2 & -4 & -6 \\ -1 & 2 & 3 \\ 1 & -2 & -3 \end{bmatrix}$

    1×3 3×1 1×1 3×1 1×3 3×3

✔**CHECKPOINT** Now try Exercise 45.

In Example 9, note that the two products are different. Even when both *AB* and *BA* are defined, matrix multiplication is not, in general, commutative. That is, for most matrices,

$$AB \neq BA.$$

This is one way in which the algebra of real numbers and the algebra of matrices differ.

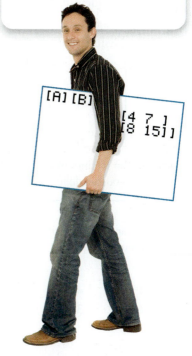

Explore the Concept

 Use the following matrices to find *AB*, *BA*, (*AB*)*C*, and *A*(*BC*). What do your results tell you about matrix multiplication and commutativity and associativity?

$$A = \begin{bmatrix} 1 & 2 \\ 3 & 4 \end{bmatrix}$$

$$B = \begin{bmatrix} 0 & 1 \\ 2 & 3 \end{bmatrix}$$

$$C = \begin{bmatrix} 3 & 0 \\ 0 & 1 \end{bmatrix}$$

Example 10 Matrix Multiplication

Use a graphing utility to find the product AB using

$$A = \begin{bmatrix} 1 & 2 & 3 \\ 2 & -5 & 1 \end{bmatrix} \quad \text{and} \quad B = \begin{bmatrix} -3 & 2 & 1 \\ 4 & -2 & 0 \\ 1 & 2 & 3 \end{bmatrix}.$$

Solution

Note that the dimension of A is 2×3 and the dimension of B is 3×3. So, the product will have dimension 2×3. Use the *matrix editor* to enter A and B into the graphing utility. Then, find the product, as shown in Figure 8.24.

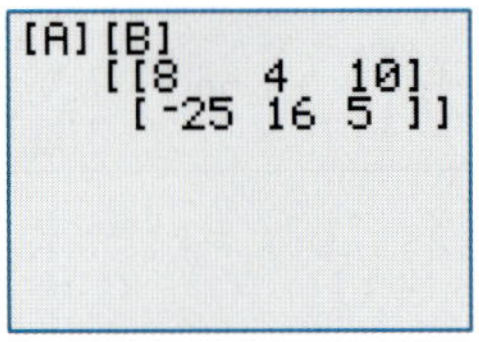

```
[A][B]
    [[8    4   10]
     [-25  16   5]]
```

Figure 8.24

✔CHECKPOINT   Now try Exercise 47.

Properties of Matrix Multiplication

Let A, B, and C be matrices and let c be a scalar.

1. $A(BC) = (AB)C$ Associative Property of Matrix Multiplication

2. $A(B + C) = AB + AC$ Left Distributive Property

3. $(A + B)C = AC + BC$ Right Distributive Property

4. $c(AB) = (cA)B = A(cB)$ Associative Property of Scalar Multiplication

---

Definition of Identity Matrix

The $n \times n$ matrix that consists of 1's on its main diagonal and 0's elsewhere is called the **identity matrix of dimension** $n \times n$ and is denoted by

$$I_n = \begin{bmatrix} 1 & 0 & 0 & \cdots & 0 \\ 0 & 1 & 0 & \cdots & 0 \\ 0 & 0 & 1 & \cdots & 0 \\ \vdots & \vdots & \vdots & & \vdots \\ 0 & 0 & 0 & \cdots & 1 \end{bmatrix}. \quad \text{Identity matrix}$$

Note that an identity matrix must be *square*. When the dimension is understood to be $n \times n$, you can denote I_n simply by I.

If A is an $n \times n$ matrix, then the identity matrix has the property that $AI_n = A$ and $I_n A = A$. For example,

$$\begin{bmatrix} 3 & -2 & 5 \\ 1 & 0 & 4 \\ -1 & 2 & -3 \end{bmatrix} \begin{bmatrix} 1 & 0 & 0 \\ 0 & 1 & 0 \\ 0 & 0 & 1 \end{bmatrix} = \begin{bmatrix} 3 & -2 & 5 \\ 1 & 0 & 4 \\ -1 & 2 & -3 \end{bmatrix} \quad AI = A$$

and

$$\begin{bmatrix} 1 & 0 & 0 \\ 0 & 1 & 0 \\ 0 & 0 & 1 \end{bmatrix} \begin{bmatrix} 3 & -2 & 5 \\ 1 & 0 & 4 \\ -1 & 2 & -3 \end{bmatrix} = \begin{bmatrix} 3 & -2 & 5 \\ 1 & 0 & 4 \\ -1 & 2 & -3 \end{bmatrix}. \quad IA = A$$

Applications

Matrix multiplication can be used to represent a system of linear equations. Note how the system

$$\begin{cases} a_{11}x_1 + a_{12}x_2 + a_{13}x_3 = b_1 \\ a_{21}x_1 + a_{22}x_2 + a_{23}x_3 = b_2 \\ a_{31}x_1 + a_{32}x_2 + a_{33}x_3 = b_3 \end{cases}$$

can be written as the matrix equation

$$AX = B$$

where A is the *coefficient matrix* of the system, B is the *constant matrix* of the system, and X is a column matrix.

$$\underbrace{\begin{bmatrix} a_{11} & a_{12} & a_{13} \\ a_{21} & a_{22} & a_{23} \\ a_{31} & a_{32} & a_{33} \end{bmatrix}}_{A} \times \underbrace{\begin{bmatrix} x_1 \\ x_2 \\ x_3 \end{bmatrix}}_{X} = \underbrace{\begin{bmatrix} b_1 \\ b_2 \\ b_3 \end{bmatrix}}_{B}$$

Example 11   Solving a System of Linear Equations

Consider the following system of linear equations.

$$\begin{cases} x_1 - 2x_2 + x_3 = -4 \\ x_2 + 2x_3 = 4 \\ 2x_1 + 3x_2 - 2x_3 = 2 \end{cases}$$

a. Write this system as a matrix equation $AX = B$.

b. Use Gauss-Jordan elimination on $[A \vdots B]$ to solve for the matrix X.

Solution

a. In matrix form $AX = B$, the system can be written as follows.

$$\begin{bmatrix} 1 & -2 & 1 \\ 0 & 1 & 2 \\ 2 & 3 & -2 \end{bmatrix} \begin{bmatrix} x_1 \\ x_2 \\ x_3 \end{bmatrix} = \begin{bmatrix} -4 \\ 4 \\ 2 \end{bmatrix}$$

b. The augmented matrix is

$$[A \vdots B] = \begin{bmatrix} 1 & -2 & 1 & \vdots & -4 \\ 0 & 1 & 2 & \vdots & 4 \\ 2 & 3 & -2 & \vdots & 2 \end{bmatrix}.$$

Using Gauss-Jordan elimination, you can rewrite this equation as

$$[I \vdots X] = \begin{bmatrix} 1 & 0 & 0 & \vdots & -1 \\ 0 & 1 & 0 & \vdots & 2 \\ 0 & 0 & 1 & \vdots & 1 \end{bmatrix}.$$

So, the solution of the system of linear equations is

$$x_1 = -1, x_2 = 2, \text{ and } x_3 = 1.$$

The solution of the matrix equation is

$$X = \begin{bmatrix} x_1 \\ x_2 \\ x_3 \end{bmatrix} = \begin{bmatrix} -1 \\ 2 \\ 1 \end{bmatrix}.$$

✔CHECKPOINT   Now try Exercise 63.

Technology Tip

 Most graphing utilities can be used to obtain the reduced row-echelon form of a matrix. The screen below shows how one graphing utility displays the reduced row-echelon form of the augmented matrix in Example 11.

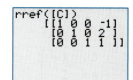

```
rref([C])
    [[1 0 0 -1]
     [0 1 0 2 ]
     [0 0 1 1 ]]
```

Evaluating an Expression In Exercises 21–24, evaluate the expression.

21. $\begin{bmatrix} -5 & 0 \\ 3 & -6 \end{bmatrix} + \begin{bmatrix} 7 & 1 \\ -2 & -1 \end{bmatrix} + \begin{bmatrix} -10 & -8 \\ 14 & 6 \end{bmatrix}$

22. $\begin{bmatrix} 6 & 9 \\ -1 & 0 \\ 7 & 1 \end{bmatrix} + \begin{bmatrix} 0 & 5 \\ -2 & -1 \\ 3 & -6 \end{bmatrix} + \begin{bmatrix} -13 & -7 \\ 4 & -1 \\ -6 & 0 \end{bmatrix}$

✓ 23. $4\left(\begin{bmatrix} -4 & 0 & 1 \\ 0 & 2 & 3 \end{bmatrix} - \begin{bmatrix} 2 & 1 & -2 \\ 3 & -6 & 0 \end{bmatrix} \right)$

24. $\frac{1}{2}([5 \quad -2 \quad 4 \quad 0] + [14 \quad 6 \quad -18 \quad 9])$

Operations with Matrices In Exercises 25–28, use the matrix capabilities of a graphing utility to evaluate the expression. Round your results to the nearest thousandths, if necessary.

25. $\frac{3}{7} \begin{bmatrix} 2 & 5 \\ -1 & -4 \end{bmatrix} + 6 \begin{bmatrix} -3 & 0 \\ 2 & 2 \end{bmatrix}$

26. $55 \begin{bmatrix} 14 & -11 \\ -22 & 19 \end{bmatrix} + \begin{bmatrix} -22 & 20 \\ 13 & 6 \end{bmatrix}$

27. $-1 \begin{bmatrix} 3.211 & 6.829 \\ -1.004 & 4.914 \\ 0.055 & -3.889 \end{bmatrix} - 1 \begin{bmatrix} 1.630 & -3.090 \\ 5.256 & 8.335 \\ -9.768 & 4.251 \end{bmatrix}$

28. $-1 \begin{bmatrix} 10 & 15 \\ -20 & 10 \\ 12 & 4 \end{bmatrix} + \frac{1}{8} \left(\begin{bmatrix} -13 & 11 \\ 7 & 0 \\ 6 & 9 \end{bmatrix} + \begin{bmatrix} -3 & 13 \\ -3 & 8 \\ -14 & 15 \end{bmatrix} \right)$

Solving a Matrix Equation In Exercises 29–32, solve for X when

$$A = \begin{bmatrix} -2 & -1 \\ 1 & 0 \\ 3 & -4 \end{bmatrix} \quad \text{and} \quad B = \begin{bmatrix} 0 & 3 \\ 2 & 0 \\ -4 & -1 \end{bmatrix}.$$

29. $X = 3A - 2B$

30. $2X = 2A - B$

✓ 31. $2X + 3A = B$

32. $2A + 4B = -2X$

Finding the Product of Two Matrices In Exercises 33–40, find AB, if possible.

33. $A = \begin{bmatrix} 2 & 1 \\ -3 & 4 \\ -1 & 6 \end{bmatrix}, \quad B = \begin{bmatrix} 0 & -3 & 0 \\ 4 & 0 & 2 \\ 8 & -2 & 7 \end{bmatrix}$

34. $A = \begin{bmatrix} 0 & -1 & 0 \\ 4 & 0 & 2 \\ 8 & -1 & 7 \end{bmatrix}, \quad B = \begin{bmatrix} 2 & 1 \\ -3 & 4 \\ 1 & 6 \end{bmatrix}$

✓ 35. $A = \begin{bmatrix} -1 & 6 \\ -4 & 5 \\ 0 & 3 \end{bmatrix}, \quad B = \begin{bmatrix} 2 & 3 \\ 0 & 9 \end{bmatrix}$

36. $A = \begin{bmatrix} 1 & 0 & 0 \\ 0 & 4 & 0 \\ 0 & 0 & -2 \end{bmatrix}, \quad B = \begin{bmatrix} 3 & 0 & 0 \\ 0 & -1 & 0 \\ 0 & 0 & 5 \end{bmatrix}$

✓ 37. $A = \begin{bmatrix} 5 & 0 & 0 \\ 0 & -8 & 0 \\ 0 & 0 & 7 \end{bmatrix}, \quad B = \begin{bmatrix} \frac{1}{5} & 0 & 0 \\ 0 & -\frac{1}{8} & 0 \\ 0 & 0 & \frac{1}{2} \end{bmatrix}$

38. $A = \begin{bmatrix} 0 & 0 & 5 \\ 0 & 0 & -3 \\ 0 & 0 & 4 \end{bmatrix}, \quad B = \begin{bmatrix} 6 & -11 & 4 \\ 8 & 16 & 4 \\ 0 & 0 & 0 \end{bmatrix}$

39. $A = \begin{bmatrix} 5 \\ 6 \end{bmatrix}, \quad B = [-3 \quad -1 \quad -5 \quad -9]$

40. $A = \begin{bmatrix} 1 & 0 & 3 & -2 \\ 6 & 13 & 8 & -17 \end{bmatrix}, \quad B = \begin{bmatrix} 1 & 6 \\ 4 & 2 \end{bmatrix}$

Operations with Matrices In Exercises 41–46, find, if possible, (a) AB, (b) BA, and (c) A^2. (*Note:* $A^2 = AA$.) Use the matrix capabilities of a graphing utility to verify your results.

41. $A = \begin{bmatrix} 1 & 2 \\ 4 & 2 \end{bmatrix}, \quad B = \begin{bmatrix} 2 & -1 \\ -1 & 8 \end{bmatrix}$

42. $A = \begin{bmatrix} 6 & 3 \\ -2 & -4 \end{bmatrix}, \quad B = \begin{bmatrix} -2 & 0 \\ 2 & 4 \end{bmatrix}$

43. $A = \begin{bmatrix} 3 & -1 \\ 1 & 3 \end{bmatrix}, \quad B = \begin{bmatrix} 1 & -3 \\ 3 & 1 \end{bmatrix}$

44. $A = \begin{bmatrix} 1 & -1 \\ 1 & 1 \end{bmatrix}, \quad B = \begin{bmatrix} 1 & 3 \\ -3 & 1 \end{bmatrix}$

✓ 45. $A = \begin{bmatrix} 7 \\ 8 \\ -1 \end{bmatrix}, \quad B = [1 \quad 1 \quad 2]$

46. $A = [3 \quad 2 \quad 1], \quad B = \begin{bmatrix} 2 \\ 3 \\ 0 \end{bmatrix}$

Matrix Multiplication In Exercises 47–50, use the matrix capabilities of a graphing utility to find AB, if possible.

✓ 47. $A = \begin{bmatrix} 7 & 5 & -4 \\ -2 & 5 & 1 \\ 10 & -4 & -7 \end{bmatrix}, \quad B = \begin{bmatrix} 2 & -2 & 3 \\ 8 & 1 & 4 \\ -4 & 2 & -8 \end{bmatrix}$

48. $A = \begin{bmatrix} 1 & -12 & 4 \\ 14 & 10 & 12 \\ 6 & -15 & 3 \end{bmatrix}, \quad B = \begin{bmatrix} 12 & 10 \\ -6 & 12 \\ 10 & 16 \end{bmatrix}$

49. $A = \begin{bmatrix} -3 & 8 & -6 & 8 \\ -12 & 15 & 9 & 6 \\ 5 & -1 & 1 & 5 \end{bmatrix}, \quad B = \begin{bmatrix} 3 & 1 & 6 \\ 24 & 15 & 14 \\ 16 & 10 & 21 \\ 8 & -4 & 10 \end{bmatrix}$

50. $A = \begin{bmatrix} -2 & 6 & 12 \\ 21 & -5 & 6 \\ 13 & -2 & 9 \end{bmatrix}, \quad B = \begin{bmatrix} 3 & 0 \\ -7 & 18 \\ 34 & 14 \\ 0.5 & 1.4 \end{bmatrix}$

Operations with Matrices In Exercises 51–54, use the matrix capabilities of a graphing utility to evaluate the expression.

51. $\begin{bmatrix} 3 & 1 \\ 0 & -2 \end{bmatrix} \begin{bmatrix} 1 & 0 \\ -2 & 2 \end{bmatrix} \begin{bmatrix} 1 & 0 \\ 2 & 4 \end{bmatrix}$

52. $-3 \begin{bmatrix} 6 & 5 & -1 \\ 1 & -2 & 0 \end{bmatrix} \begin{bmatrix} 0 & 3 \\ -1 & -3 \\ 4 & 1 \end{bmatrix}$

53. $\begin{bmatrix} 0 & 2 & -2 \\ 4 & 1 & 2 \end{bmatrix} \left(\begin{bmatrix} 4 & 0 \\ 0 & -1 \\ -1 & 2 \end{bmatrix} + \begin{bmatrix} -2 & 3 \\ -3 & 5 \\ 0 & -3 \end{bmatrix} \right)$

54. $\begin{bmatrix} 3 \\ -1 \\ 5 \\ 7 \end{bmatrix} ([5 \quad -6] + [7 \quad -1] + [-8 \quad 9])$

Matrix Multiplication In Exercises 55–58, use matrix multiplication to determine whether each matrix is a solution of the system of equations. Use a graphing utility to verify your results.

55. $\begin{cases} x + 2y = 4 \\ 3x + 2y = 0 \end{cases}$

(a) $\begin{bmatrix} 2 \\ 1 \end{bmatrix}$ (b) $\begin{bmatrix} -2 \\ 3 \end{bmatrix}$

(c) $\begin{bmatrix} -4 \\ 4 \end{bmatrix}$ (d) $\begin{bmatrix} 2 \\ -3 \end{bmatrix}$

56. $\begin{cases} 6x + 2y = 0 \\ -x + 5y = 16 \end{cases}$

(a) $\begin{bmatrix} -1 \\ 3 \end{bmatrix}$ (b) $\begin{bmatrix} 2 \\ -6 \end{bmatrix}$

(c) $\begin{bmatrix} 3 \\ -9 \end{bmatrix}$ (d) $\begin{bmatrix} -3 \\ 9 \end{bmatrix}$

57. $\begin{cases} -2x - 3y = -6 \\ 4x + 2y = 20 \end{cases}$

(a) $\begin{bmatrix} 3 \\ 0 \end{bmatrix}$ (b) $\begin{bmatrix} 6 \\ -2 \end{bmatrix}$

(c) $\begin{bmatrix} -6 \\ 6 \end{bmatrix}$ (d) $\begin{bmatrix} 4 \\ 2 \end{bmatrix}$

58. $\begin{cases} 5x - 7y = -15 \\ 3x + y = 17 \end{cases}$

(a) $\begin{bmatrix} 4 \\ 5 \end{bmatrix}$ (b) $\begin{bmatrix} 5 \\ 2 \end{bmatrix}$

(c) $\begin{bmatrix} -4 \\ -5 \end{bmatrix}$ (d) $\begin{bmatrix} 2 \\ 11 \end{bmatrix}$

Solving a System of Linear Equations In Exercises 59–66, (a) write the system of equations as a matrix equation $AX = B$ and (b) use Gauss-Jordan elimination on the augmented matrix $[A \vdots B]$ to solve for the matrix X. Use a graphing utility to check your solution.

59. $\begin{cases} -x_1 + x_2 = 4 \\ -2x_1 + x_2 = 0 \end{cases}$

60. $\begin{cases} 2x_1 + 3x_2 = 5 \\ x_1 + 4x_2 = 10 \end{cases}$

61. $\begin{cases} -2x_1 - 3x_2 = -4 \\ 6x_1 + x_2 = -36 \end{cases}$

62. $\begin{cases} -4x_1 + 9x_2 = -13 \\ x_1 - 3x_2 = 12 \end{cases}$

✓ **63.** $\begin{cases} x_1 - 2x_2 + 3x_3 = 9 \\ -x_1 + 3x_2 - x_3 = -6 \\ 2x_1 - 5x_2 + 5x_3 = 17 \end{cases}$

64. $\begin{cases} x_1 + x_2 - 3x_3 = 9 \\ -x_1 + 2x_2 = 6 \\ x_1 - x_2 + x_3 = -5 \end{cases}$

65. $\begin{cases} x_1 - 5x_2 + 2x_3 = -20 \\ -3x_1 + x_2 - x_3 = 8 \\ -2x_2 + 5x_3 = -16 \end{cases}$

66. $\begin{cases} x_1 - x_2 + 4x_3 = 17 \\ x_1 + 3x_2 = -11 \\ -6x_2 + 5x_3 = 40 \end{cases}$

Operations with Matrices In Exercises 67–72, use a graphing utility to perform the operations for the matrices A, B, and C, and the scalar c. Write a brief statement comparing the results of parts (a) and (b).

$A = \begin{bmatrix} 1 & 2 & -1 \\ 0 & -2 & 3 \\ 4 & -3 & 2 \end{bmatrix}$, $B = \begin{bmatrix} 2 & 3 & 0 \\ 4 & 1 & -2 \\ -1 & 2 & 0 \end{bmatrix}$,

$C = \begin{bmatrix} 3 & -2 & 1 \\ -4 & 0 & 3 \\ -1 & 3 & -2 \end{bmatrix}$, and $c = 3$

67. (a) $A(B + C)$ (b) $AB + AC$
68. (a) $(B + C)A$ (b) $BA + CA$
69. (a) $(A + B)^2$ (b) $A^2 + AB + BA + B^2$
70. (a) $(A - B)^2$ (b) $A^2 - AB - BA + B^2$
71. (a) $A(BC)$ (b) $(AB)C$
72. (a) $c(AB)$ (b) $(cA)B$

Operations with Matrices In Exercises 73–80, perform the operations (a) using a graphing utility and (b) by hand algebraically. If it is not possible to perform the operation(s), state the reason.

$A = \begin{bmatrix} 1 & 2 & -2 \\ -1 & 1 & 0 \end{bmatrix}$, $B = \begin{bmatrix} -1 & 4 & -1 \\ -2 & -1 & 0 \end{bmatrix}$,

$C = \begin{bmatrix} 1 & 2 \\ -2 & 3 \\ 1 & 0 \end{bmatrix}$, $c = -2$, and $d = -3$

73. $A + cB$ **74.** $A(B + C)$
75. $c(AB)$ **76.** $B + dA$
77. $CA - BC$ **78.** dAB^2
79. cdA **80.** $cA + dB$

Using a Graphing Utility In Exercises 81 and 82, use the matrix capabilities of a graphing utility to find $f(A) = a_0 I_n + a_1 A + a_2 A^2$.

81. $f(x) = x^2 - 5x + 2$, $A = \begin{bmatrix} 2 & 0 \\ 4 & 5 \end{bmatrix}$

82. $f(x) = x^2 - 7x + 6$, $A = \begin{bmatrix} 5 & 4 \\ 1 & 2 \end{bmatrix}$

83. Manufacturing A corporation has three factories, each of which manufactures acoustic guitars and electric guitars. The number of units of guitar i produced at factory j in one day is represented by a_{ij} in the matrix

$$A = \begin{bmatrix} 70 & 50 & 25 \\ 35 & 100 & 70 \end{bmatrix}.$$

Find the production levels when production is increased by 20%.

84. Manufacturing A corporation has four factories, each of which manufactures sport utility vehicles and pickup trucks. The number of units of vehicle i produced at factory j in one day is represented by a_{ij} in the matrix

$$A = \begin{bmatrix} 100 & 90 & 70 & 30 \\ 40 & 20 & 60 & 60 \end{bmatrix}.$$

Find the production levels when production is decreased by 10%.

85. Manufacturing A corporation that makes sunglasses has four factories, each of which manufactures two products. The number of units of product i produced at factory j in one day is represented by a_{ij} in the matrix

$$A = \begin{bmatrix} 100 & 120 & 60 & 40 \\ 140 & 160 & 200 & 80 \end{bmatrix}.$$

Find the production levels when production is decreased by 10%.

86. Tourism A vacation service has identified four resort hotels with a special all-inclusive package (room and meals included) at a popular travel destination. The quoted room rates are for double and family (maximum of four people) occupancy for 5 days and 4 nights. The current rates for the two types of rooms at the four hotels are represented by matrix A.

$$\begin{matrix} & \text{Hotel} & \text{Hotel} & \text{Hotel} & \text{Hotel} \\ & w & x & y & z \end{matrix}$$
$$A = \begin{bmatrix} 615 & 670 & 740 & 990 \\ 995 & 1030 & 1180 & 1105 \end{bmatrix} \begin{matrix} \text{Double} \\ \text{Family} \end{matrix} \Big\} \text{Occupancy}$$

Room rates are guaranteed not to increase by more than 12%. What is the maximum rate per package per hotel during the next year?

✓ **87. Agriculture** A fruit grower raises two crops, apples and peaches. Each of these crops is shipped to three different outlets. The number of units of crop i that are shipped to outlet j is represented by a_{ij} in the matrix

$$A = \begin{bmatrix} 125 & 100 & 75 \\ 100 & 175 & 125 \end{bmatrix}.$$

The profit per unit is represented by the matrix

$$B = [\$3.50 \quad \$6.00].$$

Find the product BA and state what each entry of the product represents.

88. Inventory Control A company sells five models of computers through three retail outlets. The inventories are given by S. The wholesale and retail prices are given by T. Compute ST and interpret the result.

$$\overbrace{}^{\text{Model}}$$
$$\begin{matrix} \text{A} & \text{B} & \text{C} & \text{D} & \text{E} \end{matrix}$$
$$S = \begin{bmatrix} 3 & 2 & 2 & 3 & 0 \\ 0 & 2 & 3 & 4 & 3 \\ 4 & 2 & 1 & 3 & 2 \end{bmatrix} \begin{matrix} 1 \\ 2 \\ 3 \end{matrix} \Big\} \text{Outlet}$$

$$\overbrace{}^{\text{Price}}$$
$$\begin{matrix} \text{Wholesale} & \text{Retail} \end{matrix}$$
$$T = \begin{bmatrix} \$840 & \$1100 \\ \$1200 & \$1350 \\ \$1450 & \$1650 \\ \$2650 & \$3000 \\ \$3050 & \$3200 \end{bmatrix} \begin{matrix} \text{A} \\ \text{B} \\ \text{C} \\ \text{D} \\ \text{E} \end{matrix} \Big\} \text{Model}$$

89. *Why you should learn it* (p. 658) A company that manufactures boats has the following labor-hour and wage requirements. Compute ST and interpret the result.

Labor per Boat

$$\overbrace{}^{\text{Department}}$$
$$\begin{matrix} \text{Cutting} & \text{Assembly} & \text{Packaging} \end{matrix}$$
$$S = \begin{bmatrix} 1.0\text{ hr} & 0.5\text{ hr} & 0.2\text{ hr} \\ 1.6\text{ hr} & 1.0\text{ hr} & 0.2\text{ hr} \\ 2.5\text{ hr} & 2.0\text{ hr} & 1.4\text{ hr} \end{bmatrix} \begin{matrix} \text{Small} \\ \text{Medium} \\ \text{Large} \end{matrix} \Big\} \begin{matrix} \text{Boat} \\ \text{size} \end{matrix}$$

Wages per Hour

$$\overbrace{}^{\text{Plant}}$$
$$\begin{matrix} \text{A} & \text{B} \end{matrix}$$
$$T = \begin{bmatrix} \$15 & \$13 \\ \$12 & \$11 \\ \$11 & \$10 \end{bmatrix} \begin{matrix} \text{Cutting} \\ \text{Assembly} \\ \text{Packaging} \end{matrix} \Big\} \text{Department}$$

90. Physical Education The numbers of calories burned by individuals of different weights performing different types of aerobic exercises for 20-minute time periods are shown in the matrix.

$$\begin{matrix} \text{120-lb} & \text{150-lb} \\ \text{person} & \text{person} \end{matrix}$$
$$B = \begin{bmatrix} 109 & 136 \\ 127 & 159 \\ 64 & 79 \end{bmatrix} \begin{matrix} \text{Bicycling} \\ \text{Jogging} \\ \text{Walking} \end{matrix}$$

(a) A 120-pound person and a 150-pound person bicycle for 40 minutes, jog for 10 minutes, and walk for 60 minutes. Organize a matrix A for the time spent exercising in units of 20-minute intervals.

(b) Find the product AB.

(c) Explain the meaning of the product AB in the context of the situation.

91. Politics The matrix

From

$$P = \begin{bmatrix} 0.6 & 0.1 & 0.1 \\ 0.2 & 0.7 & 0.1 \\ 0.2 & 0.2 & 0.8 \end{bmatrix} \begin{matrix} 1 \\ 2 \\ 3 \end{matrix} \text{ To}$$

(with column labels R D I)

is called a *stochastic matrix*. Each entry $p_{ij} (i \neq j)$ represents the proportion of the voting population that changes from party i to party j, and p_{ii} represents the proportion that remains loyal to the party from one election to the next. Compute and interpret P^2.

92. Politics Use a graphing utility to find P^3, P^4, P^5, P^6, P^7, and P^8 for the matrix given in Exercise 91. Can you detect a pattern as P is raised to higher powers?

Conclusions

True or False? In Exercises 93 and 94, determine whether the statement is true or false. Justify your answer.

93. Two matrices can be added only when they have the same dimension.

94. $\begin{bmatrix} -6 & -2 \\ 2 & -6 \end{bmatrix}\begin{bmatrix} 4 & 0 \\ 0 & -1 \end{bmatrix} = \begin{bmatrix} 4 & 0 \\ 0 & -1 \end{bmatrix}\begin{bmatrix} -6 & -2 \\ 2 & -6 \end{bmatrix}$

Think About It In Exercises 95–102, let matrices A, B, C, and D be of dimensions 2×3, 2×3, 3×2, and 2×2, respectively. Determine whether the matrices are of proper dimension to perform the operation(s). If so, give the dimension of the answer.

95. $A + 2C$ **96.** $B - 3C$

97. AB **98.** BC

99. $BC - D$ **100.** $CB - D$

101. $D(A - 3B)$ **102.** $(BC - D)A$

Think About It In Exercises 103–106, use the matrices

$$A = \begin{bmatrix} 2 & -1 \\ 1 & 3 \end{bmatrix} \text{ and } B = \begin{bmatrix} -1 & 1 \\ 0 & -2 \end{bmatrix}.$$

103. Show that $(A + B)^2 \neq A^2 + 2AB + B^2$.

104. Show that $(A - B)^2 \neq A^2 - 2AB + B^2$.

105. Show that $(A + B)(A - B) \neq A^2 - B^2$.

106. Show that $(A + B)^2 = A^2 + AB + BA + B^2$.

107. Think About It If a, b, and c are real numbers such that $c \neq 0$ and $ac = bc$, then $a = b$. However, if A, B, and C are nonzero matrices such that $AC = BC$, then A is *not necessarily* equal to B. Illustrate this using the following matrices.

$$A = \begin{bmatrix} 0 & 1 \\ 0 & 1 \end{bmatrix}, \quad B = \begin{bmatrix} 1 & 0 \\ 1 & 0 \end{bmatrix}, \quad C = \begin{bmatrix} 2 & 3 \\ 2 & 3 \end{bmatrix}$$

108. Think About It If a and b are real numbers such that $ab = 0$, then $a = 0$ or $b = 0$. However, if A and B are matrices such that $AB = O$, it is *not necessarily* true that $A = O$ or $B = O$. Illustrate this using the following matrices.

$$A = \begin{bmatrix} 3 & 3 \\ 4 & 4 \end{bmatrix}, \quad B = \begin{bmatrix} 1 & -1 \\ -1 & 1 \end{bmatrix}$$

109. Exploration Let $i = \sqrt{-1}$ and let

$$A = \begin{bmatrix} i & 0 \\ 0 & i \end{bmatrix} \quad \text{and} \quad B = \begin{bmatrix} 0 & -i \\ i & 0 \end{bmatrix}.$$

(a) Find A^2, A^3, and A^4. Identify any similarities with i^2, i^3, and i^4.

(b) Find and identify B^2.

110. Think About It Let A and B be unequal diagonal matrices of the same dimension. (A **diagonal matrix** is a square matrix in which each entry not on the main diagonal is zero.) Determine the products AB for several pairs of such matrices. Make a conjecture about a quick rule for such products.

111. Exploration Consider matrices of the form

$$A = \begin{bmatrix} 0 & a_{12} & a_{13} & a_{14} & \cdots & a_{1n} \\ 0 & 0 & a_{23} & a_{24} & \cdots & a_{2n} \\ 0 & 0 & 0 & a_{34} & \cdots & a_{3n} \\ \vdots & \vdots & \vdots & \vdots & \cdots & \vdots \\ 0 & 0 & 0 & 0 & \cdots & a_{(n-1)n} \\ 0 & 0 & 0 & 0 & \cdots & 0 \end{bmatrix}.$$

(a) Write a 2×2 matrix and a 3×3 matrix in the form of A.

(b) Use a graphing utility to raise each of the matrices to higher powers. Describe the result.

(c) Use the result of part (b) to make a conjecture about powers of A when A is a 4×4 matrix. Use the graphing utility to test your conjecture.

(d) Use the results of parts (b) and (c) to make a conjecture about powers of an $n \times n$ matrix A.

112. CAPSTONE Let matrices A and B be of dimensions 3×2 and 2×2, respectively. Answer the following questions and explain your reasoning.

(a) Is it possible that $A = B$?

(b) Is $A + B$ defined?

Cumulative Mixed Review

Condensing a Logarithmic Expression In Exercises 113 and 114, condense the expression to the logarithm of a single quantity.

113. $3 \ln 4 - \frac{1}{3}\ln(x^2 + 3)$ **114.** $\frac{3}{2}\ln 7t^4 - \frac{3}{5}\ln t^5$

8.6   The Inverse of a Square Matrix

The Inverse of a Matrix

This section further develops the algebra of matrices. To begin, consider the real number equation

$$ax = b.$$

To solve this equation for x, multiply each side of the equation by a^{-1} (provided that $a \neq 0$).

$$ax = b$$
$$(a^{-1}a)x = a^{-1}b$$
$$(1)x = a^{-1}b$$
$$x = a^{-1}b$$

The number a^{-1} is called the *multiplicative inverse* of a because

$$a^{-1}a = 1.$$

The definition of the multiplicative **inverse of a matrix** is similar.

Definition of the Inverse of a Square Matrix

Let A be an $n \times n$ matrix and let I_n be the $n \times n$ identity matrix. If there exists a matrix A^{-1} such that

$$AA^{-1} = I_n = A^{-1}A$$

then A^{-1} is called the **inverse** of A. The symbol A^{-1} is read "A inverse."

Example 1   The Inverse of a Matrix

Show that B is the inverse of A, where

$$A = \begin{bmatrix} -1 & 2 \\ -1 & 1 \end{bmatrix} \quad \text{and} \quad B = \begin{bmatrix} 1 & -2 \\ 1 & -1 \end{bmatrix}.$$

Solution

To show that B is the inverse of A, show that $AB = I = BA$, as follows.

$$AB = \begin{bmatrix} -1 & 2 \\ -1 & 1 \end{bmatrix}\begin{bmatrix} 1 & -2 \\ 1 & -1 \end{bmatrix} = \begin{bmatrix} -1+2 & 2-2 \\ -1+1 & 2-1 \end{bmatrix} = \begin{bmatrix} 1 & 0 \\ 0 & 1 \end{bmatrix}$$

$$BA = \begin{bmatrix} 1 & -2 \\ 1 & -1 \end{bmatrix}\begin{bmatrix} -1 & 2 \\ -1 & 1 \end{bmatrix} = \begin{bmatrix} -1+2 & 2-2 \\ -1+1 & 2-1 \end{bmatrix} = \begin{bmatrix} 1 & 0 \\ 0 & 1 \end{bmatrix}$$

As you can see,

$$AB = I = BA.$$

This is an example of a square matrix that has an inverse. Note that not all square matrices have inverses.

✓**CHECKPOINT**   Now try Exercise 7.

Recall that it is not always true that $AB = BA$, even when both products are defined. However, if A and B are both square matrices and $AB = I_n$, then it can be shown that $BA = I_n$. So, in Example 1, you need only check that $AB = I_2$.

What you should learn

- Verify that two matrices are inverses of each other.
- Use Gauss-Jordan elimination to find inverses of matrices.
- Use a formula to find inverses of 2×2 matrices.
- Use inverse matrices to solve systems of linear equations.

Why you should learn it

A system of equations can be solved using the inverse of the coefficient matrix. This method is particularly useful when the coefficients are the same for several systems, but the constants are different. Exercise 66 on page 680 shows how to use an inverse matrix to find a model for the number of international travelers to the United States from Europe.

Finding Inverse Matrices

When a matrix A has an inverse, A is called **invertible** (or **nonsingular**); otherwise, A is called **singular.** A nonsquare matrix cannot have an inverse. To see this, note that if A is of dimension $m \times n$ and B is of dimension $n \times m$ (where $m \neq n$), then the products AB and BA are of different dimensions and so cannot be equal to each other. Not all square matrices have inverses, as you will see later in this section. When a matrix does have an inverse, however, that inverse is unique. Example 2 shows how to use systems of equations to find the inverse of a matrix.

Example 2   Finding the Inverse of a Matrix

Find the inverse of

$$A = \begin{bmatrix} 1 & 4 \\ -1 & -3 \end{bmatrix}.$$

Solution

To find the inverse of A, try to solve the matrix equation

$$AX = I$$

for X.

$$\overset{A}{\begin{bmatrix} 1 & 4 \\ -1 & -3 \end{bmatrix}} \overset{X}{\begin{bmatrix} x_{11} & x_{12} \\ x_{21} & x_{22} \end{bmatrix}} = \overset{I}{\begin{bmatrix} 1 & 0 \\ 0 & 1 \end{bmatrix}}$$

$$\begin{bmatrix} x_{11} + 4x_{21} & x_{12} + 4x_{22} \\ -x_{11} - 3x_{21} & -x_{12} - 3x_{22} \end{bmatrix} = \begin{bmatrix} 1 & 0 \\ 0 & 1 \end{bmatrix}$$

Equating corresponding entries, you obtain the following two systems of linear equations.

$$\begin{cases} x_{11} + 4x_{21} = 1 \\ -x_{11} - 3x_{21} = 0 \end{cases} \qquad \text{Linear system with two variables, } x_{11} \text{ and } x_{21}.$$

$$\begin{cases} x_{12} + 4x_{22} = 0 \\ -x_{12} - 3x_{22} = 1 \end{cases} \qquad \text{Linear system with two variables, } x_{12} \text{ and } x_{22}.$$

Solve the first system using elementary row operations to determine that

$$x_{11} = -3 \quad \text{and} \quad x_{21} = 1.$$

From the second system you can determine that

$$x_{12} = -4 \quad \text{and} \quad x_{22} = 1.$$

Therefore, the inverse of A is

$$A^{-1} = X$$

$$= \begin{bmatrix} -3 & -4 \\ 1 & 1 \end{bmatrix}.$$

You can use matrix multiplication to check this result.

Check

$$AA^{-1} = \begin{bmatrix} 1 & 4 \\ -1 & -3 \end{bmatrix} \begin{bmatrix} -3 & -4 \\ 1 & 1 \end{bmatrix} = \begin{bmatrix} 1 & 0 \\ 0 & 1 \end{bmatrix} \checkmark$$

$$A^{-1}A = \begin{bmatrix} -3 & -4 \\ 1 & 1 \end{bmatrix} \begin{bmatrix} 1 & 4 \\ -1 & -3 \end{bmatrix} = \begin{bmatrix} 1 & 0 \\ 0 & 1 \end{bmatrix} \checkmark$$

✓CHECKPOINT   Now try Exercise 11.

Explore the Concept

Most graphing utilities are capable of finding the inverse of a square matrix. Try using a graphing utility to find the inverse of the matrix

$$A = \begin{bmatrix} 2 & -3 & 1 \\ -1 & 2 & -1 \\ -2 & 0 & 1 \end{bmatrix}.$$

After you find A^{-1}, store it as $[B]$ and use the graphing utility to find $[A] \times [B]$ and $[B] \times [A]$. What can you conclude?

In Example 2, note that the two systems of linear equations have the *same coefficient matrix A*. Rather than solve the two systems represented by

$$\begin{bmatrix} 1 & 4 & \vdots & 1 \\ -1 & -3 & \vdots & 0 \end{bmatrix}$$

and

$$\begin{bmatrix} 1 & 4 & \vdots & 0 \\ -1 & -3 & \vdots & 1 \end{bmatrix}$$

separately, you can solve them *simultaneously* by *adjoining* the identity matrix to the coefficient matrix to obtain

$$\begin{matrix} A & & & I \\ \begin{bmatrix} 1 & 4 & \vdots & 1 & 0 \\ -1 & -3 & \vdots & 0 & 1 \end{bmatrix} \end{matrix}.$$

This "doubly augmented" matrix can be represented as

$$[A \ \vdots \ I].$$

By applying Gauss-Jordan elimination to this matrix, you can solve *both* systems with a single elimination process.

$$\begin{bmatrix} 1 & 4 & \vdots & 1 & 0 \\ -1 & -3 & \vdots & 0 & 1 \end{bmatrix}$$

$$\begin{matrix} R_1 + R_2 \rightarrow \end{matrix} \begin{bmatrix} 1 & 4 & \vdots & 1 & 0 \\ 0 & 1 & \vdots & 1 & 1 \end{bmatrix}$$

$$\begin{matrix} -4R_2 + R_1 \rightarrow \end{matrix} \begin{bmatrix} 1 & 0 & \vdots & -3 & -4 \\ 0 & 1 & \vdots & 1 & 1 \end{bmatrix}$$

So, from the "doubly augmented" matrix $[A \ \vdots \ I]$, you obtained the matrix $[I \ \vdots \ A^{-1}]$.

$$\begin{matrix} A & & I \\ \begin{bmatrix} 1 & 4 & \vdots & 1 & 0 \\ -1 & -3 & \vdots & 0 & 1 \end{bmatrix} \end{matrix} \implies \begin{matrix} I & & A^{-1} \\ \begin{bmatrix} 1 & 0 & \vdots & -3 & -4 \\ 0 & 1 & \vdots & 1 & 1 \end{bmatrix} \end{matrix}$$

This procedure (or algorithm) works for any square matrix that has an inverse.

Explore the Concept

Select two 2×2 matrices A and B that have inverses. Enter them into your graphing utility and calculate $(AB)^{-1}$. Then calculate $B^{-1}A^{-1}$ and $A^{-1}B^{-1}$. Make a conjecture about the inverse of the product of two invertible matrices.

Finding an Inverse Matrix

Let A be a square matrix of dimension $n \times n$.

1. Write the $n \times 2n$ matrix that consists of the given matrix A on the left and the $n \times n$ identity matrix I on the right to obtain

   $$[A \ \vdots \ I].$$

2. If possible, row reduce A to I using elementary row operations on the *entire* matrix

   $$[A \ \vdots \ I].$$

   The result will be the matrix

   $$[I \ \vdots \ A^{-1}].$$

   If this is not possible, then A is not invertible.

3. Check your work by multiplying to see that

   $$AA^{-1} = I = A^{-1}A.$$

Example 3   Finding the Inverse of a Matrix

Find the inverse of

$$A = \begin{bmatrix} 1 & -1 & 0 \\ 1 & 0 & -1 \\ 6 & -2 & -3 \end{bmatrix}.$$

Solution

Begin by adjoining the identity matrix to A to form the matrix

$$[A \ \vdots \ I] = \begin{bmatrix} 1 & -1 & 0 & \vdots & 1 & 0 & 0 \\ 1 & 0 & -1 & \vdots & 0 & 1 & 0 \\ 6 & -2 & -3 & \vdots & 0 & 0 & 1 \end{bmatrix}.$$

Use elementary row operations to obtain the form $[I \ \vdots \ A^{-1}]$, as follows.

$$\begin{bmatrix} 1 & 0 & 0 & \vdots & -2 & -3 & 1 \\ 0 & 1 & 0 & \vdots & -3 & -3 & 1 \\ 0 & 0 & 1 & \vdots & -2 & -4 & 1 \end{bmatrix}$$

Therefore, the matrix A is invertible and its inverse is

$$A^{-1} = \begin{bmatrix} -2 & -3 & 1 \\ -3 & -3 & 1 \\ -2 & -4 & 1 \end{bmatrix}.$$

Confirm this result by multiplying A by A^{-1} to obtain I, as follows.

Check

$$AA^{-1} = \begin{bmatrix} 1 & -1 & 0 \\ 1 & 0 & -1 \\ 6 & -2 & -3 \end{bmatrix}\begin{bmatrix} -2 & -3 & 1 \\ -3 & -3 & 1 \\ -2 & -4 & 1 \end{bmatrix} = \begin{bmatrix} 1 & 0 & 0 \\ 0 & 1 & 0 \\ 0 & 0 & 1 \end{bmatrix} = I$$

✓**CHECKPOINT**   Now try Exercise 17.

The algorithm shown in Example 3 applies to any $n \times n$ matrix A. When using this algorithm, if the matrix A does not reduce to the identity matrix, then A does not have an inverse. For instance, the following matrix has no inverse.

$$A = \begin{bmatrix} 1 & 2 & 0 \\ 3 & -1 & 2 \\ -2 & 3 & -2 \end{bmatrix}$$

To see why matrix A above has no inverse, begin by adjoining the identity matrix to A to form

$$[A \ \vdots \ I] = \begin{bmatrix} 1 & 2 & 0 & \vdots & 1 & 0 & 0 \\ 3 & -1 & 2 & \vdots & 0 & 1 & 0 \\ -2 & 3 & -2 & \vdots & 0 & 0 & 1 \end{bmatrix}.$$

Then use elementary row operations to obtain

$$\begin{bmatrix} 1 & 2 & 0 & \vdots & 1 & 0 & 0 \\ 0 & -7 & 2 & \vdots & -3 & 1 & 0 \\ 0 & 0 & 0 & \vdots & -1 & 1 & 1 \end{bmatrix}.$$

At this point in the elimination process you can see that it is impossible to obtain the identity matrix I on the left. Therefore, A is not invertible.

Technology Tip

Most graphing utilities can find the inverse of a matrix by using the inverse key $\boxed{x^{-1}}$. For instructions on how to use the inverse key to find the inverse of a matrix, see Appendix A; for specific keystrokes, go to this textbook's *Companion Website*.

The Inverse of a 2 × 2 Matrix

Using Gauss-Jordan elimination to find the inverse of a matrix works well (even as a computer technique) for matrices of dimension 3×3 or greater. For 2×2 matrices, however, many people prefer to use a formula for the inverse rather than Gauss-Jordan elimination. This simple formula, which works *only* for 2×2 matrices, is explained as follows. If A is the 2×2 matrix given by

$$A = \begin{bmatrix} a & b \\ c & d \end{bmatrix}$$

then A is invertible if and only if

$$ad - bc \neq 0.$$

If $ad - bc \neq 0$, then the inverse is given by

$$A^{-1} = \frac{1}{ad - bc} \begin{bmatrix} d & -b \\ -c & a \end{bmatrix}.$$ Formula for inverse of matrix A

The denominator

$$ad - bc$$

is called the *determinant* of the 2×2 matrix A. You will study determinants in the next section.

Example 4   Finding the Inverse of a 2 × 2 Matrix

If possible, find the inverse of each matrix.

a. $A = \begin{bmatrix} 3 & -1 \\ -2 & 2 \end{bmatrix}$

b. $B = \begin{bmatrix} 3 & -1 \\ -6 & 2 \end{bmatrix}$

Solution

a. For the matrix A, apply the formula for the inverse of a 2×2 matrix to obtain

$$ad - bc = (3)(2) - (-1)(-2)$$

$$= 4.$$

Because this quantity is not zero, the inverse is formed by interchanging the entries on the main diagonal, changing the signs of the other two entries, and multiplying by the scalar $\frac{1}{4}$, as follows.

$$A^{-1} = \frac{1}{4} \begin{bmatrix} 2 & 1 \\ 2 & 3 \end{bmatrix}$$ Substitute for a, b, c, d, and the determinant.

$$= \begin{bmatrix} \frac{1}{2} & \frac{1}{4} \\ \frac{1}{2} & \frac{3}{4} \end{bmatrix}$$ Multiply by the scalar $\frac{1}{4}$.

b. For the matrix B, you have

$$ad - bc = (3)(2) - (-1)(-6)$$

$$= 0$$

which means that B is not invertible.

✓**CHECKPOINT**   Now try Exercise 29.

Explore the Concept

 Use a graphing utility to find the inverse of the matrix

$$A = \begin{bmatrix} 1 & -3 \\ -2 & 6 \end{bmatrix}.$$

What message appears on the screen? Why does the graphing utility display this message?

Systems of Linear Equations

You know that a system of linear equations can have exactly one solution, infinitely many solutions, or no solution. If the coefficient matrix A of a *square* system (a system that has the same number of equations as variables) is invertible, then the system has a unique solution, which is defined as follows.

A System of Equations with a Unique Solution

If A is an invertible matrix, then the system of linear equations represented by $AX = B$ has a unique solution given by

$$X = A^{-1}B.$$

The formula $X = A^{-1}B$ is used on most graphing utilities to solve linear systems that have invertible coefficient matrices. That is, you enter the $n \times n$ coefficient matrix $[A]$ and the $n \times 1$ column matrix $[B]$. The solution X is given by $[A]^{-1}[B]$.

Example 5   Solving a System of Equations Using an Inverse

Use an inverse matrix to solve the system.

$$\begin{cases} 2x + 3y + z = -1 \\ 3x + 3y + z = 1 \\ 2x + 4y + z = -2 \end{cases}$$

Solution

Begin by writing the system as $AX = B$.

$$\begin{bmatrix} 2 & 3 & 1 \\ 3 & 3 & 1 \\ 2 & 4 & 1 \end{bmatrix} \begin{bmatrix} x \\ y \\ z \end{bmatrix} = \begin{bmatrix} -1 \\ 1 \\ -2 \end{bmatrix}$$

Then, use Gauss-Jordan elimination to find A^{-1}.

$$A^{-1} = \begin{bmatrix} -1 & 1 & 0 \\ -1 & 0 & 1 \\ 6 & -2 & -3 \end{bmatrix}$$

Finally, multiply B by A^{-1} on the left to obtain the solution.

$$X = A^{-1}B$$

$$= \begin{bmatrix} -1 & 1 & 0 \\ -1 & 0 & 1 \\ 6 & -2 & -3 \end{bmatrix} \begin{bmatrix} -1 \\ 1 \\ -2 \end{bmatrix}$$

$$= \begin{bmatrix} 2 \\ -1 \\ -2 \end{bmatrix}$$

So, the solution is

$$x = 2, \quad y = -1, \quad \text{and} \quad z = -2.$$

Use a graphing utility to verify A^{-1} for the system of equations.

✓ CHECKPOINT   Now try Exercise 51.

Study Tip

 Remember that matrix multiplication is not commutative. So, you must multiply matrices in the correct order. For instance, in Example 5, you must multiply B by A^{-1} on the left.

Vocabulary and Concept Check

In Exercises 1 and 2, fill in the blank(s).

1. If there exists an $n \times n$ matrix A^{-1} such that $AA^{-1} = I_n = A^{-1}A$, then A^{-1} is called the _____ of A.

2. If a matrix A has an inverse, then it is called invertible or _____ ; if it does not have an inverse, then it is called _____ .

3. Do all square matrices have inverses?

4. Given that A and B are square matrices and $AB = I_n$, does $BA = I_n$?

Procedures and Problem Solving

The Inverse of a Matrix In Exercises 5–10, show that B is the inverse of A.

5. $A = \begin{bmatrix} 2 & 1 \\ 5 & 3 \end{bmatrix}$, $B = \begin{bmatrix} 3 & -1 \\ -5 & 2 \end{bmatrix}$

6. $A = \begin{bmatrix} 1 & -1 \\ -1 & 2 \end{bmatrix}$, $B = \begin{bmatrix} 2 & 1 \\ 1 & 1 \end{bmatrix}$

✓ 7. $A = \begin{bmatrix} 1 & 2 \\ 3 & 4 \end{bmatrix}$, $B = \begin{bmatrix} -2 & 1 \\ \frac{3}{2} & -\frac{1}{2} \end{bmatrix}$

8. $A = \begin{bmatrix} 1 & -1 \\ 2 & 3 \end{bmatrix}$, $B = \begin{bmatrix} \frac{3}{5} & \frac{1}{5} \\ -\frac{2}{5} & \frac{1}{5} \end{bmatrix}$

9. $A = \begin{bmatrix} 2 & -17 & 11 \\ -1 & 11 & -7 \\ 0 & 3 & -2 \end{bmatrix}$, $B = \begin{bmatrix} 1 & 1 & 2 \\ 2 & 4 & -3 \\ 3 & 6 & -5 \end{bmatrix}$

10. $A = \begin{bmatrix} 1 & 0 & -1 \\ -1 & 1 & 0 \\ 1 & 2 & 0 \end{bmatrix}$, $B = \frac{1}{3}\begin{bmatrix} 0 & -2 & 1 \\ 0 & 1 & 1 \\ -3 & -2 & 1 \end{bmatrix}$

Finding the Inverse of a Matrix In Exercises 11–20, find the inverse of the matrix (if it exists).

✓ 11. $\begin{bmatrix} 2 & 0 \\ 0 & 3 \end{bmatrix}$

12. $\begin{bmatrix} 1 & 2 \\ 3 & 7 \end{bmatrix}$

13. $\begin{bmatrix} 1 & -2 \\ 2 & -3 \end{bmatrix}$

14. $\begin{bmatrix} -7 & 33 \\ 4 & -19 \end{bmatrix}$

15. $\begin{bmatrix} 2 & 7 & 1 \\ -3 & -9 & 2 \end{bmatrix}$

16. $\begin{bmatrix} -2 & 5 \\ 6 & -15 \\ 0 & 1 \end{bmatrix}$

✓ 17. $\begin{bmatrix} 1 & 1 & 1 \\ 3 & 5 & 4 \\ 3 & 6 & 5 \end{bmatrix}$

18. $\begin{bmatrix} 1 & 2 & 2 \\ 3 & 7 & 9 \\ -1 & -4 & -7 \end{bmatrix}$

19. $\begin{bmatrix} 1 & 0 & 0 \\ 3 & 4 & 0 \\ 2 & 5 & 5 \end{bmatrix}$

20. $\begin{bmatrix} 1 & 0 & 0 \\ 3 & 5 & 0 \\ 2 & 5 & 0 \end{bmatrix}$

Finding the Inverse of a Matrix In Exercises 21–28, use the matrix capabilities of a graphing utility to find the inverse of the matrix (if it exists).

21. $\begin{bmatrix} 1 & 1 & 2 \\ 3 & 1 & 0 \\ -2 & 0 & 3 \end{bmatrix}$

22. $\begin{bmatrix} 1 & 2 & -1 \\ 3 & 7 & -10 \\ -5 & -7 & -15 \end{bmatrix}$

23. $\begin{bmatrix} -\frac{1}{2} & \frac{3}{4} & \frac{1}{4} \\ 1 & 0 & -\frac{3}{2} \\ 0 & -1 & \frac{1}{2} \end{bmatrix}$

24. $\begin{bmatrix} -\frac{5}{6} & \frac{1}{3} & \frac{11}{6} \\ 0 & \frac{2}{3} & 2 \\ 1 & -\frac{1}{2} & -\frac{5}{2} \end{bmatrix}$

25. $\begin{bmatrix} 0.1 & 0.2 & 0.3 \\ -0.3 & 0.2 & 0.2 \\ 0.5 & 0.4 & 0.4 \end{bmatrix}$

26. $\begin{bmatrix} 0.6 & 0 & -0.3 \\ 0.7 & -1 & 0.2 \\ 1 & 0 & -0.9 \end{bmatrix}$

27. $\begin{bmatrix} -1 & 0 & 1 & 0 \\ 0 & 2 & 0 & -1 \\ 2 & 0 & -1 & 0 \\ 0 & -1 & 0 & 1 \end{bmatrix}$

28. $\begin{bmatrix} 1 & -2 & -1 & -2 \\ 3 & -5 & -2 & -3 \\ 2 & -5 & -2 & -5 \\ -1 & 4 & 4 & 11 \end{bmatrix}$

Finding the Inverse of a 2 × 2 Matrix In Exercises 29–34, use the formula on page 676 to find the inverse of the 2 × 2 matrix.

✓ 29. $\begin{bmatrix} 5 & 1 \\ -2 & -2 \end{bmatrix}$

30. $\begin{bmatrix} 7 & 12 \\ -8 & -5 \end{bmatrix}$

31. $\begin{bmatrix} \frac{7}{2} & -\frac{3}{4} \\ \frac{1}{5} & \frac{4}{5} \end{bmatrix}$

32. $\begin{bmatrix} -\frac{1}{4} & -\frac{2}{3} \\ \frac{1}{3} & \frac{8}{9} \end{bmatrix}$

33. $\begin{bmatrix} 2 & 3 \\ -1 & 5 \end{bmatrix}$

34. $\begin{bmatrix} 1 & -2 \\ -3 & 2 \end{bmatrix}$

Finding a Matrix Entry In Exercises 35 and 36, find the value of the constant k such that $B = A^{-1}$.

35. $A = \begin{bmatrix} 1 & 2 \\ -2 & 0 \end{bmatrix}$, $B = \begin{bmatrix} k & -\frac{1}{2} \\ \frac{1}{2} & \frac{1}{4} \end{bmatrix}$

36. $A = \begin{bmatrix} -1 & 1 \\ 2 & 1 \end{bmatrix}$, $B = \begin{bmatrix} -\frac{1}{3} & \frac{1}{3} \\ k & \frac{1}{3} \end{bmatrix}$

Solving a System of Linear Equations In Exercises 37–40, use the inverse matrix found in Exercise 13 to solve the system of linear equations.

37. $\begin{cases} x - 2y = 5 \\ 2x - 3y = 10 \end{cases}$ 38. $\begin{cases} x - 2y = 0 \\ 2x - 3y = 3 \end{cases}$

39. $\begin{cases} x - 2y = 4 \\ 2x - 3y = 2 \end{cases}$ 40. $\begin{cases} x - 2y = 1 \\ 2x - 3y = -2 \end{cases}$

Solving a System of Linear Equations In Exercises 41 and 42, use the inverse matrix found in Exercise 17 to solve the system of linear equations.

41. $\begin{cases} x + y + z = 0 \\ 3x + 5y + 4z = 5 \\ 3x + 6y + 5z = 2 \end{cases}$ 42. $\begin{cases} x + y + z = -1 \\ 3x + 5y + 4z = 2 \\ 3x + 6y + 5z = 0 \end{cases}$

Solving a System of Linear Equations In Exercises 43 and 44, use the inverse matrix found in Exercise 28 and the matrix capabilities of a graphing utility to solve the system of linear equations.

43. $\begin{cases} x_1 - 2x_2 - x_3 - 2x_4 = 0 \\ 3x_1 - 5x_2 - 2x_3 - 3x_4 = 1 \\ 2x_1 - 5x_2 - 2x_3 - 5x_4 = -1 \\ -x_1 + 4x_2 + 4x_3 + 11x_4 = 2 \end{cases}$

44. $\begin{cases} x_1 - 2x_2 - x_3 - 2x_4 = 1 \\ 3x_1 - 5x_2 - 2x_3 - 3x_4 = -2 \\ 2x_1 - 5x_2 - 2x_3 - 5x_4 = 0 \\ -x_1 + 4x_2 + 4x_3 + 11x_4 = -3 \end{cases}$

Solving a System of Equations Using an Inverse In Exercises 45–52, use an inverse matrix to solve (if possible) the system of linear equations.

45. $\begin{cases} 3x + 4y = -2 \\ 5x + 3y = 4 \end{cases}$ 46. $\begin{cases} 18x + 12y = 13 \\ 30x + 24y = 23 \end{cases}$

47. $\begin{cases} -0.4x + 0.8y = 1.6 \\ 2x - 4y = 5 \end{cases}$ 48. $\begin{cases} 0.2x - 0.6y = 2.4 \\ -x + 1.4y = -8.8 \end{cases}$

49. $\begin{cases} -\frac{1}{4}x + \frac{3}{8}y = -2 \\ \frac{3}{2}x + \frac{3}{4}y = -12 \end{cases}$ 50. $\begin{cases} \frac{5}{6}x - y = -20 \\ \frac{4}{3}x - \frac{7}{2}y = -51 \end{cases}$

✓ 51. $\begin{cases} 4x - y + z = -5 \\ 2x + 2y + 3z = 10 \\ 5x - 2y + 6z = 1 \end{cases}$ 52. $\begin{cases} 4x - 2y + 3z = -2 \\ 2x + 2y + 5z = 16 \\ 8x - 5y - 2z = 4 \end{cases}$

Solving a System of Linear Equations In Exercises 53–56, use the matrix capabilities of a graphing utility to solve (if possible) the system of linear equations.

53. $\begin{cases} 5x - 3y + 2z = 2 \\ 2x + 2y - 3z = 3 \\ -x + 7y - 8z = 4 \end{cases}$ 54. $\begin{cases} 2x + 3y + 5z = 4 \\ 3x + 5y - 9z = 7 \\ 5x + 9y + 17z = 13 \end{cases}$

55. $\begin{cases} 7x - 3y + 2w = 41 \\ -2x + y - w = -13 \\ 4x + z - 2w = 12 \\ -x + y - w = -8 \end{cases}$

56. $\begin{cases} 2x + 5y + w = 11 \\ x + 4y + 2z - 2w = -7 \\ 2x - 2y + 5z + w = 3 \\ x - 3w = -1 \end{cases}$

Investment Portfolio In Exercises 57–60, consider a person who invests in AAA-rated bonds, A-rated bonds, and B-rated bonds. The average yields are 6.5% on AAA bonds, 7% on A bonds, and 9% on B bonds. The person invests twice as much in B bonds as in A bonds. Let x, y, and z represent the amounts invested in AAA, A, and B bonds, respectively.

$$\begin{cases} x + y + z = \text{(total investment)} \\ 0.065x + 0.07y + 0.09z = \text{(annual return)} \\ 2y - z = 0 \end{cases}$$

Use the inverse of the coefficient matrix of this system to find the amount invested in each type of bond.

	Total Investment	Annual Return
57.	$10,000	$705
58.	$10,000	$760
59.	$12,000	$835
60.	$500,000	$38,000

Electrical Engineering In Exercises 61 and 62, consider the circuit in the figure. The currents I_1, I_2, and I_3, in amperes, are given by the solution of the system of linear equations

$$\begin{cases} 2I_1 + 4I_3 = E_1 \\ I_2 + 4I_3 = E_2 \\ I_1 + I_2 - I_3 = 0 \end{cases}$$

where E_1 and E_2 are voltages. Use the inverse of the coefficient matrix of this system to find the unknown currents for the given voltages.

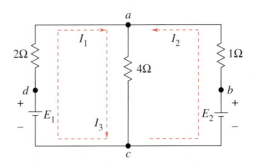

61. $E_1 = 15$ volts, $E_2 = 17$ volts
62. $E_1 = 10$ volts, $E_2 = 10$ volts

Horticulture In Exercises 63 and 64, consider a company that specializes in potting soil. Each bag of potting soil for seedlings requires 2 units of sand, 1 unit of loam, and 1 unit of peat moss. Each bag of potting soil for general potting requires 1 unit of sand, 2 units of loam, and 1 unit of peat moss. Each bag of potting soil for hardwood plants requires 2 units of sand, 2 units of loam, and 2 units of peat moss. Find the numbers of bags of the three types of potting soil that the company can produce with the given amounts of raw materials.

63. 500 units of sand
500 units of loam
400 units of peat moss

64. 500 units of sand
750 units of loam
450 units of peat moss

65. Floral Design A florist is creating 10 centerpieces for the tables at a wedding reception. Roses cost \$2.50 each, lilies cost \$4 each, and irises cost \$2 each. The customer has a budget of \$300 allocated for the centerpieces and wants each centerpiece to contain 12 flowers, with twice as many roses as the number of irises and lilies combined.

(a) Write a system of linear equations that represents the situation.

(b) Write a matrix equation that corresponds to your system.

(c) Solve your system of linear equations using an inverse matrix. Find the number of flowers of each type that the florist can use to create the 10 centerpieces.

66. *Why you should learn it* (p. 672) The table shows the numbers of international travelers y (in thousands) to the United States from Europe from 2006 through 2008. (Source: U.S. Department of Commerce)

Year	Travelers, y (in thousands)
2006	10,136
2007	11,406
2008	12,783

(a) The data can be approximated by a parabola. Create a system of linear equations for the data. Let t represent the year, with $t = 6$ corresponding to 2006.

(b) Use the matrix capabilities of a graphing utility to find an inverse matrix to solve the system in part (a) and find the least squares regression parabola $y = at^2 + bt + c$.

(c) Use the graphing utility to graph the parabola with the data points.

(d) Use the result of part (b) to estimate the numbers of international travelers to the United States from Europe in 2009, 2010, and 2011.

(e) Are your estimates from part (d) reasonable? Explain.

Conclusions

True or False? In Exercises 67 and 68, determine whether the statement is true or false. Justify your answer.

67. Multiplication of an invertible matrix and its inverse is commutative.

68. When the product of two square matrices is the identity matrix, the matrices are inverses of one another.

69. Writing Explain how to determine whether the inverse of a 2×2 matrix exists. If so, explain how to find the inverse.

70. If A is a 2×2 matrix given by $A = \begin{bmatrix} a & b \\ c & d \end{bmatrix}$, then A is invertible if and only if $ad - bc \neq 0$. If $ad - bc \neq 0$, verify that the inverse is $A^{-1} = \dfrac{1}{ad - bc} \begin{bmatrix} d & -b \\ -c & a \end{bmatrix}$.

71. Exploration Consider matrices of the form

$$A = \begin{bmatrix} a_{11} & 0 & 0 & 0 & \ldots & 0 \\ 0 & a_{22} & 0 & 0 & \ldots & 0 \\ 0 & 0 & a_{33} & 0 & \ldots & 0 \\ \vdots & \vdots & \vdots & \vdots & \ldots & \vdots \\ 0 & 0 & 0 & 0 & \ldots & a_{nn} \end{bmatrix}.$$

(a) Write a 2×2 matrix and a 3×3 matrix in the form of A. Find the inverse of each.

(b) Use the result of part (a) to make a conjecture about the inverse of a matrix in the form of A.

72. CAPSTONE Let A be a 2×2 matrix given by

$$A = \begin{bmatrix} x & 0 \\ 0 & y \end{bmatrix}.$$

Use the determinant of A to determine the conditions under which A^{-1} exists.

Cumulative Mixed Review

Solving an Equation In Exercises 73–76, solve the equation algebraically. Round your result to three decimal places.

73. $e^{2x} + 2e^x - 15 = 0$

74. $e^{2x} - 10e^x + 24 = 0$

75. $7 \ln 3x = 12$

76. $\ln(x + 9) = 2$

77. *Make a Decision* To work an extended application analyzing the numbers of U.S. households with televisions from 1985 through 2008, visit this textbook's *Companion Website*. (Data Source: Nielsen Media Research)

The Determinant of a 2 × 2 Matrix

Every *square* matrix can be associated with a real number called its **determinant.** Determinants have many uses, and several will be discussed in this and the next section. Historically, the use of determinants arose from special number patterns that occur when systems of linear equations are solved. For instance, the system

$$\begin{cases} a_1 x + b_1 y = c_1 \\ a_2 x + b_2 y = c_2 \end{cases}$$

has a solution

$$x = \frac{c_1 b_2 - c_2 b_1}{a_1 b_2 - a_2 b_1}$$

and

$$y = \frac{a_1 c_2 - a_2 c_1}{a_1 b_2 - a_2 b_1}$$

provided that $a_1 b_2 - a_2 b_1 \neq 0$. Note that each fraction has the same denominator. This denominator is called the *determinant* of the coefficient matrix of the system.

Coefficient Matrix   *Determinant*

$$A = \begin{bmatrix} a_1 & b_1 \\ a_2 & b_2 \end{bmatrix} \qquad \det(A) = a_1 b_2 - a_2 b_1$$

The determinant of the matrix A can also be denoted by vertical bars on both sides of the matrix, as indicated in the following definition.

Definition of the Determinant of a 2 × 2 Matrix

The **determinant** of the matrix

$$A = \begin{bmatrix} a_1 & b_1 \\ a_2 & b_2 \end{bmatrix}$$

is given by

$$\det(A) = |A|$$

$$= \begin{vmatrix} a_1 & b_1 \\ a_2 & b_2 \end{vmatrix}$$

$$= a_1 b_2 - a_2 b_1.$$

In this text, $\det(A)$ and $|A|$ are used interchangeably to represent the determinant of A. Although vertical bars are also used to denote the absolute value of a real number, the context will show which use is intended.

A convenient method for remembering the formula for the determinant of a 2 × 2 matrix is shown in the following diagram.

$$\det(A) = \begin{vmatrix} a_1 & b_1 \\ a_2 & b_2 \end{vmatrix} = a_1 b_2 - a_2 b_1$$

Note that the determinant is the difference of the products of the two diagonals of the matrix.

What you should learn

● Find the determinants of 2 × 2 matrices.
● Find minors and cofactors of square matrices.
● Find the determinants of square matrices.

Why you should learn it

Determinants and Cramer's Rule can be used to find the least squares regression parabola that models retail sales of family clothing stores, as shown in Exercise 27 on page 696 of Section 8.8.

The Determinant of a Square Matrix

The following definition is called *inductive* because it uses determinants of matrices of dimension $(n - 1) \times (n - 1)$ to define determinants of matrices of dimension $n \times n$.

Determinant of a Square Matrix

If A is a square matrix (of dimension 2×2 or greater), then the determinant of A is the sum of the entries in any row (or column) of A multiplied by their respective cofactors. For instance, expanding along the first row yields

$$|A| = a_{11}C_{11} + a_{12}C_{12} + \cdots + a_{1n}C_{1n}.$$

Applying this definition to find a determinant is called **expanding by cofactors.**

Try checking that for a 2×2 matrix

$$A = \begin{bmatrix} a_1 & b_1 \\ a_2 & b_2 \end{bmatrix}$$

this definition of the determinant yields $|A| = a_1 b_2 - a_2 b_1$, as previously defined.

Example 3 The Determinant of a Matrix of Dimension 3 × 3

Find the determinant of $A = \begin{bmatrix} 0 & 2 & 1 \\ 3 & -1 & 2 \\ 4 & 0 & 1 \end{bmatrix}$.

Solution

Note that this is the same matrix that was in Example 2. There you found the cofactors of the entries in the first row to be

$$C_{11} = -1, \quad C_{12} = 5, \quad \text{and} \quad C_{13} = 4.$$

So, by the definition of the determinant of a square matrix, you have

$$
\begin{aligned}
|A| &= a_{11}C_{11} + a_{12}C_{12} + a_{13}C_{13} \qquad \text{First-row expansion} \\
&= 0(-1) + 2(5) + 1(4) \\
&= 14.
\end{aligned}
$$

✔**CHECKPOINT** Now try Exercise 23.

In Example 3, the determinant was found by expanding by the cofactors in the first row. You could have used any row or column. For instance, you could have expanded along the second row to obtain

$$
\begin{aligned}
|A| &= a_{21}C_{21} + a_{22}C_{22} + a_{23}C_{23} \qquad \text{Second-row expansion} \\
&= 3(-2) + (-1)(-4) + 2(8) \\
&= 14.
\end{aligned}
$$

When expanding by cofactors, you do not need to find cofactors of zero entries, because zero times its cofactor is zero.

$$a_{ij}C_{ij} = (0)C_{ij} = 0$$

So, the row (or column) containing the most zeros is usually the best choice for expansion by cofactors.

8.7 Exercises

See www.CalcChat.com for worked-out solutions to odd-numbered exercises.
For instructions on how to use a graphing utility, see Appendix A.

Vocabulary and Concept Check

In Exercises 1 and 2, fill in the blank.

1. Both $\det(A)$ and $|A|$ represent the _____ of the matrix A.

2. The determinant of the matrix obtained by deleting the ith row and jth column of a square matrix A is called the _____ of the entry a_{ij}.

3. For a square matrix B, the minor $M_{23} = 5$. What is the cofactor C_{23} of matrix B?

4. To find the determinant of a matrix using expanding by cofactors, do you need to find all the cofactors?

Procedures and Problem Solving

The Determinant of a Matrix In Exercises 5–12, find the determinant of the matrix.

5. $\begin{bmatrix} 4 \end{bmatrix}$

6. $\begin{bmatrix} -12 \end{bmatrix}$

7. $\begin{bmatrix} 8 & 4 \\ 2 & 3 \end{bmatrix}$

8. $\begin{bmatrix} -5 & 2 \\ 6 & 3 \end{bmatrix}$

✓ **9.** $\begin{bmatrix} 6 & 2 \\ -5 & 3 \end{bmatrix}$

10. $\begin{bmatrix} 3 & -3 \\ 4 & -8 \end{bmatrix}$

11. $\begin{bmatrix} -7 & 6 \\ \frac{1}{2} & 3 \end{bmatrix}$

12. $\begin{bmatrix} 4 & -3 \\ 0 & 0 \end{bmatrix}$

Using a Graphing Utility In Exercises 13 and 14, use the matrix capabilities of a graphing utility to find the determinant of the matrix.

13. $\begin{bmatrix} 0.3 & 0.2 & 0.2 \\ 0.2 & 0.2 & 0.2 \\ -0.4 & 0.4 & 0.3 \end{bmatrix}$

14. $\begin{bmatrix} 0.1 & 0.2 & 0.3 \\ -0.3 & 0.2 & 0.2 \\ 0.5 & 0.4 & 0.4 \end{bmatrix}$

Finding the Minors and Cofactors of a Matrix In Exercises 15–18, find all (a) minors and (b) cofactors of the matrix.

15. $\begin{bmatrix} 3 & 4 \\ 2 & -5 \end{bmatrix}$

16. $\begin{bmatrix} 11 & 0 \\ -3 & 2 \end{bmatrix}$

✓ **17.** $\begin{bmatrix} -4 & 6 & 3 \\ 7 & -2 & 8 \\ 1 & 0 & -5 \end{bmatrix}$

18. $\begin{bmatrix} -2 & 9 & 4 \\ 7 & -6 & 0 \\ 6 & 7 & -6 \end{bmatrix}$

Finding a Determinant In Exercises 19–22, find the determinant of the matrix. Expand by cofactors on each indicated row or column.

19. $\begin{bmatrix} -3 & 2 & 1 \\ 4 & 5 & 6 \\ 2 & -3 & 1 \end{bmatrix}$

(a) Row 1

(b) Column 2

20. $\begin{bmatrix} -3 & 4 & 2 \\ 6 & 3 & 1 \\ 4 & -7 & -8 \end{bmatrix}$

(a) Row 2

(b) Column 3

21. $\begin{bmatrix} 6 & 0 & -3 & 5 \\ 4 & 13 & 6 & -8 \\ -1 & 0 & 7 & 4 \\ 8 & 6 & 0 & 2 \end{bmatrix}$

(a) Row 2

(b) Column 2

22. $\begin{bmatrix} 10 & 8 & 3 & -7 \\ 4 & 0 & 5 & -6 \\ 0 & 3 & 2 & 7 \\ 1 & 0 & -3 & 2 \end{bmatrix}$

(a) Row 3

(b) Column 1

Finding a Determinant In Exercises 23–32, find the determinant of the matrix. Expand by cofactors on the row or column that appears to make the computations easiest.

✓ **23.** $\begin{bmatrix} 1 & 4 & -2 \\ 3 & 2 & 0 \\ -1 & 4 & 3 \end{bmatrix}$

24. $\begin{bmatrix} -3 & 0 & 0 \\ 7 & 11 & 0 \\ 1 & 2 & 2 \end{bmatrix}$

25. $\begin{bmatrix} 6 & 3 & -7 \\ 0 & 0 & 0 \\ 4 & -6 & 3 \end{bmatrix}$

26. $\begin{bmatrix} 1 & 1 & 2 \\ 3 & 1 & 0 \\ -2 & 0 & 3 \end{bmatrix}$

27. $\begin{bmatrix} -1 & 2 & -5 \\ 0 & 3 & -4 \\ 0 & 0 & 3 \end{bmatrix}$

28. $\begin{bmatrix} 1 & 0 & 0 \\ -1 & -1 & 0 \\ 4 & 1 & 5 \end{bmatrix}$

29. $\begin{bmatrix} 2 & 6 & 6 & 2 \\ 2 & 7 & 3 & 6 \\ 1 & 5 & 0 & 1 \\ 3 & 7 & 0 & 7 \end{bmatrix}$

30. $\begin{bmatrix} 3 & 6 & -5 & 4 \\ -2 & 0 & 6 & 0 \\ 1 & 1 & 2 & 2 \\ 0 & 3 & -1 & -1 \end{bmatrix}$

31. $\begin{bmatrix} 3 & 2 & 4 & -1 & 5 \\ -2 & 0 & 1 & 3 & 2 \\ 1 & 0 & 0 & 4 & 0 \\ 6 & 0 & 2 & -1 & 0 \\ 3 & 0 & 5 & 1 & 0 \end{bmatrix}$

32. $\begin{bmatrix} 5 & 2 & 0 & 0 & -2 \\ 0 & 1 & 4 & 3 & 2 \\ 0 & 0 & 2 & 6 & 3 \\ 0 & 0 & 3 & 4 & 1 \\ 0 & 0 & 0 & 0 & 2 \end{bmatrix}$

Using a Graphing Utility In Exercises 33–36, use the matrix capabilities of a graphing utility to evaluate the determinant.

33. $\begin{vmatrix} 1 & -1 & 8 & 4 \\ 2 & 6 & 0 & -4 \\ 2 & 0 & 2 & 6 \\ 0 & 2 & 8 & 0 \end{vmatrix}$ 34. $\begin{vmatrix} 0 & -3 & 8 & 2 \\ 8 & 1 & -1 & 6 \\ -4 & 6 & 0 & 9 \\ -7 & 0 & 0 & 14 \end{vmatrix}$

35. $\begin{vmatrix} 3 & -2 & 4 & 3 & 1 \\ -1 & 0 & 2 & 1 & 0 \\ 5 & -1 & 0 & 3 & 2 \\ 4 & 7 & -8 & 0 & 0 \\ 1 & 2 & 3 & 0 & 2 \end{vmatrix}$

36. $\begin{vmatrix} -2 & 0 & 0 & 0 & 0 \\ 0 & 3 & 0 & 0 & 0 \\ 0 & 0 & -1 & 0 & 0 \\ 0 & 0 & 0 & 2 & 0 \\ 0 & 0 & 0 & 0 & -4 \end{vmatrix}$

The Determinant of a Matrix Product In Exercises 37–40, find (a) $|A|$, (b) $|B|$, (c) AB, and (d) $|AB|$. What do you notice about $|AB|$?

37. $A = \begin{bmatrix} -1 & 0 \\ 0 & 3 \end{bmatrix}$, $B = \begin{bmatrix} 2 & 0 \\ 0 & -1 \end{bmatrix}$

38. $A = \begin{bmatrix} 4 & 0 \\ 3 & -2 \end{bmatrix}$, $B = \begin{bmatrix} -1 & 1 \\ -2 & 2 \end{bmatrix}$

39. $A = \begin{bmatrix} -1 & 2 & 1 \\ 1 & 0 & 1 \\ 0 & 1 & 0 \end{bmatrix}$, $B = \begin{bmatrix} -1 & 0 & 0 \\ 0 & 2 & 0 \\ 0 & 0 & 3 \end{bmatrix}$

40. $A = \begin{bmatrix} 2 & 0 & 1 \\ 1 & -1 & 2 \\ 3 & 1 & 0 \end{bmatrix}$, $B = \begin{bmatrix} 2 & -1 & 4 \\ 0 & 1 & 3 \\ 3 & -2 & 1 \end{bmatrix}$

Using a Graphing Utility In Exercises 41 and 42, use the matrix capabilities of a graphing utility to find (a) $|A|$, (b) $|B|$, (c) AB, and (d) $|AB|$. What do you notice about $|AB|$?

41. $A = \begin{bmatrix} 6 & 4 & 0 & 1 \\ 2 & -3 & -2 & -4 \\ 0 & 1 & 5 & 0 \\ -1 & 0 & -1 & 1 \end{bmatrix}$, $B = \begin{bmatrix} 0 & -5 & 0 & -2 \\ -2 & 4 & -1 & -4 \\ 3 & 0 & 1 & 0 \\ 1 & -2 & 3 & 0 \end{bmatrix}$

42. $A = \begin{bmatrix} -1 & 5 & 2 & 0 \\ 0 & 0 & 1 & 1 \\ 3 & -3 & -1 & 0 \\ 4 & 2 & 4 & -1 \end{bmatrix}$,

$B = \begin{bmatrix} 1 & 5 & 0 & 0 \\ 10 & -1 & 2 & 4 \\ 2 & 0 & 0 & 1 \\ -3 & 2 & 5 & 0 \end{bmatrix}$

Verifying an Equation In Exercises 43–48, evaluate the determinants to verify the equation.

43. $\begin{vmatrix} w & x \\ y & z \end{vmatrix} = - \begin{vmatrix} y & z \\ w & x \end{vmatrix}$

44. $\begin{vmatrix} w & cx \\ y & cz \end{vmatrix} = c \begin{vmatrix} w & x \\ y & z \end{vmatrix}$

45. $\begin{vmatrix} w & x \\ y & z \end{vmatrix} = \begin{vmatrix} w & x + cw \\ y & z + cy \end{vmatrix}$

46. $\begin{vmatrix} w & x \\ cw & cx \end{vmatrix} = 0$

47. $\begin{vmatrix} 1 & x & x^2 \\ 1 & y & y^2 \\ 1 & z & z^2 \end{vmatrix} = (y - x)(z - x)(z - y)$

48. $\begin{vmatrix} a + b & a & a \\ a & a + b & a \\ a & a & a + b \end{vmatrix} = b^2(3a + b)$

Solving an Equation In Exercises 49–60, solve for x.

49. $\begin{vmatrix} x & 2 \\ 1 & x \end{vmatrix} = 2$ 50. $\begin{vmatrix} x & 4 \\ -1 & x \end{vmatrix} = 20$

51. $\begin{vmatrix} 2x & -3 \\ -2 & 2x \end{vmatrix} = 3$ 52. $\begin{vmatrix} x & 2 \\ 4 & 9x \end{vmatrix} = 8$

53. $\begin{vmatrix} x & 1 \\ 2 & x - 2 \end{vmatrix} = -1$ 54. $\begin{vmatrix} x + 1 & 2 \\ -1 & x \end{vmatrix} = 4$

55. $\begin{vmatrix} x + 3 & 2 \\ 1 & x + 2 \end{vmatrix} = 0$ 56. $\begin{vmatrix} x - 2 & -1 \\ -3 & x \end{vmatrix} = 0$

57. $\begin{vmatrix} 2x & 1 \\ -1 & x - 1 \end{vmatrix} = x$ 58. $\begin{vmatrix} x - 1 & x \\ x + 1 & 2 \end{vmatrix} = -8$

59. $\begin{vmatrix} 1 & 2 & x \\ -1 & 3 & 2 \\ 3 & -2 & 1 \end{vmatrix} = 0$ 60. $\begin{vmatrix} 1 & x & -2 \\ 1 & 3 & 3 \\ 0 & 2 & -2 \end{vmatrix} = 0$

∫ **Entries Involving Expressions** In Exercises 61–66, evaluate the determinant, in which the entries are functions. Determinants of this type occur when changes of variables are made in calculus.

61. $\begin{vmatrix} 4u & -1 \\ -1 & 2v \end{vmatrix}$

62. $\begin{vmatrix} 3x^2 & -3y^2 \\ 1 & 1 \end{vmatrix}$

63. $\begin{vmatrix} e^{2x} & e^{3x} \\ 2e^{2x} & 3e^{3x} \end{vmatrix}$

64. $\begin{vmatrix} e^{-x} & xe^{-x} \\ -e^{-x} & (1 - x)e^{-x} \end{vmatrix}$

65. $\begin{vmatrix} x & \ln x \\ 1 & 1/x \end{vmatrix}$

66. $\begin{vmatrix} x & x \ln x \\ 1 & 1 + \ln x \end{vmatrix}$

Conclusions

True or False? In Exercises 67 and 68, determine whether the statement is true or false. Justify your answer.

67. If a square matrix has an entire row of zeros, then the determinant will always be zero.

68. If two columns of a square matrix are the same, then the determinant of the matrix will be zero.

69. **Exploration**   Find a pair of 3×3 matrices A and B to demonstrate that $|A + B| \neq |A| + |B|$.

70. **Think About It**   Let A be a 3×3 matrix such that $|A| = 5$. Can you use this information to find $|2A|$? Explain.

Exploration   In Exercises 71–74, (a) find the determinant of A, (b) find A^{-1}, (c) find $\det(A^{-1})$, and (d) compare your results from parts (a) and (c). Make a conjecture based on your results.

71. $A = \begin{bmatrix} 1 & 2 \\ -2 & 2 \end{bmatrix}$ **72.** $A = \begin{bmatrix} 5 & -1 \\ 2 & -1 \end{bmatrix}$

73. $A = \begin{bmatrix} 1 & -3 & -2 \\ -1 & 3 & 1 \\ 0 & 2 & -2 \end{bmatrix}$ **74.** $A = \begin{bmatrix} -1 & 3 & 2 \\ 1 & 3 & -1 \\ 1 & 1 & -2 \end{bmatrix}$

Properties of Determinants   In Exercises 75–77, a property of determinants is given (A and B are square matrices). State how the property has been applied to the given determinants and use a graphing utility to verify the results.

75. If B is obtained from A by interchanging two rows of A or by interchanging two columns of A, then $|B| = -|A|$.

(a) $\begin{vmatrix} 1 & 3 & 4 \\ -7 & 2 & -5 \\ 6 & 1 & 2 \end{vmatrix} = -\begin{vmatrix} 1 & 4 & 3 \\ -7 & -5 & 2 \\ 6 & 2 & 1 \end{vmatrix}$

(b) $\begin{vmatrix} 1 & 3 & 4 \\ -2 & 2 & 0 \\ 1 & 6 & 2 \end{vmatrix} = -\begin{vmatrix} 1 & 6 & 2 \\ -2 & 2 & 0 \\ 1 & 3 & 4 \end{vmatrix}$

76. If B is obtained from A by adding a multiple of a row of A to another row of A or by adding a multiple of a column of A to another column of A, then $|B| = |A|$.

(a) $\begin{vmatrix} 1 & -3 \\ 5 & 2 \end{vmatrix} = \begin{vmatrix} 1 & -3 \\ 0 & 17 \end{vmatrix}$

(b) $\begin{vmatrix} 5 & 4 & 2 \\ 2 & -3 & 4 \\ 7 & 6 & 3 \end{vmatrix} = \begin{vmatrix} 1 & 10 & -6 \\ 2 & -3 & 4 \\ 7 & 6 & 3 \end{vmatrix}$

77. If B is obtained from A by multiplying a row of A by a nonzero constant c or by multiplying a column of A by a nonzero constant c, then $|B| = c|A|$.

(a) $\begin{vmatrix} 1 & 5 \\ 6 & 9 \end{vmatrix} = 3\begin{vmatrix} 1 & 5 \\ 2 & 3 \end{vmatrix}$ (b) $\begin{vmatrix} 2 & 8 \\ 6 & 8 \end{vmatrix} = 8\begin{vmatrix} 1 & 2 \\ 3 & 2 \end{vmatrix}$

78. **Exploration**   A **diagonal matrix** is a square matrix with all zero entries above and below its main diagonal. Evaluate the determinant of each diagonal matrix. Make a conjecture based on your results.

(a) $\begin{bmatrix} 7 & 0 \\ 0 & 4 \end{bmatrix}$ (b) $\begin{bmatrix} -1 & 0 & 0 \\ 0 & 5 & 0 \\ 0 & 0 & 2 \end{bmatrix}$ (c) $\begin{bmatrix} 2 & 0 & 0 & 0 \\ 0 & -2 & 0 & 0 \\ 0 & 0 & 1 & 0 \\ 0 & 0 & 0 & 3 \end{bmatrix}$

79. **Exploration**   A **triangular matrix** is a square matrix with all zero entries either above or below its main diagonal. Such a matrix is **upper triangular** when it has all zeros below the main diagonal and **lower triangular** when it has all zeros above the main diagonal. A diagonal matrix is both upper and lower triangular. Evaluate the determinant of each triangular matrix. Make a conjecture based on your results.

(a) $\begin{bmatrix} 3 & -2 \\ 0 & 5 \end{bmatrix}$ (b) $\begin{bmatrix} 3 & -7 & 1 \\ 0 & -5 & -9 \\ 0 & 0 & 5 \end{bmatrix}$ (c) $\begin{bmatrix} 4 & 0 & 0 & 0 \\ 3 & -3 & 0 & 0 \\ 3 & 6 & 5 & 0 \\ 2 & -2 & 1 & 2 \end{bmatrix}$

80. **CAPSTONE**   Create a study sheet showing the methods you have learned for finding the determinant of a square matrix.

81. **Proof**   Use your results in Exercises 37–42 to make a conjecture about the value of $|AB|$ for two $m \times m$ matrices A and B. Prove your conjecture for 2×2 matrices.

82. **Exploration**   Consider square matrices in which the entries are consecutive integers. An example of such a matrix is

$$\begin{bmatrix} 4 & 5 & 6 \\ 7 & 8 & 9 \\ 10 & 11 & 12 \end{bmatrix}.$$

Use a graphing utility to evaluate four determinants of this type. Make a conjecture based on the results. Then verify your conjecture.

Cumulative Mixed Review

Factoring a Quadratic Expression   In Exercises 83–86, factor the expression.

83. $x^2 - 3x + 2$ **84.** $x^2 + 5x + 6$

85. $4y^2 - 12y + 9$ **86.** $4y^2 - 28y + 49$

Solving a System of Equations   In Exercises 87 and 88, solve the system of equations using the method of substitution or the method of elimination.

87. $\begin{cases} 3x - 10y = 46 \\ x + y = -2 \end{cases}$ **88.** $\begin{cases} 5x + 7y = 23 \\ -4x - 2y = -4 \end{cases}$

8.8 Applications of Matrices and Determinants

Area of a Triangle

In this section, you will study some additional applications of matrices and determinants. The first involves a formula for finding the area of a triangle whose vertices are given by three points on a rectangular coordinate system.

Area of a Triangle

The area of a triangle with vertices (x_1, y_1), (x_2, y_2), and (x_3, y_3) is

$$\text{Area} = \pm\frac{1}{2}\begin{vmatrix} x_1 & y_1 & 1 \\ x_2 & y_2 & 1 \\ x_3 & y_3 & 1 \end{vmatrix}$$

where the symbol $(\pm)$ indicates that the appropriate sign should be chosen to yield a positive area.

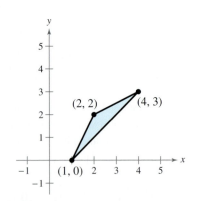

What you should learn

- Use determinants to find areas of triangles.
- Use determinants to decide whether points are collinear.
- Use Cramer's Rule to solve systems of linear equations.
- Use matrices to encode and decode messages.

Why you should learn it

Matrices can be used to decode a message, as shown in Exercise 36 on page 697.

Example 1 Finding the Area of a Triangle

Find the area of the triangle whose vertices are

$$(1, 0), \quad (2, 2), \quad \text{and} \quad (4, 3)$$

as shown in Figure 8.28.

Solution

Begin by letting

$$(x_1, y_1) = (1, 0), \quad (x_2, y_2) = (2, 2),$$

and

$$(x_3, y_3) = (4, 3).$$

Then, to find the area of the triangle, evaluate the determinant by expanding along row 1.

$$\begin{vmatrix} x_1 & y_1 & 1 \\ x_2 & y_2 & 1 \\ x_3 & y_3 & 1 \end{vmatrix} = \begin{vmatrix} 1 & 0 & 1 \\ 2 & 2 & 1 \\ 4 & 3 & 1 \end{vmatrix}$$

$$= 1(-1)^2\begin{vmatrix} 2 & 1 \\ 3 & 1 \end{vmatrix} + 0(-1)^3\begin{vmatrix} 2 & 1 \\ 4 & 1 \end{vmatrix} + 1(-1)^4\begin{vmatrix} 2 & 2 \\ 4 & 3 \end{vmatrix}$$

$$= 1(-1) + 0 + 1(-2)$$

$$= -3$$

Using this value, you can conclude that the area of the triangle is

$$\text{Area} = -\frac{1}{2}\begin{vmatrix} 1 & 0 & 1 \\ 2 & 2 & 1 \\ 4 & 3 & 1 \end{vmatrix}$$

$$= -\frac{1}{2}(-3)$$

$$= \frac{3}{2} \text{ square units.}$$

Figure 8.28

 CHECKPOINT Now try Exercise 5.

Collinear Points

What if the three points in Example 1 had been on the same line? What would have happened had the area formula been applied to three such points? The answer is that the determinant would have been zero. Consider, for instance, the three collinear points $(0, 1)$, $(2, 2)$, and $(4, 3)$, as shown in Figure 8.29. The area of the "triangle" that has these three points as vertices is

$$\frac{1}{2}\begin{vmatrix} 0 & 1 & 1 \\ 2 & 2 & 1 \\ 4 & 3 & 1 \end{vmatrix} = \frac{1}{2}\left[0(-1)^2\begin{vmatrix} 2 & 1 \\ 3 & 1 \end{vmatrix} + 1(-1)^3\begin{vmatrix} 2 & 1 \\ 4 & 1 \end{vmatrix} + 1(-1)^4\begin{vmatrix} 2 & 2 \\ 4 & 3 \end{vmatrix}\right]$$

$$= \frac{1}{2}[0 - 1(-2) + 1(-2)]$$

$$= 0$$

This result is generalized as follows.

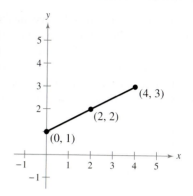

Figure 8.29

Test for Collinear Points

Three points (x_1, y_1), (x_2, y_2), and (x_3, y_3) are **collinear** (lie on the same line) if and only if

$$\begin{vmatrix} x_1 & y_1 & 1 \\ x_2 & y_2 & 1 \\ x_3 & y_3 & 1 \end{vmatrix} = 0.$$

Example 2   Testing for Collinear Points

Determine whether the points

$$(-2, -2), \quad (1, 1), \quad \text{and} \quad (7, 5)$$

are collinear. (See Figure 8.30.)

Solution

Begin by letting

$$(x_1, y_1) = (-2, -2), \quad (x_2, y_2) = (1, 1),$$

and

$$(x_3, y_3) = (7, 5).$$

Then by expanding along row 1, you have

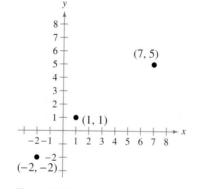

Figure 8.30

$$\begin{vmatrix} x_1 & y_1 & 1 \\ x_2 & y_2 & 1 \\ x_3 & y_3 & 1 \end{vmatrix} = \begin{vmatrix} -2 & -2 & 1 \\ 1 & 1 & 1 \\ 7 & 5 & 1 \end{vmatrix}$$

$$= -2(-1)^2\begin{vmatrix} 1 & 1 \\ 5 & 1 \end{vmatrix} + (-2)(-1)^3\begin{vmatrix} 1 & 1 \\ 7 & 1 \end{vmatrix} + 1(-1)^4\begin{vmatrix} 1 & 1 \\ 7 & 5 \end{vmatrix}$$

$$= -2(-4) + 2(-6) + 1(-2)$$

$$= -6.$$

Because the value of this determinant is *not* zero, you can conclude that the three points are not collinear.

✓**CHECKPOINT**   Now try Exercise 13.

Example 4 Using Cramer's Rule for a 3 × 3 System

Use Cramer's Rule and a graphing utility, if possible, to solve the system of linear equations.

$$\begin{cases} -x & + & z = & 4 \\ 2x - y + & z = & -3 \\ & y - 3z = & 1 \end{cases}$$

Solution

Using a graphing utility to evaluate the determinant of the coefficient matrix A, you find that Cramer's Rule cannot be applied because $|A| = 0$.

✓CHECKPOINT   Now try Exercise 21.

Example 5 Using Cramer's Rule for a 3 × 3 System

Use Cramer's Rule, if possible, to solve the system of linear equations.

$$\qquad\qquad\qquad \textit{Coefficient Matrix}$$

$$\begin{cases} -x + 2y - 3z = 1 \\ 2x \qquad + z = 0 \\ 3x - 4y + 4z = 2 \end{cases} \implies \begin{bmatrix} -1 & 2 & -3 \\ 2 & 0 & 1 \\ 3 & -4 & 4 \end{bmatrix}$$

Solution

The coefficient matrix above can be expanded along the second row, as follows.

$$D = 2(-1)^3 \begin{vmatrix} 2 & -3 \\ -4 & 4 \end{vmatrix} + 0(-1)^4 \begin{vmatrix} -1 & -3 \\ 3 & 4 \end{vmatrix} + 1(-1)^5 \begin{vmatrix} -1 & 2 \\ 3 & -4 \end{vmatrix}$$

$$= -2(-4) + 0 - 1(-2)$$

$$= 10$$

Because this determinant is not zero, you can apply Cramer's Rule.

$$x = \frac{D_x}{D} = \frac{\begin{vmatrix} 1 & 2 & -3 \\ 0 & 0 & 1 \\ 2 & -4 & 4 \end{vmatrix}}{10} = \frac{8}{10} = \frac{4}{5}$$

$$y = \frac{D_y}{D} = \frac{\begin{vmatrix} -1 & 1 & -3 \\ 2 & 0 & 1 \\ 3 & 2 & 4 \end{vmatrix}}{10} = \frac{-15}{10} = -\frac{3}{2}$$

$$z = \frac{D_z}{D} = \frac{\begin{vmatrix} -1 & 2 & 1 \\ 2 & 0 & 0 \\ 3 & -4 & 2 \end{vmatrix}}{10} = \frac{-16}{10} = -\frac{8}{5}$$

The solution is

$$\left(\tfrac{4}{5}, -\tfrac{3}{2}, -\tfrac{8}{5} \right).$$

Check this in the original system.

✓CHECKPOINT   Now try Exercise 23.

Remember that Cramer's Rule does not apply when the determinant of the coefficient matrix is zero. This would create division by zero, which is undefined.

Technology Tip

Try using a graphing utility to evaluate D_x/D from Example 4. You should obtain the error message shown below.

```
ERR:DIVIDE BY 0
1▮Quit
2:Goto
```

Cryptography

A **cryptogram** is a message written according to a secret code. (The Greek word *kryptos* means "hidden.") Matrix multiplication can be used to encode and decode messages. To begin, you need to assign a number to each letter in the alphabet (with 0 assigned to a blank space), as follows.

0 = _	9 = I	18 = R
1 = A	10 = J	19 = S
2 = B	11 = K	20 = T
3 = C	12 = L	21 = U
4 = D	13 = M	22 = V
5 = E	14 = N	23 = W
6 = F	15 = O	24 = X
7 = G	16 = P	25 = Y
8 = H	17 = Q	26 = Z

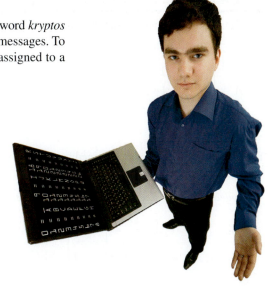

Applied Cryptography Researcher

Then the message is converted to numbers and partitioned into **uncoded row matrices,** each having n entries, as demonstrated in Example 6.

Example 6   Forming Uncoded Row Matrices

Write the uncoded row matrices of dimension 1×3 for the message

MEET ME MONDAY.

Solution

Partitioning the message (including blank spaces, but ignoring punctuation) into groups of three produces the following uncoded row matrices.

$$\begin{bmatrix} 13 & 5 & 5 \end{bmatrix} \quad \begin{bmatrix} 20 & 0 & 13 \end{bmatrix} \quad \begin{bmatrix} 5 & 0 & 13 \end{bmatrix} \quad \begin{bmatrix} 15 & 14 & 4 \end{bmatrix} \quad \begin{bmatrix} 1 & 25 & 0 \end{bmatrix}$$

M   E   E   T   M   E   M   O   N   D   A   Y

Note that a blank space is used to fill out the last uncoded row matrix.

✓**CHECKPOINT**   Now try Exercise 29(a).

To encode a message, use the techniques demonstrated in Section 8.6 to choose an $n \times n$ invertible matrix such as

$$A = \begin{bmatrix} 1 & -2 & 2 \\ -1 & 1 & 3 \\ 1 & -1 & -4 \end{bmatrix}$$

and multiply the uncoded row matrices by A (on the right) to obtain **coded row matrices.** Here is an example.

Uncoded Matrix   Encoding Matrix A   Coded Matrix

$$\begin{bmatrix} 13 & 5 & 5 \end{bmatrix} \begin{bmatrix} 1 & -2 & 2 \\ -1 & 1 & 3 \\ 1 & -1 & -4 \end{bmatrix} = \begin{bmatrix} 13 & -26 & 21 \end{bmatrix}$$

This technique is further illustrated in Example 7.

Example 7 Encoding a Message

Use the matrix A to encode the message

MEET ME MONDAY.

$$A = \begin{bmatrix} 1 & -2 & 2 \\ -1 & 1 & 3 \\ 1 & -1 & -4 \end{bmatrix}$$

Solution

The coded row matrices are obtained by multiplying each of the uncoded row matrices found in Example 6 by the matrix A, as follows.

Uncoded Matrix   *Encoding Matrix A*   *Coded Matrix*

$$[13 \quad 5 \quad 5] \begin{bmatrix} 1 & -2 & 2 \\ -1 & 1 & 3 \\ 1 & -1 & -4 \end{bmatrix} = [13 \quad -26 \quad 21]$$

$$[20 \quad 0 \quad 13] \begin{bmatrix} 1 & -2 & 2 \\ -1 & 1 & 3 \\ 1 & -1 & -4 \end{bmatrix} = [33 \quad -53 \quad -12]$$

$$[5 \quad 0 \quad 13] \begin{bmatrix} 1 & -2 & 2 \\ -1 & 1 & 3 \\ 1 & -1 & -4 \end{bmatrix} = [18 \quad -23 \quad -42]$$

$$[15 \quad 14 \quad 4] \begin{bmatrix} 1 & -2 & 2 \\ -1 & 1 & 3 \\ 1 & -1 & -4 \end{bmatrix} = [5 \quad -20 \quad 56]$$

$$[1 \quad 25 \quad 0] \begin{bmatrix} 1 & -2 & 2 \\ -1 & 1 & 3 \\ 1 & -1 & -4 \end{bmatrix} = [-24 \quad 23 \quad 77]$$

So, the sequence of coded row matrices is

$$[13 \ -26 \ 21][33 \ -53 \ -12][18 \ -23 \ -42][5 \ -20 \ 56][-24 \ 23 \ 77].$$

Finally, removing the matrix notation produces the following cryptogram.

13 −26 21 33 −53 −12 18 −23 −42 5 −20 56 −24 23 77

✓CHECKPOINT   Now try Exercise 29(b).

For those who do not know the encoding matrix A, decoding the cryptogram found in Example 7 is difficult. But for an authorized receiver who knows the encoding matrix A, decoding is simple. The receiver need only multiply the coded row matrices by

$$A^{-1}$$

(on the right) to retrieve the uncoded row matrices. Here is an example.

$$\underbrace{[13 \ -26 \ 21]}_{\text{Coded}} \underbrace{\begin{bmatrix} -1 & -10 & -8 \\ -1 & -6 & -5 \\ 0 & -1 & -1 \end{bmatrix}}_{A^{-1}} = \underbrace{[13 \quad 5 \quad 5]}_{\text{Uncoded}}$$

This technique is further illustrated in Example 8.

Technology Tip

An efficient method for encoding the message at the left with your graphing utility is to enter A as a 3×3 matrix. Let B be the 5×3 matrix whose rows are the uncoded row matrices

$$B = \begin{bmatrix} 13 & 5 & 5 \\ 20 & 0 & 13 \\ 5 & 0 & 13 \\ 15 & 14 & 4 \\ 1 & 25 & 0 \end{bmatrix}.$$

The product BA gives the coded row matrices.

```
[B] [A]
[[13   -26  21 ]
 [33   -53  -12]
 [18   -23  -42]
 [5    -20  56 ]
 [-24  23   77 ]]
```

Example 8   Decoding a Message

Use the inverse of the matrix

$$A = \begin{bmatrix} 1 & -2 & 2 \\ -1 & 1 & 3 \\ 1 & -1 & -4 \end{bmatrix}$$

to decode the cryptogram

$$13 \ -26 \ 21 \ 33 \ -53 \ -12 \ 18 \ -23 \ -42 \ 5 \ -20 \ 56 \ -24 \ 23 \ 77.$$

Solution

First, find the decoding matrix A^{-1} by using the techniques demonstrated in Section 8.6. Next partition the message into groups of three to form the coded row matrices. Then multiply each coded row matrix by A^{-1} (on the right).

Coded Matrix   Decoding Matrix A^{-1} Decoded Matrix

$$\begin{bmatrix} 13 & -26 & 21 \end{bmatrix} \begin{bmatrix} -1 & -10 & -8 \\ -1 & -6 & -5 \\ 0 & -1 & -1 \end{bmatrix} = \begin{bmatrix} 13 & 5 & 5 \end{bmatrix}$$

$$\begin{bmatrix} 33 & -53 & -12 \end{bmatrix} \begin{bmatrix} -1 & -10 & -8 \\ -1 & -6 & -5 \\ 0 & -1 & -1 \end{bmatrix} = \begin{bmatrix} 20 & 0 & 13 \end{bmatrix}$$

$$\begin{bmatrix} 18 & -23 & -42 \end{bmatrix} \begin{bmatrix} -1 & -10 & -8 \\ -1 & -6 & -5 \\ 0 & -1 & -1 \end{bmatrix} = \begin{bmatrix} 5 & 0 & 13 \end{bmatrix}$$

$$\begin{bmatrix} 5 & -20 & 56 \end{bmatrix} \begin{bmatrix} -1 & -10 & -8 \\ -1 & -6 & -5 \\ 0 & -1 & -1 \end{bmatrix} = \begin{bmatrix} 15 & 14 & 4 \end{bmatrix}$$

$$\begin{bmatrix} -24 & 23 & 77 \end{bmatrix} \begin{bmatrix} -1 & -10 & -8 \\ -1 & -6 & -5 \\ 0 & -1 & -1 \end{bmatrix} = \begin{bmatrix} 1 & 25 & 0 \end{bmatrix}$$

So, the message is as follows.

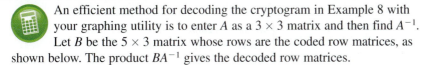

$$\begin{bmatrix} 13 & 5 & 5 \end{bmatrix} \ \begin{bmatrix} 20 & 0 & 13 \end{bmatrix} \ \begin{bmatrix} 5 & 0 & 13 \end{bmatrix} \ \begin{bmatrix} 15 & 14 & 4 \end{bmatrix} \ \begin{bmatrix} 1 & 25 & 0 \end{bmatrix}$$
M E E   T   M E   M   O N D   A Y

✓CHECKPOINT   Now try Exercise 35.

Technology Tip

An efficient method for decoding the cryptogram in Example 8 with your graphing utility is to enter A as a 3×3 matrix and then find A^{-1}.

Let B be the 5×3 matrix whose rows are the coded row matrices, as shown below. The product BA^{-1} gives the decoded row matrices.

$$B = \begin{bmatrix} 13 & -26 & 21 \\ 33 & -53 & -12 \\ 18 & -23 & -42 \\ 5 & -20 & 56 \\ -24 & 23 & 77 \end{bmatrix}$$

8.8   Exercises

See www.CalcChat.com for worked-out solutions to odd-numbered exercises.
For instructions on how to use a graphing utility, see Appendix A.

Vocabulary and Concept Check

In Exercises 1 and 2, fill in the blank.

1. _____ is a method for using determinants to solve a system of linear equations.

2. A message written according to a secret code is called a _____ .

In Exercises 3 and 4, consider three points (x_1, y_1), (x_2, y_2), and (x_3, y_3), and the determinant shown at the right.

$$\begin{vmatrix} x_1 & y_1 & 1 \\ x_2 & y_2 & 1 \\ x_3 & y_3 & 1 \end{vmatrix}$$

3. Suppose the three points are vertices of a triangle and the value of the determinant is -6. What number do you multiply -6 by to find the area of the triangle?

4. Suppose the value of the determinant is 0. What can you conclude?

Procedures and Problem Solving

Finding an Area   In Exercises 5–10, use a determinant to find the area of the figure with the given vertices.

 5. $(-2, 4), (2, 3), (-1, 5)$ **6.** $(-3, 5), (2, 6), (3, -5)$

7. $\left(0, \frac{1}{2}\right), \left(\frac{5}{2}, 0\right), (4, 3)$ **8.** $\left(\frac{9}{2}, 0\right), (2, 6), \left(0, -\frac{3}{2}\right)$

9.

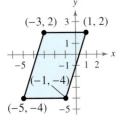

10.

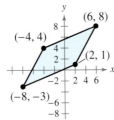

Finding a Coordinate   In Exercises 11 and 12, find x or y such that the triangle has an area of 4 square units.

11. $(-1, 5), (-2, 0), (x, 2)$ **12.** $(-4, 2), (-3, 5), (-1, y)$

Testing for Collinear Points   In Exercises 13–16, use a determinant to determine whether the points are collinear.

13. $(3, -1), (0, -3), (12, 5)$ **14.** $(3, -5), (6, 1), (4, 2)$

15. $\left(2, -\frac{1}{2}\right), (-4, 4), (6, -3)$ **16.** $\left(0, \frac{1}{2}\right), (2, -1), \left(-4, \frac{7}{2}\right)$

Finding a Coordinate   In Exercises 17 and 18, find x or y such that the points are collinear.

17. $(1, -2), (x, 2), (5, 6)$ **18.** $(-6, 2), (-5, y), (-3, 5)$

Using Cramer's Rule   In Exercises 19–24, use Cramer's Rule to solve (if possible) the system of equations.

19. $\begin{cases} -7x + 11y = -1 \\ 3x - 9y = 9 \end{cases}$ **20.** $\begin{cases} 4x - 3y = -10 \\ 6x + 9y = 12 \end{cases}$

21. $\begin{cases} 3x + 2y = -2 \\ 6x + 4y = 4 \end{cases}$ **22.** $\begin{cases} 6x - 5y = 17 \\ -13x + 3y = -76 \end{cases}$

23. $\begin{cases} 4x - y + z = -5 \\ 2x + 2y + 3z = 10 \\ 5x - 2y + 6z = 1 \end{cases}$ **24.** $\begin{cases} 4x - 2y + 3z = -2 \\ 2x + 2y + 5z = 16 \\ 8x - 5y - 2z = 4 \end{cases}$

Comparing Solution Methods   In Exercises 25 and 26, solve the system of equations using (a) Gaussian elimination and (b) Cramer's Rule. Which method do you prefer, and why?

25. $\begin{cases} 3x + 3y + 5z = 1 \\ 3x + 5y + 9z = 2 \\ 5x + 9y + 17z = 4 \end{cases}$ **26.** $\begin{cases} 2x + 3y - 5z = 1 \\ 3x + 5y + 9z = -16 \\ 5x + 9y + 17z = -30 \end{cases}$

27. **Why you should learn it**   *(p. 681)*   The retail sales (in billions of dollars) of family clothing stores in the United States from 2004 through 2008 are shown in the table. (Source: U.S. Census Bureau)

Year	Sales (in billions of dollars)
2004	72.0
2005	77.4
2006	82.0
2007	84.2
2008	83.2

The coefficients of the least squares regression parabola $y = at^2 + bt + c$, where y represents the retail sales (in billions of dollars) and t represents the year, with $t = 4$ corresponding to 2004, can be found by solving the system

$$\begin{cases} 8674a + 1260b + 190c = 15{,}489.6 \\ 1260a + 190b + 30c = 2422.0 \\ 190a + 30b + 5c = 398.8 \end{cases}$$

(a) Use Cramer's Rule to solve the system and write the least squares regression parabola for the data.

(b) Use a graphing utility to graph the parabola with the data. How well does the model fit the data?

28. MODELING DATA

The retail sales (in billions of dollars) of stores selling auto parts, accessories, and tires in the United States from 2004 through 2008 are shown in the table.
(Source: U.S. Census Bureau)

Year	Sales (in billions of dollars)
2004	67.2
2005	71.2
2006	74.5
2007	76.7
2008	78.6

The coefficients of the least squares regression parabola $y = at^2 + bt + c$, where y represents the retail sales (in billions of dollars) and t represents the year, with $t = 4$ corresponding to 2004, can be found by solving the system

$$\begin{cases} 8674a + 1260b + 190c = 14{,}325.9 \\ 1260a + 190b + 30c = 2237.5 \, . \\ 190a + 30b + 5c = 368.2 \end{cases}$$

(a) Use Cramer's Rule to solve the system and write the least squares regression parabola for the data.

(b) Use a graphing utility to graph the parabola with the data. How well does the model fit the data?

(c) Is this a good model for predicting retail sales in future years? Explain.

Encoding a Message   In Exercises 29 and 30, (a) write the uncoded 1×3 row matrices for the message, and then (b) encode the message using the encoding matrix.

Message	*Encoding Matrix*

✓ **29.** TEXT ME AT WORK
$$\begin{bmatrix} 1 & -1 & 0 \\ 1 & 0 & -1 \\ -6 & 2 & 3 \end{bmatrix}$$

30. PLEASE SEND MONEY
$$\begin{bmatrix} 4 & 2 & 1 \\ -3 & -3 & -1 \\ 3 & 2 & 1 \end{bmatrix}$$

Encoding a Message   In Exercises 31 and 32, write a cryptogram for the message using the matrix A.

$$A = \begin{bmatrix} 1 & 2 & 2 \\ 3 & 7 & 9 \\ -1 & -4 & -7 \end{bmatrix}$$

31. KEY UNDER RUG   **32.** HAPPY BIRTHDAY

Yuri Arcurs 2010/used under license from Shutterstock.com

Decoding a Message   In Exercises 33–35, use A^{-1} to decode the cryptogram.

33. $A = \begin{bmatrix} 1 & 2 \\ 3 & 5 \end{bmatrix}$ 11 21 64 112 25 50 29
53 23 46 40 75 55 92

34. $A = \begin{bmatrix} 2 & 3 \\ 3 & 4 \end{bmatrix}$ 85 120 6 8 10 15 84 117 42
56 90 125 60 80 30 45 19 26

✓ **35.** $A = \begin{bmatrix} 1 & 2 & 1 \\ -1 & 0 & 2 \\ 1 & -1 & -2 \end{bmatrix}$ 3 18 21 31 29 13 -2 -1
4 -6 28 54 -3 4 12 16
8 1 6 6 0 27 -12 -39
15 -19 -27 5 10 5

36. *Why you should learn it*   (p. 688)   The following cryptogram was encoded with a 2×2 matrix.

8 21 -15 -10 -13 -13 5 10 5 25
5 19 -1 6 20 40 -18 -18 1 16

The last word of the message is _RON. What is the message?

Conclusions

True or False?   In Exercises 37 and 38, determine whether the statement is true or false. Justify your answer.

37. Cramer's Rule cannot be used to solve a system of linear equations when the determinant of the coefficient matrix is zero.

38. In a system of linear equations, when the determinant of the coefficient matrix is zero, the system has no solution.

39. **Think About It**   Describe a way to use an invertible $n \times n$ matrix to encode a message that is converted to numbers and partitioned into uncoded column matrices.

40. **CAPSTONE**   Consider the system of linear equations

$$\begin{cases} a_1x + b_1y = c_1 \\ a_2x + b_2y = c_2 \end{cases}$$

where a_1, b_1, c_1, a_2, b_2, and c_2 represent real numbers. What must be true about the lines represented by the equations when

$$\begin{vmatrix} a_1 & b_1 \\ a_2 & b_2 \end{vmatrix} = 0?$$

Cumulative Mixed Review

Equation of a Line   In Exercises 41–44, find the general form of the equation of the line that passes through the two points.

41. $(-1, 5), (7, 3)$ **42.** $(0, -6), (-2, 10)$

43. $(3, -3), (10, -1)$ **44.** $(-4, 12), (4, 2)$

8   Chapter Summary

		What did you learn?	**Explanation and Examples**	**Review Exercises**
8.1		Use the methods of substitution and graphing to solve systems of equations in two variables (*p. 610*), and use systems of equations to model and solve real-life problems (*p. 615*).	**Substitution:** (1) Solve one of the equations for one variable. (2) Substitute the expression found in Step 1 into the other equation to obtain an equation in one variable and (3) solve the equation. (4) Back-substitute the value(s) obtained in Step 3 into the expression obtained in Step 1 to find the value(s) of the other variable. (5) Check the solution(s). **Graphing:** (1) Solve both equations for y in terms of x. (2) Use a graphing utility to graph both equations. (3) Use the *intersect* feature or the *zoom* and *trace* features of the graphing utility to approximate the point(s) of intersection of the graphs. (4) Check the solution(s).	1–22
8.2		Use the method of elimination to solve systems of linear equations in two variables (*p. 620*), and graphically interpret the number of solutions of a system of linear equations in two variables (*p. 622*).	**Elimination:** (1) Obtain coefficients for x (or y) that differ only in sign by multiplying all terms of one or both equations by suitably chosen constants. (2) Add the equations to eliminate one variable and solve the resulting equation. (3) Back-substitute the value obtained in Step 2 into either of the original equations and solve for the other variable. (4) Check the solution(s).	23–38
		Use systems of linear equations in two variables to model and solve real-life problems (*p. 624*).	A system of linear equations in two variables can be used to find the airspeed of an airplane and the speed of the wind. (See Example 6.)	39–42
8.3		Use back-substitution to solve linear systems in row-echelon form (*p. 629*).	$$\begin{cases} x - 2y + 3z = 9 \\ -x + 3y = -4 \\ 2x - 5y + 5z = 17 \end{cases} \rightarrow \begin{cases} x - 2y + 3z = 9 \\ y + 3z = 5 \\ z = 2 \end{cases}$$	43, 44
		Use Gaussian elimination to solve systems of linear equations (*p. 630*).	Use elementary row operations to convert a system of linear equations to row-echelon form. (1) Interchange two equations. (2) Multiply one of the equations by a nonzero constant. (3) Add a multiple of one equation to another equation.	45–50
		Solve nonsquare systems of linear equations (*p. 633*).	In a nonsquare system, the number of equations differs from the number of variables. A system of linear equations cannot have a unique solution unless there are at least as many equations as there are variables.	51, 52
		Graphically interpret three-variable linear systems (*p. 634*).	The graph of a system of three linear equations in three variables consists of three planes. When they intersect in a single point, the system has exactly one solution. When they have no point in common, the system has no solution. When they intersect in a line or a plane, the system has infinitely many solutions (see Figures 8.16–8.20).	53, 54
		Use systems of linear equations to write partial fraction decompositions of rational expressions (*p. 635*).	$$\frac{9}{x^3 - 6x^2} = \frac{9}{x^2(x - 6)} = \frac{A}{x} + \frac{B}{x^2} + \frac{C}{x - 6}$$	55–60
		Use systems of linear equations in three or more variables to model and solve real-life problems (*p. 638*).	A system of linear equations in three variables can be used to find the position equation of an object that is moving in a (vertical) line with constant acceleration. (See Example 8.)	61–64

What did you learn?		Explanation and Examples	Review Exercises
8.4	**Write matrices and identify their dimensions** (*p. 644*), **and perform elementary row operations on matrices** (*p. 646*).	**Elementary Row Operations** **1.** Interchange two rows. **2.** Multiply a row by a nonzero constant. **3.** Add a multiple of a row to another row.	65–80
	Use matrices and Gaussian elimination to solve systems of linear equations (*p. 647*).	Write the augmented matrix of the system. Use elementary row operations to rewrite the augmented matrix in row-echelon form. Write the system of linear equations corresponding to the matrix in row-echelon form and use back-substitution to find the solution.	81–88
	Use matrices and Gauss-Jordan elimination to solve systems of linear equations (*p. 651*).	Gauss-Jordan elimination continues the reduction process on a matrix in row-echelon form until a reduced row-echelon form is obtained. (See Example 8.)	89–98
8.5	**Decide whether two matrices are equal** (*p. 658*).	Two matrices are equal when they have the same dimension and all of their corresponding entries are equal.	99–102
	Add and subtract matrices and multiply matrices by scalars (*p. 659*), **multiply two matrices** (*p. 662*), **and use matrix operations to model and solve real-life problems** (*p. 665*).	**1.** Let $A = [a_{ij}]$ and $B = [b_{ij}]$ be matrices of dimension $m \times n$ and let c be a scalar. $A + B = [a_{ij} + b_{ij}] \qquad cA = [ca_{ij}]$ **2.** Let $A = [a_{ij}]$ be an $m \times n$ matrix and let $B = [b_{ij}]$ be an $n \times p$ matrix. The product AB is an $m \times p$ matrix given by $AB = [c_{ij}]$, where $c_{ij} = a_{i1}b_{1j} + a_{i2}b_{2j} + a_{i3}b_{3j} + \cdots + a_{in}b_{nj}$.	103–126
8.6	**Verify that two matrices are inverses of each other** (*p. 672*), **and use Gauss-Jordan elimination to find inverses of matrices** (*p. 673*).	Write the $n \times 2n$ matrix $[A \ \vdots \ I]$. Row reduce A to I using elementary row operations. The result will be the matrix $[I \ \vdots \ A^{-1}]$.	127–136
	Use a formula to find inverses of 2×2 matrices (*p. 676*).	If $A = \begin{bmatrix} a & b \\ c & d \end{bmatrix}$ and $ad - bc \neq 0$, then $A^{-1} = \dfrac{1}{ad - bc}\begin{bmatrix} d & -b \\ -c & a \end{bmatrix}$.	137–140
	Use inverse matrices to solve systems of linear equations (*p. 677*).	If A is an invertible matrix, then the system of linear equations represented by $AX = B$ has a unique solution given by $X = A^{-1}B$.	141–150
8.7	**Find the determinants of 2×2 matrices** (*p. 681*).	$\det(A) = |A| = \begin{vmatrix} a_1 & b_1 \\ a_2 & b_2 \end{vmatrix} = a_1b_2 - a_2b_1$	151–154
	Find minors and cofactors of square matrices (*p. 683*), **and find the determinants of square matrices** (*p. 684*).	The determinant of a square matrix A (of dimension 2×2 or greater) is the sum of the entries in any row or column of A multiplied by their respective cofactors.	155–166
8.8	**Use determinants to find areas of triangles** (*p. 688*) **and to decide whether points are collinear** (*p. 689*), **use Cramer's Rule to solve systems of linear equations** (*p. 690*), **and use matrices to encode and decode messages** (*p. 693*).	If a system of n linear equations in n variables has a coefficient matrix A with a nonzero determinant $|A|$, then the solution of the system is $x_1 = \dfrac{|A_1|}{|A|},\ x_2 = \dfrac{|A_2|}{|A|},\ \ldots,\ x_n = \dfrac{|A_n|}{|A|}$ where the ith column of A_i is the column of constants in the system of equations.	167–193

8   Review Exercises

See www.CalcChat.com for worked-out solutions to odd-numbered exercises.
For instructions on how to use a graphing utility, see Appendix A.

8.1

Solving a System by Substitution In Exercises 1–10, solve the system by the method of substitution.

1. $\begin{cases} x + y = 2 \\ x - y = 0 \end{cases}$

2. $\begin{cases} 2x - 3y = 3 \\ x - y = 0 \end{cases}$

3. $\begin{cases} 4x - y = 1 \\ 8x + y = 17 \end{cases}$

4. $\begin{cases} 10x + 6y = -14 \\ x + 9y = -7 \end{cases}$

5. $\begin{cases} 0.5x + y = 0.75 \\ 1.25x - 4.5y = -2.5 \end{cases}$

6. $\begin{cases} -x + \frac{2}{5}y = \frac{3}{5} \\ -x + \frac{1}{5}y = -\frac{4}{5} \end{cases}$

7. $\begin{cases} x^2 - y^2 = 9 \\ x - y = 1 \end{cases}$

8. $\begin{cases} x^2 + y^2 = 169 \\ 3x + 2y = 39 \end{cases}$

9. $\begin{cases} y = 2x^2 \\ y = x^4 - 2x^2 \end{cases}$

10. $\begin{cases} x = y + 3 \\ x = y^2 + 1 \end{cases}$

Solving a System of Equations Graphically In Exercises 11–18, use a graphing utility to approximate all points of intersection of the graphs of the equations in the system. Verify your solutions by checking them in the original system.

11. $\begin{cases} 5x + 6y = 7 \\ -x - 4y = 0 \end{cases}$

12. $\begin{cases} 8x - 3y = -3 \\ 2x + 5y = 28 \end{cases}$

13. $\begin{cases} y^2 - 4x = 0 \\ x + y = 0 \end{cases}$

14. $\begin{cases} y^2 - x = -1 \\ y + 2x = 5 \end{cases}$

15. $\begin{cases} y = 3 - x^2 \\ y = 2x^2 + x + 1 \end{cases}$

16. $\begin{cases} y = 2x^2 - 4x + 1 \\ y = x^2 - 4x + 3 \end{cases}$

17. $\begin{cases} y = 2(6 - x) \\ y = 2^{x-2} \end{cases}$

18. $\begin{cases} y = \ln(x + 2) + 1 \\ x + y = 0 \end{cases}$

19. **Finance** You invest $5000 in a greenhouse. The planter, potting soil, and seed for each plant cost $6.43, and the selling price of each plant is $12.68. How many plants must you sell to break even?

20. **Finance** You are offered two sales jobs. One company offers an annual salary of $55,000 plus a year-end bonus of 1.5% of your total sales. The other company offers an annual salary of $52,000 plus a year-end bonus of 2% of your total sales. How much would you have to sell in a year to make the second offer the better offer?

21. **Geometry** The perimeter of a rectangle is 480 meters and its length is 1.5 times its width. Find the dimensions of the rectangle.

22. **Geometry** The perimeter of a rectangle is 68 feet and its width is $\frac{8}{9}$ times its length. Find the dimensions of the rectangle.

8.2

Solving a System by Elimination In Exercises 23–32, solve the system by the method of elimination.

23. $\begin{cases} 2x - y = 2 \\ 6x + 8y = 39 \end{cases}$

24. $\begin{cases} 40x + 30y = 24 \\ 20x - 50y = -14 \end{cases}$

25. $\begin{cases} 0.2x + 0.3y = 0.14 \\ 0.4x + 0.5y = 0.20 \end{cases}$

26. $\begin{cases} 12x + 42y = -17 \\ 30x - 18y = 19 \end{cases}$

27. $\begin{cases} \frac{1}{5}x + \frac{3}{10}y = \frac{7}{50} \\ \frac{2}{5}x + \frac{1}{2}y = \frac{1}{5} \end{cases}$

28. $\begin{cases} \frac{5}{12}x - \frac{3}{4}y = \frac{25}{4} \\ -x + \frac{7}{8}y = -\frac{38}{5} \end{cases}$

29. $\begin{cases} 3x - 2y = 0 \\ 3x + 2(y + 5) = 10 \end{cases}$

30. $\begin{cases} 7x + 12y = 63 \\ 2x + 3y = 15 \end{cases}$

31. $\begin{cases} 1.25x - 2y = 3.5 \\ 5x - 8y = 14 \end{cases}$

32. $\begin{cases} 1.5x + 2.5y = 8.5 \\ 6x + 10y = 24 \end{cases}$

Solving a System Graphically In Exercises 33–38, use a graphing utility to graph the lines in the system. Use the graphs to determine whether the system is consistent or inconsistent. If the system is consistent, determine the solution. Verify your results algebraically.

33. $\begin{cases} 3x + 2y = 0 \\ x - y = 4 \end{cases}$

34. $\begin{cases} x + y = 6 \\ -2x - 2y = -12 \end{cases}$

35. $\begin{cases} \frac{1}{4}x - \frac{1}{5}y = 2 \\ -5x + 4y = 8 \end{cases}$

36. $\begin{cases} \frac{7}{2}x - 7y = -1 \\ -x + 2y = 4 \end{cases}$

37. $\begin{cases} 2x - 2y = 8 \\ 4x - 1.5y = -5.5 \end{cases}$

38. $\begin{cases} -x + 3.2y = 10.4 \\ -2x - 9.6y = 6.4 \end{cases}$

Supply and Demand In Exercises 39 and 40, find the point of equilibrium of the demand and supply equations.

	Demand	Supply
39.	$p = 37 - 0.0002x$	$p = 22 + 0.00001x$
40.	$p = 120 - 0.0001x$	$p = 45 + 0.0002x$

41. **Aerodynamics** Two planes leave Pittsburgh and Philadelphia at the same time, each going to the other city. One plane flies 25 miles per hour faster than the other. Find the airspeed of each plane given that the cities are 275 miles apart and the planes pass each other after 40 minutes of flying time.

42. **Economics** A total of $46,000 is invested in two corporate bonds that pay 6.75% and 7.25% simple interest. The investor wants an annual interest income of $3245 from the investments. What is the most that can be invested in the 6.75% bond?

8.3

Using Back-Substitution In Exercises 43 and 44, use back-substitution to solve the system of linear equations.

43. $\begin{cases} x - 4y + 3z = 3 \\ \quad\;\; - y + z = -1 \\ \quad\qquad\quad z = -5 \end{cases}$
44. $\begin{cases} x - 7y + 8z = -14 \\ \qquad y - 9z = 26 \\ \qquad\qquad z = -3 \end{cases}$

Solving a System of Linear Equations In Exercises 45–52, solve the system of linear equations and check any solution algebraically.

45. $\begin{cases} x + 3y - z = 13 \\ 2x \qquad - 5z = 23 \\ 4x - y - 2z = 14 \end{cases}$
46. $\begin{cases} x + y + z = 2 \\ -x + 3y + 2z = 8 \\ 4x + y \qquad = 4 \end{cases}$

47. $\begin{cases} x - 2y + z = -6 \\ 2x - 3y \qquad = -7 \\ -x + 3y - 3z = 11 \end{cases}$
48. $\begin{cases} 2x \qquad + 6z = -9 \\ 3x - 2y + 11z = -16 \\ 3x - y + 7z = -11 \end{cases}$

49. $\begin{cases} x - 2y + 3z = -5 \\ 2x + 4y + 5z = 1 \\ x + 2y + z = 0 \end{cases}$
50. $\begin{cases} x - 2y + z = 5 \\ 2x + 3y + z = 5 \\ x + y + 2z = 3 \end{cases}$

51. $\begin{cases} 5x - 12y + 7z = 16 \\ 3x - 7y + 4z = 9 \end{cases}$
52. $\begin{cases} 2x + 5y - 19z = 34 \\ 3x + 8y - 31z = 54 \end{cases}$

Sketching a Plane In Exercises 53 and 54, sketch the plane represented by the linear equation. Then list four points that lie in the plane.

53. $2x - 4y + z = 8$
54. $3x + 3y - z = 9$

Partial Fraction Decomposition In Exercises 55–60, write the partial fraction decomposition for the rational expression. Check your result algebraically by combining the fractions, and check your result graphically by using a graphing utility to graph the rational expression and the partial fractions in the same viewing window.

55. $\dfrac{4 - x}{x^2 + 6x + 8}$
56. $\dfrac{-x}{x^2 + 3x + 2}$

57. $\dfrac{x^2 + 2x}{x^3 - x^2 + x - 1}$
58. $\dfrac{3x^3 + 4x}{x^4 + 2x^2 + 1}$

59. $\dfrac{x^2 + 3x - 3}{x^3 + 2x^2 + x + 2}$
60. $\dfrac{2x^2 - x + 7}{x^4 + 8x^2 + 16}$

Data Analysis: Curve-Fitting In Exercises 61 and 62, find the equation of the parabola $y = ax^2 + bx + c$ that passes through the points. To verify your result, use a graphing utility to plot the points and graph the parabola.

61. $(-1, -4), (1, -2), (2, 5)$
62. $(-1, 0), (1, 4), (2, 3)$

63. **Physical Education** Pebble Beach Golf Links in Pebble Beach, California is an 18-hole course that consists of par-3 holes, par-4 holes, and par-5 holes. There are two more par-4 holes than twice the number of par-5 holes, and the number of par-3 holes is equal to the number of par-5 holes. Find the number of par-3, par-4, and par-5 holes on the course. *(Source: Pebble Beach Resorts)*

64. **Economics** An inheritance of $40,000 is divided among three investments yielding $3500 in interest per year. The interest rates for the three investments are 7%, 9%, and 11%. Find the amount of each investment if the second and third are $3000 and $5000 less than the first, respectively.

8.4

Dimension of a Matrix In Exercises 65–68, determine the dimension of the matrix.

65. $\begin{bmatrix} -3 \\ 1 \\ 10 \end{bmatrix}$
66. $\begin{bmatrix} 3 & -1 & 0 & 6 \\ -2 & 7 & 1 & 4 \end{bmatrix}$

67. $[3]$
68. $[6 \quad 7 \quad -5 \quad 0 \quad -8]$

Writing an Augmented Matrix In Exercises 69–72, write the augmented matrix for the system of linear equations.

69. $\begin{cases} 6x - 7y = 11 \\ -2x + 5y = -1 \end{cases}$
70. $\begin{cases} -x + y = 12 \\ 10x - 4y = -90 \end{cases}$

71. $\begin{cases} 8x - 7y + 4z = 12 \\ 3x - 5y + 2z = 20 \\ 5x + 3y - 3z = 26 \end{cases}$
72. $\begin{cases} 3x - 5y + z = 25 \\ -4x \qquad - 2z = -14 \\ 6x + y \qquad = 15 \end{cases}$

Writing a System of Equations In Exercises 73 and 74, write the system of linear equations represented by the augmented matrix. (Use the variables x, y, z, and w, if applicable.)

73. $\begin{bmatrix} 5 & 1 & 7 & \vdots & -9 \\ 4 & 2 & 0 & \vdots & 10 \\ 9 & 4 & 2 & \vdots & 3 \end{bmatrix}$

74. $\begin{bmatrix} 13 & 16 & 7 & 3 & \vdots & 2 \\ 1 & 21 & 8 & 5 & \vdots & 12 \\ 4 & 10 & -4 & 3 & \vdots & -1 \end{bmatrix}$

Using Gaussian Elimination In Exercises 75 and 76, write the matrix in row-echelon form. Remember that the row-echelon form of a matrix is not unique.

75. $\begin{bmatrix} 0 & 1 & 1 \\ 1 & 2 & 3 \\ 2 & 2 & 2 \end{bmatrix}$
76. $\begin{bmatrix} 4 & 8 & 16 \\ 3 & -1 & 2 \\ -2 & 10 & 12 \end{bmatrix}$

Using a Graphing Utility In Exercises 77–80, use the matrix capabilities of a graphing utility to write the matrix in reduced row-echelon form.

77. $\begin{bmatrix} 3 & -2 & 1 & 0 \\ 4 & -3 & 0 & 1 \end{bmatrix}$

78. $\begin{bmatrix} 1 & 1 & 2 & 1 & 0 & 0 \\ -1 & 0 & 3 & 0 & 1 & 0 \\ 1 & 2 & 8 & 0 & 0 & 1 \end{bmatrix}$

135. $\begin{bmatrix} 1 & 2 & 0 \\ -1 & 1 & 1 \\ 0 & -1 & 0 \end{bmatrix}$ **136.** $\begin{bmatrix} 1 & -1 & -2 \\ 0 & 1 & -2 \\ 1 & 2 & -4 \end{bmatrix}$

Finding the Inverse of a 2 × 2 Matrix In Exercises 137–140, use the formula on page 676 to find the inverse of the 2 × 2 matrix.

137. $\begin{bmatrix} -7 & 2 \\ -8 & 2 \end{bmatrix}$ **138.** $\begin{bmatrix} 10 & 4 \\ 7 & 3 \end{bmatrix}$

139. $\begin{bmatrix} -1 & 10 \\ 2 & 20 \end{bmatrix}$ **140.** $\begin{bmatrix} -6 & -5 \\ 3 & 3 \end{bmatrix}$

Solving a System of Equations Using an Inverse In Exercises 141–146, use an inverse matrix to solve (if possible) the system of linear equations.

141. $\begin{cases} -x + 4y = 8 \\ 2x - 7y = -5 \end{cases}$ **142.** $\begin{cases} 2x + 3y = -10 \\ 4x - y = 1 \end{cases}$

143. $\begin{cases} 3x + 2y - z = 6 \\ x - y + 2z = -1 \\ 5x + y + z = 7 \end{cases}$ **144.** $\begin{cases} -x + 4y - 2z = 12 \\ 2x - 9y + 5z = -25 \\ -x + 5y - 4z = 10 \end{cases}$

145. $\begin{cases} x + 2y + z - w = -2 \\ 2x + y + z + w = 1 \\ x - y - 3z = 0 \\ z + w = 1 \end{cases}$

146. $\begin{cases} x + y + z + w = 1 \\ x - y + 2z + w = -3 \\ y + w = 2 \\ x + w = 2 \end{cases}$

Solving a System of Linear Equations In Exercises 147–150, use the matrix capabilities of a graphing utility to solve (if possible) the system of linear equations.

147. $\begin{cases} x + 2y = -1 \\ 3x + 4y = -5 \end{cases}$ **148.** $\begin{cases} x + 3y = 23 \\ -6x + 2y = -18 \end{cases}$

149. $\begin{cases} -3x - 3y - 4z = 2 \\ y + z = -1 \\ 4x + 3y + 4z = -1 \end{cases}$

150. $\begin{cases} 2x + 3y - 4z = 1 \\ x - y + 2z = -4 \\ 3x + 7y - 10z = 0 \end{cases}$

8.7

The Determinant of a Matrix In Exercises 151–154, find the determinant of the matrix.

151. $\begin{bmatrix} 8 & 5 \\ 2 & -4 \end{bmatrix}$ **152.** $\begin{bmatrix} -9 & 11 \\ 7 & -4 \end{bmatrix}$

153. $\begin{bmatrix} 50 & -30 \\ 10 & 5 \end{bmatrix}$ **154.** $\begin{bmatrix} 14 & -24 \\ 12 & -15 \end{bmatrix}$

Finding the Minors and Cofactors of a Matrix In Exercises 155–158, find all (a) minors and (b) cofactors of the matrix.

155. $\begin{bmatrix} 2 & -1 \\ 7 & 4 \end{bmatrix}$ **156.** $\begin{bmatrix} 3 & 6 \\ 5 & -4 \end{bmatrix}$

157. $\begin{bmatrix} 3 & 2 & -1 \\ -2 & 5 & 0 \\ 1 & 8 & 6 \end{bmatrix}$ **158.** $\begin{bmatrix} 8 & 3 & 4 \\ 6 & 5 & -9 \\ -4 & 1 & 2 \end{bmatrix}$

Finding a Determinant In Exercises 159–166, find the determinant of the matrix. Expand by cofactors on the row or column that appears to make the computations easiest.

159. $\begin{bmatrix} -2 & 4 & 1 \\ -6 & 0 & 2 \\ 5 & 3 & 4 \end{bmatrix}$ **160.** $\begin{bmatrix} 4 & 7 & -1 \\ 2 & -3 & 4 \\ -5 & 1 & -1 \end{bmatrix}$

161. $\begin{bmatrix} -2 & 0 & 0 \\ 2 & -1 & 0 \\ -1 & 1 & 3 \end{bmatrix}$ **162.** $\begin{bmatrix} 0 & 1 & -2 \\ 0 & 1 & 2 \\ -1 & -1 & 3 \end{bmatrix}$

163. $\begin{bmatrix} 1 & 0 & -2 \\ 0 & 1 & 0 \\ -2 & 0 & 1 \end{bmatrix}$ **164.** $\begin{bmatrix} 0 & 3 & 1 \\ 5 & -2 & 1 \\ 1 & 6 & 1 \end{bmatrix}$

165. $\begin{bmatrix} 3 & 0 & -4 & 0 \\ 0 & 8 & 1 & 2 \\ 6 & 1 & 8 & 2 \\ 0 & 3 & -4 & 1 \end{bmatrix}$ **166.** $\begin{bmatrix} -5 & 6 & 0 & 0 \\ 0 & 1 & -1 & 2 \\ -3 & 4 & -5 & 1 \\ 1 & 6 & 0 & 3 \end{bmatrix}$

8.8

Finding the Area of a Figure In Exercises 167–174, use a determinant to find the area of the figure with the given vertices.

167. $(1, 0), (5, 0), (5, 8)$

168. $(-4, 0), (4, 0), (0, 6)$

169. $\left(\frac{1}{2}, 1\right), \left(2, -\frac{5}{2}\right), \left(\frac{3}{2}, 1\right)$

170. $\left(\frac{3}{2}, 1\right), \left(4, -\frac{1}{2}\right), (4, 2)$

171. $(2, 4), (5, 6), (4, 1)$

172. $(-3, 2), (2, -3), (-4, -4)$

173. $(-2, -1), (4, 9), (-2, -9), (4, 1)$

174. $(-4, 8), (4, 0), (-4, 0), (4, -8)$

Testing for Collinear Points In Exercises 175 and 176, use a determinant to determine whether the points are collinear.

175. $(-1, 7), (3, -9), (-3, 15)$

176. $(0, -5), (2, 1), (4, 7)$

Using Cramer's Rule In Exercises 177–184, use Cramer's Rule to solve (if possible) the system of equations.

177. $\begin{cases} x + 2y = 5 \\ -x + y = 1 \end{cases}$ **178.** $\begin{cases} 2x - y = -10 \\ 3x + 2y = -1 \end{cases}$

179. $\begin{cases} 5x - 2y = 6 \\ -11x + 3y = -23 \end{cases}$ **180.** $\begin{cases} 3x + 8y = -7 \\ 9x - 5y = 37 \end{cases}$

181. $\begin{cases} -2x + 3y - 5z = -11 \\ 4x - y + z = -3 \\ -x - 4y + 6z = 15 \end{cases}$

182. $\begin{cases} 5x - 2y + z = 15 \\ 3x - 3y - z = -7 \\ 2x - y - 7z = -3 \end{cases}$

183. $\begin{cases} x - 3y + 2z = 2 \\ 2x + 2y - 3z = 3 \\ x - 7y + 8z = -4 \end{cases}$

184. $\begin{cases} 14x - 21y - 7z = 10 \\ -4x + 2y - 2z = 4 \\ 56x - 21y + 7z = 5 \end{cases}$

Comparing Solution Methods **In Exercises 185 and 186, solve the system of equations using (a) Gaussian elimination and (b) Cramer's Rule. Which method do you prefer, and why?**

185. $\begin{cases} x - 3y + 2z = 5 \\ 2x + y - 4z = -1 \\ 2x + 4y + 2z = 3 \end{cases}$ **186.** $\begin{cases} x + 2y - z = -3 \\ 2x - y + z = -1 \\ 4x - 2y - z = 5 \end{cases}$

Encoding a Message **In Exercises 187 and 188, (a) write the uncoded 1×3 row matrices for the message, and then (b) encode the message using the encoding matrix.**

Message	Encoding Matrix

187. LOOK OUT BELOW $\begin{bmatrix} 2 & -2 & 0 \\ 3 & 0 & -3 \\ -6 & 2 & 3 \end{bmatrix}$

188. JUST DO IT $\begin{bmatrix} 2 & 1 & 0 \\ -6 & -6 & -2 \\ 3 & 2 & 1 \end{bmatrix}$

Decoding a Message **In Exercises 189–192, use A^{-1} to decode the cryptogram.**

189. $A = \begin{bmatrix} 1 & 0 & -1 \\ -1 & -2 & 0 \\ 1 & -2 & 2 \end{bmatrix}$ 32 −46 37 9 −48
15 3 −14 10 −1
−6 2 −8 −22 −3

190. $A = \begin{bmatrix} 1 & 2 & 0 \\ -1 & 1 & 2 \\ 2 & -1 & 2 \end{bmatrix}$ 30 −7 30 5 10 80 37
34 16 40 −7 38 −3 8
36 16 −1 58 23 46 0

191. $A = \begin{bmatrix} 1 & -1 & 0 \\ 0 & 1 & 2 \\ 1 & 1 & -2 \end{bmatrix}$ 21 −11 14 29 −11 −18
32 −6 −26 31 −19 −12
10 6 26 13 −11 −2 37
28 −8 5 13 36

192. $A = \begin{bmatrix} 1 & 1 & 0 \\ -1 & 2 & -2 \\ 1 & -1 & -2 \end{bmatrix}$ 9 15 −54 13 32 −26
8 −6 −14 −4 26 −70
−1 56 −38 28 27 −46
−13 27 −30 26 23 −48
25 4 −26 −11 31 −58
13 39 −34

193. MODELING DATA

The populations (in millions) of Florida for selected years from 2002 through 2008 are shown in the table. (Source: U.S. Census Bureau)

Year	Population (in millions)
2002	16.7
2004	17.3
2006	18.0
2008	18.3

The coefficients of the least squares regression line $y = at + b$, where y is the population (in millions) and t is the year, with $t = 2$ corresponding to 2002, can be found by solving the system

$$\begin{cases} 4b + 20a = 70.3 \\ 20b + 120a = 357 \end{cases}.$$

(a) Use Cramer's Rule to solve the system and find the least squares regression line.

(b) Use a graphing utility to graph the line from part (a).

(c) Use the graph from part (b) to estimate when the population of Florida will exceed 20 million.

(d) Use your regression equation to find algebraically when the population will exceed 20 million.

Conclusions

True or False? **In Exercises 194 and 195, determine whether the statement is true or false. Justify your answer.**

194. Solving a system of equations graphically will always give an exact solution.

195. $\begin{vmatrix} a_{11} & a_{12} & a_{13} \\ a_{21} & a_{22} & a_{23} \\ a_{31} + c_1 & a_{32} + c_2 & a_{33} + c_3 \end{vmatrix} =$

$\begin{vmatrix} a_{11} & a_{12} & a_{13} \\ a_{21} & a_{22} & a_{23} \\ a_{31} & a_{32} & a_{33} \end{vmatrix} + \begin{vmatrix} a_{11} & a_{12} & a_{13} \\ a_{21} & a_{22} & a_{23} \\ c_1 & c_2 & c_3 \end{vmatrix}$

196. What is the relationship between the three elementary row operations performed on an augmented matrix and the operations that lead to equivalent systems of equations?

197. Under what conditions does a matrix have an inverse?

8 Chapter Test

See www.CalcChat.com for worked-out solutions to odd-numbered exercises.
For instructions on how to use a graphing utility, see Appendix A.

Take this test as you would take a test in class. After you are finished, check your work against the answers in the back of the book.

In Exercises 1–3, solve the system by the method of substitution. Check your solution graphically.

1. $\begin{cases} x - y = 6 \\ 3x + 5y = 2 \end{cases}$

2. $\begin{cases} y = x - 1 \\ y = (x - 1)^3 \end{cases}$

3. $\begin{cases} 4x - y^2 = 7 \\ x - y = 3 \end{cases}$

In Exercises 4–6, solve the system by the method of elimination.

4. $\begin{cases} 2x + 5y = -11 \\ 5x - y = 19 \end{cases}$

5. $\begin{cases} 3x - 2y + z = 0 \\ 6x + 2y + 3z = -2 \\ 3x - 4y + 5z = 5 \end{cases}$

6. $\begin{cases} x - 4y - z = 3 \\ 2x - 5y + z = 0 \\ 3x - 3y + 2z = -1 \end{cases}$

7. Find the equation of the parabola $y = ax^2 + bx + c$ that passes through the points $(0, 6)$, $(-2, 2)$, and $\left(3, \frac{9}{2}\right)$.

In Exercises 8 and 9, write the partial fraction decomposition for the rational expression.

8. $\dfrac{5x - 2}{(x - 1)^2}$

9. $\dfrac{x^3 + x^2 + x + 2}{x^4 + x^2}$

$\begin{cases} -2x + 2y + 3z = 7 \\ x - y = -5 \\ y + 4z = -1 \end{cases}$

System for 13

In Exercises 10 and 11, use matrices to solve the system of equations, if possible.

10. $\begin{cases} 2x + y + 2z = 4 \\ 2x + 2y = 5 \\ 2x - y + 6z = 2 \end{cases}$

11. $\begin{cases} 2x + 3y + z = 10 \\ 2x - 3y - 3z = 22 \\ 4x - 2y + 3z = -2 \end{cases}$

12. If possible, find (a) $A - B$, (b) $3A$, (c) $3A - 2B$, and (d) AB.

$$A = \begin{bmatrix} 5 & 4 & 4 \\ -4 & -4 & 0 \\ 1 & 2 & 0 \end{bmatrix}, \quad B = \begin{bmatrix} 4 & 4 & 0 \\ 3 & 2 & 1 \\ 1 & -2 & 0 \end{bmatrix}$$

13. Find A^{-1} for $A = \begin{bmatrix} -2 & 2 & 3 \\ 1 & -1 & 0 \\ 0 & 1 & 4 \end{bmatrix}$ and use A^{-1} to solve the system at the right.

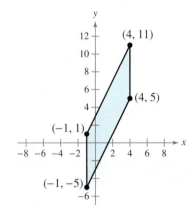

Figure for 16

In Exercises 14 and 15, find the determinant of the matrix.

14. $\begin{bmatrix} -25 & 18 \\ 6 & -7 \end{bmatrix}$

15. $\begin{bmatrix} 4 & 0 & 3 \\ 1 & -8 & 2 \\ 3 & 2 & 2 \end{bmatrix}$

16. Use determinants to find the area of the parallelogram shown at the right.

17. Use Cramer's Rule to solve (if possible) $\begin{cases} 2x - 2y = 3 \\ x + 4y = -1 \end{cases}$.

18. The flow of traffic (in vehicles per hour) through a network of streets is shown at the right. Solve the system for the traffic flow represented by x_i, $i = 1, 2, 3, 4,$ and 5.

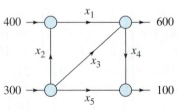

Figure for 18

Proofs in Mathematics

An **indirect proof** can be useful in proving statements of the form "p implies q." Recall that the conditional statement $p \rightarrow q$ is false only when p is true and q is false. To prove a conditional statement indirectly, assume that p is true and q is false. If this assumption leads to an impossibility, then you have proved that the conditional statement is true. An indirect proof is also called a **proof by contradiction.**

You can use an indirect proof to prove the conditional statement

"If a is a positive integer and a^2 is divisible by 2, then a is divisible by 2"

as follows. First, assume that p, "a is a positive integer and a^2 is divisible by 2," is true and q, "a is divisible by 2," is false. This means that a is not divisible by 2. If so, then a is odd and can be written as

$$a = 2n + 1$$

where n is an integer.

$a = 2n + 1$	Definition of an odd integer
$a^2 = 4n^2 + 4n + 1$	Square each side.
$a^2 = 2(2n^2 + 2n) + 1$	Distributive Property

So, by the definition of an odd integer, a^2 is odd. This contradicts the assumption, and you can conclude that a is divisible by 2.

Example   Using an Indirect Proof

Use an indirect proof to prove that $\sqrt{2}$ is an irrational number.

Solution

Begin by assuming that $\sqrt{2}$ is *not* an irrational number. Then $\sqrt{2}$ can be written as the quotient of two integers a and b ($b \neq 0$) that have no common factors.

$\sqrt{2} = \dfrac{a}{b}$	Assume that $\sqrt{2}$ is a rational number.
$2 = \dfrac{a^2}{b^2}$	Square each side.
$2b^2 = a^2$	Multiply each side by b^2.

This implies that 2 is a factor of a^2. So, 2 is also a factor of a, and a can be written as $2c$, where c is an integer.

$2b^2 = (2c)^2$	Substitute $2c$ for a.
$2b^2 = 4c^2$	Simplify.
$b^2 = 2c^2$	Divide each side by 2.

This implies that 2 is a factor of b^2 and also a factor of b. So, 2 is a factor of both a and b. This contradicts the assumption that a and b have no common factors. So, you can conclude that $\sqrt{2}$ is an irrational number.

Proofs without words are pictures or diagrams that give a visual understanding of why a theorem or statement is true. They can also provide a starting point for writing a formal proof. The following proof shows that a 2×2 determinant is the area of a parallelogram.

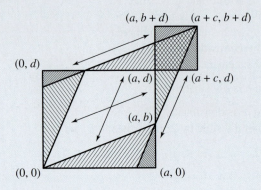

$$\begin{vmatrix} a & b \\ c & d \end{vmatrix} = ad - bc = \| \square \| - \| \square \| = \| \square \|$$

The following is a color-coded version of the proof along with a brief explanation of why this proof works.

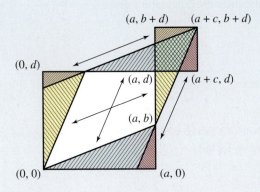

$$\begin{vmatrix} a & b \\ c & d \end{vmatrix} = ad - bc = \| \square \| - \| \square \| = \| \square \|$$

Area of $\square$ = Area of orange $\triangle$ + Area of yellow $\triangle$ + Area of blue $\triangle$ +
               Area of pink $\triangle$ + Area of white quadrilateral

Area of $\square$ = Area of orange $\triangle$ + Area of pink $\triangle$ + Area of green quadrilateral

Area of $\square$ = Area of white quadrilateral + Area of blue $\triangle$ + Area of yellow $\triangle$ −
               Area of green quadrilateral
       = Area of $\square$ − Area of $\square$

From "Proof Without Words" by Solomon W. Golomb, *Mathematics Magazine*, March 1985. Vol. 58, No. 2, pg. 107. Reprinted with permission.

9 Sequences, Series, and Probability

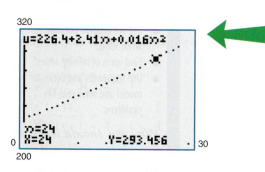

$u=226.4+2.41n+0.016n^2$

n=24
X=24   Y=293.456

Section 9.1, Example 10
Population of the United States

9.1   **Sequences and Series**

9.2   **Arithmetic Sequences and Partial Sums**

9.3   **Geometric Sequences and Series**

9.4   **The Binomial Theorem**

9.5   **Counting Principles**

9.6   **Probability**

709

Some sequences are defined **recursively.** To define a sequence recursively, you need to be given one or more of the first few terms. All other terms of the sequence are then defined using previous terms.

Example 4 A Recursive Sequence

A sequence is defined recursively as follows.

$$a_1 = 3, \ a_k = 2a_{k-1} + 1, \quad \text{where } k \geq 2$$

Write the first five terms of this sequence.

Solution

$a_1 = 3$ 1st term is given.

$a_2 = 2a_{2-1} + 1 = 2a_1 + 1 = 2(3) + 1 = 7$ Use recursion formula.

$a_3 = 2a_{3-1} + 1 = 2a_2 + 1 = 2(7) + 1 = 15$ Use recursion formula.

$a_4 = 2a_{4-1} + 1 = 2a_3 + 1 = 2(15) + 1 = 31$ Use recursion formula.

$a_5 = 2a_{5-1} + 1 = 2a_4 + 1 = 2(31) + 1 = 63$ Use recursion formula.

✓**CHECKPOINT** Now try Exercise 53.

In the next example you will study a well-known recursive sequence, the Fibonacci sequence.

Example 5 The Fibonacci Sequence: A Recursive Sequence

The Fibonacci sequence is defined recursively as follows.

$$a_0 = 1, \ a_1 = 1, \ a_k = a_{k-2} + a_{k-1}, \quad \text{where } k \geq 2$$

Write the first six terms of this sequence.

Solution

$a_0 = 1$ 0th term is given.

$a_1 = 1$ 1st term is given.

$a_2 = a_{2-2} + a_{2-1} = a_0 + a_1 = 1 + 1 = 2$ Use recursion formula.

$a_3 = a_{3-2} + a_{3-1} = a_1 + a_2 = 1 + 2 = 3$ Use recursion formula.

$a_4 = a_{4-2} + a_{4-1} = a_2 + a_3 = 2 + 3 = 5$ Use recursion formula.

$a_5 = a_{5-2} + a_{5-1} = a_3 + a_4 = 3 + 5 = 8$ Use recursion formula.

You can check this result using the *table* feature of a graphing utility, as shown in Figure 9.2.

Figure 9.2

✓**CHECKPOINT** Now try Exercise 57.

Technology Tip

To graph a sequence using a graphing utility, set the mode to *dot* and *sequence* and enter the sequence. Try graphing the sequence in Example 4 and using the *trace* feature to identify the terms. For instructions on how to use the sequence mode, see Appendix A; for specific keystrokes, go to this textbook's *Companion Website*.

```
Plot1 Plot2 Plot3
nMin=1
\u(n)⊟2u(n−1)+1
 u(nMin)⊟(3)
\v(n)=
 v(nMin)=
\w(n)=
 w(nMin)=
```

Factorial Notation

Some very important sequences in mathematics involve terms that are defined with special types of products called **factorials**.

> **Definition of Factorial**
>
> If n is a positive integer, then **n factorial** is defined as
>
> $$n! = 1 \cdot 2 \cdot 3 \cdot 4 \cdots (n-1) \cdot n.$$
>
> As a special case, zero factorial is defined as $0! = 1$.

Notice that $0! = 1$ and $1! = 1$. Here are some other values of $n!$.

$$2! = 1 \cdot 2 = 2 \quad 3! = 1 \cdot 2 \cdot 3 = 6 \quad 4! = 1 \cdot 2 \cdot 3 \cdot 4 = 24$$

The value of n does not have to be very large before the value of $n!$ becomes huge. For instance, $10! = 3,628,800$.

Factorials follow the same conventions for order of operations as do exponents. For instance,

$$2n! = 2(n!) = 2(1 \cdot 2 \cdot 3 \cdot 4 \cdots n)$$

whereas $(2n)! = 1 \cdot 2 \cdot 3 \cdot 4 \cdots 2n$.

Example 6   Writing the Terms of a Sequence Involving Factorials

Write the first five terms of the sequence given by $a_n = \dfrac{2^n}{n!}$. Begin with $n = 0$.

Algebraic Solution

$$a_0 = \frac{2^0}{0!} = \frac{1}{1} = 1 \qquad \text{0th term}$$

$$a_1 = \frac{2^1}{1!} = \frac{2}{1} = 2 \qquad \text{1st term}$$

$$a_2 = \frac{2^2}{2!} = \frac{4}{2} = 2 \qquad \text{2nd term}$$

$$a_3 = \frac{2^3}{3!} = \frac{8}{6} = \frac{4}{3} \qquad \text{3rd term}$$

$$a_4 = \frac{2^4}{4!} = \frac{16}{24} = \frac{2}{3} \qquad \text{4th term}$$

✓**CHECKPOINT**   Now try Exercise 63.

Graphical Solution

Using a graphing utility set to *dot* and *sequence* modes, enter the sequence. Next, graph the sequence, as shown in Figure 9.3.

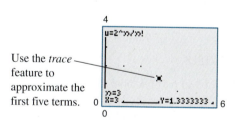

Use the *trace* feature to approximate the first five terms.

Figure 9.3

You can estimate the first five terms of the sequence as follows.

$$u_0 = 1, \quad u_1 = 2, \quad u_2 = 2,$$

$$u_3 \approx 1.333 \approx \tfrac{4}{3}, \quad u_4 \approx 0.667 \approx \tfrac{2}{3}$$

Example 7   Simplifying Factorial Expressions

a. $\dfrac{8!}{2! \cdot 6!} = \dfrac{1 \cdot 2 \cdot 3 \cdot 4 \cdot 5 \cdot 6 \cdot 7 \cdot 8}{1 \cdot 2 \cdot 1 \cdot 2 \cdot 3 \cdot 4 \cdot 5 \cdot 6} = \dfrac{7 \cdot 8}{2} = 28$

b. $\dfrac{n!}{(n-1)!} = \dfrac{1 \cdot 2 \cdot 3 \cdots (n-1) \cdot n}{1 \cdot 2 \cdot 3 \cdots (n-1)} = n$

✓**CHECKPOINT**   Now try Exercise 73.

Summation Notation

There is a convenient notation for the sum of the terms of a finite sequence. It is called **summation notation** or **sigma notation** because it involves the use of the uppercase Greek letter sigma, written as Σ.

Definition of Summation Notation

The sum of the first n terms of a sequence is represented by

$$\sum_{i=1}^{n} a_i = a_1 + a_2 + a_3 + a_4 + \cdots + a_n$$

where i is called the **index of summation**, n is the **upper limit of summation**, and 1 is the **lower limit of summation**.

Example 8   Sigma Notation for Sums

a. $\displaystyle\sum_{i=1}^{5} 3i = 3(1) + 3(2) + 3(3) + 3(4) + 3(5)$

$\qquad\quad = 3(1 + 2 + 3 + 4 + 5)$

$\qquad\quad = 3(15)$

$\qquad\quad = 45$

b. $\displaystyle\sum_{k=3}^{6} (1 + k^2) = (1 + 3^2) + (1 + 4^2) + (1 + 5^2) + (1 + 6^2)$

$\qquad\qquad\quad = 10 + 17 + 26 + 37$

$\qquad\qquad\quad = 90$

c. $\displaystyle\sum_{n=0}^{8} \frac{1}{n!} = \frac{1}{0!} + \frac{1}{1!} + \frac{1}{2!} + \frac{1}{3!} + \frac{1}{4!} + \frac{1}{5!} + \frac{1}{6!} + \frac{1}{7!} + \frac{1}{8!}$

$\qquad\qquad = 1 + 1 + \frac{1}{2} + \frac{1}{6} + \frac{1}{24} + \frac{1}{120} + \frac{1}{720} + \frac{1}{5040} + \frac{1}{40,320}$

$\qquad\qquad \approx 2.71828$

For the summation in part (c), note that the sum is very close to the irrational number $e \approx 2.718281828$. It can be shown that as more terms of the sequence whose nth term is $1/n!$ are added, the sum becomes closer and closer to e.

✓**CHECKPOINT**   Now try Exercise 87.

In Example 8, note that the lower limit of a summation does not have to be 1. Also note that the index of summation does not have to be the letter i. For instance, in part (b), the letter k is the index of summation.

Properties of Sums   (See the proofs on page 772.)

1. $\displaystyle\sum_{i=1}^{n} c = cn,$ c is a constant.

2. $\displaystyle\sum_{i=1}^{n} ca_i = c\sum_{i=1}^{n} a_i,$ c is a constant.

3. $\displaystyle\sum_{i=1}^{n} (a_i + b_i) = \sum_{i=1}^{n} a_i + \sum_{i=1}^{n} b_i$

4. $\displaystyle\sum_{i=1}^{n} (a_i - b_i) = \sum_{i=1}^{n} a_i - \sum_{i=1}^{n} b_i$

Series

Many applications involve the sum of the terms of a finite or an infinite sequence. Such a sum is called a **series.**

Definition of a Series

Consider the infinite sequence $a_1, a_2, a_3, \ldots, a_i, \ldots$

1. The sum of the first n terms of the sequence is called a **finite series** or the **partial sum** of the sequence and is denoted by
$$a_1 + a_2 + a_3 + \cdots + a_n = \sum_{i=1}^{n} a_i.$$

2. The sum of all the terms of the infinite sequence is called an **infinite series** and is denoted by
$$a_1 + a_2 + a_3 + \cdots + a_i + \cdots = \sum_{i=1}^{\infty} a_i.$$

Example 9 Finding the Sum of a Series

For the series
$$\sum_{i=1}^{\infty} \frac{3}{10^i}$$

find (a) the third partial sum and (b) the sum.

Solution

a. The third partial sum is
$$\sum_{i=1}^{3} \frac{3}{10^i} = \frac{3}{10^1} + \frac{3}{10^2} + \frac{3}{10^3}$$

$$= \frac{3}{10} + \frac{3}{100} + \frac{3}{1000}$$

$$= 0.3 + 0.03 + 0.003$$

$$= 0.333.$$

b. The sum of the series is
$$\sum_{i=1}^{\infty} \frac{3}{10^i} = \frac{3}{10^1} + \frac{3}{10^2} + \frac{3}{10^3} + \frac{3}{10^4} + \frac{3}{10^5} + \cdots$$

$$= \frac{3}{10} + \frac{3}{100} + \frac{3}{1000} + \frac{3}{10,000} + \frac{3}{100,000} + \cdots$$

$$= 0.3 + 0.03 + 0.003 + 0.0003 + 0.00003 + \cdots$$

$$= 0.33333 \ldots$$

$$= \frac{1}{3}.$$

✓CHECKPOINT Now try Exercise 117.

Technology Tip

Most graphing utilities are able to sum the first n terms of a sequence. Try using a graphing utility to confirm the results in Example 8 and Example 9(a).

```
sum(seq(1/n!,n,0
,8))
            2.71827877
```

Notice in Example 9(b) that the sum of an infinite series can be a finite number.

Application

Sequences have many applications in situations that involve recognizable patterns. One such model is illustrated in Example 10.

Example 10   Population of the United States

From 1980 through 2008, the resident population of the United States can be approximated by the model

$$a_n = 226.4 + 2.41n + 0.016n^2, \quad n = 0, 1, \ldots, 28$$

where a_n is the population (in millions) and n represents the year, with $n = 0$ corresponding to 1980. Find the last five terms of this finite sequence.   (Source: U.S. Census Bureau)

Algebraic Solution

The last five terms of this finite sequence are as follows.

$a_{24} = 226.4 + 2.41(24) + 0.016(24)^2$

≈ 293.5 2004 population

$a_{25} = 226.4 + 2.41(25) + 0.016(25)^2$

≈ 296.7 2005 population

$a_{26} = 226.4 + 2.41(26) + 0.016(26)^2$

≈ 299.9 2006 population

$a_{27} = 226.4 + 2.41(27) + 0.016(27)^2$

≈ 303.1 2007 population

$a_{28} = 226.4 + 2.41(28) + 0.016(28)^2$

≈ 306.4 2008 population

✓CHECKPOINT   Now try Exercise 121.

Graphical Solution

Using a graphing utility set to *dot* and *sequence* modes, enter the sequence. Next, graph the sequence, as shown in Figure 9.4.

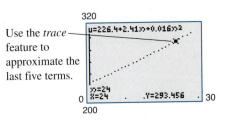

Use the *trace* feature to approximate the last five terms.

Figure 9.4

You can estimate the last five terms of the sequence as follows.

$$a_{24} \approx 293.5, \quad a_{25} \approx 296.7, \quad a_{26} \approx 299.9,$$

$$a_{27} \approx 303.1, \quad a_{28} \approx 306.4$$

Explore the Concept

A $3 \times 3 \times 3$ cube is created using 27 unit cubes (a unit cube has a length, width, and height of 1 unit), and only the faces of the cubes that are visible are painted blue (see Figure 9.5). Complete the table below to determine how many unit cubes of the $3 \times 3 \times 3$ cube have no blue faces, one blue face, two blue faces, and three blue faces. Do the same for a $4 \times 4 \times 4$ cube, a $5 \times 5 \times 5$ cube, and a $6 \times 6 \times 6$ cube, and record your results in the table below. What type of pattern do you observe in the table? Write a formula you could use to determine the column values for an $n \times n \times n$ cube.

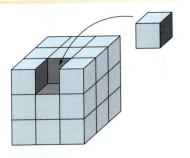

Figure 9.5

Cube \ Number of blue faces	0	1	2	3
$3 \times 3 \times 3$				

9.1 Exercises

Vocabulary and Concept Check

In Exercises 1–4, fill in the blank(s).

1. The function values $a_1, a_2, a_3, a_4, \ldots, a_n, \ldots$ are called the _____ of a sequence.

2. If you are given one or more of the first few terms of a sequence, and all other terms of the sequence are defined using previous terms, then the sequence is defined _____ .

3. For the sum $\sum_{i=1}^{n} a_i$, i is called the _____ of summation, n is the _____ of summation, and 1 is the _____ of summation.

4. The sum of the terms of a finite or an infinite sequence is called a _____ .

5. Which describes an infinite sequence? a finite sequence?
   (a) The domain consists of the first n positive integers.
   (b) The domain consists of the set of positive integers.

6. Write $1 \cdot 2 \cdot 3 \cdot 4 \cdot 5 \cdot 6$ in factorial notation.

Procedures and Problem Solving

Writing the Terms of a Sequence In Exercises 7–16, write the first five terms of the sequence. (Assume n begins with 1.)

✓ 7. $a_n = 2n + 5$
8. $a_n = 4n - 7$
9. $a_n = 3^n$
10. $a_n = \left(\frac{1}{2}\right)^n$
11. $a_n = \left(-\frac{1}{2}\right)^n$
12. $a_n = (-2)^n$
13. $a_n = \dfrac{n+1}{n}$
14. $a_n = \dfrac{n}{n+1}$
15. $a_n = \dfrac{n}{n^2+1}$
16. $a_n = \dfrac{2n}{n+1}$

Writing the Terms of a Sequence In Exercises 17–26, write the first five terms of the sequence (a) using the *table* feature of a graphing utility and (b) algebraically. (Assume n begins with 1.)

✓ 17. $a_n = \dfrac{1 + (-1)^n}{n}$
18. $a_n = \dfrac{1 + (-1)^n}{2n}$
19. $a_n = 1 - \dfrac{1}{2^n}$
20. $a_n = \dfrac{3^n}{4^n}$
21. $a_n = \dfrac{1}{n^{3/2}}$
22. $a_n = \dfrac{1}{\sqrt{n}}$
23. $a_n = \dfrac{(-1)^n}{n^2}$
24. $a_n = (-1)^n\left(\dfrac{n}{n+1}\right)$
25. $a_n = (2n-1)(2n+1)$
26. $a_n = n(n-1)(n-2)$

Using a Graphing Utility In Exercises 27–32, use the *table* feature of a graphing utility to find the first 10 terms of the sequence. (Assume n begins with 1.)

27. $a_n = 2(3n - 1) + 5$
28. $a_n = 2n(n+1)(n+2)$
29. $a_n = 1 + \dfrac{n+1}{n}$
30. $a_n = \dfrac{4n^2}{n+2}$
31. $a_n = (-1)^n + 1$
32. $a_n = (-1)^{n+1} + 1$

Writing an Indicated Term In Exercises 33–38, find the indicated term of the sequence.

33. $a_n = \dfrac{n^2}{n^2+1}$
   $a_{10} = $ ▮
34. $a_n = \dfrac{n^2}{2n+1}$
   $a_5 = $ ▮
35. $a_n = (-1)^n(3n - 2)$
   $a_{25} = $ ▮
36. $a_n = (-1)^{n-1}[n(n-1)]$
   $a_{16} = $ ▮
37. $a_n = \dfrac{2^n}{2^n+1}$
   $a_6 = $ ▮
38. $a_n = \dfrac{2^{n+1}}{2^n+1}$
   $a_7 = $ ▮

Finding the nth Term of a Sequence In Exercises 39–52, write an expression for the *apparent* nth term of the sequence. (Assume n begins with 1.)

✓ 39. $1, 4, 7, 10, 13, \ldots$
40. $3, 7, 11, 15, 19, \ldots$
41. $0, 3, 8, 15, 24, \ldots$
42. $1, \frac{1}{4}, \frac{1}{9}, \frac{1}{16}, \frac{1}{25}, \ldots$
43. $\frac{2}{3}, \frac{3}{4}, \frac{4}{5}, \frac{5}{6}, \frac{6}{7}, \ldots$
44. $\frac{2}{1}, \frac{3}{3}, \frac{4}{5}, \frac{5}{7}, \frac{6}{9}, \ldots$
45. $\frac{1}{2}, \frac{-1}{4}, \frac{1}{8}, \frac{-1}{16}, \ldots$
46. $\frac{1}{3}, -\frac{2}{9}, \frac{4}{27}, -\frac{8}{81}, \ldots$
47. $1 + \frac{1}{1}, 1 + \frac{1}{2}, 1 + \frac{1}{3}, 1 + \frac{1}{4}, 1 + \frac{1}{5}, \ldots$
48. $1 + \frac{1}{2}, 1 + \frac{3}{4}, 1 + \frac{7}{8}, 1 + \frac{15}{16}, 1 + \frac{31}{32}, \ldots$
49. $1, \frac{1}{2}, \frac{1}{6}, \frac{1}{24}, \frac{1}{120}, \ldots$
50. $1, 2, \frac{2^2}{2}, \frac{2^3}{6}, \frac{2^4}{24}, \frac{2^5}{120}, \ldots$
51. $1, 3, 1, 3, 1, \ldots$
52. $1, -1, 1, -1, 1, \ldots$

Conclusions

True or False?   In Exercises 125 and 126, determine whether the statement is true or false. Justify your answer.

125. $\sum_{i=1}^{4}(i^2 + 2i) = \sum_{i=1}^{4}i^2 + 2\sum_{i=1}^{4}i$

126. $\sum_{j=1}^{4}2^j = \sum_{j=3}^{6}2^{j-2}$

Fibonacci Sequence   In Exercises 127 and 128, use the Fibonacci sequence. (See Example 5.)

127. Write the first 12 terms of the Fibonacci sequence a_n and the first 10 terms of the sequence given by

$$b_n = \frac{a_{n+1}}{a_n}, \quad n > 0.$$

128. Using the definition of b_n given in Exercise 127, show that b_n can be defined recursively by

$$b_n = 1 + \frac{1}{b_{n-1}}.$$

Exploration   In Exercises 129–132, let

$$a_n = \frac{\left(1 + \sqrt{5}\right)^n - \left(1 - \sqrt{5}\right)^n}{2^n\sqrt{5}}$$

be a sequence with nth term a_n.

129. Use the *table* feature of a graphing utility to find the first five terms of the sequence.

130. Do you recognize the terms of the sequence in Exercise 129? What sequence is it?

131. Find expressions for a_{n+1} and a_{n+2} in terms of n.

132. Use the result from Exercise 131 to show that $a_{n+2} = a_{n+1} + a_n$. Is this result the same as your answer to Exercise 129? Explain.

∫ A Sequence Involving x   In Exercises 133–142, write the first five terms of the sequence.

133. $a_n = \dfrac{x^n}{n!}$

134. $a_n = \dfrac{x^2}{n^2}$

135. $a_n = \dfrac{(-1)^n x^{2n+1}}{2n + 1}$

136. $a_n = \dfrac{(-1)^n x^{n+1}}{n + 1}$

137. $a_n = \dfrac{(-1)^n x^{2n}}{(2n)!}$

138. $a_n = \dfrac{(-1)^n x^{2n+1}}{(2n + 1)!}$

139. $a_n = \dfrac{(-1)^n x^n}{n!}$

140. $a_n = \dfrac{(-1)^n x^{n+1}}{(n + 1)!}$

141. $a_n = \dfrac{(-1)^{n+1}(x + 1)^n}{n!}$

142. $a_n = \dfrac{(-1)^n(x - 1)^n}{(n + 1)!}$

Writing Partial Sums   In Exercises 143–146, write the first five terms of the sequence. Then find an expression for the nth partial sum.

143. $a_n = \dfrac{1}{2n} - \dfrac{1}{2n + 2}$

144. $a_n = \dfrac{1}{n} - \dfrac{1}{n + 1}$

145. $a_n = \dfrac{1}{n + 1} - \dfrac{1}{n + 2}$

146. $a_n = \dfrac{1}{n} - \dfrac{1}{n + 2}$

147. Think About It   Does every finite series whose terms are integers have a finite sum? Explain.

148. CAPSTONE   The graph represents the first six terms of a sequence.

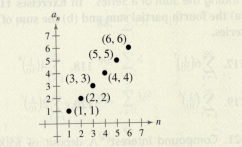

(a) Write the first six terms of the sequence.

(b) Write an expression for the apparent nth term a_n of the sequence.

(c) Use sigma notation to represent the partial sum of the first 50 terms of the sequence.

Cumulative Mixed Review

Operations with Matrices   In Exercises 149–152, find, if possible, (a) $A - B$, (b) $2B - 3A$, (c) AB, and (d) BA.

149. $A = \begin{bmatrix} 6 & 5 \\ 3 & 4 \end{bmatrix}$, $B = \begin{bmatrix} -2 & 4 \\ 6 & -3 \end{bmatrix}$

150. $A = \begin{bmatrix} 10 & 7 \\ -4 & 6 \end{bmatrix}$, $B = \begin{bmatrix} 0 & -12 \\ 8 & 11 \end{bmatrix}$

151. $A = \begin{bmatrix} -2 & -3 & 6 \\ 4 & 5 & 7 \\ 1 & 7 & 4 \end{bmatrix}$, $B = \begin{bmatrix} 1 & 4 & 2 \\ 0 & 1 & 6 \\ 0 & 3 & 1 \end{bmatrix}$

152. $A = \begin{bmatrix} -1 & 4 & 0 \\ 5 & 1 & 2 \\ 0 & -1 & 3 \end{bmatrix}$, $B = \begin{bmatrix} 0 & 4 & 0 \\ 3 & 1 & -2 \\ -1 & 0 & 2 \end{bmatrix}$

9.2   Arithmetic Sequences and Partial Sums

Arithmetic Sequences

A sequence whose consecutive terms have a common difference is called an **arithmetic sequence.**

> **Definition of Arithmetic Sequence**
>
> A sequence is **arithmetic** when the differences between consecutive terms are the same. So, the sequence
>
> $$a_1, a_2, a_3, a_4, \ldots, a_n, \ldots$$
>
> is arithmetic when there is a number d such that
>
> $$a_2 - a_1 = a_3 - a_2 = a_4 - a_3 = \cdots = d.$$
>
> The number d is the **common difference** of the sequence.

What you should learn

- Recognize, write, and find the nth terms of arithmetic sequences.
- Find nth partial sums of arithmetic sequences.
- Use arithmetic sequences to model and solve real-life problems.

Why you should learn it

Arithmetic sequences can reduce the amount of time it takes to find the sum of a sequence of numbers with a common difference. In Exercise 83 on page 727, you will use an arithmetic sequence to find the number of bricks needed to lay a brick patio.

Example 1   Examples of Arithmetic Sequences

a. The sequence whose nth term is

$$4n + 3$$

is arithmetic. The common difference between consecutive terms is 4.

$$7, 11, 15, 19, \ldots, 4n + 3, \ldots \qquad \text{Begin with } n = 1.$$

$$11 - 7 = 4$$

b. The sequence whose nth term is

$$7 - 5n$$

is arithmetic. The common difference between consecutive terms is -5.

$$2, -3, -8, -13, \ldots, 7 - 5n, \ldots \qquad \text{Begin with } n = 1.$$

$$-3 - 2 = -5$$

c. The sequence whose nth term is

$$\tfrac{1}{4}(n + 3)$$

is arithmetic. The common difference between consecutive terms is $\frac{1}{4}$.

$$1, \frac{5}{4}, \frac{3}{2}, \frac{7}{4}, \ldots, \frac{n + 3}{4}, \ldots \qquad \text{Begin with } n = 1.$$

$$\tfrac{5}{4} - 1 = \tfrac{1}{4}$$

✔CHECKPOINT   Now try Exercise 11.

The sequence $1, 4, 9, 16, \ldots$, whose nth term is n^2, is *not* arithmetic. The difference between the first two terms is

$$a_2 - a_1 = 4 - 1 = 3$$

but the difference between the second and third terms is

$$a_3 - a_2 = 9 - 4 = 5.$$

The nth term of an arithmetic sequence can be derived from the pattern below.

$a_1 = a_1$ 1st term

$a_2 = a_1 + d$ 2nd term

$a_3 = a_1 + 2d$ 3rd term

$a_4 = a_1 + 3d$ 4th term

$a_5 = a_1 + 4d$ 5th term

1 less

$a_n = a_1 + (n-1)d$ nth term

1 less

The result is summarized in the following definition.

> **The nth Term of an Arithmetic Sequence**
>
> The nth term of an arithmetic sequence has the form
>
> $$a_n = a_1 + (n-1)d$$
>
> where d is the common difference between consecutive terms of the sequence and a_1 is the first term of the sequence.

Example 2   Finding the nth Term of an Arithmetic Sequence

Find a formula for the nth term of the arithmetic sequence whose common difference is 3 and whose first term is 2.

Solution

You know that the formula for the nth term is of the form $a_n = a_1 + (n-1)d$. Moreover, because the common difference is $d = 3$ and the first term is $a_1 = 2$, the formula must have the form

$a_n = 2 + 3(n-1)$ Substitute 2 for a_1 and 3 for d.

or $a_n = 3n - 1$. The sequence therefore has the following form.

$2, 5, 8, 11, 14, \ldots, 3n - 1, \ldots$

A graph of the first 15 terms of the sequence is shown in Figure 9.6. Notice that the points lie on a line. This makes sense because a_n is a linear function of n. In other words, the terms "arithmetic" and "linear" are closely connected.

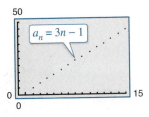

Figure 9.6

✔CHECKPOINT   Now try Exercise 21.

Example 3   Writing the Terms of an Arithmetic Sequence

The fourth term of an arithmetic sequence is 20, and the 13th term is 65. Write the first several terms of this sequence.

Solution

You know that $a_4 = 20$ and $a_{13} = 65$. So, you must add the common difference d nine times to the fourth term to obtain the 13th term. Therefore, the fourth and 13th terms of the sequence are related by

$a_{13} = a_4 + 9d.$ a_4 and a_{13} are nine terms apart.

Using $a_4 = 20$ and $a_{13} = 65$, you have

$65 = 20 + 9d.$

So, you can conclude that $d = 5$, which implies that the sequence is as follows.

$a_1 \quad a_2 \quad a_3 \quad a_4 \quad a_5 \quad a_6 \quad a_7 \quad a_8 \quad a_9 \quad a_{10} \quad a_{11} \quad a_{12} \quad a_{13} \quad \cdots$

$5, \quad 10, \quad 15, \quad 20, \quad 25, \quad 30, \quad 35, \quad 40, \quad 45, \quad 50, \quad 55, \quad 60, \quad 65, \quad \cdots$

 CHECKPOINT   Now try Exercise 35.

When you know the nth term of an arithmetic sequence *and* you know the common difference of the sequence, you can find the $(n + 1)$th term by using the *recursion formula*

$a_{n+1} = a_n + d.$ Recursion formula

With this formula, you can find any term of an arithmetic sequence, *provided* that you know the preceding term. For instance, when you know the first term, you can find the second term. Then, knowing the second term, you can find the third term, and so on.

Example 4   Using a Recursion Formula

Find the ninth term of the arithmetic sequence whose first two terms are 2 and 9.

Solution

You know that the sequence is arithmetic. Also, $a_1 = 2$ and $a_2 = 9$. So, the common difference for this sequence is

$d = 9 - 2 = 7.$

There are two ways to find the ninth term. One way is simply to write out the first nine terms (by repeatedly adding 7).

$2, \ 9, \ 16, \ 23, \ 30, \ 37, \ 44, \ 51, \ 58$

Another way to find the ninth term is to first find a formula for the nth term. Because the common difference is $d = 7$ and the first term is $a_1 = 2$, the formula must have the form

$a_n = 2 + 7(n - 1).$ Substitute 2 for a_1 and 7 for d.

Therefore, a formula for the nth term is

$a_n = 7n - 5$

which implies that the ninth term is

$a_9 = 7(9) - 5 = 58.$

 CHECKPOINT   Now try Exercise 43.

Technology Tip

Most graphing utilities have a built-in function that will display the terms of an arithmetic sequence. For instructions on how to use the *sequence* feature, see Appendix A; for specific keystrokes, go to this textbook's *Companion Website*.

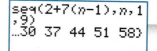

seq(2+7(n-1),n,1
,9)
...30 37 44 51 58}

The Sum of a Finite Arithmetic Sequence

There is a simple formula for the *sum* of a finite arithmetic sequence.

> **The Sum of a Finite Arithmetic Sequence**   (See the proof on page 773.)
>
> The sum of a finite arithmetic sequence with *n* terms is given by
>
> $$S_n = \frac{n}{2}(a_1 + a_n).$$

Be sure you see that this formula works only for *arithmetic* sequences.

Example 5   Finding the Sum of a Finite Arithmetic Sequence

Find the sum: $1 + 3 + 5 + 7 + 9 + 11 + 13 + 15 + 17 + 19$.

Solution

To begin, notice that the sequence is arithmetic (with a common difference of 2). Moreover, the sequence has 10 terms. So, the sum of the sequence is

$$S_n = 1 + 3 + 5 + 7 + 9 + 11 + 13 + 15 + 17 + 19$$

$$= \frac{n}{2}(a_1 + a_n) \qquad \text{Formula for sum of an arithmetic sequence}$$

$$= \frac{10}{2}(1 + 19) \qquad \text{Substitute 10 for } n, \text{ 1 for } a_1, \text{ and 19 for } a_n.$$

$$= 5(20) = 100. \qquad \text{Simplify.}$$

✔**CHECKPOINT**   Now try Exercise 57.

The sum of the first *n* terms of an infinite sequence is called the **nth partial sum.** The *n*th partial sum of an arithmetic sequence can be found by using the formula for the sum of a finite arithmetic sequence.

Example 6   Finding a Partial Sum of an Arithmetic Sequence

Find the 150th partial sum of the arithmetic sequence 5, 16, 27, 38, 49, ...

Solution

For this arithmetic sequence, you have $a_1 = 5$ and $d = 16 - 5 = 11$. So,

$$a_n = 5 + 11(n - 1)$$

and the *n*th term is $a_n = 11n - 6$. Therefore, $a_{150} = 11(150) - 6 = 1644$, and the sum of the first 150 terms is

$$S_n = \frac{n}{2}(a_1 + a_n) \qquad \text{\textit{n}th partial sum formula}$$

$$= \frac{150}{2}(5 + 1644) \qquad \text{Substitute 150 for } n, \text{ 5 for } a_1, \text{ and 1644 for } a_n.$$

$$= 75(1649) \qquad \text{Simplify.}$$

$$= 123{,}675. \qquad \text{\textit{n}th partial sum}$$

✔**CHECKPOINT**   Now try Exercise 65.

Applications

Example 7   Total Sales

A small business sells $20,000 worth of sports memorabilia during its first year. The owner of the business has set a goal of increasing annual sales by $15,000 each year for 19 years. Assuming that this goal is met, find the total sales during the first 20 years this business is in operation.

Algebraic Solution

The annual sales form an arithmetic sequence in which $a_1 = 20,000$ and $d = 15,000$. So,

$$a_n = 20,000 + 15,000(n - 1)$$

and the nth term of the sequence is

$$a_n = 15,000n + 5000.$$

This implies that the 20th term of the sequence is

$$a_{20} = 15,000(20) + 5000$$
$$= 300,000 + 5000$$
$$= 305,000.$$

The sum of the first 20 terms of the sequence is

$$S_n = \frac{n}{2}(a_1 + a_n) \qquad \text{\textit{n}th partial sum formula}$$

$$= \frac{20}{2}(20,000 + 305,000) \qquad \text{Substitute 20 for } n, \\ 20,000 \text{ for } a_1, \text{ and} \\ 305,000 \text{ for } a_n.$$

$$= 10(325,000) \qquad \text{Simplify.}$$

$$= 3,250,000. \qquad \text{Simplify.}$$

So, the total sales for the first 20 years are $3,250,000.

✓CHECKPOINT   Now try Exercise 85.

Numerical Solution

The annual sales form an arithmetic sequence in which $a_1 = 20,000$ and $d = 15,000$. So,

$$a_n = 20,000 + 15,000(n - 1)$$

and the nth term of the sequence is

$$a_n = 15,000n + 5000.$$

Use the *list editor* of a graphing utility to create a table that shows the sales for each of the first 20 years and the total sales for the first 20 years, as shown in Figure 9.7.

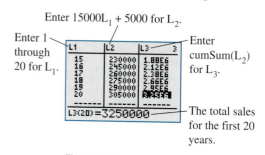

Figure 9.7

So, the total sales for the first 20 years are $3,250,000.

Figure 9.8 shows the annual sales for the business in Example 7. Notice that the annual sales for the business follow a *linear growth* pattern. In other words, saying that a quantity increases arithmetically is the same as saying that it increases linearly.

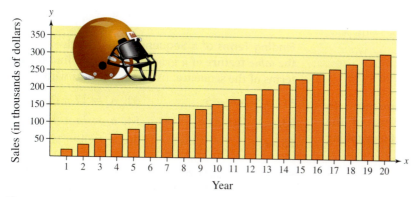

Figure 9.8

If you go on to take a course in calculus, you will study sequences and series in detail. You will learn that sequences and series play a major role in the study of calculus.

See www.CalcChat.com for worked-out solutions to odd-numbered exercises.
For instructions on how to use a graphing utility, see Appendix A.

9.2   Exercises

Vocabulary and Concept Check

In Exercises 1 and 2, fill in the blank.

1. The nth term of an arithmetic sequence has the form _____ .

2. The formula $S_n = \dfrac{n}{2}(a_1 + a_n)$ can be used to find the sum of the first n terms of an arithmetic sequence, called the _____ .

3. How do you know when a sequence is arithmetic?

4. Is 4 or 1 the common difference of the arithmetic sequence $a_n = 4n + 1$?

Procedures and Problem Solving

Determining Whether a Sequence Is Arithmetic   In Exercises 5–10, determine whether or not the sequence is arithmetic. If it is, find the common difference.

5. $10, 8, 6, 4, 2, \ldots$

6. $4, 9, 14, 19, 24, \ldots$

7. $3, \frac{5}{2}, 2, \frac{3}{2}, 1, \ldots$

8. $\frac{1}{3}, \frac{2}{3}, \frac{4}{3}, \frac{8}{3}, \frac{16}{3}, \ldots$

9. $3.7, 4.3, 4.9, 5.5, 6.1, \ldots$

10. $1^2, 2^2, 3^2, 4^2, 5^2, \ldots$

Writing the Terms of a Sequence   In Exercises 11–20, write the first five terms of the sequence. Determine whether or not the sequence is arithmetic. If it is, find the common difference. (Assume n begins with 1.)

✓ 11. $a_n = 8 + 13n$

12. $a_n = 2^n + n$

13. $a_n = \dfrac{1}{n + 1}$

14. $a_n = 1 + (n - 1)4$

15. $a_n = 150 - 7n$

16. $a_n = 2^{n-1}$

17. $a_n = 3 + 2(-1)^n$

18. $a_n = 3 - 4(n + 6)$

19. $a_n = (-1)^n$

20. $a_n = (-1)^{2n+1}$

Finding the nth Term of an Arithmetic Sequence   In Exercises 21–30, find a formula for a_n for the arithmetic sequence.

✓ 21. $a_1 = 1, d = 3$

22. $a_1 = 15, d = 4$

23. $a_1 = 100, d = -8$

24. $a_1 = 0, d = -\frac{2}{3}$

25. $4, \frac{3}{2}, -1, -\frac{7}{2}, \ldots$

26. $10, 5, 0, -5, -10, \ldots$

27. $a_1 = 5, a_4 = 15$

28. $a_1 = -4, a_5 = 16$

29. $a_3 = 94, a_6 = 85$

30. $a_5 = 190, a_{10} = 115$

Writing the Terms of an Arithmetic Sequence   In Exercises 31–38, write the first five terms of the arithmetic sequence. Use the *table* feature of a graphing utility to verify your results.

31. $a_1 = 5, d = 6$

32. $a_1 = 5, d = -\frac{3}{4}$

33. $a_1 = -10, d = -12$

34. $a_4 = 16, a_{10} = 46$

✓ 35. $a_8 = 26, a_{12} = 42$

36. $a_6 = -38, a_{11} = -73$

37. $a_3 = 19, a_{15} = -1.7$

38. $a_5 = 16, a_{14} = 38.5$

Writing Terms of an Arithmetic Sequence   In Exercises 39–42, write the first five terms of the arithmetic sequence. Find the common difference and write the nth term of the sequence as a function of n.

39. $a_1 = 15, \quad a_{k+1} = a_k + 4$

40. $a_1 = 6, \quad a_{k+1} = a_k + 5$

41. $a_1 = \frac{3}{5}, \quad a_{k+1} = -\frac{1}{10} + a_k$

42. $a_1 = 1.5, \quad a_{k+1} = a_k - 2.5$

Using a Recursion Formula   In Exercises 43–46, the first two terms of the arithmetic sequence are given. Find the missing term. Use the *table* feature of a graphing utility to verify your results.

✓ 43. $a_1 = 5, \quad a_2 = 11, \quad a_{10} = $ ▢

44. $a_1 = 3, \quad a_2 = 13, \quad a_9 = $ ▢

45. $a_1 = 4.2, \quad a_2 = 6.6, \quad a_7 = $ ▢

46. $a_1 = -0.7, \quad a_2 = -13.8, \quad a_8 = $ ▢

Graphing Terms of a Sequence   In Exercises 47–50, use a graphing utility to graph the first 10 terms of the sequence. (Assume n begins with 1.)

47. $a_n = 15 - \frac{3}{2}n$

48. $a_n = -5 + 2n$

49. $a_n = 0.4n - 2$

50. $a_n = -1.3n + 7$

Finding Terms of a Sequence   In Exercises 51–56, use the *table* feature of a graphing utility to find the first 10 terms of the sequence. (Assume n begins with 1.)

51. $a_n = 4n - 5$

52. $a_n = 17 + 3n$

53. $a_n = 20 - \frac{3}{4}n$

54. $a_n = \frac{4}{5}n + 12$

55. $a_n = 1.5 + 0.05n$

56. $a_n = 8 - 12.5n$

Finding the Sum of a Finite Arithmetic Sequence   In Exercises 57–64, find the sum of the finite arithmetic sequence.

✓ 57. $2 + 4 + 6 + 8 + 10 + 12 + 14 + 16 + 18 + 20$

58. $1 + 4 + 7 + 10 + 13 + 16 + 19$

59. $-1 + (-3) + (-5) + (-7) + (-9)$

60. $-5 + (-3) + (-1) + 1 + 3 + 5$

61. Sum of the first 100 positive integers

62. Sum of the first 50 negative integers

63. Sum of the integers from -100 to 30

64. Sum of the integers from -10 to 50

Finding a Partial Sum of an Arithmetic Sequence   In Exercises 65–70, find the indicated nth partial sum of the arithmetic sequence.

✓ **65.** $8, 20, 32, 44, \ldots, \quad n = 10$

66. $2, 8, 14, 20, \ldots, \quad n = 25$

67. $0.5, 1.3, 2.1, 2.9, \ldots, \quad n = 10$

68. $4.2, 3.7, 3.2, 2.7, \ldots, \quad n = 12$

69. $a_1 = 100, \ a_{25} = 220, \quad n = 25$

70. $a_1 = 15, \ a_{100} = 307, \quad n = 100$

Finding a Partial Sum of an Arithmetic Sequence   In Exercises 71–76, find the partial sum without using a graphing utility.

71. $\displaystyle\sum_{n=1}^{50} n$

72. $\displaystyle\sum_{n=1}^{100} 2n$

73. $\displaystyle\sum_{n=11}^{30} n - \sum_{n=1}^{10} n$

74. $\displaystyle\sum_{n=51}^{100} n - \sum_{n=1}^{50} n$

75. $\displaystyle\sum_{n=1}^{500} (n + 8)$

76. $\displaystyle\sum_{n=1}^{100} \frac{8 - 3n}{16}$

Finding a Partial Sum Using a Graphing Utility   In Exercises 77–82, use a graphing utility to find the partial sum.

77. $\displaystyle\sum_{n=1}^{20} (2n + 1)$

78. $\displaystyle\sum_{n=0}^{50} (50 - 2n)$

79. $\displaystyle\sum_{n=1}^{100} \frac{n + 1}{2}$

80. $\displaystyle\sum_{n=0}^{100} \frac{4 - n}{4}$

81. $\displaystyle\sum_{i=1}^{60} \left(250 - \tfrac{2}{5}i\right)$

82. $\displaystyle\sum_{j=1}^{200} (10.5 + 0.025j)$

83. *Why you should learn it* (p. 721)   A brick patio has the approximate shape of a trapezoid, as shown in the figure. The patio has 18 rows of bricks. The first row has 14 bricks and the 18th row has 31 bricks. How many bricks are in the patio?

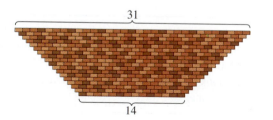

84. Performing Arts   An auditorium has 20 rows of seats. There are 20 seats in the first row, 21 seats in the second row, 22 seats in the third row, and so on. How many seats are there in all 20 rows?

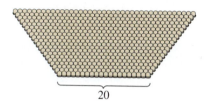

20

✓ **85. Sales**   A hardware store makes a profit of $30,000 during its first year. The store owner sets a goal of increasing profits by $5000 each year for 4 years. Assuming that this goal is met, find the total profit during the first 5 years of business.

86. Physics   An object with negligible air resistance is dropped from a plane. During the first second of fall, the object falls 16 feet; during the second second, it falls 48 feet; during the third second, it falls 80 feet; and during the fourth second, it falls 112 feet. Assume this pattern continues. How many feet will the object fall in 8 seconds?

87. MODELING DATA

The table shows the sales a_n (in billions of dollars) for Coca-Cola Enterprises from 2001 through 2008. (Source: Coca-Cola Enterprises, Inc.)

Year	Sales, a_n (in billions of dollars)
2001	15.7
2002	16.9
2003	17.3
2004	18.2
2005	18.7
2006	19.8
2007	20.9
2008	21.8

(a) Use the *regression* feature of a graphing utility to find an arithmetic sequence for the data. Let n represent the year, with $n = 1$ corresponding to 2001.

(b) Use the sequence from part (a) to approximate the annual sales for Coca-Cola Enterprises for the years 2001 through 2008. How well does the model fit the data?

(c) Use the sequence to find the total sales for Coca-Cola Enterprises over the period from 2001 through 2008.

(d) Use the sequence to predict the total sales over the period from 2009 through 2016. Is your total reasonable? Explain.

88. MODELING DATA

The table shows the numbers a_n (in thousands) of master's degrees conferred in the United States from 2000 through 2007. (Source: U.S. National Center for Education Statistics)

Year	Master's degrees conferred, a_n (in thousands)
2000	457
2001	468
2002	482
2003	513
2004	559
2005	575
2006	594
2007	605

(a) Use the *regression* feature of a graphing utility to find an arithmetic sequence for the data. Let n represent the year, with $n = 0$ corresponding to 2000.

(b) Use the sequence from part (a) to approximate the numbers of master's degrees conferred for the years 2000 through 2007. How well does the model fit the data?

(c) Use the sequence to find the total number of master's degrees conferred over the period from 2000 through 2007.

(d) Use the sequence to predict the total number of master's degrees conferred over the period from 2008 through 2018. Is your total reasonable? Explain.

Conclusions

True or False? **In Exercises 89 and 90, determine whether the statement is true or false. Justify your answer.**

89. Given the nth term and the common difference of an arithmetic sequence, it is possible to find the $(n + 1)$th term.

90. If the only known information about a finite arithmetic sequence is its first term and its last term, then it is possible to find the sum of the sequence.

Finding Terms of a Sequence **In Exercises 91 and 92, find the first 10 terms of the sequence.**

91. $a_1 = x, d = 2x$

92. $a_1 = -y, d = 5y$

93. Think About It The sum of the first 20 terms of an arithmetic sequence with a common difference of 3 is 650. Find the first term.

94. Writing Explain how to use the first two terms of an arithmetic sequence to find the nth term.

95. Think About It The sum of the first n terms of an arithmetic sequence with first term a_1 and common difference d is S_n. Determine the sum when each term is increased by 5. Explain.

96. Think About It In each sequence, decide whether it is possible to fill in the blanks to form an arithmetic sequence. If so, find a recursion formula for the sequence. Explain how you found your answers.

(a) -7, ▮, ▮, ▮, ▮, ▮, 11

(b) 17, ▮, ▮, ▮, ▮, ▮, ▮, 59

(c) $2, 6$, ▮, ▮, 162

(d) $4, 7.5$, ▮, ▮, ▮, ▮, ▮, 28.5

(e) $8, 12$, ▮, ▮, ▮, 60.75

97. Writing Carl Friedrich Gauss, a famous nineteenth century mathematician, was a child prodigy. It was said that when Gauss was 10 he was asked by his teacher to add the numbers from 1 to 100. Almost immediately, Gauss found the answer by mentally finding the summation. Write an explanation of how he arrived at his conclusion, and then find the formula for the sum of the first n natural numbers.

98. CAPSTONE Describe the characteristics of an arithmetic sequence. Give an example of a sequence that is arithmetic and one that is not arithmetic.

Finding the Sum of a Finite Arithmetic Sequence **In Exercises 99–102, find the sum using the method from Exercise 97.**

99. The first 200 natural numbers

100. The first 100 even natural numbers from 2 to 200, inclusive

101. The first 51 odd natural numbers from 1 to 101, inclusive

102. The first 100 multiples of 4 from 4 to 400, inclusive

Cumulative Mixed Review

Gauss-Jordan Elimination **In Exercises 103 and 104, use Gauss-Jordan elimination to solve the system of equations.**

103. $\begin{cases} 2x - y + 7z = -10 \\ 3x + 2y - 4z = 17 \\ 6x - 5y + z = -20 \end{cases}$

104. $\begin{cases} -x + 4y + 10z = 4 \\ 5x - 3y + z = 31 \\ 8x + 2y - 3z = -5 \end{cases}$

105. *Make a Decision* To work an extended application analyzing the amount of municipal waste recovered in the United States from 1983 through 2008, visit this textbook's *Companion Website*. (Data Source: U.S. Environmental Protection Agency)

9.3   Geometric Sequences and Series

Geometric Sequences

In Section 9.2, you learned that a sequence whose consecutive terms have a common *difference* is an arithmetic sequence. In this section, you will study another important type of sequence called a **geometric sequence.** Consecutive terms of a geometric sequence have a common *ratio*.

> ### Definition of Geometric Sequence
>
> A sequence is **geometric** when the ratios of consecutive terms are the same. So, the sequence $a_1, a_2, a_3, a_4, \ldots, a_n, \ldots$ is geometric when there is a number r such that
>
> $$\frac{a_2}{a_1} = \frac{a_3}{a_2} = \frac{a_4}{a_3} = \cdots = r, \qquad r \neq 0.$$
>
> The number r is the **common ratio** of the sequence.

What you should learn

- Recognize, write, and find the *n*th terms of geometric sequences.
- Find *n*th partial sums of geometric sequences.
- Find sums of infinite geometric series.
- Use geometric sequences to model and solve real-life problems.

Why you should learn it

Geometric sequences can reduce the amount of time it takes to find the sum of a sequence of numbers with a common ratio. For instance, Exercise 109 on page 738 shows how to use a geometric sequence to find the total vertical distance traveled by a bouncing ball.

Example 1   Examples of Geometric Sequences

a. The sequence whose nth term is 2^n is geometric. For this sequence, the common ratio between consecutive terms is 2.

$$2, 4, 8, 16, \ldots, 2^n, \ldots \qquad \text{Begin with } n = 1.$$

$$\frac{4}{2} = 2$$

b. The sequence whose nth term is $4(3^n)$ is geometric. For this sequence, the common ratio between consecutive terms is 3.

$$12, 36, 108, 324, \ldots, 4(3^n), \ldots \qquad \text{Begin with } n = 1.$$

$$\frac{36}{12} = 3$$

c. The sequence whose nth term is $\left(-\frac{1}{3}\right)^n$ is geometric. For this sequence, the common ratio between consecutive terms is $-\frac{1}{3}$.

$$-\frac{1}{3}, \frac{1}{9}, -\frac{1}{27}, \frac{1}{81}, \ldots, \left(-\frac{1}{3}\right)^n, \ldots \qquad \text{Begin with } n = 1.$$

$$\frac{1/9}{-1/3} = -\frac{1}{3}$$

✔**CHECKPOINT**   Now try Exercise 7.

The sequence

$$1, 4, 9, 16, \ldots$$

whose nth term is n^2, is *not* geometric. The ratio of the second term to the first term is

$$\frac{a_2}{a_1} = \frac{4}{1} = 4$$

but the ratio of the third term to the second term is

$$\frac{a_3}{a_2} = \frac{9}{4}.$$

Study Tip

 In Example 1, notice that each of the geometric sequences has an nth term of the form ar^n, where r is the common ratio of the sequence.

The Sum of a Finite Geometric Sequence

The formula for the sum of a *finite* geometric sequence is as follows.

> **The Sum of a Finite Geometric Sequence**   (See the proof on page 773.)
>
> The sum of the finite geometric sequence
>
> $$a_1, \ a_1r, \ a_1r^2, \ a_1r^3, \ a_1r^4, \ . \ . \ . \ , \ a_1r^{n-1}$$
>
> with common ratio $r \neq 1$ is given by
>
> $$S_n = \sum_{i=1}^{n} a_1 r^{i-1} = a_1 \left(\frac{1 - r^n}{1 - r} \right).$$

Example 6   Finding the Sum of a Finite Geometric Sequence

Find the following sum.

$$\sum_{n=1}^{12} 4(0.3)^n$$

Solution

By writing out a few terms, you have

$$\sum_{n=1}^{12} 4(0.3)^n = 4(0.3)^1 + 4(0.3)^2 + 4(0.3)^3 + \cdots + 4(0.3)^{12}.$$

Now, because

$$a_1 = 4(0.3), \qquad r = 0.3, \qquad \text{and} \qquad n = 12$$

you can apply the formula for the sum of a finite geometric sequence to obtain

$$\sum_{n=1}^{12} 4(0.3)^n = a_1 \left(\frac{1 - r^n}{1 - r} \right) \qquad \text{Formula for sum of a finite geometric sequence}$$

$$= 4(0.3) \left[\frac{1 - (0.3)^{12}}{1 - 0.3} \right] \qquad \text{Substitute } 4(0.3) \text{ for } a_1, \ 0.3 \text{ for } r, \text{ and } 12 \text{ for } n.$$

$$\approx 1.71. \qquad \text{Use a calculator.}$$

✔CHECKPOINT   Now try Exercise 55.

When using the formula for the sum of a geometric sequence, be careful to check that the index begins at $i = 1$. For an index that begins at $i = 0$, you must adjust the formula for the nth partial sum. For instance, if the index in Example 6 had begun with $n = 0$, then the sum would have been

$$\sum_{n=0}^{12} 4(0.3)^n = 4(0.3)^0 + \sum_{n=1}^{12} 4(0.3)^n$$

$$= 4 + \sum_{n=1}^{12} 4(0.3)^n$$

$$\approx 4 + 1.71$$

$$= 5.71.$$

Technology Tip

Using the *sum sequence* feature of a graphing utility, you can calculate the sum of the sequence in Example 6 to be about 1.7142848, as shown below.

```
sum(seq(4*0.3^n,
n,1,12))
         1.714284803
```

Calculate the sum beginning at $n = 0$. You should obtain a sum of about 5.7142848.

? What's Wrong?

You use a graphing utility to find the sum

$$\sum_{n=1}^{6} 7(0.5)^n$$

as shown in the figure. What's wrong?

```
7*(1-.5^6)/(1-.5
)
         13.78125
```

Geometric Series

The sum of the terms of an infinite geometric sequence is called an **infinite geometric series** or simply a **geometric series.**

The formula for the sum of a *finite geometric sequence* can, depending on the value of r, be extended to produce a formula for the sum of an *infinite geometric series.* Specifically, if the common ratio r has the property that

$$|r| < 1$$

then it can be shown that r^n becomes arbitrarily close to zero as n increases without bound. Consequently,

$$a_1\left(\frac{1 - r^n}{1 - r}\right) \longrightarrow a_1\left(\frac{1 - 0}{1 - r}\right) \quad \text{as} \quad n \longrightarrow \infty.$$

This result is summarized as follows.

The Sum of an Infinite Geometric Series

If $|r| < 1$, then the infinite geometric series

$$a_1 + a_1 r + a_1 r^2 + a_1 r^3 + \cdots + a_1 r^{n-1} + \cdots$$

has the sum

$$S = \sum_{i=0}^{\infty} a_1 r^i = \frac{a_1}{1 - r}.$$

Note that when

$$|r| \geq 1$$

the series does not have a sum.

Example 7   Finding the Sum of an Infinite Geometric Series

Find each sum.

a. $\displaystyle\sum_{n=0}^{\infty} 4(0.6)^n$

b. $3 + 0.3 + 0.03 + 0.003 + \cdots$

Solution

a. $\displaystyle\sum_{n=0}^{\infty} 4(0.6)^n = 4 + 4(0.6) + 4(0.6)^2 + 4(0.6)^3 + \cdots + 4(0.6)^n + \cdots$

$$= \frac{4}{1 - 0.6} \qquad \frac{a_1}{1 - r}$$

$$= 10$$

b. $3 + 0.3 + 0.03 + 0.003 + \cdots = 3 + 3(0.1) + 3(0.1)^2 + 3(0.2)^3 + \cdots$

$$= \frac{3}{1 - 0.1} \qquad \frac{a_1}{1 - r}$$

$$= \frac{10}{3}$$

$$\approx 3.33$$

✓**CHECKPOINT**   Now try Exercise 75.

Explore the Concept

Notice that the formula for the sum of an infinite geometric series requires that $|r| < 1$. What happens when $r = 1$ or $r = -1$? Give examples of infinite geometric series for which $|r| > 1$ and convince yourself that they do not have finite sums.

Explore the Concept

On most graphing utilities, it is not possible to find the sum of the series in Example 7(a) because you cannot enter ∞ as the upper limit of summation. Can you still find the sum using a graphing utility without entering ∞? If so, which partial sum will result in 10, the exact sum of the series?

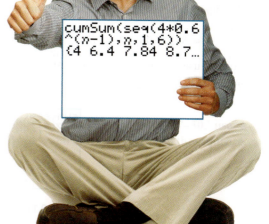

```
cumSum(seq(4*0.6
^(n-1),n,1,6))
{4 6.4 7.84 8.7…
```

Finding a Sequence of Partial Sums In Exercises 51 and 52, find the sequence of the first five partial sums S_1, S_2, S_3, S_4, and S_5 of the geometric sequence by adding terms.

51. $8, -4, 2, -1, \frac{1}{2}, \ldots$ **52.** $8, 12, 18, 27, \frac{81}{2}, \ldots$

Finding a Sequence of Partial Sums In Exercises 53 and 54, use a graphing utility to create a table showing the sequence of the first 10 partial sums S_1, S_2, S_3, $\ldots$, and S_{10} for the series.

53. $\sum_{n=1}^{\infty} 16\left(\frac{1}{2}\right)^{n-1}$ **54.** $\sum_{n=1}^{\infty} 4(0.2)^{n-1}$

Finding the Sum of a Finite Geometric Sequence In Exercises 55–64, find the sum. Use a graphing utility to verify your result.

✓ **55.** $\sum_{n=1}^{9} 2^{n-1}$ **56.** $\sum_{n=1}^{9} (-2)^{n-1}$

57. $\sum_{i=1}^{7} 64\left(-\frac{1}{2}\right)^{i-1}$ **58.** $\sum_{i=1}^{6} 32\left(\frac{1}{4}\right)^{i-1}$

59. $\sum_{n=0}^{20} 3\left(\frac{3}{2}\right)^n$ **60.** $\sum_{n=0}^{15} 2\left(\frac{4}{3}\right)^n$

61. $\sum_{i=1}^{10} 8\left(-\frac{1}{4}\right)^{i-1}$ **62.** $\sum_{i=1}^{10} 5\left(-\frac{1}{3}\right)^{i-1}$

63. $\sum_{n=0}^{5} 300(1.06)^n$ **64.** $\sum_{n=0}^{6} 500(1.04)^n$

Using Summation Notation In Exercises 65–68, use summation notation to write the sum.

65. $5 + 15 + 45 + \cdots + 3645$

66. $7 + 14 + 28 + \cdots + 896$

67. $2 - \frac{1}{2} + \frac{1}{8} - \cdots + \frac{1}{2048}$

68. $15 - 3 + \frac{3}{5} - \cdots - \frac{3}{625}$

Finding the Sum of an Infinite Geometric Series In Exercises 69–82, find the sum of the infinite geometric series, if possible. If not possible, explain why.

69. $\sum_{n=0}^{\infty} 10\left(\frac{4}{5}\right)^n$ **70.** $\sum_{n=0}^{\infty} 6\left(\frac{2}{3}\right)^n$

71. $\sum_{n=0}^{\infty} 5\left(-\frac{1}{2}\right)^n$ **72.** $\sum_{n=0}^{\infty} 9\left(-\frac{2}{3}\right)^n$

73. $\sum_{n=1}^{\infty} 2\left(\frac{7}{3}\right)^{n-1}$ **74.** $\sum_{n=1}^{\infty} 8\left(\frac{5}{3}\right)^{n-1}$

✓ **75.** $\sum_{n=0}^{\infty} 10(0.11)^n$ **76.** $\sum_{n=0}^{\infty} 5(0.45)^n$

77. $\sum_{n=0}^{\infty} -3(-0.9)^n$ **78.** $\sum_{n=0}^{\infty} -10(-0.2)^n$

79. $9 + 6 + 4 + \frac{8}{3} + \cdots$ **80.** $8 + 6 + \frac{9}{2} + \frac{27}{8} + \cdots$

81. $3 - 1 + \frac{1}{3} - \frac{1}{9} + \cdots$

82. $-6 + 5 - \frac{25}{6} + \frac{125}{36} - \cdots$

Writing a Repeating Decimal as a Rational Number In Exercises 83–86, find the rational number representation of the repeating decimal.

83. $0.\overline{36}$ **84.** $0.\overline{297}$

85. $1.2\overline{5}$ **86.** $1.3\overline{8}$

Identifying a Sequence In Exercises 87–94, determine whether the sequence associated with the series is arithmetic or geometric. Find the common difference or ratio and find the sum of the first 15 terms.

87. $8 + 16 + 32 + 64 + \cdots$

88. $17 + 14 + 11 + 8 + \cdots$

89. $90 + 30 + 10 + \frac{10}{3} + \cdots$

90. $\frac{5}{4} + \frac{7}{4} + \frac{9}{4} + \frac{11}{4} + \cdots$

91. $\sum_{n=1}^{\infty} 6n$ **92.** $\sum_{n=1}^{\infty} 2^{n-1}$

93. $\sum_{n=0}^{\infty} 6(0.8)^n$ **94.** $\sum_{n=0}^{\infty} \frac{n+4}{4}$

95. Compound Interest A principal of $1000 is invested at 3% interest. Find the amount after 10 years if the interest is compounded (a) annually, (b) semiannually, (c) quarterly, (d) monthly, and (e) daily.

96. Annuity A deposit of $100 is made at the beginning of each month in an account that pays 3% interest, compounded monthly. The balance A in the account at the end of 5 years is given by

$$A = 100\left(1 + \frac{0.03}{12}\right)^1 + \cdots + 100\left(1 + \frac{0.03}{12}\right)^{60}.$$

Find A.

97. Annuity A deposit of P dollars is made at the beginning of each month in an account earning an annual interest rate r, compounded monthly. The balance A after t years is given by

$$A = P\left(1 + \frac{r}{12}\right) + P\left(1 + \frac{r}{12}\right)^2 + \cdots$$

$$+ P\left(1 + \frac{r}{12}\right)^{12t}.$$

Show that the balance is given by

$$A = P\left[\left(1 + \frac{r}{12}\right)^{12t} - 1\right]\left(1 + \frac{12}{r}\right).$$

98. Annuity A deposit of P dollars is made at the beginning of each month in an account earning an annual interest rate r, compounded continuously. The balance A after t years is given by $A = Pe^{r/12} + Pe^{2r/12} + \cdots + Pe^{12tr/12}$. Show that the balance is given by

$$A = \frac{Pe^{r/12}(e^{rt} - 1)}{e^{r/12} - 1}.$$

Annuities In Exercises 99–102, consider making monthly deposits of P dollars in a savings account earning an annual interest rate r. Use the results of Exercises 97 and 98 to find the balances A after t years if the interest is compounded (a) monthly and (b) continuously.

99. $P = \$50$, $r = 7\%$, $t = 20$ years

100. $P = \$75$, $r = 4\%$, $t = 25$ years

✓ **101.** $P = \$100$, $r = 5\%$, $t = 40$ years

102. $P = \$20$, $r = 6\%$, $t = 50$ years

103. Geometry The sides of a square are 16 inches in length. A new square is formed by connecting the midpoints of the sides of the original square, and two of the resulting triangles are shaded. After this process is repeated five more times, determine the total area of the shaded region.

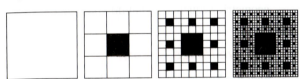

104. Geometry The sides of a square are 27 inches in length. New squares are formed by dividing the original square into nine squares. The center square is then shaded (see figure). After this process is repeated three more times, determine the total area of the shaded region.

105. Physics The temperature of water in an ice cube tray is 70°F when it is placed in a freezer. Its temperature n hours after being placed in the freezer is 20% less than 1 hour earlier.

(a) Find a formula for the nth term of the geometric sequence that gives the temperature of the water n hours after it is placed in the freezer.

(b) Find the temperatures of the water 6 hours and 12 hours after it is placed in the freezer.

(c) Use a graphing utility to graph the sequence to approximate the time required for the water to freeze.

106. Fractals In a *fractal*, a geometric figure is repeated at smaller and smaller scales. The sphereflake shown is a computer-generated fractal that was created by Eric Haines. The radius of the large sphere is 1. Attached to the large sphere are nine spheres of radius $\frac{1}{3}$. Attached to each of the smaller spheres are nine spheres of radius $\frac{1}{9}$. This process is continued infinitely.

Eric Haines

(a) Write a formula in series notation that gives the surface area of the sphereflake.

(b) Write a formula in series notation that gives the volume of the sphereflake.

(c) Is the surface area of the sphereflake finite or infinite? Is the volume finite or infinite? If either is finite, find the value.

107. MODELING DATA

The table shows the mid-year populations a_n of China (in millions) from 2002 through 2008.   (Source: U.S. Census Bureau)

Year	Population, a_n
2002	1284.3
2003	1291.5
2004	1298.8
2005	1306.3
2006	1314.0
2007	1321.9
2008	1330.0

(a) Use the *exponential regression* feature of a graphing utility to find a geometric sequence that models the data. Let n represent the year, with $n = 2$ corresponding to 2002.

(b) Use the sequence from part (a) to describe the rate at which the population of China is growing.

(c) Use the sequence from part (a) to predict the population of China in 2015. The U.S. Census Bureau predicts the population of China will be 1393.4 million in 2015. How does this value compare with your prediction?

(d) Use the sequence from part (a) to determine when the population of China will reach 1.35 billion.

Proofs in Mathematics

Properties of Sums (p. 714)

1. $\sum_{i=1}^{n} c = cn$, $\quad c$ is a constant.

2. $\sum_{i=1}^{n} ca_i = c\sum_{i=1}^{n} a_i$, $\quad c$ is a constant.

3. $\sum_{i=1}^{n} (a_i + b_i) = \sum_{i=1}^{n} a_i + \sum_{i=1}^{n} b_i$

4. $\sum_{i=1}^{n} (a_i - b_i) = \sum_{i=1}^{n} a_i - \sum_{i=1}^{n} b_i$

Proof

Each of these properties follows directly from the properties of real numbers.

1. $\sum_{i=1}^{n} c = c + c + c + \cdots + c = cn$ $\qquad$ *n* terms

The Distributive Property is used in the proof of Property 2.

2. $\sum_{i=1}^{n} ca_i = ca_1 + ca_2 + ca_3 + \cdots + ca_n$

$\phantom{\sum_{i=1}^{n} ca_i} = c(a_1 + a_2 + a_3 + \cdots + a_n)$

$\phantom{\sum_{i=1}^{n} ca_i} = c\sum_{i=1}^{n} a_i$

The proof of Property 3 uses the Commutative and Associative Properties of Addition.

3. $\sum_{i=1}^{n} (a_i + b_i) = (a_1 + b_1) + (a_2 + b_2) + (a_3 + b_3) + \cdots + (a_n + b_n)$

$\phantom{\sum_{i=1}^{n} (a_i + b_i)} = (a_1 + a_2 + a_3 + \cdots + a_n) + (b_1 + b_2 + b_3 + \cdots + b_n)$

$\phantom{\sum_{i=1}^{n} (a_i + b_i)} = \sum_{i=1}^{n} a_i + \sum_{i=1}^{n} b_i$

The proof of Property 4 uses the Commutative and Associative Properties of Addition and the Distributive Property.

4. $\sum_{i=1}^{n} (a_i - b_i) = (a_1 - b_1) + (a_2 - b_2) + (a_3 - b_3) + \cdots + (a_n - b_n)$

$\phantom{\sum_{i=1}^{n} (a_i - b_i)} = (a_1 + a_2 + a_3 + \cdots + a_n) + (-b_1 - b_2 - b_3 - \cdots - b_n)$

$\phantom{\sum_{i=1}^{n} (a_i - b_i)} = (a_1 + a_2 + a_3 + \cdots + a_n) - (b_1 + b_2 + b_3 + \cdots + b_n)$

$\phantom{\sum_{i=1}^{n} (a_i - b_i)} = \sum_{i=1}^{n} a_i - \sum_{i=1}^{n} b_i$

Infinite Series

The study of infinite series was considered a novelty in the fourteenth century. Logician Richard Suiseth, whose nickname was Calculator, solved this problem.

If throughout the first half of a given time interval a variation continues at a certain intensity; throughout the next quarter of the interval at double the intensity; throughout the following eighth at triple the intensity and so ad infinitum; The average intensity for the whole interval will be the intensity of the variation during the second subinterval (or double the intensity).

This is the same as saying that the sum of the infinite series

$$\frac{1}{2} + \frac{2}{4} + \frac{3}{8} + \cdots + \frac{n}{2^n} + \cdots$$

is 2.

> **The Sum of a Finite Arithmetic Sequence**   (p. 724)
>
> The sum of a finite arithmetic sequence with n terms is given by
>
> $$S_n = \frac{n}{2}(a_1 + a_n).$$

Proof

Begin by generating the terms of the arithmetic sequence in two ways. In the first way, repeatedly add d to the first term to obtain

$$S_n = a_1 + a_2 + a_3 + \cdots + a_{n-2} + a_{n-1} + a_n$$
$$= a_1 + [a_1 + d] + [a_1 + 2d] + \cdots + [a_1 + (n-1)d].$$

In the second way, repeatedly subtract d from the nth term to obtain

$$S_n = a_n + a_{n-1} + a_{n-2} + \cdots + a_3 + a_2 + a_1$$
$$= a_n + [a_n - d] + [a_n - 2d] + \cdots + [a_n - (n-1)d].$$

When you add these two versions of S_n, the multiples of d subtract out and you obtain

$$2S_n = (a_1 + a_n) + (a_1 + a_n) + (a_1 + a_n) + \cdots + (a_1 + a_n) \qquad n \text{ terms}$$
$$2S_n = n(a_1 + a_n)$$
$$S_n = \frac{n}{2}(a_1 + a_n).$$

> **The Sum of a Finite Geometric Sequence**   (p. 732)
>
> The sum of the finite geometric sequence
>
> $$a_1,\ a_1 r,\ a_1 r^2,\ a_1 r^3,\ a_1 r^4,\ \ldots,\ a_1 r^{n-1}$$
>
> with common ratio $r \neq 1$ is given by
>
> $$S_n = \sum_{i=1}^{n} a_1 r^{i-1} = a_1\left(\frac{1 - r^n}{1 - r}\right).$$

Proof

$$S_n = a_1 + a_1 r + a_1 r^2 + \cdots + a_1 r^{n-2} + a_1 r^{n-1}$$
$$rS_n = a_1 r + a_1 r^2 + a_1 r^3 + \cdots + a_1 r^{n-1} + a_1 r^n \qquad \text{Multiply by } r.$$

Subtracting the second equation from the first yields

$$S_n - rS_n = a_1 - a_1 r^n.$$

So, $S_n(1 - r) = a_1(1 - r^n)$, and, because $r \neq 1$, you have

$$S_n = a_1\left(\frac{1 - r^n}{1 - r}\right).$$

Example 1   Finding the Standard Equation of an Ellipse

Find the standard form of the equation of the ellipse having foci at

   (0, 1)   and   (4, 1)

and a major axis of length 6, as shown in Figure 10.19.

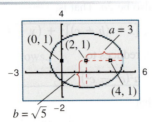

Figure 10.19

Solution

By the Midpoint Formula, the center of the ellipse is (2, 1) and the distance from the center to one of the foci is $c = 2$. Because $2a = 6$, you know that $a = 3$. Now, from $c^2 = a^2 - b^2$, you have

$$b = \sqrt{a^2 - c^2} = \sqrt{9 - 4} = \sqrt{5}.$$

Because the major axis is horizontal, the standard equation is

$$\frac{(x - 2)^2}{3^2} + \frac{(y - 1)^2}{\left(\sqrt{5}\right)^2} = 1.$$

✓CHECKPOINT   Now try Exercise 31.

Example 2   Sketching an Ellipse

Sketch the ellipse given by

   $4x^2 + y^2 = 36$

and identify the center and vertices.

Algebraic Solution

$4x^2 + y^2 = 36$	Write original equation.
$\dfrac{4x^2}{36} + \dfrac{y^2}{36} = \dfrac{36}{36}$	Divide each side by 36.
$\dfrac{x^2}{3^2} + \dfrac{y^2}{6^2} = 1$	Write in standard form.

The center of the ellipse is (0, 0). Because the denominator of the y^2-term is larger than the denominator of the x^2-term, you can conclude that the major axis is vertical. Moreover, because $a = 6$, the vertices are (0, −6) and (0, 6). Finally, because $b = 3$, the endpoints of the minor axis are (−3, 0) and (3, 0), as shown in Figure 10.20.

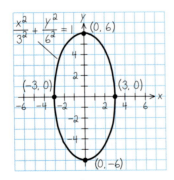

Figure 10.20

✓CHECKPOINT   Now try Exercise 37.

Graphical Solution

Solve the equation of the ellipse for y as follows.

$$4x^2 + y^2 = 36$$
$$y^2 = 36 - 4x^2$$
$$y = \pm\sqrt{36 - 4x^2}$$

Then use a graphing utility to graph

$$y_1 = \sqrt{36 - 4x^2}$$

and

$$y_2 = -\sqrt{36 - 4x^2}$$

in the same viewing window, as shown in Figure 10.21. Be sure to use a square setting.

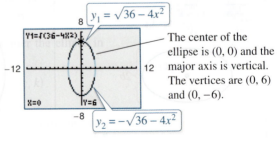

The center of the ellipse is (0, 0) and the major axis is vertical. The vertices are (0, 6) and (0, −6).

Figure 10.21

Example 3   Graphing an Ellipse

Graph the ellipse given by $x^2 + 4y^2 + 6x - 8y + 9 = 0$.

Solution

Begin by writing the original equation in standard form. In the third step, note that 9 and 4 are added to *both* sides of the equation when completing the squares.

$$x^2 + 4y^2 + 6x - 8y + 9 = 0 \qquad \text{Write original equation.}$$

$$\left(x^2 + 6x + \boxed{}\right) + 4\left(y^2 - 2y + \boxed{}\right) = -9 \qquad \begin{array}{l}\text{Group terms and factor}\\ \text{4 out of } y\text{-terms.}\end{array}$$

$$(x^2 + 6x + 9) + 4(y^2 - 2y + 1) = -9 + 9 + 4(1) \quad \text{Complete the square.}$$

$$(x + 3)^2 + 4(y - 1)^2 = 4 \qquad \begin{array}{l}\text{Write in completed}\\ \text{square form.}\end{array}$$

$$\frac{(x + 3)^2}{2^2} + \frac{(y - 1)^2}{1^2} = 1 \qquad \text{Write in standard form.}$$

Now you see that the center is $(h, k) = (-3, 1)$. Because the denominator of the x-term is $a^2 = 2^2$, the endpoints of the major axis lie two units to the right and left of the center. Similarly, because the denominator of the y-term is $b^2 = 1^2$, the endpoints of the minor axis lie one unit up and down from the center. The graph of this ellipse is shown in Figure 10.22.

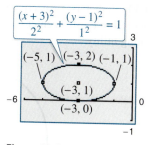

Figure 10.22

✔CHECKPOINT   Now try Exercise 41.

Example 4   Analyzing an Ellipse

Find the center, vertices, and foci of the ellipse $4x^2 + y^2 - 8x + 4y - 8 = 0$.

Solution

By completing the square, you can write the original equation in standard form.

$$4x^2 + y^2 - 8x + 4y - 8 = 0 \qquad \text{Write original equation.}$$

$$4\left(x^2 - 2x + \boxed{}\right) + \left(y^2 + 4y + \boxed{}\right) = 8 \qquad \begin{array}{l}\text{Group terms and factor}\\ \text{4 out of } x\text{-terms.}\end{array}$$

$$4(x^2 - 2x + 1) + (y^2 + 4y + 4) = 8 + 4(1) + 4 \quad \text{Complete the square.}$$

$$4(x - 1)^2 + (y + 2)^2 = 16 \qquad \begin{array}{l}\text{Write in completed}\\ \text{square form.}\end{array}$$

$$\frac{(x - 1)^2}{2^2} + \frac{(y + 2)^2}{4^2} = 1 \qquad \text{Write in standard form.}$$

So, the major axis is vertical, where $h = 1$, $k = -2$, $a = 4$, $b = 2$, and

$$c = \sqrt{a^2 - b^2} = \sqrt{16 - 4} = \sqrt{12} = 2\sqrt{3}.$$

Therefore, you have the following.

Center: $(1, -2)$

Vertices: $(1, -6)$

　　　　　 $(1, 2)$

Foci: $\left(1, -2 - 2\sqrt{3}\right)$

　　　 $\left(1, -2 + 2\sqrt{3}\right)$

The graph of the ellipse is shown in Figure 10.23.

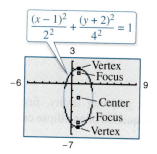

Figure 10.23

✔CHECKPOINT   Now try Exercise 43.

10.2   Exercises

See www.CalcChat.com for worked-out solutions to odd-numbered exercises.
For instructions on how to use a graphing utility, see Appendix A.

Vocabulary and Concept Check

In Exercises 1–4, fill in the blank(s).

1. An _____ is the set of all points (x, y) in a plane, the sum of whose distances from two distinct fixed points called _____ is constant.

2. The chord joining the vertices of an ellipse is called the _____ , and its midpoint is the _____ of the ellipse.

3. The chord perpendicular to the major axis at the center of an ellipse is called the _____ of the ellipse.

4. The eccentricity e of an ellipse is given by the ratio $e =$ _____ .

In Exercises 5–8, consider the ellipse given by $\dfrac{x^2}{2^2} + \dfrac{y^2}{8^2} = 1$.

5. Is the major axis horizontal or vertical?

6. What is the length of the major axis?

7. What is the length of the minor axis?

8. Is the ellipse elongated or nearly circular?

Procedures and Problem Solving

Identifying the Equation of an Ellipse   In Exercises 9–12, match the equation with its graph. [The graphs are labeled (a), (b), (c), and (d).]

(a)
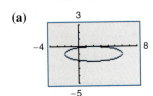

(b)
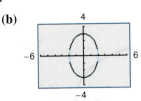

(c)
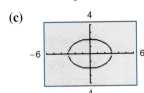

(d)
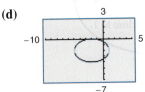

9. $\dfrac{x^2}{4} + \dfrac{y^2}{9} = 1$

10. $\dfrac{x^2}{9} + \dfrac{y^2}{4} = 1$

11. $\dfrac{(x - 2)^2}{16} + (y + 1)^2 = 1$

12. $\dfrac{(x + 2)^2}{9} + \dfrac{(y + 2)^2}{4} = 1$

Using the Standard Equation of an Ellipse   In Exercises 13–18, find the center, vertices, foci, and eccentricity of the ellipse, and sketch its graph. Use a graphing utility to verify your graph.

13. $\dfrac{x^2}{64} + \dfrac{y^2}{9} = 1$

14. $\dfrac{x^2}{16} + \dfrac{y^2}{81} = 1$

15. $\dfrac{(x - 4)^2}{16} + \dfrac{(y + 1)^2}{25} = 1$

16. $\dfrac{(x + 3)^2}{12} + \dfrac{(y - 2)^2}{16} = 1$

17. $\dfrac{(x + 5)^2}{\frac{9}{4}} + (y - 1)^2 = 1$

18. $(x + 2)^2 + \dfrac{(y + 4)^2}{\frac{1}{4}} = 1$

An Ellipse Centered at the Origin   In Exercises 19–26, find the standard form of the equation of the ellipse with the given characteristics and center at the origin.

19.
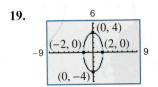

20.
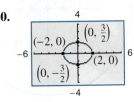

21. Vertices: $(\pm 3, 0)$; foci: $(\pm 2, 0)$

22. Vertices: $(0, \pm 8)$; foci: $(0, \pm 4)$

23. Foci: $(\pm 5, 0)$; major axis of length 14

24. Foci: $(\pm 2, 0)$; major axis of length 10

25. Vertices: $(0, \pm 5)$; passes through the point $(4, 2)$

26. Vertical major axis; passes through points $(0, 6)$ and $(3, 0)$

Finding the Standard Equation of an Ellipse   In Exercises 27–36, find the standard form of the equation of the ellipse with the given characteristics.

27.
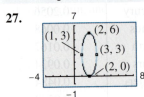

28. $(0, -1)$
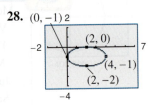

29. Vertices: $(0, 2), (8, 2)$; minor axis of length 2

30. Foci: $(0, 0), (4, 0)$; major axis of length 6

✓ **31.** Foci: $(0, 0), (0, 8)$; major axis of length 36

32. Center: $(2, -1)$; vertex: $\left(2, \frac{1}{2}\right)$; minor axis of length 2

33. Vertices: $(3, 1), (3, 9)$; minor axis of length 6

34. Center: $(3, 2)$; $a = 3c$; foci: $(1, 2), (5, 2)$

35. Center: $(0, 4)$; $a = 2c$; vertices: $(-4, 4), (4, 4)$

36. Vertices: $(5, 0), (5, 12)$; endpoints of the minor axis: $(0, 6), (10, 6)$

Using the Standard Equation of an Ellipse   In Exercises 37–48, (a) find the standard form of the equation of the ellipse, (b) find the center, vertices, foci, and eccentricity of the ellipse, and (c) sketch the ellipse. Use a graphing utility to verify your graph.

✓ **37.** $x^2 + 9y^2 = 36$ **38.** $16x^2 + y^2 = 16$

39. $49x^2 + 4y^2 - 196 = 0$

40. $4x^2 + 49y^2 - 196 = 0$

✓ **41.** $9x^2 + 4y^2 + 36x - 24y + 36 = 0$

42. $9x^2 + 4y^2 - 54x + 40y + 37 = 0$

✓ **43.** $6x^2 + 2y^2 + 18x - 10y + 2 = 0$

44. $x^2 + 4y^2 - 6x + 20y - 2 = 0$

45. $16x^2 + 25y^2 - 32x + 50y + 16 = 0$

46. $9x^2 + 25y^2 - 36x - 50y + 61 = 0$

47. $12x^2 + 20y^2 - 12x + 40y - 37 = 0$

48. $36x^2 + 9y^2 + 48x - 36y + 43 = 0$

Finding Eccentricity   In Exercises 49–52, find the eccentricity of the ellipse.

49. $\dfrac{x^2}{4} + \dfrac{y^2}{9} = 1$ **50.** $\dfrac{x^2}{25} + \dfrac{y^2}{49} = 1$

51. $x^2 + 9y^2 - 10x + 36y + 52 = 0$

52. $4x^2 + 3y^2 - 8x + 18y + 19 = 0$

53. Using Eccentricity   Find an equation of the ellipse with vertices $(\pm 5, 0)$ and eccentricity $e = \frac{3}{5}$.

54. Using Eccentricity   Find an equation of the ellipse with vertices $(0, \pm 8)$ and eccentricity $e = \frac{1}{2}$.

55. Using Eccentricity   Find an equation of the ellipse with foci $(\pm 3, 0)$ and eccentricity $e = \frac{4}{5}$.

56. Architecture   Statuary Hall is an elliptical room in the United States Capitol Building in Washington, D.C. The room is also referred to as the Whispering Gallery because a person standing at one focus of the room can hear even a whisper spoken by a person standing at the other focus. Given that the dimensions of Statuary Hall are 46 feet wide by 97 feet long, find an equation for the shape of the floor surface of the hall. Determine the distance between the foci.

57. Architecture   A fireplace arch is to be constructed in the shape of a semiellipse. The opening is to have a height of 2 feet at the center and a width of 6 feet along the base (see figure). The contractor draws the outline of the ellipse on the wall by the method discussed on page 786. Give the required positions of the tacks and the length of the string.

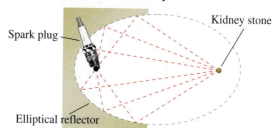

58. *Why you should learn it*   *(p. 786)*   A lithotripter machine uses an elliptical reflector to break up kidney stones nonsurgically. A spark plug in the reflector generates energy waves at one focus of an ellipse. The reflector directs these waves toward the kidney stone positioned at the other focus of the ellipse with enough energy to break up the stone, as shown in the figure. The lengths of the major and minor axes of the ellipse are 280 millimeters and 160 millimeters, respectively. How far is the spark from the kidney stone?

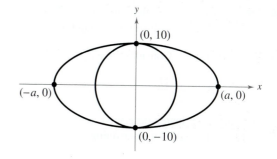

✓ **59. Astronomy**   Halley's comet has an elliptical orbit with the sun at one focus. The eccentricity of the orbit is approximately 0.97. The length of the major axis of the orbit is about 35.88 astronomical units. (An astronomical unit is about 93 million miles.) Find the standard form of the equation of the orbit. Place the center of the orbit at the origin and place the major axis on the x-axis.

60. Geometry   The area of the ellipse in the figure is twice the area of the circle. How long is the major axis? (*Hint:* The area of an ellipse is given by $A = \pi ab$.)

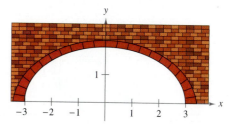

61. Aeronautics The first artificial satellite to orbit Earth was Sputnik I (launched by the former Soviet Union in 1957). Its highest point above Earth's surface was 947 kilometers, and its lowest point was 228 kilometers. The center of Earth was a focus of the elliptical orbit, and the radius of Earth is 6378 kilometers (see figure). Find the eccentricity of the orbit.

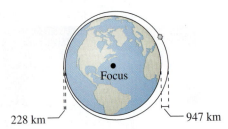

62. Geometry A line segment through a focus with endpoints on an ellipse, perpendicular to the major axis, is called a **latus rectum** of the ellipse. Therefore, an ellipse has two latera recta. Knowing the length of the latera recta is helpful in sketching an ellipse because this information yields other points on the curve (see figure). Show that the length of each latus rectum is $2b^2/a$.

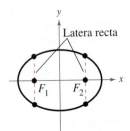

Using Latera Recta In Exercises 63–66, sketch the ellipse using the latera recta (see Exercise 62).

63. $\dfrac{x^2}{4} + \dfrac{y^2}{1} = 1$ **64.** $\dfrac{x^2}{9} + \dfrac{y^2}{16} = 1$

65. $9x^2 + 4y^2 = 36$

66. $5x^2 + 3y^2 = 15$

Conclusions

True or False? In Exercises 67 and 68, determine whether the statement is true or false. Justify your answer.

67. It is easier to distinguish the graph of an ellipse from the graph of a circle when the eccentricity of the ellipse is large (close to 1).

68. The area of a circle with diameter $d = 2r = 8$ is greater than the area of an ellipse with major axis $2a = 8$.

69. Think About It Consider the ellipse

$$\frac{x^2}{328} + \frac{y^2}{327} = 1.$$

Is this ellipse better described as *elongated* or *nearly circular*? Explain your reasoning.

70. CAPSTONE Consider the ellipse shown.

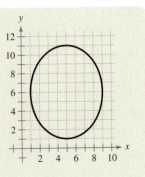

(a) Identify the center, vertices, and foci of the ellipse.

(b) Write the standard form of the equation of the ellipse.

(c) Find the eccentricity e of the ellipse.

71. Think About It At the beginning of this section, it was noted that an ellipse can be drawn using two thumbtacks, a string of fixed length (greater than the distance between the two tacks), and a pencil (see Figure 10.16). When the ends of the string are fastened at the tacks and the string is drawn taut with a pencil, the path traced by the pencil is an ellipse.

(a) What is the length of the string in terms of a?

(b) Explain why the path is an ellipse.

72. Error Analysis Describe the error in finding the distance between the foci.

$c^2 = a^2 + b^2$

$= 64 + 36$

$= 100$

So, $c = 10$ and the distance between the foci is 20 inches.

73. Think About It Find the equation of an ellipse such that for any point on the ellipse, the sum of the distances from the points $(2, 2)$ and $(10, 2)$ is 36.

74. Proof Show that $a^2 = b^2 + c^2$ for the ellipse

$$\frac{x^2}{a^2} + \frac{y^2}{b^2} = 1$$

where $a > 0, b > 0$, and the distance from the center of the ellipse $(0, 0)$ to a focus is c.

Cumulative Mixed Review

Identifying a Sequence In Exercises 75–78, determine whether the sequence is arithmetic, geometric, or neither.

75. 66, 55, 44, 33, 22, . . .

76. 80, 40, 20, 10, 5, . . .

77. $\frac{1}{4}, \frac{1}{2}, 1, 2, 4, . . .$

78. $-\frac{1}{2}, \frac{1}{2}, \frac{3}{2}, \frac{5}{2}, \frac{7}{2}, . . .$

Finding the Sum of a Finite Geometric Sequence In Exercises 79 and 80, find the sum.

79. $\displaystyle\sum_{n=0}^{6} 3^n$ **80.** $\displaystyle\sum_{n=1}^{10} 4\left(\frac{3}{4}\right)^{n-1}$

10.3   Hyperbolas

Introduction

The definition of a **hyperbola** is similar to that of an ellipse. The difference is that for an ellipse, the *sum* of the distances between the foci and a point on the ellipse is constant; whereas for a hyperbola, the *difference* of the distances between the foci and a point on the hyperbola is constant.

What you should learn
- Write equations of hyperbolas in standard form.
- Find asymptotes of and graph hyperbolas.
- Use properties of hyperbolas to solve real-life problems.
- Classify conics from their general equations.

Why you should learn it
Hyperbolas can be used to model and solve many types of real-life problems. For instance, in Exercise 50 on page 803, hyperbolas are used to locate the position of an explosion that was recorded by three listening stations.

> **Definition of a Hyperbola**
>
> A **hyperbola** is the set of all points (x, y) in a plane, the difference of whose distances from two distinct fixed points, the **foci,** is a positive constant. [See Figure 10.27(a).]

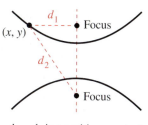

$d_2 - d_1$ is a positive constant.

(a)

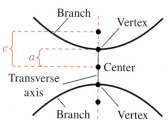

(b)

Figure 10.27

The graph of a hyperbola has two disconnected parts called the **branches.** The line through the two foci intersects the hyperbola at two points called the **vertices.** The line segment connecting the vertices is the **transverse axis,** and the midpoint of the transverse axis is the **center** of the hyperbola [see Figure 10.27(b)]. The development of the **standard form of the equation of a hyperbola** is similar to that of an ellipse. Note, however, that a, b, and c are related differently for hyperbolas than for ellipses. For a hyperbola, the distance between the foci and the center is greater than the distance between the vertices and the center.

> **Standard Equation of a Hyperbola**
>
> The **standard form of the equation of a hyperbola** with center at (h, k) is
>
> $$\frac{(x - h)^2}{a^2} - \frac{(y - k)^2}{b^2} = 1 \qquad \text{Transverse axis is horizontal.}$$
>
> $$\frac{(y - k)^2}{a^2} - \frac{(x - h)^2}{b^2} = 1. \qquad \text{Transverse axis is vertical.}$$
>
> The vertices are a units from the center, and the foci are c units from the center. Moreover, $c^2 = a^2 + b^2$. If the center of the hyperbola is at the origin $(0, 0)$, then the equation takes one of the following forms.
>
> $$\frac{x^2}{a^2} - \frac{y^2}{b^2} = 1 \qquad \text{Transverse axis is horizontal.}$$
>
> $$\frac{y^2}{a^2} - \frac{x^2}{b^2} = 1 \qquad \text{Transverse axis is vertical.}$$

Figure 10.28 shows both the horizontal and vertical orientations for a hyperbola.

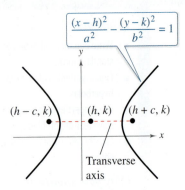

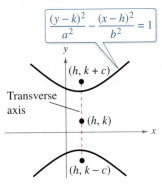

Transverse axis is horizontal.
Figure 10.28

Transverse axis is vertical.

Example 1   Finding the Standard Equation of a Hyperbola

Find the standard form of the equation of the hyperbola with foci $(-1, 2)$ and $(5, 2)$ and vertices $(0, 2)$ and $(4, 2)$.

Solution

By the Midpoint Formula, the center of the hyperbola occurs at the point $(2, 2)$. Furthermore, $c = 3$ and $a = 2$, and it follows that

$$b = \sqrt{c^2 - a^2}$$
$$= \sqrt{3^2 - 2^2}$$
$$= \sqrt{9 - 4}$$
$$= \sqrt{5}.$$

So, the hyperbola has a horizontal transverse axis, and the standard form of the equation of the hyperbola is

$$\frac{(x - 2)^2}{2^2} - \frac{(y - 2)^2}{(\sqrt{5})^2} = 1.$$

This equation simplifies to

$$\frac{(x - 2)^2}{4} - \frac{(y - 2)^2}{5} = 1.$$

Figure 10.29 shows the hyperbola.

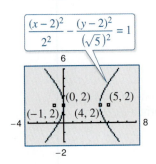

Figure 10.29

✔CHECKPOINT   Now try Exercise 39.

Technology Tip

You can use a graphing utility to graph a hyperbola by graphing the upper and lower portions in the same viewing window. To do this, you must solve the equation for y before entering it into the graphing utility. When graphing equations of conics, it can be difficult to solve for y, which is why it is very important to know the algebra used to solve equations for y.

Asymptotes of a Hyperbola

Each hyperbola has two **asymptotes** that intersect at the center of the hyperbola. The asymptotes pass through the corners of a rectangle of dimensions $2a$ by $2b$, with its center at (h, k), as shown in Figure 10.30.

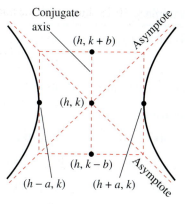

> ### Asymptotes of a Hyperbola
>
> $$y = k \pm \frac{b}{a}(x - h)$$ Asymptotes for horizontal transverse axis
>
> $$y = k \pm \frac{a}{b}(x - h)$$ Asymptotes for vertical transverse axis

Figure 10.30

The **conjugate axis** of a hyperbola is the line segment of length $2b$ joining $(h, k + b)$ and $(h, k - b)$ when the transverse axis is horizontal, and the line segment of length $2b$ joining $(h + b, k)$ and $(h - b, k)$ when the transverse axis is vertical.

Example 2   Sketching a Hyperbola

Sketch the hyperbola whose equation is

$$4x^2 - y^2 = 16.$$

Algebraic Solution

$$4x^2 - y^2 = 16 \qquad \text{Write original equation.}$$

$$\frac{4x^2}{16} - \frac{y^2}{16} = \frac{16}{16} \qquad \text{Divide each side by 16.}$$

$$\frac{x^2}{2^2} - \frac{y^2}{4^2} = 1 \qquad \text{Write in standard form.}$$

Because the x^2-term is positive, you can conclude that the transverse axis is horizontal. So, the vertices occur at $(-2, 0)$ and $(2, 0)$, the endpoints of the conjugate axis occur at $(0, -4)$ and $(0, 4)$, and you can sketch the rectangle shown in Figure 10.31. Finally, by drawing the asymptotes

$$y = 2x \quad \text{and} \quad y = -2x$$

through the corners of this rectangle, you can complete the sketch, as shown in Figure 10.32.

Graphical Solution

Solve the equation of the hyperbola for y, as follows.

$$4x^2 - y^2 = 16$$

$$4x^2 - 16 = y^2$$

$$\pm\sqrt{4x^2 - 16} = y$$

Then use a graphing utility to graph

$$y_1 = \sqrt{4x^2 - 16} \quad \text{and} \quad y_2 = -\sqrt{4x^2 - 16}$$

in the same viewing window, as shown in Figure 10.33. Be sure to use a square setting.

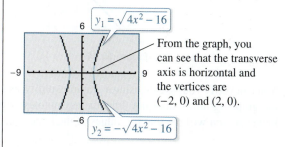

From the graph, you can see that the transverse axis is horizontal and the vertices are $(-2, 0)$ and $(2, 0)$.

Figure 10.33

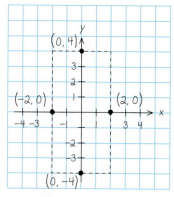

Figure 10.31

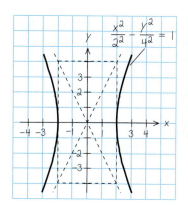

Figure 10.32

✓**CHECKPOINT**   Now try Exercise 21.

Example 3   Finding the Asymptotes of a Hyperbola

Sketch the hyperbola given by

$$4x^2 - 3y^2 + 8x + 16 = 0$$

and find the equations of its asymptotes.

Solution

$$4x^2 - 3y^2 + 8x + 16 = 0$$ Write original equation.

$$4(x^2 + 2x) - 3y^2 = -16$$ Subtract 16 from each side and factor.

$$4(x^2 + 2x + 1) - 3y^2 = -16 + 4(1)$$ Complete the square.

$$4(x + 1)^2 - 3y^2 = -12$$ Write in completed square form.

$$\frac{y^2}{2^2} - \frac{(x + 1)^2}{(\sqrt{3})^2} = 1$$ Write in standard form.

From this equation you can conclude that the hyperbola has a vertical transverse axis, is centered at $(-1, 0)$, has vertices $(-1, 2)$ and $(-1, -2)$, and has a conjugate axis with endpoints $\left(-1 - \sqrt{3}, 0\right)$ and $\left(-1 + \sqrt{3}, 0\right)$. To sketch the hyperbola, draw a rectangle through these four points. The asymptotes are the lines passing through the corners of the rectangle, as shown in Figure 10.34. Finally, using $a = 2$ and $b = \sqrt{3}$, you can conclude that the equations of the asymptotes are

$$y = \frac{2}{\sqrt{3}}(x + 1) \quad \text{and} \quad y = -\frac{2}{\sqrt{3}}(x + 1).$$

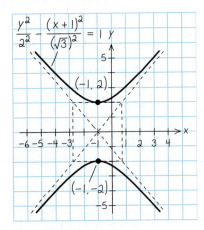

Figure 10.34

You can verify your sketch using a graphing utility, as shown in Figure 10.35. Notice that the graphing utility does not draw the asymptotes. When you trace along the branches, however, you will see that the values of the hyperbola approach the asymptotes.

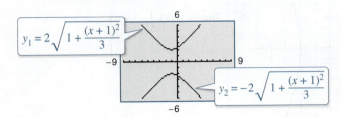

Figure 10.35

✓CHECKPOINT   Now try Exercise 25.

Example 4   Using Asymptotes to Find the Standard Equation

Find the standard form of the equation of the hyperbola having vertices $(3, -5)$ and $(3, 1)$ and having asymptotes

$$y = 2x - 8 \quad \text{and} \quad y = -2x + 4$$

as shown in Figure 10.36.

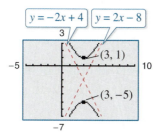

Figure 10.36

Solution

By the Midpoint Formula, the center of the hyperbola is $(3, -2)$. Furthermore, the hyperbola has a vertical transverse axis with $a = 3$. From the original equations, you can determine the slopes of the asymptotes to be

$$m_1 = 2 = \frac{a}{b} \quad \text{and} \quad m_2 = -2 = -\frac{a}{b}$$

and because $a = 3$, you can conclude that $b = \frac{3}{2}$. So, the standard form of the equation of the hyperbola is

$$\frac{(y + 2)^2}{3^2} - \frac{(x - 3)^2}{\left(\dfrac{3}{2}\right)^2} = 1.$$

✔**CHECKPOINT**   Now try Exercise 45.

As with ellipses, the *eccentricity* of a hyperbola is

$$e = \frac{c}{a} \qquad \text{Eccentricity}$$

and because $c > a$ it follows that $e > 1$. When the eccentricity is large, the branches of the hyperbola are nearly flat, as shown in Figure 10.37(a). When the eccentricity is close to 1, the branches of the hyperbola are more pointed, as shown in Figure 10.37(b).

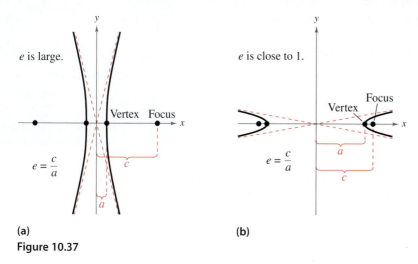

(a)

(b)

Figure 10.37

Applications

The following application was developed during World War II. It shows how the properties of hyperbolas can be used in radar and other detection systems.

Example 5   An Application Involving Hyperbolas

Two microphones, 1 mile apart, record an explosion. Microphone A receives the sound 2 seconds before microphone B. Where did the explosion occur?

Solution

Assuming sound travels at 1100 feet per second, you know that the explosion took place 2200 feet farther from B than from A, as shown in Figure 10.38. The locus of all points that are 2200 feet closer to A than to B is one branch of the hyperbola

$$\frac{x^2}{a^2} - \frac{y^2}{b^2} = 1$$

where

$$c = \frac{5280}{2} = 2640$$

and

$$a = \frac{2200}{2} = 1100.$$

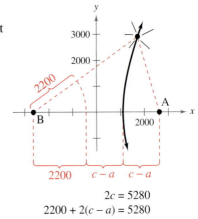

$$2c = 5280$$
$$2200 + 2(c - a) = 5280$$

Figure 10.38

So, $b^2 = c^2 - a^2 = 2640^2 - 1100^2 = 5{,}759{,}600$, and you can conclude that the explosion occurred somewhere on the right branch of the hyperbola

$$\frac{x^2}{1{,}210{,}000} - \frac{y^2}{5{,}759{,}600} = 1.$$

✔**CHECKPOINT**   Now try Exercise 49.

Another interesting application of conic sections involves the orbits of comets in our solar system. Of the 610 comets identified prior to 1970, 245 have elliptical orbits, 295 have parabolic orbits, and 70 have hyperbolic orbits. The center of the sun is a focus of each of these orbits, and each orbit has a vertex at the point where the comet is closest to the sun, as shown in Figure 10.39. Undoubtedly, there are many comets with parabolic or hyperbolic orbits that have not been identified. You get to see such comets only *once*. Comets with elliptical orbits, such as Halley's comet, are the only ones that remain in our solar system.

If p is the distance between the vertex and the focus in meters, and v is the velocity of the comet at the vertex in meters per second, then the type of orbit is determined as follows.

1. Ellipse: $v < \sqrt{2GM/p}$

2. Parabola: $v = \sqrt{2GM/p}$

3. Hyperbola: $v > \sqrt{2GM/p}$

In each of the above, $M \approx 1.989 \times 10^{30}$ kilograms (the mass of the sun) and $G \approx 6.67 \times 10^{-11}$ cubic meter per kilogram-second squared (the universal gravitational constant).

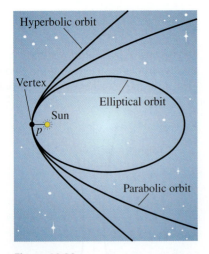

Figure 10.39

General Equations of Conics

Classifying a Conic from Its General Equation

The graph of $Ax^2 + Bxy + Cy^2 + Dx + Ey + F = 0$ is one of the following.

1. **Circle:** $A = C$ $A \neq 0$

2. **Parabola:** $AC = 0$ $A = 0$ or $C = 0$, but not both.

3. **Ellipse:** $AC > 0$ A and C have like signs.

4. **Hyperbola:** $AC < 0$ A and C have unlike signs.

The test above is valid when the graph is a *conic*. The test does not apply to equations such as

$$x^2 + y^2 = -1$$

whose graphs are not conics.

Example 6   Classifying Conics from General Equations

Classify the graph of each equation.

a. $4x^2 - 9x + y - 5 = 0$

b. $4x^2 - y^2 + 8x - 6y + 4 = 0$

c. $2x^2 + 4y^2 - 4x + 12y = 0$

d. $2x^2 + 2y^2 - 8x + 12y + 2 = 0$

Solution

a. For the equation $4x^2 - 9x + y - 5 = 0$, you have

$$AC = 4(0) = 0.   \text{Parabola}$$

So, the graph is a parabola.

b. For the equation $4x^2 - y^2 + 8x - 6y + 4 = 0$, you have

$$AC = 4(-1) < 0.   \text{Hyperbola}$$

So, the graph is a hyperbola.

c. For the equation $2x^2 + 4y^2 - 4x + 12y = 0$, you have

$$AC = 2(4) > 0.   \text{Ellipse}$$

So, the graph is an ellipse.

d. For the equation $2x^2 + 2y^2 - 8x + 12y + 2 = 0$, you have

$$A = C = 2.   \text{Circle}$$

So, the graph is a circle.

✔**CHECKPOINT**   Now try Exercise 55.

Study Tip

Notice in Example 6(a) that there is no y^2-term in the equation. Therefore, $C = 0$.

54. Photography A panoramic photo can be taken using a hyperbolic mirror. The camera is pointed toward the vertex of the mirror and the camera's optical center is positioned at one focus of the mirror (see figure). An equation for the cross section of the mirror is

$$\frac{x^2}{25} - \frac{y^2}{16} = 1.$$

Find the distance from the camera's optical center to the mirror.

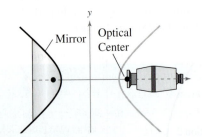

Classifying a Conic from a General Equation In Exercises 55–64, classify the graph of the equation as a circle, a parabola, an ellipse, or a hyperbola.

✓ **55.** $9x^2 + 4y^2 - 18x + 16y - 119 = 0$

56. $x^2 + y^2 - 4x - 6y - 23 = 0$

57. $16x^2 - 9y^2 + 32x + 54y - 209 = 0$

58. $x^2 + 4x - 8y + 20 = 0$

59. $y^2 + 12x + 4y + 28 = 0$

60. $4x^2 + 25y^2 + 16x + 250y + 541 = 0$

61. $x^2 + y^2 + 2x - 6y = 0$

62. $y^2 - x^2 + 2x - 6y - 8 = 0$

63. $x^2 - 6x - 2y + 7 = 0$

64. $9x^2 + 4y^2 - 90x + 8y + 228 = 0$

Conclusions

True or False? In Exercises 65–68, determine whether the statement is true or false. Justify your answer.

65. In the standard form of the equation of a hyperbola, the larger the ratio of b to a, the larger the eccentricity of the hyperbola.

66. In the standard form of the equation of a hyperbola, the trivial solution of two intersecting lines occurs when $b = 0$.

67. If $D \neq 0$ and $E \neq 0$, then the graph of

$$x^2 - y^2 + Dx + Ey = 0$$

is a hyperbola.

68. If the asymptotes of the hyperbola

$$\frac{x^2}{a^2} - \frac{y^2}{b^2} = 1, \text{ where } a, b > 0$$

intersect at right angles, then $a = b$.

69. Think About It Consider a hyperbola centered at the origin with a horizontal transverse axis. Use the definition of a hyperbola to derive its standard form.

70. Exploration Use the figure to show that

$$|d_2 - d_1| = 2a.$$

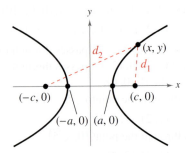

71. Think About It Find the equation of the hyperbola for any point on which the difference between its distances from the points $(2, 2)$ and $(10, 2)$ is 6.

72. Proof Show that $c^2 = a^2 + b^2$ for the equation of the hyperbola

$$\frac{x^2}{a^2} - \frac{y^2}{b^2} = 1$$

where the distance from the center of the hyperbola $(0, 0)$ to a focus is c.

73. Proof Prove that the graph of the equation

$$Ax^2 + Cy^2 + Dx + Ey + F = 0$$

is one of the following (except in degenerate cases).

Conic	Condition
(a) Circle	$A = C$
(b) Parabola	$A = 0$ or $C = 0$ (but not both)
(c) Ellipse	$AC > 0$
(d) Hyperbola	$AC < 0$

74. CAPSTONE Describe any similarities and differences in the graphs of the hyperbolas

$$\frac{x^2}{16} - \frac{y^2}{9} = 1 \quad \text{and} \quad \frac{y^2}{9} - \frac{x^2}{16} = 1.$$

Verify your results by using a graphing utility to graph both hyperbolas in the same viewing window.

Cumulative Mixed Review

Factoring a Polynomial In Exercises 75–80, factor the polynomial completely.

75. $x^3 - 16x$

76. $x^2 + 14x + 49$

77. $2x^3 - 24x^2 + 72x$

78. $6x^3 - 11x^2 - 10x$

79. $16x^3 + 54$

80. $4 - x + 4x^2 - x^3$

10.4   Parametric Equations

Plane Curves

Up to this point, you have been representing a graph by a single equation involving *two* variables such as x and y. In this section, you will study situations in which it is useful to introduce a *third* variable to represent a curve in the plane.

To see the usefulness of this procedure, consider the path of an object that is propelled into the air at an angle of 45°. When the initial velocity of the object is 48 feet per second, it can be shown that the object follows the parabolic path

$$y = -\frac{x^2}{72} + x \qquad \text{Rectangular equation}$$

as shown in Figure 10.40. However, this equation does not tell the whole story. Although it does tell you *where* the object has been, it does not tell you *when* the object was at a given point (x, y) on the path. To determine this time, you can introduce a third variable t, called a **parameter.** It is possible to write both x and y as functions of t to obtain the **parametric equations**

$$x = 24\sqrt{2}\,t \qquad \text{Parametric equation for } x$$

$$y = -16t^2 + 24\sqrt{2}\,t. \qquad \text{Parametric equation for } y$$

From this set of equations you can determine that at time $t = 0$, the object is at the point $(0, 0)$. Similarly, at time $t = 1$, the object is at the point

$$\left(24\sqrt{2},\ 24\sqrt{2} - 16\right)$$

and so on.

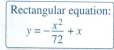
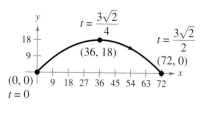

Rectangular equation:
$$y = -\frac{x^2}{72} + x$$

Parametric equations:
$$x = 24\sqrt{2}\,t$$
$$y = -16t^2 + 24\sqrt{2}\,t$$

Curvilinear motion: two variables for position, one variable for time
Figure 10.40

For this particular motion problem, x and y are continuous functions of t, and the resulting path is a **plane curve.** (Recall that a *continuous function* is one whose graph can be traced without lifting the pencil from the paper.)

> ### Definition of a Plane Curve
>
> If f and g are continuous functions of t on an interval I, then the set of ordered pairs
>
> $(f(t), g(t))$
>
> is a **plane curve** C. The equations given by
>
> $x = f(t)$ and $y = g(t)$
>
> are **parametric equations** for C, and t is the **parameter.**

Eliminating the Parameter

Many curves that are represented by sets of parametric equations have graphs that can also be represented by rectangular equations (in x and y). The process of finding the rectangular equation is called **eliminating the parameter.**

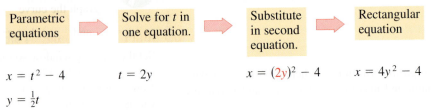

$$x = t^2 - 4 \qquad t = 2y \qquad x = (2y)^2 - 4 \qquad x = 4y^2 - 4$$

$$y = \tfrac{1}{2}t$$

Now you can recognize that the equation $x = 4y^2 - 4$ represents a parabola with a horizontal axis and vertex at $(-4, 0)$.

When converting equations from parametric to rectangular form, you may need to alter the domain of the rectangular equation so that its graph matches the graph of the parametric equations. This situation is demonstrated in Example 3.

Study Tip

It is important to realize that eliminating the parameter is primarily an aid to curve sketching. When the parametric equations represent the path of a moving object, the graph alone is not sufficient to describe the object's motion. You still need the parametric equations to determine the *position, direction,* and *speed* at a given time.

Example 3 Eliminating the Parameter

Identify the curve represented by the equations

$$x = \frac{1}{\sqrt{t + 1}} \quad \text{and} \quad y = \frac{t}{t + 1}.$$

Solution

Solving for t in the equation for x produces

$$x^2 = \frac{1}{t + 1} \quad \Longrightarrow \quad \frac{1}{x^2} = t + 1 \quad \Longrightarrow \quad \frac{1}{x^2} - 1 = t.$$

Substituting in the equation for y, you obtain the rectangular equation

$$y = \frac{t}{t + 1} = \frac{\dfrac{1}{x^2} - 1}{\dfrac{1}{x^2} - 1 + 1} = \frac{\dfrac{1 - x^2}{x^2}}{\dfrac{1}{x^2}} = \frac{1 - x^2}{x^2} \cdot \frac{x^2}{1} = 1 - x^2.$$

From the rectangular equation, you can recognize that the curve is a parabola that opens downward and has its vertex at $(0, 1)$, as shown in Figure 10.52. The rectangular equation is defined for all values of x. The parametric equation for x, however, is defined only when $t > -1$. From the graph of the parametric equations, you can see that x is always positive, as shown in Figure 10.53. So, you should restrict the domain of x to positive values, as shown in Figure 10.54.

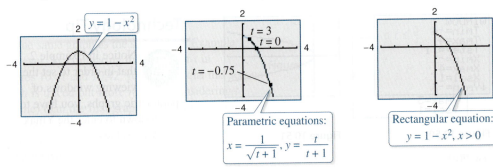

Figure 10.52   Figure 10.53   Figure 10.54

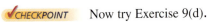 **CHECKPOINT**   Now try Exercise 9(d).

Finding Parametric Equations for a Graph

You have been studying techniques for sketching the graph represented by a set of parametric equations. Now consider the *reverse* problem—that is, how can you find a set of parametric equations for a given graph or a given physical description? From the discussion following Example 1, you know that such a representation is not unique. That is, the equations

$$x = 4t^2 - 4 \quad \text{and} \quad y = t, \ -1 \le t \le \frac{3}{2}$$

produced the same graph as the equations

$$x = t^2 - 4 \quad \text{and} \quad y = \frac{t}{2}, \ -2 \le t \le 3.$$

This is further demonstrated in Example 4.

Example 4   Finding Parametric Equations for a Given Graph

Find a set of parametric equations to represent the graph of $y = 1 - x^2$ using the parameters (a) $t = x$ and (b) $t = 1 - x$.

Solution

a. Letting $t = x$, you obtain the following parametric equations.

$\quad x = t$ Parametric equation for x

$\quad y = 1 - t^2$ Parametric equation for y

The graph of these equations is shown in Figure 10.55.

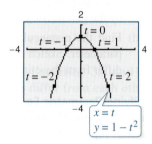

Figure 10.55

b. Letting $t = 1 - x$, you obtain the following parametric equations.

$\quad x = 1 - t$ Parametric equation for x

$\quad y = 1 - (1 - t)^2 = 2t - t^2$ Parametric equation for y

The graph of these equations is shown in Figure 10.56. Note that the graphs in Figures 10.55 and 10.56 have opposite orientations.

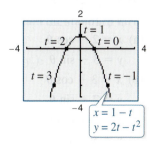

Figure 10.56

✔CHECKPOINT   Now try Exercise 47.

? What's Wrong?

You use a graphing utility in *parametric* mode to graph the parametric equations in Example 4(a). You use a standard viewing window and expect to obtain a parabola similar to Figure 10.55. Your result is shown below. What's wrong?

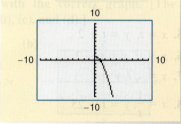

Projectile Motion   In Exercises 61 and 62, consider a projectile launched at a height of h feet above the ground at an angle of θ with the horizontal. The initial velocity is v_0 feet per second and the path of the projectile is modeled by the parametric equations

$$x = (v_0 \cos \theta)t \quad \text{and} \quad y = h + (v_0 \sin \theta)t - 16t^2.$$

61. MODELING DATA

The center field fence in Yankee Stadium is 7 feet high and 408 feet from home plate. A baseball is hit at a point 3 feet above the ground. It leaves the bat at an angle of θ degrees with the horizontal at a speed of 100 miles per hour (see figure).

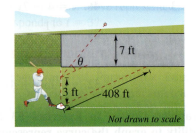

Not drawn to scale

(a) Write a set of parametric equations that model the path of the baseball.

(b) Use a graphing utility to graph the path of the baseball when $\theta = 15°$. Is the hit a home run?

(c) Use the graphing utility to graph the path of the baseball when $\theta = 23°$. Is the hit a home run?

(d) Find the minimum angle required for the hit to be a home run.

62. *Why you should learn it* (p. 805)   The quarterback 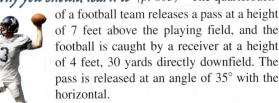 of a football team releases a pass at a height of 7 feet above the playing field, and the football is caught by a receiver at a height of 4 feet, 30 yards directly downfield. The pass is released at an angle of 35° with the horizontal.

(a) Write a set of parametric equations for the path of the football.

(b) Find the speed of the football when it is released.

(c) Use a graphing utility to graph the path of the football and approximate its maximum height.

(d) Find the time the receiver has to position himself after the quarterback releases the football.

Conclusions

True or False?   In Exercises 63–66, determine whether the statement is true or false. Justify your answer.

63. The two sets of parametric equations $x = t$, $y = t^2 + 1$ and $x = 3t$, $y = 9t^2 + 1$ correspond to the same rectangular equation.

64. Because the graphs of the parametric equations $x = t^2$, $y = t^2$ and $x = t$, $y = t$ both represent the line $y = x$, they are the same plane curve.

65. If y is a function of t and x is a function of t, then y must be a function of x.

66. The parametric equations $x = at + h$ and $y = bt + k$, where $a \neq 0$ and $b \neq 0$, represent a circle centered at (h, k) when $a = b$.

67. Think About It   The graph of the parametric equations $x = t^3$ and $y = t - 1$ is shown below. Would the graph change for the equations $x = (-t)^3$ and $y = -t - 1$? If so, how would it change?

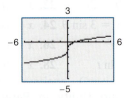

68. CAPSTONE   The curve shown is represented by the parametric equations

$$x = 6 \cos \theta \quad \text{and} \quad y = 6 \sin \theta, \quad 0 \leq \theta \leq 6.$$

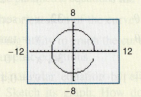

(a) Describe the orientation of the curve.

(b) Determine a range of θ that gives the graph of a circle.

(c) Write a set of parametric equations representing the curve so that the curve traces from the same point as the original curve, but in the opposite direction.

(d) How does the original curve change when cosine and sine are interchanged?

Cumulative Mixed Review

Testing for Evenness and Oddness   In Exercises 69–72, check for symmetry with respect to both axes and to the origin. Then determine whether the function is even, odd, or neither.

69. $f(x) = \dfrac{4x^2}{x^2 + 1}$

70. $f(x) = \sqrt{x}$

71. $y = e^x$

72. $(x - 2)^2 = y + 4$

10.5   Polar Coordinates

Introduction

So far, you have been representing graphs of equations as collections of points (x, y) in the rectangular coordinate system, where x and y represent the directed distances from the coordinate axes to the point (x, y). In this section, you will study a second coordinate system called the **polar coordinate system.**

To form the polar coordinate system in the plane, fix a point O, called the **pole** (or **origin**), and construct from O an initial ray called the **polar axis,** as shown in Figure 10.57. Then each point P in the plane can be assigned **polar coordinates** (r, θ) as follows.

1. $r = directed\ distance$ from O to P

2. $\theta = directed\ angle$, counterclockwise from the polar axis to segment $\overline{OP}$

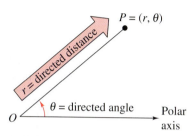

Figure 10.57

What you should learn

- Plot points and find multiple representations of points in the polar coordinate system.
- Convert points from rectangular to polar form and vice versa.
- Convert equations from rectangular to polar form and vice versa.

Why you should learn it

Polar coordinates offer a different mathematical perspective on graphing. For instance, in Exercises 9–16 on page 817, you will see that a polar coordinate can be written in more than one way.

Example 1   Plotting Points in the Polar Coordinate System

a. The point $(r, \theta) = \left(2, \dfrac{\pi}{3}\right)$ lies two units from the pole on the terminal side of the angle $\theta = \dfrac{\pi}{3}$, as shown in Figure 10.58.

b. The point $(r, \theta) = \left(3, -\dfrac{\pi}{6}\right)$ lies three units from the pole on the terminal side of the angle $\theta = -\dfrac{\pi}{6}$, as shown in Figure 10.59.

c. The point $(r, \theta) = \left(3, \dfrac{11\pi}{6}\right)$ coincides with the point $\left(3, -\dfrac{\pi}{6}\right)$, as shown in Figure 10.60.

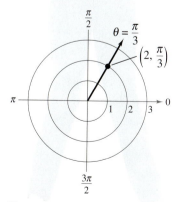

Figure 10.58

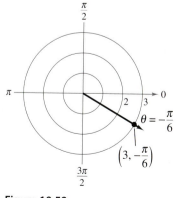

Figure 10.59

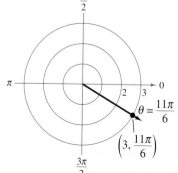

Figure 10.60

✓**CHECKPOINT**   Now try Exercise 9.

Using a Graphing Utility to Find Polar Coordinates In Exercises 45–50, use a graphing utility to find one set of polar coordinates for the point given in rectangular coordinates. (There are many correct answers.)

45. $(3, -2)$ **46.** $(-5, 2)$

47. $(\sqrt{3}, 2)$ **48.** $(3\sqrt{2}, 3\sqrt{2})$

49. $(\frac{5}{2}, \frac{4}{3})$ **50.** $(\frac{7}{4}, \frac{3}{2})$

Converting a Rectangular Equation to Polar Form In Exercises 51–68, convert the rectangular equation to polar form. Assume $a > 0$.

51. $x^2 + y^2 = 9$ **52.** $x^2 + y^2 = 16$

53. $y = 4$ **54.** $y = x$

55. $x = 8$ **56.** $x = a$

57. $3x - y + 2 = 0$ **58.** $3x + 5y - 2 = 0$

59. $xy = 4$ **60.** $2xy = 1$

61. $(x^2 + y^2)^2 = 9(x^2 - y^2)$ **62.** $y^2 - 8x - 16 = 0$

63. $x^2 + y^2 - 6x = 0$ **64.** $x^2 + y^2 - 8y = 0$

65. $x^2 + y^2 - 2ax = 0$ **66.** $x^2 + y^2 - 2ay = 0$

67. $y^2 = x^3$ **68.** $x^2 = y^3$

Converting a Polar Equation to Rectangular Form In Exercises 69–88, convert the polar equation to rectangular form.

69. $r = 4 \sin \theta$ **70.** $r = 2 \cos \theta$

71. $\theta = \dfrac{2\pi}{3}$ **72.** $\theta = \dfrac{5\pi}{3}$

73. $\theta = \dfrac{5\pi}{6}$ **74.** $\theta = \dfrac{11\pi}{6}$

75. $\theta = \dfrac{\pi}{2}$ **76.** $\theta = \pi$

77. $r = 4$ **78.** $r = 10$

79. $r = -3 \csc \theta$ **80.** $r = 2 \sec \theta$

81. $r^2 = \cos \theta$ **82.** $r^2 = \sin 2\theta$

83. $r = 2 \sin 3\theta$ **84.** $r = 3 \cos 2\theta$

85. $r = \dfrac{1}{1 - \cos \theta}$ **86.** $r = \dfrac{2}{1 + \sin \theta}$

87. $r = \dfrac{6}{2 - 3 \sin \theta}$ **88.** $r = \dfrac{6}{2 \cos \theta - 3 \sin \theta}$

Converting a Polar Equation to Rectangular Form In Exercises 89–94, describe the graph of the polar equation and find the corresponding rectangular equation. Sketch its graph.

89. $r = 6$ **90.** $r = 8$

✓ **91.** $\theta = \dfrac{\pi}{4}$ **92.** $\theta = \dfrac{7\pi}{6}$

93. $r = 3 \sec \theta$ **94.** $r = 2 \csc \theta$

Conclusions

True or False? In Exercises 95 and 96, determine whether the statement is true or false. Justify your answer.

95. If (r_1, θ_1) and (r_2, θ_2) represent the same point in the polar coordinate system, then $|r_1| = |r_2|$.

96. If (r, θ_1) and (r, θ_2) represent the same point in the polar coordinate system, then $\theta_1 = \theta_2 + 2\pi n$ for some integer n.

97. Think About It

(a) Show that the distance between the points (r_1, θ_1) and (r_2, θ_2) is
$$\sqrt{r_1^2 + r_2^2 - 2r_1 r_2 \cos(\theta_1 - \theta_2)}.$$

(b) Describe the positions of the points relative to each other for $\theta_1 = \theta_2$. Simplify the Distance Formula for this case. Is the simplification what you expected? Explain.

(c) Simplify the Distance Formula for $\theta_1 - \theta_2 = 90°$. Is the simplification what you expected? Explain.

(d) Choose two points in the polar coordinate system and find the distance between them. Then choose different polar representations of the same two points and apply the Distance Formula again. Discuss the result.

98. CAPSTONE In the rectangular coordinate system, each point (x, y) has a unique representation. Explain why this is not true for a point (r, θ) in the polar coordinate system.

99. Think About It Convert the polar equation $r = \cos \theta + 3 \sin \theta$ to rectangular form and identify the graph.

100. Think About It Convert the polar equation $r = 2(h \cos \theta + k \sin \theta)$ to rectangular form and verify that it is the equation of a circle. Find the radius of the circle and the rectangular coordinates of the center of the circle.

Cumulative Mixed Review

Solving a Triangle Using the Law of Sines or Cosines In Exercises 101–104, use the Law of Sines or the Law of Cosines to solve the triangle.

101. $a = 13, b = 19, c = 25$

102. $A = 24°, a = 10, b = 6$

103. $A = 56°, C = 38°, c = 12$

104. $B = 71°, a = 21, c = 29$

10.6   Graphs of Polar Equations

Introduction

What you should learn
- Graph polar equations by point plotting.
- Use symmetry and zeros as sketching aids.
- Recognize special polar graphs.

Why you should learn it
Several common figures, such as the circle in Exercise 10 on page 825, are easier to graph in the polar coordinate system than in the rectangular coordinate system.

In previous chapters you sketched graphs in rectangular coordinate systems. You began with the basic point-plotting method. Then you used sketching aids such as a graphing utility, symmetry, intercepts, asymptotes, periods, and shifts to further investigate the natures of the graphs. This section approaches curve sketching in the polar coordinate system similarly.

Example 1   Graphing a Polar Equation by Point Plotting

Sketch the graph of the polar equation $r = 4 \sin \theta$ by hand.

Solution

The sine function is periodic, so you can get a full range of r-values by considering values of θ in the interval $0 \le \theta \le 2\pi$, as shown in the table.

θ	0	$\dfrac{\pi}{6}$	$\dfrac{\pi}{3}$	$\dfrac{\pi}{2}$	$\dfrac{2\pi}{3}$	$\dfrac{5\pi}{6}$	π	$\dfrac{7\pi}{6}$	$\dfrac{3\pi}{2}$	$\dfrac{11\pi}{6}$	2π
r	0	2	$2\sqrt{3}$	4	$2\sqrt{3}$	2	0	-2	-4	-2	0

By plotting these points, as shown in Figure 10.68, it appears that the graph is a circle of radius 2 whose center is the point $(x, y) = (0, 2)$.

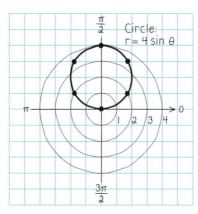

Figure 10.68

 CHECKPOINT   Now try Exercise 27.

You can confirm the graph found in Example 1 in three ways.

1. *Convert to Rectangular Form*   Multiply each side of the polar equation by r and convert the result to rectangular form.

2. *Use a Polar Coordinate Mode*   Set your graphing utility to *polar* mode and graph the polar equation. (Use $0 \le \theta \le \pi$, $-6 \le x \le 6$, and $-4 \le y \le 4$.)

3. *Use a Parametric Mode*   Set your graphing utility to *parametric* mode and graph $x = (4 \sin t) \cos t$ and $y = (4 \sin t) \sin t$.

Most graphing utilities have a *polar* graphing mode. If yours doesn't, you can rewrite the polar equation $r = f(\theta)$ in parametric form, using t as a parameter, as follows.

$$x = f(t) \cos t \qquad \text{and} \qquad y = f(t) \sin t$$

Symmetry and Zeros

In Figure 10.68, note that as θ increases from 0 to 2π the graph is traced out twice. Moreover, note that the graph is *symmetric with respect to the line* $\theta = \pi/2$. Had you known about this symmetry and retracing ahead of time, you could have used fewer points. The three important types of symmetry to consider in polar curve sketching are shown in Figure 10.69.

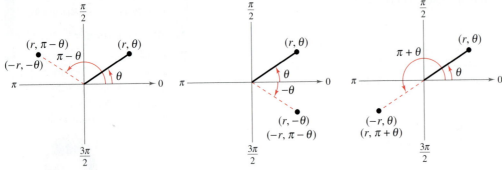

Symmetry with Respect to the Line $\theta = \dfrac{\pi}{2}$

Symmetry with Respect to the Polar Axis

Symmetry with Respect to the Pole

Figure 10.69

Testing for Symmetry in Polar Coordinates

The graph of a polar equation is symmetric with respect to the following when the given substitution yields an equivalent equation.

1. The line $\theta = \dfrac{\pi}{2}$:   Replace (r, θ) by $(r, \pi - \theta)$ or $(-r, -\theta)$.

2. The polar axis: Replace (r, θ) by $(r, -\theta)$ or $(-r, \pi - \theta)$.

3. The pole: Replace (r, θ) by $(r, \pi + \theta)$ or $(-r, \theta)$.

You can determine the symmetry of the graph of $r = 4 \sin \theta$ (see Example 1) as follows.

1. Replace (r, θ) by $(-r, -\theta)$:
$$-r = 4 \sin(-\theta) \quad \Longrightarrow \quad r = -4 \sin(-\theta) = 4 \sin \theta$$

2. Replace (r, θ) by $(r, -\theta)$:
$$r = 4 \sin(-\theta) = -4 \sin \theta$$

3. Replace (r, θ) by $(-r, \theta)$:
$$-r = 4 \sin \theta \quad \Longrightarrow \quad r = -4 \sin \theta$$

So, the graph of $r = 4 \sin \theta$ is symmetric with respect to the line $\theta = \pi/2$.

Study Tip

Recall from Section 5.3 that the sine function is odd. That is,
$$\sin(-\theta) = -\sin \theta.$$

Example 2 Using Symmetry to Sketch a Polar Graph

Use symmetry to sketch the graph of

$$r = 3 + 2 \cos \theta$$

by hand.

Solution

Replacing (r, θ) by $(r, -\theta)$ produces

$$r = 3 + 2 \cos(-\theta) = 3 + 2 \cos \theta. \qquad \cos(-u) = \cos u$$

So, by using the even trigonometric identity, you can conclude that the curve is symmetric with respect to the polar axis. Plotting the points in the table and using polar axis symmetry, you obtain the graph shown in Figure 10.70. This graph is called a **limaçon**.

θ	0	$\dfrac{\pi}{6}$	$\dfrac{\pi}{3}$	$\dfrac{\pi}{2}$	$\dfrac{2\pi}{3}$	$\dfrac{5\pi}{6}$	π
r	5	$3 + \sqrt{3}$	4	3	2	$3 - \sqrt{3}$	1

Use a graphing utility to confirm this graph.

✓CHECKPOINT Now try Exercise 31.

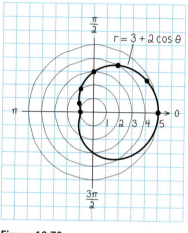

Figure 10.70

The three tests for symmetry in polar coordinates on page 820 are sufficient to guarantee symmetry, but they are not necessary. For instance, Figure 10.71 shows the graph of

$$r = \theta + 2\pi. \qquad \text{Spiral of Archimedes}$$

From the figure, you can see that the graph is symmetric with respect to the line $\theta = \pi/2$. Yet the tests on page 820 fail to indicate symmetry because neither of the following replacements yields an equivalent equation.

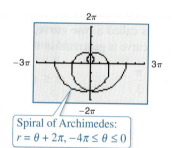

Spiral of Archimedes:
$r = \theta + 2\pi, \ -4\pi \le \theta \le 0$

Figure 10.71

Original Equation	Replacement	New Equation
$r = \theta + 2\pi$	(r, θ) by $(-r, -\theta)$	$-r = -\theta + 2\pi$
$r = \theta + 2\pi$	(r, θ) by $(r, \pi - \theta)$	$r = -\theta + 3\pi$

The equations discussed in Examples 1 and 2 are of the form

$$r = f(\sin \theta) \qquad \text{Example 1}$$

and

$$r = g(\cos \theta). \qquad \text{Example 2}$$

The graph of the first equation is symmetric with respect to the line $\theta = \pi/2$, and the graph of the second equation is symmetric with respect to the polar axis. This observation can be generalized to yield the following *quick tests for symmetry*.

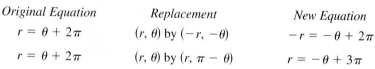

Quick Tests for Symmetry in Polar Coordinates

1. The graph of $r = f(\sin \theta)$ is symmetric with respect to the line $\theta = \dfrac{\pi}{2}$.

2. The graph of $r = g(\cos \theta)$ is symmetric with respect to the polar axis.

Graphing a Rotated Conic In Exercises 31–36, use a graphing utility to graph the rotated conic.

31. $r = \dfrac{3}{1 - \cos(\theta - \pi/4)}$ (See Exercise 15.)

32. $r = \dfrac{7}{7 + \sin(\theta - \pi/3)}$ (See Exercise 18.)

33. $r = \dfrac{4}{4 - \cos(\theta + 3\pi/4)}$ (See Exercise 17.)

34. $r = \dfrac{9}{3 - 2\cos(\theta + \pi/2)}$ (See Exercise 20.)

35. $r = \dfrac{8}{4 + 3\sin(\theta + \pi/6)}$ (See Exercise 19.)

36. $r = \dfrac{5}{-1 + 2\cos(\theta + 2\pi/3)}$ (See Exercise 22.)

Finding the Polar Equation of a Conic In Exercises 37–52, find a polar equation of the conic with its focus at the pole.

	Conic	Eccentricity	Directrix
✓ **37.**	Parabola	$e = 1$	$x = -1$
38.	Parabola	$e = 1$	$y = -4$
39.	Ellipse	$e = \frac{1}{2}$	$y = 1$
40.	Ellipse	$e = \frac{3}{4}$	$y = -4$
41.	Hyperbola	$e = 2$	$x = 1$
42.	Hyperbola	$e = \frac{3}{2}$	$x = -1$

	Conic	Vertex or Vertices
43.	Parabola	$\left(1, -\dfrac{\pi}{2}\right)$
44.	Parabola	$(8, 0)$
45.	Parabola	$(5, \pi)$
46.	Parabola	$\left(10, \dfrac{\pi}{2}\right)$
47.	Ellipse	$(2, 0), (10, \pi)$
48.	Ellipse	$\left(2, \dfrac{\pi}{2}\right), \left(4, \dfrac{3\pi}{2}\right)$
49.	Ellipse	$(20, 0), (4, \pi)$
50.	Hyperbola	$\left(1, \dfrac{3\pi}{2}\right), \left(9, \dfrac{3\pi}{2}\right)$
51.	Hyperbola	$(2, 0), (8, 0)$
52.	Hyperbola	$\left(4, \dfrac{\pi}{2}\right), \left(1, \dfrac{\pi}{2}\right)$

53. Astronomy The planets travel in elliptical orbits with the sun at one focus. Assume that the focus is at the pole, the major axis lies on the polar axis, and the length of the major axis is $2a$ (see figure). Show that the polar equation of the orbit of a planet is

$$r = \frac{(1 - e^2)a}{1 - e\cos\theta}$$

where e is the eccentricity.

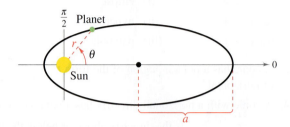

54. Astronomy Use the result of Exercise 53 to show that the minimum distance (*perihelion*) from the sun to a planet is $r = a(1 - e)$ and that the maximum distance (*aphelion*) is $r = a(1 + e)$.

Astronomy In Exercises 55–58, use the results of Exercises 53 and 54 to find the polar equation of the orbit of the planet and the perihelion and aphelion distances.

✓ **55.** Earth $a = 9.2956 \times 10^7$ miles, $e = 0.0167$

56. Mercury $a = 3.5983 \times 10^7$ miles, $e = 0.2056$

57. Venus $a = 6.7283 \times 10^7$ miles, $e = 0.0068$

58. Jupiter $a = 7.7841 \times 10^8$ kilometers, $e = 0.0484$

59. Astronomy Use the results of Exercises 53 and 54, where for the planet Neptune, $a = 4.498 \times 10^9$ kilometers and $e = 0.0086$ and for the dwarf planet Pluto, $a = 5.906 \times 10^9$ kilometers and $e = 0.2488$.

(a) Find the polar equation of the orbit of each planet.

(b) Find the perihelion and aphelion distances for each planet.

(c) Use a graphing utility to graph both Neptune's and Pluto's equations of orbit in the same viewing window.

(d) Is Pluto ever closer to the sun than Neptune? Until recently, Pluto was considered the ninth planet. Why was Pluto called the ninth planet and Neptune the eighth planet?

(e) Do the orbits of Neptune and Pluto intersect? Will Neptune and Pluto ever collide? Why or why not?

60. *Why you should learn it* (p. 827) On November 27, 1963, the United States launched a satellite named *Explorer 18*. Its low and high points above the surface of Earth were about 119 miles and 122,800 miles, respectively (see figure). The center of Earth is at one focus of the orbit.

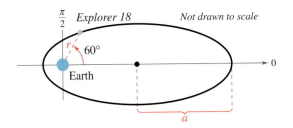

(a) Find the polar equation of the orbit (assume the radius of Earth is 4000 miles).

(b) Find the distance between the surface of Earth and the satellite when $\theta = 60°$.

(c) Find the distance between the surface of Earth and the satellite when $\theta = 30°$.

Conclusions

True or False? In Exercises 61–64, determine whether the statement is true or false. Justify your answer.

61. The graph of $r = 4/(-3 - 3 \sin \theta)$ has a horizontal directrix above the pole.

62. The conic represented by the following equation is an ellipse.

$$r^2 = \frac{16}{9 - 4 \cos\left(\theta + \dfrac{\pi}{4}\right)}$$

63. For values of $e > 1$ and $0 \le \theta \le 2\pi$, the graphs of the following equations are the same.

$$r = \frac{ex}{1 - e \cos \theta} \quad \text{and} \quad r = \frac{e(-x)}{1 + e \cos \theta}$$

64. The graph of $r = \dfrac{5}{1 - \sin[\theta - (\pi/4)]}$ can be obtained

by rotating the graph of $r = \dfrac{5}{1 + \sin \theta}$ about the pole.

65. Verifying a Polar Equation Show that the polar equation of the ellipse

$$\frac{x^2}{a^2} + \frac{y^2}{b^2} = 1 \quad \text{is} \quad r^2 = \frac{b^2}{1 - e^2 \cos^2 \theta}.$$

66. Verifying a Polar Equation Show that the polar equation of the hyperbola

$$\frac{x^2}{a^2} - \frac{y^2}{b^2} = 1 \quad \text{is} \quad r^2 = \frac{-b^2}{1 - e^2 \cos^2 \theta}.$$

Writing a Polar Equation In Exercises 67–72, use the results of Exercises 65 and 66 to write the polar form of the equation of the conic.

67. $\dfrac{x^2}{169} + \dfrac{y^2}{144} = 1$

68. $\dfrac{x^2}{9} - \dfrac{y^2}{16} = 1$

69. $\dfrac{x^2}{25} + \dfrac{y^2}{16} = 1$

70. $\dfrac{x^2}{36} - \dfrac{y^2}{4} = 1$

71. Hyperbola   One focus: $(5, 0)$
Vertices: $(4, 0), (4, \pi)$

72. Ellipse   One focus: $(4, 0)$
Vertices: $(5, 0), (5, \pi)$

73. Exploration Consider the polar equation

$$r = \frac{4}{1 - 0.4 \cos \theta}.$$

(a) Identify the conic without graphing the equation.

(b) Without graphing the following polar equations, describe how each differs from the given polar equation. Use a graphing utility to verify your results.

$$r = \frac{4}{1 + 0.4 \cos \theta}, \quad r = \frac{4}{1 - 0.4 \sin \theta}$$

74. Exploration The equation

$$r = \frac{ep}{1 \pm e \sin \theta}$$

is the equation of an ellipse with $0 < e < 1$. What happens to the lengths of both the major axis and the minor axis when the value of e remains fixed and the value of p changes? Use an example to explain your reasoning.

75. Think About It What conic does the polar equation given by $r = a \sin \theta + b \cos \theta$ represent?

76. CAPSTONE In your own words, define the term *eccentricity* and explain how it can be used to classify conics. Then explain how you can use the values of b and c to determine whether a polar equation of the form

$$r = \frac{a}{b + c \sin \theta}$$

represents an ellipse, a parabola, or a hyperbola.

Cumulative Mixed Review

Evaluating a Trigonometric Expression In Exercises 77–80, find the value of the trigonometric function given that u and v are in Quadrant IV and $\sin u = -\frac{3}{5}$ and $\cos v = 1/\sqrt{2}$.

77. $\cos(u + v)$

78. $\sin(u + v)$

79. $\sin(u - v)$

80. $\cos(u - v)$

10   Review Exercises

See www.CalcChat.com for worked-out solutions to odd-numbered exercises.
For instructions on how to use a graphing utility, see Appendix A.

10.1

Forming a Conic Section   In Exercises 1 and 2, state the type of conic formed by the intersection of the plane and the double-napped cone.

1.

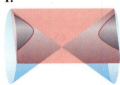

2.

Finding the Standard Equation of a Circle   In Exercises 3–6, find the standard form of the equation of the circle with the given characteristics.

3. Center at origin; point on the circle: $(-3, -4)$

4. Center at origin; point on the circle: $(8, -15)$

5. Endpoints of a diameter: $(-1, 2)$ and $(5, 6)$

6. Endpoints of a diameter: $(-2, 3)$ and $(6, -5)$

Writing the Equation of a Circle in Standard Form   In Exercises 7–10, write the equation of the circle in standard form. Then identify its center and radius.

7. $\frac{1}{2}x^2 + \frac{1}{2}y^2 = 18$ **8.** $\frac{3}{4}x^2 + \frac{3}{4}y^2 = 1$

9. $16x^2 + 16y^2 - 16x + 24y - 3 = 0$

10. $4x^2 + 4y^2 + 32x - 24y + 51 = 0$

Sketching a Circle   In Exercises 11 and 12, sketch the circle. Identify its center and radius.

11. $x^2 + y^2 + 4x + 6y - 3 = 0$

12. $x^2 + y^2 + 8x - 10y - 8 = 0$

Finding the Intercepts of a Circle   In Exercises 13 and 14, find the x- and y-intercepts of the graph of the circle.

13. $(x - 3)^2 + (y + 1)^2 = 7$

14. $(x + 5)^2 + (y - 6)^2 = 27$

Finding the Vertex, Focus, and Directrix of a Parabola   In Exercises 15–18, find the vertex, focus, and directrix of the parabola, and sketch its graph.

15. $4x - y^2 = 0$ **16.** $y = -\frac{1}{8}x^2$

17. $\frac{1}{2}y^2 + 18x = 0$ **18.** $\frac{1}{4}y - 8x^2 = 0$

Finding the Standard Equation of a Parabola   In Exercises 19–22, find the standard form of the equation of the parabola with the given characteristics.

19. Vertex: $(0, 0)$ **20.** Vertex: $(2, 0)$

    Focus: $(4, 0)$ Focus: $(0, 0)$

21.

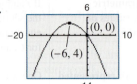

22.

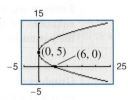

Finding the Tangent Line at a Point on a Parabola   In Exercises 23 and 24, find an equation of the tangent line to the parabola at the given point and find the x-intercept of the line.

23. $x^2 = -2y, \ (2, -2)$ **24.** $y^2 = -2x, \ (-8, -4)$

25. Architecture   A parabolic archway (see figure) is 12 meters high at the vertex. At a height of 10 meters, the width of the archway is 8 meters. How wide is the archway at ground level?

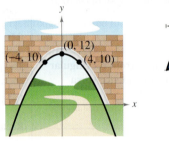

Figure for 25

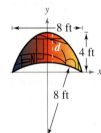

Figure for 26

26. Architecture   A church window (see figure) is bounded on top by a parabola and below by the arc of a circle.

(a) Find equations of the parabola and the circle.

(b) Use a graphing utility to create a table showing the vertical distances d between the circle and the parabola for various values of x.

x	0	1	2	3	4
d					

10.2

Using the Standard Equation of an Ellipse   In Exercises 27–30, find the center, vertices, foci, and eccentricity of the ellipse and sketch its graph. Use a graphing utility to verify your graph.

27. $\dfrac{x^2}{4} + \dfrac{y^2}{16} = 1$

28. $\dfrac{x^2}{9} + \dfrac{y^2}{8} = 1$

29. $\dfrac{(x + 1)^2}{25} + \dfrac{(y - 2)^2}{49} = 1$

30. $\dfrac{(x - 5)^2}{1} + \dfrac{(y + 3)^2}{36} = 1$

Using the Standard Equation of an Ellipse In Exercises 31–34, (a) find the standard form of the equation of the ellipse, (b) find the center, vertices, foci, and eccentricity of the ellipse, and (c) sketch the ellipse. Use a graphing utility to verify your graph.

31. $16x^2 + 9y^2 - 32x + 72y + 16 = 0$

32. $4x^2 + 25y^2 + 16x - 150y + 141 = 0$

33. $3x^2 + 8y^2 + 12x - 112y + 403 = 0$

34. $x^2 + 20y^2 - 5x + 120y + 185 = 0$

Finding the Standard Equation of an Ellipse In Exercises 35–38, find the standard form of the equation of the ellipse with the given characteristics.

35. Vertices: $(\pm 5, 0)$; foci: $(\pm 4, 0)$

36. Vertices: $(0, \pm 6)$; passes through the point $(2, 2)$

37. Vertices: $(-3, 0), (7, 0)$; foci: $(0, 0), (4, 0)$

38. Vertices: $(2, 0), (2, 4)$; foci: $(2, 1), (2, 3)$

39. Architecture A semielliptical archway is to be formed over the entrance to an estate. The arch is to be set on pillars that are 10 feet apart and is to have a height (atop the pillars) of 4 feet. Where should the foci be placed in order to sketch the arch?

40. Architecture You are building a wading pool that is in the shape of an ellipse. Your plans give an equation for the elliptical shape of the pool measured in feet as

$$\frac{x^2}{324} + \frac{y^2}{196} = 1.$$

Find the longest distance across the pool, the shortest distance, and the distance between the foci.

41. Astronomy Saturn moves in an elliptical orbit with the sun at one focus. The least distance and the greatest distance of the planet from the sun are 1.3495×10^9 and 1.5045×10^9 kilometers, respectively. Find the eccentricity of the orbit, defined by $e = c/a$.

42. Astronomy Mercury moves in an elliptical orbit with the sun at one focus. The eccentricity of Mercury's orbit is $e = 0.2056$. The length of the major axis is 72 million miles. Find the standard equation of Mercury's orbit. Place the center of the orbit at the origin and the major axis on the x-axis.

10.3

Finding the Standard Equation of a Hyperbola In Exercises 43–46, find the standard form of the equation of the hyperbola with the given characteristics.

43. Vertices: $(\pm 4, 0)$; foci: $(\pm 6, 0)$

44. Vertices: $(0, \pm 1)$; foci: $(0, \pm 2)$

45. Foci: $(0, 0), (8, 0)$; asymptotes: $y = \pm 2(x - 4)$

46. Foci: $(3, \pm 2)$; asymptotes: $y = \pm 2(x - 3)$

Sketching a Hyperbola In Exercises 47–52, (a) find the standard form of the equation of the hyperbola, (b) find the center, vertices, foci, and asymptotes of the hyperbola, and (c) sketch the hyperbola.

47. $5y^2 - 4x^2 = 20$ **48.** $x^2 - y^2 = \frac{9}{4}$

49. $9x^2 - 16y^2 - 18x - 32y - 151 = 0$

50. $-4x^2 + 25y^2 - 8x + 150y + 121 = 0$

51. $y^2 - 4x^2 - 2y - 48x + 59 = 0$

52. $9x^2 - y^2 - 72x + 8y + 119 = 0$

Finding the Asymptotes of a Hyperbola In Exercises 53–56, find the equations of the asymptotes of the hyperbola and then sketch its graph.

53. $x^2 - y^2 - 4 = 0$ **54.** $25y^2 - 9x^2 - 225 = 0$

55. $5y^2 - x^2 + 6x - 34 = 0$

56. $9x^2 - 4y^2 - 36x - 8y - 4 = 0$

57. Marine Navigation Radio transmitting station A is located 200 miles east of transmitting station B. A ship is in an area to the north and 40 miles west of station A. Synchronized radio pulses transmitted to the ship at 186,000 miles per second by the two stations are received 0.0005 second sooner from station A than from station B. How far north is the ship?

58. Physics Two of your friends live 4 miles apart on the same "east-west" street, and you live halfway between them. You are having a three-way phone conversation when you hear an explosion. Six seconds later your friend to the east hears the explosion, and your friend to the west hears it 8 seconds after you do. Find equations of two hyperbolas that would locate the explosion. (Assume that the coordinate system is measured in feet and that sound travels at 1100 feet per second.)

Classifying a Conic from a General Equation In Exercises 59–62, classify the graph of the equation as a circle, a parabola, an ellipse, or a hyperbola.

59. $3x^2 + 2y^2 - 12x + 12y + 29 = 0$

60. $4x^2 + 4y^2 - 4x + 8y - 11 = 0$

61. $5x^2 - 2y^2 + 10x - 4y + 17 = 0$

62. $-4y^2 + 5x + 3y + 7 = 0$

10.4

Sketching the Graph of Parametric Equations In Exercises 63 and 64, complete the table for the set of parametric equations. Plot the points (x, y) and sketch a graph of the parametric equations.

63. $x = 3t - 2$
     $y = 7 - 4t$

t	-2	-1	0	1	2	3
x						
y						

10   Chapter Test

See www.CalcChat.com for worked-out solutions to odd-numbered exercises.
For instructions on how to use a graphing utility, see Appendix A.

Take this test as you would take a test in class. After you are finished, check your work against the answers in the back of the book.

In Exercises 1–3, graph the conic and identify any vertices and foci.

1. $y^2 - 8x = 0$
2. $y^2 - 4x + 4 = 0$
3. $x^2 - 4y^2 - 4x = 0$

4. Find the standard form of the equation of the parabola with focus $(8, -2)$ and directrix $x = 4$, and sketch the parabola.

5. Find the standard form of the equation of the ellipse shown at the right.

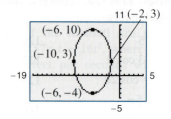

Figure for 5

6. Find the standard form of the equation of the hyperbola with vertices $(0, \pm 3)$ and asymptotes $y = \pm \frac{3}{2} x$.

7. Use a graphing utility to graph the conic $x^2 - \dfrac{y^2}{4} = 1$. Describe your viewing window.

In Exercises 8–10, sketch the curve represented by the parametric equations. Then eliminate the parameter and write the corresponding rectangular equation whose graph represents the curve.

8. $x = t^2 - 6$
   $y = \dfrac{1}{2} t - 1$

9. $x = \sqrt{t^2 + 2}$
   $y = \dfrac{t}{4}$

10. $x = 2 + 3 \cos \theta$
    $y = 2 \sin \theta$

In Exercises 11–13, find a set of parametric equations to represent the graph of the given rectangular equation using the parameters (a) $t = x$ and (b) $t = 2 - x$.

11. $4x + y = 7$
12. $y = \dfrac{3}{x}$
13. $y = x^2 + 10$

14. Convert the polar coordinates $\left(-2, \dfrac{5\pi}{6} \right)$ to rectangular form.

15. Convert the rectangular coordinates $(2, -2)$ to polar form and find two additional polar representations of this point. (There are many correct answers.)

16. Convert the rectangular equation $x^2 + y^2 - 3x = 0$ to polar form.

17. Convert the polar equation $r = 2 \sin \theta$ to rectangular form.

In Exercises 18–20, identify the conic represented by the polar equation algebraically. Then use a graphing utility to graph the polar equation.

18. $r = 2 + 3 \sin \theta$

19. $r = \dfrac{1}{1 - \cos \theta}$

20. $r = \dfrac{4}{2 + 3 \sin \theta}$

21. Find a polar equation of an ellipse with its focus at the pole, an eccentricity of $e = \frac{1}{4}$, and directrix at $y = 4$.

22. Find a polar equation of a hyperbola with its focus at the pole, an eccentricity of $e = \frac{5}{4}$, and directrix at $y = 2$.

23. For the polar equation $r = 8 \cos 3\theta$, find the maximum value of $|r|$ and any zeros of r. Verify your answers numerically.

Take this test to review the material in Chapters 8–10. After you are finished, check your work against the answers in the back of the book.

In Exercises 1–4, use any method to solve the system of equations.

1. $\begin{cases} -x - 3y = 5 \\ 4x + 2y = 10 \end{cases}$ 2. $\begin{cases} 2x - y^2 = 0 \\ x - y = 4 \end{cases}$

3. $\begin{cases} 2x - 3y + z = 13 \\ -4x + y - 2z = -6 \\ x - 3y + 3z = 12 \end{cases}$ 4. $\begin{cases} x - 4y + 3z = 5 \\ 5x + 2y - z = 1 \\ -2x - 8y = 30 \end{cases}$

In Exercises 5–8, perform the matrix operations given

$$A = \begin{bmatrix} -3 & 0 & -4 \\ 2 & 4 & 5 \\ -4 & 8 & 1 \end{bmatrix} \quad \text{and} \quad B = \begin{bmatrix} -1 & 5 & 2 \\ 6 & -3 & 3 \\ 0 & 4 & -2 \end{bmatrix}.$$

5. $3A - 2B$ 6. $5A + 3B$ 7. AB 8. BA

9. Find (a) the inverse of A (if it exists) and (b) the determinant of A.

$$A = \begin{bmatrix} 1 & 2 & -1 \\ 3 & 7 & -10 \\ -5 & -7 & -15 \end{bmatrix}$$

10. Use a determinant to find the area of the triangle with vertices $(0, 0)$, $(6, 2)$, and $(8, 10)$.

11. Write the first five terms of each sequence a_n. (Assume that n begins with 1.)

   (a) $a_n = \dfrac{(-1)^{n+1}}{2n + 3}$ (b) $a_n = 3(2)^{n-1}$

In Exercises 12–15, find the sum. Use a graphing utility to verify your result.

12. $\displaystyle\sum_{k=1}^{6} (7k - 2)$ 13. $\displaystyle\sum_{k=1}^{4} \frac{2}{k^2 + 4}$ 14. $\displaystyle\sum_{n=0}^{10} 9\left(\frac{3}{4}\right)^n$ 15. $\displaystyle\sum_{n=0}^{50} 100\left(-\frac{1}{2}\right)^n$

In Exercises 16–18, find the sum of the infinite geometric series.

16. $\displaystyle\sum_{n=0}^{\infty} 3\left(-\frac{3}{5}\right)^n$ 17. $\displaystyle\sum_{n=1}^{\infty} 5(-0.02)^n$ 18. $4 - 2 + 1 - \dfrac{1}{2} + \dfrac{1}{4} - \cdots$

19. Find each binomial coefficient.

   (a) $_{20}C_{18}$ (b) $\dbinom{20}{2}$

In Exercises 20–23, use the Binomial Theorem to expand and simplify the expression.

20. $(x + 3)^4$ 21. $(2x + y^2)^5$

22. $(x - 2y)^6$ 23. $(3a - 4b)^8$

In Exercises 24–27, find the number of distinguishable permutations of the group of letters.

24. L, I, O, N, S   25. S, E, A, B, E, E, S

26. B, O, B, B, L, E, H, E, A, D   27. I, N, T, U, I, T, I, O, N

In Exercises 28–31, identify the conic and sketch its graph.

28. $\dfrac{(y+3)^2}{36} - \dfrac{(x-5)^2}{121} = 1$ **29.** $\dfrac{(x-2)^2}{4} + \dfrac{(y+1)^2}{9} = 1$

30. $y^2 - x^2 = 16$ **31.** $x^2 + y^2 - 2x - 4y + 1 = 0$

In Exercises 32–34, find the standard form of the equation of the conic.

32.

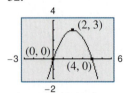

33.

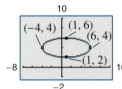

34.

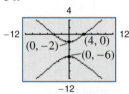

In Exercises 35–37, (a) sketch the curve represented by the parametric equations, (b) use a graphing utility to verify your graph, and (c) eliminate the parameter and write the corresponding rectangular equation whose graph represents the curve. Adjust the domain of the resulting rectangular equation, if necessary.

35. $x = 2t + 1$ **36.** $x = \cos\theta$ **37.** $x = 4\ln t$

$\quad\ y = t^2$ $\quad\ y = 2\sin^2\theta$ $\quad\ y = \frac{1}{2}t^2$

In Exercises 38–41, find a set of parametric equations to represent the graph of the given rectangular equation using the parameters (a) $t = x$ and (b) $t = \dfrac{x}{2}$.

38. $y = 3x - 2$ **39.** $x^2 - y = 16$ **40.** $y = \dfrac{2}{x}$ **41.** $y = \dfrac{e^{2x}}{e^{2x} + 1}$

In Exercises 42–45, plot the point given in polar coordinates and find three additional polar representations of the point, using $-2\pi < \theta < 2\pi$.

42. $\left(8, \dfrac{5\pi}{6}\right)$ **43.** $\left(5, -\dfrac{3\pi}{4}\right)$ **44.** $\left(-2, \dfrac{5\pi}{4}\right)$ **45.** $\left(-3, -\dfrac{11\pi}{6}\right)$

46. Convert the rectangular equation $4x + 4y + 1 = 0$ to polar form.

47. Convert the polar equation $r = 4\cos\theta$ to rectangular form.

48. Convert the polar equation $r = \dfrac{2}{4 - 5\cos\theta}$ to rectangular form.

In Exercises 49–51, identify the type of polar graph represented by the polar equation. Then use a graphing utility to graph the polar equation.

49. $r = -\dfrac{\pi}{6}$ **50.** $r = 3 - 2\sin\theta$ **51.** $r = 2 + 5\cos\theta$

52. The salary for the first year of a job is \$32,500. During the next 14 years, the salary increases by 5% each year. Determine the total compensation over the 15-year period.

53. On a game show, the digits 3, 4, and 5 must be arranged in the proper order to form the price of an appliance. If they are arranged correctly, then the contestant wins the appliance. What is the probability of winning if the contestant knows that the price is at least \$400?

54. A parabolic archway is 16 meters high at the vertex. At a height of 14 meters, the width of the archway is 12 meters, as shown in the figure at the right. How wide is the archway at ground level?

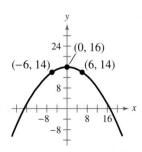

Figure for 54

Proofs in Mathematics

> **Standard Equation of a Parabola**   (p. 779)
>
> The standard form of the equation of a parabola with vertex at (h, k) is as follows.
>
> $(x - h)^2 = 4p(y - k), \quad p \neq 0$ Vertical axis, directrix: $y = k - p$
>
> $(y - k)^2 = 4p(x - h), \quad p \neq 0$ Horizontal axis, directrix: $x = h - p$
>
> The focus lies on the axis p units (*directed distance*) from the vertex. If the vertex is at the origin $(0, 0)$, then the equation takes one of the following forms.
>
> $x^2 = 4py$ Vertical axis
>
> $y^2 = 4px$ Horizontal axis

Parabolic Paths

There are many natural occurrences of parabolas in real life. For instance, the famous astronomer Galileo discovered in the 17th century that an object that is projected upward and obliquely to the pull of gravity travels in a parabolic path. Examples of this are the center of gravity of a jumping dolphin and the path of water molecules in a drinking fountain.

Proof

For the case in which the directrix is parallel to the x-axis and the focus lies above the vertex, as shown in the top figure, if (x, y) is any point on the parabola, then, by definition, it is equidistant from the focus

$(h, k + p)$

and the directrix

$y = k - p.$

So, you have

$$\sqrt{(x - h)^2 + [y - (k + p)]^2} = y - (k - p)$$

$$(x - h)^2 + [y - (k + p)]^2 = [y - (k - p)]^2$$

$$(x - h)^2 + y^2 - 2y(k + p) + (k + p)^2 = y^2 - 2y(k - p) + (k - p)^2$$

$$(x - h)^2 + y^2 - 2ky - 2py + k^2 + 2pk + p^2 = y^2 - 2ky + 2py + k^2 - 2pk + p^2$$

$$(x - h)^2 - 2py + 2pk = 2py - 2pk$$

$$(x - h)^2 = 4p(y - k).$$

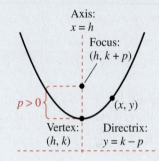

Parabola with vertical axis

For the case in which the directrix is parallel to the y-axis and the focus lies to the right of the vertex, as shown in the bottom figure, if (x, y) is any point on the parabola, then, by definition, it is equidistant from the focus

$(h + p, k)$

and the directrix

$x = h - p.$

So, you have

$$\sqrt{[x - (h + p)]^2 + (y - k)^2} = x - (h - p)$$

$$[x - (h + p)]^2 + (y - k)^2 = [x - (h - p)]^2$$

$$x^2 - 2x(h + p) + (h + p)^2 + (y - k)^2 = x^2 - 2x(h - p) + (h - p)^2$$

$$x^2 - 2hx - 2px + h^2 + 2ph + p^2 + (y - k)^2 = x^2 - 2hx + 2px + h^2 - 2ph + p^2$$

$$-2px + 2ph + (y - k)^2 = 2px - 2ph$$

$$(y - k)^2 = 4p(x - h).$$

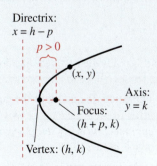

Parabola with horizontal axis

Note that when the vertex of a parabola is at the origin, the two equations above simplify to $x^2 = 4py$ and $y^2 = 4px$, respectively.

Polar Equations of Conics   (p. 827)

The graph of a polar equation of the form

$$\textbf{1. } r = \frac{ep}{1 \pm e \cos \theta} \qquad \text{or} \qquad \textbf{2. } r = \frac{ep}{1 \pm e \sin \theta}$$

is a conic, where $e > 0$ is the eccentricity and $|p|$ is the distance between the focus (pole) and the directrix.

Proof

A proof for

$$r = \frac{ep}{1 + e \cos \theta}$$

with $p > 0$ is shown here. The proofs of the other cases are similar. In the figure, consider a vertical directrix, p units to the right of the focus $F(0, 0)$. If $P(r, \theta)$ is a point on the graph of

$$r = \frac{ep}{1 + e \cos \theta}$$

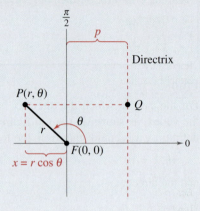

then the distance between P and the directrix is

$$
\begin{aligned}
PQ &= |p - x| \\
&= |p - r \cos \theta| \\
&= \left| p - \left(\frac{ep}{1 + e \cos \theta} \right) \cos \theta \right| \\
&= \left| p \left(1 - \frac{e \cos \theta}{1 + e \cos \theta} \right) \right| \\
&= \left| \frac{p}{1 + e \cos \theta} \right| \\
&= \left| \frac{r}{e} \right|.
\end{aligned}
$$

Moreover, because the distance between P and the pole is simply

$$PF = |r|$$

the ratio of PF to PQ is

$$
\begin{aligned}
\frac{PF}{PQ} &= \frac{|r|}{\left| \dfrac{r}{e} \right|} \\
&= |e| \\
&= e
\end{aligned}
$$

and, by definition, the graph of the equation must be a conic.

Progressive Summary (Chapters P–10)

This chart outlines the topics that have been covered so far in this text. Progressive Summary charts appear after Chapters 2, 4, 7, and 10. In each Progressive Summary, new topics encountered for the first time appear in red.

ALGEBRAIC FUNCTIONS

Polynomial, Rational, Radical

■ Rewriting

Polynomial form $\leftrightarrow$ Factored form
Operations with polynomials
Rationalize denominators
Simplify rational expressions
Exponent form $\leftrightarrow$ Radical form
Operations with complex numbers

■ Solving

Equation	*Strategy*
Linear	Isolate variable
Quadratic	Factor, set to zero
	Extract square roots
	Complete the square
	Quadratic Formula
Polynomial	Factor, set to zero
	Rational Zero Test
Rational	Multiply by LCD
Radical	Isolate, raise to power
Absolute value	Isolate, form two equations

■ Analyzing

Graphically	*Algebraically*
Intercepts	Domain, Range
Symmetry	Transformations
Slope	Composition
Asymptotes	Standard forms
End behavior	of equations
Minimum values	Leading Coefficient
Maximum values	Test
	Synthetic division
	Descartes's Rule of Signs

Numerically
Table of values

TRANSCENDENTAL FUNCTIONS

Exponential, Logarithmic, Trigonometric, Inverse Trigonometric

■ Rewriting

Exponential form $\leftrightarrow$ Logarithmic form
Condense/expand logarithmic expressions
Simplify trigonometric expressions
Prove trigonometric identities
Use conversion formulas
Operations with vectors
Powers and roots of complex numbers

■ Solving

Equation	*Strategy*
Exponential	Take logarithm of each side
Logarithmic	Exponentiate each side
Trigonometric	Isolate function
	Factor, use inverse function
Multiple angle	Use trigonometric
or high powers	identities

■ Analyzing

Graphically	*Algebraically*
Intercepts	Domain, Range
Asymptotes	Transformations
Minimum values	Composition
Maximum values	Inverse Properties
	Amplitude, Period
	Reference angles

Numerically
Table of values

OTHER TOPICS

Systems, Sequences, Series, Conics, Parametric and Polar Equations

■ Rewriting

Row operations for systems of equations
Partial fraction decomposition
Operations with matrices
Matrix form of a system of equations
nth term of a sequence
Summation form of a series
Standard forms of conics
Rectangular form $\leftrightarrow$ Polar, Parametric

■ Solving

Equation	*Strategy*
System of	Substitution
linear equations	Elimination
	Gaussian
	Gauss-Jordan
	Inverse matrices
	Cramer's Rule
Conics	Convert to standard form
	Convert to polar form

■ Analyzing

Systems:
  Intersecting, parallel, and coincident lines, determinants
Sequences:
  Graphing utility in *dot* mode, nth term, partial sums, summation formulas
Conics:
  Table of values, vertices, foci, axes, symmetry, asymptotes, translations, eccentricity
Parametric forms:
  Point plotting, eliminate parameters
Polar forms:
  Point plotting, special equations, symmetry, zeros, eccentricity, maximum r-values, directrix

Appendix A: Technology Support

Introduction

Graphing utilities such as graphing calculators and computers with graphing software are very valuable tools for visualizing mathematical principles, verifying solutions of equations, exploring mathematical ideas, and developing mathematical models. Although graphing utilities are extremely helpful in learning mathematics, their use does not mean that learning algebra is any less important. In fact, the combination of knowledge of mathematics and the use of graphing utilities enables you to explore mathematics more easily and to a greater depth. If you are using a graphing utility in this course, it is up to you to learn its capabilities and to practice using this tool to enhance your mathematical learning.

In this text, there are many opportunities to use a graphing utility, some of which are described below.

Uses of a Graphing Utility

1. Check or validate answers to problems obtained using algebraic methods.

2. Discover and explore algebraic properties, rules, and concepts.

3. Graph functions, and approximate solutions of equations involving functions.

4. Efficiently perform complicated mathematical procedures such as those found in many real-life applications.

5. Find mathematical models for sets of data.

In this appendix, the features of graphing utilities are discussed from a generic perspective and are listed in alphabetical order. To learn how to use the features of a specific graphing utility, consult your user's manual or go to this textbook's *Companion Website*. Additional keystroke guides are available for most graphing utilities, and your college library may have a resource on how to use your graphing utility.

Many graphing utilities are designed to act as "function graphers." In this course, functions and their graphs are studied in detail. You may recall from previous courses that a function can be thought of as a rule that describes the relationship between two variables. These rules are frequently written in terms of x and y. For instance, the equation

$$y = 3x + 5$$

represents y as a function of x.

Many graphing utilities have an *equation editor* feature that requires that an equation be written in "$y =$" form in order to be entered, as shown in Figure A.1. (You should note that your *equation editor* screen may not look like the screen shown in Figure A.1.)

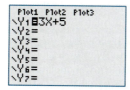

Figure A.1

Cumulative Sum Feature

The *cumulative sum* feature finds partial sums of a series. For instance, to find the first four partial sums of the series

$$\sum_{k=1}^{4} 2(0.1)^k$$

choose the *cumulative sum* feature, which is found in the *operations* menu of the *list* feature (see Figure A.2). To use this feature, you will also have to use the *sequence* feature (see Figure A.2 and page A15). You must enter an expression for the sequence, a variable, the lower limit of summation, and the upper limit of summation, as shown in Figure A.3. After pressing ENTER, you can see that the first four partial sums are 0.2, 0.22, 0.222, and 0.2222. You may have to scroll to the right in order to see all the partial sums.

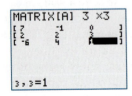

Figure A.2

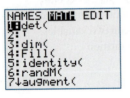

Figure A.3

Technology Tip

 As you use your graphing utility, be aware of how parentheses are inserted in an expression. Some graphing utilities automatically insert the left parenthesis when certain calculator buttons are pressed. The placement of parentheses can make a difference between a correct answer and an incorrect answer.

Determinant Feature

The *determinant* feature evaluates the determinant of a square matrix. For instance, to evaluate the determinant of

$$A = \begin{bmatrix} 7 & -1 & 0 \\ 2 & 2 & 3 \\ -6 & 4 & 1 \end{bmatrix}$$

enter the 3×3 matrix in the graphing utility using the *matrix editor*, as shown in Figure A.4. Then choose the *determinant* feature from the *math* menu of the *matrix* feature, as shown in Figure A.5. Once you choose the matrix name, A, press ENTER and you should obtain a determinant of -50, as shown in Figure A.6.

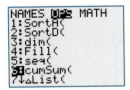

Figure A.4

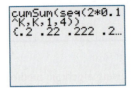

Figure A.5

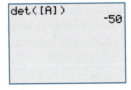

Figure A.6

Draw Inverse Feature

The *draw inverse* feature graphs the inverse function of a *one-to-one* function. For instance, to graph the inverse function of $f(x) = x^3 + 4$, first enter the function in the *equation editor* (see Figure A.7) and graph the function (using a square viewing window), as shown in Figure A.8. Then choose the *draw inverse* feature from the *draw* feature menu, as shown in Figure A.9. You must enter the function you want to graph the inverse function of, as shown in Figure A.10. Finally, press ENTER to obtain the graph of the inverse function of $f(x) = x^3 + 4$, as shown in Figure A.11. This feature can be used only when the graphing utility is in *function* mode.

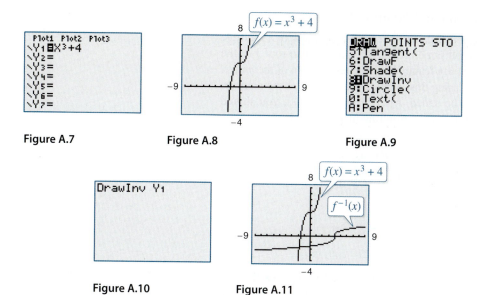

Figure A.7   Figure A.8   Figure A.9

Figure A.10   Figure A.11

Elementary Row Operations Features

Most graphing utilities can perform elementary row operations on matrices.

Row Swap Feature

The *row swap* feature interchanges two rows of a matrix. To interchange rows 1 and 3 of the matrix

$$A = \begin{bmatrix} -1 & -2 & 1 & 2 \\ 2 & -4 & 6 & -2 \\ 1 & 3 & -3 & 0 \end{bmatrix}$$

first enter the matrix in the graphing utility using the *matrix editor*, as shown in Figure A.12. Then choose the *row swap* feature from the *math* menu of the *matrix* feature, as shown in Figure A.13. When using this feature, you must enter the name of the matrix and the two rows that are to be interchanged. After pressing ENTER, you should obtain the matrix shown in Figure A.14. Because the resulting matrix will be used to demonstrate the other elementary row operation features, use the *store* feature to copy the resulting matrix to [A], as shown in Figure A.15.

Figure A.12   Figure A.13

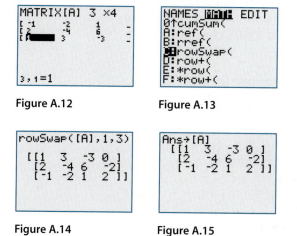

Figure A.14   Figure A.15

Technology Tip

The *store* feature of a graphing utility is used to store a value in a variable or to copy one matrix to another matrix. For instance, as shown at the left, after performing a row operation on a matrix, you can copy the answer to another matrix (see Figure A.15). You can then perform another row operation on the copied matrix. To continue performing row operations to obtain a matrix in row-echelon form or reduced row-echelon form, you must copy the resulting matrix to a new matrix before each operation.

Row Addition and Row Multiplication and Addition Features

The *row addition* and *row multiplication and addition* features add a row or a multiple of a row of a matrix to another row of the same matrix. To add row 1 to row 3 of the matrix stored in [A], choose the *row addition* feature from the *math* menu of the *matrix* feature, as shown in Figure A.16. When using this feature, you must enter the name of the matrix and the two rows that are to be added. After pressing (ENTER), you should obtain the matrix shown in Figure A.17. Copy the resulting matrix to [A].

Figure A.16

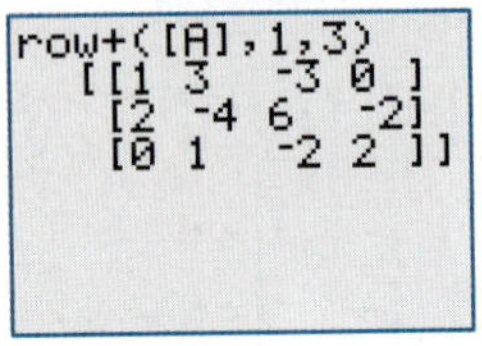

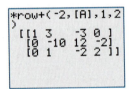

Figure A.17

To add -2 times row 1 to row 2 of the matrix stored in [A], choose the *row multiplication and addition* feature from the *math* menu of the *matrix* feature, as shown in Figure A.18. When using this feature, you must enter the constant, the name of the matrix, the row the constant is multiplied by, and the row to be added to. After pressing (ENTER), you should obtain the matrix shown in Figure A.19. Copy the resulting matrix to [A].

Figure A.18

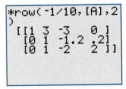

Figure A.19

Row Multiplication Feature

The *row multiplication* feature multiplies a row of a matrix by a nonzero constant. To multiply row 2 of the matrix stored in [A] by

$$-\frac{1}{10}$$

choose the *row multiplication* feature from the *math* menu of the *matrix* feature, as shown in Figure A.20. When using this feature, you must enter the constant, the name of the matrix, and the row to be multiplied. After pressing (ENTER), you should obtain the matrix shown in Figure A.21.

Figure A.20

Figure A.21

Intersect Feature

The *intersect* feature finds the point(s) of intersection of two graphs. The *intersect* feature is found in the *calculate* menu (see Figure A.22). To find the point(s) of intersection of the graphs of $y_1 = -x + 2$ and $y_2 = x + 4$, first enter the equations in the *equation editor*, as shown in Figure A.23. Then graph the equations, as shown in Figure A.24. Next, use the *intersect* feature to find the point of intersection. Trace the cursor along the graph of y_1 near the intersection and press (ENTER) (see Figure A.25). Then trace the cursor along the graph of y_2 near the intersection and press (ENTER) (see Figure A.26). Marks are then placed on the graph at these points (see Figure A.27). Finally, move the cursor near the point of intersection and press (ENTER). In Figure A.28, you can see that the coordinates of the point of intersection are displayed at the bottom of the window. So, the point of intersection is $(-1, 3)$.

Figure A.22

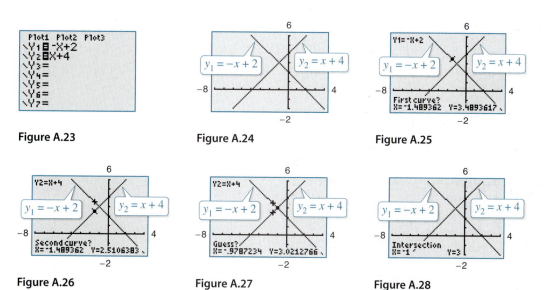

Figure A.23   **Figure A.24**   **Figure A.25**

Figure A.26   **Figure A.27**   **Figure A.28**

List Editor

Most graphing utilities can store data in lists. The *list editor* can be used to create tables and to store statistical data. The *list editor* can be found in the *edit* menu of the *statistics* feature, as shown in Figure A.29. To enter the numbers 1 through 10 in a list, first choose a list (L_1) and then begin entering the data into each row, as shown in Figure A.30.

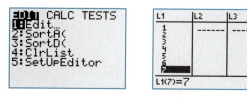

Figure A.29   **Figure A.30**

You can also attach a formula to a list. For instance, you can multiply each of the data values in L_1 by 3. First, display the *list editor* and move the cursor to the top line.

Then move the cursor onto the list to which you want to attach the formula (L_2). Finally, enter the formula 3* L_1 (see Figure A.31) and then press (ENTER). You should obtain the list shown in Figure A.32.

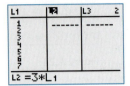
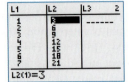

Figure A.31   **Figure A.32**

Matrix Feature

The *matrix* feature of a graphing utility has many uses, such as evaluating a determinant and performing row operations. (For instructions on how to perform row operations, see Elementary Row Operations Features on page A3.)

Matrix Editor

You can define, display, and edit matrices using the *matrix editor*. The *matrix editor* can be found in the *edit* menu of the *matrix* feature. For instance, to enter the matrix

$$A = \begin{bmatrix} 6 & -3 & 4 \\ 9 & 0 & -1 \end{bmatrix}$$

first choose the matrix name [A], as shown in Figure A.33. Then enter the dimension of the matrix (in this case, the dimension is 2×3) and enter the entries of the matrix, as shown in Figure A.34. To display the matrix on the home screen, choose the *name* menu of the *matrix* feature and select the matrix [A] (see Figure A.35), then press (ENTER). The matrix A should now appear on the home screen, as shown in Figure A.36.

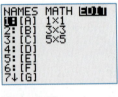

Figure A.33 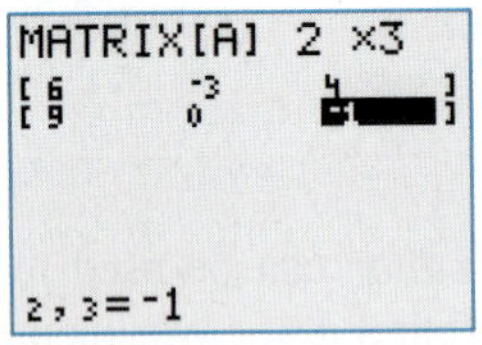   **Figure A.34**

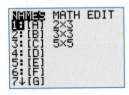
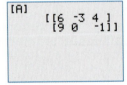

Figure A.35   **Figure A.36**

Matrix Operations

Most graphing utilities can perform matrix operations. To find the sum $A + B$ of the matrices shown at the right, first enter the matrices in the *matrix editor* as [A] and [B]. Then find the sum, as shown in Figure A.37. Scalar multiplication can be performed in a similar manner. For instance, you can evaluate $7A$, where A is the matrix at the right, as shown in Figure A.38. To find the product AB of the matrices A and B at the right, first be sure that the product is defined. Because the number of columns of A (2 columns) equals the number of rows of B (2 rows), you can find the product AB, as shown in Figure A.39.

$$A = \begin{bmatrix} -3 & 5 \\ 0 & 4 \end{bmatrix}$$

$$B = \begin{bmatrix} 7 & -2 \\ -1 & 2 \end{bmatrix}$$

[A]+[B]
```
        [[4   3]
         [-1  6]]
```

Figure A.37

7[A]
```
        [[-21 35]
         [0   28]]
```

Figure A.38

[A][B]
```
        [[-26 16]
         [-4  8 ]]
```

Figure A.39

Inverse Matrix

Some graphing utilities may not have an *inverse matrix* feature. However, you can find the inverse of a square matrix by using the inverse key $\boxed{x^{-1}}$. To find the inverse of the matrix

$$A = \begin{bmatrix} 1 & -2 & 1 \\ -1 & 3 & 0 \\ 2 & 4 & 5 \end{bmatrix}$$

enter the matrix in the *matrix editor* as [A]. Then find the inverse, as shown in Figure A.40.

[A]⁻¹
```
    [[-3  -2.8  .6 ]
     [-1  -.6   .2 ]
     [2   1.6   -.2]]
```

Figure A.40

Maximum and Minimum Features

The *maximum* and *minimum* features find relative extrema of a function. For instance, the graph of

$$y = x^3 - 3x$$

is shown in Figure A.41. In the figure, the graph appears to have a relative maximum at $x = -1$ and a relative minimum at $x = 1$. To find the exact values of the relative extrema, you can use the *maximum* and *minimum* features found in the *calculate* menu (see Figure A.42). First, to find the relative maximum, choose the *maximum* feature and trace the cursor along the graph to a point left of the maximum and press $\boxed{\text{ENTER}}$ (see Figure A.43). Then trace the cursor along the graph to a point right of the maximum and press $\boxed{\text{ENTER}}$ (see Figure A.44). Note the two arrows near the top of the display marking the left and right bounds, as shown in Figure A.45. Next, trace the cursor along the graph between the two bounds and as close to the maximum as you can (see Figure A.45) and press $\boxed{\text{ENTER}}$. In Figure A.46, you can see that the coordinates of the maximum point are displayed at the bottom of the window. So, the relative maximum is $(-1, 2)$.

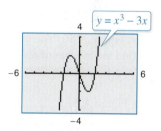

Figure A.41

Figure A.42

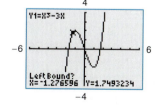

Figure A.43

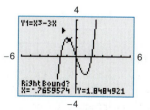

Figure A.44

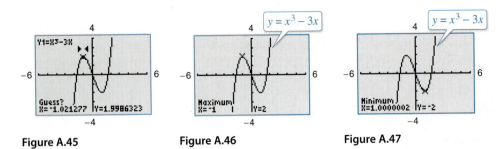

Figure A.45   **Figure A.46**   **Figure A.47**

You can find the relative minimum in a similar manner. In Figure A.47, you can see that the relative minimum is $(1, -2)$.

Mean and Median Features

In real-life applications, you often encounter large data sets and want to calculate statistical values. The *mean* and *median* features calculate the mean and median of a data set. For instance, in a survey, 100 people were asked how much money (in dollars) per week they withdraw from an automatic teller machine (ATM). The results are shown in the table below. The frequency represents the number of responses.

Amount	10	20	30	40	50	60	70	80	90	100
Frequency	3	8	10	19	24	13	13	7	2	1

To find the mean and median of the data set, first enter the data in the *list editor*, as shown in Figure A.48. Enter the amount in L_1 and the frequency in L_2. Then choose the *mean* feature from the *math* menu of the *list* feature, as shown in Figure A.49. When using this feature, you must enter a list and a frequency list (if applicable). In this case, the list is L_1 and the frequency list is L_2. After pressing (ENTER), you should obtain a mean of

$49.80

as shown in Figure A.50. You can follow the same steps (except choose the *median* feature) to find the median of the data. You should obtain a median of

$50

as shown in Figure A.51.

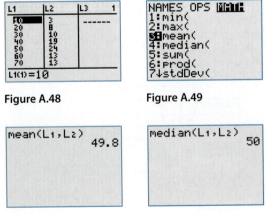

Figure A.48   **Figure A.49**

Figure A.50   **Figure A.51**

Mode Settings

Mode settings of a graphing utility control how the utility displays and interprets numbers and graphs. The default mode settings are shown in Figure A.52.

Figure A.52

Radian and Degree Modes

The trigonometric functions can be applied to angles measured in either radians or degrees. When your graphing utility is in *radian* mode, it interprets angle values as radians and displays answers in radians. When your graphing utility is in *degree* mode, it interprets angle values as degrees and displays answers in degrees. For instance, to calculate $\sin(\pi/6)$, make sure the calculator is in *radian* mode. You should obtain an answer of 0.5, as shown in Figure A.53. To calculate $\sin 45°$, make sure the calculator is in *degree* mode, as shown in Figure A.54. You should obtain an approximate answer of 0.7071, as shown in Figure A.55. Without changing the mode of the calculator before evaluating $\sin 45°$, you would obtain an answer of approximately 0.8509, which is the sine of 45 radians.

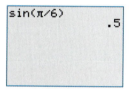

Figure A.53

Figure A.54

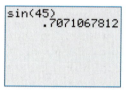

Figure A.55

Function, Parametric, Polar, and Sequence Modes

Most graphing utilities can graph using four different modes.

Function Mode   The *function* mode is used to graph standard algebraic and trigonometric functions. For instance, to graph $y = 2x^2$, use the *function* mode, as shown in Figure A.52. Then enter the equation in the *equation editor*, as shown in Figure A.56. Using a standard viewing window (see Figure A.57), you obtain the graph shown in Figure A.58.

Figure A.56

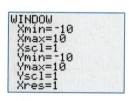

Figure A.57

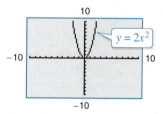

Figure A.58

Parametric Mode   To graph parametric equations such as

$$x = t + 1 \quad \text{and} \quad y = t^2$$

use the *parametric* mode, as shown in Figure A.59. Then enter the equations in the *equation editor*, as shown in Figure A.60. Using the viewing window shown in Figure A.61, you obtain the graph shown in Figure A.62.

Figure A.59

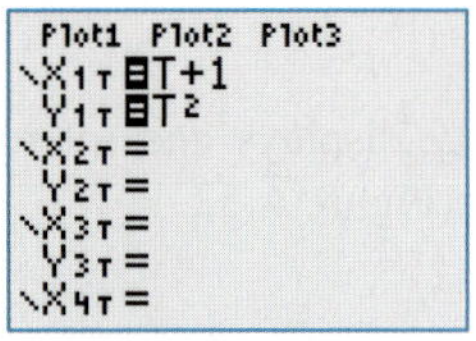

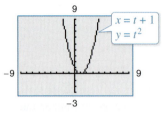

Figure A.60

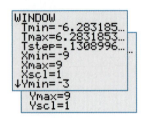

Figure A.61

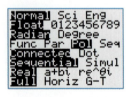

Figure A.62

Polar Mode   To graph polar equations of the form $r = f(\theta)$, you can use the *polar* mode of a graphing utility. For instance, to graph the polar equation

$$r = 2 \cos \theta$$

use the *polar* mode (and *radian* mode), as shown in Figure A.63. Then enter the equation in the *equation editor*, as shown in Figure A.64. Using the viewing window shown in Figure A.65, you obtain the graph shown in Figure A.66.

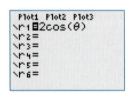

Figure A.63

Figure A.64

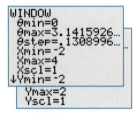

Figure A.65

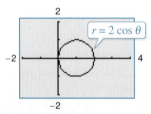

Figure A.66

Sequence Mode   To graph the first five terms of a sequence such as

$$a_n = 4n - 5$$

use the *sequence* mode, as shown in Figure A.67. Then enter the sequence in the *equation editor*, as shown in Figure A.68 (assume that *n* begins with 1). Using the viewing window shown in Figure A.69, you obtain the graph shown in Figure A.70.

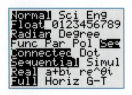

Figure A.67

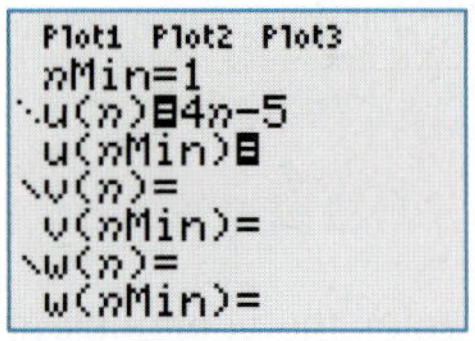

Figure A.68

Technology Tip

 Note that when using the different graphing modes of a graphing utility, the utility uses different variables. When the utility is in *function* mode, it uses the variables x and y. In *parametric* mode, the utility uses the variables x, y, and t. In *polar* mode, the utility uses the variables r and θ. In *sequence* mode, the utility uses the variables u (instead of a) and n.

Figure A.69

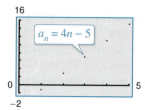

Figure A.70

Connected and Dot Modes

Graphing utilities use the point-plotting method to graph functions. When a graphing utility is in *connected* mode, the utility connects the points that are plotted. When the utility is in *dot* mode, it does not connect the points that are plotted. For instance, the graph of

$$y = x^3$$

in *connected* mode is shown in Figure A.71. To graph this function using *dot* mode, first change the mode to *dot* mode (see Figure A.72) and then graph the equation, as shown in Figure A.73. As you can see in Figure A.73, the graph is a collection of dots.

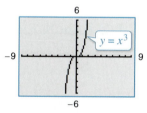

Figure A.71

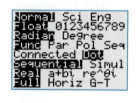

Figure A.72

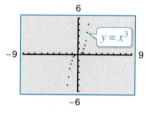

Figure A.73

A problem arises in some graphing utilities when the *connected* mode is used. Graphs with vertical asymptotes, such as rational functions and tangent functions, appear to be connected. For instance, the graph of

$$y = \frac{1}{x + 3}$$

is shown in Figure A.74. Notice how the two portions of the graph appear to be connected with a vertical line at $x = -3$. From your study of rational functions, you know that the graph has a vertical asymptote at $x = -3$ and therefore is undefined when $x = -3$. When using a graphing utility to graph rational functions and other functions that have vertical asymptotes, you should use the *dot* mode to eliminate extraneous vertical lines. Because the *dot* mode of a graphing utility displays a graph as a collection of dots rather than as a smooth curve, in this text, a blue curve is placed behind the graphing utility's display to indicate where the graph should appear, as shown in Figure A.75.

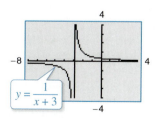

Figure A.74

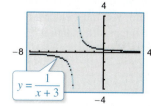

Figure A.75

$_nC_r$ Feature

The $_nC_r$ feature calculates binomial coefficients and the number of combinations of n elements taken r at a time. For instance, to find the number of combinations of eight elements taken five at a time, enter 8 (the n-value) on the home screen and choose the $_nC_r$ feature from the *probability* menu of the *math* feature (see Figure A.76). Next, enter 5 (the r-value) on the home screen and press (ENTER). You should obtain 56, as shown in Figure A.77.

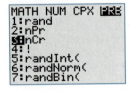

Figure A.76

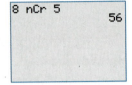

Figure A.77

$_nP_r$ Feature

The $_nP_r$ feature calculates the number of permutations of n elements taken r at a time. For instance, to find the number of permutations of six elements taken four at a time, enter 6 (the n-value) on the home screen and choose the $_nP_r$ feature from the *probability* menu of the *math* feature (see Figure A.78). Next enter 4 (the r-value) on the home screen and press (ENTER). You should obtain 360, as shown in Figure A.79.

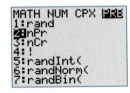

Figure A.78

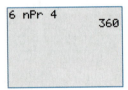

Figure A.79

One-Variable Statistics Feature

Graphing utilities are useful in calculating statistical values for a set of data. The *one-variable statistics* feature analyzes data with one measured variable. This feature outputs the mean of the data, the sum of the data, the sum of the data squared, the sample standard deviation of the data, the population standard deviation of the data, the number of data points, the minimum data value, the maximum data value, the first quartile of the data, the median of the data, and the third quartile of the data. Consider the following data, which show the hourly earnings (in dollars) of 12 retail sales associates.

8.05, 10.25, 8.45, 9.15, 8.90, 8.20, 9.25, 10.30, 8.60, 9.60, 10.05, 11.35

You can use the *one-variable statistics* feature to determine the mean and standard deviation of the data. First, enter the data in the *list editor*, as shown in Figure A.80. Then choose the *one-variable statistics* feature from the *calculate* menu of the *statistics* feature, as shown in Figure A.81. When using this feature, you must enter a list. In this case, the list is L_1. In Figure A.82, you can see that the mean of the data is

$$\bar{x} \approx 9.35$$

and the standard deviation of the data is

$$\sigma x \approx 0.95.$$

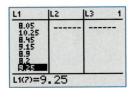

Figure A.80

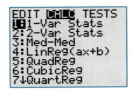

Figure A.81

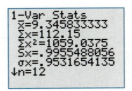

Figure A.82

Regression Feature

Throughout the text, you are asked to use the *regression* feature of a graphing utility to find models for sets of data. Most graphing utilities have built-in regression programs for the following.

Regression	*Form of Model*
Linear	$y = ax + b$ or $y = a + bx$
Quadratic	$y = ax^2 + bx + c$
Cubic	$y = ax^3 + bx^2 + cx + d$
Quartic	$y = ax^4 + bx^3 + cx^2 + dx + e$
Logarithmic	$y = a + b \ln x$
Exponential	$y = ab^x$
Power	$y = ax^b$
Logistic	$y = \dfrac{c}{1 + ae^{-bx}}$
Sine	$y = a \sin(bx + c) + d$

For instance, you can find a linear model for the data in the table.

x	y
6	222.8
7	228.7
8	235.0
9	240.3
10	245.0
11	248.2
12	254.4
13	260.2
14	268.3
15	287.0

Enter the data in the *list editor*, as shown in Figure A.83. Note that L_1 contains the *x*-values, and L_2 contains the *y*-values. Now choose the *linear regression* feature from the *calculate* menu of the *statistics* feature, as shown in Figure A.84. In Figure A.85, you can see that a linear model for the data is given by

$$y = 6.22x + 183.7.$$

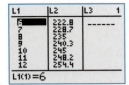

Figure A.83

Figure A.84

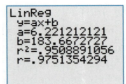

Figure A.85

When you use the *regression* feature of a graphing utility, you will notice that the program may also output an "*r*-value." (For some calculators, make sure you select the *diagnostics on* feature before you use the *regression* feature. Otherwise, the calculator will not output an *r*-value.) The *r*-value or *correlation coefficient* measures how well the linear model fits the data. The closer the value of $|r|$ is to 1, the better the fit. For the data above,

$$r \approx 0.97514$$

which implies that the model is a good fit for the data.

Row-Echelon and Reduced Row-Echelon Features

Some graphing utilities have features that can automatically transform a matrix to row-echelon form and reduced row-echelon form. These features can be used to check your solutions to systems of equations.

Row-Echelon Feature

Consider the system of equations and the corresponding augmented matrix shown below.

Linear System

$$\begin{cases} 2x + 5y - 3z = 4 \\ 4x + y = 2 \\ -x + 3y - 2z = -1 \end{cases}$$

Augmented Matrix

$$\begin{bmatrix} 2 & 5 & -3 & \vdots & 4 \\ 4 & 1 & 0 & \vdots & 2 \\ -1 & 3 & -2 & \vdots & -1 \end{bmatrix}$$

You can use the *row-echelon* feature of a graphing utility to write the augmented matrix in row-echelon form. First, enter the matrix in the graphing utility using the *matrix editor*, as shown in Figure A.86. Next, choose the *row-echelon* feature from the *math* menu of the *matrix* feature, as shown in Figure A.87. When using this feature, you must enter the name of the matrix. In this case, the name of the matrix is [A]. You should obtain the matrix shown in Figure A.88. You may have to scroll to the right in order to see all the entries of the matrix.

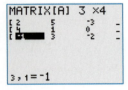

Figure A.86

Figure A.87

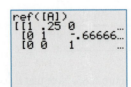

Figure A.88

Reduced Row-Echelon Feature

To write the augmented matrix in reduced row-echelon form, follow the same steps used to write a matrix in row-echelon form except choose the *reduced row-echelon* feature, as shown in Figure A.89. You should obtain the matrix shown in Figure A.90. From Figure A.90, you can conclude that the solution to the system is $x = 3$, $y = -10$, and $z = -16$.

Figure A.89

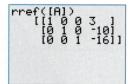

Figure A.90

Sequence Feature

The *sequence* feature is used to display the terms of sequences. For instance, to determine the first five terms of the arithmetic sequence

$$a_n = 3n + 5 \qquad \text{Assume } n \text{ begins with 1.}$$

set the graphing utility to *sequence* mode. Then choose the *sequence* feature from the *operations* menu of the *list* feature, as shown in Figure A.91. When using this feature, you must enter the sequence, the variable (in this case n), the beginning value (in this case 1), and the end value (in this case 5). The first five terms of the sequence are 8, 11, 14, 17, and 20, as shown in Figure A.92. You may have to scroll to the right in order to see all the terms of the sequence.

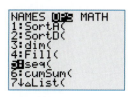

Figure A.91

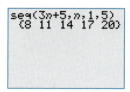

Figure A.92

Shade Feature

Most graphing utilities have a *shade* feature that can be used to graph inequalities. For instance, to graph the inequality

$$y \le 2x - 3$$

first enter the equation $y = 2x - 3$ in the *equation editor*, as shown in Figure A.93. Next, using a standard viewing window (see Figure A.94), graph the equation, as shown in Figure A.95.

Figure A.93

Figure A.94

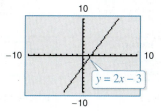

Figure A.95

Because the inequality sign is $\leq$, you want to shade the region below the line $y = 2x - 3$. Choose the *shade* feature from the *draw* feature menu, as shown in Figure A.96. You must enter a lower function and an upper function. In this case, the lower function is -10 (this is the least y-value in the viewing window) and the upper function is Y_1 ($y = 2x - 3$), as shown in Figure A.97. Then press (ENTER) to obtain the graph shown in Figure A.98.

Figure A.96

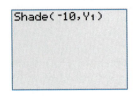

Figure A.97

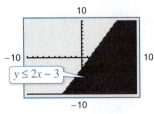

Figure A.98

To graph the inequality

$$y \geq 2x - 3$$

(using a standard viewing window), you would enter the lower function as Y_1 ($y = 2x - 3$) and the upper function as 10 (the greatest y-value in the viewing window).

Statistical Plotting Feature

The *statistical plotting* feature plots data that are stored in lists. Most graphing utilities can display the following types of plots.

Plot Type	Variables
Scatter plot	x-list, y-list
xy line graph	x-list, y-list
Histogram	x-list, frequency
Box-and-whisker plot	x-list, frequency
Normal probability plot	data list, data axis

For instance, use a box-and-whisker plot to represent the following set of data. Then use the graphing utility plot to find the least number, the lower quartile, the median, the upper quartile, and the greatest number.

17, 19, 21, 27, 29, 30, 37, 27, 15, 23, 19, 16

First, use the *list editor* to enter the values in a list, as shown in Figure A.99. Then go to the *statistical plotting editor*. In this editor you will turn the plot on, select the box-and-whisker plot, select the list you entered in the *list editor*, and enter the frequency of each item in the list, as shown in Figure A.100. Now use the *zoom* feature and choose the *zoom stat* option to set the viewing window and plot the graph, as shown in Figure A.101. Use the *trace* feature to find that the least number is 15, the lower quartile is 18, the median is 22, the upper quartile is 28, and the greatest number is 37.

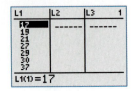

Figure A.99

Figure A.100

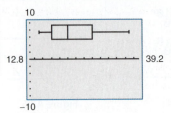

Figure A.101

Sum Feature

The *sum* feature finds the sum of a list of data. For instance, the data below represent a student's quiz scores on 10 quizzes throughout an algebra course.

22, 23, 19, 24, 20, 15, 25, 21, 18, 24

To find the total quiz points the student earned, enter the data in the *list editor*, as shown in Figure A.102. To find the sum, choose the *sum* feature from the *math* menu of the *list* feature, as shown in Figure A.103. You must enter a list. In this case the list is L_1. You should obtain a sum of 211, as shown in Figure A.104.

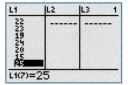

Figure A.102

Figure A.103

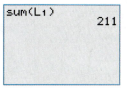

Figure A.104

Sum Sequence Feature

The *sum* feature and the *sequence* feature can be combined to find the sum of a sequence or series. For instance, to find the sum

$$\sum_{k=0}^{10} 5^{k+1}$$

first choose the *sum* feature from the *math* menu of the *list* feature, as shown in Figure A.105. Then choose the *sequence* feature from the *operations* menu of the *list* feature, as shown in Figure A.106. You must enter an expression for the sequence, a variable, the lower limit of summation, and the upper limit of summation. After pressing (ENTER), you should obtain the sum 61,035,155, as shown in Figure A.107.

Figure A.105

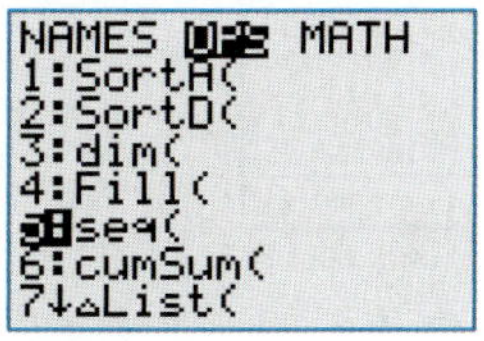

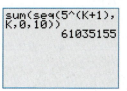

Figure A.106

Figure A.107

Table Feature

Most graphing utilities are capable of displaying a table of values with *x*-values and one or more corresponding *y*-values. These tables can be used to check solutions of an equation and to generate ordered pairs to assist in graphing an equation by hand.

To use the *table* feature, enter an equation in the *equation editor*. The table may have a setup screen, which allows you to select the starting *x*-value and the table step or *x*-increment. You may then have the option of automatically generating values for *x* and *y* or building your own table using the *ask* mode (see Figure A.108).

Figure A.108

For instance, enter the equation

$$y = \frac{3x}{x + 2}$$

in the *equation editor*, as shown in Figure A.109. In the table setup screen, set the table to start at $x = -4$ and set the table step to 1, as shown in Figure A.110. When you view the table, notice that the first x-value is -4 and that each value after it increases by 1. Also notice that the Y_1 column gives the resulting y-value for each x-value, as shown in Figure A.111. The table shows that the y-value for $x = -2$ is ERROR. This means that the equation is undefined when $x = -2$.

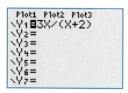

Figure A.109

Figure A.110

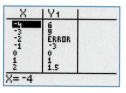

Figure A.111

With the same equation in the *equation editor*, set the independent variable in the table to *ask* mode, as shown in Figure A.112. In this mode, you do not need to set the starting x-value or the table step because you are entering any value you choose for x. You may enter any real value for x—integers, fractions, decimals, irrational numbers, and so forth. When you enter

$$x = 1 + \sqrt{3}$$

the graphing utility may rewrite the number as a decimal approximation, as shown in Figure A.113. You can continue to build your own table by entering additional x-values in order to generate y-values, as shown in Figure A.114.

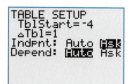

Figure A.112

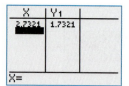

Figure A.113

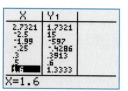

Figure A.114

When you have several equations in the *equation editor*, the table may generate y-values for each equation.

Tangent Feature

Some graphing utilities have the capability of drawing a tangent line to a graph at a given point. For instance, consider the equation

$$y = -x^3 + x + 2.$$

To draw the line tangent to the point $(1, 2)$ on the graph of y, enter the equation in the *equation editor*, as shown in Figure A.115. Using the viewing window shown in Figure A.116, graph the equation, as shown in Figure A.117. Next, choose the *tangent* feature from the *draw* feature menu, as shown in Figure A.118. You can either move the cursor to select a point or enter the x-value at which you want the tangent line to be drawn.

Because you want the tangent line to the point

$$(1, 2)$$

enter 1 (see Figure A.119) and then press (ENTER). The *x*-value you entered and the equation of the tangent line are displayed at the bottom of the window, as shown in Figure A.120.

Figure A.115

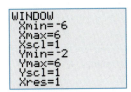

Figure A.116

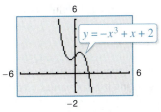

Figure A.117

Figure A.118

Figure A.119

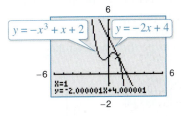

Figure A.120

Trace Feature

For instructions on how to use the *trace* feature, see Zoom and Trace Features on page A23.

Value Feature

The *value* feature finds the value of a function *y* for a given *x*-value. To find the value of a function such as

$$f(x) = 0.5x^2 - 1.5x$$

at $x = 1.8$, first enter the function in the *equation editor* (see Figure A.121) and then graph the function (using a standard viewing window), as shown in Figure A.122. Next, choose the *value* feature from the *calculate* menu, as shown in Figure A.123. You will see "X= " displayed at the bottom of the window. Enter the *x*-value, in this case $x = 1.8$, as shown in Figure A.124. When entering an *x*-value, be sure it is between the Xmin and Xmax values you entered for the viewing window. Then press (ENTER). In Figure A.125, you can see that when $x = 1.8$,

$$y = -1.08.$$

Figure A.121

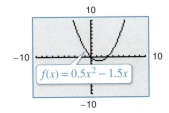

Figure A.122

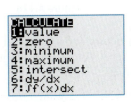

Figure A.123

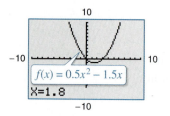

Figure A.124

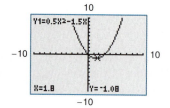

Figure A.125

Viewing Window

A viewing window for a graph is a rectangular portion of the coordinate plane. A viewing window is determined by the following six values (see Figure A.126).

Xmin = the minimum value of x
Xmax = the maximum value of x
Xscl = the number of units per tick mark on the x-axis
Ymin = the minimum value of y
Ymax = the maximum value of y
Yscl = the number of units per tick mark on the y-axis

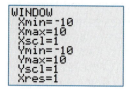

Figure A.126

When you enter these six values in a graphing utility, you are setting the viewing window. On some graphing utilities there is a seventh value for the viewing window labeled Xres. This sets the pixel resolution (1 through 8). For instance, when

Xres = 1

functions are evaluated and graphed at each pixel on the x-axis. Some graphing utilities have a standard viewing window, as shown in Figure A.127. To initialize the standard viewing window quickly, choose the *standard viewing window* feature from the *zoom* feature menu (see page A23), as shown in Figure A.128.

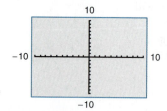

Figure A.127

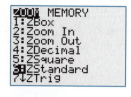

Figure A.128

By choosing different viewing windows for a graph, it is possible to obtain different impressions of the graph's shape. For instance, Figure A.129 shows four different viewing windows for the graph of

$$y = 0.1x^4 - x^3 + 2x^2.$$

Of these viewing windows, the one shown in part (a) is the most complete.

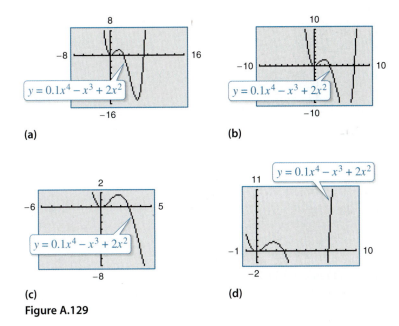

(a)

(b)

(c)

(d)

Figure A.129

On most graphing utilities, the display screen is two-thirds as high as it is wide. On such screens, you can obtain a graph with a true geometric perspective by using a square setting—one in which

$$\frac{\text{Ymax} - \text{Ymin}}{\text{Xmax} - \text{Xmin}} = \frac{2}{3}.$$

One such setting is shown in Figure A.130. Notice that the x and y tick marks are equally spaced on a square setting, but not on a standard setting (see Figure A.127). To initialize the square viewing window quickly, choose the *square viewing window* feature from the *zoom* feature menu (see page A23), as shown in Figure A.131.

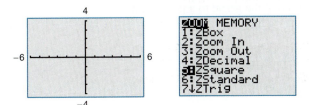

Figure A.130   **Figure A.131**

To see how the viewing window affects the geometric perspective, graph the semicircles

$$y_1 = \sqrt{9 - x^2}$$

and

$$y_2 = -\sqrt{9 - x^2}$$

using a standard viewing window, as shown in Figure A.132. Notice how the circle appears elliptical rather than circular. Now graph y_1 and y_2 using a square viewing window, as shown in Figure A.133. Notice how the circle appears circular. (Note that when you graph the two semicircles, your graphing utility may not connect them. This is because some graphing utilities are limited in their resolution. So, in this text, a blue curve is placed behind the graphing utility's display to indicate where the graph should appear.)

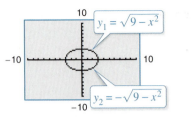

Figure A.132

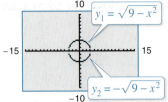

Figure A.133

Zero or Root Feature

The *zero* or *root* feature finds the real zeros of the various types of functions studied in this text. To find the zeros of a function such as

$$f(x) = 2x^3 - 4x$$

first enter the function in the *equation editor*, as shown in Figure A.134. Now graph the equation (using a standard viewing window), as shown in Figure A.135. From the graph you can see that the graph of the function crosses the *x*-axis three times, so the function has three zeros.

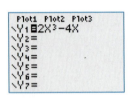

Figure A.134

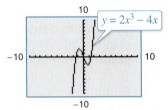

Figure A.135

To find these zeros, choose the *zero* feature found in the *calculate* menu (see Figure A.136). Next, trace the cursor along the graph to a point left of one of the zeros and press (ENTER) (see Figure A.137). Then trace the cursor along the graph to a point right of the zero and press (ENTER) (see Figure A.138). Note the two arrows near the top of the display marking the left and right bounds, as shown in Figure A.139. Now trace the cursor along the graph between the two bounds and as close to the zero as you can (see Figure A.140) and press (ENTER). In Figure A.141, you can see that one zero of the function is $x \approx -1.414214$.

Figure A.136

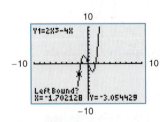

Figure A.137

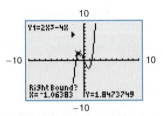

Figure A.138

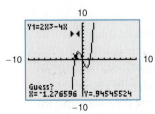

Figure A.139

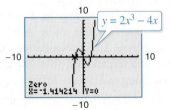

Figure A.140

Figure A.141

Repeat this process to determine that the other two zeros of the function are $x = 0$ (see Figure A.142) and $x \approx 1.414214$ (see Figure A.143).

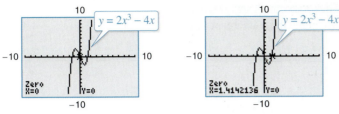

Figure A.142   **Figure A.143**

Zoom and Trace Features

The *zoom* feature enables you to adjust the viewing window of a graph quickly (see Figure A.144). For example, the *zoom box* feature allows you to create a new viewing window by drawing a box around any part of the graph.

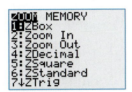

Figure A.144

The *trace* feature moves from point to point along a graph. For instance, enter the equation

$$y = 2x^3 - 3x + 2$$

in the *equation editor* (see Figure A.145) and graph the equation, as shown in Figure A.146. To activate the *trace* feature, press (TRACE); then use the arrow keys to move the cursor along the graph. As you trace the graph, the coordinates of each point are displayed, as shown in Figure A.147.

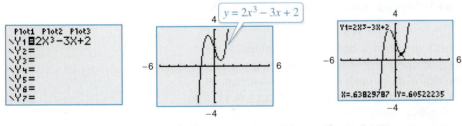

Figure A.145   **Figure A.146**   **Figure A.147**

The *trace* feature combined with the *zoom* feature enables you to obtain better and better approximations of desired points on a graph. For instance, you can use the *zoom* feature to approximate the *x*-intercept of the graph of

$$y = 2x^3 - 3x + 2.$$

From the viewing window shown in Figure A.146, the graph appears to have only one *x*-intercept. This intercept lies between -2 and -1. To zoom in on the *x*-intercept, choose the *zoom-in* feature from the *zoom* feature menu, as shown in Figure A.148. Next, trace the cursor to the point you want to zoom in on, in this case the *x*-intercept (see Figure A.149). Then press (ENTER). You should obtain the graph shown in Figure A.150.

Now, using the *trace* feature, you can approximate the *x*-intercept to be

$$x \approx -1.468085$$

as shown in Figure A.151. Use the *zoom-in* feature again to obtain the graph shown in Figure A.152. Using the *trace* feature, you can approximate the *x*-intercept to be

$$x \approx -1.476064$$

as shown in Figure A.153.

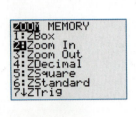

Figure A.148

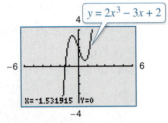

Figure A.149

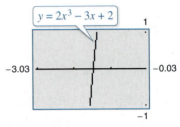

Figure A.150

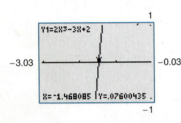

Figure A.151

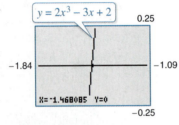

Figure A.152

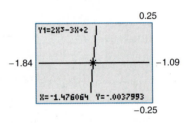

Figure A.153

Here are some suggestions for using the *zoom* feature.

1. With each successive zoom-in, adjust the *x*-scale so that the viewing window shows at least one tick mark on each side of the *x*-intercept.

2. The error in your approximation will be less than the distance between two scale marks.

3. The *trace* feature can usually be used to add one more decimal place of accuracy without changing the viewing window.

You can adjust the scale in Figure A.153 to obtain a better approximation of the *x*-intercept. Using the suggestions above, change the viewing window settings so that the viewing window shows at least one tick mark on each side of the *x*-intercept, as shown in Figure A.154. From Figure A.154, you can determine that the error in your approximation will be less than 0.001 (the Xscl value). Then, using the *trace* feature, you can improve the approximation, as shown in Figure A.155. To three decimal places, the *x*-intercept is

$$x \approx -1.476.$$

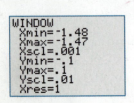

Figure A.154

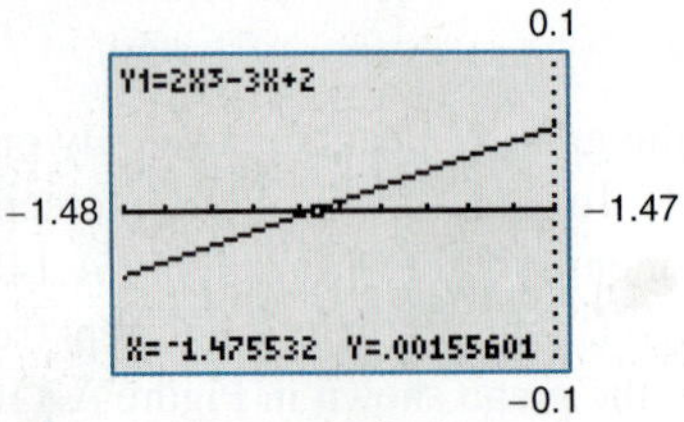

Figure A.155

Answers to Odd-Numbered Exercises and Tests

Chapter P

Section P.1   (page 9)

1. rational   **3.** prime   **5.** terms   **7.** c
8. d   **9.** e   **10.** b   **11.** f   **12.** a
13. (a) 5, 1   (b) 5, 0, 1   (c) $-9, 5, 0, 1, -4, -1$
    (d) $-9, -\frac{7}{2}, 5, \frac{2}{3}, 0, 1, -4, -1$ (e) $\sqrt{2}$
15. (a) 1, 20   (b) 1, 20   (c) $-13, 1, -10, 20$
    (d) $2.01, 0.666\ldots, -13, 1, -10, 20$
    (e) $0.010110111\ldots$
17. (a) $\frac{6}{3}, 3$ (b) $\frac{6}{3}, 3$ (c) $\frac{6}{3}, -2, 3, -3$
    (d) $-\frac{1}{3}, \frac{6}{3}, -7.5, -2, 3, -3$ (e) $-\pi, \frac{1}{2}\sqrt{2}$
19. 0.3125   **21.** $0.\overline{123}$ **23.** $-9.\overline{09}$ **25.** $\frac{23}{5}$
27. $-\frac{13}{2}$ **29.** $-1 < 2.5$
31.
$-4 > -8$
33.
$\frac{3}{2} < 7$
35.
$\frac{5}{6} > \frac{2}{3}$
37. (a) $x \le 5$ is the set of all real numbers less than or equal to 5.
    (b)
    (c) Unbounded
39. (a) $x < 0$ is the set of all negative real numbers.
    (b)
    (c) Unbounded
41. (a) $-2 < x < 2$ is the set of all real numbers greater than -2 and less than 2.
    (b)
    (c) Bounded
43. (a) $-1 \le x < 0$ is the set of all negative real numbers greater than or equal to -1.
    (b)
    (c) Bounded
45. $x < 0; (-\infty, 0)$ **47.** $y \ge 0; [0, \infty)$
49. $9 \le t \le 24; [9, 24]$ **51.** $-2 < x \le 5; (-2, 5]$
53. $[-1, 5]$ **55.** $[-3, 8)$ **57.** $[-a, a+4]$
59. The set of all real numbers greater than -6
61. The set of all real numbers less than or equal to 2
63. 10   **65.** -6
67. 1 for $x > -2$; undefined for $x = -2$; -1 for $x < -2$
69. 1   **71.** 0   **73.** $\frac{7}{2}$ **75.** $|-3| > -|-3|$
77. $-5 = -|5|$ **79.** $-|-2| = -|2|$
81. 51   **83.** $\frac{5}{2}$ **85.** $\frac{128}{75}$ **87.** $|x - 5| \le 3$
89. $|y - 0| \ge 6$ **91.** 1722.0 billion; 69.3 billion
93. 1853.4 billion; 157.8 billion
95. 2407.3 billion; 248.1 billion
97. $|\$113{,}356 - \$112{,}700| = \$656 > \$500$
    $0.05(\$112{,}700) = \$5635$
    Because the actual expenses differ from the budget by more than $500, there is failure to meet the "budget variance test."

99. $|\$37{,}335 - \$37{,}600| = \$265 < \$500$
    $0.05(\$37{,}600) = \$1880$
    Because the difference between the actual expenses and the budget is less than $500 and less than 5% of the budgeted amount, there is compliance with the "budget variance test."
101. Terms: $7x$, 4; coefficient: 7
103. Terms: $\sqrt{3}x^2, -8x, -11$; coefficients: $\sqrt{3}, -8$
105. Terms: $4x^3, \frac{x}{2}, -5$; coefficients: $4, \frac{1}{2}$
107. (a) 3   (b) -6 **109.** (a) 0   (b) 0
111. Commutative Property of Addition
113. Multiplicative Inverse Property
115. Distributive Property
117. Associative Property of Addition
119. $\frac{1}{2}$ **121.** $\frac{3}{8}$ **123.** $\frac{x}{2}$ **125.** $\frac{96}{x}$ **127.** $-\frac{7}{5}$
129. -36.00 **131.** 1.56   **133.** 13.33   **135.** 179 mi
137. False. Two numbers with different signs will always have a product less than 0.
139. (a)

n	1	0.5	0.01	0.0001	0.000001
$5/n$	5	10	500	50,000	5,000,000

    (b) $5/n$ approaches ∞ as n approaches 0.
141. When u and v have the same sign, $|u + v| = |u| + |v|$. When u and v have different signs, $|u + v| < |u| + |v|$.
143. $a \le 0$; If the original value of a is negative, then $|a|$ results in a positive number. Because a is negative, the expression $|a| = a$ states that $|a|$ is equal to a negative number, which can never happen. So, if a is originally negative, $|a|$ must equal $-a$, which is a positive value.

Section P.2   (page 20)

1. exponent, base   **3.** principal nth root
5. rationalizing   **7.** $2 - 3\sqrt{5}$ **9.** No
11. (a) 48   (b) 81   **13.** (a) 1   (b) -9
15. (a) 243   (b) $-\frac{3}{4}$ **17.** (a) $\frac{5}{6}$ (b) 27   **19.** $\frac{7}{4}$
21. -54 **23.** 1   **25.** (a) x^5 (b) $3x^9$
27. (a) $9x^2$ (b) $16x^6$
29. (a) $\frac{7}{x}$ (b) $\frac{4}{3}(x + y)^2, x + y \ne 0$
31. (a) $\frac{x^2}{y^2}, x \ne 0$ (b) $\frac{b^5}{a^5}, b \ne 0$ **33.** -1600
35. 2.125   **37.** 5103   **39.** 9.735×10^2
41. 1.0252484×10^4 **43.** -1.11025×10^3
45. 2.485×10^{-4} **47.** -2.5×10^{-6} **49.** 5.73×10^7
51. 8.99×10^{-5} **53.** 125,000   **55.** $-481,600,000$
57. 76,500,000   **59.** -0.009001 **61.** 493
63. 0.0000000000000000016022   **65.** 60,000
67. 175,000,000   **69.** 5×10^4 or 50,000
71. (a) 4.907×10^{17} (b) 1.479
73. (a) 67,082.039   (b) 39.791   **75.** 11   **77.** 4

79. -125 **81.** -7.225 **83.** 21.316 **85.** 14.499

87. 2.035 **89.** 2.709 **91.** $1,010,000.128$

93. (a) 3 (b) $2x\sqrt[5]{3}$ **95.** (a) $3y^2\sqrt{6x}$ (b) $2\sqrt[3]{4a^2/b^2}$

97. (a) $34\sqrt{2}$ (b) $22\sqrt{2}$

99. (a) $13\sqrt{x+1}$ (b) $18\sqrt{5x}$

101. $\sqrt{5}+\sqrt{3} > \sqrt{5+3}$ **103.** $5 > \sqrt{3^2+2^2}$

105. $\dfrac{\sqrt{3}}{3}$ **107.** $\dfrac{\sqrt{14}+2}{2}$ **109.** $\dfrac{3}{\sqrt{3}}$

111. $\dfrac{2}{3(\sqrt{5}-\sqrt{3})}$ **113.** (a) $\sqrt{3}$ (b) $\sqrt[3]{(x+1)^2}$

115. $64^{1/3}$ **117.** $\sqrt[5]{32}$ **119.** $(-216)^{1/3}$ **121.** $81^{3/4}$

123. $\dfrac{2}{|x|}$ **125.** $\dfrac{1}{x^3}$, $x > 0$ **127.** $\dfrac{1}{8}$ **129.** -3

131. (a) $2\sqrt[4]{2}$ (b) $\sqrt[8]{2x}$ **133.** 0.026 in.

135. Brazil: 1.02×10^4 Ireland: 3.86×10^4

China: 6.66×10^3 India: 3.07×10^3

Germany: 3.41×10^4 Mexico: 1.33×10^4

Iran: 1.152×10^4

137. True. $x^{k+1}/x = x^k x/x = x^k$, $(x \neq 0)$

139. $1 = \dfrac{a^n}{a^n} = a^{n-n} = a^0$

141. When any positive integer is squared, the unit's digit is 0, 1, 4, 5, 6, or 9. Therefore, $\sqrt{5233}$ is not an integer.

Section P.3   (page 31)

1. n, a_n **3.** First, Outer, Inner, Last

5. When each of its factors is prime

7. d   **8.** e   **9.** b   **10.** a   **11.** f   **12.** c

13. Answers will vary, but first term is $-2x^3$.

15. Answers will vary, but first term has the form $-ax^4$, $a > 0$.

17. $4x^2 + 3x + 2$

Degree: 2; leading coefficient: 4

19. $x^7 - 8$

Degree: 7; leading coefficient: 1

21. $-2x^5 + 6x^4 - x + 1$

Degree: 5; leading coefficient: -2

23. Polynomial: $4x^3 + 3x - 5$ **25.** Not a polynomial

27. $7t^2 - 4t - 3$ **29.** $2x + 17$ **31.** $x^2 - 20$

33. $8.1x^3 + 29.7x^2 + 11$ **35.** $-15z^2 + 3z$

37. $-6y^2 + 20y$ **39.** $3x^3 - 6x^2 + 3x$

41. $x^2 + 7x + 12$ **43.** $6x^2 - 7x - 5$

45. $25x^2 - 20x + 4$ **47.** $x^2 - 81$ **49.** $x^2 - 4y^2$

51. $4r^4 - 25$ **53.** $x^3 + 3x^2 + 3x + 1$

55. $8x^3 - 12x^2y + 6xy^2 - y^3$ **57.** $\frac{1}{16}x^2 - 9$

59. $5.76x^2 + 14.4x + 9$ **61.** $-3x^4 - x^3 - 12x^2 - 19x - 5$

63. $x^2 + 2xz + z^2 - 25$ **65.** $x^2 + 2xy + y^2 - 6y - 6x + 9$

67. $2x^2 + 2x$ **69.** $5(x-8)$ **71.** $2x(x^2-3)$

73. $(x-5)(3x+8)$ **75.** $(x+8)(x-8)$

77. $3(4y+3)(4y-3)$ **79.** $\left(2x-\frac{1}{3}\right)\left(2x+\frac{1}{3}\right)$

81. $[(x-1)-2][(x-1)+2] = (x-3)(x+1)$

83. $(x-2)^2$ **85.** $\left(x+\frac{1}{2}\right)^2$ **87.** $(2x-3)^2$

89. $\left(2x-\frac{1}{3}\right)^2$ **91.** $(x-4)(x^2+4x+16)$

93. $(x+1)(x^2-x+1)$ **95.** $\left(x-\frac{2}{3}\right)\left(x^2+\frac{2}{3}x+\frac{4}{9}\right)$

97. $(2x-1)(4x^2+2x+1)$

99. $(x+2-y)(x^2+4x+4+xy+2y+y^2)$

101. $(x-1)(x+2)$ **103.** $(s-2)(s-3)$

105. $-(y-4)(y+5)$ **107.** $(3x-2)(x-1)$

109. $(2x+1)(x-1)$ **111.** $(5x+1)(x+5)$

113. $-(5u-2)(u+3)$ **115.** $(x-1)(x^2+2)$

117. $(3x+2)(2x-1)$ **119.** $(x^2+1)(x-5)$

121. $10(x-2)(x+2)$ **123.** $y(y+1)(y-1)$

125. $(x-1)^2$ **127.** $(2x-1)^2$

129. $-2x(x-2)(x+1)$ **131.** $(9x+1)(x+1)$

133. $\frac{1}{96}(3x+2)(4x-3)$ **135.** $(3x+1)(x^2+5)$

137. $(u+2)(3-u^2)$ **139.** $(x-2)(x+2)(2x+1)$

141. $(x+1)^2(x-1)^2$ **143.** $3(t+2)(t^2-2t+4)$

145. $2(2x-1)(4x-1)$ **147.** $-(x+1)(x-3)(x+9)$

149. (a) $1000r^2 + 2000r + 1000$

(b)

r	$2\frac{1}{2}\%$	3%	4%
$1000(1+r)^2$	1050.63	1060.90	1081.60

r	$4\frac{1}{2}\%$	5%
$1000(1+r)^2$	1092.03	1102.50

(c) The amount increases with increasing r.

151. (a) $T(x) = 0.0475x^2 + 1.099x + 0.23$

(b)

x (mi/h)	30	40	55
T (ft)	75.95	120.19	204.36

(c) Stopping distance increases as speed increases.

153. a

155.

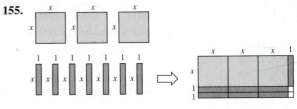

157.

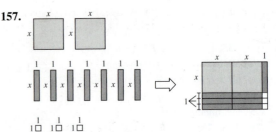

159. $4\pi(r+1)$ **161.** $4(6-x)(6+x)$

163. $(4x^3)(2x+1)^3(2x^2+2x+1)$

165. $(2x-5)^3(5x-4)^2(70x-107)$ **167.** $\dfrac{-8}{(5x-1)^2}$

169. $-14, 14, -2, 2$ **171.** $-51, 51, -15, 15, -27, 27$

173. $2, -3$ **175.** $3, -8$

177. (a) $V = \pi h(R-r)(R+r)$

(b) $V = 2\pi\left[\left(\dfrac{R+r}{2}\right)(R-r)\right]h$

179. False. $(x^2-1)(x^2+1)$ becomes a fourth-degree polynomial.

181. Degree: $m + n$

183. Answers will vary. Sample answer: To cube a binomial difference, cube the first term. Next subtract 3 times the square of the first term multiplied by the second term. Then add 3 times the first term multiplied by the square of the second term. Lastly, subtract the cube of the second term $(x - y)^3 = x^3 - 3x^2y + 3xy^2 - y^3$.

185. No. $(3x - 6)(x + 1)$ is not completely factored because 3 can be factored out of $(3x - 6)$.

187. A 3 was not factored out of the second binomial.

Section P.4   (page 42)

1. domain   **3.** complex fractions

5. When its numerator and denominator have no factors in common aside from ± 1

7. All real numbers x **9.** All nonnegative real numbers x

11. All real numbers x except $x = 3$

13. All real numbers x except $x = 1$

15. All real numbers x except $x = 3$

17. All real numbers x such that $x \geq -7$

19. All real numbers x such that $x \geq \frac{5}{2}$

21. All real numbers x such that $x > 3$

23. $3x, \ x \neq 0$ **25.** $(x + 1), \ x \neq -1$

27. $x + 2, \ x \neq -2$ **29.** $\dfrac{3x}{2}, \ x \neq 0$ **31.** $\dfrac{3y}{y + 1}, \ x \neq 0$

33. $-\dfrac{4y}{5}, \ y \neq \dfrac{1}{2}$ **35.** $-\dfrac{1}{2}, \ x \neq 5$ **37.** $y - 4, \ y \neq -4$

39. $\dfrac{x(x + 3)}{x - 2}, \ x \neq -2$ **41.** $\dfrac{y - 4}{y + 6}, \ y \neq 3$

43. $-(x^2 + 1), \ x \neq 2$ **45.** $z - 2$

47.

x	0	1	2	3	4	5	6
$\dfrac{x^2 - 2x - 3}{x - 3}$	1	2	3	Undef.	5	6	7
$x + 1$	1	2	3	4	5	6	7

The expressions are equivalent except at $x = 3$.

49. Only common factors of the numerator and denominator can be canceled. In this case, factors of terms were incorrectly canceled.

51. $\dfrac{\pi}{4}$ **53.** $\dfrac{1}{5(x - 2)}, \ x \neq 1$ **55.** $\dfrac{r + 1}{r}, \ r \neq 1, -1$

57. $\dfrac{t - 3}{(t + 3)(t - 2)}, \ t \neq -2$ **59.** $\dfrac{3}{2}, \ x \neq -y$

61. $\dfrac{x + 5}{x - 1}$ **63.** $-\dfrac{2x^2 - 5x - 18}{(2x + 1)(x + 3)}$ **65.** $-\dfrac{2}{x - 2}$

67. $-\dfrac{x^2 + 3}{(x + 1)(x - 2)(x - 3)}$ **69.** $-\dfrac{x^2 - 2x + 2}{x(x^2 + 1)}$

71. A common denominator was not found before the fractions were added.

73. $\dfrac{1}{2}, \ x \neq 2$ **75.** $x(x + 1), \ x \neq -1, 0$

77. $-\dfrac{2x + h}{x^2(x + h)^2}, \ h \neq 0$ **79.** $\dfrac{2x - 1}{2x}, \ x > 0$

81. $x^{-2}(x^7 - 2) = \dfrac{x^7 - 2}{x^2}$ **83.** $-\dfrac{1}{(x^2 + 1)^5}$

85. $\dfrac{2x^3 - 2x^2 - 5}{(x - 1)^{1/2}}$ **87.** $\dfrac{2x^2 - 1}{x^{5/2}}$

89. $-\dfrac{(x - 1)(x^2 + x + 2)}{x^2(x^2 + 1)^{3/2}}$ **91.** $\dfrac{-2(3x^2 + 3x - 5)}{\sqrt{4x + 3}(x^2 + 5)^2}$

93. $\dfrac{1}{\sqrt{x + 2} + \sqrt{x}}$ **95.** $\dfrac{1}{\sqrt{x + 2} + \sqrt{2}}, \ x \neq 0$

97. $\dfrac{1}{\sqrt{x + 9} + 3}, \ x \neq 0$ **99.** $\dfrac{x}{2(2x + 1)}$

101. (a) 6.39%   (b) $\dfrac{288(MN - P)}{N(MN + 12P)}$; 6.39%

103. (a) $\dfrac{1}{50}$ min   (b) $\dfrac{x}{50}$ min   (c) 2.4 min

105. (a)

t	0	2	4	6	8	10
T	75	55.9	48.3	45	43.3	42.3

t	12	14	16	18	20	22
T	41.7	41.3	41.1	40.9	40.7	40.6

(b) 40

107. $2x + h, \ h \neq 0$ **109.** $\dfrac{-2x - h}{x^2(x + h)^2}, \ h \neq 0$

111. $\dfrac{4n^2 + 6n + 26}{3}, \ n \neq 0$

113. False. The domain of the left-hand side is $x^n \neq 1$.

115. The expression will still have a radical in the denominator.

117. Answers will vary. Counterexample: Let $x = y = 1$. $\sqrt{1 + 1} = \sqrt{2} \neq \sqrt{1} + \sqrt{1}$

119. $\dfrac{2x^2 - 1}{x^4}$; Answers will vary.

Section P.5   (page 53)

1. Cartesian   **3.** Midpoint Formula

5. c   **6.** f   **7.** a   **8.** d   **9.** e   **10.** b

11. A: $(2, 6)$; B: $(-6, -2)$; C: $(4, -4)$; D: $(-3, 2)$

13.

15.

17. $(-5, 4)$ **19.** $(0, -6)$ **21.** Quadrant IV

23. Quadrant II   **25.** Quadrant III or IV

27. Quadrant III   **29.** Quadrant I or III

31.

33. 8   **35.** 5   **37.** 13   **39.** $\dfrac{\sqrt{277}}{6}$ **41.** $\sqrt{71.78}$

43. (a) $3, 4, 5$ (b) $3^2 + 4^2 = 5^2$
45. (a) $10, 3, \sqrt{109}$ (b) $10^2 + 3^2 = \left(\sqrt{109}\right)^2$
47. $\left(\sqrt{5}\right)^2 + \left(\sqrt{45}\right)^2 = \left(\sqrt{50}\right)^2$
49. Two equal sides of length $\sqrt{29}$
51. Opposite sides have equal lengths of $2\sqrt{5}$ and $\sqrt{85}$.
53. The diagonals are of equal length $\left(\sqrt{58}\right)$. The slope of the line between $(-5, 6)$ and $(0, 8)$ is $\frac{2}{5}$. The slope of the line between $(-5, 6)$ and $(-3, 1)$ is $-\frac{5}{2}$. The slopes are negative reciprocals, indicating perpendicular lines, which form a right angle.

55. (a)   (b) 10
    (c) $(4, 3)$

57. (a)   (b) 17
    (c) $\left(0, \frac{5}{2}\right)$

59. (a)   (b) $2\sqrt{10}$
    (c) $(2, 3)$

61. (a)   (b) $\dfrac{\sqrt{82}}{3}$
    (c) $\left(-1, \dfrac{7}{6}\right)$

63. (a)   (b) $\sqrt{110.97}$
    (c) $(1.25, 3.6)$

65. \$614.5 million   **67.** $x^2 + y^2 = 25$
69. $(x - 2)^2 + (y + 1)^2 = 16$
71. $(x + 1)^2 + (y - 2)^2 = 5$

73. $(x - 3)^2 + (y - 4)^2 = 25$
75. $(x + 2)^2 + (y - 1)^2 = 1$
77. $(x - 3)^2 + (y + 6)^2 = 16$
79. $(x - 2)^2 + (y + 1)^2 = 16$
81. Center: $(0, 0)$ **83.** Center: $(1, -3)$
    Radius: 3   Radius: 2

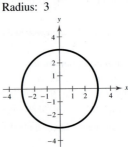

85. Center: $\left(\frac{1}{2}, \frac{1}{2}\right)$
    Radius: $\frac{3}{2}$

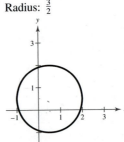

87. $(0, 1), (4, 2), (1, 4)$
89. $(-1, 5), (-4, 8), (-6, 5), (-3, 2)$ **91.** 65
93. Exactly 5 performers have been elected in each of the last 6 years; 5
95. $5\sqrt{74} \approx 43$ yd
97. True. The lengths of the sides from $(-8, 4)$ to $(2, 11)$ and from $(2, 11)$ to $(-5, 1)$ are both $\sqrt{149}$.
99. False. You would have to use the Midpoint Formula 15 times.
101. No. The scales depend on the magnitudes of the quantities measured.
103. $\left(\dfrac{3x_1 + x_2}{4}, \dfrac{3y_1 + y_2}{4}\right), \left(\dfrac{x_1 + x_2}{2}, \dfrac{y_1 + y_2}{2}\right),$
    $\left(\dfrac{x_1 + 3x_2}{4}, \dfrac{y_1 + 3y_2}{4}\right)$
    (a) $\left(\dfrac{7}{4}, -\dfrac{7}{4}\right), \left(\dfrac{5}{2}, -\dfrac{3}{2}\right), \left(\dfrac{13}{4}, -\dfrac{5}{4}\right)$
    (b) $\left(-\dfrac{3}{2}, -\dfrac{9}{4}\right), \left(-1, -\dfrac{3}{2}\right), \left(-\dfrac{1}{2}, -\dfrac{3}{4}\right)$
105. Proof

Section P.6   (page 62)

1. Line plots   **3.** frequency distribution
5. c   **6.** d   **7.** a   **8.** b   **9.** (a) 2.979   (b) 0.19
11.

```
                        ×
            ×           × × ×
            × × ×       × × × ×
            × × ×       × × × ×
  ×   ×     × × ×       × × × × ×     ×
  +---+---+---+---+---+---+---+---+---+
  10  12  14  16  18  20  22  24
            Quiz Scores
```

15

13. Sample answer:

Interval	Tally
$[7, 10)$	卌 IIII
$[10, 13)$	卌 卌 卌 卌
$[13, 16)$	卌 卌 II
$[16, 19)$	卌 III
$[19, 22)$	I

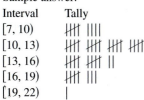

15.

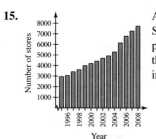

Answers will vary.
Sample answer: As time progresses from 1995 to 2008, the number of Wal-Mart stores increases at a fairly constant rate.

17.

Year	2003	2004	2005
Difference in tuition charges (in dollars)	13,480	13,996	14,525

Year	2006	2007	2008
Difference in tuition charges (in dollars)	14,988	15,946	17,171

19.

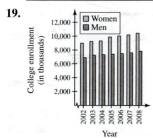

21. The price decreased slightly from 2000 to 2002.
23. About 40%   **25.** $1.85; January   **27.** From April to May
29.

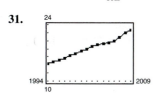

Answers will vary.
Sample answer: As time progresses from 1998 to 2009, the number of women in the workforce increases at a fairly constant rate.

31.

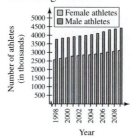

33. Answers will vary. Sample answer: A double bar graph is best because there are two different sets of data within the same time interval that do not deal primarily with increasing or decreasing behavior.

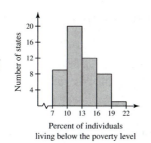

35. Answers will vary.
37. Line plots are useful for ordering small sets.
Histograms or bar graphs can be used to organize larger sets.
Line graphs are used to show trends over periods of time.

Review Exercises   (page 68)

1. (a) 11   (b) 11   (c) 11, -14
(d) 11, -14, $-\frac{8}{9}$, $\frac{5}{2}$, 0.4   (e) $\sqrt{6}$
3. (a) $0.8\overline{3}$ (b) 0.875

$$\frac{17}{24} \quad \frac{3}{4} \quad \frac{19}{24} \quad \frac{5}{6} \quad \frac{7}{8} \quad \frac{11}{12} \quad \frac{23}{24}$$

$\frac{5}{6} < \frac{7}{8}$

5. (a) The set consists of all real numbers x less than or equal to 7.
(b) (number line: 5 6 7 8 9)
(c) Unbounded
7. $[-2, -1)$ **9.** 122   **11.** $|x - 7| \geq 6$
13. (a) -11 (b) 25   **15.** (a) -18 (b) -12
17. Commutative Property of Addition
19. (a) $-8z^3$ (b) $3a^3b^2$ **21.** (a) $\dfrac{3u^5}{v^4}$ (b) m^{-2}
23. 2.585×10^9 **25.** 1.25×10^{-7} **27.** 128,000
29. 0.000018   **31.** 78   **33.** $5|a|$ **35.** $\dfrac{3}{4}$ **37.** $\dfrac{x}{3}\sqrt[3]{2}$
39. $\sqrt{3}$ **41.** $3\sqrt{3x}$ **43.** $\sqrt{2x}(2x + 1)$
45. $\dfrac{3 + \sqrt{5}}{4}$ **47.** $\dfrac{5}{2\sqrt{5}}$ **49.** 32,768   **51.** $6x^{9/10}$
53. $-2x^5 - x^4 + 3x^3 + 15x^2 + 5$; degree: 5;
leading coefficient: -2
55. $3x^2 + 7x - 1$ **57.** $2x^3 - x^2 + 3x - 9$
59. $-2a^3 - 2a^2 + 6a$ **61.** $x^2 + 13x + 36$ **63.** $x^2 - 64$
65. $x^3 - 12x^2 + 48x - 64$ **67.** $m^2 - 14m - n^2 + 49$
69. $(x + 3)(x + 5) = 5(x + 3) + x(x + 3)$
Distributive Property
71. $7(x + 5)$ **73.** $x(x^2 - 1) = x(x - 1)(x + 1)$
75. $2x(x^2 + 9x - 2)$ **77.** $(x - 13)(x + 13)$
79. $(x + 3)^2$ **81.** $(x + 6)(x^2 - 6x + 36)$
83. $(x - 9)(x + 3)$ **85.** $(2x + 1)(x + 10)$
87. $(x - 4)(x^2 - 3)$ **89.** $(x - 3)(2x + 5)$

CHAPTER P

91. All real numbers x

93. All real numbers x except $x = \frac{3}{2}$

95. $\dfrac{x}{x^2 + 7},\ x \neq 0$ **97.** $\dfrac{x - 6}{x - 5},\ x \neq -5$ **99.** $\dfrac{1}{x^2},\ x \neq \pm 2$

101. $\frac{1}{5}x(5x - 6),\ x \neq 0,\ -\frac{3}{2}$

103. $\dfrac{x^3 - x + 3}{(x - 1)(x + 2)}$ **105.** $\dfrac{x + 1}{x(x^2 + 1)}$

107. $-\dfrac{1}{xy(x + y)},\ x \neq y$

109.

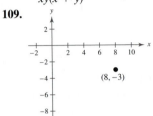

Quadrant IV

111.

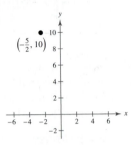

Quadrant II

113.

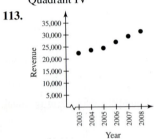

115. (a)

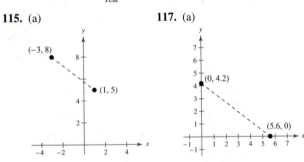

(b) Distance: 5

(c) Midpoint: $(-1, 6.5)$

117. (a)

(b) Distance: 7

(c) Midpoint: $(2.8, 2.1)$

119. $(x - 3)^2 + (y + 1)^2 = 68$

121. $(2, 5),\ (4, 5),\ (2, 0),\ (4, 0)$

123.

The price with the greatest frequency is \$100.

125.

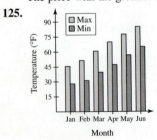

Chapter Test (page 71)

1. $-\frac{10}{3} > -|-4|$ **2.** 54 **3.** Additive Identity Property

4. (a) -18 (b) $\frac{4}{27}$ (c) $-\frac{8}{343}$ (d) $\frac{8}{729}$

5. (a) 25 (b) 6 (c) 1.8×10^5 (d) 2.7×10^{13}

6. (a) $12z^8$ (b) $\dfrac{1}{(u - 2)^7}$ (c) $\dfrac{3x^2}{y^2}$

7. (a) $15z\sqrt{2z}$ (b) $-10\sqrt{y}$ (c) $\dfrac{2}{v}\sqrt[3]{\dfrac{2}{v^2}}$

8. $-2x^5 - x^4 + 3x^3 + 3$

Degree: 5; leading coefficient: -2

9. $2x^2 - 3x - 5$ **10.** $8x^3 - 20x^2 + 12x - 30$

11. $8,\ x \neq 3$ **12.** $\dfrac{x - 1}{2x},\ x \neq \pm 1$ **13.** $x^2 - 5$

14. $x^3 - 6x^2 + 12x - 8$ **15.** $x^2 + 2xy + y^2 - z^2$

16. $x^2(2x + 1)(x - 2)$ **17.** $(x - 2)(x + 2)^2$

18. $8(x - 2)(x^2 + 2x + 4)$

19. (a) $\dfrac{\sqrt[3]{16^5}}{16} = 4\sqrt[3]{4}$ (b) $-3 - 3\sqrt{3}$

(c) $\dfrac{\sqrt{x + 2} + \sqrt{2}}{x}$

20. $\frac{5}{6}\sqrt{3x^2}$

21.

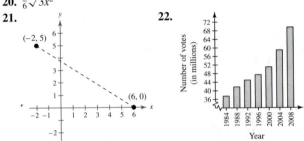

Midpoint: $\left(2, \frac{5}{2}\right)$

Distance: $\sqrt{89}$

22.

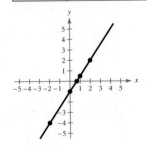

Chapter 1

Section 1.1 (page 81)

1. solution point **3.** Algebraic, graphical, numerical

5. (a) Yes (b) Yes **7.** (a) No (b) Yes

9. (a) No (b) Yes

11.

x	-2	0	$\frac{2}{3}$	1	2
y	-4	-1	0	$\frac{1}{2}$	2
Solution point	$(-2, -4)$	$(0, -1)$	$\left(\frac{2}{3}, 0\right)$	$\left(1, \frac{1}{2}\right)$	$(2, 2)$

13.

x	-1	0	1	2	3
y	3	0	-1	0	3
Solution point	$(-1, 3)$	$(0, 0)$	$(1, -1)$	$(2, 0)$	$(3, 3)$

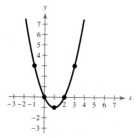

15. b   **16.** d   **17.** c   **18.** a

19.    **21.**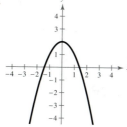

23. 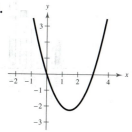   **25.**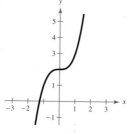

27. 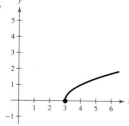   **29.**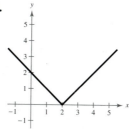

31. 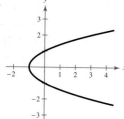   **33.**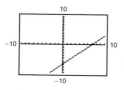

Intercepts: $(7, 0), (0, -7)$

35. 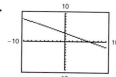   **37.**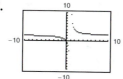

Intercepts: $(6, 0), (0, 3)$ Intercept: $(0, 0)$

39. 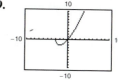   **41.**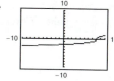

Intercepts: $(-3, 0), (0, 0)$ Intercepts: $(0, -2), (8, 0)$

43. 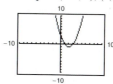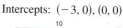   **45.**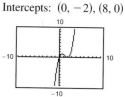

Intercepts:
$(0, 3), (1, 0), (3, 0)$ Intercepts: $(0, 0), (2, 0)$

47.

Xmin = -10
Xmax = 10
Xscl = 2
Ymin = -50
Ymax = 100
Yscl = 25

49. The graphs are identical. Distributive Property
51. The graphs are identical. Associative Property of Multiplication

53. 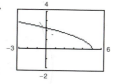   **55.**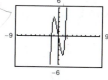

(a) $(3, 1.41)$ (a) $(-0.5, 2.47)$
(b) $(-4, 3)$ (b) $(1, -4), (-1.65, -4)$

57. $y_1 = \sqrt{16 - x^2}$ **59.** $y_1 = 2 + \sqrt{9 - (x - 1)^2}$
$y_2 = -\sqrt{16 - x^2}$ $y_2 = 2 - \sqrt{9 - (x - 1)^2}$

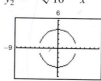   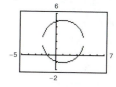

61. b, c, d
63. (a) 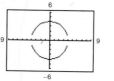   (b) $109,000
(c) $178,000

65. (a)

Year	2000	2001	2002	2003	2004
Median sales price (in thousands of dollars)	150.9	151.7	159.3	171.3	185.0

Year	2005	2006	2007	2008
Median sales price (in thousands of dollars)	198.0	207.7	211.6	207.2

The model fits the data well.

CHAPTER 1

(b) 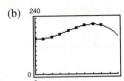   The model fits the data well.

(c) 2012: \$55,231; 2014: −\$136,682; No. There is a large decrease in the price over a short period of time, leading to negative values.

(d) 2000, 2010

67. False. $y = x^2 - 1$ has two x-intercepts.

69. Use the equations $y = 3400 + 0.05x$ and $y = 3000 + 0.07x$ to model the situation. When the sales level x equals \$20,000, both equations yield \$4400.

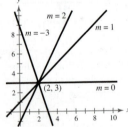

71. $2x^2 + 8x + 11$

Section 1.2   (page 92)

1. (a) iii   (b) i   (c) v   (d) ii   (e) iv   **3.** parallel

5. 0   **7.** (a) L_2 (b) L_3 (c) L_1 **9.** $\frac{3}{2}$

11.

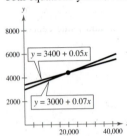

13. $m = -\frac{5}{2}$

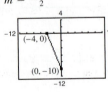

15. m is undefined.

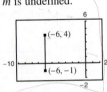

17. $(0, 1), (3, 1), (-1, 1)$ **19.** $(1, 4), (1, 6), (1, 9)$

21. $(-1, -7), (-2, -5), (-5, 1)$

23. $(3, -4), (5, -3), (9, -1)$

25. $y = 3x - 2$

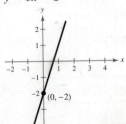

27. $y = -\frac{1}{2}x - 2$

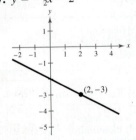

29. $x = 6$

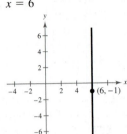

31. $y = \frac{3}{2}$

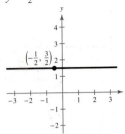

33. $y = 0.45x + 1.15$; \$8.8 million

35. $m = \frac{1}{2}$; y-intercept: $(0, -2)$; a line that rises from left to right

37. $m = \frac{2}{5}$; y-intercept: $(0, 2)$; a line that rises from left to right

39. Slope is undefined; no y-intercept; a vertical line at $x = -6$

41. $m = 0$; y-intercept: $\left(0, -\frac{2}{3}\right)$; a horizontal line at $y = -\frac{2}{3}$

43. (a) $m = 5$;
y-intercept: $(0, 3)$

(b)

45. (a) Slope is undefined;
there is no y-intercept.

(b)

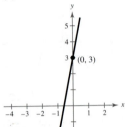

47. (a) $m = 0$;
y-intercept: $\left(0, -\frac{5}{3}\right)$

(b)

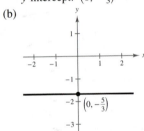

49. $y = 2x - 5$

51. $y = -\frac{3}{5}x + 2$

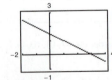

53. $x + 8 = 0$

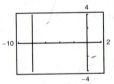

55. $y = -\frac{1}{2}x + \frac{3}{2}$

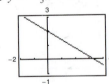

57. $y = -\frac{6}{5}x - \frac{18}{25}$

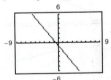

59. $y = \frac{2}{5}x + \frac{1}{5}$

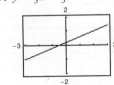

61.

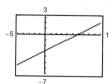

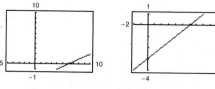

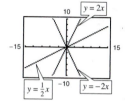

The first graph does not show both intercepts. The third graph is best because it shows both intercepts and gives the most accurate view of the slope by using a square setting.

63. Perpendicular   **65.** Parallel

67. (a) $y = 2x - 3$ (b) $y = -\frac{1}{2}x + 2$

69. (a) $y = -\frac{3}{4}x + \frac{3}{8}$ (b) $y = \frac{4}{3}x + \frac{127}{72}$

71. (a) $y = -3x - 13.1$ (b) $y = \frac{1}{3}x - \frac{1}{10}$

73. (a) $x = 3$ (b) $y = -2$ **75.** (a) $y = 1$ (b) $x = -4$

77. $y = 2x + 1$ **79.** $y = -\frac{1}{2}x + 1$

81. The lines $y = \frac{1}{2}x$ and $y = -2x$ are perpendicular.

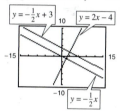

83. The lines $y = -\frac{1}{2}x$ and $y = -\frac{1}{2}x + 3$ are parallel. Both are perpendicular to $y = 2x - 4$.

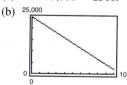

85. 12 ft

87. (a) The greatest increase was from 2001 to 2002, and the greatest decrease was from 2002 to 2003.

(b) $y = 58.625x + 995$

(c) There is an increase of about $58.625 million per year.

(d) $1581.25 million; Answers will vary.

89. $V = 125t + 1415$ **91.** $V = -2000t + 38,400$

93. (a) $V = 25,000 - 2300t$

(b)

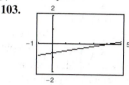

t	0	1	2	3	4
V	25,000	22,700	20,400	18,100	15,800

t	5	6	7	8	9	10
V	13,500	11,200	8900	6600	4300	2000

95. (a) $C = 27.75t + 36,500$ (b) $R = 65t$

(c) $P = 37.25t - 36,500$ (d) $t \approx 979.9$ h

97. (a) Increase of about 621 students per year

(b) 78,470; 81,575; 84,680

(c) $y = 621x + 75,365$, where $x = 0$ corresponds to 1990; $m = 621$; The slope determines the average increase in enrollment.

99. False. The slopes $\left(\frac{2}{7} \text{ and } -\frac{11}{7}\right)$ are not equal.

101.

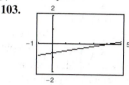

103.

a and b represent the x- and y-intercepts. a and b represent the x- and y-intercepts.

105. $3x + 2y - 6 = 0$ **107.** $12x + 3y + 2 = 0$

109. a   **111.** c

113. No. Answers will vary. Sample answer: The line $y = 2$ does not have an x-intercept.

115. Yes. Once a line parallel to the given line is established, there are an infinite number of distances away from that line, and thus an infinite number of parallel lines.

117. Yes; $x + 20$ **119.** No   **121.** No

123. $(x - 9)(x + 3)$ **125.** $(2x - 5)(x + 8)$

127. Answers will vary.

Section 1.3   (page 105)

1. domain, range, function   **3.** No   **5.** No

7. Yes. Each element of the domain is assigned to exactly one element of the range.

9. No. The National League, an element in the domain, is assigned to three items in the range, the Cubs, the Pirates, and the Dodgers; the American League, an element in the domain, is also assigned to three items in the range, the Orioles, the Yankees, and the Twins.

11. Yes. Each input value is matched with one output value.

13. (a) Function

(b) Not a function because the element 1 in A corresponds to two elements, -2 and 1, in B.

(c) Function

15. Yes; yes; Each input value (year) is matched with exactly one output value (average price) for both name brand and generic drug prescriptions.

17. Not a function   **19.** Function   **21.** Function

23. Not a function   **25.** Function   **27.** Not a function

29. (a) 7   (b) -11 (c) $3t + 7$

31. (a) 0   (b) -0.75 (c) $x^2 + 2x$

33. (a) 1   (b) 2.5   (c) $3 - 2|x|$

35. (a) Undefined   (b) $-\frac{1}{5}$ (c) $\frac{1}{y^2 + 6y}$

37. (a) 1   (b) -1 (c) $\frac{|t|}{t}$

39. (a) -1 (b) 2   (c) 6   **41.** (a) 6   (b) 3   (c) 10

43. (a) 0   (b) 4   (c) 17

45. $\{(-2, 4), (-1, 1), (0, 0), (1, 1), (2, 4)\}$

47. $\{(-2, 4), (-1, 3), (0, 2), (1, 3), (2, 4)\}$

49.

t	-5	-4	-3	-2	-1
$h(t)$	1	$\frac{1}{2}$	0	$\frac{1}{2}$	1

51. 5   **53.** $\frac{4}{3}$ **55.** All real numbers x

57. All real numbers t except $t = 0$ **59.** All real numbers x

61. All real numbers x except $x = 0, -2$

63. All real numbers y such that $y > 10$

65.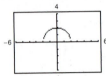

Domain: $[-2, 2]$
Range: $[0, 2]$

67.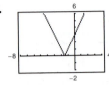

Domain: $(-\infty, \infty)$
Range: $[0, \infty)$

69. $A = \dfrac{C^2}{4\pi}$

71. (a) 1024 cm^3

(b) 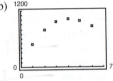   Yes, it is a function.

(c) $V = x(24 - 2x)^2, \ 0 < x < 12$

(d)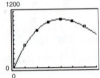

The function fits the data points; Answers will vary.

73. $A = 2xy = 2x\sqrt{36 - x^2}, \ 0 < x < 6$

75. (a) $C = 68.20x + 248,000$ (b) $R = 98.98x$

(c) $P = 30.78x - 248,000$

77. (a) The independent variable is t and represents the year. The dependent variable is n and represents the numbers of miles traveled.

(b)

t	0	1	2	3	4	5	6
$n(t)$	581	645	699	742	775	798	809

t	7	8	9	10	11	12	13
$n(t)$	844	870	895	921	947	972	998

t	14	15	16	17
$n(t)$	1024	1050	1075	1101

(c) The model fits the data well.

79. No; The ball is 9.6 feet high when it reaches the first baseman.

81. $2, c \neq 0$ **83.** $3 + h, h \neq 0$

85. $-\dfrac{1}{t}, t \neq 1$ **87.** False. The range is $[-1, \infty)$.

89. $f(x) = \sqrt{x} + 2$
Domain: $x \geq 0$
Range: $[2, \infty)$

91. No, f is not the independent variable because the value of f depends on the value of x.

93. $\dfrac{12x + 20}{x + 2}$ **95.** $\dfrac{(x + 6)(x + 10)}{5(x + 3)}, x \neq 0, \dfrac{1}{2}$

Section 1.4   (page 118)

1. decreasing   **3.** $[1, 4]$ **5.** Relative maximum

7. Domain: $(-\infty, \infty)$ **9.** Domain: $[-4, 4]$
Range: $(-\infty, 1]$ Range: $[0, 4]$
$f(0) = 1$ $f(0) = 4$

11.

Domain: $(-\infty, \infty)$
Range: $[3, \infty)$

13.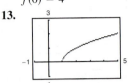

Domain: $[1, \infty)$
Range: $[0, \infty)$

15.

Domain: $(-\infty, \infty)$
Range: $[0, \infty)$

17. (a) $(-\infty, \infty)$ (b) $[-2, \infty)$ (c) $-1, 3$
(d) x-intercepts   (e) -1 (f) y-intercept
(g) $-2; (1, -2)$ (h) $0; (-1, 0)$ (i) 2

19. Function. Graph the given function over the window shown in the figure.

21. Not a function. Solve for y and graph the two resulting functions.

23. Increasing on $(-\infty, \infty)$

25. Increasing on $(-\infty, 0), (2, \infty)$
Decreasing on $(0, 2)$

27. (a)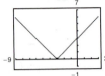

(b) Constant on $(-\infty, \infty)$

29. (a)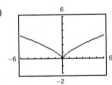

(b) Decreasing on $(-\infty, 0)$
Increasing on $(0, \infty)$

31. (a)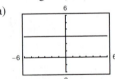

(b) Increasing on $(-2, \infty)$
Decreasing on $(-3, -2)$

33. (a)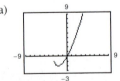

(b) Decreasing on $(-\infty, -1)$
Constant on $(-1, 1)$
Increasing on $(1, \infty)$

35.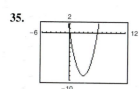

Relative minimum:
$(3, -9)$

37.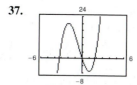

Relative minimum: $(1, -7)$
Relative maximum:
$(-2, 20)$

39.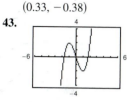

Relative minimum:
$(0.33, -0.38)$

41.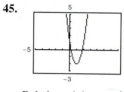

Relative minimum:
$(2, -9)$

43.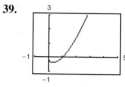

Relative minimum: $(1, -2)$
Relative maximum: $(-1, 2)$

45.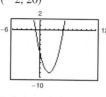

Relative minimum: $(1, -2)$

47.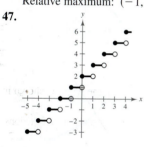

49.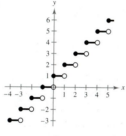

51.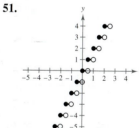

53.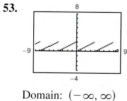

Domain: $(-\infty, \infty)$
Range: $[0, 2)$
Sawtooth pattern

55.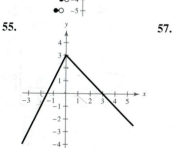

57.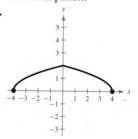

59.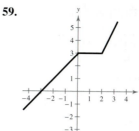

61.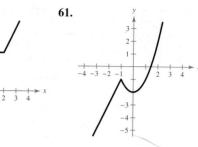

63. Even
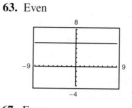

65. Neither
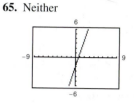

67. Even
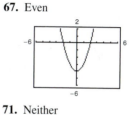

69. Neither
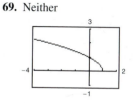

71. Neither
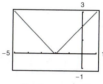

73. (a) $\left(\frac{3}{2}, 4\right)$ (b) $\left(\frac{3}{2}, -4\right)$
75. (a) $(-4, 9)$ (b) $(-4, -9)$
77. (a) $(-x, -y)$ (b) $(-x, y)$
79. (a)–(c) Neither   **81.** (a)–(c) Odd
83. (a)–(c) Odd   **84.** (a)–(c) Neither
85. (a)–(c) Even   **86.** (a)–(c) Neither
87.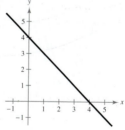

89.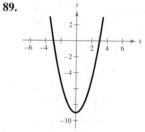

$(-\infty, 4]$ $(-\infty, -3], [3, \infty)$

91. (a) C_2 is the appropriate model. The cost of the first minute is $1.05 and the cost increases $0.08 when the next minute begins, and so on.

(b) 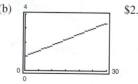 $2.49

93. $h = -x^2 + 4x - 3, \; 1 \le x \le 3$

95. (a) 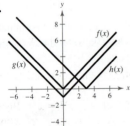   (b) Increasing from 2000 to 2005; decreasing from 2005 to 2008
(c) About 850

97. False. Counterexample: $f(x) = \sqrt{1 + x^2}$

99. c   **100.** d   **101.** b   **102.** e   **103.** a   **104.** f

105. Yes. To check whether x is a function of y, determine whether any horizontal line can be drawn to intersect the graph more than once.

107. Yes

109. (a) Even. g is a reflection in the x-axis.
(b) Even. g is a reflection in the y-axis.
(c) Even. g is a vertical shift downward.
(d) Neither. g is shifted to the right and reflected in the x-axis.

111. Proof   **113.** Terms: $-2x^2, 8x$; coefficients: $-2, 8$

115. Terms: $\dfrac{x}{3}, -5x^2, x^3$; coefficients: $\dfrac{1}{3}, -5, 1$

117. (a) -17 (b) 1 (c) $-x^2 + 3x + 1$

119. $h + 4, h \neq 0$

Section 1.5   (page 128)

1. Horizontal shifts, vertical shifts, reflections

3. $-f(x), f(-x)$

5. 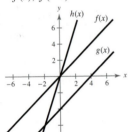   **7.**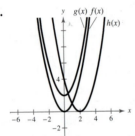

9. 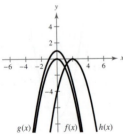   **11.**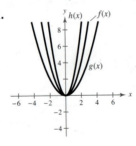

13. 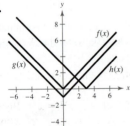   **15.**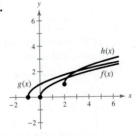

17.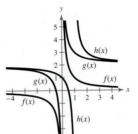

19. (a) 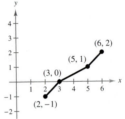   (b)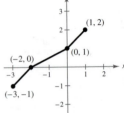

(c)   (d)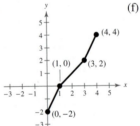

(e)   (f)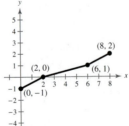

(g)

21. The graph of $f(x) = x^2$ should have been shifted one unit to the left instead of one unit to the right.

23. Vertical shift two units upward

25. Horizontal shift four units to the right

27. Vertical shift two units downward

29. Vertical shift of $y = x$; $y = x + 3$

31. Vertical shift of $y = x^2$; $y = x^2 - 1$

33. Reflection in the x-axis and a vertical shift one unit upward of $y = \sqrt{x}$; $y = 1 - \sqrt{x}$

35. Reflection in the x-axis

37. Reflection in the y-axis (identical)

39. Reflection in the x-axis   **41.** Vertical stretch

43. Vertical shrink   **45.** Horizontal shrink

47.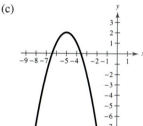

g is a horizontal shift and
h is a vertical shrink.

49.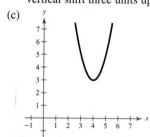

g is a vertical shrink and a
reflection in the x-axis and
h is a reflection in the y-axis.

51. (a) $f(x) = x^2$
 (b) Horizontal shift five units to the left, reflection in the x-axis, and vertical shift two units upward
 (c) 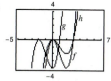   (d) $g(x) = 2 - f(x + 5)$

53. (a) $f(x) = x^2$
 (b) Horizontal shift four units to the right, vertical stretch, and vertical shift three units upward
 (c) 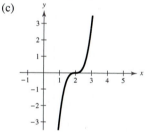   (d) $g(x) = 3 + 2f(x - 4)$

55. (a) $f(x) = x^3$
 (b) Horizontal shift two units to the right and vertical stretch
 (c) 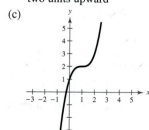   (d) $g(x) = 3f(x - 2)$

57. (a) $f(x) = x^3$
 (b) Horizontal shift one unit to the right and vertical shift two units upward
 (c)   (d) $g(x) = f(x - 1) + 2$

59. (a) $f(x) = \dfrac{1}{x}$
 (b) Horizontal shift eight units to the left and vertical shift nine units downward
 (c) 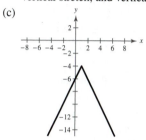   (d) $g(x) = f(x + 8) - 9$

61. (a) $f(x) = |x|$
 (b) Horizontal shift one unit to the right, reflection in the x-axis, vertical stretch, and vertical shift four units downward
 (c)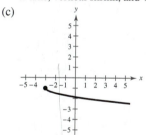

 (d) $g(x) = -2f(x - 1) - 4$
63. (a) $f(x) = \sqrt{x}$
 (b) Horizontal shift three units to the left, reflection in the x-axis, vertical shrink, and vertical shift one unit downward
 (c)

 (d) $g(x) = -\frac{1}{2}f(x + 3) - 1$
65. (a) Horizontal shift 24.7 units to the right, vertical shift 183.4 units upward, reflection in the x-axis, and a vertical shrink
 (b)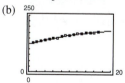

 (c) $G(t) = F(t + 10) = -0.099[(t + 10) - 24.7]^2 + 183.4$
 $= -0.099(t - 14.7)^2 + 183.4$

 To make a horizontal shift 10 years backward (10 units left), add 10 to t.
67. False. When $f(x) = x^2$, $f(-x) = (-x)^2 = x^2$. Because $f(x) = f(-x)$ in this case, $y = f(-x)$ is not a reflection of $y = f(x)$ across the x-axis in all cases.
69. $x = -2$ and $x = 3$
71. Cannot be determined because it is a vertical shift
73. c   **75.** c   **77.** Answers will vary.
79. (a) The graph of g is a vertical shrink of the graph of f.
 (b) The graph of g is a vertical stretch of the graph of f.

CHAPTER 1

81. Neither   **83.** All real numbers x except $x = 9$
85. All real numbers x such that $-10 \le x \le 10$

Section 1.6   (page 137)

1. addition, subtraction, multiplication, division
3. $g(x)$ **5.** $2x$
7.

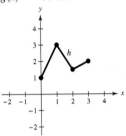

9.

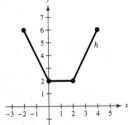

11. (a) $2x$
  (b) 6
  (c) $x^2 - 9$
  (d) $\dfrac{x + 3}{x - 3}$
  All real numbers x,
  except $x = 3$

13. (a) $x^2 - x + 1$
  (b) $x^2 + x - 1$
  (c) $x^2 - x^3$
  (d) $\dfrac{x^2}{1 - x}$
  All real numbers x,
  except $x = 1$

15. (a) $x^2 + 5 + \sqrt{1 - x}$ (b) $x^2 + 5 - \sqrt{1 - x}$
  (c) $(x^2 + 5)\sqrt{1 - x}$ (d) $\dfrac{x^2 + 5}{\sqrt{1 - x}}, \ x < 1$

17. (a) $\dfrac{x + 1}{x^2}$ (b) $\dfrac{x - 1}{x^2}$ (c) $\dfrac{1}{x^3}$ (d) $x, \ x \ne 0$

19. 9   **21.** 1   **23.** 140   **25.** $-\dfrac{24}{7}$ **27.** $4t^2 - 2t + 1$

29. $-125t^3 - 50t^2 + 5t + 2$ **31.** $\dfrac{t^2 - 1}{-t - 2}$

33.

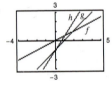

35.

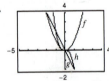

37.

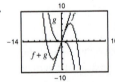

$f(x), 0 \le x \le 2;$
$g(x), x > 6$

39.

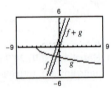

$f(x), 0 \le x \le 2;$
$f(x), x > 6$

41. (a) $(x - 1)^2$ (b) $x^2 - 1$ (c) 1
43. (a) $20 - 3x$ (b) $-3x$ (c) 20
45. (a) All real numbers x such that $x \ge -4$
  (b) All real numbers x (c) All real numbers x
47. (a) All real numbers x
  (b) All real numbers x such that $x \ge 0$
  (c) All real numbers x such that $x \ge 0$
49. (a) All real numbers x except $x = 0$ (b) All real numbers x
  (c) All real numbers x except $x = -3$
51. (a) All real numbers x (b) All real numbers x
  (c) All real numbers x
53. (a) All real numbers x
  (b) All real numbers x except $x = \pm 2$
  (c) All real numbers x except $x = \pm 2$

55. (a) $(f \circ g)(x) = \sqrt{x^2 + 4}$; $(g \circ f)(x) = x + 4, \ x \ge -4$;
  Domain of $f \circ g$: all real numbers x
  (b)

  $f \circ g \ne g \circ f$

57. (a) $(f \circ g)(x) = x$; $(g \circ f)(x) = x$;
  Domain of $f \circ g$: all real numbers x
  (b)

  $f \circ g = g \circ f$

59. (a) $(f \circ g)(x) = x^4$; $(g \circ f)(x) = x^4$;
  Domain of $f \circ g$: all real numbers x
  (b)

  $f \circ g = g \circ f$

61. (a) $(f \circ g)(x) = 24 - 5x$; $(g \circ f)(x) = -5x$
  (b) $24 - 5x \ne -5x$
  (c)

x	0	1	2	3
$g(x)$	4	3	2	1
$(f \circ g)(x)$	24	19	14	9

x	0	1	2	3
$f(x)$	4	9	14	19
$(g \circ f)(x)$	0	-5	-10	-15

63. (a) $(f \circ g)(x) = \sqrt{x^2 + 1}$; $(g \circ f)(x) = x + 1, \ x \ge -6$
  (b) $x + 1 \ne \sqrt{x^2 + 1}$
  (c)

x	0	1	2	3
$g(x)$	-5	-4	-1	4
$(f \circ g)(x)$	1	$\sqrt{2}$	$\sqrt{5}$	$\sqrt{10}$

x	0	1	2	3
$f(x)$	$\sqrt{6}$	$\sqrt{7}$	$\sqrt{8}$	3
$(g \circ f)(x)$	1	2	3	4

65. (a) $(f \circ g)(x) = |2x^3|$; $(g \circ f)(x) = 2|x|^3$ (b) $|2x^3| = 2|x|^3$
  (c)

x	-2	-1	0	1	2
$g(x)$	-16	-2	0	2	16
$(f \circ g)(x)$	16	2	0	2	16

x	-2	-1	0	1	2
$f(x)$	2	1	0	1	2
$(g \circ f)(x)$	16	2	0	2	16

67. (a) 3   (b) 0   **69.** (a) 0   (b) 4

71. $f(x) = x^2$, $g(x) = 2x + 1$

73. $f(x) = \sqrt[3]{x}$, $g(x) = x^2 - 4$

75. $f(x) = \dfrac{1}{x}$, $g(x) = x + 2$

77. $f(x) = x^2 + 2x$, $g(x) = x + 4$

79. (a) $T = \frac{3}{4}x + \frac{1}{15}x^2$

(b)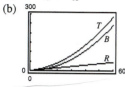

(c) B. For example, $B(60) = 240$, whereas $R(60)$ is only 45.

81. (a) $r(x) = \dfrac{x}{2}$ (b) $A(r) = \pi r^2$ (c) $(A \circ r)(x) = \pi\left(\dfrac{x}{2}\right)^2$

$(A \circ r)(x)$ represents the area of the circular base of the tank with radius $x/2$.

83. (a) $T = 1.2t^2 + 75.4t + 1220$

(b)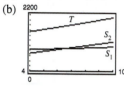

85. (a) $N(T(t))$ or $(N \circ T)(t) = 40t^2 + 590$; $N(T(t))$ or $(N \circ T)(t)$ represents the number of bacteria after t hours outside the refrigerator.

(b) $(N \circ T)(6) = 2030$; There are 2030 bacteria in a refrigerated food product after 6 hours outside the refrigerator.

(c) About 2.3 h

87. $s(t) = \sqrt{(150 - 450t)^2 + (200 - 450t)^2}$
$= 50\sqrt{162t^2 - 126t + 25}$

89. False. $g(x) = x - 3$

91. (a) $O(M(Y)) = 2\left(6 + \frac{1}{2}Y\right) = 12 + Y$; Answers will vary.

(b) Middle child is 8 years old, youngest child is 4 years old.

93. Proof   **95.** Proof   **97.** a, c

99. $(0, -5)$, $(1, -5)$, $(2, -7)$

101. $\left(0, 2\sqrt{6}\right)$, $\left(1, \sqrt{23}\right)$, $\left(2, 2\sqrt{5}\right)$ **103.** $y = 10x + 38$

105. $y = -\frac{30}{11}x + \frac{34}{11}$

Section 1.7   (page 148)

1. inverse, f^{-1} **3.** $y = x$ **5.** At most once

7. $f^{-1}(x) = \dfrac{x}{6}$ **9.** $f^{-1}(x) = x - 7$

11. $f^{-1}(x) = \frac{1}{2}(x - 1)$ **13.** $f^{-1}(x) = x^3$

15. c   **16.** b   **17.** a   **18.** d

19. $f(g(x)) = f\left(\sqrt[3]{x}\right) = \left(\sqrt[3]{x}\right)^3 = x$
$g(f(x)) = g(x^3) = \sqrt[3]{x^3} = x$

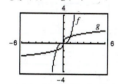

Reflections in the line $y = x$

21. $f(g(x)) = f(x^2 + 4)$, $x \geq 0$
$= \sqrt{(x^2 + 4) - 4} = x$
$g(f(x)) = g\left(\sqrt{x - 4}\right)$
$= \left(\sqrt{x - 4}\right)^2 + 4 = x$

Reflections in the line $y = x$

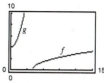

23. $f(g(x)) = f\left(\sqrt[3]{1 - x}\right) = 1 - \left(\sqrt[3]{1 - x}\right)^3 = x$
$g(f(x)) = g(1 - x^3) = \sqrt[3]{1 - (1 - x^3)} = x$

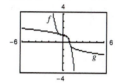

Reflections in the line $y = x$

25. (a) $f(g(x)) = f\left(-\dfrac{2x + 6}{7}\right)$
$= -\dfrac{7}{2}\left(-\dfrac{2x + 6}{7}\right) - 3 = x$
$g(f(x)) = g\left(-\dfrac{7}{2}x - 3\right)$
$= -\dfrac{2\left(-\frac{7}{2}x - 3\right) + 6}{7} = x$

(b)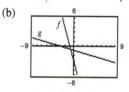

(c)

x	0	2	-2	6
$f(x)$	-3	-10	4	-24

x	-3	-10	4	-24
$g(x)$	0	2	-2	6

27. (a) $f(g(x)) = f\left(\sqrt[3]{x - 5}\right) = \left(\sqrt[3]{x - 5}\right)^3 + 5 = x$
$g(f(x)) = g(x^3 + 5) = \sqrt[3]{(x^3 + 5) - 5} = x$

(b)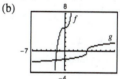

(c)

x	0	1	-1	-2	4
$f(x)$	5	6	4	-3	69

x	5	6	4	-3	69
$g(x)$	0	1	-1	-2	4

CHAPTER 1

29. (a) $f(g(x)) = f(8 + x^2)$
$$= -\sqrt{(8 + x^2) - 8}$$
$$= -\sqrt{x^2} = -(-x) = x, \ x \le 0$$
$$g(f(x)) = g\left(-\sqrt{x - 8}\right)$$
$$= 8 + \left(-\sqrt{x - 8}\right)^2$$
$$= 8 + (x - 8) = x, \ x \ge 8$$

(b)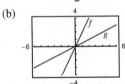

(c)

x	8	9	12	15
$f(x)$	0	-1	-2	$-\sqrt{7}$

x	0	-1	-2	$-\sqrt{7}$
$g(x)$	8	9	12	15

31. (a) $f(g(x)) = f\left(\dfrac{x}{2}\right)$
$$= 2\left(\dfrac{x}{2}\right) = x$$
$$g(f(x)) = g(2x)$$
$$= \dfrac{2x}{2} = x$$

(b)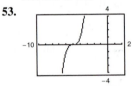

(c)

x	-4	-2	0	2	4
$f(x)$	-8	-4	0	4	8

x	-8	-4	0	4	8
$g(x)$	-4	-2	0	2	4

33. (a) $f(g(x)) = f\left(-\dfrac{5x + 1}{x - 1}\right)$
$$= \dfrac{\left(-\dfrac{5x + 1}{x - 1}\right) - 1}{\left(-\dfrac{5x + 1}{x - 1}\right) + 5} = \dfrac{\dfrac{6x}{x - 1}}{\dfrac{6}{x - 1}} = x, \ x \ne 1$$

$$g(f(x)) = g\left(\dfrac{x - 1}{x + 5}\right)$$
$$= -\dfrac{5\left(\dfrac{x - 1}{x + 5}\right) + 1}{\left(\dfrac{x - 1}{x + 5}\right) - 1} = \dfrac{\dfrac{6x}{x + 5}}{\dfrac{6}{x + 5}} = x, \ x \ne -5$$

(b)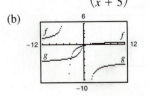

(c)

x	-3	-2	-1	0	2	3	4
$f(x)$	-2	-1	$-\frac{1}{2}$	$-\frac{1}{5}$	$\frac{1}{7}$	$\frac{1}{4}$	$\frac{1}{3}$

x	-2	-1	$-\frac{1}{2}$	$-\frac{1}{5}$	$\frac{1}{7}$	$\frac{1}{4}$	$\frac{1}{3}$
$g(x)$	-3	-2	-1	0	2	3	4

35. Yes. No two elements in the domain of f correspond to the same element in the range of f.

37. No. -3 and 0 both correspond to 6, so f is not one-to-one.

39. Not a function   **41.** Function; one-to-one

43. Function; one-to-one

45. 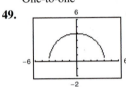   **47.**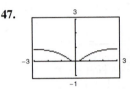

One-to-one   Not one-to-one

49. 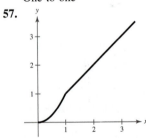   **51.**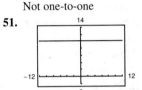

Not one-to-one   Not one-to-one

53. 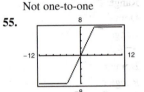   **55.**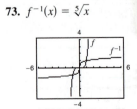

One-to-one   Not one-to-one

57.

An inverse function exists.

59. Not one-to-one   **61.** $f^{-1}(x) = \dfrac{5x - 4}{3}$

63. Not one-to-one   **65.** $f^{-1}(x) = \sqrt{x} - 3$

67. $f^{-1}(x) = \dfrac{x^2 - 3}{2}, \ x \ge 0$ **69.** $f^{-1}(x) = 2 - x, \ x \ge 0$

71. $f^{-1}(x) = \dfrac{x + 3}{2}$ **73.** $f^{-1}(x) = \sqrt[5]{x}$

Reflections in the line   Reflections in the line
$y = x$ $y = x$

75. $f^{-1}(x) = x^{5/3}$

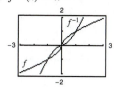

Reflections in the line $y = x$

77. $f^{-1}(x) = \sqrt{4 - x^2}$,
   $0 \le x \le 2$

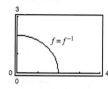

The graphs are the same.

79. $f^{-1}(x) = \dfrac{4}{x}$

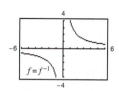

The graphs are the same.

81. $f^{-1}(x) = \sqrt{x} + 2$

Domain of f: all real numbers x such that $x \ge 2$
Range of f: all real numbers y such that $y \ge 0$
Domain of f^{-1}: all real numbers x such that $x \ge 0$
Range of f^{-1}: all real numbers y such that $y \ge 2$

83. $f^{-1}(x) = x - 2$

Domain of f: all real numbers x such that $x \ge -2$
Range of f: all real numbers y such that $y \ge 0$
Domain of f^{-1}: all real numbers x such that $x \ge 0$
Range of f^{-1}: all real numbers y such that $y \ge -2$

85. $f^{-1}(x) = \sqrt{x} - 3$

Domain of f: all real numbers x such that $x \ge -3$
Range of f: all real numbers y such that $y \ge 0$
Domain of f^{-1}: all real numbers x such that $x \ge 0$
Range of f^{-1}: all real numbers y such that $y \ge -3$

87. $f^{-1}(x) = \dfrac{\sqrt{-2(x - 5)}}{2}$

Domain of f: all real numbers x such that $x \ge 0$
Range of f: all real numbers y such that $y \le 5$
Domain of f^{-1}: all real numbers x such that $x \le 5$
Range of f^{-1}: all real numbers y such that $y \ge 0$

89. $f^{-1}(x) = x + 3$

Domain of f: all real numbers x such that $x \ge 4$
Range of f: all real numbers y such that $y \ge 1$
Domain of f^{-1}: all real numbers x such that $x \ge 1$
Range of f^{-1}: all real numbers y such that $y \ge 4$

91.

x	-4	-2	2	3
$f^{-1}(x)$	-2	-1	1	3

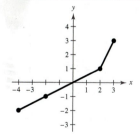

93. $\frac{1}{2}$ **95.** -2 **97.** 0 **99.** 2

101. (a) and (b)

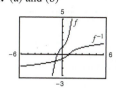

(c) Inverse function because it satisfies the Vertical Line Test.

103. (a) and (b)

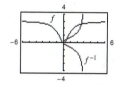

(c) Not an inverse function because it does not satisfy the Vertical Line Test.

105. 32 **107.** 600 **109.** $2\sqrt[3]{x + 3}$

111. $\dfrac{x + 1}{2}$ **113.** $\dfrac{x + 1}{2}$

115. (a) f is one-to-one because no two elements in the domain (men's U.S. shoe sizes) correspond to the same element in the range (men's European shoe sizes).

   (b) 45 (c) 10 (d) 41 (e) 13

117. (a) 19   9   64   64   4   69   29   4   64   9   104   29   94

   (b) $f^{-1}(x) = \dfrac{x - 4}{5}$; What time

119. False. For example, $y = x^2$ is even, but does not have an inverse.

121. This situation could be represented by a one-to-one function. The inverse function would represent the number of miles completed in terms of time in hours.

123. This function could not be represented by a one-to-one function because it oscillates.

125. The graph of f^{-1} is a reflection of the graph of f in the line $y = x$.

127. (a) The function will be one-to-one because no two values of x will produce the same value for $f(x)$.

   (b) $f^{-1}(50)$ represents the value of 50 degrees Fahrenheit in degrees Celsius.

129. Constant function   **131.** Proof   **133.** $9x$, $x \ne 0$

135. $-(x + 6)$, $x \ne 6$ **137.** Not a function

139. Not a function

Review Exercises   (page 154)

1.

x	-2	0	2	3	4
y	3	2	1	$\frac{1}{2}$	0
Solution point	$(-2, 3)$	$(0, 2)$	$(2, 1)$	$\left(3, \frac{1}{2}\right)$	$(4, 0)$

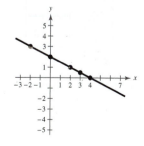

3.

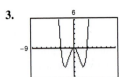

Intercepts:
$(0, 0), (\pm 2\sqrt{2}, 0)$

5.

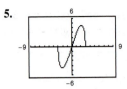

Intercepts: $(0, 0), (\pm 3, 0)$

7. (a)

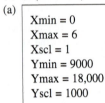

Xmin = 0
Xmax = 6
Xscl = 1
Ymin = 9000
Ymax = 18,000
Yscl = 1000

(b)

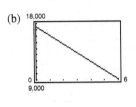

(c) 4

9.

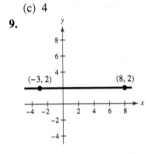

$m = 0$

11.

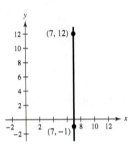

m is undefined.

13.

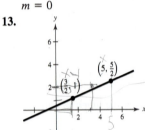

$m = \frac{3}{7}$

15.

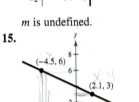

$m = -\frac{5}{11}$

17. $x - 4y - 6 = 0; (6, 0), (10, 1), (-2, -2)$

19. $3x - 2y - 10 = 0; (4, 1), (2, -2), (-2, -8)$

21. $y - 6 = 0; (0, 6), (1, 6), (-1, 6)$

23. $y = -1$

25. $y = \frac{11}{3}$

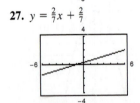

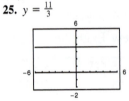

27. $y = \frac{2}{7}x + \frac{2}{7}$

29. $y = -\frac{1}{2}x + \frac{1}{2}$

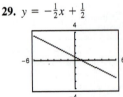

31. (a) $y = \frac{5}{4}x - \frac{23}{4}$ (b) $y = -\frac{4}{5}x + \frac{2}{5}$

33. (a) Not a function because element 20 in A corresponds to two elements, 4 and 6, in B.

(b) Function

35. Not a function   **37.** Function   **39.** Function

41. (a) 2   (b) 10   (c) $b^6 + 1$ (d) $x^2 - 2x + 2$

43. All real numbers x except $x = -2$

45. All real numbers x such that $-5 \le x \le 5$

47. (a) $C = 5.35x + 16,000$ (b) $P = 2.85x - 16,000$

49. $2h + 4x + 3, \; h \ne 0$

51.

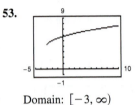

Domain: all real numbers x
Range: $(-\infty, 3]$

53.

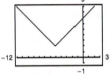

Domain: $[-3, \infty)$
Range: $[4, \infty)$

55.

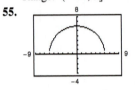

Domain: $[-6, 6]$
Range: $[0, 6]$

57.

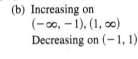

Domain: all real numbers x
Range: $[2, \infty)$

59. Function. Solve for y and graph the resulting function.

61. (a)

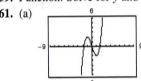

(b) Increasing on
$(-\infty, -1), (1, \infty)$
Decreasing on $(-1, 1)$

63. (a)

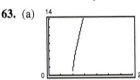

(b) Increasing on $(6, \infty)$

65. Relative minima: $(-2, 0), (2, 0)$
Relative maximum: $(0, 16)$

67.

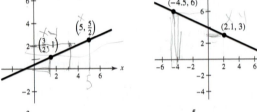

69.

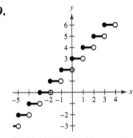

71. Even   **73.** Even   **75.** Neither   **77.** Even

79. Rational function $f(x) = \dfrac{1}{x}$

Horizontal shift three units right

$g(x) = \dfrac{1}{x - 3}$

81. Absolute value function $f(x) = |x|$
Vertical shift three units upward
$g(x) = |x| + 3$

83. 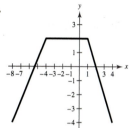   **85.**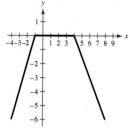

87. (a) Rational function $f(x) = \dfrac{1}{x}$

(b) Vertical shift six units downward

(c) 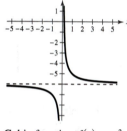   (d) $h(x) = f(x) - 6$

89. (a) Cubic function $f(x) = x^3$

(b) Horizontal shift two units right, vertical shift five units upward

(c) 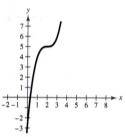   (d) $h(x) = f(x - 2) + 5$

91. (a) Square root function $f(x) = \sqrt{x}$

(b) Reflection in the x-axis, vertical shift six units upward

(c) 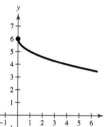   (d) $h(x) = -f(x) + 6$

93. (a) Absolute value function $f(x) = |x|$

(b) Vertical shift nine units upward

(c) 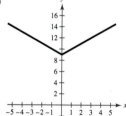   (d) $h(x) = f(x) + 9$

95. -7 **97.** -42 **99.** 5   **101.** 17

103. $f(x) = x^2$, $g(x) = x + 3$

105. $f(x) = \sqrt{x}$, $g(x) = 4x + 2$

107. $f(x) = \dfrac{4}{x}$, $g(x) = x + 2$

109.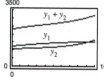

111. $f^{-1}(x) = \dfrac{x}{6}$ **113.** $f^{-1}(x) = 2x - 6$

115. Answers will vary.

117.    **119.**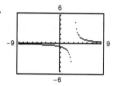

One-to-one   One-to-one

121. $f^{-1}(x) = 2x + 10$ **123.** $f^{-1}(x) = \sqrt[3]{\dfrac{x + 3}{4}}$

125. $f^{-1}(x) = x^2 - 10$, $x \geq 0$ **127.** $f^{-1}(x) = \sqrt{4x - 4}$

Chapter Test   (page 157)

1. 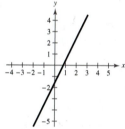   Intercepts:
$(0, -1)$, $\left(-\dfrac{1}{2}, 0\right)$, $\left(\dfrac{1}{2}, 0\right)$

2. 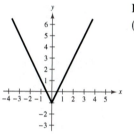   **3.**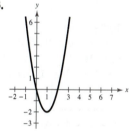

Intercepts: $\left(0, -\dfrac{8}{5}\right)$, $\left(\dfrac{4}{5}, 0\right)$ Intercepts: $(0, 0)$, $(2, 0)$

4. 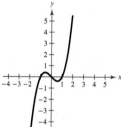   **5.**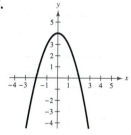

Intercepts: Intercepts:
$(-1, 0)$, $(0, 0)$, $(1, 0)$ $(-2, 0)$, $(0, 4)$, $(2, 0)$

CHAPTER 1

6.

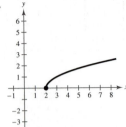

Intercept: $(2, 0)$

7. (a) $5x + 2y - 8 = 0$ (b) $2x - 5y + 20 = 0$

8. $y = -x + 1$

9. No. For some values of x there correspond more than one value of y.

10. (a) -9 (b) 1 (c) $|t - 4| - 15$ **11.** $(-\infty, 3]$

12. $C = 25.60x + 24,000$
$P = 73.9x - 24,000$

13. Odd   **14.** Even

15. Increasing: $(-2, 0), (2, \infty)$
Decreasing: $(-\infty, -2), (0, 2)$

16. Increasing: $(-2, 2)$
Constant: $(-\infty, -2), (2, \infty)$

17.

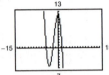

Relative minimum: $(-3.33, -6.52)$
Relative maximum: $(0, 12)$

18.

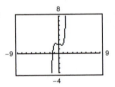

Relative minimum: $(0.77, 1.81)$
Relative maximum: $(-0.77, 2.19)$

19. (a) $f(x) = x^3$
(b) Horizontal shift five units to the right, reflection in the x-axis, vertical stretch, and vertical shift three units upward
(c)

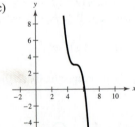

20. (a) $f(x) = \sqrt{x}$
(b) Reflection in the y-axis and horizontal shift seven units to the left
(c)

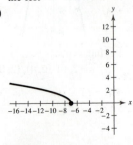

21. (a) $f(x) = |x|$
(b) Reflection in the y-axis (no effect), vertical stretch, and vertical shift seven units downward
(c)

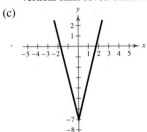

22. (a) $x^2 - \sqrt{2 - x}$, $(-\infty, 2]$ (b) $\dfrac{x^2}{\sqrt{2 - x}}$, $(-\infty, 2)$
(c) $2 - x$, $(-\infty, 2]$ (d) $\sqrt{2 - x^2}$, $\left[-\sqrt{2}, \sqrt{2}\right]$

23. $f^{-1}(x) = \sqrt[3]{x - 8}$ **24.** No inverse

25. $f^{-1}(x) = \left(\frac{8}{3}x\right)^{2/3}$, $x \geq 0$

Chapter 2

Section 2.1   (page 166)

1. equation   **3.** extraneous   **5.** Conditional equation

7. (a) Yes   (b) No   (c) No   (d) No

9. (a) No   (b) No   (c) No   (d) Yes

11. Identity   **13.** Contradiction   **15.** Conditional equation

17. $-\frac{96}{23}$ **19.** $\frac{20}{81}$ **21.** 12 **23.** -9 **25.** -10

27. $\frac{17}{48}$ **29.** 10 **31.** 4 **33.** 5 **35.** $\frac{5}{3}$

37. No solution   **39.** No solution   **41.** $h = \dfrac{2A}{b}$

43. $P = A\left(1 + \dfrac{r}{n}\right)^{-nt}$ **45.** $h = \dfrac{V}{\pi r^2}$ **47.** 61.2 in.

49. (a)

(b) $l = 1.5w$; $P = 5w$
(c) 7.5 m long $\times$ 5 m wide

51. (a) Test average $= \dfrac{\text{test 1} + \text{test 2} + \text{test 3} + \text{test 4}}{4}$
(b) 97

53. 2.5 h   **55.** 46.3 mi/h

57. 30 ft   **59.** 2.5%

61. 50 lb of each kind

63. $24,000 in notebook computers; $16,000 in desktop computers

65. $h = 27$ ft

67. (a)

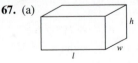

(b) 24 in. $\times$ 12 in. $\times$ 8 in.

69.

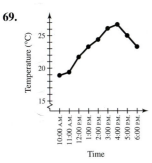

71. 3 h **73.** $x = 6$ ft
75. False. It is quadratic; $x(3 - x) = 10 \Rightarrow 3x - x^2 = 10$.
77. $9x + 27 = 0$ **79.** $2x + \frac{7}{2} = 4$
81. There is no solution because $x = 1$ is an extraneous solution.
83. $\frac{5}{8}$

85.

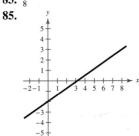

87.

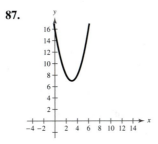

89.

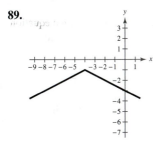

91. -28 **93.** -2580 **95.** -357

Section 2.2 (page 176)

1. x-intercept, y-intercept **3.** $(1, 0), (-1, 0)$ **5.** $-1, 1$
7. $(5, 0), (0, -5)$ **9.** $(0, 2)$ **11.** $(-2, 0), (0, 0)$
13. No intercepts **15.** $(1, 0), \left(0, \frac{1}{2}\right)$

17.

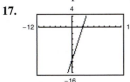

$(4, 0), (0, -12)$

19.

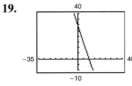

$(10, 0), (0, 30)$

21–25. Answers will vary. **27.** $\frac{12}{23}$; $f(x) = 2.3x - 1.2 = 0$
29. $\frac{85}{3}$; $f(x) = 3x - 85 = 0$ **31.** 6; $f(x) = 7x - 42 = 0$
33. -194; $f(x) = 0.20x + 38.8 = 0$
35. $\frac{9}{7}$; $f(x) = 9 - 7x = 0$ **37.** 15; $f(x) = 3x - 45 = 0$
39. $-\frac{3}{10}$; $f(x) = 10x + 3 = 0$ **41.** -1.379
43. $2.172, 7.828$ **45.** $0.5, -3, 3$ **47.** $-0.717, 2.107$
49. -1.333 **51.** ± 3.162 **53.** $5, -1$ **55.** $-1, 2.333$
57. 11 **59.** 21

61. (a)

x	-1	0	1	2	3	4
$3.2x - 5.8$	-9	-5.8	-2.6	0.6	3.8	7

$1 < x < 2$; Answers will vary.

(b)

x	1.5	1.6	1.7	1.8	1.9	2
$3.2x - 5.8$	-1	-0.68	-0.36	-0.04	0.28	0.60

$1.8 < x < 1.9$; Answers will vary. Sample answer: Use smaller intervals of x to increase accuracy.
63. $(1, 1)$ **65.** $(2, 2)$ **67.** $(8, -2)$ **69.** $(-1, 3)$
71. $(4, 1)$ **73.** $(1.45, 1.90), (-3.45, -7.90)$
75. $(0, 0), (-2, 8), (2, 8)$
77. (a) 6.46
(b) $\frac{1.73}{0.27} \approx 6.41$
Yes, the more rounding performed, the less accurate the result.
79. (a) $t(x) = \frac{x}{63} + \frac{280 - x}{54}$ **81.** (a) $A(x) = 12x$

(b)

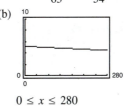

(b)

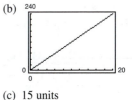

$0 \le x \le 280$ (c) 15 units
(c) 164.5 mi
83. (a) $T = 10,000 + \frac{1}{2}x$ (b) \$6800
(c) \$7600 (d) \$7500
85. (a)

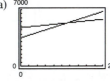

$(11.4, 5367.0)$; In 2001, both states had the same population.
(b) $(11.4, 5367.0)$; In 2001, both states had the same population.
(c) Change in population per year; Arizona's population is growing faster.
(d) Maryland: 6,011,400; Arizona: 7,410,000
Answers will vary.
87. True **89.** Answers will vary. **91.** 3 **93.** 1
95. $\frac{4\sqrt{3}}{5}$ **97.** $\frac{3(8 - \sqrt{11})}{53}$
99. $3x^2 + 13x - 30$ **101.** $4x^2 - 81$

Section 2.3 (page 184)

1. (a) ii (b) iii (c) i **3.** complex, $a + bi$
5. $-2 + 4i$ **7.** $a = -9, b = 4$ **9.** $a = 6, b = 5$
11. $5 + 4i$ **13.** -6 **15.** $-1 - 5i$ **17.** -75
19. $0.3i$ **21.** $-3 + 3i$ **23.** $7 - 3\sqrt{2}i$
25. $-14 + 20i$ **27.** $\frac{19}{6} + \frac{37}{6}i$ **29.** $-4.2 + 7.5i$
31. $-2\sqrt{3}$ **33.** -10 **35.** $12 + 20i$ **37.** $5 + i$
39. $-20 + 32i$ **41.** 24 **43.** $-13 + 84i$ **45.** $80i$
47. $4 - 3i$; 25 **49.** $-6 + \sqrt{5}i$; 41

CHAPTER 2

51. $-\sqrt{20}\,i;\ 20$ **53.** $3 + \sqrt{2}\,i;\ 11$ **55.** $-6i$

57. $\frac{8}{41} + \frac{10}{41}i$ **59.** $\frac{3}{5} + \frac{4}{5}i$ **61.** $-\frac{40}{1681} - \frac{9}{1681}i$

63. $-\frac{1}{2} - \frac{5}{2}i$ **65.** $\frac{62}{949} + \frac{297}{949}i$ **67.** $-1 + 6i$

69. $-375\sqrt{3}\,i$ **71.** i

73. (a) 8   (b) 8   (c) 8

The results are the same.

75. (a) 1   (b) i (c) $-i$ (d) -1

77. False. Any real number is equal to its conjugate.

79. False. Example: $(1 + i) + (1 - i) = 2$, which is not an imaginary number.

81. True. Answers will vary.

83. $\sqrt{-6}\sqrt{-6} = \left(\sqrt{6}\,i\right)\left(\sqrt{6}\,i\right) = 6i^2 = -6$

85. $16x^2 - 25$ **87.** $3x^2 + \frac{23}{2}x - 2$

Section 2.4   (page 196)

1. quadratic equation

3. factoring, extracting square roots, completing the square, Quadratic Formula

5. $2x^2 + 5x - 3 = 0$ **7.** $3x^2 - 60x - 10 = 0$

9. $0, -\frac{1}{2}$ **11.** $4, -2$ **13.** -5 **15.** $3, -\frac{1}{2}$

17. $2, -6$ **19.** $-a - b, -a + b$ **21.** ± 7 **23.** $16, 8$

25. $\dfrac{1 \pm \sqrt{6}\,i}{3};\ 0.33 \pm 0.82i$ **27.** 2   **29.** $-8, 4$

31. $-3 \pm \sqrt{7}$ **33.** $1 \pm \dfrac{\sqrt{6}}{3}$ **35.** $1 \pm \sqrt{5}\,i$

37. $-\dfrac{5}{4} \pm \dfrac{\sqrt{89}}{4}$ **39.** $2 \pm 3i$ **41.** $-1 \pm 4i$

43. (a)

45. (a)

(b) and (c)   (b) and (c)
$(-1, 0), (-5, 0)$ $(-0.5, 0), (1.5, 0)$

47. One real solution   **49.** One real solution

51. No real solutions   **53.** $1 \pm \sqrt{3}$ **55.** $\frac{2}{7}$

57. $-\dfrac{3}{2} \pm \dfrac{\sqrt{23}}{2}i$ **59.** $-2 \pm \dfrac{1}{2}i$ **61.** $1 \pm \sqrt{2}$

63. $6, -12$ **65.** $1 \pm \dfrac{3}{2}i$ **67.** $-\dfrac{1}{2}$ **69.** $\dfrac{-7 \pm \sqrt{73}}{4}$

71. $\frac{3}{2}$ **73.** $x^2 + x - 30 = 0;\ 2x^2 + 2x - 60 = 0$

75. $21x^2 + 31x - 42 = 0;\ 3x^2 + \frac{31}{7}x - 6 = 0$

77. $x^2 - 75 = 0;\ \dfrac{x^2}{5} - 15 = 0$

79. $x^2 - 2x - 11 = 0;\ 5x^2 - 10x - 55 = 0$

81. $x^2 - 4x + 5 = 0;\ -x^2 + 4x - 5 = 0$

83. (a)

w

$w + 14$

(b) $1632 = w^2 + 14w$

(c) Width: 34 ft; length: 48 ft

85. (a) $A(x) = -\frac{8}{3}x^2 + \frac{200}{3}x$

(b)

x	y	Area
2	$\frac{92}{3}$	$\frac{368}{3} \approx 123$
4	28	224
6	$\frac{76}{3}$	304
8	$\frac{68}{3}$	$\frac{1088}{3} \approx 363$
10	20	400
12	$\frac{52}{3}$	416
14	$\frac{44}{3}$	$\frac{1232}{3} \approx 411$

Approximate dimensions for maximum area: $24 \text{ m} \times \frac{52}{3} \text{ m}$

(c)

Approximate dimensions for maximum area: $25 \text{ m} \times \frac{50}{3} \text{ m}$

(d) and (e) $15 \times 23\frac{1}{3}$ or 35×10 m

87. (a) $s = -16t^2 + 1815$

(b)

t	0	2	4	6	8	10	12
s	1815	1751	1559	1239	791	215	-489

(c) $(10, 12);\ 10.65$ sec

89. (a) 22.36 sec   (b) 3.73 mi

91. (a) 2003

(b) Answers will vary.

(c)

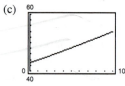

(d) 2013

(e) Answers will vary.

93. (a)

(b) $16.8°C$ (c) 2.5

95. Eastbound plane: About 550 mi/h

Northbound plane: About 600 mi/h

97. False. Both solutions are complex.

99. False. Imaginary solutions are always complex conjugates of each other.

101. (a) $0, -\dfrac{b}{a}$ (b) $0, 1$ **103.** Proof   **105.** e

107. $x^2(x - 3)(x^2 + 3x + 9)$

109. $(x + 5)\left(x - \sqrt{2}\right)\left(x + \sqrt{2}\right)$ **111.** Answers will vary.

Section 2.5   (page 207)

1. polynomial   **3.** Square both sides of the equation.
5. $0, \pm 2$ **7.** $-3, 0$ **9.** $\pm 1, 5$ **11.** $\pm 1, \pm\sqrt{3}$
13. $\pm\frac{1}{2}, \pm 4$
15. (a)

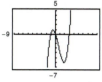

(b) and (c)
$(-1, 0), (0, 0), (3, 0)$
(d) They are the same.

17. (a)

(b) and (c)
$(-3, 0), (-1, 0),$
$(1, 0), (3, 0)$
(d) They are the same.

19. $\frac{100}{9}$ **21.** 26   **23.** -16 **25.** -256.5 **27.** 6, 7
29. 0   **31.** 0   **33.** $\frac{1}{4}$ **35.** 9   **37.** $\frac{11}{2}$ **39.** $1, -\frac{125}{8}$
41. $-59, 69$ **43.** $-116, 134$ **45.** 2, 3   **47.** 1
49. (a)

(b) and (c) $x = 5, 6$
(d) They are the same.

51. (a)

(b) and (c) $x = 0, 4$
(d) They are the same.

53. $2, -\frac{3}{2}$ **55.** $\dfrac{-3 \pm \sqrt{21}}{6}$ **57.** $4, -5$ **59.** $\dfrac{1 \pm \sqrt{31}}{3}$
61. $-\frac{1}{5}, -\frac{1}{3}$ **63.** $2, -\frac{3}{5}$ **65.** $-3, 4$ **67.** $\sqrt{3}, -3$
69. 10
71. (a)

(b) and (c) $x = -1$
(d) They are the same.

73. (a)

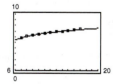

(b) and (c) $x = 1, -3$
(d) They are the same.

75. $900 total   **77.** About 4%   **79.** 14.806 lb/in.^{2}
81. (a) and (b)

The model is a good fit for the data.
(c) 2009

83. (a) $C = 126,720(8 - x) + 39,600\sqrt{16x^2 + 9}$
(b) $1,123,424.95   (c) 2 mi or 0.382 mi
(d)

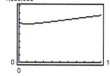

(e) $x = 1$

85. False. See Example 7.   **87.** $x^3 - 5x^2 - 2x + 10 = 0$
89. $x^4 - 3x^2 - 4 = 0$ **91.** $x = 6, x = -4$
93. $a = b = 9$, or $a = 0, b = 18$
95. Dividing by x loses the solution $x = 0$.
97. $\dfrac{25}{6x}$ **99.** $\dfrac{-3z^2 - 2z + 4}{z(z + 2)}$ **101.** 11

Section 2.6   (page 219)

1. double   **3.** $x \le -a, x \ge a$
5. No   **7.** f   **8.** a   **9.** d   **10.** b
11. e   **12.** c
13. (a) Yes   (b) No   (c) Yes   (d) No
15. (a) No   (b) Yes   (c) Yes   (d) No
17. $x > -4$

19. $x < -2$

21. $x < -\frac{1}{2}$

23. $x \ge 4$

25. $-1 < x < 3$

27. $-2 < x \le 5$

29. $-\frac{9}{2} < x < \frac{15}{2}$

31.

33.

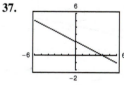

$x \le 2$

$x < 2$

35.

37.

(a) $x \ge 2$ (b) $x \le \frac{3}{2}$ (a) $-2 \le x \le 4$ (b) $x \le 4$
39. $x < -2, x > 2$ **41.** $1 < x < 13$

43. $x < -28, x > 0$ **45.** $\frac{1}{2} < x < \frac{3}{2}$

47.

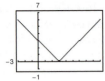

(a) $1 \le x \le 5$ (b) $x \le -1, x \ge 7$

49. $|x| \le 3$ **51.** $|x| > 3$ **53.** $|x - 7| \le 10$

55. $|x - 3| \ge 5$

57. Positive on: $(-\infty, -1) \cup (5, \infty)$

Negative on: $(-1, 5)$

59. Positive on: $\left(-\infty, \dfrac{2 - \sqrt{10}}{2}\right) \cup \left(\dfrac{2 + \sqrt{10}}{2}, \infty\right)$

Negative on: $\left(\dfrac{2 - \sqrt{10}}{2}, \dfrac{2 + \sqrt{10}}{2}\right)$

61. Positive on: $(-\infty, \infty)$

63. $(-7, 3)$

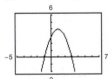

65. $(-\infty, -5], [1, \infty)$

67. $[-2, 0], [2, \infty)$

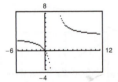

69. $(-3, -1), \left(\tfrac{3}{2}, \infty\right)$

71. $(-1, 1), (3, \infty)$

73. No solution

75. All real numbers x

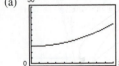

77. (a) $x = 1$ (b) $x \ge 1$ (c) $x > 1$

79.

(a) $x \le -1, \ x \ge 3$ (b) $0 \le x \le 2$

81. $(-\infty, -1), (0, 1)$ **83.** $(-\infty, -1), (4, \infty)$

85.

(a) $0 \le x < 2$ (b) $2 < x \le 4$

87. $[5, \infty)$ **89.** $(-\infty, \infty)$ **91.** $(-\infty, -2], [2, \infty)$

93. (a) 2004   (b) $(2000, 2004); (2004, 2008)$

95. (a) 10 sec   (b) $(4, 6)$

97. (a)

(b) $(2000, 2004)$

(c) $10.102 < x < 14.2749$

99. $t \ge 6.19$; In the year 2006, there were at least 900 Bed Bath & Beyond stores.

101. $t \approx 1.17$; In 2001, there were the same number of Bed Bath & Beyond stores as Williams-Sonoma stores.

103. $333\tfrac{1}{3}$ vibrations/sec   **105.** $1.2 < t < 2.4$

107. False. $10 \ge -x$ **109.** a, b **111.** iv, ii, iii, i

113. $y^{-1} = \dfrac{x}{12}$ **115.** $y^{-1} = \sqrt[3]{x - 7}$

117. Answers will vary.

Section 2.7   (page 228)

1. positive   **3.** Negative correlation

5. (a)

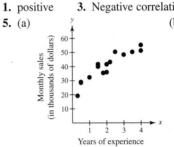

(b) Yes, the data appear somewhat linear. More experience, x, corresponds to higher sales, y.

7. Negative correlation   **9.** No correlation

11. (a) and (b)   **13.** (a) and (b)

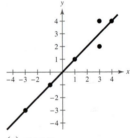   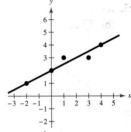

(c) $y = x$ (c) $y = \tfrac{1}{2}x + 2$

15. $y = 0.46x + 1.6$

(a)

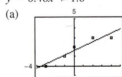

(b)

x	-3	-1	0	2	4
Linear equation	0.22	1.14	1.6	2.52	3.44
Given data	0	1	2	3	3

The model fits the data well.

17. (a)

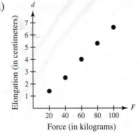

(b) $d = 0.07F - 0.3$ (c) $d = 0.066F$ (d) 3.63 cm

19. (a) and (c)

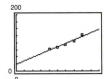

(b) $T = 12.37t + 24.04$
(d) 2010: \$147.74 million;
2015: \$209.59 million;
Answers will vary.

The model fits the data well.

(e) 12.37; The slope represents the average annual increase in salaries (in millions of dollars).

21. (a) and (c)

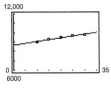

(b) $P = 38.98t + 8655.4$

(d)

Year	2010	2015	2020	2025	2030
Actual	9018	9256	9462	9637	9802
Model	9045.2	9240.1	9435	9629.9	9824.8

The model fits the data well.

(e) 10,604,400 people; Answers will vary.

23. (a) $y = 47.77x + 103.8$

(b)

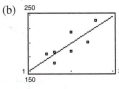

(c) The slope represents the increase in sales due to increased advertising.

(d) \$175,455

25. (a) $T = -0.019t + 4.92$
$r \approx -0.886$

(b) The negative slope means that the winning times are generally decreasing over time.

(c)

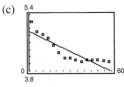

(d)

Year	1952	1956	1960	1964	1968
Actual	5.20	4.91	4.84	4.72	4.53
Model	4.88	4.81	4.73	4.65	4.58

Year	1972	1976	1980	1984	1988
Actual	4.32	4.16	4.15	4.12	4.06
Model	4.50	4.43	4.35	4.27	4.20

Year	1992	1996	2000	2004	2008
Actual	4.12	4.12	4.10	4.09	4.05
Model	4.12	4.05	3.97	3.89	3.82

The model does not fit the data well.

(e) The closer $|r|$ is to 1, the better the model fits the data.

(f) No; The winning times have leveled off in recent years, but the model values continue to decrease to unrealistic times.

27. True. To have positive correlation, the y-values tend to increase as x increases.

29. Answers will vary.

31. (a) 10 (b) $2w^2 + 5w + 7$

33. $-\frac{3}{5}$ **35.** $-\frac{1}{4}, \frac{3}{2}$

Review Exercises (page 234)

1. (a) No (b) No (c) No (d) Yes

3. $x = 9$ **5.** $x = \frac{11}{3}$

7. $x = \frac{1}{2}$ **9.** $x = 6$ **11.** $x = \frac{7}{3}$

13. September: \$325,000; October: \$364,000

15. (a)

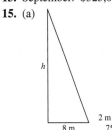

(b) $h = \frac{64}{3}$ m

17. $-3.5°C$ **19.** $(-3, 0), (0, 3)$

21. $(1, 0), (8, 0), (0, 8)$ **23.** $x = 2.2$

25. $x = -1.301$ **27.** $x = 0.338, 1.307$

29. $(1, -2)$ **31.** $(4.5, -3.125), (-3, 2.5)$

33. $6 + 5i$ **35.** $2 + 7i$ **37.** $3 + 7i$

39. $40 + 65i$ **41.** $-26 + 7i$ **43.** $3 + 9i$

45. $-4 - 46i$ **47.** -80 **49.** $1 - 6i$

51. $\frac{17}{26} + \frac{7}{26}i$ **53.** $-3, \frac{1}{2}$ **55.** $\frac{2}{3}, 5$ **57.** $0, 2$

59. $-1, 5$ **61.** $-1, 4$ **63.** $-1, \frac{3}{2}$ **65.** $-\frac{5}{2}, 3$

67. $-4 \pm 3\sqrt{2}$ **69.** $6 \pm \sqrt{6}$ **71.** $\frac{1}{2}, -5$

73. $\frac{-1 \pm \sqrt{61}}{2}$ **75.** $-2 \pm \sqrt{6}i$ **77.** $\frac{3 \pm \sqrt{33}i}{2}$

79. (a)

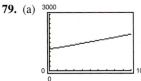

(b) and (c) 2001
(d) 2013; 2016
(e) Sample answer: The answers seem reasonable because the answers follow the graph of the model.

81. $0, \frac{2}{3}, 8$ **83.** $0, \frac{12}{5}$ **85.** $\pm 2, \pm \sqrt{3}i$ **87.** $\pm 2, \pm \sqrt{7}$

89. 5 **91.** $\frac{25}{4}$ **93.** No solution **95.** $-124, 126$

97. $-2 \pm \frac{\sqrt{95}}{5}, -4$ **99.** $-4, 1$ **101.** $\frac{1}{5}$ **103.** 2, 6

105. $-5, 15$ **107.** 1, 3 **109.** 4 investors

111. 191.5 mi/h **113.** 3%

CHAPTER 2

115. (a)

Year	2000	2001	2002	2003	2004
Population (in millions)	18.39	18.47	18.55	18.63	18.71

Year	2005	2006	2007	2008
Population (in millions)	18.78	18.86	18.94	19.02

(b)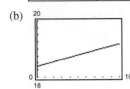

(c) and (d) 2001

(e) 2014; The answer seems reasonable.

(f) Answers will vary.

117. $(-\infty, 9)$

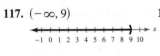

119. $\left(-\frac{5}{3}, \infty\right)$

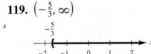

121. $[-3, 9)$

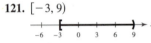

123. $(1, 3)$

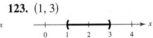

125. $(-\infty, 0], [3, \infty)$

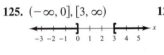

127. $\left[-\frac{1}{2}, \frac{7}{2}\right]$

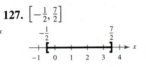

129. $(-\infty, -1], [3, \infty)$

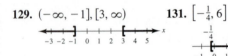

131. $\left[-\frac{1}{4}, 6\right]$

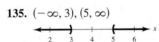

133. $[-4, 0], [4, \infty)$

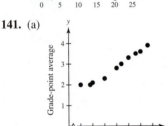

135. $(-\infty, 3), (5, \infty)$

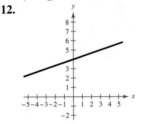

137. $(-\infty, 3), [20, \infty)$

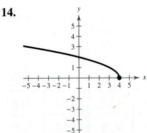

139. $0.28

141. (a)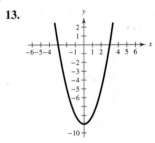

(b) Yes. Answers will vary.

143. (a)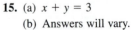

(b) Answers will vary. Sample answer: $S = 10t - 0.4$

(c) $s = 9.70t + 0.4$; This model fits the data better.

(d) 24.7 m/sec

145. False. A graph with two distinct y-intercepts is not a function.

147. False. A regression line can have a positive or negative slope.

149. Answers will vary.   **151.** $\sqrt{-6}\sqrt{-6} = 6i^2 = -6$

153. (a) 1   (b) i (c) -1 (d) $-i$

Chapter Test   (page 238)

1. $x = 3$ **2.** $x = \frac{2}{15}$ **3.** $-9 - 18i$

4. $6 + \left(2\sqrt{5} + \sqrt{14}\right)i$ **5.** $13 + 4i$ **6.** $-17 + 14i$

7. $\frac{43}{37} + \frac{38}{37}i$ **8.** $\frac{4}{13} + \frac{7}{13}i$ **9.** ± 1.414 **10.** ± 0.5

11. 0   **12.** $\pm 1, 0$ **13.** $1, 9$ **14.** $-6 \pm \sqrt{38}$

15. $\pm \frac{9}{2}$ **16.** $-3, \frac{1}{5}$ **17.** $\pm 2, \frac{4}{3}$ **18.** 2

19. $\pm \sqrt{58}$ **20.** $-\frac{5}{2}, \frac{11}{4}$

21. $\left(-\frac{9}{5}, \infty\right)$ **22.** $(3, 13)$

23. $\left(-\infty, -\frac{1}{2}\right], \left[-\frac{1}{3}, \infty\right)$ **24.** $\left(-7, -\frac{2}{3}\right)$

25. $S = 22.94t + 91.7; r \approx 0.9970; 2009$

Cumulative Test for Chapters P–2 (page 239)

1. $\dfrac{7x^3}{16y^5}, x \neq 0$ **2.** $9\sqrt{15}$ **3.** $2x^2y\sqrt{7y}$ **4.** $7x - 10$

5. $x^3 - x^2 - 5x + 6$ **6.** $\dfrac{x - 1}{(x + 1)(x + 3)}$ **7.** $(3 + x)(7 - x)$

8. $x(1 + x)(1 - 6x)$ **9.** $2(3 - 2x)(9 + 6x + 4x^2)$

10. Midpoint: $\left(-\frac{1}{2}, -2\right); d = 6\sqrt{5} \approx 13.42$

11. $\left(x + \frac{1}{2}\right)^2 + (y + 8)^2 = 16$

12.   **13.**

14.   **15.** (a) $x + y = 3$

                                    (b) Answers will vary. Sample answer: $(-1, 4), (0, 3), (1, 2)$

16. (a) $2x + y = 0$

(b) Answers will vary. Sample answer: $(0, 0), (1, -2), (2, -4)$

17. (a) $x = -\frac{3}{7}$

(b) Answers will vary. Sample answer: $\left(-\frac{3}{7}, 0\right), \left(-\frac{3}{7}, 1\right), \left(-\frac{3}{7}, -3\right)$

18. (a) $y - 6x = -9$ (b) $y + \frac{1}{6}x = \frac{10}{3}$

19. (a) $\dfrac{5}{3}$ (b) Undefined   (c) $\dfrac{5+4s}{3+4s}$

20. (a) -32 (b) 4 (c) 20 **21.** $(-\infty, \infty)$

22. $\left[-\dfrac{5}{7}, \infty\right)$ **23.** $[-3, 3]$ **24.** $\left(-\infty, -\dfrac{2}{5}\right) \cup \left(-\dfrac{2}{5}, \infty\right)$

25. Odd   **26.** No. It doesn't pass the Vertical Line Test.

27.

Decreasing on $(-\infty, 5)$
Increasing on $(5, \infty)$

28. (a) Vertical shrink   (b) Vertical shift two units up
(c) Horizontal shift two units left, reflection in x-axis

29. $x^2 + 4x + 3$ **30.** $-x^2 + 4x - 1$ **31.** $4x^2 + 9$

32. $4x^3 + x^2 + 8x + 2$ **33.** $0, 1, 2$

34. -3 **35.** $-0.667, -2$ **36.** 4.444

37. $\left(-\infty, \dfrac{120}{7}\right]$ **38.** $(-\infty, -3], \left[\dfrac{5}{2}, \infty\right)$

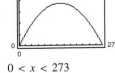

39. $\left(-\infty, -\dfrac{3}{2}\right), \left(-\dfrac{1}{4}, \infty\right)$ **40.** $(-1, 2]$

41. 4.456 in.

42. (a) $A = x(273 - x)$

(b)

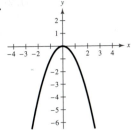

$0 < x < 273$

(c) 76.2 ft $\times$ 196.8 ft

43. (a) $R = 169.69t + 1638.9$; $r \approx 0.9860$

(b)

(c) 3675.2; 4014.6
(d) Answers will vary.

Chapter 3

Section 3.1   (page 250)

1. nonnegative integer, real   **3.** Yes; $(2, 3)$ **5.** c

6. d   **7.** b   **8.** a

9.

Reflection in the x-axis

11.

Horizontal shift three units to the left

13.

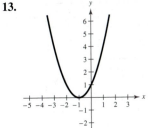

Horizontal shift one unit to the left

15.

Horizontal shift three units to the right

17. Parabola opening downward
Vertex: $(0, 25)$

19. Parabola opening upward
Vertex: $(0, -4)$

21. Parabola opening upward
Vertex: $(-4, -3)$

23. Parabola opening upward
Vertex: $(4, 0)$

25. Parabola opening upward
Vertex: $\left(\dfrac{1}{2}, 1\right)$

27. Parabola opening upward
Vertex: $(1, 6)$

29. Parabola opening upward
Vertex: $\left(\dfrac{1}{2}, 20\right)$

31. Parabola opening downward
Vertex: $(-1, 4)$
x-intercepts: $(1, 0), (-3, 0)$

33. Parabola opening upward
Vertex: $(-4, -5)$
x-intercepts: $\left(-4 \pm \sqrt{5}, 0\right)$

35. Parabola opening downward
Vertex: $(4, 1)$
x-intercepts: $\left(4 \pm \dfrac{1}{2}\sqrt{2}, 0\right)$

37. $y = -(x + 1)^2 + 4$ **39.** $f(x) = (x + 2)^2 + 5$

41. $y = 4(x - 1)^2 - 2$ **43.** $y = -\dfrac{104}{125}\left(x - \dfrac{1}{2}\right)^2 + 1$

45. $(5, 0), (-1, 0)$ **47.** $(-4, 0)$

49.

$(0, 0), (4, 0)$

51.

$\left(-\dfrac{5}{2}, 0\right), (6, 0)$

53.

$(7, 0), (-1, 0)$

55. $f(x) = x^2 - 2x - 3$
$g(x) = -x^2 + 2x + 3$

57. $f(x) = 2x^2 + 7x + 3$
$g(x) = -2x^2 - 7x - 3$

59. $55, 55$ **61.** $12, 6$

63. (a)

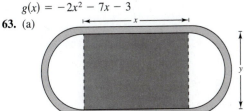

(b) $r = \dfrac{1}{2}y$; $d = y\pi$ (c) $y = \dfrac{200 - 2x}{\pi}$

CHAPTER 3

(d) $A = x\left(\dfrac{200 - 2x}{\pi}\right)$

(e)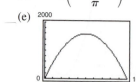

$x = 50$ m, $y = \dfrac{100}{\pi}$ m

65. (a) 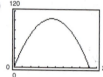   (b) $\frac{3}{2}$ ft

(c) About 104 ft   (d) About 228.6 ft

67. (a) $A = -2x^2 + 112x - 600$ (b) $x = 28$ in.

69. (a) \$54,000; \$61,600; \$61,200   (b) \$79

(c) \$62,410   (d) Answers will vary.

71. (a)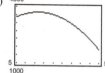

(b) 1966; 4155 cigarettes; Yes, the warning had an effect because the maximum consumption occurred in 1966.

(c) 1852 cigarettes per year; 5 cigarettes per day

73. True. The vertex is $(0, -1)$ and the parabola opens down.

75. c, d   **77.** Horizontal shift z units to the right

79. Vertical stretch $(z > 1)$ or shrink $(0 < z < 1)$ and horizontal shift three units to the right

81. $b = \pm 20$ **83.** $b = \pm 8$ **85.** Proof

87. $y = -x^2 + 5x - 4$; Answers will vary.

89. $(1.2, 6.8)$ **91.** $(2, 5), (-3, 0)$ **93.** Answers will vary.

Section 3.2   (page 262)

1. continuous   **3.** (a) solution   (b) $(x - a)$ (c) $(a, 0)$

5. No   **7.** $f(x_2) > 0$ **9.** f   **10.** h   **11.** c

12. a   **13.** e   **14.** d   **15.** g   **16.** b

17. 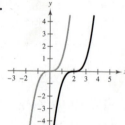   **19.**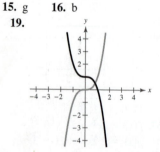

Horizontal shift two units to the right

Reflection in the x-axis and vertical shift one unit upward

21.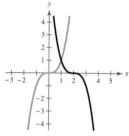

Reflection in the x-axis and horizontal shift two units to the right

23. 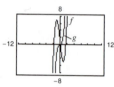   **25.**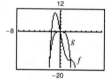

Yes, because both graphs have the same leading coefficient.

Yes, because both graphs have the same leading coefficient.

27.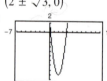

No, because the graphs have different leading coefficients.

29. Rises to the left, rises to the right

31. Falls to the left, falls to the right

33. Falls to the left, rises to the right

35. Falls to the left, falls to the right

37. (a) $\left(2 \pm \sqrt{3}, 0\right)$

(b)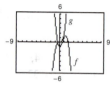

(c) $(0.27, 0), (3.73, 0)$; Answers are approximately the same.

39. (a) $(-1, 0), (1, 0)$

(b)

(c) $(-1, 0), (1, 0)$; Answers are the same.

41. (a) $(0, 0), \left(\pm\sqrt{2}, 0\right)$

(b)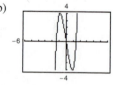

(c) $(-1.41, 0), (0, 0), (1.41, 0)$; Answers are approximately the same.

43. (a) $\left(\pm\sqrt{5}, 0\right)$

(b)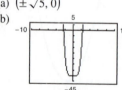

(c) $(-2.236, 0), (2.236, 0)$; Answers are approximately the same.

45. (a) $(4, 0), (\pm 5, 0)$

(b)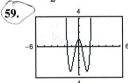

(c) $(-5, 0), (4, 0), (5, 0)$;
Answers are the same.

47. (a) $(0, 0), \left(\frac{5}{2}, 0\right)$

(b)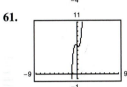

(c) $(0, 0), (2.5, 0)$;
Answers are the same.

49. ± 5 (multiplicity 1)   **51.** 3 (multiplicity 2)

53. $1, -2$ (multiplicity 1)

55. 2 (multiplicity 2), 0 (multiplicity 1)

57. $\dfrac{-5 \pm \sqrt{37}}{2}$ (multiplicity 1)

59.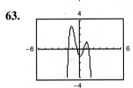

Zeros: $\pm 1.680, \pm 0.421$
Relative minima: $(\pm 1.225, -3.500)$
Relative maximum: $(0, 1)$

61.

Zero: -1.178
Relative minimum: $(0.324, 5.782)$
Relative maximum: $(-0.324, 6.218)$

63.

Zeros: $-1.618, -0.366, 0.618, 1.366$
Relative minimum: $(0.101, -1.050)$
Relative maxima: $(-1.165, 3.267),$
$(1.064, 1.033)$

65. $f(x) = x^2 - 4x$ **67.** $f(x) = x^3 + 5x^2 + 6x$

69. $f(x) = x^4 - 4x^3 - 9x^2 + 36x$ **71.** $f(x) = x^2 - 2x - 2$

73. $f(x) = x^3 - 10x^2 + 27x - 22$

75. $f(x) = x^3 + 5x^2 + 8x + 4$

77. $f(x) = x^4 + 2x^3 - 23x^2 - 24x + 144$

79. $f(x) = -x^3 - 4x^2 - 5x - 2$

81.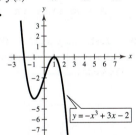

$y = -x^3 + 3x - 2$

83.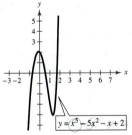

$y = x^3 - 5x^2 - x + 2$

85. (a) Falls to the left, rises to the right

(b) $(0, 0), (3, 0), (-3, 0)$

(c) and (d)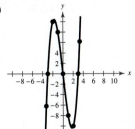

87. (a) Falls to the left, rises to the right   (b) $(0, 0), (3, 0)$

(c) and (d)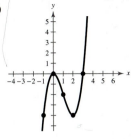

89. (a) Falls to the left, falls to the right

(b) $(\pm 2, 0), (\pm \sqrt{5}, 0)$

(c) and (d)

91. (a) Falls to the left, rises to the right   (b) $(\pm 3, 0)$

(c) and (d)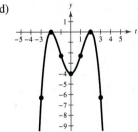

93. (a) Falls to the left, falls to the right   (b) $(-2, 0), (2, 0)$

(c) and (d)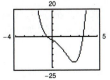

95. (a)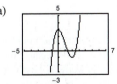

$(-1, 0), (1, 2), (2, 3)$

(b) $-0.879, 1.347, 2.532$

97. (a)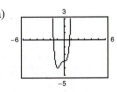

$(-2, -1), (0, 1)$

(b) $-1.585, 0.779$

99. (a)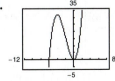

$(-1, 0), (3, 4)$

(b) $-0.578, 3.418$

101.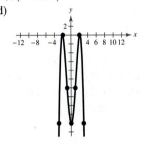

Two x-intercepts

CHAPTER 3

103.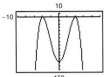

y-axis symmetry
Two x-intercepts

105.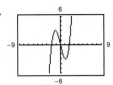

Origin symmetry
Three x-intercepts

107.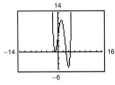

Three x-intercept

109. (a) Answers will vary.

(c)

Height, x	Volume, V
1	1156
2	2048
3	2700
4	3136
5	3380
6	3456
7	3388

$5 < x < 7$

(b) Domain: $0 < x < 18$

(d)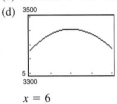

$x = 6$

111. $(200, 160)$

113. (a)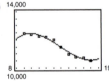

The model fits the data well.

(b)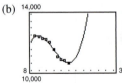

Answers will vary. Sample answer: You could use the model to estimate production in 2010 because the result is somewhat reasonable, but you would not use the model to estimate the 2020 production because the result is unreasonably high.

(c)

The model fits the data well.

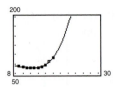

Answers will vary. Sample answer: You could use the model to estimate production in 2010 because the result is somewhat reasonable, but you would not use the model to estimate the 2020 production because the result is unreasonably high.

115. True. The degree is odd and the leading coefficient is -1.

117. False. The graph crosses the x-axis at $x = -3$ and $x = 0$.

119.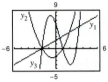

121. 69   **123.** $-\frac{1408}{49} \approx -28.73$ **125.** 109

127. $x > -8$ **129.** $-26 \le x < 7$

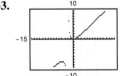

Section 3.3   (page 277)

1. $f(x)$ is the dividend, $d(x)$ is the divisor, $q(x)$ is the quotient, and $r(x)$ is the remainder.

3. constant term, leading coefficient   **5.** upper, lower

7. 7   **9.** $2x + 4,\ x \ne -3$ **11.** $x^3 + 3x^2 - 1,\ x \ne -2$

13. $x^2 - 3x + 1,\ x \ne -\frac{5}{4}$ **15.** $7x^2 - 14x + 28 - \dfrac{53}{x + 2}$

17. $3x + 5 - \dfrac{2x - 3}{2x^2 + 1}$ **19.** $x - \dfrac{x + 9}{x^2 + 1}$

21. $2x - \dfrac{17x - 5}{x^2 - 2x + 1}$ **23.** $3x^2 - 2x + 5,\ x \ne 5$

25. $6x^2 + 25x + 74 + \dfrac{248}{x - 3}$ **27.** $9x^2 - 16,\ x \ne 2$

29. $x^2 - 8x + 64,\ x \ne -8$ **31.** $4x^2 + 14x - 30,\ x \ne -\frac{1}{2}$

33.
[graph: -15 to 15, -10 to 10]

35. [graph: -9 to 9, -6 to 6]

37. $f(x) = (x - 4)(x^2 + 3x - 2) + 3,\ f(4) = 3$

39. $f(x) = \left(x - \sqrt{2}\right)\left[x^2 + \left(3 + \sqrt{2}\right)x + 3\sqrt{2}\right] - 8,$
$f\left(\sqrt{2}\right) = -8$

41. $f(x) = \left(x - 1 + \sqrt{3}\right)\left[4x^2 - \left(2 + 4\sqrt{3}\right)x - \left(2 + 2\sqrt{3}\right)\right],$
$f\left(1 - \sqrt{3}\right) = 0$

43. (a) -2 (b) 1   (c) $-\frac{1}{4}$ (d) 5

45. (a) -35 (b) -22 (c) -10 (d) -211

47. $(x - 2)(x + 3)(x - 1)$ **49.** $(2x - 1)(x - 5)(x - 2)$
Zeros: $2, -3, 1$ Zeros: $\frac{1}{2}, 5, 2$

51. (a) Answers will vary.   (b) $(2x - 1), (x - 1)$
(c) $(x + 2)(x - 1)(2x - 1)$ (d) $-2, 1, \frac{1}{2}$

53. (a) Answers will vary.   (b) $(x - 1), (x - 2)$
(c) $(x - 5)(x + 4)(x - 1)(x - 2)$ (d) $-4, 1, 2, 5$

55. (a) Answers will vary.   (b) $(x + 7), (3x - 2)$
(c) $(2x + 1)(3x - 2)(x + 7)$ (d) $-7, -\frac{1}{2}, \frac{2}{3}$

57. $\pm 1, \pm 3; \pm 1, -3$

59. $\pm 1, \pm 3, \pm 5, \pm 9, \pm 15, \pm 45, \pm\frac{1}{2}, \pm\frac{3}{2}, \pm\frac{5}{2}, \pm\frac{9}{2}, \pm\frac{15}{2}, \pm\frac{45}{2}$;
$-1, \frac{3}{2}, 3, 5$

61. 4, 2, or 0 positive real zeros, no negative real zeros

63. 2 or 0 positive real zeros, 1 negative real zero

65. (a) 1 positive real zero, 2 or 0 negative real zeros
(b) $\pm 1, \pm 2, \pm 4$
(c) 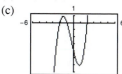   (d) $-2, -1, 2$

67. (a) 3 or 1 positive real zeros, 1 negative real zero
(b) $\pm 1, \pm 2, \pm 4, \pm 8, \pm\frac{1}{2}$
(c) 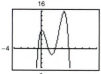   (d) $-\frac{1}{2}, 1, 2, 4$

69. (a) 2 or 0 positive real zeros, 1 negative real zero
(b) $\pm 1, \pm 3, \pm\frac{1}{2}, \pm\frac{3}{2}, \pm\frac{1}{4}, \pm\frac{3}{4}, \pm\frac{1}{8}, \pm\frac{3}{8}, \pm\frac{1}{16}, \pm\frac{3}{16}, \pm\frac{1}{32}, \pm\frac{3}{32}$
(c) 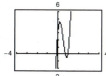   (d) $-\frac{1}{8}, \frac{3}{4}, 1$

71. Answers will vary; 1.937, 3.705

73. Answers will vary; ± 2 **75.** $\pm 2, \pm\frac{3}{2}$ **77.** $\pm 1, \frac{1}{4}$

79. d   **80.** a   **81.** b   **82.** c   **83.** $-\frac{1}{2}, 2 \pm \sqrt{3}, 1$

85. $-1, \frac{3}{2}, 4 \pm \sqrt{17}$ **87.** $-2, 0, 1 \pm \frac{\sqrt{15}}{3}$ **89.** $-1, 2$

91. $-6, \frac{1}{2}, 1$ **93.** $-3, -\frac{3}{2}, \frac{1}{2}, 4$ **95.** $-3, 0, \frac{-1 \pm \sqrt{7}}{4}$

97. $-\frac{5}{2}, -2, \pm 1, \frac{3}{2}$

99. (a) $-2, 0.268, 3.732$ (b) -2
(c) $h(t) = (t + 2)(t - 2 + \sqrt{3})(t - 2 - \sqrt{3})$

101. (a) $0, 3, 4, -1.414, 1.414$ (b) $0, 3, 4$
(c) $h(x) = x(x - 3)(x - 4)(x + \sqrt{2})(x - \sqrt{2})$

103. (a)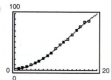
(b) The model fits the data well.
(c) About 116 subscriptions; No, because you cannot have more subscriptions than people.

105. (a) Answers will vary.
(b) 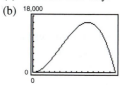 $20 \times 20 \times 40$

(c) $15, \dfrac{15 \pm 15\sqrt{5}}{2}$;
$\dfrac{15 - 15\sqrt{5}}{2}$ represents a negative volume.

107. False. If $(7x + 4)$ is a factor of f, then $-\frac{4}{7}$ is a zero of f.

109. $-2(x - 1)^2(x + 2)$

111. $-(x - 2)(x + 2)(x + 1)(x - 1)$

113. (a) $x + 1, \ x \neq 1$ (b) $x^2 + x + 1, \ x \neq 1$
(c) $x^3 + x^2 + x + 1, \ x \neq 1$
$\dfrac{x^n - 1}{x - 1} = x^{n-1} + x^{n-2} + \cdots + x^2 + x + 1, \ x \neq 1$

115. $\pm\dfrac{5}{3}$ **117.** $\dfrac{-3 \pm \sqrt{3}}{2}$

Section 3.4   (page 286)

1. Fundamental Theorem, Algebra   **3.** n linear factors

5. c   **6.** a   **7.** d   **8.** b

9 and 11. Answers will vary.

13. Zeros: $4, -i, i$. One real zero; they are the same.

15. Zeros: $\sqrt{2}i, \sqrt{2}i, -\sqrt{2}i, -\sqrt{2}i$. No real zeros; they are the same.

17. $2 \pm \sqrt{3}$
$(x - 2 - \sqrt{3})(x - 2 + \sqrt{3})$

19. $6 \pm \sqrt{10}$
$(x - 6 - \sqrt{10})(x - 6 + \sqrt{10})$

21. $\pm 5i$
$(x + 5i)(x - 5i)$

23. $\pm\frac{3}{2}, \pm\frac{3}{2}i$
$(2x - 3)(2x + 3)(2x - 3i)(2x + 3i)$

25. $\dfrac{1 \pm \sqrt{223}i}{2}$
$\left(z - \dfrac{1 - \sqrt{223}i}{2}\right)\left(z - \dfrac{1 + \sqrt{223}i}{2}\right)$

27. $\pm i, \pm 3i$
$(x + i)(x - i)(x + 3i)(x - 3i)$

29. $\frac{5}{3}, \pm 4i$
$(3x - 5)(x - 4i)(x + 4i)$

31. $-5, 4 \pm 3i$
$(t + 5)(t - 4 + 3i)(t - 4 - 3i)$

33. $1 \pm \sqrt{5}i, -\frac{1}{5}$
$(5x + 1)(x - 1 + \sqrt{5}i)(x - 1 - \sqrt{5}i)$

35. $2, 2, \pm 2i$
$(x - 2)^2(x + 2i)(x - 2i)$

37. (a) $7 \pm \sqrt{3}$ (b) $(x - 7 - \sqrt{3})(x - 7 + \sqrt{3})$
(c) $(7 \pm \sqrt{3}, 0)$

39. (a) $\frac{3}{2}, \pm 2i$ (b) $(2x - 3)(x - 2i)(x + 2i)$ (c) $(\frac{3}{2}, 0)$

41. (a) $-6, 3 \pm 4i$ (b) $(x + 6)(x - 3 - 4i)(x - 3 + 4i)$
(c) $(-6, 0)$

43. (a) $\pm 4i, \pm 3i$ (b) $(x + 4i)(x - 4i)(x + 3i)(x - 3i)$
(c) None

45. $f(x) = x^3 - 2x^2 + x - 2$

47. $f(x) = x^4 - 12x^3 + 53x^2 - 100x + 68$

49. $f(x) = x^4 + 3x^3 - 7x^2 + 15x$

51. (a) $-(x - 1)(x + 2)(x - 2i)(x + 2i)$
(b) $f(x) = -(x^4 + x^3 + 2x^2 + 4x - 8)$

53. (a) $-2(x + 1)(x - 2 + \sqrt{5}i)(x - 2 - \sqrt{5}i)$

CHAPTER 3

(b) $f(x) = -2x^3 + 6x^2 - 10x - 18$

55. (a) $(x^2 + 1)(x^2 - 7)$ (b) $(x^2 + 1)(x + \sqrt{7})(x - \sqrt{7})$
  (c) $(x + i)(x - i)(x + \sqrt{7})(x - \sqrt{7})$

57. (a) $(x^2 - 6)(x^2 - 2x + 3)$
  (b) $(x + \sqrt{6})(x - \sqrt{6})(x^2 - 2x + 3)$
  (c) $(x + \sqrt{6})(x - \sqrt{6})(x - 1 - \sqrt{2}i)(x - 1 + \sqrt{2}i)$

59. $-\frac{3}{2}, \pm 5i$ **61.** $-3, 5 \pm 2i$ **63.** $-\frac{2}{3}, 1 \pm \sqrt{3}i$

65. $\frac{3}{4}, \frac{1}{2}(1 \pm \sqrt{5}i)$ **67.** (a) $1.000, 2.000$ (b) $-3 \pm \sqrt{2}i$

69. (a) 0.750 (b) $\dfrac{1}{2} \pm \dfrac{\sqrt{5}}{2}i$

71. No. Setting $h = 50$ and solving the resulting equation yields imaginary roots.

73. False. A third-degree polynomial must have at least one real zero.

75. Answers will vary.

77. Parabola opening upward   **79.** Parabola opening upward
  Vertex: $\left(\frac{7}{2}, -\frac{81}{4}\right)$ Vertex: $\left(-\frac{5}{12}, -\frac{169}{24}\right)$

Section 3.5   (page 293)

1. rational functions   **3.** vertical asymptote

5. (a) Domain: all real numbers x except $x = 1$
(b)

x	$f(x)$	x	$f(x)$
0.5	-2	1.5	2
0.9	-10	1.1	10
0.99	-100	1.01	100
0.999	-1000	1.001	1000

x	$f(x)$	x	$f(x)$
5	0.25	-5	$-0.\overline{16}$
10	$0.\overline{1}$	-10	$-0.\overline{09}$
100	$0.\overline{01}$	-100	$-0.\overline{0099}$
1000	$0.\overline{001}$	-1000	$-0.\overline{000999}$

(c) f approaches $-\infty$ from the left and ∞ from the right of $x = 1$.

7. (a) Domain: all real numbers x except $x = 1$
(b)

x	$f(x)$	x	$f(x)$
0.5	3	1.5	9
0.9	27	1.1	33
0.99	297	1.01	303
0.999	2997	1.001	3003

x	$f(x)$	x	$f(x)$
5	3.75	-5	-2.5
10	$3.\overline{33}$	-10	-2.727
100	$3.\overline{03}$	-100	-2.97
1000	$3.\overline{003}$	-1000	-2.997

(c) f approaches ∞ from both the left and the right of $x = 1$.

9. (a) Domain: all real numbers x except $x = \pm 1$
(b)

x	$f(x)$	x	$f(x)$
0.5	-1	1.5	5.4
0.9	-12.79	1.1	17.29
0.99	-147.8	1.01	152.3
0.999	-1498	1.001	1502.3

x	$f(x)$	x	$f(x)$
5	3.125	-5	3.125
10	$3.\overline{03}$	-10	$3.\overline{03}$
100	$3.\overline{0003}$	-100	$3.\overline{0003}$
1000	3	-1000	3.000003

(c) f approaches ∞ from the left and $-\infty$ from the right of $x = -1$. f approaches $-\infty$ from the left and ∞ from the right of $x = 1$.

11. a   **12.** d   **13.** c   **14.** e   **15.** b   **16.** f

17. Vertical asymptote: $x = 0$
  Horizontal asymptote: $y = 0$

19. Vertical asymptote: $x = -3, 2$
  Horizontal asymptote: $y = 2$

21. Vertical asymptote: $x = 2$
  Horizontal asymptote: $y = -1$
  Hole at $x = 0$

23. Vertical asymptote: $x = 0$
  Horizontal asymptote: $y = 1$
  Hole at $x = -5$

25. (a) Domain: all real numbers x (b) Continuous
  (c) Horizontal asymptote: $y = 3$

27. (a) Domain: all real numbers x except $x = 3$
  (b) Not continuous
  (c) Vertical asymptote: $x = 3$
     Horizontal asymptote: $y = 0$

29. (a) Domain of f: all real numbers x except $x = 4$
    Domain of g: all real numbers x
  (b) Vertical asymptote: none; Hole in f at $x = 4$
  (c)

x	1	2	3	4	5	6	7
$f(x)$	5	6	7	Undef.	9	10	11
$g(x)$	5	6	7	8	9	10	11

(d)

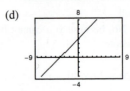

(e) Graphing utilities are limited in their resolution and therefore may not show a hole in a graph.

31. (a) Domain of f: all real numbers x except $x = -1, 3$
Domain of g: all real numbers x except $x = 3$
(b) Vertical asymptote: $x = 3$; Hole in f at $x = -1$
(c)

x	-2	-1	0	1	2	3	4
$f(x)$	$\frac{3}{5}$	Undef.	$\frac{1}{3}$	0	-1	Undef.	3
$g(x)$	$\frac{3}{5}$	$\frac{1}{2}$	$\frac{1}{3}$	0	-1	Undef.	3

(d)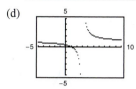

(e) Graphing utilities are limited in their resolution and therefore may not show a hole in a graph.
33. 4; less than; greater than   **35.** 2; greater than; less than
37. ± 2 **39.** 7   **41.** $-1, 3$ **43.** 2
45. (a) $28.33 million; $170 million; $765 million
(b)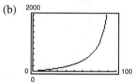

Answers will vary.
(c) No. The function is undefined at the 100% level.
47. (a)

M	200	400	600	800	1000
t	0.472	0.596	0.710	0.817	0.916

M	1200	1400	1600	1800	2000
t	1.009	1.096	1.178	1.255	1.328

The greater the mass, the more time required per oscillation.
(b) $M \approx 1306$ g
49. (a) 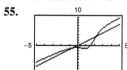   The model fits the data well.

(b) 1412 thousand; 1414 thousand; 1416 thousand; Answers will vary.
(c) $y = 1482.6$; Answers will vary.
51. False. $\dfrac{1}{x^2 + 1}$ has no vertical asymptote.
53. No. If $x = c$ is also a zero in the denominator, then f is undefined at $x = c$.
55. 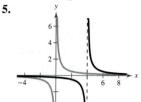   **57.** $x - y - 1 = 0$
59. $3x - y + 1 = 0$

Both graphs have the same slope.
61. $x + 9 + \dfrac{42}{x - 4}$ **63.** $2x^2 - 9 + \dfrac{34}{x^2 + 5}$

Section 3.6   (page 303)

1. slant, asymptote   **3.** Yes
5. 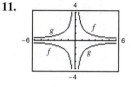   **7.**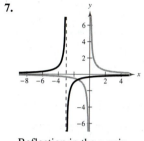

Horizontal shift four units to the right   Reflection in the x-axis, horizontal shift three units to the left

9. 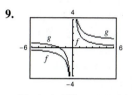   **11.**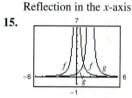

Vertical shift   Reflection in the x-axis

13. 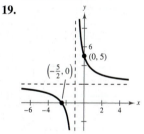   **15.**

Vertical shift   Horizontal shift

17. 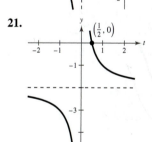   **19.**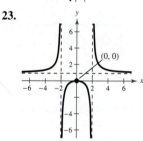

21.   **23.**

25. 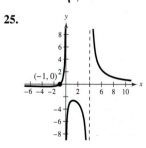   **27.**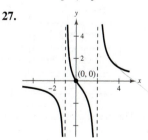

29.

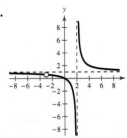

There is a hole at $x = -3$.

31.

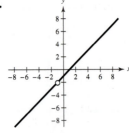

There is a hole at $x = -1$.

33.

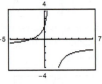

Domain: $(-\infty, 1), (1, \infty)$
Vertical asymptote: $x = 1$
Horizontal asymptote:
  $y = -1$

35.

Domain: $(-\infty, 0), (0, \infty)$
Vertical asymptote: $t = 0$
Horizontal asymptote:
  $y = 3$
Domain: $(-\infty, \infty)$
Horizontal asymptote: $y = 0$

37.

39.

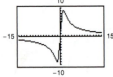

Domain:
  $(-\infty, -2), (-2, 3), (3, \infty)$
Vertical asymptotes:
  $x = -2, x = 3$
Horizontal asymptote: $y = 0$

41.

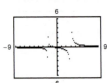

Domain: $(-\infty, 0), (0, \infty)$
Vertical asymptote: $x = 0$
Horizontal asymptote: $y = 0$

43.

There are two horizontal
asymptotes, $y = \pm 6$.

45.

There are two horizontal
asymptotes, $y = \pm 4$, and
one vertical asymptote,
$x = -1$.

47.

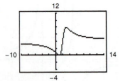

The graph crosses the horizontal asymptote, $y = 4$.

49.

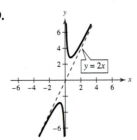

51.

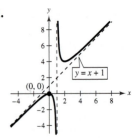

53.

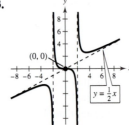

55.

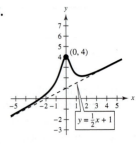

57. $(-1, 0)$ **59.** $(1, 0), (-1, 0)$

61.

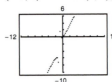

Domain: $(-\infty, -1), (-1, \infty)$
Vertical asymptote: $x = -1$
Slant asymptote: $y = 2x - 1$

63.

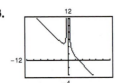

Domain: $(-\infty, 0), (0, \infty)$
Vertical asymptote: $x = 0$
Slant asymptote: $y = -x + 3$

65. Vertical asymptotes: $x = \pm 2$; horizontal asymptote: $y = 1$; slant asymptote: none; holes: none

67. Vertical asymptote: $x = -\frac{3}{2}$; horizontal asymptote: $y = 1$; slant asymptote: none; hole at $x = 2$

69. Vertical asymptote: $x = -2$; horizontal asymptote: none; slant asymptote: $y = 2x - 7$; hole at $x = -1$

71.

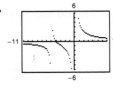

$(-4, 0)$

73.

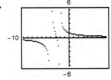

$\left(-\frac{8}{3}, 0\right)$

75.

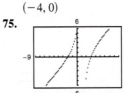

$(3, 0), (-2, 0)$

77.

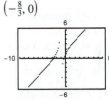

$\left(\dfrac{-3 \pm \sqrt{5}}{2}, 0\right)$

79.

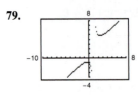

None

81.

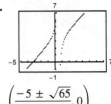

$\left(\dfrac{-5 \pm \sqrt{65}}{4}, 0\right)$

83. (a) Answers will vary.   (b) $[0, 950]$

(c)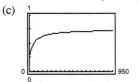

The concentration increases more slowly; the concentration approaches 75%.

85. (a) Answers will vary.

(b) $(2, \infty)$

(c)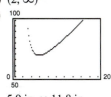

5.9 in. $\times$ 11.8 in.

87.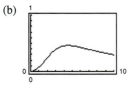

$x \approx 40$

89. (a) $C = 0$. The chemical will eventually dissipate.

(b)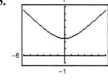

$t \approx 4.5$ h

(c) Before about 2.6 hours and after about 8.3 hours

91. (a) $A = -0.2182t + 5.665$ (b) $A = \dfrac{1}{0.0302t - 0.020}$

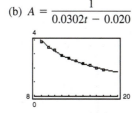

(c)

Year	1999	2000	2001	2002
Original data, A	3.9	3.5	3.3	2.9
Model from (a), A	3.7	3.5	3.3	3.0
Model from (b), A	4.0	3.5	3.2	2.9

Year	2003	2004	2005	2006
Original data, A	2.7	2.5	2.3	2.2
Model from (a), A	2.8	2.6	2.4	2.2
Model from (b), A	2.7	2.5	2.3	2.2

Year	2007	2008
Original data, A	2.0	1.9
Model from (a), A	2.0	1.7
Model from (b), A	2.0	1.9

Answers will vary.

93. False. The graph of a rational function is continuous when the polynomial in the denominator has no real zeros.

95.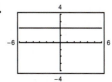

The denominator is a factor of the numerator.

97. *Horizontal asymptotes:*

If the degree of the numerator is greater than the degree of the denominator, then there is no horizontal asymptote.

If the degree of the numerator is less than the degree of the denominator, then there is a horizontal asymptote at $y = 0$.

If the degree of the numerator is equal to the degree of the denominator, then there is a horizontal asymptote at the line given by the ratio of the leading coefficients.

Vertical asymptotes:

Set the denominator equal to zero and solve.

Slant asymptotes:

If there is no horizontal asymptote and the degree of the numerator is exactly one greater than the degree of the denominator, then divide the numerator by the denominator. The slant asymptote is the result, not including the remainder.

99. $\dfrac{512}{x^3}$ **101.** 3

103.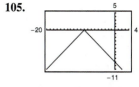

Domain: $(-\infty, \infty)$

Range: $[\sqrt{6}, \infty)$

105.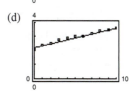

Domain: $(-\infty, \infty)$

Range: $(-\infty, 0]$

107. Answers will vary.

Section 3.7   (page 311)

1. Quadratic   **3.** Quadratic   **5.** Linear   **7.** Neither

9. (a)

(b) Linear

(c) $y = 0.14x + 2.2$

(d)

(e)

x	0	1	2	3	4	5
Actual, y	2.1	2.4	2.5	2.8	2.9	3.0
Model, y	2.2	2.3	2.5	2.6	2.8	2.9

x	6	7	8	9	10
Actual, y	3.0	3.2	3.4	3.5	3.6
Model, y	3.0	3.2	3.3	3.5	3.6

11. (a)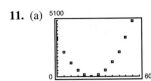

(b) Quadratic
(c) $y = 5.55x^2 - 277.5x + 3478$

(d)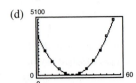

(e)

x	0	5	10	15	20	25
Actual, y	3480	2235	1250	565	150	12
Model, y	3478	2229	1258	564	148	9

x	30	35	40	45	50	55
Actual, y	145	575	1275	2225	3500	5010
Model, y	148	564	1258	2229	3478	5004

13. (a)

(b) Linear
(c) $y = 2.48x + 1.1$

(d)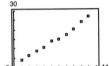

(e)

x	1	2	3	4	5
Actual, y	4.0	6.5	8.8	10.6	13.9
Model, y	3.6	6.1	8.5	11.0	13.5

x	6	7	8	9	10
Actual, y	15.0	17.5	20.1	24.0	27.1
Model, y	16.0	18.5	20.9	23.4	25.9

15. (a)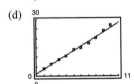

(b) Quadratic
(c) $y = 0.14x^2 - 9.9x + 591$

(d)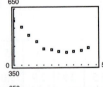

(e)

x	0	5	10	15	20	25
Actual, y	587	551	512	478	436	430
Model, y	591	545	506	474	449	431

x	30	35	40	45	50
Actual, y	424	420	423	429	444
Model, y	420	416	419	429	446

17. (a)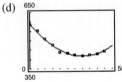

(b) $P = 0.1323t^2 - 1.893t + 6.85$
(c)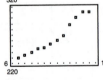

(d) July

19. (a)

(b) $P = -0.5638t^2 + 9.690t + 32.17$
(c)

(d) 2014; No. By 2021, the model gives negative values for the number of Internet users.
Answers will vary.

21. (a)

(b) $T = 7.97t + 166.1$
$r^2 \approx 0.9469$

(c)

(d) $T = 0.459t^2 - 3.51t + 232.4$
$r^2 \approx 0.9763$

(e)

(f) Quadratic
(g) Linear: 2014
Quadratic: 2011

23. True. See "Basic Characteristics of Quadratic Functions" on page 245.

25. The model is consistently above the data points.

27. (a) $(f \circ g)(x) = 2x^2 + 5$ (b) $(g \circ f)(x) = 4(x^2 - x + 1)$

29. (a) $(f \circ g)(x) = x$ (b) $(g \circ f)(x) = x$

31. $f^{-1}(x) = \dfrac{x - 5}{2}$ **33.** $f^{-1}(x) = \sqrt{x - 5}$

35. $1 + 3i$; 10   **37.** $5i$; 25

Review Exercises   (page 316)

1.
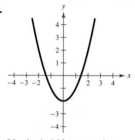
Vertical shift two units downward

3.
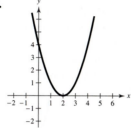
Horizontal shift two units to the right

5.
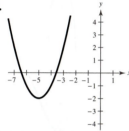
Horizontal shift five units to the left, vertical shift two units downward

7. Parabola opening upward
Vertex: $\left(-\frac{3}{2}, 1\right)$
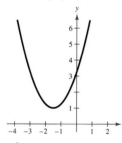
x-intercept: none

9. Parabola opening upward
Vertex: $\left(-\frac{5}{2}, -\frac{41}{12}\right)$
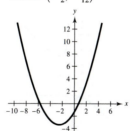
x-intercepts:
$\left(\dfrac{-5 \pm \sqrt{41}}{2}, 0\right)$

11. $f(x) = (x - 1)^2 - 4$

13. (a) $A = x\left(\dfrac{8 - x}{2}\right), 0 < x < 8$

(b)
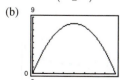
$x = 4, y = 2$

(c) $A = -\frac{1}{2}(x - 4)^2 + 8$; $x = 4, y = 2$
They are the same.

15.
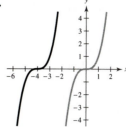
Horizontal shift four units to the left

17.
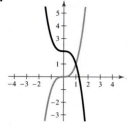
Reflection in the x-axis, vertical shift two units upward

19.
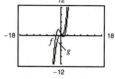
Yes. Both functions are of the same degree and have the same degree and have positive leading coefficients.

21. Falls to the left, falls to the right

23. (a) $x = -1, 0, 0, 2$
(b)
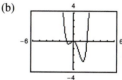
(c) $x = -1, 0, 0, 2$;
They are the same.

25. (a) $t = 0, \pm\sqrt{3}$
(b)
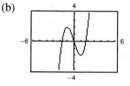
(c) $t = 0, \pm1.73$;
They are the same.

27. (a) $x = -3, -3, 0$
(b)
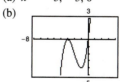
(c) $x = -3, -3, 0$; They are the same.

29. $f(x) = x^4 - 5x^3 - 3x^2 + 17x - 10$

31. $f(x) = x^3 - 7x^2 + 13x - 3$

33. (a) Rises to the left, rises to the right
(b) $x = -3, -1, 3, 3$
(c) and (d)
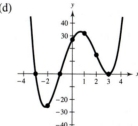

35. (a) $(-3, -2), (-1, 0), (0, 1)$ (b) $x = -2.25, -0.56, 0.80$

37. (a) $(-3, -2), (2, 3)$ (b) $x = -2.57, 2.57$

39. $8x + 5 + \dfrac{2}{3x - 2}$ **41.** $x^2 - 2, x \neq \pm1$

43. $5x + 2, x \neq \dfrac{3 \pm \sqrt{5}}{2}$

45. $0.25x^3 - 4.5x^2 + 9x - 18 + \dfrac{36}{x + 2}$

CHAPTER 3

25. Right shift of two units and downward shift of three units

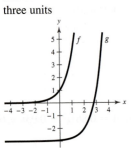

27. (a)

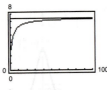

(b) $e^2 \approx 7.3891$

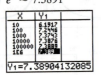

29. 9897.129   **31.** 54.164

33.

x	-2	-1	0	1	2
$f(x)$	0.16	0.4	1	2.5	6.25

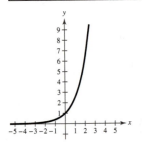

Asymptote: $y = 0$

35.

x	-2	-1	0	1	2
$f(x)$	0.03	0.17	1	6	36

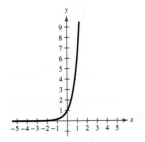

Asymptote: $y = 0$

37.

x	-3	-2	-1	0	1
$f(x)$	0.33	1	3	9	27

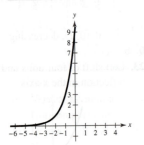

Asymptote: $y = 0$

39.

x	-1	0	1	2	3	4
y	1.04	1.11	1.33	2	4	10

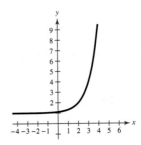

Asymptote: $y = 1$

41.

x	-2	-1	0	1	2
$f(x)$	7.39	2.72	1	0.37	0.14

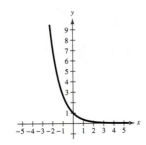

Asymptote: $y = 0$

43.

x	-6	-5	-4	-3	-2
$f(x)$	0.41	1.10	3	8.15	22.17

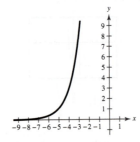

Asymptote: $y = 0$

45.

x	3	4	5	6	7
$f(x)$	2.14	2.37	3	4.72	9.39

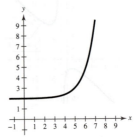

Asymptote: $y = 2$

47.

t	−2	−1	0	1	2
s(t)	1.57	1.77	2	2.26	2.54

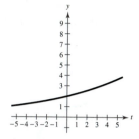

Asymptote: $y = 0$

49. (a)

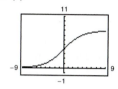

(b) $y = 0, y = 8$

51. (a)

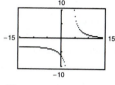

(b) $y = -3, y = 0,$
    $x \approx 3.47$

53. (86.350, 1500)

55. (a)

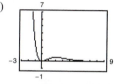

(b) Decreasing on $(-\infty, 0), (2, \infty)$
    Increasing on $(0, 2)$

(c) Relative minimum: $(0, 0)$
    Relative maximum: $(2, 0.54)$

57.

n	1	2	4	12
A	$3047.49	$3050.48	$3051.99	$3053.00

n	365	Continuous
A	$3053.49	$3053.51

59.

n	1	2	4	12
A	$5477.81	$5520.10	$5541.79	$5556.46

n	365	Continuous
A	$5563.61	$5563.85

61.

t	1	10	20
A	$12,489.73	$17,901.90	$26,706.49

t	30	40	50
A	$39,841.40	$59,436.39	$88,668.67

63.

t	1	10	20
A	$12,427.44	$17,028.81	$24,165.03

t	30	40	50
A	$34,291.81	$48,662.40	$69,055.23

65. $1530.57   **67.** $17,281.77

69. (a) $y_1 = 500(1 + 0.07)^t$
    $y_2 = 500\left(1 + \dfrac{0.07}{4}\right)^{4t}$
    $y_3 = 500e^{0.07t}$

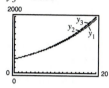

(b) y_3 yields the highest return after 20 years.
    $y_2 - y_1 = \$68.36$
    $y_3 - y_2 = \$24.40$
    $y_3 - y_1 = \$92.76$

71. (a) 10 g   (b) 7.85 g
(c)

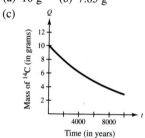

73. (a)

(b)

t	15	16	17	18	19
P	40.14	40.53	40.93	41.33	41.73

t	20	21	22	23	24
P	42.14	42.55	42.96	43.38	43.80

t	25	26	27	28	29	30
P	44.23	44.66	45.10	45.54	45.98	46.43

(c) 2037

75. True. The definition of an exponential function is $f(x) = a^x$,
    $a > 0, a \neq 1$.

77. d

CHAPTER 4

79.

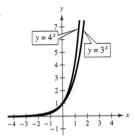

$3^x < 4^x$ when $x > 0$

81. > **83.** > **85.** $f^{-1}(x) = \dfrac{x + 7}{5}$

87. $f^{-1}(x) = x^3 - 8$ **89.** Answers will vary.

Section 4.2 (page 343)

1. logarithmic function **3.** $a^{\log_a x} = x$ **5.** $b = a^c$
7. $4^3 = 64$ **9.** $7^{-2} = \frac{1}{49}$ **11.** $32^{2/5} = 4$
13. $2^{1/2} = \sqrt{2}$ **15.** $\log_5 125 = 3$ **17.** $\log_{81} 3 = \frac{1}{4}$
19. $\log_6 \frac{1}{36} = -2$ **21.** $\log_g 4 = a$ **23.** 4 **25.** -3
27. 2.538 **29.** 7.022 **31.** 9 **33.** 2 **35.** $\frac{1}{10}$
37. $3x$ **39.** -3

41.

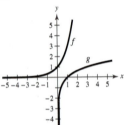

43.

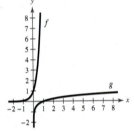

45.

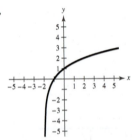

Domain: $(-2, \infty)$
Vertical asymptote: $x = -2$
x-intercept: $(-1, 0)$

47.

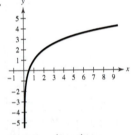
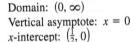

Domain: $(0, \infty)$
Vertical asymptote: $x = 0$
x-intercept: $\left(\frac{1}{2}, 0\right)$

49.

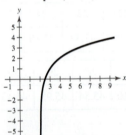

Domain: $(2, \infty)$
Vertical asymptote: $x = 2$
x-intercept: $\left(\frac{5}{2}, 0\right)$

51. b **52.** c **53.** d **54.** a **55.** Reflection in the x-axis
57. Reflection in the x-axis, vertical shift four units upward

59. Horizontal shift three units to the left and vertical shift two units downward
61. $e^0 = 1$ **63.** $e^1 = e$ **65.** $e^{1/2} = \sqrt{e}$
67. $e^{2.1972\cdots} = 9$ **69.** $\ln 20.0855\ldots = 3$
71. $\ln 3.6692\ldots = 1.3$ **73.** $\ln 1.3956\ldots = \frac{1}{3}$
75. $\ln 4.4816\ldots = \frac{3}{2}$ **77.** 1.869 **79.** 0.693 **81.** 2
83. 1.8 **85.** 0 **87.** 1
89.

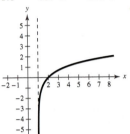

91.

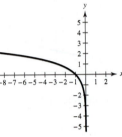

Domain: $(1, \infty)$
Vertical asymptote: $x = 1$
x-intercept: $(2, 0)$

Domain: $(-\infty, 0)$
Vertical asymptote: $x = 0$
x-intercept: $(-1, 0)$

93. Horizontal shift three units to the left
95. Vertical shift five units downward
97. Horizontal shift one unit to the right and vertical shift two units upward

99. (a)

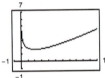

(b) Domain: $(0, \infty)$

(c) Decreasing on $(0, 2)$; increasing on $(2, \infty)$
(d) Relative minimum: $(2, 1.693)$

101. (a)

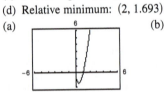

(b) Domain: $(0, \infty)$

(c) Decreasing on $(0, 0.368)$; increasing on $(0.368, \infty)$
(d) Relative minimum: $(0.368, -1.472)$

103. (a)

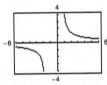

(b) Domain: $(-\infty, -2), (1, \infty)$
(c) Decreasing on $(-\infty, -2); (1, \infty)$
(d) No relative maxima or minima

105. (a)

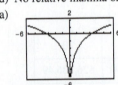

(b) Domain: $(-\infty, 0), (0, \infty)$
(c) Decreasing on $(-\infty, 0)$; increasing on $(0, \infty)$
(d) No relative maxima or minima

107. (a)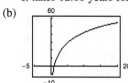

(b) Domain: $[1, \infty)$
(c) Increasing on $(1, \infty)$
(d) Relative minimum: $(1, 0)$

109. (a) 80   (b) 68.12   (c) 62.30

111. (a)

K	1	2	4	6
t	0	12.60	25.21	32.57

K	8	10	12
t	37.81	41.87	45.18

It takes 12.60 years for the principal to double.

(b)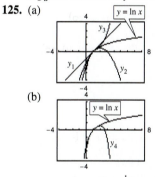

113. (a) 30 yr, 10 yr

(b) Total amount: \$323,179.20; interest: \$173,179.20
Total amount: \$199,108.80; interest: \$49,108.80

115. False. Reflect $g(x)$ about the line $y = x$.

117. 2   **119.** $\frac{1}{4}$ **121.** b

123. $\log_a x$ is the inverse of a^x only if $0 < a < 1$ and $a > 1$, so $\log_a x$ is defined only for $0 < a < 1$ and $a > 1$.

125. (a)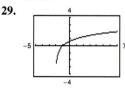

(b)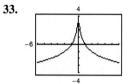

$y_4 = (x - 1) - \frac{1}{2}(x - 1)^2 + \frac{1}{3}(x - 1)^3 - \frac{1}{4}(x - 1)^4$
Answers will vary.

127. (a)

x	1	5	10	10^2
f(x)	0	0.32	0.23	0.046

x	10^4	10^6
f(x)	0.00092	0.0000138

(b) 0

129. $(x + 3)(x - 1)$ **131.** $(4x + 3)(3x - 1)$

133. $(4x + 5)(4x - 5)$ **135.** $x(2x - 9)(x + 5)$ **137.** 15

139. 2.75   **141.** 27.67

Section 4.3 (page 351)

1. change-of-base   **3.** $\log_3 24 = \dfrac{\ln 24}{\ln 3}$

5. (a) $\dfrac{\log_{10} x}{\log_{10} 5}$ (b) $\dfrac{\ln x}{\ln 5}$ **7.** (a) $\dfrac{\log_{10} x}{\log_{10} \frac{1}{5}}$ (b) $\dfrac{\ln x}{\ln \frac{1}{5}}$

9. (a) $\dfrac{\log_{10} \frac{3}{10}}{\log_{10} a}$ (b) $\dfrac{\ln \frac{3}{10}}{\ln a}$ **11.** (a) $\dfrac{\log_{10} x}{\log_{10} 2.6}$ (b) $\dfrac{\ln x}{\ln 2.6}$

13. 1.771   **15.** -2 **17.** -0.059 **19.** 2.691

21. $\ln 5 + \ln 4$ **23.** $2 \ln 5 - \ln 4$ **25.** 1.6542

27. 0.2823

29. 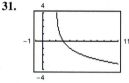   **31.**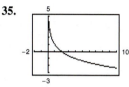

33.   **35.**

37. $\frac{3}{2}$ **39.** $4 + 4 \log_2 3$ **41.** $6 + \ln 5$ **43.** $\ln 6 - 2$

45. Answers will vary.   **47.** $\log_{10} 5 + \log_{10} x$

49. $\log_{10} t - \log_{10} 8$ **51.** $4 \log_8 x$ **53.** $\frac{1}{2} \ln z$

55. $\ln x + \ln y + \ln z$ **57.** $\log_6 a + 3 \log_6 b + 2 \log_6 c$

59. $\frac{1}{3} \ln x - \frac{1}{3} \ln y$ **61.** $\ln(x + 1) + \ln(x - 1) - 3 \ln x,\ x > 1$

63. $4 \ln x + \frac{1}{2} \ln y - 5 \ln z$

65. (a)

(b)

x	1	2	3	4	5	6
y_1	1.61	3.87	5.24	6.24	7.03	7.68
y_2	1.61	3.87	5.24	6.24	7.03	7.68

x	7	8	9	10	11
y_1	8.24	8.72	9.16	9.55	9.90
y_2	8.24	8.72	9.16	9.55	9.90

(c) $y_1 = y_2$

67. (a)   (c) $y_1 = y_2$

(b)

x	0	3	4	5
y_1	Error	4.39	4.85	5.34
y_2	Error	4.39	4.85	5.34

x	6	7	8
y_1	5.78	6.17	6.53
y_2	5.78	6.17	6.53

CHAPTER 4

69. $\ln 4x$ **71.** $\log_4 \dfrac{z}{y}$ **73.** $\log_2(x + 3)^2$

75. $\ln\sqrt{x^2 + 4}$ **77.** $\ln\dfrac{x}{(x + 1)^3}$ **79.** $\ln\dfrac{x - 2}{x + 2}$

81. $\ln\dfrac{x}{(x^2 - 4)^2}$ **83.** $\ln\sqrt[3]{\dfrac{x(x + 3)^2}{x^2 - 1}}$

85. (a)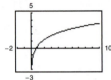

(b)

x	-5	-4	-3	-2	-1
y_1	-2.36	-1.51	-0.45	0.94	2.77
y_2	-2.36	-1.51	-0.45	0.94	2.77

x	0	1	2	3	4	5
y_1	4.16	2.77	0.94	-0.45	-1.51	-2.36
y_2	4.16	2.77	0.94	-0.45	-1.51	-2.36

(c) $y_1 = y_2$

87. (a)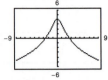

(b)

x	-1	0	1	2
y_1	Error	Error	0.35	1.24
y_2	Error	Error	0.35	1.24

x	3	4	5
y_1	1.79	2.19	2.51
y_2	1.79	2.19	2.51

(c) $y_1 = y_2$

89. (a)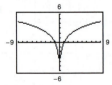

(b)

x	-5	-4	-3	-2	-1
y_1	3.22	2.77	2.20	1.39	0
y_2	Error	Error	Error	Error	Error

x	0	1	2	3	4	5
y_1	Error	0	1.39	2.20	2.77	3.22
y_2	Error	0	1.39	2.20	2.77	3.22

(c) No. The domains differ.

91. (a)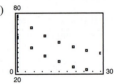

(b)

x	-4	-3	0	3
y_1	Error	Error	Error	1.61
y_2	2.48	1.61	Error	1.61

x	4	5	6
y_1	2.48	3.04	3.47
y_2	2.48	3.04	3.47

(c) No. The domains differ.

93. 2   **95.** 6.8

97. Not possible; -4 is not in the domain of $\log_2 x$.

99. 2   **101.** -4 **103.** 8   **105.** $-\dfrac{1}{2}$

107. (a) $\beta = 120 + 10\log_{10} I$

(b)

I	10^{-4}	10^{-6}	10^{-8}	10^{-10}	10^{-12}	10^{-14}
β	80	60	40	20	0	-20

109. (a)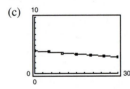

(b) $T = 54.4(0.964)^t + 21$

(c)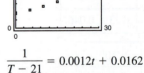

$\ln(T - 21) = -0.037t + 3.997$
$T = e^{(-0.037t + 3.997)} + 21$

(d)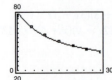

$\dfrac{1}{T - 21} = 0.0012t + 0.0162$

$T = \dfrac{1}{0.0012t + 0.0162} + 21$

111. True   **113.** False. $f(\sqrt{x}) = \frac{1}{2}f(x)$

115. True. When $f(x) = 0$, $x = 1 < e$.

117. The error is an improper use of the Quotient Property of logarithms.

$$\ln \frac{x^2}{\sqrt{x^2 + 4}} = \ln x^2 - \ln \sqrt{x^2 + 4}$$

$$= 2 \ln x - \frac{1}{2} \ln(x^2 + 4)$$

119.
$\ln 1 = 0$	$\ln 9 \approx 2.1972$
$\ln 2 \approx 0.6931$	$\ln 10 \approx 2.3025$
$\ln 3 \approx 1.0986$	$\ln 12 \approx 2.4848$
$\ln 4 \approx 1.3862$	$\ln 15 \approx 2.7080$
$\ln 5 \approx 1.6094$	$\ln 16 \approx 2.7724$
$\ln 6 \approx 1.7917$	$\ln 18 \approx 2.8903$
$\ln 8 \approx 2.0793$	$\ln 20 \approx 2.9956$

121. No. Domain of y_1: $(-\infty, 0), (2, \infty)$
Domain of y_2: $(2, \infty)$

123. $\dfrac{3x^4}{2y^3}$ **125.** $18x^3y^4$ **127.** $3 \pm \sqrt{7}$ **129.** $\pm 4, \pm \sqrt{3}$

Section 4.4   (page 361)

1. (a) $x = y$ (b) $x = y$ (c) x (d) x **3.** 7
5. Subtract 3 from both sides.   **7.** (a) Yes   (b) No
9. (a) No   (b) Yes   (c) Yes, approximate
11. (a) Yes, approximate   (b) No   (c) Yes
13. (a) Yes   (b) Yes, approximate   (c) No
15. 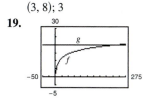   **17.**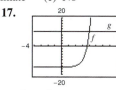

(3, 8); 3   (4, 10); 4
19. 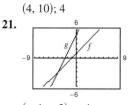   **21.**

(243, 20); 243 $(-4, -3)$; -4
23. 2   **25.** -4 **27.** -2 **29.** -4 **31.** $\ln 14$
33. $\log_{10} 36$ **35.** 5   **37.** 5   **39.** e^{-9} **41.** 5
43. 0.1   **45.** $\dfrac{e^5 + 1}{2}$ **47.** x^2 **49.** x^2
51. $2x - 1$ **53.** $x^2 + 6$ **55.** 0.944   **57.** 2
59. 184.444   **61.** 0.511   **63.** 0   **65.** -1.498
67. 6.960   **69.** -277.951 **71.** 1.609   **73.** $-1, 2$
75. 0.586, 3.414   **77.** 1.946   **79.** 183.258
81. (a)
x	0.6	0.7	0.8	0.9	1.0
e^{3x}	6.05	8.17	11.02	14.88	20.09

(0.8, 0.9)
(b) 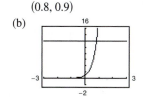   (c) 0.828

83. 21.330   **85.** 3.656

87. 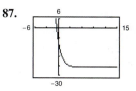   **89.**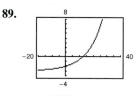

$x = -0.427$ $x = 12.207$
91. 0.050   **93.** 2.042   **95.** 1   **97.** 4453.242
99. 17.945   **101.** 103   **103.** 5.389   **105.** 1.718, -3.718
107. 2   **109.** No real solution   **111.** 180.384
113. $y = \log_a x, \quad a^y = x$
$\log_e a^y = \log_e x$
$y \log_e a = \log_e x$
$y = \dfrac{\log_e x}{\log_e a}$
$\log_a x = \dfrac{\log_e x}{\log_e a} = \dfrac{\ln x}{\ln a}$

115. (a)
x	2	3	4	5	6
$\ln 2x$	1.39	1.79	2.08	2.30	2.48

(5, 6)
(b) 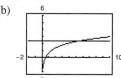   (c) 5.512

5.512
117. (a)
x	12	13	14	15	16
$6 \log_3(0.5x)$	9.79	10.22	10.63	11.00	11.36

(14, 15)
(b) 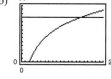   (c) 14.988

14.988
119. 1.469, 0.001   **121.** 2.928   **123.** 3.423
125. 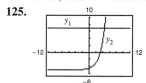   **127.**

(4.585, 7) $(-14.979, 80)$
129. 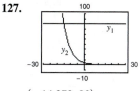   **131.** $-1, 0$ **133.** 2, 0

(663.142, 3.25)

135. $e^{-1/2} \approx 0.607$ **137.** $e^{-1} \approx 0.368$ **139.** $\dfrac{\ln y - \ln a}{b}$
141. $b \pm \sqrt{c(\ln y - \ln a)}$ **143.** (a) 9.24 yr   (b) 14.65 yr
145. (a) 27.73 yr   (b) 43.94 yr

147. (a) 682 units   (b) 779 units   **149.** 2008

151. (a)
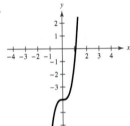
(b) $y = 20$. The object's temperature cannot cool below the room's temperature.
(c) 0.81 h

153. False; $e^x = 0$ has no solutions.

155. The error is that both sides of the equation should be divided by 2 before taking the natural log of both sides.

$$2e^x = 10$$
$$e^x = 5$$
$$\ln e^x = \ln 5$$
$$x = \ln 5$$

157. Inverse Property. You would take the natural log of both sides, which would give you $x \ln 5 = \ln 34$. So,
$$x = \frac{\ln 34}{\ln 5}.$$

159. Yes. The investment will double every $\dfrac{\ln 2}{r}$ years.

161.
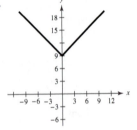

163.

165.
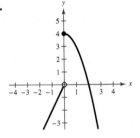

Section 4.5 (page 372)

1. (a) iv   (b) i   (c) iii   (d) vi   (e) ii   (f) v

3. sigmoidal   **5.** Exponential decay

7. c   **8.** e   **9.** b   **10.** a   **11.** d   **12.** f

	Initial Investment	Annual % Rate	Time to Double	Amount After 10 Years
13.	$10,000	3.5%	19.8 yr	$14,190.68
15.	$7500	3.30%	21 yr	$10,432.26
17.	$5000	1.25%	55.45 yr	$5665.74
19.	$63,762.82	4.5%	15.40 yr	$100,000.00

21.

r	2%	4%	6%	8%	10%	12%
t	54.93	27.47	18.31	13.73	10.99	9.16

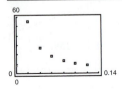

23.
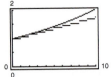
Continuous compounding

Isotope	Half-Life (years)	Initial Quantity	Amount After 1000 Years
25. ^{226}Ra	1599	10 g	6.48 g
27. ^{14}C	5700	3 g	2.66 g

29. $y = e^{0.768x}$ **31.** $y = 4e^{-0.2773x}$

33. (a) Decreasing. The negative exponent indicates that the model is decreasing.
(b) 333,680 people; 317,565 people; 308,272 people
(c) 2014

35. (a) 0.0189   (b) About 1,534,104 people

37. About 15,601 yr ago

39. (a) $V = -8305t + 49,200$ (b) $V = 49,200e^{-0.2059t}$
(c)
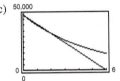
(d) Exponential model
(e) $0 < t < 2; t \geq 2$

41. (a)
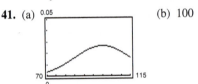
(b) 100

43. (a) About 203 animals   (b) About 13 mo
(c)
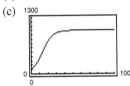

The horizontal asymptotes occur at $p = 1000$ and $p = 0$. The asymptote at $p = 1000$ means there will not be more than 1000 animals in the preserve.

45. (a) 10,000,000   (b) 125,892,541   (c) 1,258,925

47. (a) 20 dB   (b) 70 dB   (c) 120 dB

49. 97.49%   **51.** 4.64   **53.** About 31,623 times

55. (a)
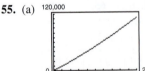
(b) 21.20 yr; yes

57. 3:00 A.M.   **59.** False. The domain can be all real numbers.

61. No. Any x-value in the Gaussian model will give a positive y-value.

63. Gaussian model   **65.** Exponential growth model

67. a; $(0, -3), \left(\frac{9}{4}, 0\right)$ **68.** b; $(0, 2), (5, 0)$

69. d; $(0, 25), \left(\frac{100}{9}, 0\right)$ **70.** c; $(0, 4), (2, 0)$

71. Falls to the left and rises to the right

73. Rises to the left, falls to the right   **75.** $2x^2 + 3 + \dfrac{3}{x - 4}$

77. Answers will vary.

Section 4.6 (page 382)

1. $y = ax^b$ **3.** Scatter plot **5.** Logarithmic model
7. Quadratic model **9.** Exponential model
11. Quadratic model
13.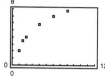

Logarithmic model

15.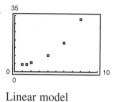

Linear model

17.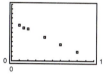

Linear model

19. $y = 4.752(1.2607)^x$; 0.96773

21. $y = 8.463(0.7775)^x$; 0.86639

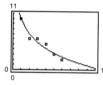

23. $y = 2.083 + 1.257 \ln x$; 0.98672

25. $y = 9.826 - 4.097 \ln x$; 0.93704

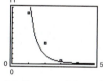

27. $y = 1.985x^{0.760}$; 0.99686

29. $y = 16.103x^{-3.174}$; 0.88161

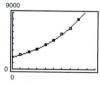

31. (a) Exponential model: $S = 1876.645(1.1980)^t$
Power model: $S = 1905.844t^{0.6018}$

(b) Exponential model: Power model:

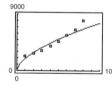

(c) Exponential model

33. (a) Linear model: (b) Power model:
$P = 2.89t + 252.9$; $P = 222.94t^{0.1048}$;
$r^2 \approx 0.9987$ $r^2 \approx 0.9850$

  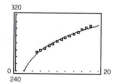

(c) Exponential model: (d) Logarithmic model:
$P = 254.445(1.0102)^t$; $P = 29.813 \ln t + 215.36$;
$r^2 \approx 0.9972$ $r^2 \approx 0.9803$

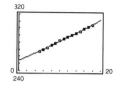

(e) The linear model is the best fit because its coefficient of determination is closest to 1.

(f) Linear:

Year	2009	2010	2011	2012	2013	2014
Population (in millions)	307.8	310.7	313.6	316.5	319.4	322.3

Power:

Year	2009	2010	2011	2012	2013	2014
Population (in millions)	303.5	305.2	306.7	308.2	309.7	311.1

Exponential:

Year	2009	2010	2011	2012	2013	2014
Population (in millions)	308.6	311.7	314.9	318.1	321.3	324.6

Logarithmic:

Year	2009	2010	2011	2012	2013	2014
Population (in millions)	303.1	304.7	306.1	307.5	308.8	310.1

(g) and (h) Answers will vary.

35. (a) $N = 1315.584(1.0644)^t$ (b) $N = 1315.584e^{0.0624t}$
(c) About 2307 stores; Answers will vary.

37. (a) $P = \dfrac{91.3686}{1 + 765.5440e^{-0.2547x}}$

(b) 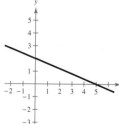 The model fits the data well.

39. True. See page 365. **41.** Answers will vary.
43. Slope: $-\frac{2}{5}$ **45.** Slope: $-\frac{12}{35}$
y-intercept: $(0, 2)$ y-intercept: $(0, 3)$

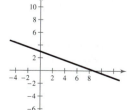

47. $y = -(x + 1)^2 + 2$ **49.** $y = -2(x - 3)^2 + 2$

Review Exercises (page 388)

1. 10.3254 **3.** 0.0001 **5.** c **6.** d **7.** b **8.** a

9.

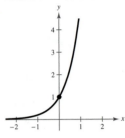

11.

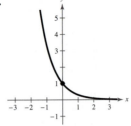

Horizontal asymptote:
  $y = 0$
y-intercept: $(0, 1)$
Increasing on $(-\infty, \infty)$

Horizontal asymptote:
  $y = 0$
y-intercept: $(0, 1)$
Decreasing on $(-\infty, \infty)$

13.

x	0	1	2	3	4
$h(x)$	0.37	1	2.72	7.39	20.09

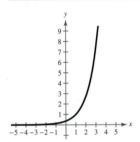

Horizontal asymptote: $y = 0$

15.

x	-2	-1	0	1	2
$h(x)$	-0.14	-0.37	-1	-2.72	-7.39

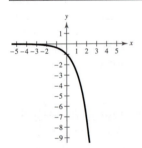

Horizontal asymptote: $y = 0$

17.

x	-1	0	1	2	3	4
$f(x)$	6.59	4	2.43	1.47	0.89	0.54

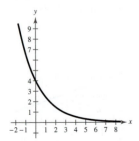

Horizontal asymptote: $y = 0$

19.

t	1	10	20
A	\$10,832.87	\$22,255.41	\$49,530.32

t	30	40	50
A	\$110,231.76	\$245,325.30	\$545,981.50

21. (a)

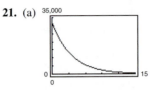

(b) \$18,000
(c) When it is first sold; Yes; Answers will vary.

23. $5^3 = 125$ **25.** $64^{1/6} = 2$ **27.** $\log_4 64 = 3$
29. $\log_{125} 25 = \frac{2}{3}$ **31.** $\log_{1/2} 8 = -3$ **33.** 3 **35.** -1
37. Domain: $(0, \infty)$ **39.** Domain: $(1, \infty)$
  Vertical asymptote: $x = 0$ Vertical asymptote: $x = 1$
  x-intercept: $(32, 0)$ x-intercept: $(1.016, 0)$

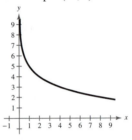
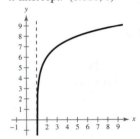

41. 3.068 **43.** 0.896
45.

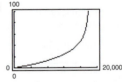

47.

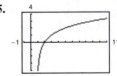

Domain: $(0, \infty)$
Vertical asymptote: $x = 0$
x-intercept: $(0.05, 0)$

Domain: $(0, \infty)$
Vertical asymptote: $x = 0$
x-intercept: $(1, 0)$

49. (a) $0 \le h < 18,000$
(b)

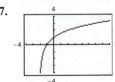

Asymptote: $h = 18,000$
(c) The time required to increase its altitude further increases.
(d) 5.46 min
51. 1.585 **53.** 2.235
55.

57.

59. 1.13 **61.** 0.41 **63.** $\ln 5 - 2$ **65.** $2 + \log_{10} 2$
67. $1 + 2 \log_5 x$ **69.** $\log_{10} 5 + \frac{1}{2} \log_{10} y - 2 \log_{10} x$
71. $\ln(x + 3) - \ln x - \ln y$ **73.** $\log_2 9x$

75. $\ln \dfrac{\sqrt{2x-1}}{(x+1)^2}$ **77.** $\ln \dfrac{3\sqrt[3]{4-x^2}}{x}$

79. (a)

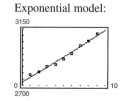

(b)

h	4	6	8	10	12	14
s	38	33	30	27	25	23

(c) The decrease in productivity starts to level off.

81. 4   **83.** −3   **85.** −5   **87.** 4096   **89.** 9

91. e^4 **93.** $e^2 + 1$ **95.** −0.757   **97.** 4.459

99. 1.760   **101.** 3.916   **103.** 1.609, 0.693

105. 200.615   **107.** 36.945   **109.** 53.598

111. No solution   **113.** 0.9   **115.** −1   **117.** 0.368

119. 10.05 yr   **121.** e   **122.** b   **123.** f   **124.** d

125. a   **126.** c   **127.** $k = 0.0177$; 11,407,330

129. (a) 9.52 weeks   (b) 21.20 weeks

131. Logistic model

133. (a) Linear model: $N = 41.5t + 2722.1$; $r^2 \approx 0.9785$
  Exponential model: $N = 2728(1.0142)^t$; $r^2 \approx 0.9818$
  Power model: $N = 2727.6t^{0.0497}$; $r^2 \approx 0.8398$

(b) Linear model: Exponential model:

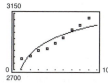

Power model:

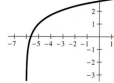

(c) The exponential model is the best fit because its coefficient of determination is closest to 1.

(d) 3,141,090   (e) 2021–2022

135. True. $e^{x-1} = e^x \cdot e^{-1} = \dfrac{e^x}{e}$ **137.** False. $x > 0$

139. Because $1 < \sqrt{2} < 2$, then $2^1 < 2^{\sqrt{2}} < 2^2$.

Chapter Test   (page 392)

1.

x	−2	−1	0	1	2
f(x)	100	10	1	0.1	0.01

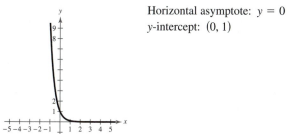

Horizontal asymptote: $y = 0$
y-intercept: $(0, 1)$

2.

x	0	2	3	4
f(x)	−0.03	−1	−6	−36

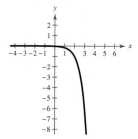

Horizontal asymptote: $y = 0$
y-intercept: $\left(0, -\frac{1}{36}\right)$

3.

x	−2	−1	0	1	2
f(x)	0.9817	0.8647	0	−6.3891	−53.5982

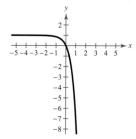

Horizontal asymptote: $y = 1$
Intercept: $(0, 0)$

4. −0.89   **5.** 9.2   **6.** 2

7.

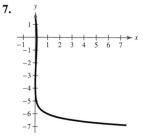

Domain: $(0, \infty)$
Vertical asymptote: $x = 0$
x-intercept: $(10^{-6}, 0)$

8.

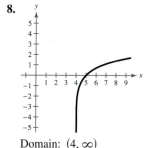

Domain: $(4, \infty)$
Vertical asymptote: $x = 4$
x-intercept: $(5, 0)$

9.

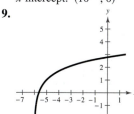

Domain: $(-6, \infty)$
Vertical asymptote: $x = -6$
x-intercept: $(-5.63, 0)$

10. 1.945   **11.** 0.115   **12.** 1.674

13. $\log_2 3 + 4\log_2 a$ **14.** $\ln 5 + \frac{1}{2}\ln x - \ln 6$

15. $\ln x + \frac{1}{2}\ln(x+1) - \ln 2 - 4$ **16.** $\log_3 13y$

17. $\ln \dfrac{x^4}{y^4}$ **18.** $\ln \dfrac{x(2x-3)}{x+2}$ **19.** 4   **20.** 2.431

21. 343   **22.** 100,004   **23.** 1.321   **24.** 1   **25.** 1.597

26. 1.649   **27.** 54.96%

28. (a) Logarithmic model: $R = 200.7 \ln t + 57.835$
Exponential model: $R = 115.47(1.227)^t$
Power model: $R = 119.22t^{0.6703}$

(b) Logarithmic model: Exponential model:

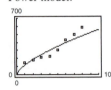   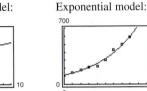

Power model:

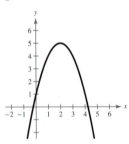

(c) Exponential model; $2483.9 million

Cumulative Test for Chapters 3–4 (page 393)

1.   **2.**

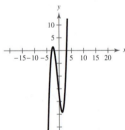   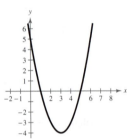

3.   **4.** $-2, \pm 2i$ **5.** 1.424

6. $4x + 2 - \dfrac{15}{x + 3}$ **7.** $2x^2 + 7x + 48 + \dfrac{268}{x - 6}$

8. Answers will vary. Sample answer: $f(x) = x^4 + x^3 + 18x$

9.   **10.**

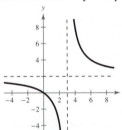   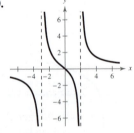

Asymptotes: Asymptotes:
$x = 3, y = 2$ $x = -3, x = 2, y = 0$

11.

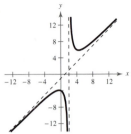

Asymptotes: $x = 2, y = x - 1$

12. Answers will vary. Sample answer: $y = \dfrac{4x^2}{x^2 + 1}$

13. Horizontal shift one unit left, vertical shift five units down

14. Reflection in the x-axis, horizontal shift three units left

15. (a) All real numbers x (b) $(0, 10)$ (c) $y = 2$

16. (a) All real numbers x (b) $(0, 2)$ (c) $y = 0$

17. (a) All real numbers x such that $x > 0$
(b) $(1, 0)$ (c) $x = 0, y = 0$

18. (a) All real numbers x such that $x > \frac{1}{2}$
(b) $\left(\frac{7}{12}, 0\right)$ (c) $x = \frac{1}{2}$

19. 6   **20.** -4 **21.** 10   **22.** -3 **23.** 1.723

24. 0.872   **25.** -7.484 **26.** -3.036

27. $\ln(x + 2) + \ln(x - 2) - \ln(x^2 + 1)$

28. $\ln \dfrac{x^2(x + 1)}{(x - 1)}$ **29.** $x \approx 1.242$ **30.** $x \approx 6.585$

31. $x = 12.8$ **32.** $x \approx 152.018$ **33.** $x = 0, 1$

34. No solution   **35.** $50, $9.50, $5.45; $5

36. (a) $1204.52   (b) $1209.93   **37.** $108.63

38. (a) 300   (b) 570   (c) 9 yr

39. (a) $A = x(273 - x)$

(b)
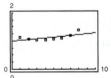 $0 < x < 273$

(c) 76.23 ft $\times$ 196.77 ft

40. (a) Quadratic model: $y = 0.0178t^2 - 0.130t + 1.26$; $r^2 \approx 0.9778$
Exponential model: $y = 1.002(1.025)^t$; $r^2 \approx 0.4009$
Power model: $y = 1.041t^{0.0564}$; $r^2 \approx 0.1686$

(b) Quadratic model: Exponential model:

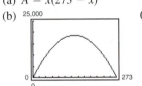   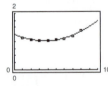

Power model:

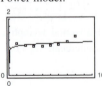

(c) The quadratic model is the best fit because its coefficient of determination is closest to 1.

(d) $1.74; Answers will vary.

Chapter 5

Section 5.1   (page 405)

1. Trigonometry   **3.** standard position   **5.** radian
7. 180°   **9.** No   **11.** 210°
13. (a) Quadrant I   (b) Quadrant III
15. (a) Quadrant II   (b) Quadrant IV
17. (a) Quadrant III   (b) Quadrant I
19. (a)   (b)

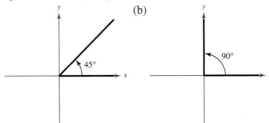

21. (a)   (b)

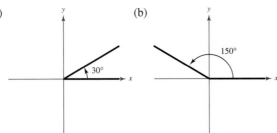

23. (a)   (b)

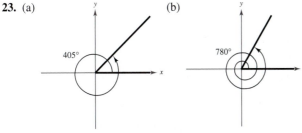

25. (a) 412°, −308°   (b) 324°, −396°
27. (a) 660°, −60°   (b) 340°, −380°
29. 64.75°   **31.** 85.308°   **33.** −125.01°   **35.** 280° 36′
37. −345° 7′ 12″   **39.** −20° 20′ 24″
41. Complement: 66°; supplement: 156°
43. Complement: 3°; supplement: 93°   **45.** 2
47. (a) Quadrant I   (b) Quadrant III
49. (a) Quadrant IV   (b) Quadrant II
51. (a) Quadrant IV   (b) Quadrant III
53. (a)   (b)

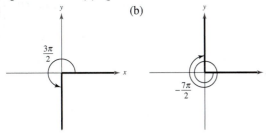

55. (a)   (b)

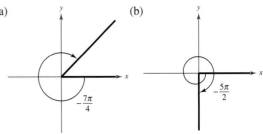

57. (a)   (b)

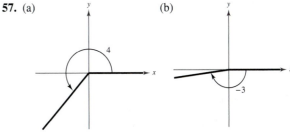

59. (a) $\dfrac{\pi}{6}$ (b) $\dfrac{5\pi}{6}$ **61.** (a) $-\dfrac{\pi}{9}$ (b) $-\dfrac{4\pi}{3}$
63. (a) 270°   (b) −210°   **65.** (a) 420°   (b) −39°
67. 2.007   **69.** −3.776   **71.** −0.014   **73.** 25.714°
75. 1170°   **77.** −114.592°
79. (a) $\dfrac{13\pi}{6}, -\dfrac{11\pi}{6}$ (b) $\dfrac{8\pi}{3}, -\dfrac{4\pi}{3}$
81. (a) $\dfrac{7\pi}{4}, -\dfrac{\pi}{4}$ (b) $\dfrac{28\pi}{15}, -\dfrac{32\pi}{15}$
83. Complement: $\dfrac{\pi}{6}$; supplement: $\dfrac{2\pi}{3}$
85. Complement: $\dfrac{\pi}{3}$; supplement: $\dfrac{5\pi}{6}$
87. $\frac{6}{5}$ rad   **89.** $\frac{8}{15}$ rad   **91.** $\frac{70}{29}$ rad   **93.** 14π in.
95. 18π m   **97.** 22.92 ft   **99.** 34.80 mi   **101.** 591.32 mi
103. 4° 2′ 33″   **105.** 275.02°   **107.** 436.97 km/min
109. (a) 80π rad/sec   (b) 25π ft/sec
111. (a) 400π rad/min to 1000π rad/min   (b) 6000π cm/min
113. False. A radian is larger: 1 rad ≈ 57.3°.
115. True. The sum of the angles of a triangle must equal
$$180° = \pi \text{ radians, and } \dfrac{2\pi}{3} + \dfrac{\pi}{4} + \dfrac{\pi}{12} = \pi.$$
117. $\dfrac{50\pi}{3}$ m²
119. (a) $A = 0.4r^2, \; r > 0; \; s = 0.8r, \; r > 0$

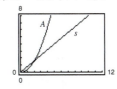

The area function changes more rapidly for $r > 1$ because it is quadratic and the arc length function is linear.
(b) $A = 50\theta, \; 0 < \theta < 2\pi; \; s = 10\theta, \; 0 < \theta < 2\pi$

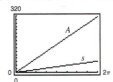

CHAPTER 5

121. Answers will vary.

123. 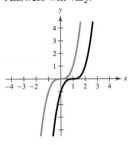   **125.**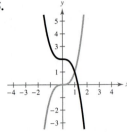

Horizontal shift one unit   Reflection in the x-axis,
to the right   vertical shift two units
                                   upward

127.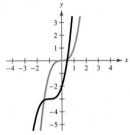

Horizontal shift one unit to the left, vertical shift three units
downward

Section 5.2   (page 416)

1. (a) iii   (b) vi   (c) ii   (d) v   (e) i   (f) iv

3. elevation, depression   **5.** 12

7. $\sin \theta = \frac{9}{41}$ $\csc \theta = \frac{41}{9}$

$\cos \theta = \frac{40}{41}$ $\sec \theta = \frac{41}{40}$

$\tan \theta = \frac{9}{40}$ $\cot \theta = \frac{40}{9}$

9. $\sin \theta = \frac{8}{17}$ $\csc \theta = \frac{17}{8}$

$\cos \theta = \frac{15}{17}$ $\sec \theta = \frac{17}{15}$

$\tan \theta = \frac{8}{15}$ $\cot \theta = \frac{15}{8}$

11. $\sin \theta = \frac{3}{5}$ $\csc \theta = \frac{5}{3}$

$\cos \theta = \frac{4}{5}$ $\sec \theta = \frac{5}{4}$

$\tan \theta = \frac{3}{4}$ $\cot \theta = \frac{4}{3}$

The triangles are similar and corresponding sides are
proportional.

13. $\cos \theta = \frac{\sqrt{11}}{6}$

$\tan \theta = \frac{5\sqrt{11}}{11}$

$\csc \theta = \frac{6}{5}$

$\sec \theta = \frac{6\sqrt{11}}{11}$

$\cot \theta = \frac{\sqrt{11}}{5}$

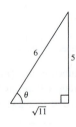

15. $\sin \theta = \frac{\sqrt{15}}{4}$

$\cos \theta = \frac{1}{4}$

$\tan \theta = \sqrt{15}$

$\csc \theta = \frac{4\sqrt{15}}{15}$

$\cot \theta = \frac{\sqrt{15}}{15}$

17. $\sin \theta = \frac{3\sqrt{10}}{10}$

$\cos \theta = \frac{\sqrt{10}}{10}$

$\csc \theta = \frac{\sqrt{10}}{3}$

$\sec \theta = \sqrt{10}$

$\cot \theta = \frac{1}{3}$

19. $\sin \theta = \frac{2\sqrt{13}}{13}$

$\cos \theta = \frac{3\sqrt{13}}{13}$

$\tan \theta = \frac{2}{3}$

$\sec \theta = \frac{\sqrt{13}}{3}$

$\csc \theta = \frac{\sqrt{13}}{2}$

21. $\frac{\pi}{6}, \frac{1}{2}$ **23.** $60°, \sqrt{3}$ **25.** $60°, \frac{\pi}{3}$ **27.** $30°, \frac{\sqrt{3}}{2}$

29. $45°, \frac{\pi}{4}$ **31.** (a) 0.1736   (b) 0.1736

33. (a) 1.3499   (b) 1.3432   **35.** (a) 5.0273   (b) 0.4142

37. $\csc \theta$ **39.** $\cot \theta$ **41.** $\cos \theta$ **43.** $\frac{\sin \theta}{\cos \theta}$

45. 1   **47.** $\cos \theta$ **49.** $\cot \theta$ **51.** $\csc \theta$

53. (a) $\sqrt{3}$ (b) $\frac{1}{2}$ (c) $\frac{\sqrt{3}}{2}$ (d) $\frac{\sqrt{3}}{3}$

55. (a) $\frac{1}{3}$ (b) $\frac{2\sqrt{2}}{3}$ (c) $\frac{\sqrt{2}}{4}$ (d) 3

57. (a) 3   (b) $\frac{2\sqrt{2}}{3}$ (c) $\frac{\sqrt{2}}{4}$ (d) $\frac{1}{3}$

59–65. Answers will vary.

67. (a) $30° = \frac{\pi}{6}$ **69.** (a) $60° = \frac{\pi}{3}$ **71.** (a) $60° = \frac{\pi}{3}$

(b) $30° = \frac{\pi}{6}$ (b) $45° = \frac{\pi}{4}$ (b) $45° = \frac{\pi}{4}$

73. $y = 35\sqrt{3}, r = 70\sqrt{3}$ **75.** $x = 8, y = 8\sqrt{3}$

77. (a)   (b) $\tan \theta = \frac{h}{21}$

(c) $h = 25.2$ ft

79. $30° = \pi/6$ **81.** 160 ft

83. (a)

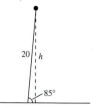

(b) $\sin 85° = \dfrac{h}{20}$

(c) 19.9 m

(d) The side of the triangle labeled h will become shorter.

(e)

Angle, θ	80°	70°	60°	50°
Height	19.7	18.8	17.3	15.3

Angle, θ	40°	30°	20°	10°
Height	12.9	10.0	6.8	3.5

(f) As $\theta \to 0°$, $h \to 0$.

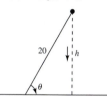

85. False. $\dfrac{\sqrt{2}}{2} + \dfrac{\sqrt{2}}{2} \neq 1$

87. Yes, with the Pythagorean Theorem. Answers will vary.

89. (a)

θ	0°	20°	40°	60°	80°
$\sin \theta$	0	0.3420	0.6428	0.8660	0.9848
$\cos \theta$	1	0.9397	0.7660	0.5	0.1736
$\tan \theta$	0	0.3640	0.8391	1.7321	5.6713

(b) Sine: increasing
Cosine: decreasing
Tangent: increasing

(c) Answers will vary.

91.

x	-1	0	1	2
$f(x)$	0.05	1	20.09	403.43

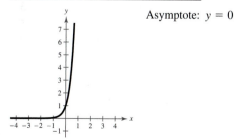

Asymptote: $y = 0$

93.

x	-1	0	1	2
$f(x)$	2.05	3	22.09	405.43

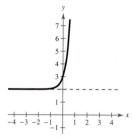

Asymptote: $y = 2$

95.

Domain: $(0, \infty)$
Vertical asymptote: $x = 0$
x-intercept: $(1, 0)$

97.

Domain: $(0, \infty)$
Vertical asymptote: $x = 0$
x-intercept: $(1, 0)$

Section 5.3   (page 428)

1. $\dfrac{y}{r}$ **3.** $\dfrac{y}{x}$ **5.** $\cos \theta$ **7.** odd

9. Reference angle   **11.** $0, \pi$

13. $\sin \theta = \dfrac{3}{5}$
$\cos \theta = \dfrac{4}{5}$
$\tan \theta = \dfrac{3}{4}$
$\csc \theta = \dfrac{5}{3}$
$\sec \theta = \dfrac{5}{4}$
$\cot \theta = \dfrac{4}{3}$

15. $\sin \theta = -\dfrac{5}{13}$
$\cos \theta = \dfrac{12}{13}$
$\tan \theta = -\dfrac{5}{12}$
$\csc \theta = -\dfrac{13}{5}$
$\sec \theta = \dfrac{13}{12}$
$\cot \theta = -\dfrac{12}{5}$

17. $\sin \theta = -\dfrac{1}{2}$ $\csc \theta = -2$
$\cos \theta = -\dfrac{\sqrt{3}}{2}$ $\sec \theta = -\dfrac{2\sqrt{3}}{3}$
$\tan \theta = \dfrac{\sqrt{3}}{3}$ $\cot \theta = \sqrt{3}$

19. $\sin \theta = \dfrac{24}{25}$
$\cos \theta = \dfrac{7}{25}$
$\tan \theta = \dfrac{24}{7}$
$\csc \theta = \dfrac{25}{24}$
$\sec \theta = \dfrac{25}{7}$
$\cot \theta = \dfrac{7}{24}$

21. $\sin \theta = -\dfrac{12}{13}$
$\cos \theta = \dfrac{5}{13}$
$\tan \theta = -\dfrac{12}{5}$
$\csc \theta = -\dfrac{13}{12}$
$\sec \theta = \dfrac{13}{5}$
$\cot \theta = -\dfrac{5}{12}$

23. $\sin \theta = \dfrac{5\sqrt{29}}{29}$
$\cos \theta = -\dfrac{2\sqrt{29}}{29}$
$\tan \theta = -\dfrac{5}{2}$
$\csc \theta = \dfrac{\sqrt{29}}{5}$
$\sec \theta = -\dfrac{\sqrt{29}}{2}$
$\cot \theta = -\dfrac{2}{5}$

25. $\sin \theta = \dfrac{4\sqrt{41}}{41}$
$\cos \theta = -\dfrac{5\sqrt{41}}{41}$
$\tan \theta = -\dfrac{4}{5}$
$\csc \theta = \dfrac{\sqrt{41}}{4}$
$\sec \theta = -\dfrac{\sqrt{41}}{5}$
$\cot \theta = -\dfrac{5}{4}$

27. Quadrant III   **29.** Quadrant I

31. $\sin \theta = \frac{3}{5}$

$\cos \theta = -\frac{4}{5}$

$\tan \theta = -\frac{3}{4}$

$\csc \theta = \frac{5}{3}$

$\sec \theta = -\frac{5}{4}$

$\cot \theta = -\frac{4}{3}$

33. $\sin \theta = -\frac{15}{17}$

$\cos \theta = \frac{8}{17}$

$\tan \theta = -\frac{15}{8}$

$\csc \theta = -\frac{17}{15}$

$\sec \theta = \frac{17}{8}$

$\cot \theta = -\frac{8}{15}$

35. $\sin \theta = \frac{\sqrt{3}}{2}$

$\cos \theta = -\frac{1}{2}$

$\tan \theta = -\sqrt{3}$

$\csc \theta = \frac{2\sqrt{3}}{3}$

$\sec \theta = -2$

$\cot \theta = -\frac{\sqrt{3}}{3}$

37. $\sin \theta = 0$

$\cos \theta = -1$

$\tan \theta = 0$

$\csc \theta$ is undefined.

$\sec \theta = -1$

$\cot \theta$ is undefined.

39. $\sin \theta = \frac{\sqrt{2}}{2}$

$\cos \theta = -\frac{\sqrt{2}}{2}$

$\tan \theta = -1$

$\csc \theta = \sqrt{2}$

$\sec \theta = -\sqrt{2}$

$\cot \theta = -1$

41. $\sin \theta = -\frac{2\sqrt{5}}{5}$

$\cos \theta = -\frac{\sqrt{5}}{5}$

$\tan \theta = 2$

$\csc \theta = -\frac{\sqrt{5}}{2}$

$\sec \theta = -\sqrt{5}$

$\cot \theta = \frac{1}{2}$

43. -1 **45.** 0   **47.** 1   **49.** Undefined

51. $\theta' = 60°$

53. $\theta' = 30°$

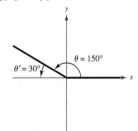

55. $\theta' = 45°$

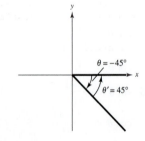

57. $\theta' = \frac{\pi}{3}$

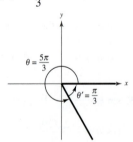

59. $\theta' = \frac{\pi}{6}$

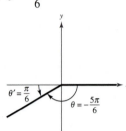

61. $\theta' = \frac{\pi}{6}$

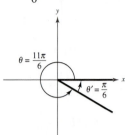

63. $\theta' = 28°$

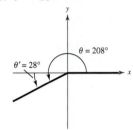

65. $\theta' = 68°$

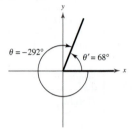

67. $\theta' = \frac{\pi}{5}$

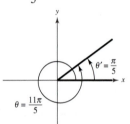

69. $\theta' \approx 1.342$

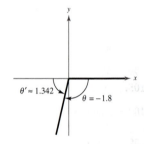

71. $\sin 225° = -\frac{\sqrt{2}}{2}$

$\cos 225° = -\frac{\sqrt{2}}{2}$

$\tan 225° = 1$

73. $\sin(-750°) = -\frac{1}{2}$

$\cos(-750°) = \frac{\sqrt{3}}{2}$

$\tan(-750°) = -\frac{\sqrt{3}}{3}$

75. $\sin \frac{5\pi}{3} = -\frac{\sqrt{3}}{2}$

$\cos \frac{5\pi}{3} = \frac{1}{2}$

$\tan \frac{5\pi}{3} = -\sqrt{3}$

77. $\sin\left(-\frac{\pi}{6}\right) = -\frac{1}{2}$

$\cos\left(-\frac{\pi}{6}\right) = \frac{\sqrt{3}}{2}$

$\tan\left(-\frac{\pi}{6}\right) = -\frac{\sqrt{3}}{3}$

79. $\sin \frac{11\pi}{4} = \frac{\sqrt{2}}{2}$

$\cos \frac{11\pi}{4} = -\frac{\sqrt{2}}{2}$

$\tan \frac{11\pi}{4} = -1$

81. $\sin\left(-\frac{17\pi}{6}\right) = -\frac{1}{2}$

$\cos\left(-\frac{17\pi}{6}\right) = -\frac{\sqrt{3}}{2}$

$\tan\left(-\frac{17\pi}{6}\right) = \frac{\sqrt{3}}{3}$

83. $\dfrac{4}{5}$ **85.** $-\sqrt{3}$ **87.** $\dfrac{\sqrt{65}}{4}$

89. $\tan 110° \approx -2.748$
$\csc 110° \approx 1.064$
$\sec 110° \approx -2.924$
$\cot 110° \approx -0.364$

91. $\tan 350° \approx -0.176$
$\csc 350° \approx -5.760$
$\sec 350° \approx 1.015$
$\cot 350° \approx -5.673$

93. $\cos \theta = -\dfrac{\sqrt{21}}{5}$

$\tan \theta = -\dfrac{2\sqrt{21}}{21}$

$\csc \theta = \dfrac{5}{2}$

$\sec \theta = -\dfrac{5\sqrt{21}}{21}$

$\cot \theta = -\dfrac{\sqrt{21}}{2}$

95. $\sin \theta = \dfrac{4\sqrt{17}}{17}$

$\cos \theta = -\dfrac{\sqrt{17}}{17}$

$\csc \theta = \dfrac{\sqrt{17}}{4}$

$\sec \theta = -\sqrt{17}$

$\cot \theta = -\dfrac{1}{4}$

97. $\sin \theta = -\dfrac{2}{3}$

$\cos \theta = \dfrac{\sqrt{5}}{3}$

$\tan \theta = -\dfrac{2\sqrt{5}}{5}$

$\sec \theta = \dfrac{3\sqrt{5}}{5}$

$\cot \theta = -\dfrac{\sqrt{5}}{2}$

99. 0.1736 **101.** -1.1918 **103.** 0.8391
105. -2.9238

107. (a) $30° = \dfrac{\pi}{6}, 150° = \dfrac{5\pi}{6}$ (b) $210° = \dfrac{7\pi}{6}, 330° = \dfrac{11\pi}{6}$

109. (a) $60° = \dfrac{\pi}{3}, 120° = \dfrac{2\pi}{3}$ (b) $135° = \dfrac{3\pi}{4}, 315° = \dfrac{7\pi}{4}$

111. (a) $150° = \dfrac{5\pi}{6}, 210° = \dfrac{7\pi}{6}$ (b) $120° = \dfrac{2\pi}{3}, 240° = \dfrac{4\pi}{3}$

113. (a) -1 (b) -0.4
115. (a) $0.25, 2.89$ (b) $1.82, 4.46$
117. (a) $-\dfrac{1}{3}$ (b) -3 **119.** (a) $-\dfrac{1}{5}$ (b) -5

121. (a) $\dfrac{1+\sqrt{3}}{2}$ (b) $\dfrac{\sqrt{3}-1}{2}$ (c) $\dfrac{3}{4}$

(d) $\dfrac{\sqrt{3}}{4}$ (e) $\dfrac{\sqrt{3}}{2}$ (f) $\dfrac{\sqrt{3}}{2}$

123. (a) 0 (b) $\sqrt{2}$ (c) $\dfrac{1}{2}$ (d) $-\dfrac{1}{2}$ (e) -1 (f) $\dfrac{\sqrt{2}}{2}$

125. (a) $\dfrac{1-\sqrt{3}}{2}$ (b) $-\dfrac{1+\sqrt{3}}{2}$ (c) $\dfrac{3}{4}$

(d) $-\dfrac{\sqrt{3}}{4}$ (e) $-\dfrac{\sqrt{3}}{2}$ (f) $-\dfrac{\sqrt{3}}{2}$

127. (a) $-\dfrac{1+\sqrt{3}}{2}$ (b) $\dfrac{1-\sqrt{3}}{2}$ (c) $\dfrac{3}{4}$

(d) $\dfrac{\sqrt{3}}{4}$ (e) $\dfrac{\sqrt{3}}{2}$ (f) $-\dfrac{\sqrt{3}}{2}$

129. (a) $-\dfrac{1+\sqrt{3}}{2}$ (b) $\dfrac{-1+\sqrt{3}}{2}$ (c) $\dfrac{1}{4}$

(d) $\dfrac{\sqrt{3}}{4}$ (e) $\dfrac{\sqrt{3}}{2}$ (f) $-\dfrac{1}{2}$

131. (a) -1 (b) 1 (c) 0 (d) 0 (e) 0 (f) 0
133. (a) -1 (b) 1 (c) 0 (d) 0 (e) 0 (f) 0
135. (a) $60.4°F$ (b) $92.3°F$ (c) $76.35°F$
137. (a) 12 mi   (b) 6 mi   (c) 6.93 mi
139. True. $0 < \cos \theta < 1$ in Quadrant I, so

$$\sin \theta < \dfrac{\sin \theta}{\cos \theta} = \tan \theta.$$

141. False. Sine is positive in Quadrant II.
143. (a)

θ	0°	20°	40°
$\sin \theta$	0	0.3420	0.6428
$\sin(180° - \theta)$	0	0.3420	0.6428

θ	60°	80°
$\sin \theta$	0.8660	0.9848
$\sin(180° - \theta)$	0.8660	0.9848

(b) $\sin \theta = \sin(180° - \theta)$

145. (a)

θ	0	0.3	0.6	0.9
$\cos\left(\dfrac{3\pi}{2} - \theta\right)$	0	-0.296	-0.565	-0.783
$-\sin \theta$	0	-0.296	-0.565	-0.783

θ	1.2	1.5
$\cos\left(\dfrac{3\pi}{2} - \theta\right)$	-0.932	-0.997
$-\sin \theta$	-0.932	-0.997

(b) $\cos\left(\dfrac{3\pi}{2} - \theta\right) = -\sin \theta$

147. As θ increases from $0°$ to $90°$, x will decrease from 12 to 0 centimeters and y will increase from 0 to 12 centimeters. Therefore, $\sin \theta = y/12$ will increase from 0 to 1 and $\cos \theta = x/12$ will decrease from 1 to 0. So, $\tan \theta = y/x$ will increase without bound. When $\theta = 90°$, $\tan \theta$ will be undefined.
149. The calculator mode is in degrees instead of radians.
151. 7 **153.** -1.752 **155.** $4.908, -5.908$

Section 5.4   (page 439)

1. amplitude   **3.** $\dfrac{2\pi}{b}$ **5.** 2π

7. It vertically shifts the graph d units.
9. (a) $x = -2\pi, -\pi, 0, \pi, 2\pi$ (b) $y = 0$

(c) Increasing: $\left(-2\pi, -\dfrac{3\pi}{2}\right), \left(-\dfrac{\pi}{2}, \dfrac{\pi}{2}\right), \left(\dfrac{3\pi}{2}, 2\pi\right)$

Decreasing: $\left(-\dfrac{3\pi}{2}, -\dfrac{\pi}{2}\right), \left(\dfrac{\pi}{2}, -\dfrac{3\pi}{2}\right)$

(d) Relative maxima: $\left(-\dfrac{3\pi}{2}, 1\right), \left(\dfrac{\pi}{2}, 1\right)$

Relative minima: $\left(-\dfrac{\pi}{2}, -1\right), \left(\dfrac{3\pi}{2}, -1\right)$

11. Period: π
Amplitude: 3

13. Period: 4π
Amplitude: $\dfrac{5}{2}$

15. Period: 2
Amplitude: $\dfrac{2}{3}$

17. Period: 2π
Amplitude: 2

19. Period: 3π
Amplitude: $\dfrac{1}{4}$

21. g is a shift of f π units to the right.

23. g is a reflection of f in the x-axis.

25. g is a reflection of f in the x-axis and has five times the amplitude of f.

27. g is a shift of f three units upward.

29. g has twice the amplitude of f.

31. g is a horizontal shift of f π units to the right.

33.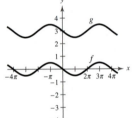

35.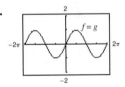

37.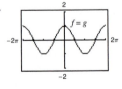

39.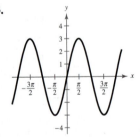

41.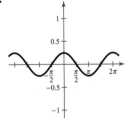

43.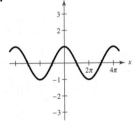

45.

47.

49.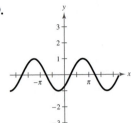

51.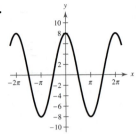

53.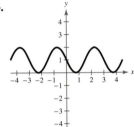

55.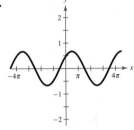

57.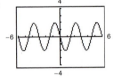
Amplitude: 2
Period: 3

59.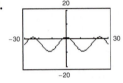
Amplitude: 5
Period: 24

61.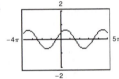
Amplitude: $\dfrac{2}{3}$
Period: 4π

63.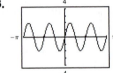
Amplitude: 2
Period: $\dfrac{\pi}{2}$

65.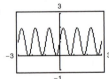
Amplitude: 1
Period: 1

67.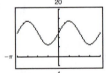
Amplitude: 5
Period: π

69.
Amplitude: $\dfrac{1}{100}$
Period: $\dfrac{1}{60}$

71. $a = -4, d = 4$ **73.** $a = -6, d = 1$

75. $a = -3, b = 2, c = 0$ **77.** $a = 1, b = 1, c = \dfrac{\pi}{4}$

79.

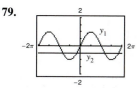

$$x = -\frac{\pi}{6}, -\frac{5\pi}{6}, \frac{7\pi}{6}, \frac{11\pi}{6}$$

81. (a)

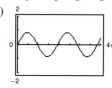

    (b) 6 sec
    (c) 10 cycles/min
    (d) The period of the model would decrease because the time for a respiratory cycle would decrease.

83. (a)

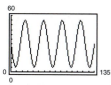

    (b) Minimum height: 5 ft
        Maximum height: 55 ft

85. (a) 365 days. The cycle is 1 year.
    (b) 30.3 gallons per day. The average is the constant term of the model.
    (c)

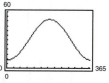

    Consumption exceeds 40 gallons per day from the beginning of May through part of September.

87. (a) and (c)

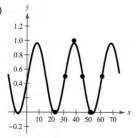

    The model fits the data well.
    (b) $y = 0.493 \sin(0.209x - 0.114) + 0.472$
    (d) 30 days   (e) 12.9%

89. True. The period of $\sin x$ is 2π. Adding 2π moves the graph one period to the right.

91. False. The function $y = \frac{1}{2}\cos 2x$ has an amplitude that is one-half that of the function $y = \cos x$.

93.

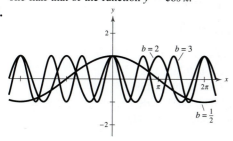

The value of b affects the period of the graph.
$b = \frac{1}{2} \to \frac{1}{2}$ cycle; $b = 2 \to 2$ cycles; $b = 3 \to 3$ cycles

95. e   **97.** c

99. (a)

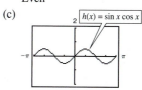

    (b)

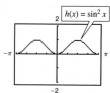

       Even   Even

    (c)

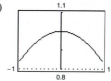

    Odd

101. (a)

x	-1	-0.1	-0.01
$\dfrac{\sin x}{x}$	0.8415	0.9983	1.0000

x	-0.001	0	0.001
$\dfrac{\sin x}{x}$	1.0000	Undefined	1.0000

x	0.01	0.1	1
$\dfrac{\sin x}{x}$	1.0000	0.9983	0.8415

    (b)

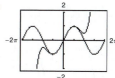

    $f \to 1$ as $x \to 0$
    (c) The ratio approaches 1 as x approaches 0.

103. (a)

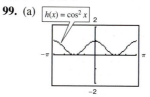

    The polynomial function is a good approximation of the sine function when x is close to 0.
    (b)

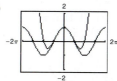

    The polynomial function is a good approximation of the cosine function when x is close to 0.

CHAPTER 5

(c) $\sin x \approx x - \dfrac{x^3}{3!} + \dfrac{x^5}{5!} - \dfrac{x^7}{7!}$

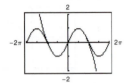

$\cos x \approx 1 - \dfrac{x^2}{2!} + \dfrac{x^4}{4!} - \dfrac{x^6}{6!}$

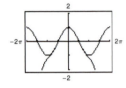

The accuracy increased.

105.

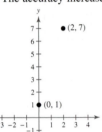

$m = 3$

107. $487.014°$

109. Answers will vary.

Section 5.5 (page 451)

1. vertical   **3.** tangent, cotangent

5. (a) $x = -2\pi, -\pi, 0, \pi, 2\pi$ (b) $y = 0$

  (c) Increasing on $\left(-2\pi, -\dfrac{3\pi}{2}\right), \left(-\dfrac{3\pi}{2}, -\dfrac{\pi}{2}\right), \left(-\dfrac{\pi}{2}, \dfrac{\pi}{2}\right),$

    $\left(\dfrac{\pi}{2}, \dfrac{3\pi}{2}\right), \left(\dfrac{3\pi}{2}, 2\pi\right)$

  (d) No relative extrema   (e) $x = -\dfrac{3\pi}{2}, -\dfrac{\pi}{2}, \dfrac{\pi}{2}, \dfrac{3\pi}{2}$

7. (a) No x-intercepts   (b) $y = 1$

  (c) Increasing on $\left(-2\pi, -\dfrac{3\pi}{2}\right), \left(-\dfrac{3\pi}{2}, -\pi\right),$

    $\left(0, \dfrac{\pi}{2}\right), \left(\dfrac{\pi}{2}, \pi\right)$

    Decreasing on $\left(-\pi, -\dfrac{\pi}{2}\right), \left(-\dfrac{\pi}{2}, 0\right),$

    $\left(\pi, \dfrac{3\pi}{2}\right), \left(\dfrac{3\pi}{2}, 2\pi\right)$

  (d) Relative minima: $(-2\pi, 1), (0, 1), (2\pi, 1)$

    Relative maxima: $(-\pi, -1), (\pi, -1)$

  (e) $x = -\dfrac{3\pi}{2}, -\dfrac{\pi}{2}, \dfrac{\pi}{2}, \dfrac{3\pi}{2}$

9.

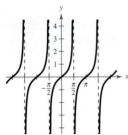

11.

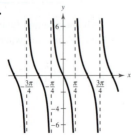

13.

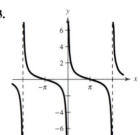

15.

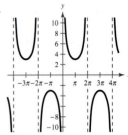

17.

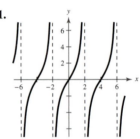

19.

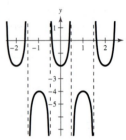

21.

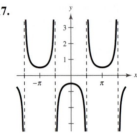

23.

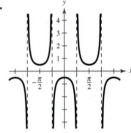

25.

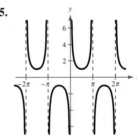

27.

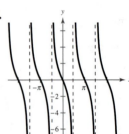

29.

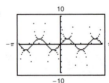

Answers will vary.

31.

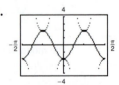

Answers will vary.

33.

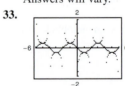

Answers will vary.

35. $-5.498, -2.356, 0.785, 3.927$

37. $-4.189, -2.094, 2.094, 4.189$

39. $-5.236, -2.094, 1.047, 4.189$

41. Even   **43.** Odd   **45.** Odd

47. 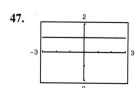 **49.**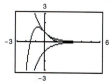

Not equivalent;   Equivalent

y_1 is undefined at $x = 0$.

51. d; as x approaches 0, $f(x)$ approaches 0.
52. a; as x approaches 0, $f(x)$ approaches 0.
53. b; as x approaches 0, $g(x)$ approaches 0.
54. c; as x approaches 0, $g(x)$ approaches 0.

55.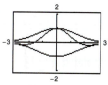

$-e^{-x} \le e^{-x} \cos x \le e^{-x}$

Touches $y = \pm e^{-x}$ at $x = n\pi$

Intercepts at $x = \dfrac{\pi}{2} + n\pi$

57.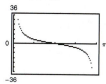

$-e^{-x^2/4} \le e^{-x^2/4} \cos x \le e^{-x^2/4}$

Touches $y = \pm e^{-x}$ at $x = n\pi$

Intercepts at $x = \dfrac{\pi}{2} + n\pi$

59. (a) $f \to -\infty$ (b) $f \to \infty$ (c) $f \to -\infty$ (d) $f \to \infty$
61. (a) $f \to \infty$ (b) $f \to -\infty$ (c) $f \to \infty$ (d) $f \to -\infty$
63. $d = 5 \cot x$ **65.** (a)

(b) Not periodic and damped; approaches 0 as t increases.

67. (a) Yes. To each t there corresponds one and only one value of y.
(b) 1.3 oscillations/sec   (c) $y = 12(0.221)^t \cos(8.2t)$
(d) $y = 12e^{-1.5t} \cos(8.2t)$
(e)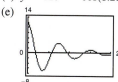

Answers will vary.
69. True. The sine function is damped.

71. True. $\sec x = \csc\left(x - \dfrac{\pi}{2}\right) = \dfrac{1}{\sin\left(x - \dfrac{\pi}{2}\right)}$

73. (a)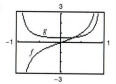

(b) $\left(-1, \frac{1}{3}\right)$ (c) $\left(-1, \frac{1}{3}\right)$; The intervals are the same.

75. (a)

x	-1	-0.1	-0.01
$\dfrac{\tan x}{x}$	1.5574	1.0033	1.0000

x	-0.001	0	0.001
$\dfrac{\tan x}{x}$	1.0000	Undefined	1.0000

x	0.01	0.1	1
$\dfrac{\tan x}{x}$	1.0000	1.0033	1.5574

(b)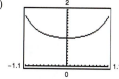

$f \to 1$ as $x \to 0$

(c) The ratio approaches 1 as x approaches 0.

77.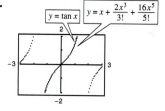

The polynomial function is a good approximation of the tangent function when x is close to 0.
79. Distributive Property   **81.** Additive Identity Property
83. Not one-to-one   **85.** One-to-one. $f^{-1}(x) = \dfrac{x^2 + 14}{3}, x \ge 0$
87. Domain: all real numbers x
Intercepts: $(-4, 0), (1, 0), (0, -4)$
No asymptotes
89. Domain: all real numbers x
Intercept: $(0, 5)$
Asymptote: $y = 2$

Section 5.6   (page 462)

1. $y = \sin^{-1} x$, $-1 \le x \le 1$ **3.** $\sin^{-1} x$ or arcsin x

5. (a) $\dfrac{\pi}{6}$ (b) 0   **7.** (a) $\dfrac{\pi}{2}$ (b) 0

9. (a) $\dfrac{\pi}{6}$ (b) $-\dfrac{\pi}{4}$ **11.** (a) $-\dfrac{\pi}{3}$ (b) $\dfrac{\pi}{3}$

13. (a) $\dfrac{\pi}{3}$ (b) $-\dfrac{\pi}{6}$

CHAPTER 5

15. (a)

x	-1	-0.8	-0.6	-0.4	-0.2
y	-1.571	-0.927	-0.644	-0.412	-0.201

x	0	0.2	0.4	0.6	0.8	1
y	0	0.201	0.412	0.644	0.927	1.571

(b) 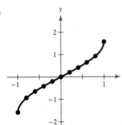 (c)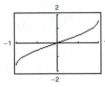

They are the same.

(d) Intercept: $(0, 0)$; symmetric about the origin

17. $\left(-\sqrt{3}, -\dfrac{\pi}{3}\right), \left(-\dfrac{\sqrt{3}}{3}, -\dfrac{\pi}{6}\right), \left(1, \dfrac{\pi}{4}\right)$ **19.** 0.47

21. 1.50   **23.** 0.72   **25.** -0.85 **27.** -1.41

29. 0.85   **31.** 1.29

33. 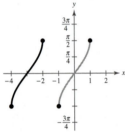   **35.**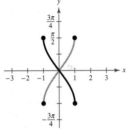

g is a horizontal shift of f three units to the left. g is a reflection of f in the y-axis.

37. 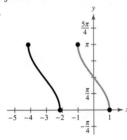   **39.**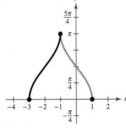

g is a horizontal shift of f π units to the left. g is a reflection of f in the y-axis and a horizontal shift of f two units to the left.

41. 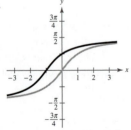   **43.**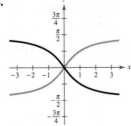

g is a horizontal shift of f one unit to the left. g is a reflection of f in the y-axis.

45. $\theta = \arctan \dfrac{x}{8}$ **47.** $\theta = \arcsin \dfrac{x+2}{5}$

49. $\sqrt{4 - x^2}$; $\theta = \arcsin \dfrac{x}{2}$, $\theta = \arccos \dfrac{\sqrt{4 - x^2}}{2}$, $\theta = \arctan \dfrac{x}{\sqrt{4 - x^2}}$

51. $\sqrt{x^2 + 2x + 5}$; $\theta = \arcsin \dfrac{x+1}{\sqrt{x^2 + 2x + 5}}$, $\theta = \arccos \dfrac{2}{\sqrt{x^2 + 2x + 5}}$, $\theta = \arctan \dfrac{x+1}{2}$

53. 0.7   **55.** -0.3 **57.** 0   **59.** $-\dfrac{\pi}{6}$ **61.** $\dfrac{\pi}{2}$

63. $\dfrac{\pi}{2}$ **65.** $\dfrac{4}{5}$ **67.** $\dfrac{7}{25}$ **69.** $\dfrac{\sqrt{34}}{5}$ **71.** $\dfrac{\sqrt{5}}{3}$

73. $\dfrac{1}{x}$ **75.** $\sqrt{-x^2 - 4x - 3}$ **77.** $\dfrac{\sqrt{25 - x^2}}{x}$

79. $\dfrac{\sqrt{x^2 + 7}}{x}$ **81.** $\dfrac{14}{\sqrt{x^2 + 196}}$ **83.** $\dfrac{|x - 1|}{\sqrt{x^2 - 2x + 10}}$

85. 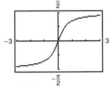   **87.**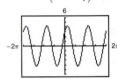

89. 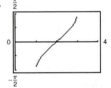   **91.** $3\sqrt{2} \sin\left(2t + \dfrac{\pi}{4}\right)$

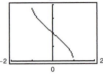

The two forms are equivalent.

93. $\dfrac{\pi}{2}$ **95.** $\dfrac{\pi}{2}$ **97.** π

99. (a)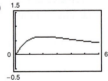

11 ft   34 ft

(b) 0.574 rad   (c) 12.94 ft

101. (a)

(b) 2 ft

(c) $\beta = 0$; As the camera moves farther from the picture, the angle subtended by the camera approaches zero.

103. (a) $\theta = \arctan \dfrac{x}{20}$ (b) 0.245 rad, 0.540 rad

105. False. $\arctan 1 = \dfrac{\pi}{4}$

107.   **109.**

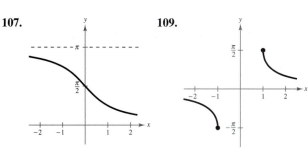

111. $\dfrac{\pi}{4}$ **113.** $\dfrac{5\pi}{6}$ **115 and 117.** Proofs

119. $\dfrac{\sqrt{2}}{2}$ **121.** $\dfrac{\sqrt{3}}{3}$

123. $\cos\theta = \dfrac{\sqrt{11}}{6}$

$\tan\theta = \dfrac{5\sqrt{11}}{11}$

$\csc\theta = \dfrac{6}{5}$

$\sec\theta = \dfrac{6\sqrt{11}}{11}$

$\cot\theta = \dfrac{\sqrt{11}}{5}$

125. $\cos\theta = \dfrac{\sqrt{7}}{4}$

$\tan\theta = \dfrac{3\sqrt{7}}{7}$

$\csc\theta = \dfrac{4}{3}$

$\sec\theta = \dfrac{4\sqrt{7}}{7}$

$\cot\theta = \dfrac{\sqrt{7}}{3}$

Section 5.7   (page 472)

1. harmonic motion   **3.** No
5. $B = 60°$ **7.** $A = 19°$ **9.** $A \approx 26.57°$
  $a \approx 5.77$ $a \approx 4.82$ $B \approx 63.43°$
  $c \approx 11.55$ $c \approx 14.81$ $c \approx 13.42$
11. $A \approx 72.76°$ **13.** $B = 77°45'$
  $B \approx 17.24°$ $a \approx 91.34$
  $a \approx 51.58$ $b \approx 420.70$
15. 5.12 in.   **17.** 8.21 ft   **19.** 19.70 ft   **21.** 109.63 ft
23. 2089.99 ft
25. (a)

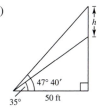

  (b) $h = 50(\tan 47°40' - \tan 35°)$ (c) 19.87 ft
27. (a) $l = \sqrt{h^2 + 28h + 10{,}196}$ **29.** 70.35°

  (b) $\theta = \arccos\dfrac{100}{l}$ (c) $h \approx 56$ ft

31. 75.97°   **33.** 5098.78 ft   **35.** 0.66 mi
37. 104.95 nm south, 58.18 nm west

39. (a) N 58° E   (b) 68.82 m   **41.** N 56.31° W
43. 1933.32 ft   **45.** 3.23 mi
47. (a) 61.82°; 15.64°   (b) 31.10 ft   **49.** 78.69°
51. 35.26°   **53.** $y = \sqrt{3}\,r$ **55.** $d = 8\sin\pi t$

57. $d = 3\cos\dfrac{4\pi t}{3}$ **59.** (a) 4   (b) 4   (c) 4   (d) $\dfrac{1}{16}$

61. (a) $\frac{1}{16}$ (b) 70   (c) 0   (d) $\frac{1}{140}$ **63.** $\omega = 528\pi$

65. (a)   (b) $\dfrac{\pi}{8}$ sec   (c) $\dfrac{\pi}{32}$ sec

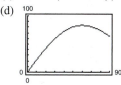

67. (a)

Base 1	Base 2	Altitude	Area
8	$8 + 16\cos 10°$	$8\sin 10°$	22.06
8	$8 + 16\cos 20°$	$8\sin 20°$	42.46
8	$8 + 16\cos 30°$	$8\sin 30°$	59.71
8	$8 + 16\cos 40°$	$8\sin 40°$	72.65
8	$8 + 16\cos 50°$	$8\sin 50°$	80.54
8	$8 + 16\cos 60°$	$8\sin 60°$	83.14
8	$8 + 16\cos 70°$	$8\sin 70°$	80.71

(b)

Base 1	Base 2	Altitude	Area
8	$8 + 16\cos 56°$	$8\sin 56°$	82.73
8	$8 + 16\cos 58°$	$8\sin 58°$	83.04
8	$8 + 16\cos 59°$	$8\sin 59°$	83.11
8	$8 + 16\cos 60°$	$8\sin 60°$	83.14
8	$8 + 16\cos 61°$	$8\sin 61°$	83.11
8	$8 + 16\cos 62°$	$8\sin 62°$	83.04

  83.14 ft²
(c) $A = 64(1 + \cos\theta)(\sin\theta)$
(d)

  83.14 ft²; They are the same.
69. (a)   (b) 12 mo; Yes, there are
         12 months in a year.
      (c) 2.77; The maximum change
         in the number of hours of
         daylight

71. True. In $a\sin\omega t$ or $a\cos\omega t$, a is a real number.
73. False. The amplitude is equal to a.
75. Answers will vary.
77. $4x - y + 6 = 0$ **79.** $4x + 5y - 22 = 0$

81. All real numbers x **83.** All real numbers x

Review Exercises (page 480)

1. $40°$

3. (a)

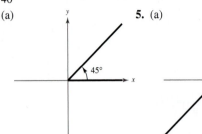

5. (a)

(b) Quadrant I
(c) $405°, -315°$

(b) Quadrant III
(c) $225°, -495°$

7. Complement: $85°$; supplement: $175°$

9. Complement: none; supplement: $23°$ **11.** $135.279°$

13. $6.572°$ **15.** $135° 17' 24''$ **17.** $-85° 21' 36''$

19. (a)

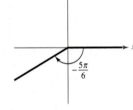

(b) Quadrant III

(c) $\dfrac{10\pi}{3}, -\dfrac{2\pi}{3}$

21. (a)

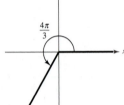

(b) Quadrant III

(c) $\dfrac{7\pi}{6}, -\dfrac{17\pi}{6}$

23. Complement: $\dfrac{3\pi}{8}$; supplement: $\dfrac{7\pi}{8}$

25. Complement: $\dfrac{\pi}{5}$; supplement: $\dfrac{7\pi}{10}$ **27.** 1.641

29. 7.243 **31.** $128.571°$ **33.** $-200.535°$ **35.** $\frac{25}{12}$ rad

37. $\dfrac{46\pi}{3}$ m   **39.** 6000π cm/min

41. $\sin\theta = \dfrac{\sqrt{65}}{9}$

$\cos\theta = \dfrac{4}{9}$

$\tan\theta = \dfrac{\sqrt{65}}{4}$

$\csc\theta = \dfrac{9\sqrt{65}}{65}$

$\sec\theta = \dfrac{9}{4}$

$\cot\theta = \dfrac{4\sqrt{65}}{65}$

43. $\sin\theta = \dfrac{5\sqrt{61}}{61}$

$\cos\theta = \dfrac{6\sqrt{61}}{61}$

$\tan\theta = \dfrac{5}{6}$

$\csc\theta = \dfrac{\sqrt{61}}{5}$

$\sec\theta = \dfrac{\sqrt{61}}{6}$

$\cot\theta = \dfrac{6}{5}$

45. $\sin\theta = \dfrac{7}{24}$

$\cos\theta = \dfrac{\sqrt{527}}{24}$

$\tan\theta = \dfrac{7\sqrt{527}}{527}$

$\csc\theta = \dfrac{24}{7}$

$\sec\theta = \dfrac{24\sqrt{527}}{527}$

$\cot\theta = \dfrac{\sqrt{527}}{7}$

47. $\sin\theta = \dfrac{\sqrt{17}}{17}$

$\cos\theta = \dfrac{4\sqrt{17}}{17}$

$\tan\theta = \dfrac{1}{4}$

$\csc\theta = \sqrt{17}$

$\sec\theta = \dfrac{\sqrt{17}}{4}$

$\cot\theta = 4$

49. (a) 0.1045 (b) 0.1045 **51.** (a) 0.7071 (b) 1.4142

53. Answers will vary.   **55.** 235 ft

57. $\sin\theta = \dfrac{4}{5}$

$\cos\theta = \dfrac{3}{5}$

$\tan\theta = \dfrac{4}{3}$

$\csc\theta = \dfrac{5}{4}$

$\sec\theta = \dfrac{5}{3}$

$\cot\theta = \dfrac{3}{4}$

59. $\sin\theta = \dfrac{2\sqrt{53}}{53}$

$\cos\theta = -\dfrac{7\sqrt{53}}{53}$

$\tan\theta = -\dfrac{2}{7}$

$\csc\theta = \dfrac{\sqrt{53}}{2}$

$\sec\theta = -\dfrac{\sqrt{53}}{7}$

$\cot\theta = -\dfrac{7}{2}$

61. $\sin\theta = \dfrac{15\sqrt{481}}{481}$

$\cos\theta = \dfrac{16\sqrt{481}}{481}$

$\tan\theta = \dfrac{15}{16}$

$\csc\theta = \dfrac{\sqrt{481}}{15}$

$\sec\theta = \dfrac{\sqrt{481}}{16}$

$\cot\theta = \dfrac{16}{15}$

63. $\sin\theta = -\dfrac{\sqrt{11}}{6}$

$\cos\theta = \dfrac{5}{6}$

$\tan\theta = -\dfrac{\sqrt{11}}{5}$

$\csc\theta = -\dfrac{6\sqrt{11}}{11}$

$\cot\theta = -\dfrac{5\sqrt{11}}{11}$

65. $\cos\theta = -\dfrac{\sqrt{55}}{8}$ $\sec\theta = -\dfrac{8\sqrt{55}}{55}$

$\tan\theta = -\dfrac{3\sqrt{55}}{55}$ $\cot\theta = -\dfrac{\sqrt{55}}{3}$

$\csc\theta = \dfrac{8}{3}$

67. $\theta' = 30°$

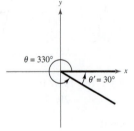

69. $\theta' = \dfrac{\pi}{4}$

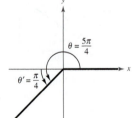

71. $\theta' = 84°$

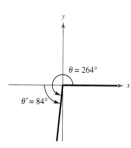

73. $\theta' = \dfrac{\pi}{5}$

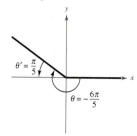

103.

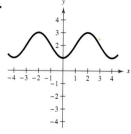

105.

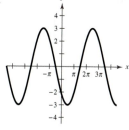

107. $a = -2,\ b = 1,\ c = \dfrac{\pi}{4}$ **109.** $a = -4,\ b = 2,\ c = \dfrac{\pi}{2}$

111.

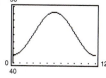

Maximum sales: June
Minimum sales: December

75. $\sin 240° = -\dfrac{\sqrt{3}}{2}$

$\cos 240° = -\dfrac{1}{2}$

$\tan 240° = \sqrt{3}$

77. $\sin(-210°) = \dfrac{1}{2}$

$\cos(-210°) = -\dfrac{\sqrt{3}}{2}$

$\tan(-210°) = -\dfrac{\sqrt{3}}{3}$

79. $\sin\left(-\dfrac{9\pi}{4}\right) = -\dfrac{\sqrt{2}}{2}$

$\cos\left(-\dfrac{9\pi}{4}\right) = \dfrac{\sqrt{2}}{2}$

$\tan\left(-\dfrac{9\pi}{4}\right) = -1$

81. $\sin 4\pi = 0$
$\cos 4\pi = 1$
$\tan 4\pi = 0$

83. 0.6494   **85.** 3.2361

87.

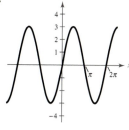

89.

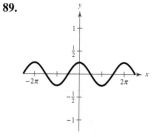

113.

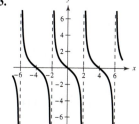

115.

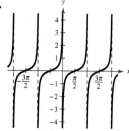

117.

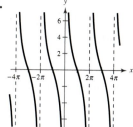

119.

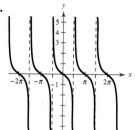

91. Period: 2; amplitude: 5   **93.** Period: π; amplitude: 3.4

95.

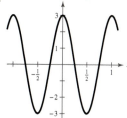

97.

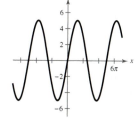

121.

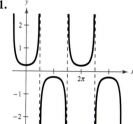

123.

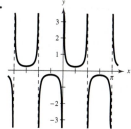

99.

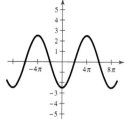

101.

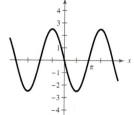

125.

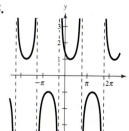

127.

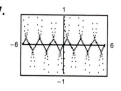

Answers will vary.

CHAPTER 5

121. $\cos \theta = \pm \sqrt{1 - \sin^2 \theta}$

$\tan \theta = \pm \dfrac{\sin \theta}{\sqrt{1 - \sin^2 \theta}}$

$\csc \theta = \dfrac{1}{\sin \theta}$

$\sec \theta = \pm \dfrac{1}{\sqrt{1 - \sin^2 \theta}}$

$\cot \theta = \pm \dfrac{\sqrt{1 - \sin^2 \theta}}{\sin \theta}$

The sign depends on the choice of θ.

123. $\dfrac{x^2 + 6x - 8}{(x + 5)(x - 8)}$ **125.** $\dfrac{-5x^2 + 8x + 28}{(x^2 - 4)(x + 4)}$

127.   **129.**

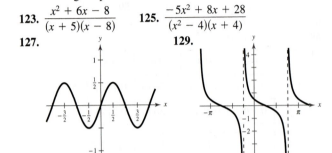

Section 6.2   (page 502)

1. $\cot u$ **3.** $\tan u$ **5.** $\cos^2 u$ **7.** $-\sin u$ **9.** No

11–19. Answers will vary.

21.

x	0.2	0.4	0.6	0.8
y_1	4.8348	2.1785	1.2064	0.6767
y_2	4.8348	2.1785	1.2064	0.6767

x	1.0	1.2	1.4
y_1	0.3469	0.1409	0.0293
y_2	0.3469	0.1409	0.0293

23.

x	0.2	0.4	0.6	0.8
y_1	4.8348	2.1785	1.2064	0.6767
y_2	4.8348	2.1785	1.2064	0.6767

x	1.0	1.2	1.4
y_1	0.3469	0.1409	0.0293
y_2	0.3469	0.1409	0.0293

25.

x	0.2	0.4	0.6	0.8
y_1	5.0335	2.5679	1.7710	1.3940
y_2	5.0335	2.5679	1.7710	1.3940

x	1.0	1.2	1.4
y_1	1.1884	1.0729	1.0148
y_2	1.1884	1.0729	1.0148

27.

x	0.2	0.4	0.6	0.8
y_1	5.1359	2.7880	2.1458	2.0009
y_2	5.1359	2.7880	2.1458	2.0009

x	1.0	1.2	1.4
y_1	2.1995	2.9609	5.9704
y_2	2.1995	2.9609	5.9704

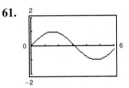

29. $\cot(-x) = -\cot(x)$ **31.** $(\tan^2 x)^2$

33–57. Answers will vary.

59.   **61.**

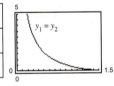

$y = 1$ $y = \sin x$

63. Answers will vary.   **65.** 1   **67.** 2

69–75. Answers will vary.

77. (a) Answers will vary.

(b)

θ	15°	30°	45°	60°	75°	90°
s	18.66	8.66	5	2.88	1.34	0

(c) Maximum: 15°; minimum: 90°   (d) Noon

79. True. For instance, $(\sec^2 \theta - 1)/\sec^2 \theta = \sin^2 \theta$ was verified two different ways on page 498.

81. False. $\sin^2\left(\dfrac{\pi}{4}\right) + \cos^2\left(\dfrac{\pi}{4}\right) \neq 1 + \tan^2\left(\dfrac{\pi}{4}\right)$

83. (a) Answers will vary.

(b) No. Division by zero.

85. $a \cos \theta$ **87.** $a \sec \theta$ **89.** $\sqrt{\tan^2 x} = |\tan x|; \dfrac{3\pi}{4}$

91. $|\tan \theta| = \sqrt{\sec^2 \theta - 1}; \dfrac{3\pi}{4}$ **93.** Answers will vary.

95.

x	-4	-2	0	2	3
y	3.0625	3.25	4	7	11

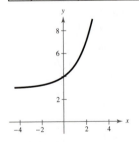

Horizontal asymptote at $y = 3$

97.

x	-3	-2	0	2	4
y	9	5	2	1.25	1.0625

Horizontal asymptote at $y = 1$

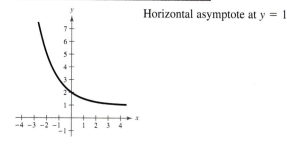

Section 6.3 (page 513)

1. general **3.** No **5–9.** Answers will vary.

11. $30°, 150°$ **13.** $120°, 240°$ **15.** $45°, 225°$

17. $\dfrac{5\pi}{6}, \dfrac{7\pi}{6}$ **19.** $\dfrac{3\pi}{4}, \dfrac{7\pi}{4}$ **21.** $\dfrac{5\pi}{6}, \dfrac{11\pi}{6}$

23. $\dfrac{7\pi}{6}, \dfrac{11\pi}{6}$ **25.** $\dfrac{\pi}{6}, \dfrac{7\pi}{6}$ **27.** $\dfrac{3\pi}{4}, \dfrac{7\pi}{4}$

29. $\dfrac{7\pi}{6} + 2n\pi, \dfrac{11\pi}{6} + 2n\pi$ **31.** $\dfrac{\pi}{6} + 2n\pi, \dfrac{11\pi}{6} + 2n\pi$

33. $\dfrac{\pi}{6} + n\pi, \dfrac{5\pi}{6} + n\pi$ **35.** $\dfrac{\pi}{3} + n\pi, \dfrac{2\pi}{3} + n\pi$

37. $\dfrac{2\pi}{3}, \dfrac{5\pi}{3}$ **39.** $\dfrac{\pi}{4}, \dfrac{3\pi}{4}, \dfrac{5\pi}{4}, \dfrac{7\pi}{4}$ **41.** $0, \dfrac{\pi}{2}, \pi, \dfrac{3\pi}{2}$

43. $\dfrac{\pi}{3}, \pi, \dfrac{5\pi}{3}$ **45.** No solution

47. $\dfrac{\pi}{6}, \dfrac{5\pi}{6}, \dfrac{7\pi}{6}, \dfrac{11\pi}{6}$ **49.** 3.6652, 4.7124, 5.7596

51. 0.8614, 5.4218 **53.** 1.5708 **55.** 0.5236, 2.6180

57. (a)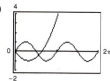
(b) $\sin 2x = x^2 - 2x$
(c) $(0, 0), (1.7757, -0.3984)$

59. (a)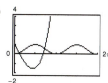
(b) $\sin^2 x = e^x - 4x$
(c) $(0.3194, 0.0986),$
$(2.2680, 0.5878)$

61. $2\pi + 4n\pi$ **63.** $\dfrac{\pi}{8} + \dfrac{n\pi}{2}$ **65.** $\dfrac{2\pi}{3} + n\pi, \dfrac{5\pi}{6} + n\pi$

67. $\dfrac{\pi}{2} + 4n\pi, \dfrac{7\pi}{2} + 4n\pi$ **69.** $x = -1, 3$ **71.** $x = \pm 2$

73. 1.1071, 4.2487 **75.** 0.8603, 3.4256

77. 0, 2.6779, 3.1416, 5.8195

79. 0.3398, 0.8481, 2.2935, 2.8018

81. $\dfrac{\pi}{4}, \dfrac{5\pi}{4}$, arctan 5, arctan $5 + \pi$ **83.** $-1.154, 0.534$

85. 1.110

87. (a)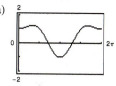
Maxima: $(0.7854, 1), (3.9270, 1)$
Minima: $(2.3562, -1),$
$(5.4978, -1)$

(b) $\dfrac{\pi}{4}, \dfrac{3\pi}{4}, \dfrac{5\pi}{4}, \dfrac{7\pi}{4}$

89. (a)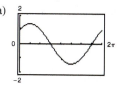
Maxima: $(1.0472, 1.25), (5.2360, 1.25)$
Minima: $(0, 1), (3.1416, -1), (6.2832, 1)$

(b) $0, \dfrac{\pi}{3}, \pi, \dfrac{5\pi}{3}, 2\pi$

91. (a)

Maximum: $(0.7854, 1.4142)$
Minimum: $(3.9270, -1.4142)$

(b) $\dfrac{\pi}{4}, \dfrac{5\pi}{4}$

93. 1 **95.** May, June, July

97. (a) All real numbers x except $x = 0$
(b) y-axis symmetry; horizontal asymptote: $y = 1$
(c) Oscillates (d) Infinite number of solutions
(e) Yes. 0.6366

99. 0.04 sec, 0.43 sec, 0.83 sec **101.** $36.87°, 53.13°$

103. (a)

$x \approx 0.86, A \approx 1.12$

(b) $0.6100 < x < 1.0980$

105. 1 **107.** False. $\sin x - x = 0$ has one solution, $x = 0$.

109. False. The range of the sine function does not include 3.4.

111. Answers will vary. **113.** 2.164 rad

115. -0.007 rad **117.** Answers will vary.

Section 6.4 (page 521)

1. $\sin u \cos v - \cos u \sin v$ **3.** $\dfrac{\tan u + \tan v}{1 - \tan u \tan v}$

5. $\cos u \cos v + \sin u \sin v$

7. Sample answer: $\sin(45° + 60°)$ **9.** (a) $-\dfrac{1}{2}$ (b) $-\dfrac{3}{2}$

11. (a) $\dfrac{\sqrt{2} - \sqrt{6}}{4}$ (b) $\dfrac{1 + \sqrt{2}}{2}$

13. (a) $\dfrac{\sqrt{2} + \sqrt{6}}{4}$ (b) $\dfrac{\sqrt{2} - 1}{2}$

15. $\sin 105° = \dfrac{\sqrt{2} + \sqrt{6}}{4}$ **17.** $\sin 195° = \dfrac{\sqrt{2} - \sqrt{6}}{4}$

$\cos 105° = \dfrac{\sqrt{2} - \sqrt{6}}{4}$ $\cos 195° = -\dfrac{\sqrt{2} + \sqrt{6}}{4}$

$\tan 105° = -2 - \sqrt{3}$ $\tan 195° = 2 - \sqrt{3}$

109. $\sin \dfrac{u}{2} = \dfrac{\sqrt{10}}{10}$

$\cos \dfrac{u}{2} = \dfrac{3\sqrt{10}}{10}$

$\tan \dfrac{u}{2} = \dfrac{1}{3}$

111. $\sin \dfrac{u}{2} = \dfrac{3\sqrt{14}}{14}$

$\cos \dfrac{u}{2} = \dfrac{\sqrt{70}}{14}$

$\tan \dfrac{u}{2} = \dfrac{3\sqrt{5}}{5}$

113. $-|\cos 4x|$ **115.** $\tan 5x$ **117.** $V = \sin \dfrac{\theta}{2} \cos \dfrac{\theta}{2}$ m^3

119. $3\left(\sin \dfrac{\pi}{2} + \sin 0\right)$ **121.** $\dfrac{1}{2}(\cos \alpha - \cos 9\alpha)$

123. $2 \cos\left(\dfrac{9\theta}{2}\right) \cos\left(\dfrac{\theta}{2}\right)$ **125.** $2 \cos x \sin \dfrac{\pi}{4}$

127. $y = \dfrac{1}{2}\sqrt{10} \sin\left(8t - \arctan \dfrac{1}{3}\right)$

129. $\dfrac{\sqrt{10}}{2}$ ft   **131.** False. $\cos \dfrac{\theta}{2} > 0$

133. True. Answers will vary.   **135.** Answers will vary.

137. No. $\sin \theta = \dfrac{1}{2}$ has an infinite number of solutions but is not an identity.

139. $y_3 = y_2 + 1$

Chapter Test   (page 539)

1. $\sin \theta = \dfrac{-3\sqrt{13}}{13}$ **2.** 1   **3.** 1   **4.** $\csc \theta \sec \theta$

$\cos \theta = \dfrac{-2\sqrt{13}}{13}$

$\csc \theta = \dfrac{-\sqrt{13}}{3}$

$\sec \theta = \dfrac{-\sqrt{13}}{2}$

$\cot \theta = \dfrac{2}{3}$

5. $0, \dfrac{\pi}{2} < \theta \le \pi, \dfrac{3\pi}{2} < \theta < 2\pi$

6.

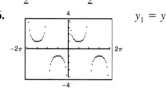

$y_1 = y_2$

7–12. Answers will vary.   **13.** $\sqrt{3} + 2$

14. $\dfrac{1}{16}\left(\dfrac{10 - 15 \cos 2x + 6 \cos 4x - \cos 6x}{1 + \cos 2x}\right)$ **15.** $\tan 2\theta$

16. $2(\sin 6\theta + \sin 2\theta)$ **17.** $-2 \cos \dfrac{7\theta}{2} \sin \dfrac{\theta}{2}$

18. $0, \dfrac{3\pi}{4}, \pi, \dfrac{7\pi}{4}$ **19.** $\dfrac{\pi}{6}, \dfrac{\pi}{2}, \dfrac{5\pi}{6}, \dfrac{3\pi}{2}$ **20.** $\dfrac{\pi}{6}, \dfrac{5\pi}{6}, \dfrac{7\pi}{6}, \dfrac{11\pi}{6}$

21. $\dfrac{\pi}{6}, \dfrac{5\pi}{6}, \dfrac{3\pi}{2}$ **22.** $-2.938, -2.663, 1.170$

23. $\sin 2u = \dfrac{4}{5}$ **24.** $76.52°$

$\cos 2u = -\dfrac{3}{5}$

$\tan 2u = -\dfrac{4}{3}$

Chapter 7

Section 7.1   (page 550)

1. oblique   **3.** $\dfrac{1}{2}bc \sin A; \dfrac{1}{2}ab \sin C; \dfrac{1}{2}ac \sin B$

5. AAS (or ASA) and SSA

7. $C = 95°, b \approx 24.59$ in., $c \approx 28.29$ in.

9. $A = 40°, a \approx 15.69$ cm, $b \approx 6.32$ cm

11. $C = 74° 15', a \approx 6.41$ km, $c \approx 6.26$ km

13. $B \approx 21.55°, C \approx 122.45°, c \approx 11.49$

15. $B = 60.9°, b \approx 19.32, c \approx 6.36$

17. $B = 18° 13', C \approx 51° 32', c \approx 40.05$

19. $B \approx 48.74°, C \approx 21.26°, c \approx 48.23$

21. $A = 48°, b \approx 2.29, c \approx 4.73$

23. $A = 35°, a \approx 36.50, b \approx 11.05$

25. $A \approx 44° 14', B \approx 50° 26', b \approx 38.67$ **27.** No solution

29. Two solutions

$B \approx 72.21°, C \approx 49.79°, c \approx 10.27$

$B \approx 107.79°, C \approx 14.21°, c \approx 3.30$

31. Given: $A = 36°, a = 5$

(a) One solution if $b \le 5$ or $b = \dfrac{5}{\sin 36°}$.

(b) Two solutions if $5 < b < \dfrac{5}{\sin 36°}$.

(c) No solution if $b > \dfrac{5}{\sin 36°}$.

33. Given: $A = 10°, a = 10.8$

(a) One solution if $b \le 10.8$ or $b = \dfrac{10.8}{\sin 10°}$.

(b) Two solutions if $10.8 < b < \dfrac{10.8}{\sin 10°}$.

(c) No solution if $b > \dfrac{10.8}{\sin 10°}$.

35. 28.19 square units   **37.** 1782.32 square units

39. 2888.57 square units

41. (a)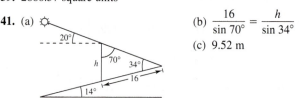

(b) $\dfrac{16}{\sin 70°} = \dfrac{h}{\sin 34°}$

(c) 9.52 m

43. $240.03°$

45. 15.53 km from Colt Station; 42.43 km from Pine Knob

47. $\theta \approx 16.08°$

49. (a) $\alpha \approx 5.36°$

(b) $\beta = \arcsin\left(\dfrac{d \sin \theta}{58.36}\right)$ (c) $d = \sin(84.64 - \theta)\left[\dfrac{58.36}{\sin \theta}\right]$

(d)

θ	10°	20°	30°	40°	50°	60°
d	324.08	154.19	95.19	63.80	43.30	28.10

51. False. The triangle cannot be solved if only three angles are known.

53. False. AAS and ASA cases have unique solutions.

55. (a) Answers will vary; Sample answer: $b = 4$

(b) Answers will vary; Sample answer: $b = 7$

(c) Answers will vary; Sample answer: $b = 10$

57. $\tan \theta = -\dfrac{12}{5}; \csc \theta = -\dfrac{13}{12}; \sec \theta = \dfrac{13}{5}; \cot \theta = -\dfrac{5}{12}$

59. $3(\sin 11\theta + \sin 5\theta)$ **61.** $\dfrac{3}{2}\left(\sin \dfrac{11\pi}{6} + \sin \dfrac{3\pi}{2}\right)$

Section 7.2   (page 557)

1. $c^2 = a^2 + b^2 - 2ab \cos C$ **3.** No   **5.** Yes

7. $A \approx 40.80°, B \approx 60.61°, C \approx 78.59°$

9. $A \approx 49.51°, B \approx 55.40°, C \approx 75.09°$

11. $A \approx 31.40°, C \approx 128.60°, b \approx 6.56$ mm

13. $A \approx 26.38°, B \approx 36.34°, C \approx 117.28°$

15. $B \approx 29.44°, C \approx 100.56°, a \approx 23.38$

17. $A \approx 36.87°, B \approx 53.13°, C = 90°$

19. $A \approx 103.52°, B \approx 38.24°, C \approx 38.24°$

21. $A \approx 154° \, 14\,', C \approx 17° \, 31\,', b \approx 8.58$

23. $A \approx 37° \, 6'\, 7'', C \approx 67° \, 33'\, 53'', b \approx 9.94$

	a	b	c	d	θ	ϕ
25.	4	8	11.64	4.96	30°	150°
27.	10	14	20	13.86	68.20°	111.80°
29.	15	16.96	25	20	77.22°	102.78°

31. Law of Cosines; $A \approx 102.44°, C \approx 37.56°, b \approx 5.26$

33. Law of Sines; no solution

35. Law of Sines; $C = 103°, a \approx 0.82, b \approx 0.71$

37. 104.57   **39.** 19.81   **41.** 0.27 ft²   **43.** 15.52

45. 35.19   **47.** 483.40 m

49. (a) N 59.7° E   (b) N 72.8° E   **51.** 72.28°

53. $PQ \approx 9.43, QS = 5, RS \approx 12.81$ **55.** 18,617.66 ft²

57. (a) $49 = 2.25 + x^2 - 3x \cos \theta$

(b) $x = \frac{1}{2}\left(3 \cos \theta + \sqrt{9 \cos^2 \theta + 187}\right)$

(c) 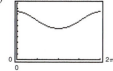   (d) 6 in.

59. True   **61.** Proof

63. To solve the triangle using the Law of Cosines, substitute values into $a^2 = b^2 + c^2 - 2bc \cos A$.

Simplify the equation so that you have a quadratic equation in terms of c. Then, find the two values of c, and find the two triangles that model the given information.

Using the Law of Sines will give the same result as using the Law of Cosines.

Sample answer: An advantage of using the Law of Cosines is that it is easier to choose the correct value to avoid the ambiguous case, but its disadvantage is that there are more computations. The opposite is true for the Law of Sines.

65. Proof   **67.** $-\dfrac{\pi}{2}$ **69.** $\dfrac{\pi}{3}$

Section 7.3   (page 569)

1. directed line segment   **3.** magnitude

5. standard position   **7.** resultant

9. magnitude and direction   **11.** Answers will vary.

13. $\langle 1, 3 \rangle, \|\mathbf{v}\| = \sqrt{10}$ **15.** $\langle -3, 2 \rangle, \|\mathbf{v}\| = \sqrt{13}$

17. $\langle 0, 5 \rangle, \|\mathbf{v}\| = 5$ **19.** $\langle 8, 6 \rangle, \|\mathbf{v}\| = 10$

21. $\left\langle \dfrac{3}{5}, -\dfrac{3}{5} \right\rangle, \|\mathbf{v}\| = \dfrac{3\sqrt{2}}{5}$ **23.** $\left\langle \dfrac{7}{6}, \dfrac{9}{5} \right\rangle, \|\mathbf{v}\| = \dfrac{\sqrt{4141}}{30}$

25.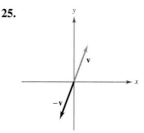

27.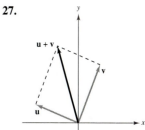

29.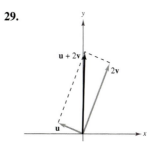

31.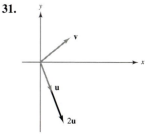

33.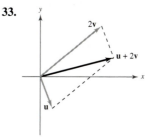

35.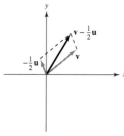

37. (a) $\langle 11, 3 \rangle$

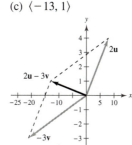

(b) $\langle -3, 1 \rangle$

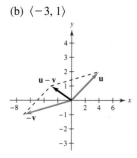

(c) $\langle -13, 1 \rangle$

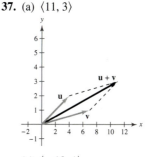

(d) $\langle 23, 9 \rangle$

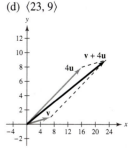

39. (a) $\langle -4, -4 \rangle$

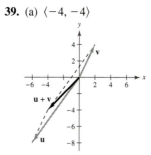

(b) $\langle -8, -12 \rangle$

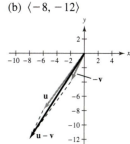

11.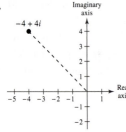

$4\sqrt{2}$

13.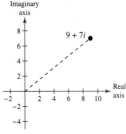

$\sqrt{130}$

15. $2\left(\cos\dfrac{\pi}{2} + i\sin\dfrac{\pi}{2}\right)$ **17.** $4(\cos\pi + i\sin\pi)$

19. $3\sqrt{2}\left(\cos\dfrac{5\pi}{4} + i\sin\dfrac{5\pi}{4}\right)$ **21.** $2\left(\cos\dfrac{11\pi}{6} + i\sin\dfrac{11\pi}{6}\right)$

23.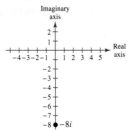

$8\left(\cos\dfrac{3\pi}{2} + i\sin\dfrac{3\pi}{2}\right)$

25.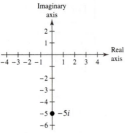

$5\left(\cos\dfrac{3\pi}{2} + i\sin\dfrac{3\pi}{2}\right)$

27.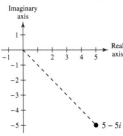

$5\sqrt{2}\left(\cos\dfrac{7\pi}{4} + i\sin\dfrac{7\pi}{4}\right)$

29.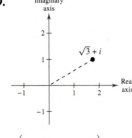

$2\left(\cos\dfrac{\pi}{6} + i\sin\dfrac{\pi}{6}\right)$

31.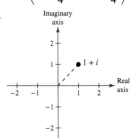

$\sqrt{2}\left(\cos\dfrac{\pi}{4} + i\sin\dfrac{\pi}{4}\right)$

33.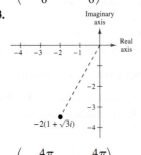

$4\left(\cos\dfrac{4\pi}{3} + i\sin\dfrac{4\pi}{3}\right)$

35.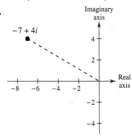

$\sqrt{65}(\cos 2.622 + i\sin 2.622)$

37.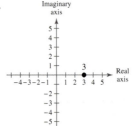

$3(\cos 0 + i\sin 0)$

39.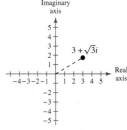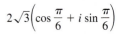

$2\sqrt{3}\left(\cos\dfrac{\pi}{6} + i\sin\dfrac{\pi}{6}\right)$

41.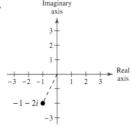

$\sqrt{5}(\cos 4.249 + i\sin 4.249)$

43.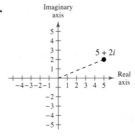

$\sqrt{29}(\cos 0.381 + i\sin 0.381)$

45.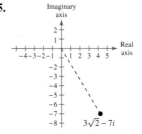

$\sqrt{67}(\cos 5.257 + i\sin 5.257)$

47.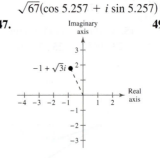

$-1 + \sqrt{3}\,i$

49.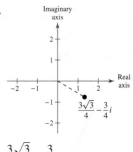

$\dfrac{3\sqrt{3}}{4} - \dfrac{3}{4}i$

51.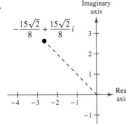

$$-\frac{15\sqrt{2}}{8} + \frac{15\sqrt{2}}{8} i$$

53.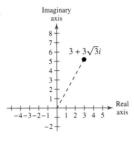

$$3 + 3\sqrt{3}\, i$$

55.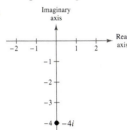

$$-4i$$

57.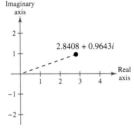

$$2.8408 + 0.9643i$$

59. $4.6985 + 1.7101i$ **61.** $4.7693 + 7.6324i$

63.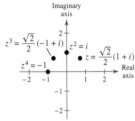

The absolute value of each is 1.

65. $10(\cos 0 + i \sin 0)$ **67.** $6(\cos \pi + i \sin \pi)$

69. $\frac{10}{9}(\cos 200° + i \sin 200°)$ **71.** $\frac{11}{50}(\cos 130° + i \sin 130°)$

73. $\cos 30° + i \sin 30°$ **75.** $\frac{1}{2}(\cos 80° + i \sin 80°)$

77. $6(\cos 312° + i \sin 312°)$

79. (a) $2\sqrt{2}\left(\cos \frac{7\pi}{4} + i \sin \frac{7\pi}{4}\right)$ (b) and (c) 4

$$\sqrt{2}\left(\cos \frac{\pi}{4} + i \sin \frac{\pi}{4}\right)$$

81. (a) $2\sqrt{2}\left(\cos \frac{\pi}{4} + i \sin \frac{\pi}{4}\right)$ (b) and (c) 4

$$\sqrt{2}\left(\cos \frac{7\pi}{4} + i \sin \frac{7\pi}{4}\right)$$

83. (a) $2\left(\cos \frac{3\pi}{2} + i \sin \frac{3\pi}{2}\right)$ (b) and (c) $2 - 2i$

$$\sqrt{2}\left(\cos \frac{\pi}{4} + i \sin \frac{\pi}{4}\right)$$

85. (a) $2\left(\cos \frac{3\pi}{2} + i \sin \frac{3\pi}{2}\right)$ (b) and (c) $-2 - 2\sqrt{3}i$

$$2\left(\cos \frac{11\pi}{6} + i \sin \frac{11\pi}{6}\right)$$

87. (a) $2(\cos 0 + i \sin 0)$ (b) and (c) $2 - 2i$

$$\sqrt{2}\left(\cos \frac{7\pi}{4} + i \sin \frac{7\pi}{4}\right)$$

89. (a) $5(\cos 0.93 + i \sin 0.93)$

$$2\left(\cos \frac{5\pi}{3} + i \sin \frac{5\pi}{3}\right)$$

(b) and (c) $\left(\frac{3}{4} - \sqrt{3}\right) + \left(\frac{3\sqrt{3}}{4} + 1\right)i$

91. (a) $5(\cos 0 + i \sin 0)$ (b) and (c) $\frac{5}{4} - \frac{5}{4}i$

$$2\sqrt{2}\left(\cos \frac{\pi}{4} + i \sin \frac{\pi}{4}\right)$$

93. (a) $4\left(\cos \frac{\pi}{2} + i \sin \frac{\pi}{2}\right)$ (b) and (c) $2 - 2i$

$$\sqrt{2}\left(\cos \frac{3\pi}{4} + i \sin \frac{3\pi}{4}\right)$$

95.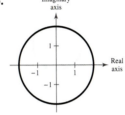

97.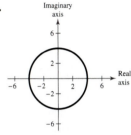

99.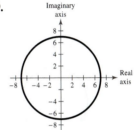

101.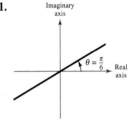

103.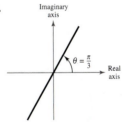

105.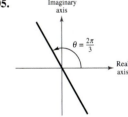

107. $-2 + 2i$ **109.** $8i$ **111.** $-32\sqrt{3} + 32i$

113. $\frac{125}{2} + \frac{125\sqrt{3}}{2}i$ **115.** i **117.** $4.5386 - 15.3428i$

119. 256 **121.** $-597 - 122i$

123. $2048 + 2048\sqrt{3}i$ **125.** $\frac{9\sqrt{2}}{2} + \frac{9\sqrt{2}}{2}i$

127. Answers will vary.   **129.** $1 + i, -1 - i$

131. $-\frac{\sqrt{6}}{2} + \frac{\sqrt{6}}{2}i, \frac{\sqrt{6}}{2} - \frac{\sqrt{6}}{2}i$

133. $-1.5538 + 0.6436i, 1.5538 - 0.6436i$

135. $\frac{\sqrt{6}}{2} + \frac{\sqrt{2}}{2}i, -\frac{\sqrt{6}}{2} - \frac{\sqrt{2}}{2}i$

137. (a) $\sqrt{5}(\cos 60° + i \sin 60°)$

$$\sqrt{5}(\cos 240° + i \sin 240°)$$

CHAPTER 7

41. $\langle 30, 9 \rangle$ **43.** $\langle 74, -5 \rangle$

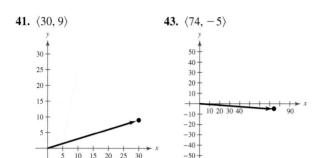

45. $\langle 0, -1 \rangle$ **47.** $\dfrac{\sqrt{29}}{29}\langle 5, -2 \rangle$ **49.** $9\mathbf{i} - 8\mathbf{j}$

51. $\|\mathbf{v}\| = 7;\ \theta = 60°$ **53.** $\|\mathbf{v}\| = \sqrt{41};\ \theta = 38.7°$

55. $\|\mathbf{v}\| = 3\sqrt{2};\ \theta = 225°$

57. 133.92 lb, 5.55° from the 85-lb force   **59.** 115.47 lb

61. -20 **63.** 7   **65.** 25   **67.** -40 **69.** 2.802

71. $\dfrac{11\pi}{12}$

73.   **75.**

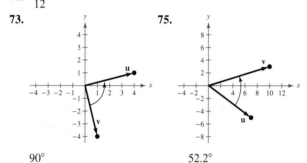

90°   52.2°

77. Parallel   **79.** Neither   **81.** $\frac{1}{2}$ **83.** -1

85. $\frac{13}{17}\langle -4, -1 \rangle,\ \left\langle -\frac{52}{17}, -\frac{13}{17} \right\rangle + \left\langle -\frac{16}{17}, \frac{64}{17} \right\rangle$

87. $\frac{5}{2}\langle -1, 1 \rangle,\ \left\langle -\frac{5}{2}, \frac{5}{2} \right\rangle + \left\langle \frac{9}{2}, \frac{9}{2} \right\rangle$ **89.** 72,000 ft-lb

91.   **93.**

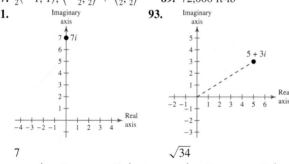

7    $\sqrt{34}$

95. $2\sqrt{2}\left(\cos\dfrac{7\pi}{4} + i\sin\dfrac{7\pi}{4}\right)$ **97.** $2\left(\cos\dfrac{7\pi}{6} + i\sin\dfrac{7\pi}{6}\right)$

99. $10\left(\cos\dfrac{3\pi}{4} + i\sin\dfrac{3\pi}{4}\right)$ **101.** $4(\cos 240° + i\sin 240°)$

103. (a) $2\sqrt{2}\left(\cos\dfrac{7\pi}{4} + i\sin\dfrac{7\pi}{4}\right)$ (b) and (c) 12

$3\sqrt{2}\left(\cos\dfrac{\pi}{4} + i\sin\dfrac{\pi}{4}\right)$

105. (a) $3\sqrt{2}\left(\cos\dfrac{7\pi}{4} + i\sin\dfrac{7\pi}{4}\right)$ (b) and (c) $-\dfrac{3}{2}i$

$2\sqrt{2}\left(\cos\dfrac{\pi}{4} + i\sin\dfrac{\pi}{4}\right)$

107. $\dfrac{625}{2} + \dfrac{625\sqrt{3}}{2}i$ **109.** $2035 - 828i$

111. $\pm(0.3660 + 1.3660i)$ **113.** $-1 + i,\ 1 - i$

115. (a) $3\left(\cos\dfrac{\pi}{4} + i\sin\dfrac{\pi}{4}\right)$ (b)

$3\left(\cos\dfrac{7\pi}{12} + i\sin\dfrac{7\pi}{12}\right)$

$3\left(\cos\dfrac{11\pi}{12} + i\sin\dfrac{11\pi}{12}\right)$

$3\left(\cos\dfrac{5\pi}{4} + i\sin\dfrac{5\pi}{4}\right)$

$3\left(\cos\dfrac{19\pi}{12} + i\sin\dfrac{19\pi}{12}\right)$

$3\left(\cos\dfrac{23\pi}{12} + i\sin\dfrac{23\pi}{12}\right)$

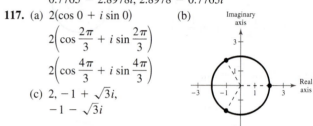

(c) $2.1213 + 2.1213i,\ -0.7765 + 2.8978i,$
$-2.8978 + 0.7765i,\ -2.1213 - 2.1213i,$
$0.7765 - 2.8978i,\ 2.8978 - 0.7765i$

117. (a) $2(\cos 0 + i\sin 0)$ (b)

$2\left(\cos\dfrac{2\pi}{3} + i\sin\dfrac{2\pi}{3}\right)$

$2\left(\cos\dfrac{4\pi}{3} + i\sin\dfrac{4\pi}{3}\right)$

(c) $2,\ -1 + \sqrt{3}i,$
$-1 - \sqrt{3}i$

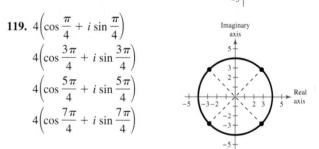

119. $4\left(\cos\dfrac{\pi}{4} + i\sin\dfrac{\pi}{4}\right)$

$4\left(\cos\dfrac{3\pi}{4} + i\sin\dfrac{3\pi}{4}\right)$

$4\left(\cos\dfrac{5\pi}{4} + i\sin\dfrac{5\pi}{4}\right)$

$4\left(\cos\dfrac{7\pi}{4} + i\sin\dfrac{7\pi}{4}\right)$

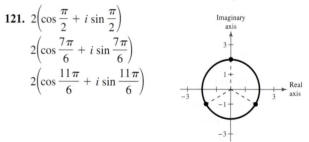

121. $2\left(\cos\dfrac{\pi}{2} + i\sin\dfrac{\pi}{2}\right)$

$2\left(\cos\dfrac{7\pi}{6} + i\sin\dfrac{7\pi}{6}\right)$

$2\left(\cos\dfrac{11\pi}{6} + i\sin\dfrac{11\pi}{6}\right)$

123. True. $\sin 90°$ is defined in the Law of Sines.

Chapter Test   (page 601)

1. $C = 46°,\ a \approx 13.07,\ b \approx 22.03$

2. $A \approx 22.33°,\ B \approx 49.46°,\ C \approx 108.21°$

3. $B \approx 40.11°,\ C \approx 104.89°,\ a \approx 7.12$

4. Two solutions
$B \approx 41.10°,\ C \approx 113.90°,\ c \approx 38.94$
$B \approx 138.90°,\ C \approx 16.10°,\ c \approx 11.81$

5. No solution   **6.** $B \approx 14.79°,\ C \approx 15.21°,\ c \approx 4.93$

7. 675 ft   **8.** 2337 m²

9. $\mathbf{w} = \langle 12, 13 \rangle,\ \|\mathbf{w}\| \approx \sqrt{313}$

10. (a) $\langle 8, 8 \rangle$ (b) $\langle -12, -22 \rangle$ (c) $\langle -4, -26 \rangle$

11. (a) $\langle -7, -18 \rangle$ (b) $\langle -2, 32 \rangle$ (c) $\langle -24, 20 \rangle$

12. (a) $13\mathbf{i} + 17\mathbf{j}$ (b) $-17\mathbf{i} - 28\mathbf{j}$ (c) $-\mathbf{i} - 14\mathbf{j}$

13. (a) $-\mathbf{j}$ (b) $5\mathbf{i} + 9\mathbf{j}$ (c) $11\mathbf{i} + 17\mathbf{j}$

14. $\left\langle \dfrac{3\sqrt{13}}{13}, -\dfrac{2\sqrt{13}}{13} \right\rangle$ **15.** $\left\langle \dfrac{18\sqrt{34}}{17}, -\dfrac{30\sqrt{34}}{17} \right\rangle$

16. $\theta \approx 14.87°, 250.15$ lb   **17.** -1 **18.** $105.95°$

19. Yes. $\mathbf{u} \cdot \mathbf{v} = 0$ **20.** $\left\langle \frac{185}{26}, \frac{37}{26} \right\rangle$, $\mathbf{u} = \left\langle \frac{185}{26}, \frac{37}{26} \right\rangle + \left\langle -\frac{29}{26}, \frac{145}{26} \right\rangle$

21. $z = 6\sqrt{2}\left(\cos \dfrac{3\pi}{4} + i \sin \dfrac{3\pi}{4} \right)$ **22.** $-50 - 50\sqrt{3}\,i$

23. $-\dfrac{6561}{2} + \dfrac{6561\sqrt{3}}{2}i$ **24.** $5832i$

25. $4\left(\cos \dfrac{\pi}{12} + i \sin \dfrac{\pi}{12} \right)$

$4\left(\cos \dfrac{7\pi}{12} + i \sin \dfrac{7\pi}{12} \right)$

$4\left(\cos \dfrac{13\pi}{12} + i \sin \dfrac{13\pi}{12} \right)$

$4\left(\cos \dfrac{19\pi}{12} + i \sin \dfrac{19\pi}{12} \right)$

26. $5\left(\cos \dfrac{\pi}{8} + i \sin \dfrac{\pi}{8} \right)$

$5\left(\cos \dfrac{5\pi}{8} + i \sin \dfrac{5\pi}{8} \right)$

$5\left(\cos \dfrac{9\pi}{8} + i \sin \dfrac{9\pi}{8} \right)$

$5\left(\cos \dfrac{13\pi}{8} + i \sin \dfrac{13\pi}{8} \right)$

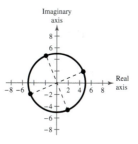

Cumulative Test for Chapters 5–7 (page 602)

1. (a)

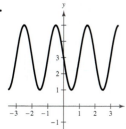

$\theta = -150°$

(b) $210°$

(c) $-\dfrac{5\pi}{6}$

(d) $30°$

(e) $\sin(-150°) = -\dfrac{1}{2}$

$\cos(-150°) = -\dfrac{\sqrt{3}}{2}$

$\tan(-150°) = \dfrac{\sqrt{3}}{3}$

$\csc(-150°) = -2$

$\sec(-150°) = -\dfrac{2\sqrt{3}}{3}$

$\cot(-150°) = \sqrt{3}$

2. $146.1°$ **3.** $\cos \theta = -\frac{5}{13}$

4.

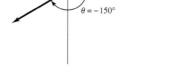

5.

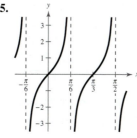

6.

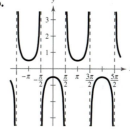

7. $a = 3, b = \pi, c = \pi$

8. $\dfrac{3}{5}$ **9.** $-\dfrac{\sqrt{3}}{3}$ **10.** $\dfrac{2x}{\sqrt{4x^2 + 1}}$ **11.** $2 \tan \theta$

12–14. Answers will vary.   **15.** $\dfrac{3\pi}{2} + 2n\pi$

16. $\dfrac{\pi}{6} + n\pi, \dfrac{5\pi}{6} + n\pi$ **17.** $1.7646, 4.5186$

18.

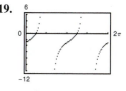

$\dfrac{\pi}{3}, \dfrac{5\pi}{3}$

19.

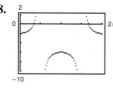

$\dfrac{\pi}{4}, \dfrac{5\pi}{4}$

20. $\dfrac{16}{63}$ **21.** $\dfrac{4}{3}$ **22.** $\dfrac{2\sqrt{5}}{5}$

23. $2 \cos 6x \cos 2x$ **24–27.** Answers will vary.

28. $B \approx 14.89°$ **29.** $B \approx 52.82°$
   $C \approx 119.11°$ $C \approx 95.18°$
   $c \approx 17.00$ $a \approx 5.32$

30. $B = 55°$ **31.** $A \approx 26.07°$
   $b \approx 20.14$ $B \approx 33.33°$
   $c \approx 24.13$ $C \approx 120.60°$

32. 131.71 in.^{2} **33.** 94.10 in.^{2} **34.** $3\mathbf{i} + 5\mathbf{j}$

35. $\dfrac{\sqrt{5}}{5}\mathbf{i} - \dfrac{2\sqrt{5}}{5}\mathbf{j}$ **36.** -5 **37.** 1

38. $\left\langle -\frac{1}{13}, -\frac{5}{13} \right\rangle$; $\mathbf{u} = \left\langle \frac{105}{13}, -\frac{21}{13} \right\rangle + \left\langle -\frac{1}{13}, -\frac{5}{13} \right\rangle$

39. $3\sqrt{2}\left(\cos \dfrac{3\pi}{4} + i \sin \dfrac{3\pi}{4} \right)$ **40.** $-9 + 3\sqrt{3}\,i$

41. $-12\sqrt{3} + 12i$

42. $1.4553 + 0.3436i, -1.4553 - 0.3436i$

43. $1, -\dfrac{1}{2} + \dfrac{\sqrt{3}}{2}i, -\dfrac{1}{2} - \dfrac{\sqrt{3}}{2}i$

44. $5\left(\cos \dfrac{\pi}{4} + i \sin \dfrac{\pi}{4} \right)$

$5\left(\cos \dfrac{3\pi}{4} + i \sin \dfrac{3\pi}{4} \right)$

$5\left(\cos \dfrac{5\pi}{4} + i \sin \dfrac{5\pi}{4} \right)$

$5\left(\cos \dfrac{7\pi}{4} + i \sin \dfrac{7\pi}{4} \right)$

45. 5 ft

46. $d = 7 \sin \dfrac{\pi}{4}t$

47. $54.34°; 489.45$ km/h

48. $80.28°$

CHAPTER 7

Chapter 8

Section 8.1   (page 616)

1. system, equations   **3.** substitution

5. Break-even point

7. (a) No   (b) No   (c) No   (d) Yes

9. (a) No   (b) Yes   (c) No   (d) No

11. $(2, 2)$ **13.** $(2, 6), (-1, 3)$

15. $(0, 2), \left(\sqrt{3}, 2 - 3\sqrt{3}\right), \left(-\sqrt{3}, 2 + 3\sqrt{3}\right)$ **17.** $(4, 4)$

19. $(5, 5)$ **21.** $\left(\frac{1}{2}, 3\right)$ **23.** $(1, 1)$ **25.** $\left(\frac{20}{3}, \frac{40}{3}\right)$

27. No solution   **29.** \$4000 at 4%, \$14,000 at 6%

31. \$3500 at 7.6%, \$14,500 at 8.8%   **33.** $(-2, 0), (3, 5)$

35. No real solution   **37.** $(0, 0), (1, 1), (-1, -1)$

39. $(4, 3)$ **41.** $\left(\frac{5}{2}, \frac{3}{2}\right)$ **43.** No real solution

45. $(3, 6), (-3, 0)$ **47.** $(4, -0.5)$ **49.** $(8, 3), (3, -2)$

51. $(\pm 1.540, 2.372)$ **53.** $(0, 1)$ **55.** $(2.318, 2.841)$

57. $(2.25, 5.5)$ **59.** $(0, -13), (\pm 12, 5)$ **61.** $(1, 2)$

63. $(-2, 0), \left(\frac{29}{10}, \frac{21}{10}\right)$ **65.** No real solution   **67.** $(0.25, 1.5)$

69. $(0.287, 1.751)$ **71.** $(0, 1), (1, 0)$ **73.** $\left(-4, -\frac{1}{4}\right), \left(\frac{1}{2}, 2\right)$

75.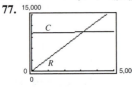

192 units; \$1,910,400

77.

3133 units; \$10,308

79. 6 m × 9 m

81. (a)

Week	Animated	Horror
1	336	42
2	312	60
3	288	78
4	264	96
5	240	114
6	216	132
7	192	150
8	168	168
9	144	186
10	120	204
11	96	222
12	72	240

(b) and (c) $x = 8$

(d) The answers are the same.

(e) During week 8 the same number of animated and horror films were rented.

83. (a) $C = 9.45x + 16,000$ **85.** 8 mi × 12 mi

$R = 55.95x$

(b) 344 units

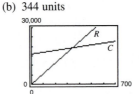

87. (a) $\begin{cases} x + \quad\quad y = 20,000 \\ 0.055x + 0.075y = \quad 1300 \end{cases}$

(b)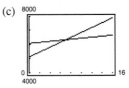

(c) \$10,000. The solution is $(10,000, 10,000)$.

89. (a)

t	Year	Arizona	Indiana
0	2000	5118	6080
1	2001	5289.9	6115.7
2	2002	5461.8	6151.4
3	2003	5633.7	6187.1
4	2004	5805.6	6222.8
5	2005	5977.5	6258.5
6	2006	6149.4	6294.2
7	2007	6321.3	6329.9
8	2008	6493.2	6365.6

(b) 2008

(c)

(d) $(7.06, 6332.15)$

$(7.06, 6332.15)$

(e) At one point in 2007, the populations of Arizona and Indiana were equal.

91. False. You can solve for either variable before back-substituting.

93. For a linear system, the result will be a contradictory equation such as $0 = N$, where N is a nonzero real number. For a nonlinear system, there may be an equation with imaginary roots.

95. (a) $\begin{cases} 3x + y = 3 \\ 3x + y = 5 \end{cases}$ (b) $\begin{cases} 3x + y = 4 \\ 2x + y = 2 \end{cases}$ (c) $\begin{cases} 6x + 3y = 9 \\ 2x + y = 3 \end{cases}$

97. (a)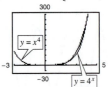

(b) There are three points of intersection when b is even.

99. $y = -\frac{2}{7}x + \frac{45}{7}$ **101.** $y = 3$ **103.** $y = \frac{30}{17}x - \frac{18}{17}$

105. Domain: All real numbers x except $x = 6$
Asymptotes: $y = 0$, $x = 6$

107. Domain: All real numbers x except $x = \pm 4$
Asymptotes: $y = 1$, $x = \pm 4$

109. Domain: All real numbers x
Asymptote: $y = 0$

Section 8.2   (page 625)

1. method, elimination   **3.** Inconsistent   **5.** Yes

7. $(2, 1)$ **9.** $(1, -1)$

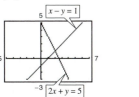

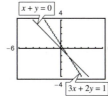

11. Inconsistent

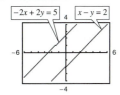

13. $\left(2, \frac{1}{2}\right)$ **15.** $(3, 4)$ **17.** $(4, -1)$ **19.** $\left(\frac{12}{7}, \frac{18}{7}\right)$

21. Inconsistent   **23.** b; One solution, consistent

24. a; Infinitely many solutions, consistent

25. c; One solution, consistent

26. d; No solutions, inconsistent   **27.** $\left(\frac{3}{2}, -\frac{1}{2}\right)$

29. Inconsistent   **31.** All points on $6x + 8y - 1 = 0$

33. $(5, -2)$ **35.** All points on $-5x + 6y = -3$

37. All points on $5x - 6y - 3 = 0$ **39.** $(101, 96)$

41. $\left(\frac{90}{31}, -\frac{67}{31}\right)$ **43.** $(-1, 1)$ **45.** $\left(1, \frac{1}{2}\right)$

47.

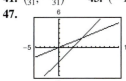

Consistent; $(5, 2)$

49.

Inconsistent

51.

Consistent; all points on $8x - 14y = 5$

53.

$(3.833, 7)$

55.

$(6, 5)$

57.

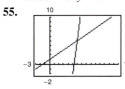

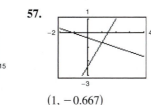

$(1, -0.667)$

59.

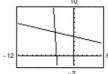

$(-4, 5)$

61. $(4, 1)$ **63.** $(2, -1)$ **65.** $(6, -3)$ **67.** $\left(\frac{49}{4}, \frac{33}{4}\right)$

69. $\begin{cases} 3x + \frac{1}{2}y = 4 \\ x + 3y = 24 \end{cases}$ **71.** $\begin{cases} 2x + 2y = 11 \\ x - 4y = -7 \end{cases}$

73. $(240, 404)$ **75.** $(2{,}000{,}000, 100)$

77. Plane: 550 mi/h; wind: 50 mi/h

79. (a) $\begin{cases} 5.00A + 3.50C = 5087.50 \\ A + C = 1175 \end{cases}$

(b) $A = 650$, $C = 525$; Answers will vary.

(c) $A = 650$, $C = 525$

81. 9 oranges, 7 grapefruits

83. 185 movies, 125 video games

85. $y = 0.97x + 2.1$ **87.** $y = -2.5x + 5.54$

89. (a) and (b) $y = 14x + 19$

(c)

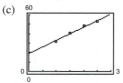

(d) 41.4 bushels per acre

91. True. A linear system can have only one solution, no solution, or infinitely many solutions.

93. False. Sometimes you will be able to get only a close approximation.

95. (a) $\begin{cases} x + y = 10 \\ x + y = 20 \end{cases}$ (b) $\begin{cases} x + y = 4 \\ 3x + 3y = 12 \end{cases}$

97. $u = 1$; $v = -\tan x$

99. $x \leq -\frac{22}{3}$

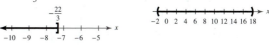

101. $-2 < x < 18$

103. $-5 < x < \frac{7}{2}$ **105.** $\ln 6x$

107. $\log_9 \dfrac{12}{x}$ **109.** $\ln \dfrac{x^2}{x + 2}$ **111.** Answers will vary.

Section 8.3   (page 639)

1. row-echelon   **3.** Gaussian   **5.** three-dimensional

7. Independent   **9.** (a) No   (b) Yes   (c) No   (d) No

11. (a) No   (b) No   (c) Yes   (d) No

13. $(2, -2, 2)$ **15.** $(3, 10, 2)$ **17.** $\left(\frac{11}{4}, 7, 11\right)$

19. $\begin{cases} x - 2y + 3z = 5 \\ y - 2z = 9 \\ 2x - 3z = 0 \end{cases}$

It removed the x-term from Equation 2.

21. $(1, 2, 3)$ **23.** $(-4, 8, 5)$ **25.** $(2, -3, -2)$

27. Inconsistent   **29.** $\left(1, -\frac{3}{2}, \frac{1}{2}\right)$ **31.** $(-a + 3, a + 1, a)$

33. Inconsistent   **35.** Inconsistent   **37.** $(-1, 1, 0)$

39. $(2a, 21a - 1, 8a)$ **41.** $\left(-\frac{3}{2}a + \frac{1}{2}, -\frac{2}{3}a + 1, a\right)$

43. $\begin{cases} x + y + z = 1 \\ 2x + y + z = 4 \\ x + y - 3z = -7 \end{cases}$ **45.** $\begin{cases} x + y + 2z = -10 \\ -x + 12y + 8z = -14 \\ x + 14y - 4z = -6 \end{cases}$

47. 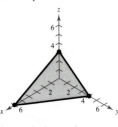   **49.**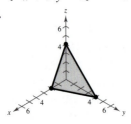

$(6, 0, 0), (0, 4, 0),$ $(2, 0, 0), (0, 4, 0),$
$(0, 0, 3), (4, 0, 1)$ $(0, 0, 4), (0, 2, 2)$

51. $\dfrac{A}{x} + \dfrac{B}{x - 14}$ **53.** $\dfrac{A}{x} + \dfrac{B}{x^2} + \dfrac{C}{x - 10}$

55. $\dfrac{A}{x - 5} + \dfrac{B}{(x - 5)^2} + \dfrac{C}{(x - 5)^3}$

57. $\dfrac{1}{2}\left(\dfrac{1}{x - 1} - \dfrac{1}{x + 1}\right)$ **59.** $\dfrac{1}{x} - \dfrac{1}{x + 1}$

61. $\dfrac{3}{2x - 1} - \dfrac{2}{x + 1}$ **63.** $-\dfrac{3}{x} - \dfrac{1}{x + 2} + \dfrac{5}{x - 2}$

65. $\dfrac{3}{x} - \dfrac{1}{x^2} + \dfrac{1}{x + 1}$ **67.** $2x - 7 + \dfrac{17}{x + 2} + \dfrac{1}{x + 1}$

69. $x + 3 + \dfrac{6}{x - 1} + \dfrac{4}{(x - 1)^2} + \dfrac{1}{(x - 1)^3}$

71. $\dfrac{3}{x} - \dfrac{2}{x - 4}$

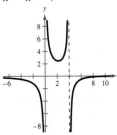 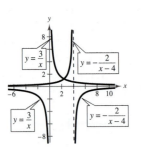

The vertical asymptotes are the same.

73. $s = -16t^2 + 144$ **75.** $s = -16t^2 - 32t + 400$

77. $y = \frac{1}{2}x^2 - 2x$ **79.** $y = x^2 - 6x + 8$

81. $x^2 + y^2 - 10x = 0$ **83.** $x^2 + y^2 + 6x - 8y = 0$

85. \$300,000 at 8%, \$400,000 at 9%, and \$75,000 at 10%

87. $187,500 + s$ in certificates of deposit
    $187,500 - s$ in municipal bonds
    $125,000 - s$ in blue-chip stocks
    s in growth stocks

89. 13 two-point baskets, 6 three-point baskets, 9 free throws

91. $I_1 = 1, I_2 = 2, I_3 = 1$ **93.** $y = -\frac{5}{24}x^2 - \frac{3}{10}x + \frac{41}{6}$

95. $y = x^2 - x$

97. (a) $\begin{cases} 900a + 30b + c = 55 \\ 1600a + 40b + c = 105 \\ 2500a + 50b + c = 188 \end{cases}$

    $y = 0.165x^2 - 6.55x + 103$

(b) 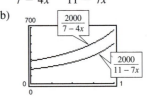   (c) 453 ft

99. (a) $\dfrac{2000}{7 - 4x} - \dfrac{2000}{11 - 7x}, \ 0 \le x \le 1$

(b)

101. False. The leading coefficients are not all 1.

103. The student did not work the problem correctly. Because $\dfrac{x^2 + 1}{x(x - 1)}$ is an improper fraction, the student should have divided before decomposing.

105. No. There are two arithmetic errors. The constant in the second equation should be -11 and the coefficient of z in the third equation should be 2.

107. $x = 5, y = 5, \lambda = -5$

109. (a) $-4, 0, 3$ **111.** (a) $-4, -\frac{3}{2}, 3$

(b) 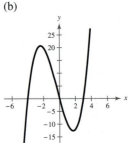   (b)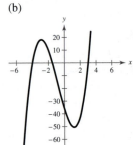

113. $\dfrac{\pi}{6} + n\pi$ **115.** Answers will vary.

Section 8.4   (page 653)

1. matrix   **3.** Gauss-Jordan elimination   **5.** No

7. 1×2 **9.** 3×1 **11.** 2×2

13. $\begin{bmatrix} 4 & -3 & \vdots & -5 \\ -1 & 3 & \vdots & 12 \end{bmatrix}$ 2×3

15. $\begin{bmatrix} 1 & 10 & -2 & \vdots & 2 \\ 5 & -3 & 4 & \vdots & 0 \\ 2 & 1 & 0 & \vdots & 6 \end{bmatrix}$ 3×4

17. $\begin{bmatrix} 7 & -5 & 1 & \vdots & 13 \\ 19 & 0 & -8 & \vdots & 10 \end{bmatrix}$ 2×4

19. $\begin{cases} 3x + 4y = 9 \\ x - y = -3 \end{cases}$

21. $\begin{cases} 9x + 12y + 3z = 0 \\ -2x + 18y + 5z = 10 \\ x + 7y - 8z = -4 \end{cases}$

23. Add -3 times R_2 to R_1.   **25.** Interchange R_1 and R_2.

27. $\begin{bmatrix} 1 & 4 & 3 \\ 0 & 2 & -1 \end{bmatrix}$

29. $\begin{bmatrix} 1 & 1 & 4 & -1 \\ 0 & 5 & -2 & 6 \\ 0 & 3 & 20 & 4 \end{bmatrix}, \begin{bmatrix} 1 & 1 & 4 & -1 \\ 0 & 1 & -\frac{2}{5} & \frac{6}{5} \\ 0 & 3 & 20 & 4 \end{bmatrix}$

31. (a) i) $\begin{bmatrix} 3 & 0 & \vdots & -6 \\ 6 & -4 & \vdots & -28 \end{bmatrix}$ ii) $\begin{bmatrix} 3 & 0 & \vdots & -6 \\ 0 & -4 & \vdots & -16 \end{bmatrix}$

iii) $\begin{bmatrix} 3 & 0 & \vdots & -6 \\ 0 & 1 & \vdots & 4 \end{bmatrix}$ iv) $\begin{bmatrix} 1 & 0 & \vdots & -2 \\ 0 & 1 & \vdots & 4 \end{bmatrix}$

(b) $x = -2, y = 4$ (c) Answers will vary.

33. i)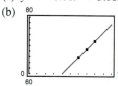

```
row+([A],2,1)→[B
]
  [[3 0 -6 ]
   [6 -4 -28]]
```

ii)
```
*row+(-2,[B],1,2
)→[C]
  [[3 0 -6 ]
   [0 -4 -16]]
```

iii)
```
*row(-1/4,[C],2)
→[D]
  [[3 0 -6]
   [0 1 4 ]]
```

iv)
```
*row(1/3,[D],1)→
[E]
  [[1 0 -2]
   [0 1 4 ]]
```

35. Reduced row-echelon form   **37.** Not in row-echelon form
39. Not in row-echelon form

41. $\begin{bmatrix} 1 & 3 & \frac{3}{2} & 5 \\ 0 & 1 & \frac{3}{14} & 0 \\ 0 & 0 & 1 & -\frac{35}{12} \end{bmatrix}$ **43.** $\begin{bmatrix} 1 & -1 & -1 & 1 \\ 0 & 1 & 6 & 3 \\ 0 & 0 & 0 & 0 \end{bmatrix}$

45. $\begin{bmatrix} 1 & 0 & 0 \\ 0 & 1 & 0 \\ 0 & 0 & 1 \end{bmatrix}$ **47.** $\begin{bmatrix} 1 & 0 & -\frac{3}{7} & -\frac{8}{7} \\ 0 & 1 & -\frac{12}{7} & \frac{10}{7} \end{bmatrix}$

49. $\begin{cases} x - 2y = 4 \\ \quad y = -3 \end{cases}$ **51.** $\begin{cases} x - y + 2z = 4 \\ \quad y - z = 2 \\ \quad\quad z = -2 \end{cases}$

$(-2, -3)$ $(8, 0, -2)$

53. $(7, -5)$ **55.** $(-4, -8, 2)$ **57.** $(3, 2)$
59. Inconsistent   **61.** $(3, -2, 5, 0)$ **63.** $(4, -3, 2)$
65. $(2a + 1, 3a + 2, a)$ **67.** $(7, -3, 4)$
69. $(0, 2 - 4a, a)$ **71.** $(-5a, a, 3)$
73. Yes; $(-1, 1, -3)$ **75.** No   **77.** $y = x^2 + 2x + 5$
79. $y = 2x^2 - x + 1$ **81.** $f(x) = -9x^2 - 5x + 11$
83. $f(x) = x^3 - 2x^2 - 4x + 1$
85. $I_1 = \frac{13}{10}, I_2 = \frac{11}{5}, I_3 = \frac{9}{10}$
87. $\begin{cases} x + 5y + 10z + 20w = 95 \\ x + y + z + w = 26 \\ \quad y - 4z = 0 \\ x - 2y = -1 \end{cases}$

15 \$1 bills, 8 \$5 bills, 2 \$10 bills, 1 \$20 bill

89. $\dfrac{8x^2}{(x-1)^2(x+1)} = \dfrac{2}{x+1} + \dfrac{6}{x-1} + \dfrac{4}{(x-1)^2}$

91. (a) $y = -0.01t^2 + 3.08t + 47.7$

(b)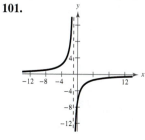

(c) 2010: \$77.50,
2015: \$91.65,
2020: \$105.30

(d) Answers will vary.

93. (a) $x_1 = s, x_2 = t, x_3 = 600 - s,$
$x_4 = s - t, x_5 = 500 - t, x_6 = s, x_7 = t$

(b) $x_1 = 0, x_2 = 0, x_3 = 600, x_4 = 0, x_5 = 500,$
$x_6 = 0, x_7 = 0$
(c) $x_1 = 500, x_2 = 100, x_3 = 100, x_4 = 400,$
$x_5 = 400, x_6 = 500, x_7 = 100$
95. True. See Example 7.
97. $\begin{cases} x + y + 7z = -1 \\ x + 2y + 11z = 0 \\ 2x + y + 10z = -3 \end{cases}$ **99.** No; Answers will vary.

(Answer is not unique.)

101. 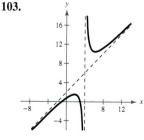   **103.**

Asymptotes: Asymptotes:
$x = -1, y = 0$ $x = 4, y = x + 2$

Section 8.5   (page 667)

1. equal   **3.** zero, O
5. (a) iii   (b) i   (c) iv   (d) v   (e) ii
7. No, not in general.   **9.** $x = 5, y = -8$
11. $x = -1, y = 4, z = 6$

13. (a) $\begin{bmatrix} 8 & -1 \\ 1 & 7 \end{bmatrix}$ (b) $\begin{bmatrix} 2 & -3 \\ 5 & -5 \end{bmatrix}$

(c) $\begin{bmatrix} 15 & -6 \\ 9 & 3 \end{bmatrix}$ (d) $\begin{bmatrix} 9 & -8 \\ 13 & -9 \end{bmatrix}$

15. (a) $\begin{bmatrix} 9 & 5 \\ 1 & -2 \\ -3 & 15 \end{bmatrix}$ (b) $\begin{bmatrix} 7 & -7 \\ 3 & 8 \\ -5 & -5 \end{bmatrix}$

(c) $\begin{bmatrix} 24 & -3 \\ 6 & 9 \\ -12 & 15 \end{bmatrix}$ (d) $\begin{bmatrix} 22 & -15 \\ 8 & 19 \\ -14 & -5 \end{bmatrix}$

17. (a) $\begin{bmatrix} 5 & 5 & -2 & 4 & 4 \\ -5 & 10 & 0 & -4 & -7 \end{bmatrix}$

(b) $\begin{bmatrix} 3 & 5 & 0 & 2 & 4 \\ 7 & -6 & -4 & 2 & 7 \end{bmatrix}$

(c) $\begin{bmatrix} 12 & 15 & -3 & 9 & 12 \\ 3 & 6 & -6 & -3 & 0 \end{bmatrix}$

(d) $\begin{bmatrix} 10 & 15 & -1 & 7 & 12 \\ 15 & -10 & -10 & 3 & 14 \end{bmatrix}$

19. (a) Not possible   (b) Not possible

(c) $\begin{bmatrix} 18 & 0 & 9 \\ -3 & -12 & 0 \end{bmatrix}$ (d) Not possible

21. $\begin{bmatrix} -8 & -7 \\ 15 & -1 \end{bmatrix}$

23. $\begin{bmatrix} -24 & -4 & 12 \\ -12 & 32 & 12 \end{bmatrix}$ **25.** $\begin{bmatrix} -17.143 & 2.143 \\ 11.571 & 10.286 \end{bmatrix}$

27. $\begin{bmatrix} -4.841 & -3.739 \\ -4.252 & -13.249 \\ 9.713 & -0.362 \end{bmatrix}$ **29.** $\begin{bmatrix} -6 & -9 \\ -1 & 0 \\ 17 & -10 \end{bmatrix}$

31. $\begin{bmatrix} 3 & 3 \\ -\frac{1}{2} & 0 \\ -\frac{13}{2} & \frac{11}{2} \end{bmatrix}$ **33.** Not possible

35. $\begin{bmatrix} -2 & 51 \\ -8 & 33 \\ 0 & 27 \end{bmatrix}$ **37.** $\begin{bmatrix} 1 & 0 & 0 \\ 0 & 1 & 0 \\ 0 & 0 & \frac{7}{2} \end{bmatrix}$

39. $\begin{bmatrix} -15 & -5 & -25 & -45 \\ -18 & -6 & -30 & -54 \end{bmatrix}$

41. (a) $\begin{bmatrix} 0 & 15 \\ 6 & 12 \end{bmatrix}$ (b) $\begin{bmatrix} -2 & 2 \\ 31 & 14 \end{bmatrix}$ (c) $\begin{bmatrix} 9 & 6 \\ 12 & 12 \end{bmatrix}$

43. (a) $\begin{bmatrix} 0 & -10 \\ 10 & 0 \end{bmatrix}$ (b) $\begin{bmatrix} 0 & -10 \\ 10 & 0 \end{bmatrix}$ (c) $\begin{bmatrix} 8 & -6 \\ 6 & 8 \end{bmatrix}$

45. (a) $\begin{bmatrix} 7 & 7 & 14 \\ 8 & 8 & 16 \\ -1 & -1 & -2 \end{bmatrix}$ (b) $[13]$ (c) Not possible

47. $\begin{bmatrix} 70 & -17 & 73 \\ 32 & 11 & 6 \\ 16 & -38 & 70 \end{bmatrix}$ **49.** $\begin{bmatrix} 151 & 25 & 48 \\ 516 & 279 & 387 \\ 47 & -20 & 87 \end{bmatrix}$

51. $\begin{bmatrix} 5 & 8 \\ -4 & -16 \end{bmatrix}$ **53.** $\begin{bmatrix} -4 & 10 \\ 3 & 14 \end{bmatrix}$

55. (a) No (b) Yes (c) No (d) No
57. (a) No (b) Yes (c) No (d) No

59. (a) $\begin{bmatrix} -1 & 1 \\ -2 & 1 \end{bmatrix}\begin{bmatrix} x_1 \\ x_2 \end{bmatrix} = \begin{bmatrix} 4 \\ 0 \end{bmatrix}$ (b) $\begin{bmatrix} 4 \\ 8 \end{bmatrix}$

61. (a) $\begin{bmatrix} -2 & -3 \\ 6 & 1 \end{bmatrix}\begin{bmatrix} x_1 \\ x_2 \end{bmatrix} = \begin{bmatrix} -4 \\ -36 \end{bmatrix}$ (b) $\begin{bmatrix} -7 \\ 6 \end{bmatrix}$

63. (a) $\begin{bmatrix} 1 & -2 & 3 \\ -1 & 3 & -1 \\ 2 & -5 & 5 \end{bmatrix}\begin{bmatrix} x_1 \\ x_2 \\ x_3 \end{bmatrix} = \begin{bmatrix} 9 \\ -6 \\ 17 \end{bmatrix}$ (b) $\begin{bmatrix} 1 \\ -1 \\ 2 \end{bmatrix}$

65. (a) $\begin{bmatrix} 1 & -5 & 2 \\ -3 & 1 & -1 \\ 0 & -2 & 5 \end{bmatrix}\begin{bmatrix} x_1 \\ x_2 \\ x_3 \end{bmatrix} = \begin{bmatrix} -20 \\ 8 \\ -16 \end{bmatrix}$ (b) $\begin{bmatrix} -1 \\ 3 \\ -2 \end{bmatrix}$

67. (a) $\begin{bmatrix} 7 & -2 & 5 \\ -6 & 13 & -8 \\ 16 & 11 & -3 \end{bmatrix}$ (b) $\begin{bmatrix} 7 & -2 & 5 \\ -6 & 13 & -8 \\ 16 & 11 & -3 \end{bmatrix}$

The answers are the same.

69. (a) $\begin{bmatrix} 26 & 11 & 0 \\ 11 & 20 & -3 \\ 11 & 14 & 0 \end{bmatrix}$ (b) $\begin{bmatrix} 26 & 11 & 0 \\ 11 & 20 & -3 \\ 11 & 14 & 0 \end{bmatrix}$

The answers are the same.

71. (a) $\begin{bmatrix} 25 & -34 & 28 \\ -53 & 34 & -7 \\ -76 & 30 & 21 \end{bmatrix}$ (b) $\begin{bmatrix} 25 & -34 & 28 \\ -53 & 34 & -7 \\ -76 & 30 & 21 \end{bmatrix}$

The answers are the same.

73. (a) and (b) $\begin{bmatrix} 3 & -6 & 0 \\ 3 & 3 & 0 \end{bmatrix}$

75. Not possible, undefined   **77.** Not possible, undefined

79. (a) and (b) $\begin{bmatrix} 6 & 12 & -12 \\ -6 & 6 & 0 \end{bmatrix}$ **81.** $\begin{bmatrix} -4 & 0 \\ 8 & 2 \end{bmatrix}$

83. $\begin{bmatrix} 84 & 60 & 30 \\ 42 & 120 & 84 \end{bmatrix}$ **85.** $\begin{bmatrix} 90 & 108 & 54 & 36 \\ 126 & 144 & 180 & 72 \end{bmatrix}$

87. $[\$1037.50 \quad \$1400.00 \quad \$1012.50]$

The entries represent the total profits made at the three outlets.

89. $\begin{bmatrix} \$23.20 & \$20.50 \\ \$38.20 & \$33.80 \\ \$76.90 & \$68.50 \end{bmatrix}$ The entries represent labor costs at the two plants for the three boat sizes.

91. $\begin{bmatrix} 0.40 & 0.15 & 0.15 \\ 0.28 & 0.53 & 0.17 \\ 0.32 & 0.32 & 0.68 \end{bmatrix}$ P^2 represents the proportion of changes in party affiliations after two elections.

93. True. To add two matrices, you add corresponding entries.
95. Not possible   **97.** Not possible   **99.** 2×2
101. 2×3 **103 and 105.** Answers will vary.

107. $AC = BC = \begin{bmatrix} 2 & 3 \\ 2 & 3 \end{bmatrix}, A \neq B$

109. (a) $A^2 = \begin{bmatrix} -1 & 0 \\ 0 & -1 \end{bmatrix}, A^3 = \begin{bmatrix} -i & 0 \\ 0 & -i \end{bmatrix}, A^4 = \begin{bmatrix} 1 & 0 \\ 0 & 1 \end{bmatrix}$

The entries on the main diagonal are i^2 in A^2, i^3 in A^3, and i^4 in A^4.

(b) $B^2 = \begin{bmatrix} 1 & 0 \\ 0 & 1 \end{bmatrix}$

B^2 is the identity matrix.

111. (a) $A = \begin{bmatrix} 0 & 2 \\ 0 & 0 \end{bmatrix}, B = \begin{bmatrix} 0 & 2 & 3 \\ 0 & 0 & 4 \\ 0 & 0 & 0 \end{bmatrix}$

(Answers are not unique.)

(b) A^2 and B^3 are zero matrices.

(c) $A = \begin{bmatrix} 0 & 2 & 3 & 4 \\ 0 & 0 & 5 & 6 \\ 0 & 0 & 0 & 7 \\ 0 & 0 & 0 & 0 \end{bmatrix}$

A^4 is the zero matrix.

(d) A^n is the zero matrix.

113. $\ln \dfrac{64}{\sqrt[3]{x^2 + 3}}$

Section 8.6   (page 678)

1. inverse   **3.** No   **5–9.** Answers will vary.

11. $\begin{bmatrix} \frac{1}{2} & 0 \\ 0 & \frac{1}{3} \end{bmatrix}$ **13.** $\begin{bmatrix} -3 & 2 \\ -2 & 1 \end{bmatrix}$ **15.** Does not exist

17. $\begin{bmatrix} 1 & 1 & -1 \\ -3 & 2 & -1 \\ 3 & -3 & 2 \end{bmatrix}$ **19.** $\begin{bmatrix} 1 & 0 & 0 \\ -\frac{3}{4} & \frac{1}{4} & 0 \\ \frac{7}{20} & -\frac{1}{4} & \frac{1}{5} \end{bmatrix}$

21. $\begin{bmatrix} -\frac{3}{2} & \frac{3}{2} & 1 \\ \frac{9}{2} & -\frac{7}{2} & -3 \\ -1 & 1 & 1 \end{bmatrix}$ **23.** $\begin{bmatrix} -12 & -5 & -9 \\ -4 & -2 & -4 \\ -8 & -4 & -6 \end{bmatrix}$

25. $\dfrac{5}{11}\begin{bmatrix} 0 & -4 & 2 \\ -22 & 11 & 11 \\ 22 & -6 & -8 \end{bmatrix}$ **27.** $\begin{bmatrix} 1 & 0 & 1 & 0 \\ 0 & 1 & 0 & 1 \\ 2 & 0 & 1 & 0 \\ 0 & 1 & 0 & 2 \end{bmatrix}$

29. $\begin{bmatrix} \frac{1}{4} & \frac{1}{8} \\ -\frac{1}{4} & -\frac{5}{8} \end{bmatrix}$ **31.** $\frac{1}{59}\begin{bmatrix} 16 & 15 \\ -4 & 70 \end{bmatrix}$ **33.** $\begin{bmatrix} \frac{5}{13} & -\frac{3}{13} \\ \frac{1}{13} & \frac{2}{13} \end{bmatrix}$

35. $k = 0$ **37.** $(5, 0)$ **39.** $(-8, -6)$ **41.** $(3, 8, -11)$

43. $(2, 1, 0, 0)$ **45.** $(2, -2)$

47. Not possible, because A is not invertible.   **49.** $(-4, -8)$

51. $(-1, 3, 2)$ **53.** $(0.3125t + 0.8125, 1.1875t + 0.6875, t)$

55. $(5, 0, -2, 3)$

57. \$7000 in AAA-rated bonds, \$1000 in A-rated bonds, and \$2000 in B-rated bonds

59. \$9000 in AAA-rated bonds, \$1000 in A-rated bonds, and \$2000 in B-rated bonds

61. $I_1 = \frac{1}{2}$ ampere, $I_2 = 3$ amperes, $I_3 = 3.5$ amperes

63. 100 bags for seedlings, 100 bags for general potting, 100 bags for hardwood plants

65. (a) $\begin{cases} 2.5r + 4l + 2i = 300 \\ -r + 2l + 2i = 0 \\ r + l + i = 120 \end{cases}$

(b) $\begin{bmatrix} 2.5 & 4 & 2 \\ -1 & 2 & 2 \\ 1 & 1 & 1 \end{bmatrix}\begin{bmatrix} r \\ l \\ i \end{bmatrix} = \begin{bmatrix} 300 \\ 0 \\ 120 \end{bmatrix}$

(c) 80 roses, 10 lilies, 30 irises

67. True, $AA^{-1} = I = A^{-1}A$.

69. Answers will vary.

71. (a) Answers will vary.

(b) $A^{-1} = \begin{bmatrix} \frac{1}{a_{11}} & 0 & 0 & 0 & \cdots & 0 \\ 0 & \frac{1}{a_{22}} & 0 & 0 & \cdots & 0 \\ 0 & 0 & \frac{1}{a_{33}} & 0 & \cdots & 0 \\ \vdots & \vdots & \vdots & \vdots & \cdots & \vdots \\ 0 & 0 & 0 & 0 & \cdots & \frac{1}{a_{nn}} \end{bmatrix}$

73. $\ln 3 \approx 1.099$ **75.** $\dfrac{e^{12/7}}{3} \approx 1.851$

77. Answers will vary.

Section 8.7   (page 685)

1. determinant   **3.** -5 **5.** 4 **7.** 16 **9.** 28

11. -24 **13.** -0.002

15. (a) $M_{11} = -5, M_{12} = 2, M_{21} = 4, M_{22} = 3$

(b) $C_{11} = -5, C_{12} = -2, C_{21} = -4, C_{22} = 3$

17. (a) $M_{11} = 10, M_{12} = -43, M_{13} = 2, M_{21} = -30,$
$M_{22} = 17, M_{23} = -6, M_{31} = 54, M_{32} = -53,$
$M_{33} = -34$

(b) $C_{11} = 10, C_{12} = 43, C_{13} = 2, C_{21} = 30, C_{22} = 17,$
$C_{23} = 6, C_{31} = 54, C_{32} = 53, C_{33} = -34$

19. (a) -75 (b) -75 **21.** (a) 170 (b) 170

23. -58 **25.** 0 **27.** -9 **29.** -168 **31.** 412

33. -336 **35.** 410

37. (a) -3 (b) -2

(c) $\begin{bmatrix} -2 & 0 \\ 0 & -3 \end{bmatrix}$ (d) $6; |AB| = |A| \cdot |B|$

39. (a) 2 (b) -6

(c) $\begin{bmatrix} 1 & 4 & 3 \\ -1 & 0 & 3 \\ 0 & 2 & 0 \end{bmatrix}$ (d) $-12; |AB| = |A| \cdot |B|$

41. (a) -25 (b) -220

(c) $\begin{bmatrix} -7 & -16 & -1 & -28 \\ -4 & -14 & -11 & 8 \\ 13 & 4 & 4 & -4 \\ -2 & 3 & 2 & 2 \end{bmatrix}$ (d) $5500; |AB| = |A| \cdot |B|$

43–47. Answers will vary.   **49.** $x = \pm 2$ **51.** $x = \pm\frac{3}{2}$

53. $x = 1 \pm \sqrt{2}$ **55.** $x = -4, -1$ **57.** $x = 1, \frac{1}{2}$

59. $x = 3$ **61.** $8uv - 1$ **63.** e^{5x} **65.** $1 - \ln x$

67. True. Expansion by cofactors on a row of zeros is zero.

69. Answers will vary. Sample answer:

$A = \begin{bmatrix} 1 & 0 & -3 \\ 6 & -2 & 7 \\ 9 & 5 & -1 \end{bmatrix}, \quad B = \begin{bmatrix} 3 & 1 & 5 \\ -8 & 1 & 0 \\ -7 & 6 & -2 \end{bmatrix}$

$|A + B| = -328, |A| + |B| = -404$

71. (a) 6 (b) $\begin{bmatrix} \frac{1}{3} & -\frac{1}{3} \\ \frac{1}{3} & \frac{1}{6} \end{bmatrix}$ (c) $\frac{1}{6}$ (d) They are reciprocals.

73. (a) 2 (b) $\begin{bmatrix} -4 & -5 & 1.5 \\ -1 & -1 & 0.5 \\ -1 & -1 & 0 \end{bmatrix}$

(c) $\frac{1}{2}$ (d) They are reciprocals.

75. (a) Columns 2 and 3 are interchanged.

(b) Rows 1 and 3 are interchanged.

77. (a) 3 is factored from the second row.

(b) 2 and 4 are factored from the first and second columns, respectively.

79. (a) 15 (b) -75 (c) -120

The determinant of a triangular matrix is the product of the numbers along the main diagonal.

81. Answers will vary.   **83.** $(x - 2)(x - 1)$

85. $(2y - 3)^2$ **87.** $(2, -4)$

Section 8.8   (page 696)

1. Cramer's Rule   **3.** $-\frac{1}{2}$ **5.** $\frac{5}{2}$ **7.** $\frac{33}{8}$ **9.** 24

11. $x = 0, -\frac{16}{5}$ **13.** Collinear   **15.** Not collinear

17. $x = 3$ **19.** $(-3, -2)$ **21.** Not possible

23. $(-1, 3, 2)$ **25.** (a) and (b) $\left(0, -\frac{1}{2}, \frac{1}{2}\right)$

27. (a) $y = -1.086t^2 + 15.949t + 25.326$

(b) 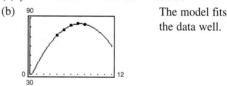   The model fits the data well.

29. (a) $\begin{bmatrix} 20 & 5 & 24 \end{bmatrix}, \begin{bmatrix} 20 & 0 & 13 \end{bmatrix}, \begin{bmatrix} 5 & 0 & 1 \end{bmatrix},$
$\begin{bmatrix} 20 & 0 & 23 \end{bmatrix}, \begin{bmatrix} 15 & 18 & 11 \end{bmatrix}$

(b) $-119 \;\; 28 \;\; 67 \;\; -58 \;\; 6 \;\; 39 \;\; -1 \;\; -3 \;\; 3$
$-118 \;\; 26 \;\; 69 \;\; -33 \;\; 7 \;\; 15$

31. $1 \;\; -43 \;\; -108 \;\; 49 \;\; 91 \;\; 91 \;\; 1 \;\; -29 \;\; -73 \;\; 33 \;\; 42 \;\; 15 \;\; 7 \;\; 14 \;\; 14$

33. HAPPY NEW YEAR

35. IF YOU CANT BE KIND BE VAGUE

37. True. Cramer's Rule divides by the determinant.

39. Answers will vary.   **41.** $x + 4y - 19 = 0$

43. $2x - 7y - 27 = 0$

Review Exercises　(page 700)

1. $(1, 1)$　　**3.** $\left(\frac{3}{2}, 5\right)$　　**5.** $(0.25, 0.625)$　　**7.** $(5, 4)$

9. $(0, 0), (2, 8), (-2, 8)$　　**11.** $(2, -0.5)$

13. $(0, 0), (4, -4)$　　**15.** $(-1, 2), (0.67, 2.56)$

17. $(4, 4)$　　**19.** 800 plants　　**21.** 96 m $\times$ 144 m

23. $\left(\frac{5}{2}, 3\right)$　　**25.** $(-0.5, 0.8)$　　**27.** $\left(-\frac{1}{2}, \frac{4}{5}\right)$　　**29.** $(0, 0)$

31. $\left(\frac{14}{5} + \frac{8}{5}a, a\right)$

33.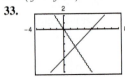
Consistent; $(1.6, -2.4)$

35.
Inconsistent

37.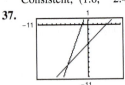
Consistent; $(-4.6, -8.6)$

39. $\left(\frac{500{,}000}{7}, \frac{159}{7}\right)$

41. 218.75 mi/h; 193.75 mi/h　　**43.** $(2, -4, -5)$

45. $\left(\frac{38}{17}, \frac{40}{17}, -\frac{63}{17}\right)$　　**47.** $(3a + 4, 2a + 5, a)$

49. $\left(-\frac{19}{6}, \frac{17}{12}, \frac{1}{3}\right)$　　**51.** $(a - 4, a - 3, a)$

53.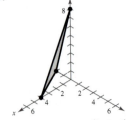
Sample answer: $(0, 0, 8), (0, -2, 0), (4, 0, 0), (1, -1, 2)$

55. $\dfrac{3}{x + 2} - \dfrac{4}{x + 4}$　　**57.** $\dfrac{1}{2}\left(\dfrac{3}{x - 1} - \dfrac{x - 3}{x^2 + 1}\right)$

59. $\dfrac{2x - 1}{x^2 + 1} + \dfrac{-1}{x + 2}$　　**61.** $y = 2x^2 + x - 5$

63. 4 par-3 holes, 10 par-4 holes, 4 par-5 holes

65. 3×1　　**67.** 1×1

69. $\begin{bmatrix} 6 & -7 & \vdots & 11 \\ -2 & 5 & \vdots & -1 \end{bmatrix}$

71. $\begin{bmatrix} 8 & -7 & 4 & \vdots & 12 \\ 3 & -5 & 2 & \vdots & 20 \\ 5 & 3 & -3 & \vdots & 26 \end{bmatrix}$

73. $\begin{cases} 5x + y + 7z = -9 \\ 4x + 2y = 10 \\ 9x + 4y + 2z = 3 \end{cases}$

75. $\begin{bmatrix} 1 & 1 & 1 \\ 0 & 1 & 2 \\ 0 & 0 & 1 \end{bmatrix}$　　**77.** $\begin{bmatrix} 1 & 0 & 3 & -2 \\ 0 & 1 & 4 & -3 \end{bmatrix}$

79. $\begin{bmatrix} 1 & 0 & 0 \\ 0 & 1 & 0 \\ 0 & 0 & 1 \end{bmatrix}$　　**81.** $(10, -12)$　　**83.** $(-0.2, 0.7)$

85. $\left(\frac{1}{2}, -\frac{1}{3}, 1\right)$　　**87.** $(3a + 1, -a, a)$　　**89.** $(2, -3, 3)$

91. $\left(1, 2, \frac{1}{2}\right)$　　**93.** $(1, 2, 2)$　　**95.** $(3, 0, -4)$

97. $(2, 6, -10, -3)$　　**99.** $x = 12, y = -7$

101. $x = 1, y = 11$

103. (a) $\begin{bmatrix} 17 & -17 \\ 13 & 2 \end{bmatrix}$　(b) $\begin{bmatrix} -3 & 23 \\ -15 & 8 \end{bmatrix}$

(c) $\begin{bmatrix} 14 & 6 \\ -2 & 10 \end{bmatrix}$　(d) $\begin{bmatrix} 37 & -57 \\ 41 & -4 \end{bmatrix}$

105. (a) $\begin{bmatrix} 6 & 5 & 8 \\ 1 & 7 & 8 \\ 5 & 1 & 4 \end{bmatrix}$　(b) $\begin{bmatrix} 6 & -5 & 6 \\ 9 & -9 & -4 \\ 1 & 3 & 2 \end{bmatrix}$

(c) $\begin{bmatrix} 12 & 0 & 14 \\ 10 & -2 & 4 \\ 6 & 4 & 6 \end{bmatrix}$　(d) $\begin{bmatrix} 6 & 15 & 10 \\ -7 & 23 & 20 \\ 9 & -1 & 6 \end{bmatrix}$

107. $\begin{bmatrix} -13 & -8 & 18 \\ 0 & 11 & -19 \end{bmatrix}$　**109.** $\begin{bmatrix} 9 & -7 \\ -9 & 4 \end{bmatrix}$

111. $\begin{bmatrix} 48 & -18 & -3 \\ 15 & 51 & 33 \end{bmatrix}$　**113.** $\begin{bmatrix} -8 & -4 \\ 7 & -17 \\ -17 & -2 \end{bmatrix}$

115. $\dfrac{1}{3}\begin{bmatrix} 6 & 2 \\ -4 & 11 \\ 10 & 0 \end{bmatrix}$　**117.** $\begin{bmatrix} 14 & -2 & 8 \\ 14 & -10 & 40 \\ 36 & -12 & 48 \end{bmatrix}$

119. $[30]$　**121.** $\begin{bmatrix} 14 & -22 & 22 \\ 19 & -41 & 80 \\ 42 & -66 & 66 \end{bmatrix}$　**123.** $\begin{bmatrix} 1 & 17 \\ 12 & 36 \end{bmatrix}$

125. (a) $\begin{bmatrix} 525.88 & 47.40 \\ 734.94 & 66.20 \\ 861.76 & 77.20 \end{bmatrix}$

The entries represent the dairy mart's sales and profits on milk for Friday, Saturday, and Sunday.

(b) \$190.80

127. Answers will vary.

129. $\begin{bmatrix} 4 & -5 \\ 5 & -6 \end{bmatrix}$　**131.** $\begin{bmatrix} \frac{1}{2} & -1 & -\frac{1}{2} \\ \frac{1}{2} & -\frac{2}{3} & -\frac{5}{6} \\ 0 & \frac{2}{3} & \frac{1}{3} \end{bmatrix}$　**133.** $\begin{bmatrix} \frac{1}{5} & \frac{1}{5} \\ \frac{1}{10} & -\frac{1}{15} \end{bmatrix}$

135. $\begin{bmatrix} 1 & 0 & 2 \\ 0 & 0 & -1 \\ 1 & 1 & 3 \end{bmatrix}$　**137.** $\begin{bmatrix} 1 & -1 \\ 4 & -\frac{7}{2} \end{bmatrix}$　**139.** $\begin{bmatrix} -\frac{1}{2} & \frac{1}{4} \\ \frac{1}{20} & \frac{1}{40} \end{bmatrix}$

141. $(36, 11)$　　**143.** $(2, -1, -2)$　　**145.** $(1, -2, 1, 0)$

147. $(-3, 1)$　　**149.** $(1, 1, -2)$　　**151.** -42　　**153.** 550

155. (a) $M_{11} = 4, M_{12} = 7, M_{21} = -1, M_{22} = 2$

(b) $C_{11} = 4, C_{12} = -7, C_{21} = 1, C_{22} = 2$

157. (a) $M_{11} = 30, M_{12} = -12, M_{13} = -21, M_{21} = 20,$
$M_{22} = 19, M_{23} = 22, M_{31} = 5, M_{32} = -2,$
$M_{33} = 19$

(b) $C_{11} = 30, C_{12} = 12, C_{13} = -21, C_{21} = -20,$
$C_{22} = 19, C_{23} = -22, C_{31} = 5, C_{32} = 2, C_{33} = 19$

159. 130　　**161.** 6　　**163.** -3　　**165.** 279　　**167.** 16

169. 1.75　　**171.** $\frac{13}{2}$　　**173.** 48　　**175.** Collinear

177. $(1, 2)$　　**179.** $(4, 7)$　　**181.** $(-1, 4, 5)$

183. $(0, -2.4, -2.6)$　　**185.** (a) and (b) $\left(\frac{53}{33}, -\frac{17}{33}, \frac{61}{66}\right)$

187. (a) $[12 \ 15 \ 15], [11 \ 0 \ 15], [21 \ 20 \ 0],$
$[2 \ 5 \ 12], [15 \ 23 \ 0]$

(b) $-21 \ 6 \ 0 \ -68 \ 8 \ 45 \ 102 \ -42 \ -60 \ -53$
$20 \ 21 \ 99 \ -30 \ -69$

189. I WILL BE BACK

191. THAT IS MY FINAL ANSWER

193. (a) $y = 0.275t + 16.2$

(b)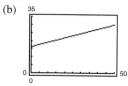

(c) and (d) $t \approx 13.8$ or 2013

195. True. This is the correct sum of the two determinants.

197. An $n \times n$ matrix A has an inverse A^{-1} if $\det(A) \neq 0$.

Chapter Test   (page 706)

1. $(4, -2)$ **2.** $(0, -1), (1, 0), (2, 1)$

3. $(8, 5), (2, -1)$ **4.** $\left(\frac{28}{9}, -\frac{31}{9}\right)$ **5.** $\left(-\frac{2}{3}, -\frac{1}{2}, 1\right)$

6. $(1, 0, -2)$ **7.** $y = -\frac{1}{2}x^2 + x + 6$

8. $\dfrac{5}{x - 1} + \dfrac{3}{(x - 1)^2}$ **9.** $\dfrac{1}{x} + \dfrac{2}{x^2} - \dfrac{1}{x^2 + 1}$

10. $(-2a + 1.5, 2a + 1, a)$ **11.** $(5, 2, -6)$

12. (a) $\begin{bmatrix} 1 & 0 & 4 \\ -7 & -6 & -1 \\ 0 & 4 & 0 \end{bmatrix}$

(b) $\begin{bmatrix} 15 & 12 & 12 \\ -12 & -12 & 0 \\ 3 & 6 & 0 \end{bmatrix}$

(c) $\begin{bmatrix} 7 & 6 & 12 \\ -18 & -16 & -2 \\ 1 & 10 & 0 \end{bmatrix}$

(d) $\begin{bmatrix} 36 & 20 & 4 \\ -28 & -24 & -4 \\ 10 & 8 & 2 \end{bmatrix}$

13. $\begin{bmatrix} -\frac{4}{3} & -\frac{5}{3} & 1 \\ -\frac{4}{3} & -\frac{8}{3} & 1 \\ \frac{1}{3} & \frac{2}{3} & 0 \end{bmatrix}$, $(-2, 3, -1)$ **14.** 67   **15.** -2

16. 30   **17.** $\left(1, -\frac{1}{2}\right)$

18. $x_1 = 700 - s - t, x_2 = 300 - s - t,$
$x_3 = s, x_4 = 100 - t, x_5 = t$

Chapter 9

Section 9.1   (page 717)

1. terms   **3.** index, upper limit, lower limit

5. (a) Finite sequence   (b) Infinite sequence

7. 7, 9, 11, 13, 15   **9.** 3, 9, 27, 81, 243

11. $-\frac{1}{2}, \frac{1}{4}, -\frac{1}{8}, \frac{1}{16}, -\frac{1}{32}$ **13.** $2, \frac{3}{2}, \frac{4}{3}, \frac{5}{4}, \frac{6}{5}$ **15.** $\frac{1}{2}, \frac{2}{5}, \frac{3}{10}, \frac{4}{17}, \frac{5}{26}$

17. (a) 0, 1, 0, 0.5, 0   (b) 0, 1, 0, $\frac{1}{2}$, 0

19. (a) 0.5, 0.75, 0.875, 0.938, 0.969   (b) $\frac{1}{2}, \frac{3}{4}, \frac{7}{8}, \frac{15}{16}, \frac{31}{32}$

21. (a) 1, 0.354, 0.192, 0.125, 0.089   (b) $1, \dfrac{1}{2^{3/2}}, \dfrac{1}{3^{3/2}}, \dfrac{1}{8}, \dfrac{1}{5^{5/2}}$

23. (a) $-1, 0.25, -0.111, 0.063, -0.04$

(b) $-1, \frac{1}{4}, -\frac{1}{9}, \frac{1}{16}, -\frac{1}{25}$

25. (a) and (b) 3, 15, 35, 63, 99

27. 9, 15, 21, 27, 33, 39, 45, 51, 57, 63

29. $3, \frac{5}{2}, \frac{7}{3}, \frac{9}{4}, \frac{11}{5}, \frac{13}{6}, \frac{15}{7}, \frac{17}{8}, \frac{19}{9}, \frac{21}{10}$ **31.** 0, 2, 0, 2, 0, 2, 0, 2, 0, 2

33. $\frac{100}{101}$ **35.** -73 **37.** $\frac{64}{65}$ **39.** $a_n = 3n - 2$

41. $a_n = n^2 - 1$ **43.** $a_n = \dfrac{n + 1}{n + 2}$ **45.** $a_n = \dfrac{(-1)^{n+1}}{2^n}$

47. $a_n = 1 + \dfrac{1}{n}$ **49.** $a_n = \dfrac{1}{n!}$

51. $a_n = (-1)^n + 2(1)^n = (-1)^n + 2$

53. 28, 24, 20, 16, 12   **55.** 3, 4, 6, 10, 18

57. 1, 3, 4, 7, 11   **59.** 6, 8, 10, 12, 14; $a_n = 2n + 4$

61. 81, 27, 9, 3, 1; $a_n = \dfrac{243}{3^n}$

63. (a) 1, 1, 0.5, 0.167, 0.042   (b) $1, 1, \frac{1}{2}, \frac{1}{6}, \frac{1}{24}$

65. (a) 1, 0.333, 0.4, 0.857, 2.667

(b) $1, \frac{1}{3}, \frac{2}{5}, \frac{6}{7}, \frac{8}{3}$

67. (a) 1, 0.5, 0.042, 0.001, 2.480×10^{-5}

(b) $1, \frac{1}{2}, \frac{1}{24}, \frac{1}{720}, \frac{1}{40,320}$

69. $\frac{1}{12}$ **71.** 495   **73.** $n + 1$ **75.** $\dfrac{1}{2n(2n + 1)}$

77. c   **78.** b   **79.** d   **80.** a

81.

83.

85.

87. 35   **89.** 40   **91.** 30   **93.** $\frac{9}{5}$ **95.** 238

97. 30   **99.** 81   **101.** $\frac{47}{60}$

103. $\displaystyle\sum_{i=1}^{9} \dfrac{1}{3i} \approx 0.94299$ **105.** $\displaystyle\sum_{i=1}^{8} \left[2\left(\dfrac{i}{8}\right) + 3\right] = 33$

107. $\displaystyle\sum_{i=1}^{6} (-1)^{i+1} 3i = -546$ **109.** $\displaystyle\sum_{i=1}^{20} \dfrac{(-1)^{i+1}}{i^2} \approx 0.821$

111. $\displaystyle\sum_{i=1}^{5} \dfrac{2^i - 1}{2^{i+1}} \approx 2.0156$ **113.** $\frac{75}{16}$ **115.** $-\frac{3}{2}$

117. (a) $\frac{3333}{5000}$ (b) $\frac{2}{3}$ **119.** (a) $\frac{1111}{10,000}$ (b) $\frac{1}{9}$

121. (a) $A_1 = \$5037.50$, $A_2 = \$5075.28$,
$A_3 = \$5113.35$, $A_4 = \$5151.70$,
$A_5 = \$5190.33$, $A_6 = \$5229.26$,
$A_7 = \$5268.48$, $A_8 = \$5307.99$

(b) $6741.74

123. $72,443 million

125. True by the Properties of Sums

127. 1, 1, 2, 3, 5, 8, 13, 21, 34, 55, 89, 144;
$1, 2, \frac{3}{2}, \frac{5}{3}, \frac{8}{5}, \frac{13}{8}, \frac{21}{13}, \frac{34}{21}, \frac{55}{34}, \frac{89}{55}$

129. 1, 1, 2, 3, 5

131. $a_{n+1} = \dfrac{1}{2}a_n + \dfrac{\left(1 + \sqrt{5}\right)^n + \left(1 - \sqrt{5}\right)^n}{2^{n+1}}$

$a_{n+2} = \dfrac{3}{2}a_n + \dfrac{\left(1 + \sqrt{5}\right)^n + \left(1 - \sqrt{5}\right)^n}{2^{n+1}}$

133. $x, \dfrac{x^2}{2}, \dfrac{x^3}{6}, \dfrac{x^4}{24}, \dfrac{x^5}{120}$ **135.** $-\dfrac{x^3}{3}, \dfrac{x^5}{5}, -\dfrac{x^7}{7}, \dfrac{x^9}{9}, -\dfrac{x^{11}}{11}$

137. $-\dfrac{x^2}{2}, \dfrac{x^4}{24}, -\dfrac{x^6}{720}, \dfrac{x^8}{40{,}320}, -\dfrac{x^{10}}{3{,}628{,}800}$

139. $-x, \dfrac{x^2}{2}, -\dfrac{x^3}{6}, \dfrac{x^4}{24}, -\dfrac{x^5}{120}$

141. $x+1, -\dfrac{(x+1)^2}{2}, \dfrac{(x+1)^3}{6}, -\dfrac{(x+1)^4}{24}, \dfrac{(x+1)^5}{120}$

143. $\dfrac{1}{4}, \dfrac{1}{12}, \dfrac{1}{24}, \dfrac{1}{40}, \dfrac{1}{60}; \dfrac{1}{2} - \dfrac{1}{2n+2}$

145. $\dfrac{1}{6}, \dfrac{1}{12}, \dfrac{1}{20}, \dfrac{1}{30}, \dfrac{1}{42}; \dfrac{1}{2} - \dfrac{1}{n+2}$

147. Yes, if there is a finite number of integer terms, you can always find a sum.

149. (a) $\begin{bmatrix} 8 & 1 \\ -3 & 7 \end{bmatrix}$ (b) $\begin{bmatrix} -22 & -7 \\ 3 & -18 \end{bmatrix}$

(c) $\begin{bmatrix} 18 & 9 \\ 18 & 0 \end{bmatrix}$ (d) $\begin{bmatrix} 0 & 6 \\ 27 & 18 \end{bmatrix}$

151. (a) $\begin{bmatrix} -3 & -7 & 4 \\ 4 & 4 & 1 \\ 1 & 4 & 3 \end{bmatrix}$ (b) $\begin{bmatrix} 8 & 17 & -14 \\ -12 & -13 & -9 \\ -3 & -15 & -10 \end{bmatrix}$

(c) $\begin{bmatrix} -2 & 7 & -16 \\ 4 & 42 & 45 \\ 1 & 23 & 48 \end{bmatrix}$ (d) $\begin{bmatrix} 16 & 31 & 42 \\ 10 & 47 & 31 \\ 13 & 22 & 25 \end{bmatrix}$

Section 9.2 (page 726)

1. $a_n = a_1 + (n-1)d$

3. A sequence is arithmetic when the differences between consecutive terms are the same.

5. Arithmetic sequence, $d = -2$

7. Arithmetic sequence, $d = -\frac{1}{2}$

9. Arithmetic sequence, $d = 0.6$

11. 21, 34, 47, 60, 73
Arithmetic sequence, $d = 13$

13. $\frac{1}{2}, \frac{1}{3}, \frac{1}{4}, \frac{1}{5}, \frac{1}{6}$
Not an arithmetic sequence

15. 143, 136, 129, 122, 115
Arithmetic sequence, $d = -7$

17. 1, 5, 1, 5, 1
Not an arithmetic sequence

19. $-1, 1, -1, 1, -1$
Not an arithmetic sequence

21. $a_n = -2 + 3n$ **23.** $a_n = 108 - 8n$

25. $a_n = \frac{13}{2} - \frac{5}{2}n$ **27.** $a_n = \frac{10}{3}n + \frac{5}{3}$

29. $a_n = 103 - 3n$ **31.** 5, 11, 17, 23, 29

33. $-10, -22, -34, -46, -58$ **35.** $-2, 2, 6, 10, 14$

37. 22.45, 20.725, 19, 17.275, 15.55

39. 15, 19, 23, 27, 31; $d = 4$; $a_n = 11 + 4n$

41. $\frac{3}{5}, \frac{1}{2}, \frac{2}{5}, \frac{3}{10}, \frac{1}{5}$; $d = -\frac{1}{10}$; $a_n = -\frac{1}{10}n + \frac{7}{10}$

43. 59 **45.** 18.6

47. 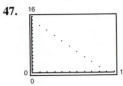   **49.**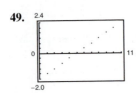

51. $-1, 3, 7, 11, 15, 19, 23, 27, 31, 35$

53. 19.25, 18.5, 17.75, 17, 16.25, 15.5, 14.75, 14, 13.25, 12.5

55. 1.55, 1.6, 1.65, 1.7, 1.75, 1.8, 1.85, 1.9, 1.95, 2

57. 110 **59.** -25 **61.** 5050 **63.** -4585

65. 620 **67.** 41 **69.** 4000 **71.** 1275 **73.** 355

75. 129,250 **77.** 440 **79.** 2575 **81.** 14,268

83. 405 bricks **85.** $200,000

87. (a) $a_n = 0.84n + 14.9$

(b)

Year	2001	2002	2003	2004
Sales (in billions of dollars)	15.7	16.6	17.4	18.3

Year	2005	2006	2007	2008
Sales (in billions of dollars)	19.1	19.9	20.8	21.6

The model fits the data well.

(c) $149.5 billion (d) $203.2 billion; Answers will vary.

89. True. Use the recursion formula, $a_{n+1} = a_n + d$.

91. $x, 3x, 5x, 7x, 9x, 11x, 13x, 15x, 17x, 19x$ **93.** 4

95. $S_n + 5n$

97. Answers will vary. Sample answer: Gauss saw that the sum of the first and last numbers was 101, the sum of the second and second-last numbers was 101, and so on. Seeing that there were 50 such pairs of numbers, Gauss simply multiplied 50 by 101 to get the summation 5050.

$a_n = (n+1)\left(\dfrac{n}{2}\right)$, where n is the total number of natural numbers.

99. 20,100 **101.** 2601 **103.** $(1, 5, -1)$

105. Answers will vary.

Section 9.3 (page 735)

1. geometric, common **3.** geometric series **5.** $|r| < 1$

7. Geometric sequence, $r = 3$

9. Not a geometric sequence

11. Geometric sequence, $r = -\frac{1}{2}$

13. Geometric sequence, $r = 2$

15. Not a geometric sequence

17. 6, 18, 54, 162, 486 **19.** $1, \frac{1}{2}, \frac{1}{4}, \frac{1}{8}, \frac{1}{16}$

21. $5, -\frac{1}{2}, \frac{1}{20}, -\frac{1}{200}, \frac{1}{2000}$ **23.** $1, e, e^2, e^3, e^4$

25. 64, 32, 16, 8, 4; $r = \frac{1}{2}$; **27.** 9, 18, 36, 72, 144; $r = 2$;
$a_n = 64\left(\frac{1}{2}\right)^{n-1}$ $a_n = 9(2)^{n-1}$

29. $6, -9, \frac{27}{2}, -\frac{81}{4}, \frac{243}{8}$; $r = -\frac{3}{2}$;
$a_n = 6\left(-\frac{3}{2}\right)^{n-1}$

31. (a) 0.000034 (b) $\frac{2}{59{,}049}$ **33.** (a) 44.949 (b) $\frac{32{,}768}{729}$

35. (a) and (b) -243 **37.** (a) and (b) -646.803

39. $a_n = 7(3)^{n-1}$; 45,927 **41.** $a_n = 5(6)^{n-1}$; 50,388,480

43. $\frac{1}{128}$ **45.** $-\frac{2}{9}$

47. 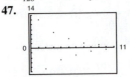   **49.**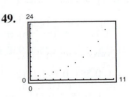

51. $8, 4, 6, 5, \frac{11}{2}$

53.

n	S_n
1	16
2	24
3	28
4	30
5	31
6	31.5
7	31.75
8	31.875
9	31.9375
10	31.96875

55. 511   **57.** 43   **59.** 29,921.31   **61.** 6.4

63. 2092.60   **65.** $\displaystyle\sum_{n=1}^{7} 5(3)^{n-1}$ **67.** $\displaystyle\sum_{n=1}^{7} 2\left(-\frac{1}{4}\right)^{n-1}$

69. 50   **71.** $\frac{10}{3}$

73. Series does not have a finite sum because $\left|\frac{7}{3}\right| > 1$.

75. $\frac{1000}{89}$ **77.** $-\frac{30}{19}$ **79.** 27   **81.** $\frac{9}{4}$ **83.** $\frac{4}{11}$ **85.** $\frac{113}{90}$

87. Geometric; $r = 2$; 262,136   **89.** Geometric; $r = \frac{1}{3}$; 135

91. Arithmetic; $d = 6$; 720   **93.** Geometric; $r = 0.8$; 28.944

95. (a) \$1343.92   (b) \$1346.86   (c) \$1348.35
(d) \$1349.35   (e) \$1349.84

97. Answers will vary.

99. (a) \$26,198.27   (b) \$26,263.88

101. (a) \$153,237.86   (b) \$153,657.02   **103.** 126 in.^{2}

105. (a) $T_n = 70(0.8)^n$ (b) 18.4°F; 4.8°F
(c)

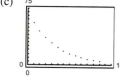

3.5 h

107. (a) $a_n = 1269.10(1.006)^n$
(b) The population is growing at a rate of 0.6% per year.
(c) 1388.2 million. This value is close to the prediction.
(d) 2010

109. 42 ft   **111.** True. The terms all equal a_1.

113. $3, \dfrac{3x}{2}, \dfrac{3x^2}{4}, \dfrac{3x^3}{8}, \dfrac{3x^4}{16}$ **115.** $100e^{8x}$

117. (a)

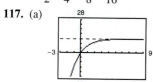

(b)

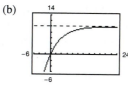

Horizontal asymptote: Horizontal asymptote:
$y = 12$ $y = 10$
Corresponds to the sum   Corresponds to the sum of
of the series   the series.

119. Divide the second term by the first to obtain the common ratio. The nth term is the first term times the common ratio raised to the $(n-1)$th power.

121. 45.65 mi/h   **123.** -102 **125.** Answers will vary.

Section 9.4 (page 744)

1. $_nC_r$ or $\dbinom{n}{r}$ **3.** Binomial Theorem, Pascal's Triangle

5. 21   **7.** 15,504   **9.** 14   **11.** 1   **13.** 210

15. 4950   **17.** 749,398   **19.** 1225   **21.** 31,125

23. $x^4 + 8x^3 + 24x^2 + 32x + 16$

25. $a^3 + 9a^2 + 27a + 27$ **27.** $y^3 - 12y^2 + 48y - 64$

29. $x^5 + 5x^4y + 10x^3y^2 + 10x^2y^3 + 5xy^4 + y^5$

31. $r^6 + 18r^5s + 135r^4s^2 + 540r^3s^3 + 1215r^2s^4$
$\quad + 1458rs^5 + 729s^6$

33. $x^5 - 5x^4y + 10x^3y^2 - 10x^2y^3 + 5xy^4 - y^5$

35. $1 - 12x + 48x^2 - 64x^3$

37. $x^8 + 8x^6 + 24x^4 + 32x^2 + 16$

39. $x^{10} - 25x^8 + 250x^6 - 1250x^4 + 3125x^2 - 3125$

41. $x^8 + 4x^6y^2 + 6x^4y^4 + 4x^2y^6 + y^8$

43. $x^{18} - 6x^{15}y + 15x^{12}y^2 - 20x^9y^3 + 15x^6y^4 - 6x^3y^5 + y^6$

45. $\dfrac{1}{x^5} + \dfrac{5y}{x^4} + \dfrac{10y^2}{x^3} + \dfrac{10y^3}{x^2} + \dfrac{5y^4}{x} + y^5$

47. $\dfrac{16}{x^4} - \dfrac{32y}{x^3} + \dfrac{24y^2}{x^2} - \dfrac{8y^3}{x} + y^4$

49. $-512x^4 + 576x^3 - 240x^2 + 44x - 3$

51. $2x^4 - 24x^3 + 113x^2 - 246x + 207$

53. $-4x^6 - 24x^5 - 60x^4 - 83x^3 - 42x^2 - 60x + 20$

55. $61,440x^7$ **57.** $360x^3y^2$ **59.** $1,259,712x^2y^7$

61. $-4,330,260,000x^3y^9$ **63.** 1,737,104   **65.** 180

67. $-489,888$ **69.** 210   **71.** 35   **73.** 6

75. $81t^4 - 216t^3v + 216t^2v^2 - 96tv^3 + 16v^4$

77. $32x^5 - 240x^4y + 720x^3y^2 - 1080x^2y^3 + 810xy^4 - 243y^5$

79. $x^5 + 10x^4y + 40x^3y^2 + 80x^2y^3 + 80xy^4 + 32y^5$

81. $x^{3/2} + 15x + 75\sqrt{x} + 125$

83. $x^2 - 3x^{4/3}y^{1/3} + 3x^{2/3}y^{2/3} - y$

85. $3x^2 + 3xh + h^2, \ h \neq 0$

87. $6x^5 + 15x^4h + 20x^3h^2 + 15x^2h^3 + 6xh^4 + h^5, \ h \neq 0$

89. $\dfrac{\sqrt{x+h} - \sqrt{x}}{h} = \dfrac{1}{\sqrt{x+h} + \sqrt{x}}, \ h \neq 0$

91. -4 **93.** $161 + 240i$ **95.** $2035 + 828i$

97. $-115 + 236i$ **99.** $-23 + 208\sqrt{3}\,i$ **101.** 1

103. $-\frac{1}{8}$ **105.** 1.172   **107.** 510,568.785

109.

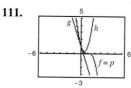

g is shifted four units to the left of f.
$g(x) = x^3 + 12x^2 + 44x + 48$

111.

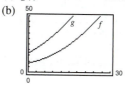

113. 0.273   **115.** 0.171

$p(x)$ is the expansion of $f(x)$.

117. (a) $g(t) = 0.044t^2 + 1.32t + 17.1$
(b)

119. True. Pascal's Triangle is made up of binomial coefficients.

121. False. The correct term is $126,720x^4y^8$.

123. $n + 1$ terms

125. (a) $5(2x)^4(-3y)^1 = -240x^4y$

(b) $_6C_3(\frac{1}{2}x)^3(7y)^3 = 857.5x^3y^3$

127 and 129. Answers will vary. **131.** $\begin{bmatrix} 1 & 2 \\ -0.5 & -0.5 \end{bmatrix}$

Section 9.5 (page 753)

1. Fundamental Counting Principle **3.** Permutation

5. 8 **7.** 6 **9.** 11 **11.** 10 **13.** 120 **15.** 1024

17. (a) 900 (b) 648 (c) 180 **19.** 16,000,000

21. (a) 35,152 (b) 3902 **23.** (a) 100,000 (b) 20,000

25. (a) 720 (b) 48 **27.** 24 **29.** 336 **31.** 120

33. 27,907,200 **35.** 197,149,680 **37.** 120

39. 362,880 **41.** 11,880 **43.** 50,653

45. ABCD, ABDC, ACBD, ACDB, ADBC, ADCB,
BACD, BADC, CABD, CADB, DABC, DACB,
BCAD, BDAC, CBAD, CDAB, DBAC, DCAB,
BCDA, BDCA, CBDA, CDBA, DBCA, DCBA

47. 420 **49.** 2520 **51.** 10 **53.** 4 **55.** 1

57. 15,504 **59.** 850,668

61. AB, AC, AD, AE, AF, BC, BD, BE, BF, CD, CE, CF, DE, DF, EF

63. 2.53×10^{17} **65.** 195,249,054

67. (a) 17,550 (b) 1053 (c) 27,378 **69.** 24

71. 462,000 **73.** 5 **75.** 20 **77.** $n = 5$ or $n = 6$

79. $n = 10$ **81.** $n = 3$ **83.** $n = 2$ **85.** False.

87. For some calculators the answer is too large.

89. $_nP_r$ represents the number of ways to choose and order r elements out of a collection of n elements.

91 and 93. Answers will vary. **95.** 35 **97.** $(-2, -8)$

Section 9.6 (page 762)

1. sample space **3.** mutually exclusive

5. $0 \le P(E) \le 1$ **7.** $P(E) = 1$

9. $\{(H, 1), (H, 2), (H, 3), (H, 4), (H, 5), (H, 6),$
$(T, 1), (T, 2), (T, 3), (T, 4), (T, 5), (T, 6)\}$

11. $\{ABC, ACB, BAC, BCA, CAB, CBA\}$ **13.** $\frac{3}{8}$ **15.** $\frac{7}{8}$

17. $\frac{3}{13}$ **19.** $\frac{3}{52}$ **21.** $\frac{5}{36}$ **23.** $\frac{11}{12}$ **25.** $\frac{3}{100}$ **27.** $\frac{9}{25}$

29. $\frac{1}{5}$ **31.** $\frac{2}{5}$ **33.** 0.25 **35.** $\frac{1}{3}$ **37.** 0.88 **39.** $\frac{7}{20}$

41. $\frac{4}{13}$ **43.** $\frac{4}{13}$ **45.** $\frac{9}{16}$ **47.** $\frac{3}{32}$ **49.** $\frac{1}{8}$ **51.** $\frac{2}{125}$

53. $P(\{\text{Taylor wins}\}) = \frac{1}{2}$
$P(\{\text{Moore wins}\}) = P(\{\text{Perez wins}\}) = \frac{1}{4}$

55. (a) 20.22 million (b) 0.294 (c) 0.866

57. (a) $\frac{1}{120}$ (b) $\frac{1}{24}$ **59.** (a) $\frac{14}{55}$ (b) $\frac{12}{55}$ (c) $\frac{54}{55}$

61. 0.1024 **63.** (a) $\frac{1}{15,625}$ (b) $\frac{4096}{15,625}$ (c) $\frac{11,529}{15,625}$

65. (a) $\frac{\pi}{4}$ (b) Answers will vary.

67. True. The sum of the probabilities of all outcomes must be 1.

69. (a) As you consider successive people with distinct birthdays, the probabilities must decrease to take into account the birth dates already used. Because the birth dates of people are independent events, multiply the respective probabilities of distinct birthdays.

(b) $\frac{365}{365} \cdot \frac{364}{365} \cdot \frac{363}{365} \cdot \frac{362}{365}$ (c) Answers will vary.

(d) Q_n is the probability that the birthdays are *not* distinct, which is equivalent to at least two people having the same birthday.

(e)

n	10	15	20	23	30	40	50
P_n	0.88	0.75	0.59	0.49	0.29	0.11	0.03
Q_n	0.12	0.25	0.41	0.51	0.71	0.89	0.97

(f) 23

71. (a) No. $P(A) + P(B) = 0.76 + 0.58 = 1.34 > 1$. The sum of the probabilities is greater than 1, so A and B cannot be mutually exclusive.

(b) Yes. $A' = 0.24$, $B' = 0.42$, and $A' + B' = 0.66 < 1$, so A' and B' can be mutually exclusive.

(c) $0.76 \le P(A \cup B) \le 1$

73. 15 **75.** 165

Review Exercises (page 768)

1. $\frac{2}{3}, \frac{4}{5}, \frac{8}{9}, \frac{16}{17}, \frac{32}{33}$ **3.** $a_n = 5n$ **5.** $a_n = \frac{2}{2n - 1}$

7. $9, 5, 1, -3, -7$ **9.** $\frac{1}{380}$ **11.** $(n + 1)(n)$ **13.** 30

15. $\frac{205}{24}$ **17.** $\sum_{k=1}^{20} \frac{1}{2k} \approx 1.799$ **19.** $\sum_{k=1}^{9} \frac{k}{k + 1} \approx 7.071$

21. (a) $\frac{1111}{2000}$ (b) $\frac{5}{9}$ **23.** (a) $\frac{15}{8}$ (b) 2

25. (a) 3015.00, 3030.08, 3045.23, 3060.45, 3075.75, 3091.13, 3106.59, 3122.12

(b) $3662.38

27. Arithmetic sequence, $d = -2$

29. Arithmetic sequence, $d = \frac{1}{2}$

31. $3, 7, 11, 15, 19$ **33.** $1, 4, 7, 10, 13$

35. $35, 32, 29, 26, 23; d = -3; a_n = 38 - 3n$

37. $a_n = 103 - 3n; 1600$ **39.** 80 **41.** 6375

43. (a) $45,000 (b) $202,500

45. Geometric sequence, $r = 2$

47. Geometric sequence, $r = -\frac{1}{3}$

49. $4, -1, \frac{1}{4}, -\frac{1}{16}, \frac{1}{64}$ **51.** $9, 6, 4, \frac{8}{3}, \frac{16}{9}$ or $9, -6, 4, -\frac{8}{3}, \frac{16}{9}$

53. $120, 40, \frac{40}{3}, \frac{40}{9}, \frac{40}{27}; r = \frac{1}{3}; a_n = 120(\frac{1}{3})^{n-1}$

55. (a) $-\frac{1}{2}$ (b) -0.5 **57.** 127 **59.** 3277 **61.** 32

63. 12 **65.** (a) $a_t = 130,000(0.7)^t$ (b) $21,849.10

67. 45 **69.** 126 **71.** $x^4 + 20x^3 + 150x^2 + 500x + 625$

73. $a^5 - 20a^4b + 160a^3b^2 - 640a^2b^3 + 1280ab^4 - 1024b^5$

75. 20 **77.** 70 **79.** 10

81. (a) 216 (b) 108 (c) 36 **83.** 239,500,800

85. 5040 **87.** $n = 3$ **89.** 28 **91.** 479,001,600

93. $\frac{1}{9}$ **95.** (a) 0.416 (b) 0.8 (c) 0.074

97. True. $\frac{(n + 2)!}{n!} = \frac{(n + 2)(n + 1)n!}{n!} = (n + 2)(n + 1)$

99. (a) Each term is obtained by adding the same constant (common difference) to the preceding term.

(b) Each term is obtained by multiplying the same constant (common ratio) by the preceding term.

Chapter Test (page 771)

1. $1, -\frac{2}{3}, \frac{4}{9}, -\frac{8}{27}, \frac{16}{81}$ **2.** $12, 16, 20, 24, 28$

3. $-x, \frac{x^2}{2}, -\frac{x^3}{3}, \frac{x^4}{4}, -\frac{x^5}{5}$

4. $-\dfrac{x^3}{6}, -\dfrac{x^5}{120}, -\dfrac{x^7}{5040}, -\dfrac{x^9}{362,880}, -\dfrac{x^{11}}{39,916,800}$ **5.** 7920

6. $\dfrac{1}{n+1}$ **7.** $2n$ **8.** $a_n = n^2 + 1$

9. $a_n = 5100 - 100n$ **10.** $a_n = 4\left(\frac{1}{2}\right)^{n-1}$

11. $\displaystyle\sum_{n=1}^{12} \dfrac{2}{3n+1}$ **12.** $\displaystyle\sum_{n=1}^{\infty} 2\left(\frac{1}{4}\right)^{n-1}$ **13.** 189   **14.** 28.80

15. $\frac{25}{7}$ **16.** $16a^4 - 160a^3b + 600a^2b^2 - 1000ab^3 + 625b^4$

17. 84   **18.** 1140   **19.** 72   **20.** 328,440

21. $n = 3$ **22.** 26,000   **23.** 12,650   **24.** $\frac{3}{26}$

25. $\frac{1}{462}$ **26.** (a) $\frac{1}{4}$ (b) $\frac{121}{3600}$ (c) $\frac{1}{60}$ **27.** 0.25

Chapter 10

Section 10.1   (page 782)

1. conic section   **3.** circle, center

5. The standard form of the equation of a circle; (h, k) represents the center of the circle, r represents the radius of the circle.

7. $x^2 + y^2 = 16$ **9.** $(x - 3)^2 + (y - 7)^2 = 53$

11. $(x + 3)^2 + (y + 1)^2 = 7$

13. Center: $(0, 0)$ **15.** Center: $(-2, 7)$
Radius: 7   Radius: 4

17. Center: $(1, 0)$ **19.** $x^2 + y^2 = 4$
Radius: $\sqrt{15}$ Center: $(0, 0)$
                Radius: 2

21. $x^2 + y^2 = \dfrac{3}{4}$
Center: $(0, 0)$
Radius: $\dfrac{\sqrt{3}}{2}$

23. $(x - 1)^2 + (y + 3)^2 = 1$ **25.** $\left(x + \frac{3}{2}\right)^2 + (y - 3)^2 = 1$
Center: $(1, -3)$ Center: $\left(-\frac{3}{2}, 3\right)$
Radius: 1   Radius: 1

27. Center: $(0, 0)$ **29.** Center: $(-2, -2)$
Radius: 4   Radius: 3

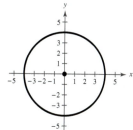

31. Center: $(7, -4)$ **33.** Center: $(-1, 0)$
Radius: 5   Radius: 6

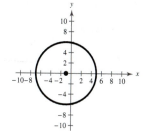

35. x-intercept: $(2, 0)$ **37.** x-intercepts:
y-intercepts: $\left(1 \pm 2\sqrt{7}, 0\right)$
$\left(0, -3 \pm \sqrt{5}\right)$ y-intercepts:
                    $(0, 9), (0, -3)$

39. x-intercepts: $\left(6 \pm \sqrt{7}, 0\right)$
y-intercept: none

41. (a) $x^2 + y^2 = 2704$ (b) Yes   (c) 2 mi

43. e   **44.** b   **45.** d   **46.** f   **47.** a   **48.** c

49. $x^2 = \frac{3}{2}y$ **51.** $x^2 = -6y$ **53.** $y^2 = -8x$

55. $x^2 = -4y$ **57.** $y^2 = -8x$ **59.** $y^2 = 9x$

61. Vertex: $(0, 0)$ **63.** Vertex: $(0, 0)$
Focus: $\left(0, \frac{1}{2}\right)$ Focus: $\left(-\frac{3}{2}, 0\right)$
Directrix: $y = -\frac{1}{2}$ Directrix: $x = \frac{3}{2}$

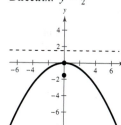

65. Vertex: $(0, 0)$ **67.** Vertex: $(-1, -2)$
Focus: $\left(0, -\frac{3}{2}\right)$ Focus: $(-1, -4)$
Directrix: $y = \frac{3}{2}$ Directrix: $y = 0$

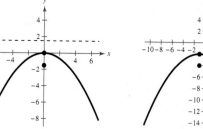

69. Vertex: $(-2, -3)$ **71.** Vertex: $\left(-\frac{3}{2}, 2\right)$
Focus: $(-4, -3)$ Focus: $\left(-\frac{3}{2}, 3\right)$
Directrix: $x = 0$ Directrix: $y = 1$

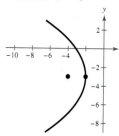
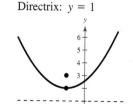

73. Vertex: $(1, 1)$ **75.** Vertex: $(-2, 1)$
Focus: $(1, 2)$ Focus: $\left(-2, -\frac{1}{2}\right)$
Directrix: $y = 0$ Directrix: $y = \frac{5}{2}$

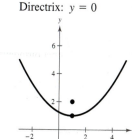
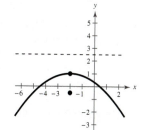

77. Vertex: $\left(\frac{1}{4}, -\frac{1}{2}\right)$
Focus: $\left(0, -\frac{1}{2}\right)$
Directrix: $x = \frac{1}{2}$

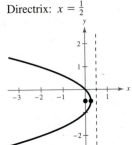

79. $(x - 3)^2 = -(y - 1)$ **81.** $y^2 = 4(x + 4)$
83. $y^2 = 2(x + 2)$ **85.** $(y - 2)^2 = -8(x - 5)$
87. $x^2 = 8(y - 4)$ **89.** $(y - 2)^2 = 8x$
91.

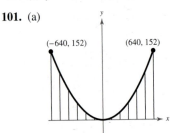

(2, 4)
93. $4x - y - 8 = 0$; $(2, 0)$ **95.** $4x - y + 2 = 0$; $\left(-\frac{1}{2}, 0\right)$
97. (a) $x^2 = 12{,}288y$ (in feet)   (b) 22.6 ft
99. (a) $y^2 = 6x$ (b) 2.67 in.
101. (a)

(−640, 152)   (640, 152)

(b) $x^2 = \dfrac{51{,}200}{19}y$

(c)

x	0	200	400	500	600
y	0	14.844	59.375	92.773	133.59

103. $y^2 = 640x$ **105.** (a) $x^2 = -49(y - 100)$ (b) 70 ft
107. $y = \frac{3}{4}x - \frac{25}{4}$ **109.** $y = \frac{\sqrt{2}}{2}x - 3\sqrt{2}$
111. False. $x^2 + (y + 5)^2 = 25$ represents a circle with its center at $(0, -5)$ and a radius of 5.
113. False. A circle is a conic section.
115. True. The vertex is the closest point to the directrix or focus.
117. True. If the axis is horizontal, then the directrix must be vertical.
119.

The intersection results in a point.

121. $y = -\sqrt{2(x - 2)} - 1$
123. Minimum: $(-0.75, -1.13)$
125. Minimum: $(0.88, -3.11)$; maximum: $(-0.88, 1.11)$

Section 10.2   (page 792)

1. ellipse, foci   **3.** minor axis   **5.** Vertical   **7.** 4
9. b   **10.** c   **11.** a   **12.** d
13. Center: $(0, 0)$
Vertices: $(\pm 8, 0)$
Foci: $(\pm\sqrt{55}, 0)$
Eccentricity: $\dfrac{\sqrt{55}}{8}$

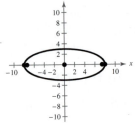

15. Center: $(4, -1)$
Vertices: $(4, 4), (4, -6)$
Foci: $(4, 2), (4, -4)$
Eccentricity: $\frac{3}{5}$

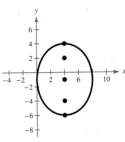

17. Center: $(-5, 1)$
Vertices: $\left(-\frac{7}{2}, 1\right), \left(-\frac{13}{2}, 1\right)$
Foci: $\left(-5 \pm \dfrac{\sqrt{5}}{2}, 1\right)$
Eccentricity: $\dfrac{\sqrt{5}}{3}$

19. $\dfrac{x^2}{4} + \dfrac{y^2}{16} = 1$ **21.** $\dfrac{x^2}{9} + \dfrac{y^2}{5} = 1$ **23.** $\dfrac{x^2}{49} + \dfrac{y^2}{24} = 1$
25. $\dfrac{x^2}{400/21} + \dfrac{y^2}{25} = 1$ **27.** $\dfrac{(x - 2)^2}{1} + \dfrac{(y - 3)^2}{9} = 1$
29. $\dfrac{(x - 4)^2}{16} + \dfrac{(y - 2)^2}{1} = 1$ **31.** $\dfrac{x^2}{308} + \dfrac{(y - 4)^2}{324} = 1$
33. $\dfrac{(x - 3)^2}{9} + \dfrac{(y - 5)^2}{16} = 1$ **35.** $\dfrac{x^2}{16} + \dfrac{(y - 4)^2}{12} = 1$
37. (a) $\dfrac{x^2}{36} + \dfrac{y^2}{4} = 1$
(b) Center: $(0, 0)$ (c)
Vertices: $(\pm 6, 0)$
Foci: $(\pm 4\sqrt{2}, 0)$
Eccentricity: $\dfrac{2\sqrt{2}}{3}$

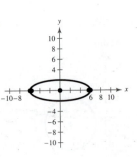

39. (a) $\dfrac{x^2}{4} + \dfrac{y^2}{49} = 1$ (c)

(b) Center: $(0, 0)$
Vertices: $(0, \pm 7)$
Foci: $\left(0, \pm 3\sqrt{5}\right)$
Eccentricity: $\dfrac{3\sqrt{5}}{7}$

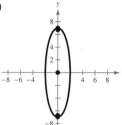

41. (a) $\dfrac{(x + 2)^2}{4} + \dfrac{(y - 3)^2}{9} = 1$ (c)

(b) Center: $(-2, 3)$
Vertices: $(-2, 6), (-2, 0)$
Foci: $\left(-2, 3 \pm \sqrt{5}\right)$
Eccentricity: $\dfrac{\sqrt{5}}{3}$

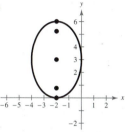

43. (a) $\dfrac{\left(x + \frac{3}{2}\right)^2}{4} + \dfrac{\left(y - \frac{5}{2}\right)^2}{12} = 1$ (c)

(b) Center: $\left(-\dfrac{3}{2}, \dfrac{5}{2}\right)$
Vertices: $\left(-\dfrac{3}{2}, \dfrac{5 \pm 4\sqrt{3}}{2}\right)$
Foci: $\left(-\dfrac{3}{2}, \dfrac{5}{2} \pm 2\sqrt{2}\right)$
Eccentricity: $\dfrac{\sqrt{6}}{3}$

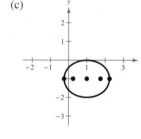

45. (a) $\dfrac{(x - 1)^2}{25/16} + (y + 1)^2 = 1$ (c)

(b) Center: $(1, -1)$
Vertices: $\left(\dfrac{9}{4}, -1\right), \left(-\dfrac{1}{4}, -1\right)$
Foci: $\left(\dfrac{7}{4}, -1\right), \left(\dfrac{1}{4}, -1\right)$
Eccentricity: $\dfrac{3}{5}$

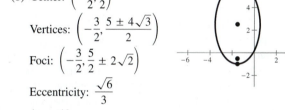

47. (a) $\dfrac{\left(x - \frac{1}{2}\right)^2}{5} + \dfrac{(y + 1)^2}{3} = 1$

(b) Center: $\left(\dfrac{1}{2}, -1\right)$ (c)
Vertices: $\left(\dfrac{1}{2} \pm \sqrt{5}, -1\right)$
Foci: $\left(\dfrac{1}{2} \pm \sqrt{2}, -1\right)$
Eccentricity: $\dfrac{\sqrt{10}}{5}$

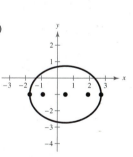

49. $\dfrac{\sqrt{5}}{3}$ **51.** $\dfrac{2\sqrt{2}}{3}$ **53.** $\dfrac{x^2}{25} + \dfrac{y^2}{16} = 1$

55. $\dfrac{x^2}{225/16} + \dfrac{y^2}{81/16} = 1$ **57.** $\left(\pm\sqrt{5}, 0\right);$ 6 ft

59. $\dfrac{x^2}{321.84} + \dfrac{y^2}{19.02} = 1$ **61.** $e = \dfrac{c}{a} \approx 0.052$

63.   **65.**

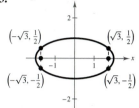

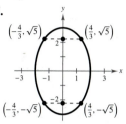

67. True. The ellipse is more elongated when e is close to 1.
69. Nearly circular because its eccentricity is about 0.055, which is close to zero.
71. (a) $2a$
(b) The sum of the distances from the two fixed points is constant.
73. $\dfrac{(x - 6)^2}{324} + \dfrac{(y - 2)^2}{308} = 1$ **75.** Arithmetic
77. Geometric   **79.** 1093

Section 10.3   (page 802)

1. hyperbola   **3.** (h, k) **5.** Center
7. b   **8.** c   **9.** a   **10.** d
11. Center: $(0, 0)$ **13.** Center: $(0, 0)$
Vertices: $(\pm 1, 0)$ Vertices: $(0, \pm 1)$
Foci: $\left(\pm\sqrt{2}, 0\right)$ Foci: $\left(0, \pm\sqrt{5}\right)$
Asymptotes: $y = \pm x$ Asymptotes: $y = \pm\frac{1}{2}x$

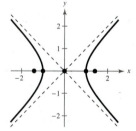

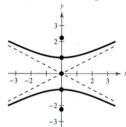

15. Center: $(0, 0)$ **17.** Center: $(1, -2)$
Vertices: $(0, \pm 5)$ Vertices:
Foci: $\left(0, \pm\sqrt{106}\right)$ $(3, -2), (-1, -2)$
Asymptotes: $y = \pm\frac{5}{9}x$ Foci: $\left(1 \pm \sqrt{5}, -2\right)$
Asymptotes:
$y = -2 \pm \frac{1}{2}(x - 1)$

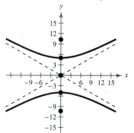

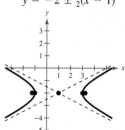

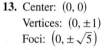

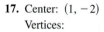

19. Center: $(1, -5)$

Vertices: $\left(1, -5 \pm \dfrac{1}{3}\right)$

Foci: $\left(1, -5 \pm \dfrac{\sqrt{13}}{6}\right)$

Asymptotes: $y = -5 \pm \dfrac{2}{3}(x - 1)$

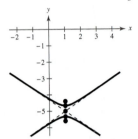

21. (a) $\dfrac{x^2}{9} - \dfrac{y^2}{4} = 1$ (c)

(b) Center: $(0, 0)$

Vertices: $(\pm 3, 0)$

Foci: $\left(\pm \sqrt{13}, 0\right)$

Asymptotes: $y = \pm \dfrac{2}{3}x$

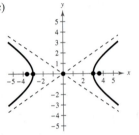

23. (a) $\dfrac{x^2}{3} - \dfrac{y^2}{2} = 1$ (c)

(b) Center: $(0, 0)$

Vertices: $\left(\pm \sqrt{3}, 0\right)$

Foci: $\left(\pm \sqrt{5}, 0\right)$

Asymptotes: $y = \pm \dfrac{\sqrt{6}}{3}x$

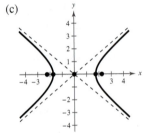

25. (a) $(x - 2)^2 - \dfrac{(y + 3)^2}{9} = 1$ (c)

(b) Center: $(2, -3)$

Vertices: $(3, -3), (1, -3)$

Foci: $\left(2 \pm \sqrt{10}, -3\right)$

Asymptotes:

$y = -3 \pm 3(x - 2)$

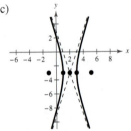

27. (a) $(x + 1)^2 - 9(y + 3)^2 = 0$

(b) It is a degenerate conic.   (c)
The graph of this equation
is two lines intersecting
at $(-1, -3)$.

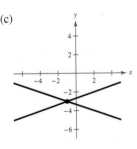

29. (a) $\dfrac{(y + 3)^2}{2} - \dfrac{(x - 1)^2}{18} = 1$ (c)

(b) Center: $(1, -3)$

Vertices: $\left(1, -3 \pm \sqrt{2}\right)$

Foci: $\left(1, -3 \pm 2\sqrt{5}\right)$

Asymptotes:

$y = -3 \pm \dfrac{1}{3}(x - 1)$

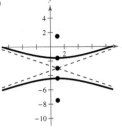

31. $\dfrac{y^2}{4} - \dfrac{x^2}{12} = 1$ **33.** $\dfrac{x^2}{1} - \dfrac{y^2}{25} = 1$

35. $\dfrac{17y^2}{1024} - \dfrac{17x^2}{64} = 1$ **37.** $\dfrac{(x - 4)^2}{4} - \dfrac{y^2}{12} = 1$

39. $\dfrac{(y - 5)^2}{16} - \dfrac{(x - 4)^2}{9} = 1$ **41.** $\dfrac{y^2}{9} - \dfrac{4(x - 2)^2}{9} = 1$

43. $\dfrac{(y - 2)^2}{4} - \dfrac{x^2}{4} = 1$ **45.** $\dfrac{(x - 2)^2}{1} - \dfrac{(y - 2)^2}{1} = 1$

47. $\dfrac{(x - 3)^2}{9} - \dfrac{(y - 2)^2}{4} = 1$

49. $\dfrac{x^2}{98,010,000} - \dfrac{y^2}{13,503,600} = 1$

51. (a) $x^2 - \dfrac{y^2}{27} = 1$ (b) 1.89 ft = 22.68 in.

53. $\left(12\sqrt{5} - 12, 0\right) \approx (14.83, 0)$ **55.** Ellipse

57. Hyperbola   **59.** Parabola   **61.** Circle

63. Parabola

65. True. For a hyperbola, $c^2 = a^2 + b^2$. The larger the ratio of b to a, the larger the eccentricity of the hyperbola, $e = c/a$.

67. False. If $D = E$ or $D = -E$, the graph is two intersecting lines. For example, the graph of $x^2 - y^2 - 2x + 2y = 0$ is two intersecting lines.

69. Answers will vary.   **71.** $\dfrac{(x - 6)^2}{9} - \dfrac{(y - 2)^2}{7} = 1$

73. Proof   **75.** $x(x + 4)(x - 4)$ **77.** $2x(x - 6)^2$

79. $2(2x + 3)(4x^2 - 6x + 9)$

Section 10.4   (page 810)

1. plane curve, parametric equations, parameter

3. Eliminate the parameter.   **5.** c   **6.** d

7. b   **8.** a

9. (a)

t	0	1	2	3	4
x	0	1	$\sqrt{2}$	$\sqrt{3}$	2
y	2	1	0	-1	-2

(b)

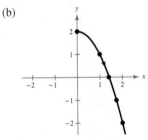

The curve starts at $(0, 2)$ and moves along the right half of the parabola.

(c)

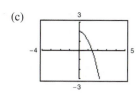

(d) $y = 2 - x^2$

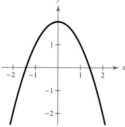

The graph is an entire parabola rather than just the right half.

11. b

13.

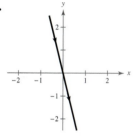

$y = -4x$

15.

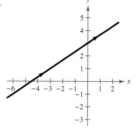

$y = \frac{2}{3}x + 3$

17.

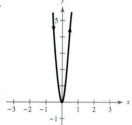

$y = 16x^2$

19.

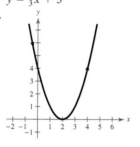

$y = (x - 2)^2$

21.

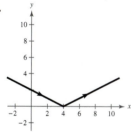

$y = \frac{1}{2}|x - 4|$

23.

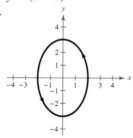

$\frac{x^2}{4} + \frac{y^2}{9} = 1$

25.

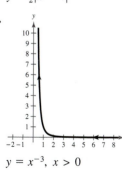

$y = x^{-3}, \ x > 0$

27.

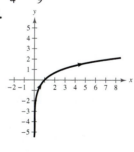

$y = \ln x$

29.

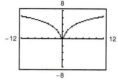

31.

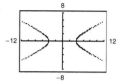

33.

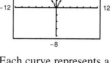

35. Each curve represents a portion of the line $y = 2x + 1$.

Domain	Orientation
(a) $(-\infty, \infty)$	Left to right
(b) $[-1, 1]$	Depends on θ
(c) $(0, \infty)$	Right to left
(d) $(0, \infty)$	Left to right

37. $y - y_1 = \dfrac{y_2 - y_1}{x_2 - x_1}(x - x_1)$ **39.** $\dfrac{(x - h)^2}{a^2} + \dfrac{(y - k)^2}{b^2} = 1$

41. $x = 3 - 5t, y = 1 + 5t$ **43.** $x = 5\cos\theta, y = 4\sin\theta$

45. (a) $x = t, y = 5t - 3$ (b) $x = 2 - t, y = -5t + 7$

47. (a) $x = t, y = \dfrac{1}{t}$ (b) $x = 2 - t, y = \dfrac{1}{2 - t}$

49. (a) $x = t, y = 6t^2 - 5$
   (b) $x = 2 - t, y = 6t^2 - 24t + 19$

51. (a) $x = t, y = e^t$ (b) $x = 2 - t, y = e^{2-t}$

53.

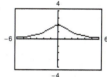

55.

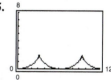

57. b   **58.** c   **59.** d   **60.** a

61. (a) $x = (146.67\cos\theta)t$
   $y = 3 + (146.67\sin\theta)t - 16t^2$

   (b)

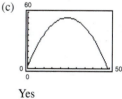

   No

   (c)

   Yes

   (d) About $19.4°$

63. True. Both sets of parametric equations correspond to $y = x^2 + 1$.

65. False. The set $x = t^2$, $y = t$ does not correspond to y as a function of x.

67. Yes, the orientation would be reversed.

69. Even   **71.** Neither

Section 10.5   (page 817)

1. pole

3. $x = r\cos\theta, y = r\sin\theta$ and $\tan\theta = \dfrac{y}{x}, r^2 = x^2 + y^2$

5. $(0, 4)$ **7.** $\left(\dfrac{\sqrt{2}}{2}, \dfrac{\sqrt{2}}{2}\right)$

9.

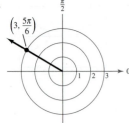

$\left(3, -\dfrac{7\pi}{6}\right), \left(-3, \dfrac{11\pi}{6}\right), \left(-3, -\dfrac{\pi}{6}\right)$

11.

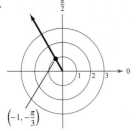

$\left(-1, \dfrac{5\pi}{3}\right), \left(1, \dfrac{2\pi}{3}\right), \left(1, -\dfrac{4\pi}{3}\right)$

13.

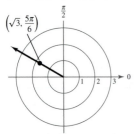

$\left(-\sqrt{3}, \dfrac{11\pi}{6}\right), \left(\sqrt{3}, -\dfrac{7\pi}{6}\right), \left(-\sqrt{3}, -\dfrac{\pi}{6}\right)$

15.

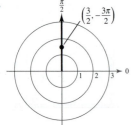

$\left(\dfrac{3}{2}, \dfrac{\pi}{2}\right), \left(-\dfrac{3}{2}, \dfrac{3\pi}{2}\right), \left(-\dfrac{3}{2}, -\dfrac{\pi}{2}\right)$

17.

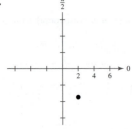

$\left(2, -2\sqrt{3}\right)$

19.

$\left(\dfrac{\sqrt{2}}{2}, \dfrac{\sqrt{2}}{2}\right)$

21.

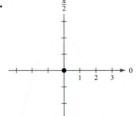

$(0, 0)$

23.

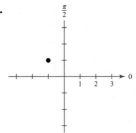

$(-1.004, 0.996)$

25.

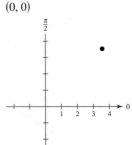

$(3.549, 3.522)$

27. $(1.53, 1.29)$ **29.** $(-1.20, -4.34)$
31. $(-0.02, 2.50)$ **33.** $(-3.60, 1.97)$

35.

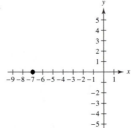

$(7, \pi), (-7, 0)$

37.

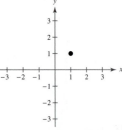

$\left(\sqrt{2}, \dfrac{\pi}{4}\right), \left(-\sqrt{2}, \dfrac{5\pi}{4}\right)$

39.

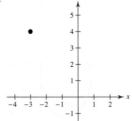

$(5, 2.214), (-5, 5.356)$

41.

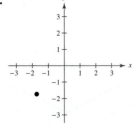

$\left(\sqrt{6}, \dfrac{5\pi}{4}\right), \left(-\sqrt{6}, \dfrac{\pi}{4}\right)$

43.

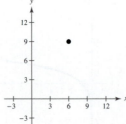

$(10.82, 0.98), (-10.82, 4.12)$
45. $(3.61, -0.59)$ **47.** $(2.65, 0.86)$ **49.** $(2.83, 0.49)$
51. $r = 3$ **53.** $r = 4 \csc \theta$ **55.** $r = 8 \sec \theta$

57. $r = -\dfrac{2}{3 \cos \theta - \sin \theta}$ **59.** $r^2 = 8 \csc 2\theta$

61. $r^2 = 9 \cos 2\theta$ **63.** $r = 6 \cos \theta$ **65.** $r = 2a \cos \theta$

67. $r = \tan^2 \theta \sec \theta$ **69.** $x^2 + y^2 - 4y = 0$

71. $y = -\sqrt{3}x$ **73.** $y = -\dfrac{\sqrt{3}}{3}x$ **75.** $x = 0$

77. $x^2 + y^2 = 16$ **79.** $y = -3$ **81.** $(x^2 + y^2)^3 = x^2$

83. $(x^2 + y^2)^2 = 6x^2y - 2y^3$ **85.** $y^2 = 2x + 1$

87. $4x^2 - 5y^2 = 36y + 36$

89. The graph is a circle centered at the origin with a radius of 6; $x^2 + y^2 = 36$.

91. The graph consists of all points on the line that makes an angle of $\pi/4$ with the positive x-axis; $x - y = 0$.

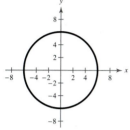

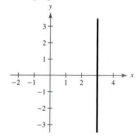

93. The graph is a vertical line through $(3, 0)$; $x - 3 = 0$.

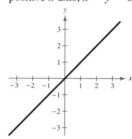

95. True. Because r is a directed distance, (r, θ) can be represented by $(-r, \theta \pm (2n + 1)\pi)$, so $|r| = |-r|$.

97. (a) Answers will vary.

(b) The points lie on a line passing through the pole.
$$d = \sqrt{r_1^2 + r_2^2 - 2r_1 r_2} = |r_1 - r_2|$$

(c) $d = \sqrt{r_1^2 + r_2^2}$ (Pythagorean Theorem)
Answers will vary.

(d) Answers will vary. The Distance Formula should give the same result in both cases.

99. $\left(x - \dfrac{1}{2}\right)^2 + \left(y - \dfrac{3}{2}\right)^2 = \dfrac{5}{2}$; A circle centered at $\left(\dfrac{1}{2}, \dfrac{3}{2}\right)$ with a radius of $\dfrac{\sqrt{10}}{2}$.

101. $A \approx 30.68°$
$B \approx 48.23°$
$C \approx 101.09°$

103. $a \approx 16.16$
$b \approx 19.44$
$B \approx 86°$

Section 10.6   (page 825)

1. convex limaçon   **3.** lemniscate

5. When (r, θ) can be replaced with $(r, \pi - \theta)$ or $(-r, -\theta)$ and yield an equivalent equation

7. Rose curve   **9.** Lemniscate   **11.** Rose curve

13. a   **15.** c

17. Polar axis   **19.** $\theta = \dfrac{\pi}{2}$ **21.** $\theta = \dfrac{\pi}{2}$ **23.** Pole

25.

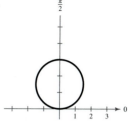

27.

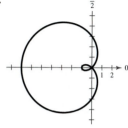

29.

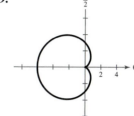

31.

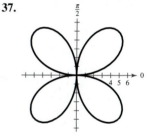

33.

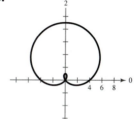

35.

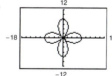

Symmetry: polar axis

Zeros: $\dfrac{\pi}{6}, \dfrac{\pi}{2}, \dfrac{5\pi}{6}$

37.

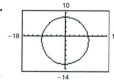

Symmetry:

$\theta = \dfrac{\pi}{2}$, polar axis, pole

Zeros: $0, \dfrac{\pi}{2}, \pi, \dfrac{3\pi}{2}, 2\pi$

39.

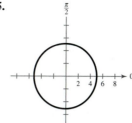

Answers will vary.

41.

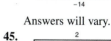

Answers will vary.

43.

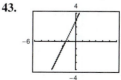

Answers will vary.

45.

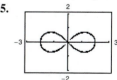

Answers will vary.

47.

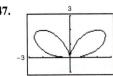

Answers will vary.

49.

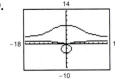

Answers will vary.

51.

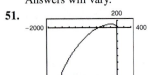

Answers will vary.

53.

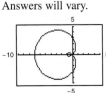

$0 \le \theta < 2\pi$

55.

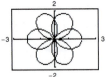

$0 \le \theta < 4\pi$

57.

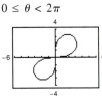

$0 \le \theta < \dfrac{\pi}{2}$

59.

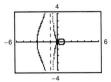

61.

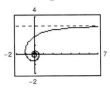

63. True. $n = 5$ **65.** Answers will vary.

67. (a) $r = 2 - \sin\left(\theta - \dfrac{\pi}{4}\right)$ (b) $r = 2 + \cos \theta$

(c) $r = 2 + \sin \theta$ (d) $r = 2 - \cos \theta$

69.

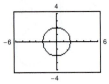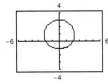

$k = 0$; circle $k = 1$; convex limaçon

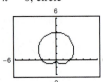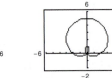

$k = 2$; cardioid $k = 3$; limaçon with inner loop

71. $x = -3, 3$ **73.** $x = \dfrac{13}{5}$

Section 10.7 (page 831)

1. conic   **3.** vertical

5.

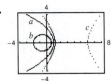

(a) Parabola
(b) Ellipse
(c) Hyperbola

7.

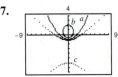

(a) Parabola
(b) Ellipse
(c) Hyperbola

9. b   **10.** c   **11.** f   **12.** e   **13.** d   **14.** a
15. Parabola   **17.** Ellipse   **19.** Ellipse   **21.** Ellipse
23. Hyperbola
25. Parabola   **27.** Hyperbola

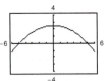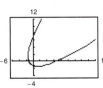

29. Ellipse   **31.**

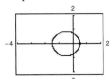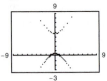

33.   **35.**

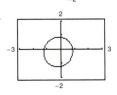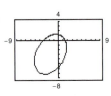

37. $r = \dfrac{1}{1 - \cos \theta}$ **39.** $r = \dfrac{1}{2 + \sin \theta}$

41. $r = \dfrac{2}{1 + 2 \cos \theta}$ **43.** $r = \dfrac{2}{1 - \sin \theta}$

45. $r = \dfrac{10}{1 - \cos \theta}$ **47.** $r = \dfrac{10}{3 + 2 \cos \theta}$

49. $r = \dfrac{20}{3 - 2 \cos \theta}$ **51.** $r = \dfrac{16}{3 + 5 \cos \theta}$

53. Answers will vary.

55. $r = \dfrac{9.2930 \times 10^7}{1 - 0.0167 \cos \theta}$ **57.** $r = \dfrac{6.7280 \times 10^7}{1 - 0.0068 \cos \theta}$

Perihelion: Perihelion:
9.1404×10^7 mi 6.6781×10^7 mi
Aphelion: Aphelion:
9.4508×10^7 mi 6.7695×10^7 mi

59. (a) $r_{\text{Neptune}} = \dfrac{4.4977 \times 10^9}{1 - 0.0086 \cos \theta}$

$r_{\text{Pluto}} = \dfrac{5.5404 \times 10^9}{1 - 0.2488 \cos \theta}$

(b) Neptune: Perihelion: 4.4593×10^9 km
                Aphelion: 4.5367×10^9 km
    Pluto: Perihelion: 4.4366×10^9 km
              Aphelion: 7.3754×10^9 km

(c)

(d) Yes; because on average, Pluto is farther from the sun than Neptune.

(e) Using a graphing utility, it would appear that the orbits intersect. No, Pluto and Neptune will never collide because the orbits do not intersect in three-dimensional space.

61. False. The equation can be rewritten as $r = \dfrac{-4/3}{1 + \sin\theta}$.

Because ep is negative, p must be negative, and because p represents the distance between the pole and the directrix, the directrix has to be below the pole.

63. True. The graphs represent the same hyperbola.

65. Answers will vary.   **67.** $r^2 = \dfrac{24{,}336}{169 - 25\cos^2\theta}$

69. $r^2 = \dfrac{400}{25 - 9\cos^2\theta}$ **71.** $r^2 = \dfrac{144}{25\sin^2\theta - 16}$

73. (a) Ellipse

(b) $r = \dfrac{4}{1 + 0.4\cos\theta}$ is reflected about the line $\theta = \dfrac{\pi}{2}$.

$r = \dfrac{4}{1 - 0.4\sin\theta}$ is rotated 90° counterclockwise.

75. Circle   **77.** $\dfrac{\sqrt{2}}{10}$ **79.** $\dfrac{\sqrt{2}}{10}$

Review Exercises   (page 836)

1. Hyperbola   **3.** $x^2 + y^2 = 25$

5. $(x - 2)^2 + (y - 4)^2 = 13$

7. $x^2 + y^2 = 36$ **9.** $\left(x - \frac{1}{2}\right)^2 + \left(y + \frac{3}{4}\right)^2 = 1$

Center: $(0, 0)$ Center: $\left(\frac{1}{2}, -\frac{3}{4}\right)$

Radius: 6   Radius: 1

11.   **13.** $\left(3 \pm \sqrt{6}, 0\right)$

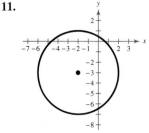

Center: $(-2, -3)$

Radius: 4

15. Vertex: $(0, 0)$ **17.** Vertex: $(0, 0)$

Focus: $(1, 0)$ Focus: $(-9, 0)$

Directrix: $x = -1$ Directrix: $x = 9$

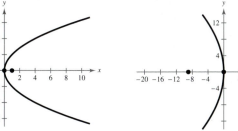

19. $y^2 = 16x$ **21.** $(x + 6)^2 = -9(y - 4)$

23. $2x + y - 2 = 0; (1, 0)$ **25.** $8\sqrt{6}$ m

27. Center: $(0, 0)$

Vertices: $(0, \pm 4)$

Foci: $\left(0, \pm 2\sqrt{3}\right)$

Eccentricity: $\dfrac{\sqrt{3}}{2}$

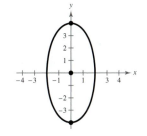

29. Center: $(-1, 2)$

Vertices: $(-1, 9), (-1, -5)$

Foci: $\left(-1, 2 \pm 2\sqrt{6}\right)$

Eccentricity: $\dfrac{2\sqrt{6}}{7}$

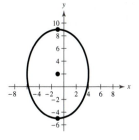

31. (a) $\dfrac{(x - 1)^2}{9} + \dfrac{(y + 4)^2}{16} = 1$ (c)

(b) Center: $(1, -4)$

Vertices: $(1, 0), (1, -8)$

Foci: $\left(1, -4 \pm \sqrt{7}\right)$

Eccentricity: $\dfrac{\sqrt{7}}{4}$

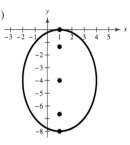

33. (a) $\dfrac{(x + 2)^2}{1/3} + \dfrac{(y - 7)^2}{1/8} = 1$ (c)

(b) Center: $(-2, 7)$

Vertices: $\left(-2 \pm \dfrac{\sqrt{3}}{3}, 7\right)$

Foci: $\left(-2 \pm \dfrac{\sqrt{30}}{12}, 7\right)$

Eccentricity: $\dfrac{\sqrt{10}}{4}$

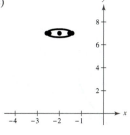

35. $\dfrac{x^2}{25} + \dfrac{y^2}{9} = 1$ **37.** $\dfrac{(x - 2)^2}{25} + \dfrac{y^2}{21} = 1$

39. The foci should be placed 3 feet on either side of the center at the same height as the pillars.

41. $e \approx 0.0543$ **43.** $\dfrac{x^2}{16} - \dfrac{y^2}{20} = 1$

45. $\dfrac{(x - 4)^2}{16/5} - \dfrac{y^2}{64/5} = 1$

47. (a) $\dfrac{y^2}{4} - \dfrac{x^2}{5} = 1$ (c)

(b) Center: $(0, 0)$

Vertices: $(0, \pm 2)$

Foci: $(0, \pm 3)$

Asymptotes: $y = \pm\dfrac{2\sqrt{5}}{5}x$

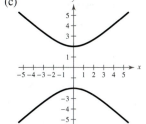

49. (a) $\dfrac{(x - 1)^2}{16} - \dfrac{(y + 1)^2}{9} = 1$ (c)

(b) Center: $(1, -1)$

Vertices:

$(5, -1), (-3, -1)$

Foci: $(6, -1), (-4, -1)$

Asymptotes:

$-1 \pm \dfrac{3}{4}(x - 1)$

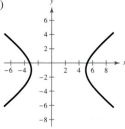

CHAPTER 10

51. (a) $\dfrac{(x+6)^2}{101/2} - \dfrac{(y-1)^2}{202} = 1$

(b) Center: $(-6, 1)$

Vertices: $\left(-6 \pm \dfrac{\sqrt{202}}{2}, 1\right)$

Foci: $\left(-6 \pm \dfrac{\sqrt{1010}}{2}, 1\right)$

Asymptotes: $y = 1 \pm 2(x+6)$

(c)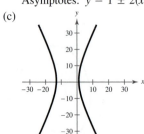

53. $y = \pm x$ **55.** $y = \pm \dfrac{\sqrt{5}}{5}(x-3)$

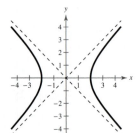 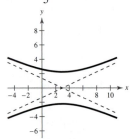

57. About 72 mi   **59.** Ellipse   **61.** Hyperbola

63.

t	-2	-1	0	1	2	3
x	-8	-5	-2	1	4	7
y	15	11	7	3	-1	-5

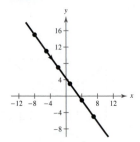

65.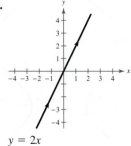

$y = 2x$

67.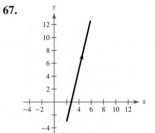

$y = 4x - 11, \; x \geq 2$

69.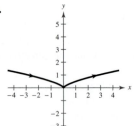

$y = \dfrac{1}{2}x^{2/3}$

71.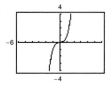

73. 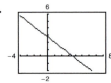   **75.**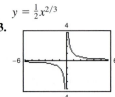

77. 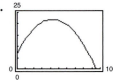   **79.**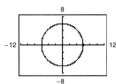

81.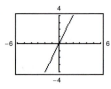

83. (a) $x = t, \; y = 6t + 2$

(b) $x = 1 - t, \; y = 8 - 6t$

85. (a) $x = t, \; y = t^2 + 2$ (b) $x = 1 - t, \; y = t^2 - 2t + 3$

87. $x = t, \; y = 5$ **89.** $x = -1 + 11t, \; y = 6 - 6t$

91. 54.22 ft/sec

93.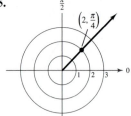

21.93 ft

95.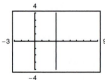

$\left(2, -\dfrac{7\pi}{4}\right), \left(-2, \dfrac{5\pi}{4}\right), \left(-2, -\dfrac{3\pi}{4}\right)$

97.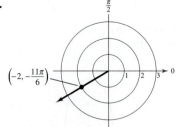

$\left(-2, \dfrac{\pi}{6}\right), \left(2, \dfrac{7\pi}{6}\right), \left(2, -\dfrac{5\pi}{6}\right)$

99.

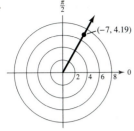

$(-7, -2.09), (7, -5.23), (7, 1.05)$

101.

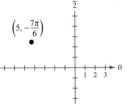

$\left(5, -\dfrac{7\pi}{6}\right)$

$\left(-\dfrac{5\sqrt{3}}{2}, \dfrac{5}{2}\right)$

103.

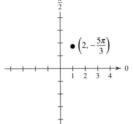

$\left(2, -\dfrac{5\pi}{3}\right)$

$(1, \sqrt{3})$

105.

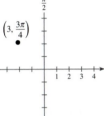

$\left(3, \dfrac{3\pi}{4}\right)$

$\left(-\dfrac{3\sqrt{2}}{2}, \dfrac{3\sqrt{2}}{2}\right)$

107.

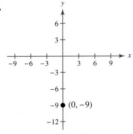

$(0, -9)$

$\left(-9, \dfrac{\pi}{2}\right), \left(9, \dfrac{3\pi}{2}\right)$

109.

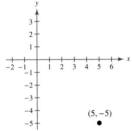

$(5, -5)$

$\left(-5\sqrt{2}, \dfrac{3\pi}{4}\right), \left(5\sqrt{2}, \dfrac{7\pi}{4}\right)$

111. $r = 9$

113. $r = 4\cos\theta$ **115.** $r^2 = 5\sec\theta\csc\theta$

117. $r^2 = \dfrac{1}{1 + 3\cos^2\theta}$ **119.** $x^2 + y^2 = 25$

121. $x^2 + y^2 = 3x$

123. $(x^2 + y^2)^2 - x^2 + y^2 = 0$ **125.** $y = -\dfrac{\sqrt{3}}{3}x$

127.

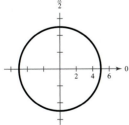

129.

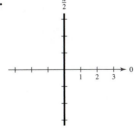

131.

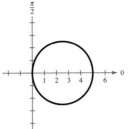

133. Dimpled limaçon   **135.** Limaçon with inner loop

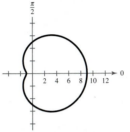
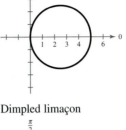

Symmetry: Polar axis   Symmetry: $\theta = \dfrac{\pi}{2}$

Zeros of r: None   Zeros of r: $\theta \approx 0.64, 2.50$

137. Rose curve   **139.** Lemniscate

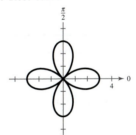
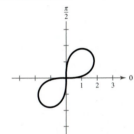

Symmetry: Pole, polar   Symmetry: Pole

axis, $\theta = \pi/2$ Zeros of r: $\theta = 0, \dfrac{\pi}{2}$

Zeros of r: $\theta = \dfrac{\pi}{4}, \dfrac{3\pi}{4}, \dfrac{5\pi}{4}, \dfrac{7\pi}{4}$

141. Hyperbola   **143.** Ellipse

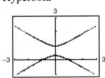
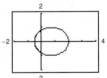

145. Ellipse

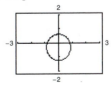

147. $r = \dfrac{4}{1 - \cos\theta}$ **149.** $r = \dfrac{5}{3 - 2\cos\theta}$

151. $r = \dfrac{1.512}{1 - 0.093\cos\theta}$

Perihelion: 1.383 astronomical units

Aphelion: 1.667 astronomical units

153. False. The equation of a hyperbola is a second-degree equation.

CHAPTER 10

155. (a) Vertical translation   (b) Horizontal translation
(c) Reflection in the y-axis   (d) Vertical shrink
157. The orientation would be reversed.

Chapter Test   (page 840)

1.

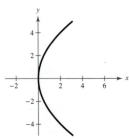

Vertex: $(0, 0)$
Focus: $(2, 0)$

2.

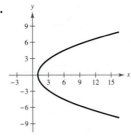

Vertex: $(1, 0)$
Focus: $(2, 0)$

3.

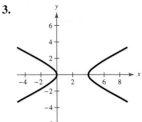

Vertices: $(0, 0), (4, 0)$
Foci: $\left(2 \pm \sqrt{5}, 0\right)$

4. $(y + 2)^2 = 8(x - 6)$

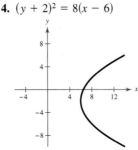

5. $\dfrac{(x + 6)^2}{16} + \dfrac{(y - 3)^2}{49} = 1$ **6.** $\dfrac{y^2}{9} - \dfrac{x^2}{4} = 1$

7.

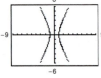

Answers will vary.

8.

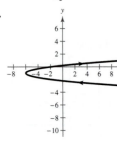

$(y + 1)^2 = \frac{1}{4}(x + 6)$

9.

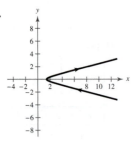

$\dfrac{x^2}{2} - \dfrac{y^2}{1/8} = 1, \ x \geq \sqrt{2}$

10.

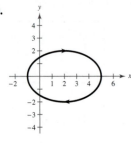

$\dfrac{(x - 2)^2}{9} + \dfrac{y^2}{4} = 1$

11. (a) $x = t, y = 7 - 4t$ (b) $x = 2 - t, y = 4t - 1$

12. (a) $x = t, y = \dfrac{3}{t}$ (b) $x = 2 - t, y = \dfrac{3}{2 - t}$

13. (a) $x = t, y = t^2 + 10$ (b) $x = t - 2, y = t^2 - 4t + 14$

14. $\left(\sqrt{3}, -1\right)$

15. Sample answer: $\left(2\sqrt{2}, \dfrac{7\pi}{4}\right); \left(2\sqrt{2}, -\dfrac{\pi}{4}\right), \left(-2\sqrt{2}, \dfrac{3\pi}{4}\right)$

16. $r = 3 \cos \theta$ **17.** $x^2 + (y - 1)^2 = 1$

18. Limaçon with inner loop   **19.** Parabola

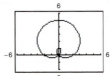

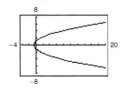

20. Hyperbola

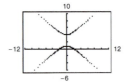

21. $r = \dfrac{4}{4 + \sin \theta}$ **22.** $r = \dfrac{10}{4 + 5 \sin \theta}$

23. Maximum: $|r| = 8$

Zeros of r: $\theta = \dfrac{\pi}{6}, \dfrac{\pi}{2}, \dfrac{5\pi}{6}$

Cumulative Test for Chapters 8–10
(page 841)

1. $(4, -3)$ **2.** $(8, 4), (2, -2)$

3. $\left(\frac{3}{5}, -4, -\frac{1}{5}\right)$ **4.** $(1, -4, -4)$

5. $\begin{bmatrix} -7 & -10 & -16 \\ -6 & 18 & 9 \\ -12 & 16 & 7 \end{bmatrix}$ **6.** $\begin{bmatrix} -18 & 15 & -14 \\ 28 & 11 & 34 \\ -20 & 52 & -1 \end{bmatrix}$

7. $\begin{bmatrix} 3 & -31 & 2 \\ 22 & 18 & 6 \\ 52 & -40 & 14 \end{bmatrix}$ **8.** $\begin{bmatrix} 5 & 36 & 31 \\ -36 & 12 & -36 \\ 16 & 0 & 18 \end{bmatrix}$

9. (a) $\begin{bmatrix} -175 & 37 & -13 \\ 95 & -20 & 7 \\ 14 & -3 & 1 \end{bmatrix}$ (b) 1

10. 22 **11.** (a) $\frac{1}{5}, -\frac{1}{7}, \frac{1}{9}, -\frac{1}{11}, \frac{1}{13}$ (b) $3, 6, 12, 24, 48$

12. 135 **13.** $\frac{47}{52}$ **14.** 34.48 **15.** 66.67 **16.** $\frac{15}{8}$

17. $-\frac{5}{51}$ **18.** $\frac{8}{3}$ **19.** (a) 190 (b) 190

20. $x^4 + 12x^3 + 54x^2 + 108x + 81$

21. $32x^5 + 80x^4y^2 + 80x^3y^4 + 40x^2y^6 + 10xy^8 + y^{10}$

22. $x^6 - 12x^5y + 60x^4y^2 - 160x^3y^3 + 240x^2y^4 - 192xy^5 + 64y^6$

23. $6561a^8 - 69,984a^7b + 326,592a^6b^2 - 870,912a^5b^3$
$\qquad + 1,451,520a^4b^4 - 1,548,288a^3b^5 + 1,032,192a^2b^6$
$\qquad - 393,216ab^7 + 65,536b^8$

24. 120 **25.** 420 **26.** $302,400$ **27.** $15,120$

28. Hyperbola

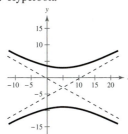

29. Ellipse

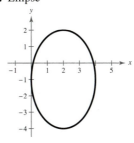

30. Hyperbola

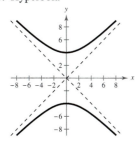

31. Circle

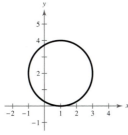

32. $(x - 2)^2 = -\frac{4}{3}(y - 3)$

33. $\dfrac{(x - 1)^2}{25} + \dfrac{(y - 4)^2}{4} = 1$

34. $\dfrac{(y + 4)^2}{4} - \dfrac{x^2}{16/3} = 1$

35. (a) and (b)   (c) $y = \dfrac{x^2 - 2x + 1}{4}$

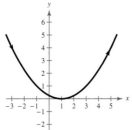

36. (a) and (b)   (c) $y = 2 - 2x^2, \ -1 \le x \le 1$

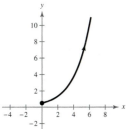

37. (a) and (b)   (c) $y = 0.5e^{0.5x}, \ x \ge 0$

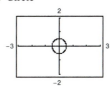

38. (a) $x = t, y = 3t - 2$ (b) $x = 2t, y = 6t - 2$

39. (a) $x = t, y = t^2 - 16$ (b) $x = 2t, y = 4t^2 - 16$

40. (a) $x = t, y = \dfrac{2}{t}$ (b) $x = 2t, y = \dfrac{1}{t}$

41. (a) $x = t, y = \dfrac{e^{2t}}{e^{2t} + 1}$ (b) $x = 2t, y = \dfrac{e^{4t}}{e^{4t} + 1}$

42.

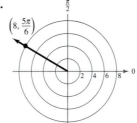

$\left(8, -\dfrac{7\pi}{6}\right), \left(-8, -\dfrac{\pi}{6}\right), \left(-8, \dfrac{11\pi}{6}\right)$

43.

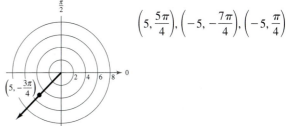

$\left(5, \dfrac{5\pi}{4}\right), \left(-5, -\dfrac{7\pi}{4}\right), \left(-5, \dfrac{\pi}{4}\right)$

44.

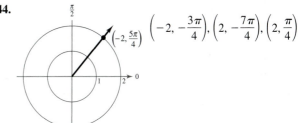

$\left(-2, -\dfrac{3\pi}{4}\right), \left(2, -\dfrac{7\pi}{4}\right), \left(2, \dfrac{\pi}{4}\right)$

45.

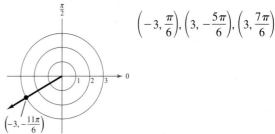

$\left(-3, \dfrac{\pi}{6}\right), \left(3, -\dfrac{5\pi}{6}\right), \left(3, \dfrac{7\pi}{6}\right)$

46. $r = -\dfrac{1}{4 \sin \theta + 4 \cos \theta}$ **47.** $(x - 2)^2 + y^2 = 4$

48. $\dfrac{\left(x + \frac{10}{9}\right)^2}{\frac{64}{81}} - \dfrac{y^2}{\frac{4}{9}} = 1$

49. Circle

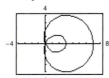

50. Dimpled limaçon

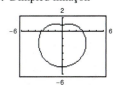

51. Limaçon with inner loop

52. $701,303.32

53. $\frac{1}{4}$ **54.** $24\sqrt{2}$ m

CHAPTER 10

Index of Selected Applications

Go to this textbook's *Companion Website* for a complete list of applications.

Index